Environmental Science and Engineering

Environmental Science

Series Editors

Rod Allan
Ulrich Förstner
Wim Salomons

For further volumes:
http://www.springer.com/series/3234

Robert Maliva · Thomas Missimer

Arid Lands Water Evaluation and Management

Springer

Robert Maliva
Schlumberger Water Services
1567 Hayley Lane
Suite 202 Fort Myers
FL 33907
USA

Thomas Missimer
Water Desalination and Reuse Center
King Abdullah University of Science
and Technology
Thuwal 23955-6900
Saudi Arabia

ISSN 1431-6250
ISBN 978-3-642-29103-6 ISBN 978-3-642-29104-3 (eBook)
DOI 10.1007/978-3-642-29104-3
Springer Heidelberg New York Dordrecht London

Library of Congress Control Number: 2012936762

Printed on acid-free paper

Springer is part of Springer Science+Business Media (www.springer.com)

Preface

Arid and semi-arid lands constitute about 30% of the Earth's land environment. The need to develop new freshwater sources in these water-short areas is rapidly increasing because of both population and economic growth. Many of these arid lands are dependent to a large degree on non-renewable resources that are slowly being mined with no real plans to replace them in the future. Therefore, considerable attention must be placed on the integrated management of existing and new water resources that will supply future potable, agricultural, industrial demands. Within the concept of sustainability, the maintenance of the natural system also must be considered and commonly, when it is ignored, negatively impacts primary water supply sources and food security.

This book is an integrated approach to evaluation and management of water resources in arid lands. Methods and management practices applied to arid lands are commonly unique to these stressed regions and principles applied to humid or semi-humid regions cannot necessarily be applied to arid or semi-arid lands. Many arid regions, such as the Middle East and parts of Africa and South Asia, require the integration of technical solution approaches with fundamental economic and sociocultural challenges. Natural system generated water supplies cannot meet future demands in most cases, therefore, the development of new water sources, such as desalination of brackish-water and seawater and increased reuse of wastewater are required.

This book is not solely a technical monograph that provides methods to evaluate primarily groundwater systems in arid lands, but contains a holistic approach to water management in arid lands. Water supply development must contain a degree of economic realism with the sophistication matching the abilities of the water users and operators to be able to pay for and manage the infrastructure. In the past decades large errors have been made in arid lands water management, wherein well-meaning investors and non-governmental organizations have funded and constructed water and wastewater supply and treatment systems that cannot be operated or paid for by the end users. For example, some poorly designed dams have aided one part of a basin and devastated another part of it; mega-groundwater supply projects have been developed that resulted in land subsidence, high evapotranspiration losses, and loss of ariable land; sophisticated, high-technology wastewater plants have been constructed that lie unused because the economy and educational levels do not allow for them to be operated and maintained; and environmental "restoration" projects have been developed to reforest farmlands with an unexpected impact of lowering aquifer water levels by several meters, causing reductions in local food production from the remaining farmlands.

We believe that much can be learned for the past practices used to sustain people in arid lands where management of the localized water resources was paramount to survival. While localized water harvesting techniques cannot produce the vast qualities of water needed to supply large cities using centralized systems, they can provide needed water for the hundreds of thousands of small villages and farms that occur throughout the arid lands of the world. The wise capture and storage of temporally rare rainfall events was the key to survival in the past and may also be so in the future.

Economic, sociocultural, and policy issues are often of greater importance in water management than technical issues. The best science and engineering in water resources management is of little value if it does not get incorporated into actual water management practices and projects. We have endeavored to objectively present basic concepts, with the understanding than men and women of good will may have different perspectives, values, and beliefs. Any opinions expressed are solely those of the authors.

Arid lands will face considerable challenges in the future as the Earth's climate warms and causes more extreme periods of drought with longer duration and also, the extreme storm events will become more common, resulting in the need to create storage using a variety of techniques including dams and reservoirs, aquifer recharge and recovery of surface and impaired waters, and a variety of smaller-scale artificial recharge projects. Large water deficits will occur in arid regions supplied by rivers and streams dependent on seasonal glacier-melt flows as these vast ice reservoirs melt in response to warmer temperatures. The issue of global climate challenge will force management of water in arid lands to become fully integrated using all available existing sources, the development of new sources (desalination and water reuse), and the expansion of storage types. New methods of controlling water demand will be needed, such as the development of new food production methods, such as crop production using saline water on genetically-engineered crops and growing food crops in graint, climate controlled greenhouses.

The assessment and management of water resources in arid and semi-arid lands is a dynamic field of research. A positive frustration in preparing this manuscript has been that after we "completed" a chapter, a new paper would later be published that includes valuable information and insights that required further revision of the chapter. Indeed, it is impossible to write a definitive text on any dynamic field as science is constantly progressing.

The intended users of this book should include university students studying engineering, hydrogeology, economics, and environmental restoration; planners that need to understand fundamental water management principles and how they apply to arid lands; practicing design engineers involved in surface-water management, water treatment design, wastewater treatment plant design, and the integration of multiple water treatment supply and distribution systems; and most of all government officials and elected leaders that need to guide the decision-making process into practical pathways based on fundamental economic principles with the needs of the people and environment being considered at every step of the development and implementation process. Many subjects covered in chapters or sections of this text could be comprehensive books by themselves. Therefore, the information presented is a general summary of the general concepts and principles involved and available the techniques to evaluate and manage water resources. Extensive references are provided to provide a greater depth of understanding and examples of the diversity of approaches that have been taken to evaluate arid lands. The focus of this book is largely on groundwater resources, because in most arid and semi-arid lands surface water resources are limited and already fully utilized. Surface water, particularly in ephemeral systems, is critical as the source of groundwater recharge. Groundwater quality issues, particularly contamination, are of increasing importance, but merit a dedicated book in itself to adequately address.

Fort Myers, FL, USA, March 2012 Robert Maliva

Thuwal, Saudi Arabia, March 2012 Thomas Missimer

Acknowledgments

The ideas and concepts presented in this book are largely the product of the work of numerous dedicated workers in the water resources field. We are particularly grateful for the insights over the years from co-workers too numerous to name. Hydrology and hydrogeology, in common with other sciences, has advanced largely in an evolutionary manner through the diverse contributions of many scientists and engineers working throughout the world. Indeed, the greatest benefit of researching and writing this book has been the discovery or rediscovery of important and valuable work performed by earlier workers in the field. The authors would particularly like to thank the following people for their thoughtful reviews of parts of the books: Vincent Amy, Weixing Guo, Lisa Latkowski, Frank Winslow, Steve Carroll, Robin Nava, Shahnawaz Sinha, Jörg Drewes, Kim Choon Ng, Christiana Hoppe-Jones, Khan Jadoon, and Nigel Merte, Mr. Samir Al-Mashharawi provided some important photographs. We also thanks the editors at Springer for their diligent work on the manuscript.

Contents

Part I Arid Regions Water Management and Issues

1 Introduction . . . 3
1.1 Introduction . . . 3
1.2 Global Water Budget and Water Stress . . . 11
1.3 Technical Issues . . . 14
1.4 Socioeconomic and Political Issues . . . 15
1.5 Appropriateness, Values, and Priorities . . . 16
References . . . 19

2 Aridity and Drought . . . 21
2.1 Definition of Aridity . . . 21
2.2 Aridity Indices . . . 23
2.3 Causes of Aridity . . . 25
2.4 Droughts . . . 27
2.4.1 Definition of Droughts . . . 27
2.4.2 Impacts of Drought . . . 30
2.4.3 Drought Indices . . . 32
2.5 Deserts and Desertification . . . 33
2.5.1 Great Plains Dust Bowl of the 1930s . . . 34
2.5.2 Causes and Management of Desertification . . . 36
References . . . 37

Part II Arid Lands Geology and Hydrogeology: An Overview

3 Geology of Arid Lands . . . 43
3.1 Introduction . . . 43
3.2 Geomorphology and Geology of Arid Lands . . . 45
3.2.1 Introduction . . . 45
3.2.2 Mountains . . . 46

3.2.3 Low-Angle Bedrock Surfaces 50
3.2.4 Alluvial Fans 54
3.2.5 River Plains (Fluvial Deposits) 64
3.2.6 Dry Watercourses (Arroyos and Wadis) 72
3.2.7 Badlands 75
3.2.8 Playas 77
3.2.9 Sand Seas (Ergs) 81
3.2.10 Desert Flats 85
3.2.11 Recent Volcanic Deposits 87
3.3 Arid Lands Geology and Hydrogeology 88
References 89

4 Aquifer Concepts in Arid Lands 95
4.1 Aquifer Types 95
4.1.1 Aquifer Hydraulic Classification 95
4.1.2 Perched Aquifers 98
4.2 Arid Lands Aquifer Types 100
4.2.1 Local Alluvial Aquifers 102
4.2.2 Sand Dune (erg) Aquifers 104
4.2.3 Basin-Fill Aquifers 105
4.2.4 Intermontane and Submontane Carbonate and Sandstone Aquifers 108
4.2.5 Fractured Rock Aquifers (Hard-Rock Aquifers) . . . 109
4.2.6 Regional and Intermediate Aquifers and Flow Systems 112
4.2.7 Non-renewable Groundwater Resources 112
References 113

5 Introduction to Aquifer Hydraulics 117
5.1 Introduction 117
5.1.1 Aquifer Heads, Potentiometric Surface, and Groundwater Flow 117
5.2 Groundwater Flow 119
5.3 Aquifer Hydraulic Properties 120
5.3.1 Transmissivity 121
5.3.2 Storativity 121
5.3.3 Leakance 122
5.3.4 Hydraulic Diffusivity 124
5.4 Porosity and Storage 124
5.5 Aquifer Heterogeneity 127
5.6 Vadose Zone Hydrology 129
5.6.1 Capillary Processes 131
5.6.2 Water Potential 133

5.6.3 Unsaturated or Vadose Zone Processes and Flow . . . 134
5.6.4 Infiltration . . . 138
5.6.5 Kinematic Wave Theory . . . 142
5.6.6 Impacts of Bedding and Heterogeneity on Vadose Zone Flow (Funnel or Finger-Flow) . . . 142
5.6.7 Impacts of Air Entrapment and Flow on Vadose Zone Processes (Lisse Effect) . . . 143
References . . . 145

Part III Water Budget and Recharge

6 Water Budget . . . 151
6.1 Introduction . . . 151
6.2 Surface Water Components . . . 152
6.3 Recharge Within the Context of the Water Budget . . . 153
6.4 Captured Discharge and Recharge . . . 154
6.5 Water Budget and Groundwater Resources Management . . . 156
6.6 Water Budget Component Evaluation . . . 160
References . . . 160

7 Precipitation and Evapotranspiration . . . 163
7.1 Introduction . . . 163
7.2 Precipitation in Arid Regions . . . 164
7.3 Precipitation Measurements . . . 165
7.3.1 Rain Gauges . . . 166
7.3.2 Radar Measurement of Rainfall . . . 167
7.3.3 Satellite Rainfall Measurement . . . 168
7.4 Precipitation Frequency Analysis . . . 169
7.5 Evapotranspiration . . . 172
7.6 Measurement of Actual ET . . . 173
7.6.1 Lysimeters . . . 174
7.6.2 Energy Balance and Micrometeorological Methods . . . 175
7.6.3 Energy Balance Bowen Ratio Method . . . 175
7.6.4 Eddy Covariance Method . . . 176
7.6.5 Geophysical Methods . . . 177
7.7 Measurements of Reference ET . . . 178
7.7.1 Modified Penman–Monteith Method . . . 178
7.7.2 Priestley–Taylor Method . . . 180
7.7.3 Blaney–Criddle Method . . . 180
7.7.4 Pan Evaporation Measurements . . . 181
References . . . 183

8 Recharge Concepts and Settings . . . 187
8.1 Introduction . . . 187
8.2 Direct, Indirect, Matrix, and Macropore Recharge . . . 189
8.3 Factors Controlling the Rate of Infiltration and Recharge in Arid Regions . . . 190
8.3.1 Effects of Vegetation and Land Use/Land Cover . . . 194
8.3.2 Effects of Temperature . . . 196
8.3.3 Effects of Soil Moisture Gradients and Changes . . . 197
8.4 Arid and Semiarid Region Recharge Location . . . 197
8.4.1 Mountain Block and Front Recharge . . . 198
8.4.2 Wadi Recharge . . . 202
8.4.3 Flat Lands Depressional Recharge . . . 204
References . . . 205

9 Sustainability and Safe Yield . . . 209
9.1 Introduction . . . 209
9.2 Sustainability Water Resources Management Concepts . . . 210
9.3 Safe Yield and Sustainability . . . 212
9.4 Acceptability of the Impacts of Groundwater Use . . . 214
9.5 Evaluation of Safe Yield and Sustainability . . . 216
9.6 Mitigation . . . 216
References . . . 218

Part IV Water Resources Assessment Methods

10 Water Resources Assessment Methods: Assessment of Groundwater Resources . . . 223
10.1 Introduction . . . 223
10.2 Desktop Investigations . . . 225
10.3 Exploratory Well and Production Well Construction . . . 226
10.4 Data Collection During Well Drilling . . . 228
10.5 Aquifer Hydraulic Testing . . . 229
10.5.1 Aquifer Performance Testing . . . 230
10.5.2 Aquifer Testing During Drilling . . . 232
10.5.3 Packer Tests . . . 234
10.6 Vadose-Zone Hydraulic Testing Methods . . . 235
10.6.1 Infiltration Tests . . . 235
10.6.2 Soil Moisture Measurement . . . 237
10.6.3 Soil Matric Potential Measurement . . . 241
References . . . 243

11 Recharge Measurement in Arid and Semiarid Regions 247
11.1 Introduction 247
11.2 Direct Physical Measurement of Infiltration and Recharge Rates. 249
11.2.1 Lysimeter. 250
11.2.2 Seepage Meters. 251
11.2.3 Darcy–Buckingham Law Based Methods 251
11.3 Water-Table Fluctuation Method 253
11.4 Aquifer Water Budget-Based Methods 254
11.4.1 Aquifer Water Budget 255
11.4.2 Spatially Distributed Water Budget Models 256
11.4.3 Darcy Flux in Aquifers (Flow Tubes) 258
11.4.4 Stream Gauging: Baseflow Method 260
11.4.5 Surface Water Budget of Channels 261
11.5 Environmental Tracers. 262
11.5.1 Chloride Mass Balance Method 262
11.5.2 Age Tracers 265
11.5.3 Chlorofluorocarbons 267
11.5.4 Tracer Versus Water Migration. 268
11.6 Neutron Probe Measurements. 269
11.7 Borehole Temperature Profiles 269
11.8 Groundwater Modeling 270
11.9 Recharges Estimates in Arid and Semiarid Environments . . . 270
References 272

12 Environmental Isotopes 277
12.1 Introduction 277
12.2 Isotope Basics 278
12.3 Stable Isotopes 278
12.4 Radioisotopes. 281
12.4.1 Radioactive Dating Concepts 281
12.4.2 Radiocarbon (Carbon 14) Dating 282
12.4.3 Tritium 284
12.4.4 Chlorine 36 286
12.4.5 Krypton 81. 287
12.5 Stable Isotopes of Water 287
References 290

13 Wadi Recharge Evaluation 293
13.1 Introduction 293
13.2 Methods for Measuring Wadi Recharge. 295
13.3 Wadi Rainfall-Runoff Relationships 298
13.3.1 Precipitation 300
13.3.2 Regression Methods 301

13.3.3 Soil Conservation Service Curve Number Method . . . 301
13.3.4 Hydrographs . . . 304
13.3.5 Geomorphic Relationship . . . 310
13.4 Transmission Losses . . . 310
13.4.1 Transmission Loss Relationships from Experimental Watersheds . . . 312
13.4.2 Transmission Loss and Tributary Flows . . . 314
13.5 Transmission Losses and Recharge . . . 315
13.6 Hydrologic Modelling . . . 315
13.6.1 HEC-HMS Model Code . . . 317
13.6.2 Soil and Water Assessment Tool . . . 317
13.6.3 The Precipitation-Runoff Modeling System Model . . . 318
13.6.4 Applications of SW Modeling for Wadi Recharge . . . 318
13.7 Wadi Recharge Seasonal Timing . . . 319
13.8 Temperature Data and Wadi Recharge . . . 320
13.8.1 Determination of Temporal and Spatial Patterns of Recharge . . . 321
13.8.2 Determination of Hydraulic Conductivity and Infiltration Rates . . . 324
References . . . 325

14 Microgravity . . . 329
14.1 Introduction . . . 329
14.2 Microgravity Use in Groundwater Investigations . . . 331
14.3 Relative Gravity Surveys . . . 332
14.4 Applications of Microgravity to Water Management . . . 334
14.4.1 Mapping the Distribution of Injected Water in an MAR System . . . 335
14.5 Gravity Recovery and Climate Experiment . . . 336
References . . . 341

15 Compaction and Land Subsidence . . . 343
15.1 Introduction . . . 343
15.2 Land Subsidence Measurement . . . 348
15.2.1 Conventional (On Ground) Surveying . . . 348
15.2.2 Extensometers . . . 349
15.2.3 Global Positioning Systems . . . 350
15.2.4 InSAR . . . 352
15.2.5 Measured Land Subsidence Amounts and Rates . . . 353
15.3 Impacts of Subsidence . . . 355
15.4 Case Studies . . . 356

15.4.1 Mexico City . . . 356
15.4.2 Antelope Valley, California . . . 359
15.5 Mitigation . . . 359
References . . . 361

16 Surface and Airborne Geophysics . . . 365
16.1 Introduction . . . 365
16.2 Electrical Resistivity and Electromagnetic Techniques . . . 366
16.3 Electrical Resistivity Method . . . 368
16.4 Electromagnetic Surveys . . . 373
16.4.1 Frequency-Domain Electromagnetic Soundings for Shallow Data Collection . . . 374
16.4.2 Transient Electromagnetic (TDEM) Sounding Methods for Deep Data Collection . . . 375
16.5 Applications of Resistivity-Based Surface Geophysics to Groundwater Investigations . . . 377
16.5.1 Mapping of Saline-Water Interface . . . 377
16.5.2 Depth to the Water Table. . . . 380
16.5.3 Mapping of Freshwater Lenses . . . 380
16.5.4 Managed Aquifer Recharge Investigations . . . 380
16.5.5 Formation and Aquifer Mapping . . . 381
16.5.6 Mapping of Recharge Areas . . . 382
16.5.7 Mapping of Regional Aquifer Flow Orientation (Fractured Rock Aquifers) . . . 382
16.6 Ground Penetrating Radar . . . 383
16.7 Surface Nuclear Magnetic Resonance . . . 385
16.8 Magnetotellurics . . . 387
16.9 Seismic Reflection . . . 388
16.10 Airborne Geophysics . . . 390
16.10.1 Mapping Aquifers in Arid Lands . . . 391
16.10.2 Mapping Coastal the Saline-Water/Fresh-Water Interface . . . 392
16.10.3 Detection of Saline-Water Contamination and Other Contamination Plumes . . . 393
16.10.4 Recharge and Salinity Mapping . . . 394
16.11 Applications of Surface Geophysics to Groundwater Investigations . . . 395
References . . . 395

17 Borehole Geophysical Techniques . . . 401
17.1 Introduction . . . 401
17.2 Quality Assurance and Control . . . 402
17.3 Caliper Logs . . . 404
17.4 Natural Gamma Ray Log . . . 406

17.5 Spontaneous Potential 406
17.6 Resistivity Logs 409
17.7 Sonic Logs. 410
17.8 Nuclear Logging. 413
17.8.1 Density Log 413
17.8.2 Neutron Log. 416
17.9 Flowmeter Logs 417
17.9.1 Introduction 417
17.9.2 Spinner 418
17.9.3 Electromagnetic Borehole Flowmeter (EBF). 419
17.9.4 Heat-Pulse Flowmeter 421
17.9.5 Interprctation of Flowmeter Log Data 421
17.10 Temperature and Fluid Resistivity Logs 423
17.11 Borehole Imaging Logs 424
17.11.1 Borehole Video Survey 424
17.11.2 Optical Televiewer 426
17.11.3 Acoustic-Televiewer Log 426
17.11.4 Microresistivity Imaging 428
17.12 Nuclear Magnetic Resonance Log. 428
17.13 Elemental Capture Spectroscopy Log 430
17.14 Development of Borehole Geophysical Logging Programs. . . 431
References 433

18 Remote Sensing 435
18.1 Introduction 435
18.2 LIDAR 437
18.3 Aerial Photograph Interpretation. 438
18.4 Applications of Remote Sensing to Groundwater Management 439
18.4.1 Lineament Mapping 439
18.4.2 Identification of Structural and Sedimentological Features of Hydrogeological Significance 445
18.4.3 Identification of Recharge and Discharge Areas . . . 446
18.4.4 Vegetation Mapping 448
18.4.5 Evapotranspiration and Recharge Rate Quantification 448
18.4.6 Soil Moisture Measurement 449
18.4.7 Interferometric Synthetic Aperture Radar (InSAR) 451
18.5 Remote Sensing and Groundwater Modeling 451
18.6 Limitations of Remote Sensing. 452
References 453

19 Geographic Information Systems 457
19.1 Introduction 457
19.2 GIS Principles 458
19.3 GIS Applications 459
19.3.1 Groundwater Model Development 460
19.3.2 Recharge Investigations 461
19.3.3 Site Evaluation Studies 464
19.3.4 Mapping of Contamination Vulnerability 467
19.4 Hydrologic Data Management 468
19.5 The Future of GIS in Arid Lands Water Management 471
References 471

20 Groundwater Flow and Solute-Transport Modeling 475
20.1 Introduction 475
20.2 Groundwater Modeling for Arid and Semiarid Lands 476
20.3 Limitations of Groundwater Modeling 477
20.4 Groundwater Model Development 478
20.4.1 Statement of Problem and Modeling Objectives . . . 478
20.4.2 Conceptual Model 479
20.4.3 Choice of Model Code 480
20.4.4 Model Development and the Population of the Model Grid 482
20.4.5 Model Calibration 483
20.4.6 Model Verification, Post Audits, and Refinement 486
20.4.7 Predictive Simulations 487
20.5 Groundwater Flow Modeling Codes 487
20.5.1 MODFLOW 488
20.5.2 MicroFEM 488
20.6 Groundwater Model Codes for Solute Transport 488
20.6.1 MT3DMS 489
20.6.2 SEAWAT 489
20.6.3 SUTRA 489
20.6.4 HST3D 490
20.6.5 FEFLOW 490
20.6.6 ECLIPSE® 490
20.6.7 Hydrus 491
20.6.8 PHT3D 491
20.7 Simulation of Compaction and Land Subsidence 491
20.8 Integrated Surface Water-Groundwater Modeling 492
References 495

Part V Water Management Techniques

21 Ancient Water Management 501
21.1 Introduction 501
21.2 Ancient Jordan 502
21.2.1 Jawa 503
21.2.2 Petra 503
21.3 Negev Desert 507
21.4 Greek and Roman Civilizations 508
21.5 Ancient China 509
21.6 Southwestern United States 510
21.6.1 Anasazi 510
21.6.2 Hohokam 512
21.7 Qanats 513
21.8 The Collapse of Ancient Cultures and Climate 517
21.8.1 Akkadian Empire, Mesopotamia 519
21.8.2 Classic Maya civilization, Central America 519
21.8.3 Tiwanaku Civilization, Bolivian and Peruvian Altiplano, Near Lake Titicaca 520
21.8.4 Mississippian Chiefdoms (Cahokia), United States 520
21.8.5 Angkor, Cambodia 521
21.8.6 Neolithic Cultures of the Central Plain of China . . . 521
21.8.7 Anasazi, Southwestern United States 521
21.9 Conclusions 522
References 524

22 Rainwater, Stormwater, and Fog Harvesting 529
22.1 Introduction 529
22.2 Rainwater Harvesting Concepts 532
22.3 Domestic Scale Rainwater Harvesting Systems 536
22.3.1 Earthwork Systems 538
22.3.2 Storage Systems 538
22.4 Roof Rainwater Harvesting Systems 539
22.5 Stormwater Harvesting 540
22.6 Legal Issues 541
22.7 Large-Scale Rainwater and Stormwater Harvesting 542
22.7.1 Rainwater Harvesting in India 543
22.7.2 Rainwater Harvesting in the Loess Plateau, China 545
22.8 Fog Harvesting 547
22.8.1 Introduction 547
22.8.2 Saudi Arabia 550
22.8.3 Western South Africa 550

22.8.4 Yemen 550
22.8.5 Valencia, Spain 551
22.8.6 Southern Peru 552
22.8.7 Other Areas 552
22.8.8 Conclusions 553
22.9 Lithic Mulch Agriculture 553
22.10 Rainwater Harvesting Economics 554
References 556

23 Managed Aquifer Recharge 559
23.1 Introduction 559
23.2 Types of MAR Systems 560
23.2.1 Injection or Recharge Wells 560
23.2.2 Surface Spreading 571
23.3 MAR by Enhancement of Natural Recharge Processes 589
23.3.1 In-Channel Recharge Methods 589
23.3.2 Wadi Dams 590
23.3.3 Sand Dams 594
23.4 MAR Systems Involving Subsurface Engineering 597
23.4.1 Groundwater Dams 598
23.4.2 Artificial Aquifers 600
23.4.3 Physical Storage ASR Using Horizontal Flow Barriers 600
23.5 Stormwater Recharge 602
23.6 MAR System Implementation 606
23.6.1 Phase I: Desktop Feasibility Assessment 607
23.6.2 Phase II: Field Investigations and Testing of Potential System Sites 607
23.6.3 MAR System Design 608
23.6.4 Phase IV: Pilot System Construction 609
23.6.5 Phase V: Project Review and Adaptive Management 609
23.6.6 Phase VI: System Expansion 610
23.7 Managed Aquifer Recharge System Risks and Risk Assessment 610
23.7.1 Microbiological Processes 613
23.7.2 Chemical Contaminants 615
23.8 Economics of MAR Projects 615
23.9 Unmanaged Aquifer Recharge 617
23.10 Groundwater Banking 620
References 623

24 Riverbank Filtration 631
24.1 Introduction 631
24.2 RBF System Design Issues 633

24.3 Evaluation of RBF System Performance: Pathogen Removal 635
24.4 Evaluation of RBF System Performance: Chemical Contaminant Removal 637
24.5 Clogging 639
24.6 Modeling of RBF Systems 640
24.7 Modeling of Contribution of Bank Filtered Water 641
24.8 Limitations and Opportunities of RBF 642
References 643

25 Coastal Water Management Issues 647
25.1 Introduction 647
25.2 Causes of Saline–Water Intrusion 648
25.3 Location and Characterization of Saline–Water Interface 652
25.4 Monitoring of Saline–Water Intrusion 653
25.5 Simulation of Saline–Water Intrusion 656
25.6 Coastal Regions and Climate Change 657
25.7 Mitigation of Saline–Water Intrusion 658
25.7.1 Positive Salinity Barriers Systems 659
25.7.2 Extractive Salinity Barriers 662
25.7.3 Injection and Extraction Barrier Systems 663
25.7.4 Physical Barrier Systems 664
25.7.5 Optimization 664
References 666

26 Domestic and Agricultural Water Conservation 669
26.1 Introduction 669
26.2 Domestic Water Conservation and Saving 672
26.2.1 Domestic Water Savings Opportunities 672
26.2.2 Domestic Water Savings Implementation 674
26.3 Irrigation and Agricultural Water Conservation 682
26.3.1 Agricultural Irrigation Primer 682
26.3.2 Soil Salinization 684
26.3.3 Agricultural Irrigation Efficiency 686
26.3.4 Irrigation Water Use Efficiency Techniques 691
26.3.5 Saline-Water Irrigation 693
26.3.6 Irrigation Water Use Efficiency Incentives 694
References 695

Part VI Desalination

27 Desalination: Desalination in Arid Lands 701
27.1 Introduction 701

27.2 Thermal Desalination Processes 704
27.2.1 Multi-Stage Flash Distillation 705
27.2.2 Multiple Effect Distillation 708
27.2.3 Vapor Compression Distillation 710
27.2.4 Adsorption Desalination (New Technology Not Yet Fully Commercial) 711
27.3 Membrane Desalination Processes 714
27.3.1 Seawater Reverse Osmosis 714
27.3.2 Brackish-Water Reverse Osmosis 718
27.3.3 Electrodialysis Reversal 719
27.3.4 Forward Osmosis 720
27.4 Hybrid Facilities 721
27.4.1 Combined Power Generation, Multistage Flash Distillation, and Reverse Osmosis 722
27.4.2 Combined Power Generation, Multistage Flash Distillation, and Nanofiltration (and Reverse Osmosis) 723
27.4.3 Combined Power Generation, Multiple-Effect Distillation, and Reverse Osmosis 723
27.4.4 Combined Multiple Effect Distillation and Adsorption Desalination 724
27.5 Alternative Energy Use in Desalination 725
27.5.1 Solar-Powered, Direct and Indirect Solar Desalination 727
27.5.2 Wind-Powered Desalination 731
27.5.3 Geothermal Powered Desalination 732
27.5.4 Nuclear-Powered Desalination 733
27.6 Future Innovations and Energy and Cost Reductions 736
27.6.1 Membrane Pretreatment of Distillation Treatment Plant Intake Water (and RO in Hybrid Systems) . . . 736
27.6.2 Use of Waste Heat, Solar Energy, and Geothermal Energy with Adsorption Desalination 737
27.6.3 Use of Subsurface Intakes for Seawater Ro Plants 737
27.6.4 Use of Operational Aquifer Storage and Recovery to Improve Efficiency 739
27.6.5 Combination of Power Plant, Distillation Plant, Reverse Osmosis Plants and Aquifer Storage and Recovery on a Single Site to Improve Operational Efficiency 740
27.7 Current Comparative Energy Use and Desalination Economics 741
27.7.1 Energy Use of Various Desalination Technologies 741

27.7.2 Comparative Costs of Current Desalination Technologies . . . 743
References . . . 744

28 Environmental Issues in Desalination . . . 749
28.1 Introduction . . . 749
28.2 Energy Consumption Issues . . . 750
28.3 Intake Issues . . . 751
28.4 Concentrate Disposal Issues . . . 752
28.4.1 Physical and Chemical Properties of Concentrate . . . 753
28.4.2 Surface Discharge to Tidal Waters . . . 753
28.4.3 Subsurface Disposal Methods . . . 759
28.4.4 Evaporation Ponds . . . 762
28.4.5 Zero Discharge/Crystallization . . . 763
28.5 Desalination Environmental Impacts with Future Innovations . . . 764
References . . . 765

Part VII Wastewater Reuse in Arid Lands

29 Wastewater Reuse . . . 773
29.1 Introduction . . . 773
29.2 Wastewater Terminology . . . 777
29.3 Wastewater Treatment Processes . . . 777
29.4 Wastewater Reuse for Agricultural Irrigation . . . 782
29.5 Potable Reuse . . . 786
29.5.1 Direct Potable Reuse . . . 787
29.5.2 Indirect Potable Reuse . . . 788
29.5.3 Wastewater MAR . . . 790
29.6 Economic Issues . . . 792
29.7 Public Acceptance of Wastewater Reuse . . . 796
29.8 Industrial Wastewater Reuse . . . 799
29.9 Environmental Concerns Over Reuse . . . 800
References . . . 801

30 Wastewater Reuse Health and Safety Issues . . . 805
30.1 Introduction . . . 805
30.2 Risks Associate with Wastewater Reuse . . . 805
30.2.1 Risk Management and Reduction . . . 807
30.2.2 Monitoring . . . 807
30.3 Pathogenic Microorganisms . . . 808
30.4 Chemical Contaminants . . . 810

30.5 Compounds of Emerging Concern 811
30.6 Epidemiological Studies and Risk Assessments 816
30.7 Microbiological Standards and Guidelines 820
30.8 Implementing Health-Based Targets and Monitoring 821
References 823

31 Managed Aquifer Recharge and Natural Water Treatment Processes in Wastewater Reuse 827
31.1 Introduction 827
31.2 Water Quality Improvements During SAT 829
31.2.1 Introduction 829
31.2.2 Nutrient Removal During SAT 829
31.2.3 Pathogen Removal During SAT 830
31.2.4 Organic Carbon Removal During SAT 830
31.2.5 CEC Removal During SAT 832
31.3 Water Quality Improvements in the Groundwater Environment 833
31.3.1 Fate and Transport of Pathogenic Microorganisms in Groundwater Environments 834
31.3.2 Microbial Risk Evaluation and Management 836
31.3.3 Organic Carbon Removal 838
31.3.4 Disinfection Byproducts 840
31.3.5 CECs in Groundwater Environment 842
References 844

Part VIII Water Policy and Management

32 Water Policy and Governance 851
32.1 Introduction 851
32.2 Tragedy of the Commons 854
32.3 Water as an Undervalued Asset 858
32.4 Water as a Human Right 859
32.5 Cultural Conservatism 861
32.6 Government Policy and Water Law 862
32.7 Water Governance Options 864
32.7.1 Absolute Ownership 865
32.7.2 Prior Appropriation Doctrine 865
32.7.3 Reasonable Use Doctrine 866
32.7.4 Correlative Rights Doctrine 867
32.7.5 Restatements of Torts 867
32.8 Closing Thoughts 868
References 868

33 Religious and Cultural Influences on Water Management 871
33.1 Introduction ... 871
33.2 Water Management in Islam ... 872
33.3 Water Management in Christianity and Judaism ... 875
33.4 Buddhism ... 876
33.5 Hinduism ... 877
33.6 Conclusions ... 877
References ... 878

34 Economics ... 881
34.1 Introduction ... 881
34.2 Water as an Economic Good ... 885
34.3 Self-Sufficiency and Food Security ... 888
34.4 Virtual Water ... 889
34.5 Water Ownership and Market-Driven Solutions ... 893
34.6 Water Markets ... 896
34.7 Chilean Water Market Experience ... 898
34.8 Third Party Effects of Water Trading ... 900
34.9 Subsidies ... 902
34.10 Privatization of Public Water Treatment and Distribution Infrastructure ... 904
References ... 907

35 Integrated Water Resources Management ... 911
35.1 Introduction ... 911
35.2 Meaning and Implementation of IWRM ... 912
35.3 Conjunctive Use ... 916
35.4 Optimization ... 918
35.5 Multiple Criteria Decision Analysis ... 919
35.6 Decision Support Systems ... 920
35.7 IWRM and Water Management Planning ... 922
35.8 IWRM in Arid and Semiarid Lands ... 923
References ... 924

36 Non-Renewable Groundwater Resources ... 927
36.1 Non-Renewable and Fossil Groundwater Definitions ... 927
36.2 Utilization of Non-Renewable Groundwater ... 928
36.3 Assessment of Non-Renewable Resources ... 930
36.4 Legal and Institutional Controls of Non-Renewable Groundwater Use ... 934
36.5 Transboundary Non-Renewable Groundwater Resources ... 935
36.6 Examples of Non-Renewable Groundwater Management ... 936
36.6.1 Saudi Arabia ... 936
36.6.2 North Western Sahara Aquifer System ... 938

36.6.3 Nubian Sandstone Aquifer System 939
36.6.4 Great Man-Made River Project 941
36.6.5 Great Artesian Basin 943
36.6.6 Northern Chile 945
References 947

37 Transboundary Water Management 953
37.1 Introduction 953
37.2 International (Transboundary) River Water Law 956
37.2.1 Helsinki Rules and Berlin Rules 957
37.2.2 1997 Convention on the Non-navigation Uses of International Watercourses 958
37.3 Conflict Resolution 959
37.4 Transboundary Aquifers 959
37.5 Transboundary Groundwater Law 961
37.5.1 Bellagio Draft Treaty 962
37.5.2 Internationally Shared Aquifer Resource Management Programme 962
37.5.3 1997 Convention on the Non-Navigation Uses of International Watercourses 963
37.5.4 Law of Transboundary Aquifers 963
37.6 Transboundary Groundwater Management: United States and Mexico 965
37.6.1 Hueco Bolson Aquifer 965
37.6.2 All-American Canal Lining 968
References 969

Part IX Global Climate Change

38 Overview and Impacts on Arid Land Water Resources 975
38.1 Climate Variability and Change: An Overview 975
38.1.1 Natural Climate Changes: Cyclic, Abrupt and General Variation on a Large Scale 975
38.1.2 Natural Climatic Variations Affecting Global Human Populations 977
38.1.3 Climate Change Induced by Man's Activities 979
38.2 Effects of Global Climate Change on Hydrologic Systems 980
38.2.1 General Effects 980
38.2.2 Effects on Freshwater Systems and Uncertainty 982
38.2.3 Modeling of Global Climate Change with Emphasis on Semi-Arid and Arid Lands Precipitation 986
38.3 Global Climate Change: Impacts to Arid Land Surface-Water Hydrology 988

38.4 Global Climate Change: Impacts to Groundwater Systems. . . 990
38.5 Regional Variability in Surface Water and Groundwater Impacts 993
38.6 Impacts of Climate Change on Agriculture and Food Supply Security 994
References 996

39 Adaptation and Mitigation of Global Climate Change in Arid Lands 1001
39.1 Adaptive Strategies in Response to Climate Change 1001
39.2 Use of Stationarity or Non-Stationarity in Future Water Resources Planning and Management with Consideration of Global Warming 1005
39.3 Adaptive Water Management Strategies 1007
39.4 Increase in Water Supply by Development of New Water Resources: Supply Side Approach. 1008
39.4.1 Development of Additional Groundwater Resources 1009
39.4.2 Increased Storage (Reservoirs and Aquifer Storage and Recovery) 1010
39.4.3 Increase Amount of Water Reuse 1012
39.4.4 Desalination 1013
39.4.5 Rainwater Harvesting 1014
39.4.6 Water Transfers and Diversions 1014
39.5 Demand Reduction 1016
39.5.1 Conservation 1017
39.5.2 Reallocation of Existing Uses (Permanent and Temporary) 1018
39.5.3 Agriculture Water Use Changes and Water Trading 1019
39.5.4 Temporarily Forgoing a Use. 1020
39.5.5 Permanently Forgoing a Use 1021
39.6 Reduction and Control of Greenhouse Gas Emissions in the Future 1021
References 1022

Part X Conclusions

40 Conclusions 1027
40.1 Introduction 1027
40.2 Importance of Groundwater 1027
40.3 Water as an Undervalued Resource. 1028
40.4 Groundwater Quality. 1029
40.5 Groundwater Resource Assessments 1030

40.6 Data Management 1031
40.7 Demand Management 1032
40.8 Virtual Water Trade 1033
40.9 Integrated Water Resources Management 1033
40.10 Wastewater Reuse 1034
40.11 MAR for Wastewater Treatment 1036
40.12 Desalination 1037
40.13 Managed Aquifer Recharge 1038
40.14 Global Climate Change 1039
40.15 Adaptation to Climate Change 1040
40.16 Closing Thoughts 1041
References 1042

Curriculum Vitae 1043

Index 1047

Part I
Arid Regions Water Management and Issues

Chapter 1
Introduction

1.1 Introduction

There is growing recognition that water supply has become an ever more difficult global challenge. Water has been described as "*the defining crisis of the twenty-first century*" (Pearce 2006). Solomon (2010, p. 367) proclaimed that

> water scarcity is cleaving an explosive fault line between freshwater Haves and Have-Nots across the political, economic, and social global landscapes of the twenty-first century.

Water scarcity problems are of obvious national and economic security importance and are receiving great government and academic attention. Water scarcity issues are also entering the general public consciousness. Recently published mass-market books provide approachable discussions of current and potential future freshwater challenges (e.g., Postel 1992; Reisner 1999; De Villiers 1999; Glennon 2002, 2009; Pearce 2006; Solomon 2010).

It is important to recognize at the start of any discussion on water management in arid and semiarid lands is that water scarcity occurs on three main levels. First, and most critical, is the supply of water to meet basic human consumption and sanitary needs, which involves the smallest volume of water, but has the greatest human impacts when not available. Second, is the supply of water needed for local food and industrial production. Third, is water used for amenities that are considered basic elements of civilized life in developed countries.

Gleick (1996) proposed that a basic water requirement (BWR) of 50 liters (13.2 gal) per person per day (l/p/d) of clean water is needed to meet the four basic needs, including water for drinking, human hygiene, sanitary services, and food preparation. The BWR does not include the water required for growing food. Water consumption in western Europe and the United States is many times the BWR. Gleick (1996, 1998) proposed that the 50 l/p/d (13.2 gal/p/d) should be considered a fundamental human right. The World Health Organization (Howard and Bartram

R. Maliva and T. Missimer, *Arid Lands Water Evaluation and Management*,
Environmental Science and Engineering, DOI: 10.1007/978-3-642-29104-3_1,

2003) derived a minimum water requirement of 100 l/p/d (26.4 gal/p/d) and above as the optimum domestic water supply to promote good health.

The use of water for non-critical activities, such as public and private lawn and landscape irrigation, may seem trivial and frivolous when compared to potable water shortages elsewhere, but these issues still have to be addressed in the management of water resources in developed areas. Different regions face different water-scarcity challenges that are not directly related. The lack of clean water to meet basic needs in parts of the South Asia and Africa, for example, cannot be resolved by reducing lawn and golf course irrigation in Las Vegas, Nevada, or Phoenix, Arizona.

Water quality is also becoming an ever important issue in water-scarce regions. Limits on the capacity of water to attenuate contaminants have rarely been recognized, and consequently pollution of surface water and groundwater is a growing problem almost everywhere (Winpenny 1994). Freshwater resources are locally under threat from processes such as saline-water intrusion in coastal areas, agricultural return flows and soil salinization, industrial waste disposal and spills, and domestic and agricultural wastewater discharges.

Agriculture is the greatest global water use by a large margin, and water scarcity and food supply are inexorably linked. Irrigated agriculture constitutes about 70% of global water use (Shiklomanov 1993). Increased food production over the past several decades, the so-called Green Revolution, has been achieved in many areas through unsustainable groundwater pumping. Fedoroff et al. (2010) made a sobering observation that global population is widely projected to increase by 3 billion people by the middle of the 21st century, but the supply of arable land has not changed for more than a half-century. New land brought into cultivation has been offset by the losses of arable land through urbanization, desertification, and soil salinization. The challenges caused by increases in population are compounded by increases in prosperity. Demand for water is directly related to prosperity; as disposable income increases, water use increases (Price 2002). The increased water use with increasing prosperity is both direct and indirect. A major indirect increase in water consumption is related to a change in diet away from grains and towards more vegetables and meats, which require greater amounts of water to produce. There is thus a great need in agriculture for both increased land productivity and water productivity.

The Comprehensive Assessment of Water Management in Agriculture (2007) reported that a fifth of the world's population, more than 1.2 billion people, live in areas of physical water scarcity that do not have enough water to meet the minimum demands of the population. Although part of the water problem is due to a physical scarcity of water related to population growth, it was also concluded that the main reason for water problems lie elsewhere, in the management of water and to a (Comprehensive Assessment of Water Management in Agriculture, p. 10):

> lack of commitment to water and poverty, inadequate and inadequately targeted investment, insufficient human capacity, ineffective institutions, and poor governance.

Solomon (2010, p. 376) similarly observed that

> Despite its growing scarcity and preciousness to life, ironically, water is also man's most misgoverned, inefficiently allocated and profligately wasted natural resource.

A fundamental question is whether freshwater is truly scarce in a physical sense on a global scale (supply problem) or whether it is available, but should be used more productively (demand problem; Rijsberman 2005). In light of the small requirements for domestic use, lack of access to water for domestic use is typically not the result of water scarcity in the physical sense. The lack of access tends to be caused by poor water delivery service or because people in developing countries often do not have access to sufficient financial resources to avail themselves of the services (Rijsberman 2005).

A combination of population growth and the migration of people into urban areas have in many developing areas wiped out progress made in supplying clean water and sanitation (Gleick 1993). Urban water management is an important discipline in its own right. The focus of this book is on water resources management. Nevertheless, it is important to recognize that inadequate or non-existent water and wastewater treatment, transmission, and distribution infrastructure profoundly affect the quality of life in urban areas of developing countries. Mays (2002, 2009) provides excellent overviews of urban water supply issues.

The primary loci of water scarcity naturally lie within arid and semiarid lands, which by definition have limited water resources. Within arid lands, water supplies are highly variable and consumption is increasing. The increase in consumption is related to both increasing population and an increase in per capita consumption that is associated with increased prosperity. Many arid nations now consume more water than is replenished naturally within their borders. Large-scale water transfers sustain some arid regions, but it is clear that this is not a significant viable option for most areas in the future, because of politics, economics, and the unavailability of unallocated water for export. Winpenny (1994) observed that projects to increase water supply are tending to encounter hydrologic limits, face increasing costs, entail increasing environmental costs and risks to which the public is increasingly sensitive, and demand growing government subsidies.

Superimposed on the population-related increase in water demands are changes in water supply associated with global climate change. The need to regulate water supplies, to store available water during periods of excess, to optimize the use of existing resources, and to exploit and develop new resources is only too apparent (Agnew and Anderson 1992). Considerable uncertainty remains concerning how global climate change will impact local water resources. Local decreases in precipitation may have the combined deleterious effects on local waters resources by increasing irrigation requirements while also decreasing both surface water supplies and groundwater recharge.

Several quotes are germane to water resources management in arid and semiarid lands:

> The significant problems we face cannot be solved at the same level of thinking we were at when we created them (Albert Einstein).

> There's no such thing as a free lunch (Economist Milton Friedman).

> Have you got a problem? Do what you can where you are with what you've got. (U.S. President Theodore Roosevelt)

> No problem is so big or so complicated that it can't be run away from. (Cartoonist Charles Schulz)

Careful, prudent, and imaginative management is especially necessary in water-scarce lands in order to enable demands to be met with the best possible water within a reasonable economic framework (Pereira et al. 2002).

Groundwater management in semiarid and arid regions involves several types of issues:

- technical issues, which involve developing an understanding of the volume of available water resources, augmentation options, water quality, and the impacts of water use on the resource itself, surface-water and groundwater users, and dependent environments;
- socioeconomic and political issues, which are related to allocation of water resources between competing sectors of society, intergenerational equity with respect to the use of non-renewable resources, and decisions on the allocation of societal economic resources with respect to the development of alternative water sources;
- international political considerations with respect to the use of transboundary water resources.

A fundamental reality is that effective water management requires a sound technical understanding of water resources, but good science does not necessarily lead to sound water-management decisions. Societal dynamics are complex, and societal and political values and traditions are often discordant with optimal management of available resources. For example, some irrigation schemes are initiated and continued because of political and cultural support for local agriculture even though the schemes may be highly inefficient uses of water and cannot be justified on economic grounds. Changing water use patterns typically require institutional strengthening and coordination, and stakeholder acceptance, which involves education. Water users must understand and accept that current water use is unsustainable and that restrictions in water use or the implementation of conservation measures are in their personal and community's best interest.

The technological side of water management cannot be fully divorced from policy and economic issues. An overview is provided of water policy, social, and economic issues associated with the management of water resources in arid and semiarid regions (Chap. 32 through Chap. 34). It is recognized that difficult choices and trade-offs must be made between environmental, economic, social, and political objectives. The choices made are to a very large degree affected by local economic resources. For example, Arnold and Arnold (2009) observed that the developed world tends to overspend in the environmental area in order to achieve health benefits that are largely illusory and that in purchasing higher levels of water treatment or restricting the acceptable uses of reclaimed water, people may be paying for peace of mind as opposed to material health improvements. Scientists cannot provide a definitive answer to value-laden policy choices, but must endeavor to provide objective and accurate information on potential options and their consequences, upon which informed policy choices can be made.

The ultimate water-management challenge in arid and semiarid lands is to bring the supply and demand of water into a long-term balance, while also meeting social and environmental goals. The Comprehensive Assessment of Water Management in Agriculture (2007) recognizes the triple water management goals of ensuring food security, reducing poverty, and conserving ecosystems. It must also be recognized that water-scarcity problems exist on many levels. The most fundamental goals are to meet the basic minimum domestic requirements for human health and for growing food to meet local needs (subsistence and local commercial agriculture). Another major challenge is providing a safe and reliable water supply for growing urban areas. Water supplies are also needed for industrial activities, electrical power generation, mining, and large-scale agricultural operations. In more developed and prosperous arid areas, water resources are employed for municipal and residential landscape irrigation and for recreational facilities.

Achieving a balance between water supply and demand will require either increasing the water supply available to a region or decreasing the local water demand. Water demand can be decreased by more efficient use of water and curtailing wasteful activities. Solutions to local water scarcity will likely require a combination of supply-side and demand-side measures. The focus of this book is on the supply side of the equation, primarily of the optimization of the use of existing water resources. This focus in no way reflects a lack of appreciation of the importance of the demand side of the equation, which consists of conservation strategies. In arid regions every drop of water counts. Indeed, it is commonly much less expensive, but not always, to decrease demands through increased water use efficiency than to develop additional water supplies. Much has already been written on water conservation methods and strategies (e.g., Postel 1992; Vickers 2001; Pereira et al. 2002) and there is little new to add. Conservation strategies are summarized in Chap. 26.

Although the occurrence of water scarcity is recognized, the failure to treat water as a scare commodity lies at the heart of the problem (Winpenny 1994). There are three main reasons why water supply and consumption demand patterns are in many places failing to adapt to increasing water demands (Winpenny 1994):

- Water is underpriced relative to its cost of provision or is provided free of cost.
- Water is regarded as a public good and is thus available to all. As no one can be denied access to water, there is no incentive for private entities to invest in its preservation and improvement.
- There is a failure to adequately consider the costs of environmental externalities associated with supply water.

Water-policy decisions are inherently value laden and men and women of goodwill often have different priorities, perspectives, and self-interests with respect to the allocation and manner of use water. Once an issue is clearly and accurately presented and there is confidence in the source of information, people will tend to make responsible decisions with respect to water use, particularly if there is the perception that any sacrifices are shared.

Much has been written on water policy. That water is vital to life is a cliché. It is also becoming widely understood and accepted that improving the efficiency of water use and reducing waste and losses are the most affordable and practical solutions to water scarcity (Wiener 1977; Lazarova et al. 2000). Improving the efficiency of water use on a national level in countries and regions facing severe water scarcity must involve a comprehensive analysis and implementation program that considers all aspects of water management (Wiener 1977). In other words, an integrated water resources management approach needs to be taken (Chap. 35).

Water supply is a volatile political issue, both domestically and between nations. There is an oft quoted saying, commonly attributed to the writer Mark Twain, that

> whiskey is for drinking, water is for fighting over

Water-management plans cannot ignore the political and economic dimensions of the problem. Solutions that are not politically acceptable are non-starters. For example, it is recognized that the main water-resources challenge in arid lands stems to a large degree from increases in demand associated with population growth, which cannot be met in the long-term with "traditional" water supplies. Reducing the population or preventing migration of new people into arid regions such as the American Southwest, has been suggested as a solution (e.g., Glennon 2002), but the reality is that such a solution is politically unacceptable and will never be implemented (Rajagopal-Durbin and Durbin 2008). Populations will increase in arid and semiarid lands and real solutions must be found to meet the associated increase in water demands.

Falkenmark and Lindh (1993) proposed that cultures undergo three main development phases with respect to water resources, which was expanded upon by Turton (1999). The first phase is the pre-industrial phase in which water resources are abundant and the human influence on water resources is characterized by relatively few encroachments into river beds. Although water resources in arid lands cannot be considered to be abundant, they are sufficient to meet the basic needs of relatively small populations.

The second phase of water-resource development involves implementation of supply-side strategies, which are designed to increase the available supply of water. Supply-side strategies include the construction of dams and conveyances for the inter-basin transfer of water. The second phase may also involve the large-scale exploitation of groundwater resources, which are still abundant. Supply-side strategies are the easiest to implement and are politically popular. Turton (1999) used the water development history on the American West as an extreme example of supply-side water management. Supply-side water management can lead to a vicious circle wherein new supply results in increased demands, and thus the need to develop more supplies (Turton 1999). Eventually a point is reached where supply-side solutions are no longer feasible and societies are forced to resort to much less popular water-demand management.

Much of the warm arid and semiarid lands of the world are in the final phase of water-resources development in which all water that can be readily mobilized is being put to beneficial use. Further supplies can only be made accessible by employing non-conventional technologies. In practice, the solutions to water scarcity involve the following four main elements:

- optimization of the use of existing water supplies (e.g., increase storage, reuse of wastewater, better management)
- creation of new sources of water (e.g., desalination)
- increasing the efficiency of water use so that less is needed for a given activity
- reallocation of resources to the most efficient and valuable uses.

The first element of water demand management is the implementation of technological changes to increase water use efficiency (i.e., doing more with what you have) (Turton 1999). Water use efficiency is relatively easy to apply and is unlikely to cause disruptions. Technology to increase water use efficiency can be imported if it is not locally available.

The second element of water demand management is the reallocation of water between sectors, particularly from agriculture to the industrial and commercial sectors (i.e., doing other things with water) (Turton 1999). The benefits of reallocation can be huge, but the process can be highly stressful, as jobs and livelihoods are lost in the agricultural sector. The social stress can be reduced if farmers and agricultural workers can find jobs in other growing sectors of the economy. In order for management strategies to be successful, the perception of water needs to be shifted, which is no easy feat as perceptions are often culturally-bound (Turton 1999).

The transition in water-demand management to the reallocation stage requires a high level of adaptive capacity, which is defined as the capacity for managing natural resources stresses (Ohlsson 1999; Turton 1999). Societies with little adaptive capacity with respect to water do not have institutions in place that can materially impact water management. Adaptive capacity involves the following elements (Turton 1999):

- willingness and ability of people who are governed to accept water-demand management
- institutional and intellectual capital to innovate and find solutions
- ability of the regime to generate water demand management strategies that are perceived as being both reasonable and legitimate by those being governed.

The adaptive-capacity requirements to address water scarcity are related to the magnitude of the problem. Clearly, some societies have more resources with which to work, allowing for more options and opportunities to find solutions. Also, the causes of water shortages (such as population growth and climate change) may be outside of the purview of water managers, which necessitates a broader societal approach to the problem.

A scale-issue exists in water management between what might be referred to as large-scale and small-scale technologies, or as Gleick (2002) described, "hard water" versus "soft water". Small-scale technologies are techniques or methods that could be used on a household or village scale to secure a source of water or to improve overall water productivity. Many small-scale technologies are simple water-management methods, such as rainwater harvesting, that have been employed in arid lands for millennia, and more efficient irrigation practices (e.g., drip irrigation). The importance of small technologies to global water-resources management should not be underestimated. One cistern will have a small impact on a societal water supply, but thousands or millions of cisterns can store a huge amount of water that might otherwise be lost to evaporation. The value of small-scale technologies is increasingly being recognized in both rural and urban regions. Rainwater harvesting (Chap. 22) has been widely embraced in rural semiarid areas of rural India and China as means to secure drinking water and irrigation water supplies. Rainwater harvesting is also being turned to in water-short urban areas. For example, the city of Tucson, located in the Sonoran Desert of Arizona, passed an ordinance in 2009 requiring new commercial buildings to have cisterns to store water to meet some of their irrigation water needs.

Inasmuch as the global population is increasingly becoming urbanized, large-scale centralized technologies are still needed to provide and manage water supplies. Urban populations require large-scale water technology to provide water that is used directly for domestic purposes and indirectly used for the production of food. Alternative water-supply development in urban areas of arid regions also increasingly involves large-scale technologies, such as desalination. The rate of urban population growth in many developing countries is unprecedented in history and there are, thus, no experiences that local governments and utilities can draw upon for the development water supply and wastewater treatment infrastructure to accommodate the growth (Falkenmark and Lindh 1993).

Water resources are divided between endogenous and exogenous supplies. Endogenous water supplies are locally derived from precipitation falling within a region. Exogenous water originates in neighboring countries and enters a region through aquifers and rivers. Management of exogenous resources is becoming increasingly contentious. Inasmuch as aquifers may extended across a broad region and may cross multiple international borders, their management increasing involves large-scale technology, including sophisticated groundwater modeling, water level and quality monitoring, and artificial recharge techniques.

The subject of water-resources management and development in arid regions is very broad and includes numerous individual topics for which separate books can and have been written (e.g., Agnew and Anderson 1992; Lowi 1993; Allan 2001;

Koundouri et al. 2006; Schneir-Madanes and Courel 2010). A single volume on the subject, therefore, necessarily must be restricted to abbreviated discussions of topics. This book is focused to a large degree on groundwater management, because in most arid and semiarid lands surface water resources are either not a significant water source or are already fully utilized. As a gross simplification, surface water use tends to be self-limiting. The amount of water that can be withdrawn from a river is largely dictated by current water flow and reservoir storage capacity. Groundwater resources are both a blessing and challenge in arid and semiarid lands because of the large storage capacities of aquifers and potential for over-exploitation. Groundwater abstraction in many arid and semiarid lands now greatly exceeds the rate of recharge. The over-exploitation allows for needed socioeconomic development, but cannot be sustained over the long term. Aquifers are increasingly also being used to provide storage and treatment for excess surface water and wastewater through managed aquifer recharge (Chap. 23). The critical water-management challenge facing many arid and semiarid lands is determining how to optimize the use of and protect groundwater resources in order to obtain the maximum long-term societal value from them.

The focus of this book is on the technologies now available for the management of water resources in arid and semiarid lands. The focus is also on groundwater resources which constitute over 99% of all liquid freshwater on earth, although most is in deep aquifers and is non-renewable (Jackson et al. 2001). Surface water is considered to the extent that it is linked to water management and groundwater systems. This book is not meant to be a "how to book" that provides step-by-step instructions on water-resources management technologies. Instead the objective is to provide an overview of the technologies now available that have applications for water-resources management and development in arid and semiarid regions. The goal is to provide the basis or theoretical background for the technologies, the type of information they can provide, and the limitations of the technologies. Water-resources development has truly become a team effort in which the project team must include specialized expects in the various disciplines involved. Professionals involved in the management of water resources projects need to be aware of and have an understanding of the different "tools that are in the box" and ensure that the project team includes professionals who are truly experts in their use.

1.2 Global Water Budget and Water Stress

Global freshwater occurrence was reviewed by Shiklomanov (1993). The total volume of freshwater in the hydrosphere is approximately 35 million km^3, which corresponds to about 2.5% of total volume of water in the hydrosphere. The distribution of freshwater is summarized in Table 1.1. Approximately 99% of all liquid freshwater is present in aquifers and greater than 75% of the groundwater is non-renewable.

Table 1.1 Distribution of freshwater in hydrosphere (after Shiklomanov 1993)

Water type	Volume (km^3)	Percent total freshwater	Percent liquid freshwater
Glaciers and permanent snow cover	24,064	68.70	–
Groundwater	10,530	30.06	98.73
Lakes	91	0.26	0.85
Atmospheric water	12.9	0.04	0.12
Ground ice/permafrost	300	0.86	–
Swamp water, river flow, biological water, and soil moisture	31.21	0.09	0.29
Total	*35,029*	*100*	*100*

The volume of freshwater present in each reservoir is only part of the story. Water is cycled and a critical issue is the amount of "new" freshwater is available for human use within the hydrologic cycle. The global freshwater cycle was reviewed by Shiklomanov (1993), Postel et al. (1996), and Jackson et al. (2001). Only water flowing through the solar-powered hydrologic cycle is renewable. The annual net surface freshwater flux (i.e., run-off) provided by the solar-powered hydrologic cycle, which is essentially precipitation on land minus evapotranspiration, is approximately 40,000 km^3/yr. The net surface freshwater flux is several times greater than the amount that is needed to sustain the world's current population. Human water use also includes the amount of the total ET on land ($\approx$70,000 km^3/yr) appropriated in crop lands, grazing lands, and trees harvested for fuel and timber. Appropriated ET involves utilization of soil moisture, which is a renewable resource, and is involved in the "virtual" water trade (Sect. 34.4; Allan 2001). Postel et al. (1996) estimated that humanity uses about 26% of the terrestrial ET.

Only 12,500 km^3/yr (31%) of the approximately 40,000 km^3/yr of total annual run-off was estimated by Postel et al. (1996) to be accessible both geographically and temporally. An estimated 19% of the run-off is geographically inaccessible because of remoteness from population centers, such as the flows of the Amazon, Congo-Zaire and northern undeveloped rivers. Temporally inaccessible flow includes floods and monsoon flows that are not captured by dams. Postel et al. (1996) calculated that humans use approximately 54% of the 12,500 km^3/yr of accessible flow, which presumably increased over the 14 years since their paper was published. The principal means for expanding the accessible flows are to store more flood-run off or desalinate seawater (Postel et al. 1996). Desalination is used mainly for developing drinking water supplies in water scarce and energy-rich or affluent countries.

On a global scale the amount of renewable freshwater resources greatly exceeds current and anticipated future withdrawals (Oki and Kanae 2006). The basic water-resources problems are caused by the high variability of water resources in time and space. The distribution of water is not well matched geographically and temporally to human needs (Postel et al. 1996). Movement of large volumes of

Table 1.2 Water stress indicators

	No stress	Low stress	Medium stress	High stress
Falkenmark index (m^3 per capita)	–	>1,700	1,700–1,000	<1,000
Relative water stress index	<0.1	0.1–0.2	0.2–0.4	>0.4

water to water-stressed areas is generally only feasible when gravity can be used (Oki and Kanae 2006). The water demands for ecosystems and navigation also need to be met.

Several quantitative indices have been proposed as means of evaluating the adequacy of the water supply of a nation to meet its needs. The indices include or consider (Raskin et al. 1997):

- use-to-resource ratio, which gauges the average overall pressure on water resources and threats to aquatic ecosystems
- coefficient of variation of precipitation, which is a measure of hydrological fluctuation
- storage-to-flow ratio, which is an indicator of the capacity of infrastructure to manage fluctuating water supplies
- import dependence, which is an index of reliance on inflows from contiguous countries.
- socioeconomic coping capacity index (e.g., average per capita income), which represents the ability of a society to endure emerging water problems and uncertainties.

Two of the most widely used water stress indicators are provided in Table 1.2. The Falkenmark indicator relates water resources to population (Falkenmark et al. 1989). It is the most widely used indicator because of its conceptual simplicity and ease of application. Water stress or water scarcity is considered to occur when the annual water available is less than 1,700 m^3/yr (448,800 gal/yr) per capita and severe water stress is considered to occur with the water supply falls below 1,000 m^3/yr (264,000 gal/yr) per capita.

The relative water stress index (RWSI) relates average annual water use (withdrawals; W) to the total renewable water supply (W_R).

$$RWSI = \frac{W}{W_R} \tag{1.1}$$

The RWSI can also be corrected for supplemental water (W_S) obtained from desalination, which is a water use that does not involve use of renewable water

$$RWSI = \frac{W - W_S}{W_R} \tag{1.2}$$

Withdrawals greater than 0.2 of the renewable water supply are considered to be indicative of water stress.

Evaluation of national vulnerability to water stress requires consideration of factors that are not incorporated into the basic water- stress indices, including (Rijsberman 2005; Arnell 1999; Raskin et al. 1997):

- national averages may hide serious water scarcity at smaller scales
- resource accessibility
- indicators do not take in account the availability of infrastructure that modifies the availability of water to users
- simple thresholds do not reflect important variations among countries due to factors such as climate and customs, and the efficiency of water use
- resource reliability—in arid and semiarid lands only a small percentage of annual runoff may be accessible because flows are concentrated in a short time
- capability to cope with changes and stresses.

Llamas (2004) noted that ratios of water resources to population, such as the Falkenmark Index, have little practical application and are misleading as most water problems are related to quality and not to relative abundance. Furthermore, a significant number of regions (e.g., Israel and parts of Spain) have a high economic and social standard of living despite water scarcity (<500 m^3/yr [132,000 gal/yr] per capita). Additional water stress indicators have been proposed that attempt to incorporate some of the these factors, but have the disadvantages of complexity, reliance on expert judgment, and not being intuitive, and thus not understandable to non-experts (Rijsberman 2005).

1.3 Technical Issues

A fundamental requirement for effective water-resources management in arid and semiarid lands is having an accurate, quantitative knowledge of the quantity, quality, and hydrogeology of all local water resources. The technical issues determine the water resources that that are potentially available to a society for its beneficial use and the impacts associated with their use. Socioeconomic considerations dictate how water resources are utilized or not utilized. Technical issues dictate what options are available to water managers, while socioeconomic issues dictate which options may exist and ultimately are chosen by those responsible for water management. It is important to appreciate that technical and socioeconomic issues are closely intertwined, because human activities profoundly impact water resources in many direct and indirect manners.

The technical water management issues broadly include all the factors that affect the physical availability of water and physiochemical responses of hydrologic systems to water use. The latter can, in turn, impact biological systems. Key technical water management parameters and issues include:

- geographic extent, storage volume, hydraulic properties, and water quality of local aquifers,
- aquifer water balance and recharge rate,

- stream (wadi) rainfall, runoff, and recharge relationships,
- effects of groundwater abstraction (e.g., saline-water intrusion, land subsidence, reductions in spring and stream flow),
- managed aquifer recharge options and associated changes in water quality,
- alternative water supply options (e.g., wastewater reuse and desalination) and
- impacts of disposal of water treatment byproducts (e.g., desalination concentrate).

1.4 Socioeconomic and Political Issues

Socioeconomic and political issues essentially involve the human element of water management and water policy. Where any resource is scarce relative to demands, decisions have to be made as to how the resource is allocated between competing entities. Natural environments are also a competing entity, which may be either ignored or given a very high priority depending upon local circumstances. Water must be allocated between sectors in a society (e.g., agriculture, industry, and municipal) and between individuals in each sector (e.g., individual farmers or landowners). Thus, basic decisions must be made as to:

- Who decides as to how water is allocated?
- How are water allocation decisions made?
- What are the priorities for water allocation?
- Who pays for the water-supply and management infrastructure and for the use of water?

Water supply and management must also compete for financial resources with other sectors of society. Inasmuch as financial resources are finite, allocation decisions approach a zero-sum game. For example, the decision to invest more money in water-supply infrastructure may result in less money being available for other sectors of society, such as transportation and education.

There are two end-member approaches for water management, which are government command and control and a free market. Under a command and control system, a government department or regulatory agency sets water policy, which is typically implemented through some type of permitting or authorization process. Under a free-market approach, water is considered to be an economic good and its use is determined through market processes. A third end-member is anarchy, in which there is essentially no control over water use. In the absence of controls over water use, a tragedy of the commons scenario (Sect. 32.2) may occur where competition for the use of a resource for personal gain eventually destroys it.

In most areas of the world, freshwater use is managed under some type of command and control system, in which a governmental regulatory agency has the authority to regulate water use. There are considerable differences between countries and states concerning how much discretion regulatory agencies have in the management of water. The rights of water users may be clearly established in law, or water managers may be given wide latitude of independent action. The enforcement authority and coercive powers that regulatory agencies have with respect to ensuring compliance with water use rules and regulations also vary considerably. Some countries lack the human and financial resources to effectively manage water use.

Ideally, governmental water policy should be in harmony with the values of most of society. Water policy should also be in harmony with hydrologic reality. The fundamental challenge facing water managers is achieving an acceptable balance between competing water demands (e.g., domestic, agricultural, industrial, and environmental), while preventing unacceptable harm to water resources and other adverse impacts. Water management becomes even more complex for transboundary aquifers and surface water bodies, particularly if there is a state of distrust or hostility between the nations involved.

Although it has been proposed that treating water purely as an economic commodity would increase the economic efficiency of its use, it is clear that there are numerous social factors that necessarily must enter water-management decisions. At the most basic level, water is essential for life, and very few would argue that people should be denied water to drink because they cannot afford to pay for it. Domestic food production in many nations is considered a national security issue, even though it may be a greatly inefficient use of water from an economic perspective.

Water-allocation decisions are ultimately value decisions. There are often no right or wrong decisions concerning socioeconomic issues, unlike technical issues, which are ultimately dictated by the physical realities of the hydrologic system. For example, abstraction from a non-renewable groundwater supply will result in the technical response of a decrease in amount of water in storage and a decline in water levels and perhaps some physical damage, such as land subsidence (Chap. 36). The socioeconomic questions include whether or not the abstraction is providing commensurate social and economic benefits relative to the value of the resource. Another difficult question is whether it is in the long-term interest of a country to use the water now rather than to reserve it for later use when the need may be more severe (i.e., balancing current needs against intergenerational equity). There is not a clear-cut answer to most socioeconomic questions.

The objective of this book with respect to water policy issues is to present the questions, and avoid interjecting the authors' personal values by proffering answers. Returning to the example of use of non-renewable groundwater resources, the important matter is that the parties involved in the decision-making process have a full understanding of the implications of various groundwater development options so that an informed decision can be made. Water policy decisions need to be based on sound science.

1.5 Appropriateness, Values, and Priorities

Recurring questions in water management, and virtually all other areas of society, is prioritization or 'right-sizing' the use of resources and management actions. Where water is scarce, priorities must be set so that the uses of limited water resources are in accordance with societal values. Prioritization decisions can be made through command-and-control systems or through the market place. The latter implicitly involves the decision that water use will be prioritized and allocated based on economic values.

A basic societal goal should be to provide everyone with an adequate supply of safe water that is obtainable in a sustainable manner without any adverse impacts. However, water management does not occur in isolation and necessarily must be performed within the constraints of societal economic, social, and human resources. There are opportunity costs associated with all economic decisions, which are often not given adequate consideration. For example, money invested in upgrading water treatment facilities may yield positive public health and associated economic benefits, but it is possible that greater public health benefits could be achieved through investing the resources in better nutrition or medical facilities. Or the opposite may be true in some circumstances, in that it may be far less expensive to improve water treatment, than to treat or bear the societal and economic costs of waterborne diseases. Very difficult decisions often need to be made on the allocation of finite societal resources.

It is critical to be cognizant that societies have different expectations and risk tolerances concerning water-supply issues and financial resources to invest in water supply and treatment infrastructure. Water and food are so fundamental and intimate to human life, that there may be hypersensitivity over any safety issues, particularly in the more developed countries. For example, the presence of trace quantities of pharmaceutical and endocrine disrupting compounds (compounds of emerging concerns) has resulted in a plethora of alarmist popular press articles and an explosive increase in research, despite the fact that there is no clear direct linkage between their presence at their detected concentrations and any human health risk. Indeed, human exposure may be much greater from the food and air pathways (Stanford et al. 2010). While very high levels of treatment to reduce the concentrations of CECs of wastewater used for aquifer recharge may be socially required for peace of mind in an affluent area, such as California (USA), such treatment would be a misallocation of resources in developing countries where much more serious health risks exist.

The focus of the book is largely on the technologies available to assess, manage, and develop water resources. The technologies and tools available to water managers have increased dramatically over the past several decades and a decision has to be frequently made between what is possible, what is necessary, and what is superfluous, and choosing between the state-of-the-art options versus what is good enough (i.e., acceptable but not optimal). The research for this book reinforced the authors' observation that there is a sizeable dichotomy between academic hydrological research and applied water management. Research that will attract grant funding and be publishable tends to focus on the development and application of new technologies and analytical tools. However, water managers and suppliers do not have unlimited budgets, often operate under tight time constraints, and may not have the human and technical resources to apply the latest state-of-the-art technology. These technical limitations are typically much more severe in developing countries. Simple technologies are not necessarily inferior to advanced technologies, particularly when the latter are unaffordable. Advanced technologies need to provide real benefits that are demonstrably commensurate with the costs of their application.

For example, groundwater modeling is playing an ever increasing role in water-resources management, while at the same time in some areas groundwater monitoring has decreased dramatically (Vörösmarty and Sahagian 2000; Alley 2006). Developing sophisticated and expensive integrated surface-water and ground-water models incorporating remote sensing and geographic information system (GIS) technology may be the current technical state of the art, but it is certainly an open question whether water resources would be better managed by improving the amount and quality of monitoring data so as to provide more hard data concerning what is actual occurring. Again, the question arises as to what is the most effective investment of finite financial resources toward achieving water-management goals. Priorities must be set to maximize the value from existing financial and human resources.

An additional water-management consideration is between efficiency and flexibility. Water use efficiency can be defined as the ratio between the amount of water required for a given activity versus the amount actually used, delivered, or abstracted from the water source (Sect. 26.3.3). Efficiency in water use is important, but Pereira et al. (2002, p. 50) pointed out a risk associated with highly efficient systems.

> in times of a drought a wasteful or inefficient water use system can often,with considerable effort, be temporarily improved to allow greater use to be made of the limited water available. On the other hand a totally water-use efficient system may contain no flexibility for temporarily increasing supplies or for surviving with a lesser supply and may fail completely.

Flexibility in water- supply system design is increasingly being recognized as an adaptation against the uncertainties associated with climate change (Chap. 39). Important elements of water-system flexibility include increasing the diversity of supplies through the development of alternative water supplies and increasing storage through managed aquifer-recharge technologies (Chap. 23).

Effective arid-lands water management requires a great diversity of knowledge of the various tools and technologies available, and their capabilities and limitations (i.e., appropriateness for the problem at hand). For example, quantification of aquifer recharge is a critical element of the management of groundwater resources in arid regions. Those charged with evaluating and managing groundwater resources need to be aware of the various techniques available for measuring aquifer recharge, and developing a program that obtains the best possible data under existing financial constraints. On a broader scale, water managers need to be aware of all the various alternative water-supply options that are available and appropriate in their region and be able to select the option or options that can most cost-effectively supply additional water at a quality suitable for the intended use.

It is clear that there is no single, universal solution to dealing with water scarcity. Solutions will vary between nations, and also evolve within a country as it economically develops. The great challenge facing those responsible for water management is setting priorities and developing water supply solutions that are technically sound, affordable, and compatible with local social, cultural, and political values.

References

Agnew, C., & Anderson, W. (1992). *Water in the arid realm*. London: Routledge.

Allan, J. A. (2001). *The middle east water question hydropolitics and the global economy*. London: I.B. Tauris.

Alley, W. M. (2006). Tracking U.S. groundwater reserves for the future. *Environment, 48*(3), 10–25.

Arnell, N. W. (1999). Climate change and global water resources. *Global Environmental Change, 9*, 532–549.

Arnold, R. G., & Arnold, K. P. (2009). Integrated urban water management in the Tucson, Arizona metropolitan area. In L. M. Mays (Ed.), *Integrated urban water management: Arid and semi-arid regions* (pp. 113–138). Paris: UNESCO, Leiden, Taylor & Francis.

Comprehensive Assessment of Water Management in Agriculture. (2007). *Water for food, water for life: A comprehensive assessment of water management in agriculture*. London: Earthscan and Colombo, International Water Management Institute.

De Villiers, M. (1999). *Water Fate of our most precious resource*. Toronto: Stoddart.

Falkenmark, M., & Lindh, G. (1993). Water and economic development. In Gleick P. H. (Ed.), *Water in crisis, a guide to the world's freshwater resources* (pp. 80–91). New York: Oxford University Press.

Falkenmark, M., Lundqvist, J., & Widstrand, C. (1989). Macro-scale water scarcity requires micro-scale approaches: Aspects of vulnerability in semi-arid development. *Natural Resources Forum, 13*, 258–267.

Fedoroff, N. V., Battisti, D. S., Beachy, R. N., Cooper, P. J. M., Fischhoff, D. A., Hodges, C. N., et al. (2010). Radically rethinking agriculture for the 21st century. *Science, 327*, 833–834.

Gleick, P. H. (1993). Water in the 21st century. In P. H. Gleick (Ed.), *Water in crisis, a guide to the world's freshwater resources* (pp. 105–113). New York: Oxford University Press.

Gleick, P. H. (1996). Basic water requirements for human activities: Meeting basic needs. *Water International, 21*, 83–92.

Gleick, P. J. (1998). The human right to water. *Water Policy, 1*, 487–503.

Gleick, P. H. (2002). Soft water paths. *Nature, 418*, 373.

Glennon, R. J. (2002). *Water follies: Groundwater pumping and the fate of America's fresh waters*. Washington, DC: Island Press.

Glennon, R. (2009). *Unquenchable America's water crisis and what to do about it*. Washington, DC: Island Press.

Howard, G., & Bartram, J. (2003) Domestic water quality, service level and health: World Health Organization document WHO/SDE/WSH/03.02.

Jackson, R. B., Carpenter, S. R., Dahm, C. N., McKnight, D. M., Naiman, R. J., Postel, S. L., et al. (2001). Water in a changing world. *Ecological Applications, 11*, 1027–1045.

Koundouri, P., Karousakis, K., Assimacopoulous, K., Jeffrey, D., & Lange, M. A. (2006). *Water management in arid and semi-arid regions: Interdisciplinary perspectives*. Cheltenham: Edward Elgar.

Lazarova, V., Cirelli, G., Jeffrey, P., Salgot, M., Icekson, N., & Brissaud, F. (2000). Enhancement of integrated water management and reuse in Europe and the middle east. *Water Science and Technology, 42*(1–2), 193–202.

Llamas, M.R. (2004). Water and ethics. Use of groundwater: UNESCO International Hydrogeological Programme.

Lowi, M. R. (1993). *Water and power*. Cambridge: Cambridge University Press.

Mays, L. W. (Ed.). (2002). *Urban water supply handbook*. New York: McGraw-Hill.

Mays, L.W (Ed.). (2009). *Integrated urban water management: Arid and semi-arid regions*. Paris: UNESCO, and Leiden, Taylor & Francis.

Ohlsson, L. (1999). *Environment, scarcity and conflict: A study of malthusian concerns*. Göteborg: Department of Peace and Development Research, University of Göteborg.

Oki, T., & Kanae, S. (2006). Global hydrological cycles and world water resources. *Science, 313*, 1068–1072.

Pearce, F. (2006). *When rivers run dry*. Massachusetts: Beacon Press.

Pereira, L.S., Cordery, I., & Iaconides, I. (2002). Coping with water scarcity: UNESCO, International Hydrological Programme, IHP-VI, Technical Documents in Hydrology No. 58, p. 269.

Postel, S. (1992). *Last oasis: Facing water scarcity*. New York: WW Norton.

Postel, S. L., Daily, G. C., & Ehrlich, P. R. (1996). Human appropriation of renewable freshwater. *Science, 271*, 785–877.

Price M. (2002). Who needs sustainability? In K. M. Hissock, M. O. Rivett & R. M. Davison (Eds.), *Sustainable groundwater development* (pp. 75–81): London: Geological Society, Special Publications 193.

Rajagopal-Durbin, A., & Durbin, T. J. (2008). Wells are not always water follies: Sustainable groundwater policies for the American West. *Water Policy, 10*(2), 145–164.

Raskin, P., Gleick, P., Kirshen, P., Pontius, G., & Strzepek, K. (1997). *Water futures: Assessment of long-range patterns and problems background document to the comprehensive assessment of the freshwater resources of the world report*. Stockholm: Stockholm Environment Institute.

Reisner, M. (1999). *Cadillac desert: The American West and its disappearing water* (2nd ed.). New York: Penguin.

Rijsberman, F. R. (2005). Water scarcity: Fact or fission? *Agricultural Water Management, 80*(1–3), 5–23.

Schneir-Madanes, G., & Courel, M. F. (2010). *Water and sustainability in arid regions bridging the gap between physical and social sciences*. Dordrecht: Springer.

Shiklomanov, I. A. (1993). World Freshwater Resources. In P. H. Gleick (Ed.), *Water in crisis: A guide to the world's freshwater resources* (pp. 13–24). Oxford: Oxford University Press.

Solomon, S. (2010). *Water: The epic struggle for wealth, power and civilization*. New York: Harper Collins.

Stanford, B. D., Snyder, S. A., Trenholm, R. A., Holandy, J. C., & Vaderford, B. J. (2010). Estrogenic activity of US drinking waters: A relative exposure comparison. *Journal American Water Works Association, 120*(11), 55–65.

Turton A.R. (1999). *Water scarcity and social adaptive capacity: Towards an understanding of the social dynamics of water demand management in developing countries*. MEWRE Occasional Paper No. 9, School of Oriental and African Studies, University of London, England.

Vickers, A. (2001). *Handbook of water use and conservation*. Amherst: WaterPlow Press.

Vörösmarty, C. J., & Sahagian, D. (2000). Anthropogenic disturbance of the terrestrial water cycle. *BioScience, 50*, 753–765.

Wiener, A. (1977). Coping with water deficiency in arid and semiarid countries through high-efficiency water management. *Ambio, 6*(1), 77–82.

Winpenny, J. (1994). *Managing water as an economic resource*. London: Routledge.

Chapter 2
Aridity and Drought

2.1 Definition of Aridity

Aridity is a term that most people conceptually understand, and it evokes images of dry, desert lands with sparse natural surface-water bodies and rainfall, and commonly only scant vegetation, which is adapted to a paucity of water. Aridity has a wide variety of landscape manifestations (Fig. 2.1) including barren rock hills and plains, sand dune fields, and vegetation dominated by cacti and other xerophytic plants. Despite their usual parched appearance, deserts can become remarkably verdant when rain does fall (Fig. 2.1d). The primary focus of this book is largely on arid and semiarid lands with warm climates in which water scarcity is more severe because of greater populations and associated water use. However, aridity also occurs in regions with cold climates in which precipitation falls mainly as snow. Much of Antarctica and the Arctic slope of Alaska are considered to be polar deserts because they receive little net precipitation each year (Smiley and Zumberge 1971).

Non-technical dictionaries typically define the word arid with respect to climate in terms such as barren, parched, dry or without moisture, or as having insufficient water, rainfall or precipitation to support vegetation or agriculture. However, it is difficult to define "arid" precisely and delineate the boundary between different degrees or levels of aridity or its opposite, humidity. For example, what is the threshold for an area to be considered arid instead of semiarid and what are the specific landscape and water resources manifestations of an arid versus semiarid climate?

Numerous climatic and biological classifications have been proposed for arid lands (McGinnies 1988). Aridity, as defined by the shortage of moisture, is essentially a climatic phenomenon that is based on average climatic conditions over a region (Agnew and Anderson 1992). A fundamental distinction exists between aridity, which is a long-term climatic phenomenon and droughts, which are a temporary phenomenon (water deficit). Thornthwaite (1948) recognized that

R. Maliva and T. Missimer, *Arid Lands Water Evaluation and Management*,
Environmental Science and Engineering, DOI: 10.1007/978-3-642-29104-3_2,

Fig. 2.1 Desert landscapes. **a** Rocky desert in southwestern Jordan (photo by Weixing Guo). **b** Sand dune field in Abu Dhabi, United Arab Emirates. **c** East coast of Aruba, near Noord. **d** Sonoran Desert near Phoenix, Arizona, USA, after rains

whether a climate is moist or dry is dependent upon both the amount of precipitation and whether or not the precipitation is greater or less than the water needed to offset evaporation and plant transpiration, which are collectively termed evapotranspiration (ET). In other words, aridity is a function of both precipitation and the potential evapotranspiration rate (ET_p). An additional factor effecting aridity is temperature and the annual timing of precipitation. Rainfall during cold seasons is more effective in areas with a sufficiently high temperature for plant growth, because less water is lost to direct evapotranspiration during cold periods than in the hot season (Walton 1969).

Potential evapotranspiration is the amount of transfer of water to the atmosphere that would be possible under ideal conditions of soil moisture and vegetation (Thornthwaite 1948). Penman (1948, 1956) more precisely defined potential evapotranspiration as "the amount of water transpired in a given time by a short green crop, completely shading the ground, of *uniform* height and with adequate water status in the soil profile". A critical issue for arid region water resources is the difference between ET_p and the actual rate of evapotranspiration (ET_a). Potential evaporation rates in some arid regions (e.g., Arabian Peninsula) may exceed 4 m (13.1 ft) per year. However, the actual rate of evapotranspiration in land areas depends upon the amount of moisture available. A characteristic feature of warm arid land areas is that ET_a rates are actually very low because of the paucity of water that is available for transfer to the atmosphere. On the contrary, the ET_a rates of surface water bodies in arid lands are very high, approaching ET_p values. The high ET_a rates results in a loss of water and the concentration of salts.

2.2 Aridity Indices

Numerous numerical indices have been proposed to quantify the degree of dryness of a climate at a given location, and thus define climatic zones. Aridity indices were reviewed by Walton (1969) and Stadler (2005). There is a lack of agreement over the approaches used to delineate the exact boundaries between lands having different levels of aridity, although there is an agreement over the general location of arid regions (Agnew and Anderson 1992). Aridity indices inherently include an element of circularity in that they are calibrated against known aridity patterns. For example, the Atacama Desert of northern Chile is widely recognized to be the driest desert in the world. Hence, the numerical thresholds for the extremely arid or hyperaridity category for aridity indices are based on the values of the indices for the Atacama Desert and similar areas.

Several commonly used aridity indices are discussed herein to illustrate basic concepts. From a practical water management perspective, aridity indices do not have much relevance. It is typically self-evident that an arid region under investigation is indeed arid. Nor is there a strong reason to prefer one index to another with respect to water management. Aridity indices have greater value for the tracking the effects of climate change on local water resources, if sufficiently accurate data are available for mapping local changes in the values of the indices over time.

The simplest aridity index is based solely on precipitation. A commonly used rainfall-based definition is that an arid region receives less than 10-in or 250 mm of precipitation per year. This criterion for aridity was used by the Intergovernmental Panel on Climate Change (IPCC 2007). Semiarid regions are commonly defined by annual rainfalls between 10 and 20-in (250 and 500 mm).

The UNESCO (1979) aridity index (AI) is based on the ratio of annual precipitation (P) and potential evapotranspiration rates as follows:

$$AI = \frac{P}{ET_p} \tag{2.1}$$

where, ET_p is calculated using the Penman (1948) formula. The UNESCO system is attractive in that it is conceptually and operationally simple and is based solely on the two main parameters that define aridity. Warm arid regions have low P and high ET_p rates and thus very low AI values. UNESCO (1979) proposed a classification of climate zones based on AI index, in which arid regions are defined by an index of less than 0.20 (Table 2.1). Alternative versions of the classification use an AI value for 0.05 for the boundary between hyperarid and arid regions.

Table 2.1 UNESCO (1979) Aridity classification

Classification	Aridity index
Hyperarid	AI < 0.03
Arid	0.03 < AI < 0.20
Semi-arid	0.20 < AI < 0.50
Dry subhumid	0.50 < AI < 0.65

Data on local ET_p rates may not be available, which constrains the use of the UNESCO index. The De Martonne (1926) aridity index (A_m) instead uses temperature as a proxy for ET_p as follows:

$$A_m = \frac{P}{T + 10} \tag{2.2}$$

where, P (cm) is the annual precipitation and T (°C) the annual mean temperature. The equation is appropriate for temperatures greater than –9.9°C. The De Martonne aridity index decreases (approaches zero) with increasing aridity.

The Thornthwaite (1948) classification scheme, and variations thereof, is also commonly used to map the distribution of non-polar dry lands. Meigs (1953) prepared a map of the distribution of non-polar arid regions for UNESCO based on the Thornthwaite classification, which still widely cited (Fig. 2.2). Thornthwaite (1948) defined indices of humidity (I_h) and aridity (I_a) as follows:

$$I_h = \frac{100\,s}{n} \tag{2.3}$$

$$I_a = \frac{100\,d}{n} \tag{2.4}$$

where,

s = surplus water, which is defined as the sum of the monthly difference between precipitation (P) and ET_p for those months when P exceeds ET_P (cm)

d = water deficiency, which is defined as the sum of the monthly difference between ET_P and precipitation (P) and for those months when ET_p exceeds P (cm)

n = water need, which is the sum of monthly values of potential evapotranspiration for the surplus or deficiency months (cm).

Thornthwaite (1948) then defined the moisture index (I_m) as follows:

$$I_m = I_h - 0.6 I_a \quad or \quad \frac{100\,s - 60\,d}{n} \tag{2.5}$$

Thornthwaite (1948) considered the semiarid climate type to have a moisture index between –20 and –40 and the arid climate type to have a moisture index between –40 and –60.

It must be emphasized that the exact numerical cut-offs between the different climate classes are arbitrary lines drawn in a climatic continuum and that they do not represent a fundamental climatic change. For example, the climatic different

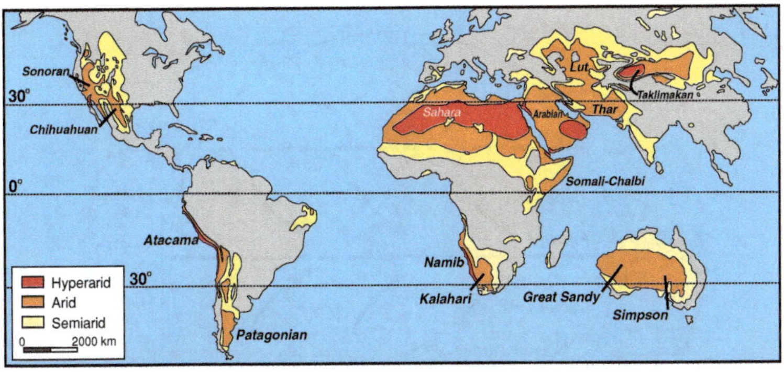

Fig. 2.2 Map of distribution of non-polar arid lands (after Meigs 1953). *Source* http://pubs.usgs.gov/gip/deserts/what/world.html

between areas with UNESCO aridity indices of 0.19 and 0.21 would not be significant yet they would be assigned to different classifications (arid and semiarid, respectively).

2.3 Causes of Aridity

Rainfall occurs when moist air is cooled through either the process of ascent, mixing, radiation cooling, or contact cooling. The processes leading to aridity tend to prevent such cooling through maintaining air stability, creating temperature inversions, or through warming of the atmosphere. Descending air tends to get warmer, which decreases its relative humidity and increases its "dryness". Most arid lands lie in the tropics and therefore, receive substantial inputs of solar radiation energy, which can be expended on heating the environment, wind generation, or be utilized for evaporation (Walton 1969; Agnew and Anderson 1992).

There are four main causes of regional aridity (Thompson 1975; Agnew and Anderson 1992):

(1) *High pressure*: Air heated at the equator rises, moves towards the poles, and then descends in the subtropics at around 20 to 30° north and south latitude (Fig. 2.3a). Compression and warming of the descending air mass leads to dry and stable atmospheric conditions (e.g., Sahara Desert). This atmospheric circulation pattern is referred to as the "Hadley cell."
(2) *Continental winds*: Winds blowing across continental interiors have a reduced opportunity to absorb moisture and are fairly stable with low humidity (e.g., SW Asia, Middle East). It is not the absolute distance from oceans that is important, but rather the distance from the ocean along the moist airstream flow paths (Walton 1969).

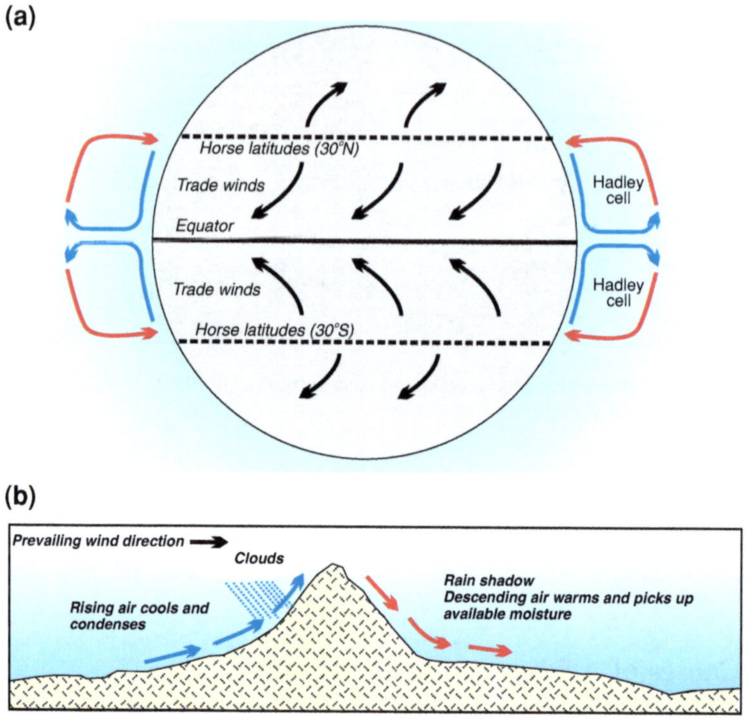

Fig. 2.3 Cause of aridity. **a** Hadley cell circulation. Air is warmed and rises to the top of troposphere near the equator. The cooled air descends in the subtropics, creating a zone of dry and stable high-pressure conditions. **b** Rain shadow effect. Rising air at the windward slope of mountain ranges cools and loses moisture. Descending air on the leeward sides of mountains warms during descent, decreasing its relative humidity

(3) *Rain shadow effect*: The cooling of air forced aloft by mountain ranges causes precipitation and a loss of moisture. The descending air on the leeward side of the ranges is warmer at a given altitude compared to the windward side because of the greater adiabatic lapse rates (rate of temperature change with elevation) of dry air compared to moist or saturated air. The windward sides of mountain ranges may have high rates of precipitation, while the leeward sides (a relatively short geographic distance away) may have semiarid or arid conditions (e.g., northern Rocky Mountains USA) (Fig. 2.3b).

(4) *Cold ocean currents*: Cold ocean currents near land can cool the lower atmospheric, but warmer air aloft creates a temperature inversion that prevents the ascent of air and thus precipitation. As the cool air moves inland it is warmed and as a result its humidity decreases (e.g., coastal Peru, Oman, and Namibia).

Aridity is mainly the result of large-scale, persistent, atmospheric and oceanic circulations patterns or regional geography and topography. The cause of aridity is not generally germane to local groundwater management. Aridity, and associated water scarcity, is a long-term, hydrologic and climatic condition with which local populations must adapt irrespective of its cause. However, the cause of aridity is significant within the context of global climate change, which may impact global atmospheric and oceanic circulation patterns (Chap. 39). For example, global warming may result in an expansion or shift of the Hadley cells, which could increase the extent and change the positions of the subtropical dry zones.

2.4 Droughts

2.4.1 Definition of Droughts

Drought is a concept that is generally understood on a basic level, but is difficult to quantify. Palmer (1965, p. 2) defined a drought as a meteorological phenomenon that is characterized by "prolonged and abnormal moisture deficiency", and more specifically (p. 3) as

> An interval of time, generally on the order of months or years in duration, during which the actual moisture supply at a given place rather consistently falls short of climatically expected or climatically appropriate moisture supply.

A drought can alternatively be broadly defined as a temporary, recurring reduction in the precipitation in an area. Key words are 'temporary', 'reduction' and 'recurring'. The term 'reduction' implies that water levels are less than what are considered normal. The term 'recurring' signifies that droughts are a normal part of the climatic cycle. Droughts are thus not 'abnormal' events, but are rather low-end extremes in the normal variation in the overall water supply.

Aridity and drought are not synonymous. Aridity is a measure of long-term average climatic conditions. Both humid and arid regions experience droughts. However, the inter-year variation in precipitation is greater in arid regions and there is a greater probability of below average precipitation in any particular year (Smakhtim and Schipper 2008). Arid regions are thus more prone to droughts and may experience more severe impacts from droughts.

The definition of drought has real world consequences. The misidentification of recurring temporary dry spells as droughts, which implies an extreme abnormal situation, can result in societies becoming maladapted and increasingly vulnerable because of unnecessary asset depletion or inappropriate mitigation measures (Smakhtin and Schipper 2008). As noted by Smatkhtin and Schipper (2008, p. 137)

> If people, agriculture and industries are adversely affected by normal climatic fluctuations, the problem is not really related to drought. Instead, this points to a high vulnerability of the population and/or economic sectors to already existing climatic variability.

Frequent droughts at a specific location suggest that a problem with the quantitative definition of a drought is likely (Smakhtin and Schipper 2008).

Frequent droughts in a region could be a temporary statistical anomaly (i.e., chance occurrence of multiple rare events within a short time period), an unrecognized cycle rooted in global climate variation (e.g., a dry phase of a multidecadal climatic cycle), a manifestation of a longer-term change in local climate to more arid conditions, or perhaps reflect a poor characterization of local climate conditions. Insufficient monitoring data may be available to properly statistically characterize the true natural variations in rainfall. Apparent frequent droughts may occur if the historic climatic record for a station coincides with a wetter than normal period, and current rainfall patterns are close to the drier norm.

Persistence of drought conditions may indicate the onset of drier conditions due to climate change, requiring the reevaluation of what is normal (Agnew and Anderson 1992), at least on a water management time scale. Cyclicity in climatic conditions occurs on multiple-time scales, ranging upwards to the 10,000's of years for the Earth orbital-driven (Milankovitch) glacial cycles (Hays et al. 1976; Muller and MacDonald 1997; Roe 2006). Shorter-term climatic variations, such as the Pacific Decadal Oscillation (PDO) and Atlantic Multidecadal Oscillation (AMO), appear to have a major influence on multidecadal drought frequency in some areas (e.g., United States; McCabe et al. 2004).

The definition of droughts needs to consider (Dracup et al. 1980):

- nature of the water deficit that is being considered (e.g., precipitation, stream flow, soil moisture)
- averaging period (e.g., days, months, years)
- selection of the truncation level (cut off) to separate droughts from the remainder of the time and
- regional aspects.

A drought is a period of less than normal water supply, which needs to be considered within the context of the temporal pattern of local uses of water and how they can be altered or redistributed to temporarily meet the demands. The nature of the water deficit that is considered should be that which has the greatest impact on man's activities in the affected area (Dracup et al. 1980). For example, if an area is dependent on hydroelectric power, streamflow may be the water-supply parameter of primary interest. On the contrary, soil moisture will be a critical parameter in areas that practice dryland agriculture. However, water-supply shortages ultimately stem from meteorological conditions.

Climatic data are typical evaluated using smoothing techniques to reduce the affects of random fluctuations and to evaluate and clarify long-term trends in a time-series of data (e.g., daily rainfall readings). A common technique is moving average smoothing, which is either the sum or average value over a specific averaging period. For example, the trailing moving six-month total rainfall for a given time period is the sum of daily rainfall values over the 6 month period immediately preceding the established starting date. Increasing the averaging period tends to reduce the variance of the data and smooth out fluctuations.

The selection of the averaging period for a particular water-supply investigation is dependent almost entirely on the purpose of the study (Dracup et al. 1980) and the sensitivity of the system to water deficits of varying durations and magnitudes. Shorter averaging periods will result in a greater number of calculated drought (extreme) events. Shorter averaging periods will also result in increased serial correlation in which conditions during a given time period are affected by previous conditions (i.e., time periods are not independent events). For example, single months in which the total monthly rainfall is 20% below normal (i.e., long-term mean) will occur more frequently than months in which the trailing 12 month total rainfall total was 20 % below normal.

The deficit period used to define a drought needs to consider the sensitivity of local populations to water deficits of different durations. Using the above example, a rainfall deficiency over a single month may not have significant adverse impacts and thus would not qualify as a drought. However, a 20 % deficiency in rainfall over a 12-month period may over-tax available water storage and supply infrastructure and, therefore, have serious societal impacts.

A critical issue for identifying and quantifying droughts is the local historic rainfall distribution (i.e., what is "normal"?). The size of the historical water supply data is important in determining the type and accuracy of the analysis that may be performed. The sample size must be large enough to guarantee that sample statistics are reasonable approximations of the corresponding population parameters (Dracup et al. 1980). The frequency of rare statistical events (e.g., 100-year drought) inherently has large uncertainties if there are limited historical data.

For a region to receive its long-term average annual precipitation in a year is a rare event. Most years will be either wetter or drier than the mean or median. The key issue concerning the definition of drought is when does a deficiency in precipitation (or stream flow or soil moisture) reach a threshold (truncation level) so that a region is experiencing a true 'drought' as opposed to just a drier than normal year (i.e., temporary dry spell). In addition to intensity (i.e., magnitude of the precipitation deficiency), droughts must also be defined in terms of duration. The severity of a drought is the product of the magnitude of the water supply deficit and its duration. It is generally understood that a drought is a prolonged water shortage relative to the climatic norm at a location, but the necessary length of time for a dry period to become a drought is not universally defined and may ultimately depend upon local circumstances and impacts.

Droughts have a relatively strong spatial component in arid regions because of the spatial variability of rainfall (Şen 2008). Nearby areas, thus, may have different exposures to a drought depending upon where infrequent rainfalls occurred. Local areas may also be impacted by droughts caused by deficits in rainfall elsewhere in a watershed. Downstream communities dependent on river flows for water supply, for example, can be impacted by rainfall deficits in the upper basin of rivers.

Although droughts are ultimately a meteorological phenomenon, from a water management perspective, they are usually defined in terms of their impacts, particularly on human activities. A myriad of definitions of droughts have been proposed in the technical literature, which involves consideration of the magnitude of the reduction in water supply, the normal variation in water supply, the impacts of the reductions, and ideally a quantitative criterion (or criteria) for when the reduction in water supply reaches the drought threshold. Smakhtin and Schipper (2008) cautioned that because of the difference in drought manifestations, there cannot and probably should not be a universal quantitative definition of a drought.

Several different general types of droughts have been recognized, which were proposed by Subrahmanyam (1967), Matthai (1979) and Wilhite and Glantz (1985), and subsequently discussed in numerous papers and reports (e.g., Agnew and Anderson 1992; Williams-Sether et al. 1994; National Drought Mitigation Center 2006). The main drought types are:

- meteorological drought, which is defined only in terms of precipitation deficiencies, in absolute amounts, for a given period
- climatological drought, which is defined in terms of precipitation deficiencies, in percentages of normal values
- atmospheric drought, which is defined not only in terms of precipitation deficiencies but also in terms of temperature, humidity, or wind speed
- agricultural drought, which is defined principally in terms of soil moisture and plant behavior
- hydrologic drought, which is defined in terms of reduction of streamflow, reduction in lake or reservoir storage, and lowering of ground-water levels
- socioeconomic drought, which occurs when the demand for an economic good exceeds supply as a result of a weather-related shortfall in water supply
- water-management drought, which characterize water-supply shortages caused by the failure of water-management practices or facilities, such as an integrated water-supply system and surface or subsurface storage, to bridge normal or abnormal dry periods and equalize the water supply throughout the year (Matthai 1979).

Regulatory droughts may be added to this list, whereby water may be physically available, but its use is precluded by regulatory policy. For example, groundwater or surface water abstractions may be limited at times in order to maintain environmental flows in streams or wetland water levels. However, every type of drought is always ultimately a meteorological drought (Smakhtin and Schipper 2008). An important distinction is between drought and drought impacts. As can be observed in the listed definitions, the term 'drought' is frequently used to refer to the adverse impacts of the deficiency in rainfall, rather than to the deficiency of rainfall as a meteorological event (Smakhtin and Schipper 2008). The local societal impacts of droughts (i.e., water deficiencies), depend upon the adaptive capacity of the affected societies. Societies vary greatly in their capacity to cope with temporary water scarcity.

2.4.2 Impacts of Drought

In economic terms, droughts are the most costly natural disaster to strike the United States (Cook et al. 2007) and many other countries. The primary impact of droughts is on food production, as agriculture is by far the largest water user.

Droughts may also have severe environmental, economic, and social impacts. The environmental and socioeconomic impacts of droughts are controlled to a large degree by the duration of droughts, rather than their severity, because recovery from the cumulative damage of consecutive drought years is more difficult (Cook et al. 2007). The impacts of drought also depend upon human and ecosystem demand for water, available water-resources management capabilities and practices, as well as the meteorological and hydrological characteristics of the drought (Loucks and Gladwell 1999).

Vulnerability to droughts depends in part on the gap between average water use and the safe yield of a system (Frederick and Schwarz 2000), and the adaptive capacity of the water system and society as a whole. The greatest vulnerability occurs when water supplies are already stretched to meet demands during normal hydrologic conditions. Many regions simply do not have the food, water, and economic resources to overcome multiple-year droughts, particularly in water-stressed regions where resources even during normal years may be barely adequate (or inadequate) to meet local needs (e.g., Ethiopia and some other African countries). Vulnerability to droughts is also related to whether the primary water source is groundwater or surface water. Areas dependent on surface water tend to be more vulnerable to droughts because the impacts of reduced precipitation are felt quicker. On the contrary, urban areas that are supplied water primarily by desalination (e.g., many major coastal cities the Middle East), have a very low vulnerability to droughts. The impact of droughts also depends upon the degree of societal development, its per capita GDP, and the density of the rural population (Le Houérou 1996). Droughts have also had severe impacts on some ancient cultures, and may have contributed to, or were the primary cause of the cultural collapse (Chap. 21).

The human and economic impacts of droughts have a strong spatial component. To local communities facing the brunt of drought conditions, the impacts can be catastrophic, including loss of income, malnutrition, starvation, and migration. However, the impacts of drought decrease at greater geographic and political scales, and some regions or economic groups may actually benefit from droughts. For example, unaffected farmers may benefit from higher prices for their crops (Frederick and Schwarz 2000).

Planning for and adapting to drought conditions are complicated by droughts usually lacking a clearly defined beginning and often a clearly defined end (Loucks and Gladwell 1999). Droughts have a slow initiation and they are usually only recognized when the drought is already well established. The slow initiation and undefined end of a drought makes it very difficult to take defensive actions (Pereira et al. 2002). A period of drier than normal conditions could be either a dry spell that is about to end, the onset of longer-term drought that will have severe impacts on water users, the beginning of an unrecognized natural climatic fluctuation, or the beginning of a climatic change to increasing aridity. Megadroughts lasting for decades or centuries have been documented in the paleoclimatic record (Sect. 21.8). Similarly, a period of rainfall during a drought may represent either the breaking of the drought condition or just a temporary, short-term event in a continuing drought. An important challenge is the development of prediction and early warning skills to provide lead time to implement policies and measures to mitigate the effects of droughts (Pereira et al. 2002).

2.4.3 Drought Indices

A variety of drought indices have been developed to quantify whether or not a region is experiencing a drought and to categorize the seriousness of the drought. Water resources need to be managed on a continuous basis, so whether or not water shortages cross a specified numerical threshold does not have great operational significance. Drought indices are useful for mapping regional water supply trends, both temporal and spatial. Drought indices are also used to define disaster conditions that qualify for government assistance and where and when emergency water restrictions may be required. Drought indices were reviewed by Hayes (2007).

2.4.3.1 Percentage of Normal Precipitation

The percentage of normal precipitation is the actual precipitation divided by the local normal precipitation (e.g., 30-year mean) and multiplied by 100%. The percentage of normal precipitation is calculated for a variety of time scales. The limitation of percentage of normal precipitation is that rainfall is not normally distributed and in arid regions the mean rainfall may be substantially greater than the median rainfall because data are skewed by infrequent major wet events. The advantage of the method is its simplicity and understandability.

2.4.3.2 Palmer Drought Severity Index

The Palmer Drought Severity Index (Palmer 1965) is a soil moisture balance method in which the primary inputs are precipitation, evapotranspiration, and the soil available water capacity (AWC). Evapotranspiration is calculated using the Thornthwaite (1948) formula. The PDSI is widely used in the United States and was derived so that a given index value has the same climatological meaning regardless of location. The limitations of the method were described by Alley (1984), which included its complexity. Application of the PDSI is difficult in many parts of the world due to the absence of the data required for its application (Şen 2008).

2.4.3.3 Standardized Precipitation Index

Standardized precipitation is defined as the difference from the mean for a specified time period divided by the standard deviation, where the mean and standard deviation are determined from past records (McKee et al. 1993). SPI values are calculated for various time scales. The long-term precipitation data are fitted to a probability distribution, which are transformed to a normal distribution so the mean SPI for the location and time period is zero. Drought occurs when the SPI continuously reaches an intensity of –1.0 or less and ends when the SPI becomes positive. The SPI is the most widely used drought index and has the advantage that it is based only on precipitation and, thus, has a lesser data requirement than the PDSI.

2.5 Deserts and Desertification

The term "desert" is usually loosely defined as an area receiving little rainfall. Deserts are often described in terms of the manifestations of their water scarcity, such as the absence of permanent surface water bodies (with the exception of rivers originating outside of the desert), an often sparse flora that is adapted to dry conditions (xerophytes), and a sandy or rocky land cover. The term "desert" is widely used synonymously with arid and hyperarid regions. The US Geological Survey defines desert as a region receiving less than 10 inches (25 cm) of rainfall per year.

Desertification is the defined by the United Nations Convention to Combat Desertification (UNCCD) to mean land degradation in arid, semiarid, and dry sub-humid areas resulting from various factors including climatic variations and human activities. Land degradation is in turn defined as the reduction or loss of the biological or economic productivity of land (Millennium Ecosystem Assessment 2005). The UNCCD definition is, in some respects, unsatisfactory in that it includes both natural and anthropogenic causes, which is an important distinction. Mitigation options are available for anthropogenic desertification, which includes discontinuing or modifying the activities responsible for land degradation, which may not be a practical alternative to cope with natural desertification. Natural desertification is a normal process resulting from climatic change. For example, the Sahara Desert has alternated between wet (savanna) and desert conditions during Pleistocene and Holocene time in response to natural global climate changes (Haynes 2001; White and Mattingly 2006).

Desertification is recognized as being one of the greatest threats to global food security. It is common in dry lands, but is by no means restricted to these areas (Le Houérou 1996). The term "desertification" is not applied to natural deserts. Desertification has been occurring for more than a thousand years, but it was not until the 20th century that it was considered to be a serious problem because the option had earlier been available to move on to new lands (Dregne 1986).

Desertification is a process, and there is no firm agreement as to what the threshold is for land to be considered "desertified". It is recognized that there is a continuum in the degree of desertification. Dregne and Nan-Ting (1992) categorize degrees of desertification, ranging from slight to very severe, based on the declines in agricultural productivity. Very severely desertified land was considered by

Dregne and Nan-Ting (1992) to have decreases in productivity of 50 % or more. The Millennium Ecosystem Assessment (2005) suggested a broader, but more difficult to measure, criterion of a decrease to the capacity of ecosystems to supply services.

2.5.1 Great Plains Dust Bowl of the 1930s

Perhaps the best known example of desertification is the Dust Bowl, which impacted the Great Plains of the Western United States in the 1930s. The Dust Bowl was one of greatest environmental catastrophes ever. It affected approximately 400,000 km^2 (100,000,000 acres) of land that was centered in parts of the states of Texas, Oklahoma, New Mexico, Colorado, and Kansas (Worster 1979). The name Dust Bowl refers to the enormous dust storms, which in some instances reached the eastern coast of the United States. Drought conditions and dust storms destroyed crops, covered lands with sand dunes, suffocated animals, caused respiratory illnesses in humans, damaged and covered structures, carried away huge quantities of valuable top soil, and resulted in enormous economic losses (Fig. 2.4). Hundreds of thousands of people were forced to migrate from the region. Numerous technical books and articles have been written on the causes and agricultural, economic, and social impacts of the Dust Bowl. The Dust Bowl also had profound cultural impacts, which are reflected in novels, movies, and music.

Droughts are a frequent occurrence in the Great Plains, but dust bowl conditions are not. The ultimate cause of the dust bowl was environmentally maladapted agricultural practices (Worster 1979). The native grasses (sod) in the Great Plains protect the soil from erosion and also minimized surface water runoff. In order to maximize wheat production, the land was stripped of its native vegetation to such an extent that there was no protection against the wind. The farmers as a matter of pride were at that time referred to as "sodbusters." Agricultural operations expanded greatly in the 1920s due to a combination of wetter than average conditions and mechanizations, which allowed for larger farms to be developed.

Worster (1979) discussed the deeper socio-economic causes of the dust bowl. First was an economic drive for greater prosperity, which drove farmers to plant more and more wheat, and thus "bust" more of the native sod. An additional important factor was a belief that farmers could overcome any deficiencies in the land and climate. For example, there was at that time what would be considered now an absurd belief that the "rains would follow the plow." Agriculture was viewed as being autonomous with respect to the environment.

The Dust Bowl revealed the need for improved agricultural practices in the Great Plains, such as terracing and planting trees to form shelter belts. As the rains returned and grain prices increased in the 1940s, there was a renewal in the maladapted agricultural practices that caused the Dust Bowl (Worster 1979).

Fig. 2.4 Images of the extreme land degradation during the Dust Bowl of American Great Plains in the 1930s. Formerly productive farm lands were devastated. (Cimarron County, Oklahoma, April, 1936, Arthur Rothstein, photographer. *Source* U.S. Library of Congress

Lessons of the Dust Bowl were quickly forgotten or ignored. The area impacted by the Dust Bowl has recovered and the Great Plains are called the "bread basket of world" for its winter wheat harvest. However, current agriculture is now largely supported by unsustainable groundwater pumping.

2.5.2 Causes and Management of Desertification

Desertification in drylands is ultimately the result of long-term failures to balance demand for and supply of ecosystem services (Millennium Ecosystem Assessment 2005). The direct causes of desertification stem largely from drastic reduction or destruction of the perennial plant cover and a simplification of the vegetation structure, which results in declines in the structure of soils. Soils become subject to erosion by wind and water, salinization by evaporation, trampling by animals, and water logging in topographic depressions (Le Houérou 1996). Anthropogenic factors that can result in desertification include over-exploitation, over-grazing, bad irrigation practices, illegal and excessive logging, bush and forest fires, and deforestation due to population increases (Le Houérou 1996; Holtz 2008). Desertification and land degradation are not sudden, spectacular events, but rather gradual processes usually resulting from continued poor land management practices.

Land areas differ in their susceptibility to degradation depending upon their characteristics and climate (Eswaran et al. 2001). However, it has been observed that lands that are kept free from human and livestock interference (e.g., uninhabited areas, parks, and preserves) usually do not undergo any kind of degradation or desertification even under drought conditions (Dregne 1986; Le Houérou 1996). Native plant and animal communities are typically adapted to survive droughts and recover once drought conditions ease. Adverse climatic conditions may not be the primary cause of desertification, but may trigger or accelerate the process (Le Houérou 1996). The "deadly combination" is land abuse during good periods and its continuation during periods of deficient rainfall (Dregne 1986).

Desertification is having profound impacts on many societies as it results in reduced land productivity, and thus the food supply, causing or aggravating famine, malnutrition, poverty, and migratory movements (Holtz 2008). It is a long-lasting process that can have irreversible consequences for the production potential of the land and the local environment. Many areas that are particularly vulnerable to desertification, such as semiarid to weakly arid regions of Africa and Asia, cannot afford losses in agricultural productivity (Eswaran et al. 2001). As lands can no longer sustain their current populations, people are forced to migrate to other lands (increasing their population) or to urban slums (Le Houérou 1996). Increases in population due to the influx of migrants can, in turn, exacerbate local desertification. Over 1 billion people in over 100 countries are directly affected by desertification or are at risk (Holtz 2008) and an estimated 250 million people in developing countries have already been affected by severe land degradation (Reynolds et al. 2007).

Desertified land can be restored. Indeed, when money is of no concern, virtually any land can be restored to something approximating its original productivity (Dregne and Nan-Ting 1992). Restoration of desertified lands is cost-effective if the income generated as a result of restoration exceeds the costs of the restoration. Economic analyses of land restoration are complicated, particularly in developing countries, because of the difficulty in quantifying local crop values and labor costs for what is considered to be subsistence farming. Nevertheless, Dregne and Nan-Ting (1992) calculated that 1,860 billon hectares (7.2 million mi^2) of desertified land could be repaired at a cost-benefit ratio of about 1:2.5. Although, the numbers will be different under present-day land and economic conditions, there is little doubt that restoration of other than very severely desertified land can be very cost effective.

There is a huge amount of literature on all aspects of desertification and land use practices, including numerous dedicated books and many hundreds of journal articles and conference proceedings papers. Desertification is largely a land management rather than water management issue, and hence, is not a major focus of this book. The discussion of desertification herein is thus necessarily superficial. Nevertheless, it is important to recognize that desertification is intertwined with water management on many levels. Increased integration of land and water management and a proactive strategy are key methods for desertification prevention (Millennium Ecosystem Assessment 2005).

A key issue is that dryland human-environmental systems are coupled, dynamic, and co-adapting, and are the product of complex interactions between biophysical and human subsystems that are constantly changing in response to both external and internal drivers (Reynolds et al. 2007). On a large-scale level, vulnerability to desertification is generally inversely related to water availability. However, increasing the availability of water can have negative local impacts on the land. For example, overgrazing was inadvertently made worse in parts of the Sahel by the drilling of additional wells that provided year-long drinking water for livestock, which resulted in the concentration of grazing around the wells, thereby causing desertification (Dregne 1986).

Poor land management and irrigation practices are also often associated. For example, land degradation caused by salinization and water-logging is often related largely to local irrigation water management practices, particularly inadequate drainage. Declines in soil structure associated with desertification can adversely impact the water holding capacity and infiltration rates through soils. As more water runs off the land surface, less is retained in the soil and available for local recharge. Countries, communities, and individuals not having the resources (physical, technical, human, governance, and economic) to sustainably manage soil resources face similar challenges with respect to water resources.

References

Agnew, C., & Anderson, W. (1992). *Water in the arid realm*. Routledge: London.

Alley, W. M. (1984). The Palmer drought severity index: Limitations and assumptions. *Journal of Climate and Applied Meteorology, 23*(5615), 1100–1109.

Cook, E. R., Seager, R., Cane, M. A., & Stahle, D. W. (2007). North American drought: Reconstructions, causes and consequences. *Earth Science Reviews, 81*, 93–134.

De Martonne, E. (1926). Aréisme et indice artidite. *Comptes Rendus de L'Acad Sci, Paris, 182*, 1395–1398.
Dracup, J. A., Lee, K. S., & Paulson, E. G., Jr. (1980). On the definition of drought. *Water Resources Research, 16*, 297–302.
Dregne, H. E. (1986). Desertification of arid lands. In F. El-Baz & M. H. A. Hassan (Eds.), *Physics of desertification* (pp. 4–34). Springer: Dordrecht.
Dregne, H. E., & Nan-Ting, C. (1992). Global desertification dimensions and costs. In H. E. Dregne (Ed.), *Dregradation and restoration of arid land: Lubbock* (pp. 249–282). Texas: Texas Tech University.
Eswaran, H., Lal, R., & Reich, P. F. (2001). Land degradation: An overview. In E. M. Bridges, I. D. Hannam, L. R. Oldeman, F. W. T. Pening de Vries, S. J. Scherr, & S. Sompatpanit (Eds.), *Responses to land degradation, Proceedings* 2nd *International Conference on Land Degradation, Khon Kaen, Thailand* (pp. 20–35). New Delhi, India: Oxford Press.
Frederick, K. D., & Schwarz, G. E. (2000). *Socioeconomic impacts of climate variability on U.S. water resources*. Resources for the Future Discussion Paper 00-21, Washington, DC.
Hayes, M. J. (2007). Drought indices: Intermountain West Climate Summary, July 2007, pp. 2–6.
Haynes, C. V. Jr (2001). Geochronology and climate change of the Pleistocene–Holocene transition in the Darb el Arba'in Desert, Eastern Sahara. *Geoarchaeology, 16*, 119–141.
Hays, J. D., Imbrie, J., & Shackleton, N. J. (1976). Variations in the Earth's orbit: Pacemaker of the ice ages. *Science, 194*(4270), 1121–1132.
Holtz, U. (2008). *The United Nations Convention to Combat Desertification (UNCCD) and its political dimensions*. Paper prepared for fifth parliamentary round table, United Nations Convention to Combat Desertification, Havana, Cuba.
IPCC. (2007). Climate change 2007: Impacts, adaptation and vulnerability. In M. L. Parry, O. F. Canziani, J. P. Palutikof, P. J. van der Linden, & C. E. Hanson (Eds.), *Contribution of working group II to the fourth assessment, report of the intergovernmental panel on climate change*. Cambridge, UK: Cambridge University Press.
Le Houérou, H. N. (1996). Climate change, drought and desertification. *Journal of Arid Environments, 34*, 133–185.
Loucks, D. P., & Gladwell, J. S. (1999). *Sustainability criteria for water resources systems*. Cambridge, UK: Cambridge University Press, International Hydrology Series.
Matthai, H. F. (1979). *Hydrologic and human aspects of the 1976–77 drought*. U.S. Geological Survey Professional Paper 1130, p. 84.
McCabe, G. J., Palecki, M. A., & Betancourt, J. L. (2004). Pacific and Atlantic ocean influences on multidecadal drought frequency in the United States. *PNAS, 101*(12), 4136–4141.
McKee, T. B., Doesken, N. J., & Klesit, J. (1993). *The relationship of drought frequency and duration to time scales: Proceedings Eighth Conference on Applied Climatology* (p. 6), 17–22 January 1993, Anahiem, CA.
Meigs, P. (1953). World distribution of arid and semi-arid homoclimates. In UNESCO (Eds.), *Reviews of research on arid zone hydrology* (pp. 203–210). Paris: UNESCO.
Millennium Ecosystem Assessment. (2005). *Ecosystem and human well-being: Desertification synthesis*. Washington, DC: World Resources Institute.
Muller, R. A., & MacDonald, G. J. (1997). Glacial cycles and astronomical forcing. *Science, 277*(5323), 215–218.
National Drought Mitigation Center. (2006). What is drought? Understanding and defining drought: http://drought.unl.edu/whatis/concept.htm.
Palmer, W. C. (1965). Meteorological drought: United States Department of Commerce, Weather Bureau, Research Paper No. 45.
Penman, H. (1948). Natural evaporation from open water, bare soil and grass. *Proceedings of the Royal Society, A193*, 120–146.
Penman, H. L. (1956). Evaporation: An introduction survey. *Netherlands Journal of Agricultural Science, 4*, 8–29.
Pereira, L. S., Cordery, I., & Iaconides, I. (2002). Coping with water scarcity: UNESCO, International Hydrological Programme, IHP-VI, Technical Documents in Hydrology No. 58.

Reynolds, J. F., Smith, D. M. S., Lambin, E. F., Turner, B. L., I. I., Mortimore, M., Batterbury, S. P. J., et al. (2007). Global desertification: Building a science for dryland development. *Science, 316*, 847–851.

Roe, G. (2006). In defense of Milankovitch. *Geophysical Research Letters, 33*(L24703), 5.

Şen, Z. (2008). *Wadi hydrolog* (p. 347). Boca Raton, FL: CRC Press.

Smakhtin, V., & Schipper, E. L. F. (2008). Droughts: The impact of semantics and perceptions. *Water Policy, 10*(2), 131–143.

Smiley, T. L., & Zumberge, J. H. (1971). Polar deserts. *Science, 174*(4004), 79–80.

Stadler, S. J. (2005). Aridity indexes. In J. E. Oliver (Ed.), *Encyclopedia of world climatology* (pp. 89–94). Heidelberg: Springer.

Subrahmanyam, V. P. (1967). Incidence and spread of continental drought: World Meteorological Organization, International Hydrological Decade, Reports on WMO/IHD Projects, no. 2, Geneva, Switzerland.

Thompson, R. D. (1975). *The climatology of arid world.* University of Reading, UK, Department of Geography Paper No. 35, p. 39.

Thornthwaite, C. W. (1948). An approach toward a rational classification of climate. *Geographical Review, 38*(1), 55–94.

United Nations Educational, Scientific and Cultural Organization (UNESCO). (1979). *Map of the world distribution of arid regions: Map at scale 1:25,000,000 with explanatory note.* MAB Technical Notes 7, UNESCO, Paris.

Walton, K. (1969). *The arid zone*. Chicago, IL: Aldine Publishing Co.

White, K., & Mattingly, D. J. (2006). Ancient lakes of the Sahara. *American Scientist, 94*(1), 58–65. ISSN 0003-0996.

Wilhite, D., & Glantz, M. (1985). Understanding the drought phenomenon: The role of definitions. *Water International, 10*(3), 111–120.

Williams-Sether, T., Macek-Rowland, K. M., & Emerson, D. G. (1994). *Climatic and hydrologic aspects of the 1988–92 drought and the effect on people and resources of North Dakota, Vol. 29*, p. 57. North Dakota State Water Commission Water Resources Investigation.

Worster, D. (1979). *Dust bowl: Southern plains in the 1930's.* New York: Oxford University Press.

Part II
Arid Lands Geology and Hydrogeology: An Overview

Chapter 3
Geology of Arid Lands

3.1 Introduction

The fundamental framework of arid lands aquifers and aquifer systems is formed by variations in the geology, which dictate subsurface hydraulic conditions. A voluminous literature exists on the geomorphology and geology of arid lands and dedicated books have been written on many aspects of these subjects. Therefore, the discussion on the geology has been necessarily condensed into an introduction of the various types of geological deposits found in arid and semi-arid lands, and how the geology of these lands controls the flow and storage of water. Of particular importance is the relationship between rainfall, surface water (run-off), and groundwater recharge.

Arid and semi-arid lands constitute between 26 and 47% of the Earth's land surface (Koppen 1931; Thornthwaite 1948; Meigs 1953; Shantz 1956; United Nations 1977; Heathcote 1983; UNEP 1992). The geology in these lands is quite diverse and includes virtually every known rock type. Arid lands are characterized by an overprint of terrestrial sedimentation atop the bedrock types. Bedrock types determine the composition of the sediments locally generated and their resistance to weathering and erosion. The flow of water and air control how the sediments are moved and deposited. The predominant surficial geology within arid and semi-arid lands consists of alluvial, aeolian, and fluvial sediments with the bedrock geology being more exposed at higher altitudes and the terrestrial sediments covering the middle and lower parts of basins.

Terrestrial sedimentary processes in arid lands differ in many ways from those in more humid regions with a wetter climate and lower actual evapotranspiration rates (Bull 1991). The effects of vegetation in controlling the runoff of rainfall are significant factors that control the development of the fundamental geomorphic and geologic features of arid lands. Within arid land areas, even small differences in average rainfall accumulations can significantly affect the patterns of sedimentation. For example in hyperarid regions, wind transport of sediment is more

R. Maliva and T. Missimer, *Arid Lands Water Evaluation and Management*, Environmental Science and Engineering, DOI: 10.1007/978-3-642-29104-3_3,

significant and leads to the deposition of mixed aeolian and alluvial sediment sequences. Whereas in semi-arid or dry-subhumid regions, the predominant sediment sequences are mixed alluvial and fluvial deposits reflecting the greater importance of surface water flow.

The low rainfall in arid lands reduces the rate of soil formation and chemical weathering. Siliciclastic deposits tend to be chemically immature in that they retain a relatively high percentage of unstable minerals such as feldspars. High evaporation rates allow for the local precipitation and persistence of evaporite minerals, which are typically absent in humid climates.

Tectonics also plays a significant role in the geology of arid lands deposits (Rendell 1997). Actively rising mountain areas produce steep altitude gradients that allow the combination of mass wastage and fluvial processes to produce coarse-grained deposits. In tectonically inactive areas, the slopes of land surface are lower and the terrestrial sediments tend to be combinations of coarse and fine-grained sediments. Aeolian deposits are significant in some stable cratonic areas. A critical aspect of the tectonic influence on arid lands geology is the timeframe and duration of tectonic events and their influence on the rates of sediment deposition and the distribution of landforms and channels. There is an important balance between the uplift rate of mountains and the erosion rate that can increase, decrease, or maintain slopes over geologic time. These slope changes affect the patterns of sedimentation, particularly in alluvial fan systems.

Geologic history of arid lands also influences the physical patterns of sedimentary sequences and current land forms. Most arid lands have experienced periods of higher humidity in the past, which resulted in the deposition of different types of sediments than those currently being deposited within a given basin. These earlier wetter (i.e., pluvial) periods have great water resources significance in some areas, where they resulted in the recharge of much of the groundwater now present in groundwater basins (Chap. 36). Naturally occurring sea level changes over geological time have greatly impacted coastal sedimentation. For example, in the eastern Arabian Gulf region, naturally-occurring global sea level changes of the past have produced a pattern of intervals of terrestrial sediments or rock alternating with marine deposits, with the latter tending to form aquifers. The cyclicity from humid to arid over extended periods of time may produce rather complex, mixed alluvial, fluvial, and marine sequences. From a hydrologic perspective, sediment and rock types vary in their hydraulic properties and, therefore, the sedimentological history of a basin determines the locations of both aquifers and low-permeability confining units.

Post-depositional chemical processes (i.e., diagenesis) within arid lands deposits also have a significant effect on the distribution of porosity and permeability within a basin at various scales (Watson and Nash 1997). In many interior basins and within periodically-wetted channel deposits, calcretes or caliche (Gouldie 1973; Wright and Tucker 1991) and silcretes (Gouldie 1973; Summerfield 1983; Watson and Nash 1997; Ullyott et al. 1998) form. These crusts or cemented layers, which form near land surface, either partially or completely, bind the sediments in layers. Calcretes and silcretes result in compositional, textural and hydraulic heterogeneities within sediment sequences, and are of particular importance in controlling groundwater flow, and cause significant variation in both vertical and horizontal permeability. The cemented areas are also less subject to wind erosion and can locally act to stabilize the surface of arid lands.

3.2 Geomorphology and Geology of Arid Lands

3.2.1 Introduction

Climatic conditions and local geology interact to create the characteristic geomorphic forms found in arid lands. There is a distinctive relationship between the dominant geomorphic processes and the resulting landscape characteristics in all geomorphic systems, which is particularly true in arid and semi-arid lands (Ritter et al. 2002). Wilson (1968) suggested that the dominant geomorphic processes in arid lands, in order of importance, are desiccation, wind action, and running water, which collectively create landscapes containing dunes, salt pans (playas), deflation basins, cavernous weathering, angular slopes, and arroyos (wadis). Wilson (1968) also concluded that in semi-arid and semi-humid regions the dominant processes, in order of importance, are running water, weathering (especially mechanical), and rapid mass movements, which produce dominant landscape features that include pediments, alluvial fans, angular slopes with coarse debris, and badlands.

An important issue affecting the formation of geomorphic and geologic features in arid lands is the rate of change within the processes that create the depositional systems. The concept of dynamic equilibrium in open systems was defined by Hack (1960) who noted "that when in equilibrium a landscape may be considered a part of an open system in a steady state of balance in which every slope and every form is adjusted to every other." Bull (1975) suggested that there are a series of independent and dependant variables that cause "allometric" changes in landforms and an equilibrium or steady-state does not really occur within these land forms. For time spans greater than 1,000 years, the independent variables are rock types and structure, climate, total relief as controlled by vertical tectonic movements, base level at the mouth of a drainage basin, and drainage-basin area. The corresponding dependent variables subject to feedback mechanisms are, in order of increasing dependency, hillslope morphology, drainage net, soil-profile development, vegetation, sediment yield, stream-channel slope and patterns, and water and sediment discharge from the basin. Over short time frames, some type of equilibrium may occur, particularly within small, closed systems in arid lands. However, in large, open arid land systems, it is unlikely that all parts of the system will ever reach an equilibrium state, although parts of it may do so. These geomorphic dynamics are quite important in the creation of the geologic structure of the landforms that form the architecture of the associated groundwater systems. Geomorphic dynamics control the location, type, and characteristics of sediment deposits, which ultimately form aquifers and confining strata.

Table 3.1 Geomorphic landscapes in arid regions (after Clements et al. 1957; Mabbutt 1976; Thomas 1997a, b)

Landscape type	SW USA	Sahara	Libya	Arabia	Australia
Mountains	38.1	43	39	47	16
Low angle bedrock surfaces	0.7	10	6	1	14
Alluvial fans	31.4	1	1	4	–
River plains	1.2	1	3	1	13
Dry watercourses	3.6	1	1	1	–
Badlands	2.6	2	8	1	–
Playas	1.1	1	1	1	1
Sand seas	0.6	28	22	26	38
Desert flats[a]	20.5	10	18	16	18
Recent volcanic deposits	0.2	3	1	2	–

The given percentages are only approximate. The errors between areas are great based on the degree of investigation

Mabbutt (1976) data were compiled in a different manner and may not have the accuracy as the others

[a] Undifferentiated and includes areas bordering playas

Thomas (1997a, b) summarized the arid regions landscapes in various global regions, based on the work of Clements et al. (1957) and Mabbutt (1976) (Table 3.1). There are significant differences between the predominant landscape types in different regions of the world and their accompanying geology. Landscape types and geology ultimately control local hydrogeology. Each of these landscapes is discussed in terms of its geologic significance and its effects on processes that produce basin sediment sequences that are important in forming the framework of aquifers and aquifer systems.

3.2.2 Mountains

Mountainous terrains are common in arid lands and are the source of the runoff and sediments that form many of the lower-altitude landscapes. The geology of mountains terrains, in particular the lithologies present, greatly affect the weathering and erosion rates, which, in turn, determine the composition and geometry of down-gradient sedimentary deposits. Tectonic activities associated with the mountain areas, such as uplift and faulting also play significant roles in creating the available accommodation space within basins in which sediments are transported and deposited. Tectonic movement creates differences in slope with the greater slopes having a greater potential for mass wastage and resulting in more rapid sediment movement into basins and higher velocity (greater energy) water flows.

Tectonism is perhaps the single most important factor that controls landforms in mountainous regions and the associated down-slope landforms, particularly in arid lands. A summary of the differences between tectonically active and inactive regions and their associated geomorphic forms is shown in Fig. 3.1. The rapid

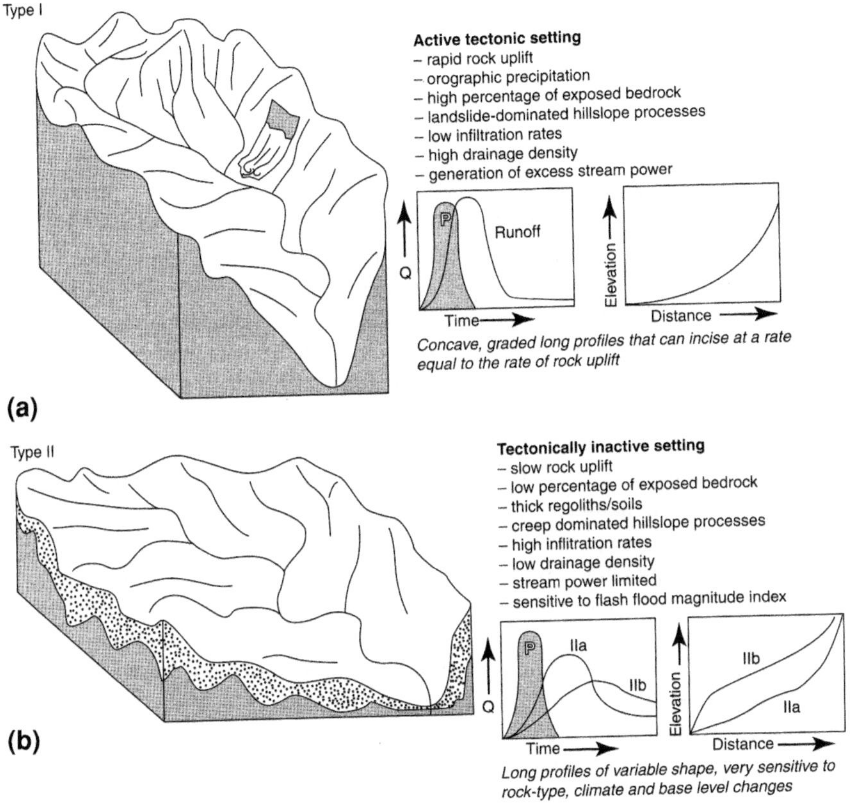

Fig. 3.1 Generalized geomorphic, hydrologic and long-profile characteristics of tectonically active (**a**) and tectonically inactive (**b**) mountainous regions (modified from Pazzaglia et al. 1998). Note that Q is discharge and P is a unit precipitation event

uplift of the mountains maintains high slopes into the piedmont and basin floor, causing predominantly down-slope landslide sediment movement (mass wastage) and a great amount of stream power during rainfall events, which allows both the downward incising of stream channels and rapid sediment transport. Landscapes in areas of rapid uplift also have a high percentage of rock exposure, low infiltration rates, and high rates of runoff.

Tectonically inactive mountainous regions, on the contrary, have a lower percentage of exposed rock with greater development of soil and sediment profiles. Infiltration in tectonically inactive mountainous areas is also greater with corresponding lower runoff rates and stream power. The altitude of the mountainous region may also affect landforms as a result of the rain shadow effect. The seaward (windward) sides of mountains may receive much greater rainfall than the leeward sides, because of the cooling of air forced aloft by mountain ranges and associated precipitation (Sect. 2.3).

Climate interaction with the bedrock geology also greatly affects the formation of various landforms and controls, and to a large degree, the supply of alluvial and fluvial sediments delivered into the basin (Bull 1991). Thermal expansion and contraction is commonly extreme in arid areas, causing rock spalling and breakage, which reduce fragments to a size that allows down-slope transport (Gray 1965; Gouldie 1997). Also, when small amounts of moisture occur within the rock, the effects of heating can be magnified, producing more rapid rates of breakage (Winkler 1975). Similarly, in both warm and cold arid lands, the different colors of the minerals within the bedrock or large rock fragments causes differential thermal expansion rates, which results in spalling or cracking (Hockman and Kessler 1950). Thermal fragmentation produces both large clasts and sand-sized sediment, which tend to be both compositionally and texturally immature.

Resistance to weathering combined with active tectonic movement is another key factor in maintaining steep slopes which affect the development of arid land geomorphic forms. Dense igneous and metamorphic rocks, shales, and dense, low porosity sandstones tend to have very high rates of runoff. Rain falling on these areas covered with dense, low-permeability rock, can only enter the rock via fractures. The corresponding runoff coefficients are quite high. Relatively small rainfall events in mountainous areas in which the land surface consists predominantly of dense rock can produce rapid flows within channels. The high-velocity and energy flows can erode (incise) channels and transport coarse sediments, which are deposited downstream in alluvial cones and fans. Alluvial fans and cones tend to be more common and best developed in arid and semi-arid lands, rather than in hyperarid regions, where rainfall events are rare or nearly non-existent (e.g., Atacama Desert). An example of an arid area that drains from igneous rock (part of ophiolite sequence) is the mountain range bordering Oman and the United Arab Emirates (Fig. 3.2). This area receives a small annual average rainfall, but large rainfall events occur on about a 5 year timeframe.

Mountainous areas that consist of carbonate rocks or other sedimentary rock types with high degrees of porosity and permeability can produce channels or wadis and alluvial fans with different geometries than those that form in areas of low porosity rock (Fig. 3.3). Note that the steep channels draining the Jabal Hafit carbonate anticline in near Al Ain, Abu Dhabi, are relatively narrow and steep sided. Highly karstic carbonate mountains show lesser degrees of alluvial fan construction because drainage of rainfall occurs in the subsurface and via collapsed karst features (Figs. 3.4a ,b). Carbonate rocks also tend to be eroded by chemical processes and generally do not produce large volumes of sediments. Carbonate rocks are chemically reactive and small amounts of moisture can allow cementation of eroded rock fragments, resulting in the stabilization and maintenance of steeper slopes in some regions. In hyperarid and arid areas, small amounts of rainfall may not run off, but can infiltrate into the bedrock, slowly forming caves and other dissolution features. Only larger-scale rainfall events will contribute flow to the channels and tend to produce narrow, shallow wadis in the middle and distal parts of alluvial fans (Fig. 3.5). The subsurface drainage also allows springs to develop near the base of the mountains, which can be important local water sources.

Fig. 3.2 View of the mountains of Oman from within the Schwaib Wadi of the United Arab Emirates. Note the wide outwash areas from the mountains transports sediment via channels into an outwash alluvium fan. Overland flow washes across the fan into the Schwaib Wadi during high rainfall events. The primary flow into the wadi comes from these mountains and enters from the left in the photograph. Additional flow is added through the fenced and screened area

Transport of sediments from mountains into the basins is also related to climate in hyperarid and arid regions. Within hyperarid regions, sand-sized and smaller grains are transported down-slope primarily by wind as illustrated in the arid region of southern Peru (Figs. 3.6, 3.7). However, as rock fragments and sands are produced, periodically unstable slope deposits collapse and the sediment is transported down-slope as a rock fall or dry or wet landslides. In both hyperarid and semi-arid regions, infrequent rainfall events transport sand and larger-size rock fragments into channel areas and ultimately down-slope into alluvial fans, wadis, or desert flats. The rate of down-slope sediment movement is controlled collectively by sediment production, slope, and the frequency of rainfall events acting together, as controlled by rock type, tectonics, and basin accommodation space. More frequent rainfall events in semi-arid or dry subhumid regions tend to cause greater amounts of sediment to be delivered into basins. Landforms created by deposition of water-transported sediments, such as alluvial fans, become more prominent and wadis become incised and filled with alluvial or fluvial sediments.

The piedmont is the slope transition between the mountains and the basin floor (Fig. 3.8) and can contain a number of geomorphic forms, including exposed rock,

Fig. 3.3 Upper wadi located high on the face of Jabal Hafit in the United Arab Emirates. This is a steep, anticlinal feature standing about 800 m above the desert floor. The limestone ranges from Eocene to Oligocene in age and is a triple-porosity carbonate that contains small to moderate-sized karst features. Drainage of rainfall occurs both as surface runoff and subsurface drainage. Natural springs and some newer wells occur in the lower part of the feature in the vegetated area shown

erosional pediments without sediment cover or minimal cover, alluvial fans, wadis, and badlands. Badlands may also occur within the basin depending upon the local hydrologic conditions (semi-arid locations). The basin floor can contain desert flats, deflation plains, wadis, rivers and river plains, and playas. Near intersections with coastal areas, the playas may merge with coastal sabkkas. There is considerable significance with regard to the relationships between geomorphic landforms, geologic facies models, and hydrogeology, because the longitudinal topographic and facies (sediment deposit stacking) profile from the mountains into the basin floor controls the subsurface flow of water into arid basins.

3.2.3 Low-Angle Bedrock Surfaces

Exposed angular bedrock surfaces occur primarily in the higher altitude areas of arid lands as mountain slopes. Low-angle bedrock surfaces also commonly occur on erosional pediments located below escarpment slopes within the piedmont

Fig. 3.4 *Top* The carbonate mountains in the Ma'aloula area of Syria are highly karstified and few channelized discharge channels occur. The drainage is primary through the ground or into cave systems or collapsed cave channels. *Bottom* Water flows during heavy rainfall events into collapsed karst features that ultimately drain to streams

physiographic component of the mountain-basin geomorphic system. These exposed rock surfaces are formed by the erosion of the escarpment into the underlying bedrock directly adjacent to mountains (Fig. 3.9). Low-angle bedrock surfaces also occur in the lower-altitude parts of arid lands, particularly in cratonic areas. These low angle bedrock exposures can contain fractured rocks with some hydraulic conductivity or can be smooth and produce high rates of runoff during rare rainfall events.

Fig. 3.5 Narrow and shallow wadis developed along the distal margin of the Jabal Hafit carbonate anticline in the United Arab Emirates. Note that many of these channels have formed on the very lowest part of the feature and may be related to the emergence of subsurface drainage

Fig. 3.6 High-altitude discharge channel containing rock fragments, boulders, and wind-blown sand. The site is located southwest of Arequipa, Peru, in a hyperarid to arid area

Fig. 3.7 Pattern of hyperarid to arid sediment transport into a fluvial valley. The rock type in the mountains consists of igneous, crystalline granitic rocks with high weathering resistance. Note that the fine-grained sands eventually are blown into drainage channels and outwash into Rio Tambo across a fault scarp bounding the southern side of the valley

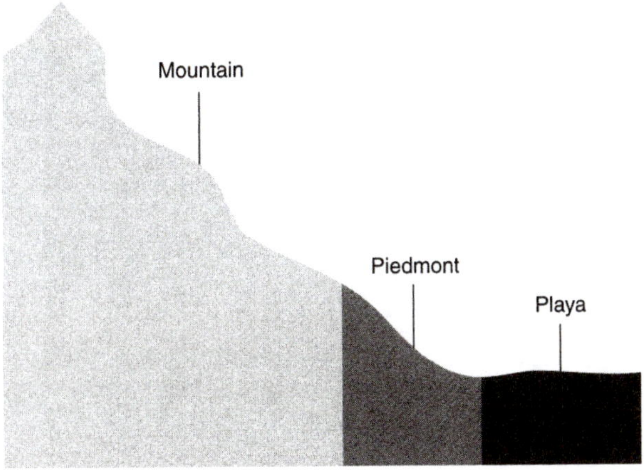

Fig. 3.8 Physiographic components of a mountain-basin geomorphic system showing the position of the piedmont zone and basin floor with associated depositional features (modified from Ritter et al. 2002)

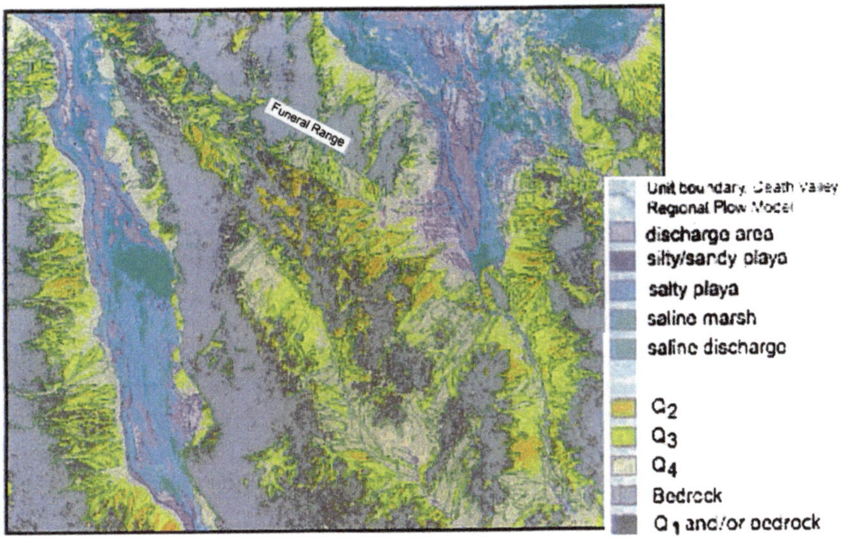

Fig. 3.9 Thematic surficial map with unit boundaries from the Death Valley Regional Groundwater Flow model overlain for comparison. Note that the low angle bedrock exposures are shown as Q1 or the dark gray color areas within the lower slopes in pediment areas (from Jayko et al. 2005)

Fig. 3.10 Alluvial fans in Death Valley National Park, California, U.S.A. *Source* National Park Service

3.2.4 Alluvial Fans

Alluvial cones and fans are common depositional geomorphic forms in all classifications of arid lands. Alluvial fan is defined as

> A low, outspread, relatively flat to gently sloping mass of loose rock material, shaped like an open fan or a segment of a cone, deposited by a stream (especially in semi-arid region) at the place where it issues from a narrow mountain valley upon a plain or broad valley, or where a tributary stream is near or at its junction with the main stream, or wherever a constriction in a valley abruptly ceases or the gradient of the stream suddenly decreases; it is steepest near the mouth of the valley where its apex points upstream, and it slopes gently and convexly outward with a gradually decreasing gradient (American Institute of Geology 1997, p. 17).

An example of an alluvial fan in an arid climate is provided in Fig. 3.10.

The overall geomorphology of alluvial fans is collectively influenced by the interactions between three fundamental sets of factors: those that influence fan "context" (tectonics, gross topography, accommodation space), those that influence water and sediment delivery to the fan and the processes operating on the fan, and those that influence the relationship between the fan and adjacent environments with base level being one of the most important (Harvey 2005). Considerable research has been performed on alluvial fan morphology and geometry based on the climate in which the fan developed (Kochel and Johnson 1984; Ritter et al. 1995; Harvey 2004; Hartley et al. 2005), the geology of the mountains feeding sediment into the fan (Bull 1964a; Hooke 1968; Hooke and Rohrer 1977; Leece 1991), the slope of the channel feeding the fan (tectonism) (Bull 1964b; 1972; Blair and McPherson 1994a; Ritter et al. 1995), the stream power (Bull 1979), and the drainage basin area of the fan (Bull 1964a, 1977; Harvey 1997). Some important general references on alluvial fan morphology, genesis, geology, and evolution over time include Bull (1972, 1991), Rust (1979), Miall (1981), Nilsen (1982), Blair and McPherson (1994a, b), and Harvey (1997).

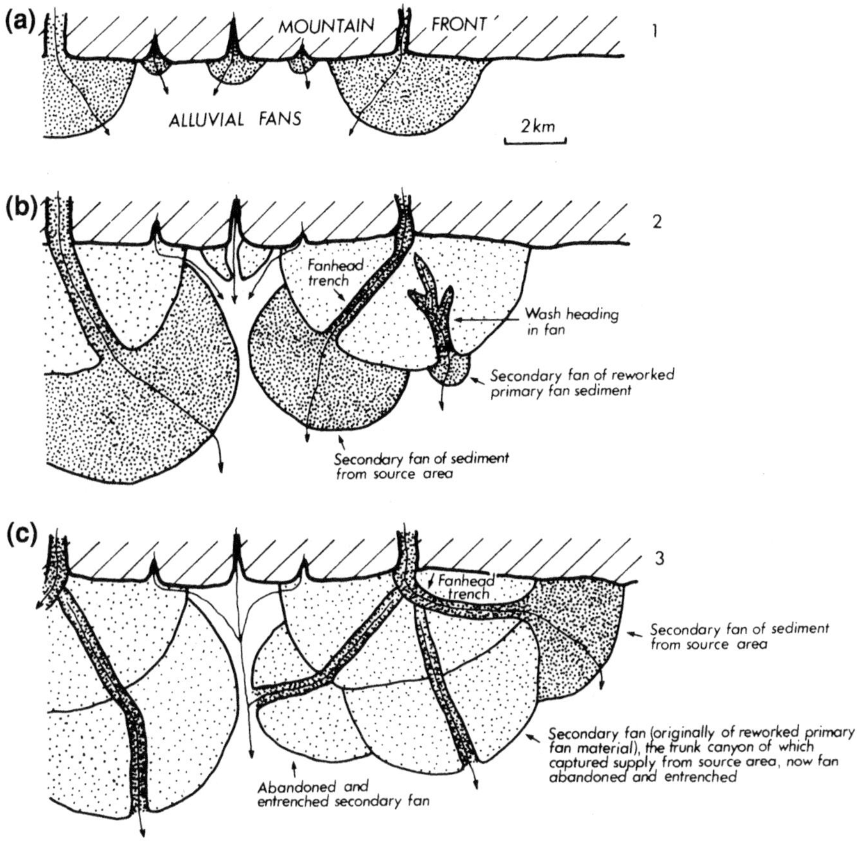

Fig. 3.11 Aerial view of various types of alluvial fans (modified from Denny 1967 and Nilsen 1982). View (**a**) shows small alluvial fans or cones at the base of a recently elevated mountain. View (**b**) shows alluvial fan dissection of an original fan by wash from a mountain with building of a new fan segments. View (**c**) shows two drainage diversions have taken place causing wash from a mountain to abandon segments and build a new segment at the mouth

Alluvial fans are found in the piedmont component of the mountain-basin geomorphic system (Fig. 3.9). The geometry of alluvial fans varies from simple small cones to larger-scale 180° aprons with either a single feeder channel or more complex multi-channel systems showing changes in direction as accommodation space fills on the fan surface and the distribution channels shift (Fig. 3.11). Alluvial fans formed in arid regions have a generally different set of characteristics compared to those that form in more humid regions (Kochel and Johnson 1984). The key differences in arid-region alluvial fans compared to humid-region alluvial fans are:

Table 3.2 Generalized Characteristics of alluvial fans formed in different environments (modified from Kochel and Johnson 1984)

Parameter	Arid fans	Humid-glacial fans	Humid-tropical fans
Fan morphology			
Plan view	Broad, fanlike, Symmetrical	Broad, fanlike Symmetrical	Broad, fanlike Symmetrical
Axial profile	Segmented (20–100 m/km)	Smooth (1–20 m/km)	Smooth
Thickness	Up to 100 s m	Up to 100 s m	Up to 100 s m
Area	Small	Very large	Large
Depositional processes			
Major processes	Debris flow Braided stream Sheet flood Sieve flood	Braided stream	Braided stream Debris flow
Return interval	1–50 years, discrete events	0 to few days, seasonally constant	Seasonally constant to discrete
Fan area activated	10–50 %	80–100 %	30–70 %
Triggering processes	Heavy rain Snow melt	Meltwater Outwash	Heavy rain Monsoon
Discharge	Flashy	Seasonal	Seasonal

(1) the axial profile is segmented and has a steep slope ranging from 20 to 100 m/km (2–40 to 200 ft/mile)
(2) the area is generally small compared to humid-region fans
(3) the number of processes forming the alluvial fans is greater and includes debris flow, braided stream, sheet flood, and sieve flood
(4) the return interval is 1–50 years per discrete event
(5) the fan activated area is 10–50%
(6) the triggering processes are heavy rain and snowmelt
(7) the discharge is virtually always flashy.

A comparison between alluvial fan characteristics in different climates is presented in Table 3.2. The primary and secondary sedimentary processes acting in the formation of alluvial fans dictate the geometry and internal structure of these features. The primary sedimentary processes involve both mass wastage (landslides) and water movement, which interact with a variety of secondary processes (Blair and McPherson 1994a) (Table 3.3).

It is quite important to understand the difference in the geology and subsurface structure between alluvial fans and other depositional environments, such as rivers and deltas. Commonly, some of these depositional environments are confused within ancient rock settings. The geology of alluvial fans differs considerably from river and river delta deposits based on a number of characteristics, including (Blair and McPherson 1994a):

Table 3.3 Primary verses secondary sedimentary processes active on alluvial fans (from Blair and McPherson 1994a)

Primary processes	Secondary processes
(A) Generated by bedrock cliff failure	Winnowing by overland flow
Sediment-gravity flows	Rilling and gullying
Rockfalls	Rainsplash erosion
Rock Slides	Wind erosion or deposition
Rock avalanches	Plant rooting
(B) Generated by colluvial slope failure	Animal burrowing
Sediment-gravity flows	Sediment weathering
Colluvial slides	Soil development
Cohesive debris flows	Groundwater flow
Noncohesive debris flows	Case hardening
Fluid-gravity flows	Subsurface cementation
Sheetfloods	Lateral erosion
Incised-channel floods	Tectonic faulting or folding

- steep radial profiles,
- relatively short radial distances,
- high degrees of radial relief,
- concave-downward sectional profile,
- location in the piedmont,
- very high propensity for unconfined or non-channelized flow and related depositional features,
- absence of floodplain deposits,
- relatively steep radical slopes ranging from 1.5 to 25°,
- coarser-grained sediments, including boulders, cobbles, and pebbles,
- very large slopes relative to grain size,
- minor effect of vegetation and no effect in hyperarid regions,
- very small to small drainage basin area,
- very high relief within drainage basin,
- high propensity for flash floods,
- very common gravity flows (landslides),
- common supercritical water flows,
- large effects of slope failures,
- flow depth, width and stream competency, flow capacity, and water discharge all decrease greatly downstream, and
- aggradational events are very infrequent.

Debris-flow and sheetflow processes control the pattern of deposition on all alluvial fans. Downslope, a change occurs from stream flow in confined channels to open areas, which results in the deposition of the large clasts first, with the finer sediments deposited on the distal parts of the fan. Debris-flow and sheetflood dominated fans show a difference in the overall pattern of sedimentation (Fig. 3.12). As alluvial fans evolve over time, changes in the slope caused by vertical tectonic movement and climate will produce alternating layering of debris-dominated

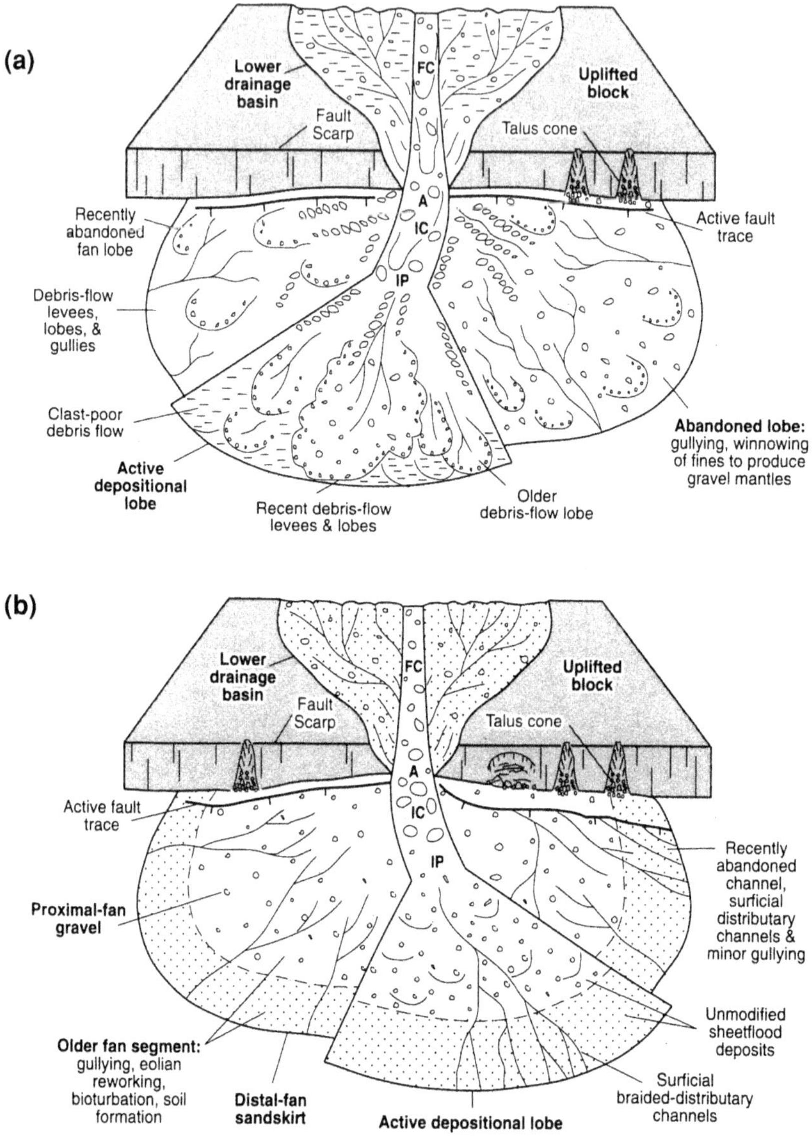

Fig. 3.12 Schematic diagrams showing the lower drainage basin, and primary and secondary depositional features of alluvial fans, including those dominated by (**a**) debris-flow processes (Type I) and (**b**) sheetflood processes (Type II). Note abbreviations: A = fan apex, FC = drainage-basin feeder channel, IC = incised channel on fan, IP = fan intersection point (modified from Blair and McPherson 1994a)

and sheetflood-dominated deposits, thereby resulting in a complex internal geology (Fig. 3.13). The internal geology of alluvial fans is extremely heterogeneous and the corresponding variation in hydraulic conductivity is quite extreme. The heterogeneity

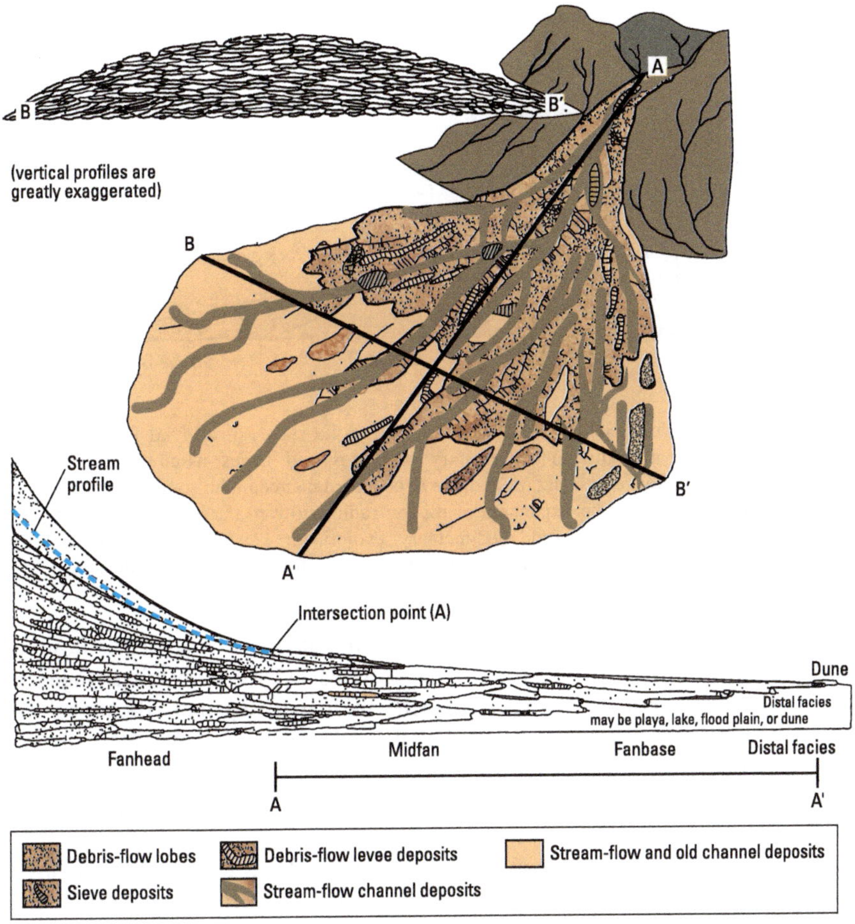

Fig. 3.13 Generalized model of alluvial fan sedimentation (modified from Spearing 1974 as presented in Nilsen 1982). The fan surface is presented in color with the longitudinal profile as A–A′ and the sectional profile as B–B′

results in creation of a layered aquifer consisting of permeable coarse-grained flow (aquifer) zones, separated by lower permeability, fine-grained, confining or semi-confining deposits.

Alluvial fans evolve over time to produce variable geometries and internal sedimentologic structures. All alluvial fans begin with a pre-existing slope on the margin of a mountain with available accommodation space in the bounding basin. According to Blair and McPherson (1994a) (Fig. 3.14), there are four fundamental stages of fan development, which are deposited on progressively decreasing slopes. The initial precursor stage consists of talus deposits at the base of mountains. Stage 1 consists of rock slide and colluvial slide debris deposited atop of the tallus deposits. Stage 2 fan deposits consist mostly of water-deposited sediments and vary in

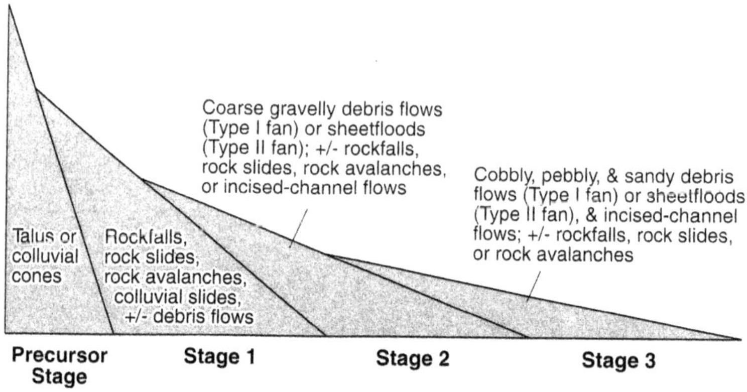

Fig. 3.14 Cross-sectional schematic diagram depicting the stages of alluvial fan evolutionary stages and their relationship to sedimentary processes and facies assemblages (vertically exaggerated). Depositions slopes increase to the *left* (toward the mountain source) and decrease to the *right* with a corresponding expansion of the fan radius (toward the basin floor). The stages refer to common morphologic and sedimentary evolutionary changes that fans and their respective drainage basins undergo as they increase in size (from Blair and McPherson 1994a)

composition depending on whether they are dominated by coarse debris-flow sediment (Type I fan) or sheetflow transported sediments (Type II fan). In a stage 2 fan deposit, various amounts of rock avalanche sediments and incised-channel flow sediments can also occur. A stage 3 alluvial fan sediment sequence is highly variable, depending on the fan type, and consists of some combination of cobbles, pebbles and sand deposited from sheetfloods, rockfalls, rock slides and rock avalanches.

The development of either a Type I or II alluvial fan and the accompanying internal sedimentological and hydrogeological characteristics are dependent on tectonism (slope and accommodation space), climate, and mountain rock type (Blair and McPherson 1994a). Type I fans differ in general character from Type II fans in that they contain coarser sediments (rock slide and debris flow dominated) and a lesser degree of depositional inter-fingering with basin-floor sediments. Type II fans contain a greater abundance of water-deposited gravels and sands and inter-finger to a greater degree with basin-floor sediment deposits (Fig. 3.15). An example of a Type II alluvial fan with sediment at a shoreline is shown in Fig. 3.16a. The fan lies between two rocky headlands and has been truncated by shoreline erosion. The truncation has occurred upgradient from the distal sand skirt and cuts across the inter-fingered gravel and sheetflood couplets. An eroded face of the alluvial fan shows sheetflood stratification consisting of predominantly sand and gravel with some cobbles and pebbles (Fig. 3.16b). Detailed characteristics of Type I and II alluvial fans are provided in Table 3.4.

Within both modern and ancient alluvial fan deposits, there are repetitive sequences of coarsening upwards deposits (Fig. 3.17) beginning with the basin floor deposits, which are overlain by sand and gravel deposited on the distal edge of the

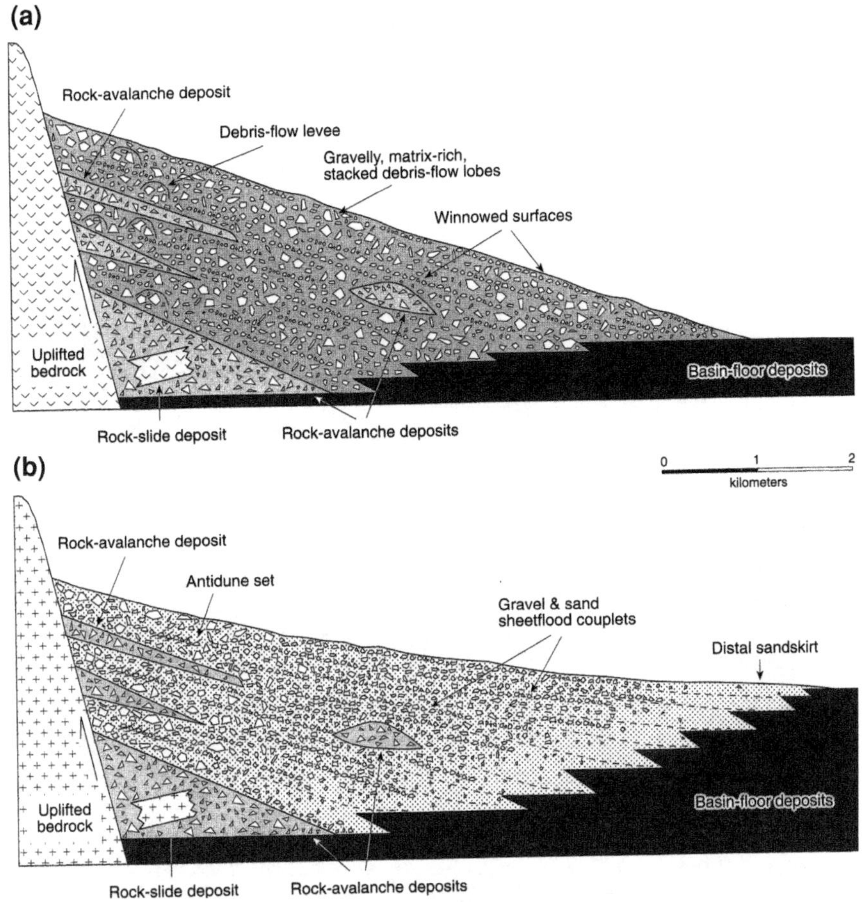

Fig. 3.15 Schematic cross-sections of sedimentary facies based on case studies for stage 2 or 3 alluvial fans, including (**a**) a debris-flow dominated fan (Type I), and (**b**) a sheetflood-dominated fan (Type II) with vertical exaggeration (from Blair and McPherson 1994a)

fan. The upper deposits in the sequence consist of boulders, cobbles, and pebbles deposited either as debris flows or within outwash channels. These stacked para sequences or sequences can be dominated by either debris flow, streamflow, or mixed deposits, as shown in an example from an ancient deposit (Fig. 3.18). Alluvial fan deposits also undergo further diagenetic processes which produce partially cemented layers within the already heterogeneous sediments (Fig. 3.19). Calcium carbonate crusts are deposited either internally within the sediments or on exposure horizons.

A critical part of the hydrogeology of alluvial fan deposits is the relationship and hydraulic connection between the high permeability alluvial sediments and the finer-grained deposits forming the basin floor (Fig. 3.20). Alluvial fan deposits often control the movement of groundwater from adjoining mountainous terrains

Fig. 3.16 *Top* An example of a Type II alluvial fan with the distal part of the fan truncated by a shoreline. The fan lies about 10 km south of Puerto Libertad, Mexico on the eastern shore of the Sea of Cortez. The fan lies between two rocky headlands and has been truncated by shoreline erosion. Note the gullies on the fan which occur near the distal part of it. *Bottom* Internal sedimentary structure of the alluvial fan. Bedding of sand and gravel is evident and the occurrence of pebbles and a few larger,angular rock fragments occurs

to the basin floor. In some areas the alluvial fan sediments grade laterally into fluvial plain or playa sediments, thereby limiting exploitable groundwater occurrence to only the mountain front and connected alluvial fan sediments. Where there is connection and inter-fingering of alluvial sediments with windblown or

Table 3.4 Characteristics of type I verses type II alluvial fans (from Blair and McPherson 1994a)

Feature	Type I alluvial fan	Type II alluvial fan
Dominant primary process and facies	Debris flows, especially lobe facies	Sheetfloods, especially couplet facies
Minor primary processes	Rockfall, rock slide, rock avalanche colluvial slide, incised channel	Rockfall, rock slide, rock avalanche, incised channel, noncohesive debris flow
Dominant secondary Processes	Winnowing by overland flows and wind to produce desert pavements, boulder mantles, gullies, and shallow channels	Winnowing by overland flows and wind to produce desert pavements, gullies, and shallow distributary channels
Typical grain sorting and size	Very poorly sorted clayey boulder, pebble, and cobble gravel	Poorly sorted sandy and boulder cobble to pebble cobble
Downfan trend in maximum clast size	Relatively constant	Typically decreases from boulders to pebbles or sand
Typical grain shape	Angular	Angular to subangular
Typical stratification style	Poorly or subtly stratified except for secondary winnowed surfaces	Well-stratified coarse gravel and sandy fine gravel couplets
Presence of granular or sandy interbeds	Rare	Common
Presence of a distal sand- skirt facies	Rare	Common
Presence of depositional clay matrix	Common	Rare
Drainage-basin size	Small to moderate	Small to large
Feeder channel length	Short to moderate	Moderate to long
Typical bedrock lithology underlying the drainage basin	Pelitic metamorphic rocks, shale, aphanitic, volcanic rocks, or mafic plutonic rocks; also granite or or gneissic rocks weathering under humid climate	Quartzite, quartz-rich conglomerate or sandstone; also granite or gneissic rocks weathering under arid conditions
Clay abundance in the drainage-basin colluvial slopes	Moderate to abundant	Rare
Common average slope value	5–15°	2–8°
Downfan slope style	Constant or straight	Distally decreasing or planoconcave
Permeability	Low	High
Porosity	Low	High
Connectivity of permeable units	Low	High

Fig. 3.17 Coarse-grained sheetflood deposit within a shallow channel located about 3 km from the shoreline near Mollendo, Peru. A lithified volcanic ash deposit divides the alluvial deposit and has been breached by a flood event. The sequence at this location is a mix of coarse alluvial deposits alternating with volcanic ash

dune deposits or coalescing alluvial fans, greater volumes of groundwater resources may occur over a greater area of the basin floor.

3.2.5 River Plains (Fluvial Deposits)

Fluvial sediment sequences in arid lands greatly differ from those in humid lands (Table 3.5). Knighton and Nanson (1997, pp. 201–202) conclude that

> The regime of arid zone rivers is characteristically unsteady. The uncertainties of hydrologic input coupled with the highly variable effects of transmission loss make their behavior much more difficult to predict than that of their humid zone counterparts. Relationships between process and form are less clear-cut, partly because discharge generally decreases downstream and partly because high-magnitude events exert such a dominant influence, both hydrologically and geomorphologically. The effects of floods can be preserved for long periods of time, especially where vegetation is very sparse, so that channel form may never become completely adjusted to coexisting processes. The crucial role of vegetation in determining the efficacy of erosional processes acting within and outside the channel/floodplain system is indicated by the sediment yield curves, whose form underlies how sensitive dryland environments can be to small changes in conditions.

Fluvial deposits are treated herein as the sum of those deposits in channels and those in the associated floodplain of the river or stream that are basinward of alluvial

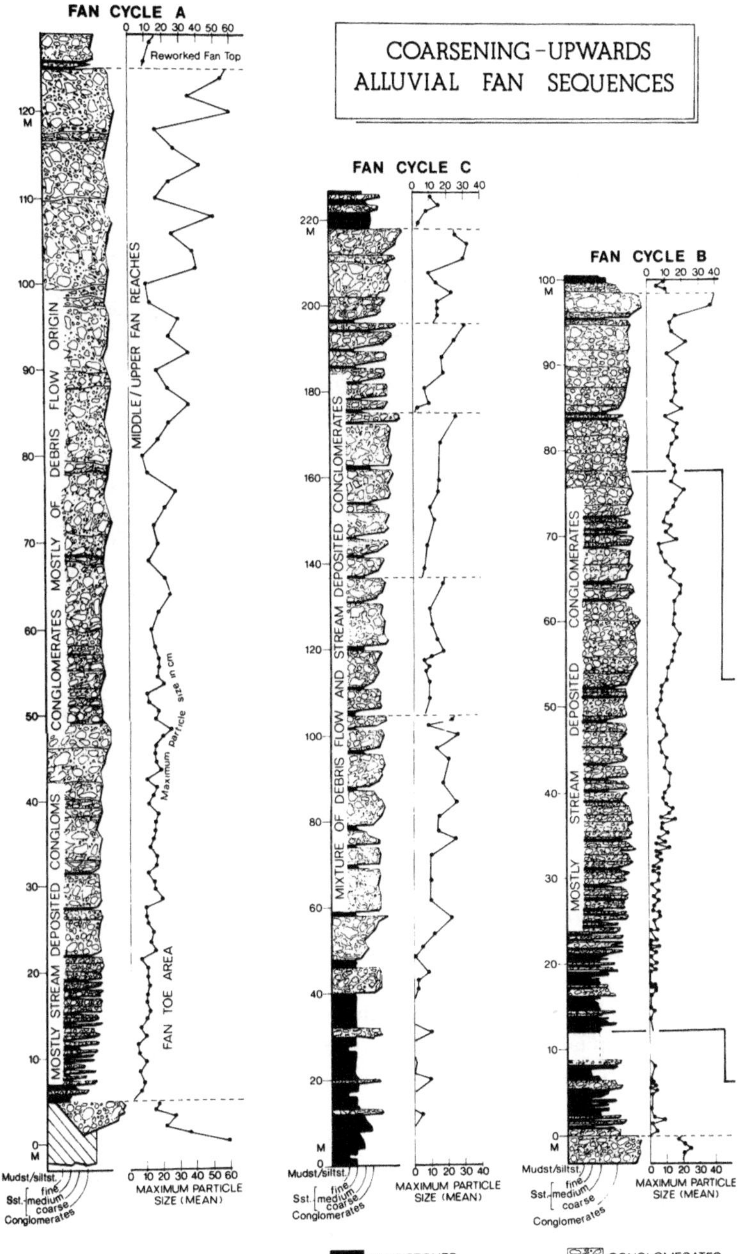

Fig. 3.18 Alluvial fan deposits tend to produce a series of coarsening upward sediment parasequences or sequences. This is an example of some ancient alluvial fan sequences from the Devonian of Hornelen Basin, Norway (from Steel et al. 1977 as modified by Rust 1979)

Fig. 3.19 Alluvial fan sheetflood deposit lying above a volcanic ash deposit with some layers cemented by calcium carbonate. The location is at the same site as shown in Fig. 3.17 (near Mollendo, Peru)

fan deposits. Wadis (ephemeral streams) share some similar sedimentary features and sequences with perennial fluvial deposits, but are confined within the channels (including the terraces within the channel boundaries) and are generally coarser-grained and less sorted, because of flashy flow and sediment transport. The horizontal relationship of alluvial fan facies and fluvial facies is shown in Fig. 3.20.

Fluvial deposits differ from alluvial fan deposits in that the latter include rock slide, colluvial slide, and wet slide deposits, which do not occur within fluvial sediment sequences. In certain cases, ephemeral streams can carry sediment downstream from mountains or lower slope high-elevation areas, without the formation of alluvial fans, producing sediment sequences that share similar characteristics to perennial stream deposits. Climate affects the relationship between drainage basin size and the bank-full width of a stream. Drier regions tend to contain streams that are very wide relative to their drainage areas in order to accommodate the flashy nature of storm events (Fig. 3.21). The relationship between the bank-full channel width and the drainage area of streams in hyperarid

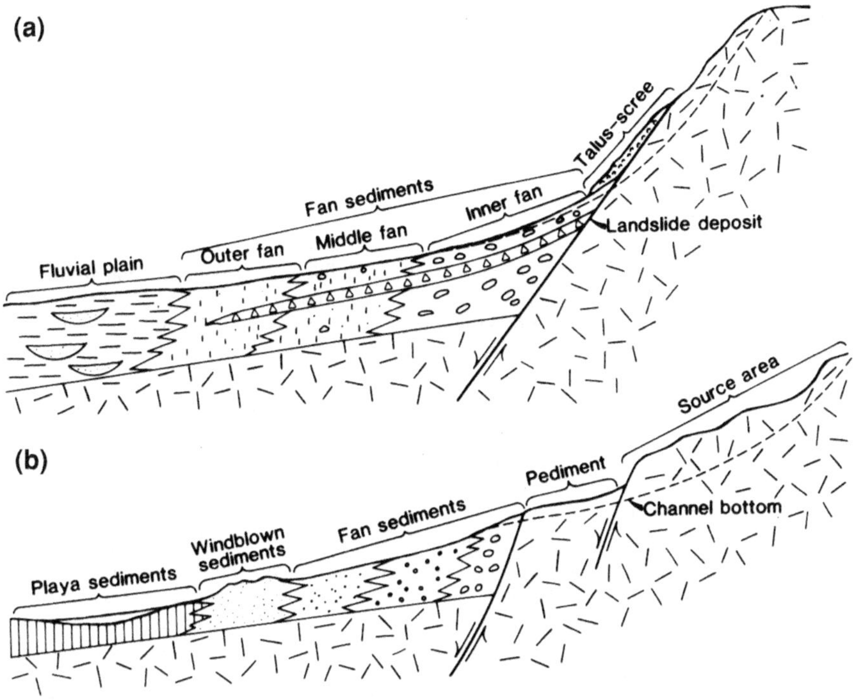

Fig. 3.20 Cross-section showing the lateral relationship of alluvial fan deposits with basin floor sediment facies. **a** Inter-fingering with fluvial plain deposits. **b** Inter-fingering with playa and aeolian deposits (from Nilsen 1982)

areas varies in the extreme because there are a large number of variables controlling the runoff characteristics of a given basin (see alluvial fans, Sect. 3.2.4).

The characteristics of fluvial sedimentary deposits are created by the threshold power of the streams as horizontal stream gradients decrease. As stream gradients decrease downgradient, flow velocity and energy decreases, coarser sediments are deposited, and the remaining sediment load becomes finer grained. The relationship between the critical power of a stream in an arid region and the resulting geometry of the valley and associated sediment deposits is illustrated in Fig. 3.22. Near the proximal segment of a stream adjacent to a mountainous area, fluvial sediment sequences will be coarser-grained and may merge with purely alluvial sediments, such as found in alluvial fans.

Within mid-range segments of streams, the sediment sequences may take on the characteristics of a braided stream. In the distal, basinal stream segment, the outwash deposits may resemble more classical fluvial sequences with a number of stacked, fining upward units. The main fluvial facies correspond to the alluvial fan edge, the proximal braided river, and the distal braided river (Miall 1978b) (Table 3.6 and Fig. 3.23). Rust (1979) described two fundamental sediment facies that relate to arid lands fluvial deposits. These facies correspond to proximal

Table 3.5 A comparison of hydrologic input, throughput, and output in arid and humid zones (from Knighton and Nanson 1997)

	Arid zone	Humid zone
Input	Low and unreliable. Limited duration but often high intensity storms. Extremely variable at the event and annual scales patially concentrated events	Relatively high and dependable (often seasonally so). Long duration (often frontal) precipitation of variably intensity. Temporal variability is much less. Large areas generally affected
Thoughput	Horton overall flow dominant. Rapid onset of surface runoff. Relatively high runoff coefficient. Decreasing discharge downstream due totransmission losses	Infiltration, throughflow and groundwater flow more significant. Longer lag between precipitation and runoff. Lower runoff coefficent. Increasing discharge downstream due to tributary inflows
Output	Mostly intermittent. Extremely flashy regime. Sharply peaked runoff hydrographic. Considerable interannual variability	Largely perennial. Relatively steady regime. Runoff hydrographs have lower amplitude. Dependable interannual flows
Channels	Drainage densities can be high but networks may not be fully integrated. Floods as major drainage controls. Long recovery time after disturbances. Transient behavior is dominant	Well integrated drainage networks. Channels adjust to a range of more frequent discharges. Channel recovery is quicker. Tendency for channels to equilibrate

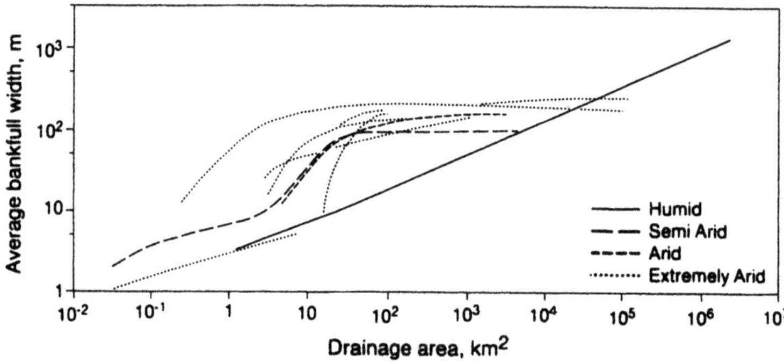

Fig. 3.21 Diagram showing the relationship between channel width and the drainage area for different climates (from Wolman and Gerson 1978, as presented in Knighton and Nanson 1997)

braided rivers and braidplains (Gp) and distal braided rivers and braidplains (Gm). Proximal braided rivers and braidplains are apparently only distinguishable in ancient rock deposits and not in more recent deposits, so they can be treated as one deposit. This deposit is characterized by horizontally bedded, clast-supported, imbricate gravels. Distal braided rivers and braidplains are characterized by fining-upward parasequences dominated by clast-supported, trough cross-stratified gravel, which is overlain by horizontally-stratified gravel or trough cross-stratified

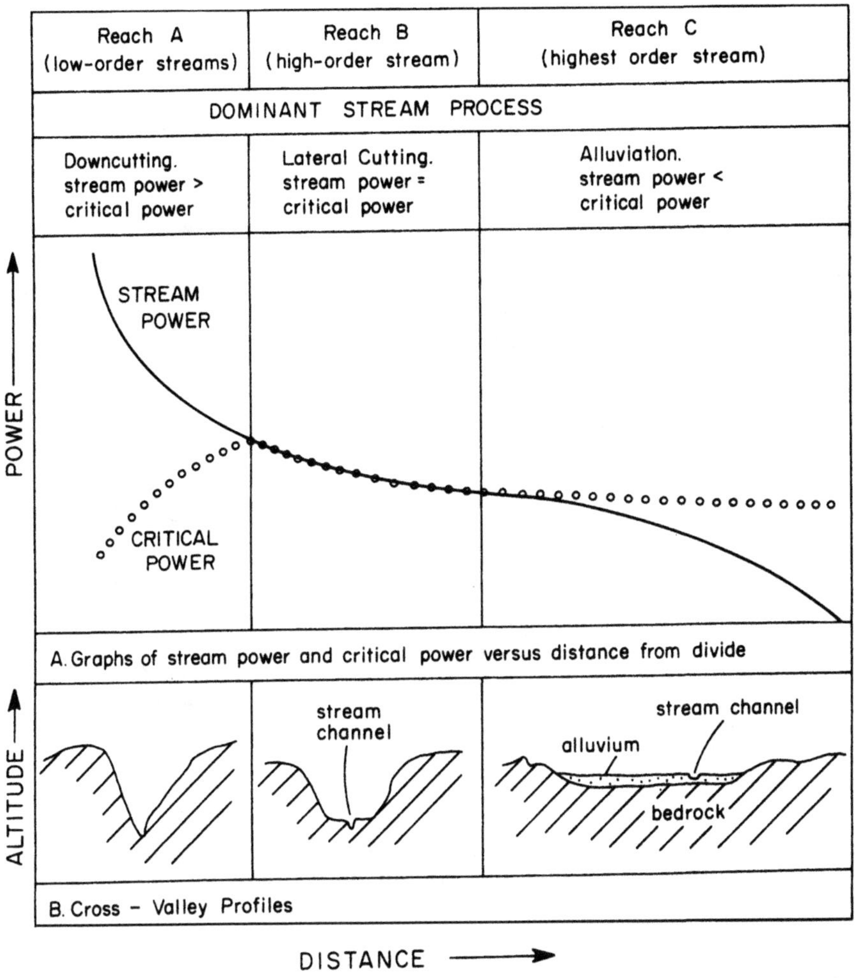

Fig. 3.22 Relationships between valley profiles and stream power and critical power for arid rocky drainage basins (after Bull 1979)

Table 3.6 Three facies assemblages in gravel- and sand-dominated braided river deposits (based on Miall 1977 and Rust 1978)

Name	Environmental setting	Main facies	Minor facies
Trollheim type (G_I)	Proximal rivers (predominantly alluvial Fans) subject to debris flows	Gms, Gm	St, Sp, Fl, Fm
Scott type (G_{II})	Proximal rivers (including alluvial fans) with stream flows	Gm	Gp, Gt, Sp, St, Sr, Fl Fm
Donjek type (G_{III})	Distal gravelly rivers (cyclic deposits)	Gm, Gt St	Gp, Sh, Sr, Sp, Fl, Fm

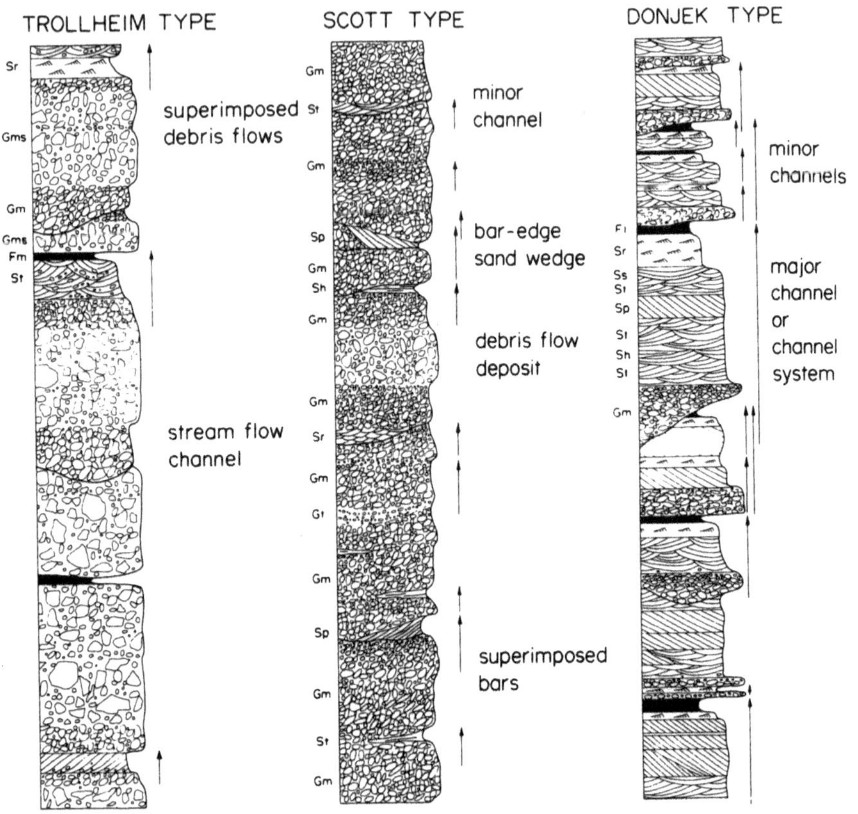

Fig. 3.23 Vertical sedimentary profile models for braided stream deposits. The Trollheim Type represents a proximal stream adjacent or merging with an alluvial fan. The Scott Type is represents a profile proximal within the stream basin, but away from alluvial fan debris flows. The Donjek Types represents distal deposits near or at the flow loss point for dry lands fluvial systems. The arrows show small-scale cyclic parasequences or sequences, depending upon the scale of the features. Conglomerate clasts are not shown to scale. The facies descriptions are given in Table 3.7 (from Miall 1978b)

and horizontally-stratified sand. Each parasequence ends at the top with a massive or laminated mud. Each parasequence is topped with a discontinuity surface or disconformity depending on the timing between depositional events. The facies symbols in the text and their descriptions are given in Table 3.7.

It can be difficult to distinguish between ephemeral and perennial stream deposits in the middle and distal parts of dry land fluvial systems. Apparently, more investigations have been conducted on ancient rock deposits than on modern arid land fluvial deposits. Reid and Frostick (1997) suggest that the distinguishing characteristics of desert or dry lands stream deposits, include generally thin (0.1–0.3 m; [0.3–1.0 ft] scour and fill deposits, horizontally-laminated sand beds

Table 3.7 Lithofacies and sedimentary structure of modern and ancient stream deposits identified in this text (modified from Miall 1977)

Facies code	Lithofacies	Sedimentary structures	Interpretations
Gms	Massive, matrix supported gravel	None	Debris flow deposits
Gm	Massive or crudely bedded gravel	Horizontal bedding imbrications	Longitudinal bars, lag deposits sieve deposits
Gt	Gravel stratified	Trough crossbeds	Minor channel fills
Gp	Gravel stratified	Planar crossbeds	Linguoid bars or deltaic growths from older bar remnants
St	Sand, medium to very coarse may be pebbly	Solitary (theta) or grouped (pi) trough crossbeds	Dunes (lower flow regime)
Sp	Sand, medium to very coarse may be pebbly	Solitary (alpha) or grouped (omicron) planar crosssets	Linguoid, traverse bars, sand waves (lower flow regime)
Sr	Sand, very fine to coarse	Ripple marks of all types	Ripples (lower flow regime)
Sh	Sand, very fine to very coarse may be pebbly	Horizontal lamination, parting or streaming lineation	Planar bed flow (l. and u. flow regime)
Ss	Sand, fine to coarse, may be pebbly	Broad, shallow scours including eta cross-stratification	Scour fills
Fl	Sand, silt, mud	Fine lamination, very small ripples	Overbank or waning flood deposits
Fm	Mud, silt	Massive, desiccation cracks	Overbank or drape deposits

with alternating coarse and fine sands, and the presence of mud drapes and interclass near the distal part of the stream basin.

Major rivers run through some of the most arid lands in the world (e.g., Nile River, Colorado River of the Western United States, Rio Tambo of Peru). The headwaters of these rivers commonly occur in humid regions or in mountains fed by glacial or snow melt, located far from arid lands. The fluvial sediment facies within these systems include virtually all fluvial system types, ranging from mixed alluvial fan and fluvial, proximal braided river, distal braided river, sand bars within heavily eroded canyons, typical overbank floodplain sands, silts, and muds, and deltaic facies. There rivers are normally perennial systems, except in some specific reaches during very dry periods. The rivers may be gaining streams in the incised canyon reaches and losing streams in the hyperarid, moderately-sloping reaches. The river-floodplain and bounding basin water interaction is controlled by the nature (hydraulic properties) of the underlying and bounding sediments and rock formations in connection with these rivers. The detailed geology of the deposits associated with perennial rivers is discussed in a number of very good references (e.g., Miall 1978a; Smith and Rodgers 1999; Blum et al. 2005).

3.2.6 Dry Watercourses (Arroyos and Wadis)

Dry watercourses or stream channels are common geomorphic features in arid lands, which are of great hydrologic importance as loci of groundwater recharge. These ephemeral streams can be either erosional or depositional features, depending upon their position (upstream versus downstream), time, and flow conditions during a specific storm event. Sediment transport commonly occurs in large quantities in these channels and produces over-bank deposits of coarse and fine-grained sediment. Dry watercourses should perhaps be termed "alluvial rivers", which are defined as rivers that occupy a broad flood-plain belt over which the depth of alluvium deposited by the river equals or exceeds the depth to which scour takes place in time of flood (American Geological Institute 1997).

Wadi is a term used in the desert regions of the Middle East, northern Africa, and southwest Asia for "a stream bed or channel, or a steep-sided and boulder ravine, gully, or valley, or a dry wash, that is usually dry except during the rainy season, and that often forms an oasis" (American Geological Institute 1997, p. 706). Wadi may refer to either only the channel or, in some usages, to the entire stream valley. A wadi is similar in nature to the arroyo of the southwestern part of the United States. Arroyos are defined as "a term applied in the arid and semi-arid regions of the southwest United States to the small, deep, flat-floored channel or gully of an ephemeral stream or of an intermittent stream, usually with vertical or steeply cut banks of unconsolidated material at least 60 cm (24 in.) high; it is usually dry, but may be transformed into a temporary watercourse or short-lived torrent after heavy rainfall" (American Geological Institute 1997, p. 36).

The terms wadi and arroyo are often used interchangeably in the technical literature for ephemeral streams, depending upon geographic location. Generally, wadis have a larger range in landscape size compared arroyos. However, there are ephemeral rivers in the southwestern United States (e.g., Salt River in Arizona) that are similar to the larger wadis of Saudi Arabia. Commonly, in the United States, larger wadi-like features are termed "rivers" or ephemeral streams. An example of a large wadi is shown in Fig. 3.24, located approximately 50 km (30 mi) northwest of Riyadh, Saudi Arabia. The illustrated wadi contains a central channel and terraces adjacent to the primary channel below the rim of the feature. The wadi is located distal from steep-slope areas. Arroyos in the southwestern United States (e.g., Fig. 3.25) are often considerably smaller features. Arroyos located proximal to steep-sloped foothills often have steep banks and contain very large, angular rock fragments, reflecting the high energy of flash floods.

Few comprehensive investigations of wadi sediments have been performed to determine the three-dimensional sedimentological and hydrogeological characteristics of the channel deposits. Al-Shaibani (2008) performed a hydrogeologic investigation of Wadi Wajj in western Sandi Arabia. More than 30 hand-dug wells were constructed to collect sediment samples and some grain size analyses were performed. The thickness of the wadi deposits ranged from a few m to 20 m (few ft to 65.6 ft) as measured from land surface to the base of the channel. Al-Shaibani

Fig. 3.24 Large wadi located approximately 50 km northwest of Riyadh in the Al-Bawada area of Saudi Arabia. Note the steep relief at the edge of the wadi and the terraces above the base channel. This is a distal erosional feature containing a mixed bedload of pebbles, sand, and some mud. The trees growing in the central channel show that no severe flood event has occurred in recent time and there is sufficient stored water to maintain the vegetative growth (Photograph was provided by Frank Winslow)

describes only two hydraulically connected units, which are a gravel and sand unit with high hydraulic conductivity and fractured and weathered crystalline rock below the sediments. Unweathered crystalline rock forms the base of the channel. The channel fill sediments are poorly sorted, consisting of mostly fine to coarse-sized sand. Unfortunately, no more detailed descriptions of the geology were reported.

Sedimentation or erosion within wadis is determined by the location of the wadi with regard to a mountain front or other high ground (which determines the slope and energy of water), and the intensity of a given flood event. Obviously, large, intense flood events can cause either erosion or deposition of coarse-grained sediment within the channel. The surface of the Schwaib Wadi in the United Arab Emirates consists of alluvial deposits, containing predominantly pebbles and sand, which were deposited during the last major flood event (Fig. 3.26a). A smaller event that carried only mud, occurring at some time in 2008 or early in 2009, and is represented by a thin mud deposit, which is present on the eastern side of the channel, near the fence dividing the United Arab Emirates from Oman (Fig. 3.26b).

The variation in flow energy and resulting sediment grain size results in wadi channel deposits that consist of pebble and sand deposits separated by thin muddy units or, in some cases, episodic scouring events that homogenize the deposits into

Fig. 3.25 Small arroyo located in White Tank Mountain Park, Arizona. The photograph was taken about 1 week after a substantial rainfall event. Note the steep sides of the feature and the large, angular rock bedload. This is a proximal feature located very close to the sediment source-the foothills in the background of the photograph

poorly-sorted mixes of pebbles, sand, and clay or mud clasts. The sedimentology of wadi deposits is quite important in producing the variable vertical and horizontal hydraulic conductivity that controls aquifer recharge and general groundwater flow. Currently, insufficient geologic information has been collected within wadi or arroyo features to truly analyze and model these features. This is a research subject of considerable importance.

Another important feature of wadis and arroyos is the relief of the bedrock underlying the channel. The bedrock altitude does not always follow the topography of land surface and the current channel. Longitudinal undulation of the bedrock surface below a wadi channel can have a significant effect on horizontal, subsurface water flow beneath a channel. Subsurface highs of the bedrock can act as subsurface groundwater dam and cause entrapment of water or cause emergence (discharge) of water as it drains toward the basin floor.

There is much to be learned concerning wadi geology and hydrogeology. These systems are very heterogeneous and the sediment sequences can be repetitive or fully mixed. Moving away from alluvial fans or the mountain front, the vertical variations in geology can have the characteristics of virtually any of the sequences shown in Fig. 3.23 or variations within these sediment types. Detailed geological investigations are needed along wadi channels from the mountain front to basin floors to gain better knowledge on the internal geology of these features.

Fig. 3.26 *Top* Floor of the Schwaib Wadi in the United Arab Emirates facing southeast. The fence line is the border with Oman. Runoff flows from the mountains in the background into the wadi. Note that the wadi surface contains a poorly-sorted deposit consisting predominantly of pebbles and sand. *Bottom* View of the floor of the Schwaib Wadi. A relatively recent (late 2008 or 2009) event produced a low flow carrying mud into the wadi. Note that the deposit has desiccated and occurs as mud clasts

3.2.7 Badlands

The geomorphic term "badland" probably was derived from the French phase "mauvais terres a traverse" or land that is "difficult to travel across" (Campbell 1997). Badlands form in horizontally-bedded, low permeability, soft-rock terrains by

Fig. 3.27 *Top* Badlands landscape from the Petrified Forest National Park, Arizona, USA. *Bottom* Badlands erosion and gullying (photographs provided by Thomas M. Scott)

fluvial erosion of the landscape (Campbell 1997). Badlands are a relatively minor part of the landscape in arid areas. Clements et al. (1957), in a rather narrow and selective review, determined that badlands constitute about 2.6% of the southwest part of the United States, 8% of the Libyan Desert, 2% of the Sahara Desert, and 1% of the Arabian Desert. Example of a badlands landscape is given in Figs. 3.27a, b.

A key aspect of badlands that influences arid land geology and hydrogeology is the relatively soft, clayey nature of the rocks and sediments in which badlands are formed. As these sediments weather and are transported downslope in rainfall events, they produce a significant load of clay and silt, which may form clay clasts or mud drapes in downstream fluvial deposits. Badlands share a common trait with pediments in that small amounts of rainfall (as little as 3 mm [0.1 in]) will result in runoff (e.g. badlands of Alberta; Byran et al. 1978).

An important part of badlands is gullies that form and convey runoff downslope. Badlands gullies are considered by some to be distinct from arroyos, which are alluvium-filled and are flow-through features (Graf 1983), while others make no distinction between badlands gullies and arroyos, such as Cooke and Reaves (1976), who term arroyos "valley-bottom gullies". Deep gullies within badlands can form the headwaters of ephemeral streams that may be arroyos or wadis. Also, badlands can lie at the headwaters of perennial streams and contribute significant runoff based on their very high runoff coefficient.

Cyclicity in the erosion of gullies within badlands can cause episodic downstream movement of muddy sediments. The deposition of mud within the ephemeral stream channels (arroyos or wadis) or overbank in the floodplain can result in the formation of internal mud beds or laminations within predominantly gravel and sand deposits. This is another source of heterogeneity within alluvial deposits that can affect local groundwater flow.

This discussion on badlands has emphasized their contribution of sediments and runoff to wadis, arroyos, and the over alluvial/fluvial systems in outwash deposits, particularly contributions that affect heterogeneity and subsurface hydraulic properties. Runoff, erosion, and geochemical processes that create geomorphic features within badlands, such as gully formation, piping, and collapse, are complex and described in greater detail by Campbell (1997), Bryan and Yair (1982), Jones (1971, 1990), and Schumm et al. (1984).

3.2.8 Playas

Playas, playa lakes, playa basins, and associated deposits constitute only about 1% of the arid land environments, but can be quite significant (Shaw and Thomas 1997). A playa is defined as "a dry, vegetation-free, flat area, the lowest part of an undrained desert basin, underlain by stratified clay, silt, or sand, and commonly by soluble salts. The term is also applied to a basin containing an expanse of playa(s) which may be marked by ephemeral lakes" (American Institute of Geology 1997). A number of other terms have been applied to the same geomorphic feature in different parts of the world. These terms are: salina, salada, salar, saline mudflat, pan, salt pan, alkali flat, and others. In the Middle East, the terms playa and sabkha are sometimes used interchangeably. In the sedimentological literature, the term "sabkha" is typically restricted to coastal sabkhas, which are a type of tidal flat characterized by the deposition of evaporite minerals (e.g., gypsum and anhydrite). Coastal sabkhas are flooded by periodic tidal events, but in certain cases may have

Table 3.8 Playa type classification (after Snyder 1962)

Author	Playa types		
Foshag (1926)	Moist playa	Dry playa	___
Thompson (1929)	Moist playa	Dry playa	Crystal-body playa
Jaeger (1942)	Salt pan	Salt-free or dry playa	Lime pan or fflalkpfannen
Stone (1956)	Moist type Salt crusted/ Clay crusted	Dry type Clay pan/Lime pan	Compound/Artificial/ Crystal body __ __ __

a groundwater inflow component and there can be a seaward extension of a playa into a coastal sabkha (Bye and Haribson 1991).

There are different types of playas that vary depending on the geological conditions in and surrounding the valley system, and whether they form in either closed-basin or open-basin systems (Snyder 1962). Playas can be classified based on their predominant moisture content and association with evaporate formation (Table 3.8). The geological characteristics of playa sediments and the salinity of the system are greatly dependent on the degree of isolation of the systems and the inputs of surface water and groundwater. The geology of playas or pans is very much controlled by the hydrology and Bowler (1986) defined the morphology of evaporation basins by the interaction of surface water and groundwater (Fig. 3.28). With increasing aridity, Bowler (1986) determined that the predominant playa sedimentation changed from carbonates to evaporites.

Within closed basins or valleys with no surface water or groundwater outflow, much of the surface water flow through alluvial fans and ephemeral streams is lost during transmission to channel infiltration. Usually only during major flood events does surface water reach the playa at the topographic low point in a basin. Groundwater discharge occurs within closed-basin playas, commonly as seeps or springs where the water table position is characteristically high. The salinity of the water in playas increases as the result of evaporation, and evaporite deposits form that commonly consist of salt (Fig. 3.29) and gypsum mixed with clay and some wind-blown sand.

Salt accumulates very slowly within playa basins if surface water is the sole source of the salts. Hardie et al. (1978) found that a major flood in Lake Eyre (Australia) in 1950, covered 8,000 km^2 (3,089 mi^2) and evaporated within 2 years leaving a thin layer halite over only one tenth of the original flood area. Holser (1979) calculated that a 200 m (656 ft) depth of surface water would produce a 3 m (98. Ft) thick deposit of halite. The shallow stratigraphy of a playa within Salar de Atacama, Chile, has a salt crust in the upper 20–25 cm (7.9–9.8 in) with mixed salt and clay (with some sand) or clay lying from 20 to 30 cm (7.9–11.8 in.) or greater below the upper crust. Mixed evaporites and fine-grained siliciclastics occur to depths over 135 cm (53.1 in.) below surface (McCartney 2001). This deposit must be very old or the groundwater input from springs contributes a significant portion of the salts to the basin. A dry playa surface can be rough and furrowed or relatively smooth like the Bonnaville salt flats in Utah.

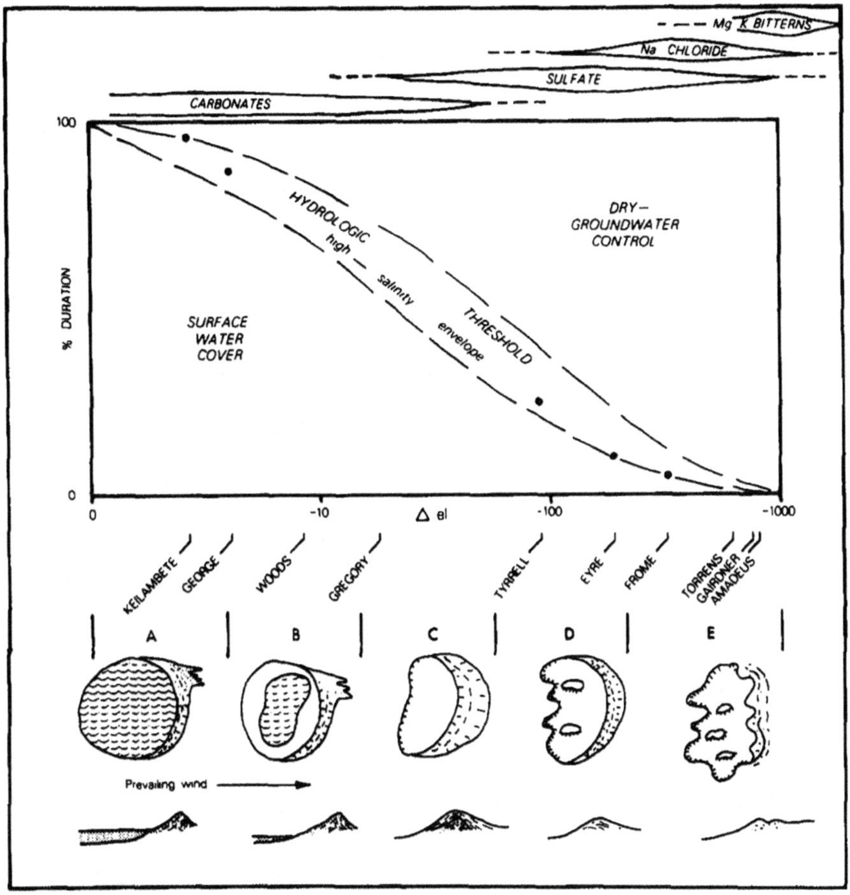

Fig. 3.28 Classification of playas (pans) based on hydrology, morphology, and surface-water/groundwater interaction. The hydrology controls the deposition of evaporate type. The disequilibrium index (Δe) is taken from data collected in Australia from humid to dry regions (Keilambete to Amadeus). (from Bowler 1986)

Open-basin playas tend to flood during periodic wet periods and be dry during dry periods. Typically, closed valleys containing drained playas do not have evaporite deposits, because the surface water tends to recharge unconfined aquifers with water levels located far below land surface, not allowing evaporative concentration (Snyder 1962). The playa geology in these valleys consists of clay, silt, and fine sand that collectively allow vertical flow of water. Classification of open-basin playas has been based on the position of the water table by Mabutt (1977), Wood and Sanford (1990) and Rosen (1994). Rosen (1994) described playas with a low water table as "recharge playas". These playa types tend to have clay or vegetated surfaces and no accumulation of evaporites.

Fig. 3.29 Salt deposits in Badwater Basin salt flats (playa), Death Valley National Park, California, U.S.A. *Source* National Park Service

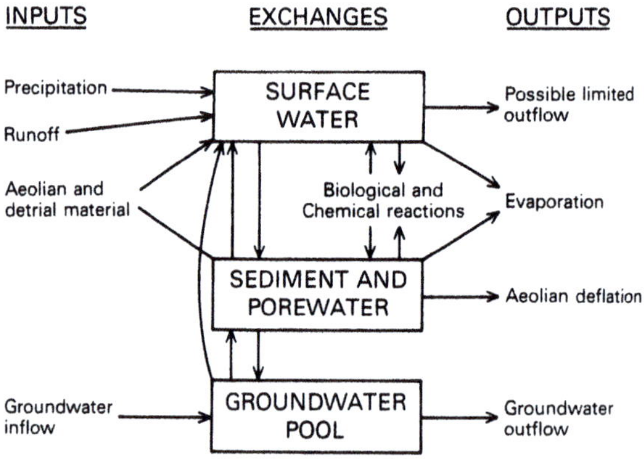

Fig. 3.30 Diagram showing the idealized exchange pathways for water and salts in a playa (pan) system (after Torgersen et al. 1986)

Playa facies are very site-specific and are controlled by the basin inflow, width of the valley, relative hydraulic gradient into the basin (tectonic controlled), and whether the valley is a fully closed or an open system with respect to groundwater flow. The characteristics of important inflows and outflows to and from playas have been defined by Torgerson et al. (1986) and are shown in Fig. 3.30. These inputs and outputs determine the sediment types and geometry of the playa

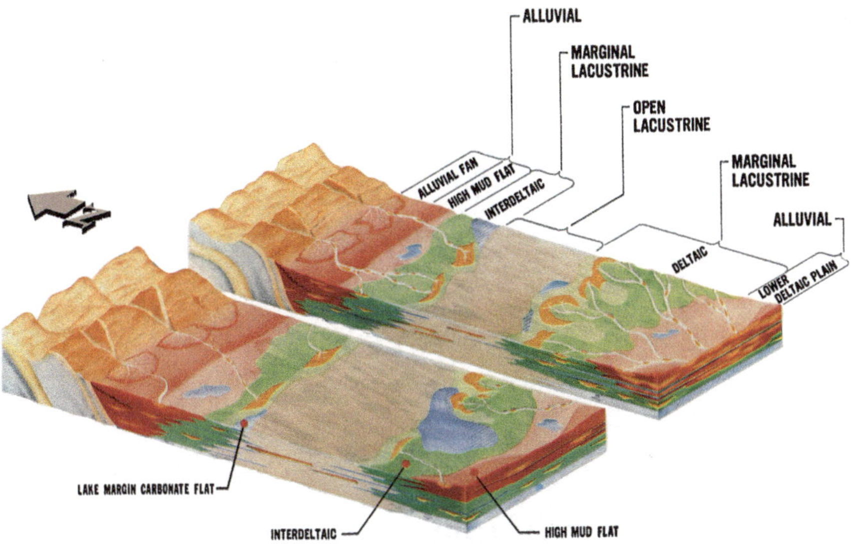

Fig. 3.31 Block diagram showing the distribution of interpreted open-circulation, marginal lacutrine, and alluvial depositional environments of the western part of Lake Uinta, Utah, as it existed during the early Eocene. Diagonally stripped blue pattern is grain-supported carbonate rock; solid blue pattern is mud-supported carbonate rock. Yellow pattern is sandstone, and claystone units are shown in their natural color. Some thin graded siliciclastic units of probable turbidite origin are present in delta fronts and in the open lacustrine facies. Much of the open lacustrine is composed of kerogen-rich carbonate units. The width of Lake Uinta in the diagram is about 40 km. Vertical exaggeration is between 15 and 20. The diagram was modified from Ryder et al. (1976) as presented in Fouch and Dean (1982)

lithofacies. An example of an ancient geologic depositional pattern of a playa/lacustrine system is the Eocene in the western part of Lake Uinta, Utah (Fig. 3.31). In this case, there is a transition or facies change from alluvial fan sediments into high mud flat deposited sediments into an inter-deltaic sediment facies and ultimately into an open lake sediment facies consisting of predominantly organic-rich carbonates. This system was likely relatively wet and hydrologically open, because the central lake or playa did not produce an evaporite sequence.

Within closed systems, the playa sediments are a combination of fine-grained clays and silts, mixed with evaporites. The overall sediment sequence has a very low hydraulic conductivity and will not transmit water out into the desert floor (where lower) or via groundwater flow. These sediment accumulations could act as groundwater dams, trapping water in alluvial fan sediments or in the sediments within feeding wadis or arroyos.

Within open playa basins, the sediments, although still having a generally fined-grained nature, have sufficient vertical and horizontal hydraulic conductivity to allow groundwater recharge to occur, with perhaps some of the groundwater entering inter-basin flow systems. Where ancient playa or lacustrine carbonate deposits occur, the initial sediments will also have a low hydraulic conductivity.

Fig. 3.32 **a** Ripple marks on a dune with lizard tracks crossing the feature. This is the smallest bed form in an erg system. The photograph was taken in the United Arab Emirates on the eastern end of the Rub al Khali erg. **b** Draa in Rub al Khali erg near Al Ain, United Arab Emirates. The draa is a sand ridge that is the largest bed form in an erg and constitutes the largest thickness of permeable sand within an erg. **c** Dune field near the southeastern margin of the Rub al Khali erg in the United Arab Emirates. The dunes are the middle-sized bed form in an erg. The mountains in the foreground are in Oman

However, the system may develop into an aquifer created by dissolution and diagenesis over an extended time period with a connection to high permeability alluvial fan sediments.

3.2.9 Sand Seas (Ergs)

Large-scale aeolian deposits are an important part of arid lands geology and hydrogeology. These predominantly sand deposits commonly inter-finger with alluvial fan, wadi, and playa deposits. Since aeolian sand deposits are generally quite permeable, they have the capability to store and transmit groundwater and therefore, can be locally important in the hydrogeology of arid lands.

Arid lands aeolian deposits or sand seas are termed "ergs", which is an Arabic term. Wilson (1973, p. 78) defined erg as "an area where wind-laid sand deposits cover at least 20% of the ground, and which is large enough to contain draas." A number of other terms have also been applied to ergs, such as "sand sea", "sand-ridge desert" (Australia), "nafud" (Arabia), "edeyen" (Libya), "issaouana"

(Algeria), and "kum" or perki" (Central Asia) (Wilson 1973). Ergs contain three repetitive bed forms, including, from small to large scale, ripples (1–20 cm [0.4–7.9 in.] apart), dunes (commonly 25–250 m [82–820 ft] apart), and draas (sand ridges 1–3 km [0.6–1.8 mi] apart) (Wilson 1972). Examples of the three bed forms from in the eastern part of the Rub al Khali (United Arab Emirates part) erg are shown in Figs. 3.32a, b, c.

There are three types of ergs, including free and active, vegetated and active, and fixed by vegetation (Wilson 1973). Active ergs may contain some vegetation and receive some rainfall, but the vegetation and rainfall do not interfere with the movement of the sand and evolution of the landform. Progressive increase in the amount of vegetation affects the bed-form pattern (Melton 1940), especially a rounding of features and the occurrence of parabolic blowout dunes (McKee 1966). When the surface vegetation becomes significant, land forms become fixed and the feature would be considered a dead erg.

A key issue regarding the relationship between the geology and hydrogeology of ergs is their geometry and relationship to mountains, alluvial fans, wadis, and playas. Ergs are virtually always confined to basins and tend to terminate at significant breaks in slope (Wilson 1973). Ergs tend to occur in the lower altitudes areas of basins, but are also controlled by small changes in relief within the basins (Wilson 1973) and rarely occur on highland massifs (Wilson 1971). Ergs require a source of the sand to form. Ergs, therefore, will generally form down-wind of sand sources, particularly areas with high deflation rates and near sand-sized fluvial deposits (Wilson 1973). Sands can be transported both into an erg and pass from one erg into another at considerable distance (Kuenen 1960).

In terms of size, they range from 1 to 1,000,000 km^2 (0.4–386,100 mi^2) in area, but most fixed ergs have areas greater than 12,000 km^2 (4,633 mi^2) (Wilson 1973). The Rub al Khali erg on the Arabian Peninsula covers 560,000 km^2 (216,216 mi^2) (see Figs. 3.32a, b, c). The shape of ergs is irregular and depends upon on the topographic conditions within the basin in which they lie. Sand occurrence within an erg is not a 100% cover, particularly along margins and between major draas and in an open erg. The sand cover generally ranges between 20 and 80% (Wilson 1973). The thickness of sand within an erg is highly variable, but there are draas that have crest to trough heights ranging from 10 to 400 m (32.8–1,312 ft) (Wilson 1973). Four ergs studied by Wilson (1973) have a mean average sand thickness of 14.5 m (48 ft). Algerian ergs have an average sand thickness of 26 m (85.3 ft) (Miall 1990).

Despite the fact that dune sands have a generally high hydraulic conductivity, variations in composition and grain size can cause heterogeneities. Also, in some dead erg systems, deposits of calcrete and silcrete can form that will tend to inhibit vertical water movement. Four types of grain size regimes have been defined from erg sands, which include: (1) very coarse-sand ergs (sheet sands without dunes), (2) coarse-sand ergs (dunes present without slip faces), (3) fine-sand ergs, and (4) mixed sand ergs (dunes with and without slip faces, intermixed) (Wilson 1973). The size distribution of the sands affects the horizontal and vertical hydraulic conductivity. Sands with a bimodal size distribution containing a coarse and fine

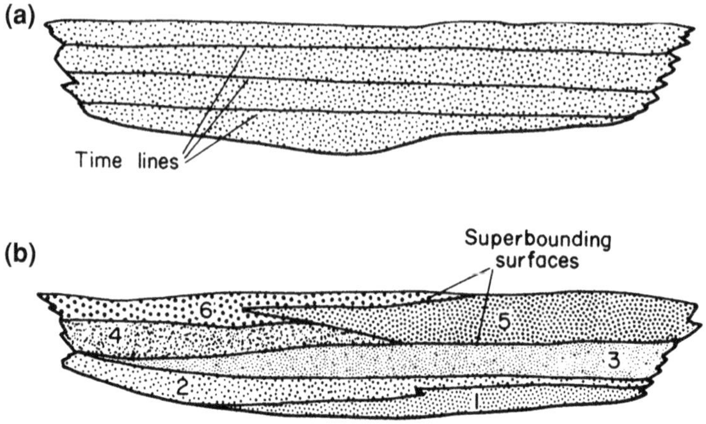

Fig. 3.33 Diagram showing two examples of basin-fill erg deposits. **a** Succession of erg blankets filling the entire basin with vertical accretion is shown in (**a**). Diachronous basin-fill deposits formed by migrating ergs is shown in (**b**). (from Kocurek 1988)

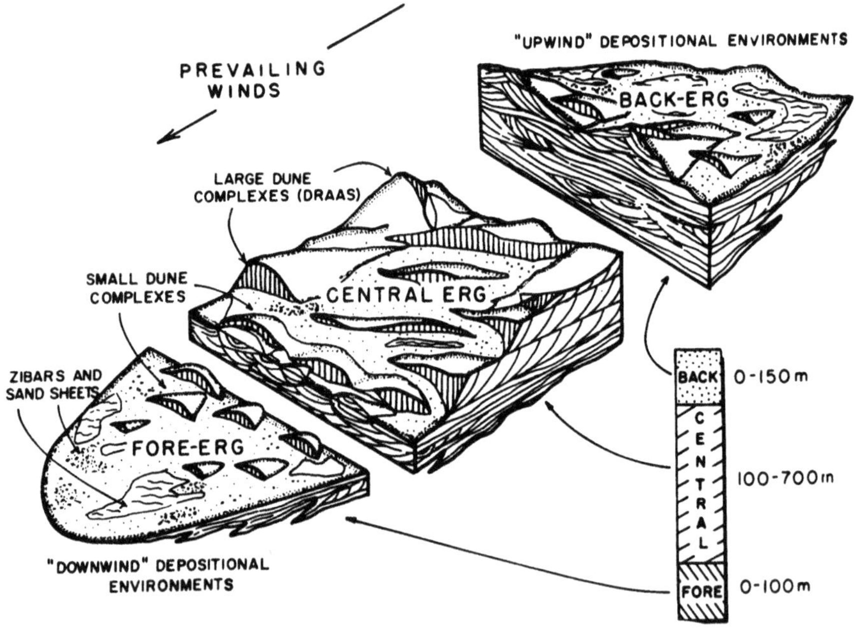

Fig. 3.34 Diagram showing a conceptual model of a migrating erg. Note the greater average thickness of the erg on the windward or proixmal end, the highest draas in the middle section, and the thinning of sand and lower density of dunes on the distal end of the erg (from Porter 1986)

component have lower hydraulic conductivities than better sorted sands. Variations in grain size characteristics also greatly affect the morphology of dunes (Bagnold 1941).

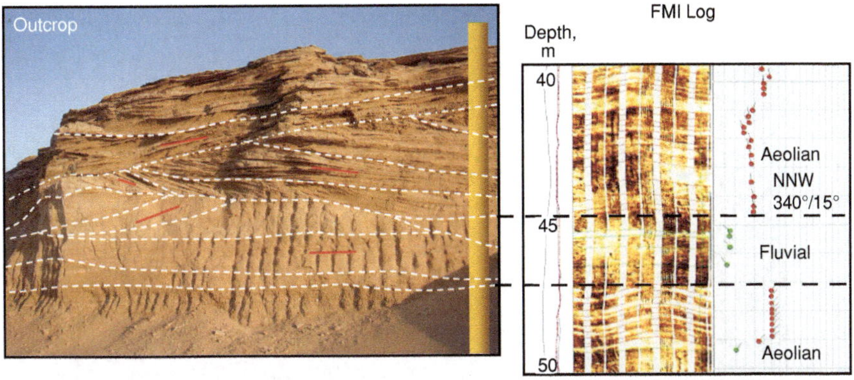

Fig. 3.35 Outcrop of a mixed erg and fluvial deposit from the Rub al Khali erg in the United Arab Emirates. Note the cross-bedding aeolian features at the top of the sequence, the flat-bedded fluvial deposit in the center, and the aeolian deposit at the base. The facies are defined visually and in an FMI geophysical log. In many cases, the fluvial facies may contain a higher percentage of silt and clay-sized particles, thereby, creating vertical heterogeneity within the sediment sequence (from Maliva and Missimer 2010)

Erg facies interact with other arid lands facies to affect the flow of water from mountain systems into the central basins. Basin infill deposits can consist solely of erg deposits, which may be either erg blankets that fill the entire basin or facies deposited by migrating ergs with variable sediment size characteristics (Fig. 3.33). The sides of the basins may be alluvial fans, desert flats, or crystalline rock types. Erg deposits are typically thicker in the upwind area or the back-erg with a marked reduction in sand thickness and density of dunes in the downwind or distal part of the erg (Fig. 3.34). Commonly, erg deposits vary vertically with interbedded fluvial or alluvial deposits (Fig. 3.35). Where an erg is hydraulically connected to sediments with high hydraulic conductivity, it can become a reservoir for fresh-water or brackish water moving downward into a basin from mountains. In erg deposits lying adjacent or coincident with playa facies, high salinity water may move into the ergs deposits. Ergs and desert sand deposits may be an important part of arid lands hydrogeology, because they can contain usable water supplies. Ergs are discussed in considerably greater detail by Wilson (1973), Cooke et al. (1993), Lancaster (1995), Thomas (1997a, b), and Laity (2008).

3.2.10 Desert Flats

Desert flat is a term used by Mabbutt (1977) and Clements et al. (1957) for an area of low topographic relief located within a dry lands basin floor. These low-relief areas are commonly eroded by aeolian action and perhaps could be termed deflation flats or plains. Deflation is specifically defined as "the sorting out, lifting,

Fig. 3.36 Photograph of a "desert flat" is the area known as Pampa de la Joya in southern Peru. The area has very low relief and contains an area covered by desert pavement and sparsely scattered, isolated barchan dunes

and removal of loose fine-grained particles (clay and silt sizes) by the turbulent eddy action of the wind, as along a sand-dune coast or in a desert, a form of wind erosion" (American Geological Institute 1997, p. 166). There are a number of depositional features that could be classified as "desert flats", which are wide, flat playas (the Booneville salt flat), wind-eroded plains (such as those from on the up-wind side of ergs), or fluvial valleys created by streams which have had no flow for extended time periods and have been eroded flat by winds, with wind-transported sediments infilling depressions. Another term applied to these feature is desert plain which is a general term used by Blackwelder (1931) for any plain commonly found in a desert, including flood plains, structural plains, playas, bajadas, and pediments (American Geological Institute 1997).

An example of a "desert flat" is the area known as Pampa de la Joya in southern Peru. This pampa is a near level plain located at an altitude of 1,262 m (4,139 ft) NGVD between the City of Arequipa and the coastal town of Mollendo (Finkel 1959). The plain is extremely arid, containing no vegetation, and contains only a few, widely separated drainage gullies that carry rare, but in some cases intense, rainfall events. The plain contains widely-spaced, individual barchan dunes that have been studied for over 50 years. The desert floor consists of gravel and small pebbles that have been polished to varying degrees by the wind (Fig. 3.36). There is a limited amount of sand on the desert flat and most of the surface is desert pavement. The surface of the flat is not commonly cemented, but there are some areas containing cohesive volcanic ash that has become cemented in dry arroyos located seaward and below the plain to the west (see Fig. 3.17, channel containing partially lithified volcanic ash). This type of desert flat or plain is common

Fig. 3.37 Inter-mountain valley located southwest of Arequipa, Peru. The valley has low relief and some rounded hills covered with a thin veneer of sand eroded by wind from the adjacent granitic mountains

throughout the world and may be considered to be both an erosional and depositional feature. Many of these features are covered by desert pavement or an armor of packed stones and gravel that is about 10–20 cm (3.9–7.9 in) thick, and is commonly uncemented and underlain by sand or silt that may be up to a meter (3.3 ft) in thickness (McFadden et al. 1987; Breed et al. 1997).

Many desert flats occur within inter-mountain basin valleys. These areas are not as flat as that occurring in Pampa de la Joya and contain a number of low relief mounds and a thin veneer of sand eroded from nearby mountains (Fig. 3.37). The desert flat areas are commonly underlain by bedrock at relatively shallow depths and these valleys are, as a result, not very good groundwater reservoirs.

Unfortunately, the geomorphic term desert flat or plain is not clearly defined with the context of arid lands. These features have varying geological origins, and overlap with other depositional features, such as playas and fluvial floodplains. In locations where the plain is underlain by permeable sediments with some hydraulic connection to alluvial fans or ephemeral streams, fresh, brackish, or high salinity water might occur within these features, especially where they occur at the very bottom of a basin.

3.2.11 Recent Volcanic Deposits

Volcanic rock deposits play an important role in the geology of many arid lands regions, particularly those that occur on active tectonic plate margins. Recent volcanic deposits can be a source of sand for aeolian deposits within ergs (Thomas

Fig. 3.38 Photograph of a river valley south of Arequipa, Peru that cuts through some volcanic ash deposits within the floodplain in the river valley. Note the location of Volcan Misti in the foreground. The last major eruption covered a large region area in Arequipa and to the west and southwest with a blanket of white ash deposits. Some of the older ash deposits from this volcano and others in the near vicinity sometimes weather to a tan to light brown color

1997a, b), may form the pediment or basin floor substrate (Berger 1997; parts of the desert in Argentina in Patagonia), or may contribute ash deposits that accumulate in wadis (Fig. 3.17) or are deposited as wind-blown deposits through a wide-area of an arid region (Fig. 3.38). Relatively recent volcanic rocks and sediments have been documented in arid areas along the eastern margin of the Red Sea in Saudi Arabia, from southern Peru to central Chile within the Atacama region Fig. 3.39, and in many other arid regions. Typically, basalt deposits may be either jointed and create fractured-rock type aquifers or may contain some matrix porosity in vesicles, lava tubes, and interflow strata. In aa lava flows, the blocky nature of the lava causes the formation of large, well-connected pores that have the potential to store and transmit water. Volcanic rocks and associated sediments have not been greatly studied with arid lands in terms of hydraulic connection and their potential use as aquifer systems.

3.3 Arid Lands Geology and Hydrogeology

A key principle is that arid regions often contain geological features that are capable of storing and transmitting large quantities of water, but often do not because water is not available (i.e., the rate of recharge is too low). For example, some lava deposits may be extremely transmissive, but are dry features because of

Fig. 3.39 Photograph of the Sababcaya Volcao, Peru. Note the lava flows into an arid area with essentially no visible vegetation. There are alluvial fans coming off of drainage features to the west of the volcano from a another mountain (from the NASA Earth Observatory)

the absence of significant local recharge. Recharge in arid regions tends to be highly episodic on different time scales (Chap. 7). Much of the groundwater resources present in some larger aquifers were deposited during earlier wetter (pluvial) times. With respect to water resources, present conditions in arid and semi-arid lands are not indicative of past conditions. The geological history of arid lands influences the location and characteristics of aquifers.

Local geological conditions are of particular significance to the extent that they allow for concentration of surface-water runoff and local infiltration and ultimately aquifer recharge. Thus, ephemeral streams (e.g., wadis and arroyos) have a greatly disproportionate influence on arid lands water resources because they are the site of surface-water accumulation and have hydraulic characteristics (coarse grain sizes and high vertical hydraulic conductivities) that are favorable for infiltration and percolation. On the contrary, recent sand dunes (ergs) may have little significance on local modern water resources, because they are the site of little, if any, current recharge. However, dune deposits (ergs) may be significant hydrogeological features because they retain water emplaced during earlier pluvial times (Sect. 4.2.2) and are potential storage locations for large aquifer storage and recovery projects.

References

Al-Shaibani, A. M. (2008). Hydrogeology and hydrochemistry of a shallow alluvial aquifer, western Saudi Arabia. *Hydrogeology Journal, 16*, 155–165.

American Geological Institute. (1997). *Glossary of geology* (4th ed., 769 pp) In J. Jackson (Ed.), Alexandria, VA.

Bagnold, R. A. (1941). *The physics of blown sand and desert dunes* (p. 265). London: Methuen.

Berger, I. A. (1997). South America. In D. S. G. Thomas (Ed.), *Arid zone geomorphology: Process, forms, and changes in drylands* (pp. 543–562). Chichester, England: Wiley.

Blackwelder, E. (1931). Pleistocene glaciation in the Sierra Nevada and basin ranges. *Geological Society of America Bulletin, 42*, 865–922.

Blair, T. C., & McPherson, J. G. (1994a). Alluvial fans and their natural distinction from rivers based on morphology, hydraulic processes, sedimentary processes, and facies assemblages. *Journal of Sedimentary Research, A64*(3), 450–489.

Blair, T. C., & McPherson, J. G. (1994b). Alluvial fan processes and forms. In A. D. Abrahams & A. J. Parsons (Eds.), *Geomorphology of desert environments* (pp. 354–402). London: Chapman & Hall.

Blum, M., Marriott, S., & Leclair, S., (Eds.), (2005). *Fluvial sedimentology, VII: IAS Special Publication 35*. New York: Wiley-Blackwell.

Bowler, J. M. (1986). Spatial variability and hydrological evolution of Australian Lake basins: Analogue for Pleistocene hydrological change and evaporate formation. *Palaeogeography, Palaeoclimatology, Palaeoecology, 54*, 21–41.

Breed, C. S., McCauley, J. F., Whitney, M. I., Tchakerian, V. P., & Laity, J. E. (1997). Winde erosion in drylands. In D. S. G. Thomas (Ed.), *Arid zone geomorphology: Process, forms, and changes in drylands* (pp. 437–464). Chichester, England: Wiley.

Bryan, R. B., & Yair, A. (Eds.). (1982). *Badlands geomorphology and piping*. Norwich, UK: GeoBooks.

Bryan, R. B., Yair, A., & Hodges, W. K. (1978). Factors controlling the initiation of runoff and piping in Dinosaur Provincial Park badlands, Alberta, Canada. *Zeitschrift fur Geomorphologie, Supplementband, 29*, 151–168.

Bull, W. B. (1964a). *Geomorphology of segmented alluvial fans in western Fresno County* (pp. 89–129). California: U. S. Geological Survey Professional Paper 352-E.

Bull, W. B. (1964b). *Alluvial fans and near-surface subsidence in western Fresno County* (pp. A1–A70). California: U. S. Geological Survey Professional Paper 437-A.

Bull, W. B. (1972). Recognition of alluvial fan deposits in the stratigraphic record. In J. K. Rigby & W. K. Hamblin (Eds.), *Recognition of ancient sedimentary environments* (pp. 63–83). Tulsa, OK: SEPM Special Publication 16.

Bull, W. R. (1975). Allometric change of landforms. *Geological Society of America Bulletin, 86*, 1489–1498.

Bull, W. B. (1979). Threshold of critical power in streams. *Geological Society of America Bulletin, 90*, 453–464.

Bull, W. B. (1991). *Geomorphic responses to climatic change*. Oxford: Oxford University Press.

Bye, J. A. T., & Harbison, I. P. (1991). Transfer of inland salts to the marine environment at the head of Spencer Gulf, South Australia. *Palaeogeography, Palaeoclimatology, Palaeoecology, 84*, 357–368.

Campbell, I. A. (1997). Badlands and badland gullies. In D. S. G. Thomas (Ed.), *Arid zone geomorphology: Process, forms, and changes in drylands* (pp. 261–291). Chichester, England: Wiley.

Clements, T., Merriam, R. H., Eyman, J. L., Stone, R. O., & Reade, H. L. (1957). *A study of desert surface conditions*. U. S. Army Environmental Protection Division Technical Report EP-53, Natick, Massochusetts, U. S. Army.

Cooke, R. U., Warren, A., & Goudie, A. S. (1993). *Desert geomorphology*. London: UCL Press.

Denny, C. S. (1967). *Alluvial fans in the death valley region*. Nevada, CA: U. S. Geological Survey Professional Paper 466.

Finkel, H. J. (1959). The barchans of southern Peru. *Journal of Geology, 67*, 614–647.

Fouch, T. D., & Dean, W. E. (1982). Lacustrine and associated clastic deposition environments. In P. A. Scholle & D. Spearing (Eds.), *Sandstone depositional environments* (pp. 87–114). Tulsa, OK: American Association of Petroleum Geologists Memoir 31.

Gouldie, A. S. (1973). *Duricrusts in tropical and subtropical landscapes*. Oxford: Clarendon Press.

Gouldie, A. S. (1997). Weathering processes. In D. S. G. Thomas (Ed.), *Arid zone geomorphology: Process, form and change in drylands* (pp. 25–39). Chichester, England: Wiley.

Graf, W. L. (1983). The Arroyo problem—Palaeohydrology and Palaeohydraulics in the Short Term. In K. Gregory (Ed.), *Background to paleohydrology* (pp. 279–302). New York: Wiley.

Gray, W. M. (1965). Surface spalling by thermal stresses in rocks. In *Rock Mechanic Symposium: Toronto, Proceedings of the Ottawa*, Canada Department of Mines and Technical Surveys.

Hack, J. T. (1960). Interpretation of erosional topography in humid temperate regions. *American Journal of Science (Bradley Volume), 258-A*, 80–97.

Hardie, L. A., Smoot, J. P., & Eugster, H. P. (1978). Saline lakes and their deposits: A sedimentological approach. In A. Matter & M. Tucker (Eds.), *Modern and ancient lake sediments: International Association of Sedimentologists* (pp. 7–41), Special Publication 2.

Hartley, A. J., Mather, A. E., Jolley, E., & Turner, P. (2005). Climatic controls on alluvial-fan activity, Coastal Cordillera, northern Chile, In A. M. Harvey, A. E. Mather & M. Stokes (Eds.), *Alluvial fans: Geomorphology, sedimentology, dynamics* (pp. 95–115). London: Geological Society Special Publications, no. 251.

Harvey, A. M. (1997). The role of alluvial fans in arid fluvial systems. In D. S. G. Thomas (Ed.), *Arid zone geomorphology: Process, forms, and changes in drylands* (pp. 231–260). Chichester, England: Wiley.

Harvey, A. M. (2004). The response of dry-region alluvial fans to late Quaternary climate change. In A. S. Alsharhan, W. W. Wood, A. S. Gouldie, A. Fowler, & E. M. Abdellatif (Eds.), *Desertification in the third millennium* (pp. 83–98). Balkema: Rotterdam.

Harvey, A. M. (2005). Differential effects of base-level, tectonic setting and climatic change on Quatnerary alluvial fans in the northern Great Basin, Nevada, USA. In A. M. Harvey, A. E. Mather & M. Stokes (Eds.), *Alluvial fans: Geomorphology, sedimentology, dynamics* (pp. 117–131). London: Geological Society Special Publications, no. 251.

Heathcote, R. L. (1983). *The arid lands: Their use and abuse*. London: Longman.

Hockman, A., & Kessler, D. (1950). Thermal and moisture expansion studies of some domestic granites: U.S Department of Commerce. *National Bureau of Standards (RP 2087), 44*, 395–410.

Holser, W. T. (1979). Mineralogy of evaporates, in Marine minerals: Review in mineralogy. *Mineralogy Society of America, 6*, 211–294.

Hooke, R. Le. B. (1968). Steady-state relationships on arid-region alluvial fans in closed basins. *American Journal of Science, 266*, 609–629.

Hooke, R. Le. B., & Rohrer, W. L. (1977). Relative erodibility of source rock types from second order variations in alluvial fan size. *Bulletin of the Geological Society of America, 88*, 1177–1182.

Jayko, A. S., Menges, C. M., & Thompson, R. A. (2005). *Digital method for regional mapping of surficial basin deposits in arid regions, example from central Death Valley, Inyo County, California* (49 pp). U.S. Geological Survey Open-File Report 2005-1445.

Jones, A. A. A. (1971). Soil piping and stream channel initiation. *Water Resources Research, 7*, 602–610.

Jones, J. A. A. (1990). Piping effects in drylands. *Geological Society of America Special Paper, 252*, 207–138.

Knighton, D., & Nanson, G. (1997). Distinctiveness, diversity and uniqueness in arid zone river systems. In D. S. G. Thomas (Ed.), *Arid zone geomorphology: Process, forms, and changes in drylands* (pp. 185–203). Chichester, England: Wiley.

Kochel, R. C., & Johnson, R. A. (1984). Geomorphology and sedimentology of humid-temperate alluvial fans, central Virginia. In E. Koster & R. Steel (Eds.), *Gravels and conglomerates* (pp. 109–122). Canada: Canadian Society of Petrology and Geology, Memoir 10.
Kocurek, G. (1988). First-order and super bounding surface in eolian sequences-bounding surfaces revisited. *Sedimentary Geology, 56*, 193–206.
Koppen, W. (1931). Die climate der Erde: Berlin.
Kuenen, P. H. (1960). Experimental abrasion 4: Eolian action. *Journal of Geology, 68*, 427–449.
Laity, J. (2008). *Deserts and desert environments*. Chichester, England: Wiley-Blackwell.
Lancaster, N. (1995). *The geomorphology of desert dunes* (p. 312). London: Routledge.
Leece, S. A. (1991). Influence of lithologic erodibility on alluvial fan area, western White Mountains, California and Nevada. *Earth Surface Processes and Landforms, 16*, 11–18.
Mabbutt, J. A. (1976). Physiographic setting as an indication of inherent resistance to desertification. *WGDAL, 1976*, 189–197.
Mabbutt, J. A. (1977). *Desert landforms*. Cambridge, MA: MIT Press.
Maliva, R. G., & Missimer, T. M. (2010). Aquifer storage and recovery and managed aquifer recharge using wells: Planning, hydrogeology, design, and operation: Houston, Schlumberger Water Services, Methods in Water Resources Evaluation Series No. 2.
McCartney, J. (2001). *Hydraulic and hydrochemical interactions in the Tilopozo groundwater zone, Salar de Atacama region II, Chile*. Unpublished Master of Science Thesis, National Centre for Groundwater Management, University of Technology, Sydney, Australia, 114 pp. and appendices.
McFadden, L. D., Wells, S. G., & Jerciovich, M. J. (1987). Influences of eolian and pedogenic processes on the origan and evolution of desert pavements. *Geology, 15*, 504–508.
McKee, E. D. (1966). Structures of dunes at White Sands National Monument, New Mexico. *Sedimentology, 7*, 1–60.
Meigs, P., 1953, World distribution of arid and semi-arid homoclimates, In UNESCO (Eds.), *Reviews of research on arid zone hydrology* (pp. 203–210). Paris: UNESCO.
Melton, F. A. (1940). A tentative classification of sand-dunes. *Journal of Geology, 48*, 113–174.
Miall, A. D. (1977). A review of the braided river depositional environment. *Earth Science Reviews, 13*, 1–62.
Miall, A. D. (Ed.) (1978a). *Fluvial sedimentology*. Calgary, AB: Canadian Society of Petroleum Geologists Memoir 5.
Miall, A. D. (1978b). Lithofacies types and vertical profile models in braided river deposits: A summary. In A. D. Miall (Ed.), *Fluvial sedimentology* (pp. 597–604). Calgary, AB: Canadian Society of Petroleum Geologists Memoir 5.
Miall, A. D. (1981). Alluvial sedimentary basins: Tectonic setting and basin architecture. In A. D. Miall (Ed.), *Sedimentation and tectonics in alluvial basins* (pp. 1–33).Canada: Geological Association of Canada Special Paper 23.
Miall, A. D. (1990). *Principals of sedimentary basin analysis* (2nd ed.). New York: Springer.
Nilsen, T. H. (1982). Alluvial fan deposits. In P. A. Scholle & D. Spearing (Eds.), *Sandstone depositional environments* (pp. 49–86). Tulsa, OK: American Association of Petroleum Geologists Memoir 31.
Pazzaglia, F., Gardner, T., & Merrits, D. (1998). Bedrock fluvial incision and longitudinal profile development over geologic timescales determined by fluvial terraces. In K. Tinkler & E. Wohl (Eds.), *Rivers over rock: Fluvial processes in bedrock channels* (pp. 207–235). Washington, DC: American Geophysical Union.
Porter, M. L. (1986). Sedimentary record of erg migration. *Geology, 14*, 497–500.
Reid, I., & Frostick, L. E. (1997). Channel form, flows and sediments in deserts. In D. S. G. Thomas (Ed.), *Arid zone geomorphology: Process, forms, and changes in drylands* (pp. 205–229). Chichester, England: Wiley.
Rendell, H. (1997). Tectonic frameworks. In D. S. G. Thomas (Ed.), *Arid zone geomorphology* (2nd ed., pp. 13–22). West Sussex, England: Wiley.
Ritter, J. B., Miller, J. R., Enzel, Y., & Wells, S. G. (1995). Reconciling the roles of tectonism and climate in Quaternary alluvial fan evolution. *Geology, 23*(3), 245–248.

Ritter, D. F., Kochel, R. C., & Miller, J. R. (2002). *Process geomorphology* (4th ed.). Long Grove, IL: Wayland Press, Inc.

Rosen, M. R. (1994). The importance of groundwater in playas: A review of playa classifications and the sedimentology and hydrology of playas. In M. R. Rosen (Ed.), *Paleoclimate and basin evolution of playa systems* (pp. 1–18). Boulder, CO: Geological Society of America Special Paper 289.

Rust, J. B. (1978). Depositional models for braided alluvium. In A. D. Miall (Ed.), *Fluvial sedimentology* (pp. 605–625). Calgary, AB: Canadian Society of Petroleum Geologists Memoir 5.

Rust, B. R. (1979). Coarse alluvial deposits. In R. G. Walker (Ed.), *Facies models* (pp. 9–22). Geoscience Canada, Reprint Series 1.

Ryder, R. T., Fouch, T. D., & Elison, J. H. (1976). Early tertiary sedimentation in the western Uinta Basin, Utah. *Geological Society of America Bulletin, 87*, 496–512.

Schumm, S. A., Harvey, M. D., & Watson, C. C. (1984). *Incised channels: Morphology, dynamics and control*. Littleton, CO: Water Resources Publications.

Shantz, H. L. (1956). History and problems of arid lands development. In G. F. White (Ed.), *The future of arid lands* (pp. 3–25). Washington, DC: American Association for the Advancement of Science.

Shaw, P. A., & Thomas, D. S. G. (1997). Pans, playas and salt lakes. In D. S. G. Thomas (Ed.), *Arid zone geomorphology: Process, form and change within drylands* (pp. 293–318). Chichester, England: Wiley.

Smith, N., & Rodgers, J. (1999). *Fluvial sedimentology: IAS Special Publication VI*. New York: Wiley-Blackwell.

Snyder, C. T. (1962). A hydrologic classification of valleys in the Great Basin, western United States. *Bulletin of the International Association of Scientific Hydrology, 7*(3), 53–59.

Spearing, D. R. (1974). Alluvial fan deposits: Summary sheets of sedimentary deposits, Sheet 1. Boulder, CO: Geological Society of America.

Steel, R. J., Maehle, S., Nilsen, H., Roe, S. L., & Spinnagr, A. (1977). Coarsening-upward cycles in the alluvium of Hornelen Basin (Devonian), Norway: Sedimentary response to tectonic events. *Geological Society of America, 88*, 1124–1134.

Summerfield, M. A. (1983). Petrography and diagenesis of silcrete from the Kalahari Basin and Cape Coastal Zone, southern Africa. *Journal of Sedimentary Petrology, 53*(3), 895–910.

Thomas, D. S. G. (1997a). Arid environments: Their nature and extent. In D. S. G. Thomas (Ed.), *Arid zone geomorphology: Process, forms, and changes in drylands* (pp. 3–12). Chichester, England: Wiley.

Thomas, D. S. G. (1997b). Sand seas and Aeolian bed forms. In D. S. G. Thomas (Ed.), *Arid zone geomorphology: Process, forms, and changes in drylands* (pp. 371–412). Chichester, England: Wiley.

Thornthwaite, C. W. (1948). An approach toward a rational classification of climate. *Geographical Review, 38*(1), 55–94.

Torgersen, T., De Dekker, P., Chivas, A. R., & Bowler, J. M. (1986). Salt lakes: A discussion of the processes influencing palaeoenvironmental interpretations and recommendations for future study. *Palaeogeography, Palaeoclimatology, Palaeoecology, 54*, 7–19.

Ullyott, J. S., Nash, D. J., & Shaw, P. A. (1998). Recent advances in silcrete research and their implications for the origin and paleoenvironmental significance of sarsens. *Proceedings of the Geologist's Association, 109*(4), 255–270.

UNEP. (1992). *World atlas of desertification*. Sevenoaks: Edward Arnold.

United Nations. (1977). *Status of desertification in the hot arid regions, climatic aridity index map and experimental world scheme of aridity and drought probability, at a scale of 1:25,000,000, explanatory note*. UN Conference on Desertification A/CONF, 74/31, New York, United Nations.

Watson, A., & Nash, D. J. (1997). Desert crusts and varnishes. In D. S. G. Thomas (Ed.), *Arid zone geomorphology* (2nd ed., pp. 69–107). West Sussex, England: Wiley.

Wilson, L. (1968). Morphogenetic classification. In R. W. Fairbridge (Ed.), *Encyclopedia of geomorphology* (pp. 717–728). New York: Reinhold Book Company.
Wilson, I. G. (1971). Desert sandflow basins and a model for the development of ergs. *Geography Journal, 137*, 180–199.
Wilson, I. G. (1972). Universal discontinuities in bedforms produced by the wind. *Journal of Sedimentary Geology, 42*, 667–669.
Wilson, I. G. (1973). Ergs. *Sedimentary Geology, 10*, 77–106.
Winkler, E. M. (1975). *Stone: Properties, durability in man's environment* (2nd ed.). New York: Springer.
Wolman, M. G., & Gerson, R. (1978). Relative scales of time and effectiveness of climate in watershed geomorphology. *Earth Surface Processes, 3*, 189–208.
Wood, W. W., & Sanford, W. E. (1990). Groundwater control of evaporate deposition. *Economic Geology, 85*, 1226–1235.
Wright, P. V., & Tucker, M. E. (1991). Calcretes: An introduction. In V. P. Wright & M. E. Tucker (Eds.), *Calcretes: International Association of Sedimentologists* (pp. 1–22). Oxford: Blackwell Scientific.

Chapter 4
Aquifer Concepts in Arid Lands

4.1 Aquifer Types

There are three main overlapping classification schemes for aquifers that are potential water sources in arid and semiarid lands. Aquifers may be classified according to their

- hydraulic properties and the properties of adjoining confining strata (e.g., confined, unconfined),
- geographical and topographic extent of their flow systems (e.g., local, regional),
- predominant lithology or deposition environment (e.g., alluvial, fractured rock).

4.1.1 Aquifer Hydraulic Classification

Using the Dutch terminology, there are four different types of aquifers that are differentiated based on the response of the aquifer to a pumping condition, which are unconfined, semi-unconfined, semi-confined, and confined (Kruseman and de Ridder 1970; Fig. 4.1). The two end-member aquifer types are unconfined and confined. Unconfined aquifers (which are also referred to as water table aquifers) are characterized by their upper boundary being the water table, which is defined as the regional surface where the fluid pressure of a porous medium is equal to the atmospheric pressure (Sect. 5.6). The water table also marks the boundary between the saturated zone and the overlying unsaturated zone, which is also referred to as the vadose zone. However, a capillary fringe is often present above the water table which is partially saturated with its moisture content increasing from the top to the

R. Maliva and T. Missimer, *Arid Lands Water Evaluation and Management*, Environmental Science and Engineering, DOI: 10.1007/978-3-642-29104-3_4,

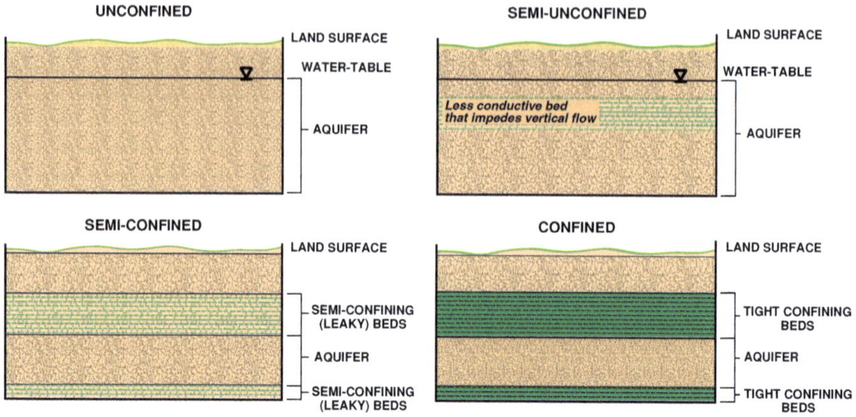

Fig. 4.1 Types of aquifers based on the classification of Kruseman and de Ridder (1970)

bottom. The base of the capillary fringe can be nearly saturated, but it is still above the water table because this water is under capillary tension and not in equilibrium with atmospheric pressure.

The Kruseman and de Ridder (1970) definition of an unconfined aquifer is idealized in that it requires that there are no confining or semi-confining strata present that would impede the rise or fall of the water table. Fully unconfined aquifers can therefore occur only in homogeneous and isotropic siliciclastic sediment bodies or carbonate units that have essentially equal vertical and horizontal hydraulic conductivities. Unconfined aquifers are hydraulically characterized by their hydraulic conductivity (transmissivity) and specific yield. There is an implicit assumption that the horizontal to vertical ratio of hydraulic conductivity is unity or close to it. When water is placed into or removed from this aquifer type, it is truly entering or leaving storage. Unconfined aquifers may have large storage volumes because their specific yield values are typically orders of magnitude greater than the storativity of confined aquifers. However, unconfined aquifers are more vulnerable to contamination and often have high hydraulic gradients, which can result in rapid groundwater movement.

Confined aquifers are the opposite end-member to an unconfined aquifer. The aquifer is confined by the presence of essentially impervious confining strata both above and below it (Theis 1935). Water is produced through the compaction of the aquifer matrix and expansion of water in response to decreases in pressure, rather than the draining of water from pores (i.e., dewatering). The potentiometric surface in a confined aquifer continuously declines during pumping and does not stabilize unless an equilibrium is reached with captured recharge or discharge (Sect. 6.4). The water level decline occurs along the Theis curve and shows a decreasing rate of decline over time for a given pumping rate (Theis 1935). Aquifer heterogeneity may affect the rate of pressure change, especially in aquifers that are thinning or thickening, or are laterally bounded. The characteristic aquifer hydraulic properties are transmissivity and storativity. Vertical leakage is assumed to be zero or very close to it.

Unconfined and confined aquifers are idealized end-members that have characteristic hydraulic conditions that are rarely met, although they may be approached in some aquifers. Aquifers described as being unconfined aquifers typically have some heterogeneity or anisotropy that influences vertical water movement. Similarly, aquifers described as being confined may experience some leakage of water from or through underlying and overlying confining units under pumping conditions. On a small localized scale, and under low pumping or injection rates and duration conditions, some aquifers may approach true unconfined or confined aquifer behavior.

The terms "semi-unconfined aquifer" and "semi-confined aquifer" are used to describe aquifers that have hydraulic conditions intermediate between unconfined and confined aquifers. A semi-unconfined aquifer occurs where there is some natural degree of variation in sediment properties that creates a disparity in the flow of water in the vertical direction from that in the horizontal direction. Although there may be no major change in sediment type within the aquifer, the vertical flow of water is inhibited by a localized reduction in vertical hydraulic conductivity. This reduction can be caused by the occurrence of simple bedding or grading within a bed. A small amount of clay or silt within the upper part of the aquifer can also create a significant anisotropy ratio in hydraulic conductivity (horizontal hydraulic conductivity divided by vertical hydraulic conductivity) with values ranging from 3 to over 100 in relatively homogeneous aquifers. Where an aquifer is composed of interbedded lithologies with large differences in hydraulic properties, anisotropy ratios may be in the 100s or 1,000s.

The characteristic hydraulic properties of semi-unconfined aquifers are hydraulic conductivity, specific yield, and a vertical drain factor. During an aquifer test, the potentiometric surface of the aquifer will decline along the Theis curve and then rapidly come to a temporary equilibrium or a "delayed yield" response (Boulton 1954a, b, 1963; Neuman 1975). As pumping continues and the stored water above the inhibiting sediment is drained, the potentiometric surface will again decline along the Theis curve. Semi-unconfined aquifers are common throughout the world, especially in coastal zones near land surface, in fluvial and alluvial systems occurring in a variety of different regions and climatic types, and in mixed siliciclastic and carbonate systems. Most aquifers that are described as being unconfined actually meet the Kruseman and de Ridder (1970) definition of a semi-unconfined aquifer.

A semi-confined or leaky aquifer occurs where the aquifer is bounded both above and below by sediments or rocks with lower vertical hydraulic conductivities than the aquifer. Some leakage of water occurs through one or both confining units when the aquifer is pumped (Hantush and Jacob 1955; Hantush 1956, 1959, 1960). The degree of confinement is directly related to the vertical hydraulic conductivity and the thickness of the bounding confining units. The characteristic hydraulic units for this aquifer type are transmissivity, storativity, and leakance (leakage coefficient). It is important to understand that leakance (as measured in an

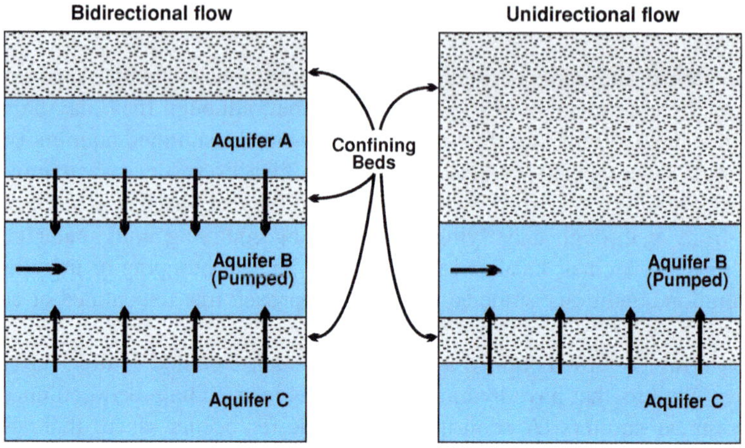

Fig. 4.2 Pumping-induced vertical flow into a leaky aquifer. Vertical flow into the pumped aquifer (B) may be either bidirectional or unidirectional depending upon the properties of the overlying and underlying confining units

aquifer performance test) is a parameter that defines the potential for movement of water into the aquifer through both underlying and overlying bounding confining units. When the upper and lower confining units have similar properties, then leakage is governed by bi-directional flow from both the top and bottom during pumping and injection (Fig. 4.2). Typically, one of the confining units is much thicker or has a much lower vertical hydraulic conductivity than the other, which creates a semi-confined or leaky aquifer controlled by essentially unidirectional vertical flow (Fig. 4.2). Semi-confined aquifers are very common in all rock types and geographic settings. They are perhaps the most common type of aquifer. When a semi-confined aquifer is pumped, the potentiometric surface declines and eventually comes into equilibrium with the pumping rate and the cone of depression ceases to expand. At equilibrium, the rate of water leaking into the aquifer is equal to the pumping rate.

4.1.2 Perched Aquifers

A perched aquifer is a local aquifer that is located above the regional or local water table. Perched aquifers occur because of the presence of a relatively impermeable layer within the vadose zone that impedes the vertical percolation of water (Fig. 4.3). A lense of groundwater accumulates within the vadose zone above the impermeable layer. Perched aquifers are most common in inland areas within arid and semiarid lands, where thick unsaturated zones may be present. Perched aquifers are usually absent or poorly developed in humid and coastal areas where the water table is located close to land surface and the vadose zone is relatively thin. Where the vadose

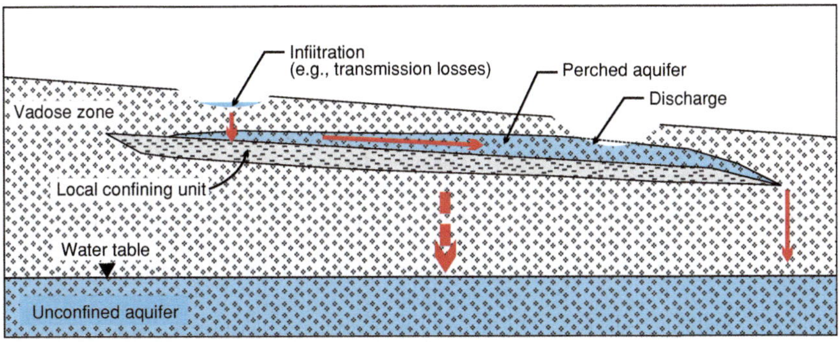

Fig. 4.3 Conceptual diagram of a perched aquifer system. Groundwater in perched aquifers may leak through the underlying confining unit or flow down-gradient if the confining unit has a slope. Water may also be lost to discharge in springs or streams or to deep-rooted vegetation

zone is thin, there typically is not enough room for a perched aquifer to be present. The less permeable strata that could be responsible for perched aquifer conditions in settings with a thick vadose zone will tend to be located below the water table in humid areas where the water table is shallow. The type of heterogeneities that cause perched aquifer conditions in areas with a deep water table would instead result in a high-degree of internal aquifer heterogeneity in areas with a shallow water table and corresponding thin vadose zone.

Perched aquifers are a minor source of water in arid and semiarid lands, but can be sufficient to meet limited local domestic demands in locations such as large wadis. Perched aquifer conditions are more significant for their impacts on local groundwater flow, solute (contaminant) transport, and aquifer recharge. The presence of a low permeability layer within the unsaturated zone may reduce the rate of local recharge to the underlying aquifer. The rate of recharge will depend on the vertical hydraulic conductivity and thickness of the layer. The reduced recharge rate may be beneficial if the percolating water contains contaminants at concentrations of concern. The retention of water at shallow depths increases its susceptibility to later loss by evapotranspiration, particularly if the water is retained in the root zone of plants.

If the low permeability strata are sloping from either depositional dynamics or structural deformation, the water in the perched aquifers will flow down gradient (Fig. 4.3). The potential for development of perched aquifers and the geometry of both flow zones and confining zones should be evaluated in the context of the geology of the aquifer of interest. For example, the strata that constitute alluvial fans (Sect. 3.2.4) tend to parallel the depositional slope, which can result in down-slope flow of water perched above inter-channel clay-rich layers. Recharge to the underlying aquifer may occur either at the boundaries of or through discontinuities in the impermeable layer. Discharge (springs) may also occur in down gradient areas where the perched aquifer is exposed.

Table 4.1 Main arid and semiarid lands aquifer types

Type	Characteristics
Local alluvial aquifers	Consist of alluvial sediments located beneath and adjacent to wadis and alluvial fans. Local recharge occurs during storm events and the aquifer storage volume is small
Intermontane and submontane basin-fill aquifers	Consist predominantly of alluvial sediments deposited within and adjacent to mountains. Recharge occurs mainly along periphery of basins (alluvial fans and wadis) or in mountain blocks. Often large storage volumes
Intermontane and submontane carbonate and sandstone aquifers	Sedimentary rock aquifers that are recharged primarily in highlands surrounding basins
Sand dune aquifers	Consist of groundwater present beneath sand dunes. Primary recharge may have occurred during earlier pluvial periods with little modern recharge
Fractured rock aquifers	Commonly crystalline rock. Local recharge in fractured zones or in areas of alluvial cover. Small storage volume and highly variable water yields
Regional renewable aquifers	Sedimentary or fractured rock aquifers of regional extent with often minimal intra-basinal recharge. Main recharge area is topographically higher wetter areas. Aquifers may have very large storage volumes
Non-renewable	Negligible current recharge. Contain water emplaced during earlier pluvial periods. Regional non-renewable aquifers may store enormous volumes of water

4.2 Arid Lands Aquifer Types

The basic hydrogeology of aquifers in arid lands does not significantly differ from that of aquifers in humid areas. Groundwater flows through transmissive strata from areas of high altitude or pressure to areas of low altitude or pressure. The types of aquifers, in terms of lithology, depend predominantly upon local geology rather than climate. The key difference concerning groundwater resources is the location and amount of recharge. Arid lands characteristically have low precipitation rates, high potential evaporation rates, and thus low rates of groundwater recharge. Recharge also tends to much more localized in arid lands than in more humid regions. Groundwater recharge in arid and semiarid lands is discussed in greater detail in Chap. 7.

Aquifers in arid and semiarid lands can be assigned to six broad categories based on their scale, geology, and amount and location of recharge (Table 4.1). Local flow systems are restricted to a single basin with boundaries that are controlled by local lithologic, tectonic, and topographic features (Tóth 1962, 1963; Maxey 1968) (Fig. 4.4). The topographic high recharge areas and topographic low discharge areas are located adjacent to each other in local flow systems

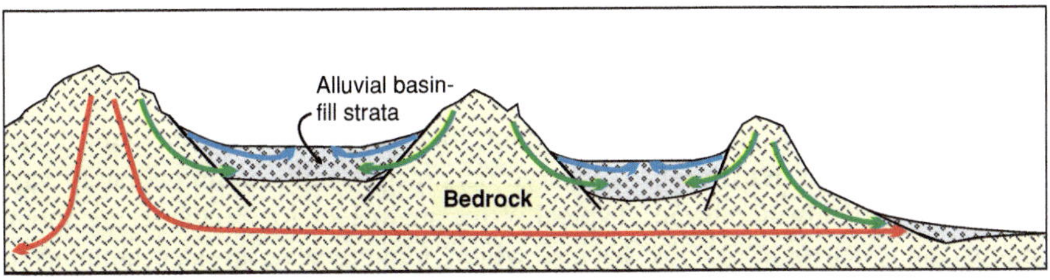

Fig. 4.4 Conceptual diagram of flow systems in mountainous regions. Regional flow systems (*red arrows*) originate in the topographically highest areas and discharge at the regional low point. Local flow systems are restricted to individual basins and may involve recharge within the bedrock mountain blocks (*green arrows*) or occur entirely within alluvial sediments (*blue arrows*)

(Tóth 1962, 1963). Typically, in arid lands recharge occurs mainly at higher altitudes in mountainous areas and discharge occurs in adjoining basins.

Intermediate and regional flow systems are characterized by one or more topographic highs located between the recharge and discharge areas, with the regional flow systems characterized by recharge areas at the major topographic high and discharge at the major topographic low (Tóth 1962, 1963). Regional aquifers extend beyond individual surface drainage (topographic) and groundwater basins and are commonly recharged primarily in more distant wetter areas (Tóth 1963; Maxey 1968). Negligible recharge may actually occur within the arid and semiarid lands in which the water is used. Water may flow over long distances through bedrock masses previously presumed to be impermeable (Maxey 1968).

Local groundwater flow systems typically have relatively short flow paths, cool temperatures (within several degrees of average atmospheric temperature), and low total dissolved solids concentrations (Maxey 1968; Mifflin 1988; Hibbs and Darling 2005). Regional flow systems on the contrary often have warmer temperatures, higher total dissolved solids concentrations, and more mature hydrochemical facies caused by greater water–rock interaction (Maxey 1968; Hibbs and Darling 2005). Environmental tracers (Sect. 11.5) may also be used to differentiate between local and regional (intermediate and deep) groundwater systems. For example, water in regional groundwater systems (far from its recharge area) will usually be considerably older, as determined using groundwater dating techniques, than groundwater in local systems.

Groundwater may flow to depths of hundreds of meters (hundreds of feet) below land surface before reaching regional flow paths. A key point is that basin hydrogeology may be misinterpreted if only data from shallow wells are considered (Mifflin 1988; Hibbs and Darling 2005). Head pressure data are needed at both various depths and geographic locations in order to determine if hydraulically separate aquifers are present and to determine the three-dimensional groundwater flow directions within the system.

Depending on the local geology, intermediate and regional (inter-basin) flow systems may underlie local basin-fill aquifers. Within the basin and range province of the Western United States, inter-basin flow systems appear to be best developed in carbonate strata that contain fracture and solution secondary porosity (Mifflin 1988). Interbasin flow is less well developed where the bedrock has a low permeability.

Wadi and alluvial fan aquifers, and intermontane and submontane aquifers are characterized by primarily local recharge occurring within or immediately adjacent to the groundwater basin. Typically, the main loci for recharge are mountainous areas surrounding the basins, which may receive greater amounts of precipitation and have lower evapotranspiration rates because of lower temperatures.

4.2.1 Local Alluvial Aquifers

Local aquifers of limited areal extent may form in arid and semiarid lands at locations where permeable sediments are present and surface water flow (run-off) is focused. Local alluvial aquifers are intrabasinal aquifers composed of sands, gravels, and silts deposited in stream and river beds and alluvial fans. The aquifers may also include some underlying, weathered bedrock that can be hydraulically conducted to fractured rock aquifers. As discussed in Sect. 3.2.6, ephemeral river systems are commonly referred to using the Arabic name "wadis" or the Spanish name "arroyos". The term "wadi", which is used herein, may also more be broadly used to describe the entire river valley and catchment rather than just the stream channel. Individual wadi and alluvial fan deposits may contain aquifers, which can be important local water sources.

The geology of wadi and alluvial fan deposits is discussed in Chap. 3. Alluvial aquifers occur in three main contiguous geomorphic and sedimentological environments:

- wadis in bedrock-bounded valleys in upland areas (Fig. 4.5),
- alluvial fans,
- basin floor (flatland wadi channels).

Streams in arid and semiarid regions are typically loosing streams in that water flows from the stream into the underlying and bordering aquifer. Streams and rivers located in the wetter end of the arid climate spectrum may have reaches that are episodically or perennially gaining (Newman et al. 2006). Run-off in typical loosing streams is progressively lost to streambed infiltration as it flows downstream, which is referred to as transmission loss. Depending upon the intensity and duration of a rainfall event, and local watershed and channel geometry and characteristics, surface flow in wadis may be restricted because of transmission losses to upland areas, or may continue to flow downstream to alluvial fans and channels on valley floors. Wadi channel transmission losses may not fit the classic definition of a losing stream in reaches upstream of the basin compared to perennial streams that flow through basins with a water table consistently below the stream stage. In these locations, recharge from the stream can become part of

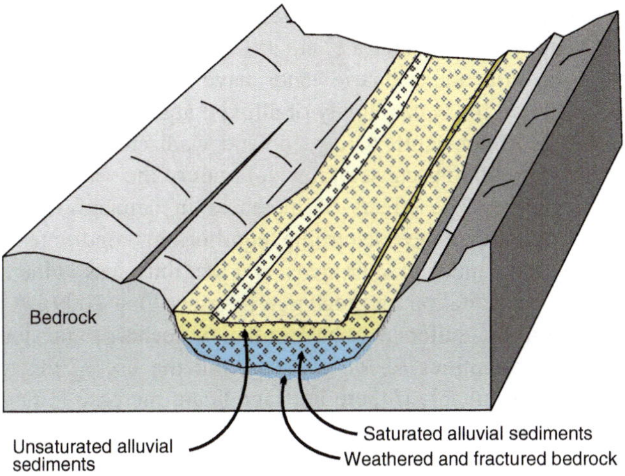

Fig. 4.5 Block diagram of a wadi aquifer. Water is stored in alluvial sediment and, in some systems, in an underlying zone of weathered and fractured bedrock

recharge to a regional aquifer, whereas wadi recharge is commonly limited only to the alluvium within the wadi channel (local aquifer) until water reaches the basin floor and aquifers connected to the basin edge.

Wadi and alluvial deposits share some basic features. The slope, flow velocities (and energy), and corresponding grain size of sediment in wadi channels tends to decrease downstream. Similarly, grain-size and hydraulic conductivity decrease downstream in alluvial fan deposits. The thickness of the vadose zone also increases in the downstream direction (Al-Shaibani 2008; Şen 2008). Alluvial aquifers deposited by ephemeral flows have highly episodic recharge, whose rate will depend upon the duration of the stream flow event (and thus time available for recharge), stream channel geometry, and the characteristics of the channel bed sediments, particularly their vertical hydraulic conductivity.

Wadis play a major role in arid lands aquifer recharge because they are the site of the accumulation of runoff during storms and their coarse-grained bed sediments usually have a relatively high hydraulic conductivity. Wadi alluvial aquifers differ in the extent to which they are connected to underlying aquifers (Newman et al. 2006). Recharge in wadis fills both local alluvial aquifers located along basin margins and may also be important water source for larger-scale, basin-fill aquifers. Wadi recharge is discussed in detail in Chap. 13.

Fluvial sediments hydraulically-connected to perennial rivers may constitute small aquifers in situations where the surrounding strata are relatively impermeable. Perennial rivers may act as either drains or recharge sites depending on the position of the water table with respect to the river stage. Commonly, rivers located at the low point in a basin acted as a drain (gaining streams), but may latter become recharge sources (loosing streams) after groundwater development lowers the water table.

Communities and settlements were often established based on proximity to minor aquifer water sources (Pereira et al. 2002). Highly variable and intermittent rainfall patterns in arid and semiarid lands have the potential to impose great temporal variability on the water supply of alluvial aquifers of limited areal extent (Love et al. 2007). Minor aquifers (e.g., upland wadi channel aquifers) are vulnerable to over-exploitation because of low volumes of the water reserves and their unpredictable recharge. The effects of increases in demands and decreases in recharge would be immediate, disturbing equilibrium conditions. Alluvial fan aquifers generally have much greater areas, and are, thus, less vulnerable to short-term variations in precipitation and recharge (van der Gun 2009). A fundamental issue is that local minor aquifers may have sufficient recharge and storage capacity to sustain small communities for generations, but the use of the resource may quickly become unsustainable if there is a significant increase in populations and concomitant water demand.

4.2.2 Sand Dune (erg) Aquifers

Sand dune aquifers are local or regional aquifers that contain groundwater at the base of aeolian sand deposits. A notable example is the sand dune aquifer in the United Arab Emirates, which is aerially the largest aquifer in the country. Dunes cover 74% of the total area of the country and local or regional aquifers occur within this giant dune or erg system (Murad et al. 2007). Geochemical data indicate that the groundwater was emplaced approximately 6,000 years ago during wetter conditions and is essentially a non-renewable resource (Wood et al. 2003; Wood 2011). Berger (1992) proposed, based on deep percolation modeling results, that dunefields (ergs) can provide a conduit for groundwater recharge because of their relatively high hydraulic conductivity and absence of vegetation.

Field evidence for modern recharge rates in dune aquifers in arid lands is sparse. Scanlon et al. (2010) determined that the recharge rate in a sparsely vegetated dune/interdune area of the Thar Desert region of Rajasthan, India, is 2.7–5.6 mm/year (0.11–0.22 in/year), which is 2–3% of the annual precipitation of 165 mm/year (6.5 in/year). An analysis of historical recharge rates in sand dunes located in the Badain Jaran Desert of northwestern China indicates that the average recharge rate over the last 2,000 years has been 0.9 mm/year (0.035 in/year) (Ma et al. 2009). Other investigations in this area showed historical recharge rates ranging from 0.9 to 1.4 mm/year (0.035–0.055 in/year) (Ma and Edmunds 2006; Ma et al. 2007; Gates et al. 2008). Edmunds and Tyler (2002) suggest that in desert dune areas with an annual precipitation range of 100–200 mm/year (3.9–7.9 in) or less, the direct recharge rate is close to zero, which supports the contention of Wood et al. (2003) and Wood (2011) that water in dune aquifers was emplaced during earlier pluvial periods. In some desert dune aquifers, recharge is not direct, but results from flood losses originating from ephemeral streams that pass through the dune fields, such as occurs in the Kuiseb Dune area in Namibia (Klaus et al. 2008).

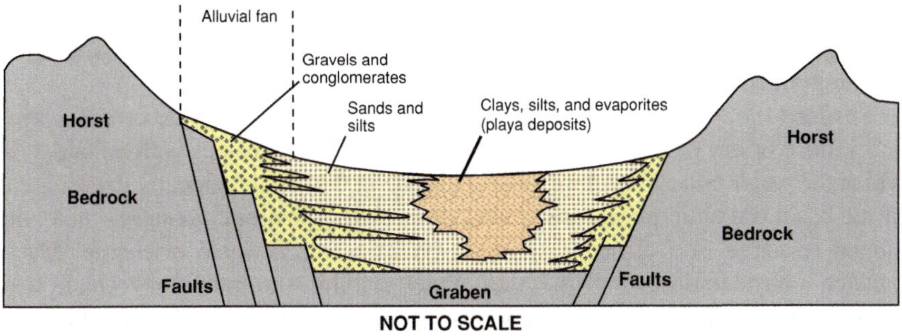

Fig. 4.6 Basin fill aquifers are located between bedrock mountain blocks. Coarse-grained, relatively high permeability sediments were deposited near the basin margins, primarily as alluvial fans and fluvial deposits. Finer-grained sediments occur near the center of the basin

4.2.3 Basin-Fill Aquifers

Basin-fill aquifers are composed predominantly of siliciclastic sediments that fill intermontane basins (Fig. 4.6). Basin-fill aquifers are sometimes referred to as alluvial aquifers. However, the term "alluvial aquifer" is also used to refer specifically to aquifers composed of "alluvium", which is defined as sediments deposited by flowing water. Thus, basin-fill aquifers, and particularly the most permeable coarser-grained units, are often composed predominantly of alluvium. The term alluvial aquifer is also applied to local intrabasinal aquifers.

In terms of depositional environments, basin-fill aquifers are typically composed predominantly of alluvial fan, fluvial, aeolian, and lacustrine deposits. Some basin-fill aquifers may contain interbedded volcanic deposits (e.g., Basin of Mexico; Mazari and Mackay 1993; Carrera-Hernández and Garskin 2007). As discussed in Chap. 3, alluvial deposits in basin-fill aquifers consist of interbedded strata of highly variable grain size, and silt and clay content, which results in a high degree of aquifer heterogeneity. Groundwater flow occurs predominantly in coarse-grained alluvial and channel-fill deposits. Finer-grained overbank, lacustrine, and caliche deposits can form confining or semiconfining units that impede vertical flow.

Sediment-grain size decreases downgradient and inward from the basin margins due to a decrease in hydraulic energy. Gravels and conglomerates are deposited near the basin margins and the finest sediments are deposited in playas and lakes at the topographic low points of the basin (Sect. 3.2.8). Variations in sedimentation rate and hydraulic energy, related to pulses of tectonic activity or climate variation, result in coarser-grained, more permeable sediments being transported further into the basin. Well-sorted river and delta deposits and fine-grained lacustrine sediments may be deposited during wetter, pluvial periods (Mifflin 1988). Basins having thick deposits of alluvium often contain stacked aquifers (or subaquifers) consisting of intervals with abundant coarse-grained sediments (sand and gravels) separated by intervals composed mostly of fine-grained, low permeability sediments. The degree of hydraulic separation commonly increases towards the basin center as fine-grained confining strata become more abundant.

Snyder (1962) distinguished two main types of playas: phreatic or wet playas and vadose or dry playas. Phreatic playas tend to occur in basins with no outlet, in which the water table rises close to land surface in the topographically lowest areas of the basin. Phreatic playas are groundwater discharge areas. At steady state, the annual recharge in a closed basin must be balanced by local discharge. Mass-balance considerations indicate that under natural (no pumping) conditions, groundwater levels in a closed basin must rise towards the basin floor until a long-term equilibrium is reached between recharge and discharge.

Salts commonly accumulate on and below the surface of phreatic playas caused by the evaporation of surface waters that flow into the playas during major rainfall events and upwards flowing groundwater. Although phreatic playas are the site of evaporation, the primary discharge method may actually be transpiration by phreatophyte vegetation rather than direct evaporation (Mifflin 1988; Robson and Banta 1995). The groundwater below and in the vicinity of phreatic playas tends to be of poor quality because of elevated salinity. Examples of phreatic playas are the Great Salt Lake Desert in Utah (Snyder 1962), the Atacama playas, and the salt flats in Death Valley National Park, California (Fig. 3.29).

The water table occurs at a substantial depth below the base of the vadose playas so that the upwards flow of groundwater towards land surface does not occur to a significant degree. A groundwater outlet from the basin is present, which prevents the water table from rising to land surface (Snyder 1962). Dry playas may also be local depressions in an open groundwater basin. Vadose playas are not discharge areas, but may instead be recharge areas during wet periods.

The hydrologic characteristics of basin-fill aquifers vary depending on whether the basins are open or closed with respect to surface water and groundwater. Basin-fill aquifers are classified based on whether they are undrained, fully drained, or partially drained (Snyder 1962; Maxey 1968). An undrained basin has no interbasin groundwater flow. The flow system is entirely local and natural groundwater discharge occurs within the basin at phreatic (wet) playas (Fig. 4.7a) or in some basins in springs or base flow into streams. In a fully drained basin, recharged water leaves the basin through interbasin flow, and significant local natural (pre-development) discharge does not occur (Fig. 4.7b). The water table is located a substantial distance below land surface and any playas are dry. Fully drained basin-fill aquifers are thus part of intermediate or regional flow systems. Partially drained basins have discharge through both interbasin flow and phreatic playas (Fig. 4.7c).

An important hydrogeologic observation is that type of flow system can be inferred from some basic observations. Active intrabasin recharge, the absence of a phreatic (wet) playa, and a great depth to water in the pre-development groundwater flow system are good evidence for interbasin groundwater flow based on fluid-mass balance considerations (Maxey and Jameson 1948; Mifflin 1988;

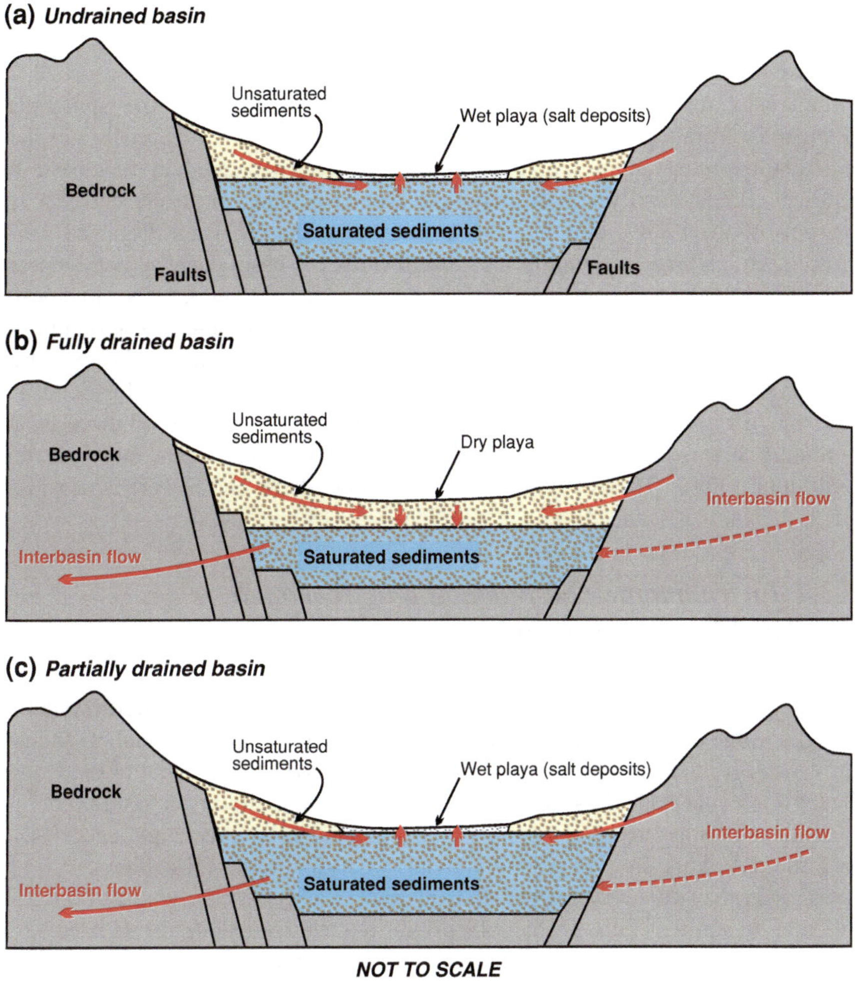

Fig. 4.7 Conceptual diagrams of basin fill aquifer types using classification of Synder (1962) and Maxey (1968). In a closed basin, the only discharge location is a wet (phreatic) playa at the low point in the basin or baseflow to a stream or spring. Fully drained basins are characterized by interbasin flow being the primary groundwater outflow. The water table is located well below land surface and the playa is dry Drainage in partially drained basin occurs both at wet playas and through interbasin flow

Hibbs and Darling 2005). Negligible groundwater discharge within a basin indicates discharge by downward flow and then deep aquifer flow into adjacent basins.

The basin-fill aquifers in the Basin and Range province of the western United States are very good examples of intermontane aquifers (Mifflin 1988). Extension faulting caused the formation of a repetitive pattern of north–south striking mountain ranges and deep basins, which were filled with thick deposits of siliciclastic deposits. Depending upon the bedrock type (carbonate versus low permeability strata), basins may be either drained or poorly drained.

Basin-fill aquifers may contain great volumes of water in storage because of the large areal extent and great thickness of sediments. In arid lands, the aquifers are prone to overdraft and associated declines in aquifer water levels because of their inherent low recharge rates. Salinity tends to increase with depth in thick basin-fill aquifers. Better quality (lower salinity) water also tends to be present near the margins of the basins, closer to the location of recharge. Total dissolved solids concentrations tend to naturally increase towards the basin center due to concentration of salts by evapotranspiration (Robson and Banta 1995). Upper freshwater aquifers may be vulnerable to increases in salinity with progressive groundwater development. For example, brackish-water intrusion occurs in the Hueco Bolson aquifer system of the states of Texas and New Mexico (U.S.A.) and Chichuahua (Mexico) from several different processes, including leakage from interbeds and confining strata, lateral encroachment, upconing of saline water, and irrigation drain return flows (Hibbs 1999; Hutchinson 2006; Sheng and Devere 2005).

4.2.4 Intermontane and Submontane Carbonate and Sandstone Aquifers

Local aquifers may be composed of carbonate, siliciclastic, or fractured crystalline (metamorphic or igneous) rock. Submontane aquifers also include blanket sand and gravel aquifers, such as the High Plains Aquifer in the midcontinent of the United States (Gutentag et al. 1984). Recharge may occur either directly in basin-bordering mountains or other topographic highs (e.g., mountain block recharge; Sect. 8.4.1), or through leakage from overlying alluvial (wadi) and basin-fill aquifers.

A basin may contain an upper alluvial aquifer and an underlying carbonate or siliciclastic aquifer, both of which are primarily recharged along the uplifted basin boundary (Fig. 4.8). Many of the primary aquifers in the eastern Arabian Peninsula (e.g., Dammam Formation and Umm Er Rhaduma Formation) are submontane aquifers in which natural recharge occurs mainly in upland areas and discharge occurs along the coast in playas (sabkhas) and offshore outcrops. Wells completed in topographically low, down-gradient areas may have been naturally flowing prior to the initiation of intense groundwater development.

Carbonate and sandstone aquifers in arid regions do not differ greatly from similar aquifers in humid regions. Carbonate aquifers tend to be highly heterogeneous with flow dominated by secondary porosity. The most extreme example of heterogeneous carbonate aquifers is karstic systems in which the bulk of groundwater flow occurs through dissolutional flow conduits. Karst systems in carbonate aquifers in arid lands formed mainly during earlier pluvial periods. Current rainfall is typically too sparse to cause significant carbonate dissolution. Aquifers in arid regions are much more prone to overdraft and associated declining aquifer water levels because of low recharge rates.

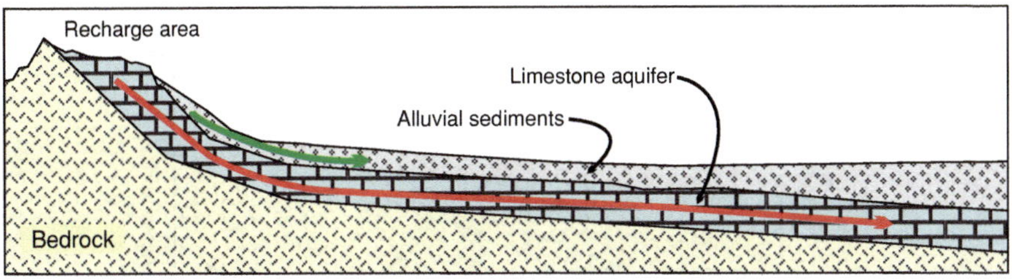

Fig. 4.8 Conceptual diagram of a regional aquifer. Recharge occurs in upland areas where aquifer strata are exposed. Groundwater may flow 10s or 100s of km from recharge area. In arid regions, there may be no significant recharge outside of the upland recharge area

4.2.5 Fractured Rock Aquifers (Hard-Rock Aquifers)

Hard rock terrain refers to an area underlain by igneous and metamorphic rock, such as granites and gneisses, which are essentially impermeable. Groundwater availability is largely restricted to fractured and weathered horizons. Fractured rock aquifers are important water sources in hard rock terrains but are unpredictable in terms of initial and sustainable yields. Well yields are usually highly variable depending on whether or not a well intercepts fractures, and the number of open fractures intercepted. Groundwater flow occurs within fractures and faults that have open apertures, which constitute only a small fraction of the total volume of the rock. Groundwater availability is limited because of the very low overall porosities (and storativities; Sect. 5.3.2) and often very low transmissivities (Howard et al. 1992). The volume of water stored in the aquifers is substantially less than that stored in sedimentary aquifers. The productive capacity (total volume of water than may be produced) of wells completed in fractured rock may be very low, unless the fracture network has a very great extent or is well connected to a recharge area. Low storativity of hard rock aquifers results in their being vulnerable to significant depletion during sustained drought periods (Wright 1992). Where high-yielding wells are drilled, questions may still remain concerning the sustainability of the resource because of the low aquifer storage volumes. Wells may initially have high production rates, which may drop off rapidly as the volume of stored water in fractures is depleted.

Hard rock aquifers contain two main zones, bed rock and regolith (Wright 1992) (Fig. 4.9). The regolith consists of weathered bed rock, alluvial deposits (transported materials), and soils. Weathering can either substantially increase fracture permeability or reduce permeability by the precipitation or formation of clay minerals within fracture apertures.

Mantles of weathered crystalline rock (regolith) are a common feature of the landscapes across equatorial Africa, South America, and Asia. The thickness and properties of the mantles reflect the results of deep weathering and stripping (colluvial and fluvial erosion), which, in turn, are controlled by climate and the

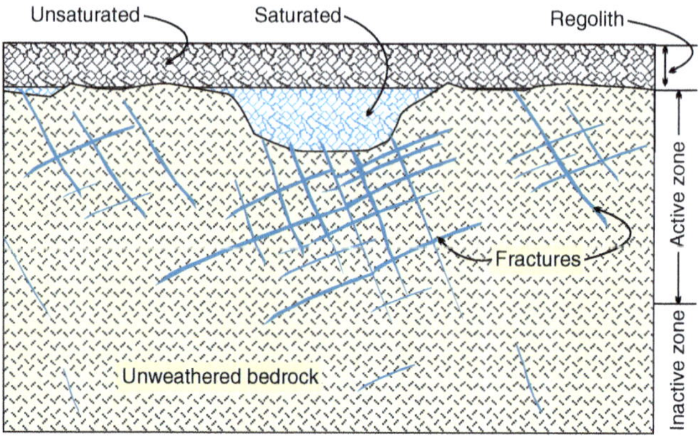

Fig. 4.9 Conceptual diagram of a fractured rock system. Groundwater flow occurs mainly in the upper active zone, which contains common open fractures. Saturated regolith may provide water storage for the fractures in the active zone

tectonic and geomorphic evolution of the region (Wright 1992; Taylor and Howard 2000). Tectonic quiescence allows for deep weathering and the development of thick regoliths. Tectonic uplift induces stripping of sediments and exposure of the rock surface.

The regolith itself may serve as a local aquifer where it has a relatively thick and permeable saturated zone. The regolith also provides storage for underlying fracture systems. Zones of high hydraulic conductivity tend to be located at the base of aquifers that occur in erosional mantles, because of the less aggressive weathering associated with saturated conditions. Coarse, partly weathered clasts tend to preferentially occur just above bedrock (McFarlane et al. 1992). Chemical weathering within the unsaturated zone tends to produce a clay-rich residuum of low hydraulic conductivity.

Groundwater is transmitted via coarse weathered debris at the base of the weathered mantle and by fractures in the underlying bed rock. The weathered mantle and fractured bedrock act as an integrated system (Taylor and Howard 2000). During pumping tests of fractured rock zones in Uganda, a leaky aquifer response and drawdowns in monitoring wells completed in the weathered mantle both indicate a hydraulic connection between the fractured rock and mantle (Taylor and Howard 2000). The groundwater storage provided by the mantle (regolith) or alluvial deposits is important for water supply because of the very low storativity of the fractured rock (McFarlane et al. 1992; Wright 1992; Taylor and Howard 2000; Limaye 2010). Flow through fractured rock is also a source of recharge into wadi aquifer systems and contributes water to the aquifers during low recharge periods (Al-Shaibani 2008).

Numerous studies have documented that groundwater flow in fractured crystalline aquifers predominantly occurs in a shallow higher permeability zone (the "active zone"), which overlies a deeper, less fractured, and thus less permeable zone ("inactive zone", Maréchal 2010). The base of the active zone depends upon local hydrogeologic conditions, but can be determined from local well drilling and testing experiences. For example, an extensive packer-testing program in a fractured rock aquifer in Uganda indicated that in most (>90%) of the studied wells, most (>75%) of the water was produced from one or two test zones, which indicates that water-producing fractures are uncommon. There was no relationship between high-yielding fractures and depths within the upper 60 m (197 ft) of bedrock (Howard et al. 1992). Fractures were much less developed at greater depths and 60 m (197 ft) appeared to represent the bottom of the aquifers (active zones). No relationship was observed between fracture permeability and physiography. Other studies suggest a tendency for fracturing to be better developed at shallower depths, which would be expected if the fracturing was due primarily to decompression (e.g., Carruthers et al. 1991; Taylor and Howard 2000; Howard et al. 1992; Maréchal 2010). Erosional unloading allows old incipient fractures to open and tensional cracks to form (Carruthers et al. 1991). Fracture aperture widths are also controlled to a significant degree by the vertical pressure load of the rock mass and its plasticity. Fractures tend to aneal (close) with depth resulting in a tendency for fractured rock aquifers to have a lower productivity with depth.

Fractured rock aquifers are characterized by high degrees of both vertical and horizontal heterogeneity, which can greatly impact local groundwater flow (Howard et al. 1992; Taylor and Howard 2000). Fracture-dominated aquifer heterogeneity is of particular importance for local solute-transport. For example, a geochemical study of managed aquifer recharge in Rajasthan, India, demonstrated the effects of fractures on local groundwater flow and solute transport in a hard rock aquifer (Stiefel et al. 2009). Wells located relatively close to the source of aquifer recharge (check dams) did not always receive a higher proportion of the artificially recharged water (as evidence by tracer concentrations) compared to more down-gradient wells. Groundwater flow from the recharge areas largely bypasses some nearby wells.

Over-draft of hard rock aquifers can result in a rapid lowering of water levels because of the low storativity of the aquifers. The common perceived solution is to drill deeper wells. However, as a general rule, the probability of increasing the yield of a given well is very low beyond the active zone (Maréchal 2010). The phenomenon of the "sunken cost fallacy" is often active in well drilling in hard rock areas, whereby increasing resources are invested in deepening dry holes in the statistically unlikely hope of avoiding the loss of a well (Maréchal 2010). The rational strategy is to instead stop drilling once the base of the active zone is reached and drill a new well elsewhere.

Development of fractured aquifers presents a number of specific challenges. The performance of wells depends upon whether or not the borehole intercepts sufficiently open fractures. Various remote sensing techniques, including lineament analysis and surface geophysics, have been applied with varying success in hard rock terrains in order to improve the success rate of wells (Chap. 18). Examination of regional relationships between terrain and well yields can assist in the location of the new wells. For example, in Malawi, high yields of wells completed in the regolith is related to very low surface relief, which in turn correlates with the total saturated regolith thicknesses (McFarlane et al. 1992).

The hydraulics of fractured rock aquifers are complex and experiences at one location cannot be translated to other areas (McFarlane et al. 1992). Drawdown behavior in pumped hard rock aquifers is usually greatly affected by heterogeneity and the scale of heterogeneity may be large relative to the scale of the test. It is therefore very difficult to get reliable information on aquifer hydraulics from standard aquifer pumping tests (Solomon and Quiel 2006). Although aquifers have a regional extent, they respond to abstraction in a "discontinuous" fashion due to either discontinues or barrier boundaries within fracture systems, or to constraints from the low permeability regolith (Wright 1992).

4.2.6 Regional and Intermediate Aquifers and Flow Systems

Regional and intermediate flow systems by definition (Tóth 1962, 1963) have a great geographic extent that may incorporate multiple local aquifer systems. Local aquifers may be hydraulically connected to regional and intermediate flow systems, and may be either a net sink or source of water to the regional and intermediate flow systems. Fully drained basin-fill aquifers are well connected to a regional or intermediate flow system (Sect. 4.2.3). Multiple local aquifers may be integrated via regional and intermediate flow systems, implying that the impact of groundwater pumping in a local basin may extend beyond the boundary of the groundwater basin, especially from a water budget perspective.

Although regional aquifers may contain large quantities of water, the long flow lengths and travel times, and elevated temperatures can result in deterioration in water quality due to fluid-rock interactions. For example, in the San Luis Potosi region of Mexico, vertical flow from a regional aquifer reduces the drawdowns from local groundwater pumping but also results in elevated, unhealthy fluoride concentrations in produced water (Carrillo-Rivera et al. 1996, 2002).

Evaluation of regional and intermediate aquifers and flow systems is greatly hampered in many areas by the paucity of data on their hydrogeology, chemistry, pressures, and flow rates. Indeed, it may not even be known in many areas if such flow systems even exist to a significant degree, because deep wells may not be present.

4.2.7 Non-renewable Groundwater Resources

Non-renewable groundwater resources are characterized by the absence of current recharge or current recharge (replenishment) rates that are very low on a human time scale and relative to abstraction rates. The recharge of non-renewable aquifers commonly occurred during earlier wetter pluvial periods. Enormous quantities of non-renewable water are present under some arid regions (e.g., Sahara Desert, parts of the Arabian Peninsula). Non-renewable aquifers are a critical source of water in some arid and semiarid lands. Utilization of non-renewable groundwater resources essentially involves the removal or mining of water from a finite storage volume, which has important implications for sustainability and inter-generational equity. Non-renewable aquifers are discussed in Chap. 36.

References

Al-Shaibani, A. M. (2008). Hydrogeology and hydrochemistry of a shallow alluvial aquifer, western Saudi Arabia. *Hydrogeology Journal, 16*, 155–165.

Berger, D. L. (1992). Ground-water recharge through active sand dunes in northwestern Nevada. *Water Resources Bulletin, 28*(5), 959–965.

Boulton, N. S. (1954a). Unsteady radial flow to a pumped well allowing for delayed yield from storage. *International Association of Scientific Hydrology Publication, 37*, 472–477.

Boulton, N. S. (1954b). The drawdown of the water table under nonsteady conditions near a pumped well in an unconfined formation. *Institute of Civil Engineers Proceedings (London), 3*, 564–579.

Boulton, N. S. (1963). Analysis of data from non-equilibrium pumping tests allowing for delayed yield from storage. *Institute of Civil Engineers Proceedings (London), 26*, 469–482.

Carrera-Hernández, J. J., & Garskin, S. J. (2007). The Basin of Mexico aquifer system: Regional groundwater level dynamics and database development. *Hydrogeology Journal, 15*, 1577–1590.

Carrillo-Rivera, J. J., Cardona, A., & Edmunds, W. M. (2002). Use of abstraction regime and knowledge of hydrogeological conditions to control high fluoride concentration in abstracted groundwater. *Journal of Hydrology, 261*, 24–47.

Carrillo-Rivera, J. J., Cardona, A., & Moss, D. (1996). Importance of the vertical component of groundwater flow: A hydrogeochemical approach in the valley of San Luis Potosi. *Mexico Journal of Hydrology, 186*, 23–44.

Carruthers, R. M., Greenbaum, D., Peart, R. J., & Herbert, R. (1991). Geophysical investigation of photolineaments in southeast Zimbabwe. *Quarterly Journal of Engineering and Hydrogeology, 24*, 437–451.

Edmunds, W. M., & Tyler, S. W. (2002). Unstaurated zones as archives of past climate: Toward a new proxy for continental regions. *Hydrogeology Journal, 10*, 216–228.

Gates, J. B., Edmunds, W. M., Ma, J., & Sheppard, P. R. (2008). A 700-year history of groundwater recharge in the drylands of NW China and links to East Asian monsoon variability. *The Holocene, 18*, 1045–1054.

Gutentag, E. D., Heimes, F. J., Krothe, N. C., Luckey, R. R., & Weeks, J. B. (1984). *Geohydrology of the high plains aquifer in parts of Colorado, Kansas, Nebraska, New Mexico, Oklahoma, South Dakota, Texas, and Wyoming*. U.S. Geological Survey Professional Paper 1400-B.

Hantush, M. S. (1956). Analysis of data from pumping tests in leaky aquifers. *American Geophysical Union Transactions, 37*(6), 702–714.

Hantush, M. S. (1959). Nonsteady flow to flowing wells in a leaky aquifer. *Journal of Geophysical Research, 64*(8), 1043–1052.

Hantush, M. S. (1960). Modification of the theory of leaky aquifers. *Journal of Geophysical Research, 65*(11), 3713–3725.

Hantush, M. S., & Jacob, C. E. (1955). Nonsteady radial flow in an infinite leaky aquifer. *Transactions of the American Geophysical Union, 36*, 95–100.

Hibbs, B. J. (1999). Hydrogeologic and water quality issues along the El Paso/Juarez Corridor; an international case study. *Environmental and Engineering Geoscience, 5*, 27–39.

Hibbs, B. J., & Darling, B. K. (2005). Revisting a classification scheme for U.S.-Mexico alluvial basin-fill aquifers. *Ground Water, 43*(5), 750–763.
Howard, K. W. F., Hughes, M., Charlesworth, D. L., & Ngobi, G. (1992). Hydrogeologic evaluation of fracture permeability in crystalline basement aquifers of Uganda. *Applied Hydrogeology (Hydrogeology Journal), 1*(1), 55–65.
Hutchinson, W. R. (2006). *Groundwater management in El Paso, Texas*. Boca Raton, FL: Dissertation.com.
Klaus, J., Kulls, C., & Dahan, O. (2008). Evaluating the recharge mechanism of the Lower Kuiseb Dune area using mixing cell modeling and residence time data. *Journal of Hydrology, 358*, 304–316.
Kruseman, G. P., & de Ridder, N. A. (1970). *Analysis and evaluation of pumping test data*. Wageningen: International Institute for Land Reclamation and Improvement, Bulletin 11.
Limaye, S. D. (2010). Review: Groundwater development and management in the Deccan Traps (basalts) of western India. *Hydrogeology Journal, 18*, 543–558.
Love, D., de Hamer, W., Owen, R. J. S., Booij, M., Uhlenbrook, S., & Hoekstra, A. Y. et al. (2007). Case studies of groundwater—surface water interactions and scale relationships in small alluvial aquifers. *8th WATERNET/WARFSA/GWP-SA Symposium, 31 October–2 November 2007, Lusaka, Zambia* (p. 6).
Ma, J., Ding, Z., Gates, J. B., & Su, Y. (2007). Chloride and the environmental isotopes as the indicators of groundwater recharge in the Gobi Desert, northwest China. *Environmental Geology, 55*(7), 1407–1419.
Ma, J., & Edmunds, W. M. (2006). Groundwater and lake evolution in the Badain Jaran desert ecosystem, Inner Mongolia. *Hydrogeology Journal, 14*, 1231–1243.
Ma, J., Edmunds, W. M., He, J., & Jia, B. (2009). A 2000 year geochemical record of paleoclimate and hydrology derived from dune sand moisture. *Paleoleography, Paleoclimatology, Paleoecology, 276*, 38–46.
Maréchal, J.-C. (2010). Editor's message: The sunk cost fallacy of deep drilling. *Hydrogeology Journal, 18*, 287–289.
Maxey, G. B. (1968). Hydrogeology of desert basins. *Hydrogeology Journal, 6*(5), 10–22.
Maxey, G. B., & Jameson, C. H., (1948) Geology and water resources of Las Vegas, Pahrump, and Indian Spring Valleys, Clark and Nye Counties, Nevada: Nevada State Engineer, Water Resources Bulletin 5.
Mazari, M., & Mackay, D. M. (1993). Potential for groundwater contamination in Mexico City. *Environmental Science and Technology, 27*(5), 794–801.
McFarlane, M. J., Chilton, P. J., & Lewis, M. A. (1992). Geomorphological controls on borehole yields: A statistical study in an area of basement rocks in central Malawi. In E. P. Wright & W. G. Burgess (Eds.), *Hydrogeology of crystalline basement aquifers in Africa* (pp. 131–154). Geological Society of London Special Publication 66.
Mifflin, M. D., (1988). Chapter 8, Region 5, Great Basin. In W. Back, J. S. Rosenshern & P. R. Seaber (Eds.), *Hydrogeology, the Geology of North America*, Vol. O-2 (pp. 69–78). Boulder, CO: Geological Society of America.
Murad, A. A., Al Nuaimi, H., & Al Hammadi, M. (2007). Comprehensive assessment of water resources in the United Arab Emirates (UAE). *Water Resources Management, 21*, 1449–1463.
Neuman, S. P. (1975). Analysis of pumping test data from anisotropic unconfined aquifers considering delayed gravity response. *Water Resources Research, 11*, 329–342.
Newman, B. D., Vivoni, E. R., & Groffman, A. R. (2006). Surface water—groundwater interactions in semiarid drainages of the American southwest. *Hydrological Process, 20*, 3371–3394.
Pereira, L. S., Cordery, I., & Iaconides, I. (2002). Coping with water scarcity: UNESCO, International Hydrological Programme, IHP-VI, Technical Documents in Hydrology No. 58.
Robson, S. G., & Banta, E. R. (1995). Ground water atlas of the United States. Arizona, Colorado, New Mexico, and Utah: U.S. Geological Survey Atlas HA 730-C.

Scanlon, B. R., Mukherjee, A., Gates, J., Reedy, R. C., & Sinha, A. K. (2010). Groundwater recharge in natural dune systems and agricultural ecosystems in the Thar Desert Region, Rajasthan, India. *Hydrogeology Journal, 18*, 959–972.

Şen, Z. (2008). *Wadi hydrology*. Boca Raton, FL: CRC Press.

Sheng, Z., & Devere, J. (2005). Understanding and managing the stressed Mexico-USA transboundary Hueco bolson aquifer in the El Paso de Norte region as a complex system. *Hydrogeology Journal, 13*, 813–825.

Snyder, C. T. (1962). A hydrologic classification of valleys in the Great Basin, western United States. *Bulletin of the International Association of Scientific Hydrology, 7*(3), 53–59.

Solomon, S., & Quiel, F. (2006). Groundwater study using remote sensing and geographic information systems (GIS) in the central highlands of Eritrea. *Hydrogeology Journal, 14*, 1029–1041.

Stiefel, J. M., Melesse, A. M., McClain, M. E., Price, R. M., Anderson, E. P., & Chauhan, N. K. (2009). Effects of rainwater-harvesting-induced artificial recharge on the groundwater of wells in Rajasthan, India. *Hydrogeology Journal, 17*, 2061–2073.

Taylor, R., & Howard, K. (2000). A tectono-geomorphic model for the hydrogeology of deeply weathered crystalline rock: Evidence from Uganda. *Hydrogeology Journal, 8*, 279–294.

Theis, C. V. (1935). The relation between the lowering of the piezometric surface and the rate and duration of discharge of a well using groundwater storage. *Transactions American Geophysical Union, 16*, 519–524.

Tóth, J. A. (1962). A theory of groundwater motion in small drainage basins in Central Alberta, Canada. *Journal of Geophysical Research, 67*, 4375–4387.

Tóth, J. (1963). A theoretical analysis of ground-water flow in small drainage basins. *Journal of Geophysical Research, 68*(16), 4795–4812.

Van der Gun, J. A. M. (2009). Climate change and alluvial aquifers in arid regions. In F. Ludwig, P. Kabat, H. van Schaik & M. van der Valk (Eds.), *Climate change adaptation in the water sector* (pp. 159–176). London: Earthscan.

Wood, W. W. (2011). Source of paleo-groundwater in the Emirate of Abu Dhabi, United Arab Emirates: Evidence from unusual oxygen and deuterium isotope data. *Hydrogeology Journal, 19*(1), 155–161.

Wood, W. W., Rizk, Z. S., & Alsharhan, A. S. (2003). Timing of recharge, and the origin, evolution and distribution of solutes in a hyperarid aquifer system. In A. S. Alsharhan & W. W. Wood (Eds.), *Water resources perspectives: Evaluation, management and policy* (pp. 295–312). Amesterdam: Elsevier.

Wright, E. P. (1992). The hydrogeology of crystalline basement aquifer in Africa. In E. P. Wright & W. G. Burgess (Eds.), *Hydrogeology of crystalline basement aquifers in Africa* (pp. 1–27). Geological Society of London Special Publication 66.

Chapter 5
Introduction to Aquifer Hydraulics

5.1 Introduction

Management of groundwater resources requires an understanding of the response of aquifers to groundwater withdrawals. Groundwater impact analyses are now usually performed by numerical groundwater flow and solute-transport modeling, which has become a specialized discipline (Chap. 20). A basic knowledge of hydraulic principles is necessary for understanding the behavior of aquifers in arid and semiarid lands. A number of excellent references on groundwater hydraulics are available including Freeze and Cherry (1979), Lohman (1979), Heath (1983), Driscoll (1986), Kruseman and de Ridder (1970, 1991), Walton (1970, 1991), Kasenow (1997, 2006), Domenico and Schwartz (1998), Fetter (2001), and Todd and Mays (2005).

5.1.1 Aquifer Heads, Potentiometric Surface, and Groundwater Flow

Perhaps the most important principle in understanding the movement of both surface water and groundwater is that water flows "downhill" or from areas of high altitude or pressure to areas of lower altitude or pressure. Hydraulic head is a measure of water pressure above a specified datum, such mean sea level or, in the United States, the National Geodetic Vertical Datum of 1929 (NGVD) or North American Vertical Datum of 1988 (NAVD 88). It is the sum of the heads related to altitude, pressure, and a velocity component as expressed in the Bernoulli equation (Todd 1959).

The term "potentiometric surface" refers to an imaginary surface that connects points of equal altitude to which water will rise in tightly cased wells completed in

R. Maliva and T. Missimer, *Arid Lands Water Evaluation and Management*, Environmental Science and Engineering, DOI: 10.1007/978-3-642-29104-3_5,

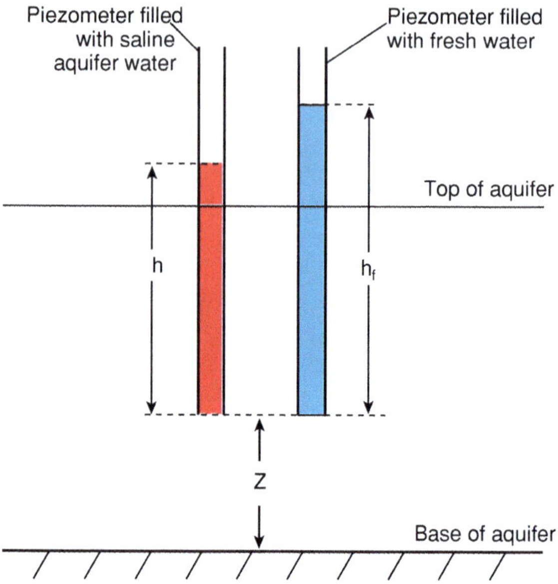

Fig. 5.1 Equivalent freshwater heads. Water levels in piezometers open to the same aquifer will differ as a function of the density of the waters

the same aquifer. Potentiometric surface maps are generated by contouring water level altitude data from an aquifer referenced to the same datum. Potentiometric surface maps are commonly used to determine both the direction of groundwater flow and the hydraulic gradient, which is rate of change of head divided by distance.

For freshwater aquifers it is the standard and an acceptable procedure to use water-level altitudes in wells to generate potentiometric surface maps and determine hydraulic gradients. However, water altitude data cannot be used directly to calculate and compare heads in an aquifer in which there are significant variations in water temperature or salinity and thus density. Hydraulic heads in an aquifer are a function of both the height of the water column in a well and the density of the water. For a given aquifer head (pressure), the altitude of water in a well will decrease with increasing salinity. In order to generate an accurate potentiometric surface map, altitude data should be converted into equivalent freshwater heads (h_f) as follows (Fig. 5.1; Guo and Langevin 2002):

$$h_f = \frac{\rho}{\rho_f} h - \frac{\rho - \rho_f}{\rho_f} Z \tag{5.1}$$

where,

h = head (water altitude) measured in well above base of aquifer datum (ft, m)
ρ = density of water in well (g/cm^3)
ρ_f = density of freshwater in well (g/cm^3, $\approx$1.0 at 4°C)
Z = elevation of the base of well above base of aquifer datum (ft, m).

Lower water altitudes in parts of the aquifer containing saline water can give the false impression that a hydraulic gradient exists between the parts of an aquifer containing freshwater to the parts of the aquifer containing saline water, when no such gradient actually exists. Conversion of all measured heads to equivalent freshwater heads is necessary for the development of a groundwater flow and solute transport models, model calibration, and interpretation of the calculated results (Guo and Langevin 2002).

5.2 Groundwater Flow

Darcy's Law forms the foundation of groundwater flow analysis by relating flow rate to hydraulic gradient, as follows:

$$Q = -KA\left(\frac{dh}{dl}\right) \quad or \tag{5.2}$$

$$Q = -KA\left(\frac{\Delta h}{\Delta l}\right) \tag{5.3}$$

where,
K = hydraulic conductivity, which has the units of length over time (e.g., m/s)
Q = flow rate or discharge (m^3/s)
dh/dl, Δh/Δl = hydraulic gradient (dimensionless, m/m)
A = the cross-sectional area of the flow path (m^2).

The average linear velocity of flow (v) can be calculated as follows:

$$v = -\frac{K}{\varphi_e}\left(\frac{dh}{dl}\right) \tag{5.4}$$

where φ_e = effective porosity.

Flow velocity is inversely proportional to effective porosity. One can conceptualize this relationship by considering a pipe with a constant flow rate (Q). If an obstruction is placed in the pipe (i.e. its effective porosity is reduced), then the flow velocity (length/time) must be increased to maintain the same flow rate (volume/time). In a layered (bedded) aquifer, the flow rate in each bed will be proportional to the hydraulic conductivity of the bed. Large differences in hydraulic conductivity may occur in heterogeneous aquifers, which will result in corresponding variations in both flow rate and average linear flow velocity. Inasmuch as travel distance is the product of hydraulic conductivity and time, water in the beds with the greatest hydraulic conductivity will also travel the greatest distance per unit of time under prevailing hydraulic gradients.

Hydraulic conductivity is an extrinsic property of a rock or sediment that depends upon properties of the fluid. In the water industry, it is assumed that the flow of freshwater through aquifers is of equal and constant density and viscosity, which make hydraulic conductivity measurements comparable. Permeability is an intrinsic property of a rock or sediment in that it is not dependent on other variables or conditions. The relationship between hydraulic conductivity (K) and intrinsic permeability (k) is as follows:

$$K = \frac{k\rho g}{\mu} \tag{5.5}$$

where,
ρ = density (kg/m^3 or g/cm^3)
g = gravitation acceleration (9.807 m/s^2)
μ = dynamic viscosity (kg m^{-1} s^{-1}).

Dynamic viscosity is also expressed as units of Pascal-seconds (Pa s), which is equivalent to 1 kg m^{-1} s^{-1}, and in centipoise units (cP), which are equal to 1×10^{-3} kg m^{-1} s^{-1}. Permeability has unit of length squared, with the SI unit being m^2. The unit for permeability commonly used in the oil and gas industry is the millidarcy (mD), which is equivalent to 9.8969×10^{-16} ($\approx 1 \times 10^{-15}$) m^2 or 9.8969×10^{-12} ($\approx 1 \times 10^{-11}$) cm^2. Permeability analyses of core samples are often reported in millidarcies, particularly by laboratories that make analyses primarily for the oil and gas industry.

The dynamic viscosity of water is a function of temperature, pressure, and salinity. The relationship between the dynamic viscosity of water and temperature (T, °K) is expressed by the equation:

$$\mu = 2.414 \times 10^{-5} \times 10^{\left(\frac{247.8}{(T - 140)}\right)} \tag{5.6}$$

A decrease in temperature from 30 to 20°C results in an increase in dynamic viscosity from 0.796×10^{-3} to 1.000×10^{-3} Pa s, a 25.6% increase. Temperature changes can thus have a large impact on hydraulic conductivity, which should be taken into consideration in managed aquifer recharge (MAR) systems that recharge water into an aquifer with a different native water temperature or at different times of the year. An increase in viscosity will result in a decrease in hydraulic conductivity, which has been documented to cause substantial seasonal and diurnal variations in infiltration rates (Sect. 13.8.1) and increases in injection pressure (Maliva and Missimer 2010). Salinity has a much lesser effect on viscosity than temperature, but may be significant in systems involving hypersaline fluids.

5.3 Aquifer Hydraulic Properties

The three main hydraulic properties of an aquifer are hydraulic conductivity (and in turn transmissivity), storativity (specific yield or storage coefficient), and leakance (or drain factor).

5.3.1 Transmissivity

Transmissivity (T) is a measure of the ability of an aquifer to transmit water. It is defined by the volume of water that will flow through a unit width (typically 1 m or 1 ft) of the cross-sectional area of an aquifer under a unit hydraulic gradient (1 m/1 m or ft/ft).Transmissivity has the units of length squared divided by time, such as m^2/day or ft^2/day. Transmissivity is also expressed in the literature using the older unit of gallons per day per ft (gpd/ft). For a given injection or recovery volume, aquifers with high transmissivities have lower injection pressures and drawdowns and are thus capable of accepting or producing larger quantities of water. Greater transmissivities will result in water migrating more rapidly under prevailing hydraulic gradients.

Transmissivity is equivalent to the product of the average aquifer hydraulic conductivity $(\overline{K})$ and the aquifer thickness (b);

$$T = \overline{K}b \tag{5.7}$$

The word "average" must be stressed. In heterogeneous aquifers, most of the transmissivity lies in the fraction of the total aquifer thickness that has the greatest hydraulic conductivity. For an aquifer divided into "n" number of beds, the transmissivity is:

$$T = \sum_{i=1}^{n} K_i b_i \tag{5.8}$$

where, K_i and b_i are the thickness and hydraulic conductivity of each bed.

It is erroneous to make inferences concerning fluid migration rates from average hydraulic conductivities calculated by dividing aquifer transmissivity by total thickness. Considering that all aquifers are heterogeneous to some extent, actual groundwater flow rates and distances traveled will be different than values estimated from the average aquifer hydraulic conductivity.

In the case of unconfined aquifers, the reduction of water levels from pumping (and associated decrease in saturated thickness) results in a decrease in transmissivity. Thus, as aquifers are dewatered, the rate of drawdown will increase because of the decrease in transmissivity, even if pumping rate remains unchanged.

5.3.2 Storativity

Storativity (S), which is also referred to as the storage coefficient, is defined as the volume of water that is released from a unit *area* of aquifer (1 m^2 or 1 ft^2) under a unit decline (1 m or 1 ft) of hydraulic head. Storativity is a dimensionless parameter. Specific storage (S_s) is defined as the volume of water that is released from a unit *volume* of aquifer under a unit decline of hydraulic head. Specific storage has the units of one divided by length (m^{-1}). The storativity of a confined aquifer is the vertically integrated specific storage value, which for a homogeneous aquifer is the product of its specific storage and the thickness of the aquifer (b),

$$S = S_s b \tag{5.9}$$

Specific yield (S_y) is the storage term for an unconfined aquifer. Specific yield is defined as the volume of water that a unit area of an unconfined aquifer releases by drainage per unit decline of the water table. Specific yield, which is also referred to as drainage porosity, is thus equivalent to the storativity of an unconfined aquifer.

The water released from storage in a confined aquifer by pumping is caused by a combination of compaction of the aquifer and expansion of the water, both caused by a decrease in water pressure. The water ultimately released from an unconfined aquifer is due to the drainage (dewatering) of the aquifer. Unconfined aquifers have a delayed yield response to pumping in which the initial water production is largely from depressuring, followed by production from drainage.

Specific yield values are usually orders of magnitude greater than the storativity values of a confined aquifer for the same rock or sediment types. Specific yield is related to the effective (interconnected) porosity of the sediment or rock. The specific yield of porous granular sediments and rock are usually in the 0.05–0.4 range whereas storativity values are often on the order of 1×10^{-3} to 1×10^{-5}. Thus, a given drop in water level or pressure in an unconfined aquifer will result in the release of a much greater volume of water than in a confined aquifer. Similarly, a given rate of pumping will induce much greater drawdowns in a confined aquifer than in an unconfined aquifer. The rate of drawdown within an aquifer may decrease dramatically when the potentiometric surface falls below the top of the aquifer and the aquifer becomes unconfined (Fig. 5.2).

Although conceptually simple, the accurate measurement of specific yield is a very difficult because the drainage of water from pore spaces can be very slow (months or longer), especially in fine-grained sediments. Specific yield may be estimated from laboratory drainage tests, aquifer pumping tests (delayed-yield curve analysis and the volume-balance method), and using microgravimetric data. Comparison of the results of various types of specific yield measurements in sand suggests that measured specific yield values depend upon the type of test, the time scale of the test, and method of data analysis (Nwankwor et al. 1984).

5.3.3 Leakance

Leakance or leakance coefficient (L) is defined as the volume of water that flows through a unit area of a semi-confining unit per unit head difference per unit of time. The leakance of a semi-confining unit is equal to the equivalent vertical hydraulic conductivity($\overline{K}_z$) of the unit divided by its thickness (b):

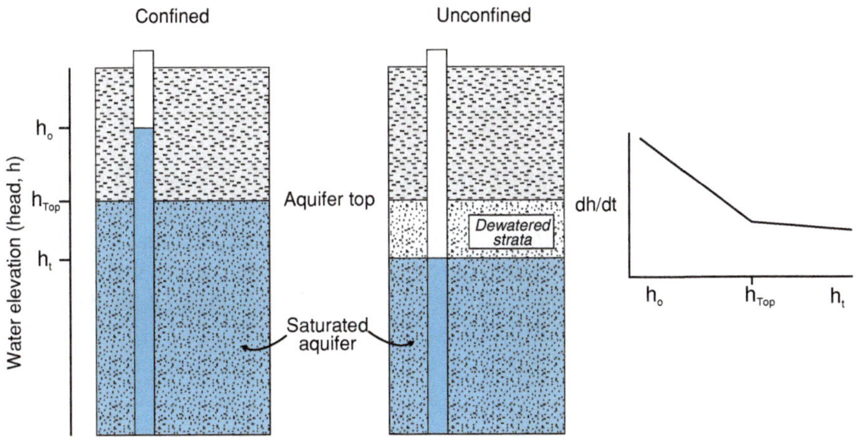

Fig. 5.2 Diagram illustrating the relationship between aquifer heads with respect to the top of a confined aquifer and the rate of decline in head at a constant abstraction rate. The rate of decline in head decreases dramatically when aquifer heads drop below the top of aquifer and water starts to be produced primarily from dewatering of the aquifer (unconfined conditions)

$$L = \frac{\overline{K}_z}{b} \tag{5.10}$$

The unit of leakance is the reciprocal of time (typically days^{-1}). Equivalent vertical hydraulic conductivity can be calculated as follows:

$$\overline{K}_z = \frac{b}{\sum_{i=1}^{n} \frac{b_i}{K_i}} \tag{5.11}$$

Equivalent vertical hydraulic conductivity, and thus leakance, is largely dependent upon the vertical hydraulic conductivity of the least permeable (tightest) beds within the semi-confining unit.

Although leakance is a property of an individual semi-confining unit, many aquifer test analyses of leaky aquifers, such as Hantush-Walton method, provide leakance values for the tested aquifer, which reflect the combined leakage into the aquifer from both the underlying and overlying semi-confining units. For groundwater modeling purposes, the aquifer leakance must be divided between the overlying and underlying semi-confining units. Often the lithologies of the semi-confining unit suggest how the leakance should be approximately divided. If one unit is particularly clay-rich, and thus likely to have a relatively low vertical hydraulic conductivity, then it is reasonable to assign the clay-rich unit a small fraction of the total aquifer leakance. The leakance distribution can be refined as part of the model calibration process.

Field testing methods are available that can be employed to measure the differences in leakance between the upper and lower confining units. For example, pressure transducers may be installed in confining strata in order to measure the pressure response of pumping (Neuman and Witherspoon 1972). Analysis of cores taken of confining units can also provide data on vertical hydraulic conductivity. In practice, methods that more directly measure the hydraulic properties of confining strata tend to be expensive and time-consuming process and, as a result, and are uncommonly performed.

Field measurement of leakance can be quite difficult because the length of pumping time required to obtain a clear departure from the Theis curve can be lengthy. Also, in layered systems, there is an inherent assumption that the heads in the adjacent aquifers are constant during the term of the test which may not actually occur. Therefore, the analysis of aquifer test data used to determine leakance must be very carefully done.

5.3.4 Hydraulic Diffusivity

Hydraulic diffusivity (α), which is also referred to as aquifer diffusivity, is defined as the transmissivity divided by storativity.

$$\alpha = T/S \tag{5.12}$$

Hydraulic diffusivity is a not widely used parameter, but has relevance for arid and semiarid regions because it is proportional to the speed at which head (pressure) changes are propagated through porous media. Hydraulic diffusivity thus relates to the time lag between local groundwater pumping and aquifer pressure responses elsewhere in an aquifer. For example, a substantial time lag may occur between groundwater pumping in one part of a groundwater basin and adverse impacts elsewhere in the basin, such as a reduction in spring flow or river baseflow. The absence of adverse impacts to current pumping, therefore, is not conclusive evidence that continued pumping will not result in adverse impacts in the future. Present day water levels in regional aquifers in arid regions may still partially reflect past wetter conditions, rather than being in equilibrium with current conditions.

Hydraulic diffusivity is positively related to transmissivity and inversely related to storativity. Pressure changes will propagate faster across an aquifer with a relatively high transmissivity and low storativity. For a given transmissivity, confined aquifers will have much greater hydraulic diffusivities than unconfined aquifers because of their much lower storativity values. Drawdowns from groundwater pumping will tend to be propagated much more slowly in unconfined aquifers than in confined aquifers.

5.4 Porosity and Storage

Porosity is the pore space in the volume of aquifer of interest divided by its total volume. The pores within an aquifer that are connected and contribute to the space through which water flows constitute what is termed effective porosity. Effective

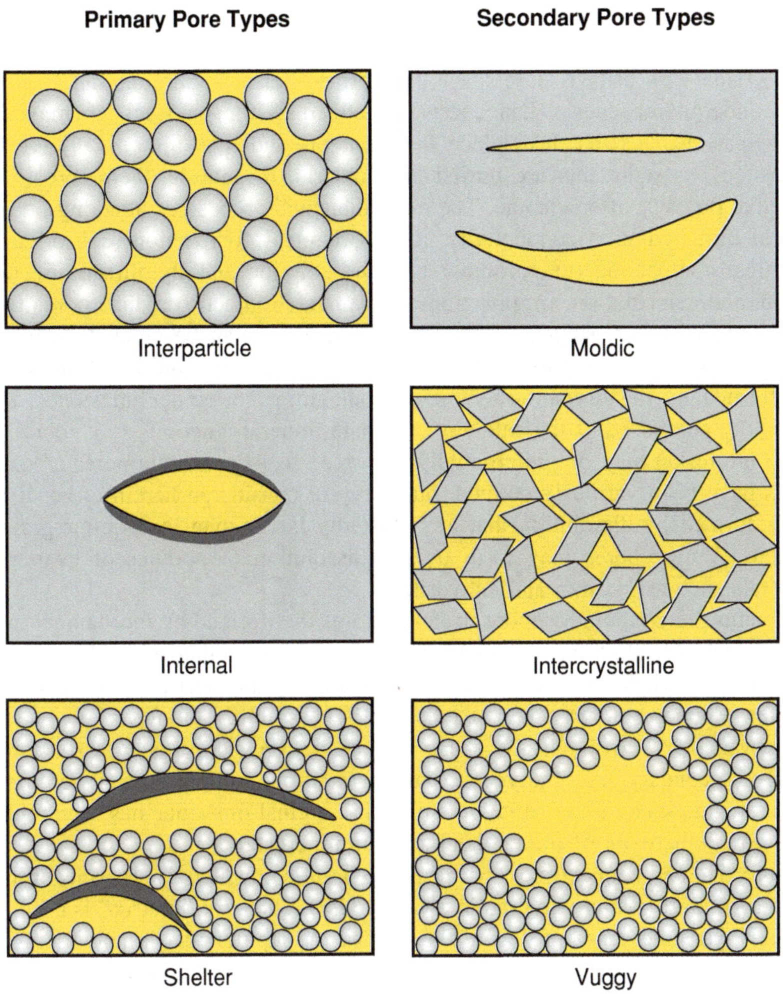

Fig. 5.3 Diagrams of common pore type (*yellow*) in sedimentary rock Primary pores are present in the unaltered sediments. Secondary pores form after deposition as the result of dissolution and replacement processes

porosity typically is less than total porosity. Effective porosity may be only a small fraction of the total porosity of some rocks in which most of the pores are not interconnected.

The main pore types found in sedimentary rocks were described by Choquette and Pray (1970) and are schematically illustrated in Fig. 5.3. Porosity can be subdivided into primary porosity and secondary porosity. Primary porosity is the original "fabric" or pattern of pores that is present immediately after the sediments are deposited. Primary porosity includes intergranular (interparticle) porosity which is the space between the grains, and intragranular porosity which is pore spaces that occur within grains. The interior of a hollow shell, such as an intact bivalve or gastropod, is an example of an intragranular pore.

Secondary porosity is pore space that forms after deposition as sediments and rock undergo diagenesis. Diagenesis is defined as the physical, chemical, and biological changes that take place in sediments and rock after they have been deposited. As sediments are buried and undergo diagenesis, their porosity and effective porosity often change. The main diagenetic processes affecting porosity, and in turn hydraulic conductivity, are mechanical compaction (consolidation), chemical compaction (intergranular pressure solution and stylolitization), dissolution, and cementation. Compaction and cementation act to reduce porosity.

Secondary pores, such as fossil molds and vugs, can be created during diagenesis, and tend to be particularly important in carbonate aquifers. Moldic pores are formed by the dissolution of individual grains. In some limestones, shells originally composed of the calcium carbonate mineral aragonite (e.g., corals and some mollusks) may be preferentially dissolved to form moldic pores. "Vuggy" pores have been variously defined, but they are considered herein to be discrete voids formed by dissolution that are markedly larger than the aquifer grains or crystals. Vuggy pores may form by the dissolution of nodules of evaporative minerals such as gypsum and anhydrite.

Fractures are tabular or more irregular pores that formed by mechanical failure of the rock. Conduits are elongate dissolution features that allow for the rapid (sometimes turbulent) flow of water, and are a characteristic feature of karst aquifer systems. Fractures may be enlarged by dissolution and form flow conduits. Intercrystalline porosity is the space present between crystals in a rock and is usually secondary. A relatively common example of intercrystalline porosity occurs within sucrosic dolomite, in which the original limestone has been replaced by a porous network of rhombohedral dolomite crystals.

Pore size and connectivity control the hydraulic conductivity (permeability) of the rock matrix (primary hydraulic conductivity). The term "matrix" is used in an aquifer hydraulic sense to denote the part of the aquifer that is not fractured or conduit porosity. A basic relationship in granular sediments and rocks (e.g., sands and sandstones) is that pore size and thus hydraulic conductivity increases as grain size increases (with consideration of sorting) (Shepard 1989; Alyamani and Sen 1993). At the fine end of the grain-size spectrum (fine silts and clays), hydraulic conductivity becomes exceedingly low with the water becoming essentially immobile.

Hydraulic conductivity is primarily controlled by the size of the pore throats that connect pores rather than the size of pores themselves. Pore throats are the constrictions that connect adjoining pores and can be conceptualized as being equivalent to the stem of a funnel. For a given pressure differential, flow rate is proportional to the cross-sectional area (and shape) of the pore throats. The presence of small quantities of clay (or suspended solids in water recharged by injection or surface spreading) can greatly reduce the hydraulic conductivity of a sediment or rock if it clogs the pore throats.

There is an overall tendency for primary porosity to decrease with progressive burial and diagenesis with a concomitant reduction of pore-size, connectivity, and hydraulic conductivity. Sediment types differ in their susceptibility to processes that result in reductions in porosity and hydraulic conductivity. Sediments composed of soft, clay-rich rock fragments may quickly loose nearly all of their effective porosity through mechanical compaction. Clean quartz sands on the contrary are much more resistant to mechanical compaction.

5.5 Aquifer Heterogeneity

All aquifers are heterogeneous to some degree with respect to porosity and hydraulic conductivity. There are two main types of aquifer heterogeneity, (1) layered or matrix heterogeneity, and (2) multiple porosity systems. It is not uncommon for an aquifer to have both types of heterogeneity. Layered heterogeneity is formed by differences in the composition of sediment at the time of deposition, diagenetic history, or both. Sedimentary deposits often have variations in grain size, shape, and sorting both within a bed (horizontal heterogeneity) and between beds (vertical heterogeneity). For example, river (fluvial) deposits typically have grain-size variations at a given location in a channel reflecting spatial differences in current velocity and horizontal variations between channel and non-channel deposits (over-bank deposition). The variation in sediment properties leads to heterogeneity in pore size and shape, and hydraulic conductivity. Diagenesis can either mute or accentuate differences in primary porosity. The effects of sedimentology on the distribution of hydraulic conductivity (permeability) are discussed by Castle et al. (2004).

When an aquifer develops secondary porosity by either dissolution or fracturing, specific parts of the aquifer (flow conduits or fractures) can have greater effective porosities and associated hydraulic conductivities. Karstic limestones are an example of a dual-porosity aquifer, which consists of matrix porosity and conduit porosity (Fig. 5.4). Triple-porosity (or tri-porosity) aquifers also occur, which include additional fracture porosity or a second generation of dissolution porosity. The Cenozoic carbonates of the Arabian Gulf that are primary oil reservoirs are a good example of a triple-porosity system or aquifer. The Biscayne Aquifer of South Florida is also considered to be either a dual- or triple-porosity aquifer (Cunningham et al. 2004, 2006, 2009). Dual- or tri-porosity aquifers may have extreme heterogeneity with flow occurring predominantly in the fractures or conduits.

Aquifer heterogeneity affects the bulk hydraulic properties of aquifers. The presence of flow zones with enhanced permeability can result in aquifers having greater transmissivities than would occur in homogenous aquifers dominated by matrix-flow. The presence of secondary porosity dominated flow may be detected by transmissivities calculated from pumping test results that are substantially

Fig. 5.4 (*Top*) Karstic limestone (Carboniferous age) in the Yorkshire Dales, near Malham, England. Extreme example of a dual-porosity system in which fracture apertures have been greatly widened by limestone dissolution. (*Bottom*) Karstic limestone exposure on Curacao. Cave is about 1 m high

(orders of magnitude) greater than the values that would be expected based on the matrix hydraulic conductivity of aquifer rock types.

Aquifer heterogeneity is particularly important for solute transport, which is strongly influenced by fluid flow paths rather than bulk aquifer hydraulic parameters. Solute transport is of critical concern for assessing and remediating contaminated sites and also for managed aquifer recharge (MAR) systems in which there are significant differences in water quality between the recharged water and native groundwater (Chap. 23). For example, aquifer heterogeneity can impact both the chemistry and flow direction of recharged treated wastewater (reclaimed water), which may have public health significance if there are nearby sensitive

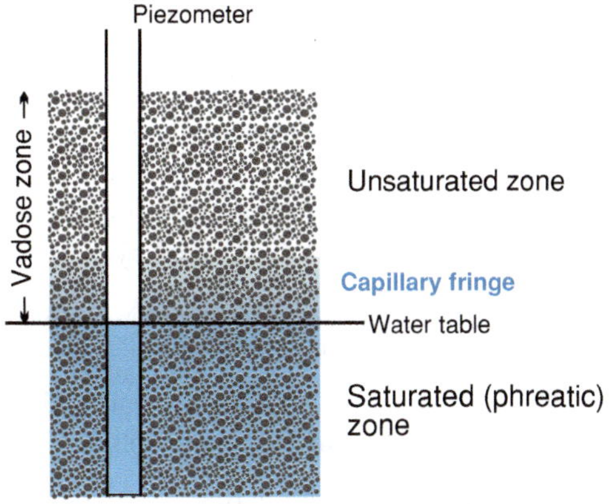

Fig. 5.5 Diagram of the relationship between the vadose zone, unsaturated zone, and water table. The lower part of the capillary fringe may be saturated, but is above the water table

receptors. Similarly, if the native groundwater is of lesser quality, aquifer heterogeneity could result in the rapid migration of stored water away from its injection point and adversely impact its recoverability.

Aquifer heterogeneity may also control aquifer recharge in arid and semiarid lands. The great degree of variation in hydraulic conductivity present in some arid region deposits results in very large (often > 100:1) ratios of horizontal to vertical hydraulic conductivity, which favors horizontal flow over vertical flow. Zones of enhanced vertical hydraulic conductivity, such a fractures zones or root zones, may be the primary loci for focused groundwater recharge as they provide a conduit through low permeability strata (Chap. 8).

5.6 Vadose Zone Hydrology

Unconfined groundwater systems are subdivided into two main hydrologic zones; the vadose zone and the underlying phreatic zone (Fig. 5.5). The vadose zone is defined as the interval extending from land surface to the top of the regional water table. The term "regional" is used to exclude local perched aquifers. The terms "vadose zone" and "unsaturated zone" are often used interchangeably but they are technically not the synonymous (Stephens 1996). The top of the phreatic zone, which is also the top of an unconfined aquifer, is usually referred to as the water table. The water table is technically defined as the level or surface at which the static groundwater pressure is equal to the atmospheric pressure. As the water table is not in motion, the upward pressure exerted by the groundwater must be equal to the downward pressure from the weight of the atmosphere.

The vadose zone is separated from the phreatic zone by the capillary fringe, which is transitional between the two zones. The moisture content of capillary fringe increases downwards as the water table is approached. The capillary fringe can be largely (and at its base fully) saturated with water that is under surface tension. Hence, the difference between the vadose zone and unsaturated zone is that the former contains that saturated part of the capillary fringe. The thickness of the capillary fringe is a function of pore size. Very fine grained sediments have greater capillary pressures and thus can have thick (>1 m; >3.3 ft) capillary fringes. Small quantities of clay occurring within predominantly sandy sediments can also affect the maximum height and distribution of the capillary fringe.

The vadose zone was described by Harter and Hopmans (2004, p. 180) as being

> an intrinsic part of the hydrologic cycle, essentially controlling interrelationships between precipitation, infiltration, surface runoff, evapotranspiration, and groundwater recharge.

The vadose zone serves several main functions related to regional-scale hydrology (Harter and Hopmans 2004)

- partitioning of precipitation (and applied irrigation water) into infiltration, runoff, evapotranspiration, interflow, and groundwater recharge,
- storing and transferring water in the root zone between the atmosphere and deep vadose zone including interflow,
- storing and transferring water between the root zone and groundwater below,
- storing, transferring, filtering, adsorbing, retarding, and attenuating solutes and contaminants before they reach the groundwater.

The vadose zone is considered in distinctly different spatial and temporal scales by watershed and water-resources engineers and scientists, and soil physicists (Harter and Hopmans 2004). The former often consider the vadose zone as a thin skin that separates surface-water resources from groundwater resources, while the latter are much more concerned with fine-scale and more local processes within the vadose zone.

Vadose zones processes are often given little consideration in water resources evaluations in humid areas because groundwater is commonly present a short distance below land surface and, as a result, the thickness of the vadose zone is minimal. On the contrary, in arid regions, the top of the phreatic zone is often a great distance below land surface [10's of meters (or feet) or more]. Vadose zone processes thus have a much greater importance in arid and semiarid regions as precipitation has a much longer vertical flow path before it reaches the underlying aquifer.

There is an extensive literature on vadose zone hydrological process. Several dedicated textbooks are available (e.g., Stephens 1996; Selker et al. 1999) and the subject is of great importance in the soil sciences, agriculture, and contaminant hydrology. The mechanism and controls of the movement of contaminants from the unsaturated soil zone to the groundwater is of fundamental importance in management and remediation of contamination. Overviews of vadose (unsaturated) zone hydrology are provided in most groundwater text and reference books (e.g., Fetter 2001; Schwartz and Zhang 2003; Todd and Mays 2005). The basic principles of vadose zone hydrology are herein reviewed. A key issue concerning the management of arid and semiarid groundwater resources is the effect of vadose zone processes on aquifer recharge.

Groundwater recharge or basin runoff has been traditionally treated by hydrologists in a lumped, "black box" fashion (Harter and Hopmans 2004). An important issue for groundwater resources evaluations is how much water passes through the vadose zone to recharge the underlying aquifer. Specific vadose zone processes are often not considered. For example, average recharge rates for an aquifer may be obtained from environmental tracers or from the response of the water table to rainfall events (Chap. 10). These recharge rates may be applied to groundwater models directly as a net recharge to the water-table aquifer, as opposed to incorporating recharge into model by simulating precipitation, evapotranspiration, and run-off.

The "black box" approach is often adequate for practical groundwater resources management. Sufficient data are often not available for populating detailed distributed parameter models of the vadose zone. While such models have the attraction of being perceived as being more scientific and technically rigorous, the reality is that as additional parameters are added to a model, and thus degrees of freedom, the uncertainty in the model results may actually increase. The fundamental questions that must be addressed at the onset of any water resources investigation are what are the (1) specific objectives, (2) available data, and (3) resources available to obtain additional data and for model development. It is, therefore, important to have an understanding of vadose zone processes, so an informed decision can be made concerning the scale and detail in which such processes are to be considered in the investigation.

Salts, nutrients (e.g., nitrate), and contaminants tend to accumulate in the vadose zone. Changes in recharge rates and aquifer water levels can result in the mobilization of salts and chemicals accumulated in the vadose, with associated impacts on groundwater quality and surface waters (Newman et al. 2006). Vadose zone process, therefore, need to be considered in arid and semiarid region water quality studies.

5.6.1 Capillary Processes

Water above the water table is under a negative pressure or tension caused by capillary pressure. Capillary pressure is defined as the difference in pressure across the interface between two immiscible fluids, which are referred to as wetting and non-wetting phases, as follows:

$$P_c = P_{nw} - P_w \tag{5.13}$$

where,

P_c = capillary pressure (Pa or cm of water)

P_{nw} = pressure of the non-wetting phase (Pa or cm of water), and

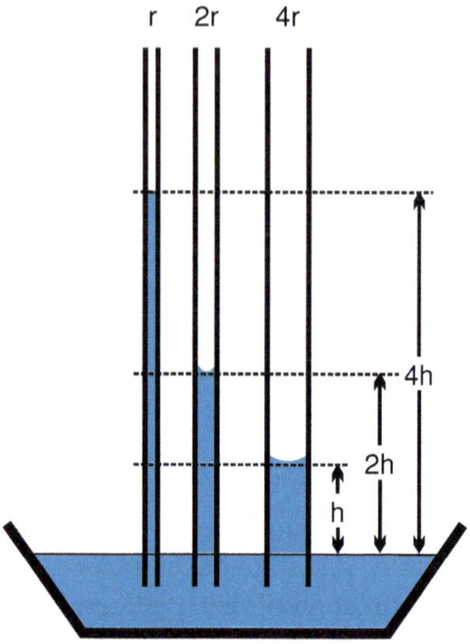

Fig. 5.6 Inverse relationship between capillary tube radius (r) and the height of water column (h)

P_w = pressure of the wetting phase (Pa or cm of water).

In the case of the vadose zone, the wetting phase is water and the non-wetting phase is air. Capillary pressure (P_c) can be calculated using the Young–Laplace equation as follows:

$$P_c = \frac{2\gamma \cos\theta}{r} \tag{5.14}$$

where,

γ = surface tension (N/M, J/m^2)

θ = wetting angle (degrees)

r = effective radius of the interface (m).

Using a vertical capillary tube as an example, in which water is the wetting phase (Fig. 5.6) the height (h) at which water would rise above the static water level in a tube with a radius (r) is calculated as:

$$h = \frac{2\gamma \cos\theta}{\rho g r} \tag{5.15}$$

where,

h = height of water column (m)

g = gravitational acceleration (9.807 m/s^2)

ρ = density of water ($\approx$1000 kg/m^3).

The water in the capillary tube is under lesser pressure than atmospheric pressure and is "pulled" upwards. Both γ and θ are properties of the fluids and solids. With respect to the vadose zone, capillary pressure results in water being pulled into the formation. From this analysis it can be seen that capillary pressure is inversely proportional to the radius of the interface, which in porous rock corresponds to pore throat size. The pore throats that connect larger pores behave in essence as small capillary tubes.

There are two important implications of the inverse relationship between capillary pressure and interface radius (pore throat diameters). Inasmuch as coarse-grained sediments have larger pore and pore throat diameters, they have corresponding lesser capillary pressures. Unsaturated, very-fined grained materials (silts and clays) will tend to pull in and strongly hold water. Secondly, capillary pressure decreases with increasing water content. The water that initially enters a sediment or rock will be prefentially drawn into the smallest pores and pore throats with the greatest capillary pressures. Water that subsequently enters the unit will fill progressive larger pores and pore throats and will have a lower capillary pressure.

Under a given surface tension, fine-grained materials may have more pores filled with water, and can have a higher hydraulic conductivity than a coarse-grained deposit. The relationship between hydraulic conductivity and grain size in unsaturated sediments may be the opposite of the relationship under saturated conditions. This relationship has been demonstrated by Heilig et al. (2003) in a set of field experiments, showing that at a matric pressure of −280 cm (9.16 ft), coarse sand had a hydraulic conductivity of 1.7×10^{-8} cm s^{-1} (4.8×10^{-4} ft/day) and fine sand had a corresponding hydraulic conductivity of 1.2×10^{-4} cm s^{-1} (3.4 ft/day).

The amount of water that can be held in a sediment or rock by capillary pressure is referred to as its "specific retention" (S_r), which is defined as the ratio of the volume of water that will be retained in a saturated sample against gravity (V_r) to the total volume of the sample (V).

$$S_r = \frac{V_r}{V} \tag{5.16}$$

Some of the water retention in vadose zone rock or sediment is caused by processes other than capillary pressure, such as molecular absorption. The ratio of the volume of water that can be drained from a sample by gravity (V_g) to the total volume is referred to as the specific yield (S_y),

$$S_y = \frac{V_g}{V} \tag{5.17}$$

The sum of the specific retention and specific yield is equal to the total porosity.

5.6.2 Water Potential

Water potential ψ (g, m, o), which is also referred to as the soil–water potential, is the sum of gravitational potential, matric potential, and osmotic potential. Water potential has the units of pressure (bars, atmospheres, Kpa, or height of water). Typically, osmotic potential (ψ_0) is very small relative to the matric potential (ψ_m).

The gravitation potential (ψ_g) (is often also small relative to the matric potential differences in dry vadose soils.

Matric potential is the negative tension with which water is held in the soil and is inversely proportional to the soil moisture content. It is strongly related to soil texture and is inversely proportional to grain size. Matric potential is the sum of capillary and adsorptive forces and is always negative. Osmotic potential is the pressure related to differences in salinity and is insignificant in most situations. Osmosis is the process by which water flows across semi-permeable membranes from areas of low solute concentrations (high water potential) to areas of higher solute concentration (lower water potential). Within soils, clay layers may act as semi-permeable membranes. Gravitational potential is related to differences in elevation (elevation head). Assuming osmotic and gravitational potentials are insignificant relative to matric potentials, water within the vadose zone will move from areas of high matric potential to areas of low (more negative) matric potential.

Water potential (ψ) is inversely related to the volumetric water content of the soil, sediment, or rock. The relationship between water potential and volumetric water content is illustrated by soil–water retention curves (Fig. 5.7). Water potential is zero where the soil is saturated and volumetric water content is equal to the effective (interconnected) porosity. The shape of the soil–water retention curve is dependent upon the pore size distribution.

Water potential increases as water is drained from the pores. In sands and gravels, the drainage of water from large pores results in a relatively minor associated increase in matric potential, which is reflected as a flat to gently sloping segment on the soil–water retention curve. Drainage of the remaining water from the very small pores and pore throats results in a large increase in matric potential.

The water content of a soil at a given water potential depends upon the wetting history of the soil, which is referred to as hysteresis. The soil–water retention curves differ depending on whether the sample is being wetted or dried (Fig. 5.8). At a given water potential, the water content will be greater for drainage than for wetting. During drainage water is retained in some relatively large pores by smaller pore throats until the water potential decreases (tension increases) sufficiently at pore throats to allow the water to drain.

5.6.3 Unsaturated or Vadose Zone Processes and Flow

The transport of water and solutes through the vadose zone includes some of the most complex and least predictable processes in groundwater hydrology. There are many processes occurring within the vadose zone starting at land surface, where water must infiltrate into the unsaturated soil or surficial rock. Water entering the vadose zone then percolates downward, displacing air, which moves upward. The percolating water flow is subject to a variety of forces, including gravity and suction caused by capillary forces. Water must pass through the full thickness of

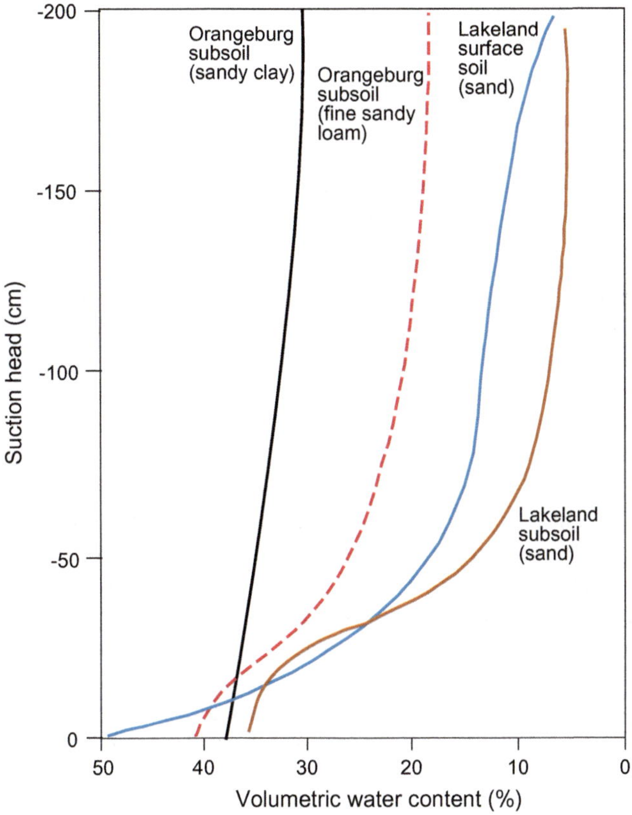

Fig. 5.7 Examples of soil water retention curves for some Florida, USA, soils. *Source*: Bouma et al. (1982)

Fig. 5.8 Conceptual diagram illustrating the difference in soil–water retention curves between soils that undergoing wetting and drying, which is referred to as hystersis

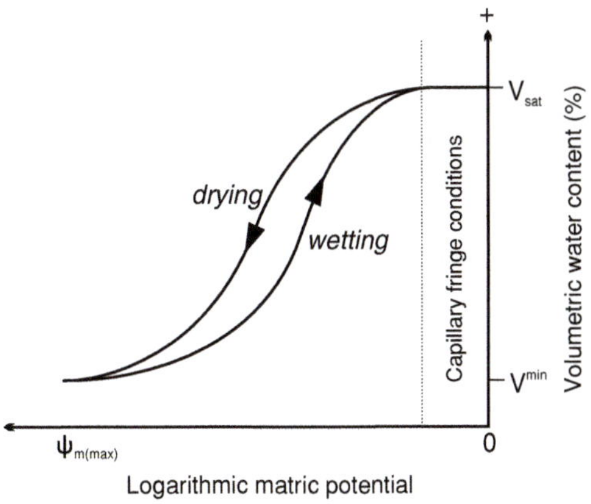

the vadose zone before reaching the water table. Upon reaching the water table (and becoming recharge), the water enters the groundwater realm and its flow is controlled by hydraulic head, which, in turn, is primarily controlled by gravity.

Rainfall that reaches land surface can take three primary pathways

- It can infiltrate into the vadose zone,
- It can run off the land and enter streams, rivers, or other surface water bodies, or
- It can pond, be intercepted (stored above groundwater surface) or otherwise locally remain at land surface to either evaporate or later infiltrate.

A number of factors control the infiltration and subsequent percolation of water through the unsaturated zone. The condition of the land surface and the physical properties of the soil, in particular, affect the ability of rainfall to infiltrate past the soil–water interface.

Once rain water passes through the soil-atmosphere interface, it becomes a vertically-moving "wetted front", sometimes termed a gravity-induced "kinematic wave" (Charbeneau 1984, 2000; Morel-Seytoux 1987). The downward movement of the wetted front is greatly influenced by the moisture gradient or the "suction" pressure caused by capillary action above the water table. The wetted front is pulled down by both gravity and capillary pressure. As the wetted front moves deeper in the unsaturated zone, there is a tendency for the overall moisture content to increase as the water table is approached. This increase in moisture content reduces capillary pressure-driven flow. The downward flowing water also displaces air trapped in the unsaturated zone, which must move upwards and can significantly inhibit the percolation process. Finally, water becomes recharge to the underlying aquifer after it passes through unsaturated zone and crosses the water table.

A form of Darcy's Law is used to describe the flow of water through the vadose zone. One-dimensional vertical flow through the vadose zone can be expressed using the modification of Darcy's law referred to as Darcy-Buckingham law:

$$q_z = -K(\theta)\frac{\partial h}{\partial z} \tag{5.18}$$

where,

q_z = specific flow rate in the vertical (z) direction (m/d)
$K(\theta)$ = hydraulic conductivity (m/d) at volumetric soil moisture θ
h = total head (m)
z = distance (m)
θ = volumetric soil moisture content (unitless, m^3/m^3).

Volumetric soil moisture content is the ratio of the volume of water in a soil sample (V_w) to the total volume of the sample (V_t)

$$\theta = \frac{V_w}{V_t} \tag{5.19}$$

The Darcy-Buckingham equation differs from Darcy's equation (Eq. 5.2) in that hydraulic conductivity is a function of the volumetric soil moisture content rather than being a constant. Thus, as the moisture content of soil at a given depth increases as the result of infiltration (for example, during and after a rainfall event), its hydraulic conductivity progressively changes. Therein is the fundamental difficulty in applying the Darcy-Buckingham equation, in that the quantitative relationship between hydraulic conductivity and soil moisture content is poorly constrained.

Within the vadose zone, total head is a function of both elevation (z) and the water potential [$\psi(\theta)$]:

$$h = z + \psi(\theta) \tag{5.20}$$

The water potential, which consists mainly of matric potential (suction head) is expressed in units of pressure (KPa, bars) or in equivalent head units (m, ft).

Within the vadose zone, the water potential [$\psi(\theta)$] is a negative value and is a function of water content and sediment type (Sect. 5.6.5). Water potentials can be measured in the field using tensiometers and other techniques (Sect. 10.5.6). Equations 5.18 and 5.20 can be combined as follows:

$$q_z = -K(\theta)\frac{\partial(z + \psi(\theta)}{\partial z} \quad or \tag{5.21}$$

$$q_z = -K(\theta)(1 + \frac{\partial\psi(\theta)}{\partial z}) \tag{5.22}$$

Both hydraulic conductivity and soil–water potential are non-linear functions of water content (θ) and Eqs. 5.21 and 5.22 cannot be directly solved. Water potential is inversely related to water content (Fig. 5.7). Unsaturated hydraulic conductivity increases with water content (Fig. 5.9). At low soil moisture contents, water is preferentially drawn into the smallest pores where it is tightly held and thus flow is minimal. As soil moisture content increases, capillary pressure decreases (i.e., water is less tightly heald), and greater flow rates occur under a given hydraulic gradient.

A great amount of effort has been made in the soil sciences to obtain solutions to these equations, including development of a predictive capability for hydraulic conductivity. The relationship between unsaturated hydraulic conductivity and soil moisture content is dependent on the soil properties, particularly the grain types, pore size distribution, and texture (Bouwer 1964). Soils are heterogeneous both in matrix properties and the presence and types of macroporosity. Unsaturated hydraulic conductivities can be estimated from the soil–water moisture retention function with at least one measurement of hydraulic conductivity at some water content (e.g., Brooks and Corey 1964; Campbell 1974; Van Genuchten 1980). Empirical data are required to determine the soil-type specific values for coefficients in hydraulic conductivity versus soil-moisture equations.

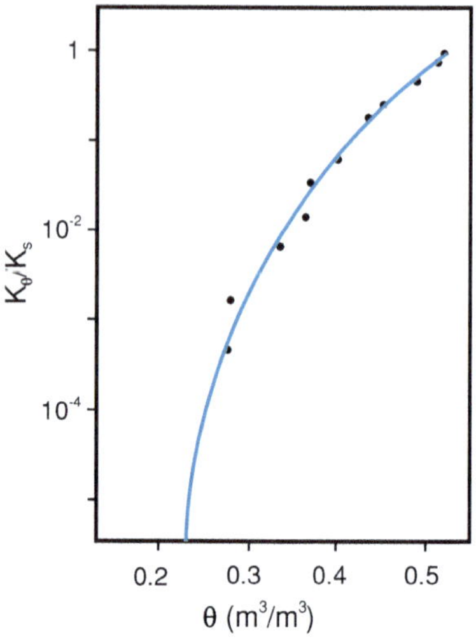

Fig. 5.9 Example of the relationship between soil moisture and relative hydraulic conductivity (ratio of measured hydraulic conductivity at soil moisture θ to saturated hydraulic conductivity for a loam soil (after Van Genuchten 1980)

5.6.4 Infiltration

Infiltration is the process by which surface water enters the soil (vadose zone). The term "percolation" is used to describe the downward movement of water through the vadose zone by gravity. Deep percolation refers to water the percolates past the root zone of plants (Bouwer 1978). Precipitation infiltrates into the vadose zone and usually percolates in a general downward direction, where some it will enter the groundwater realm at the water table. The flux of soil water across the water table into the phreatic zone is called recharge. Recharge processes are discussed in detail in Chap. 8.

The process of infiltration is controlled by a large number of conditions, which include

- type of soil, such as sand, loam, clay, and its associated hydraulic properties,
- soil vertical hydraulic conductivity,
- moisture content of the uppermost part of the soil,
- soil moisture gradient with depth in the unsaturated or vadose zone,
- slope of land surface,
- vegetation or organic debris located at land surface,
- presence of physical pathways in the soil surface (e.g., deep mud cracks),
- presence of impervious crusts or coatings at the land surface,
- rate of rainfall verses the ability of the water to pass through the interface,
- degree that a given soil traps air between the wetted front and the water table,
- pore size distribution of the soil, which controls capillarity,
- heterogeneity of the soil profile,
- depth of ponding of water above land surface, and
- time.

There are several terms that apply to infiltration. Infiltration capacity is the rate at which rainfall can be absorbed by a given soil type at land surface conditions. Field moisture capacity is the quantity of water that is retained in a freely drained soil after the excess water has drained away and the rate of downward movement has materially decreased. The term "field capacity" has different meanings and Stephens (1996) recommended that the term be avoided by hydrogeologists. Field moisture capacity in the sense used by Horton (1933) is equivalent to specific retention. Rainfall excess is the amount of rain that falls at intensities exceeding the infiltration capacity of the soil and thus becomes runoff.

Perhaps the most influential and still relevant paper on infiltration was Robert Horton's (1933) paper "The role of infiltration on the hydrologic cycle". Horton (1933) recognized that soil acts as a separating surface between infiltration and surface runoff. Rainfall will infiltrate into the soil at its infiltration capacity with the rainfall excess being available for surface runoff. The initial infiltrated water will first make up any soil moisture-deficiency (e.g., soil moisture content below the field moisture capacity). Once the field moisture-capacity is reached, infiltrated water will be available for groundwater recharge. Horton (1933) also noted that only a fraction of the rainfall excess will actually become runoff into streams. A variable amount of the runoff will be captured in transit as surface detention, and will either infiltrate into the soil or later evaporate.

A key observation is that after a rainfall event, the rate of infiltration decreases over time in a regular, cyclical manner (Horton 1933; Jury and Horton 2004). The decrease in infiltration is caused by the packing of the soil surface, swelling of the soil, inwashing of fine materials into soil surface openings, and the progressive filling of soil pores with water.

Horton's infiltration model is expressed by the equation (Horton 1940):

$$f_t = f_c + (f_0 - f_c)e^{-\mathrm{kt}} \tag{5.23}$$

where,

f_t = infiltration rate at time t (mm/h)

f_0 = initial infiltration rate or maximum infiltration rate (mm/h)

f_c = final or equilibrium infiltration rate after the soil has been saturated (mm/h)

t = time (h)

k = decay constant specific to the soil (h^{-1}).

The total volume of infiltration (Ft; units = mm) after time t can be calculated as

$$F_t = f_c t + \frac{(f_0 - f_c)}{k}\left(1 - e^{-kt}\right) \tag{5.24}$$

Horton's equation provides a useful conceptual understanding of the infiltration process, but has the practical limitation that it is difficult to obtain accurate values for the parameters. Horton's model also does not consider conduit flow.

Infiltration rate is most rapid in dry soils, because of their greater water potential (Sect. 5.6.5). The capillary pressure of dry sediments acts to draw in

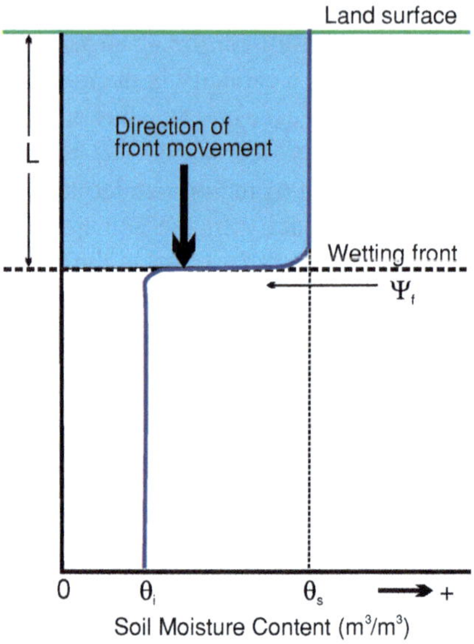

Fig. 5.10 Diagram of Green-Amp equation parameters

water. However, the recharge rate in dry sediments will be less than the rate in pre-wetted sediments as water is retained under capillary pressure and the pore spaces must first be filled before significant percolation to the water table occurs.

Chow et al. (1988) defined cumulative infiltration (F(t)) as the accumulated depth of water infiltrated, with the mathematical definition of:

$$F(t) = \int_0^t f(t)dt \tag{5.25}$$

The Green and Ampt (1911) method is commonly used for estimation of infiltration in most soil types. Green and Ampt developed a concept termed the "wetting front", which is a sharp boundary between the upper saturated soil with a soil moisture content (θ_s) and the underlying soil with a moisture content (θ_i) (Fig. 5.10). The main assumptions underlying the Green-Ampt equation are that the wetting front advances at the same rate with depth, soil in the wetted regions has constant properties, and the soil–water potential below the wetting front does not change with time and depth (Jury and Horton 2004). The moisture content of the soil will increase by ($\theta_s - \theta_i$) or ($\Delta\theta$) as the front passes. Hence, the infiltration volume is equal to the product of $\Delta\theta$ and the distance the front moved since the start of infilitration (L).

After mathematical derivation by coupling Darcy's Law with the model, the Green-Ampt equations for instantaneous infiltration rate and cumulative infiltration under conditions of negligible surface-water ponding depth become

$$f(t) = K_s \left[\frac{\psi_f \Delta\theta}{F(t)} + 1 \right] \tag{5.26}$$

Table 5.1 Green-Ampt infiltration parameters for various soil types

Soil type	Porosity	Effective porosity	Wetting front suction head (mm)	Hydraulic Conductivity (mm/h)
Sand	0.437	0.417	49.5	117.8
Loamy sand	0.437	0.401	61.3	29.9
Sandy loam	0.453	0.412	110.1	10.9
Loam	0.463	0.434	88.9	3.4
Silt Loam	0.501	0.486	166.8	6.5
Sandy clay loam	0.398	0.330	218.5	1.5
Clay loam	0.464	0.309	208.8	1.0
Silty clay loam	0.471	0.432	273.0	1,0
Sandy clay	0.403	0.321	239.0	0.6
Silty clay	0.479	0.423	292.2	0.5
Clay	0.475	0.385	316.3	0.3

Source Rawls et al. (1983)

$$F(t) = K_s t + \psi_f \Delta\theta \ln\left(1 + \frac{F(t)}{\psi_f \Delta\theta}\right) \tag{5.27}$$

where,
f(t) = instantaneous infiltration rate at time t (mm/h)
F(t) = cumulative infiltration at time t (mm)
K_s = saturated hydraulic conductivity (mm/h)
ψ_f = soil–water potential immediately below the wetting front (mm)
$\Delta\theta$ = change in soil moisture across the wetting front (dimensionless).

The change in soil moisture across the wetting front is referred to as the soil moisture deficit. In order to solve the Green Ampt equation for the instantaneous infilitration rate or cumulative infilitration after a given time (t), data are needed for the variables K_s, $\Delta\theta$, and ψ_f. Examination of Eq. 5.27 reveals that the variable being solved for, F(t), is present in the equation itself, which precludes direct solution by algebraic manipulation. However, Eq. 5.27 can be solved by numerical techniques. Different values of F(t) can be tried until the equation converges.

Saturated hydraulic conductivity is a property of the soil composition and texture, which can be either measured directly or typical values for the soil type may be used. Soil–water potential (suction head) is as function of soil type and volumetric water content and is graphically illustrated in water retention curves (Sect. 5.6.2). The Green-Ampt infiltration parameters for a large number of soil types were determined by Rawls et al. (1983), (1993) (Table 5.1).

Infiltration in arid lands has some unusual aspects in certain areas. In many arid and hyperarid areas, the infiltration capacity of the soils varies by orders of magnitude. Wadi sediment tends to allow rapid infiltration of rainfall and flood events compared to terrace and desert "pavement" areas, where rainfall and runoff pond in depressions. Terrace and desert pavement areas commonly contain either some clay or other fine-grained sediment mixed in the surface soil layer or is partially cemented with calcium carbonate. In many cases, infiltration only penetrates the upper 10 cm

(4 in.) of the soil profile and the ponded water evaporates before significant infiltration can occur. Another issue in the western area of Saudi Arabia near Jeddah is that intense rainfall events are sometimes preceded by easterly winds that result in sandstorms, which mobilize considerable quantities of dust into the atmosphere with corresponding deposits on land. The rainfall event that follows washes the dust into depressional areas and as the turbid water infiltrates, the fine clays in the dust plug the pores, reducing the infiltration capacity.

5.6.5 Kinematic Wave Theory

Kinematics is defined as the study of motion without references to mass and force. The underlying concept of the kinematic wave theory in general is that there is a functional relationship between flux, concentration, and position, which can be mathematically expressed by the law of conservation of mass through the continuity equation and a flux-concentration relationship (Singh 1997; Singh 2002). The flux relationship used for groundwater flow is Darcy's law. With respect to unsaturated flow, the kinematic wave theory is based on the redistribution of water, assuming flow is only by gravity drainage (Charbeneau 1984, 2000; Morel-Seytoux 1987). After a period of infiltration, there are three parts to the descending wave, which are: (1) a draining part with increasing water content with depth, (2) a constant or vertical plateau part, and (3) the wetting front. The basic assumption of this theory is that pressure gradients within the unsaturated zone are negligible and the front suction head is constant. Therefore, the downward flow is described by

$$\frac{\partial \theta}{\partial t} = -\frac{dK(\theta)}{d\theta}\frac{\partial \theta}{\partial z} \tag{5.28}$$

where,
θ = residual water content
K = the hydraulic conductivity.

The analytical solution of the kinematic wave model can be used to model soil water distribution and recharge (Charbeneau 1984, 2000). The kinematic wave model for soil moisture movement is applicable at high moisture contents at which gravity flow dominates. The model is not applicable to low soil moisture conditions at which capillary flow dominates (Singh 2002). A more detailed discussion of this process is contained in Todd and Mays (2005).

5.6.6 Impacts of Bedding and Heterogeneity on Vadose Zone Flow (Funnel or Finger-Flow)

Heterogeneity within the vadose zone, particularly, layering of fine-grained and coarse-grained soils, commonly causes funneling or "funnel" flow to occur during downward movement of the wetting front. This issue is particularly important in arid lands, because the funneling flow effect can change the pathway of water movement, affecting infiltration, recharge rate, and solute transport.

Water flow is commonly diverted along lithology changes in the soil because of changes in capillary pressure along an interface. Heilig et al. (2003) suggests that if the textural interface between fine-grained sediment and underlying coarser sediment is sufficiently long and the flow is continuous, at some location and at a time, the matric potential may be high enough to allow water to break through into the underlying coarse unit. There is some conflict between differing research results on the matric pressure difference required to cause the breakthrough (Heilig et al. 2003).

Soil heterogeneity caused by vertical differences in sediment grain size, stratification, and disturbances, such as roots, mud infilled mud cracks, and other sedimentary structures, has a pronounced effect on the pattern of moisture movement along the wetting front. The occurrence of funneling flow or finger flow has been demonstrated in the field and in laboratory experiments (Ross 1990; Kung 1990a, b, 1993; Ju and Kung 1993; Sililo and Tellum 2000; Heilig et al. 2003). Finger flow and the unevenness of unsaturated zone moisture distribution can be measured using time domain reflectometry or ground penetrating radar (Evett 2003; Robinson et al. 2003). It is also being simulated using models such as HYDRUS or STOMP, and stochastic methods, which allow recharge to be determined in regional models (Yoon et al. 2007 Foussereau et al. 2000a, b).

Because of the significant heterogeneity in arid lands soils, particularly in wadis, terrace soils, and desert pavement areas, the vadose zone and infiltration processes are inherently affected by heterogeneities. Both the infiltration rate and the recharge rate to the aquifer system are affected (Kuhn and Yair 2004; Mugabe et al. 2007). There is a considerable need for more detailed characterization of arid lands soil and vadose zone hydraulic properties.

5.6.7 Impacts of Air Entrapment and Flow on Vadose Zone Processes (Lisse Effect)

The process of infiltration occurs as essentially a two-phase flow system, wherein water is displacing pore air (Faybishenko 1999; Hillel 1998). Direct observations have shown that air can inhibit infiltration when it is entrapped ahead of the wetting front or is trapped as bubbles within the transmission zone (Constantz et al. 1988; Faybishenko 1999; Wang et al. 1998; Weir and Kissling 1992). As air pressure increases in the vadose zone below the wetting front, air will "vent" upward to allow water to flow deeper into the system. This uneven purging of air is controlled by the continuity of the wetting front, the grain size distribution of the soil, heterogeneities within the soil profile, and corresponding variations in the distribution of hydraulic conductivity (Bouwer 1966). Air venting can cause a "finger flow" pattern within the vadose zone that is independent of soil heterogeneity, but usually occurs in concert with physical heterogeneities. Non-uniform

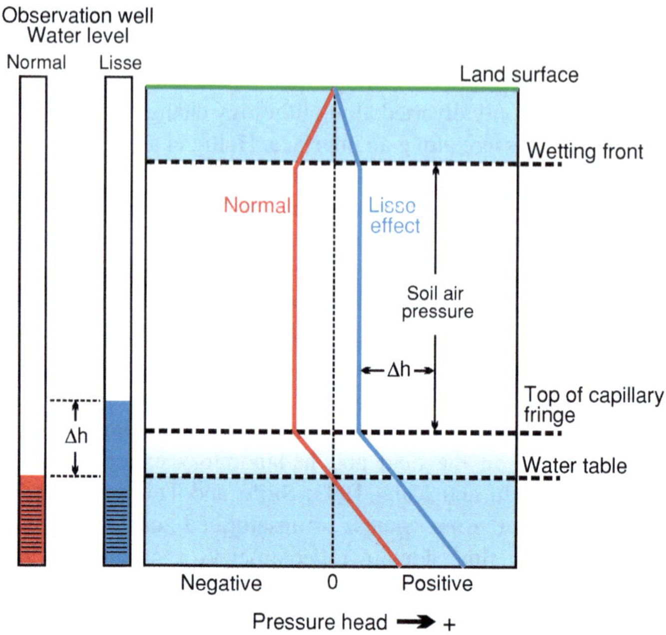

Fig. 5.11 Diagram illustrating the Lisse effect. Downward movement of a wetting front increases the pressure of air trapped in the underlying vadose zone to values greater than atmospheric pressure. The increase in air pressure causes a rise in water levels (Δh) in observation wells completed in the water table aquifer (after Weeks 2002; Guo et al. 2008)

early infiltration can isolate parts of the pore network by air encapsulation, thereby, altering recharge rates and the sorptive capacity of the soil (Carrick et al. 2010).

A more extreme effect of trapped air during infiltration is known as the "Lisse effect". Weeks (2002, p. 652) defined the Lisse Effect as

> a rarely noted phenomenon occurring when infiltration caused by intense rain seals the surface soil layer to airflow, trapping and compressing air under pressure in the unsaturated zone.

The Lisse effect was first observed by Thal Larsen in 1932, in water level data from a well located in the Village of Lisse, Holland (Hooghoudt 1947). The effect causes water levels in a well to rise abnormally high after a rainfall event, well above what would be predicted by solely filling the connected pores (specific yield). The cause of this additional pressure head in wells is that the wetted front is driven by gravity and the weight of the water and is pulled downward by capillary pressure head (negative). The compression of trapped air causes water pressure at the water table to become greater than atmospheric. The compressed air pushes water into a well, causing the water level to rise above atmospheric pressure (Fig. 5.11). The change in head produced by the Lisse Effect is expressed as

$$\Delta h = P_{wc}\left(\frac{m}{h - m}\right) \tag{5.29}$$

where,
Δh = rise in head caused by the Lisse Effect (m)
P_{wc} = atmospheric pressure expressed in terms of water column height (m)
m = depth of the wetting front penetration (m)
h = distance from the top of the capillary fringe to land surface (m).

Weeks (2002) concluded that the Lisse Effect is likely more common than supposed and can produce an abnormal rise in water levels of about 0.10–0.55 m (0.3–1.8 ft). It is very important to recognize this effect, if it locally occurs, when analyzing groundwater recharge by using hydrograph water levels (Healy and Cook 2002) and rainfall accumulations (i.e., the water-table fluctuation method; Sect. 8.3), because there could be a substantial over-estimation of groundwater recharge. This is particularly important in arid lands, where the soil types and conditions, and infrequent and intense rainfall events, can cause trapped air to be a significant issue in aquifer recharge.

References

Alyamani, M. S., & Sen, Z. (1993). Determination of hydraulic conductivity from complete grain-size distribution curves. *Ground Water, 31*(4), 551–555.

Bouma, J., Brown, R. B., & Rao, P. S. C. (1982). *Retention of water: Basics of soil-water relationships–Part II*. Gaineville, Soil and Water Science Department, Florida Cooperative Extension Service, Institute of Food and Agricultural Sciences, University of Florida, Document.

Bouwer, H. (1964). Unsaturated flow in ground-water hydraulics. *Journal of Hydraulics Division, American Society of Civil Engineers, 90*(HY5), 121–144.

Bouwer, H. (1966). Rapid field measurement of air entry value and hydraulic conductivity of soil as significant parameters in flow system analysis. *Water Resources Research, 2*(4), 729–738.

Bouwer, H. (1978). *Groundwater hydrology*. New York: McGraw-Hill.

Brooks, R. H., & Corey, A. T. (1964). *Hydraulic properties of porous media*. Hydrology Papers, no. 3, Colorado State University, Fort Collins, CO.

Campbell, G. S. (1974). A simple method for determining unsaturated conductivity from moisture retention data. *Soil Science, 117*(6), 311–314.

Carrick, S., Buchan, G., Almond, P., & Smith, N. (2010). Atypical early-time infiltration into a structured soil near field capacity: The dynamic interplay between sorptivity, hydrophobicity, and air encapsulation. *Geoderma, 160*, 579–589.

Castle, J. W., Molz, F. J., Lu, S., & Dinwiddie, C. L. (2004). Sedimentology and fractal-based analysis of permeability data, John Henry Member, Straight Cliffs Formation (Upper Cretaceous), Utah, U.S.A. *Journal of Sedimentary Research, 74*, 270–284.

Charbeneau, R. J. (1984). Kinematic models for soil moisture and solute transport. *Water Resources Research, 17*(3), 699–706.

Charbeneau, R. J. (2000). *Groundwater hydraulics and pollutant transport*. Upper Saddle River, NJ: Prentice Hall.

Choquette, P. W., & Pray, L. C. (1970). Geologic nonclature and classification of porosity in sedimentary carbonates. *American Association of Petroleum Geologists Bulletin, 54*, 207–250.

Chow, V. T., Maidment, D. R., & Mays, L. W. (1988). *Applied hydrology*. New York: McGraw-Hill.
Constantz, J., Herhelrath, W. N., & Murphy, F. (1988). Air encapsulation during infiltration. *Soil Science Society of America Journal, 52*, 10–16.
Cunningham, K. J., Carlson, J. L., Wingard, G. L., Robinson, E, & Wacker, M. A. (2004). Characterization of aquifer heterogeneity using cyclostratigraphy and geophysical methods in the upper part of the karstic Biscayne Aquifer, southeastern Florida. *U.S. Geological Survey Water-Resources Investigations Report* 03-4208.
Cunningham, K. J., Sukop, M. C., Huang, H., Alvarez, P. F., Curran, H. A., Renken, R. A., et al. (2009). Prominence of ichnologically influenced macroporosity in the karst Biscayne aquifer: Stratiform "super-K" zones. *Geological Society of American Bulletin, 121*, 164–180.
Cunningham, K. J., Wacker, M. A., Robinson, E., Dixon, J. F., & Wingard, G. L. (2006). A cyclostratigraphic and borehole-geophysical approach to development of a three-dimensional conceptual hydrogeologic model of the karstic Biscayne Aquifer, southeastern Florida. *U.S. Geological Survey Scientific Investigations Report* 2005-5235.
Domenico, P. A., & Schwartz, F. W. (1998). *Physical and Chemical Hydrogeology* (2nd ed.,). New York: Wiley.
Driscoll, F. G. (1986). *Groundwater and wells* (2nd ed.,). St. Paul, MN: Johnson Filtration Systems.
Evett, S. R. (2003). Soil water measurement by time domain reflectometry. In B. A. Stewart & T. A. Howell (Eds.), *Encyclopedia of water science* (pp. 894–898). New York: Marcel Dekker.
Faybishenko, B. (1999). Comparison of laboratory and field methods for determining the quasi-saturated conductivity of soils. In M. T. Van Genuchten, F. J. Leij, & L. Wu (Eds.), *Proceedings of Characterization of the Hydraulic Properties of Unsaturated Porous Media. Proceedings of the International Worship on Characterization and Measurement of Unsaturated Porous Media, Riverside, California*, October 22–24, 1997, University of California, pp. 279–292.
Fetter, C. W. (2001). *Applied hydrogeology* (4th ed.,). Englewood Cliffs: Prentice Hall.
Foussereau, X., Graham, W. D., & Rao, P. S. C. (2000a). Stochastic analysis of transient flow in unsaturated heterogeneous soils. *Water Resources Research, 36*(4), 891–910.
Foussereau, X., Graham, W. D., Akpoji, G. A., Destouni, G., & Rao, P. S. C. (2000b). Stochastic analysis of transport in unsaturated heterogeneous soils under transient flow regimes. *Water Resources Research, 36*(4), 911–921.
Freeze, R. A., & Cherry, J. A. (1979). *Groundwater*. Englewood Cliffs: Prentice-Hall.
Green, W H., & Ampt, G. A. (1911). Studies on soil physics: The flow of air and water through soils. *Journal of Agricultural Science, 4*, 1–24.
Guo, H., Jiao, J. J. & Weeks, E. P. (2002). Rain-induced subsurface airflow and Lisse effects. *Water Resources Research, 44*, W07409.
Guo, W., & Langevin, C. D. (2002). User's Guide to SEAWAT: A computer program for simulation of three-dimensional variable-density ground-water flow. *U.S. Geological Survey Open-File Report* 01-434.
Harter, T., & Hopmans, J. W. (2004). Role of vadose-zone processes in regional-scale hydrology: review of opportunities and challenges. In R. A. Feddes & J. C. Van Dam (Eds.), *Unsaturated zone modeling: programs, applications and challenges* (pp. 179–208). Doordrecht: Kluwer.
Healy, R. W., & Cook, P. G. (2002). Using ground water levels to estimate recharge. *Hydrogeology Journal, 10*(1), 91–109.
Heath, R. C. (1983). Basic ground-water hydrology. *U.S. Geological Survey Water-Supply Paper 2220.*
Heilig, A., Steenhuis, T. S., Walter, M. T., & Herbert, S. J. (2003). Funneled flow mechanisms in layers soil: Field investigations. *Journal of Hydrology, 279*, 210–223.
Hillel, D. (1998). *Environmental soil physics*. San Diego: Academic Press.
Hooghoudt, S. B. (1947). Waarnemingen van grondwaterstanden voor de landbouw: Comisse voor Hydrologisch TNO. *Verslagen Techische Bijeenkomsten, 1–6*, 185–201.

Horton, R. E. (1933). The role of infiltration in the hydrologic cycle. *Transactions American Geophysical Union, 14*, 446–460.

Horton, R. E. (1940). An approach towards a physical interpretation of infiltration capacity. *Proceedings Soil Society of America, 5*, 399–417.

Ju, S. H., & Kung, K.-J. S. (1993). Simulating funnel-type preferential flow and overall flow property induced by multiple soil layers. *Journal of Environmental Quality, 22*(3), 432–442.

Jury, W. A., & Horton, R. (2004). *Soil physics*. Hoboken: Wiley.

Kasenow, M. (1997). *Applied ground-water hydrology and well hydraulics*. Highlands Ranch, CO: Water Resources Publications.

Kasenow, M. (2006). *Aquifer test data: Evaluation and analysisp*. Highlands Ranch, CO: Water Resources Publications.

Kruseman, G. P., & de Ridder, N. A. (1970). *Analysis and evaluation of pumping test data Bulletin, 11*. Wageningen International Institute for Land Reclamation and Improvement: The Netherlands.

Kruseman, G. P. & de Ridder, N. A. (1991). *Analysis and evaluation of pumping test data*. Wageningen International Institute for Land Reclamation and Improvement: Publication 47.

Kuhn, N. J., & Yair, A. (2004). Spatial distribution of surface conditions and runoff generation in small arid watersheds, Zin Valley Badlands, Israel. *Geomorphology, 57*, 183–200.

Kung, K.-J. S. (1990a). Preferential flow in sandy vadose zone: I. Field observations. *Geoderma, 46*, 51–58.

Kung, K.-J. S. (1990b). Preferential flow in sandy vadose zone: II. Mechanism and implications. *Geoderma, 46*, 59–71.

Kung, K.-J. S. (1993). Laboratory observations of the funnel flow mechanism and its influence on solute transport. *Journal of Environmental Quality, 22*, 91–102.

Lohman, S. W. (1979). *Ground-water hydraulics*. U.S. Geological Survey Professional Paper 708.

Maliva, R. G., & Missimer, T. M. (2010). *Aquifer storage and recovery and managed aquifer recharge using wells: Planning, hydrogeology, design, and operation*. Houston, Schlumberger Water Services: Methods in Water Resources Evaluation Series No. 2.

Morel-Seytoux, H. J. (1987). Multiphase flow in porous media, 4, 103–174. In P. Novak (Ed.), *Developments in hydraulic engineering*. New York: Elsevier.

Mugabe, F. T., Hodnett, M. G., & Senzanje, A. (2007). Effect of temporal rainfall distribution and soil type on soil moisture and runoff generation in semi-arid Zimbabwe. *Nordic Hydrology, 38*(3), 249–263.

Neuman, S. P., & Witherspoon, P. A. (1972). Field determination of the hydraulic properties of leaky multiple aquifer systems. *Water Resources Research, 8*(5), 1284–1298.

Newman, B. D., Vivoni, E. R., & Groffman, A. R. (2006). Surface water—groundwater interactions in semiarid drainages of the American southwest. *Hydrological Process, 20*, 3371–3394.

Nwankwor, G. I., Cherry, J. A., & Gillham, R. W. (1984). A comparative study of specific yield determinations for a shallow sand aquifer. *Ground Water, 22*(6), 764–772.

Rawls, W. J, Ahuja, L. R., Brakensiek, D. L., & Shirmohammadi, A. (1993). Infiltration and soil movement (pp. 5.1–5.51). In D. R. Maidment (Ed.), *Handbook of hydrology*. New York: McGraw-Hill, Inc.

Rawls, W. J., Brakensiek, D. L., & Miller, N. (1983). Green-Ampt infiltration parameters from soils data. *Journal of Hydraulic Engineering, 109*, 62–70.

Robinson, D. A., Jones, S. B., Wraith, J. M., Or, D., & Friedman, S. P. (2003). A review of advances in dielectric and electrical conductivity measurement in soils using time domain reflectometry. *Vadose Zone Journal, 2*, 444–475.

Ross, B. (1990). The diversion capacity of capillary barriers. *Water Resources Research, 26*(10), 2625–2659.

Schwartz, F. W., & Zhang, H. (2003). *Fundamentals of ground water*. New York: Wiley.

Selker, J. S., McCord, J. T., & Keller, C. K. (1999). *Vadose zone processes*. Boca Raton, FL: Lewis.

Shepard, R. G. (1989). Correlations of permeability and grain size. *Ground Water, 27*(5), 633–638.
Sililo, O. T. N., & Tellum, J. H. (2000). Fingering in unsaturated zone flow: A qualitative review with laboratory experiments on heterogeneous systems. *Ground Water, 38*(6), 864–871.
Singh, V. P. (1997). *Kinematic wave modeling in water resources: Environmental Hydrology*. New York: Wiley.
Singh, V. P. (2002). Is hydrology kinematic? *Hydrologic Processes, 16*, 667–716.
Stephens, D. B. (1996). *Vadose zone hydrology*. Boca Raton FL: CRC Press.
Todd, D. K. (1959). *Groundwater hydrology*. New York: Wiley.
Todd, D. K., & Mays, L. W. (2005). *Groundwater hydrology* (3rd ed). New York: Wiley.
Van Genuchten, M. T. (1980). A closed-form equation for predicting the hydraulic conductivity of unsaturated soils. *Soil Science Society of America Journal, 44*, 892–898.
Walton, W. C. (1970). Groundwater resource evaluation. New York: McGraw-Hill.
Walton, W. C. (1991). *Principles of groundwater engineering*. New York: CRC Press.
Wang, Z., Feyen, J., van Genuchten, M. T., & Nielsen, D. R. (1998). Air entrapment effects on infiltration rate and flow instability. *Water Resources Research, 34*, 213–222.
Weeks, E. P. (2002). The Lisse effect revisited. *Ground Water, 40*(6), 652–656.
Weir, G. J., & Kissling, W. M. (1992). The influence of airflow on the vertical infiltration of water into soil. *Water Resources Research, 28*(10), 2765–2772.
Yoon, H., Valocchi, A. J., & Werth, C. J. (2007). Effect of soil moisture dynamics on dense nonaqueous phase liquid (DNAPL) spill zone architecture in heterogeneous porous media. *Journal of Contaminant Hydrology, 90*(3–4), 159–184.

Part III
Water Budget and Recharge

Chapter 6
Water Budget

6.1 Introduction

A quantitative understanding of aquifer water budgets is critical for effective, long-term water management. Water budgets can be directly calculated assuming that sufficient data are available. However, water budgets are now much more commonly implicitly evaluated as part of numerical surface-water and groundwater models or integrated surface-water/groundwater models. Nevertheless, development of accurate conceptual models, and in turn, numerical models of hydrologic systems requires knowledge of the main components of their water budgets and their magnitudes. The basic water budget of either a groundwater or surface water system can be expressed as:

$$\sum \text{Output} = \sum \text{Inputs} \; - \; \text{Change in Storage} \tag{6.1}$$

Both the water inputs and outputs include a number of components, with importance varying between hydrologic systems (Fig. 6.1). The general concepts of the water budget as applied to arid lands are discussed in this chapter. More detailed technical discussion of precipitation and evapotranspiration and the recharge of groundwater systems are provided in Chaps. 7 and 8.

In general, the main water inputs into a surface water and groundwater basin include local precipitation, surface-water inflows, groundwater inflows, and any anthropogenic inflows, such as recharged treated wastewaters, agricultural return flows, and desalinated waters. The main outputs are evapotranspiration, groundwater pumping, and surface and groundwater flows that leave the basin. Local water budgets may also be modified by various-types of managed (and unplanned) aquifer recharge methods or processes (Chap. 23). For example, dams constructed across wadis decrease the local outflow of surface water from basins and increase the amount of groundwater recharge. Leakage from water and wastewater transmission mains and the operation of septic systems may greatly increase groundwater recharge in urban areas, with associated rising groundwater levels (Sect. 23.9).

R. Maliva and T. Missimer, *Arid Lands Water Evaluation and Management*, Environmental Science and Engineering, DOI: 10.1007/978-3-642-29104-3_6,

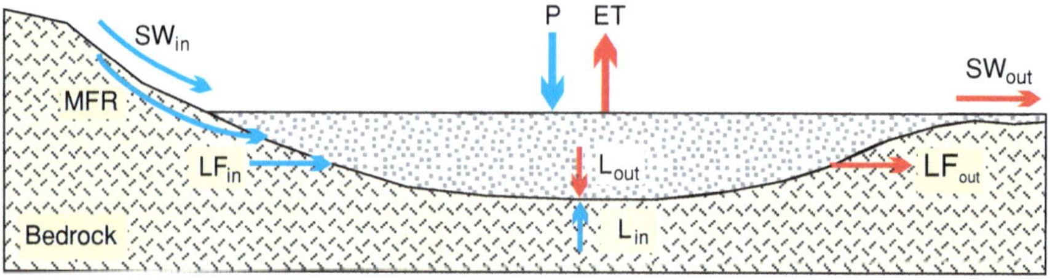

Fig. 6.1 Conceptual diagram of the main elements of arid lands water budgets, which are precipitation, evapotranspiration, surface water flows (*SW*), mountain front recharge (*MFR*), lateral groundwater flow (*LF*), and vertical leakage (*L*)

Table 6.1 Main elements of hydraulic water budgets

Inputs		Outputs	
Parameter	Symbol	Parameter	Symbol
Recharge to water table	R	Discharges (natural)	D
Precipitation	P	Evapotranspiration	ET
Injection (artificial recharge)	R_a	Pumping	Q
Lateral flow into aquifer	LF_{in}	Lateral flow out of aquifer	LF_{out}
Vertical flow (leakage) into aquifer	VF_{in}	Vertical flow (leakage) out of aquifer	VF_{out}
Surface water inflows	SW_{in}	Surface water outflows	SE_{out}
Unplanned anthropogeneic recharge	AR	Baseflow to streams	BF

Groundwater and surface-water systems are linked to varying degrees. Groundwater and surface-water budgets may be analyzed separately or together in an integrated model. For groundwater systems in arid regions, where the predominant direction of flow is downwards from surface water into underlying aquifers, surface-water interaction can be incorporated into groundwater budgets through the recharge term. Similarly, groundwater interactions can be incorporated into surface-water budgets through transmission loss and baseflow terms.

The main water budget components are listed in Table 6.1. Individual water budget components may be further divided into subcomponents. For example, groundwater components of vertical flow or leakage into or out of an aquifer may occur through both the top and bottom of the aquifer. Not all of the components listed in Table 6.1 may have a material impact on local budgets. Recharge and surface water interactions are not significant, by definition, in aquifers containing non-renewable groundwater.

6.2 Surface Water Components

Surface-water flows in arid lands pose some difficult issues in terms of their integration into overall water management. There are two fundamental components of surface water; inflows and outflows of perennial streams or rivers and the interaction of ephemeral streams with the groundwater systems based on episodic and temporally irregular flood events. Both issues may influence the water budget of regions, countries, or basins.

The headwaters of perennial streams or rivers that pass through arid lands typically occur in mountainous areas that have higher precipitation rates, and are commonly quite distant from arid lands. These streams tend to be losing streams as they pass through semi-arid to hyper-arid lands and add water directly into underlying local aquifer systems or indirectly provide recharge via agricultural irrigation practices. Changes in stream flows (and thus water budgets) occur as the result of local precipitation, transmission losses (recharge to underlying aquifers), wastewater discharges (including agricultural drainage waters), and surface water withdrawals for agricultural, industrial, and municipal uses. Groundwater pumping may impact surface water budgets by inducing additional recharge (transmission losses). Notable examples of perennial streams that greatly impact the water budgets of broad arid and semiarid regions include the Nile River, Tigris-Euphrates River System, and the Colorado River.

Ephemeral streams (wadis and arroyos) convey water from coupled precipitation-runoff-flood events that primarily impact the water budget of the basins into which they discharge. In a natural condition, flash floods within these ephemeral streams rapidly convey water into the downstream areas, either tidal waters or the basin floor. The flashy nature of ephemeral stream flow in arid regions, and common paucity of stream gauging and aquifer water level data, makes calculation of water budgets within small, arid surface-water basins quite complex. The hydrology of wadis and ephemeral streams is covered in greater detail in Chap. 13.

6.3 Recharge Within the Context of the Water Budget

Assuming net lateral and vertical flows into an aquifer are negligible, under steady state conditions (i.e., there is no change in the volume of water in storage) then natural recharge (R_o) and discharge (D_o) are in a dynamic equilibrium as

$$R_o = D_o \tag{6.2}$$

Natural recharge is equal to the difference between precipitation and evapotranspiration minus any net surface-water flow (SW) or interflow (i.e., flow within the vadose zone; IF) out of the aquifer or study area, calculated as

$$R = P - ET - (SW_{out} - SW_{in}) - (IF_{out} - IF_{in}) \tag{6.3}$$

Theis (1940) in his classic paper on the source of water derived from wells noted that:

> Discharge by wells is thus a new discharge superimposed upon a previously stable system, and it must be balanced by an increase in the recharge of the aquifer, or by a decrease in the old natural discharge, or by the loss of storage in the aquifer, or by a combination of these.

Theis (1940) concepts were expanded upon by Lohman (1979), Bredehoeft et al. (1982), and Bredehoeft (2002), as summarized herein. After the start of pumping (Q_p), the aquifer water budget can be expressed as:

$$(R_o + \Delta R) = (D_o + \Delta D) - Q_p + \frac{\Delta S}{\Delta t} = 0 \qquad (6.4)$$

where,
ΔR = change in recharge rate
ΔD = change in natural discharge
$(\Delta S/\Delta t)$ = rate of change in storage.

Since R_o and D_o are equal, Eq. 6.3 can be simplified and rearranged to:

$$Q_p = \Delta R - \Delta D + \frac{\Delta S}{\Delta t} \qquad (6.5)$$

All groundwater use initially involves the removal of water from storage. The reduction in the volume of stored water can occur by either dewatering of an unconfined aquifer or depressuring a confined aquifer with an associated compression of the aquifer matrix. Once a steady state is reached and no change in storage is occurring ($\Delta S/\Delta t = 0$), pumping is balanced by either an increase in recharge or a decrease in discharge, or a combination of both. The adjustments in recharge and discharge rates as a result of groundwater pumping are referred to as "capture". Recharge may be captured in a situation where excess water exists at times so that the aquifer becomes overfilled and available recharge is rejected (Theis 1940).

The Surficial Aquifer System of South Florida provides an excellent example of captured recharge. During the summer wet season of most years, the aquifer fills to its capacity and excess precipitation becomes surface water that runs off mainly as sheet flow or through drainage systems. Any pumping induced drawdown of the water table during the dry season, and thus reduction in storage, is reversed once the wet season begins. On annual basis, groundwater pumping and other abstractions are balanced by captured recharged during the wet season.

6.4 Captured Discharge and Recharge

Another key observation of Theis (1940) with respect to arid and semiarid regions is that:

> no matter how great the normal recharge, if under natural conditions none of it was rejected by the aquifer, then there is no possibility of balancing the well discharge by increased recharge, except by the use of artificial processes such as water spreading

Arid and semiarid regions in which the water table is located well below land surface do not have rejected recharge as defined by Theis (1940). Bredehoeft (2002) noted that in many, if not most, groundwater situations, the rate of recharge cannot be impacted by pumping. The absence of a significant impact of pumping on the natural recharge is especially true in arid environments because the aquifers do not become filled. Aquifer recharge is controlled by vertical flow rate (hydraulic conductivity) limitations in the unsaturated sediments, lack of vadose zone moisture, (infiltrated water is retained in the vadose rather than percolating to the water table and becoming recharge) or air entrapment, rather than water being rejected because the aquifer is full. In the absence of capture of rejected recharge water, the pumping rate (Q_p) must equal the captured discharge in order for a new steady state to be established, as follows:

$$Q_p = \Delta D \quad \text{if} \quad \frac{\Delta S}{\Delta t} = 0 \tag{6.6}$$

Bredehoeft and Durbin (2009, p. 508) noted that

> Capture is an all-important concept in managing ground water; a ground water system can only be maintained indefinitely if the pumping is equaled by capture – a combined decrease in the undeveloped discharge and increase in the undeveloped recharge. If pumping continually exceeds capture, then water levels in the system can never stabilize, and the system will continue to be depleted.

Captured discharge may be stream baseflow or a reduction in evapotranspiration by vegetation caused by the lowering of the water table or physical removal of phreatophytic vegetation (i.e., control of salt cedar growth in the Western United States; Pattison et al. 2011). Evapotranspiration is the principal, and may be the sole, mechanism of groundwater discharge in some arid and semiarid regions (Nichols 1994). Phreatophyte vegetation in arid regions is very efficient in extracting available moisture and may occur in areas where the water table is located 15 m (50 ft) or more below land surface. Substantial water savings may be possible by removing phreatophytes or replacing them with shallowly rooted vegetation (Bouwer 1978). However, phreatophyte removal needs to be carefully planned as the vegetation may have wildlife and scenic values that should be preserved (Bouwer 1978; Pattison et al. 2011). Although relationships have been developed relating evapotranspiration rates to plant density, leaf area, and depth to groundwater (e.g., Nichols 1994), accurate quantification of basin-scale changes in evapotranspiration and thus discharge caused by groundwater pumping are not practical or quantitative.

Capture may impact surface-water users (Theis 1940) and sensitive environments. Captured discharge may reduce the baseflow to a surface water bodies, spring flow, or may induce flow from the surface water to the groundwater (i.e., induced recharge). Induced recharge from surface water bodies may adversely impact water users who have rights to the surface water bodies (Balleau 1988). Thus, neither groundwater nor surface water systems should be evaluated in isolation.

Decreases in stream and springs flows can harm organisms dependent upon these water sources. However, captured discharge may otherwise be leaked out to tide or not used by humans or be critical for preservation of the environment. As is discussed in Sect. 9.2, the acceptability of impacts from captured discharge is an important consideration in the assessment of the sustainability of water supplies.

To emphasize a key point, if the rate of capture is less than the pumping rate, then an equilibrium condition cannot be reached and some of the pumped water will be produced from storage (i.e., the aquifer will be "mined"). An important implication of the this analysis is that knowledge of natural "virgin" recharge (R_o) and discharge (D_o) rates have little importance in assessing sustainable yield of an aquifer (Bredehoeft et al. 1982; Bredehoeft 2002; Bredehoeft and Durbin 2009). Bredehoeft (2002) noted that it is a persistent myth that the virgin recharge rate is important in determining the magnitude of sustainable groundwater development. It is, instead, the rates of change of recharge and discharge that are the critical variables. Inasmuch as ΔD cannot exceed D_o, the latter parameter still has significance in water budget analysis as being the upper limit value. The value of ΔR depends upon surface-water-groundwater interactions and bears no relationship to R_o.

From Eq. 6.4, as is typically the case in arid regions where $\Delta R = 0$ (i.e., there are no surface water bodies from which recharge may be induced), then the maximum steady state ($\Delta S/\Delta t = 0$) pumping rate occurs when the entire discharge is captured ($\Delta D = D_0$). Also, greater pumping rates may be sustainable if pumping induces inter-aquifer leakage into the pumped aquifer or there is significant lateral flow into the basin. Lateral flow through a groundwater basin can be considered in an analogous manner as discharge and recharge. Some or all of the lateral flow through the basin may be captured by pumping without causing a decrease in the total volume of water in storage.

6.5 Water Budget and Groundwater Resources Management

Bredehoeft et al. (1982) identified the following implications concerning the relationships between the water budget and water supply:

1. Some water must be removed from storage (i.e., mined) in order to develop a groundwater supply. The removal of water from the vicinity of production wells is necessary to create hydraulic gradients towards the well.
2. Time delays occur in the response of aquifers to pumping, which results in aquifers not reaching steady-state (equilibrium) conditions or that it may take decades or longer for an equilibrium condition to be reached.
3. The location of pumping wells significantly changes the dynamic response of the aquifer to pumping and the rate at which natural discharge can be captured.
4. The magnitude of groundwater development depends upon the hydrologic effects that can be tolerated ultimately or at any given time.
5. To calculate hydrologic impacts it is necessary to know the hydraulic properties and boundaries of the aquifer.

The rate at which aquifers approach equilibrium to a new regime of pumping will depend upon the aquifer type and the magnitude of the pumping impact.

Confined aquifers will approach equilibrium orders of magnitude of time quicker than unconfined aquifers because of their greater diffusivity (Sect. 5.3.4), which needs to be considered in the planning of water-supply development (Theis 1940). It may take a very long time (centuries) for a large unconfined aquifer to reach equilibrium with a new pumping regime, if such an equilibrium is possible. Confining layer storage and leakance, and boundary conditions also effect the time required for an aquifer to reach equilibrium with a new pumping regime.

The water balance in humid regions is usually dominated by local processes, such as precipitation, evapotranspiration, and recharge. In groundwater basins in arid and semiarid regions, other elements of the water balance may have proportionally greater significance. Local aquifers may also be replenished by the leakage of water from regional flow systems (Maxey 1968). Leakage to and from the pumped aquifer from adjoining aquifers must also be considered in the aquifer water budget. The standard groundwater-balance equation for a groundwater basin can be expressed as (Carrillo-Rivera 2000):

$$LF_{in} - LF_{out} + VF_a + VF_b - Q_p = \Delta S/\Delta t \tag{6.7}$$

where, VF_a and VF_b are the net vertical flow (leakage) into the aquifer from above (a) and below (b). Captured recharge and discharge also need to be considered if they are significant components of the water budget.

Carrillo-Rivera (2000) documented that vertical flow from below was an important component in the water budget of two studied groundwater basins in Mexico. A key observation was that the change in hydraulic heads are two orders of magnitude less than the decline calculated based on the rates of abstraction and lateral flow into the basin, and the aquifer storativity value determined from pumping tests. The drawdown rate declined despite no corresponding reduction in aquifer pumping. There was also a pronounced increase in groundwater temperature and chemistry signature (increased F, B, Li, and Na concentrations) that cannot be explained by lateral flow. The hydraulic, temperature and water quality data all indicate the entry of vertical groundwater flow from deep regional circulation via vertical faults and fractures into the system. Failure to incorporate regional flow into water budgets may result in an underestimation of groundwater resources (Maxey 1968; Carrillo-Rivera 2000). Groundwater management plans should also consider the chemical response of captured regional flow with regard to elements such as fluoride (Carrillo-Rivera et al. 1996, 2002; Carrillo-Rivera 2000). Vertical flow can cause an unacceptable deterioration in the quality of produced water.

An additional factor that can be very important in the water budget of shallow unconfined aquifers is recharge from irrigation return flows (Alley et al. 1999; Kendy 2003; Devlin and Sophocleous 2005), and other unplanned recharge, such canal seepage, septic system recharge, and pipe leakage (Sect. 23.9). Once used, water may re-enter the groundwater system and be used again, and thus, needs to be included in the water budget as additional recharge in order to avoid over-estimating net extractions. Quantification of unplanned recharge is inherently challenging because of its unplanned and uncontrolled nature. Little effort has been made in most areas to accurately measure unplanned recharge and its occurrence may not even be fully appreciated.

A fundamental concept concerning the impact of agricultural or other development on water resources is that actual change in water use at a site is predominantly the change in evapotranspiration between the existing and new land cover (Stewart and Mills 1967) provided that there is no change in net runoff (Maliva and Hopfensperger 2007). Another factor that needs to be considered is the ET rate of native vegetation, which is usually largely eliminated as land is developed. Land development needs to be viewed as essentially a change in crop. The pumping of water for irrigation results in a net loss of water to the groundwater system to the degree that it results in higher ET rates. Return flow can be incorporated into the water budget calculations by either using net pumping (pumping rate − return flow rate) for the term "Q_p" in the equations introduced or by including it in the recharge term. The construction of buildings and pavement can have a net beneficial impact on water resources by resulting in a dramatic reduction in ET as existing vegetation is removed, provided that the increased runoff is retained.

Failure to consider returns flow can result in an over-estimation of the impacts of groundwater pumping on the shallow unconfined aquifer. If the pumped water is derived from deeper aquifers, then irrigation can result in a net recharge to shallow aquifers with the water loss being from the source aquifer. Dichotomies may exist between modeled drawdowns and measured drawdowns if all components of the water budget, such as return flows, are not considered (Maliva and Hopfensperger 2007). Attempts to reduce aquifer drawdowns by adopting more efficient irrigation methods may also fail to achieve desired goals (Kendy 2003). If the rate of crop evapotranspiration is unchanged, then the benefits of more efficient irrigation may be largely cancelled out by a reduction in agricultural return flows (Kendy 2003). Continued drawdowns may occur if the irrigated area is expanded.

Changes in the local water budget are not necessarily either beneficial or harmful. Alley (2006, p. 16) observed that

> The amount of groundwater available for use depends upon how the changes in inflow and outflow affect the surrounding environment and upon the extent to which society is willing to accept the resultant environmental changes

Evaluation of the impacts of groundwater use thus requires groundwater modeling and it is not a matter of simply evaluating the aquifer overall water budget (Bredehoeft et al. 1982; Bredehoeft 2002; Devlin and Sophocleous 2005). It is important to understand how a particular aquifer system works so that the impacts resulting from its use can be predicted and to allow for informed resource planning decisions to be made (Kalf and Wooley 2005). The modeling must incorporate all aspects (inputs and outputs) of the groundwater system.

An additional consideration is the transient response of aquifers, particularly deep aquifers in arid and semiarid regions (Seiler et al. 2008). The transient behavior of aquifers (and other reservoirs) is related to the mean turn-over time, which in turn, relates to the age of the water. Current aquifer behavior, such as discharge rates and local aquifer heads, may not be in equilibrium with current recharge, much less recently introduced groundwater pumping. Current (pre-development) hydraulic gradients may reflect at least in part an on-going decline in a fossil hydraulic gradient that formed during earlier pluvial periods (Bourdon 1977; Houston and Hart 2004). Not considering transient responses in the development of groundwater management strategies in arid region aquifers with long mean turn-over times can result in either an over-estimation or under-estimation of sustainable yield (Seiler et al. 2008).

Aquifers may also take a very long time to achieve a new steady state after the initiation of new pumping, if such equilibrium can ever be reached. Walton (2011) defined the aquifer response time as the time it takes for water levels and storage changes throughout the aquifer system to become negligible after a change in withdrawals. The aquifer response time depends primarily on aquifer diffusivity (and, in turn transmissivity and storativity), confining layer storage and leakance, and boundary conditions (Walton 2011). Aquifer response time may range from days to centuries (or longer for very large regional systems). A related concept is the time lag between the start of pumping and hydraulic response at a given location (e.g., surface water body, wetland), which is also a function of the distance from the withdrawal point.

Consider the interaction of a well with a stream or other surface water body (Bredehoeft et al. 1982; Balleau 1988; Sophocleous 2002; Bredehoeft and Durbin 2009). Groundwater will initially be produced from storage (groundwater depletion). It may take a very long time (10–1,000s of years) until the cone of depression reaches the surface-water body and a steady state is achieved in which pumping is balanced by induced recharge. Groundwater is thus first produced entirely from storage, but over time the amount of water pumped approaches a balance with induced recharge. Therefore, during the initial stages of groundwater usage, an increase in total availability of water in a basin is achieved because of the production of groundwater from storage. However, over time, the total available water (combined surface water and groundwater) decreases due to the reduction in surface water caused by induced recharge (Balleau 1988).

From a water management perspective, aquifer response times and time lags are a function of the distance between pumped wells and surface water bodies or other environments or features of concern. Bredehoeft and Durbin (2009) concluded that "it makes a big difference in the response of a system where the wells are located". Bredehoeft and Durbin (2009) also pointed out the limitations of monitoring in water management. Current water levels may not represent the maximum drawdowns that will result from pumping that occurred to date. Drawdowns may continue to increase after pumping is stopped as the system slowly approaches a new equilibrium. Maximum drawdowns (impacts) will be larger than those observed at the time pumping stops, especially if monitoring is at some distance from the pumping wells or center (Bredehoeft and Durbin 2009).

6.6 Water Budget Component Evaluation

Water budgets are conceptually simple, but can be very challenging to actually calculate because of the paucity of acceptable quality data. Water budgets can be evaluated through the development and calibration of a groundwater flow model, but model results are not unique.

Hence, it is important to obtain model-independent values for at least some of the more important water-budget parameters, particularly recharge and related hydraulic parameters. Independently obtained data places constraints on the range of possible values and thus reduces model uncertainties. Recharge and natural discharge are the most critical variables for evaluating groundwater resources in arid and semiarid regions because the recharge rate is typically small and it ultimately controls to a large degree the amount of renewable groundwater resources locally available. More detailed analysis of water budget elements are provided in chapters addressing precipitation and evapotranspiration (Chap. 7), recharge concepts (Chap. 8), and techniques for measuring recharge rates (Chap. 11).

Unconfined aquifers in large basins pose a great challenge because of the long aquifer response times. Very long-term simulations are required to determine the eventual effects of groundwater abstractions and to determine whether or not a new equilibrium (state steady state) condition is achievable and the time that will elapse until such a steady state is achieved.

It must be emphasized that aquifers in arid regions generally differ from humid regions in terms of the much greater time required to reach a new equilibrium state because of their low recharge rates. Returning to the South Florida example, unconfined aquifers usually reach a new equilibrium to changes in pumping within a year because of the capture of rejected recharge. Confined aquifers in humid regions also tend reach new equilibriums quicker because of greater and more regular (less temporally variable) recharge rates.

References

Alley, W. M. (2006). Tracking U.S. groundwater reserves for the future. *Environment, 48*(3), 10–25.

Alley, W. M., Reilly, T. E., & Franke, O. L. (1999). Sustainability of ground-water resources. U.S. Geological Survey Circular 1186.

Balleau, W. P. (1988). Water approximation and transfer in a general hydrogeologic system. *Natural Resources Journal, 29*(2), 269–291.

Bourdon, D. J. (1977). Flow of fossil groundwater. *Quarterly Journal of Engineering Geology and Hydrogeology, 10*, 97–124.

Bouwer, H. (1978). *Groundwater hydrology*. New York: McGraw-Hill.

Bredehoeft, J. D. (2002). The water budget myth revisited: Why hydrogeologists model. *Ground Water, 40*, 340–345.

Bredehoeft, J. D., & Durbin, T. (2009). Ground water development—the time of capture problem. *Ground Water, 47*(4), 506–514.

Bredehoeft, J. D., Papadopulos, S. S., & Cooper, H. H., Jr. (1982). *The water-budget myth: Studies in Geophysics Scientific Basis of Water Resources Management*. (pp. 51–57). Washington, DC: National Academy Press.

Carrillo-Rivera, J. J. (2000). Application of the groundwater-balance equation to indicate interbasin and vertical flow in two semi-arid drainage basins, Mexico. *Hydrogeology Basin, 8*, 503–520.

Carrillo-Rivera, J. J., Cardona, A., & Edmunds, W. M. (2002). Use of abstraction regime and knowledge of hydrogeological conditions to control high fluoride concentration in abstracted groundwater. *Journal of Hydrology, 261*, 24–47.

Carrillo-Rivera, J. J., Cardona, A., & Moss, D. (1996). Importance of the vertical component of groundwater flow: A hydrogeochemical approach in the valley of San Luis Potosi Mexico. *Journal of Hydrology, 186*, 23–44.

Devlin, J. F., & Sophocleous, M. (2005). The persistence of the water budget myth and its relationship to sustainability. *Hydrogeology Journal, 13*, 549–554.

Houston, J., & Hart, D. (2004). Theoretical head decay in closed basin aquifers: An insight into fossil groundwater and recharge events in the Andes of northern Chile. *Quarterly Journal of Engineering Geology and Hydrogeology, 37*, 131–139.

Kalf, F. R. P., & Wooley, D. R. (2005). Application and methodology for determining sustainable yield in groundwater systems. *Hydrogeology Journal, 13*, 295–312.

Kendy, E. (2003). The false promises of sustainable pumping rates. *Ground Water, 41*, 2–4.

Lohman, S. W. (1979). Ground-water hydraulics. U.S. Geological Survey Professional Paper 708.

Maliva, R. G., & Hopfensperger, K. P. (2007). Impacts of residential development on humid tropical freshwater resources: Southwest Florida experience. *Journal of the American Water Resources Association, 43*, 1540–1549.

Maxey, G. B. (1968). Hydrogeology of desert basins. *Hydrogeology Journal, 6*(5), 10–22.

Nichols, W. D. (1994). Groundwater discharge by phreatophyte shrubs in the Great Basin as related to depth to groundwater. *Water Resources Research, 30*, 3265–3274.

Pattison, R. R., D'Antonio, C. M., Dudley, T. L., Allander, K. K., & Rice, B. (2011). Early impacts of biological control on canopy cover and water use of the invasive saltcedar tree (Tamarix spp.) in western Nevada. *Physiological Ecology, 165*, 605–616.

Seiler, K.-P., Gu, W.-Z. & Stichler, W. (2008). Transient response of groundwater systems to climate changes. In W. Dragoni & B. S. Sukhija (Eds.), *Climate change and groundwater* (pp. 111–119). Geological Society of London Special Publications No. 288.

Sophocleous, M. (2002). Interactions between groundwater and surface water: The state of the science. *Hydrogeology Journal, 10*, 52–67.

Stewart, E. H., & Mills, W. C. (1967). Effect of depth to water table and plant density on evapotranspiration rate in Southern Florida. *Transactions of the American Society of Agricultural Engineers, 10*, 746–747.

Theis, C. V. (1940). The source of water derived from wells. Essential factors controlling the response of an aquifer to development. *Civil Engineering, 10*, 277–280.

Walton, W. C. (2011). Aquifer system response time and groundwater supply management. *Ground Water, 49*, 126–127.

Chapter 7
Precipitation and Evapotranspiration

7.1 Introduction

The primary water input into local hydrologic systems is precipitation. Under long-term, steady-state conditions precipitation (P) is primarily either lost to evapotranspiration (ET), recharges underlying aquifers (R), runs off as surface water (SW), or infiltrates into the vadose zone and flows down-gradient as soil moisture (which is sometimes termed interflow; IF) as follows:

$$P = ET + R + SW + IF \tag{7.1}$$

Local water budgets in arid regions are dominated by precipitation and evapotranspiration. Vadose zone interflow is a minor component of local water budgets and typically recharge rates are only a small percentage of rainfall. However, the paucity of vegetation and the occurrence of surface crusts in arid and semi-arid regions results in low soil-infiltration rates, which combined with high-intensity, short-duration convective rainfall, favor runoff (Wheater 2002). Overland flow, concentrated by topography, converges on wadi channel networks and tends to result in flood flow. Arid lands are, thus, prone to flash floods, which can he highly destructive and frequently result in loss of life. In flat-lying areas virtually all precipitation that falls in inter-channel areas may be lost to evapotranspiration.

Quantification of precipitation and evapotranspiration are important elements for water management in arid and semiarid regions. However, in many countries, particularly developing nations, there is a profound paucity of accurate long-term climatological data.

R. Maliva and T. Missimer, *Arid Lands Water Evaluation and Management*, Environmental Science and Engineering, DOI: 10.1007/978-3-642-29104-3_7,

7.2 Precipitation in Arid Regions

Rain in hot deserts (e.g., Arabian, Sahara, Libyan, Great Australian) usually falls in convectional showers of short duration and limited areal extent on a small number of rainy days each year (Walton 1969). Such torrential rainfall events are often associated with intense run-off and erosion. The intense rainfall events replenish groundwater supplies and introduce fluvial processes into the desert landscape. By moistening the soil, they give an opportunity for short-lived annual plants to grow, which is critical to the overall desert ecology.

The amount of water vapor that air can hold is related to temperature. Rainfall occurs when moist air is cooled through ascent, mixing, or radiation, and, as a result, water vapor condenses. Processes leading to aridity tend to prevent such cooling by maintaining air stability, creating temperature inversions, or warming of the atmosphere. Commonly, rainfall can occur at high altitudes and the actual moisture never reaches the ground because of dry air at lower altitudes. When rainfall does reach land surface, it can occur as a dramatic, intense, localized downpours. Arid lands differ in their rainfall regimes and several patterns are recognized (McGinnies 1979, cited by Agnew and Anderson 1992; Almazroui 2011a, b):

- two rainy seasons (e.g., Venezuela)
- winter rains (e.g., north African Coast)
- summer rains (e.g., Sahel)
- seldom rains (e.g., Sahara, Atacama)
- autumn orographic effects and wind convergence (e.g., Saudi Arabia, Red Sea)

Some arid regions, such as the southwestern United States and northern Mexico receive sparse, sporadic rainfall throughout the year, with the greatest rates occurring during the summer "monsoon" season.

Convective precipitation in arid environments is characterized by extreme spatial and temporal variability (i.e., spottiness). Fluctuation in precipitation in arid lands means that annual average values are less useful in water management in these areas than perhaps anywhere else in the world (Heathcote 1983). Annual rainfall histograms in arid regions have a positive skew with more years having annual rainfall below the annual average (Şen 2008). The mode and median annual rainfall are smaller than the average. A histogram of historic annual rainfall for Phoenix, Arizona, shows the typical positive skew (Fig. 7.1).

The range of annual rainfall in arid and semiarid regions is generally less than the range in humid regions when expressed in absolute values (i.e., cm; in) because of the overall small values. However, the range when expressed as the ratio of highest to lowest annual amount tends to be much greater in arid regions than in more humid regions. Arid regions may experience many years or decades with rainfall totals near or below average, followed by an extreme rainfall event that may cause severe or catastrophic flooding. An example is the 2009 Jeddah

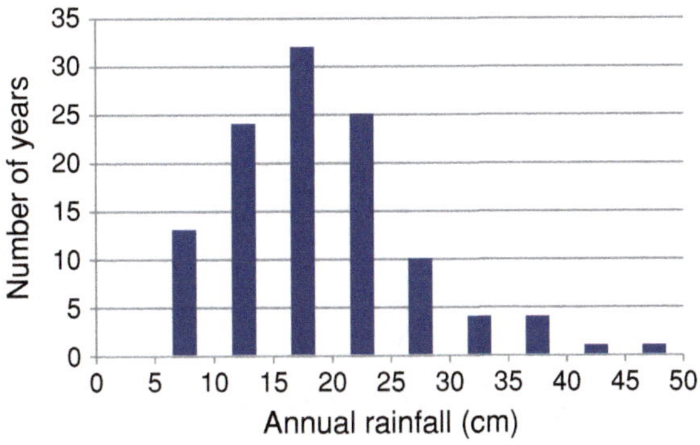

Fig. 7.1 Histogram of annual rainfall in Phoenix, Arizona (1896–2009). Mean = 19.3 cm, median = 18.0 cm. *Source* National Weather service

flood, wherein 90 mm (3.5 in.) of rainfall occurred in a 4-h period, which is nearly twice the average total annual rainfall accumulation in that region of Saudi Arabia.

Water management in arid regions is also hampered by the paucity of long-term monitoring data. The period of record may not provide an adequate representation of the historic variation in rainfall if the period happened not to include less frequent extreme rainfall events. Local populations that settled in the flood plains of ephemeral streams may, as a result, not be prepared for periodic flood events that are a normal part of the local hydrology. Major rainfall events that have a disproportionate impact on local aquifer recharge may also not be included in short periods of record.

7.3 Precipitation Measurements

The two main methods for measuring rainfall are rain gauges and radar-based remote sensing. Rain gauges provide point measurements of rainfall. The great spatial and temporal variation in rainfall in arid regions (particularly for convective precipitation) creates large monitoring challenges when conventional rain gauges are used. Typically, the density of rain gauges is much too coarse to obtain an accurate record of precipitation within catchments. Convective storm cells are often smaller than catchment areas and, therefore, may or may not reach any given rain gauge within the catchment area during a storm event. However, it is unlikely that a rain gauge centrally located within a catchment would have a long-term bias in recorded rainfall unless there are topographic or geographic controls over local rainfall. Rain gauge data may provide an inaccurate assessment of the total rainfall in a catchment for individual storms, but would be expected to have a satisfactory accuracy for long-term average rainfall values provided that the tracks of storms are random with respect to the location of the gauge.

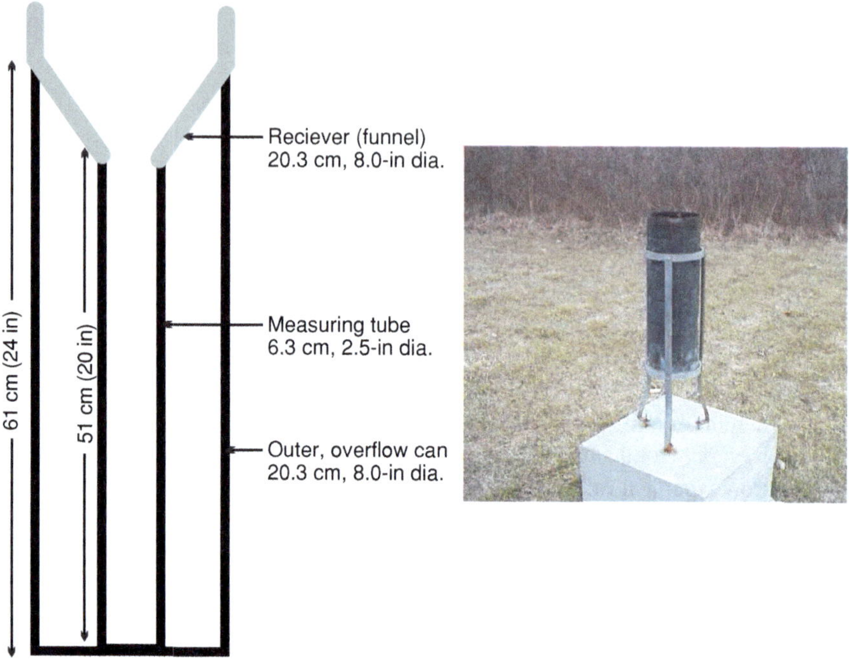

Fig. 7.2 Diagram and photograph of the standard U.S. National Weather Service non-recording rain gauge. (*Photo source* National Weather Service)

7.3.1 Rain Gauges

Rain gauges are the simplest and most accurate means to measure rainfall. The basic elements of a rain gauge is a collection container with an opening of known area (A) and a means for measuring the volume of rain (V) collected over a given time period. The precipitation rate (P) is calculated as the collected volume divided by the opening area (V/A). Snowfall can be measured as equivalent rainfall by allowing the snow to melt and then measuring its volume, but most commonly a specially designed "board" is used to make these measurements.

Although rain gauges are conceptually very simple and easy to construct, their design can affect recorded rainfall amounts. It is, therefore, important to use standardized equipment to provide uniformity and continuity of precipitation data. The U.S. National Weather Service (NWS) standard 8-in. (20.3 cm) non-recording rain gauge is used throughout the world for official rainfall measurements (Fig. 7.2). The standard 8-in. (20.3 cm) diameter gauge has four main components: an 8-in. (20.3 cm) diameter overflow can, an 8-in. (20.3 cm) diameter collector funnel, a 2.5-in. (6.4 cm) diameter measuring tube, and a measuring stick. The collector funnel and measuring tube are sized so that the cross-section area of the collector is 10 times the area of the measuring tube. Thus, 1 cm

(2.5 in.) of water in the measuring tube corresponds to 0.1 cm (0.04 in.) of rainfall. The gauge usually is installed at least 24 cm (9.4 in.) above land surface. Measurements are performed manually using the measuring stick, ideally daily.

A limitation of rain gauges in general is that they are not reliable under windy conditions in which there is a strong horizontal component to the rainfall. Non-recording rain gauges are also labor intensive, requiring frequent manual measurements, which can be time consuming (and thus expensive) in remote locations. The gauges also do not provide data on the intensity of rainfall events, as the only rate information is the average value between successive readings. In hot dry areas, evaporative losses can impact data, especially if readings are performed infrequently. Commonly, powdered cork is added to some gauges to record rainfall events as a cut line between readings and thereby, reducing the evaporative loss error.

Alternatives to the standard non-recording rain gauges are systems that automatically record water volume. Tipping bucket systems collect rainfall using a funnel that channels water into a small seesaw-like container (bucket) of known capacity. After a design specific volume of water fills the bucket, the lever tips, dumping the collected water and sending an electronic signal that is recorded. Accuracy is limited by the bucket size. Tipping bucket rain gauges tend to slightly underestimate rainfall as small rain events may not trigger the bucket tip and may experience losses of water during high intensity rains. The great advantage of tipping bucket rain gauges is they are less labor intensive as only infrequent trips to the station are required to download the collected data. The rain gauges (and other weather data) can also be transmitted in real time or periodically downloaded to a central office using remote telemetry units (RTUs) that can be wireless transmitters or use a satellite feed or telephone line. Another important advantage of tipping bucket systems is that the rate of rainfall can be determined from the frequency of tips.

The weighing rain gauge is another example of an automatic rain gauge. Water is collected in a storage bin in which the weight of the water is continuously recorded. Alternatively, a pressure transducer can be used to determine the height of water within a measuring tube.

7.3.2 Radar Measurement of Rainfall

Radar measurement of rainfall can provide data over large areas as opposed to the point measurements of rain gauges, and is thus potentially more useful for evaluation of catchment or basin-wide water budgets. Weather radar involves the transmittance of directional pulses of microwave radiation and recording the return signal from reflections off of droplets of water or ice particles in the atmosphere. The intensity of the return pulse is related to the concentration of droplets or particles in the atmosphere, i.e., the precipitation rate. Rain accumulation is determined by multiplying the average value over a point or area by the time between images. Total accumulations are calculated by adding up the accumulations from all images over the time period of interest. The Next Generation Radar (NEXRAD) weather radar system (technically named the WSR-88D Doppler radar system) is used for rainfall measurements in the United States and employs a 10-cm (3.9 in.) wavelength radar wave.

The radar return, which is expressed as a reflectivity factor (Z), is converted to a radar rainfall estimate (P_r) through an empirical Z-P relationship. The Z-P relationship varies as a function of many factors (including drop size) and it is not possible to derive a single equation that is accurate for every point in a given radar domain and for every storm type and intensity (Hunter 1996; Hardegree et al. 2008; and references therein). A correction factor also needs to be applied to convert radar rainfall (Rr) to gauge rainfall (R_g). The errors associated with NEXRAD rainfall estimates are discussed by Hunter (1996). A comparison of NEXRAD and gauge precipitation by Hardegree et al. (2008) indicates that radar is more likely to detect high intensity rainfall events than light rains and is, therefore, more appropriate for flood forecasting than for long-term water balance and natural rainfall modeling applications.

Skinner et al. (2009) compared NEXRAD and rain gauge precipitation data in South Florida and found that NEXARD tended to underestimate rainfall, particularly large rainfall amounts, relative to the gauge network. Use of NEXRAD data in arid regions may also produce another type of error which involves the detection of high altitude moisture that does not reach the ground as actual rainfall. Radar rainfall measurements are increasingly being used in surface water and groundwater modeling because it provides spatial and temporal data that can be readily incorporated into models using a GIS framework. Particularly where rainfall is spatially heterogeneous within the study area and time period, radar rainfall data may be more accurate than rainfall maps obtained by interpolations between small numbers of scattered rain gauges. However, it is critical that radar rainfall data be ground-truthed and calibrated against rain gauge data. The effects of uncertainty in rainfall rates on groundwater models should be evaluated as part of the sensitivity analysis.

Processing of the radar rain data is a complex exercise, and in many areas (particularly developing countries) such data and expertise may not available. However, radar rainfall data are commercially available in some areas. For example, the OneRain Company provides gauge-adjusted radar rainfall estimates in the United Sates based on the NWS NEXARD data. Radar rainfall estimates can be calibrated against user provided rain gauge data and publically available gauge data.

7.3.3 Satellite Rainfall Measurement

Satellite rainfall estimates offer the potential of much greater areal coverage compared to land-based radar measurements. The current state of the art is a combination of passive microwave (PMW) data from low-earth orbiting satellites and infrared (IR) data from geostationary satellites (Gebremichael et al. 2010; Pereira Filho et al. 2010; and references therein). PMW data provide more direct information on rainfall, but has the limitation of low sampling frequency. IR data has a weak physical relationship to surface rainfall, but has a high spatial and temporal frequency. The combination of PMW and IR data can overcome some of the limitations of each individual data source for estimating rainfall. Several different PMW/IR methods have been developed to estimate rainfall, which were summarized by Gebremichael et al. (2010) and Pereira Filho et al. (2010).

The Climate Precipitation Center Morphing (CMORPH) technique (Joyce et al. 2004) estimates rainfall exclusively from PMW-derived data. IR-derived data are used to derive a motion field and in essence act as a spatial and temporal interpolator of the PMW-derived rainfall data. The PERSIANN-CCS technique (Hong et al. 2004) obtains estimates of precipitation from the IR data, but uses the PMW data for training the neural network that assigns precipitation estimates to the IR temperatures (Gebremichael et al. 2010).

Pereira Filho et al. (2010) compared CMORPH rainfall estimates with rain gauge data in the Amazon Basin of Brazil. A large amount of scatter ($R^2 = 0.1$) occurred in daily data, which was attributed to errors in both the gauge and CMORPH data. Much less scatter ($R^2 = 0.7$) occurred in the yearly precipitation data as local errors tend to be reduced by the larger time integration. Satellite rainfall was found to have much potential for detecting spatial and temporal patterns of rainfall. The CMORPH data typically overestimate rainfall in continental areas and under estimate rainfall in coastal areas.

Gebremichael et al. (2010) compared CMORPH and PERSIANN-CCS rainfall estimates with data from a dense network of gauges in the Nile River Basin. Both satellite estimates underestimated heavy rainfall events. The 39-day rainfall totals for a study periods were 483 mm (19 in.) for the rain gauges, 328 mm (12.9 in.) for CMORPH, and 244 mm (9.6 in.) for PERSIANN-CCS. Based on the results of the study, Gebremichael et al. (2010) recommended extreme caution with using satellite rainfall estimates because the estimates could be subject to significant errors. The study results also demonstrated the need for establishing high-quality, dense, ground-based rainfall monitoring systems in selected sites to serve as validation sites.

7.4 Precipitation Frequency Analysis

Water management in arid and semiarid regions is often impacted by extreme events such as droughts, severe storms, and floods. The magnitude of extreme events is inversely related to the frequency of their occurrence, with severe events (on either side of the spectrum) occurring infrequently. Frequency analysis of low precipitation events (i.e., droughts) is important for planning supplemental water supplies. Flood frequency is of obvious importance in the design of storm-water control infrastructure. Infrequent, high magnitude rainfall events are also important in that they may be responsible for most of the groundwater recharge within a basin.

Statistical methods are routinely applied to precipitation and storm flow data sets to determine both average conditions and the magnitude of infrequent severe events, such as the 1 in 100 year flood or drought and the return period of a drought or flood of a given magnitude. Frequency analyses of hydrologic data typically assume that the data are independent and identically distributed and the hydrologic system producing the data is considered to be stochastic, space-independent, and time independent (Chow et al. 1988). A major challenge in many arid and semiarid regions is that there is a great long-term temporal variation in precipitation, but continuous long-term monitoring data are often not available. Relatively short-term data sets (e.g., <30 years) may not be representative of a region if the data happened to have been collected in a wetter or drier period compared to the long-term norm. Global climate change (Chaps. 38 and 39) further complicates the interpretation of frequency data if past data are not representative of future climate conditions.

Statistical analysis of precipitation data is a fundamental issue in surface-water hydrology and is addressed in general texts on the subject (e.g., Chow et al. 1988). Rainfall data from arid and semiarid regions data are best fit to one of several probability distribution functions as such as normal (Gaussian), log-normal, gamma, Weibul, and Gumbel, which describe the shape of the frequency versus magnitude distribution (e.g., Abdullah and Al-Mazouri 1998; Tilahun 2006). Rainfall frequency analysis can be performed using widely available statistical software packages (e.g., SPSS and SAS), which allow for the evaluation of the probability distribution function that best fits field data.

An older, but still commonly used, technique is the probability paper method, which is presented herein because of its simplicity and because it well illustrates basic concepts. Annual precipitation data from Phoenix, Arizona (Phoenix WSFO AP station obtained from the Western Climate Research Center) are used as an example. Frequency analysis can also be performed using shorter time intervals (monthly or less), provided that such data are available.

The first step in the probability paper method is the reordering and ranking of the data in descending order, with the greatest magnitude value assigned a rank of one. The reordering and ranking of the data can be easily performed using a spreadsheet (e.g., Excel) sort function. A graph of the Phoenix annual rainfall versus rank is provided in Fig. 7.3. The probability of annual rainfall equaling or exceeding the value "x" of rank "m" (which is designated "$P(X_m)$") is calculated using plotting formulas.

A number of plotting formulas have been proposed or used for rainfall data over the past century, which are summarized by Makkonen (2006). The widely used Hazen (1914) and Weibull (1939) formulas are, in order,

$$P(x_m) = (m - 0.5)/n \tag{7.2}$$

$$P(x_m) = m/(n + 1) \tag{7.3}$$

where "n" is the total number of data points.

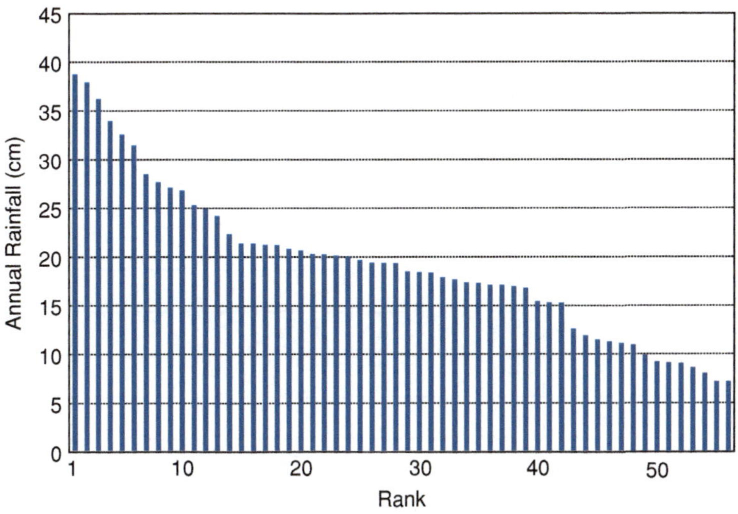

Fig. 7.3 Annual rainfall in Phoenix, Arizona (1950–2006) plotted versus rank

Critchley and Siegert (1991) in the FAO Water Harvesting Manual recommended the equation

$$P(x_m) = \frac{(m - 0.375)}{(n + 0.25)} \tag{7.4}$$

which has a similar form as some other plotting formulas reviewed by Chow et al. (1988) and Makkonen (2006). From a practical perspective, the plotting formulas differ in the calculated return frequencies, with the Weibull formula being more conservative in terms of estimated flood frequencies in that it gives shorter return frequencies (Makkonen 2006). Equations 7.2 through 7.4 can be converted to probability as a percentage by multiplying by 100.

The rank of each value is entered into the selected plotting formula and the data plotted on normal probability paper. The probability paper plot for the Phoenix data using the Weibull formula is provided in Fig. 7.4. A best fit line is drawn through the data points. If the data are normally distributed, then the data should plot on a straight line. The patterns of departure from a straight line can provide insights into the deviations from a normal distribution (Sokol and Rohlf 1981). For example, the shape of the Phoenix data plot is indicative of a platykurtic distribution, in which there are more years near the mean value and fewer values at the tails (i.e., there is negative excess kurtosis).

Probability refers to the likelihood that a specific event will occur. It is possible (but unlikely) that a 1 in 1,000 year annual rainfall or drought may occur next year or that two 100 year events may occur in successive years, which is a concept often not understood by the lay public. The occurrence of what are considered rare events, could be due either to chance (i.e., probability density function is correct

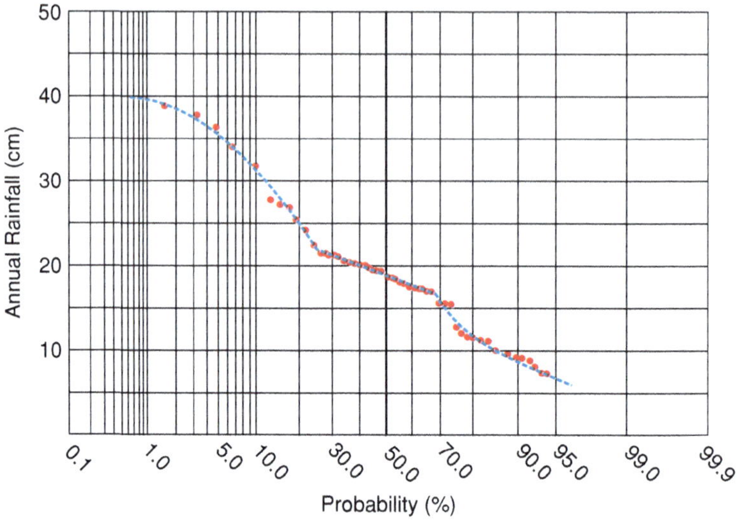

Fig. 7.4 Probability paper plot of the Phoenix, Arizona, annual rainfall data. Plot represents the probability that annual rainfall will equal or exceed a given value

and a rare event just happened to have occurred) or that the probability density function is incorrect (i.e., it is based on unrepresentative data). For example, if the data used to generate a probability density function were from an abnormally dry period, then greater than the predicted annual rainfall years would be more frequent once normal climatic conditions return.

7.5 Evapotranspiration

Evapotranspiration (ET) is sum of evaporation and plant transpiration from land surface. It is typically expressed as a rate in terms of length of a water column per time (e.g., mm/d, cm/yr, in/yr). ET is of fundamental importance in water resources as it is by far the major outflux of water from land surface. There are three main types of ET that are referred to in the literature, actual ET (ET_a), reference ET (ET_o), and potential ET (PET). Actual ET, as the name implies, is the actual flux of water to the atmosphere from land surface over a defined period of time. The similar parameters, ET_o and PET, are the ET rates that will occur from a reference surface that is not short of water, and are dependent on local meteorological conditions. Actual ET is also a function of water availability. Thus, in arid and semiarid regions, ET_a rates are usually much lower than ET_o and PET rates.

Relating ET to standard specific surfaces provides a reference to which ET from other crops or surfaces can be related (Allen et al. 1998). Actual ET rates are related to ET_o rates by a crop coefficient (K_c) as follows:

$$ET_a = ET_oK_c \tag{7.5}$$

Thus, once ET_o in an area of interest is determined and the K_c values are known, then ET_a rates can be calculated.

There are two generally recognized reference surfaces; short crop and long crop. The short crop is a hypothetical reference crop with an approximate height of 0.12 m (4.72 in.), a fixed surface resistance of 70 s m^{-1}, and an albedo of 0.23 (Allen et al. 1998). The short crop is similar to an extensive surface of clipped green grass, actively growing, completely shading the ground, and with adequate water (Allen et al. 1998; Task Committee on Standardization of Reference Evapotranspiration 2005). The tall crop has an approximate height of 0.5 m (1.6 ft) and is similar to full cover alfalfa. Reference ET has largely superseded potential ET because the latter is imprecisely defined. Penman (1948) defined PET in terms of a "short green crop", which could correspond to many types of horticultural and agronomic crops (Irmak and Haman 2003). Reference ET is very widely used in agriculture, but the term potential ET is still often used in the water resources literature, especially when assessing full landscape ET losses.

Allen et al. (1998) and the American Society of Civil Engineers' Task Committee on Standardization of Reference Evapotranspiration (2005) have become the standard guides for evaluating ET especially with respect to agriculture. ET is typically measured in terms of length of the water column per unit time (e.g., m/d). ET can also be evaluated through the latent heat of vaporization (λ), which is the energy of heat required to vaporize water. Latent heat of vaporization is a function of temperature. ET rates in units of length/time units and latent heat fluxes (MJ m^{-2} day^{-1} or Wm^{-2} day^{-1}; MJ ft^{-2} day^{-1} or W ft^{-2} day^{-1}) are convertible.

Reference ET rate (ET_o) is a climatic parameter that expresses the evaporative power of the atmosphere, and is independent of biological and non-weather related variables such as crop type, crop development, crop roughness, management practices, ground cover, plant density, and soil moisture (Allen et al. 1998). The above factors all impact the actual ET rates. As a climatic parameter, ET_o can be computed from weather data, with the recommended standard procedure being the FAO Penman–Monteith method (Allen et al. 1998).

7.6 Measurement of Actual ET

Actual evapotranspiration rate is a critical variable for evaluations of local water budgets, but is also a very difficult parameter to accurately determine and requires expensive equipment and well-trained personnel. Commonly used methods and their underlying concepts are herein discussed, but it must be strongly emphasized that the techniques require specialized equipment and expertise. Determination of actual and reference ET rates from meteorological data is a specialized scientific subdiscipline that has its own extensive literature. A meteorologist trained in the

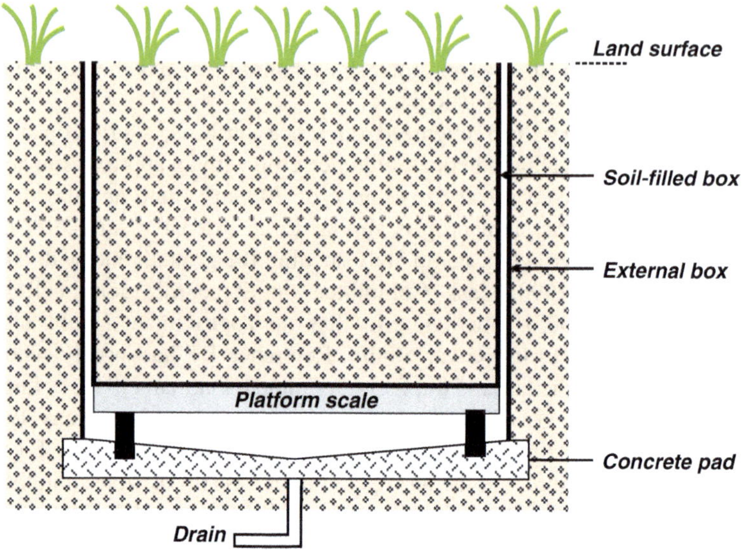

Fig. 7.5 Schematic diagram of a weighing lysimeter. The soil-filled sample box may also be equipped with tensiometers and a neutron probe access tube (After Gee and Jones 1985)

applicable techniques to measure actual or reference ET needs to be on the project team if meteorological based estimates of ET are contemplated.

7.6.1 Lysimeters

ET rates can be measured directly using lysimeters, which can also be used to measure recharge rates (Sect. 11.2.1). A lysimeter essentially consists of cylinder or box of soil and vegetation that is installed flush to land surface (Fig. 7.5). Weighing lysimeters are constructed on balances or other types of scales that allow for the measurement of slight changes in the mass of the soil column. Water losses from evapotranspiration can be estimated from the decrease in the mass of the soil column, corrected for any drainage and precipitation that occurred during the measurement period. ET_a is calculated in units of length by dividing the corrected water loss by the surface area of the lysimeter.

Lysimeters have the great advantage in that they can provide a direct measurement of ET_a. However, lysimeters are expensive to construct and operate, and provide only a point measurement of ET. Lysimeters are also physically limited as to the types of vegetative covers that the can be tested, and work best for grasses and small plant covers. Lysimeters are not effective for measuring the ET rates of forested areas because mature trees cannot normally be grown within a lysimeter.

7.6.2 Energy Balance and Micrometeorological Methods

ET_a rates can be estimated from latent heat fluxes, which are evaluated using energy balance and micrometeorological methods. The basic underlying principle is the conservation of energy. The energy arriving at an evaporating surface must be balanced by energy leaving the surface for the same time period (Allen et al. 1998). The conservation of energy equation for a system in which only vertical fluxes are considered is (Allen et al. 1998; Payero et al. 2003):

$$R_n - G - L - H = 0 \tag{7.6}$$

$$R_n = G + L + H \tag{7.7}$$

where,
R_n = net radiation to the surface ($Jm^{-2}s^{-1}$ or Wm^{-2})
G = soil heat flux ($Jm^{-2}s^{-1}$ or Wm^{-2})
H = sensible heat flux (i.e., heat transferred to the air during a change of temperature that is not accompanied by a change of state ($Jm^{-2}s^{-1}$ or Wm^{-2})
L = latent heat flux from the surface ($Jm^{-2}s^{-1}$ or Wm^{-2}).

7.6.3 Energy Balance Bowen Ratio Method

If the values of Rn, G, and H can be measured, then the latent heat flux, and in turn, ET_a rate can be calculated. The Energy Balance Bowen Ratio (EBBR) method is widely used to measure latent heat flux and thus actual ET rates. The Bowen Ratio (β) relates sensible to latent heat flux as follows:

$$\beta = \frac{H}{L} = \gamma\left(\frac{K_h}{K_w}\right)\left(\frac{\frac{\partial T}{\partial z}}{\frac{\partial e}{\partial z}}\right) \tag{7.8}$$

where,
K_h = eddy diffusivity for heat (m^2s^{-1})
K_W = eddy diffusivity for water vapor (m^2s^{-1})
T = temperature (°C)
Z = elevation above surface (m)
e = vapor pressure (kPa)
γ = psychrometric constant ($kP°C^{-1}$).

The pyschrometric constant is defined as

$$\gamma = \frac{C_p P}{\lambda_v \varepsilon} \tag{7.9}$$

where,
C_p = specific heat of air at constant pressure ($Jkg^{-1}°C^{-1}$)

P = atmospheric pressure (kPa)
λ_v = latent heat of water vaporization (Wm^{-2})
ε = ratio of the molecular weight of water vapor to dry air ($\approx$ 0.622).

The eddy diffusivities for heat and water vapor are approximately equal. For two measurements at different vertical elevations, Eq. (7.8) results in:

$$\beta = \gamma\left(\frac{\Delta T}{\Delta e}\right) \tag{7.10}$$

Equations 7.6 and 7.8 rearrange to:

$$L = \frac{R_n - G}{1 + \beta} \tag{7.11}$$

In order to estimate the actual ET rate using the EBBR method, the following parameters need to be measured: difference in temperature (ΔT) and vapor pressure (Δe) between two elevations, atmospheric pressure, net radiation to the surface (R_n), and soil heat flux (G).

EBBR instrumentation is commercially available, but it must be stressed that trained personnel are required for its operation and more importantly for the interpretation of the data. Published comparisons of EBBR and lysimeter data indicated differences of less than 10% between the two methods (Prueger et al. 1997).

Fritschen (1965) tested the validity of Bowen ratio method in hot arid lands. ET measurements were performed in an irrigated valley near Phoenix, Arizona using lysimeters and the Bowen ratio method. The measured ET rates from both methods were in agreement (±5%) during daylight hours. There was rather poor agreement during the nighttime hours, which were not material to the overall ET rates as nocturnal ET is relatively unimportant under most conditions. Prueger et al. (1997) reported that in a comparative study of the EBBR and lysimeter techniques in the semiarid northern Great Plains of the United States, the measured ET rates were similar. The lysimeter measurements yielded greater ET rates following large rainfall events. This problem was attributed to inadequate drainage from the lysimeter. The results of the comparative studies confirm that the EBBR method can provide good estimations of ET over large areas, which is not practically possible with lysimeters (Prueger et al. 1997).

7.6.4 Eddy Covariance Method

The eddy covariance method is a statistical micrometeorological technique for measuring actual ET rates based on measurements of the vertical component of turbulence or eddies of air flow and the relationship (covariance) of wind velocity with scalar variables such as humidity, gas concentration (e.g., carbon dioxide) or temperature (Tanner and Greene 1989). The underlying principle is that the evaporation rate can be calculated by measurements of fast fluctuations in the vertical wind speed and water vapor. Conceptually, the vertical components of wind eddy transport pulses of water vapor upwards or downwards with the sum of the pulses (positive for upwards, negative for downwards) corresponding to the actual ET rate. The upward components of eddies tend to have higher water vapor content from evapotranspiration at land surface or in the plant canopy than the downward components, so the net sum is positive. The eddy covariance method thus provides a direct estimate of the transfer of water vapor from land surface or plant canopy to the overlying atmosphere.

The eddy covariance method is most commonly performed using stationary towers that are equipped at a minimum with a sonic anemometer (i.e., instrument to measure wind speed), infrared gas analyzer and temperature and humidity probes. The anemometer and infrared gas analyzer perform the fast measurements of wind speed and humidity (water vapor). All of the instruments are connected to a data logger. Eddy covariance stations are often equipped with other meteorological instruments such as a pyranometer or radiometer (i.e., devices used measure solar radiation). Eddy covariance stations are commercially available

7.6.5 Geophysical Methods

Surface geophysical techniques are increasingly being used by agricultural researchers and farm managers to assess soil moisture gradients and, in turn, ET_a loss. Measurements of changes in the electrical permittivity of soils are used to estimate associated changes in the soil water content, and in turn ET_a The geophysical techniques used to measure soil moisture include impedance probes (Bell 1987; Bell et al. 1987), time domain reflectometry (TDR) (Gregory et al. 1995; Robinson et al. 2003; Timlin et al. 2007; Graeff et al. 2010), ground-penetrating radar (GPR) (Huismann et al. 2003), and active microwave remote sensing.

Perhaps the two techniques showing the most promise for large-scale spatial and temporal assessment of evapotranspiration determination are TDR and GPR. A detailed review of TDR was made by Robinson et al. (2003). TDR has been demonstrated to allow accurate measurements of changes in soil moisture and thus ET. Comparison with micro-lysimeter data showed that the lysimeter measurements produced results accurate to ±0.5 mm day (±0.02 in/day), whereas the TDR measurements showed accuracies to ±1.3 mm per day (±0.05 in/day) (Plauborg 1995). Soil moisture profiles have been monitored in arid regions using the TDR method (Laurent et al. 2005). Considerable recent research using GPR shows that this technique can be used to accurately measure soil moisture, and thus ET losses, and can produce results comparable to TDR estimates (Huisman et al. 2003). The GPR technique is being modified to allow measurements to be made of both soil hydraulic properties and moisture content, which will considerably increase its utility (Jadoon et al. 2008; Lambot et al. 2009; Jadoon et al. 2010, 2011).

As a general limitation, TDR and GPR techniques can only be applied to soils containing freshwater. The presence of elevated salinities in soil waters affects the electrical properties of the soils, and makes interpretations of the data difficult or impossible. These techniques are further discussed in Chap. 10.

7.7 Measurements of Reference ET

A variety of methods have been developed for the measurement of reference ET rates (ET_o). The accuracy of the methods may vary between locations and climatic conditions. Calibration to local climatic conditions may be necessary depending on the use of the data and the accuracy required. It behooves any project team in which ET_o measurements are considered to include a specialist who is intimately familiar with the different methods, the requirements for data collection, and the advantages and limitations of the different methods under project site conditions. The following methods are commonly used to determine ET_o rates and are summarized herein:

- Modified Penman–Monteith,
- Priestly-Taylor Method,
- Pan evaporation, and
- Blaney–Criddle.

Allen et al. (1998) provide a thorough discussion of the theory and methods for calculated ET_o rates using the Penman–Monteith and pan evaporation methods.

7.7.1 *Modified Penman–Monteith Method*

Modified Penman–Monteith method and derivations thereof are meteorological techniques for computing evaporation from vegetated surfaces (Penman 1948; Monteith 1965). The Penman–Monteith method allows calculation of ET_o, expressed as mm/day as follows (e.g., Allen et al. 1998):

$$ET_o = \left(\frac{1}{\lambda}\right) \frac{\Delta(R_n - G) + \frac{86{,}400 \rho_a C_P (e_s^o - e_a)}{r_a}}{\Delta + \gamma\left(1 + \frac{r_s}{r_{av}}\right)} \tag{7.12}$$

where,
λ = latent heat of vaporization (MJ kg^{-1})
Δ = slope of the saturation vapor pressure–temperature relationship (kPa $°C^{-1}$)
R_n = net radiation (MJ m^{-2} day^{-1})
G = soil heat flux (MJ m^{-2} day^{-1})
ρ_a = air density (kg m^{-3})

C_p = specific heat of dry air (MJ $kg^{-1}{}^{\circ}C^{-1}$)
$(e_s^o - e_a)$ = vapor pressure deficit of the air (saturated minus actual vapor pressure; kPa)
γ = psychrometric constant (kPa $^{\circ}C^{-1}$)
r_c = canopy (surface) resistance (s m^{-1})
r_a = aerodynamic resistance (s m^{-1}).

Canopy and aerodynamic resistances are functions of surface roughness, leaf area, and stomatal resistance and are thus dependent on factors such as crop type, height, and degree of land cover.

Allen et al. (1998) proposed a FAO Penman–Monteith method as the sole standard method for calculation of reference ET, as it shows relatively accurate and consistent performance under both arid and humid conditions. The FAO Penman–Monteith equation for daily ET (mm/day) from a short (0.12 m) crop reference is as follows (Allen et al. 1998):

$$ET_o = \frac{0.408\Delta(R_n - G) + \gamma \frac{900}{T+273} u_2(e_s - e_a)}{\Delta + \gamma(1 + 0.34u_2)} \tag{7.13}$$

where,
T = mean daily temperature at 2 m of height (°C)
u_2 = wind speed at 2 m of height (m s^{-1}).

The weather measurements should be made at 2 m (6.6 ft) above an extensive surface of green grass that completely shades the ground, and is not short of water (Allen et al. 1998). The American Society of Civil Engineers (Task Committee on Standardization of Reference Evapotranspiration 2005) adopted a more generalized version of Eq. 7.13, which also allows for the calculation of hourly ET_o rates.

From the these equations it can be seen that ET_o rates are calculated from meteorological parameters that provide energy for the vaporization of water and the control the movement of water vapor from the evaporating surface. Key parameters are solar radiation, air temperature, air humidity, and wind speed. The former two parameters control the energy inputs required for vaporization. Air humidity is critical variable because that rate of evaporation decreases as air approaches saturation with respect to water vapor. Thus, the rate of evaporation decreases as humidity increases at a given net radiation flux and air temperature. Wind speed and air turbulence control the transport of water vapor from evaporating surfaces. Under low wind and turbulence conditions, the water vapor concentration in the air above an evaporating surface increases and thus the evaporation rate decreases. Air turbulence replaces the water vapor-rich area above the evaporating surface with drier air, which allows for higher evaporation rates. The persistence of a high-humidity layer above a free water surface tends to significantly reduce diffusive loss of water to the atmosphere. The turbulent action of wind breaks up this layer and greatly increases the diffusive water loss.

The methods for determining the required meteorological parameters were reviewed by Allen et al. (1998) and Task Committee on Standardization of Reference Evapotranspiration (2005). Some data can be measured directly, while other values can be derived from empirical relationships. ET_o rates can be estimated from just temperature (minimum and maximum) and solar radiation data using empirical relationships for the other variables. Meteorological data have traditionally been measured from on-site weather stations. There are increasing efforts now to use remote sensing data, such as from satellites, to obtain areally extensive measurements of ET_o rates (Chap. 18).

Missing climatic data can be estimated using regional data or from other parameters. The soil heat flux, for example, is typically much less than the net solar radiation and may be ignored if no data are available. Values estimated from stations with missing data should be compared against ET_o values calculated from stations with full data sets and a sensitivity analysis should be performed to determine the potential errors associated with assumed or estimated data (Allen et al. 1998).

7.7.2 Priestley–Taylor Method

The Priestley-Taylor method (Priestley and Taylor 1972) is essentially the energy part of the Penman–Monteith equation in which the aerodynamic terms are replaced by a constant "α" referred to as the Priestley-Taylor coefficient. Reference ET (ET_o) in mm/d is calculated as follows:

$$ET_o = \frac{1}{\lambda}\frac{\Delta(R_n - G)}{\Delta + \gamma}\alpha \tag{7.14}$$

where,
λ = latent heat of vaporization (MJ/kg)
γ = psychromatic constant (kPa °C^{-1}).

The Priestley–Taylor method is useful for calculating ET_o rates where data are not available for aerodynamic variables such as relative humidity and wind speed. The key variables are net radiation and temperature as the soil heat flux is commonly insignificant. Temperature data are necessary to calculate the slope of the saturation vapor pressure–temperature relationship (Δ). The Priestley–Taylor method is suitable for remote sensing investigations as R_n and T can be estimated from satellite data. The Priestley–Taylor coefficient has a value of approximately 1.26 for free water and moist vegetated surfaces (Priestley and Taylor 1972). Higher values up to 1.74 are recommended for arid regions (Jensen et al. 1990).

7.7.3 Blaney–Criddle Method

The Blaney–Criddle Method (Blaney and Criddle 1950; Blaney 1959) was developed as a means of estimating monthly crop consumptive use (evapotranspiration) from monthly temperature data. The original Blaney–Criddle formula is as follows:

$$ET_o = KTP/100 \tag{7.15}$$

where,
ET_o = consumptive use or evapotranspiration (inches/month)
K = empirical coefficient for crop and for month
T = mean monthly temperature (°F)
P = percentage of the daylight hours of the year occurring during a given month.

The Blaney–Criddle can alternatively be expressed in SI units as

$$ET_o = K(32 + 1.8T)P/3.94 \tag{7.16}$$

where,
ET_o = consumptive use or evapotranspiration (mm/month)
T = mean monthly temperature (°C).

Several modifications of the Blaney–Criddle have been developed. The FAO modified Blaney–Criddle formula developed by Doorenbos and Pruit (1977) incorporates relative humidity and wind speed:

$$ET_o = a + b(P(0.46T + 8)) \tag{7.17}$$

where,
ET_o = consumptive use or evapotranspiration (mm/month)
T = mean monthly temperature (°C)
a, b = adjustment factors for minimum relative humidity, daylight duration, and wind speed.

The Blaney–Criddle Method is thus simple to use because it is computationally simple, uses relatively accessible climate data (temperature), and a large volume of data on crop coefficient values are available. The Blaney–Criddle method has a relatively low accuracy and provides only a rough or "order of magnitude" estimate, and is particularly inaccurate under extreme conditions (Brouwer and Heibloem 1986). The method is used primarily for estimating agricultural irrigation requirements rather than the ET_o of native land covers.

7.7.4 Pan Evaporation Measurements

The measurement of the actual evaporation rates in pans is very commonly used to estimate evaporation rates in surface-water bodies (e.g., lakes) and reference evapotranspiration rates. Measured evaporation rates are a function of pan design, so standardized pan designs are typically used to allow for comparisons between sites. The most commonly used pan design in the United States is the Class A evaporation pan (Fig. 7.6), which is a steel cylinder with a diameter of 47.5 in. (120.7 cm) and a depth of 10 in. (25 cm). The pans are placed in open ground upon a level slatted wooden platform. Class A evaporation pans are commercially available.

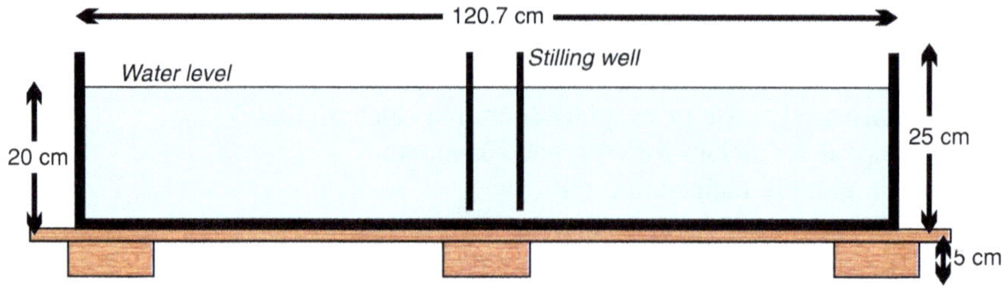

Fig. 7.6 Diagram of a Class A evaporation pan

An alternative pan design is the Colorado sunken pan, which is 92 cm (3 ft) square pan that is 46 cm (18 in.) deep. The Colorado sunken pan is placed in the ground with the rim 5 cm (2 in.) above soil level (Allen et al. 1998).

Evaporation is typically measured every 24 h by either the change in water level or the amount of water that needs to be added to bring the water up to a reference level. The change in water level can be impacted by extraneous factors. Precipitation will obviously impact pan water levels and must be recorded. Animals (birds in particular) drinking from and splashing the water can also impact readings so the pans are usually placed inside chain-link or wire mesh cages. Meteorological data such as temperatures (maximum and minimum), humidity, and wind speed are commonly also recorded as evaporation pan stations, as these data can assist in the interpretation of the data.

Class A pan evaporation stations are operated by state, local, and national weather and water management agencies, so in many locations current and historical data may be available. In order to estimate (ET_o) rates, pan evaporation rates must be multiplied by a pan coefficient (K_p),

$$ET_o = K_p(ET_{pan}) \tag{7.18}$$

Pan coefficients are a function of a number of variables, particularly local meteorological conditions (e.g., Synder 1992; Synder et al. 2005). The metal construction of the pans results in a greater heating of contained water and thus higher evaporation rates. The storage of heat in pans can cause significant evaporation at night, whereas most crops have significant transpiration only during the daytime (Allen et al. 1998). Station maintenance and operator skill and attention can also impact measurements. Pan evaporation rates, and thus pan coefficients, may change over time because of aging, deterioration, and repainting of the pan systems (Allen et al. 1998). It is recommended that ET_{pan} values be calibrated against ET_o values computed with the Penman–Monteith method (Allen et al. 1998).

Pan evaporation rates (ET_{pan}) are not directly equivalent to actual surface-water evaporation rates. Evaporation rates from surface-water bodies depend on local conditions, including size (area and shape), depth, vegetation, wind fetch, wind speed, and albedo, which vary between surface-water bodies. Local meteorological conditions at an isolated pan station on land will often be different from conditions at the surface of a lake. A pan evaporation coefficient must be applied to the pan evaporation rates to obtain an estimate of lake evaporation rates. Values in the range of 0.7–0.8 are commonly used to estimate free surface-water evaporation rates.

References

Abdullah, M. A., & Al-Mazroui, M. A. (1998). Climatological study of the southwestern region of Saudi Arabia. I. Rainfall analysis. *Climate Research, 9*, 213–233.

Agnew, C., & Anderson, W. (1992). *Water in the arid realm* (329 pp.). London: Routledge.

Allen, R. G., Pereira, L. S., Raes, D., & Smith, M. (1998). *Crop evapotranspiration—Guidelines for computing crop water requirements*. Rome, Italy, FAO Irrigation and Drainage Paper 56, Food and Agricultural Organization of the United Nations.

Almazroui, M. (2011a). Calibration of TRMM rainfall climatology over Saudi Arabia during 1998–2009. *Atmospheric Research, 99*(3–4):155–165.

Almazroui, M. (2011b). Sensitivity of a regional climate model on the simulation of high intensity rainfall events over the Arabian Peninsula and around Jeddah (Saudi Arabia). *Theoretical and Applied Climatology, 104*(1), 261–276.

Bell, J. P. (1987). *Neutron probe practice* (3rd ed.). Wallingford, UK: Institute of Hydrology Report 19: Institute of Hydrology.

Bell, J. P., Dean, T. J., & Hodnett, M. G. (1987). Soil moisture measurement by an improved capacitance technique: II. Field techniques, evaluation, and calibration. *Journal of Hydrology (Amsterdam), 93*, 79–90.

Blaney, H. F. (1959). Monthly consumptive use requirements for irrigated crops. *Journal of the Irrigation and Drainage Division, 85*(1), 1–12.

Blaney, H.F., & Criddle, W.D. (1950). Determining water requirements in irrigated areas fromclimatological and irrigation data: U.S. Department of Agriculture, Soil Conservation Survey, SCS-TP 96,(p. 44).

Brouwer, C, & Heibloem, M. (1986). *Irrigation water management: Irrigation water needs*. United Nations Food and Agriculture Organization (FAO) Irrigation Water Management, Training Manual No. 3.

Chow V. T., Maidment D. R., & Mays L.W. (1988). *Applied hydrology*. New York: McGraw- Hill.

Critchley, W., & Siegert, K. (1991). *Manual for the design and construction of water harvesting schemes for plant production*. Rome: Food and Agriculture Organization of the United Nations

Doorenbos, J., & Pruitt, W. O. (1977). *Guidelines for predicting crop water requirements, FAO Irrigation and Drainage Paper 24* (2nd ed.). Rome: Food and Agriculture Organization of the United Nations, Rome.

Fritschen, L. J. (1965). Accuracy of evapotranspiration determinations by the Bowen ratio method. *Bulletin International Association of Scientific Hydrology, 10*, 38–48.

Gebremichael, M., Anagnostou, E. N., & Bitew, M. M. (2010). Critical steps for continuing advancement of satellite rainfall applications for surface hydrology in the Nile River Basin. *Journal American Water Resources Association, 46*(2), 360–366.

Gee, G. W., & Jones, T. L. (1985). *Lysimeters at the Hanford site: present use and future needs*. Report prepared for the U.S. Department of Energy, Pacific Northwest Laboratory, Richland, WA.

Graeff, T., Zehe, E., Schlaeger, S., Morgner, M., Bauer, A., Becker, R., et al. (2010). *Hydrology and Earth System Sciences Discussions, 7*, 269–311.

Gregory, P. J., Poss, R., & Micin, S. (1995). Use of time-domain reflectometry (TDR) to measure the water-content of sandy soils. *Australian Journal of Soil Research, 33*(2), 265–275.

Hardegree, S. P., Van Vactor, S. S., Levinson, D. H., & Winstall, A. H. (2008). Evaluation of NEXRAD radar precipitation products for natural resource applications. *Rangeland Ecology & Managament, 61*(3), 346–353.
Hazen, A. (1914). Storage to be provided in impounding reservoirs for municipal water supply. *Transactions American Society of Civil Engineers, 77*, 1539–1640.
Heathcote, R. L. (1983). *The arid lands: Their use and abuse*. London: Longman.
Hong, Y., Hsu, K. L., Sorooshian, S., & Gao, X. (2004). Precipitation estimation from remotely sensed imagery using an artificial neural network cloud classification system. *Journal of Applied Meteorology, 43*, 1834–1852.
Huisman, J. A., Hubbard, S. S., Redman, J. D., & Annan, A. P. (2003). Measuring soil water content with ground-penetrating radar: A review. *Vadose Zone Journal, 2*, 476–491.
Hunter, S. M. (1996). WSD-88D rainfall estimation: Capabilities, limitations and potential improvements. *National Weather Digest, 20*, 26–38.
Irmak, S., & Haman, D. Z. (2003). *Evapotranspiration: potential or reference?* Gainesville, FL: University of Florida Institute of Food and Agricultural Sciences Document ABE 343.
Jadoon, K. Z., Slob, E. C., Vanclooster, M., Vereecken, H., & Lambot, S. (2008). Uniqueness and stability analysis of hydrophysical inversion for time-lapse ground-penetrating radar estimates of shallow soil hydraulic properties. *Water Resources Research, 44*, W09 421, 639 doi:10.1029/2007WR006.
Jadoon, K. Z., Lambot, S., Scharnagl, B., van der Klug, J., Slob, E., & Vereechen, H. (2010). Quantifying field scale soil hydrogeophysical properties using full-waveform inversion of proximal zero-offset ground-penetrating radar. *Near Surface Geophysics, 8*(6), 483–491.
Jadoon, K. Z., Slob, E., Vereecken, H., & Lambot, S. (2011). Analysis of antenna transfer functions and phase center position for modeling off-ground GPR. *IEEE Transactions on Geosciences and Remote Sensing, 48*(5), 1949–1662.
Jensen, M. E., Burman, R. D., & Allen, R. G. (1990). *Evaporation and irrigation water requirement*. American Society of Civil Engineers Manuals and Reports on Engineering Practice No. 70.
Joyce, R. J., Janowiak, J. E., Arkin, P. A., & Xie, P. (2004). CMORPH: A method that produces global precipitation estimates from passive microwave and infrared data at high spatial and temporal resolution. *Journal of Hydrometeorology, 5*, 487–503
Lambot, S., Rhebergen, J., Slob, E., Lopera, O., Jadoon, K. Z., & Vereecken, H. (2009). Remote estimation of the hydraulic properties of a sandy soil using full-waveform integrated hydrogeophysical inversion of time-lapse, off-ground GPR data. *Vadose Zone Journal, 8*(3), 743–754.
Laurent, J.-P., Ruelle, P., Delage, L., Zairi, A., Nouna, B. B., & Adjmi, T. (2005). Monitoring soil water content profiles with a commercial TDR system: Comparison field tests and laboratory calibration. *Vadose Zone Journal, 4*, 1030–1036.
Makkonen, L. (2006). Plotting positions in extreme value analysis. *Journal of Applied Meteorology and Climatology, 45*, 334–340.
McGinnies, W. G. (1979). Arid land ecosystems, common features throughout the world. In D. W. Goodall, R. A. Perry, & K. M. W. Howes (Eds.), *Arid land ecosystems: Structure, functioning and management* (pp. 299–318). Cambridge: Cambridge University Press.
Monteith, J. L. (1965). Evaporation and environment. In Fogg, G. E. (Ed.), *Symposium of the society for experimental biology, The state and movement of water in living organisms* (Vol. 19, pp. 205–234). New York: Academic Press.
Payero, J. O., Neale, C. M. U., Wright, J. L., & Allen, R. G. (2003). Guidelines for validating Bowen ratio data. *Transactions of the American Society of Agricultural Engineers, 46*(4), 1051–1060.
Penman, H. (1948). Natural evaporation from open water, bare soil and grass. *Proceedings of the Royal Society, A193*, 120–146.
Pereira Filho, A. J., Carbone, R. E., Janowiak, J. E., Arkin, P., Joyce, R., Hallak, R. et al. (2010) Satellite rainfall estimates over South America—possible applicability to the water management of large watersheds. *Journal American Water Resources Association, 46*(2), 344–360.

Plauborg, F. (1995). Evaporation from bare soil in a temperate humid climate-measurement using micro-lysimeters and time domain reflectometry. *Agricultural and Forestry Meteorology, 76*, 1–17.

Priestley, C. H. B., & Taylor, R. J. (1972). On the assessment of surface heat flux and evaporation using large-scale parameters. *Monthly Weather Review, 100*(2), 81–92.

Prueger, J. H., Hatfield, J. L., Aase, J. K., & Pikul, J. L., Jr. (1997). Bowen-ratio comparsions with lysimeter evapotranspiration. *Agronomy Journal, 84*, 730–736.

Robinson, D. A., Jones, S. B., Wraith, J. M., Or, D., & Friedman, S. P. (2003). Vadose Zone Journal. *A review of advances in dielectric and electrical conductivity measurement in soils using time domain reflectometry, 2*, 444–475.

Şen, Z. (2008). *Wadi hydrology*. Boca Raton, FL: CRC Press.

Skinner, C., Bloetscher, F., & Pathak, C. S. (2009). Comparison of NEXRAD and rain gauge precipitation measurements in South Florida. *Journal of Hydrologic Engineering, 14*(3), 248–260.

Sokol, R. R., & Rohlf, F. J. (1981). Biometry (2nd ed.). New York: W.H. Freeman.

Synder, R. L. (1992). Equation for evaporation pan to evapotranspiration conversions. *Journal of Irrigation and Drainage Engineering, 118*, 977–980.

Synder, R. L., Orang, M., Matyac, S., & Grisner, M. E. (2005). Simplified estimation of reference evapotranspiration from pan evaporation data in California. *Journal of Irrigation and Drainage Engineering, 131*, 249–253.

Tanner, B. D., & Greene, J. P. (1989). Measurements of sensible heat and water vapor fluxes using eddy correlation methods. Final report prepared for the U.S. Army Dugway Proving Grounds, Dugway, Utah.

Task Committee on Standardization of Reference Evapotranspiration. (2005).*The ASCE standardized reference evapotranspiration equation*. Reston, VA: Environmental and Water Resources Institute, American Society of Civil Engineers.

Tilahun, K. (2006). The characterization of rainfall in the arid and semi-arid regions of Ethiopia. *Water SA, 32*(3), 429–436.

Timlin, D., Fleisher, D., Kim, S.-H., Reddy, V., & Baker, J. (2007). Evaporation measurement in controlled environment chambers: A comparison between time domain reflectometry and accumulation of condensate from cooling coils. *Agronomy Journal, 99*, 166–173.

Walton, K. (1969). *The arid zone*. Chicago, IL: Aldine Publishing Co.

Weibull, W. (1939). A statistical theory of the strength of materials. *Ingenoirs Vetenskaps Akadanien Handlinger, 151*, 1–47.

Wheater, H. S. (2002). Hydrological processes in arid and semi arid area. In Wheater, H., & Al-Weshah, R. A. (Eds.), *Hydrology of wadi systems, IHP-V, Technical documents in hydrology, no. 55* (pp. 5–22). Paris: UNESCO.

Chapter 8
Recharge Concepts and Settings

8.1 Introduction

Groundwater recharge is the process by which water moves from the Earth's surface into the groundwater system. It is the ultimate source of renewable groundwater and is thus a fundamentally critical parameter for water-resources investigations. A tremendous volume of technical literature on groundwater recharge has been published over the past 30 years, with numerous dedicated books on the subject (e.g., Simmers 1988, 1997; Sharma 1989; Lerner et al. 1990; Kinzelbach et al. 2002; Seiler and Gat 2007; Healy 2010). It is the intent in this chapter to provide an overview of the basic recharge concepts as applied to arid lands. Techniques available to quantify recharge rates in arid and semiarid environments are discussed in Chaps. 11 through 13. The focus of this chapter is on groundwater recharge on a long-term basis within a groundwater basin or catchment, as this ultimately controls the supply of groundwater available for use. It is recognized that small-scale variations in recharge rate may also be important for contaminant migration.

Aquifers may be recharged by both natural and anthropogenic water sources. The main natural recharge process begins with infiltration of precipitation and surface waters. Anthropogenic sources include irrigation losses from canals and farm fields (i.e., return flows, deep percolation), leaked water from water and wastewater pipes, wastewater discharges to the ground (e.g., septic systems and infiltration basins), and intentional recharge activities (managed aquifer recharge; Chaps. 23). Irrigation recharge is especially important in arid regions. The quantity of irrigation water that recharges the groundwater in arid areas is usually large relative to the natural recharge from precipitation because large irrigation systems are commonly located in regions of low precipitation and low natural recharge (Winter et al. 1998) and often derive their water from external sources.

Natural recharge in arid lands is limited by the amount of water that the climate provides and the rate at which water infiltrates into the soil and percolates

R. Maliva and T. Missimer, *Arid Lands Water Evaluation and Management*, Environmental Science and Engineering, DOI: 10.1007/978-3-642-29104-3_8,

downwards to the water table. In arid and semiarid regions, the rate of evapotranspiration typically exceeds the precipitation rate most of the time, therefore, limiting the potential for groundwater recharge. However, in some areas of basins, precipitation may exceed ET during parts of the year and recharge of surface-water runoff may occur. For example, Flint et al. (2004) determined that recharge is temporally and spatially restricted in the Great Basin of the United States and occurs within only 5% of the basin. Most of the local recharge and runoff occurs at or is generated from the mountainous boundaries of the basin, which experience greater precipitation and lower temperatures (Flint et al. 2004). This same pattern is common in semi-arid and arid lands of the Middle East, particularly in Lebanon, Syria, Morocco, and western Saudi Arabia, and in the Western United States.

Small, temporally isolated rainfall events result in little or no recharge because the shallowly infiltrated water is quickly lost to evapotranspiration. During minor rainfall events, virtually all of the water that is retained in the upper part of the vadose zone or shallowly ponded on impervious surfaces is lost to ET. Most groundwater recharge in arid and semi-arid regions occurs instead during infrequent wet periods. Natural climate variability of various time scales is a fundamental control on groundwater resources of some aquifers (Gurdak et al. 2007). Present-day climatic conditions may not be representative of the conditions under which most recharge into a given aquifer occurred. The combination of low recharge rates and great spatial and temporal variability results in groundwater recharge being a very difficult parameter to accurately estimate, especially in arid and semiarid lands. The difficulty in assessing recharge is further greatly complicated in many areas by a paucity of climatic, soil, and geologic data.

A clear distinction needs to be made, both conceptually and for modeling purposes, between the potential amount of water available for recharge from the soil zone and the amount of water that actually reaches the water table and replenishes the aquifer (Simmers 1998). A high percentage, and in some cases, all of the water that locally infiltrates into the shallow vadose zone (i.e., soil) may subsequently be lost to evapotranspiration (Wheater 2002).

Net infiltration is defined as the percolation flux that passes below the depth at which the rate of removal by evapotranspiration becomes insignificant (Flint et al. 2002). It thus consists of water that percolates downwards pass the root zone of plants. Net infiltration is a function of precipitation, air temperature, root zone depth and root density, soil hydraulic properties and depths, and bedrock permeability (Flint et al. 2004). Under steady state conditions, the net infiltration flux will be equal to the unconfined aquifer recharge rate, unless some of the infiltrating water discharges before reaching the water table. For example, the net infiltration could enter a perched aquifer that discharges at a spring or a seep (Flint et al. 2002).

Where the vadose zone is thick, heterogeneity can greatly impact local groundwater flow. Thin, areally extensive clay layers, for example, can result in the lateral spreading of percolating water (as opposed to downward flow) and the stranding of water in the vadose zone before it reaches that water table (Izbicki et al. 2007). Slope of the vadose zone also forces infiltrating water to take a longer pathway, within both vertical and horizontal components, to the water table. Thus, within arid and semiarid environments, the volume of water that infiltrates into the vadose zone may be substantially greater than the volume of water that eventually reaches the water table, particularly where the vadose zone is thick. The mechanics of unsaturated zone flow are discussed in Chap. 5.

8.2 Direct, Indirect, Matrix, and Macropore Recharge

There are two main types of recharge, direct and indirect. Direct recharge, also referred to as diffuse and distributed recharge, is water that enters the groundwater system by vertical infiltration and percolation at the site of precipitation or application. Indirect or focused recharge involves the runoff water from the area of precipitation and its infiltration and percolation to the water table at another location. Indirect recharge sites may be depressions into which runoff flows and ponds (e.g., dry or vadose playas; Wood and Sanford 1995), ephemeral stream channels, flood plains, or alluvial fans. Direct recharge is known to be of decreasing significance with increasing aridity (Simmers 1990, 1998; De Vries and Simmers 2002). Direct or diffuse recharge is also much slower than indirect recharge. For example, in a study of the High Plains of the Western United States, the landscape is dominated by diffuse (slow path recharge), whereas the bulk of the recharge occurs by focused (fast path) recharge in a much smaller fraction of the area (Gurdak et al. 2007). Dry playas may be an important recharge location in some arid and semiarid basins.

Recharge is also subdivided into matrix and macropore recharge. Matrix or interstitial recharge occurs through the intergranular pore spaces of the sediment or rock. Macropore recharge, on the contrary, occurs through zones of enhanced permeability. Macropores include features such as deep desiccation cracks, animal burrows, root tubes, pipes, fissures, and fractures (Stephens 1996; Wood et al. 1997; Wood 1999). The higher permeability of macropores results in much more rapid rates of recharge compared to matrix recharge. Rain water is more likely to escape evapotranspiration if it rapidly percolates through macropores.

Sediment or rock with both matrix porosity and macroporosity are referred to as dual-porosity systems. Multiple generations or types of macroporosity may be present and some aquifers have been documented as having triple porosity systems. Total recharge or the recharge flux is the sum of the matrix and macropore recharge. Recharge often occurs simultaneously through both matrix porosity and macroporosity. Water percolating through macropores may be completely absorbed by the surrounding matrix before it reaches the water table (Stephens 1996). Macropores may also become clogged over time, which can result in the spatial distribution of deep soil–water movement gradually changing over time (Stephens 1996).

8.3 Factors Controlling the Rate of Infiltration and Recharge in Arid Regions

Surface permeability is a critical variable in determining the rate at which precipitation infiltrates into the soil and whether or not it either runs-off or locally ponds. More permeable soils allow for the more rapid infiltration and downward percolation of rainwater. In order for a significant fraction of precipitation to recharge the groundwater, the infiltrating water must escape evapotranspiration by passing rapidly below land surface and through the root zone of vegetation and the zone of diffusive loss. Key issues are the

- composition and hydraulic properties of soils,
- intensity, duration, and magnitude of rainfall events,
- the presence and abundance of macropores present at land surface that are capable of accepting infiltration,
- the degree to which the macropores are connected to underlying aquifers,
- the condition of the land surface, including roughness, vegetative cover, and moisture content,
- the soil moisture gradient within the vadose zone, and
- the hydraulic conductivity of the aquifer.

Secondary macroporosity features (e.g., fractures and solution conduits) can be locally critical for groundwater recharge because they may allow for rapid downward movement of water to the underlying aquifer, which would not otherwise occur. In mountainous areas, a major source of recharge to adjoining basin-fill or alluvial aquifers is infiltration into fractured bedrock, which is essentially impermeable where not fractured. Materials within the unsaturated zone rarely display homogenous properties and often consist of beds of sediment or rock with widely varying saturated hydraulic conductivities and a strong potential for lateral rather than vertical flow, especially above lithological discontinuities (Simmers 1990). Vertical hydraulic conductivity, both unsaturated and saturated, can be reduced by caliche layers, which can greatly retard vertical water movement. A combination of minimal vegetation cover, permeable soils or fractured porous bedrock, and high-intensity rainfall events are most favorable for recharge in arid environments (De Vries and Simmers 2002).

Recharge rate often varies spatially within a groundwater basin with the highest rates occurring in coarser grained sediments near water sources, such as river and wadi channels and alluvial fans. Topographic depressions in which surface water accumulates are often predominant locations of infiltration and unsaturated flow in arid regions (Scanlon et al. 1997). Variation in infiltration capacity within a basin will control the location of recharge. Precipitation or snow melt in areas underlain by low permeability rock (e.g., crystalline bedrock) will largely run-off and may infiltrate into and recharge downstream areas underlain by fractured rock or coarse-grained alluvial deposits (Fig. 8.1).

Fig. 8.1 White Tank Mountains, near Phoenix, Arizona. Upper part of catchment consists mostly of crystalline rock that favors runoff over recharge (*top*). Recharge occurs mainly in downstream areas where wadis are underlain by coarse-grained alluvial deposits (*bottom*)

The rate of net infiltration and recharge depends mainly on climate, soil type, moisture gradient, and vegetation. Areas with little rainfall naturally have low recharge rates. Local recharge rates are dependent on local topography, and soil and aquifer characteristics. Factors that impact local recharge rates include (Simmers 1990; Stonestrom and Harril 2007):

- orientation of mountains with respect to prevailing storm tracks,
- scale, altitude, and local relief of mountains,
- exposure and thus amount of solar radiation and evapotranspiration,

- shading from mountains, which also affects ET,
- slope, which effects the rate of run-off,
- soil properties (porosity, hydraulic conductivity, thickness, heterogeneity, soil moisture),
- presence and characteristics of rivers,
- land use and land cover, and
- irrigation practices.

The difficulty of estimating recharge is compounded where land use and land cover has recently changed. Present-day recharge conditions may not be representative of past conditions. Longer-term variations in recharge rates must also be considered when evaluating recharge rates in arid regions. Natural flow regimes in deep unsaturated zones of arid interfluvial environments are rarely in hydraulic equilibrium with near-surface boundary conditions imposed by present day plant-soil–water-atmospheric dynamics (Walvoord et al. 2004). The very long travel time (sometimes decades or longer) from land surface to the deeper parts of the unsaturated zone results in deep vadose-zone water flow tending not to be closely tied to current near-surface conditions.

In areas with a deep water table (large unsaturated zone thickness), it is often assumed that all the water that passes through the root zone has escaped ET and could recharge the groundwater. However, it is important to recognize that some arid region vegetative types, for example saltcedar and other phreatophytes, have very deep roots and that infiltrated water can be lost to vapor transport and root collection at considerable deep below land surface (de Vries and Simmers 2002). The direction of groundwater flow in the vadose zone may also not be vertical. The presence of relatively impermeable layers within the vadose zone may result in some percolated water becoming perched and never reaching the water table. Whether or not infiltrating water will eventually contribute to recharge depends on the retention storage in the shallow vadose zone (root zone) and evapotranspiration (Tuinhof et al. 2003).

The storage capacity of soil, which is related to both porosity and thickness, is also an important factor controlling net infiltration and groundwater recharge, especially in areas subject to high ET rates. Shallow soil with a high storage capacity will tend to retain water near land surface, which will be subject to later loss through evapotranspiration. The more deeply water penetrates into the soil, the less likely it will be subsequently lost to ET (Flint and Flint 1995; Tuinhof et al. 2003). All other factors being equal, fractured rock and exposed karst surfaces with minimal storage capacities and high vertical hydraulic conductivities will have relatively high recharge rates because the percolating water is more likely to escape ET.

The results of an investigation at the proposed Yucca Mountain nuclear repository in Nevada indicate that in environments with high evaporative demands, the depth-of-water penetration is more important for producing a higher net infiltration (recharge) rate than the volume of water entering the soil profile (i.e., total infiltration) (Flint and Flint 1995). A key conclusion of the Yucca

Mountain study was that thinner soil covers with lesser storage capacities favors deeper infiltration, especially where surface runoff is concentrated in areas underlain by fractured bedrock (Flint and Flint 1995). Ridge tops had the deepest penetration of infiltrating water compared to alluvial channels and terraces. Channels can have deep penetration but the net infiltration was not volumetrically significant because events that generate runoff are rare and channels occupy a relative small area (Flint and Flint 1995). Recharge in the channels is also inhibited by the short duration of flooding (flash floods), the presence of high concentrations of mud in the flowing water, and temporary air entrapment within the upper part of the alluvial aquifer (Missimer et al. 2012; see Chap. 5).

The depth of penetration of potential recharge water is influenced by (Flint and Flint 1995):

- potential for surface storage (depth of soils, layering, impermeable layers—caliche, slope, and aspect),
- timing of precipitation (storm location and duration),
- presence of fractures,
- relative saturation of the wetting front when it reaches fractured bedrock, and
- existing moisture profile within the unsaturated zone.

Groundwater recharge rates in arid and semiarid regions are thus not necessarily proportional to annual rainfall. The timing and distribution of precipitation events are also important considerations in groundwater recharge. In order for groundwater recharge to occur, the water must pass through the vadose zone, and particularly the root zone of vegetation. Small precipitation events that are dispersed in time will be lost entirely to evapotranspiration or to wetting and moisture storage within the vadose zone. Precipitation events must be sufficient to allow for drainage through the full thickness of the vadose zone. U. S. Geological Survey (USGS) Basin Characterization Model simulation results for the interior southwestern United States suggest that snowmelt is the dominant factor in converting excess water into runoff and in-place groundwater recharge. Accumulation of snow delays the delivery of liquid water to the land surface, and thus increases the possibility that subsequent combinations of precipitation and snowmelt in ensuing months will exceed the soil–water storage capacity and result in drainage past the root zone (Flint and Flint 2007). These conditions also increase the possibility of runoff.

More precipitation at Yucca Mountain also infiltrated during the winter, when ET is low and runoff is rare, because of the occurrence of lower intensity storms and slowly melting snows on the ridgetops for several weeks each winter (Flint and Flint 1995). Large volumes of water penetrated below the root zone and avoided the high ET demands of the following summer. Oxygen isotope data from the San Pedro Basin, Arizona, indicate that mountain block recharge occurs mostly in the winter (Wahi et al. 2008). The summer monsoonal rain produces intense events in which most of the water runs off or is lost to ET. Deuterium and oxygen isotopes and hydrograph data from Spring Mountain, southern Nevada, also indicate that late spring snowmelt is the primary source of recharge (as determined from spring flow) and that summer, short-duration convective storms contribute little to the total annual recharge (Winograd et al. 1998).

Significant recharge may occur only once every one or more decades (Llamas 2004). The infrequency and irregularity of major recharge events are major complexities to recharge estimation. Climatic variations on decadal and century time scales affect precipitation and, in turn, surface-water flow and groundwater recharge. The El Niño Southern Oscillation (ENSO) and Pacific Decadal Oscillation (PDO) have large impacts on regional precipitation and groundwater recharge. Groundwater recharge in ephemeral streams (wadis) in southeastern Arizona, USA, varies by about a factor of three between El Niño and La Niña years (Pool 2005). The increased recharge appears to occur only in channels, which experience increased flows during wet periods. The effect is not observed in inter-channel areas because the increased precipitation results in a corresponding increase in vegetation productivity. Considerable difficulty may occur in recharge estimation if long-term values are required while only short-term data are available, particularly in areas of high or increasing groundwater exploitation (Simmers 1990).

8.3.1 Effects of Vegetation and Land Use/Land Cover

Vegetation cover can have a profound effect on recharge rates. Deep-rooted vegetation can result in dramatically lower recharge rates compared to ground covered with shallow-rooted vegetation and bare ground (e.g., Gee et al. 1992). Plants utilize soil moisture from precipitation and return it to the atmosphere via transpiration. Rainwater must pass through the root zone before it can recharge the groundwater system. Vegetation adapted to water scarce conditions (i.e., xerophytes) is highly adapted to extracting soil moisture (Stonestrom and Harril 2007). Both field data and modeling results have demonstrated that a change in land cover from deep-rooted native xerophytic vegetation (trees and shrubs) to shallower-rooted agricultural crops and landscaping may actual increase groundwater recharge rates as the latter are less efficient in extracting soil moisture (Keese et al. 2005; Scanlon et al. 2005). Recharge rates have increased by one or two orders of magnitude in some areas with a natural or man-caused change in vegetative type (Scanlon et al. 2006).

The impacts of land use changes on water resources depend on numerous factors including the original vegetation to be replaced, the vegetation that is replacing it, whether the change is permanent or temporary, and associated land management practices involving alteration of drainage (Scanlon et al. 2007). Natural forests have greater ET rates than other types of vegetation, and the reduced ET in cultivated areas from converted forests provides more water for groundwater recharge and streamflow. The latter can result in increased stream baseflow. The conversion of native vegetation to rain-fed agriculture generally increases water quantity but decreases water quality (Scanlon et al. 2007). Water quality effects include the mobilization into the groundwater of natural salts and nutrients (e.g., nitrates) that have accumulated in the vadose zone, and soil salinization caused by the raising of the water table to close to land surface (Scanlon et al. 2007). Proposed or implemented forestation (reforestation) projects could reduce groundwater recharge, decrease runoff, and decrease stream sediment loads, the effects of which need to be considered in water resources management (Scanlon et al. 2007).

In the Amargosa Desert Basin (Mojave Desert, South Nevada, East California) the gradient of temperature, water (matric) potential (Sect. 5.6.2), and stable isotopes beneath the native vegetation indicate a net upward flow of heat and water. The extraction of soil moisture within the root zone creates an upwards water potential gradient. On the contrary, profiles beneath irrigated field and channels sites indicate a long-established deep percolation in which the sediments have water potentials associated with approximately steady state downward flow (Stonestrom et al. 2007). Studies at a nearby waste disposal site in the Mojave Desert of Nevada also indicated no evidence of continued downward penetration of precipitation in vegetated areas (Andraski 1997). Rangeland systems in the southwestern United States share similar characteristics, particularly low matric potentials and upward potential gradients, indicative of discharge rather than recharge conditions (Scanlon et al. 2005).

Precipitation may penetrate more rapidly in areas with natural vegetation, but a rapid decrease in water potential subsequently occurs reflecting the efficiency of root-zone water depletion by native plants (Andraski 1997). The root-zones of xerophytic plants tend to have lower soil moisture content than occurs in the overlying and underlying soils. The lower soil moisture results in a greater (more negative) water (matric) potential (Sect. 5.6.2). As a result, the direction of groundwater flow in the unsaturated soils both overlying and underlying the root zone is toward the root zone. This peak in water potential in the root zone creates a barrier against the movement of water to the water table.

Direct (diffuse) recharge may occur in settings where vegetation is absent. Houston (2009) reported that low rates of diffuse recharge may occur during wetter than average periods (particularly La Niña years) in the high Andes above 4000 m (13,000 ft), where vegetation is sparse or absent and the coarse soils have high infiltration capacities.

Agricultural conversion affects key vegetation parameters, which, in turn, affect evapotranspiration rates and thus recharge, including fractional vegetation coverage, wilting point, and root depth (Scanlon et al. 2005, 2006). Wilting point is the minimum matric potential at which plants can take up water. The wilting point of native arid and semiarid region rangeland vegetation is typically much lower than that of typical agricultural crops (Scanlon et al. 2005). Native rangeland vegetation can draw much more water out of the soil than typical agricultural crops, which is an adaptation to water scarcity. The transition in land use and land cover (LU/LC) from rangeland vegetation to cultivated crops can result in an increase in groundwater recharge. The initiation of recharge is related in part to the addition of water through irrigation. However, the change from upwards to downwards water potential gradient can also occur as a result of the change in

LU/LC to dryland agriculture. The recharge associated with dryland agriculture appears to be associated with reduced interception, reduced evapotranspiration, shallow rooting depths, fallow periods and increased soil permeability caused by plowing (Scanlon et al. 2005, 2006).

Irrigation can contribute to higher groundwater levels in unconfined aquifers if either surface water or groundwater pumped from a deep confined aquifer is used as the water source, because it results in a net additional influx of water to the soil-unconfined aquifer system. Where irrigation is performed using groundwater from the local unconfined aquifer, the drawdowns from groundwater withdrawals greatly exceeds any potential rise in water levels from changes in LU/LC (Scanlon et al. 2006).

The change from discharge to recharge associated with the change in LU/LC from native rangeland to agriculture raises the possibility that removal of native vegetation can be used as a water-management tool (Scanlon et al. 2005, 2006). However, disruption of native ecosystems has adverse ecological implications. The initiation of groundwater recharge can also have adverse impacts on groundwater quality. Recharge can increase the solute concentration of groundwater by leaching of salts from the vadose zone that have accumulated naturally in the soils over thousands of years (Scanlon et al. 2005, 2006). Rising groundwater levels can also contribute to salinization of soils. Therefore, land cover and use conversions need to be evaluated within the context of overall water management, especially with consideration of water quality.

8.3.2 Effects of Temperature

The warmer temperatures during the summer can result in significant reductions in infiltration rates because of the more rapid loss of standing and shallowly infiltrated water in the upper vadose zone. Infiltration rates are also related to water temperature because of the temperature dependency of hydraulic conductivity, which is a function of the kinematic viscosity of water. Field and modeling studies have demonstrated that large diurnal variations in infiltration rates occur at some sites, which can be attributed to the temperature-dependence of the hydraulic conductivity (Jaynes 1990; Constantz et al. 1994; Ronan et al. 1998). There is a strong relationship between temperature and the viscosity of water and thus hydraulic conductivity. As temperature increases, the viscosity of water decreases and hydraulic conductivity increases, which for a given hydraulic gradient will result in more rapid infiltration. During a field study near Carson City, Nevada, both field data and modeling results indicate the infiltration rates (and thus stream losses) are greatest in the late afternoon when temperatures are the greatest and least in the morning when temperatures are lowest (Ronan et al. 1998). Infiltration rates in the Dan Region Project (Israel), a soil-aquifer-treatment system for municipal wastewater, also have been observed to have temperature dependence with lower rates occurring during the winter when the viscosity of water is greater (Icekson-Tal et al. 2003).

The magnitude of the temperature effect on recharge will depend to a large degree of the magnitude of the diurnal and seasonal variation in temperature. The magnitude of diurnal stream temperature variations is inversely proportional to discharge, since the thermal mass of streams increases with increasing discharge (Constantz et al. 1994). Temperature effects on hydraulic conductivity need to be considered relative to other temperature-related phenomena, such as evapotranspiration. In the cited studies, the effect of temperature on stream losses was much greater than the potential variations in ET rates.

8.3.3 *Effects of Soil Moisture Gradients and Changes*

Soil moisture and the associated soil-moisture gradient from land surface to the water table greatly influence the vertical hydraulic conductivity of unsaturated soils and sediments (Chap. 5). The amount of moisture affects every aspect of recharge from the initial infiltration of water at land surface to the rate of percolation of water through the vadose zone. The hydraulic conductivity of soils is dependent on the volumetric soil moisture content (i.e., fraction of the porosity that is filled with water). As the volumetric soil content increases, so does the unsaturated hydraulic conductivity. Dry surficial soils will initially have high infiltration rates because of their high water (matric) potential. Water is essentially pulled into the soil by capillary pressure. However, the water drawn into the fine pores by capillary pressure tends to be retained in the soil, rather percolate downwards. Thus, although a dry soil surface can result in an initially greater infiltration rate, the amount of recharge is typically reduced. The recharge from a given rainfall event will actually be greater if the soil is pre-wetted from previous rainfall.

The retention of water in dry sediments (in the absence of focused flow) results in the water staying near land surface, and passing neither through the zone of diffuse evaporative loss (0.5–1.5 m [1.6–4.9 ft] depending on the soil characteristics) or below the maximum root penetration depth. The water will thus be largely or entirely lost to ET before it has a chance to reach the water table. Hence, minor rainfall events are generally not hydrologically significant in arid lands because they do not significantly contribute to recharge. However, they are of ecological importance to indigenous flora and fauna that are adapted to take advantage of whatever sparse ephemeral water may be available.

8.4 Arid and Semiarid Region Recharge Location

A critical concept is that recharge occurs only in limited areas, and is non-existent or negligible over most of the semiarid and arid landscape (Newman et al. 2006). Natural recharge is predominantly indirect, occurring at locations where surface water (runoff) accumulates or flows. Pilgrim et al. (1988) noted that there are three broad classifications of hydrologic systems in arid regions, which have a bearing on recharge processes:

- sloping regions with an integrated steam (wadi) network,
- plain lands with primitive or no stream network, instead repetitive microhydrology,
- regions with major inputs of surface water and groundwater from more humid areas, frequently with extensive irrigated agriculture (e.g., Nile, Colorado, Indus river basins, Lebanon and Syria karst aquifers).

In sloping regions with integrated stream networks, recharge occurs in upland areas by direct or indirect infiltration into bedrock and small catchments (mountain block recharge) and within perennial or, more commonly, ephemeral stream (wadi) channels. Wadi recharge is also important in the reaches of the channels located on the mountain front (alluvial fans) and basin floor. In areas without a well-developed integrated stream network, recharge often occurs mainly in vadose (dry) playas and other depressions. Vadose playas can be considered as the terminal element of many wadi systems. Anthropogenic recharge, such as canal, pipe losses and irrigation return flows, may dominate local groundwater budgets, particularly where natural recharge is very low. These unplanned and unmanaged recharge mechanisms (Sect. 22.9) are predominant in some arid lands municipalities (e.g., Riyadh and Jeddah, Saudi Arabia; Kuwait City) and locally result in rising water tables, which necessitates drainage for protection of building foundations, basements, and underground structures.

8.4.1 Mountain Block and Front Recharge

Mountain Front Recharge (MFR) is generally used to describe the contribution of mountain regions to the recharge of aquifers in adjacent basins (Wilson and Guan 2004). There are several possible definitions of the "mountain front". Wilson and Guan (2004) proposed that the piedmont zone is the best definition of the mountain front (Fig. 8.2), and that definition is used herein. Piedmont is formally defined as meaning "the foot of the mountains" and in semiarid lands is termed "bajada' (Fairbridge 1968; Wooldridge and Morgan 1937). The piedmont zone is the sedimentary apron that occurs between the mountains and basin floor, and in arid regions includes alluvial fan deposits, the bajada and the pediment (Balchin and Pye 1956).

Mountain front recharge is an encompassing term loosely defined to include all water that enters the basin aquifer *through* the mountain front and includes water that infiltrates into both fractured rock and alluvium in mountain stream beds. The term "mountain block recharge" (MBR) is used herein to describe water that specifically enters a groundwater basin from adjacent mountain bedrock via subsurface fractures and fissures. There is been some inconsistency in the literature concerning the precise definition of MFR and MBR. Under the Wilson and Guan (2004) definition, MFR includes water entering the basin aquifer from both the

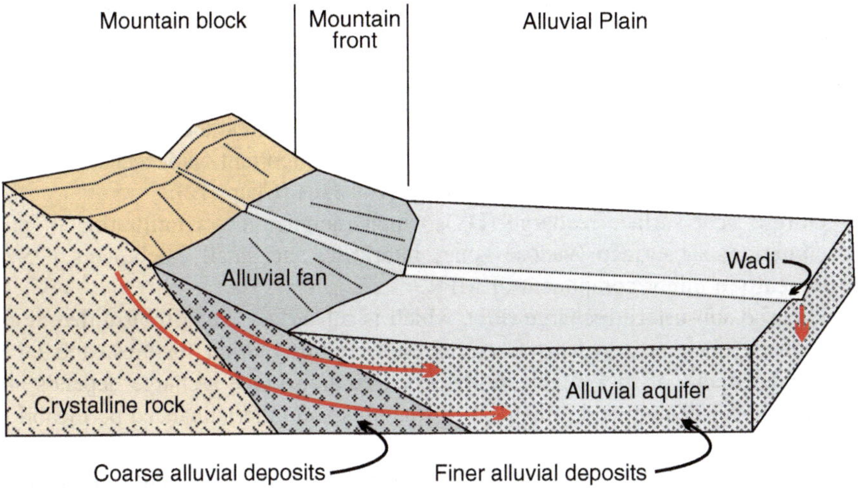

Fig. 8.2 Conceptual diagram of a mountain block and front showing principal recharge types

mountain block (MBR) and recharge in alluvial fans, perennial streams and reaches of ephemeral streams (i.e., wadis) within the mountains and mountain front.

Basin floor recharge (BFR) consists of both focused stream recharge and distributed recharge on the basin floor. Wadi recharge (Sect. 8.4.2), thus, includes both MRF and BFR depending upon the position of stream reach with respect to the lower boundary of the mountain front. MFR is a very important in arid and semiarid lands because basin floor recharge is often negligible (Chap. 4 and Chap. 11). Basins instead may contain local discharge areas. MFR is, therefore, the only natural (non-anthropogenic) source of water for basin aquifers in many arid lands.

Greater recharge tends to occur in mountains because of a combination of greater precipitation due to orogenic effects, lower temperatures (and thus reduced evapotranspiration), and thinner soils that can store less water (Wilson and Guan 2004). Total precipitation in the mountain block is typically much greater than that in the mountain front because of the much larger area of the former.

MFR has been historically studied from two perspectives; basin-centered and mountain-centered (Wilson and Guan 2004). For the basin-centered perspective, the mountains are viewed as a boundary condition for the basin. Recharge to the basin aquifer from adjacent mountains is estimated through the model calibration process. The models can be improved with geochemical data to help estimate recharge rates, flow velocity, and groundwater age. Mountain-centered methods estimate recharge from local precipitation and evapotranspiration. Both the basin-centered and mountain-centered approaches have their limitations, which were discussed by Wilson and Guan (2004). As is the case for estimation of recharge rates in general, multiple methods should used as often as possible (Chap. 11).

MFR contains four main components that were summarized by Wilson and Guan (2004):

- Focused near surface flow recharge (FS), which is the contribution from surface runoff and shallow subsurface water transmitted by streambed sediments at the mountain front. FS includes the volumetrically important wadi recharge within stream reaches in the mountains and alluvial fans (Chap. 12).
- Diffuse near surface recharge (DS), which includes direct infiltration of precipitation, subsurface (vadose zone) interflow, and small catchments. DS is typically a minor component of MFR.
- Focused subsurface recharge (FR), which is subsurface water transmitted along bedrock openings, such as fractures and karstic solution features, and other enhancements of rock permeability. Focused subsurface recharge depends largely on the density and properties of fractures and other enhanced permeability features.
- Diffuse subsurface recharge (DR), which is water that flows across the contact zone between the bedrock of the mountain block and the sediments of the basin aquifer.

MFR can be evaluated using two mass balance approaches (Wilson and Guan 2004):

$$\mathrm{MFR} = \mathrm{FS} + \mathrm{DS} + \mathrm{FR} + \mathrm{DR} \tag{8.1}$$

$$\mathrm{MFR} = \mathrm{P_b} + \mathrm{P_f} - \mathrm{ET_b} - \mathrm{ET_f} - \mathrm{DRO} \tag{8.2}$$

where,

P_b, P_f = precipitation on the mountain block and mountain front, respectively

ET_b, ET_f = evapotranspiration on the mountain block and mountain front, respectively

DRO = streamflow at the downstream end of the mountain front.

Depending on the setting, some of the variables in Eqs. 8.1 and 8.2 may be eliminated, if the value of the variable is negligible. The methods for estimating recharge are discussed in Chap. 11. The difficulty with the use of Eq. 8.1 is accurately determining the values of each of the recharge components, which inherently will have a great uncertainty due to the heterogeneity of mountainous areas. Precipitation can be estimated from secondary variables (terrain and atmospheric characteristics) in addition to direct measurements. Remote sensing and GIS technologies have been employed to map the distribution of precipitation in mountainous terrains (Chap. 19). Determination of ET rates in complex mountainous terrains is much more complex and thus results in great uncertainties in recharge estimates.

An alternative approach is a modification of the flow tube method (Sect. 11.4.3) to determine MFR (e.g., Wahi et al. 2008). Groundwater flow velocity is determined by dating groundwater along the flow paths using radioisotopes, such as ^{14}C or ^{3}H. Flow velocity is calculated from the difference in the age of water between two points along a flow path and the distance between the two points. Recharge rate can estimated from the cross-sectional area of the flow tube and average porosity. The flow tube method is subject to errors associated with the age dating, heterogeneities in groundwater flow, and general tortuosity on varying scales that affect the length of the flow path.

A key point emphasized by Wilson and Guan (2004) is that estimation of MFR must start with a proper understanding of the local hydrologic processes (i.e., correct conceptual model). An important technical issue is the partitioning of infiltrating precipitation between vadose interflow and deep penetration. If the bedrock is fractured it may be able to accept water at rates high enough to lead to significant deep penetration and active deep groundwater recharge into basin aquifers. Conversely, strong soil layering on hillsides can prevent the downward infiltration to the water table and divert water downslope as shallow vadose interflow, causing it to be lost to evapotranspiration.

Faults and fracture zones can be loci of recharge or can be low-permeability zones that impede groundwater flow depending upon their properties. Open-fractures can have high permeability, whereas cement-filled or pressure-annealed fractures may be essentially impermeable barriers. Sigda and Wilson (2003) studied the hydraulic properties of faults in poorly lithified sands and sandstone in the Rio Grande Rift of New Mexico. The faults contained deformation bands formed by cataclasis (grinding and breaking of grains) and compaction, which dramatically reduced mean pore size, increased the clay-size fraction, and decreased sorting. The deformation bands, as a result of the formation structural processes, have low porosity and saturated hydraulic conductivity compared to the parent sandstone. As the formation dries under unsaturated conditions, the hydraulic conductivity of the deformation band material decreases much less rapidly than does the hydraulic conductivity of the undeformed sands. Water tends to be retained in finer grained material because of its greater matric (water) potential. A cross-over occurs with increasing drying at which the hydraulic conductivity of the deformation bands becomes greater than that of the undeformed sandstone. Deformation bands can thus be sufficiently conductive to enhance downward moisture transport through the vadose zone, which could increase recharge (Sigda and Wilson 2003).

Water chemistry data from wells and natural discharges (e.g., springs) may be diagnostic of the origin of the water, time of water entry into the aquifer system, and precipitation rate and transmittal of the water. Hydrogen and oxygen stable isotope ratios of precipitation vary both seasonally and with elevation in the mountains, thus, providing a tool in some circumstances to determine when and where a water sample recharged into the aquifer system (Chap. 12).

Noble gas concentrations have also been demonstrated to be useful tracers for the source of recharge waters (Manning and Solomon 2003, 2005; Cederberg et al. 2009; Manning 2011). The concentration of noble gases (e.g., helium, neon, argon, and krypton) in recharging water is a function of the temperature, barometric pressure, and salinity at the water table during recharge. After the recharged water enters the aquifer, it becomes isolated from the atmosphere and its dissolved gas concentrations become fixed. Noble gases are by definition, non-reactive, and therefore their concentrations do not change along flow paths. Noble gasses can thus provide a record of temperature and pressure at the recharge location, which is related to the altitude of recharge. MBR water may be distinguishable from basin and other lower altitude recharge by its colder temperature of recharge, which can be determined from noble gas data measured in water samples collected from mountain springs and other discharge points with constrained altitudes. The fraction of higher-altitude MBR water can be evaluated using mixing equations. However, stable isotope and noble gas data provide information on the source of recharge, not the amount of recharge. The total recharge needs to be obtained using other methods.

The importance of focused mountain block recharge (FR, MBR) versus wadi recharge (FS) for the recharge of basin aquifers in arid and semiarid environments varies depending on local hydrogeology. Faults may be effective hydraulic barriers to horizontal groundwater flow from mountain blocks, in which case basin recharge occurs mainly through stream channels (e.g., Chowdhury et al. 2008). Cederberg et al. (2009) determined that MBR and stream and canal recharge in the Utah Valley, Utah, were roughly equal. Basins thus need to be independently evaluated with regard to the source of recharge water, assuming that this issue is locally relevant to water resources management.

8.4.2 Wadi Recharge

Ephemeral stream (river) systems are major sites of recharge in semiarid and arid regions. In some arid lands they are the sole geographic locations of significant recharge. Alluvial aquifers recharged by wadi systems are a critical water source in many arid lands upon which local communities are often dependent for domestic and agricultural water supplies. Long-term management of alluvial aquifers depends on the establishment of a balance between the amounts of recharge, discharge, and withdrawals. Local alluvial aquifers are susceptible to over-exploitation because of their relative low storage volumes and recharge rates. Ephemeral river systems also can present a threat to local communities because of the periodic occurrence of destructive flash floods. Salih and Ghanem (2002, p. 146) noted that

> It is ironic that, in countries that usually suffer from severe water shortage, floods are responsible for appreciable loss of life and property.

Wadis serve other functions in arid and semiarid communities. The channels and sands are often used as source of building materials. Wadis are often used as a convenient location for the disposal of sewage effluent, agricultural drainage, dewatering effluent, urban stormwater, and industrial wastewater (Alhamid et al. 2007). The combined flows entering some wadis in the Middle East exceed the natural capacity of the alluvium to absorb them, which has resulted in the establishment of local permanent flows (Alhamid et al. 2007). Wadis may also serve a water-treatment function as the quality of groundwater in wadi aquifers may be significantly better than surface-water quality (Alhamid et al. 2007).

Wadi hydrology is more complex than the hydrology of humid fluvial systems because of the great spatial and temporal variation of precipitation. Hydrologic methods developed for humid regions may, therefore, be inappropriate or subject to large errors when applied to arid systems (Şen 2008). Assessment and effective management of groundwater resources in wadi systems require the following basic information and analysis:

- spatial and temporal distribution of precipitation (both within and between years),
- partitioning of precipitation between run-off and infiltration, including the geographic location of infiltration (transmission losses) and the relationship between precipitation rate (intensity) and peak discharge and stream (flood) stage,
- fate of infiltrated water, particularly the percentage of the infiltrated water that actually recharges freshwater aquifers and provides exploitable water,
- impacts of groundwater withdrawals on both water levels (drawdowns) and water quality.

Actual groundwater recharge may be a small fraction of transmission losses (i.e., infiltrated water).

Evaluation and management of alluvial aquifers in arid regions is often greatly complicated by a paucity of long-term, site-specific hydrologic data. A fundamental paradox of arid lands water management in general is that a greater density of gaging stations and long recording periods are required to assess the spatial and temporal heterogeneity in precipitation and recharge, but such data are seldom available in arid lands. Expansion of rainfall and groundwater monitoring networks in a study area or the use of remote sensing techniques can address spatial heterogeneity, but many decades of monitoring may be necessary to characterize the temporal variability.

An additional concern in the sustainable management of wadi systems is water quality. Wadis were often preferential geographic sites for location of communities in arid lands because of the availability of water. Wadi aquifers by their nature are unconfined and are particularly vulnerable to pollution from a variety of anthropogenic sources (Salih and Ghanem 2002). The quality of water in wadi aquifers can be adversely impacted by a variety of activities and processes, such as the discharges of wastewaters, drainage water, and agricultural return flows, and the evaporative concentration of salts.

Methods for determining wadi recharge are discussed in Chap. 13. There are several end-member approaches that have been taken. Vadose zone processes may not be explicitly considered. Recharge can be estimated using geochemical or physical methods that consider the aquifer response to rainfall events, without quantifying rainfall-runoff relations and channel transmission losses. Alternatively, sequential approaches have been taken which consider rainfall-runoff relationships, transmission losses in channels, and then percolation of the infiltrated water (transmission losses) to the water table.

8.4.3 Flat Lands Depressional Recharge

Flat-land recharge refers to relatively flat lying areas in which integrated perennial or ephemeral stream systems are either absent or poorly developed. Instead, runoff from small catchments accumulates in low-lying areas (depressions), which are the primary sites for local recharge. For example, a pronounced surface feature of Qatar is the large number ($\approx$ 850) of shallow depressions, which are the surface expression of subsurface collapse caused by the dissolution of underlying calcium sulfate (anhydrite) and calcium carbonate (limestone) formations (Eccleston and Harhash 1982). After rainfall events that are sufficient to satisfy the surface soil moisture deficit, runoff reaches the depressions by incipient drainage channels or overland sheet flow. These depressions are significant recharge sites as evidenced by the underlying groundwater having relatively low salinities (Eccleston and Harhash 1982). However, depression floor silts may inhibit recharge, which is evidenced by a slow decline in water levels that approaches the rate of open-water evaporation (Eccleston and Harhash 1982). The government of Qatar implemented a program of installing recharge wells in depressions in order to increase the rate of recharge and, thus, decreasing evaporative losses (Al-Rashed and Sherif 2000; United Nations Environment Programme 2001).

Another example of depressional recharge is the southern High Plains Aquifer of the United States. There are approximately 66,000 playas in the southern Great Plains physiographic province of the United States. The playas in the High Plains Aquifer area are dry playas in which the water table is located tens of meters (more than 30 ft) below the playa floors (Gurdak and Rose 2010). Gurdak and Rose (2010) reviewed the role of playas in aquifer recharge. Historically, opinion as to the role of playas in groundwater recharge ranged from them being evaporation pans, in which little recharge occurred, to a main locus of focused recharge. The cumulative data obtained by many workers indicate that playas are not exclusively evaporation pans and that some recharge occurs in playas through their clay floors.

The recharge rate in playas appears to be substantially greater (1 to 2 orders of magnitude) than the rate in interplaya regions (land areas between playas). Data collected from soils between playas shows that the moisture water (matric) potential gradient is commonly directed upward from the water table to the root zone, which indicates that substantial recharge is not likely to occur in these areas under the current climate regime (McMahon et al. 2006; Gurdak and Rose 2010). Playas thus serve the same role as wadis in being areas of focused recharge. A positive correlation has been documented between the clay content of the playa floor deposits and infiltration rate, which appears to be due to desiccation cracks being enhanced flow paths. The recharge rates are, in part, controlled by spatial and temporal patterns in the physical characteristics of the playas, climate, and surrounding land use practices (Gurdak and Rose 2010).

References

Alhamid, A. A., Alfayzi, S. A., & Hamadto, M. A. (2007). A sustainable water resources management plan for Wadi Hanifa in Saudi Arabia. *Journal King Saud University, Engineering Sciences, 19*(2), 209–222.

Al-Rashed, M. F., & Sherif, M. M. (2000). Water resources in the GCC countries: An overview. *Water Resources Management, 14*, 59–75.

Andraski, B. J. (1997). Soil-water movement under natural-site and waste-site conditions: A multiple-year field study in the Mojave Desert, Nevada. *Water Resources Research, 33*, 1901–1916.

Balchin, W. G. V., & Pye, N. (1956). Piedmont profiles in the arid cycle. *Proceedings of the Geologists Association of London, 66*, 167–188.

Cederberg, J. R., Gardner, P. M., & Thiros, S. A. (2009). *Hydrology of Northern Utah Valley, Utah County, Utah, 1975–2005*. U.S. Geological Survey Scientific Investigations report 2008-5197.

Chowdhury, A. H., Uliana, M., & Wade, S. (2008). Ground water recharge and flow characterization using multiple isotopes. *Ground Water, 46*, 426–436.

Constantz, J., Thomas, C. L., & Zellweger, G. (1994). Influence of diurnal variations in stream temperature on streamflow loss and groundwater recharge. *Water Resources Research, 30*, 3253–3264.

De Vries, J. J., & Simmers, I. (2002). Groundwater recharge: An overview of processes and challenges. *Hydrogeology Journal, 10*, 5–17.

Eccleston, B. L., & Harhash, I. E. (1982). The Hydro-Geology of Qatar: Water Resources and Agricultural Development Project Phase III, Ministry of Industry and Agriculture.

Fairbridge, R. W. (1968). Piedmont landforms. In Fairbridge, R. W. (Ed.), *The encyclopedia of geomorphology* (pp. 842–845). New York: Reinhold Book Corporation.

Flint, A. L., Flint, L. E., Kwicklis, E. M., Fabryka-Martin, J. T., & Bodvarson, G. S. (2002). Estimating recharge at Yucca Mountain, Nevada, USA, comparison of methods. *Hydrogeology Journal, 10*, 180–204.

Flint, A. L., Flint, L. E., Hevesi, J. A., & Blainey, J. B. (2004). Fundamental concepts of recharge in the desert Southwest: A regional modeling perspective. In J. F. Hogan, F. M. Phillips, & B. R. Scanlon (Eds.), *Groundwater recharge in a desert environment: The Southwestern United States, Water Science and Application Series* (Vol. 9, pp. 159–184). Washington, DC: American Geophysical Union.

Flint, L. E., & Flint, A. L. (1995). *Shallow infiltration processes at Yucca Mountain, Nevada—neutron logging data, 1984–1993*. U.S. Geological Survey Water-Resources Investigations Report 95-4035.

Flint, L. E., & Flint, A. L. (2007). Regional analysis of groundwater recharge. In D. A. Stonestrom, J. Constantz, T. P. A. Ferré, & S. A. Leake (Eds.), *Ground-water recharge in the arid and semiarid Southwestern United States* (pp. 29–60). U.S. Geological Survey Professional Paper 1703.

Gee, G. W., Fayer, M. J., Rockhold, M. L., & Campbell, M. D. (1992). Variations in recharge at the Hanford Site. *Northwest Science, 66*, 237–250.

Gurdak, J. J., Hanson, R. T., McMahon, P. B., Bruce, B. W., NcCray, J. E., Thyne, G. D., et al. (2007). Climate variability controls on unsaturated water and chemical movement, High Plains Aquifer, USA. *Vadose Zone Journal, 6*, 533–547.

Gurdak, J. J., & Rose, C. D. (2010). Recharge rates and chemistry beneath playas of the High Plains aquifer. *Hydrogeology Journal, 18*(8), 1747–1772.

Healy, R. W. (2010). *Estimating groundwater recharge*. Cambridge: Cambridge University Press.

Houston, J. (2009). A recharge model for high altitude, arid, Andean aquifers. *Hydrologic Processes, 23*, 2383–2393.

Icekson-Tal, N., Avraham, O., Sack, J., & Cikurel, H. (2003). Water reuse in Israel—the Den Region Project: Evaluation of water quality and reliability of plant's operation. *Water Science and Technology: Water Supply, 3*(4), 231–237.

Izbicki, J. A., Johnson, R. U., Kulongoski, J., & Predmore, S. (2007). Ground-water recharge from small intermittent streams in the western Mojave Desert, California. In D. A. Stonestrom, J. Constantz, T. P. A. Ferré, & S. A. Leake (Eds.), *Ground-water recharge in the arid and semiarid southwestern United States* (pp. 157–184). U.S. Geological Survey Professional Paper 1703.

Jaynes, D. B. (1990). Temperature variations effects on field-measured infiltration. *Soil Science Society of America Journal, 54*, 305–311.

Keese, K. E., Scanlon, B. R., & Reedy, R. C. (2005). Assessing controls on diffuse groundwater recharge using unsaturated flow modeling. *Water Resources Research, 41*, W06010, doi:10:1029/2004WR003841.

Kinzelbach, W., Aeschbach, W., Alberich, C., Goni, I. B., Beyerle, U., Brunner, P., et al. (2002). *A survey of methods from groundwater recharge in arid and semi arid regions: Early Warning and Assessment Report Series UNEP/DEWA/RS.02-2*. Nairobi, Kenya: United Nations Environment Programme,

Lerner, D. N., Issar, A. S., & Simmers, I. (1990). *Groundwater recharge, a guide to understanding and estimating natural recharge*. Kennilworth: International Association of Hydrogeologists.

Llamas, M. R. (2004) Water and ethics. Use of groundwater: UNESCO International Hydrogeological Programme.

Manning, A. H. (2011). Mountain-block recharge, present and past, in the eastern Española, New Mexico, USA. *Hydrogeology Journal, 19*, 379–397.

Manning, A. H., & Solomon, D. K. (2003). Using noble gases to investigate mountain front recharge. *Journal of Hydrology, 275*, 194–207.

Manning, A. H., & Solomon, D. K. (2005). An integrated environmental tracer approach to characterizing groundwater circulation in a mountain block. *Water Resources Research, 41*, W12412, 18, doi:10.1029/2005WR004178.

Missimer, T. M., Drewes, J., Amy, G., Maliva, R., & Keller, S. (2012). Restoration of wadi aquifers by artificial recharge with treated wastewater. *Ground Water* (in press).

McMahon, P. B., Dennehy, K. F., Bruce, B. W., Böhlke, J. K., Michel, R. L., Gurdak, J. J. et al. (2006). Storage and transit time of chemicals in thick unsaturated zones under rangeland and irrigated cropland, High Plains, United States. *Water Resources Research, 42*, W03413, doi:10.1029/2005WR004417.

Newman, B. D., Vivoni, E. R., & Groffman, A. R. (2006). Surface water—groundwater interactions in semiarid drainages of the American southwest. *Hydrological Process, 20*, 3371–3394.

Pilgrim, D. H., Chapman, T. E., & Doran, D. G. (1988). Problems of rainfall-runoff modeling in arid and semiarid regions. *Hydrological Sciences Journal, 33*(4), 379–400.

Pool, D. R. (2005). Variations in climate and ephemeral channel recharge in southeastern Arizona, United States. *Water Resources Research, 41*, W11403, doi: 10.1029/2004WR003255.

Ronan, A. D., Prudic, D. E., Thodal, C. E., & Constantz, J. (1998). Field study and simulation of diurnal temperature effects on infiltration and variably saturated flow beneath an ephemeral stream. *Water Resources Research, 34*, 2137–2153.

Salih, A. M. A., & Ghanem, A. H. M. (2002). Sustainable management of wadi systems. In H. Wheater & R. A. Al-Weshah (Eds.), *Hydrology of wadi systems* (pp. 147–156). IHP-V Technical Documents in Hydrology. Paris: UNESCO

Scanlon, B. R., Jolly, I., Sophocleous, M., & Zhang, L. (2007). Global impacts of conervsions from natural to agricultural ecosystems on water resources: quantity versus quality. *Water Resources Research, 43*, W03437, doi:10.1029/2006WR005486.

Scanlon, B. R., Keese, K. E., Flint, A. L., Flint, L. E., Gaye, C. B., Edmunds, W. M., et al. (2006). Global synthesis of groundwater recharge in semiarid and arid regions. *Hydrological Processes, 20*, 3335–3379.

Scanlon, B. R., Reedy, R. C., Stonestrom, D. A., Prudic, D. E., & Dennehy, K. F. (2005). Impact of land use and land cover change on groundwater recharge and quality in the southwestern US. *Global Change Biology, 11*, 1577–1593.

Scanlon, B. R., Tyler, S. W., & Wierenga, P. J. (1997). Hydrologic issues in arid, unsaturated systems and implications for contaminant transport. *Review of Geophysics, 35*, 461–490.

Seiler, K. P., & Gat, J. R. (2007). *Groundwater recharge from run-off, infiltration and percolation*. Berlin: Springer.

Şen, Z. (2008). *Wadi hydrology*. Boca Raton, FL: CRC Press.

Sharma, M. L. (1989). *Groundwater recharge*. London: Taylor and Francis.

Sigda, J. M., & Wilson, J. L. (2003). Are faults preferential flow paths through semiarid and arid vadose zones? *Water Resources Research, 39*(8), 1225, 14.

Simmers, I. (1988). *Estimation of natural groundwater recharge*. North Atlantic Treaty Organization, Scientific Affairs Division.

Simmers, I. (1990). Aridity, groundwater recharge and water resources management. In D. N. Lerner, A. S. Issar, & I. Simmers (Eds.), *Groundwater recharge, a guide to understanding and estimating natural recharge (Contributions to Hydrogeology 8)* (pp. 1–20). Kennilworth, UK: International Associations of Hydrogeologists.

Simmers, I. (1997). *Recharge of phreatic aquifers in (semi-)arid areas (IAH International Contributions to Hydrogeology 19)*. London: Taylor & Francis.

Simmers, I. (1998). Groundwater recharge: an overview of estimation "problems" and recent developments. In N. S. Robins (Ed.), *Groundwater pollution, aquifer recharge and vulnerability* (pp. 107–115). London: Geological Society, Special Publication 130.

Stephens, D. B. (1996). *Vadose zone hydrology*. Boca Raton, FL: CRC Press.

Stonestrom, D. A., & Harril, J. R. (2007). Ground-water recharge in the arid and semiarid southwestern United States—climate and geologic framework, In D. A. Stonestrom, J. Constantz, T. P. A. Ferré, & S. A. Leake (Eds.), *Ground-water recharge in the arid and semiarid southwestern United States* (pp. 1–27). U.S. Geological Survey Professional Paper 1703

Stonestrom, D. A., Prudic, D. E., Walvoord, M. A., Abraham, J. D., Stewart-Deaker, A. E., Glancy, P. A., et al. (2007). Focused ground-water recharge in the Amargosa desert basin. In D. A. Stonestrom, J. Constantz, T. P. A. Ferré, & S. A. Leake (Eds.), *Ground-water recharge in the arid and semiarid southwestern United States* (pp. 107–136). U.S. Geological Survey Professional Paper 1703.

Tuinhof, A., Olsthoorn, T., Heederick, J. P., & De Vries, J. (2003). Management of aquifer recharge and subsurface storage, a promising option for coping with increasing storage needs. In A. Tuinhof, & J. P. Heederick (Eds.), *Management of aquifer recharge and subsurface storage* (pp. 3–18). Netherlands National Committee—International Association of Hydrogeologists (NNC-IAH) Publication No. 4.

United Nations Environment Programme (2001). *Sourcebook of alternative technologies for freshwater augmentation in West Asia*. IETC Technical Publications Series Issue 8F, 341 pp.

Wahi, A. K., Hogan, Ekwurzel, B., Baillie, M. N., & Eastoe, C. J. (2008). Geochemical quantification of semiarid recharge. *Ground Water, 46*, 414–425.

Walvoord, M. A., Stonestrom, D. A., Andraski, B. J., & Striegl, R. G. (2004). Constraining the inferred paleohydrologic evolution of a deep unsaturated zone in the Amaragosa Desert. *Vadose Zone Journal, 3*, 502–512.

Wheater, H. S. (2002). Hydrological processes in arid and semi arid area. In H. Wheater, & R. A. Al-Weshah (Eds.), *Hydrology of wadi systems, IHP-V, Technical documents in hydrology, no. 55* (pp. 5–22). Paris: UNESCO.

Wilson, J. L., & Guan, H. (2004). Mountain-block hydrology and mountain-front recharge. In F. M. Phillips, J. Hogan, & B. Scanlon (Eds.), *Groundwater recharge in a desert environment: The Southwestern United States* (pp. 113–127). Washington, DC: American Geophysical Union.

Winograd, I. J., Riggs, A. C., & Coplen, T. B. (1998). The relative contributions of summer and cool-season precipitation to groundwater recharge, Springs Mountains, Nevada, USA. *Hydrogeology Journal, 6*, 77–93.

Winter, T. C., Harvey, J. W., Franke, O. L., & Alley, W. M. (1998). Ground water and surface water. A single resource. *U.S. Geological Survey Circular, 1139.*

Wood, W. W. (1999). Use and misuse of the chloride-mass balance method in estimating ground water recharge. *Ground Water, 37*, 2–3.

Wood, W. W., & Sanford, W. E. (1995). Chemical and isotopic methods for quantifying ground-water recharge in a regional, semiarid environment. *Ground Water, 33*, 458–468.

Wood, W. W., Rainwater, K. A., & Thompson, D. B. (1997). Quantifying macropore recharge: Examples from an semi-arid area. *Ground Water, 35*, 1097–1106.

Wooldridge, S. W., & Morgan, R. S. (1937). *The physical basis of geography: An outline of geomorphology*. London: Longmans, Green and Company.

World Nuclear Association. (2011). Nuclear desalination: www.world-nuclear.org/info/in71.html.

Chapter 9
Sustainability and Safe Yield

9.1 Introduction

No other concept has gained greater currency in planning and development policy as sustainability. The concept of sustainability can be generally understood as implying that an activity can be continued into the indefinite future. However, considerable disagreement still exists as to the precise definition of sustainability and the requirements or implications of sustainability, particularly with respect to water resources. The New Oxford American Dictionary (McKean 2004) defines "sustainable" and "sustained yield" as:

> able to be maintained at a certain rate or level

> a level of exploitation or crop production that is maintained by restricting the quantity harvested to avoid long term depletion

A plethora of definitions of the term sustainability has been proposed with respect to development activities. The most widely used definition is that of the Brundtland Commission (World Commission of Development and the Environment 1987):

> Sustainable development is development that meets the needs of the present without compromising the ability of future generations to meet their own needs.

It contains within it two key concepts:

- the concept of 'needs', in particular the essential needs of the world's poor, to which overriding priority should be given; and
- the idea of limitations imposed by the state of technology and social organization on the environment's ability to meet present and future needs.

A fundamental aspect of sustainable development is the belief that future generations have a right to the same, if not a better, economy, environment, and quality of life than we enjoy today (Loucks and Gladwell 1999). Clearly,

R. Maliva and T. Missimer, *Arid Lands Water Evaluation and Management*, Environmental Science and Engineering, DOI: 10.1007/978-3-642-29104-3_9,

anticipating the needs and desires of future generations and how current resource management decisions will impact them is fraught will difficulty. A fundamental challenge to sustainable development is "discounting", which in economics means that people value something more today than they value having the same item in the future. People will tend to be inclined to utilize a resource today, even though it may have greater value in the future.

Sustainable development is an overall goal, and does not imply that all aspects of development and resource management must be sustainable everywhere and at every time. Sustainability is a function of various economic, environmental, ecological, social, and physical goals and objective analysis must inevitably involve multi-objective tradeoffs in a multi-disciplinary and multi-participatory decision-making process (Loucks and Gladwell 1999). For example, the tradeoff of allowing some limited degradation of the natural environment (for example, land clearance for a mine) that is important for local or national economic development can be part of sustainable development. The important issue is that informed decisions should be made considering all the benefits and costs (impacts) of proposed development plans.

9.2 Sustainability Water Resources Management Concepts

On a broad-scale, sustainable use of water implies resource conservation, environmental friendliness, technological appropriateness, economic viability, and social acceptability of development issues (Pereira et al. 2002). Loucks and Gladwell (1999, p. 30) proposed a definition of sustainability with respect to water:

> Sustainable water resources systems are those designed and managed to fully contribute to the objectives of society, now and in the future, while maintaining their ecological, environmental, and hydrologic integrity

Defining sustainability with respect water resources must consider three important aspects of water resources, particularly in arid and semi-arid lands:

(1) Groundwater and surface-water resources are part of integrated hydrologic systems in which any changes in water input or output necessarily affects other parts of the system. For example, groundwater withdrawals must result in either a decrease in groundwater discharge, increased recharge, reduction of water in storage, or another compensating change.
(2) Water cycles within the hydrologic system and it is important to consider all parts of the system. Water pumped from the ground does not disappear and its fate needs to be considered.
(3) The response of hydrologic systems to perturbations, whether they are caused by humans (e.g., changes in the extraction rate) or natural, can be slow because of transient and delayed responses. For example, an increase in pumping this year that does not cause a concurrent measurable reduction in spring flow provides no assurance that a pumping-induced reduction in spring flow will not occur in the future.

Sustainability requires management of water resources with a long-term perspective. In planning for sustainability, it is also critical to recognize that local and global climate is subject to change from both natural and anthropogenic causes. Sustainable water-resources systems are thus designed and operated in ways that make them more adaptive, robust, and resilient to future uncertainties (Loucks and Gladwell 1999). Systems need to be able to adapt to wide range of future plausible changes. The sustainability concept would exclude the use of non-renewable groundwater resources (Chap. 37), the use of which is inherently non-sustainable. The current sustainability paradigm has largely evolved to the concept that we should live within the renewable supply of groundwater and that stored groundwater should be treated as a "revolving account" in which none of it, or not much more of it, is permanently depleted (Shomaker 2007). In other words, groundwater is to be managed as a reservoir with the function of smoothing the irregularities in the surface water supply, but not as a reservoir full of water to be withdrawn and put to use (Shomaker 2007).

The concept that water use should be sustainable is on the surface, an intuitive conclusion and is philosophically and emotional appealing. However, it should be recognized that humans have advanced at times by a series of unsustainable developments (Price 2002; Alley and Leake 2004). Unsustainable water uses are not inherently bad. It is critical for all involved in water-resources management to understand what is mean by terms such as 'sustainability', 'safe yield' and 'overexploitation', how are they evaluated, and what are their real-world implications. Indeed, Wood (2003) noted that sustainability is ultimately not possible for some resources, and is fundamentally flawed as both a goal and operational policy. At some point population and associated demands will exceed supply, and it may not be possible to meet all demands of a finite resource (i.e., local water resources) because of exponentially increasing population and per capita water use (Wood 2003).

Water-management and policy decisions regarding sustainability have both scientific and political dimensions. The key challenge for water resources professionals is to evaluate the hydrologic effects of various development strategies in such a way that their long-term implications can be properly assessed (Alley and Leake 2004). In other words, it is the job of scientists to provide objective answers to the question 'what would happen if this course of action is taken?' and 'what are the impacts of current water policies?' The political dimension of the question is deciding which option should be taken. The word 'political' is used in sense that policy decisions are being made by government officials, stakeholders, or the general population. Although there is a desire to make water-policy decisions on a purely objective scientific or economic basis, the reality is that the decision-making process is often ultimately driven by personal values and opinions as to what is best for society. Decisions concerning economic development versus environmental preservation, in particular, are, to large degrees, subjective, value-laden choices, commonly underpinned within an economic framework.

It is also very important to note that identification of aspects that make water resource development unsustainable have been more successful than the development and implementation of remedial measures that reduce or eliminate those undesirable effects (Attia 2001). In the management of water resources, and other economic and development issues, it is often politically expedient to continue popular unsustainable practices rather than to make difficult decisions that require unpopular sacrifices necessary to place society on a sustainable course.

9.3 Safe Yield and Sustainability

Discussions of groundwater resources management in arid lands now invariably involve consideration of the topics termed "safe yield" and "sustainability". It has long been understood that unlimited pumping of an aquifer will eventually result in its depletion or will otherwise result in the degradation of the water quality to the point where the water can no longer be put to its intended use. There is a natural desire to determine the rate of aquifer pumping that could occur (or be allowed) on a long-term basis without causing significant harm to the resource. The fundamental idea behind the safe or sustainable yield concept, quantifying the desirable development of a groundwater basin, remains relevant today (Alley and Leake 2004). It is important that the concepts be unambiguously understood and suitable for quantitative application (Kalf and Wooley 2005).

There is no universally accepted definition of either of the fundamental terms, "safe yield" and "sustainability," as they apply to groundwater, much less guidelines for their incorporation into water management plans. The history of the concepts and definitions of these terms was reviewed recently by Kalf and Woolley (2005). Numerous definitions of sustainability, safe yield, and related concepts (e.g., sustainable yield, sustainable pumping rate) have been proposed followed by papers that sought to clarify the situation, which lead to even more disagreement and uncertainty.

As Freeze and Cherry (1979) noted, that although the concept of safe yield has been widely used in groundwater evaluations, there has always been widespread dissatisfaction with it. Custodio (2002) noted that the terms "overexploitation", "over-draft", "over-use", "over-extraction", "over-development", and "unsustainable use" have been also increasingly used since the 1970s, but often without a precise meaning and definition. The definition of safe yield and sustainability with respect to water resources is not just a semantic question. Definitions of terms may ultimately dictate policy. The water-management response to a mandate that water use be sustainable or not exceed an aquifer safe yield will depend upon how the terms are defined.

The term "safe yield" appears to have been first used by Lee (1915) and the early definitions of the term focused on the amount of water than could be withdrawn from an aquifer perennially without depleting the supply (Kalf and Wooley 2005). The term "groundwater mining" came into use to denote a situation where groundwater withdrawals cause a progressive reduction in the total volume of water in storage in a groundwater basin. Groundwater mining results if the total outputs of water from the groundwater basin (drainage and pumping) exceed the total inputs (recharge) on a continuous long-term basis with a resulting decrease in volume of water stored in the basin.

The definition of safe yield was subsequently expanded by some workers to include consideration of water quality changes and other impacts. Todd (1959, p. 200) proposed the following concise definition:

> The safe yield of a groundwater basin is the amount of water which can be withdrawn from it annually without producing an undesired result. Any draft in excess of safe yield is overdraft.

Undesirable results can include a host of potential impacts. An improvement on the Todd (1959) definition would be to change the word "undesired" to "unacceptable" in recognition the some undesired results typically occur in all water projects. All development projects and virtually all of man's activities impact the environment in one manner or another. The more important issue is whether adverse impacts reach a threshold for which they are considered to not be acceptable. Undesired or unacceptable results of groundwater pumping can include (Winter et al. 1998; Sophocleous 2002; Zektser et al. 2005):

- declining aquifer water levels and associated increasing pumping costs,
- a progressive decrease in spring discharge, river base flow, lake levels, wetland hydroperiods, and associated ecological impacts,
- changes in groundwater flow patterns and mixing, and concomitant deterioration of water quality, such as increases in salinity (saline-water intrusion), which is a critical issue in some arid lands and coastal aquifers,
- pore pressure decreases and associated land subsidence and fissuring, and
- damage to plant communities as the water table drops below root zones, including both natural systems and agricultural crops.

Huisman and Olsthoorn (1983) made the important observation that safe yield (i.e., maximum permissible abstraction) depends on subjective considerations of a judicial, economic, political, aesthetic, etc. nature, and that determination of safe yield is always a compromise between conflicting interests. Llamas (2001) further observed that "undesirability" depends mainly on the social perception of the issue, which is related more to the legal, cultural and economic background of a region than to hydrogeologic facts. He also noted a critical point in that the social relevance of the conflict between natural system conservation and groundwater development and its solutions will differ from country to country and may also change with time and economic development.

Devlin and Sophocleous (2005) using the term "sustainable pumping" instead of "safe yield" and defined the concept as the pumping rate that can be maintained without indefinitely dewatering or mining an aquifer. With reference to the similar term "sustainable yield", Kalf and Wooley (2005) emphasized that a groundwater basin must reach a new equilibrium in which total inputs and total outputs are balanced. Devlin and Sophocleous (2005) excluded any consideration of environmental impacts in consideration of sustainable pumping rate. The concept of sustainability usually includes consideration of issues such as ecology, water quality, and human and environmental welfare (Devlin and Sophocleous 2005).

Groundwater sustainability is broadly defined as (Alley et al. 1999; Alley and Leake 2004, p. 13) as:

> the development and use of ground water resources in a manner that can be maintained for an indefinite time without causing unacceptable environmental, economic, or social consequences.

The term 'indefinite' is also problematic as it essentially means forever. A water environment that is "sustainable" under one climate regime may become unsustainable when climate inevitably changes from natural and anthropogenic sources (Price 2002). An important question is "what is the planning horizon for assessments of sustainability?" "Forever" is a superficially logical answer but may not be practical in light of a changing climate caused by natural variation or anthropogenic impacts. The real world policy questions involve, for example, whether groundwater use should be limited now when impacts to surface water bodies may occur 100, 200, or 2,000 years into the future. When one considers the tremendous social, political, economic, and technologic changes humankind has experienced over the past century and will likely occur at an accelerated pace in the future, it is folly to believe that the use of any resource can be planned today for centuries to come. Instead, any discussion of sustainability should define a time period over which sustainability is being considered (Price 2002).

Safe yield differs from sustainability, as commonly defined, in that it almost always is considered in terms of an annual water withdrawal, whereas the temporal pattern of sustainability is more open-ended (Alley and Leake 2004). Walton (2011) commented that sustainable yield needs to consider aquifer response time, which is defined as the time that it takes for water level and storage changes throughout the aquifer to become negligible after an increase or decrease in withdrawals. The impact of groundwater withdrawals can be estimated if the duration of model simulations is longer than aquifer response times.

Overexploitation can be considered to be long-term groundwater extractions that exceed the aquifer or groundwater basin safe or sustainable yield. Custodio (2002, p. 260) proposed that overexploitation:

> remains essentially a loose concept since no unique, constant value of sustainable use can be attached. As such, overexploitation may be a useful term to point out that some negative effects of aquifer development are of concern to some social groups.

9.4 Acceptability of the Impacts of Groundwater Use

Safe yield and sustainability, and related water-resources concepts (e.g., over exploitation), are not purely scientific concepts, and necessarily involve social, economic, and political considerations. The term 'unacceptable' is subject to personal and thus societal interpretation. All groundwater pumping in arid lands will in the long run result in declines of natural discharges with consequent environmental impacts and some time lag will occur between the pumping and the onset of the impacts (Sophocleous 2005). It is therefore inherently impossible to have a no impact groundwater use.

It is generally understood that the use of groundwater inevitably has some impacts on surface environments. Rajagopal-Durbin and Durbin (2008) noted that:

> Population growth and environmental changes are inextricably intertwined. Along with the former, which is inevitable, comes the latter, which, although unfortunate, must be embraced as an unavoidable consequence (p. 146)

and

> human beings cannot be introduced into the groundwater equation without taking something else out of the equation (p. 157)

It is not realistic to adopt a no impact criterion to the development of water or other resources, and it is not responsible or ethical to just ignore the impacts. The criteria for what is considered overexploitation could be so restrictive that it would become impossible for any use of an aquifer to occur (Custodio 2002). Water use inherently involves trade-offs in order to try to balance the direct and indirect costs and the not easily evaluated environmental damage on one side, with the benefits to direct users and society on the other side (Custodio 2002). Sustainable groundwater development, thus, involves balancing economic, environmental and human (social) requirements (Hiscock et al. 2002). With respect to environmental issues, Price (2002) noted that is important to:

- be clear about what we are trying to protect and why,
- be sensible about the way we achieve it, and
- realize that most things in life are a compromise and that conflicts of interest exist, in which each interest alone may be perfectly reasonable.

In other words, decisions should be made in an informed manner and that the nature of trade-offs and the reasons and basis for selecting an option should be logically understood.

Rajagopal-Durbin and Durbin (2008) proposed that a pragmatic approach should be taken in which it is recognized that development associated with population growth cannot occur without some environmental impacts and a conscious decision needs to be made as to which environmental resources have the greatest value and should receive the greatest protection. The final step in the groundwater management process is designing groundwater systems that impact lower-value resources over high-value resources. Rajagopal-Durbin and Durbin (2008) also noted that management of groundwater resources means minimizing the need to use it.

Unresolved questions concern how environmental resources are ranked both amongst themselves and against non-environmental resources or activities (e.g., agriculture) and who gets to decide. Such decisions are already being made, but usually not in explicit terms.

9.5 Evaluation of Safe Yield and Sustainability

Evaluation of the safe or sustainable yield of a groundwater basin must consider the basin water budget, which includes all inputs (sources) and outputs (sinks). The evaluation must also include consideration of the size and boundaries, hydraulic properties, and water-quality distribution of groundwater basins. Groundwater modeling is now the standard tool for evaluating water-development strategies (Chap. 20).

Evaluations of safe yield must also consider the time delay between the start of pumping and adverse impacts, which depends upon aquifer diffusivity and the distance between the wells and the surface water (or sensitive environments) in question. There may be long time-lags between the start of a groundwater abstraction and when significant captured recharge or discharge becomes detectable at the surface water bodies located proximal and distal to the pumping center (Balleau 1988; Custodio 2002; Sophocleous 2002, 2005; Shomaker 2007). If pumping exceeds the rate of captured recharge or discharge, an aquifer may never reach a new equilibrium and water levels will continue to decline as water is continuously removed from storage (Chap. 6).

Short-term monitoring data may, therefore, be inadequate to determine whether a water use is safe or sustainable, if the criteria for determination of safe yield or sustainability is the perpetual absence of impacts to a particular surface-water body or the environment in general. Monitoring may also under-estimate the eventual impacts of abstractions to date, as water levels may continue to decline after pumping stops, especially for monitoring wells distant from the abstraction location (Bredehoeft and Durbin 2009). Assuming it is practically feasible to reduce groundwater pumping in response to monitored impacts exceeding a sustainability criterion (e.g., allowable decrease in stream discharge), there may be a large time lag after the termination or reduction in pumping for the aquifer water levels and local discharge to recover to target levels (Bredehoeft and Durbin 2009).

9.6 Mitigation

Once it is determined that current water-use practices in a region are not sustainable, then the issue arises as to how to respond. Water-resources policies focused on safe yield and sustainability also need to consider mitigation options. Mitigation, with respect to environment issues, is defined as the elimination or reduction of the frequency, magnitude, or severity of verified impacts to the environment. An integral part of water-resources management is identifying potential risks and mitigation opportunities.

A legally constituted oversight program that has the authority to implement changes in the water-use patterns or cause a change of sources can be used through the creation of a formal governmental agency or some type of legally-constituted user group. Even though groundwater use is clearly unsustainable in some arid and semiarid regions, response to the unsustainable water use is often not fully effective or non-existent. The most obvious solution to unsustainable water use is to reduce groundwater pumping. However, well owners and landowners may resist water use reductions and restrictions to address groundwater mining (overdraft) because they have no alternative sources of water (Schlager 2006). Restrictions also threaten future economic growth and development (Schlager 2006). Unsustainable groundwater use is self-correcting in the long-term in the sense that lowering of aquifer water levels (and associated need for well deepening and increased pumping costs) and deterioration in water quality will ultimately force a reduction in pumping.

Mitigation may also involve reducing impacts to sensitive environments and surface waters resulting from groundwater use. Shomaker (2007) provides an insightful discussion of ground water use in New Mexico, where water use is governed by the prior appropriation doctrine. Water policy in New Mexico is driven to a large degree by surface-water rights, because they are often senior to groundwater rights. Surface-water rights often control and limit access to groundwater, where there is a link between surface-water and groundwater. Groundwater pumping is limited by concerns that pumping diminishes river flows by intercepting groundwater flow that would otherwise discharge to streams or by inducing flow from the stream into an aquifer. The impact to surface water flow by increased groundwater pumping could be mitigated by the return flow of the additional treated municipal wastewater generated from the additional water use and by directly pumping some groundwater into the river (Shomaker 2007).

If dehydration of sensitive environments is an important environmental concern, then in some circumstances pumping of groundwater (or water from other sources) into the sensitive environments could be used as a mitigation measure. The authors have worked on water-supply projects involving wetlands or lake levels, in which the potential impacts caused by pumping could be mitigated by diverting a very small fraction of the total groundwater withdrawal to the wetland area or lake in question. Therefore, small-scale, practical means of providing mitigation measure may be available to groundwater users.

The impacts of unsustainable groundwater use may also be mitigated by managed aquifer recharge (Chap. 23). Seasonal surface (storm) water flows and treated wastewater can be used to increase the total recharge to the aquifer and bring the water use closer into balance with replenishment. The link between sustainability and integrated water resources management can provide a framework to both mitigate groundwater impacts and increase the overall water supply (Chap. 36). Such issues as indirect and direct reuse of domestic wastewater can become critical issues in the overall maintenance of water supplies (Chaps. 29 through Chap. 31).

References

Alley, W. M., & Leake, S. A. (2004). A journey from safe yield to sustainability. *Ground Water, 42*(1), 12–16.

Alley, W. M., Reilly, T. E., & Franke, O. L. (1999). *Sustainability and ground-water resources*. U.S. Geological Survey Circular 1186.

Attia, F. A. P. (2001). National and regional policies concerning sustainable water use. In *Proceedings, Regional Aquifer Systems in Arid Zones—Managing Non-renewable Resources, Tripoli, Libya, 20–24 November 1999* (pp. 439–450): Paris: UNESCO.

Balleau, W. P. (1988). Water approximation and transfer in a general hydrogeologic system. *Natural Resources Journal, 29*(2), 269–291.

Bredehoeft, J. D., & Durbin, T. (2009). Ground water development—the time of capture problem. *Ground Water, 47*(4), 4, 506–514.

Custodio, E. (2002). Aquifer exploitation: What does it mean? *Hydrogeology Journal, 10*, 254–277.

Devlin, J. F., & Sophocleous, M. (2005). The persistence of the water budget myth and its relationship to sustainability. *Hydrogeology Journal, 13*, 549–554.

Freeze, R. A., & Cherry, J. A. (1979). *Groundwater* (604 pp). Englewood Cliffs, NJ: Prentice-Hall.

Hiscock, K. M., Rivett, M. O., & Davison, R. M. (2002). Sustainable groundwater development. In K. M. Hissock, M. O. Rivett, & R. M. Davison (Eds.), *Sustainable groundwater development* (pp. 1–14). Geological Society, London Special Publications 193.

Huisman, L., & Olsthoorn, T. N. (1983). *Artificial groundwater recharge*. Boston: Pitman Advanced Publishing.

Kalf, F. R. P., & Wooley, D. R. (2005). Application and methodology for determining sustainable yield in groundwater systems. *Hydrogeology Journal, 13*, 295–312.

Lee, C. H. (1915). The determination of safe yield of underground reservoirs of the closed basin type. *Transactions of the American Society of Civil Engineers, 78*, 148–151.

Llamas, M. R. (2001). Considerations on ethical issues in relation to groundwater development and/or mining. In *Proceedings, Regional Aquifer Systems in Arid Zones—Managing Non-renewable Resources, Tripoli, Libya, 20–24 November 1999* (pp. 475–487). Paris: UNESCO.

Loucks, D. P., & Gladwell, J. S. (1999). *Sustainability criteria for water resources systems*. Cambridge: Cambridge University Press, International Hydrology Series.

McKean, E. (2004). *The New Oxford American Dictionary* (2nd ed.). New York: Oxford University Press.

Pereira, L. S., Cordery, I., & Iaconides, I. (2002). *Coping with water scarcity*. UNESCO, International Hydrological Programme, IHP-VI, Technical Documents in Hydrology No. 58.

Price, M. (2002). Who needs sustainability? In K. M. Hissock, M. O. Rivett, & R. M. Davison (Eds.), *Sustainable groundwater development* (pp. 75–81). Geological Society, London Special Publications 193.

Rajagopal-Durbin, A., & Durbin, T. J. (2008). Wells are not always water follies: Sustainable groundwater policies for the American West. *Water Policy, 10*(2), 145–164.

Schlager, E. (2006). Challenges of governing groundwater in U.S. Western states. *Hydrogeology Journal, 14*(3), 350–360.

Shomaker, J. (2007). What shall we do with all this ground water? *Natural Resources Journal, 47*, 781–791.

Sophocleous, M. (2002). Interactions between groundwater and surface water: The state of the science. *Hydrogeology Journal, 10*, 52–67.

Sophocleous, M. (2005). Groundwater recharge and sustainability in the High Plains aquifer in Kansas, USA. *Hydrogeology Journal, 13*, 351–365.

Todd, D. K. (1959). *Groundwater hydrology* (336 pp). New York: Wiley

Walton, W. C. (2011). Aquifer system response time and groundwater supply management. *Ground Water, 49*, 126–127

Winter, T. C., Harvey, J. W., Franke, O. L., & Alley, W. M. (1998). *Ground water and surface water* (79 pp). A single resource: U.S. Geological Survey Circular 1139.

Wood, W. W. (2003). Water sustainability: Science of science fiction? Perspective from one scientist. In A. S. Alsharhan, & W. W. Wood (Eds.), *Water resources perspectives: Evaluation, management and policy* (pp. 45–51). Amsterdam: Elsevier.

World Commission of Development and the Environment. (1987). Report of the World Commission on Environment and Development: Our common future: Annex to General Assembly document A/42/427, Development and International Co-operation: Environment, August 2, 1987 (http://www.un-documents.net/ocf-02.html).

Zektser, S., Loáieiga, H. A., & Wolf, J. T. (2005). Environmental impacts of groundwater overdraft: selected case studies in the southwestern United States. *Environmental Geology, 47*, 396–404.

Part IV
Water Resources Assessment Methods

Chapter 10
Water Resources Assessment Methods: Assessment of Groundwater Resources

10.1 Introduction

The fundamental challenge for managing groundwater resources in arid and semiarid lands is evaluating the response of aquifers to future short-term and long-term abstractions. This technical knowledge is then used as the basis for water-policy and management decisions. For example, groundwater pumping may be restricted to quantities that are considered sustainable, recognizing that sustainable or safe yields are, in practice, values that are difficult to quantify (Chap. 9). Information is, therefore, needed on the impacts of various pumping rates and well distributions on aquifer water budgets, surface-water resources, and sensitive environments. In the case of non-renewable groundwater resources, a strategy of planned depletion may be employed (Chap. 36), in which case information is needed on the likely rate of future declines in aquifer water levels under different pumping scenarios and on various water resource replacement strategies (e.g., managed aquifer recharge) that may be required to meet future human and natural system requirements.

Numerical groundwater modeling is a standard tool for making groundwater resource assessments and projections (Chap. 20). However, accurate groundwater flow and solute transport modeling requires data on aquifer hydrogeology, which may sparse or not readily available in many arid and semiarid lands. Thus, groundwater resource assessments must start with aquifer characterization on both regional and local scales. The basic data required include:

- volume of water in storage,
- aquifer hydrogeology (e.g., three-dimensional distribution of conductive and confining strata),
- values of hydraulic parameters (e.g., transmissivity, storativity, leakance) of each aquifer and confining unit,

R. Maliva and T. Missimer, *Arid Lands Water Evaluation and Management*, Environmental Science and Engineering, DOI: 10.1007/978-3-642-29104-3_10,

- water chemistry and quality (e.g., salinity, major and minor ion concentrations, presence and concentration of chemical and biological contaminants) variations with depth and geographic locations,
- aquifer water levels (heads),
- aquifer water budgets, including recharge and discharge location and rates, and well pumping data and,
- aquifer boundary conditions.

Aquifer characterization methods employed in semiarid and arid lands do not significantly differ from those employed in more humid regions. However, characterization programs (i.e., choice of testing methods) for groundwater investigations in more arid lands may vary from those in humid regions because of the much lower rates and greater spatial variability of aquifer recharge and the greater importance of vadose-zone processes.

Aquifer characterizations should start with a review of available existing data (i.e., a desktop investigation), which may include government agency reports and databases, academic publications, and reports prepared by private consultants. Field testing programs are then designed and implemented to obtain additional required information. The details of the characterization program must be developed with consideration of the needs to meet the overall objectives of the groundwater investigation. For example, a reconnaissance investigation to screen for the possible presence of developable water sources will require less detail than is needed for the development of a groundwater model to be used to make resource allocation assessments.

A wide-variety of tools of varying sophistication is now available to assist in aquifer characterization and data management. Aquifer characterization programs need to optimize the amount and quality of data obtained under the project financial and time constraints. Characterization programs should also focus on the end uses of the data. In particular, if a major objective of a water resources management or development project is the building of a groundwater model (for example, to be used to assess the impacts of a proposed groundwater scheme), then the aquifer characterization program should focus on the main, specific data requirements for development of the model. More complex water management projects, such as managed aquifer recharge systems, have additional data requirements relevant to the need in some systems to recharge and recover water without significant deterioration in water quality (Maliva and Missimer 2010).

Design of aquifer testing programs and the selection of well drilling methods must also be compatible with each other. Well testing methods often require specific borehole conditions. For example, pumping tests required a stabilized borehole, which necessitates installation of some type of screen in strata composed of unconsolidated sediments. Another example is that borehole geophysical logging and packer testing tools have minimum and maximum borehole diameter constraints and many logs require an uncased borehole. Ideally, well construction procedures can be tailored to the requirements of project-specific testing plans. However, in some instances, testing plans may have to be altered to accommodate less flexible well drilling methods caused by the lack of the equipment in a given region or the high cost of using the optimal method.

It is imperative that those involved in groundwater resources assessment projects be aware of the tools that are available, the type of data they can provide, the costs and time requirements, and their advantages and limitations. Any chapter on groundwater resources assessment techniques can only provide a cursory description of each technique since entire books have been written on some of the specific techniques. The intent of Chaps. 10–20 is to present basic concepts, some primary references, and examples that illustrate the applications of the techniques.

10.2 Desktop Investigations

There are very few areas of the world in which at least some data are not already available on local and regional geology and hydrogeology. The first phase of any groundwater resources investigation should be a desktop assessment, which involves a review of available existing data. The term "desktop" refers to the investigation being performed in the office without any fieldwork, other than perhaps a site reconnaissance or well inventory. Information that may be included in desktop investigations includes:

- technical reports and maps prepared by governmental agencies,
- governmental databases (e.g., monitoring data, well records),
- technical publications in scientific and engineering journals,
- books,
- reports prepared by consultants,
- governmental records and datasets (e.g., permit applications and well completion reports), and
- remote sensing data (Chap. 18).

The principal value of a desktop investigation is avoidance of unnecessarily duplication of existing data and information. Existing data can also be very valuable as a guide to the development of exploratory well programs and can augment data collected during the programs.

Over the past decade, the volume of available hydrogeological and geological data has increased dramatically through the internet, particularly for investigators who do not have ready access to a major university or government library. Search engines (e.g., Scopus, GeoRef, Geobase) allow for the identification of papers in areas of interest and copies of journal articles can be accessed in real time via the internet, but often for a charge. Obtaining copies of older papers and books may still be a challenge. Unfortunately, much historic data on groundwater resources is not readily accessible. The term "gray data" refers to books, reports, maps, notes and datasets, which are unpublished or were published in very limited quantities and are difficult to access (Cobbing and Davies 2011). Gray data includes proprietary information that is not intended for public access and distribution. Fragmentation is also a problem, where data on the groundwater resources of a region may be scattered between numerous locations. An important concern is that few people may know what data are available and where they are located, and that once they retire, the data may be lost to current scientists and policy makers (Cobbing and Davies 2011).

Cobbing and Davies (2011) described the South African Development Community (SADC) Groundwater Grey Literature Archive, the purpose of which is to promote and support groundwater exploration, development, and management, in southern Africa and reduce the temptation for expensive duplication of work that may already have been done. The archive involves the scanning of data and making it openly available on the internet. In the SADC database, many of the earlier studies were performed by European geological societies and the only paper copies of the reports may exist in Europe, not in Africa where they are most needed (Cobbing and Davies 2011).

The inaccessibility and loss of gray data are certainly not restricted to developing countries. For example, in Florida, in which groundwater is the primary water source, an enormous volume of data on local hydrogeology and groundwater resources were collected in support of obtaining water resources and regulatory approval for new land developments and water supply projects. A limited number of paper copies of the reports were prepared, which are becoming very difficult to obtain as time passes. Engineering companies that prepared the reports commonly have document retention policies, in which copies of reports are destroyed after a specified period of time. Fortunately, the transition to electronic publishing of reports is facilitating long-term document retention and sharing.

10.3 Exploratory Well and Production Well Construction

Aquifer hydraulic testing is performed using either existing production or monitoring wells, or wells newly constructed as part of an aquifer characterization program. Well drilling represents a major cost for groundwater resource assessments. There is a clear cost savings in the use of existing wells for aquifer hydraulic testing, as opposed to drilling new wells. However, the savings may turn out to be illusory in the long-term if the existing wells are not located in areas where data are most needed or are improperly constructed for the testing program objectives.

Three of the best references on water well construction and testing are: Groundwater and Wells (Driscoll 1986; Sterrett 2007), Handbook of Ground Water Development (Roscoe Moss Company 1990), and Water Wells and Boreholes (Misstear et al. 2006). Well design and the selection of well drilling methods are not influenced by aridity per se. However, well drilling and construction plans need to consider the great depths to water in some aquifers in semiarid and arid lands. Great depths to water can result in increased pressures on casings and screens, which necessitate more robust materials and care during construction (particularly well grouting) compared to wells with shallow depths to water. Casing tensile strength and screen collapse strengths are important design considerations. In areas that are prone to compaction and land subsidence, compression couplings are recommended in order to avoid damage to wells (Sect. 15.5).

The basic best design principles for wells to be used as long-term production wells should be followed in both arid and more humid lands. Wells should be designed to maximize well efficiency, which involves consideration of both open area and length of open holes or screens, and casing and screen diameters. Proper well development is also critical for well performance. Poorly constructed and developed wells have low efficiencies (greater head losses inside the well during pumping) and, as a result, greater operational costs (energy costs for pumping) and often shortened operational lives.

A key issue for the development of production or injection wells is to construct a well with a high efficiency at the target design capacity. Well efficiency (E, %) is defined in terms of theoretical and actual drawdowns (or head increases during injection) as follows:

$$E = \frac{s_{theoretical}}{s_{measured}} * 100 \tag{10.1}$$

The theoretical drawdown (s) is that which would occur in a 100% efficient well (i.e., a well in which no head losses occurs). The theoretical drawdown is essentially the drawdown in the aquifer adjacent to the well. The measured drawdown is measured inside the well casing. As a generalization, well efficiencies of 70% or greater are considered acceptable in screened wells. Open-hole constructed wells should have a higher efficiency, close to 100%.

Well inefficiency is the result of head losses within the well. The main causes of head losses within wells are (1) frictional head loss within the well casing, (2) head losses across the well screen and filter pack, and (3) head losses caused by formation damage (clogging) in the vicinity of the borehole (skin effect). Formation damage may be caused by drilling mud or fine formation materials that are not removed from the well during development or from mineral precipitation (scaling) and biological growth after well construction. Well construction, drilling, and testing methods should all be chosen so as to minimize or prevent all three of the main head losses to the extent practicable.

Head losses within a well may be reduced by increasing the diameter of the well, which decreases the flow velocity and associated frictional head losses. However, for a given flow rate, the frictional head loss asymptotically approaches zero with increasing diameter. A casing diameter can be selected above which additional reductions in head loss become negligible. The frictional head losses within casings of different material (roughness), diameters, and lengths, can be estimated for different flow velocities using the Hazen and Williams equation (Maliva and Missimer 2010). Usually, as casing diameters exceed 40 cm (15.7 in.), the amount of efficiency improvement with increasing diameter decreases as frictional head losses become inconsequential. In no case should the flow velocity in the casing exceed about 3 m/s (10 ft/s) because damage could be done to the formation at the casing set point in open-hole wells or the screen entrance velocities would be too high, causing excessive entrainment of sediment during well operation.

Minimization of head loss is an important factor in screen and filter pack design. Selecting the proper filter pack, screen slot size and corresponding screen-slot velocity are critical for maximizing well efficiency.

Unfortunately, water wells still often have sub-optimal construction because of a lack of resources or failure of the owner or consultant to specify a high-efficiency well design. The total life-cycle costs of poorly designed and constructed wells will usually exceed those of properly designed and developed high-efficiency wells. For example, the construction cost differential between high-efficiency (high open area) well screens and much less expensive slotted screens (in some instances field constructed using a circular saw) are commonly much less than the difference in operational costs over the life of the well.

10.4 Data Collection During Well Drilling

Valuable geological and hydrogeological data can be collected during well drilling, which is particularly important for exploratory well programs. The fundamental formation analysis task during well drilling is to determine the site-specific geology and hydrogeology through a description of well cuttings (and core samples, if obtained). Well cuttings are fragments of rock or sediment recovered from the borehole during drilling. Formation testing procedures were reviewed by Maliva and Missimer (2010) for managed aquifer recharge (MAR) systems and are applicable to aquifer characterization in general. Basic well cuttings analysis should allow for the determination of:

- stratigraphic formations and the location of formation boundaries,
- main lithologies (rock and sediment types),
- location of potential transmissive aquifers or zones (flow zones) and confining strata,
- presence and location of mineral phases that could cause clogging or impact the quality of stored water MAR system wells, and
- location of strata suitable for the seating of casings, placement of screens, and other critical construction issues.

The sample descriptions or characterization are used to determine the need for additional analysis and to choose the samples or intervals for testing. Well cuttings and core descriptions, along with geophysical logging, form the foundation of formation analysis, but are often poorly performed. Inadequate attention is paid to determining the depth in the well from which samples were derived. Depending upon the drilling method and well depth, there may a long time lag (5 min or more) for cuttings to travel from the bottom of the borehole to the collection point at land surface. The lag time must be considered in determining sample depth. When drilling is rapid and the travel time is great, there may be a large (1 m; 3 ft) or greater difference between the current depth of the drill bit and depth from which cuttings were derived. The field inspector must also determine the lithologic material being penetrated by observing the bit penetration rate and rig noise so if geologic materials are mixed at surface, an actual assessment of the lithology at the bottom-hole depth during drilling can be made. Mixed samples without detailed field notes leads to inaccurate geological assessments.

Poor sample collection is regrettably often followed by inadequate or inaccurate descriptions, which were made by poorly trained field geologists, hydrogeologists, engineers, or technicians. It is not uncommon for well construction to be overseen by engineering construction supervisors rather than geologists or hydrogeologists or for the only sample description to be made by the well drillers. Unfortunately adequate sample description is become a dying art, as it is no longer emphasized in university undergraduate and graduate programs. It is recommended that samples be collected and described by trained and experienced geologists in the following manner or order: basic lithology, color, textural and compositional description, hardness, apparent porosity and permeability, accessory minerals, fossils, and any additional observed features. The Shell Oil Company Sample Examination Manual (Swanson 1981) distributed by the American Association for Petroleum Geologists is an excellent, widely used reference on sample description.

10.5 Aquifer Hydraulic Testing

The main source of data on the primary hydraulic parameters of aquifers (transmissivity, storativity, and leakance; Chap. 2) is aquifer performance testing (pumping tests). Other testing procedures, such as borehole and surface geophysics (Chaps. 16 and 17) and drill stem (packer) tests, can provide information on heterogeneity in aquifer hydraulics and water quality. Aquifer testing procedures and data interpretation are discussed in general hydrogeology texts and some specialized volumes (e.g., Stallman 1969; Lohman 1979; Driscoll 1986; Kruseman and de Ridder 1991; Dawson and Istok 1991; Walton 1991, 1997, 2006; Kasenow 1997, 2006; Sterrett 2007; Weight 2008). Maliva and Missimer (2010) discuss aquifer testing procedures with respect to managed aquifer recharge.

An important issue in the design and interpretation of aquifer hydraulic testing is the volume of the aquifer investigated. Test methods differ in the volume of the aquifer from which data are obtained (Fig. 10.1). At the low end of the volume-of-investigation spectrum are core plug analyses for which data is obtained only on the small cylinder of material being tested, which is commonly 2.5 cm (1 in.) in diameter. At the high end of the spectrum are multiple well pumping tests, tidal-response analysis, and tracer tests. The results (e.g., hydraulic conductivity values) from methods with a relatively small volume of investigation may not be representative of the aquifer as a whole, especially in aquifers with a high degree of heterogeneity.

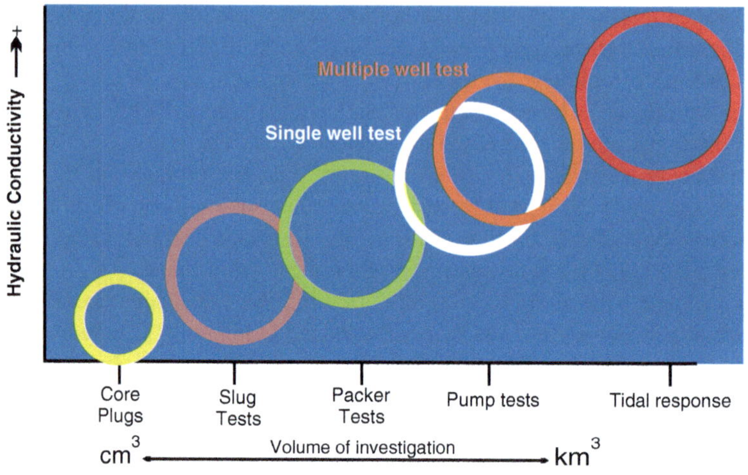

Fig. 10.1 Scaling factor: measured hydraulic conductivity tends to increase with the volume of investigation

As a generalization, a scale effect exists in which calculated transmissivity (or average hydraulic conductivity) value increases with the volume of investigation. Hydraulic conductivity values from core plugs tend to reflect the properties of the rock itself (i.e., matrix permeability). Methods with larger volumes of investigation tend to be strongly influenced by secondary pore systems and other aquifer heterogeneity. In particular, as the volume of investigation increases, there is a greater likelihood that flow zones with high hydraulic conductivity will be encountered. An extreme example of the scale effect occurs in karstic limestones in which transmissivity values estimated from core plug data may be orders of magnitude lower than values obtained from multiple well pump tests that encountered flow conduits.

All methods and scales of measurement provide important information. The scale factor can be taken advantage of to obtain insights onto aquifer heterogeneity and thus patterns of groundwater flow and solute transport. It is, however, quite important to understand the scaling factor when developing groundwater models and in regional groundwater resource assessments so that the error range in hydraulic values, particularly hydraulic conductivity, can be evaluated. In the case of large potable-supply wellfield and managed aquifer recharge projects the volume of investigation for aquifer testing should of a comparable size as the reservoir or plume of stored or produced water (Maliva and Missimer 2010).

10.5.1 Aquifer Performance Testing

Aquifer performance tests (APTs), also referred to as just pumping tests, are an integral component of groundwater resources investigations. APT tests involve the recording of the aquifer pressure response to controlled groundwater pumping. The simplest APT is a one-well test in which the drawdown in a production well is recorded upon the initiation of pumping. Single well testing can only yield transmissivity and estimated storativity values and give results less representative of the aquifer as a whole compared to multiple-well tests. The preferred test method is a multiple-well test in which usually one production well is pumped and drawdown is recorded in the production well and one or more observation wells. Drawdown is now normally recorded using electronic pressure transducers and dataloggers. Recovery of water levels after the pump is turned off is usually also recorded. The collected time-drawdown data are interpreted to provide data on aquifer hydraulic properties (transmissivity, storativity, leakance). APTs can also provide important information on well yields, well performance (efficiency), and the stability of water quality.

Proper performance of an APT and the collection and interpretation of the data requires attention to detail. It is imperative that pumping rates and changes in water levels (pressure) in the production well and observation wells be accurately measured and recorded and that the water level data be corrected for extraneous (non-pumping test-related) influences such as barometric pressure changes, tidal influences, rainfall and changes in aquifer pressure caused by other aquifer users. Background water level data must be collected before the start of the pumping test to allow for the calculation of drawdowns. APT tests are commonly analyzed using graphical procedures in which time is plotted on a logarithmic axis or by some computer hydraulic modeling methods. Frequent readings (1 s) are, therefore, needed at the start of the test, whereas 10 or 20 min readings are sufficient for the later part of the test.

A critical aspect of aquifer testing is to maintain constant, known pumping rates within a small tolerance, commonly at about 5% or less of the initial rate. The pumping system should include a means for adjusting and accurately measuring flow rates, such as a recently calibrated flow meter. It is good practice to have a secondary method for confirming the flow rate, which could be an orifice assembly with a manometer tube or recording the time required to fill a container of known volume. APTs should be performed for as long a period as is practicably possible (at least several days) in order to allow for the determination of leakance values, observe the delayed yield in unconfined aquifers, and detect pumping-induced changes in water quality. However, it is recognized that in some circumstances only short duration tests may be performed because of water disposal issues.

Pump tests can be performed using either existing production wells, newly installed production wells or test wells constructed at a smaller size and cost. Another option is to use an existing well, as a production well and installing one or more observation wells. Use of existing wells for aquifer testing offers an obvious cost savings, but consideration should be given to whether the well is properly constructed for testing purposes. Existing wells may be open to only part of the aquifer in question (i.e., are partially penetrating wells), which can complicate interpretation of test results. An existing well may also be open to more than one aquifer. Partially penetrating wells are common in thick aquifer systems in which water quality deteriorates with depth (e.g., some basin fill aquifers in the southwestern United States, Nubian Aquifer System of North Africa). In the absence of data on deeper parts of aquifer, data from the tested shallower wells (e.g., average hydraulic conductivity, specific storage) are often extrapolated to the untested deeper parts of the aquifers or corrected for the vertical component of the flow path using various mathematical methods. However, the hydraulic properties of upper production zones may not be representative of the aquifer as a whole. The greater burial depths of the deep parts of aquifers may result in increased compaction and diagenesis with associated lower porosities and hydraulic conductivities. Vertical variation in hydraulic properties in some aquifers does not allow any reasonable partial-penetration corrections to be made.

Analytical methods have been developed for the interpretation of aquifer performance tests performed on partially penetrating wells. However, the underlying assumptions of the analytical methods are typically not met, particularly that the aquifer is homogeneous and isotropic. Valid criticism has been raised that analytical methods that have been traditionally used for the analysis of aquifer performance test data are becoming obsolete because they are based on simplified conceptual models that are not representative of actual aquifer hydrogeology, particular with respect to heterogeneity (Yeh and Lee 2007; Walton 2006, 2008).

The recommended alternatives include numerical inversion techniques, such as programs based on the MODFLOW code. Numerical methods have the great advantage that they are much more flexible for the incorporation of aquifer heterogeneity. Numerical methods can also more readily accommodate other deviations from ideal conditions, such as variations in pumping rates and the corrections for external water level changes during testing, such as from a rainfall event. A limitation of numerical techniques in general is that they do not provide unique solutions. This issue can be addressed by calculating the hydraulic conductivity value at one or more observation wells independently and holding those values constant during the modeling exercise. This method helps limit the number of potential hydraulic value combinations, but does not fully eliminate the non-uniqueness issue.

10.5.2 Aquifer Testing During Drilling

APTs are usually performed after the completion of well construction. Single-well pumping tests can be performed after the drilling of a borehole to various depths in order to obtain coarse-scale information or down-hole changes in hydraulic conductivity (transmissivity), specific capacity (well yield divided by drawdown), and water quality.

Performance of a series of pumping tests during drilling is most easily done on wells completed in competent (lithified) rock in which the borehole is stable. After drilling a specified depth below the lowest casing, the drill string is removed and a short-term (several hours to one day) pumping test performed. A transmissivity value (T_1) for the test interval ($D_1 - D_0$) (Fig. 10.2) is calculated from the time-

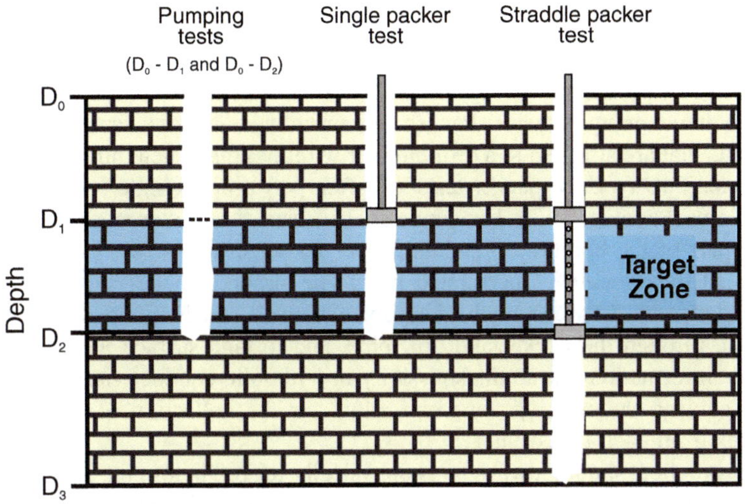

Fig. 10.2 Testing options for a target aquifer zone. Pumping tests can be performed both after drilling to the *top* and *bottom* of the zone. A single (*off-bottom*) packer test may be performed after drilling to bottom of the zone. Alternatively, a straddle packer test may be performed some time after drilling through the target zone

drawdown data. After further drilling to a greater depth (D_2), another pumping test is performed and a transmissivity value (T_2) is obtained for the interval ($D_2 - D_0$). The transmissivity value for the interval ($D_2 - D_1$) is the difference between the two calculated transmissivity values ($T_2 - T_1$). If there is no significant increase in transmissivity between two progressive pumping tests, then it can be concluded that last drilled interval has a very low transmissivity (i.e., the interval is not productive). Water samples should be collected during each test to evaluate down-hole changes in water chemistry such as increases in salinity.

Data collection options are more limited in aquifers composed of unconsolidated sand, gravel, or other material. The borehole must be stabilized during both drilling and the installation of casings and screens. The mud-rotary method is widely used when drilling through sands. The weight of the column of drilling mud results in sufficient pressure to prevent caving of the borehole wall. The principle disadvantage of the mud-rotary method is that it is more difficult to perform pumping tests during drilling. Once the mud is pumped out, the borehole is prone to collapse.

Aquifer pumping tests can be performed when the mud-rotary-mud method is used by installing a temporary screen. Temporary screens may also be installed when drilling using other methods in which the borehole is stabilized using a casing (e.g., cable tool and dual-tube methods). A sand or gravel filter pack, sealed at top with bentonite, is usually installed around the temporary screen. Upon completion of testing, the temporary screen is removed and drilling resumed.

10.5.3 Packer Tests

Packer tests (also referred to as drill stem tests) are an alternative or supplement to full borehole pumping tests. Packers are installed with one or more inflatable elements that are attached to a drill pipe (or wireline) that when inflated seal off the borehole. Two types of packer tests are often performed; single (off bottom) and straddle packer (Fig. 10.2). For single packer tests, the packer is set to isolate the lower part of the borehole. Straddle-packer tests consist of two packers that are separated by a length of perforated pipe. The straddle-packer system isolates a selected interval between the packers for testing. Water is pumped from the test interval, through the packer and drill pipe, and to land surface. Packer tests can provide information on the transmissivity or hydraulic conductivity of an interval of an aquifer or confining zone and allow for the collection of discrete water samples from each tested interval.

The borehole should always be geophysically logged before the performance of packer tests. Packers are designed to be inflated inside boreholes with specified maximum and minimum diameters. At a minimum, an X−Y caliper log should always be performed prior to running a packer test on an open borehole. The caliper log is necessary to determine the actual borehole diameter. Packers should be set at depth intervals in which the borehole is close to being circular, has a near constant diameter, and has a low degree of roughness. Sharp-edged features in a borehole or high degrees of irregularity, such as a large solution cavity, can damage the packer or allow leakage of water around the packer, thereby causing inaccurate data to be obtained.

Both single and straddle-packer tests have their advantages. Leakage around the packer, and thus production of spurious data, is an important concern for packer tests. Single-packer tests have an advantage in that only one packer element has to be properly seated, and thus there is a lesser potential for significant leakage during the test. Single-packer tests are also usually performed shortly after the interval to be tested has been drilled, so there is less time for invasion of borehole fluids into the tested intervals to occur. Single-packer tests may therefore require less development time before representative formation water samples can be obtained. A disadvantage of single-packer tests is that the packer assembly usually has to be tripped in and out of the borehole for each test, which increases testing time and cost.

Straddle-pack tests have the advantage in that multiple tests can be performed without having to remove the packer assembly from the well for each test. The packer assembly can be raised or lowered to the next test interval. A borehole can be drilled to total depth and then a full-suite of geophysical logs can be run. Multiple tests can then be conducted on intervals determined from analysis of the logs to be of hydrogeological interest and have suitable borehole conditions. Multiple packer tests are most easily run on boreholes drilled using the reverse-air rotary method because there is minimal invasion of drilling fluids into the formation and clogging of the formation pores and the borehole wall (i.e., mudcake development).

Short-term pumping tests are performed after packer instillation, and the time-drawdown data evaluated to determine the transmissivity (and in turn average hydraulic conductivity) of the tested interval. Head losses within the small-diameter drilling pipe may be substantial and should be corrected as part of the data analysis. It is also good practice to set a pressure transducer above the packer in the annulus between the drill pipe and borehole in order to detect leakage. Pressure (water level) above the packer should not materially change when the packed-off interval is pumped.

10.6 Vadose-Zone Hydraulic Testing Methods

Vadose-zone processes are not directly addressed in most arid and semiarid region water resources investigations, which are primarily focused on characterizing large-scale aquifer processes. Data on vadose zone properties such as infiltration rates, moisture content, and matric potential may be useful for evaluations of recharge processes (Chaps. 11–13) and managed artificial recharge systems involving land application. Vadose zone processes are also important for contamination assessments and remediation. Some of the basic methods for hydraulic testing of vadose zone sediments are discussed herein. More detailed discussions are provided in soils texts (e.g., Hillel 1998; Dane et al. 2002; Anderson 2005). A variety of geophysical methods can be used to measure vadose zone hydraulic properties, such as time domain reflectometry and ground penetrating radar (Chap. 16).

10.6.1 Infiltration Tests

The double-ring infiltrometer test is the most widely used method to measure vertical infiltration rates in unconsolidated sediment. The method is described by Bouwer (1986); Gregory et al. (2005), the American Society of Testing and Materials (ASTM 2009), and many environmental geology textbooks. ASTM Standard D3385-09 provides a standard guideline for the test, but in practice a variety of methodologies are used depending in local circumstances.

The double-ring infiltrometer consists of two concentric metal cylinders that are driven into the soil (Fig. 10.3). There is flexibility in the choice of cylinder diameters. The ASTM Standard D3385-9 apparatus consists of two rings of 12 in. (30 cm) and 24 (60 cm) inches in diameter. Double-ring infiltrometers are commercially available. Water is added into the inner ring and the rate of infiltration is measured. The objective of the test is to measure the vertical infiltration rate and thus vertical hydraulic conductivity. The second, outer ring minimizes errors associated with lateral flow. The annulus between the inner and outer rings is kept full of water to reduce leakage between the rings. Single-ring infiltrometer tests have greater errors associated with lateral flow.

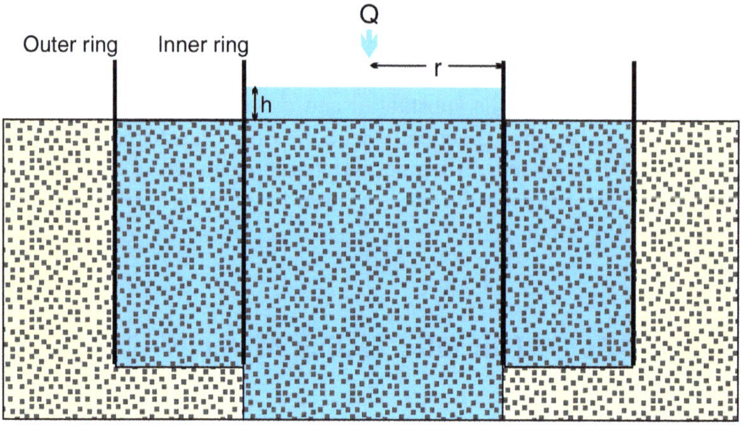

Fig. 10.3 Schematic diagram of a double-ring infiltrometer. During a constant-rate test, the flow rate (Q, m^3/s) into the inner ring is adjusted until a constant water level (h) and thus steady-state infiltration rate is achieved. Hydraulic conductivity can be estimated by dividing the flow rate by the area of the inner ring (πr^2)

Double-ring infiltrometer tests can be performed in either a constant-rate or falling-head manner. In general, the constant-rate test method is preferred because it provides a lower error range. The water level in the inner ring is kept constant and volume of flow is recorded. Commonly, a Mariotte bottle (siphon) is used to maintain a constant flow rate. A flow regulating valve (float valve) could also be used to maintain a constant head. The test is continued until the flow rate is constant and a steady state is achieved. In practice, tests have durations of about 2 h unless the sediments have a very low vertical hydraulic conductivity. Falling head tests involve the recording of the time required for a decrease in head. Falling-head tests tend to underestimate hydraulic conductivity in coarse-grained sediments, but have the advantage of requiring less water and time to conduct.

Assuming a unit hydraulic gradient, hydraulic conductivity can be calculated directly from the infiltration rate (volume/time) by dividing the flow rate by the area of inner ring. An operational source of error in infiltrometer tests is leakage between the ring and soil, which would result in an overestimation of hydraulic conductivity. There is still an error associated with lateral flow even with double-ring tests. Both single and double-ring infiltrometer tests have great accuracy when larger diameter rings are used. However, the greater diameters increase the volume of water required, which may be an operational constraint in sandy soils that have relative high hydraulic conductivities, especially in arid lands where water may have to be transported large distances to test sites. Numerical simulations indicate the infiltration rates from double-ring tests using a 30 cm (12 in.) diameter outer ring may be 56–150% greater than the actual vertical infiltration rate (Wu et al. 1997). The error (correction factor) is reduced to 20–33% when the outer ring diameter is increased to 120 cm (47.2 in.).

Double-ring infiltrometer tests are also point measurements of infiltration at a site, and thus, the results may not be representative of the area of interest. Multiple tests should be performed to better characterize the infiltration area. Layered soils can also significantly affect infiltration rates, depending upon the position of the wetting front relative to the textural discontinuity and the time of measurement (Wu et al. 1997). Infiltration rates vary depending on the antecedent soil moisture with more rapid rates occurring in dry soils. Soils should be pre-wetted or the infiltration tests should have a long enough duration so that steady-state saturation conditions are achieved. A time-series of rates recorded during a test can provide insights into the relationship between antecedent soil moisture and infiltration rate.

It is important to achieve a good seal between the rings and sediments in order to ensure that the flow of the water is through the tested sediments. Larger areas (diameters) are preferred in order to reduce effects of lateral divergence of flow beneath the infiltrometer. Infiltrometers can be either circular or square. In the case of the design of infiltration basins, standard diameter double-ring tests are often performed for initial site screening and testing. Further testing may then be performed using larger area test trenches or basins.

10.6.2 Soil Moisture Measurement

Soil moisture is a critical parameter for agriculture. Various soil-moisture measurement techniques have been developed for agriculture with the objective of optimizing irrigation-water application rates. However, agricultural soil-moisture measurement methods may not have the accuracy, precision, and operational range needed for scientific applications. Soil-moisture data are used in some methods for evaluating infiltration rates and groundwater recharge.

10.6.2.1 Gravimetric Method

The gravimetric method is the simplest and most accurate means to measure the water content of a soil or sediment sample. The mass of water in the sample is the difference in the mass of the sample before and after drying in an oven. The volumetric water content θ is calculated as

$$\theta = \frac{m_{wet} - m_{dry}}{\rho_w V_b} \tag{10.2}$$

where,

m_{wet} = mass of wet sample (g)
m_{dry} = mass of dried sample (g)
ρ_w = density of water (g/ml)
V_b = sample volume (ml).

The gravimetric method is performed on grab (not in situ) samples. It has a high accuracy, if performed correctly, because mass can be measured very accurately. Care must be taken to avoid evaporative loss of water from samples between the time of collection and measurement. Care must also be taken to accurately determine the undisturbed volume of the sample, which can be reduced by compaction during sample collection.

10.6.2.2 Soil-Moisture Blocks

Soil moisture blocks consist of two electrodes embedded in a porous media, commonly gypsum. The principle is that the moisture content of the block will tend to equilibrate with the moisture content of the adjoining soil. Changes in the moisture content of the blocks results in a change in its resistivity, which is measured using a resistivity meter. Resistivity is either measured manually or a data logging system is used. The resistivity values are then converted into soil moisture values using instrument specific conversion coefficients. For more accurate measurements of soil moisture, the soil-moisture blocks should be calibrated against local soils. Soil-moisture blocks are a relatively inexpensive means to measure soil-moisture, but they have a relatively low accuracy and have a predominantly agricultural use. The blocks also have a limited operation life, as gypsum will dissolve over time.

10.6.2.3 Neutron-Moisture Probe

The neutron-moisture meter (or probe) was developed in the 1950s as a field tool for measuring soil moisture (Gardner et al. 1952; Stone et al. 1955). The theory and applications of neutron moisture probes are reviewed by Kramer et al. (1992). Neutron moisture probes use a low-strength radioactivity source to measure soil moisture. The neutron moisture probe is similar to the neutron borehole geophysical log that has long been used in the oil and gas industry (Chap. 17).

The basic concept behind nuclear logging is that the probe contains a radioactive source that bombards the adjoining formation with fast neutrons. The radioactive source typically consists of a radioactive emitter and beryllium (e.g., americium beryllium). The fast neutrons interact (collide) with atoms in the surrounding formation and are slowed down, scattered, and loose energy (i.e., moderated). Neutron moderation is dominated by collisions with hydrogen atoms. Collisions of fast neutrons with heavier atoms results in little loss of neutron velocity. Sources that emit higher energy neutrons sample a larger volume of soil, but the effect is minor (Ziemer et al. 1967). A 'cloud' of slow neutrons forms in the formation surrounding the probe.

The interaction or collision of fast neutrons with hydrogen atoms results in the emission of gamma rays. When neutrons are slowed down to an energy level of 0.025 eV they are described as 'thermal neutrons'. Neutrons with an intermediate energy level are referred to as 'epithermal' neutrons. The neutron interactions may be measured by detectors that measure the released gamma rays, epithermal neutrons, or thermal neutrons. Probes that record thermal neutrons are most commonly used in modern tools. The neutron moisture probe differs from the borehole logging tool in that lower strength sources are used and the probe has a single detector (Kramer et al. 1992). Neutron-moisture probes are, therefore, smaller, less expensive, and simpler to use compared to a neutron borehole geophysical logging tool. Neutron moisture probes are portable and can be carried and operated by a single person. Hydrogen atoms are present in sediments and rocks in pore waters, mineral phases (especially clay minerals), solid organic matter, and hydrocarbons. In saturated rock, the neutron borehole geophysical log is used to measure porosity, as the concentration of hydrogen atoms in the formation is correlated with total porosity. In unsaturated sediments and rock, the neutron log response is proportional to the total water concentration in the formation. The neutron-moisture probe thus provides a means for measuring the water content in the vadose zone, which is a function of the porosity of the formation and the degree of saturation.

The raw neutron log response (count rate) must be processed to obtain a measurement of porosity or soil moisture. The count rate will also reflect hydrogen atoms present in phases other than water in pore spaces, such as hydrogen atoms within clay minerals. Neutron counts may be reduced by slow neutron absorbers, such as chlorine and boron. Neutron counts may, therefore, be significantly reduced in arid solids with a buildup of salts (Kramer et al. 1992). Neutron-moisture probes may be used in grouted wells after the grout has stabilized. The thickness of the grout should not be more than 7.6 cm (3 in.) (Kramer et al. 1992).

The precision, or reproducibility of successive neutron measurements, depends upon instrument reliability, background stability, and count time per measurement. Longer count times result in greater precision (Kramer et al. 1992). For quantitative measurement of porosity or soil moisture content, it is critical that the tool be adequately calibrated. Calibration of soil moisture probes has traditionally been performed in the laboratory or field by measuring the response of a given probe in a target soil types at two or more known soil moisture contents and regressing these to a linear model (Douglass 1966; Kramer et al. 1992). The soil moisture contents should encompass the moisture contents of interest in the natural system.

Laboratory calibration can be performed using a 0.2 m^3 (55-gallon) drum with a centered access tube (of identical construction as the field access tubes) and packed with target soils of different moisture contents (Kramer et al. 1992). Field calibration can be performed by collecting soil samples next to wells or access tubes of the same construction as study stations and measuring the soil moisture content by other means such a gravimetrically (Kramer et al. 1992; Andraski 1997). Ideally, separate calibration curves should be prepared for each soil type present in the strata of interest. However, where numerous soil types are present, performance of separate calibrations may not be practically possible. Grouping of similar soil types may therefore be necessary (Kramer et al. 1992). The effort made in calibration should be commensurate with the accuracy requirement for a given project.

Time-series profiles at stations can be used to evaluate both changes in the volumetric water content and the depth of penetration of infiltrating water, particularly when a precipitation event or seasonal infiltration reaches the water table (Flint and Flint 1995). A key issue for time series analyses is assessing and correcting for any long-term drift in the readings caused by decay of the radioactive source in the probe, use of different instruments, or other causes.

10.6.2.4 Time-Domain Reflectometry

The dielectric constant (permittivity) of soils is a function of the volumetric soil moisture content. The time-domain reflectometry (TDR) method measures soil moisture by determining the value of the dielectric constant of soils from the velocity of propagation of high-frequency electromagnetic pulses. TDR systems consist of a pulse generator, sampling oscilloscope and coaxial cable with two or more parallel soil (i.e., needle, nail) probes (i.e., waveguides). The TDR technique was reviewed by Topp et al. (1980), Ledieu et al. (1986), Plauborg (1995), Gregory et al. (1995), Ferre and Topp (2002), Robinson et al. (2003), Laurent et al. (2005), Timlin et al. (2007), and Graeff et al. (2010).

An electromagnetic pulse is generated at the top of the probes, which is propagated to the bottom the probes. The technique records the time between the reflections from the top and base of the soil probe. The underlying equation is

$$K = \left(\frac{ct}{2L}\right)^2 \tag{10.3}$$

where,
K = dielectric constant (dimensionless)
c = speed of light in free space (m/s)
t = time between reflections (s)
L = length of probe needles (L).

The main advantages of TDR are that it is a quick and non-destructive technique, and can be used for monitoring spatial and temporal variations of soil moisture through the installation of numerous probes. The main disadvantages are a small volume of investigation and high costs of instruments. TDR is rapidly becoming the technology of choice to monitor soil moisture and to assess moisture gradients (Plauborg 1995).

10.6.2.5 Resistivity-Based Geophysical Methods

The electrical resistivity of unsaturated sediments is a function of water content. Resistivity based geophysical methods are thus useful for assessing the degree of water saturation within the vadose zone. Resistivity is also a function of the salinity of the water. Surface and borehole geophysical methods are discussed in

Chaps. 16 and 17. Surface geophysical methods used to evaluate soil moisture include DC resistivity (electrical resistivity tomography), time-domain reflectometry (TDR) and ground penetrating radar (GPR).

The application of GPR for the measurement of soil moisture is discussed by Jadoon et al. (2008, 2010, 2011), Jonard et al. (2011) and Lambot et al. (2009). Quantitative resistivity-based techniques for the measurement of water saturation have the basic limitation that the bulk resistivity of sediments also depends upon other factors such as porosity and mineralogy. Detailed site-specific calibration is needed for accurate quantification of soil moisture content. Surface resistivity methods are particularly useful for evaluating spatial and temporal changes in water saturation such as related to local recharge.

10.6.3 Soil Matric Potential Measurement

Soil matric (water) potential is a very important parameter in agriculture and soil science because it is directly related to the ability of plants to extract water from soils. Soil matric potential can also be used to estimate soil moisture content using soil–water retention curves. Matric potential is used in some arid region groundwater studies to investigate local groundwater recharge. Matric potential may be used to estimate infiltration rates using the Darcy–Buckingham relationship (Sect. 5.6.3). Matric potential can also be qualitatively used to determine the local direction of water flow in the vadose zone.

Soil water potential measurement techniques were recently reviewed by Druner and Or (2005). The underlying principle of the techniques is that a hydraulic equilibrium is created between the soil water in the porous media around the instrument and the measuring device. Although measurement of matric potential is conceptually simple, care must be taken in order to obtain accurate data for quantitative interpretations. It is critical to understand the limitations of the techniques, operation and maintenance requirements, and the various processes than can affect measured potential values.

10.6.3.1 Tensiometers

Tensiometers are widely used to measure matric potential of soils and to determine irrigation schedules. Tensiometers can be automated to control irrigation water applications when the soil water potential decreases to a pre-determined critical value (Smajstrla and Harrison 1998) and are thus, a useful tool for water conservation by allowing for the reduction of over-irrigation.

A tensiometer consists of a porous cup, connected through a rigid plastic or glass body, to a vacuum measuring device, such as a pressure transducer, bourdon tube or manometer (Fig. 10.4). The porous cup is typically constructed of ceramic material because of its structural strength. The tensiometer is filled with water. The

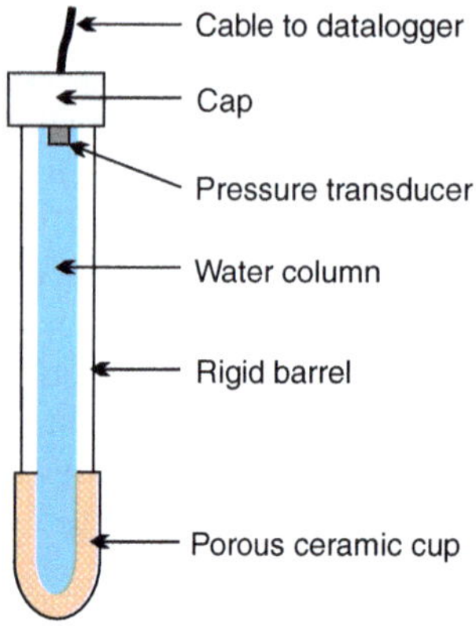

Fig. 10.4 Basic tensiometer construction diagram

porous cup is placed in close hydraulic contact with the soil to be tested. The basic principle is that water will flow from the tensiometer into the soil, which will create a partial vacuum inside the tensiometer. The flow of water will continue until the negative pressure inside the tensiometer is equal to the negative pressure in the soil. The negative pressure measured inside the tensiometers is equal to the soil matric potential (plus any osmotic pressure, which is usually minor).

There are a wide variety of tensiometers designs that are commercially available, with the least expensive varieties being manually-read instruments. Tensiometers that use pressure transducers have the great advantage for vadose zone studies in that continuous readings can be taken and recorded.

Tensiometers have the limitation that their range of operation is limited to 0 to about 85 centibars (J/kg), which is a useful range for plant water studies, but is limited compared to the range in values observed in field soils (Campbell 1988; Smajstria and Harrision 1998). Tensiometers are not suitable for drier conditions. They must also be periodically maintained by expelling air that enters the instruments and must be refilled with water.

10.6.3.2 Thermocouple Pyschrometry

Thermocouple psychrometry is based on an equilibrium being established between liquid soil water and water vapor in the ambient soil atmosphere. It does not require a continuous liquid phase for measurement and has the great advantage that it works over a wide range of soil moistures. It can be used within relatively dry soils where tensiometers are unsuitable. Thermocouple psychrometry is discussed in detail by Andraski and Scanlon (2002). The underlying principle of thermocouple psychrometry is that the partial pressure of water vapor in the soil atmosphere is equal to the vapor pressure of the soil water. Water pressure (ψ; sum of matric and osmotic potential) can be related to relative humidity using the Kelvin equation (Andraski and Scanlon 2002):

$$\psi = (RT/V_{mw})\ln\left(\frac{p}{p_o}\right) \tag{10.4}$$

where,
ψ = water potential (MPa)
R = Universal gas constant (8.314×10^{-6} $MJmol^{-1}K^{-1}$)
T = temperature (K)
V_{mw} = molar volume of water (1.8×10^{-5} m^3 mol^{-1}), and
p/p_o = relative humidity
p = actual vapor pressure of air in equilibrium with the liquid phase (Mpa) and
p_o = the saturation vapor pressure (MPa) at temperature (T).

Psychrometers are used to determine relative humidity from the difference between wet-bulb and dry-bulb temperatures, which in thermocouple pyschrometry is measured using a thermocouple. In situ thermocouple psychrometer sensors have a screened or porous ceramic cap that allows heat and vapor transfer between the soil and sensor chamber in which the thermocouple assembly is located. Data are recorded and stored using a datalogging system. Meticulous instrument maintenance and calibration are essential to instrument performance (Andraski and Scanlon 2002).

References

Anderson, M. G. (Ed.). (2005). *Encyclopedia of hydrological sciences*. Chichester. UK: Wiley.

Andraski, B. J. (1997). Soil-water movement under natural-site and waste-site conditions. *A multiple-year field study in the Mojave Desert, Nevada, Water Resources Research, 33*, 1901–1916.

Andraski, B. J., & Scanlon, B. R. (2002). Thermocouple psychrometry. In J. H. Dane, & G. C. Topp (Eds.), *Methods of soil analyses, Part 4, Physical Methods* (pp. 609–648). Madison, WI: Soil Science Society of America.

ASTM. (2009). *D3385–09 Standard test method for infiltration rate in soils in field using double-ring infiltrometer: West Conshohocken*. Pennsylvania: ASTM International.

Bouwer, H. (1986). Intake rate: Cylindrical infiltrometer. In A. Klute (Ed.) Methods of soil analysis, Part 1: Physical and mineralogical methods (pp. 825–844). Madison, WI: American Society of Agronomy and Soil Science Society of America, Agronomy Monograph No. 9.

Campbell, G. S. (1988). Soil water potential measurement and overview. *Irrigation Science, 9*. 265–273.

Cobbing, J. E., & Davies, J. (2011). Improving access to southern Africa's groundwater "grey data": *Hydrogeology Journal 19*(6). 1117–1120.

Dane, J. H., & Topp, G. C. (Eds.). (2002). *Methods of soil analyses, Part 4*. Madison, WL: Soil Science Society of America: Physical Methods.

Dawson, K. J., & Isotak, J. D. (1991). *Aquifer testing: Design and analysis of pumping and slug tests*. Boca Raton, FL: Lewis Publishers.

Douglass, J. E. (1966). Volumetric calibration of neutron moisture probe *Soil Science Society of America Proceedings 30*, 541–544.

Driscoll, F. G. (1986). Groundwater and Wells (2nd ed.). St. Paul, MN: Johnson Filtration Systems.

Durner, W., & Or, D. (2005). Soil water potential measurement. In M. G. Anderson (Ed.) *Encyclopedia of hydrological sciences*. Chichester: Wiley.

Ferre, T. P .A., and Topp, G. C. (2002). Water content, In Dane, J. H., & Topp, G. C. (Eds.), *Methods of soil analysis. Part 4*. Madison, WI: Soil Science Society of America, pp. 434–446.

Flint, L. E., & Flint, A. L. (1995). Shallow infiltration processes at Yucca Mountain, Nevada—neutron logging data, 1984–1993: U.S. Geological Survey Water-Resources Investigations Report 95-4035, 46 p.

Gardner, W., & Kirkham, D. (1952). Determination of soil moisture by neutron scattering. *Soil Science, 73*, 391–402.

Graeff, T., Zehe, E., Schlaeger, S., Morgner, M., Bauer, A., & Becker, R., et al. (2010). *Hydrology and Earth System Sciences Discussions, 7*, 269–311.

Gregory, P. J., Poss, R., & Micin, S. (1995). Use of time-domain reflectometry (TDR) to measure the water-content of sandy soils. *Australian Journal of Soil Research, 33*(2), 265–275.

Gregory, J. H., Dukes, M. D., Miller, G. L., & Jones, P. H. (2005). Analysis of double-ring infiltration techniques and development of a simple automatic water delivery system: Applied Turfgrass. *Science*, doi:10.1094/ATS-2005-0531-01-MG.

Hillel, D. (1998). *Environmental soil physics*, San Diego, California: Academic Press.

Jadoon, K. Z., Slob, E. C., Vanclooster, M., Vereecken, H., & Lambot, S. (2008). Uniqueness and stability analysis of hydrophysical inversion for time-lapse ground-penetrating radar estimates of shallow soil hydraulic properties. *Water Resources Research*, 44 p. W09 421, doi:10.1029/2007WR006639

Jadoon, K. Z., Lambot, S., Scharnagl, B., van der Klug, Slob, E., & Vereechen, H. (2010). Quantifying field scale soil hydrogeophysical properties using full-waveform inversion of proximal zero-offset ground-penetrating radar: Near Surface. *Geophysics, 8*(6), 483–491.

Jadoon, K. Z., Slob, E., Vereecken, H., & Lambot, S. (2011). Analysis of antenna transfer functions and phase center position for modeling off-ground GPR: In *IEEE Transactions on Geosciences and Remote Sensing, 48*(5), 1949–1662.

Jonard, F., Weihermuller, L., Jadoon, K. Z., Schwank, M., Vereecken, H., & Lambot, S. (2011) Mapping field-scale soil moisture with L-band radiometer and off-ground GPR over a bare soil. *IEEE Transactions on Geosciences and Remote Sensing, 49*, 2863–2875.

Kasenow, M. (1997). *Applied ground-water hydrology and well hydraulics*, Highlands Ranch. Colorado: Water Resources Publications.

Kasenow, M. (2006). *Aquifer test data: evaluation and analysis*, Highlands Ranch. Colorado: Water Resources Publications.

Kramer, J. H., Cullen, S.J ., & Everett, L.G. (1992). Vadose zone monitoring with neutron moisture probe: *Ground Water Monitoring and Remediation, 12*(3), 177–187.

Kruseman, G. P. & de Ridder, N. A. (1991). Analysis and evaluation of pumping test data: International Institute for Land Reclamation and Improvement. Wageningen, The Netherlands, Publication 47.

Lambot, S., Rhebergen, J., Slob, E., Lopera, O., Jadoon, K. Z., & Vereecken, H. (2009). Remote estimation of the hydraulic properties of a sandy soil using full-waveform integrated hydrogeophysical inversion of time-lapse, off-ground GPR data: *Vadose Zone Journal, 8*(3), 743–754.

Laurent, J.-P., Ruelle, P., Delage, L., Zairi, A., Nouna, B. B., & Adjmi, T. (2005). Monitoring soil water content profiles with a commercial TDR system. *Comparison field tests and laboratory calibration, Vadose Zone Journal, 4*, 1030–1036.

Ledieu, J., De Ridder, P., De Clerck, P., & Dautrebande, S. (1986). A method of measuring soil moisture by time-domain reflectometry. *Journal of Hydrology, 88*, 319–328.

Lohman, S. W. (1979). Ground-water hydraulics. *U.S. Geological Survey Professional Paper 708*.

Maliva, R. G., & Missimer, T. M., (2010). Aquifer storage and recovery and managed aquifer recharge using wells: Planning, hydrogeology, design, and operation: Houston, Schlumberger Water Services.*Methods in Water Resources Evaluation Series No. 2*.

Misstear, B., Banks, D., & Clark, L. (2006). *Water Wells and Boreholes*. Chichester, UK: Wiley.

Plauborg, F. (1995). Evaporation from bare soil in a temperate humid climate-measurement using micro-lysimeters and time domain reflectometry. *Agricultural and Forest Meteorology*, *76*(1), 1–17.

Robinson, D. A., Jones, S. B., Wraith, J. M., Or, D., & Friedman, S. P. (2003). A review of advances in dielectric and electrical conductivity measurement in soils using time domain reflectometry. *Vadose Zone Journal, 2*, 444–475.

Roscoe Moss Company. (1990). *Handbook of ground water development*, New York: Wiley.

Smajstria, A. G., Harrision, D. S. (1998). Tensiometers for soil moisture measurement and irrigation scheduling: University of Florida IFAS Extension Publication #CIR487.

Stallman, R. W. (1969). Aquifer testing design, observation, and data analysis, *U. S. Geological Survey Techniques of Water Resources Investigations*, Chapter B1, book 3.

Sterrett, R. J. (2007). Groundwater and wells (3rd ed.). St. Paul, MN: Johnson Screens.

Stone, J. F., Kirkham, D., & Read, A. A. (1955). Soil moisture determination by a portable neutron scattering moisture meter. *Soil Science Society of America Journal, 19*, 419–423.

Swanson, R., (1981). Sample examination manual, *AAPG Methods in Exploration 1.*

Timlin, D., Fleisher, D., Kim, S.-H., Reddy, V., & Baker, J. (2007). Evaporation measurement in controlled environment chambers. *A comparison between time domain reflectometry and accumulation of condensate from cooling coils: Agronomy Journal, 99*, 166–173.

Topp, G. C., Davis, J. L., & Annan, A. P. (1980). Electromagnetic determination of soil water content: measurements in coaxial transmission lines. *Water Resources Research, 16*, 574–582.

Walton, W. C. (1991). *Principles of groundwater engineering*. New York: CRC Press.

Walton, W. C. (1997). Practical aspects of ground water modeling (3rd ed.), Westerville, OH: National Ground Water Association.

Walton, W. C. (2006). *Aquifer test modeling*, Boca Raton, FL: CRC Press.

Walton, W.C. (2008). *Upgrading aquifer test analysis: Ground water*, v. 46, p. 660-662.

Weight, W. D. (2008). *Hydrogeology field manual*, (2nd ed.). New York: McGraw-Hill.

World Nuclear Association. (2011). Nuclear desalination: www.world-nuclear.org/info/in71.html.

Wu, L., Pan, L., Roberson, M.J., & Shouse, P. J. (1997). Numerical evaluations of ring-infiltrometers under various soil conditions: *Soil Science, 162*(11). 771–777.

Yeh, T.-C., & Lee, C. H. (2007). Time to change the way we collect and analyze data for aquifer characterization. *Ground Water, 45*, 116–118.

Ziemer, R. R., Goldberg, I., & MacGillivray, N. A. (1967). *Measuring moisture near surface—minor difference due to neutron source type*. Berkley, CA: U.S. Department of Agriculture Forest Service, Pacific Southwest Forest and Range Experiment Station, Research Note PSW-158.

Chapter 11
Recharge Measurement in Arid and Semiarid Regions

11.1 Introduction

Measuring the rate of groundwater recharge is particularly difficult in arid and semiarid lands, which are characterized by great spatial and temporal variability in recharge. Measurements of a limited number of local (point) recharge rates over a short (several years) period does not provide data representative of basin-wide rates. Most of the recharge tends to occur during rare large rainfall events and the recharge may be concentrated to certain geographic areas such as wadis or depressions that capture runoff. It is therefore highly misleading to assess and express recharge rates in terms of mean annual recharge or recharge as a proportion of the mean annual rainfall (Simmers 1990, 1998). The recharge measurement program must be capable of capturing infrequent and localized pulses of recharge. An additional consideration is that recharge rates in arid and semiarid lands are usually small relative to the resolution and errors of the measurement methods. The uncertainty introduced by errors in recharge rate measurement or calculation is an important consideration for regional groundwater flow models that are used for water management.

When evaluating groundwater recharge it is essential to start with a good conceptualization of different recharge mechanisms that may be important in the study area and tailor the choice of techniques or methods used to the conceptual model. Vital elements in the development of a conceptual model for recharge within a groundwater catchment or basin include (Sophocleous 2004):

- frequency and magnitude of rainfall events,
- importance of overland flow (runoff into and out of basin),
- location of flow ways and catchments,
- pattern and amount of evaporation with respect to land cover,
- potential geographic variation in recharge rates related to soil or rock type, land cover, and topography and,
- presence of macroporosity recharge features.

R. Maliva and T. Missimer, *Arid Lands Water Evaluation and Management*, Environmental Science and Engineering, DOI: 10.1007/978-3-642-29104-3_11,

A variety of methods has been used to estimate recharge rates. The word "estimate" is emphasized. All measurement techniques have inherent uncertainties, which may be substantial. The choice of techniques employed in a water-resources investigation should be based on the appropriateness of the technique for the study site conditions (as evaluated in the conceptual modeling process), the accuracy and uncertainties associated with the method, and economic considerations. Simmers (1990, p. 7) noted that:

> Successful groundwater recharge estimation depends on first identifying the probable flow mechanism and important features influencing recharge for a given locality, since it cannot be assumed that a procedure successfully developed for one area will prove equally reliable for another.

and (p. 20),

> The need to proceed from a well-defined conceptualization of different recharge processes is thus necessary, as is the need to use more than one technique for result verification—recharge estimation should thus be viewed as an iterative process, not as a once-and-for-all calculation.

All water-resources investigations have financial constraints and the development of a program to evaluate recharge needs to optimize the amount and quality of data that can be obtained within the project budget. The investigation of infiltration at the Yucca Mountain, Nevada, site of a proposed high-level nuclear waste repository, as summarized by Flint et al. (2002), is by far the most intensive and expensive investigation of local recharge ever performed, which could be justified based on the overall project budget (over $9 billion U.S. dollars spent to date). However, the budget for most recharge studies is a very small fraction of that available for Yucca Mountain. The challenge for groundwater professionals is to choose amongst the recharge estimation methods available to develop a program that most cost-effectively provides the data needed for an investigation.

A number of review papers have been published on methods available for the estimation of groundwater recharge rates in general and with respect to arid and semiarid environments (e.g., Allison 1988; Gee and Hillel 1988; Lerner et al. 1990, 1997; Simmers 1990; Allison et al. 1994; Stephens 1996; Scanlon et al. 1997, 2002; Flint et al. 2002; Sophocleous 2004; Herczeg and Leaney 2011).

There are three basic types of methods for estimating recharge rates:

- analysis of the rate of water movement downwards through the unsaturated zone (i.e., 'from above', Allison 1988),
- inferring recharge rates from water-table changes (i.e., 'from below', Allison 1988), and
- as a residual in water budget estimates (including modeling investigations).

A wide variety of specific methods has been developed in these three categories. However, most of them have not been widely used in actual applied water-resources investigations, particularly in water-scarce areas. Some techniques involve the use of very specialized equipment, involve a great amount of effort and cost, or just have uncertain accuracy. Some published academic studies utilized techniques that are not widely (commercially) available.

Consideration must also be given to the scale of measurement. Many techniques, particularly those involving hydraulic and tracer analyses of the vadose zone, are point measurements of recharge. A fundamental question is whether or not and how recharge rates obtained from point measurements relate to values over the area of interest and are truly representative in the necessary spatial scale (Simmers 1990, 1998). For example, is the recharge rate calculated from one or several point analyses representative of a groundwater basin recharge rate? For projects of a regional scale, groundwater-based (as opposed to vadose zone-based) methods are relevant and small-scale variability in local recharge ceases to be an issue (Simmers 1998). Local recharge is, however, very important for the transport of surficial contaminants into the groundwater system.

All methods are based on assumptions concerning boundary conditions and transport processes, and require accurate data for the various parameters involved. There are inherent uncertainties in all methods used to estimate recharge rates. Some of the uncertainties related to environmental tracers used in arid environments were discussed by Scanlon (2000). In light of the uncertainties with each individual recharge estimation method, it is generally recommended that multiple techniques should be used to increase the reliability of recharge estimates (e.g., Lerner et al. 1990; Ekwurzel et al. 1994; Stephens 1996; Simmers 1998; Scanlon 2000; Flint et al. 2002; Scanlon et al. 2002; Sibanda et al. 2009). Data from several different techniques can provide insights into the recharge processes that have been operating at a site, which cannot be obtained from a single technique.

Consistency of groundwater ages obtained from different independent data sources increases the confidence in calculated ages. Discrepancies between different methods may provide insights into the hydrological and geochemical process that are active in a groundwater system (see Sect. 11.5.4). It is important to also recognize that different methods, even if applied correctly and based on high-quality data, may yield different recharge fluxes because the methods may represent different temporal and spatial scales (Flint et al. 2002).

There are some general misunderstandings concerning some terms commonly used within the context of infiltration and recharge calculation methods. Net infiltration is sometimes used interchangeably with recharge, but is clearly not recharge. Net infiltration is water that has entered the vadose zone and has passed the root zone. Infiltrated water must pass through the full thickness of the vadose zone and enter the water-table before it becomes recharge. This issue is particularly important in arid lands because the vadose zone tends to be quite thick in inland areas and the wetted front must travel a significant vertical distance before recharge can occur.

11.2 Direct Physical Measurement of Infiltration and Recharge Rates

Direct physical measurement techniques involve the determination of the aquifer water flux to the water-table. Their principal advantage is that they directly measure the parameter(s) of interest. Their major disadvantage is that they are point measurements, and the calculated recharge rates may not be representative of basin-wide recharge rates.

11.2.1 Lysimeter

A lysimeter is the only direct physical method to measure matrix recharge. A lysimeter consists of a large circular or square container of disturbed or undisturbed soil, usually covered with vegetation, that is similar or identical to the soil in the study area (Fig. 7.5). The top of the lysimeter is open and flush with the surrounding soil surface. The walls of the cylinder isolate the enclosed soil from the surrounding soil so the only direction of water flow is vertical. Conceptually, lysimeters are quite simple. The volume of water that infiltrates during a rainfall event can be calculated from the increase in the mass of the soil (corrected for evapotranspiration). The volume of water that flows through the lysimeter and drains out the bottom corresponds to the recharge rate. From measured rainfall rates, the percentage of the total rainfall that becomes recharge can be calculated. In general, the larger diameter and greater depth of the lysimeter and the closer the soil surface and vegetation are to the surroundings, the better the estimate of recharge (Gee et al. 1992). By using large lysimeters (surface area ~100 m^2 [1,076 ft^2]) recharge rates of about 1 mm/year (0.04 in/year) can be resolved (Scanlon et al. 2002).

Lysimeters also allow for the direct measurement of evapotranspiration. Weighing lysimeters (Fig. 7.5) are constructed on balances that allow for the measurement of slight changes in the mass of the soil column. Evapotranspiration can also be measured from non-weighing and floating lysimeters. Evapotranspiration is estimated from the decrease in mass of the soil column corrected for any drainage and precipitation event. Lysimeters can thus provide direct measurements of two of the most critical parameters in the water balance.

Lysimeters are uncommonly used in water resources investigations, particularly in arid regions, because they are expensive to construct and operate. Lysimeters also provide data at a single point, which inherently would not be expected to be representative of a groundwater basin. Construction and operation of multiple lysimeters is cost prohibitive for most investigations. Lysimeters have been used for some academic and government studies, where accurate local vertical recharge information has great importance. For example, multiple lysimeters were constructed at the Hanford nuclear reservation site in the state of Washington as part of the assessment of contamination from nuclear weapons production activities (Gee et al. 1992). In addition, installation of lysimeters without disturbing the soil column can be challenging. Disturbed soils could affect infiltration and percolation rates, and therefore, measured recharge rates.

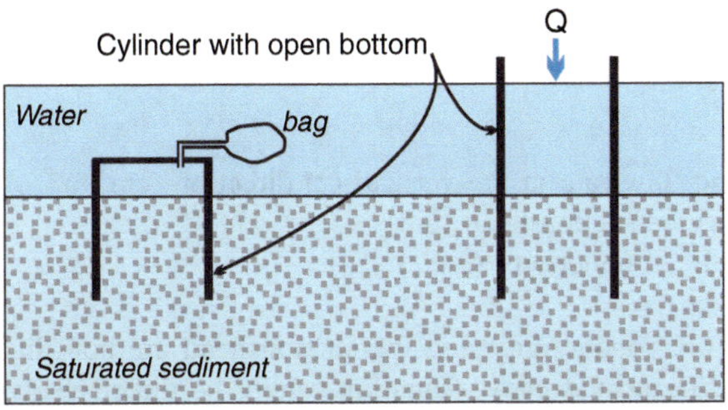

Fig. 11.1 Conceptual diagram of seepage meters. An open-bottomed cylinder is driven into saturated sediments and the downward flow of water is measured. An early design measured flow from the difference in the mass of a bag of water. Seepage rate can be estimated from the flow of water (Q) needed to maintain a constant head

11.2.2 Seepage Meters

Seepage meters allow for the measurement of the seepage rate across the sediment–water interface such as a canal or stream bed. The basic design is an open cylinder that is driven into the sediment and a means for measuring the flux of water across the cylinder. A readily available, common design is the use of a standard 55-gallon (200 l) drum. The seepage rate through the barrel can be measured using either the falling head or the constant head method. An early design by Lee (1977) uses a plastic bag filled with water to measure seepage volume (Fig. 11.1). The seepage volume was determined from the difference in the weight of the bag, and therefore volume of water, before and after the test (ΔV). More sophisticated methods have since been developed to measure seepage volumes. The seepage flux (Q; m/d) can be calculated as:

$$Q = \frac{\Delta V}{tA} \tag{11.1}$$

where
A = cross-sectional area of the cylinder (m^2), and
t = elapsed time (day).
The advantages of seepage meters (Sophocleous 2004) are:

- relative lightweight and easy to operate,
- conceptually simple and rapid measurements can be made, and
- provides direct measurement of seepage flux.

11.2.3 Darcy–Buckingham Law Based Methods

One-dimensional vertical flow through the vadose zone can be expressed using the modification of Darcy's law referred to as Darcy–Buckingham law or equation (Sect. 5.6.3)

$$q_z = -K(\theta)\left(1 + \frac{\partial\psi(\theta)}{\partial z}\right) \tag{11.2}$$

where

q_z = volumetric flow rate in the vertical (z) direction (cm^3/s)

$\psi(\theta)$ = capillary or suction head (cm)

$K(\theta)$ = hydraulic conductivity (cm/s)

z = distance (cm).

Both hydraulic conductivity and the suction head are non-linear functions of water content (θ) and the above equation cannot be directly solved. In practice, estimating recharge rates using the Darcy–Buckingham Law involves a considerable amount of work to obtain data on suction heads and unsaturated hydraulic conductivities, which have a high degree of heterogeneity. Functional relationships can be used to estimate hydraulic conductivity, but these too have significant uncertainties, particularly if general (as opposed to site-specific) relationships are used.

An assumed linear hydraulic gradient can be estimated ($\Delta h/\Delta z$) from tensiometer readings obtained at two or more points in the vadose zone. Using an average unsaturated hydraulic conductivity value for the interval between the tensiometer measurements, $K(\overline{\theta})$, the vertical flux can estimated as

$$q \approx -K(\overline{\theta})\left(\frac{\Delta h}{\Delta z}\right) \tag{11.3}$$

However, it must be understood, that the estimated recharge rates will have a high degree of uncertainty. The calculated recharge rates may not be representative of basin wide rates, because of the small volume of investigation and much of the recharge may occur through macropores.

Another strategy for using the Darcy–Buckingham equation is to estimate recharge rates using a numerical model that incorporates Darcy's law. The recharge rate can be estimated through the model calibration process. Modeling-based recharge estimation is discussed further in Sects. 11.4 and 11.8.

Modeling could be used to calculate water saturations and water potentials for different assumed infiltration fluxes given the estimated or measured hydraulic properties for the strata in the study area. Water saturation profiles can be determined using neutron probe or surface geophysical data (Chap. 16). Percolation fluxes can then be estimated through the calibration process by minimizing the difference between the measured and simulated water saturations and potentials (Flint et al. 2002).

The Darcian method has the advantage in that it is a direct means of calculating percolation and recharge rates, but is also very laborious and expensive, requiring specialized field equipment and personnel. It is also a point measurement and a very large number of measurements would be required in order to obtain a statistically valid estimation of basin-wide recharge rates (Sophocleous 1991).

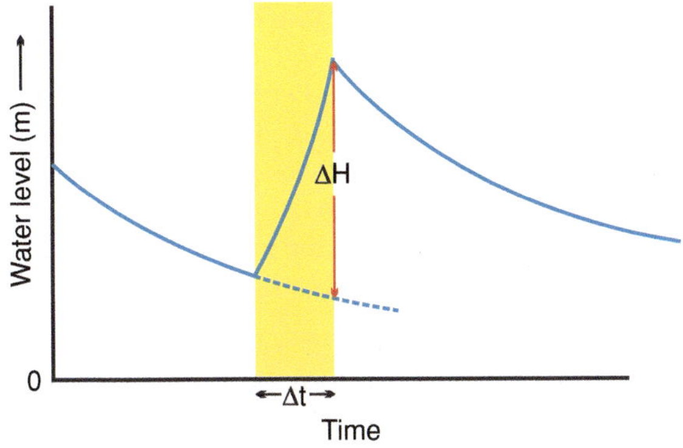

Fig. 11.2 Water-table fluctuation method. The recharge rate for the event that occurred during the time interval Δt is the product of the increase in head (ΔH) and aquifer specific yield

11.3 Water-Table Fluctuation Method

The water-table fluctuation (WTF) method is one of the most widely used methods to estimate recharge rates. Recharge rates are estimated from the aquifer response to rainfall events. The water-table fluctuation method was reviewed by Healy and Cook (2002). For an unconfined aquifer, the recharge from an individual rainfall event can be estimated from the rise in the water-table (ΔH) in response to the rainfall event (Fig. 11.2) based on the aquifer specific yield (S_y):

$$R = S_y \Delta H \quad (11.4)$$

Cumulative recharge is the sum of the responses to all rain events over a period of time. The fundamental requirements of the water-table fluctuation method are that the water-table level responds to rainfall events and that there is no net transport of water from the aquifer. The water-table response must fully reflect the water added into storage. There may be a considerable delay between the rainfall event and water-table response depending on the thickness and hydraulic properties of the vadose zone sediments or rock. Other factors may impact local aquifer water levels, such as changes in barometric pressure, groundwater flow (saturated zone and interflow), diurnal fluctuation in evapotranspiration, entrapped air, and groundwater pumping (Sophocleous 1991; Healy and Cook 2002). Significant corrections to the hydrograph might be required if there are abnormal water level responses caused by trapped air or the Lisse Effect (Weeks 2002; Chap. 5)

A critical parameter in the water-table fluctuation method is specific yield of the water-table aquifer. Specific yield is the volumetric percentage of water that can be drained from a sediment or rock under gravity. An important fact that is often not appreciated is that gravity drainage from fine-grained sediment may take years to be completed. The methods used to estimate specific yield include (Neuman 1987; Healy and Cook 2002):

- Column drainage approach—measurement of the volume of water that drains out of a saturated column of sediment,

- Aquifer performance testing (e.g., Boulton 1963; Neuman and Witherspoon 1972),
- Volume balance (pumped volume divided by volume of cone of depression)
- Water budget (to calculate change in storage),
- Surface geophysical methods (neutron, microgravity) used to estimate change in water storage (Chap. 16),
- Estimates based on soil types (Manghi et al. 2009), and
- Estimates based on grain size analysis (size and sorting) and grain shape.

The main advantages of the water-table fluctuation method are its simplicity, ease of use, low cost, and absence of assumptions on the mechanism by which water flows through the vadose zone (Healy and Cook 2002). It is best applied to shallow unconfined aquifers that exhibit rapid water-level rises and declines in response to rainfall and drainage. Deep aquifers may not show rapid responses to recharge events because wetting fronts tend to disperse over long vertical distances (Healy and Cook 2002). The method also works best for short-term water level rises in response to individual storms, as opposed to long-duration, low-intensity precipitation (Sophocleous 1991; Healy and Cook 2002). The application of the method is also much more complex in dual-porosity systems, particularly because of difficulties in obtaining a reliable estimate of the specific yield (Healy and Cook 2002).

A disadvantage of the water-table fluctuation method is that significant withdrawals are taking place in many shallow aquifers in arid lands. Estimation of withdrawal quantities can have significant challenges as water use is often not metered. One method that has been applied to characterize withdrawals is performance of detailed well inventories, including direct measurements of flow rates. However, there may be large uncertainties in the period of operation of wells. In addition, recharge due to irrigation return flows will also complicate assessment of recharge using the water-table fluctuation method.

The water-table fluctuation method can also be used to calculate recharge rates through the calibration of groundwater flow models, which can incorporate other inflows and outflows of water into an unconfined aquifer. GIS has been be used to estimate basin recharge rates using data on basin-wide withdrawal rates, and point measurements of groundwater levels and specific yields (e.g., Manghi et al. 2009; Sect. 19.3.2).

11.4 Aquifer Water Budget-Based Methods

Recharge rates can be estimated from the aquifer water budget using a variety of different available approaches. Recharge can be estimated as the residual from aquifer water budget, if the values for all of the other main parameters in the budget can be determined or estimated with a high degree of certainty. Under steady state (or near steady state) conditions, recharge can be estimated from discharge rates or the rate of flow of groundwater from recharge to discharge areas.

11.4.1 Aquifer Water Budget

Recharge can be estimated as the residual in aquifer water-budget calculations. The calculations can either be made directly from measured or estimated values of the main water-budget parameters or, commonly, recharge is estimated during the calibration of groundwater flow models (Chap. 20). Recharge is often the least well constrained element of groundwater models and is commonly a primary parameter varied to achieve model calibration. Model-derived recharge estimates should, therefore, be constrained by other independent estimates of local recharge rates.

The basic water budget equation for a specific period of time is:

$$\text{Total inputs} = \text{total outputs} + \text{change in storage}$$

or

$$P + Q_{In} = E_a + Q_{OUT} + \Delta S \tag{11.5}$$

where,
P = precipitation (including anthropogenic inputs such as irrigation)
Q_{In} = total inputs (surface water, groundwater, and vadose zone flow)
Q_{OUT} = total outputs (surface water, groundwater, vadose zone flow, and well withdrawals)
E_a = actual evapotranspiration
ΔS = change in storage.

Equation 11.5 can be expanded into include individual components of the total inputs and outputs. Components that are insignificant for a given investigation can be dropped. Where the other terms in the water budget are negligible (e.g., net surface water and groundwater flow into or out of the study area are not significant), recharge can be estimated as the difference between precipitation and evapotranspiration. The water-budget method is more difficult to accurately apply in arid and semiarid regions than in humid regions because the precipitation rate is frequently only slightly different from the actual evapotranspiration rate, with both components dominating the water budget. Small errors in either or both components can cause proportionally large errors in calculated recharge rates (Stephens 1996; De Vries and Simmers 2002; Scanlon et al. 2002; Sophocleous 2004). The actual evapotranspiration rate value is subject to the greatest uncertainty and, thus, can have a large impact on calculated recharge rates. The water budget may also be difficult to apply because of a lack of accurate data on abstractions (e.g., Al-Shaibani 2008).

Consider an example of a closed basin (no runoff) that receives on average 22 cm (8.7 in) of rainfall per year and has an average annual actual ET rate of 20 cm (7.9 in). Under steady state conditions (no change in storage), the estimated recharge rate would be the 2 cm (0.8 in). A 10 % error in ET rates would be 2 cm (0.8 in), which would result in a 100 % error in the estimated recharge rate. Water budget calculations are also hampered by uncertainty within the values for the hydrogeological parameters, particularly specific yield, which determine the storage change associated with changes in groundwater levels (Scanlon et al. 2002; Manghi et al. 2009).

There is current interest in developing the ability to calculate recharge rates from precipitation and evapotranspiration data obtained using satellite remote sensing data, which are discussed in Chap. 18. However, estimates of recharge currently obtained using remote sensing data may be inaccurate in terms of absolute value, but may still provide useful information of the spatial distribution of recharge.

11.4.2 Spatially Distributed Water Budget Models

GIS-based spatially distributed water budget models have been developed to map and quantify the distribution of groundwater recharge within basins and to evaluate total basin water budgets. The objective of these models is to use georeferenced data to calculate or estimate the values of the critical components in the water budget. The water budget is estimated on a cell-by-cell basis (a grid cell is a defined spatial area). Estimates of recharge using spatially distributed water budget models have the same limitations as estimates from basin-wide water budgets; the results are dependent on the accuracy of input variables, which may be poorly constrained. In arid and semiarid regions, the errors associated with P and ET values are substantially greater than the recharge rate. Quantification of indirect recharge, which predominates in arid regions, is particularly difficult even with the fine areal discretization possible using GIS.

Spatially distributed water budgets share some common features:

- ET rates are estimated from hydrometeorological data and/or vegetative cover information.
- Available water is estimated as the difference between precipitation and ET.
- Available water is portioned between surface runoff and recharge, with recharge rate being a function of soil and bedrock properties.

Flint et al. (2004) developed a GIS based Basin Characterization Model (BCM) to determine the spatial and temporal variability of recharge in basins. The BCM uses spatially distributed estimates of monthly precipitation, average temperature, potential evapotranspiration, and soil–water storage, and bedrock permeability. Depending on soil and bedrock permeability, excess water is partitioned as either (1) in place recharge, and (2) runoff that can become mountain front recharge or stream channel recharge.

The BCM model involves calculation of the available water (AW) in each grid cell as follows (all units in mm/month):

$$\mathrm{AW} = \mathrm{P} + \mathrm{SN_m} - \mathrm{ET_P} - \mathrm{SN_a} + \mathrm{SW_s} \qquad (11.6)$$

where,
P = precipitation,
$\mathrm{SN_m}$ = snowmelt

ET_P = potential evapotranspiration [calculated using the Priestley–Taylor method (Chap. 7)

SN_a = accumulated snow

SW_s = stored soil water carried over from the previous month.

The potential runoff (RO_P) and potential recharge (R_P) are estimated by

$$RO_P = AW - SW_T \tag{11.7}$$

$$R_P = AW - RO_P - FC \tag{11.8}$$

where,

SW_T = total water storage capacity of the soil

FC = field capacity of the soil.

The maximum recharge rate of the BCM model is the bedrock permeability, expressed as cm^3 (in^3) of water per cm^2 (in^2) of grid cell area per month (or cm/month [in/month] within each grid cell). The BCM model does not simulate the flow of runoff between model cells. Instead, an assumed percentage of the runoff is calculated as eventually recharging the aquifer. For example, Flint et al. (2004) assumed 10% of the runoff recharged the aquifer in their Great Basin study.

Batelaan and De Smedt (2007) documented WetSpass, a GIS-based spatially-distributed water budget model for estimating groundwater recharge, actual ET, and surface runoff. WetSpass is an acronym for Water and Energy Transfer between Soil, Plants and Atmosphere under quasi Steady State. WetSpass can be linked to a groundwater flow model (e.g., MODFLOW). The groundwater flow model provides the position of the water-table and WetSPass returns a recharge estimate.

Spatially-distributed recharge is estimated as a function of vegetation, soil textures, slope, groundwater depth, and long-term average hydrometeorological parameters (Batelaan and De Smedt 2007). Each raster cell in WetSpass is sub-divided into a vegetated, bare soil, open water, and impervious fraction for which independent water budgets are maintained. For the vegetated fraction of the raster cell, the seasonal water budget is expressed as

$$P = I + RO_v + T_V + R_v \tag{11.9}$$

$$ET_{tot} = I + T_V + ET_S \tag{11.10}$$

where,

P = precipitation

I = interception

RO_v = surface run off from vegetated area

T_v = actual transpiration

R_v = groundwater recharge in vegetated area

ET_{tot} = actual total ET

ET_s = ET from bare soil.

Evapotranspiration estimations are based on the Penman–Monteith equation (Sect. 7.6.1).

Surface runoff potential ($RO_{v\text{-}POT}$) is estimated using a runoff factor (f_i), which is an empirical function found in published tables. It is dependent on vegetation type (V), soil texture (ST), slope (SL), and groundwater saturated area (D).

$$RO_{v-POT} = f_1(V, ST, SL, D)(P - I) \tag{11.11}$$

Actual runoff (RO_v) is estimated using a second runoff factor (f_2) that is a function of seasonal precipitation intensity distribution (P_i), soil infiltration capacity (I_C), and groundwater saturated area (D):

$$RO_v = f_2(P_i, I_c, D)\ RO_{v-POT} \tag{11.12}$$

The water budget in bare soil, surface water bodies, and impervious surfaces is evaluated in a comparable manner. WetSpass models have large data requirements and a large number of degrees of freedom. Successful model development involves a substantial literature review for parameter estimation, assuming the necessary data are available for the study area. Tihahun and Merkel (2009), for example, populated input raster maps for a groundwater basin in Ethiopia using linear regressions of precipitation, temperature, and evapotranspiration versus altitude.

As is the case for recharge estimation in general, the results of spatially-distributed water budget models should be compared with estimates made using other methods. The large number of variables (and thus degrees of freedom) leads to non-unique solutions. This concern applies to all model-based solutions, because models may converge under many potential multivariate solutions. Therefore, each variable should be constrained to the degree possible to ensure that the modeling results are reasonable.

11.4.3 Darcy Flux in Aquifers (Flow Tubes)

Under steady-state flow conditions, the groundwater flow rate downgradient from the recharge area is equal to the recharge rate (volume/time), if vadose zone percolation is the only source of water and the aquifer has impermeable boundaries (minimal vertical leakage; Stephens 1996). The groundwater flow and recharge rates can be defined by

$$Q = -T\frac{\Delta h}{\Delta l}w \tag{11.13}$$

$$R = \frac{Q}{A} + S\frac{\Delta h}{\Delta t} \tag{11.14}$$

where,

Q = flow rate (m^3/d) (ft^3/d)

T = transmissivity (m^2/d) (ft^3/d)

$\Delta h/\Delta l$ = hydraulic gradient (m/m) (ft/ft)

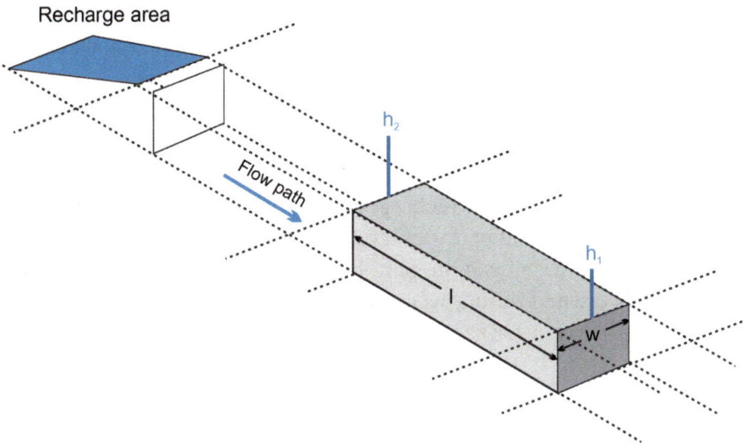

Fig. 11.3 Flow-tube method in a confined aquifer. Under steady state conditions with no significant leakage, the recharge flux (volume/time) is equal to the flow rate through the tube, which is the product of transmissivity, flow tube width (w) and hydraulic gradient ($[h_2-h_1]/l$)

w = aquifer width (m) (ft)
R = recharge rate (m/d) (ft/d)
A = area of recharge (m^2) (ft^3)
S = storativity or specific yield depending on aquifer type (confined versus unconfined, dimensionless)
$\Delta h/\Delta t$ = change in head over time.

As a simplification, the aquifer can divided into flow tubes in which the long axes are oriented parallel to the flow direction (Fig. 11.3). There is thus no lateral flow between adjacent flow tubes. The term "w" in Eq. 11.13 becomes the width of the flow tube and the term "A" in Eq. 11.14 refers to the surface of the recharge surface of the flow tube. The Darcy flux method is conceptually simple, but can be very difficult to apply because of the complexity of aquifer and recharge geometries and the need to account for all other fluxes of water into and out of the aquifer (e.g., vertical leakage into and out of aquifer, and groundwater abstractions or additions).

For a confined aquifer at steady state, the recharge flux (m^3/d; ft^3/d) is equal to the flow rate between the recharge and discharge area, which can be calculated from the transmissivity (m^2/d; ft^2/d), width (m; ft), and hydraulic gradient at one point along the flow path of the aquifer. The recharge rate (m/d; ft/day) can be obtained by dividing the recharge flux by the recharge area of the flow tube. Recharge could also be estimated from measured values of the discharge from an aquifer, as steady state recharge must equal discharge (with corrections made for pumping and other sources and sinks of water). Measurement of discharge can be quite complicated if there is large change in aquifer geometry from the recharge to discharge areas, and where there are multiple discharge points (e.g., springs) or the discharged water is quickly lost to ET. For example, it can be very difficult to accurately measure discharge rates in places with high evaporation rates or where discharge occurs primarily by plant transpiration. The errors associated with discharge estimations may be large relative to the recharge rates.

However, in regional aquifers in arid lands, the assumption that the aquifer is in a steady state condition may not be valid. For example, the flow volumes estimated from hydraulic gradients in North African regional aquifers are too large based on reasonable estimates of current recharge (Bourdon 1977; Lloyd and Farag 1978). The current hydraulic gradients observed in regional basins, such as in Saudi Arabia and North Africa, can in part be attributed to the on-going decay of recharge mounds created during wetter periods during Pleistocene time (Bourdon 1977; Lloyd and Farag 1978).

11.4.4 Stream Gauging: Baseflow Method

In the case of gaining streams (which receive water from groundwater discharge), basin-wide recharge rates can be estimated from the stream baseflow (i.e., flow not associated with storms). When considered at a watershed scale into which there is negligible net groundwater flow, Eq. 11.5 reduces to

$$R - Q_S - Q_w = \Delta S \tag{11.15}$$

where,
R = recharge to the aquifer
Q_S = discharge of groundwater to streams (baseflow)
Q_w = groundwater withdrawals
ΔS = change in storage.

Under steady-state conditions in which there is no change in storage (i.e., aquifer heads have not changed), Eq. 11.17 can be further reduced to:

$$R = Q_S + Q_w \tag{11.16}$$

or

$$R = Q_S, \text{ if well withdrawals are negligible} \tag{11.17}$$

The recharge rate under steady-state conditions is equal to the baseflow out of the watershed. Baseflow can be determined by removing the effects of short-term storm events from stream hydrographs. Baseflow measurements thus require stream gauging data, which is often unavailable in semiarid and arid lands, and especially for ephemeral streams. Changes in baseflow over time reflect the removal of water from storage (Horton 1933). Where a stream has flow during a substantial wet period (rare in arid lands), recharge can be determined using the Meyboom (1961) or the Rorabaugh (1964) methods. The Meyboom method involves determination of the seasonal stage recession rates to assess baseflow which originates from an aquifer in hydraulic connection to a stream. The

Rorabaugh method uses recession curve displacement. There are also a variety of automated methods to determine baseflow from stream hydrographs (Rutledge 1993, 1998; Rutledge and Daniel 1994), which can be performed either manually or, more recently, using computer programs (i.e., Rutledge 1998). Seepage into channel streams may also quickly evaporate, so the baseflow method often is not appropriate in arid and semiarid lands.

11.4.5 Surface Water Budget of Channels

Recharge within surface-water channels can also be estimated from channel water budgets. Wadi recharge is discussed in detail in Chap. 13. Critical data are flow rates from stations at the upstream and downstream end of the channel reach (studied segment). Recharge would be reflected by a decrease in flow between the upstream and downstream gauges, which is the referred to as a 'transmission loss'. The transmission loss needs to be corrected for any additional inputs or outputs of water between the gauging stations, such as tributary flow, evapotranspiration, precipitation, withdrawals, and interflow (i.e., flow within the vadose zone). The channel water budget can be expressed as (Lerner et al. 1997; Scanlon et al. 2002):

$$R = Q_{UP} - Q_{DOWN} - \sum Q_{IN} - \sum Q_{OUT} - E_a - \frac{\Delta S}{\Delta t} \tag{11.18}$$

where,

Q_{UP}, Q_{DOWN} = flow rates at upstream and downstream ends of reach

$\sum Q_{IN}$, $\sum Q_{OUT}$ = rates of tributary outflows and inflows

E_a = evaporation from surface water or stream bed (including transpiration from vegetation)

$\Delta S/\Delta t$ = rate of change of channel and unsaturated zone storage.

Precipitation into the channel can also be included, but would normally be a very small fraction of the total stream flow. Interflow (i.e., lateral vadose zone flow) may also need to be considered if it results in a significant net inflow or outflow from the investigated stream reach. The surface-water budget analyses have the same limitation as basin or catchment-scale analyses in that uncertainties in the accuracy of the stream flow rate and tributary flows result in a proportionately much larger uncertainty in the calculated recharge rate.

The method is simpler to apply to ephemeral streams in semiarid and arid regions, where the stream has loosing conditions all of the time (negligible baseflow into the stream over the reach of interest). Infiltration volume over an interval of time would be equal to the stream loss, minus local evapotranspiration. However, the volume of water that actually reaches the water-table (i.e., aquifer recharge) may be a small fraction of the transmission loss because of its entrapment within the vadose zone as soil moisture and later loss to evapotranspiration

(Chap. 13). Depth to the water-table is an important factor in determining whether or not transmission losses over a stream reach can be appropriately used to estimate recharge from the stream.

11.5 Environmental Tracers

Environmental tracers include naturally occurring elements, anthropogenic compounds, and stable and radioactive isotopes. Environmental isotopes are discussed in Chap. 12. There are four different kinds of methods using to calculate recharge rates from environmental tracers (Beyerle 2002):

- methods based on a time dependent input function (e.g., tritium dating),
- methods based on the radioactive decay of isotopes whose initial concentrations are known (e.g., Carbon-14),
- methods based on the increase in daughter isotopes of radioactive decay over time (e.g., Helium 3), and
- methods based on the concentration of a (largely) time independent input (e.g., chloride mass balance).

Tracer measurements should be obtained below the root zone in order to estimate recharge rates. Otherwise, the tracer is a measure of the soil–water flux, which may significantly exceed the actual recharge rate (i.e., the water that actual reaches the water-table, Stephens 1996).

11.5.1 Chloride Mass Balance Method

The chloride mass balance method (Ericksson and Khunakasem 1969) is a widely used technique that is appropriate for measuring the low recharge rates in arid and semiarid environments. The concept behind the chloride mass balance (CMB) method is quite simple and it involves relatively inexpensive analyses. The basic principle of the CMB method is that rainfall contains low concentrations of chloride (and other ions), which are concentrated by evaporation to a degree inversely proportionally to the recharge rate. The total mass of the ions is the product of the volume of rainfall and ion concentration. If an ion is conservative, in that its concentration does not change as a result of reactions with the soil, aquifer solids or other ions in solution, then the total mass of the ion in the recharge water will remain the same. As the result of evaporation, the volume of water decreases and the concentration of the ion increases. Chloride is the preferred ion to use because it usually behaves in a conservative manner and can be accurately and relatively inexpensively measured.

A recharge rate (R) can be calculated if the precipitation rate (P) and concentrations of chloride in the rainfall (Cl_p) and shallow groundwater (Cl_{GW}) are known:

$$R = \frac{Cl_P}{Cl_{GW}} P \tag{11.19}$$

However, the soil typically contains some chloride deposited as dryfall, which is mobilized by infiltrating precipitation. The total chloride (Cl_T) is thus the sum of the amount in rainfall (Cl_p) and the contribution from dryfall (DF) multiplied by precipitation rate:

$$R = \frac{Cl_P + DF}{Cl_{GW}} P = \frac{Cl_T}{Cl_{GW}} P \tag{11.20}$$

Dry fall is dissolved atmospheric chloride or wind-blown particulate salt in coastal or salt pan areas that was deposited and retained on the ground surface between precipitation events. Dry fall is often not explicitly considered. It is combined in the Cl_p term or a regional mean rainfall to dryfall ratio is used (e.g., Krulikas and Giese 1995).

The accuracy of the calculated recharge rates depends upon on the accuracy of the measurement of each parameter. The greatest uncertainties relate to the total chloride input and the degree to which site conditions meet method assumptions with regard to transport processes and boundary conditions.

Precipitation rates can typically be readily measured using rain gauges. The chloride concentration in the shallow groundwater can be measured from samples obtained from monitoring wells. The chloride concentration of rainwater is a function of local geography and climatic conditions, such as proximity to the coast and the prevailing wind direction and speed. It is therefore important to obtain chloride concentration data that is representative of the local rainwater. Ideally, chloride concentration should be measured in rain samples collected near the recharge station site rather than using regional estimates. Reported rainfall chloride concentrations are often in the range of 0.4–0.7 mg/L, but may be higher in some locations, particularly near the coast. Uncertainty in long-term chloride deposition rates remain a limitation of the CMB method. There is need for the development of long-term records of wet and dry chloride deposition in arid and semiarid regions worldwide (Scanlon et al. 2006).

Equations 11.19 and 11.20 use arithmetic and weighted average values for the precipitation rate and chloride concentration of precipitation. Consideration is typically not given to deviations around the average and the possibility of covariation of precipitation and chloride concentrations, which can be incorporated into the CMB analysis (Subyani and Sen 2006; Sen 2008). An error would be introduced into the recharge estimations if the precipitation chloride concentration of major recharge events is significantly different from the average rate. In arid regions where the downward flow rate and rate of recharge are very low, the water in shallow aquifers may be very old. Current average precipitation rates and average rainfall chloride concentration data may not be representative of values at the time the precipitation water entered the aquifer.

For the CMB method to give valid results several assumptions and boundary conditions must be met (Wood and Sanford 1995; Wood 1999; Goni 2002):

- The chloride concentration in the groundwater originates only from precipitation that recharges the aquifer.
- Chloride is conservative in the system.
- The chloride mass flux has not changed over time.
- One-dimensional downward piston flow of groundwater occurs (i.e., minimal lateral flow of water in the vadose zone).
- There is no recycling or concentration of chloride within the aquifer.

Recycling of chloride could occur, for example, if groundwater is pumped from the aquifer and the irrigation return flow is evaporatively concentrated.

The CMB method assumes that a groundwater basin is closed with respect to chloride in that all chloride present in the precipitation enters the local groundwater system. Any chloride that enters the aquifer from runoff (surface water flow) into the groundwater basin must be quantified. Similarly, chloride that leaves a groundwater basin in surface water outflow must also be considered. The chloride concentration of runoff may be greater than that of precipitation because local runoff may become enriched in chloride from interaction with surficial sediments and rock, or through evaporation. Failure to consider the enrichment of runoff with chloride could result in an underestimation of local infiltration rates (Flint et al. 2002).

The CMB method can also be applied using chloride concentration data from the vadose (unsaturated) zone obtained from cores (Cl_{UZ}):

$$R = \frac{Cl_P}{Cl_{UZ}} P \tag{11.21}$$

The water retained in cores will occur primarily within the sediment or rock matrix rather than macropores (i.e., fractures and vertical flow conduits). Therefore, recharge rates calculated using the CMB method from unsaturated zone chloride concentrations will represent the matrix or interstitial recharge rate (Wood et al. 1997). Macropore recharge water will bypass the matrix porosity and its chloride concentration will not be represented by the chloride concentration data from the vadose-zone sediment or rock matrix sampled from a core. Estimated recharge rates from the matrix would thus be biased towards the slower moving matrix flow. Macropore recharge rates can be estimated as the difference between the total recharge rates (determined from saturated zone sample) and matrix recharge rate.

Several methods have been used to obtain chloride concentration data from vadose-zone cores. Soluble cation concentrations may be obtained from core samples by first obtaining the water content by drying and then extracting (redissolving) the cations. Gravimetric water content is usually obtained by drying at 105°C. Oven-dried samples are mixed with an equal mass of deionized water and shaken periodically for 24 h. The supernatant is filtered and analyzed for ion concentrations using standard analytical methods. Pore-water concentration is determined by dividing the supernatant concentration by the gravimetric water content and multiplying by the density of water. (Scanlon et al. 2005; Stonestrom et al. 2007).

Interstitial water samples can also be extracted using centrifuge methods (Goni 2002) or directly collected by gravity drainage. Alternatively, samples may be obtained by immiscible liquid displacement. A dense immiscible liquid is added to the sample and the displaced water is collected once it floats to the top.

Chloride concentration data in arid regions may show a pronounced anomaly (interval of relatively high concentration) in the vadose zone, which indicates that recharge does not occur at a given location. In this case, chloride is accumulating in the soil zone and is not reaching the water-table (Moore 2007; Stonestrom et al. 2007).

The chloride mass balance method can also be used to estimate the amount of recharge from percolation ponds. Sukhija (2008) used the CMB method to estimate the percentage of recharge from water stored in several percolation ponds (locally referred to as "tanks") in India after monsoon rains. Water is lost from the ponds by a combination of infiltration into the underlying aquifer and evaporation. Assuming that infiltration is the only significant mechanism by which chloride leaves the pond, the decrease in the total mass of chloride can be used to calculate infiltration rate. The data requirements are estimates of the total volume of water in the ponds at different times, which can be estimated from pond area and depths, and the average chloride concentration. The infiltration rate between precipitation events is calculated from the change in the total mass of chloride in the pond and the average chloride concentration of the water.

11.5.2 Age Tracers

Various tracers are available to date the age of water within the unsaturated or saturated zone. Tracers are used in three main manners:

- identification of an anthropogenic peak in tracer concentration of known age.
- dating through decay process (radiometric decay of radioactive isotopes).
- matching concentration variations to known natural temporal variations in concentration.

Once a point date of groundwater age is obtained, the recharge rate can be estimated from the volume of water between the measuring point and a datum, which would normally be the water-table for groundwater recharge or land surface for infiltration through the vadose zone.

For saturated flow, the corresponding equation is:

$$R = \phi \Delta Z/t \tag{11.22}$$

where,
R = recharge rate (m/year)
ΔZ = distance from source area to dated point (m)
ϕ = average porosity (m^3/m^3)
t = time (difference in age between measuring point and datum) (years).
Equation 11.22 is valid for only cylindrical flow. Alternative equations would need to be used for non-cylindrical flow, such as radially diverging flow or constricted flow (area of recharge area is greater than the cross-sectional area of the aquifer). Consideration needs to also be given to leakage into or out of the aquifer along the flow path.

For downward percolation with the vadose zone, the recharge rate (R) can be calculated as follows:

$$R = \Delta Z \overline{V_W}/t \qquad (11.23)$$

where,
ΔZ = distance between dated horizon and land surface (m)
$\overline{V_W}$ = average volumetric water content of strata above the dated horizon (m^3/m^3).

Age tracers used for groundwater investigations include tritium, chlorine 36 (^{36}Cl), carbon 14 (^{14}C), and chlorofluorocarbons. The choice of tracers depends upon the anticipated age of the water to be dated and the behavior of the tracer within the study area. Tracer dates can be impacted by fluid-rock interaction, volatilization, and other process, which affect tracer concentration. Cost is also a key consideration. The one-time cost of performing tracer analyses may be ultimately less than long-term monitoring of hydrologic data (Scanlon et al. 2002). Environmental isotopic dating techniques are discussed in Chap. 12.

Recharge rates (R) can also be measured from the rate of migration of a tracer front. Fronts could be a change in water quality related to a change in land use or land cover, such as the presence of elevated nutrient or salt concentrations (Scanlon et al. 2005). The basic equation is

$$R = \theta v = \theta(z_2 - z_1)/(t_2 - t_1) \qquad (11.24)$$

where,
z_1, z_2 = position of fronts at times t_1, t_2.
θ = average volumetric water content of interval above the front
v = trace front velocity.

The practical limitation on the use of tracer front migration for recharge estimation is that a time series of data are required. The data requirements may be reduced if the time at which the tracer first reached the water-table is known or can be accurately estimated from historical data.

11.5.3 Chlorofluorocarbons

Chlorofluorocarbons are a class of chemicals that were introduced into the atmosphere that have no natural sources. Two main chlorofluorocarbons are of interest for determining the age of waters and thus recharge rates; CCl_3F (CFC-11, Freon-11) and CCl_2F_2 (CFC-12, Freon-12). Industrial production and atmospheric introduction of CFC-12 began in the 1931 and that of CFC-11 in the 1940s. Chlorofluorocarbons are attractive environmental tracers because they are detectable at very low concentrations, extremely stable, and behave essentially as inert compounds (Thompson et al. 1974). The presence or absence of CFCs can be used as a rough indictor of groundwater age. When significant quantities of $ClCl_3F$ are present in a water sample, then the sample must contain some post-1931 recharge (Thompson and Hayers 1979).

Industrial production and atmospheric measurements provide a relatively accurate CFC-11 and CFC-12 source function for the last 70 years (Ekwurzel et al. 1994). The concentration of CFCs in a water sample reflects the concentration in the atmosphere and temperature at the time the water sample became isolated from the vadose zone. The partitioning of gases between water and the atmosphere is a function of temperature, and can be predicted based on the compound's Henry Law coefficient. Atmospheric concentrations at the time of isolation from the vadose zone can be calculated from the measured concentration and recharge temperature, and then compared to reconstructed atmospheric CFC concentrations to determine the CFC-model age (Busenberg and Plummer 1992). Age can be determined from both CFC-11 and CFC-12 data, which serve as a cross-check on the method (Ekwurzel et al. 1994).

Busenberg and Plummer (1992) provide a detailed discussion of CFC sampling and analytical procedures. A critical issue is avoiding contamination of samples with atmospheric CFCs during sample collection. The contamination issue is particularly acute for older samples with low CFC concentrations that would be materially changed by even a small interaction with recent air containing much higher CFC concentrations. CFC-model ages reflect the time since the water became isolated from the unsaturated zone atmosphere, which may be significantly less than the time since the water entered the aquifer from precipitation. The presence of thick vadose zones can substantially modify the concentrations of CFCs and their ratios (Busenberg and Plummer 1992). The concentrations of CFCs may vary over a period of groundwater use for a variety of reasons discussed by Busenberg and Plummer (1992). Drawdown during groundwater pumping, for example, may introduce younger water with high CFC concentrations into a well. Improperly grouted wells may also allow entry of younger water with high CFC concentrations from an adjacent shallower aquifer. Groundwater may also become contaminated with CFCs particularly near urban areas, and therefore, the groundwater concentrations may not be directly related to atmospheric concentrations.

CFCs are a useful tracer for measuring groundwater age, but the method has the limitation that meticulous sample collection procedures must be followed to obtain accurate data. CFC concentrations may also change in the vadose zone and groundwater environments by various physical and chemical processes, which need to be considered in the data analysis. Busenberg and Plummer (1992) recommended that other tracers (e.g., tritium) and other hydrological information should be used to confirm CFC-model recharge ages, particularly near urban areas where the shallow groundwater could be contaminated with anomalous CFCs concentrations.

11.5.4 Tracer Versus Water Migration

The flux of water during infiltration does not necessarily coincide with flux of solutes, such as chloride, particularly on a short-term transient basis. The transport of anions may be faster than the average water advective flow velocity due to the anion exclusion effect. Electrostatic repulsion by negatively charged solid surfaces may force anions into the pore centers where the flow velocity is greater. Smaller pores may be effectively cut-off for anion transport while still being open to water transport (Gvirtzman and Gorelick 1991). Gvirtzman and Gorelick (1991) proposed that the exclusions from small pores may result in more tortuous flow paths and greater dispersion. The anion exclusion effect would tend to result in an overestimation of the recharge rate.

The opposite tracer migration relationship was observed in a study of nuclear fallout tracer migration in desert soils in New Mexico by Phillips et al. (1988) using ^{36}Cl and tritium as tracers. The ^{36}Cl peak was found to occur at shallower depth than the tritium peak even though the ^{36}Cl entered the soil earlier (Phillips et al. 1988). Chloride transport was thus evidently slower than that or the tritium-enriched water. This is opposite what would be expected from the anion exclusion effect. Phillips et al. (1988) speculated that the retarded downward solute transport was caused by 'vapor gaps', which are unevenly distributed pores within the unsaturated zone, across which the liquid flow paths dead end, but the transport of water continues in the vapor phase. Eventually equilibrium is reached between solute-transport into and out of the vapor gaps and net accumulation of solute no longer takes place. The vapor gap phenomenon would not impact the chloride concentration profile but would retard the movement of a transient pulse, such as the ^{36}Cl fallout pulse.

A detailed analysis of all the factors than can affect tracer movement is beyond the scope of this book. The key point made herein is that tracer movement does not necessarily closely coincide with advective flow, which introduces a potential error in tracer-based recharge estimations. Hence, it is recommended that multiple independent techniques be used to estimate recharge rates. Significant differences between methods may indicate that the underlying assumptions of one or more of the methods are not being met.

11.6 Neutron Probe Measurements

Neutron (moisture) probe/meter (Sect. 10.6.5.3) measurements can be used to quantify soil moisture. A time series of analyses of a borehole can be used to determine changes in soil moisture volume, and thus infiltration volumes, over time.

A detailed neutron logging program was performed at the Yucca Mountain nuclear repository site in Nevada, as documented by Flint and Flint (1995) and Flint et al. (2002). Time-series analyses were performed on a monthly basis in 98 boreholes located at different topographic settings. Time series analyses can be performed to determine both the volumetric water change and the depth of water penetration. A key issue for recharge studies is whether or not (and how much) infiltration occurs and whether the infiltrated water reaches the water-table and thus becomes aquifer recharge.

Net infiltration can be calculated as the increase in water content below the zone of ET. However, neutron log-based estimations of ET may be under-estimated if fracture flow bypasses the soil zone (Flint et al. 2002). The Yucca Mountain study demonstrated that depending upon soil storage capacity, most of the infiltrated water is held near surface. Lower near-surface storage capacity favors deep infiltration and recharge.

A key technical issue for neutron probes is instrument calibration, which is necessary to obtain accurate quantitative water moisture measures. Calibration curves are generated relating neutron counts to samples of known moisture content. It is also critical to assess and correct for any long-term drift in the instrument readings.

11.7 Borehole Temperature Profiles

Flowing groundwater can transfer heat. The deviation of a measured temperature profile from a theoretical profile considering only heat-conduction can be used to estimate the percolation flux (i.e., advective flow; Flint et al. 2002). Advantages of using temperature data is that it can be readily collected with a high sensitivity and accuracy and the thermal conductivity of sediments and rock is relatively insensitive to variations in saturation and water potential (Flint et al. 2002).

The percolation flux can be estimated through groundwater modeling by matching temperature profiles generated with various percolation rates to measured profiles. Boundary conditions are the average surface temperature and the groundwater temperature near the top of the saturated zone. The downward movements of heat or cold pulses from diurnal or seasonal temperature variations are potential targets for the calibration of a heat-flow model. In arid environments, rainfall commonly occurs only seasonally. Therefore, seasonal temperature variation associated with precipitation periods should be considered while evaluating temperature data for the assessment of recharge.

11.8 Groundwater Modeling

Numerical (computer) modeling has become the most important tool for managing water resources. The goal of groundwater modeling is to develop the ability to predict the response of land development, use of surface water connected to groundwater, and groundwater abstractions on groundwater resources and environments dependent on the resources. The groundwater model development process in itself provides an opportunity to develop a better understanding of the processes active in a groundwater system and to test conceptual models. The techniques for modeling recharge were reviewed by Sanford (2002). Groundwater modeling in arid lands is discussed in Chap. 20.

The model calibration process provides an opportunity to refine estimates for hydraulic parameters, such as recharge rates. If values of the other model parameters are known well enough, the model can be used to constrain the recharge rate (Sanford 2002). Recharge is often estimated through calibration of groundwater flow models. A great advantage of the use of groundwater modeling to evaluate groundwater recharge is than once a calibrated model has been developed it can be used as a predictive tool to assess groundwater management options (Scanlon et al. 2006).

However, groundwater model calibration does not provide unique solutions, and a well-calibrated model may be inaccurate. As is the case for water budget estimates in general, large errors in model determined recharge rates may occur if there are errors in the dominant parameters of precipitation and evapotranspiration or groundwater abstraction. The accuracy of model-derived recharge rates may be increased by inverse modeling using chemical tracers (Sanford 2002). A critical point is that modeling is not a substitute for field investigations of recharge, but rather should be viewed as tool for better interpretation of the collected data. Commonly, groundwater model results (i.e., simulated heads within the model domain) have a high sensitivity to recharge rates.

Spatial variations in recharge can also be evaluated in greater detail using integrated surface water-groundwater models. An advantage of integrated surface water/groundwater modeling in semiarid and arid lands is that the water flow direction is predominantly downwards. Stream baseflow is typically absent or negligible. Recharge rates calculated in surface-water models may be used as an input to groundwater flow models. A complication of integrated surface water/groundwater modeling in semiarid and arid lands is accurately incorporating the long time delay between water entering the vadose zone and it reaching the underlying aquifer (Chung et al. 2010). Also, fully integrated groundwater-surface water models require extensive data input which may not be available in sufficient detail to allow accurate model development.

11.9 Recharge Estimates in Arid and Semiarid Environments

A key lesson that needs to be emphasized for developing a program to evaluate recharge rates is that it is desirable to use as many different approaches as possible because of the uncertainties associated with each approach (Scanlon et al. 2002).

Table 11.1 Estimated recharge rates from some arid and semiarid region studies

	Mean annual precipitation (mm)	Mean annual recharge (mm)	Recharge/ precipitation	Method	References
Wadi Yalamlam, Saudi Arabia	231	22.8	10%	CMB	Subyani and Sen (2006)
Southern High Plains Aquifer, USA	485	11 ± 2	2%	CMB	Wood and Sanford (1995)
South and central New Mexico	230	8.4–9.4, 1.5–2.5	3.7–4.1% 0.7–1.1%	Tritium CMB	Phillips et al. (1988)
Interior Southwestern US (7 basins)	106–303	0.9–4.4	0.7–1.8%	BCM modeling	Flint and Flint (2007)
Upper Virgin Basin, SW Utah	303	17.3	5.7%	BCM modeling	Flint and Flint (2007)
Wadi Tharad, Saudi Arabia	57.5	6.1	11%	CMB	Subyani (2004)
Wadi Wajj, Saudi Arabia	268	11.5	4.8%	CMN	Al-Shaibani (2008)
Wadi, Marwani, KSA	60	10.9	18%	CMB	Al-ahmadi and El-fiky (2009)
Nyamandhlovu aquifer, Zimbawe	555	15–20	2.7–3.6%	CBM, WTF flow net, dating, modeling	Sibanda et al. (2009)
Amargosa Basin,Southern Nevada and Californa	180	2.9	1.6%	Hydrologic models	Osterkamp et al. (1994)

Use of multiple techniques allows for the narrowing of the window of possible recharge rates and processes. The CMB method is generally recommended for studies of recharge in arid lands because it is sensitive to low recharge rates, is relatively accurate, easy to perform, and inexpensive (e.g., Sibanda et al. 2009).

Other tracers may also be used based on study-specific conditions. Water budget methods have considerable uncertainty, but are useful for constraining the range of possible recharge rates and are suited for evaluating both local (e.g., wadi) and basin-wide recharge. Recharge rates should be further refined during the development of a calibrated groundwater model. The use of multiple techniques allows for the identification of outlying or hydrogeologically unreasonable values, which might indicate unexpected local conditions. Strategies to cope with quantitative estimation and prediction of spatially variable recharge rates have still yet to reach a stage of practical application for actual water resources management. The ultimate solution will likely lie in a combination of selective ground-truth data collection, geostatistics, and remote sensing (Simmers 1990).

Groundwater recharge rates in semiarid and arid environments were reviewed by Scanlon et al. (2006). Recharge rates obtained using the CMB method from approximately 140 study sites ranged between 0.2 and 35 mm/year (0.008 and 1.4 in/year), which represents 0.1–5 % of the long-term average annual precipitation. Estimated recharge rates from some studies are presented in Table 11.1 to illustrate the range of estimated values and techniques used.

References

Al-ahmadi, M. E., & El-Fiky, A. A. (2009). Hydrogeochemical evaluation of shallow alluvial aquifer of Wadi Marwani Western Saudi Arabia. *Journal of King Saud University (Science), 21*, 179–190.

Allison, G. B., Gee, G. W., & Tyler, S. W. (1994). *Soil Sciences Society America Journal, 58*, 6–14.

Allison, G. B. (1988). A review of some of the physical chemical and isotopic techniques available for estimating groundwater recharge. In I. Simmers (Ed.), *Estimation of natural groundwater recharge* (pp. 49–72). North Atlantic Treaty Organization, Scientific Affairs Division

Al-Shaibani, A. M. (2008). Hydrogeology and hydrochemistry of a shallow alluvial aquifer. *western Saudi Arabia: Hydrogeology Journal, 16*, 155–165.

Batelaan, O., & de Smedt, F. (2007). GIS-based recharge estimation by coupling surface-subsurface water balances. *Journal of Hydrology, 337*, 337–355.

Beyerle, U. (2002). Groundwater dating using environmental tracers and black box models. In W. Kinzelbach, W. Aeschbach, C. Alberich, I. B. Goni, U. Beyerle, P. Brunner, W.-H. Chiang, J. Rueedi & K. Zoellmann (Eds.), *A survey of methods from groundwater recharge in arid and semiarid regions: Early Warning and Assessment Report Series UNEP/DEWA/RS.02-2* (pp. 32–37). United Nations Environment Programme, Nairobi, Kenya.

Boulton, N. S. (1963). Analysis of data from non-equilibrium pumping tests allowing for delayed yield from storage. *Institute of Civil Engineers Proceedings (London), 26*, 469–482.

Bourdon, D. J. (1977). Flow of fossil groundwater. *Quarterly Journal of Engineering Geology and Hydrogeology, 10*, 97–124.

Busenberg, E., & Plummer, L. N. (1992). Use of chlorofluorocarbons (CCl_3F and CCl_2F_2) as hydrologic tracers and age-dating tools: The alluvium and terrace system of Central Oklahoma. *Water Resources Research, 28*, 2257–2283.

Chung, H.-M., Kim, N.-W., Lee, J., & Sophocleous, M. (2010). Assessing distributed groundwater recharge using integrated surface water-groundwater modeling. *Hydrogeology Journal, 18*, 1253–1264.

De Vries, J. J., & Simmers, I. (2002). Groundwater recharge: An overview of processes and challenges. *Hydrogeology Journal, 10*, 5–17.

Ekwurzel, B., Schlosser, P., Smethie, W. M., Plummer, L.N., Busenberg, E., Michel, R.L., Weppernig, R., & Stute, M. (1994). Dating of shallow groundwater: Comparison of the transient tracers $^3H/^3He$, chlorofluorocarbons, and ^{85}Kr. *Water Resources Research, 30*, 1693–1708.

Ericksson, E., & Khunakasem, V. (1969). Chloride concentration in groundwater, recharge rate and rate of deposition of chloride in Israel Coastal Plain. *Journal of Hydrology*, 7, 178–197.

Flint, L. E., & Flint, A. L. (1995). *Shallow infiltration processes at Yucca Mountain, Nevada—neutron logging data, 1984–1993*. U.S. Geological Survey Water-Resources Investigations Report 95-4035.

Flint, L. E., & Flint, A. L. (2007) Regional analysis of groundwater recharge. In D. A. Stonestrom, J. Constantz, T. P. A. Ferré & S. A. Leake (Eds). *Ground-water recharge in the arid and semiarid southwestern United States*. (pp. 29–60). U.S. Geological Survey Professional Paper 1703.

Flint, A. L., Flint, L. E., Kwicklis, E. M., Fabryka-Martin, J. T., & Bodvarson, G. S. (2002). Estimating recharge at Yucca Mountain Nevada, USA, comparison of methods. *Hydrogeology Journal, 10*, 180–204.

Flint, A. L., Flint, L. E., Hevesi, J. A., & Blainey, J. B. (2004). Fundamental concepts of recharge in the desert Southwest: A regional modeling perspective. In J. F. Hogan, F. M. Phillips & B. R. Scanlon (Eds.), *Groundwater recharge in a desert environment: The Southwestern United States, water science and application series* (Vol. 9, pp. 159–184). Washington DC: American Geophysical Union.

Gee, G. W., & Hillel, D. (1988). Groundwater recharge of arid regions: Review and critique of estimation methods. *Hydrological Processes, 2*, 255–266.

Gee, G. W., Fayer, M. J., Rockhold, M. L., & Campbell, M. D. (1992). Variations in recharge at the Hanford Site. *Northwest Science, 66*, 237–250.

Goni, I. B. (2002). Chloride method in the unsaturated zone. In W. Kinzelbach, W. Aeschbach, C. Alberich, I. B. Goni, U. Beyerle, P. Brunner, W.-H. Chiang, J. Rueedi & K. Zoellmann (Eds.), *A survey of methods from groundwater recharge in arid and semiarid regions: Early warning and assessment report series UNEP/DEWA/RS.02-2* (pp. 22–31). Nairobi, Kenya: United Nations Environment Programme.

Gvirtzman, H., & Gorelick, S. M. (1991). Dispersion and advection on unsaturated porous media enhanced by anion exchange. *Nature, 352*, 793–795.

Healy, R. W., & Cook, P. G. (2002). Using ground water levels to estimate recharge. *Hydrogeology Journal, 10*(1), 91–109.

Herczeg, A. L., & Leaney, F. W. (2011). Review: Environmental tracers in arid-zone hydrology. *Hydrogeology Journal, 19*(1), 17–30.

Horton, R. E. (1933). The role of infiltration in the hydrologic cycle. *Transactions American Geophysical Union, 14*, 446–460.

Krulikas, R. K., & Giese, G. L. (1995) *Recharge to the surficial aquifer system in Lee and Hendry Counties. Florida*: U.S. Geological Survey Water-Resources Investigations report 95-4003.

Lee, D. R. (1977). A device for measuring seepage flux in lakes and estuaries. *Limnology and Oceanography, 22*(1), 140–147.

Lerner, D. N., Issar, A. S., & Simmers, I. (1990). Groundwater recharge, a guide to understanding and estimating natural recharge. *International Associations of Hydrogeologists, Contributions to Hydrogeology* (vol. 8). Kennilworth.

Lerner, D. N., Issar, A. S., & Simmers, I. (1997). Groundwater recharge. In O. M. Saether & P. de Caritat (Eds.), *Geochemical processes, weathering and groundwater recharge in catchments* (pp. 109–150). Rotterdam: AA Balkema.

Lloyd, J. W., & Farag, M. H. (1978). Fossil ground-water gradients in arid sedimentary basins. *Ground Water, 16*(6), 388–393.

Manghi, F., Mortazavi, B., Crother, C., & Hamdi, M. R. (2009). Estimating regional groundwater recharge using a hydrological budget method. *Water Resources Management, 23*, 2475–2489.

Meyboom, P. (1961). Estimating ground water recharge from stream hydrographs. *Journal of Geophysical Research, 66*, 1203–1214.

Moore, S. J. (2007). Streamflow, infiltration, and recharge in Arroyo Hondo, New Mexico. In D. A. Stonestrom, J. Constantz, T. P. A. Ferré & S. A. Leake (Eds.), *Ground-water recharge in the arid and semiarid southwestern United States* (pp. 137–155). U.S. Geological Survey Professional Paper 1703.

Neuman, S. P., & Witherspoon, P. A. (1972). Field determination of the hydraulic properties of leaky multiple aquifer systems. *Water Resources Research, 8*(5), 1284–1298.

Neuman, S. P. (1987). On methods of determining specific yield. *Ground Water, 25*, 679–684.

Osterkamp, W. R., Lane, L. J., & Savard, C. S. (1994). Recharge estimates using a geomorphic/distributed parameter simulation approach. *Amargosa River Basin: Water Resources Bulletin, 30*(3), 493–507.

Phillips, F. M., Mattick, J. L., Duval, T. A., Elmore, D., & Kubik, P. W. (1988). Chlorine 36 and tritium from nuclear weapons fallout as tracers for long-term liquid and vapor movement in desert soils. *Water Resources Research, 24*, 1877–1891.

Rorabaugh, M. I. (1964). *Estimating changes in bank storage and ground water contribution to streamflow* (Vol. 63, pp. 432–441). International Association of Scientific Hydrology Publication.

Rutledge, A. T., & Daniel, C. C, I. I. I. (1994). Testing an automated method to estimate ground-water recharge from streamflow records. *Ground Water, 32*(2), 180–189.

Rutledge, A. T. (1993).*Computer programs for describing the recession of ground-water discharge and for estimating mean ground-water recharge and discharge from stream records*. U.S. Geological Survey Water-Resources Investigations Report 93-4121.

Rutledge, A. T. (1998). *Computer programs for describing the recession of ground-water discharge for estimating mean groundwater-recharge and discharge from streamflow records*. U.S. Geological Survey Water Resources Investigations Report 98-4148.

Sanford, W. (2002). Recharge and groundwater models: An overview. *Hydrogeology Journal, 10*, 110–120.

Scanlon, B. R. (2000). Uncertainties in estimating water fluxes and residence times using environmental tracers in an arid unsaturated zone. *Water Resources Research, 36*, 395–409.

Scanlon, B. R., Keese, K. E., Flint, A. L., Flint, L. E., Gaye, C. B., Edmunds, W. M., et al. (2006). Global synthesis of groundwater recharge in semiarid and arid regions. *Hydrological Processes, 20*, 3335–3379.

Scanlon, B. R., Tyler, S. W., & Wierenga, P. J. (1997). Hydrologic issues in arid unsaturated systems and implications for contaminant transport. *Review of Geophysics, 35*, 461–490.

Scanlon, B. R., Healy, R. W., & Cook, P. G. (2002). Choosing appropriate techniques for quantifying groundwater recharge. *Hydrogeology Journal, 10*, 18–39.

Scanlon, B. R., Reedy, R. C., Stonestrom, D. A., Prudic, D. E., & Dennehy, K. F. (2005). Impact of land use and land cover change on groundwater recharge and quality in the southwestern US. *Global Change Biology, 11*, 1577–1593.

Şen, Z. (2008). *Wadi hydrology*. Boca Raton: CRC Press.

Sibanda, T., Nonner, J. C., & Uhlenbrook, S. (2009). Comparison of groundwater recharge estimation methods for the semi-arid Nyamandblvu area, Zimbabwe. *Hydrogeology Journal, 17*, 1427–1441.

Simmers, I. (1990). Aridity, groundwater recharge and water resources management. In D. N. Lerner, A. S. Issar & I. Simmers (Eds.), *Groundwater recharge, a guide to*

understanding and estimating natural recharge (Contributions to Hydrogeology 8) (pp. 1–20). Kennilworth: International Associations of Hydrogeologists.
Simmers, I. (1998). Groundwater recharge: An overview of estimation "problems" and recent developments, In N. S. Robins (Ed.), *Groundwater pollution, aquifer recharge and vulnerability* (Vol. 130, pp. 107–115) London: Geological Society (Special Publication).
Sophocleous, M. (1991). Combining the soilwater balance and water-level fluctuation methods to estimate natural ground-water recharge: Practical aspects. *Journal of Hydrology, 124*, 229–241.
Sophocleous, M. (2004). Groundwater recharge. In L. Silveira, S. Wohnlich & E. J. Usunoff (Eds.), *Encyclopedia of life support systems (EOLSS)*. Oxford: Eolss Publishers. Retrieved from http://www.eolss.net.
Stephens, D. B. (1996). *Vadose zone hydrology*. Boca Raton: CRC Press.
Stonestrom, D. A., Prudic, D. E., Walvoord, M. A., Abraham, J. D., Stewart-Deaker, A. E. & Glancy, P. A., et al. (2007). Focused ground-water recharge in the Amargosa Desert Basin. In D. A. Stonestrom, J. Constantz, T. P. A. Ferré & S. A. Leake (Eds.), *Ground-water recharge in the arid and semiarid southwestern United States* (pp. 107–136) U.S. Geological Survey Professional Paper 1703.
Subyani, A. M. (2004). Use of chloride-mass balance and environmental isotopes for evaluation of groundwater recharge in the alluvial aquifer, Wadi Tharad, Western Saudi Arabia. *Environmental Geology, 46*, 741–769.
Subyani, A., & Sen, Z. (2006). Refined chloride mass-balance method and its application in Saudi Arabia. *Hydrological Processes, 20*, 4373–4380.
Sukhija, B. S. (2008). Adaptation to climate change: Strategies for sustaining groundwater resources during droughts. In W. Dragoni & B. S. Sukhija (Eds.), *Climate change and groundwater* (Vol. 288, pp. 169–181). Geological Society of London Special Publication.
Thompson, G. M., Hayes, J. M., & Davis, S. N. (1974). Fluorocarbon tracers in hydrology. *Geophysical Research Letters, 1*, 177–180.
Thompson, G. M., & Hayers, J. M. (1979). Trochlorofluoromethane in groundwater—as possible tracer and indicator of groundwater age. *Water Resources Research, 15*, 546–556.
Tilahun, K., & Merkel, B. J. (2009). Estimation of groundwater recharge using a GIS-based distributed water balance model in Dire Dawa, Ethiopia. *Hydrogeology Journal, 17*, 1443–1457.
Weeks, E. P. (2002). The Lisse effect revisited. *Ground Water, 40*(6), 652–656.
Wood, W. W., & Sanford, W. E. (1995). Chemical and isotopic methods for quantifying ground-water recharge in a regional, semiarid environment. *Ground Water, 33*, 458–468.
Wood, W. W., Rainwater, K. A., & Thompson, D. B. (1997). Quantifying macropore recharge: examples from an semi-arid area. *Ground Water, 35*, 1097–1106.
Wood, W. W. (1999). Use and misuse of the chloride-mass balance method in estimating ground water recharge. *Ground Water, 37*, 2–3.

Chapter 12
Environmental Isotopes

12.1 Introduction

Environmental isotopes are useful tools for evaluating the source and time of precipitation of groundwater and geochemical reactions during groundwater flow (i.e., fluid–rock interactions). Such data can be used to estimate recharge rates, regional groundwater flow rates, and the time of emplacement of fossil groundwater. Isotopic analyses are not a routine element of water resources investigations largely because of their costs and the nature of the information they can provide. Environmental isotopic analyses have been typically applied to large (basin) scale investigations. Some of the applications of environmental isotopes have been of more academic than applied hydrogeological interest. For example, determination of the age of non-renewable groundwater resources usually does not have much bearing on current water management, so long as it is recognized that the groundwater is in fact non-renewable on a human time scale. From a practical perspective, it matters little if water is 10,000 years old or 1,000,000 years old.

The employment of environmental isotopic tools, as is the case for all technologies applied to water resources investigations, must be justified in terms of the type of information they can provide, the potential quality (accuracy) of the data, and their cost effectiveness relative to other techniques available for obtaining the same or similar information. The most important application of environmental isotopes in the management of groundwater resources in arid lands is as a tool for the development or confirmation of conceptual models and the evaluation of recharge rates (Sect. 11.5). Environmental tracers, including isotopes, can be used to identify sources of recharge water and salinity, and thus, allow refinement of conceptual models of groundwater flow and to test the reliability of groundwater modeling results (e.g., Eastoe et al. 2008; Bassett et al. 2008; Druhan et al. 2008; Hutchinson and Hibbs 2008).

Another application of environmental isotopes is as a forensic tool for assessing point-to-point connection between wells, recharge sources, and springs (Bassett

R. Maliva and T. Missimer, *Arid Lands Water Evaluation and Management*, Environmental Science and Engineering, DOI: 10.1007/978-3-642-29104-3_12,

et al. 2008). The age of groundwater, as determined from environmental isotopes, can also provide important information on the vulnerability of the water to contamination. Aquifers that contain old water (and thus have a high degree of isolation from land surface processes) are less vulnerable to contamination. Environmental isotopes are also very useful for assessing emergency groundwater supplies as they can provide unique information on the dynamics, origin and environmental conditions of recharge, and mixing of waters (Vrba and Verhagen 2006).

12.2 Isotope Basics

Environmental isotopes are discussed in most isotope geochemistry and groundwater chemistry books and are reviewed in detail by Clark and Fritz (1997), Mazor (2003), and Mook (2005). Isotopes are atoms of an element that have different numbers of neutrons, and thus, atom weights. The term 'nuclide' is essentially synonymous with 'isotope', with the former being used more commonly in nuclear physics, whereas the latter is typically used in chemistry applications. By definition, all atoms of an element have the same number of protons (i.e., atomic number). For example, two isotopes of oxygen that are of interest in hydrogeology are ^{16}O and ^{18}O. The former, which is by far the most common isotope in water, has eight protons and eight neutrons. The latter isotope of oxygen has eight protons and ten neutrons. The main isotopes of hydrogen are ^{1}H (H; protium) and the much less abundant ^{2}H (D, deuterium) and ^{3}H (T, tritium). Other isotopes discussed in this section include those of chlorine, carbon, and krypton.

Isotopes are either stable or radioactive. Stable isotopes do not undergo radioactive decay, which is the spontaneous disintegration of an atom into daughter nuclides with the associated emission of ionizing radiation. Radioactive isotopes (radionuclides) have an unstable nucleus and undergo radioactive decay. Some radionuclides have such a long half-life that can be considered to be essentially stable, at least on the time scales typically of interest in groundwater investigations. The decay rate of a radionuclide is a constant value, which is usually expressed as its half-life. A half-life is the period of time in which the number of radionuclide atoms in a substance decreases by half. Thus, the age of a substance (e.g., water, fossil, mineral crystal) can be determined from the change in the concentration of the parent nuclide from the time of the substance formed to the present.

12.3 Stable Isotopes

Stable isotopes of an element have the same atomic number, and chemical behavior, but different atomic weight. The difference in atomic weight results in the fractionation of isotopes, from which lies the principal usefulness of stable isotopes. Fractionation is express in terms of a fractionation factor (α), defined as:

$$\propto = \frac{R_{products}}{R_{reactants}} \quad (12.1)$$

where, R is ratio of the heavy isotope to the light isotope (e.g., $^{13}C/^{12}C$, $^{18}O/^{16}O$).

Fractionation results in changes in the isotopic ratios of both the reactant and the remaining product. For example, during evaporation water molecules containing the lighter ^{16}O isotope preferentially enter the gas phase, leaving behind water that is enriched in the heavier ^{18}O isotope. The water that evaporates from seawater will be isotopically lighter than seawater. Meteoric water that precipitates from evaporated seawater as either rain or snow will also be isotopically lighter than the seawater source. The preferential incorporation of water depleted in both ^{18}O and D in polar ice caps during glacial periods resulted in a shift in the seawater isotopic ratios towards heavier values. Similarly, warm (interglacial or greenhouse) periods, when the ice caps were reduced in area or absent, are marked by isotopically lighter seawater. Changes in the oxygen isotopic ratios of seawater resulting from climate change are recorded in the oxygen isotopic ratios of ancient calcites (e.g., fossil shells) that precipitated in equilibrium with seawater.

The degree of fractionation is temperature dependent with greater fractionation rates occurring at lower temperatures. Empirical equations have been developed to express the temperature dependence of fractionation factors, such as between calcite and water. Thus, if the oxygen isotopic ratio of the source water is known (or can be independently determined), then the oxygen isotopic ratio of calcite (inorganic and biologically precipitated) can be used to determine the temperature at the time of precipitation. Oxygen isotope thermometry using marine fossils is a very active area of research to evaluate paleoclimate changes. If the temperature at the time of precipitation is known or the range of temperatures can be constrained from other data sources, then the oxygen isotopic ratios of calcite cement can be used to evaluate the salinity of pore waters.

Biological processes result in the fractionation of carbon isotopes. The lighter ^{12}C isotope is preferentially taken up by plants and incorporated into organic matter. The oxidation of organic matter in soil zones releases the isotopic lighter carbon and decreases the $^{13}C/^{12}C$ ratio of the dissolved inorganic carbon. Nitrogen isotopes have been used for some time to evaluate the source and fate of nitrates and other nitrogen contaminants in groundwater (Clark and Fritz 1997).

Individual stable isotope concentrations are not separately measured. Instead, the ratio of the trace isotope to the more common isotope is measured using a mass spectrometer. For consistency, trace isotope ratios are reported relative to the ratio of a standard and expressed using delta (δ) notation. Using oxygen isotopes as an example:

$$\delta^{18}O = \frac{(^{18}O/^{16}O)\text{sample-}(^{18}O/^{16}O)\text{standard}}{(^{18}O/^{16}O)\text{standard}} \times 1000 \quad (12.2)$$

Hydrogen isotopes are similarly expressed as δD values using the D/H ratio. The standard for oxygen and hydrogen isotopes in water is Vienna Standard Mean Ocean Water (VSMOW or V-SMOW). Delta or 'del' values have units of parts per thousand or "per mil" (‰). Stable isotope ratios were previously measured relative to the Standard Mean Ocean Water (SMOW) standard. Some confusion may exist in the literature with respect to samples run relative to the old SMOW standard, samples that were run relative to VSMOW and reported as being relative to SMOW, and samples run and reported relative to VSMOW. In any event, VSMOW was developed and calibrated to have the same isotopic ratios as the SMOW standard.

Carbon isotopes and oxygen isotopes in carbonate minerals were originally measured relative to the ratio of the calcite of a belemnite fossil from the Pee Dee Formation of South Carolina, which is referred to as the PDB standard. The actual supply of the PDB standard has long been exhausted. A VPDB standard has now been established as the international standard for all $\delta^{13}C$ measurements and for the measurement of $\delta^{18}O$ in carbonate minerals. The VPDB standard is considered to be isotopically identical to the original PDB standard. The $\delta^{18}O$ values for carbonate minerals may also be reported relative to VSMOW. A conversion equation can be used to convert between $\delta^{18}O$ values reported relative to the VSMOW and VPDB standards.

The stable isotope ratios of water cannot be practically directly measured. The oxygen isotope ratios of water are measured by first equilibrating the water sample with carbon dioxide gas, and then measuring the isotope ratio of the carbon dioxide gas. A conversion factor is then applied to obtain the $\delta^{18}O$ value for the water. The oxygen and carbon isotope ratios of carbonate minerals are determined by reacting the sample with phosphoric acid, and measuring the ratio of the evolved gas and applying a fractionation factor. The deuterium content of water is measured by reducing the water to elemental hydrogen gas and measuring the D/H ratio. Inasmuch as all hydrogen in water is converted to hydrogen gas no isotopic fractionation occurs.

From a practical perspective, water samples are typically provided to a stable isotope laboratory and a report is provided with delta values (e.g., $\delta^{18}O$, $\delta^{13}C$, δD) reported with respect to a standard. Stable isotope data should be subject to the same quality control and quality assurance practices as other chemical data. Duplicate samples should ideally be run and, if a project budget allows, blind standards (samples of known isotopic ratios) should be given to the laboratory for analysis to confirm the accuracy of analyses.

Oxygen and hydrogen isotopes are most useful for water resources investigations because water is composed of these elements and their fractionation results in changes in their ratios that can be related to hydrological processes and water sources. Since the mass of oxygen and hydrogen present as water in a given volume of an aquifer is typically much larger than the mass present as reactive minerals, oxygen and hydrogen isotopic ratios have a relatively low tendency to significantly change as the result of fluid–rock interaction. On the contrary, the mass of carbon present as dissolved inorganic carbon in groundwater is small fraction of the mass of carbon that is present in carbonate minerals in a limestone or dolostone. Carbon isotope ratios in groundwater may therefore significantly change as a result of fluid–rock interaction.

Other stable isotope systems, such as nitrogen, sulfur, chlorine, and strontium, have value for evaluations of biogeochemical and geological processes, but have limited value for groundwater resources evaluations. Where differences in isotopic ratios exist between two water sources, the ratios can be used as tracers to assess the origins of waters and the mixing of different waters. However, non-isotopic tracers, such as elemental concentrations, are often less expensive options to use.

12.4 Radioisotopes

12.4.1 Radioactive Dating Concepts

The decay of radioactive nuclide occurs at a constant rate that is expressed by the equation

$$N = N_0 e^{-\lambda t} \tag{12.3}$$

where,

N = number of atoms at time "t"
N_0 = number of atoms at time t = 0
λ = decay constant
t = time.

Equation 12.3 can be rewritten in terms of half-life ($T_{1/2}$) as follows:

$$N = N_0 e^{-0.693t/T_{1/2}} \tag{12.4}$$

where,

$$T_{1/2} = 0.693/\lambda \tag{12.5}$$

The elapsed time (Δt) can be calculated from the change in the number of atoms (or concentration) of a radionuclide:

$$\Delta t = \frac{1}{\lambda} ln\left(\frac{N_0}{N}\right) \tag{12.6}$$

Dating can also be performed using the present number of atoms (or concentration) of a radionuclide (parent, N_P) and its decay product (daughter, N_D), because the original number of atoms of the parent radionuclide (N_0) are the sum of the present number of atoms of the measured nuclide and its daughter:

$$N_0 = N_P + N_D \tag{12.7}$$

$$\Delta t = \frac{1}{\lambda} ln\left(1 + \frac{N_D}{N_P}\right) \tag{12.8}$$

12.4.2 Radiocarbon (Carbon 14) Dating

Carbon 14 (^{14}C) dating (also referred to a radiocarbon or carbon dating) is a well-known technique for dating carbonaceous material up to about 50,000 to 60,000 years old. The half-life of ^{14}C is 5,730 years. Carbon 14 forms in the atmosphere by the interaction of gamma rays with ^{14}N atoms. The basis of radiocarbon dating is that the ratio of ^{14}C to total carbon atoms in the atmosphere has been nearly constant. However, some variations in ^{14}C have occurred, particularly in post-industrial times. The burning of fossil fuels has added older carbon with essentially no ^{14}C to the atmosphere, which has decreased the fraction of ^{14}C in the atmosphere. Operating in the opposite direction, atmospheric nuclear weapons testing added ^{14}C to the atmosphere before the atmospheric test ban. Additionally, cosmic ray intensity has fluctuated over time. Radiocarbon dates must therefore be converted to calendar years using calibration curves that are based on radiocarbon analyses of materials of independently determined age (e.g., tree rings).

Radiocarbon dating of groundwater samples is performed on dissolved inorganic carbon (DIC). The standard processing procedure is to precipitate the DIC using an ammoniacal-strontium (or barium) chloride solution (U.S. Geological Survey 1996). The precipitate is then filtered and washed with CO_2 free distilled water under an inert atmosphere and dried. The CO_2 is released by reaction with phosphoric acid. The isotopic analyses are performed using an accelerator with mass spectrometry (AMS), fluid scintillation counters or proportional counters. Radiocarbon data can be performed using AMS on one liter (0.26 gal) samples and in some samples as small as 250 ml (0.07 gal) (U.S. Geological Survey 1996). The easiest way to sample a large volume of water is to precipitate the dissolved inorganic carbon in the field (Rueedi 2002), which eliminates the need to transport large volumes of water.

Activities of ^{14}C are referenced to an international standard known as 'modern carbon' (MC). The activity of modern carbon is defined as 95% of the ^{14}C activity in 1950 of the United States National Bureau of Standards (NBS; now National Institute of Standards and Technology) oxalic acid standard (Clark and Fritz 1997). Measured ^{14}C activities are expressed as a percent of modern carbon (pmC). Carbon 14, like ^{13}C, is fractionated during organic and inorganic phase changes and reactions, and as a result ^{14}C activities must be normalized to a common $\delta^{13}C$ value of −25‰. Without the normalization to a common $\delta^{13}C$ value, ^{14}C ages would vary depending on the degree and direction of fractionation. Samples with $\delta^{13}C$ greater than −25‰ will have elevated ^{14}C activities gained through isotopic fractionation and would, therefore, have artificially lower ages if a correction for fractionation is not made. The lack of this correction is a common error in some groundwater investigations of the past.

The basis of radiocarbon dating is that ^{14}C is incorporated into organic matter at an activity that is proportional to the atmospheric concentration. Once an organism dies, ^{14}C is no longer incorporated and the activity of ^{14}C starts to decrease due to radioactive decay. Samples of carbonaceous organic matter can be dated by the fraction of the original ^{14}C that remains using Eq. 12.6. If the standard initial ^{14}C of 100 pmC is categorically used, then ages are expressed in radiocarbon years, which may be significantly different from calendar years, because of variations in atmospheric ^{14}C concentration over time (Clark and Fritz 1997).

Although radiocarbon dating is conceptually very simple, complexity lies in the details. Radiocarbon data are subject to error by the incorporation of older, radioactively dead carbon. There have been numerous studies using ^{14}C to date groundwater and to estimate groundwater flow rates. The half-life of ^{14}C (5,730 years) makes it suitable for dating older water in confined aquifers. A limitation of the ^{14}C dating is that the $^{14}C/^{12}C$ ratio in water may be greatly impacted by other sources of carbon, such as soil gases, pedogenic carbon, and aquifer minerals. Inasmuch as the volume of dissolved inorganic carbon in groundwater is a minute fraction of the carbon that may be present in the aquifer minerals (particularly limestones and dolostones), only a small amount of fluid–rock interaction can greatly impact ^{14}C dates. The carbon in aquifer minerals is typically "dead" with respect to ^{14}C, so the result of any addition of mineral carbon tends to increase ^{14}C ages. For this reason, aquifers with a non-carbonate matrix would be much preferred to an aquifer with a carbonate matrix for radiocarbon dating.

A voluminous body of literature exists concerning correction of ^{14}C ages, many of which were listed by Zhu (2000). The basic established practice is to use major ion and $\delta^{13}C$ mass balance to reconstitute the proportions of carbon from different sources. The algebraic solutions include several presumptions including (Zhu and Murphy 2000)

- The mass transfer reactions active in the aquifer apply.
- Only two carbon sources are typically considered.
- Only the dominant aqueous species for major groundwater constituents are evaluated.
- Isotopic exchange does not alter the ratios caused by natural decay.

Inverse modeling techniques, such as using the NETPATH program (Plummer et al. 1994), can be used to determine the reactions or processes responsible for the changes in groundwater chemistry between precipitation and the observation point. The main advantage of inverse modeling is that it allows for considerable flexibility to evaluate complex hydrogeological conditions, particularly if the flow path is divided into segments that presumably represent relatively uniform geochemical conditions (Zhu and Murphy 2000). Inverse modeling also involves several assumptions that need to be evaluated including (Zhu and Murphy 2000):

- Two water analyses from 'initial' and 'final' wells represent packets of water that flow along the same path.
- Dispersion and diffusion do not significantly affect solution chemistry.
- Chemical steady-state conditions occurred during the time period of interest.
- Inter-aquifer mixing is insignificant (no vertical leakage affecting age of water).

Some geochemical processes that are active in carbonate aquifers can significantly impact the ^{14}C of DIC but yet may not leave a distinct geochemical signature. The recrystallization of metastable carbonates (aragonite, high-magnesium calcite) to low-magnesium calcite may leave a very subtle or no water chemistry signature if the groundwater has a similar $\delta^{13}C$ as the carbonate rock. In aquifers that are being actively pumped, the induced leakage of water through confining beds can greatly affect the C-14 values in the aquifer being studied.

Zhu (2000) demonstrated a two-step methodology for determining recharge rates, which first involves geochemical modeling to correct groundwater ^{14}C ages, followed by ^{14}C transport and groundwater flow modeling. The model was calibrated against both corrected ^{14}C ages and aquifer heads. Zhu (2002) further assessed the modeled recharge rates against rates obtained by the CMB method. There are still uncertainties in the modeled recharge rates related to other parameters, such as effective porosity. Effective porosity is positively correlated with simulated recharge rate.

A key lesson is that radiocarbon dating can be a very useful tool for evaluating groundwater age and flow, but that the interpretation of the data involves a number of assumptions that may not hold for a given study area. There is no "one size fits all" formula (Zhu and Murphy 2000). Detailed knowledge of the aquifer hydrological and geochemical processes is necessary for accurate modeling and interpretation of ^{14}C data.

12.4.3 Tritium

Tritium is a radioactive isotope of hydrogen (symbol T or ^{3}H) that contains one proton and two neutrons. Tritium, being incorporated into the water molecule, is an excellent tracer of the movement of water itself. The half-life of tritium is 12.32 years, which makes tritium dating a useful method for dating very young groundwater. Tritium is measured by counting the beta minus electron (β^-) emissions in a liquid scintillation counter.

Tritium occurs in nature in very small quantities. Concentration of tritium is expressed in tritium units (TU) in which one TU is equal to 1 tritium atom per 1×10^{18} hydrogen atoms. Tritium is naturally produced in the upper atmosphere (stratosphere) by the action of cosmic radiation with atmospheric gases, especially ^{14}N. One very notable source of tritium was atmospheric nuclear weapons testing. Large quantities of tritium were added to the atmosphere during above-ground nuclear testing performed in the 1950's and early 1960's. The released tritium entered the hydrosphere, and precipitation became enriched in tritium. The concentration of tritium in precipitation peaked in 1963 (Clark and Fritz 1997).

The 1963 tritium peak is a very useful marker for hydrologic studies because it provides a precise age date for water. Groundwater present a position above the 1963 peak can be interpreted as having been emplaced after 1963. The difficulty of using the 1963 tritium peak as a marker is that a large number of closely spaced samples are required to locate the peak. The tritium peak can also not be used as a marker for waters younger than the middle 1960s, after the peak. An additional challenge is that the nuclear peak is progressively becoming less distinct as the concentration of nuclear testing tritium is decreasing due to radioactive decay.

Tritium can also be used to date groundwater samples using the $^3H/^3He$ dating technique, which utilizes daughter nuclide of tritium decay, tritogenic helium ($^3He_{trit}$). An advantage of $^3H/^3He$ dating is that it is independent of ^3H input because the ratio of the nuclides is considered, not the amount of each nuclide (Solomon and Sudicky 1995). The $^3H/^3He$ age of samples is calculated as follows (Schlosser et al. 1988):

$$\Delta t = 1/\lambda\left(1 + \frac{^3He_{trit}}{^3H}\right) \qquad (12.9)$$

Helium is a noble gas and is, therefore, nonreactive in the groundwater environment. Its local total concentration and isotopic ratio can be changed as a result of dispersion. Isotopic analyses also measure the total ^3He ($^3He_{total}$) concentration which is the sum of $^3He_{trit}$ and ^3He from other sources, such as the atmosphere ($^3He_{atm}$), excess air trapped above equilibrium solubility ($^3He_{exc}$), and mantle and nuclear (i.e., terrigenic) sources ($^3He_{terr}$) (Solomon and Sudicky 1995):

$$^3He_{total} = {}^3He_{tritr} + {}^3He_{exc} + {}^3He_{atm} + {}^3He_{terr} \qquad (12.10)$$

A critical element of the $^3H/^3He$ dating technique is the correction of measured $^3He_{total}$ for non-tritigenic ^3He. If the only source of non-tritigenic ^3He is atmospheric, then the correction for $^3He_{atm}$ can be made based on the ^4He content of the sample and the $^3He/^4He$ ratios of the sample and air (Schlosser et al. 1988). Terrigenic ^3He is derived primarily from the decay of ^6Li and uranium/thorium series elements. Where significant amounts of terrigenic ^3He are present, a correction for $^3He_{terr}$ can be made using neon concentration data (Schlosser et al. 1989). However, in most groundwater systems the amount of terrigenic ^3He is insignificant.

Groundwater ages determined from the $^3H/^3He$ ratio will approximate true groundwater travel times provided that dispersive mixing between the water table (point of water intake) and point of measurement is a weak process and the ^3H input has been constant over time (Solomon and Sudicky 1995). Groundwater in low recharge settings may lose significant amounts of $^3He_{trit}$ through diffusion across the water table which could materially impact calculated ages (Schlosser et al. 1989; Ekwurzel et al. 1994).

Under transient conditions where the ^3H input has varied, such as near the middle 1960s bomb testing peak, the effects of dispersion on $^3H/^3He$ ages must be examined using a specified time-dependent ^3H input function (Solomon and Sudicky 1995). Solomon et al. (1993) documented that $^3H/^3He$ groundwater ages agree with average velocities computed using the position of the bomb peak. If the 1963 tritium peak can be identified in the unsaturated zone of a core or borehole, then the seepage rate can be calculated by dividing the depth to the peak by the number of years since 1963. As time progresses the tritium bomb peak is becoming less distinctive because of the radioactive decay of the tritium. The results of a comparative study of tracers performed at the Delmarva Peninsula on the U.S. Atlantic Coast demonstrated that the sum of $^3H + {}^3He_{trit}$ defines the bomb peak much more distinctly that the ^3H data alone (Ekwurzel et al. 1994).

12.4.4 Chlorine 36

A related environmental tracer from nuclear weapons fallout is chlorine 36 (^{36}Cl). The ^{36}Cl peak occurred in about 1955, 8 years before the tritium peak. The $^{36}Cl/Cl$ ratio in water is in parts per 10^{15}. Chlorine 36 can provide information on the rate of movement of chloride and thus solutes in general.

Chlorine 36 has been used to date pore water in three different manners (Scanlon 2000):

(1) Water fluxes over the past 50 years can be estimated from the fallout peak.
(2) Temporal changes in the $^{36}Cl/Cl$ ratio during the past 40,000 years may be used in some instances to trace water movement (Plummer et al. 1997).
(3) The radioactive decay of ^{36}Cl can used to date water up to 1,000,000 years.

Chlorine 36 has seldom been used in water resources investigations because it is a non-standard analysis and thus expensive.

The primary natural source of ^{36}Cl in the atmosphere is the interaction of cosmic rays with argon. The $^{36}Cl/Cl$ ratio data can be used to date groundwater provided that (Cresswell et al. 1999):

- the only sink for ^{36}Cl in the aquifer is radioactive decay,
- the only source of additional ^{36}Cl is normal deep subsurface production or that additional sources can be identified and quantified, and
- the current production rate for ^{36}Cl is the same at the time of recharge.

The time elapsed since recharge can be determined using the radioactive decay equation. The half-life of ^{36}Cl is 301,000 years, so the method is appropriate for dating old groundwater. Cresswell et al. (1999), for example, used $^{36}Cl/Cl$ data to determine that aquifers in the southwestern part of the Northern Territory, Australia, underwent substantial recharge during a favorable, wetter interglacial climatic regime. The implication is the groundwater resources that are being relied upon in the region are non-renewable, and therefore, not a sustainable source.

Another application of the ^{36}Cl method is that it has been used as a tracer to identify potential origins of chloride in groundwater. Murad and Krishnamurthy (2003) used ^{36}Cl and ion balance data to determine that chloride in groundwater in an aquifer in the Eastern United Arab Emirates had a source other that seawater and evaporative concentration. The elevated chloride concentrations were attributed to sources such as fertilizers, pesticides, and animal wastes.

12.4.5 Krypton 81

Krypton-81 is a natural radioactive tracer that can be used to date old groundwater systems. Krypton 81 (^{81}Kr) has a half life of 2.29×10^5 years and is produced in the upper atmosphere by cosmic-ray induced spallation and neutron activation of stable Kr isotopes (Sturchio et al. 2004). The major advantages of using the ^{81}Kr method are that its rate of formation is believed to have been constant over time and there are negligible subsurface sources and sinks of Kr. However, it is difficult to measure ^{81}Kr because of its low isotopic abundance and solubility in water (Sturchio et al. 2004). Sturchio et al. (2004) successfully dated water samples from the Nubian Aquifer in Egypt using the laser-based atom trap trace analysis (ATTA) method. The ^{81}Kr dates were consistent with dates obtained using the ^{36}Cl method and estimated travel times from the source area.

12.5 Stable Isotopes of Water

Both oxygen and hydrogen isotopes are fractionated by various hydrological processes including evaporation of seawater and freshwater, and precipitation. A fundamental discovery in stable isotope hydrology is that the $\delta^{18}O$ and δD of modern meteoric waters (i.e., rainfall) throughout the world are related by the equation (Craig 1961):

$$\delta D = 8.0\delta^{18}O + 10‰\,(SMOW) \tag{12.11}$$

The relationship was subsequently modified based on additional data to (Rozanski et al. 1993):

$$\delta D = 8.17\delta^{18}O + 11.27.8‰\,(VSMOW) \tag{12.12}$$

Equation (12.11), when plotted (Fig. 12.1), is referred to as the “Global Meteoric Water Line” (GMWL). The $\delta^{18}O$ and δD values of rainfall in both warm and cold regions plot approximately on the GMWL, but in different areas. As the degree of fractionation decreases with increasing temperature, meteoric waters in warmer regions plot closer to the seawater values of 0‰ (VSMOW). Meteoric

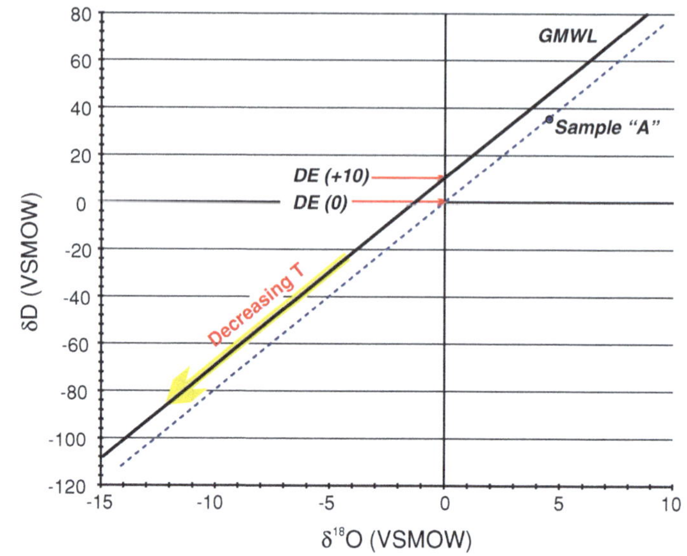

Fig. 12.1 Plot of the global meteoric water line (GMWL). Precipitation in colder climate plots towards the lower left. Deuterium excess (DE) of a water sample is defined as the y-intercept of a line with a slope of 8.0 through a sample's composition. The DE of the GMWL is +10. The DE of sample "A" is 0

waters in colder regions tend to be isotopically lighter because of the greater fractionation. Lighter isotopes are preferentially incorporated into the gas phase during evaporation.

The global meteoric water line is a global average. Local waters may depart from the GMWL. A series of local or regional water lines have been defined, whose slope and δD intercept vary from the GMWL because of differences in local climatic conditions, particularly humidity. Differences in climatic conditions can be spatial (geographic) or temporal.

Under low humidity conditions, a kinetic isotope effect occurs, which results in the enrichment of the isotopically light water molecules into the atmosphere $^{1}H_2^{16}O$, and the enrichment of the heavier molecules, $DH^{16}O$ and $H_2^{18}O$, in the evaporating water (Clark and Fritz 1997). The humidity-related kinetic isotope effect results in meteoric waters in drier regions being enriched in deuterium and thus, plotting to the left of the GMWL. The deuterium excess (DE) in a water sample is defined as:

$$\text{DE} = \delta\text{D} - 8\delta^{18}\text{O} \tag{12.13}$$

The DE is the y-intercept of a line with a slope of 8.0.

Comparison of Eq. 12.13 with the GMWL reveals that average global meteoric water (i.e., water that plots on the GMWL) will have a deuterium excess of 10. The value of the deuterium excess depends upon variations in humidity, wind speed, and sea surface temperature (Gonfiantini 1986; Clark and Fritz 1997). The deuterium excess and slope of the local meteoric line are also influenced by secondary evaporation in which a fraction of the precipitation evaporates before it reaches the ground. Secondary evaporation results in a lesser slope (<8) of the local meteoric line and greater deuterium excesses. Secondary evaporation occurs to a greater degree in drier regions and has a greater relative impact during lighter rain events (Clark and Fritz 1997).

Deuterium excess is of climatological value for identifying the sources of precipitation. From a groundwater resources management perspective, deuterium excess, local meteoric lines, and position of meteoric water lines can be useful for determining the source of groundwater. Groundwater that plots off the current local meteoric water line and has distinctly different deuterium excess can be inferred to have been emplaced under different climatic conditions than current precipitation. Clark and Fitz (1997) cautioned that it is important to use a large precipitation data set in order to improve the confidence limits of the local meteoric water line.

Deuterium excess has been used as one tool for determining the age of fossil groundwater. Groundwater that was emplaced in cooler and more humid climatic conditions during Pleistocene glacial maxima would be expected to have different deuterium excesses than water that recharged under current more arid conditions. For example, in the MENA countries, the modern MWL is characterized by a deuterium excess of 15–30‰, whereas groundwater that recharged during more cooler and humid periods in the past have a MWL that plots closer to the GMWL and has lesser deuterium excesses (Clark and Fritz 1997).

Differences in isotopic ratios between groundwater and precipitation may also be caused by physical and chemical processes that occurred during infiltration (Clark and Fritz 1997). A strong enrichment of groundwater with heavy isotopes may occur in arid regions because of high evaporation rates and the large fractionation associated with evaporation (Clark and Fritz 1997). Stable isotopes may also be used as tracers for evaluation of mixing using simple algebraic mixing curves. However, $\delta^{18}O$ and δD values are often not conservative and may be affected by fluid–rock interactions. Oxygen isotopes are particularly vulnerable to modification in carbonate aquifers because of exchange with oxygen in carbonate minerals.

Stable isotopes can also provide insights to the recharge process. For example, the $\delta^{18}O$ and δD values of shallow groundwater may differ from the values of average annual precipitation. The stable isotope ratios of groundwater samples can be used to determine the season of its recharge (e.g., Simpson et al. 1970; Cunningham et al. 1998; Blasch and Bryson 2007; Wahi et al. 2008). Differences in stable isotopic ratios of shallow groundwater may reflect the isotopic composition of precipitation during the main recharge period, which may differ from the average annual composition (Clark and Fitz 1997). For example, in Arizona, there are two main precipitation periods, the summer monsoon and winter wet season, in which the precipitation has different deuterium and oxygen isotope ratios. Summer monsoon rainfall occurs under warmer temperatures and thus tends to have higher (heavier) $\delta^{2}H$ and $\delta^{18}O$ values.

The deuterium and oxygen isotope ratios of precipitation also vary with altitude ($\delta^{18}O$ decreases with increasing altitude), which can be used as a tool to determine the location and type of recharge (Clark and Fritz 1997; Wahi et al. 2008). However, the variation in isotope ratios with altitude does not occur in all locations (Winograd et al. 1998; Cunningham et al. 1998) and the magnitude of the seasonal and altitude effects may be similar, which can make it difficult to distinguish between the two effects (Wahi et al. 2008).

References

Bassett, R. L., Steinward, A., Jorat, S., Peterson, C., & Jackson, R. (2008). Forensic isotope analysis to refute a hydrologic conceptual model. *Ground Water, 46*, 372–383.

Blasch, K. W., & Bryson, J. R. (2007). Distinguishing sources of groundwater using δ^2H and $\delta^{18}O$. *Ground Water, 45*, 294–308.

Clark, I., & Fritz, P. (1997). *Environmental isotopes in hydrogeology*. Boca Raton: Lewis Publishers.

Craig, H. (1961). Isotopic variations in meteoric waters. *Science, 133*, 1702–1703.

Cresswell, R., Wischusen, J., Jacobson, G., & Fifield, K. (1999). Assessment of recharge to groundwater systems in the arid southwestern part of Northern Territory, Australia, using chlorine-36. *Hydrogeology Journal, 7*, 393–404.

Cunningham, E. E. B., Long, A., Eastoe, C., & Bassett, R. L. (1998). Migration of recharge waters downgradient from the Santa Catalina Mountains into the Tucson basin aquifer, Arizona. *Hydrogeology Journal, 6*, 94–103.

Druhan, J. L., Hogan, J. F., Eastoe, C. J., Hibbs, B. J., & Hutchinson, W. R. (2008). Hydrogeologic controls on groundwater recharge and salinization: A geochemical analysis of the northern Hueco Bolson aquifer, Texas, USA. *Hydrogeology Journal, 16*, 281–296.

Eastoe, C. J., Hibbs, B. J., Granados Olivas, A., Hogan, J. F., Hawley, J., & Hutchinson, W. R. (2008). Isotopes in the Hueco Bolson aquifer, Texas (USA) and Chihuahua (Mexico): Local and general implications for recharge sources in alluvial basins. *Hydrogeology Journal, 16*, 737–747.

Ekwurzel, B., Schlosser, P., Smethie, W. M., Plummer, L. N., Busenberg, E., & Michel, R. L., (1994). Dating of shallow groundwater: Comparison of the transient tracers 3H/3He, chlorofluorocarbons, and 85Kr. *Water Resources Research. 30,* 1693–1708.

Gonfiantini, R. (1986). Environmental isotopes in lake studies. In P. Fritz & J. C. Fontes (Eds.), *Handbook of environmental isotope geochemistry* (Vol. 2, pp. 113–168). New York: Elsevier.

Hutchinson, W. R., & Hibbs, B. J. (2008). Ground Water budget analysis and cross-formational leakage in an arid basin. *Ground Water, 46*, 384–395.

Mazor, E., (2003). *Chemical and isotopic groundwater hydrology* (3rd ed.). New York: Marcel Dekkar.

Mook, W. G. (2005). *Introduction to isotope hydrology: stable and radioactive isotopes of hydrogen, carbon, and oxygen: IAH International Contributions to Hydrogeology 25*. London: Taylor & Francis.

Murad, A. A., & Krishnamurthy, R. V. (2003). Factors controlling groundwater quality in Eastern United Arab Emirates: A chemical and isotopic approach. *Journal of Hydrology, 286*, 227–235.

Plummer, M. A., Phillips, F. M., Fabryka-Martin, J., Turin, H. P., Wigand, P. E., & Sharma, P. (1997). Chlorine-36 in fossil rat urine: an archive of cosmogenic nuclide deposition during the past 40,000 years. *Science, 277*, 538–541.

Rozanski, K., Araguás-Araguás, L., & Gonfiantini, R. (1993). Isotopic patterns in modern global precipitation. In P. K. Swart et al. (Ed.), *Climate change in continental isotopic records: Geophysical. Monograph Series 78* (pp. 1–36). Washington, DC: American Geophysical Union.

Rueedi, U., (2002). Groundwater dating by ^{14}C. In W. Kinzelbach, W. Aeschbach, C. Alberich, I. B. Goni, U. Beyerle, P. Brunner, W. -H. Chiang, J. Rueedi & K. Zoellmann (Eds.), *A survey of methods from groundwater recharge in arid and semiarid regions: Early Warning and Assessment Report Series UNEP/DEWA/RS.02-2* (pp. 44–53). Nairobi, Kenya: United Nations Environment Programme.

Scanlon, B. R. (2000). Uncertainties in estimating water fluxes and residence times using environmental tracers in an arid unsaturated zone. *Water Resources Research, 36*, 395–409.

Schlosser, P., Stute, M., Dorr, H., Sonntag, C., & Munnich, K. O. (1988). Tritium/^{3}He dating of shallow groundwater. *Earth and Planetary Science Letters, 89*, 353–362.

Schlosser, P., Stute, M., Sonntag, C., & Munnich, K. O. (1989). Tritogenic ^{3}He in shallow groundwater. *Earth and Planetary Science Letters, 94*, 245–256.

Simpson, E. S., Thorud, D. B., & Friedman, I. (1970). Distinguishing seasonal recharge to groundwater by deuterium analysis in southern Arizona. *International Association of Hydrologists Publication, 92*, 112–121.

Solomon, D. K., Schiff, S. L., Poreda, R. J., & Clarke, W. B. (1993). A validation of the $^{3}H/^{3}He$ method for determining groundwater recharge. *Water Resources Research, 29*, 2951–2962.

Solomon, D. K., & Sudicky, E. A. (1995). Tritium and helium-3 isotope ratios for direct estimation of spatial variations in groundwater recharge. *Water Resources Research, 27*, 2309–2319.

Sturchio, N. C., Du, X., Purtschert, R., Lehmann, B. E., Sultan, M., & Patterson, L. J., et al. (2004). One million year old groundwater in the Sahara revealed by krypton-81 and chlorine-36. *Geophysics Research Letters 31*, L. 05503, doi:10.1029/2003GL01923.

U.S. Geological Survey (1996). Collection, processing, and analysis of carbon isotope samples. National water quality laboratory technical memorandum 1996-05.

Vrba, J., & Verhagen, B. T. (2006). Groundwater for emergency situations. A framework document: International Hydrological Program (IHP) VI, Series on Groundwater No. 12. Paris: UNESCO.

Wahi, A. K., Hogan, Ekwurzel, B., Baillie, M. N., & Eastoe, C. J. (2008). Geochemical quantification of semiarid recharge. *Ground Water, 46*, 414–425.

Winograd, I. J., Riggs, A. C., & Coplen, T. B. (1998). The relative contributions of summer and cool-season precipitation to groundwater recharge, Springs Mountains, Nevada, USA. *Hydrogeology Journal, 6*, 77–93.

Zhu, C. (2000). Estimate of recharge from radiocarbon dating of groundwater and numerical flow and transport modeling. *Water Resources Research, 36*, 2607–2620.

Zhu, C., & Murphy, W. M. (2000). On radioactive dating of ground water. *Ground Water, 38*, 802–804.

Chapter 13
Wadi Recharge Evaluation

13.1 Introduction

Recharge in arid and semiarid regions is very low because of the paucity of precipitation. The distribution of recharge is both temporally and spatially highly heterogeneous. Most, if not all, of the water that infiltrates soils is subsequently loss to evapotranspiration. Recharge typically occurs predominantly during storm events and primarily occurs in areas in which runoff is concentrated, such as in ephemeral stream channels (i.e., wadis). Infrequent large storms can be the major source of recharge within arid and semiarid region basins, with wadi recharge playing an important role in regulating the total resources of a watershed (Besbes et al. 1978).

Studies of groundwater recharge using a variety of techniques in the Southwestern United States and elsewhere indicate that virtually no natural recharge occurs in alluvial valley floors between ephemeral stream channels (e.g., Scanlon et al. 2006; Izbicki et al. 2007). For example, valley floor sediments in the Mojave Desert of southern California have very low water contents and high negative matric potentials, which indicate that gravity drainage does not occur (Izbicki et al. 2007).

Wadi infiltration (transmission losses) plays an important role in both aquifer recharge and flood propagation. A key feature in arid lands is that flow in wadis quickly infiltrates into the coarse alluvial sediments during low to moderate intensity events (Fig. 13.1). Many world-wide studies have documented or reported that surface water reaches the center of the basin floor or, in the case of coastal wadis, the sea, only during exceptionally large floods (e.g., Onder and Abdulrazzak 1993). Much of the water occurring as runoff during large floods does not recharge the aquifer. However, records for high magnitude events in arid regions are extremely rare (Lange et al. 1999).

Transmission losses in channel alluvium are subject to several removal processes before it can reach the water table and contribute to aquifer recharge,

R. Maliva and T. Missimer, *Arid Lands Water Evaluation and Management*, Environmental Science and Engineering, DOI: 10.1007/978-3-642-29104-3_13,

Fig. 13.1 Photographs of a small ephemeral stream channel near Phoenix, Arizona (U.S.A.) after a low-intensity rainfall. Flow occurred in the upper reach of the channel (*top*), but the water quickly infiltrated into the coarse bed sediments and no flow occurred several 100 m downstream (*bottom*)

including evapotranspiration by near or in-channel vegetation, evaporation from wetted channel sediments (diffusive evaporation), and retention above impeding (semiconfining) layers within the vadose zone and subsequent downgradient discharge or loss to evapotranspiration (Fig. 13.2; Goodrich et al. 2004). Actual aquifer recharge may be a small percentage of wadi transmission losses (i.e., infiltrated water). The ratio of recharge to transmission loss is wadi specific and also varies between rainfall events.

Recharge in wadis located in developed areas may also be enhanced by human activities. Reaches of wadis downstream of urban and agricultural areas may receive additional water from sewage system leakage, wastewater treatment system discharges, dewatering effluent discharges, and irrigation return flows (Al-Shaibani 2008). The high permeability of alluvial gravels results in rapid infiltration and percolation, which makes wadi aquifers vulnerable to contamination. Al-Shaibani (2008) documented elevated nitrate, chloride, phosphate and coliform bacteria concentrations in the Wadi Wajj aquifer, Kingdom of Saudi Arabia (KSA), downstream of the city of Taif relative to up-gradient samples.

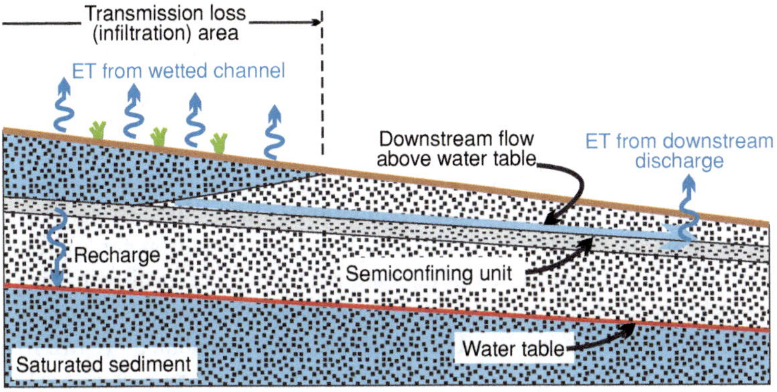

Fig. 13.2 Diagram illustrating the fate of transmission losses in ephemeral channels. Water may recharge the underlying unconfined aquifer, be lost to ET in the channel vicinity, be stored in the vadose zone, or flow downstream above semiconfining units where it is later lost to ET

Anthropogenic contamination in Wadi Marwani, KSA, is also evident by increases in salinity and nitrate concentration (Al-ahmadi and Al-Fiky 2009). The primary contaminant sources are agricultural operations, urban runoff, and septic systems. Extent of contamination can be assessed using upstream water quality data as a baseline.

13.2 Methods for Measuring Wadi Recharge

Wadi recharge is a critical issue for groundwater management in arid lands because in some groundwater basins it is the major, if not sole, source of significant modern aquifer recharge and thus supply of renewable groundwater. Quantification of wadi recharge is complicated by the fact that it occurs only during some precipitation events and rainfall in many arid regions is highly sporadic and spatially variable. The database on wadi recharge rates is quite modest, with the highest quality data obtained from heavily monitored experimental watersheds. In the vast majority of watersheds in arid and semiarid lands, monitoring data is woefully inadequate (or non-existent) to accurately estimate wadi recharge rates. Considerable effort has been expended to develop methods to estimate wadi recharge from limited datasets.

Methods of quantifying recharge in general are discussed in Chap. 11 and many could be used for the specific case of estimating recharge in wadis resulting from flood events. There are two basic types of approaches that are used to evaluate wadi recharge. Recharge can be estimated from the hydraulic response of the water table or through geochemical tracers, without implicit consideration of wadi hydraulic processes. Alternatively, wadi recharge can evaluated through an integrated approach, in which rainfall/runoff relationships, transmission losses, and infiltration are all considered (Sects. 13.3–13.6). The ultimate goal of the latter approach is the development of a calibrated, integrated surface-water and groundwater model, that can predict both the surface-water flows and aquifer response to future rainfall events, and be used to evaluate both surface-water and groundwater development options.

Aquifer-based approaches have the advantage of estimating directly the key variable of interest for groundwater evaluations, the recharge rate. An average long-term recharge rate can be used as a boundary condition for a groundwater model. If a sufficient time-series of data are available, a rainfall-recharge function may be developed that could be of use for predictive models using stochastic estimates of future rainfall. The chloride mass balance (CMB) method (Sect. 11.5.1) can be a very cost-effective tool for estimating recharge rates, if the effective rainfall (mm/yr; in/yr) and the chloride concentration of the rainfall are known. Effective rainfall is the amount of precipitation that enters a channel as direct input and as runoff, and for the CMB method must be corrected for runoff that leaves the studied watershed. Recharge can also be estimated from the water table response (Sect. 11). An important consideration in estimation of wadi recharge is that where the water table is located a great depth below the channel floor (i.e., unsaturated zone is very thick), there will be a long travel time for percolating water to reach the water table and become recharge.

As is the case for estimation of recharge in general, multiple methods should ideally be used to estimate wadi recharge. There have been relatively few comparative studies of different methods for estimating wadi recharge rates. For example, Besbes et al. (1978) discussed the application of four methods for estimating recharge in ephemeral streams in Tunisia, which are generally applicable to wadi recharge. These methods include:

- calculation from infiltration rate, wetted channel area, and duration of flood,
- water table recovery (WTF method),
- model calibration, and
- deconvolution of assumed linear relationship between stream recharge and recovery of the water table.

Estimation of recharge rates using a measured or assumed infiltration rate is problematic because it is difficult to accurately estimate hydraulic parameters. Infiltration rate, which is a function of the vertical hydraulic conductivity of alluvial sediments, may have substantial temporal and spatial variation. Furthermore, the relationship needs to be established between infiltration and aquifer recharge. It is important to emphasize that in some situations (particularly minor rainfall events and flow events) much of the water that infiltrates into wadi beds may not actually recharge the underlying unconfined aquifers.

Although the WTF method is conceptually simple, accurate implementation in an area of localized (focused) recharge has greater data requirements than is often the case in areas of diffuse (distributed) recharge. Multiple piezometers are needed to determine the spatial increase in water levels, which needs to be evaluated relative to the potentiometric surface of the aquifer that would have occurred without the recharge event. The effective porosity of the aquifer also needs to be determined in order to convert the increase in saturated volume to recharged water volume. Care must be taken in the analysis to assess the potential impacts of trapped air on the hydrograph during rainfall events and make the appropriate corrections to the water level data.

Recharge can be calculated through the calibration of a groundwater flow model, which in essence incorporates the WTF method. A limitation of the use of model calibration for recharge estimation is that it does not provide a unique calibration. The uncertainty associated with modeling decreases with data availability to narrow the envelope of possible solutions. Model calibration is a recommended approach because a calibrated groundwater flow model is a valuable tool for assessment and management of alluvial aquifers. Recharge rates determined from other methods should be used to constrain the calibration process.

Assuming a linear relationship between stream recharge and recovery of the water table, deconvolution can provide an empirical relationship between recharge and recovery data for a large flood (Besbes et al. 1978). The recharge volume for the flood V_R can be determined from model calibration or water table recovery. The linear impulse response equation is (Besbes et al. 1978):

$$\phi_i(t) = \frac{h_i(t) - d_i(t)}{V_R} \tag{13.1}$$

where,

$\phi_i(t)$ = Linear impulse response relating recharge to recovery in observation well "i" (m/m^3)

$h_i(t)$ = observed head in piezometer "i" (m)

$d_i(t)$ = Assumed head in piezometer "i" if no recharge had occurred (m)

V_R = recharge volume (m^3).

Besbes et al. (1978) noted that the impulse response function method works until the next flood in the stream modifies the behavior of the system. Impulse functions vary between piezometers and give different recharge volumes for measured piezometer responses to rainfall. This requires some type of statistical averaging technique to be applied to give adequate spatial estimates.

Goodrich et al. (2004) performed a comparative study at the Walnut Gulch Experimental Watershed, located in Southern Arizona, USA, of methods available for the estimation of recharge rates. The evaluated methods were:

- channel reach water balance,
- calibration of groundwater mounding models (aquifer response to channel flow events),
- microgravity measurements,
- geochemical tracers (CMB, stable isotopes), and
- vadose zone water and temperature modeling.

Recharges rates calculated using the five methods varied within a factor of three with the channel reach water balance method giving the greatest rate and mounding analysis the lowest. The variation was due to uncertainties associated with each method and differences in the temporal scale of the measurements (Goodrich et al. 2004).

The Goodrich et al. (2004) study was a research project conducted at an experimental watershed and involved an atypically large volume of data. It was recognized that the data requirements for some methods involved much more time and expense than is practical for most investigations. Although the channel reach water balance method is conceptually simple, quantification of all the inputs and outputs is required, which involves a considerable effort and has associated uncertainties. For example, accurately estimating evapotranspiration from near channel vegetation and wetted sediments is a complex process with large uncertainties. Microgravity measurements (Chap. 14) provide information on changes in the total water mass in both the unsaturated and saturated zones, and thus, do not directly provide information on the volume of water reaching the aquifer. Temperature data can provide information on local (point) recharge rates, but is data intensive and requires relatively complex modeling for its interpretation (Sect. 13.8).

The CMB and mounding analysis methods are the most practical in terms of data requirements (Goodrich et al. 2004). The CMB method can be applied in two manners (Goodrich et al. 2004). Aquifer recharge can be estimated from the chloride concentration of the runoff and channel transmission loss. Alternatively, a watershed-wide estimate can be made using a watershed-wide estimate of rainfall and measured chloride concentration data from the rain. Both methods have the limitations and uncertainties associated with the CMB method in general. Mounding analysis requires water level data from preferably multiple observation wells in the channel vicinity and has the general limitation of the non-uniqueness of model calibration. Goodrich et al. (2004) estimated that 15–40% of the overall recharge in the San Pedro River Basin is from ephemeral recharge during monsoonal rainfall. Channel reach water balance estimates indicated that the potential recharge was 86–92% of transmission losses.

13.3 Wadi Rainfall-Runoff Relationships

A fundamental water resources issue in wadi systems (and throughout arid and semiarid lands in general) is the fate of rainfall. Rainfall is partitioned between evapotranspiration, runoff, and infiltration (and ultimately recharge). Key differences in stream hydrology between arid and humid lands are (Pilgrim et al. 1988):

- baseflow is essentially absent in arid regions, whereas channel transmission losses are of critical importance,
- rainfall events tend to be more variable in space and time,

- plant cover is sparse and consists mainly of xerophytes and ephemeral grasses and small leafy plants, and
- phreatophytes located along stream channels may have a significant effect on groundwater recharge.

Şen (2008, p. 133) correctly cautioned that

> In general, absence of reliable long-term data and experimental research, and the tendency to employ humid zone experience and modeling, may lead to results that are highly inaccurate.

Significant differences may also exist between different arid and semiarid areas (Pilgrim et al. 1988) including:

- types of run-off producing storms (e.g., convective storms, tropical cyclones),
- effects of relief. Low relief regions often lack integrated drainage and alluvial fans are much less dominant in the landscape, and
- climatological differences (temperature and rainfall regimes).

An important hydrologic characteristic of arid regions is that rainfall events are often of short duration and intense. Stream flow in arid regions is typically dominated by surface runoff, because baseflow is usually negligible. Quick return subsurface flow, wherein infiltrated water meets a subsurface horizon of lower hydraulic conductivity, travels above the interface, and reappears as a seep or spring, is also usually not significant in arid regions, except perhaps after a long succession of storms.

The paucity of vegetation and the presence of exposed rock and soil crusts result in low infiltration rates and high runoff rates. Minor rainfall events (showers) are not significant for groundwater recharge and runoff because the water absorbed in shallow, surficial soil is quickly lost to evapotranspiration. The precipitation just wets surfaces or the uppermost few cm (in) of the vadose zone. During heavy rainfall events, large amounts of surface runoff can occur even though the underlying soil profile may remain substantially dry, due to soil crusting and the intensity of rainfall exceeding the infiltration rate. A threshold value of rainfall needs to be exceeded in order for significant runoff to be generated. Şen (2008) reported that storm rainfall needs to exceed a total of about 20 mm (0.8 in.) to generate significant runoff in wadis.

Antecedent soil moisture conditions and the depth to the water table are important factors in the rainfall-runoff relationship. In areas in which the water table is deep, groundwater mounds do not typically rise through the vadose zone to intersect channels. The channel and water table are hydraulically disconnected to variable degrees. Where the water table is shallow or storage volume is small, wadi aquifers may fill. Subsequent storms may, therefore, have significantly different rainfall-runoff relationships.

Rainfall-runoff relationships play a key role in surface water resources planning, design, and operations. Clearly, the preferred method is to empirically establish rainfall-runoff relationships from actual data on drainage basin rainfall and discharge measured at one or more gauging stations. However, in most basins such data are not available. The paucity of available data constitutes the greatest problem in arid and semiarid lands modeling, which is due to the isolation, low population density (and thus high cost), and low economic potential of many of these areas (Pilgrim et al. 1988). Even if a decision was made to obtain such data, it would take a long time in arid lands to obtain the data because of the high degree of temporal variability in rainfall.

The hydrologic parameters used to describe the rainfall-runoff processes in wadi systems must be calibrated and verified based on historic and measured rainfall and flood flow data, but such data are seldom available (Al-Weshah 2002). Thus, in arid regions, hydrologists are faced with the problem of trying to find the rainfall-runoff relationship by artificial, indirect, and synthetic ways (Şen 2008). The various methods for evaluating rainfall-runoff relationships involve empirical coefficients, which are local values that typically cannot be directly determined in ungauged catchments. In the absence of basin or watershed specific data, it is desirable, if possible, to determine the values for the coefficients from similar nearby gauged basins. Data from similar gauged catchment basins also allow the determination of the appropriateness of evaluation methods in the project site area.

A discussion of basic hydrologic principles as they may be applied to arid and semiarid stream systems follows. The Part 630 Hydrology of National Engineering Handbook prepared by the Natural Resources Conservation Service of the U.S. Department of Agriculture, is an excellent, widely-cited reference on hydrology that is available online. Individual, periodically updated chapters are cited herein. There are numerous textbooks that also provide good introductions of hydrologic principles and procedures that can be applied to arid region stream hydrology (e.g., Chow et al. 1988; Bras 1989; Ramírez 2000; Şen 2008; Soliman 2010).

13.3.1 Precipitation

Evaluation of rainfall-runoff relationships must start with data on rainfall within the catchment, watershed or basin of interest. Methods for measuring rainfall are discussed in Chap. 7. Rainfall within arid regions tends to have relatively high degrees of spatial heterogeneity, particularly with respect of convective storms. The spottiness of rainfall greatly impacts conventional analysis of rainfall-runoff relationships because of uncertainties in catchment-wide rainfall distribution. Monitoring points (rain gauges) are typically much too sparse to provide accurate data on catchment-wide rainfall. Wheater (2002) reported a situation wherein wadi flow occurred even though the rain gauges indicated no precipitation because the convective storm happened to have missed them. Radar data can provide data on the spatial distribution of rainfall, but it is not available for most arid and semiarid areas.

A major technical challenge is the determination of the areal values for rainfall rates and volumes in a study are from what is often at best a small number of point measurements from scattered rain gauges. A variety of methods are available for obtaining a real estimates of precipitation from point measurements, which were reviewed in the Natural Resources Conservation Service (1993), Şen (2008), and Soliman (2010). The methods include arithmetic mean, Kriging, Thiessen polygons, distance weighting, isohyetal analysis, and percentage weighting.

13.3.2 Regression Methods

The simplest method for quantification of the rainfall-runoff relationship is by regression analysis. Peak runoff rates or flood magnitudes (e.g., 100-year event) as measured at a stream gauging station, can be related to catchment rainfall, area, and other variables. Runoff can be related to a single variable or multiple variables and the relationship can be either linear or exponential. A relationship established for one catchment may be applied to ungauged catchments that have similar geological and hydrological conditions. However, it is important to recognize that there will always be some differences in the hydrologic behavior of different catchments, even those that appear similar.

Regression methods are based on data from gauged streams and require a long-term monitoring period to collect sufficient data to establish the statistical relationship (i.e., determine the values of the various regression coefficients). The statistical significance of the regression (correlation coefficient) should be evaluated to determine which relationships are significant. The relationship between statistically determined rainfall and runoff probabilities and the length of the monitoring period is complex, particularly in arid lands in which there is great temporal variability in rainfall (Sect. 7.3). Relatively short duration (several decades or less) monitoring records may not be representative of long-term climatic conditions, especially the frequency and magnitude of extreme events.

13.3.3 Soil Conservation Service Curve Number Method

The United States Soil Conservation Service curve number (SCS-CN) method is an empirical, lumped-parameter, water budget-based method for estimating the direct runoff from storm rainfall. The method can also be used to estimate infiltration from storm events. The SCS-CN method is documented by the Natural Resources Conservation Service (2004a) and in most hydrology texts. The SCS-CN method is presented herein because it is conceptually simple and has considerable practical applicability. Lumped-parameter, empirical methods are often considered to be less scientifically rigorous compared to distributed parameter methods based on actual soil and channel properties. However, distributed parameter methods and models have very large data requirements, which typically greatly exceed the available or practicably obtainable data. The scientific rigor of distributed-parameter methods is commonly largely lost through the use of assumed or default values for parameters.

The SCS-CN method is based on the portioning of rainfall (P) between runoff (Q), initial abstraction (I_a), and actual soil retention (F):

$$P = Q + I_a + F \text{ (units of inches or mm)} \tag{13.2}$$

Initial abstraction consists mainly of interception, infiltration during the early parts of storms, and surface depression storage. The initial abstraction ratio (IAR) is the initial abstraction (I_a) divided by the maximum potential retention or storage (S)

$$IAR = I_a/S \tag{13.3}$$

An empirically based IAR of 0.2 was originally used in the method and is a suitable default value. However, the actual site-specific value may vary from 0.2. The quantity ($P - I_a$) can be considered to be the effective rainfall that is available for surface run-off and infiltration (retention).The underlying equation of the SCS-CN method for calculating runoff (Q) is

$$Q = \frac{(P - I_a)^2}{(P - I_a) + S} \tag{13.4}$$

The parameters as previously described are in units of length (mm, in).

Equation (13.4) is not physically based, but was instead developed to obtain a rainfall-runoff curve of the proper shape (Natural Resources Conservation Service 2004a). The curves should show that minor rain events generate minimal runoff, whereas most of the precipitation of large storm events should run off once the maximum potential retention occurs (i.e., soils are saturated).

Using an initial abstraction ratio of 0.2, Eq. (13.4) becomes

$$Q = \frac{(P - 0.2S)^2}{(P + 0.8S)} \tag{13.5}$$

The curve number (CN) is a transformation of S as follows:

$$CN = \frac{1000}{10 + S} \text{(S in inches)} \tag{13.6a}$$

$$CN = \frac{1000}{10 + \frac{S}{25.4}} \text{(S in mm)} \tag{13.6b}$$

Equations 13.6a and 13.6b can be rearranged to calculate S from CN values, as follows

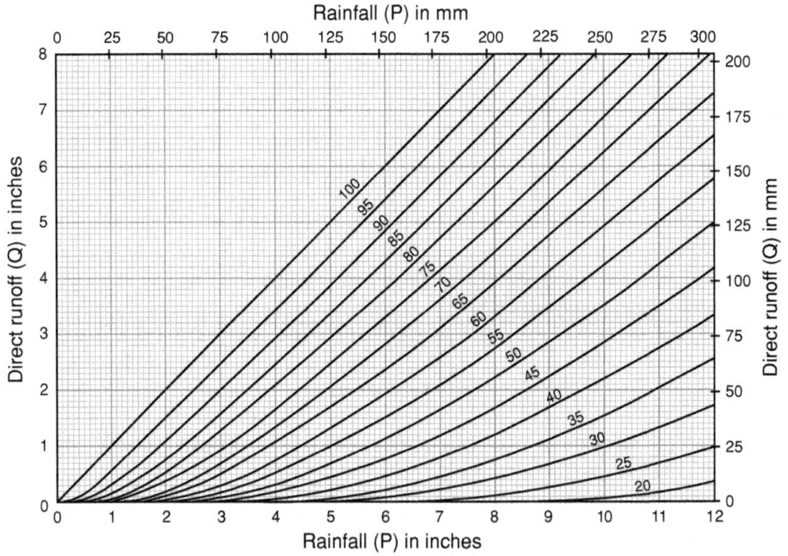

Fig. 13.3 Relationship between rainfall, runoff, and SCS-CN method curve number

$$S = \frac{1000}{CN} - 10(\text{S in inches}) \tag{13.7a}$$

$$S = \frac{25,400}{CN} - 254(\text{S in mm}) \tag{13.7b}$$

Combining Eqs. (13.5) and (13.7a, b) yields a relationship between precipitation and runoff for different curve numbers, which is illustrated in Fig. 13.3. Curve numbers close to 100 represent impervious surface conditions in which virtually all rainfall runs off. On the contrary, curve numbers close to 0 represent highly pervious conditions with large storage volumes in which rainfall results in minimal runoff.

The critical issue in the SCS-CN method is the choice of the CN. Tables are available that assign curve numbers to various combinations of hydrologic soil group (soil), land use, and treatment class (cover), which can be used as a guide to approximate local values (Natural Resources Conservation Service 2004b). Variability in CN results from rainfall intensity and duration, total rainfall, soil moisture conditions, cover density, stage of growth, and temperature, which are collectively called the "antecedent runoff condition" (Natural Resources Conservation Service 2004a). The accuracy of the method appears to be less where runoff is a small fraction of the total rainfall (i.e., where CN values are low or rainfall values are small) (Natural Resources Conservation Service 2004a).

No means exists for estimating the errors in CN of ungauged watersheds; only comparisons of estimated and actual runoffs indicate how well estimates of CN are being made. Comparisons for gauged watersheds, though not directly applicable to

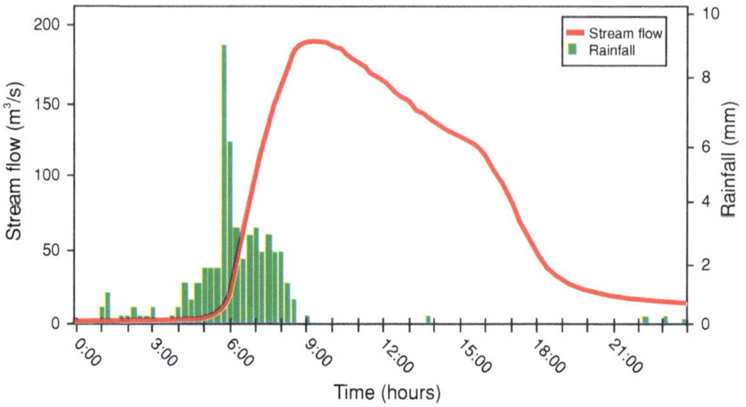

Fig. 13.4 Example of a stream hydrograph. Peachtree Creek, Georgia (U.S.A), December 24, 2002: *Source* USGS (http://ga.water.usgs.gov/edu/watercyclestreamflow.html)

ungauged watersheds, are useful as guides to judgment in estimating CN and as sources of data for reducing estimation errors (Natural Resources Conservation Service 2004a).

Şen (2008) proposed an alternative methodology for arid regions, the infinitesimal runoff coefficient (IRC), in which the rainfall-runoff relationship is a function of a parameter 'k' referred to as the saturation factor, as follows:

$$\frac{dQ}{dP} = 1 - e^{-kP} \tag{13.8}$$

$$Q = P - \frac{1}{k}\left(1 - e^{-kP}\right) \tag{13.9}$$

The IRC method has a similar limitation as the SCS-CN method in that it requires estimation of the value of the saturation parameter, which should take into account the hydrological soil group, land use and treatment classes, and antecedent moisture content (Şen 2008). The IRC method is graphically represented in a similar manner as the SCS-CN method as a series or runoff versus rainfall curves for different k values.

The runoff calculated using the SCS-CN or IRC methods is the water available for wadi recharge through transmission losses (Sect. 13.4). The water that infiltrates within the watershed may be lost entirely to ET or may be the source of mountain block recharge.

13.3.4 Hydrographs

Hydrographs are plots of discharge versus time and have a characteristic bell shape (Fig. 13.4). Hydrographs can provide and illustrate three important data:

- peak discharge,
- time to peak discharge, and
- total discharge (area under curve).

The shape of hydrographs is a function of drainage basin geometry, surficial geology, rainfall (total, and spatial and temporal distribution), and storm direction, and no two hydrographs are exactly alike. Hydrographs in arid regions tend to have relatively sharp peaks. Runoff is characterized by spates that last on average 12 h from start to finish and typically have a sharp rise and rapid regression (Şen 2008). The rise can be extraordinarily rapid, which has caused vehicles and people that attempted to cross what were dry channels at the time they entered to be swept away by flash floods.

There are four basic types of hydrographs that are used in watershed studies (Natural Resources Conservation Service 2007).

- Natural hydrographs—obtained directly from the flow records of a gauged stream.
- Synthetic hydrographs—calculated using watershed parameters and storm characteristics.
- Unit hydrographs—resulting from a unit (e.g., 1 in. or cm) of direct runoff distributed uniformly over the watershed in response to a rainfall of a specific duration.
- Dimensionless unit hydrographs—developed to represent several unit hydrographs; plotted using the ratios of time (t) to time to peak (t_p) and rate of discharge (Q) to peak rate (Q_p).

13.3.4.1 Natural Hydrographs

Natural hydrographs are generated by data from gauging stations at which the relationship between water level and flow has been established. At each gauging station, a rating curve is developed which expresses the functional relationship between the rate of discharge (Q) and water depth (D). Rating curves typically appear as a power function of the form:

$$Q = aD^b \tag{13.10}$$

where, 'a' and 'b' are empirical constants. Rating curves depend upon:

- cross-sectional area and shape (profile) of the channel,
- slope of the channel in the vicinity of the station, and
- frictional losses due to the roughness of the wetted perimeter (Manning roughness coefficient).

Inasmuch as the rating curves are a function of channel shape and bed conditions, they will change over time if stream conditions at the gauging station change. Gauging stations typically have artificial controls built into the channel to stabilize the stage-discharge relationship, such as a permanent foundation and sides to the stream. Gauging stations are often established at weirs and bridges. Hydrographs can be generated from data at natural (unmodified) channels. However, channel shape and bed-roughness may change gradually over time or very rapidly during a major flood due to scour and sediment deposition, and, as a result, the relationship between discharge and depth will also change.

13.3.4.2 Unit Hydrographs

A unit hydrograph represents the response of a watershed to one unit (inch or cm) of excess rainfall distributed uniformly over the entire watershed. Excess rainfall is used in this context as rainfall that results in runoff. Unit hydrograph theory assumes that discharge at any time is proportional to the volume of runoff and that the time factors that affect hydrograph shape are constant. Unit hydrograph theory includes the following principles (Natural Resources Conservation Service 2007):

- Principle of invariance. The hydrograph of surface runoff from a watershed resulting from a given pattern of rainfall is constant.
- Principle of superposition. The hydrograph resulting from a given pattern of excess rainfall in a watershed can be built by superimposing the unit hydrographs of subareas of the watershed. The subareas should, if possible, have a homogenous drainage pattern, homogenous land use, and approximately the same size.
- Principle of proportionality. The ordinates of the hydrograph are proportional to the volume of the excess rainfall.

The great value of the unit hydrograph theory is that once the unit hydrograph for a basin is determined, the watershed response to a given amount of excess precipitation is a multiplier of the unit hydrograph. Peak discharge can be determined for a given excess precipitation.

Dimensionless unit hydrographs (DUHs) are plots of Q/Q_p and t/t_p. Natural Resources Conservation Service (2007) developed a DUH from numerous natural unit hydrographs from watersheds varying widely in size and geographic location (Fig. 13.5), which can be considered a default DUH. However, DUHs vary from watershed to watershed based on a number of factors including size of watershed, geomorphic characteristics, geologic characteristics, watershed slope, watershed length, volume of storage, degree of channelization of the stream network, and degree of urbanization (Natural Resources Conservation Service 2007).

The DUH can be represented by an equivalent triangular hydrograph having the same units of time and discharge, and having the same percent volumes on the rising and receding sides of the triangle (Fig. 13.5). The peak discharge rate can be calculated as follows (Natural Resources Conservation Service 2007):

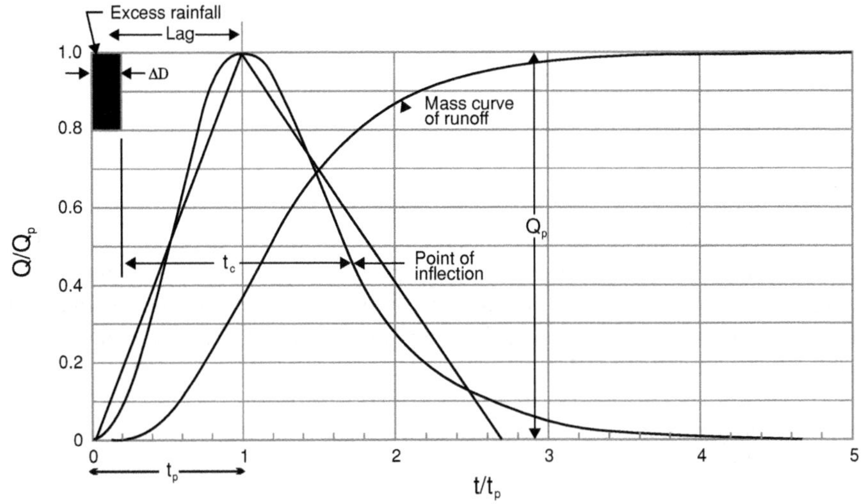

Fig. 13.5 Dimensionless unit hydrograph and equivalent triangular hydrograph. Parameters are defined in text. *Source* Natural Resources Conservation Service (2007)

$$Q_p = \frac{484AQ}{t_p} \tag{13.11}$$

where,

Q_p = peak discharge in ft^3/s
484 = peak rate factor (value for default DUH)
A = drainage area (mi^2)
Q = runoff (inches)
t_P = time to peak (hours).

The metric version of Eq. (13.11) is

$$Q_p = \frac{0.2083AQ}{t_p} \tag{13.12}$$

where,

Q_p = peak discharge in m^3/s
0.2083 = peak rate factor (value for default DUH, equivalent to 484 for English units)
A = drainage area (km^2)
Q = runoff (mm)
t_P = time to peak (hours).

The value of the peak rate factor varies depending upon topography and ranges [(in English units; Eq. (13.11)] from 600 in steep terrains to 300 in very flat swampy areas. The corresponding range for Eq. (13.12) in metric units is 0.258 to 0.129.

The time to peak is a function of the duration of unit excess rainfall (ΔD) in hours and the watershed lag (t_L) in hours as follows:

$$t_P = \frac{\Delta D}{2} + t_L \tag{13.13}$$

The watershed lag is the time from the center of mass of excess rainfall (ΔD/2) to the time of peak discharge. An important hydrologic parameter is the time of concentration (t_c), which is the time for runoff to travel from the hydraulically most distant point in the watershed to the point in question. Time of concentration is a function of channel length, slope, channel roughness, and cross-sectional characteristics of the channel (area and wetted perimeter).

The average relationship between time of concentration to watershed lag is (Natural Resources Conservation Service 2007):

$$t_L = 0.6t_c \tag{13.14}$$

Synthetic hydrographs can be generated for different excess rainfall values by multiplying the unit hydrograph by excess rainfall. The duration of the unit excess rainfall is related to the time of concentration as follows:

$$\Delta D = 0.133t_c \tag{13.15}$$

Watershed-specific DUH and unit hydrographs can be developed for gauged watersheds by calibration against local actual rainfall and runoff data. Procedures for generating DUHs are provided by the Natural Resources Conservation Service (2007). Unit hydrographs can be constructed for any location in a regularly-shaped watershed from a DUH once values for q_p and t_p are defined for a unit (e.g., 1 in. or 1 cm) rainfall event.

Most ephemeral stream systems do not have any or sufficient gaging data for the generation of unit hydrographs. However, a regional study of the DUH for gauged watersheds within the region of interest may reveal insights into what DUH to use for ungauged watersheds located within the region.

13.3.4.3 Synthetic Hydrographs in Ungauged Watersheds

It is preferable to develop unit hydrographs based on watershed-specific rainfall-runoff data. In the absence of such data, synthetic unit hydrographs can be derived from watershed characteristics. The Snyder (1938) Synthetic Unit Hydrograph is an empirical method for deriving the time to peak and peak flow of a unit hydrograph. Snyder's method is presented in most hydrology texts (e.g., Chow et al. 1988; Bras 1989; Ramírez 2001; Soliman 2010) from which the following summary was derived.

The basic lag (t_b) in hours is related to channel length as follows:

$$t_b = C_1 C_t (L L_c)^{0.3} \quad (13.16)$$

where,
C_t = lag coefficient (ranges from 1.2 to 2.2)
C_1 = coefficient, 0.75 for metric units, 1 for English units
L = main channel length (km, miles) from the outlet to upstream divide,
L_c = centroid length, distance along channel (km, miles) from outlet to centroid of watershed area.

The lag time or basin lag is the time from the center of mass of excess rainfall to the hydrograph peak.

Peak discharge of the standard unit hydrograph (q_p) is determined by

$$Q_{p=} \frac{C_2 C_p A}{t_b} \quad (13.17)$$

where,
Q_p = peak discharge (m^3/s, ft^3/s)
C_2 = 2.75 for metric units, 640 for English units
C_p = peak flow coefficient, ranges from 0.5 to 0.7
A = watershed area (km^2, mi^2).

The lag coefficient and peak flow coefficient are empirical coefficients that are derived from gauged watersheds and are dependent on watershed slopes and storage characteristics.

The time to peak (t_p) is equal to the lag time plus half the duration of the excess precipitation event

$$t_P = \frac{\Delta D}{2} + t_b \quad (13.18)$$

where, $\Delta D = t_b/5.5$.

Other methods have been developed to predict peak runoff. The Rational Method is often used for the design of urban stormwater systems and may be used for small watersheds. The core equation is

$$Q = kACi \quad (13.19)$$

where,
Q = peak discharge (ft^3/s, m^3/s)
A = watershed area (acres, hectacres)
C = run-off coefficient
i = design rainfall storm intensity (in/hr, mm/hr)
k = conversion factor (1.008 for English units, 0.00278 for metric units).

The rainfall intensity (i) is the average rainfall rate in in/hr (or mm/hr) for specific rainfall duration and a selected frequency. Peak surface runoff occurs when the entire watershed is contributing runoff, which occurs at the time of concentration. The runoff coefficient is an empirical constant which has a value approaching 1 for impervious surfaces and 0 for an extremely pervious surface from which no runoff occurs. The design rainfall intensity is a constant intensity rain with duration equal to the time of concentration. Values for the runoff coefficient for different land surfaces can be obtained from reference tables. The main technical challenge of applying the Rational Method is determining the value for the runoff coefficient (between 0 and 1).

13.3.5 Geomorphic Relationship

Rainfall-runoff relationships in ungauged basins may be estimated from relationships between geomorphic parameters and rainfall determined from a watershed considered to be "typical" of the physiographic province. Geomorphic parameters than can be readily obtained from aerial photographs include drainage area, basin shape factor, drainage density, relief ratio, and main channel length and slope (Murphey et al. 1977). Linear and multiple linear regression analyses have been used to relate geomorphic parameters to hydrographic characteristics. The Murphey et al. (1977) study of the Walnut Gulch Experimental Watershed near Tombstone, Arizona, determined that the watershed area is the most reliable parameter to use in the prediction of the watershed hydraulic response of ephemeral streams to thunderstorm precipitation. Geomorphic parameters were most accurate for predicting rise time, duration of flow, peak discharge, and peak to volume ratio.

13.4 Transmission Losses

Transmission losses to infiltration (TL) in ephemeral streams can be quantified as the difference in discharge between upstream (V_1) and downstream (V_2) gaging stations, plus any lateral (tributary) runoff (V_t) and baseflow (BF) minus evapotranspiration losses (ET).

$$TL = V_1 - V_2 + V_t + BF - ET \text{ (units = volume)} \tag{13.20}$$

Within ephemeral channels in arid or semiarid lands, baseflow will typically be negligible or non-existent. Evapotranspiration will also be minor during the time frame of an individual storm event, although significant ET loss of temporarily stored (depressions in land surface) or infiltrated water may subsequently occur. Evapotranspiration by riparian vegetation can play an important role in alluvial aquifer water budgets. To convert transmission losses from single event measurements to a representative value of water fluxes through decadal or longer time scales requires an integration of events to obtain an estimate of mean runoff and infiltration (Osterkamp et al. 1995).

The data required to quantify transmission losses are typically not available for most ephemeral streams. Stream gauges are often not practical in wadis because of the expense associated with measurement of infrequent flows in channels that may be scoured or change course during stream flow (Izbicki et al. 2007). Flow in wadis is often highly destructive and gauging equipment is prone to damage. Alternative methods have been developed to record the occurrence and quantify ephemeral stream transmission flow. For example, temperature data have been successfully used to monitor the basinward extent of flood flows and associated transmission losses (Sect. 13.8).

A water management goal is to develop the capability of determining transmission losses (and ultimately aquifer recharge) from precipitation data, upstream hydrograph parameters, and stream channel characteristics. Procedures are needed for accurately estimating transmission losses that are simple, utilize easily measured channel characteristics and flow volumes, and are transferable to ungauged areas. Simplified procedures require less information, but are general in application. On the contrary, more complex physically-based methods have greater data requirements, which are very often not available (Walters 1990). For example, a transmission loss model described by Osterkamp et al. (1995) requires inputs of contributing areas of runoff for upland and lateral areas, channel dimensions, hydraulic conductivity of channel alluvium, mean annual precipitation, and magnitudes of storm events and runoff properties for each watershed. Rarely are accurate values of such variables available, so a variety of estimates need to be made. Use of channel parameters to estimate transmission losses is further complicated by the fact that losses along a particular channel reach can be highly variable in both time and space due to the transient development and subsequent erosion of clogging layers (Besbes et al. 1978; Stephens 1996).

From a practical groundwater management perspective, the data collection requirements should be commensurate with the capabilities, resources, and budgets of the water management agencies. For example, data intensive academic or government research programs on individual watersheds may not provide a practical template for regional water resources management programs.

One approach that has been taken is to examine transmission losses in stream systems with multiple gauges in order to try to determine general transmission loss relationships that could be applied to ungauged streamed systems. Transmission losses have been quantified in some experimental watersheds in which an atypically large number of steam and rain gauges have been installed. Single or multiple regression analyses were used to relate transmission losses to hydrologic and stream characteristic parameters (Walters 1990; Sorman and Abdulrazzak 1993; Adbulrazzak and Sorman 1994; Sharma and Murthy 1994a, b; Cataldo and Pierce

2005). Transmission losses in ephemeral channels have been shown to be related to (Sorman and Abdulrazzak 1993; Stephens 1996; Cataldo and Pierce 2005)

- flow duration,
- channel length and width,
- channel geomorphology (distribution of riffle and pools sequences; Newman et al. 2006),
- antecedent moisture content (time since previous flood event)
- depth to water table,
- peak discharge,
- flow sequence,
- thickness and characteristics of alluvium (e.g., hydraulic conductivity),
- presence of clogging layers and surface crusts,
- water temperature (through its effects on hydraulic conductivity),
- suspended solids concentration of runoff,
- watershed geometry, and
- location and intensity of the isohyets (rainfall contours) in the watershed.

The most important variables are flow volume and duration, peak inflow rate, and wetted channel perimeter (e.g., Walters 1990; Lane et al. 1971; Sorman and Abdulrazzak 1993; Wheater 2002; Cataldo and Pierce 2005). These parameters are critical in that they determine the opportunity in time and space afforded for infiltration. Clearly, longer stream flow durations provide more time for infiltration to occur. Greater wetted channel perimeters results in a greater area for infiltration to occur. Flow rates and volumes affect the wetted channel perimeter and hydraulic head within the channel.

13.4.1 Transmission Loss Relationships from Experimental Watersheds

Lane et al. (1971) and Cataldo and Pierce (2005) studied transmission losses in the Walnut Gulch Experimental Watershed (WGEW) in southern Arizona, which contains multiple stream gauges and numerous rain gauges. The magnitudes of the transmission losses were determined by the difference in discharge between stations for storm events in which all the runoff originated above the upper station. In this manner, lateral flow is not significant.

Transmission losses in streams can be compared using units of volume lost per unit channel length (e.g., acre-feet/mi or m^3/km). Transmission losses (in acre-feet per mile) at the WGEW were found to be a power function involving both inflow volume and peak discharge rate as follows (Cataldo and Pierce 2005):

$$\mathrm{TL/mile} = 0.278 * \mathrm{V}_1^{0.750} : \mathrm{R}^2 = 0.825 \quad (13.21)$$

$$\mathrm{TL/mile} = 0.026 * \mathrm{Q}_{\mathrm{P1}}^{0.775} : \mathrm{R}^2 = 0.721 \quad (13.22)$$

where,

TL = transmission loss (acre feet) (experimental work not easy converted to metric scale)
V_1 = inflow volume (acre feet)
Q_{P1} = peak discharge rate (ft^3/s)
R^2 = Pearson correlation coefficient.

Significant errors occurred when the WGEW based equations were applied to another stream system. The WGEW equations overestimated transmission losses at another stream system in Arizona (Queen Creek) by 180% using the volume relationship. An average 77% difference between predicted and measured transmission losses occurred using the peak discharge relationship. The large spread in transmission losses for similar inflow conditions indicate the discharge volume and peak discharge are the primary variables controlling transmission losses, but not the only variables (Lane et al. 1971; Cataldo and Pierce 2005). Bed hydraulic conductivity would be expected to exert a strong control on transmission losses.

Cataldo et al. (2010) subsequently presented modified versions of Eqs. (13.21) and (13.22) that account for differences in stream bed hydraulic conductivity between the studied stream system and the WGEW. These equations are

$$TL/km = 1.01 * V_1^{0.75} * K/K_{WG} \quad (13.23)$$

$$TL/km = 317.29 * Q_{P1}^{0.78} K/K_{WG} \quad (13.24)$$

where,

V_1 = inflow volume (m^3)
Q_{P1} = peak discharge rate (m^3/s)
K = stream bed vertical hydraulic conductivity of study system (cm/s)
K_{WG} = vertical hydraulic conductivity of WGEW stream bed (cm/s).

Where data from multiple analyses of hydraulic conductivity are available, an average hydraulic conductivity value can be used. The main assumptions of the Cataldo et al. (2010) method is that there is no lateral tributary flow in the stream reach in question and flow is constrained within the channel banks. Cataldo et al. (2010) proposed their method as simple, rapid, planning level method that requires only a minimum amount of pre-existing data to determine transmission losses. They also noted with respect to regression and empirical equations that caution must be exercised when attempting to extrapolate specific findings to multiple sites. The WGEW may not be an appropriate baseline in other geographic areas. Equivalent Equations to 13.23 and 13.24 may be developed using gauging data from a stream more representative of local hydrogeological and climatological conditions. The key consideration is that less effort and cost is required to establish a local experimental watershed to determine the regional stream flow-transmission loss relationships than to establish a full monitoring program on all ephemeral streams.

Walters (1990) studied transmission losses in southwestern Saudi Arabia during storm events in which there were no tributary inflows in the studied reach. Multiple regression analysis was used to elucidate the relationship between transmission loss in the first mile (1.6 km) of the channel (V_{1m}) and flow and channel parameters. Transmission loss for the first mile (1.6 km) was used to standardize the data because stream losses tend to decrease over the length of the channel (Jordan 1977). The equation developed by Jordan (1977) is

$$V_{1m} = V_1 \left[1 - \left(\frac{V_2}{V_1} \right)^{\frac{1}{x_{1-2}}} \right] \tag{13.25}$$

where, flow volumes are in acre-feet and the x_{1-2} is the distance between gaging stations (miles). Walters (1990) developed three equations for estimating channel (recharge) losses:

$$V_{1m} = 0.0500V_a : R^2 = 0.91 \tag{13.26}$$

$$V_{1m} = 0.0006225W^{1.216}V_a^{0.507} : R^2 = 0.71 \tag{13.27}$$

$$V_{1m} = 0.103V_a^{0.872} : R^2 = 0.42 \tag{13.28}$$

where,
W = active channel width (ft)
V_a = flow volume at upstream gauging station (acre-feet)
R^2 = correlation coefficient.

Similar equations could be developed using metric units. The middle equation appears to be a better predictor for smaller floods, and the upper and lower tended to make better predictions for larger floods (Walters 1990). Equations 13.26 through 13.28 are specific to the studied stream system, and are presented as an example of the type of analyses that are possible in a gauged watershed. The main application is to use relationships developed in a gauged watershed to other watersheds in the study region for which adequate gauging data are not available.

From a practical water management perspective, a centrally located watershed that is believed to be representative of a regional study area could be selected for initial gauging in order to obtain general rainfall-runoff-transmission loss relationships. The gauging network could later be expanded to include other watersheds as resources become available.

13.4.2 Transmission Loss and Tributary Flows

Calculation of transmission losses becomes more complex when tributary flows are significant. Tributary flows can be quantified by independently assessing the runoff from each contributing catchment, provided that there are sufficient data available (which is typically not the case). Shentis et al. (1999) presented a method for determining lateral inflow into an ephemeral stream that is based on the following two assumptions:

(1) Runoff from small catchment areas (<300 km^2; <110 mi^2) depends on their surface lithology, and
(2) Spatial distribution of runoff reflects the spatial distribution of precipitation. Runoff depths from catchment areas can be estimated from records of representative hydrometric stations located in the same climatic subregion and that have the same basic lithostratigraphic composition.

Employment of these assumptions involves geologic (lithostratigraphic) mapping of the watershed, selection of representative stations, and association of recurrence intervals for runoff volumes at the selected stations (Ben-Zvi and Shentsis 2000)

13.5 Transmission Losses and Recharge

Transmission losses in wadis represent water that is available for recharge. An important issue is that a substantial fraction of the transmission losses may not result in net aquifer recharge. Recharge rates may be significantly less than channel infiltration rates. Alluvial channel sediments may have high infiltration rates, but much of the water may be lost to ET. A neutron logging study performed at the Yucca Mountain nuclear repository site in Nevada showed that within channels and terraces most of the infiltrated water remained near surface (within 2 m; 6.7 ft) because of the relatively high storage capacity of the shallow soils (Flint and Flint 1995; Flint et al. 2002). Near surface water is vulnerable to loss through ET during subsequent dry periods, especially when riparian phreatophyte vegetation is abundant. Independent assessment of wadi aquifer recharge is still necessary.

There are very limited data available on the ratio of aquifer recharge to transmission losses in ephemeral streams in arid lands. Some studies indicate a relatively high ratio. For example, the average ratio of estimated groundwater recharge to transmission loss was 75% in a wadi system (Tabalah basin) in southwestern Saudi Arabia (Sorman and Abdulrazzak 1993). It is clear from the available data that there is no technical basis for any generalizations on the ratio, and that stream systems need to be evaluated on an individual basis.

13.6 Hydrologic Modelling

The methods that are used for modeling of rainfall-runoff-transmission loss relationships in ephemeral stream systems incorporate three fundamental elements, which are

- simulation of the relationship between rainfall and runoff,
- simulation of the channelization of runoff from the catchment, and
- simulation of transmission losses in channels.

Hydrologic modeling programs have been developed that incorporate all of these three elements, plus additional features, such as statistical simulations of rainfall. Integrated surface water-groundwater models combine a surface-water model with a groundwater flow model. Transmission losses are simulated to result in recharge to the groundwater flow model. Hydrologic modeling is a major discipline in its own right. Provided herein is a discussion of some basic modelling concepts.

The state of the art for simulating complex hydrologic systems is distributed-parameter hydrologic modeling programs such as the U.S. Army Corps of Engineers (2009) HEC-HMS program, the U.S. Department of Agriculture SWAT program (Neitsch et al. 2005), and the U.S. Geological Survey Precipitation-Runoff Modeling System (PRMS) (Leavesley et al. 1983). These programs are current, widely used open-source codes. In general, open-source codes are preferred because they undergo testing and benefit from improvements developed by a wide pool of users. Proprietary codes however, should be considered if they have specific features that are important for a given project.

The basic concepts behind distributed-parameter, rainfall-runoff modeling is discussed by Lane (1982) and Osterkamp et al. (1994). Channel networks, including major tributary inflows, are represented by a variable number of channel reaches. Each channel reach may receive inflow from an upstream channel reach or subcatchment and lateral flow from adjacent subcatchment areas. Runoff from each subcatchment is calculated from precipitation data and runoff properties, for example, by the SCS-CN method. Transmission losses are a function of hydraulic conductivity, channel length and width, and the duration of flow. The entire watershed is represented by a simple channel network in which runoff volumes from contributing areas are computed individually, runoff is routed downstream, and total runoff summed. Outputs are hydrograph shapes and duration and water-balance calculations, including transmission losses.

Rainfall-flow models are preferably calibrated against actual storm data (i.e., downstream hydrographs). In ungauged areas, channel flows can be estimated from channel characteristics (e.g., Hedman and Osterkamp 1982).

Surface-water modeling can be much more data intensive than simple groundwater flow modeling because in surface-water models there are many more climatological and hydrologic parameters for which values must be assigned. GIS is typically used to map and input streams and sub-basins and to geographically assign input values. Surface-water modeling is thus a specialized discipline that requires considerable experience in order to obtain meaningful results.

13.6.1 HEC-HMS Model Code

The HEC-HMS model code is used to calculate the precipitation-runoff response from inputs of a meteorological model. HEC-HMS includes three components: basin model, meteorological model, and control specifications.

Basin models that are created in HEC-HMS to represent a watershed may include the following elements: sub-basins, stream reaches, junctions, sources, sinks, reservoirs, and diversions. Basin models of complex dendritic watersheds could contain hundreds of individual elements. Multiple options are provided to users for the method or mathematical models used to simulate hydrologic processes within each basin model element. The meteorological models include precipitation, evapotranspiration, and snowmelt elements, with multiple methods available for each element. Statistical methods are available for generating synthetic precipitation.

HEC-HMS is a powerful tool that has great capabilities, which have been progressively increased over time as new versions of the code are released. The limitation of HEC-HMS (and similar codes) is that it is deterministic. Values must be assigned to the boundary conditions, initial conditions, and parameters in each individual element. As model complexity increases with each additional element added, there is a corresponding increase in the data requirements, which is a major limitation in most arid and semiarid regions where there is a paucity of hydrologic and hydrogeologic data. It is thus possible to create and run highly complex models, but they could be highly inaccurate in the absence of accurate input data. HEC-HMS models can be calibrated against stream hydrographs, and the program includes optimization routines, but there is still the general issue of the non-uniqueness of model calibration.

13.6.2 Soil and Water Assessment Tool

The Soil and Water Assessment Tool (SWAT) is a physically-based model developed by the U.S. Department of Agriculture to predict the impacts of land management practices on water, sediment, and agricultural chemical yields in large complex watersheds (Neitsch et al. 2005). SWAT is designed to use readily available inputs and can be used for simulations in which no monitoring (calibration) data are available (e.g., ungauged watersheds). SWAT is a continuous time model with daily time steps, and is not designed to simulate detailed, single event storm routing.

SWAT allows for a number of different physical processes to be simulated in a watershed, which is subdivided into a number of sub-watersheds or sub-basins. The hydrology of the watershed is divided into the land phase, which controls the amount of water, sediment, and chemical loading into the main channel in each sub-basin, and the water or routing phase, which defines the movement of water and sediments through the channel network of the watershed to the outlet. Climate variables include daily precipitation, maximum and minimum air temperature, solar radiation, wind speed, and relative humidity. SWAT has weather generator capability that uses long-term climate data to generate the daily climatic time series. Infiltration and runoff can be calculated using either the SCS CN or Green-Ampt methods.

The building blocks of the SWAT land phase are hydrologic response units (HRUs), which are lumped land areas within a sub-basin that are comprised of unique land cover, soil, and management combinations. Runoff is predicted separately for each HRU based on a water balance equation and is routed to obtain the total runoff of the sub-basin. The flood routing procedure simulates evaporative losses and channel transmission losses.

13.6.3 The Precipitation-Runoff Modeling System Model

Precipitation-Runoff Modeling System (PRMS) is a modular-design, deterministic, distributed-parameter modeling system developed by the U.S. Geological Survey to evaluate the impacts of various combinations of precipitation, climate, and land use on streamflow, sediment yields, and general basin hydrology (Leavesley et al. 1983). Conceptually, is it similar to SWAT in that the basin or watershed is divided into homogeneous response units (HRU) based on the basin characteristics. Water and energy balances are computed daily for each HRU. The sum of the responses of all HRU's is weighted on a unit-area basis. The watershed is conceptualized as a series of interconnected flow planes and channel segments. Surface runoff is routed over the flow planes into the channel segments and channel flow is routed through the watershed channel system.

13.6.4 Applications of SW Modeling for Wadi Recharge

There has been limited use to date of hydrologic models to evaluate rainfall-runoff relationships in ephemeral stream systems. Some examples of modeling of wadi runoff and recharge in arid and semiarid basins are herein discussed.

Lange et al. (1999) and Lange and Leibundgut (2000) developed an uncalibrated model of Nahal (Wadi) Zin, in the Negev Desert of Israel. The simple rainfall-runoff model incorporated only measurable parameters and existing field data to simulate runoff generation on the terrain in response to local, high-intensity rainfall and transmission losses in channel alluvium. Rainfall was determined from radar data that was calibrated against local rain gauge data. Rainfall excess and runoff were calculated using the Horton method (Sect. 5.6.2). Twenty one different terrain types were identified and their infiltration processes were carefully assessed. The Muskkingum-Cunge technique (Cunge 1969) was used for streamflow routing in over 350 channel segments. Transmission losses were calculated by assigning different constant infiltration rates to inner channels and bars, which were based on previously conducted experiments.

The Nahal Zin study demonstrated that an uncalibrated model could provide reasonable predictions of peak discharges. Mapping of the different terrain types is an important part of the model procedure and requires sound field experience (Lange et al. 1999). If the various channel and terrain hydrologic properties are carefully determined, then meaningful applications may be possible for ungauged catchments. The uncertainty of the model results were assessed by "field-based" analysis of the maximum uncertainty ranges for the different input parameters.

Remote sensing employing GIS was used with HEC-HMS to evaluate runoff in Wadi Madoneh, Jordan, as part of an artificial recharge investigation (Hammouri and El-Naqa 2007). Remote sensing and GIS technologies were used to estimate the spatial variation in hydrologic parameters that are the input into HEC-HMS model (e.g., land use, soil, vegetation, drainage). The rainfall-runoff relationship was evaluated using HEC-HMS by the SCS-CN method. The model was calibrated against measured runoff data from a storm event by varying the curve numbers. The modeling results indicated that only precipitation events exceeding 14.3 mm (0.6 in.) within a 24-h period would generate runoff. The potential application of the model is that it could be used to estimate rainfall-runoff relationships in other ungauged watersheds with similar conditions.

Menking et al. (2003) used SWAT to simulate historic variations in runoff related to climate change in the Estancia Basin of New Mexico, USA. The goal of the study was to investigate the effects of changes in rainfall, and thus runoff, on water levels in the central playa, which were much higher in the past. The Menking et al. (2003) study is particularly noteworthy because SWAT was partially integrated into the groundwater flow model MODFLOW. SWAT runoff delivered to the basin flow was placed at the top of MODFLOW stream reaches and allowed to flow towards the central lake. This integration allowed for a useful assessment of the effects of climate change on basin hydrology.

Modeling tools are thus available for the simulation of wadi runoff and recharge, but their limitation is that they are data intensive. Use of remote sensing and GIS techniques has greatly demonstrated utility for interpolating and extrapolating limited field data. The greatest value of hydrologic modeling of wadi systems will likely be for the prediction of runoff and recharge for ungauged basins.

13.7 Wadi Recharge Seasonal Timing

Wadi recharge is clearly restricted to times of the year during which there is significant rainfall. The seasonal timing of rainfall is not an issue in parts of the world with a single wet season and thus precipitation source period. In areas with more than one wet season or period, the seasonal timing of precipitation may impact rainfall-runoff relationships. Significant differences in recharge rates may occur between seasonal wet periods.

In the arid and semiarid southwestern United States, precipitation occurs as rainfall during the summer monsoon and winter rain fronts, and as snowfall in mountainous areas. Stewart-Deaker et al. (2007) reported the results of a study of Abo Arroyo, central New Mexico. Storm flow accounts for nearly all the water passing through a stream gauge. Basin recharge from streambed infiltration was found to be almost entirely from storms associated with the summer monsoon.

Environmental tracers (Cl, Br) provided evidence for focused recharge beneath the active channels, with no recharge currently in adjacent terraces. A combination of the gauging data and extent of flow data from the temperature probes indicated that infiltration consumes most of water in transit. Only the largest flow passed through Abo Arroyo and reached the Rio Grande during the summer monsoon season during which most of the streambed recharge occurred.

Conversely at other sites snowmelt may be the dominant source of infiltration water. In Arroyo Hondo in north-central New Mexico snowmelt events were more likely than summer monsoon events to result in streambed infiltration because (Moore 2007):

- snowmelt is of longer duration,
- the total annual volume of snowmelt is greater although monsoon thunderstorms may have greater rates,
- almost all of the snowmelt streamflow is lost to infiltration, and
- ET rates are lower during the spring snowmelt period than during the summer and fall monsoon period.

The cited studies illustrate the difficulties of making accurate generalizations concerning the hydrology of wadi systems, particularly between systems in different geographic and geologic regions. Wadi rainfall-runoff-recharge relationships are highly specific to local geological, hydrogeological, and climatological conditions.

13.8 Temperature Data and Wadi Recharge

Streambed temperature data can provide valuable information on the pattern of stream flow, which can facilitate estimates of stream flow frequency, duration, travel time, and transmission losses (Constantz et al. 2001). Streambed temperature data can also be used to obtain estimates of vertical hydraulic conductivity, flow velocity, and recharge rates and locations.

The basic concepts behind the use of heat as a tracer of near-stream water movement are summarized by Constantz et al. (2001) and Constantz and Stonestrom (2003). Naturally-occurring variations in temperature can be used to track the heat carried by flowing water. The heat, hydraulically transported by flowing water, can be used as a tracer in the same manner as chemical tracers. There are several distinct advantages to the use of temperature as a tracer (Constantz and Stonestrom, 2003).

- Temperature is a natural signal and thus does not require the addition of chemicals.
- Temperature is a relative inexpensive parameter to measure and equipment is readily available to automatically perform long-term time series of measurements.
- Temperature data are immediately available for inspection and interpretation.

Temperature data may also compliment other data types such as water levels and pressures, and chemical tracers.

Shallow surface water and surficial soils experience diurnal variations in temperature that reflect the atmospheric variations in temperature. Groundwater temperatures on the contrary are relatively constant on a diurnal time scale. Variation in sediment temperature below the streambed reflects the balance between the diurnal oscillating transport of heat via conduction and the transport of heat by advection. Infiltrated water increases the heat capacity of sediments and thus the magnitude of the response to diurnal atmospheric temperature oscillations. The response of sediment temperatures to the advective transport of heat reflects not only the presence or absence of flowing water, but also the direction of water flow between the streambed and underlying sediments.

Streambed temperature measurements are performed by installing temperature probes within channel sediments. Recorded changes in temperatures may reflect stream flow events (with associated infiltration), as well as precipitation and the passing of cold fronts. Simulation results by Constantz et al. (2001) indicate that the optimal position for installing temperature probes for streambed temperature analyses is 10–20 cm (3.9–7.9 in.) below the streambed surface. This shallow burial depth allows for the differentiation of streambed events from precipitation and passing fronts. However, the burial depth of temperature probes may change over time due to scour and the deposition of sediments.

Temperature probes can be deployed in a series along the channel of a selected ephemeral stream reach to obtain data on the pattern of stream flow. Benchmark temperature probes installed outside of the studied channel and probes installed in the channel (above the streambed) can facilitate interpretation of the temperature data, such as the differentiation between precipitation and streamflow events (Constantz et al. 2001). Temperature probes can also be installed in nested clusters to examine temporal variation in water and heat flow with depth.

13.8.1 Determination of Temporal and Spatial Patterns of Recharge

Temporal and spatial patterns of recharge within ephemeral alluvial rivers can be evaluated using temperature data. Temporal changes in the diurnal temperature response at a given station and differences in response between stations within a channel or between a channel and off-channel benchmark may be indicative of changes in vertical water flow. The changes in vertical water flow include the initiation or cessation of gaining (discharge) or losing (infiltration) conditions. Hydrological conditions within a stream channel, such as frozen conditions, flowing water with ice, flowing water without ice, standing water, and local precipitation can produce distinct, identifiable, diurnal temperature patterns (Constantz et al. 2001).

The temperature patterns associated with gaining, loosing, and dry ephemeral stream conditions were reviewed by Constantz et al. (2001), Constantz and Stonestrom (2003), and Stewart-Deaker et al. (2007). Periods of gaining (i.e., discharge of groundwater to streams) is manifested by relatively constant temperatures in shallow streambed sediments and a dampening of the diurnal fluctuations. The upwards migration of groundwater that is buffered from the temperature fluctuations at land surface, will result in less variable near surface streambed temperatures (Fig. 13.6).

Losing (infiltration) conditions result in the downward advection of surface water that experiences diurnal oscillation in atmospheric temperatures. The downward advection results in large fluctuations in shallow sediment temperatures that will closely follow in-stream variations. Periods of infiltration will also result in increases in the amplitude of diurnal fluctuations in temperature in deeper temperature probes. Downward advection can carry diurnal fluctuations deep into the temperature profile (Stonestrom et al. 2007).

Shallow temperature probes in ephemeral streams without flow will experience relatively large diurnal temperature fluctuations. The temperature fluctuations will decrease rapidly with depth compared to saturated sediments because of the low thermal conductivity of dry sediments. The initiation of infiltration may also be observed by a change in the temperature of shallow streambeds sediments, which would reflect a difference in temperature between the flowing water and the local atmosphere.

The onset and cessation of both gaining and loosing (i.e., recharge conditions) events are evident in temperature versus time plots. For example, an abrupt decrease in diurnal temperature range within the channel streambed suggests the initiation of streamflow and an abrupt increase may suggest recession of that flow. There is greater confidence in the signal if it is not observed in off-channel stations.

Within an ephemeral stream, temperature data can used to determine how far into the basin the flow from a given storm event extended and the duration of the flow. A series of temperature probes can be installed in an ephemeral stream and the presence or absence of recharge (loosing conditions) detected from the temperature response.

Flow rate data at an up-gradient gaging station versus temperature data can be used to determine how far a given flood event extends. Classification systems can be developed based on the thermal detection of travel distances. For example, a flow of given magnitude, as measured at an up-stream gaging station, can be related to travel a given distance into the groundwater basin.

The key value of being able to relate flow magnitude to distance is that it allows for quantitative estimation of stream losses, which are due primarily to infiltration

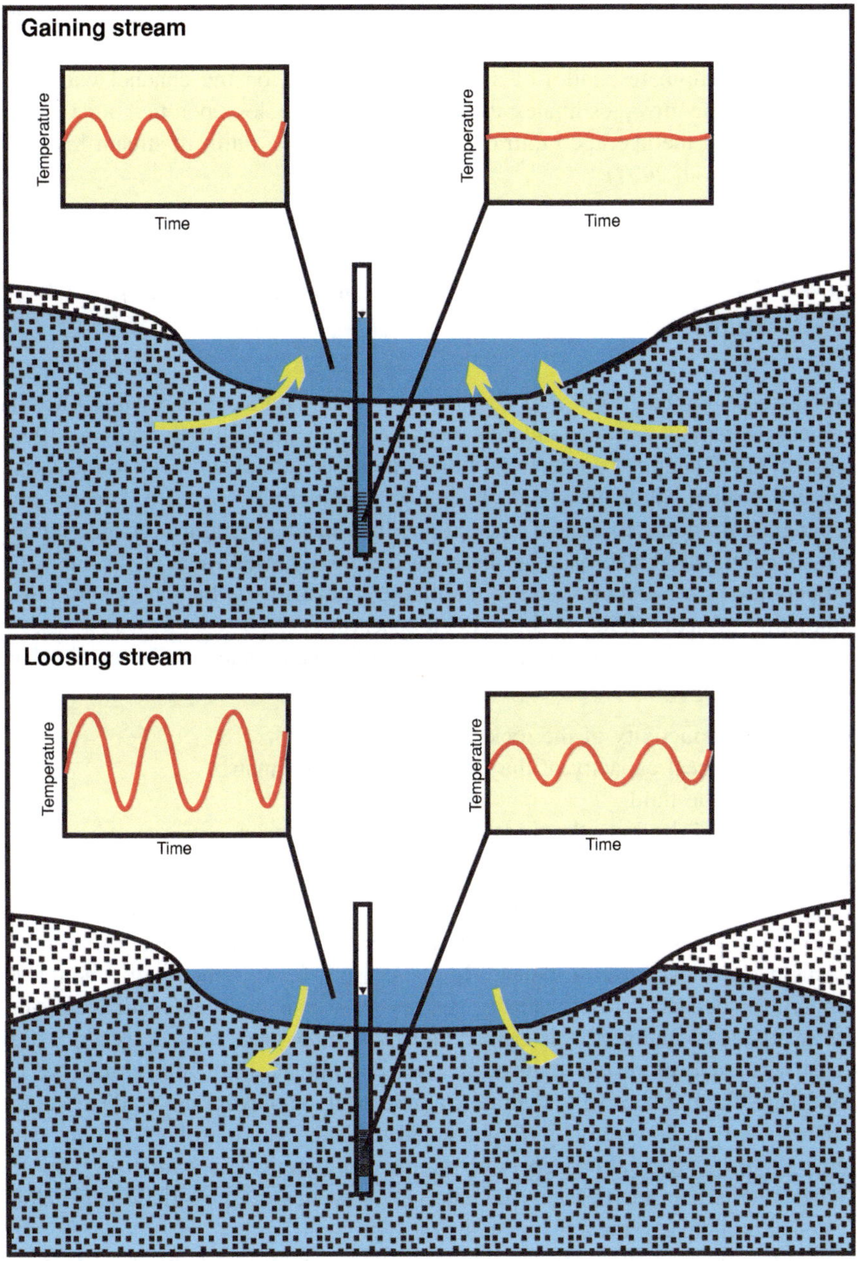

Fig. 13.6 Conceptual diagram of temperature histories of gaining and losing reaches of streams. Groundwater is buffered from temperature fluctuations at land surface. Groundwater below gaining streams will have a very muted response to atmospheric temperature fluctuations (After Constantz and Stonestrom 2003)

if lateral inflows and outflows are negligible. If the differences in flow between two points in a stream are known, then estimates can be made of the average loss per unit length (kilometer; mile) of a stream. From data on the channel width and duration of the flow, estimates can be made of the loss per unit area of the streambed and the average infiltration rate (assuming a uniform stream loss rate, Stonestrom et al. 2007).

13.8.2 Determination of Hydraulic Conductivity and Infiltration Rates

Initial studies in the 1960s proposed that temperature data can be used to trace the vertical movement of water and thus recharge rates (Suzuki 1960; Stallman 1963). Lapham (1989) applied a numerical finite-difference model approach to measuring vertical ground-water flow and hydraulic conductivity beneath streams that experience pronounced annual fluctuations in water temperature. Inverse numerical modeling has since become the most commonly used method for analyzing temperature profile data because of the inherent flexibility, which allows for closer simulation of actual field conditions and observations.

Heat transfer in sediments is a function of several main variables (Lapham 1989) including:

- thermal conductivity of the rock-fluid matrix,
- volumetric heat capacity of the fluid and rock-fluid matrix,
- density of the fluid,
- wet-bulk and dry-bulk densities, and
- vertical fluid velocity.

In order to determine the vertical fluid velocity and thus recharge rates from temperature profile data, the physical and thermal properties of the sediments in the field site must be determined. However, from measurements of a couple of properties, such as wet-bulk density and dry-bulk density, mathematical relations are available to indirectly determine the remaining properties (Lapham 1989). Sensitivity analysis results indicate that temperature profiles (temperature versus depth) are highly sensitive to hydraulic conductivity and hydraulic gradient values, and thus vertical flow velocity (Lapham 1989). Temperature profiles are somewhat sensitive to the thermal properties of saturated sediments (Lapham 1989), but the values of these properties have a relatively low degree of variation in natural systems, particularly for a given sediment type.

Temperature profiles can be obtained from depth-series of measurements obtained from a single monitoring well. A temperature probe can be slowly lowered down the well (e.g., Lapham 1989) or the well may be geophysically logged. Alternatively, a series of nested monitoring wells or piezometers can be installed at each temperature profile station. Temperature probes connected to a data logging system (or integrated probes) installed in a well can provide a continuous record of temperature change over time. Surface-water temperature must also be recorded. Nested temperature probes are the preferred strategy because of the continuous data collection, which allows for the evaluation of temperature changes on different time scales.

References

Abdulrazzak, M. J., & Sorman, A. U. (1994). Transmission losses from ephemeral stream in arid regions. *Journal of Irrigation and Drainage Engineering, 120*(3), 669–675.

Al-ahmadi, M. E., & El-Fiky, A. A. (2009). Hydrogeochemical evaluation of shallow alluvial aquifer of Wadi Marwani, Western Saudi Arabia. *Journal of King Saud University (Science), 21*, 179–190.

Al-Shaibani, A. M. (2008). Hydrogeology and hydrochemistry of a shallow alluvial aquifer, western Saudi Arabia. *Hydrogeology Journal, 16*, 155–165.

Al-Weshah, R. A. (2002). Rainfall-runoff analysis and modeling in wadi systems. In H. Wheater, & R. A. Al-Weshah (Eds.), *Hydrology of wadi systems: IHP-V, technical documents in hydrology* (Vol. 55, pp. 87–111). Paris: UNESCO.

Ben-Zvi, A., & Shentsis, S. I. (2000). Runoff events in the Negev, Israel, In M. A. Hassan, O. Slaymaker, & S. M. Berkowicz (Eds.), *The hydrology-geomorphology interface: Rainfall, floods, sedimentation, land use* (pp. 53–71). Wallingford: International Association of Hydrological Sciences Publication No. 261.

Besbes, M., Delhomme, J. P., & de Marsily, G. (1978). Estimating recharge from ephemeral streams in arid regions: A case study a Kairouan, Tunisia. *Water Resources Research, 14*, 281–290.

Bras, R. L. (1989). *Hydrology. An introduction to hydrologic science*. Reading: Addison Wesley.

Cataldo, J., & Pierce, R. J. (2005). *An analysis of transmission losses in ephemeral streams: A case study in Walnut Gulch Experimental Watershed.* Tombstone: A Report to the National Center for Housing and the Environment

Cataldo, J. C., Behr, C., Montalto, F. A., & Pierce, R. J. (2010). Prediction of transmission losses in ephemeral streams, western U.S.A. *The Open Hydrology Journal, 4*, 19–34.

Chow, V.T., Maidment, D., & Mays, L.W. (1988). *Applied Hydrology*. New York: McGraw Hill.

Constantz, J., & Stonestrom, D. A. (2003). Heat as a tracer of water movement near streams. In D. A. Stonestrom, & J. Constantz (Eds.), *Heat as a tool for studying the movement of groundwater near streams* (pp. 1–6). U.S. Geological Survey Circular 1260.

Constantz, J., Stonestrom, D. A., Stewart, A. E., Niswonger, R., & Smith, T. R. (2001). Analysis of streambed temperatures in ephemeral streams to determine streamflow frequency and duration. *Water Resources Research, 37*, 317–328.

Cunge, J. A. (1969). On the subject of flood propogation method. *Journal of Hydrologic Research, 7*(2), 205–230.

Flint, A. L., Flint, L. E., Kwicklis, E. M., Fabryka-Martin, J. T., & Bodvarsson, G. S. (2002). Estimating recharge at Yucca Mountain, Nevada, USA, comparison of methods. *Hydrogeology Journal, 10*, 180–204.

Flint, L. E., & Flint, A. L. (1995) *Shallow infiltration processes at Yucca Mountain, Nevada—neutron logging data, 1984–1993*. U.S. Geological Survey Water-Resources Investigations Report 95-4035.

Goodrich, D. C., Williams, D. G., Unkrich, C. L., Hogan, J. F., Scott, R. L., Hultine, K. R., et al. (2004). Comparison of methods to estimate channel recharge, Walnut Gulch, San Pedro River Basin. In F. M. Phillips, J. F. Hogan, & B. Scanlon (Eds.), *Recharge and vadose zone processes: Alluvial basins of the Southwestern United States, Water science and application* (Vol. 9, pp. 77–99). Washingston, D.C.: American Geophysical Union.

Hammouri, N., & El-Naqa, A. (2007). Hydrological modeling of ungauged wadis in arid environments using GIS: A case study of Wadi Madoneh in Jordan. *Revista Mexicana de Ciencias Geológicas, 24*(2), 185–196.

Hedman, E. R., & Osterkamp, W. R. (1982). *Streamfow characteristics related to channel geometry of streams in the Western United States.* U.S. Geological Survey Water-Supply Paper 2193.

Izbicki, J. A., Johnson, R. U., Kulongoski, J., & Predmore, S. (2007). Ground-water recharge from small intermittent streams in the western Mojave Desert, California. In D. A. Stonestrom, J. Constantz, T. P. A.Ferré, & S. A. Leake (Eds.), *Ground-water recharge in the arid and semiarid Southwestern United States.* (pp. 157–184). U.S. Geological Survey Professional Paper 1703.

Jordan, P. R. (1977). Streamflow transmission losses in western Kansas. *Journal of Hydraulic Division, ASCE, 103*, 905–919.

Lane, L. J. (1982). Distributed model for small semiarid watersheds. *Journal of the Hydraulics Division, American Society of Civil Engineers, 108*, 1114–1131.

Lane, L. J., Diskin, M. H., & Renard, K. G. (1971). Input-output relationship for an ephemeral stream channel system. *Journal of Hydrology, 13*, 22–40.

Lange, J., & Leibundgut, C. (2000). Non-calibrated arid zone rainfall-runoff modeling. In M. A. Hassan, O. Slaymaker, & S. M. Berkowicz (Eds.), *The hydrology-geomorphology interface: Rainfall, floods, sedimentation, land use* (pp. 45–52). Wallingford: International Association of Hydrological Sciences Publication No. 261.

Lange, J., Leibundgut, C., Greenbaum, N., & Schick, A. P. (1999). A noncalibrated rainfall-runoff model for large, arid catchmemts. *Water Resources Research, 35*(7), 2161–2172.

Lapham, W. W. (1989). *Use of temperature profiles beneath streams to determine rates of vertical ground-water flow and vertical hydraulic conductivity.* U.S. Geological Survey Water-Supply Paper 2337.

Leavesley, G. H., Lichty, R. W., Troutman, B. M., & Saindon, L. G. (1983). *Precipitation-runoff modeling system. User's manual.* U.S. Geological Survey Water-Resources Investigations Report 83-4238.

Menking, K. M., Syed, K. H., Anderson, R. Y., Shafike, N. G., & Arnold, J. G. (2003). Model estimates of runoff in the closed, semiarid Estancia Basin, Central New Mexico, USA. *Hydrological Sciences, 48*(6), 953–970.

Moore, S. J. (2007). Streamflow, infiltration, and recharge in Arroyo Hondo, New Mexico. In D. A. Stonestrom, J. Constantz, T. P. A. Ferré, & S. A. Leake (Eds.), *Ground-water recharge in the arid and semiarid southwestern United States* (pp. 137–155) .U.S. Geological Survey Professional Paper 1703.

Murphey, J. B., Wallace, D. E., & Lane, L. J. (1977). Geomorphic parameters predict hydrograph characteristics in the Southwest. *Water Resources Bulletin, 12*, 25–38.

Natural Resources Conservation Service. (1993). Chapter 4. Storm rain fall depth, Part 630, Hydrology, National Engineering Handbook, U.S. Department of Agriculture, Natural Resources Conservations Services.

Natural Resources Conservation Service. (2004a). Chapter 10. Estimation of direct runoff from storm rainfall, Part 630, Hydrology, National Engineering Handbook, U.S. Department of Agriculture, Natural Resources Conservations Services.

Natural Resources Conservation Service. (2004b). Chapter 9. Hydrologic soil-cover complexes, Part 630, Hydrology, National Engineering Handbook, U.S. Department of Agriculture, Natural Resources Conservations Services.

Natural Resources Conservation Service. (2007). Chapter 16. Hydrographs, Part 630, Hydrology, National Engineering Handbook, U.S. Department of Agriculture, Natural Resources Conservations Services.

Neitsch, S. L., Arnold, J. G., Kiniry, J. R., & Williams, J. R. (2005) *Soil water assessment tool theoretical documentation, version 2005.* U.S. Department of Agriculture, Agricultural research Center.

Newman, B. D., Vivoni, E. R., & Groffman, A. R. (2006). Surface water—groundwater interactions in semiarid drainages of the American southwest. *Hydrological Process, 20*, 3371–3394.

Onder, H., & Abdulrazzak, M. J. (1993). Analysis of salt water intrusion in coastal aquifers. *Journal King Addulaziz University, Meteorology, Environment, and Arid Land Agricultural Sciences, 4*, 43–67.

Osterkamp, W. R., Lane, L. J., & Savard, C. S. (1994). Recharge estimates using a geomorphic/distributed parameter simulation approach, Amargosa River Basin. *Water Resources Bulletin, 30*(3), 493–507.

Osterkamp, W. R., Lane, L. C., & Menges, C. M. (1995). *Techniques of ground-water recharge estimates in arid/semi-arid areas, with examples from Abu Dhabi* (Vol. 31, pp. 349–369).

Pilgrim, D. H., Chapman, T. E., & Doran, D. G. (1988). Problems of rainfall-runoff modeling in arid and semiarid regions. *Hydrological Sciences Journal, 33*(4), 379–400.

Ramírez, J. A. (2000). Prediction and modeling of flood hydrology and hydraulics. In E. E. Wohl, (Ed.), *Inland flood hazards: Human, riparian and aquatic communities* (pp. 293–333). Cambridge: Cambridge University Press.

Scanlon, B. R., Keese, K. E., Flint, A. L., Flint, L. E., Gaye, C. B., Edmunds, W. M., et al. (2006). Global synthesis of groundwater recharge in semiarid and arid regions. *Hydrological Processes, 20*, 3335–3379.

Şen, Z. (2008). *Wadi hydrology*. Boca Raton: CRC Press.

Sharma, K. D., & Murthy, J. S. R. (1994a). Estimating transmission losses in an arid region. *Journal of Arid Environments, 26*, 209–219.

Sharma, K. D., & Murthy, J. S. R. (1994b). Estimating transmission losses in an arid region—a realistic approach. *Journal of Arid Environments, 27*, 107–112.

Shentsis, I., Meirovich, L., Ben-Zvu, A., & Rosenthal, E. (1999). Assessment of transmission losses and groundwater recharge from runoff events in a wadi under shortage of data on lateral inflow, Negev, Israel. *Hydrological Processes, 13*, 1649–1663.

Snyder, F. F. (1938). Synthetic unit hydrographs. *Transactions American Geophysical Union, 19*, 447–454.

Soliman, M. M. (2010). *Engineering hydrology of arid and semi-arid regions* (p. 395).: Boca Raton: CRC Press/Taylor & Francis Group.

Sorman, A. U., & Abdulrazzak, M. J. (1993). Infiltration-recharge through wadi beds in arid regions. *Hydrological Sciences—Journal des Sciences Hydrologiques, 38*(3), 173–186.

Stallman, R. W. (1963). Computation of ground-water velocity from temperature data. In R. Bentall, (Ed.), *Methods of collecting and interpreting ground-water data*. (pp. 36–45). U.S. Geological Survey Water-Supply Paper 1544.

Stephens, D. B. (1996). *Vadose zone hydrology*. Boca Raton: CRC Press.

Stewart-Deaker, A. E., Stonestrom, D. A., & Moore, S. J. (2007). Streamflow, infiltration, and ground-water recharge at Abo Arroyo, New Mexico, In D. A., Stonestrom, J., Constantz, T. P. A., Ferré, & S. A., Leake (Eds.), *Ground-water recharge in the arid and semiarid southwestern United States* (pp. 83–105). U.S. Geological Survey Professional Paper 1703.

Stonestrom, D. A., Prudic, D. E., Walvoord, M. A., Abraham, J. D., Stewart-Deaker, A. E., Glancy, P. A., et al. (2007). Focused ground-water recharge in the Amargosa Desert Basin. In D. A. Stonestrom, J. Constantz, T. P. A. Ferré, & S. A. Leake (Eds.), *Ground-water recharge in the arid and semiarid Southwestern United States*. (pp. 107–136). U.S. Geological Survey Professional Paper 1703.

Suzuki, S. (1960). Percolation measurements based on heat flow through soil with special reference to paddy field. *Journal of Geophysical Research, 65*, 2883–2885.

U.S. Army of Corps of Engineers. (2009). *Hydrologic model system HEC-HMS, User's manual, version 3.4, August 2009*. Davis: U.S. Army Corps of Engineers, Institute for Water Resources, Hydrologic Engineering Center

Walters, M. O. (1990). Transmission losses in arid regions. *Journal of Hydraulic Engineering, 116*, 129–138.

Wheater, H. S. (2002). Hydrological processes in arid and semi arid area, In H. Wheater, & R. A. Al-Weshah (Eds.), *Hydrology of wadi systems, IHP-V, Technical documents in hydrology* (Vol. 55, pp. 5–22). Paris: UNESCO.

Chapter 14
Microgravity

14.1 Introduction

Gravity survey data can provide quantitative information on changes in the mass of water within an aquifer, which has obvious water resources value. The technique is particularly useful for unconfined aquifers that experience changes in water levels caused by pumping or recharge (natural or artificial). Unconfined alluvial aquifers are important water sources in many arid regions, so gravity surveys have applications for water-resources investigations in these regions. Combined monitoring of gravity and water levels in wells can be a useful tool for improved water storage monitoring and for developing a better understanding of aquifer storage properties (Pool 2008). Gravitational data has limited value for assessment of confined aquifers, in which changes in well water levels (potentiometric surface levels) are primarily caused by changes in aquifer pressure rather than the volume of water stored in the aquifer.

Gravitational methods are discussed in most geophysics textbooks (e.g., Telford et al. 1976; Milson 2003; Burger et al. 2006). Newton's law of universal gravitation states that every particle of matter attracts every other particle of matter with a force that is proportional to the product of the masses of the particles and inversely proportional to the square of the distance between the particles. The law is expressed mathematically as

$$F = G\frac{(m_1 m_2)}{r^2} \tag{14.1}$$

where,

F = gravitational force of either particle (N)
G = gravitational constant (6.673×10^{-11} N m^2 kg^{-2}; m^3 kg^{-1} s^{-2})
m_1, m_2 = masses of the two particles (kg)
r = distance between the two particles (m).

R. Maliva and T. Missimer, *Arid Lands Water Evaluation and Management*,
Environmental Science and Engineering, DOI: 10.1007/978-3-642-29104-3_14,

In accordance with Newton's second law (force = mass time acceleration), gravitational acceleration (g) at the earth surface is defined as:

$$g = G\frac{m}{r^2} \quad (14.2)$$

where,
m = mass of the earth (kg)
r = radius of the earth (m).

The SI unit for gravitational acceleration is the Gal, which is equal to 1 cm/s^2. The milligal (mG) is equal to 1×10^{-3} Gal and the microgal (μG) is equal to 1×10^{-6} Gal. Similar equations can be expressed in English units, but are now seldom, if ever, used in gravity surveys.

The Earth is not homogenous and local changes in the mass of buried strata will result in corresponding changes in the gravitational acceleration at locations on the Earth's surface, which may be measured using a sensitive gravimeter (also referred to as gravitometer). Depending on the areal scale and purpose of the investigation, measured (observed) gravitational accelerations (g_{obs}) require several corrections as follows:

(1) Longitudinal variation (g_n)—gravitational acceleration is greater near the equator than at the poles and a correction needs to be applied for the station latitude.

(2) Free air correction (FA_{cor})—decrease in gravitational acceleration occurs with increasing altitude (i.e., distance from the center of the earth). The altitude of the stations must therefore be accurately measured using GPS or other techniques.

(3) Bouguer correction (B_{cor})—excess material between sea level and the station, which results in an increase in gravitational acceleration.

(4) Terrain correction (T_{cor})—corrections for nearby variations in altitudes above and below the station altitude.

Gravitational acceleration data are often expressed in terms of the Bouguer anomaly (Δg_b) defined as:

$$\Delta g_b = g_{obs} - g_n + FA_{cor} - B_{cor} \quad (14.3)$$

Gravitational acceleration measurements are also effected by instrument drift and earth tides, for which corrections must be applied. The normal procedure is to establish one or more base stations at which measurements are periodically made during a gravitational survey. Earth tidal effects can be corrected by frequent measurement of the base station or determined using widely available computer programs.

Field gravimeter readings typically record the relative gravity difference between two points. Absolute gravitation acceleration is much more difficult to measure in the field. Absolute values can be determined by linking a survey to a station at which absolute gravitational acceleration has been previously determined, such as an International Gravity Standardization Net (1971) (IGSN71) observation site (Burger et al. 2006).

14.2 Microgravity Use in Groundwater Investigations

The main change in mass of interest in groundwater investigations is associated with changes in the amount of water stored in an aquifer. Saturated aquifer strata will have a greater density and mass, than unsaturated aquifer strata in which the pores are filled with air. Dewatering of an aquifer will be manifested by a decrease in the mass of the aquifer and thus a lesser gravitational acceleration measured at land surface.

The change in the gravitational acceleration (Δg) associated with changes in the density of aquifer strata caused by dewatering can be calculated using the Bouguer slab equation as follows:

$$\Delta g = 0.04193 h \Delta\rho \tag{14.4}$$

where,
Δg = change in acceleration in mGal
h = thickness of buried slab (i.e., aquifer that was dewatered) (m)
$\Delta\rho$ = density change (g/cm^3).

The density change can be calculated as

$$\Delta\rho = Sy\rho_w \tag{14.5}$$

where,
S_y = aquifer specific yield (for an unconfined aquifer)
ρ_w = density of water (≈1 g/cm^3).

The Bouguer slab equation is based on an infinite slab. However, for the purpose of gravitational surveys, a horizontally continuous aquifer is a close approximation of an infinite slab. The approximation error caused by the use of the Bouguer slab approximation is less than 5% for a horizontal disk-shaped body that has a radius of 20 times its depth (Pool and Schmidt 1997). The magnitude of the change in the measured gravitational acceleration depends only on the change in density and thickness, but not on burial depth (Turcotte and Schubert 1982). The product of h $\Delta\rho$ has the units of mass divided by length squared. Gravitational surveys can thus provide data on changes in the mass of water, which would have to be corrected for water density and porosity (storage coefficient) to estimate changes in aquifer water level (and moisture content within the vadose zone).

Equation 14.4 can be simplified as follows to express gravitational acceleration changes (μGals) as a function of specific yield (S_y) and the interval of water table fluctuation (b) (change in thickness of saturated sediments) for an unconfined aquifer (Pool and Schmidt 1997; Pool 2008):

$$\Delta g = 41.9 S_y b(m) \tag{14.6}$$

$$\Delta g = 12.77 S_y b(ft) \tag{14.7}$$

where,
b(m) = interval of water table fluctuation in meters
b(ft) = interval of water table fluctuation in feet.

Gravity and water storage changes are more complex in confined aquifers because compaction or expansion can cause a significant change in gravity (Pool and Schmidt 1997).

As an example of the change in the gravitational acceleration associated with the dewatering of an aquifer, consider the dewatering of 3 m (9.8 ft) of an aquifer with a specific yield of 0.30. The change in the gravitational acceleration can be calculated as:

$$\Delta g = 41.9 * 0.30 * 3 = 37.7 \mu\text{Gals} \tag{14.8}$$

Modern microgravimeters have sensitivity of about 1 μGal, so the hypothetical dewatering presented above should be detectable. Microgravimeters can also detect relatively large changes in the volume of water in the vadose zone. These equations can be reordered so the change in storage (Sb) can be calculated as follows (Pool and Schmidt 1997):

$$\text{Storage (m)} = \Delta g/41.9 \tag{14.9}$$

$$\text{Storage (ft)} = \Delta g/12.77 \tag{14.10}$$

If the actual change in water level in the aquifer (b') is known, for example, from monitoring well data, then the change in gravitational acceleration can be used to estimate the specific yield (S_y) of the aquifer (Pool and Schmidt 1997):

$$S_y = \frac{\Delta g}{41.9b'(m)} \tag{14.11}$$

14.3 Relative Gravity Surveys

Gravitational acceleration measurements are influenced by extraneous factors such as Earth tides, instrument drift, altitude, and regional gradients in gravity. Earth tides are an elastic deformation of the Earth caused by the gravitational fields of Sun and Moon. For groundwater investigations, the parameter of concern is commonly changes in the volume stored in an aquifer, which can be evaluated through changes in the gravitational acceleration rather than the actual value of the gravitational acceleration. Changes in aquifer water storage are thus evaluated through time-series relative-gravity surveys.

Time-series surveys performed at the same locations have the great advantage of eliminating the need for commonly applied corrections, such as for terrain and latitude, provided that there are no regional changes in non-aquifer mass or altitude. If a microgravimeter is placed at the same position for each measurement at a location, any site effects should cancel out. Relative gravity data still need to be corrected for instrument drift, Earth tides, and environmental effects (Davis et al. 2008).

The procedures for relative-gravity surveys employed in a groundwater investigation in the Tucson, Arizona area were summarized by Pool and Schmidt (1997), which serves as an example of the usefulness of microgravity data in aquifer storage assessments. Differences in gravity were measured relative to a base station located in a bedrock area that is far enough away from the study area so as to not be impacted by changes in the mass of water in the aquifer. Closed-loop gravitation surveys were performed starting and ending at the bedrock base station in order to evaluate instrument drift. Linear drifts between the starting and end readings can be readily used to correct the data. More problematic are non-linear survey drifts caused by an inaccurate approximation of Earth tides, changes in the temperature of the instrument housing, atmospheric effects, and jarring of the instrument. Pool and Schmidt (1997) used algorithms to correct for Earth tides and conducted surveys at times corresponding to linear portions of the Earth tide curves whenever possible.

The Pool and Schmidt (1997) study was performed as part of an investigation of an artificial recharge project. The initial gravity station network consisted of 43 stations in a 15.5 km^2 (6 mi^2) study area. Seven additional stations were subsequently added. The gravitational survey data allowed for the quantification of the increase in the volume of storage due to recharge, and subsequent decreases caused by groundwater withdrawal and a net outflow of groundwater from the study area. There was a good correlation of water level and gravity data in wells located in the recharge area (Rillito Creek) and flood plain where the largest water level rise and gravity change occurred (>4.6 m; >15 ft, and 30 μGals). Moderate to poor correlation occurred elsewhere, which Pool and Schmidt (1997) interpreted as being caused by storage change in perched aquifers and the unsaturated zone.

Pool (2008) discussed theoretical hydrogeology, gravity and water-level relationships. An important consideration is that there are expected and observed combined gravity and water level responses that significantly differ between different aquifer and well types and that changes in water levels within a monitoring well may not reflect actual changes in the volume of water in storage.

Storage within a confined aquifer occurs as the result of the expansion and contraction of the aquifer materials and water, rather than by the draining and filling of pores (i.e., true increases and decreases in the volume of water stored). Depressurization of the aquifer as the result of pumping can cause large changes in water levels within wells with little or no detectable change in gravity readings because there was little change in the actual mass of water. Gravitation changes might be measured that correspond to compression of the aquifer or land subsidence. Storage within unconfined aquifers is changed by the draining and filling of pores and thus results in a significant change in the mass of water. A positive correlation should therefore occur between water level and gravity, which is related to the specific yield of the aquifer (Eq. 14.11).

Complex relations between aquifer water levels and gravity may occur in areas with multiple compressible aquifers. Gravity change at land surface may result from both aquifer compaction and changes in the amount of water stored in pore spaces. Perched aquifer conditions may develop, in which an increase in gravity reflecting a recharge event may precede changes in aquifer water levels. Sites with multiple compressible aquifers may have a poor correlation between water levels and gravity, and thus the amount of water in storage. Water levels in wells completed in multiple aquifers may have heads that represent values in a single aquifer or a composite of aquifers. Gravitational data thus provides an additional constraint on the interpretation of water level data, which may be somewhat ambiguous.

The effects of loading in a multi-aquifer setting will typically allow gravitational assessment of only the unconfined aquifer at the top of the sequence. The change in gravity in this case is likely controlled overwhelmingly by the change in storage at the top of the system with little variation occurring with deeper aquifers. Loading caused by differences in the mass of the water in unconfined aquifers and the vadose zone can significantly impact pressure (head) in underlying aquifers (Maliva et al. 2011 and references therein). Data on changes in the mass of water in an unconfined aquifer and overlying vadose zone obtained from gravitational surveys may be of value in interpreting water level data from underlying confined aquifers, if there is a significant loading effect.

14.4 Applications of Microgravity to Water Management

Microgravity profiling is a non-invasive and conceptually simple method, but requires the use of highly sensitive instruments (gravimeters), strict data acquisition procedures and quality controls, and sophisticated processing to evaluate and interpret the data. The major advantages of the technique are that it is non-invasive and that an experienced professional can perform numerous readings in a day.

Time-series microgravity monitoring data can complement water level data from monitoring wells. If the technique is to be used with respect to managed aquifer recharge (MAR) projects, then a survey needs to be performed before the start of recharge in order to determine baseline conditions. The monitoring stations should be marked so that subsequent readings can be performed at the exact same spot. Davis et al. (2005, 2008) used metal spikes to mark sites, which had the advantage of being locatable using a metal detector even if covered with snow (or sand). Most of the gravity stations in the Pool and Schmidt (1997) study were concrete pads, which is the preferred option.

An important consideration is whether or not microgravity surveys will have the resolution to provide the needed information. Forward modeling based on likely changes in water mass should be performed to determine the likely change in microgravity. The modeled changes can be used to determine whether the collected data would have suitable resolution and for the design of the survey.

14.4.1 Mapping the Distribution of Injected Water in an MAR System

A principal application of time-lapse microgravity monitoring is for the monitoring of physical storage MAR systems in unconfined aquifers. Microgravity is not appropriate for confined aquifers because changes in head do not have a corresponding magnitude change in the mass of water. Microgravity has great potential for monitoring surface application systems in which vadose zone storage and transport may be significant. Differences between water level changes measured in piezometers and water mass changes obtained using microgravity would provide an estimate of the change in the volume of water in the vadose zone.

14.4.1.1 Lancaster, California Aquifer Storage and Recovery Site

Time-series microgravity data can provide information on changes in groundwater levels in unconfined aquifer as the result of managed aquifer recharge. In order to interpret the microgravity data, the specific yield of the aquifer is needed, which can be determined from microgravity and water level change data from the same location. Howle et al. (2002) demonstrated the use of microgravity data to map the mounding of the water table at the Lancaster, California, ASR site. The primary advantage of using microgravity to monitor water level changes is that it is substantially less expensive to monitor numerous microgravity stations than to install observation wells, particularly if the water table is located a great distance below land surface. The major limitations of the technique are that microgravity-determined water levels are less accurate than standard well measurements and continuous monitoring is not practical. The lower accuracy is the result of the instrument accuracy and because microgravity measurements are influenced to varying degrees by part of the injection mound not directly under the gravity station. The latter can be corrected by modeling and was found to have an inconsequential effect at the Antelope ASR site (Howle et al. 2002).

14.4.1.2 Arvada, Colorado ASR System

Microgravity was successfully used to map the location of stored water at the City of Arvada ASR project in Colorado (Davis et al. 2005, 2008). The ASR project consists of water injected into an abandoned coal mine (Leyden Mine). The mine is approximately 300 m (1,000 ft) deep and has two mined-out coal seams in the lower Laramie Formation (Cretaceous). The area of mine is 16 km^2 (10 mi^2). Abandoned coal mines, including the Leyden Mine, collapse over time, forming rubble zones. The rubble zones and any remaining open mine shafts provide water storage.

Time-lapse microgravity measurements were successfully used to identify the locations of water-filled rubble zones. The emplacement of water into porous rock results in an increase in subsurface mass, which is detectable by an increase in measured gravity between successive gravimeter readings. The performance of time-series gravimeter readings at the same locations eliminated the need for commonly applied corrections, such as for terrain and latitude. Only corrections for drift and Earth tides were required (Davis et al. 2005, 2008).

14.4.1.3 Weber River ASR Site, Utah

The Weber River Basin Aquifer Storage and Recovery (ASR) Pilot Project (WRBASR) is not strictly an ASR system using the now widely accepted concept of the ASR because it involves aquifer recharge using infiltration basins rather than wells. High precision relative microgravity surveys were conducted before, during, and after infiltration in order to track groundwater movement in lieu of installing more expensive multiple groundwater monitoring wells (Chapman et al. 2008). The storage zone is an alluvial fan system (Delta Aquifer) in which the water table is located approximately 70 m (230 ft) below land surface. A fine-grained, clay layer is present in the vadose zone, which impedes vertical water flow. The microgravity survey was able to detect the mass of infiltrated water. Mass calculations derived from Gaussian integration matched the infiltrated water mass to within 20% (Chapman et al. 2008). The microgravity survey results detected saturation of the vadose zone above the clayey layer to create a perched aquifer. The hydraulic properties of the upper aquifer above the clayey layer were estimated from the decay of the groundwater mound quantified using the microgravity time-series data (Chapman et al. 2008).

Although microgravity data obtained for the WRBASR provided useful information, the results do not indicate that microgravity surveys are a substitute for a groundwater monitoring wells. Microgravity time series and water level monitoring data are complimentary to each other. The data presented in the Chapman et al. (2008) study does not fully address the question of how much of the infiltrated water actually reaches the water table and thus is available for later use.

14.5 Gravity Recovery and Climate Experiment

Microgravity measurements made using a gravimeter can provide a point measurement of changes in water storage. Satellite measurements of the Earth's gravity field have the potential to allow the acquisition of basin-wide data on recharge and aquifer depletion. The Gravity Recovery and Climate Experiment (GRACE) space mission was launched to study processes involving changes in the Earth's mass distribution. The GRACE mission demonstrated the utility of satellite gravity measurements for measuring changes in water storage (Wahr et al. 2004), which is clearly important information for water resources management.

The GRACE mission was launched by NASA on March 17, 2002, and consists of two identical satellites orbiting in the same plane and acting in unison (NASA 2003). Global coverage is roughly every 30 days. The mission had a planned duration of 5 years, but was subsequently expended to 8 years. An even longer lifetime, estimated to be until 2016, is expected based on the actual mission status (Günter et al. 2007). The GRACE mission allows for measurements of the Earth's gravity field from space, producing precision measurements orders of magnitude greater than previous measurements. The mission has multiple purposes including producing a precise model of the geoid, monitoring changes in water and ice on the continents, and obtaining data on ocean surface and deep currents. From a water resources perspective, the GRACE mission provided an opportunity to measure changes in groundwater storage.

The basic principle behind the GRACE mission is that measurement of changes in the distance between the twin satellites are used to make gravitational field measurements (Fig. 14.1). Differences in the mass of the Earth below the satellites is caused by changes in the water content of aquifers, resulting in changes in the gravitational field, which affect the movement of the satellites. As the satellites pass over a gravity anomaly, the associated change in the speed of the lead satellite causes a change in the distance between the two satellites, which is measured using an extremely precise (within 10 μm) microwave ranging system. The distance between the satellites closes as the trailing satellite passes over the anomaly. Satellite global positioning systems (GPS) are used to determine the exact position of the satellites over the Earth to within a centimeter or less. The raw data require considerable processing to obtain changes in water total volume (e.g., Swenson and Wahr 2002) before the hydrologic parameters of interest, such as soil moisture, groundwater recharge, and evapotranspiration can be determined. The GRACE mission has a limited life. However, future projects may allow for even more accurate measurements of the gravitation field of the Earth.

At the time at which this book was being prepared, the initial results of the GRACE mission data concerning groundwater resources were starting to be published. The initial results are very encouraging as they demonstrate that satellite gravity data can provide basin-scale information on hydraulic parameters that are necessary in water resources evaluations. Satellite gravity measurements have the clear limitation of a very coarse spatial resolution (Becker 2006) and are thus not suitable for local studies. Some of the GRACE studies are particularly relevant to arid and semiarid regions.

The High Plains Aquifer of the United States Great Plains is textbook example of an aquifer that has experienced considerable depletion from groundwater pumping for irrigation. The aquifer is thus a prime candidate for evaluation of the ability of GRACE data to monitor changes in groundwater storage. Strassberg et al. (2007, 2009) performed a comparative study of GRACE-derived terrestrial water storage (TWS) and groundwater storage changes from in situ measured data.

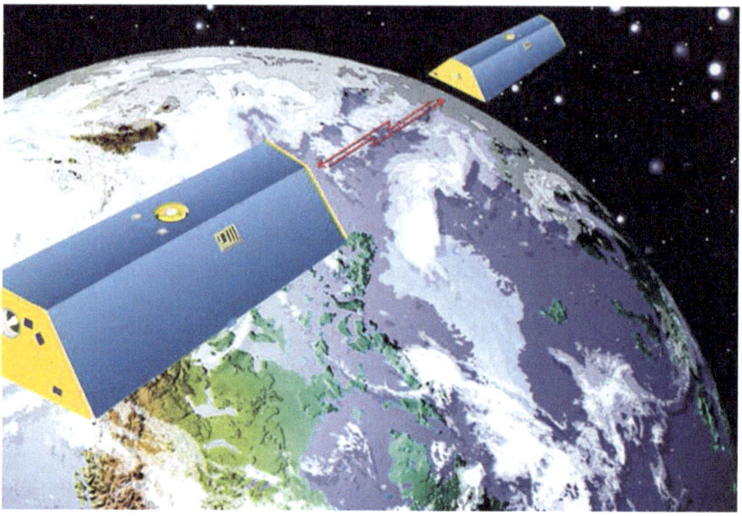

Fig. 14.1 Artist rendition of the GRACE system. *Source* NASA Jet Propulsion Laboratory

The change in TWS (ΔTWS) is equivalent to the change in soil moisture (ΔSM) and groundwater storage (ΔGWS) as:

$$\Delta TWS = \Delta SM + \Delta GWS \quad (14.12)$$

The units of Eq. (14.12) are equivalent water height (cm or mm or in), which can be converted to volume by multiplying by area.

Other factors that can contribute to ΔTWS, such as changes in surface water, ice and snow, and biomass, were judged by Strassberg et al. (2009) to not be significant in the study area. GRACE can provide an estimate of ΔTWS, but in order to calculate ΔGWS, ΔSM must be obtained by other means. A dense network of soil moisture stations are present in the study area, which allowed for direct measurements of soil moisture. Soil moisture was also simulated using the Noah land surface model (Ek et al. 2003). Groundwater water storage changes were evaluated in situ using well data. The results of the Strassberg et al. (2009) study were that ΔTWS obtained from GRACE compared well with the sum of ΔGWS and ΔSM (both in situ and modeled). There was also a good correspondence between ΔGWS calculated as GRACE ΔTWS minus the simulated ΔSM, and ΔGWS calculated from groundwater level data. The results of this study demonstrated the potential for using GRACE gravity measurements to monitor changes in GWS and SM over large semiarid and arid regions in which the groundwater monitoring network is sparse or non-existent.

The accuracy of calculations of ΔGWS using Eq. (14.12) is dependent on the accuracy of the measurement of ΔSM and other changes in the TWS. The change in ΔGWS may be small relative to the change in ΔSM (e.g., Rodell et al. 2007). The accuracy of GRACE-derived estimates of ΔGWS is limited by errors in the

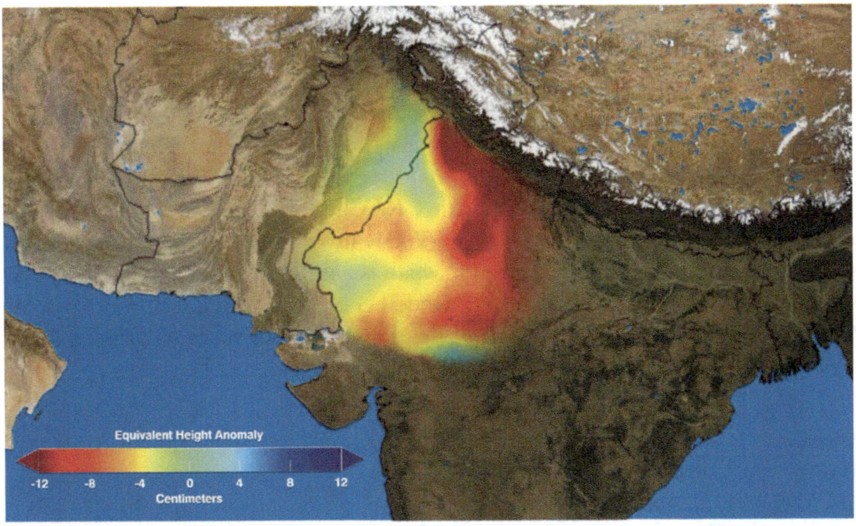

Fig. 14.2 GRACE groundwater depletion map for northern India between November 2002 and November 2008. Changes in groundwater mass are expressed as equivalent height anomalies in centimeters. *Source* NASA Goddard Space Flight Center

background models used to estimate other changes in the TWS, which accumulate in the resulting ΔGWS data (Günter et al. 2007). Areas with conditions favorable for deriving groundwater storage changes from GRACE data may be those where the ΔGWS changes are large in absolute terms and are also large relative to variation in other storage components (Günter et al. 2007).

Tiwari et al. (2009) took a similar approach for evaluating groundwater storage in northern India by subtracting estimated soil moisture changes predicted by land surface models from GRACE ΔTWS data. A negative trend was mapped over northern India (Fig. 14.2), which is largest broad-scale negative trend evident by GRACE anywhere in the world that is not caused by thinning of ice sheets and glaciers. The negative trend is located in a region that has among the world's highest population density. The trend of decreasing groundwater storage is approximately 2 cm/year (0.8 in./year), which corresponds to a 10 cm/year (3.9 in./year) lowering of the water table using an assumed porosity of 0.2. The total mass loss during the study period (April 2002 through June 2008) was 54 ± 9 km^3/year (1.9 million gal/ft^2). The decline in groundwater storage is evidence that the aquifer is not being fully recharged each rainy season.

The decline in water levels in northern Indian was also evaluated by Rodell et al. (2009) using a similar approach of combining ΔTWS data from GRACE with soil–water storage evaluated using the Global Data Assimilation Systems (GLDAS). Soil–water storage changes were not significant (i.e., no overall trend) over the August 2002 to October 2008 study period, which had about normal rainfall. Rodell et al. (2009) calculated an equivalent height depletion of

4.0 ± 1.0 cm/year (1.6 ± 0.4 in./year), which corresponds to a 0.33 m/year (1.1 ft/year) decline in the water using a specific yield of 0.12. Rodell et al. (2009) concluded that the consequences of the aquifer depletion in the area measured using GRACE (which currently has a population of 114,000,000) may include a reduction in agricultural output and shortages of potable water leading to intense socio-economic stress.

The TWS changes measurable with GRACE have great potential value for water budget calculations. Rodell et al. (2004) investigated the use of GRACE to estimate evapotranspiration rates (ET), which is a difficult variable to quantify on a regional, groundwater basin scale. The Rodell et al. (2004) approach was to use the GRACE ΔTWS data in the solution of the drainage basin water balance equation:

$$ET = P - Q - \Delta TWS \text{ (cm)} \tag{14.13}$$

where,
P = precipitation
Q = net stream flow.

All parameters are in equivalent water height (cm or in).

It was assumed that the net groundwater flow within a drainage basin would be negligible. The Mississippi River Basin was used as an example. The values of P and Q are observational data that are more accurately obtainable than the ET and ΔTWS values. The ΔTWS measurements derived from GRACE data are a perfect fit for water budget studies because they are horizontally and vertical integrated quantities (Rodell et al. 2004). The derived ΔTWS values include both changes in the water volume in the saturated and unsaturated zone. The estimated ET values based on the GRACE water balance approach were comparable to values obtained from the land-surface model and a combined atmospheric-terrestrial water budget approach.

Cesanelli and Guarracino (2011) used GRACE data to estimate ET rates in the Salado Basin of Argentina. GRACE data were used to obtain monthly estimates of the change in TWS (ΔTWS). ET rates were obtained from the ΔTWS values and precipitation data were obtained from rain gauge stations. Runoff was estimated as 5% of the precipitation.

The unprecedented heat wave of 2003 in Europe provided an opportunity to use GRACE to examine the impact of an extreme climatic event on TWS (Anderson et al. 2005). The temporal changes of TWS measured by GRACE agreed qualitatively and, at least in part, quantitatively with hydrological-based estimates of the intra-year and inter-year evolution of TWS. The difference between the datasets tended to be within the GRACE margin of error. Estimates of TWS from GRACE data were also compared to measurements from two superconducting gravimeter stations. The GRACE and gravimeter data had similar patterns for the 2001–2003 study period with differences in the magnitude of change related to different spatial extents of the measurements. The study demonstrated the potential of GRACE for investigation of variations in terrestrial water storage such as would occur in response to extreme climatic events (Anderson et al. 2005).

Gravity-based estimates of groundwater storage, such as from GRACE data may be invaluable in data-poor regions of the world (i.e., large parts of developing areas) and in regions where data are not centralized or are unobtainable due to political boundaries (Rodell et al. 2007). Although the GRACE mission has a finite life, future gravity satellite missions are expected to offer greater spatial resolution and more precise determination of water storage changes (Günter et al. 2007). Satellite gravity data is thus expected to play an expanding role in water management on a global basis in the future.

References

Anderson, O. B., Seneviratne, S. I., Hinderer, J., & Viterbo, P. (2005). GRACE-derived terrestrial water storage depletion associated with the 2003 European heat wave. *Geophysical Research Letters, 32*, L18405, 4.

Becker, M. W. (2006). Potential for satellite remote sensing of ground water. *Ground Water, 44*, 306–318.

Burger, H. R., Sheehan, A. F., & Jones, G. H. (2006). *Introduction to applied geophysics: Exploring the shallow subsurface*. New York: W. W. Norton.

Cesanelli, A., & Guarracino, L. (2011). Estimation of regional evapotranspiration in the extended Salado Basin (Argentina) from satellite gravity measurements. *Hydrogeology Journal, 19*, 629–639.

Chapman, D. S., Sahm, E., & Gettings, P. (2008). Monitoring aquifer recharge using repeated high-precision gravity measurements: A pilot study in South Weber, Utah. *Geophysics, 73*(6), WA83–WA93.

Davis, K., Li, Y., Batzle, M., & Raynolds, B. (2005). *Time-lapse gravity monitoring of an aquifer storage recovery project in Leyden, Colorado*. Golden: Colorado School of Mines, Center for Gravity, Electrical & Magnetic Studies.

Davis, K., Li, Y., & Batzle, M. (2008). Time-lapse gravity monitoring: A systematic 4D approach with application to aquifer storage and recovery. *Geophysics*, *73*(6), WA61–WA69.

Ek, M. B., Mitchell, K. E., Lin, Y., Rogers, E., Grunmann, P., Korev, V., et al. (2003). Implementation of the Noah land surface model advances in the National Centers for Environmental Prediction operational mesoscale Eta model. *Journal of Geophysical Research, 108*, 8861. doi:10.1029/2002JD003296.

Günter, A., Schmidt, R., & Döll, P. (2007). Supporting large-scale hydrogeological monitoring and modeling by time-variable gravity data. *Hydrogeology Journal, 15*, 167–170.

Howle, J. F., Phillips, S. P., & Ikehara, M. E. (2002). Estimating water-table change using microgravity surveys during an ASR program in Lancaster, California. In P. Dillon (Ed.), *Management of aquifer recharge for sustainability* (pp. 269–272). Lisse: Swets & Zeitlinger.

Maliva, R. G., Coulibaly, K., Guo, W., & Missimer, T. M. (2011). Confined aquifer loading: Implications for groundwater management. *Ground Water, 49*(3), 302–304.

Milson, J. (2003). *Field geophysics* (3rd ed.). Chichester: Wiley.

NASA. (2003). Studying the earth's gravity from space: The Gravity Recovery and Climate Experiment (GRACE). NASA Facts. The Earth System Sciences Pathfinder Series, FS-2002-1-029-GSFC (rev. December 2003), p. 6.

Pool, D. R. (2008). The utility of gravity and water-level monitoring at alluvial aquifer wells in southern Arizona. *Geophysics, 73*(6), WA49–WA59. doi:10.1190/1.2980395.

Pool, D. R., & Schmidt, W. (1997). *Measurement of ground-water storage change and specific yields using the temporal-gravity method near Rillito Creek*. Tucson, Arizona. U.S. Geological Survey Water Resources Investigations Report 97-4125.

Rodell, M., Famiglietti, J. S., Chen, J., Seneviratne, S. I., Viterbo, P., Hall, S., & Wilson, C. R. (2004). Basin scale estimates of evapotranspiration using GRACE and other observations. *Geophysical Research Letters, 31*, L20504, 4. doi:10.1029/2004gl020873.

Rodell, M., Chen, J., Kato, H., Famiglietti, J.-S., Nigro, J., & Wilson, C. R. (2007). Estimating groundwater storage changes in the Mississippi River Basin (USA) using GRACE. *Hydrogeology Journal, 15*, 159–166.

Rodell, M., Velicogna, I., & Framigletti, J. S. (2009). Satellite-based estimates of groundwater depletion in India. *Nature, 460*, 999–1002.

Strassberg, G., Scanlon, B.R., and Rodell, M., 2007, Comparison of seasonal terrestrial water storage variations from GRACE with ground-water level measurements from the High Plains Aquifer (USA). *Geophysical Research Letters, 34*, L14402, 5. doi:10.1029/2007/GL030139.

Strassberg, G., Scanlon, B. R., & Chambers, D. (2009). Evaluation of groundwater storage monitoring with the GRACE satellite: Case study of the High Plains aquifer, central United States. *Water Resources Research, 45*, W05410, 10. doi:10.1029/2008WR006892.

Swenson, S., & Wahr, J. (2002). Methods for inferring regional surface-mass anomalies from gravity recovery and climate experiment (GRACE) measurements of time-variable gravity. *Journal of Geophysical Research, 113*, B08410. doi:10.1029/2001JB000576.

Telford, W. M., Geldart, L. P., Sheriff, R. E., & Keys, D. A. (1976). *Applied geophysics*. Cambridge: Cambridge University Press.

Tiwari, V. M., Wahr, J., & Swenson, S. (2009). Dwindling groundwater resources in northern India, from satellite gravity observations. *Geophysical Research Letters*. doi:10.1029/2009GL039401.

Turcotte D. L., & Schubert, G. (1982). *Geodynamics: Applications of continuum physics to geological problems*. New York: Wiley.

Wahr, J., Swenson, S., Zlotnicki, V., & Velicogna, I. (2004). Time-variable gravity from GRACE: First results. *Geophysical Research Letters, 31*, L11501. doi:10.1029/2004GL019779.

Chapter 15
Compaction and Land Subsidence

15.1 Introduction

It is well documented that groundwater abstractions can induce land subsidence in many arid lands where groundwater is the only viable water source (e.g., Cuevas 1936; Domenico et al. 1966; Poland and Davis 1956; Poland 1961, 1970; Poland et al. 1975; Tolman and Poland 1940; Wilson and Grace 1942; Winslow and Wood 1959). Within the United States, an area of more than 26,000 km^2 (10,039 mi^2) has been permanently lowered as the result of withdrawal of underground fluids (National Research Council 1991; Holzer and Galloway 2005). Land subsidence induced by groundwater pumpage was reviewed by Galloway et al. (1999a, b), Holzer and Galloway (2005), and Galloway and Burbey (2011). The land subsidence in the San Joaquin Valley, California, USA, (Figs. 15.1 and 15.2), which was caused largely by groundwater pumping for irrigation, was observed by Galloway and Riley (1999, p. 23) to be "one of the single largest alterations of the land surface attributed to mankind".

Although there is not a direct relationship between climate and land subsidence, the greatest documented subsidence tends to have occurred in dry regions in which there are limited surface water resources, a need for intensive irrigation, and a groundwater source of supply. Land subsidence may not be obvious, especially where it occurs in a relatively uniform manner over a broad area. It is more obvious when differential subsidence occurs, which can result in damage to buildings and infrastructure, changes in stream and gravity main grades, and fissuring. Local lowering of land surface may also result in the encroachment of seawater or tidal flooding (e.g., Houston Ship Channel area, Texas). Subsidence may also be detected where land surface elevation changed with respected to a fixed vertical datum. A very obvious sign of land surface is protruding casings of deep wells, which are anchored at depth (Fig. 15.3). Horizontal movement may also be important, particularly near the margins of subsidence or uplift features (Galloway and Burbey 2011).

R. Maliva and T. Missimer, *Arid Lands Water Evaluation and Management*, Environmental Science and Engineering, DOI: 10.1007/978-3-642-29104-3_15,

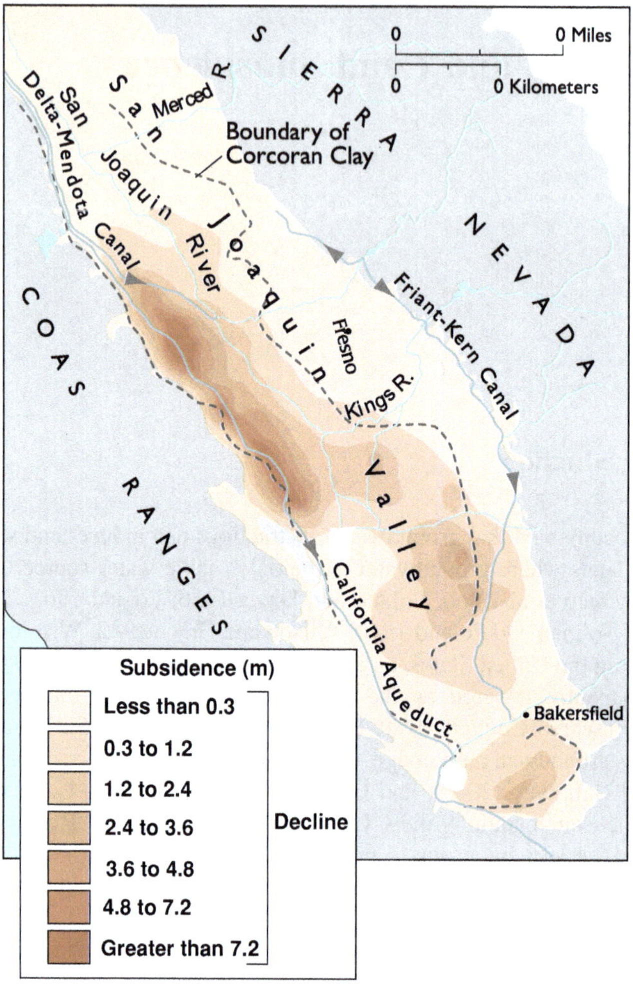

Fig. 15.1 Map of land subsidence in the San Joaquin Valley, California (U.S.A.) from 1926 to 1970 (after Galloway and Riley 1999)

The total stress acting on a horizontal plane in buried strata is due to the weight of the overlying rock and water. The total stress is borne by both the aquifer solids or skeleton (i.e., grains and cemented rock) and the pore waters. The portion of the total vertical stress borne by the aquifer skeleton is called the effective vertical stress, (σ_z') which was defined by Terzaghi (1936) in one-dimension as the difference between the total vertical stress (σ_z) and pore pressure (P), as follows:

$$\sigma_z' = \sigma_z - P \tag{15.1}$$

Fig. 15.2 Often reproduced photograpth of the approximate location of maximum land subsidence in the United States (San Joaquin Valley, California). Signs of poles show approximate atltutude of the land surface in 1925, 1955, and 1977. *Source* Galloway and Riley (1999)

According to Newton's Laws of Motion, for a system not in motion, the total downward stress from the weight of the overlying rock and water is balanced by the effective stress and pore pressure. If the pore pressure decreases as the result of pumping (or other causes), a corresponding increase in the effective stress must occur. The increase in effective stress is responsible for compaction and thus land subsidence.

For the case of vertical subsidence, the relationship between changes in effective stress and compaction is expressed by the equations:

$$\frac{db}{b} = \propto \sigma'_z \tag{15.2}$$

$$\mathrm{db} = \mathrm{b}\alpha \mathrm{d}\sigma'_{\mathrm{z}} \tag{15.3}$$

where,
db = change in thickness of interval of sediment or rock in question (m)
b = thickness of interval (m)
α = one-dimensional compressibility (Pa^{-1}, or $kg^{-2}ms$), and
$d\sigma_z{}'$ = change in effective vertical stress (Pa).

These equations are for metric units and there may be others developed using English units.

Assuming a constant geostatic load, then the change in effective stress will be equal to the change in pore pressure, which can be related to changes in head as

$$\Delta\sigma_z' = \rho_w g \Delta h \tag{15.4}$$

$$\frac{db}{dh} = b \propto g\rho_w \tag{15.5}$$

where,
ρ_w = density of water (kg/m^3)
g = gravitational acceleration constant (9.80665 m/s^2)
Δh = change in head (m).

From these equations, the amount of compaction of an interval of sediment or rock is a function of the compressibility and thickness of the interval and the change in effective stress. Beds of sediment or rock with different compressibilities will undergo different amounts of vertical compaction. In order to estimate the total compaction (and thus associated land subsidence), it is necessary to integrate Eq. (15.5) with the entire thickness of strata affected by the change in effective stress.

Silt and clay deposits that form semi-confining units tend to have high porosities but very low hydraulic conductivities. These deposits usually have much greater compressibilities than relatively clean sands and lithified aquifer strata. The silt and clay beds will typically undergo greater compaction than the more permeable strata that constitute the aquifers. Areas that have thick sequences of fine-grained sediments with high compressibilities are most vulnerable to subsidence. Reduction in effective stress associated with aquifer drawdowns may not cause significant subsidence in groundwater systems dominated by well-lithified aquifer and confining strata.

Modeling results by Hoffmann et al. (2003) indicate that spatial heterogeneities of interbed storage in the studied aquifer system (Antelope Valley, California) were responsible for the documented uneven distribution of land subsidence. A key point is that the subsidence pattern reflects both the spatial distribution of drawdown and variability of storage coefficients (and thus compressibility) of interbeds. Chen et al. (2003) similarly demonstrated that the geographic distribution of land subsidence in Suzhou City, Jiangsu Province, coastal China, was controlled by a combination of both the thickness of soft clay unit and aquifer drawdown. A similar relationship between compaction and the distribution of more highly compressible strata was also documented by Calderhead et al. (2011) in the Toluca Valley or Mexico. Land subsidence can be reduced by moving groundwater withdrawals to areas in which the compactable clays are relatively thin (Chen et al. 2003).

The response to increased effective stress (strain) may include both elastic (reversible) and inelastic (irreversible) components. For an elastic strain, the sediment or rock returns to its previous state after the stress has been removed. In the case of inelastic strain, the change (compaction) is irreversible. Grain repacking (consolidation) is an example of what is predominantly an inelastic strain. Under stress, grains are reoriented and packed together more tightly. For example, plate-shaped clay mineral flakes will become oriented perpendicular to the principle stress direction. When the stress is removed, the grains do not revert back into the original looser grain packing. Inelastic compaction of aquitards (semi-confining units) results in a one-time release of the 'water of compaction', which is essentially a non-renewable water supply that can be mined only once at the expense of land subsidence (Sneed and Galloway 2000).

Land subsidence often has both reversible and irreversible components. Some elastic deformation is inherent in the production of groundwater from confined aquifers. Water is produced by the expansion of the water and compression of the aquifer material. The storativity of an aquifer is a function of the water and aquifer matrix compressibilities. Fine-scale land surface subsidence and rebound in response to seasonal or shorter term changes in aquifer water level can be measured using sensitive instruments (e.g., extensometers). Elastic response is evidenced by a rebound occurring during periods of increased aquifer heads. Land subsidence caused by inelastic compaction and its associated impacts are irreversible.

The stress history of the deposits also plays an important role in their vulnerability to subsidence. Older deposits that were exposed to very large loads, from glaciers or seawater inundation in the past, may become over-consolidated (preconsolidated) and thus less vulnerable to further consolidation created by decreases in hydrostatic pressure. So long as the effective stress remains above the 'preconsolidation stress', subsidence will remain small and reversible. Once the preconsolidation stress is exceeded, the compressibility of the aquifer system increases dramatically and the resulting compaction is largely irreversible (Riley 1969; Holzer and Galloway 2005).

An additional consideration is that time delays occur between the reduction in aquifer heads and compaction of the interbedded or adjoining fine-grained sediments. Land subsidence will not occur immediately after the start of aquifer pumping and could continue for some after the rate of pumping is reduced and aquifer water levels recover. Such continued subsidence after groundwater levels have stabilized or started to recover is occurring in the San Joaquin Valley, where land subsidence as great as 8.8 m (29 ft) has occurred (Galloway and Riley 1999).

The time delay is caused by the low vertical hydraulic conductivity of the fine-grained sediments. Subsidence can only occur as fast as water can migrate out of the strata undergoing compaction. It may decades or centuries for pore pressures in thick aquitards (semiconfining units) to equilibrate with pressure changes in aquifers. Compaction and associated land subsidence will therefore tend to continue, even if groundwater levels have stabilized or even partially recover

Table 15.1 Methods for measuring vertical land subsidence

Method	Resolution (mm)	Spatial density (Samples/survey)
Spirit level	0.1–1	10–100
Borehole extensometer	0.01–0.1	1–3
GPS	20	10–100
InSAR	10	10,000–10,000,000

Source Galloway et al. (1999b)

(e.g., Hoffmann et al. 2003). Compaction may occur more rapidly when clay units contain interbedded sands that facilitate drainage.

Residual compaction is the most important source of persistence deformation is some areas (Galloway and Hoffmann 2007). Persistent deformation is the difference between (1) the amount of compaction that will ultimately occur given an increase in applied stress, and (2) the compaction that has occurred at a specific time. Residual compaction can be evaluated from the compaction rate that continues to occur once aquifer water levels have stabilized.

15.2 Land Subsidence Measurement

Commonly used methods to measure land subsidence are summarized in Table 15.1. The methods vary in the resolution, cost, and practical spatial density of measurements. Borehole extensometers have the finest resolution and greatest accuracy, but are expensive to construct. As a result, there is typically a low spatial density of extensometers in most study areas. Ideally, multiple methods should be used in which, for example, extensometer data may serve as a control on less accurate methods with coarser resolutions.

15.2.1 Conventional (On Ground) Surveying

Land subsidence can be measured by conventional surveying techniques. Stable benchmarks need to be established that are a well-connected to the land surface. It is also necessary to have vertically stable reference points located outside of the area affected by subsidence (Galloway et al.1999a, b). In the case of an aquifer system in an alluvial basin, reference benchmarks may be established in hard rock areas outside of the basin. It must also be emphasized that the vertical elevation changes in most areas will be small, particularly over a short (several year) time period. It is therefore imperative that a high level of care be performed in the measurements and that professional surveyors are used.

Fig. 15.3 Photographs illustrating subsidence of land surface near Las Vegas, Nevada (U.S.A.) relative to a well casing that is anchored at depth. The wellhead has been essentially stationary while the land surface has been lowered. *Source* Pavelko et al. (1999)

The widely used spirit level surveying technique can be used to measure subsidence over relative small areas. Spirit-level surveying can be very accurate and is relatively inexpensive. Professional surveyors are available in most developed regions and standard surveying equipment can be used. The limitation of the technique is that accuracy decreases with distance. The stable reference point must therefore be located relatively close (within approximately 5 mi/8 km) of the areas of interest (Galloway et al. 1999a, b).

15.2.2 Extensometers

Extensometers measure the movement of land surface with respect to a fixed datum, which is typically either a steel pipe or cable that is cemented in place at the bottom of a well. The extensometer in essence establishes a 'subsurface bench mark' at the bottom of the borehole. Compaction (or expansion) is measured as the change in distance between the bottom of the well and land surface. The extensometer well is constructed with an uncemented open hole or compression joints so that the well is not damaged during compaction. The rod or cable must also be able to freely move with respect to the well casing. With respect to a pipe extensometer, the length of the pipe does not change over time. Compaction is thus recorded by changes (increases) in the distance between the top of the pipe (or marker on the pipe) and land surface (Fig. 15.4). The sensitive recording instruments that are used in extensometers can measure changes in land surface elevations down to a fraction of a millimeter. The cable extensometer operates under the same principle as the rod extensometer and measures compaction or

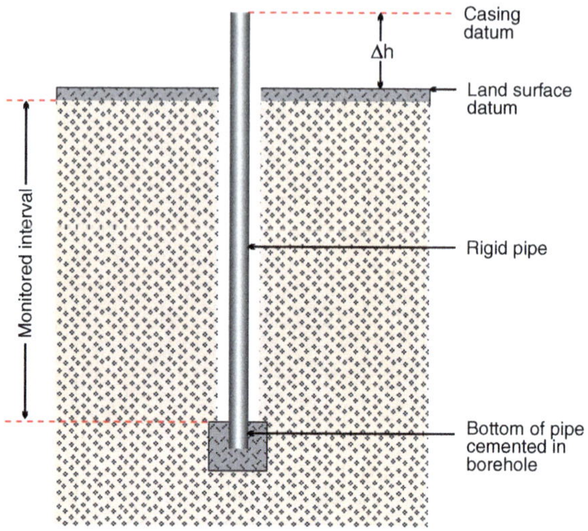

Fig. 15.4 Conceptual diagram of a extensometer. Land subsidence caused by compaction of the monitored interval is measured by the increase in distance between a datum on the pipe and land surface (Δh)

expansion by changes in the length of the cable with respect to land surface. Multiple position borehole extensometers incorporate multiple markers (e.g., magnetic) anchored to the formation borehole to measure compaction over multiple depth intervals (Galloway and Burbey 2011).

Extensometers measure compaction of the strata between the bottom of the casing and land surface, not the total amount of land surface subsidence that occurs at a site. Some compaction may occur in the strata below the base of the extensometer. The difference between subsidence measured using vertical extensometers and other methods (GPS, InSar) may be caused by compaction of strata below the extensometer.

Vertical extensometers are the most precise method for measuring land subsidence, but have the limitation that they are expensive to construct and operate, which thus limits the numbers of stations. Highly sensitive extensometers can continuously measure minute elastic compression and expansion that accompany small changes in groundwater level and relatively large-scale deformation typical of irreversible compaction of semi-confining units (aquicludes; confining beds) (Galloway et al. 1999b). Stress–strain data obtained from coupled extensometers and groundwater monitoring wells can be used to obtain data on the compressibility and hydraulic properties of semi-confining units, which are necessary for predictive modeling (Galloway et al. 1999b).

15.2.3 Global Positioning Systems

The Global Positioning Systems (GPS) uses radiowave signals from a constellation of satellites in medium Earth orbit to determine the location (latitude, longitude, and commonly also altitude) of a receiver. The GPS satellites continuously transmit messages containing the time the message was sent, precise orbital information, the general system health and rough orbits of all GPS satellites (the almanac). The GPS receiver uses the transmitted data to precisely measure the travel time of the radio signal from each satellite from which a suitable signal is received. The travel times are used to calculate the distance between each satellite and the receiver.

The distance data are then processed using the trilateration (triangulation). Conceptually, the distance of a GPS receiver from a satellite is a spherical surface that is centered on the satellite. The distances from three satellites would define three spheres that typically intersect at two points, one of which is located on the surface of the Earth.

GPS is widely used for measuring changes in land surface altitude. Consumer hand-held units do not have near the accuracy for measuring vertical altitude at the accuracy needed for surface deformation investigations. Top of the line survey-grade, dual-frequency, units are required. Static survey method is commonly employed to obtain the highest accuracy needed for land subsidence investigations. The static survey method requires the use of at least two receivers; one each at two (or more) sites (Carruth et al. 2007). Each receiver logs data simultaneously from at least four common satellites. A time series of measurements is performed at the each survey point, with each series of measurement requiring a long occupation time (0.5–8 h). The accuracy of static GPS elevation measurements is 20 mm (1 in.) or less.

GPS measurements can be impacted by a variety of system noise sources (Behr 2001):

- satellite clock error,
- receiver clock error,
- ephemeris errors (difference between the expected and actual orbital position of a GPS satellite),
- ionosphere delay,
- troposphere delay,
- poor geometry,
- multipath,
- antenna-related errors, and
- monument stability.

The listed errors can by managed through data processing and by careful site design. It is clear from the numerous factors that can impact GPS measurements that GPS surveying for land subsidence investigations requires specialized and expensive equipment and a trained professional in order to accurately measure the small changes in land surface elevation that might occur over a relatively short period of time.

15.2.4 InSAR

Satellite Interferometric Synthetic Aperture Radar (InSAR) basically involves that bouncing of radar signals off the ground. Time series of measurements are used to measure changes in the distance between the satellite and ground and thus changes in land surface altitude. The great advantages of InSAR is that it is often less expensive than obtaining numerous point measurements from labor-intensive spirit-leveling and GPS surveys, and can provide millions of data points in a region of about 10,000 Km^2 (3,860 mi^2) in area. Pixels on an InSar displacement map are typically 30–90 m^2 (323–969 ft^2) on the ground (Bawden et al. 2003).

InSAR was reviewed by Massonnet and Feigl (1998), Bürgmann et al. (2000), and (Galloway and Hoffmann 2007). Use of InSAR to map changes in land surface altitude has increased dramatically over the past decade because of the ability of technique to efficiently produce spatially complete detailed maps of altitude changes that would not be practical using land based techniques. InSAR can be used to measure subsidence over a broad area such as an entire groundwater basin or metropolitan areas.

The InSAR process measures the phase shift between synthetic aperture radar (SAR) images obtained from two satellite passes. SAR has a very narrow effective beam, which allows for fine spatial resolution. Data processing is a critical element of InSAR as several factors can affect measured phase shifts. The two images must first be georeferenced (co-registered) so that each pixel represents the same ground area in both images. InSAR data is subject to error from orbit error, atmospheric delay effects, and topographic effects.

The InSAR data are processed to generate interferograms in which changes in land surface altitude are expressed as color bands. An example of an interferogram, prepared by the U.S. Geological Survey for Las Vegas, Nevada, area is provided in Fig. 15.5. Each color cycle represents a range of change, which for the interferogram in Fig. 15.5 is 100 mm (4-in.). The color cycle of the Cabral-Cano et al. (2008) study of Mexico City represents 28 mm (1 in.). Interferograms for one satellite look direction measures only the scalar length change in the line of site and does not resolve the three orthogonal components of the displacement (Cabral-Cano et al. 2008). The scalar length of change does not necessarily correspond to vertical displacement. The length of change will correspond to change in vertical altitude for data sets with a low angle of incidence.

InSAR measures changes in land surface altitude, which can have causes other than deformation. Measured changes in land surface altitude may be caused by interference from crop canopies, land leveling, and the presence of a snow cover. As is the case for remote sensing data in general, InSAR measurements should ideally be calibrated against other land altitude change data such as GPS or extensometer.

Galloway et al. (1998) proposed that InSAR measurement of displacement could be used in conjunction with hydraulic head measurements to compute elastic storage coefficients in confined aquifer systems undergoing significant seasonal deformation. In order to measure the storage coefficients, the study area must be undergoing only elastic deformation, and, thus, there should be no net subsidence and the previous maximum stress should not be exceeded. InSAR was successfully used to measure the storage coefficient in the Las Vegas Valley, Nevada (Hoffmann et al. 2001). The InSAR-measured seasonal displacements were in good agreement with bore extensometer measurements, but were consistently larger, which was explained by InSAR measuring displacement in a greater thickness of deforming sediments. The calculated storage coefficient of 7.3×10^{-4} is well within the range of values obtained from calibrated groundwater flow models and aquifer performance tests.

Persistent scatter InSAR (PSI) is based on a stack of interferograms relative to a single master image. PSI has a higher density of measurements than standard SAR interferometry, which allows for finer spatial resolution of changes in land surface altitude (Osmanoğla et al. 2010). For example, PSI data from central Mexico City could resolve a lesser degrees of compaction along a Metro line due to its over-consolidated foundation (Osmanoğla et al. 2010). The Mexico City PSI data showed no evidence for significant uplift from recharge associated with the rainy season or any other artificial recharge process.

The limitations of InSAR include (Galloway and Hoffmann 2007):

- Thin coarse-grained, high-permeability units that are most important for groundwater production deform the least.
- Loss of coherence occurs in areas that are heavily vegetated or distributed by agricultural activities.
- Suitable SAR data may not be available for a study area for the time period of interest. The temporal coverage of InSAR data may be inadequate to constrain estimates of the time constants of compacting interbeds (Hoffmann et al. 2003).
- Data processing software is expensive and a significant investment is involved to develop the skill necessary to generate representative interferrograms. InSAR is a specialized discipline.
- A paucity of ground-truthed data is available in many areas.

15.2.5 Measured Land Subsidence Amounts and Rates

Land subsidence may not be apparent when subsidence occurs evenly over broad areas. Determining the amount of subsidence that occurred in response to groundwater pumping at some locations is complicated by the paucity or absence of baseline land surface elevation data, particularly in areas with a long history (many decades) of groundwater pumping. Features may be present that can serve in essence as crude extensometers, such as well casings anchored in underlying bedrock, particularly if the original distance from the top of casing to land surface and date of installation are approximately known.

Table 15.2 Reported local maximum groundwater pumping-induced land subsidence

Site	Subsidence (ft)	Subsidence (m)	Source
Eloy, Arizona	15	4.6	Gelt (1992), Leake (2004)
West of Phoenix, AZ	18	5.5	Leake (2004)
Las Vegas, NV	6	1.8	Leake (2004)
Lancaster, CA	6.6	2.0	Sneed & Galloway (2000) Leake (2004)
South of Mendota, CA	29	8.8	Leake (2004)
Santa Clara Valley, CA	12	3.7	Leake (2004)
El Paso, TX	1	0.3	Leake (2004)
Houston, TX	9	2.7	Leake (2004)
Tucson, AZ (NWArva Valley)	1	0.3	Carruth et al. (2007)
Mexico City	>29.5	>9	Ortega-Guerro et al. (1999)
Chalco Basin, Mexico	26	8	Ortega-Guerro et al. (1999)
Downtown Mexico City	31.8	9.7	Cabral-Cano et al. (2008)
Toluca Valley, Mexico	6.6	2	Calderhead et al. (2011)
Suzhou City, Jiangsu Province, China	3.3	1 m	Chen et al. (2003)

Note: Current subsidence values may be greater due to subsequent compaction

Table 15.3 Reported groundwater pumping-induced land subsidence rates

Site	Subsidence rate (ft/year)	Subsidence rate (m/year)	Source
Chalco Basin (near Mexico City	1.3	0.4	Ortega-Guerro et al. (1999)
Mexico City	>1.1	>0.35	Cabral-Cano et al. (2008)
Tucson, AZ	0.022	0.007	Carruth et al. (2007)
Suzhou City, Jiangsu Provicence, China	0.23	0.07 (1 m/14 years)	Chen et al. (2003)
Santa Clara Valley, California	0.72	0.22	Poland (1970)
Toluca Valley, Mexico	0.27	0.083 (37.5 cm/4.5 years)	Calderhead et al. (2011)

Documented subsidence amounts and measured subsidence rates obtained from a partial literature review are provided in Tables 15.2 and 15.3, respectively. Total subsidence amounts are a function of local geology (presence of highly compressible strata), aquifer drawdowns (increase in effective stress), and time. Notable areas of extreme land subsidence are Mexico City and some southern California agricultural areas.

15.3 Impacts of Subsidence

Land subsidence can cause a wide variety of adverse impacts to buildings, roads, and other infrastructure. Problems caused by subsidence (Galloway et al. 1999a; Leake 2004; Holzer and Galloway 2005) include:

1. Permanent inundation and vulnerability to flooding and storms,
2. Change in altitude and slope of streams, canals, and drains,
3. Damage to bridges, roads, railroad tracks, storm drains, sanitary sewers, water mains, canals and sewers,
4. Damage to private and public buildings,
5. Failure of well casings and screens,
6. Impacts to the operation of gravity sewer and storm drains,
7. Opening of earth fissures,
8. Modification of aquifer hydraulic properties, and
9. Need to raise levees and other flood control structures.

Economic consequences of subsidence, while large, are generally factored into yearly maintenance budgets rather than accounted for as unique natural disasters at a single point in time (Cabral-Cana et al. 2008). Slow, gradual, land subsidence can be ignored or accommodated for quite a while, until it reaches a state in which responsive action is required. On the contrary, the economic impacts of rapid subsidence events, such as a house or automobile falling into a sinkhole, are much easier to quantify.

It is therefore difficult to quantify the economic damage from pumping-induced long-term subsidence, especially on an annual basis. There are consequently few published studies on the economic impacts of land subsidence caused groundwater pumping. Ingebritsen and Jones (Ingebrtitsen and Jones 1999), for example, estimated than the cost of land subsidence in the Santa Clara Valley, California, has exceeded $300 Million (1998 U.S. dollars).

Compaction has significant impacts on groundwater resources. Intergranular compaction decreases porosity and thus storage capacity. The decrease in porosity also can result in a decrease in hydraulic conductivity (permeability). Water squeezed out of confining units (aquitards) by inelastic compaction is permanently lost and is essentially mined. It has been estimated that 10–30% of the water produced from a compacting aquifer was squeezed out of confining beds (Galloway et al. 1999a, b). The compaction of confining layers can decrease their vertical hydraulic conductivity and leakance value, which can decrease recharge to underlying pumped aquifers (Chen et al. 2003). The resulting decrease in recharge would accelerate the decline of water levels, and in turn compaction and associated land subsidence.

15.4 Case Studies

15.4.1 Mexico City

Mexico City, Mexico, depends mainly on groundwater from an extensive region alluvial-pyroclastic aquifer for its water supply. The production aquifer is overlain by thick lacustrine deposits that are composed of compressible and extremely porous clay-rich Quaternary deposits. Depressurization from groundwater pumping resulted in consolidation of the aquitard and associated land subsidence, which exceeds 9 m (29.5 ft) in downtown Mexico City (Ortega-Guerrero et al. 1999).

Groundwater pumping was subsequently relocated to new wellfields located outside of the city, such as the Chalco Basin, which was the subject of a modeling study by Ortega-Guerrero et al. (1999). The maximum subsidence in the Chalco Basin was reported to be 8 m (26.2 ft) and the subsidence rate the center of the basin was reported to be 0.4 m/yr (1.3 ft/yr). As reviewed and further assessed by Ortega-Guerrero et al. (1999), the aquitard shows nonparametric behavior during consolidation, as the parameters controlling groundwater flow in a high compressible aquitard can significantly change during compaction. A decrease in pore pressure and effective stress results in a decrease in porosity, which in turn leads to a decrease in hydraulic conductivity. The corresponding equations are (Ortega-Guerrero et al. 1999)

$$de(\sigma_e) = C_c \log\left(\frac{\sigma_{eo} + d\sigma_e}{\sigma_{eo}}\right) \tag{15.6}$$

$$dK'(e) = K'_0(e)(10^{de/m)} - 1) \tag{15.7}$$

where,

e = void ratio (unitless)

de = change in void ratio

$d\sigma_e$ = change in effective stress (Pa)

C_c = compression index (slope of e-logσ_e plot, determined from laboratory tests)

σ_{eo} = effective stress under current hydraulic stress (Pa)

dK′ = change in hydraulic conductivity (m/s)

K'_0 = initial hydraulic conductivity (m/s)

m = slope e-logK′ plot (generally estimated based on laboratory samples).

Ortega-Guerrero (1999) developed a one-dimensional model using field measured values of hydraulic parameters that was calibrated against current hydraulic head profiles, historic changes in land surface elevation. The model was then used for predictive and simulations at different future pumping rates.

Subsidence was investigated in the Mexico City vicinity by Cabral-Cano et al. (2008) using InSAR and GPS. The combined data indicate that current subsidence rates locally exceed 350 mm/yr (1.1 ft/yr), which are close to the maximum historic rates. The historic subsidence in downtown Mexico City was reported to be 9.7 m (31.8 ft). The results of a persistent scatter InSAR study of central Mexico City indicates a maximum subsidence rate of 300 mm/yr (1 ft/yr) with 95% of the study area having had rates of between 11.8 and 250.1 mm/yr (0.04 and 0.82 ft/yr) (Osmanoğla et al. 2010).

Analytical results suggest that the land subsidence observed can be represented with soil parameters (compressibility) that correspond to a spatial composite of silt and clay (Cabral-Cano et al. 2008). Results confirm that land subsidence is caused mainly by pressure loss in the shallow clay-rich lacustrine aquitard.

15.4.1.1 Arizona

Land subsidence has been occurring in southern and south-central Arizona in response to groundwater withdrawals for over 70 years. The impacted area includes the Phoenix and Tucson metropolitan areas, which contain a large majority of the population of the state. Maximum reported drawdowns are as much as 5.5 m (18 ft) in the vicinity of Luke Air Force Base (west of Phoenix) and 4.6 m (15 ft) in the Eloy area southeast of Phoenix (Leake 2004). The relationship between groundwater decline and subsidence in Arizona sedimentary basins is complex and varies as a function of total thickness of sediments, composition (silt and clay versus sand and gravel), and compressibility (Gelt 1992; Arizona Land Subsidence Group 2007).

A serious manifestation of land subsidence in Phoenix is the development of earth fissures, which are cracks, or seams of separation in the ground that form as a result of tensional forces related to differential subsidence (Fig. 15.5). The fissures cut across surface drainage features and form near vertical pathways to the water table. Associated erosion from water flowing into fissures can widen the fissures into fissure gullies, which can happen during a single heavy storm event, (Gelt 1992; Arizona Land Subsidence Group 2007). The tensional forces responsible for fissure development tend to occur in the transitional zone between sedimentary rock undergoing subsidence and more rigid bedrock (Gelt 1992; Arizona Land Subsidence Group 2007). Fissuring has adversely impacted buildings, canals and drainage structures, and public infrastructures. Fissures may also act as conduits for contaminants to reach the water table (Gelt 1992). Localized fissuring also impacts local property values.

Land subsidence was investigated in the Tucson area by the U.S. Geological Survey. The study is noteworthy because the data were collected using GPS, vertical extensometers and InSAR (Carruth et al. 2007). The Tucson area is located in an arid region (Sonoran Desert) in the Basin and Range physiographic province. The groundwater is contained within aquifers consisting of basin-fill coarse alluvial sediments. Land subsidence in Tucson Active Management Areas is due

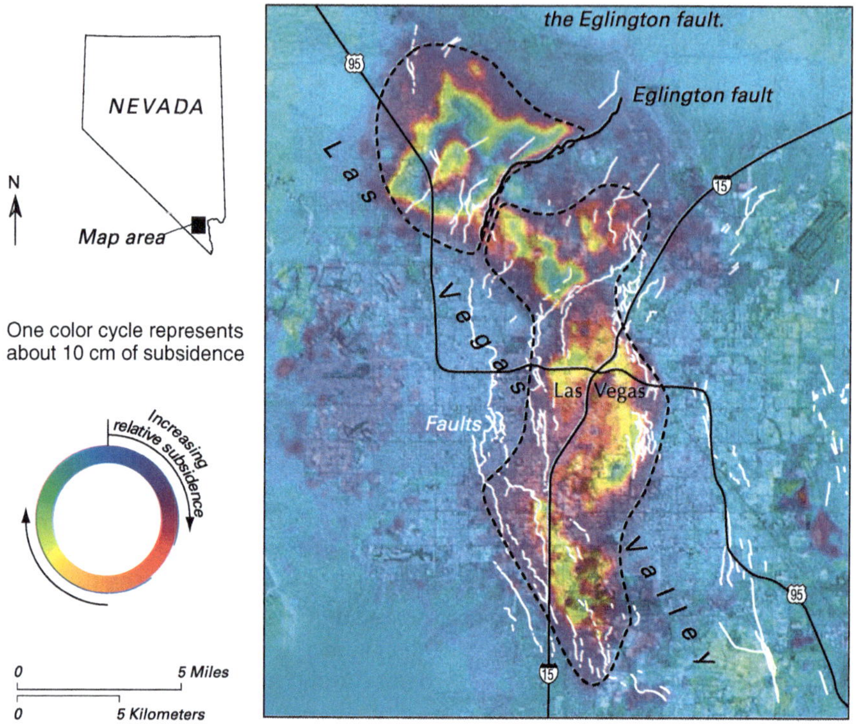

Fig. 15.5 InSar derived surface deformation map for the Las Vegas Valley showing subsidence between April 1992 and December 1997. *Source* Galloway et al. (2000)

primarily to the aquifer dewatering in response to aquifer pumping. Groundwater levels declined by as much as 13.7 m (45 ft) in the Tucson Basin and 16.8 m (55 ft) in the Arva basin between 1989 and 2005. The resulting overall long-term subsidence rates are modest. Maximum measured subsidence rates at stations in the two studied basins (Tucson and Arva) were 12.7 cm (5.0 in.) and 12.2 cm (4.8 in.), respectively, over a 17–18 year period, which corresponds to a maximum rate of about 0.007 m/year. Average subsidence over the 17–18 year period was about 3.3 cm (1.3 in.) and 7.1 cm (2.8 in.).

Reported land subsidence from the GPS measurements were somewhat less than the values obtained from the vertical extensometers. Carruth et al. (2007) interpreted the data as indicating that most of the land subsidence was caused by compaction of the strata intercepted by the extensometers, which were mostly set between 244 m to 427 m (800–1,400 ft) below land surface, and that there was subsidiary compaction of deeper strata.

15.4.2 Antelope Valley, California

Antelope Valley, southern California has experienced widespread land subsidence due to compaction of aquitards interspersed throughout the alluvial aquifer system. Throughout the Lancaster subbasin, ground water levels have declined more than 30 m (100 ft) between 1915 and 1991, and as such as 90 m (300 ft) near the southern and western margins of the subbasin. The drawdowns were primarily caused by agricultural withdrawals. Measured land subsidence in the Lancaster subbasin exceeds 2.0 m (6.6 ft) (Sneed and Galloway 2000). Differential land subsidence has caused formation of earth fissures and has altered surface drainage gradients contributing to erosion and flooding problems. The subsidence pattern reflects the areal distribution and aggregate thickness of compressible fine-grained sediment, but also may be affected by uneven bedrock subcrops (Sneed and Galloway 2000). Water levels show a clear recovery since the middle 1970s, but land subsidence is continuing, albeit at a reduced rate, because of the continued slow drainage from the compressible parts of aquifer systems (interbeds) (Sneed and Galloway 2000; Hoffmann et al. 2003).

Hoffmann et al. (2003) used inverse modeling with the MODFLOW code and SUB package (Chap. 20) to simulate land subsidence and estimate the inelastic storage coefficient and time constant. Time constants ranged for 10s to over 200 years, which indicated that delayed compaction and land subsidence will continue to occur.

15.5 Mitigation

The primary means to mitigate the impacts of land subsidence is to reduce or terminate groundwater pumping. Improved knowledge of the controls over the rate of land subsidence within a groundwater basin may also allow for the reduction of impacts by moving groundwater production to areas with a lower susceptibility to subsidence. With adequate monitoring programs and institutional controls, optimal benefits may be achieved for both subsidence mitigation and resource development (Galloway et al. 1999b).

Managed aquifer recharge (i.e., re-pressurization of the aquifer system) is also a mitigation option for land subsidence and has been successfully applied in the Las Vegas region. Replenishment of the aquifer by recharge during low demand or high supply periods can result or even eliminate the net long-term aquifer overdraft. The delay between pumping and subsidence reduces the potential for the seasonal cycle of net pumping and recharge to cause inelastic subsidence. Testing and modeling results in Lancaster, California, have also demonstrated that artificial recharge through injection can be used to manage or prevent land subsidence (Phillips et al. 2002, 2003). A key technical issue is optimizing water injection and groundwater production. Areas that are experiencing land subsidence caused by groundwater pumping may have a combination of high water demands and a lack of economically viable supply alternatives. If excess water is seasonally available, then artificial recharge or aquifer storage and recovery may be a means for controlling subsidence.

Land subsidence is an adverse impact associated with groundwater development that needs to be incorporated into the decision-making process as to what constitutes the safe or sustainable yield of any aquifer (Chap. 9). The socio-economic benefits of groundwater use must be balanced against the costs of land subsidence. The distribution of benefits and costs also needs to be considered. Who benefits from groundwater pumping and who is damaged by land subsidence?

The National Research Council (1991) addressed potential mitigation strategies for land subsidence in the United States. One option to address economic damage is a market-based approach that internalizes the subsidence costs. For example, an extraction tax could be placed on groundwater users that would be used to compensate parties adversely impact by land subsidence.

Another mitigation option is an insurance system. The National Research Council (1991, p. 50) defined an "insurable risk" as

> presenting itself to a sufficiently large and homogenous group of exposure units to permit reasonably accurate prediction of average losses, where the loss is definite but unpredictable in time and place and where there is no threat of a major catastrophe of unknown proportions.

An insurance scheme for land subsidence would likely not meet the requirement of being 'unpredictable in time and place'. With increasing monitoring and modeling capability, the areas most likely to be impacted by land subsidence related to groundwater pumping would become more predictable.

Earth fissuring (Fig. 15.6) is serious local problem in some areas that affects land development. In the construction of new facilities or infrastructure, fissures must either be avoided or mitigated by some means. Fissuring is a particular concern in parts of Arizona as home construction has been increasing in areas prone to fissuring. The most effective method for mitigating fissuring involves sealing the fissure and rerouting drainage to prevent surface flows from eroding fissures into large destructive gullies (Arizona Land Subsidence Group 2007). However, more often old fissures are backfilled with soil and debris which conceals rather than mitigates the hazard (Arizona Land Subsidence Group 2007). Commonly, homeowners are unaware of the existence of concealed fissures on their property until damage is done. Currently there are no technical guidelines or protocols in place in Arizona for defining fissure zones or for developing nearby lands, nor is the phenomenon understood well enough to forecast where or when fissures will occur (Arizona Land Subsidence Group 2007).

Other damage can be mitigated through construction than can accommodate subsidence. For example, the risk of damage to wells can be reduced through the use of compression sections. Compression sections are concentric joints of well casing, in which in the inner joints (of same diameter as main well casing), can move within the outer, larger-diameter shell casing. The compression sections should be located at depth intervals that are prone to compression failure, as determined from the investigation of the failed wells in the project area or from investigation of local geology.

Fig. 15.6 Earth fissure from land subsidence near Picacho Mountain, Arizona, USA. *Source* Carpenter (1999)

No recommendations are made concerning how to balance pumping and subsidence, which ultimately must be based on local circumstances and values. Although reduction in groundwater pumping is the ultimate solution to pumping-induced subsidence, the economic and social costs of reducing pumping may exceed the economic damage from subsidence, especially where alternative water resources are not available. Adaptation to land subsidence may turn out to be less expensive than prevention. Relocation of pumping may help resolve a local land subsidence problem, but the net result could be that subsidence is, in effect, exported to another area. The important issue for water resources management in arid and semiarid regions is the determination of the potential extent of land subsidence and its likely impacts and costs, and having the decision-making process made in full consideration of the data.

References

Arizona Land Subsidence Group. (2007). Land surface and earth fissures in Arizona. *Arizona Geological Society Contributed Report CR-07-C.*

Bawden, G. W., Sneed, N., Stork, S. V., & Galloway, D. L. (2003). Measuring human-induced land subsidence from space. *U.S. Geological Survey Fact Sheet 069-03.*

Behr, J. A. (2001). *Global positioning system (GPS) applications for deformation monitoring.* Geotechnical instrumentation for field measurements, University of Florida's Division of Continuing Education. (www.orionmonitoring.com/Assets/GPS%20Monitoring.pdf).

Bürgmann, R., Rosen, P. A., & Fielding, E. J. (2000). Synthetic aperture radar interferometry to measure earth's surface topography and its deformation. *Annual Review of Earth and Planetary Sciences, 28*, 169–209.

Cabral-Cano, E., Dixon, T. H., Miralles-Wilhelm, F., Díaz-Molina, O., Sánchez-Zamora, O., & Carande, R. E. (2008). Space geodetic imaging of rapid ground subsidence in Mexico City. *Geological Society of America Bulletin, 120*, 1556–1566.

Calderhead, A. I., Therrien, R., Rivera, A., Martel, R., & Garfias, J. (2011). Simulating pumping-induced regional land subsidence with the use of InSAR and field data in the Toluca Valley. *Mexico: Advances in Water Resources, 34*, 83–97.

Carpenter, M.C. (1999). South-Central Arizona. In D. Galloway, D. R. Jones, & S. E. Ingebritsen (Eds.), *Land subsidence in the United States: U.S. Geological survey circular 1182* (pp. 65–78).

Carruth, R. L., Pool, D. R., & Anderson, C. E. (2007). Land subsidence and aquifer-system compaction in the Tucson Active Management Area, south-central Arizona, 1987–2005. *U.S. Geological Survey Scientific Investigations Report 2007–5190.*

Chen, C., Pei, S., & Jiao, J. J. (2003). Land subsidence caused by groundwater exploitation in Suzhou City. *China: Hydrogeology Journal, 11*(2), 275–287.

Cuevas, J. A. (1936). Foundation conditions in Mexico City. *Proceedings of the International Conference on Soil Mechanics, Cambridge, MA 3*, 233–237.

Domenico, P. A., Mifflin, M. D., & Mindling, A. L. (1966). Geologic controls on land subsidence in Las Vegas Valley. In *Proceedings of the Annual Engineering Geology, Soil Engineering Symposium, Moscow, Idaho*, pp. 113–121.

Galloway, D. L., & Burbey, T. J. (2011). Regional land subsidence accompanying groundwater extraction. *Hydrogeology Journal, 19*(8), 1459–1486.

Galloway, D. & Riley, F. S. (1999). San Joaquin Valley, California. In D. Galloway, D. R. Jones & S. E. Ingebritsen (Eds.), Land subsidence in the United States: *U.S. Geological Survey Circular, 1182*, (pp. 23–34).

Galloway, D. L., & Hoffmann, J. (2007). The application of satellite differential SAR interferometry-derived groundwater displacements in hydrogeology. *Hydrogeology Journal, 15*, 133–154.

Galloway, D. L., Hudnut, K. W., Ingebritsen, S. E., Phillips, S. P., Peltzer, G., Rogez, F., et al. (1998). Detection of aquifer system compaction and land subsidence using interferometric synthetic aperture radar. *Antelope Valley, Mojave Desert, California: Water Resources Research, 34*, 2573–2585.

Galloway, D., Jones, D. R., & Ingebritsen, S. E. (1999a) Land subsidence in the United States. *U.S. Geological Survey Circular 1182.*

Galloway, D., Ingebritsen, S. E., Riley, F. S., Ikehara, M. E., & Carpenter, M. C. (1999b). Land subsidence in the United States. In D. Galloway, D. R. Jones, & S. E. Ingebritsen (Eds.), *Land subsidence in the United States: U.S. Geological Survey Circular 1182* (pp. 141–158).

Galloway, D., Jones, D. R. & Ingebritsen, S. E. (2000). Land subsidence in the United States. *U.S. Geological Survey Fact Sheet 165-00.*

Gelt, J. (1992). Land subsidence, earth fissures change. *Arizona's Landscape, Arroyo, Summer 1992, 6*(2).

Hoffmann, J., Zebker, H. A., Galloway, D. L., & Amelug, F. (2001). Seasonal subsidence and rebound in Las Vegas Valley. *Nevada, observed by synthetic aperture radar interferometry: Water Resources Research, 37*, 1551–1566.

Hoffmann, J., Galloway, D. L., & Zebker, H. A., (2003). Inverse modeling of interbed storage parameters using land subsidence observations, Antelope Valley, California. *Water Resources Research, 39*(2), 1031. doi:10.1029/2001WR001252.

Holzer, T. L. & Galloway, D.L. (2005). Impacts of land subsidence caused by withdrawal of underground fluids in the United States. In J. Ehlen, W. C. Haneberg & R. A. Larson (Eds.),

Humans as geologic agents: Geological Society of America reviews in engineering geology (Vol. XVI, pp. 87–99).

Ingebrtitsen, S. E., & Jones, D. R. (1999). Santa Clara Valley, California. In D. Galloway, D. R. Jones & S. E. Ingebritsen (Eds.), *Land subsidence in the United States: U.S. Geological Survey Circular 1182*, (pp. 15–22).

Leake, S. A. (2004). Land subsidence from ground-water pumping. *U.S. Geological Survey.* http://geochange.er.usgs.gov/sw/changes/anthropogenic/subside/.

Massonnet, D., & Feigl, K. L. (1998). Radar interferometry and its application to changes in the earth's surface. *Reviews of Geophysics, 36*(4), 441–500.

National Research Council. (1991). *Mitigation losses from land subsidence in the United States.* Washington, DC: National Academy Press.

Ortega-Guerrero, A., Rudolph, D. L., & Cherry, J. A. (1999). Analysis of long-term land subsidence near Mexico City: Field investigations and predictive modeling. *Water Resources Research, 35*, 3327–3341.

Osmanoğla, B., Dixon, T. H., Wdowinski, S., Cabral-Cano, E., & Jiang, Y. (2010). Mexico City subsidence observed with persistent scatter INSAR. *Journal of Applied Earth Observations and Geoinformation, 13*(1), 1–12.

Pavelko, M. T., Wood, D. B. & Laczniak, R. J. (1999). Las Vegas, Nevada. In D. Galloway, D. R. Jones & S. E. Ingebritsen (Eds.), *Land subsidence in the United States: U.S. Geological Survey Circular 1182* (pp. 49–64).

Phillips, S. P., Carlson, C. S., Metzger, L. F., Sneed, M., Galloway, D. L., Hudnut, K. W., et al. (2002). Optimal management of an ASR program to control land subsidence in Lancaster, California. In P. J. Dillon (Ed.), *Management of aquifer recharge for sustainability, Lisse* (pp. 361–366). A.A: Balkema.

Phillips, S. P., Carlson, C. S., Metzger, L. F., Howle, J. F., Galloway, D. L., Sneed, M., Galloway, D. L., Ikehara, M. E., Hudnut, K. W. & King, N. E. (2003). Analysis of tests of subsurface injection, storage, and recovery of freshwater in Lancaster, Antelope Valley, California. *U.S. Geological Survey Water-Resources Investigations Report 03-40611.*

Poland, J. F. (1961). The coefficient of storage in a region of major subsidence caused by compaction of an aquifer. *U. S. Geological Survey Professional Paper, 424-B,* 52–54.

Poland, J.F. (1970). Land subsidence and aquifer-system compaction, Santa Clara Valley, California, USA. *Proceedings of the Tokyo Symposium on Land Subsidence* (Vol. 1, pp. 285–294). International Association of Scientific Hydrology and UNESCO, September 1969.

Poland, J. F., et al. (1975) Land subsidence in the San Joquin Valley, California as of 1972. *U. S. Geological Professional Paper 437-H.*

Poland, J. F., & Davis, G. H. (1956). Subsidence of the land surface in Tulare-Wasco (Delano) and Los Baons-Kettleman City area. *San Joaquin Valley, California: Transactions of the American Geophysical Union, 37*, 287–296.

Riley, F. S. (1969). Analysis of borehole extensometer data from central California. In L. J. Tison (Ed.), *Land subsidence: International Association of Hydrological Sciences Publication* (Vol. 89, pp. 423–431). Wallingford: IAHS.

Sneed, M., & Galloway, D. L. (2000). Aquifer-System Compaction: Analyses and simulations-the holly site, edwards air force base, Antelope Valley, California. *U.S. Geological survey water-resources investigations report 00-4015.*

Terzaghi, K. (1936). The shear resistance of saturated soils. *Proceedings First International Conference on Soil Mechanics & Foundation Engineering* Cambridge, MA (Vol. 1, pp. 54–56).

Tolman, C. F., & Poland, J. F. (1940). Ground-water infiltration and ground-surface recession in Santa Clara Valley. *Santa Clara County, California: Transactions American Geophysical Union, 21*, 23–34.

Wilson, G., & Grace, H. (1942). The settlement of London due to underdrainage of the London clay. *Journal of the Institute of Civil Engineering (London), 19,* 100–127.

Winslow, A. G., & Wood, L. A. (1959). Relation of land subsidence to groundwater withdrawals in the upper Gulf Coastal region of Texas. *American Institute of Mining Engineers, Mining Division, Mining Engineering, 11,* 1030–1034.

Chapter 16
Surface and Airborne Geophysics

16.1 Introduction

Surface geophysical methods can provide much useful information on local and field-scale hydrogeology, which compliments, but not replaces, more traditional borehole-based field measurements. The principal value of surface geophysical techniques is that they can provide a denser data coverage at a lower cost and in a shorter time period than invasive methods (e.g., ground sampling and borehole techniques). However, there is still a need for borehole geological, water quality, hydraulic, and additional geophysical data to calibrate and validate surface geophysical data. Basic references on surface geophysics and applied geophysics in general include Zohdy et al. (1974), Telford et al. (1990), Eastern Research Group (1993), Reynolds (1997), Sharma (1997), American Society of Civil Engineers (1998), Milson (2003) and Burger et al. (2006).

Surface geophysical methods are particularly valuable in detecting contrasts on subsurface geological and hydrogeological properties, which include changes in lithology and subsurface structure, the contact between aquifer and confining strata, and the boundaries between waters with pronounced differences in salinity (e.g., interface between fresh and saline groundwater).

Some of the specific advantages of surface geophysical techniques are:

- reduction in need for invasive sampling,
- less expense than monitoring well installation and test borings, which allows for a much greater sampling densities,
- some methods (e.g., resistivity-based) measure parameters directly related to hydrogeological parameters of interest (e.g., salinity),
- may allow for interpolation between sparse borehole data,
- quicker than methods requiring drilling of new boreholes, and
- often a less expensive means to collect areally-extensive data sets.

R. Maliva and T. Missimer, *Arid Lands Water Evaluation and Management*, Environmental Science and Engineering, DOI: 10.1007/978-3-642-29104-3_16,

Surface geophysical data are typically interpreted by inversion solutions in which the cause is inferred from effects. Forward solutions, on the contrary, proceed from cause to effect. Inversion solutions essentially involve finding a best fit solution between theoretical geophysical responses generated for various earth models (e.g., configurations of layers and their properties) and the actual field (measured) data. Inversion solutions are not unique and ambiguity exists in the interpretations of all geophysical methods, which can be reduced by an understanding of the geological reality (American Society of Civil Engineers 1998). Thus, surface geophysical interpretations are more accurate if the data processing is based on a sound conceptual knowledge of the strata being evaluated. Field (borehole-based) hydrogeological data are needed for the development of conceptual models and for the calibration and validation of geophysical results. It is also advisable to use several complementary geophysical methods in an integrated exploration program rather than relying upon a single method (American Society of Civil Engineers 1998).

Forward modeling of the geophysical response to various hydrogeological systems can be used to determine in advance if the target is detectable using a given technique and to select and design the appropriate testing procedures for an investigation (Fitterman and Stewart 1986; Mills et al. 1988). It is therefore important to have some prior knowledge of local geology and the physical properties (e.g., resistivity, relative permittivity, and conductivity) of the materials that will be encountered. For example, Minsley et al. (2011) demonstrated the value of performing sensitivity analyses before the application of hydrogeophysical methods to map a subsurface plume of freshwater injected in a proposed ASR system and changes in the effective stress in the storage zone resulting from changes in pore water pressure during the operation of the system. The sensitivity analyses indicted that resistivity methods could differentiate between stored freshwater and native groundwater, but that seismic data could not provide useful information on effective stress. Failure to adequately evaluate whether or not the exploration target is detectable at a project site using various proposed surface geophysical techniques can result in costly and embarrassing failed surveys.

16.2 Electrical Resistivity and Electromagnetic Techniques

No other physical property of naturally-occurring rocks and soils displays such a wide range of values as resistivity (and its reciprocal parameter, conductivity). Hence resistivity-based methods have great demonstrated value for obtaining data on subsurface hydrogeology (Zohdy 1974). The two main classes of resistivity-based surveying methods are electrical resistivity and electromagnetic methods. Both methods are based on the resistance to the passage of an electrical current offered by soil and rock. The magnitude of the resistance depends largely upon the rock and sediment type, porosity (value and degree of interconnection), and the salinity of the pore water. Dry rock has the greatest resistivity, whereas clays and shales, and high-porosity rock filled with saline water have low resistivities.

Electrical resistivity methods provide information on underlying geoelectric layers, which do not necessarily coincide with geological layers (Zohdy 1974). Geoelectric layers are defined by resistivity and thicknesses, the boundaries of which could be caused by differences in salinity, degree of saturation, and rock or sediment type.

The relationship between electrical resistance, potential, and current is expressed by Ohm's law:

$$r = V/I \tag{16.1}$$

where,
r = resistance (ohms)
V = potential (volts)
I = current (amperes).

Resistance can be measured by applying a current to the material in question and measuring the difference in potential. Resistivity is an intrinsic property of a material, which is defined as follows:

$$R = r\,A/L \tag{16.2}$$

where,
R = resistivity in ohm-meters
r = resistance (ohms)
A = cross-sectional area normal to the flow of current (square meters)
L = length (meters).

Equation 16.2 indicates that the measured resistance values must be processed for the geometry of the investigated aquifer volume (A and L parameters) in order to determine resistivity, the parameter of interest.

The resistivity of saturated geologic formation can be quantitatively analyzed using the Archie (1942) equation as follows:

$$\frac{R_o}{R_w} = F = \frac{a}{\varphi^m} \tag{16.3}$$

where,
R_o = resistivity of a 100% water-saturated formation (ohm-m) (true resistivity)
R_w = resistivity of pore waters (ohm-m)
F = formation factor (dimensionless)
φ = porosity (m^3/m^3)
a = constant, usually equal to about 1.0. Both R_o and R_w are at formation temperature
m = constant (cementation factor), which is commonly between 1.8 and 2.0.

Archie's equation, which is most commonly used for the interpretation of borehole geophysical logs, is applicable when the electrical conductivity through interconnected pore space is much greater than all other forms of electrical conductivity in the formation (Kobr et al. 2005). The presence of intergranular clay or clay coatings on grains can invalidate Archie's law, particularly in fresh and brackish water (Kobr et al. 2005). A combination of resistivity sounding and induced polarization can allow for better resolution of groundwater salinity in clay-rich aquifers (e.g., Seara and Granda 1987).

Measured resistivity values are thus a function of both groundwater resistivity and the aquifer or water-bearing strata porosity. A given formation resistivity value can thus be produced by a range of combinations of formation water resistivity (i.e., groundwater salinity) and porosity values. Hence, the importance of having some independent information on local geology, such as approximate aquifer porosity, which can constrain the interpretation of resistivity data.

Resistivity methods are particularly useful for the location of interfaces between materials with sharply different lithologies and resistivities, such as the

- interface between saline and fresh groundwater,
- contact between porous rock and impermeable (very high resistivity) bedrock,
- location of relatively high resistivity freshwater bearing coarse sediments (e.g., channel deposits) amidst less resistive clayey deposits, and
- top of the water table.

A critical consideration is that the target must have a significant resistivity contrast in order for it to be detected. Resistivity-based methods may not be able to detect the water table, if it is not associated with a significant change in resistivity. Resistivity methods are well suited for detecting changes in salinity, and have been demonstrated to be particularly useful for determining the location and shape of coastal saline-water interfaces (Sect. 16.5.1).

16.3 Electrical Resistivity Method

The electrical resistivity method is the oldest and most widely used surface geophysical method applied to groundwater investigations, because it can provide useful information and is cost effective. The application of the electrical resistivity method to groundwater investigations is discussed in detail by Zohdy (1974). Data collection for the electrical resistivity method is relatively simple and inexpensive. An electrical current is applied to the ground using two electrodes. Either a direct current (DC) or low frequency alternating current (AC) is used. In practice DC is now seldom used, and instead most surveys are performed using very low frequency (<20 Hz) AC current, which results in measured resistivities that are essentially the same as direct current resistivities.

The difference in potential (voltage loss) is measured between two additional (potential) electrodes that do not carry current. Typically, the current electrodes and potential electrodes have a collinear arrangement. In vertical electrical sounding (VES) analyses, information on changes in the resistivity of the earth materials with depth at a test site is obtained by performing multiple analyses, in which the electrode spacing at the test site is progressively increased. Horizontal profiles or areal maps are constructed by conducting measurements using the same electrode spacing over a geographic area. Two-dimensional profiles or

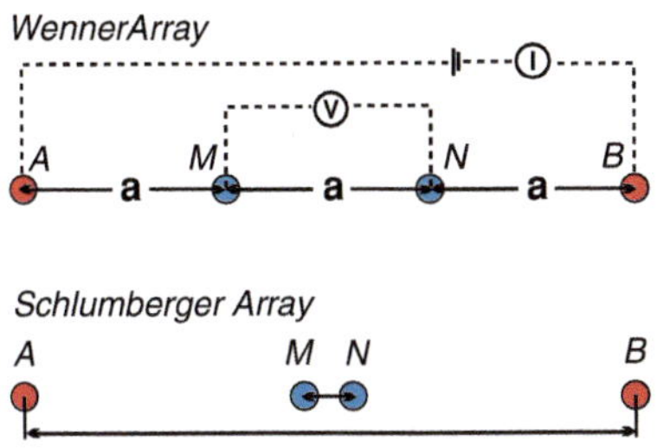

Fig. 16.1 Schematic diagram of the Wenner and Schlumberger electrical resistivity arrays. An electrical current (I) is applied to the current electrodes (A, B) and the potential difference (V) is measured between two potential electrodes (M, N)

three-dimensional diagrams can be obtained by performing vertical soundings at multiple locations along a linear transect or within a grid.

There are variations in the electrical resistivity method, which differ in the configuration (relative spacing of the electrodes). The oldest and conceptually simplest configuration is the Wenner electrode array, in which there is equal distance spacing between the four electrodes (Fig. 16.1). The Schlumberger electrode array also has a symmetric configuration, but differs from the Wenner array in that the distance between the two potential electrodes is much less than the distance between the current and potential electrodes. The dipole–dipole array consists of two closely spaced pairs of current and potential electrodes, in which the potential electrode are located outside, rather than between, the current electrodes. The Schlumberger array allows for quicker analyses because the inner (potential) electrodes are only occasionally moved during a vertical sounding. With the Wenner array, all four electrodes are moved for each analysis.

The raw data for electrical resistivity testing are the electrical current (I), potential difference between the two potential electrodes (ΔV), and the electrode spacing. Geoelectrical data are usually expressed in terms of apparent resistivities (R_a ohm-m, Ω-m) as follows:

$$R_a = K\frac{\Delta V}{I} \tag{16.4}$$

where,
K = geometrical factor (dimensionless)
ΔV = potential difference (volts)
I = current (amperes).

The general equation for the geometrical factor (K) is (Zohdy 1974)

$$K = 2\pi\frac{1}{\frac{1}{\overline{AM}} - \frac{1}{\overline{BM}} - \frac{1}{\overline{AN}} + \frac{1}{\overline{BN}}} \tag{16.5}$$

where, AM, BM, AN, and BN are the electrode spacings (m).

For the Wenner array, wherein the electrode spacings (a) are equal, apparent resistivity can be calculated as

$$R_a = 2\pi a\,\Delta V/I \tag{16.6}$$

Apparent resistivity is calculated using the Schlumberger array as (Zohdy 1974):

$$R_a = \pi \frac{\left(\frac{\overline{AB}}{2}\right)^2 - \left(\frac{\overline{MN}}{2}\right)^2}{\overline{MN}} \left(\frac{\Delta V}{I}\right) \tag{16.7}$$

Calculated apparent resistivity (R_a) values do not directly relate to any property of the underlying sediment and rock. The depth of investigation increases with electrode spacing, or more precisely the distance between the current and potential electrodes. At greater depths of investigation, the vertical resolution decreases as the electrode spacing increases. Electrode spacing is expressed as one-half the distance between the current electrodes (AB/2). There is no universal quantitative relationship between electrode spacing and depth of investigation. An often used 'rule of thumb' that assumes that electrode spacing (AB/2) is equal to the depth of investigation is incorrect (Zohdy 1974).

Electrical resistivity data are amenable to useful qualitative evaluation. The shape of the sounding curve provides general information about the relatively resistivity of underlying geoelectric layers. Maps of apparent resistivity for horizontal surveys can be used to identify variations in subsurface hydrogeology, such as the presence of coarser-grained channel deposits and structural features (e.g., Zohdy 1974; Driscoll 1986). Sharply increasing resistivity and high values (>10 Ω-m) at the bottom of a sounding may be indicative of impermeable bedrock. Cross-sections of apparent resistivity versus electrode spacing can provide insights to the location of the major lithological boundaries or salinity changes, such as the coastal saline-water freshwater interface.

The raw data for resistivity soundings is presented as a plot of apparent resistivity versus current electrode spacing (AB/2), which are normally plotted on logarithmic scales (Fig. 16.2). Quantitative analysis of surface resistivity data is performed using inverse modeling techniques, which involves the comparison of measured sounding curves to calculated theoretical sounding curves. The theoretical sounding curves are generated based on conceptual earth models that can incorporate different number of layers, layer thicknesses, and layer resistivities. Distinct types of curves have been defined with the shape depending on the number and relative resistivity of the geoelectric layers (Zohdy 1974). The curve matching was originally performed manually using catalogs of standard curves, but is now done with computers, which is much faster and accurate, and allows for evaluation of a much wide range of conceptual earth models. Commercially available inversion software for processing electrical resistivity data include EarthImager 1D, 2D, and 3D packages produced by Advanced Geosciences Inc. and RES2DINV produced by Heritage Geophysics. The inversion processes

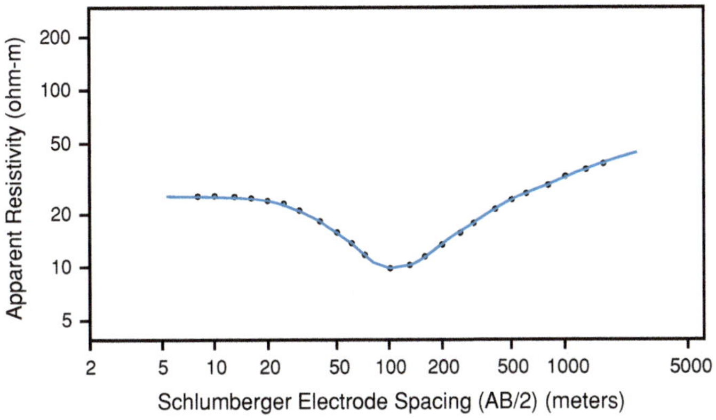

Fig. 16.2 Example of Schlumberger sounding curve (Zohdy 1974)

involves iterative schemes in which the model is continuously updated until an acceptable match (specified error level) is reached between the field data and model.

The resistivity of specific sediment or rock types can vary based on localized concentration of various mineral assemblages. To enhance accuracy of the resistivity data interpretation the electrical properties of sediment or rock samples (specifically the resistivity, dielectric permittivity and permeability), can be measured in the laboratory. Artificial mixes of sediment types from the area of investigation can be made and their properties measured to improve the accuracy of the geoelectric profile inversion interpretations. The electrical properties of unlithified sediment are affected by compaction during burial, which affects the bulk porosity. The effects of compaction can also be assessed in the laboratory by compressing the sediments in a piston apparatus and then measuring the electrical properties of the samples. Simulated synthetic profiles can then be developed for comparison with field data to improve interpretation. This information can also be used to calibrate electromagnetic and ground-penetrating radar data sets, which also are based on the geoelectrical properties of the sediment. The key point is that more information on local geology and hydrogeology and the geoelectric properties of the local sediment and rock, allows for better constrained and thus more accurate inversion solutions.

Once bulk resistivity has been interpreted using inversion techniques, groundwater resistivity and salinities can be determined using Archie's (1942) law if some data are available on porosity or groundwater salinity from boreholes or laboratory measurements. Borehole resistivity, salinity, and porosity data from points along resistivity soundings and profiles are used to calculate formation factors, which can be used to interpret bulk resistivity data, provided that the geology is relatively homogenous or well constrained (e.g., Wilson et al. 2005).

The limitations on use of the surface electrical resistivity method include non-uniqueness of inversion solutions and, in general, a shallow depth of investigation for hydrogeological investigations. However, vertical electrical soundings have been performed to depths of 500 m (1,640 ft) or greater (e.g., Worthington 1977). In order to achieve great sounding depths, very large (≥1,000 m; ≥3,280 ft) electrode spacings and increased current are required. Electromagnetic methods (e.g., time-domain electromagnetic induction (TDEM); Sect. 16.4) are usually used instead for deep soundings. Electrical resistivity analyses can be impacted by a variety of extraneous factors, which are discussed by Zohdy (1974). Some of the more important extraneous factors are buried metal objects in investigation areas, discontinuous layers, and equipment problems such as current leakage from poorly insulated cables.

Many different layered models may produce a similar apparent resistivity curve. Worthington (1977, p. 840) noted that

> A quantitative interpretation of geoelectrical sounding data in the form of horizontally stratified layers of different resistivity is generally not meaningful without some reliable stratigraphic control, or some data relating to the intrinsic resistivities of the various geoelectrical units, at several places within the survey area.

Interpretation of vertical sounding data (either DC resistivity or TDEM) is also greatly complicated if significant lateral heterogeneity in resistivity occurs near the sounding site. In order to determine which model is most representative of local hydrogeological conditions, collection of some independent data is necessary. Borehole resistivity logs from one or more boreholes in the study area, ideally near a VES, are very useful for constraining interpretations of resistivity data.

Electrical resistivity surveys were initially performed by installing pairs of current and voltage electrodes, taking manual readings, and then moving the electrodes to another survey site. A limitation of the conventional VES method is that it is labor intensive (and thus expensive) to continuously move the electrodes in order to perform an array of closely spaced soundings. Conventional VES methods are still used to obtain point resistivity versus depth profiles. However, the current state of the art is multi-electrode, continuous vertical electrical sounding (CVES) systems and the interpretation of the data in the form of electrical resistivity tomograms.

A prototype CVES system was described by Van Overmeeren and Ritsema (1988). Multicore cables are used with electrode take-outs at evenly spaced intervals, commonly 5 m; 16.4 ft. The cables are rolled out in the direction of the profiles. The electrodes are individually addressable, so it is possible through the selection of electrodes to vary the location and depth of investigation of each reading, as well as the electrode array type (e.g., Wenner, Schlumberger, dipole, and gradient). The Wenner array or a combination of Wenner and Schlumberger arrays are typically used because of the constant electrode spacing. If dedicated electrodes are used, time lapse resistivity profiles could be performed to monitor temporal changes in salinity (De Franco et al. 2009) and aquifer recharge (Venessa et al. 2011).

CVES systems are controlled by a portable computer that is programmed to control which electrodes are energized as current electrodes and which electrodes

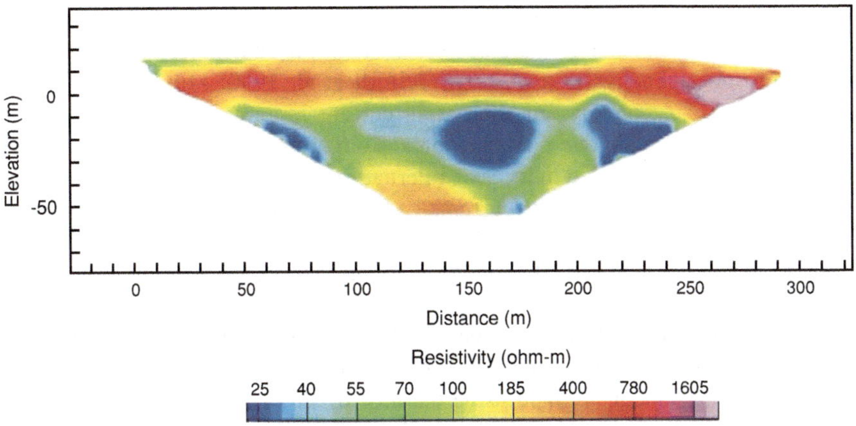

Fig. 16.3 The example of a DC-resistivity cross-section (tomogram). High-resistivity unsaturated sediments and saturated sands and gravels containing low specific conductance groundwater are underlain by lower resistivity clays and sediments containing higher specific conductance groundwater. *Source* Degnan and Brayton (2010)

collect voltage data. Rapid readings can be performed to collect data at different locations and depths along the electrode cable transect. The collected data for each reading include, at a minimum, the electrode geometry, measured current, voltages and spontaneous potentials. The data are pre-processed to obtain apparent resistivity values. Tomographic profiles (i.e., cross sections) of apparent resistivity data are processed using inversion techniques to obtain an electrical resistivity tomogram of bulk resistivity versus depth (Fig. 16.3). Three-dimensional surveys are possible if the electrodes are arranged in a grid.

Self-potential (SP) is a naturally occurring electric potential difference of the subsurface, caused by charge separation in clay or other minerals. Usually non-polarizable electrodes are used to measure SP signal. During pumping test experiments the flow of groundwater is responsible for measurable electrical field at the ground surface owing to the electrokinetic coupling between the Darcy velocity and the electrical current density (Rizzo et al. 2004). For instance, Titov et al. (2005) reported that for an unconfined aquifer the experimental hydraulic heads and SP signal associated with a pumping test can be reproduced with a numerical model based on the finite-difference method.

16.4 Electromagnetic Surveys

Electromagnetic induction methods induce currents within the Earth without having to have direct contact with the ground. The methods can thus be performed either on the ground or by airborne surveys. Electromagnetic induction methods are based on Faraday's Law, in which a changing magnetic field can generate an electrical current, and Ampere's Law, in which a changing electric field can generate a magnetic field. The basic theory of electromagnetic induction is discussed in most geophysics texts (e.g., Telford et al. 1990; Milson 2003; Burger et al. 2006). Conceptually, an alternating current loop at or above land surface induces a primary magnetic field that spreads out above and below the transmitter. The primary magnetic field induces an alternating current in subsurface conductors that are referred to as "eddy currents". The eddy currents, in turn, induce a secondary magnetic field that differs from the primary magnetic field in both amplitude and phase. The EM receiver is a loop that detects both the primary and secondary magnetic fields.

Electromagnetic methods can be divided into transient (TEM), frequency-domain (FEM) and continuous wave (CW) systems. Frequency-domain techniques induce a primary magnetic field using the flow of sinusoidal alternating current at either one or more frequencies in a wire or coil. Transient systems, on the contrary, induce electromagnetic fields through transient pulses of electric current. TEM methods tend to have greater depths of investigation and have been more widely used in groundwater investigations to obtain data on subsurface lithological and salinity changes. FEM systems are used more commonly for investigations of shallow geology and hydrogeology.

16.4.1 Frequency-Domain Electromagnetic Soundings for Shallow Data Collection

A simple, widely used type of FEM system consists of two horizontal co-planar coils (transmitter and receiver), connected by a shielded cable, that are a fixed distance apart. The cable connecting the transmitter and receiver provides a reference signal used to calculate the induced secondary field. Systems using horizontal co-planar coils are often referred to as Slingram systems. For small portable systems, the coils are attached to the ends of a rigid pole, and the entire apparatus can be carried by a single person. The depth of penetration is related to the coil separation, with a greater separation resulting in a greater effective depth of investigation. The depth of penetration of electromagnetic systems also depends on the conductivity of the surface, with a low conductivity lithology resulting in a smaller depth of investigation as compared to a high conductivity lithology. A generalization is that the maximum depth of penetration of Slingram systems is roughly equal to twice the coil separation, but in practice the depth is often less (Milson 2003).

The basic underlying principle of FEM is that the magnitude and phase of the induced electromagnetic currents are related to subsurface electrical conductivity. Subsurface electrical conductivity is, in turn, related to soil and rock lithology, degree of saturation, and water salinity. The primary advantages of FEM methods are that they do not require ground contact and the measurements can be performed quickly, which allows for the economical coverage of large areas. FEM methods are particularly suitable for mapping lateral changes in shallow subsurface geological and hydrogeological conditions. As is the case with DC resistivity data, the measured conductivity values are integrated values over a volume of soil or rock. Inversion techniques are, therefore, necessary in order to quantitatively evaluate subsurface hydrogeology. However, often spatial variations in conductivity are of primary interest, rather than the actual electromagnetic conductivity values. For example, FEM surveys can be used to map the location of subsurface channels if they are less conductive (more resistive) than surrounding clay-rich strata. FEM surveys are susceptible to interference from power lines and metallic features such as metal pipes and fences. However, FEM surveys are commonly performed specifically to locate subsurface metallic objects, such as underground storage tanks.

16.4.2 Transient Electromagnetic (TDEM) Sounding Methods for Deep Data Collection

The major technical challenge of electromagnetic methods is to separate the secondary magnetic field signal from the stronger primary field. The time-domain electromagnetic method (TDEM) is the most widely used electromagnetic technique for groundwater investigations. TDEM achieves separation of the primary and secondary field signals by using a pulsed electrical signal and measuring the decay of the secondary field after the primary field is turned off. The decay rate of the secondary field is inversely proportional to conductivity with highly conductive beds having slower decay rates. Either spatially separate transmitter and receiver coils are used or a small receiver coil is placed in the center of a larger transmittal coil to perform field soundings.

The size of the transmitter loop and applied current can be determined by forward modeling using layered Earth computer programs, which allow for the evaluation of whether the physics of TDEM soundings will allow for detection to the target depth (American Society of Civil Engineers 1998). TDEM surveys with 100 m transmitter loops have been used to obtain estimates of resistivity down to depths of several hundred meters, which would require electrode arrays several kilometers in length if conventional DC resistivity methods were to be used (Milson 2003). The main disadvantages of TDEM compared to DC resistivity are that the equipment is more expensive and the interpretation methods are more complex (Fitterman and Stewart 1986).

The raw data are processed using inversion techniques. Hypothetical simple (few layered) earth models are generated and the theoretical response for the model is calculated. The model is refined until the theoretical response matches the observed or measured field response (Fig. 16.4). As is the case for inversion techniques in general, the TDEM data processing method does not yield a unique

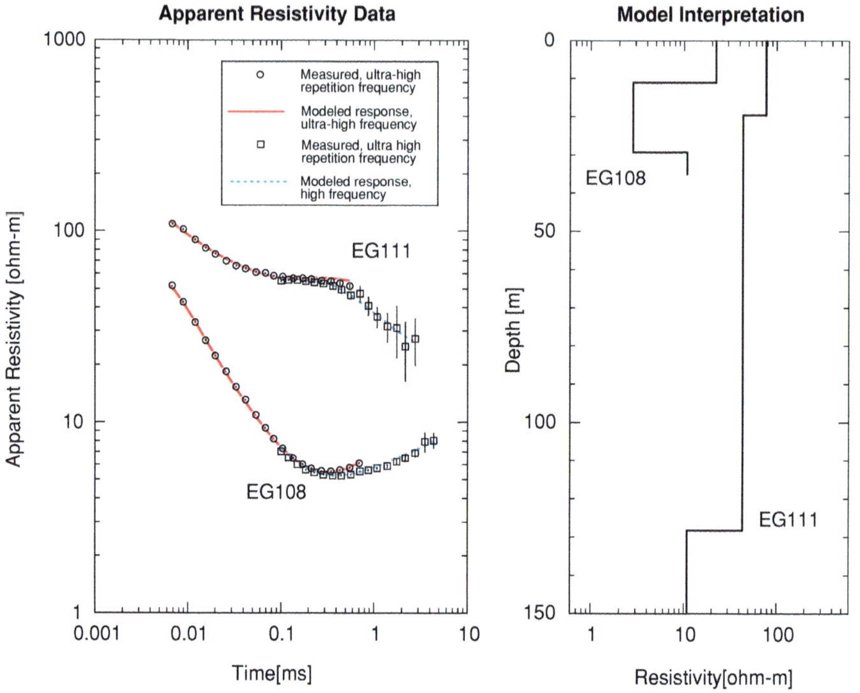

Fig. 16.4 Example of TDEM data interpretation from a survey performed in the Everglades National Park, Florida, USA (Fitterman et al. 1999). Inversion techniques are used which involve matching modeled theoretical responses for hypothetical earth models to the measured data

solution. In studies involving numerous TDEM soundings, the results should be ground-truthed against resistivity (salinity) data from one or more wells located in the study area. TDEM is also a relatively coarse technique. The minimum thickness zone that can be resolved by TDEM is several orders of magnitude greater than what can be resolved by borehole electric logs.

The great advantage of TDEM is that it is by far a less expensive means of obtaining estimates of water quality changes with depth compared to drilling and testing of a well, but the cost and time savings come at the expense of accuracy. Some of the assumptions and limitations of the TDEM method include

- The TDEM method assumes horizontal homogenous layers. Horizontal heterogeneities can result in erroneous interpretations (e.g., Taylor et al. 1992).
- Care must be taken to avoid interference from conductive items (metal) in the sounding vicinity. Localized good conductors, such as buried metallic objects or sulfide ore bodies, generate eddy currents that dominate decay curves and prevent valid depth soundings (Milson 2003).
- TDEM works best for identifying and locating sharp contrasts in resistivity, such as those occurring at major lithological changes or significant changes in salinity at the freshwater-saltwater interface. TDEM is better suited for detecting highly

conductive targets (e.g., saline water) than poorly conductive targets (Fitterman and Stewart 1986).

For example, in a TDEM survey performed in Florida, the approximate depth of the 5,000 mg/L chloride concentration interface (isochlor) could be located because it typically occurs in zone where there is a sharp change in salinity, and thus resistivity, with depth. The depth of the 250 mg/L isochlor, for which there is not an associated sharp change in resistivity, could only be estimated based on empirical relationships between the location of the 250 mg/L isochlor and the saltwater interface (Subsurface Detection Investigations, Inc. 1995) .

TDEM surveys provide an estimate of formation resistivity; the same parameter measured with greater accuracy and resolution in borehole resistivity logs. To obtain an estimate of salinity, formation water resistivity is calculated using Archie's Law (Sect. 14.5.4), which requires knowledge of porosity, the empirical constant, and the formation factor. Formation water resistivity is then converted to chloride or TDS concentration using an aquifer-specific functional relationship between water resistivity and salinity parameters. In the absence of porosity versus depth data, an assumed porosity is used, the appropriateness of which will affect the accuracy of calculated resistivity values. TDEM soundings give reasonably accurate average depths to sloping interfaces and are less sensitive to variable surface topography compared to other surface geophysical methods (American Society of Civil Engineers 1998).

16.5 Applications of Resistivity-Based Surface Geophysics to Groundwater Investigations

Resistivity-based surface geophysical techniques continue to have a wide range of applications for groundwater resources evaluation, especially in arid lands. Their greatest value has been in applications involving the mapping of differences in the salinity of groundwater and the depth to groundwater (position of the water table). Resistivity-based techniques can also provide useful information on the hydrogeology (location of confining units and aquifers) and subsurface structure. The wide variety of resistivity-based surface geophysical surveys performed to date provides guidance as to possible future applications.

16.5.1 Mapping of Saline-Water Interface

A very important application of surface geophysics is mapping vertical and horizontal saline-water/fresh-water interfaces. The exact position of the saline-water/fresh-water interface in most coastal areas is poorly constrained on a local level (Chap. 25). A combination of monitoring well water quality measurements, borehole geophysical data, and surface geophysics can be used to accurately map the position of the interface. Resistivity-based methods are well suited for the vertical and horizontal mapping of the interface because of the large difference in resistivity between freshwater and saline water.

Zohdy (1974) provides references to some early (1950s and 1960s) case histories, including documentation of a U.S. Geological Survey investigation at the White Sands Missile Range in New Mexico. Electrical soundings were successfully used to map the base of the freshwater aquifer. Investigations of the saline-water/fresh-water interface using DC resistivity methods are too numerous to document in herein. Some more recent studies which demonstrate the potential applications of the technology are discussed.

The saline-water interface was mapped in part of coastal west-central Florida using the DC resistivity method with Wenner arrays (Fretwell and Stewart 1981). The study involved 28 soundings in a karstic limestone aquifer. The Fretwell and Stewart (1981) study illustrates that multiple interpretations of resistivity data may be plausible. An intermediate depth low resistivity zone could be either (1) a more permeable zone sandwiched between less permeable zones, (2) a tongue of saline water sandwiched between freshwater water zones, or (3) a low resistivity clay lying between more resistive limestones. The study results demonstrated that the saline-water interface could be mapped in karstic limestones using DC resistivity and could provide hydrologic information between wells used for control.

TDEM was successfully used to delineate the position of the 500 mg/L chloride contours (isochlors) in coastal Monterey County, California. One hundred soundings were performed during 3 weeks of field work. TDEM was effective in determining the salinity stratification with depth, but ground-truth data were required to correlate formation resistivities to the concentrations of dissolved solids in the groundwater. Mills et al. (1988) stressed the importance of field data, such as water chemistry data from monitoring wells and borehole geophysical logs, in order to obtain the full benefits of the surface geophysical data.

VES soundings were used to evaluate the hydrogeology of the coastal plain of the Wadi Surdad area of western Yemen, for which there were limited data (Van Overmeeren 1989). A total of 219 soundings were performed using Schlumberger arrays with maximum electrode spacings (AB/2), in most instances, of between 700 and 900 m (2,300 and 2,950 ft). An important lesson of the study is that it demonstrated the principle of equivalence; many different layered models may produce practically the same resistivity curves. Two geoelectrical models were produced that could explain the sounding curves, which varied based on whether the depth to the interface between the aquifer and underlying basement was used as a fixed value in the inversion process or whether the resistivity of the aquifer material was assigned a fixed value. Borehole geophysical log data were used to differentiate between the geoelectrical models.

A combination of VES, seismic refraction surveys, and TDEM were used to map the position of the saline-water interface in Wadi Al Hawasinah, located along the Batinal coast of northern Oman (Abdalla et al. 2010). Comparison of the positions of the saline-water/fresh-water interface between 2002 and 2007 indicate that the interface had receded (migrated seawards) by about 600 m (1,970 ft) over the 5-year time period, which corresponds to an average rate of 120 m/year (394 ft/year). The recession of the interface was attributed to increased recharge induced by a wadi dam and regulation of groundwater pumping (Abdalla et al. 2010).

The profiles of the saline-water interface were generated using the DC resistivity method at the outlet of Wadi Ham along the Fujairah and Kalbha coast of the UAE. Four 2D surveys were generated using Wenner arrays. The resistivity data were processed to map areas of fresh, brackish, and saline groundwater (Sherif et al. 2006).

Vertical electrical (resistivity) soundings and seismic reflection profiling were used in northeast Malaysia to map salinity in a shallow coastal aquifer system (Samsudin et al. 2008). A Schlumberger array was used, and the apparent earth resistivity data versus electrode spacing data were interpreted using inverse resistivity software (RESIXP software developed by Interpex, Inc.). The VES and seismic survey data could discern sand aquifers and clay confining units and differentiate between fresh and brackish groundwater. CVES was also successfully used in northwest Malaysia to generate tomograms that discerned the boundary between fresh and saline groundwater (Abdul Nassir et al. 2000).

VES was used in the Mamora coastal plain of Morocco, north of Rabat, to determine geological structure and the depth to the base of the surficial aquifer, and to assess horizontal and vertical salinity variations (Benkabbour et al. 2004). The study included over 60 VES measurements using Schlumberger arrays. The studied aquifer consists of Pliocene calcarenites (cemented carbonate sands) and overlying Quaternary sandstones and sands. The VES data analysis allowed the top of the saturated zone to be mapped along with the base of the aquifer, which is marked by a downward transition to less resistive Miocene marls. The VES data allowed for the mapping of areas most impacted by saline water intrusion.

VES and CVES were used to locate the position of the saline-water interface at a study area on the Kapiti Coast on the North Island of New Zealand (Wilson et al. 2005). VES was used to determine the approximate position of the interface and CVES was used to refine the location and shape of the interface. Archie's (1942) formation factor was estimated using bulk resistivity data from the CVES profiles and borehole conductivity data. Diffuse mixing involving as little as 1% seawater could be tracked by a significant reduction in bulk resistivity.

Time-lapse resistivity tomography was performed in the Venice Lagoon area of Italy (De Franco et al. 2009). The electrical resistivity tomography (ERT) program was set up to perform 10 resistivity tomographic measurements per day over a 9 month period starting in November 2005. The time-lapse ERT program was able to detect spatial (horizontal and vertical) and temporal variations in saline-water intrusion into a vulnerable aquifer system. The technique was effective to detect relationships between saline-water intrusion and environmental parameters. Seasonal variations in water quality were evident on the tomograms. Some issues that occurred during the 9 month time lapse period were development of oxidation films on the electrode takes and vegetation causing a poor stake—soil coupling.

16.5.2 Depth to the Water Table

DC resistivity and electromagnetic methods can be used to locate the approximate position of the water table. Where there is a sharp transition from unsaturated to saturated conditions, the water table may be a distinct geoelectric boundary that is marked by a downward change from relatively highly resistive dry sediment or rock to less resistive saturated strata. In arid lands, where the water table may be quite deep, the power inputs and sensitivity of the methods should be ascertained prior to using the equipment in the field to measure the water table depth. Forward modeling can be used to test the ability of a given instrument or array to penetrate to the depth necessary to meet the objectives of the investigation. Resolution of the water table using resistivity-based methods may be poor in areas where there is a gradual transition in volumetric water content to 100% saturation at the water table.

16.5.3 Mapping of Freshwater Lenses

Resistivity methods can be used to locate freshwater lenses, particularly where there is a sharp resistivity contrast between the freshwater saturated strata, overlying desiccated strata, and underlying strata containing saline waters. Young et al. (2004) documented the use of TDEM soundings to detect and map freshwater lenses in karstic limestones in Central Oman. Freshwater lenses form within regional aquifers containing saline water as a result of episodic focused recharge. TDEM soundings allowed detection of fresh water accumulations in otherwise saline-water containing, electrically conductive limestones. Positive resistivity anomalies were detected over the freshwater lenses. Young et al. (2004), however, suggested that the anomalies may be due to the removal of clays from the formation by recharging water rather than by the recharge of fresh water itself. The TDEM data could detect the base of the freshwater lenses (i.e., contact with underlying saline waters). However, the top of the lenses and their thicknesses could not be resolved. The Young et al. (2004) Oman study documents the value of the TDEM survey data as a relatively low cost reconnaissance tool to locate potential freshwater resources, which could be further assessed using test wells.

16.5.4 Managed Aquifer Recharge Investigations

Resistivity based surface geophysical methods can be used to map the geographic extent of freshwater injected to brackish and saline-water aquifers. Minsley et al. (2011) performed an assessment of the applications of hydrogeophysical methods for ASR, based on a proposed system in Kuwait. A synthetic aquifer model was used to simulate the advection and dispersion of injected freshwater. The salinity distribution obtained from the solute-transport modeling was converted to changes in bulk resistivity values. Forward modeling was then used to estimate DC resistivity responses. The results demonstrated that resistivity based methods (DC resistivity and TDEM) are sensitive enough to detect the freshwater plume.

16.5.5 Formation and Aquifer Mapping

Changes in lithology and water saturation are often associated with substantial changes in formation resistivity, which can be detected using areal DC resistivity and electromagnetic surveys and soundings. Surface resistivity has been used to differentiate between coarse-grained aquifers (e.g., channel deposits) within or between finer-grained semi-confining strata, the contact between dry and moist deposits, and the tops and bottoms of aquifers. Information on subsurface geology obtained from DC resistivity and electromagnetic surveys can be used to make more informed decisions concerning the best locations for either test production wells.

Worthington (1977) documented the use of DC resistivity surveys to delineate the most promising sites for future groundwater development in the Kalahari Desert of what is now Namibia. The investigation included 210 soundings using Schlumberger arrays with current electrode separations of 750 m to 7 km (2,460–22,960 ft). The study is noteworthy because of the very large electrode spacings that were used, and thus depths of investigation, which were estimated to be 250–800 m (820–2,624 ft). The base of the sedimentary basin could be identified by the high resistivity of the basement rock. Correlations were observed between the geoelectrically-derived thickness, resistivity, and transverse resistance of the main aquifer (Middle Kalahari) and the yield of boreholes. The main value of the resistivity survey method is that it could allow for the optimization of the initial stages of groundwater exploration in arid and semiarid regions, in general, by increasing the probability of successful wells. Worthington (1977) reported that the success rate of wells in the Kalahari Basin was then less than 40% without use of geophysical methods.

TDEM can be used to evaluate gross aquifer lithologies if there are significant resistivity differences between the lithologic units. Fitterman et al. (1991) were able to use TDEM to map buried paleochannels in alluvial deposits in Abu Dhabi. The channel deposits are important because they serve as both storage reservoirs and recharge zones for the important local shallow aquifer systems. The TDEM data could be used to identify the base of the channels, but not the water table. The transition from partial to full saturation is gradual and not defined electrically. Fitterman et al. (1991) also demonstrated the importance of using auxiliary data, such as well logs, to a construct a geophysical model used for the interpretation of the data. Initial geophysical interpretations made without the benefit of well information failed to identify the resistive gravels that form the main body of the aquifers. The gravels were detectable using a refined model based on well data.

Taylor et al. (1992) documented the detection of faults in an alluvial piedmont aquifer north of Reno, Nevada, using closely spacing TDEM soundings. Faults could be detected by contrasts in resistivity, but not by the actual values. The TDEM survey data provided guidance concerning the extent of areas that likely have significant groundwater resources and thus reduced the number of wells required to characterize the resources.

TDEM soundings were used to broadly define aquifer zones in an alluvial aquifer in Oman (Young et al. 1998). The main aquifer zones, which are composed of clean gravels, clayey gravels, and cemented gravels, could be identified, along with the basal contact of the aquifer. The TDEM data could also be used to map the depth to the interface between freshwater and saline-water along the coast. The TDEM data could not be used to accurately map the top of the water table. The total thickness of the saturated gravel aquifer could be mapped from the base of the upper resistive TDEM layer (upper clean gravels) and water table depth data obtained from wells.

16.5.6 Mapping of Recharge Areas

In some settings, saturated shallow sediments may be distinguished from dry sediments by differences in resistivity. Resistivity profiles and soundings were used to map the location of deep percolation and recharge in the Amargosa Desert Basin of Nevada, USA (Abraham and Lucius 2004; Stonestrom et al. 2007). Numerical inversions of automated multi-electrode Schlumberger-array soundings were used to identify three categories of alluvium

1. low water content coarse gravel and high desiccated surface material ($R > 200\ \Omega$-m),
2. moist alluvium ($R < 20\ \Omega$-m), and
3. other low to low-medium water content alluvium in areas without active recharge (intermediate values of R).

The DC-resistivity data revealed areas of low resistivity, and thus high water content, beneath the ephemeral stream channels (Fig. 16.5). Deep percolation and active recharge were narrowly focused beneath the Amargosa River channel. Recent recharge appears to be negligible in the inter-channel areas.

16.5.7 Mapping of Regional Aquifer Flow Orientation (Fractured Rock Aquifers)

Groundwater flow in hardrock areas is dominated by fractures. Preferred orientation of fractures can result in hydraulic and electrical anisotropies in aquifers. Surface azimuthal resistivity (using DC resistivity and EM methods) has been used to

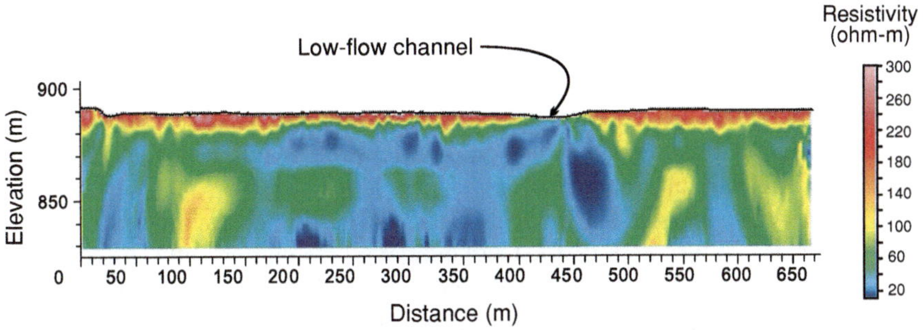

Fig. 16.5 DC-resistivity cross section from the Amargosa Desert, Nevada, in which focused recharge is evident under an ephemeral channel by lower resistivitiies, which reflect a greater moisture content (Stonestrom et al. 2007)

determine the orientation of fractures in aquifers (e.g., Skinner and Heinson 2004). Variations in electrical resistivity are measured as a function of geographic orientation, which is then used to determine preferred fracture orientation, and thus, groundwater flow direction. Using DC resistivity as an example, a surface azimuthal resistivity survey can be performed by rotating a standard collinear four-electrode array about a mid-point. Different depths can be investigated by changing the electrode spacing. The apparent resistivity data requires processing using two or three-dimensional models in order to calculate real earth resistivities and actual anisotropy direction. Skinner and Heinson (2004) demonstrated the utility of both DC and EM resistivity in providing a non-invasive determination of the orientation of maximum hydraulic connection in a fractured rock aquifer in the Clare Valley, South Australia.

16.6 Ground Penetrating Radar

Ground-penetrating radar (GPR) is a geophysical method that uses short bursts of electromagnetic radiation in the microwave (radio) band to detect subsurface structures. GPR is discussed in most surface geophysics textbooks (e.g., Telford et al. 1990; Reynolds 1997; Milson 2003; Burger et al. 2006), and is reviewed by Baker et al. (2007) and Jol (2009). The basic principle is that an electromagnetic wave generated by a transmitting antenna travels through the tested soil or rock at a velocity that is primarily a function of the permittivity (dielectric constant) of the material. When the wave hits an object or layer with a different permittivity, part of the signal is reflected to the surface and is detected by a receiving antenna. Part of the wave energy also continues to travel downward, and may be reflected back by deeper reflectors. The digital control unit of the GPR systems records the reflections as two-way travel times (usually in nanoseconds). The data are processed and plotted as two-way vertical travel time (y-axis) versus horizontal distance (x-axis).

The electrical conductivity of a material influences penetration depth. Low-conductivity materials, such as unsaturated and coarse-grained sediments and solid rock, cause little attenuation and have relatively great penetration depths. Saturated sediments, on the contrary, have relatively high electrical conductivities and permittivities. The water table may thus be a strong GPR reflector due to the difference in electrical properties.

GPR has the great advantage that it can be performed rapidly, is relatively inexpensive to run, has high vertical and horizontal resolution, and is usually close to real time interpretation. GPR has a number of applications for the evaluation of shallow (unconfined) aquifers. GPR can be used to map the depth to bedrock, depth to the water table, and the thickness of soil. It can be used to map buried channels, if there is a significant difference in resistivity between channel and adjoining sediments or rock. GPR is very commonly used to locate buried objects such as pipes, tanks, and drums. It is widely used in archaeological investigation to located subsurface structures.

Another application of GPR is the mapping of soil moisture, which is discussed by Huisman et al. (2003) and Lunt et al. (2005). Dielectric permittivity, and thus electromagnetic wave velocity, is a function of soil moisture. Dielectric permittivity values obtained from GRP can be converted to volumetric water content using either site specific or general empirical relationships. Electromagnetic wave velocities are calculated from GPR travel time if the travel distance is known. For example, cross-borehole measurements use the distance between the boreholes containing the transmitter and receiver to calculate a travel velocity from the travel time between wells.

A technique that has wider potential applicability is to use the two-way travel time associated with a geological surface. For example, Lunt et al. (2005) used the two-way travel time from land surface to a clay bed reflector to obtain the average dielectric permittivity of the soils above the reflector and, in turn, estimate the average water content of the soils. The limitation of this method is that it requires the existence of a reflector at a known depth across the study site. The great advantage of the use of GPR for the measurement of volumetric water content is that it is a noninvasive method and can cost-effectively provide a greater spatial density of measurements than is practicably possible using conventional point-measurement methods (e.g., gravimetric, neutron probe, and time-domain reflectometry techniques).

Recently, coupled hydrogeophysical inversion schemes have been tested for time-lapse GPR measurements to remotely estimate the unsaturated soil hydraulic properties (Kowalsky et al. 2005; Jadoon et al. 2008; Lambot et al. 2009). A major advantage of this approach is that it provides physics-based regularization of the geophysical model and only those permittivity and conductivity profiles are selected that obey the hydrodynamic laws.

16.7 Surface Nuclear Magnetic Resonance

Nuclear magnetic resonance (NMR) methods measure the hydrogen present in a formation as water. Surface NMR, also referred to as magnetic resonance sounding, (MRS), is a technique of growing interest because it is the only method that can directly detect freshwater in the subsurface. Within saturated formations, MRS measures the free water, and thus provides a measure of porosity. MRS has the potential to provide quantitative data on pore sizes, which in turn can provide an estimate of hydraulic conductivity. NMR is an established tool for borehole geophysical logging (Sect. 17.12). MRS has already passed the experimental stage and is evolving into a useful tool for hydrogeophysics (Yaramanci and Müller-Petke 2009). However, there is still more development to be performed in order for MRS to become a standard surface geophysical tool.

MRS theory is discussed by Schirov et al. (1991); Goldman et al. (1994), Sushakov (1996), Yaramanci (2000); Legchenko et al. (2002, 2004), Roy and Lubczynski (2003) and Yaramanci and Müller-Petke (2009). The basic concept of MRS is that a current loop at land surface is used to generate an excitation pulse that causes hydrogen protons to precess around the local magnetic field of the Earth. Various loop configurations are used including a square [commonly 50–150 m (164–492 ft) on side], a circular loop, and a figure eight. The relaxation of the protons generates a weak, time-varying magnetic field that is detected in the same loop and an induced current. The depth of investigation is a function of the excitation intensity (q), which is the product of the current intensity (I_o) and duration (τ):

$$q = I_o\tau \tag{16.8}$$

The units of q are ampere-milliseconds. The data recorded for each excitation pulse are the dead time delay, initial amplitude (E_o), relaxation or decay time (T), and phase.

Pore wall interaction is the key relaxation process for NMR. Relaxation times are correlated to pore size. Precessing protons will interact with a pore wall more rapidly in sediments or rock with small pores. Three different relaxation times are used in NMR analyses, including:

T_1 = longitudinal relaxation time,
T_2 = transverse relaxation time, and
T_3 = observed NMR relaxation time.

NMR data are interpreted using inversion techniques. Transmissivity can be estimated from MRS data as follows (Legchenko et al. 2002; Voullamoz et al. 2007; Boucher et al. 2009)

$$T_{MRS} = C_p \phi_{MRS} (T_1)^2 z \tag{16.9}$$

where,
T_{MRS} = transmissivity (m^2/s)

C_p = parametric factor (m/s/ms^2)
T_1 = T_1 relaxation constant (ms)
z = saturated thickness of aquifer (m)
ϕ_{MRS} = water content estimated from MRS (dimensionless, m^3/m^3).

Equation 16.9 involves the multiplication of an average hydraulic conductivity obtained from the NMR data by the aquifer thickness. MRS data has a large volume of investigation, so the data represents a volumetric average value rather than a point measurement. In dual-porosity systems, the data will tend to represent the bulk matrix porosity rather than volumetrically minor secondary porosity, which may provide most of the hydraulic conductivity. In order to obtain accurate measurements of transmissivity, the value of C_p needs to be determined from local pumping tests. MRS inversion solutions are not unique, so it is critical to have field data for calibration (Legchenko et al. 2002; Roy and Lucczynski et al. 2003; Boucher et al. 2009).

MRS does not detect water in very small pores and bound water, so clays have a very low or no MRS porosity (Yaramanci and Müller-Petke 2009). Both the amplitude and phase of the MRS signal are affected by electrical conductivity. MRS data cannot be accurately interpreted in isolation. The MRS inverse problem is ill posed. It is impossible, for a particular layer, to know both the layer thickness and water content (Legchenko et al. 2004). MRS and electrical or electromagnetic (resistivity) methods should always be performed together, especially at sites where no existing information on local hydrogeology is available (Yaramanci 2000). Performance of both MRS and TDEM at each station allows for a combined interpretation that may be more accurate than the separate results from each method alone (Goldman et al. 1994; Yaramanci 2000, Legchenko et al. 2004). MRS and TDEM have been proven to be effective in detecting the presence and amount of groundwater and evaluating its salinity (Goldman et al. 1994).

NMR signals are susceptible to distortion by ambient electromagnetic noise of different origins, such as nearby power lines. Other sources of noise are geological conductors (rock types, coatings on grains) and lightening. MRS may have an insufficient depth of investigation because shallow electrically conductive (e.g., clayey) layers may screen a deeper aquifer (Roy and Lubczynski 2003; Vouillamoz et al. 2007). Intensive research is still needed to determine all the parameters that influence the MRS measured decay times and to development techniques to improve the signal to noise ratio (Yaramanci and Müller-Petke 2009).

Voulillamoz et al. (2008) and Boucher et al. (2009) documented the results on NMR soundings performed on an unconfined sandstone aquifer in Niger. This investigation is notable because it includes 14 transmissivity values and six specific yield values that were obtained from sites where NMR soundings were also performed. The mean uncertainty (deviation from pump test values) of the computed T_{MRS} was 150%. The accuracy of estimation of C_p values was shown to increase with increasing amounts of calibration (i.e., pumping test) data. The MRS water constants (φ_{MRS}) were consistently greater than the specific yield values, likely caused by the measurement of total pore water content including the specific retention (trapped or adsorbed water) component of aquifer water.

Combination of MRS and VES can be used as screening tool combination for increasing the success rate of well drilling. Vouillamoz et al. (2007) investigated the potential use of MRS and VES as a tool to increase the success rate of water wells in North Rakhine State, Myanmar. Here, over 59% of more than 1,000 boreholes were unsuccessful mainly because of very low yields or poor water quality. MRS can locate productive aquifers and to quantify their storativity and transmissivity. VES can be used to estimate salinity. The key issue is whether the geophysics would save money, which depends upon the costs of the surveys and well construction, and the improvement in the well success rate (Vouillamoz et al. 2007).

MRS appears to be a promising, non-invasive method for obtaining a large data set of transmissivity and specific yields for shallow aquifers, which can aid in aquifer model parameterization (Boucher et al. 2009). MRS may provide specific yield and transmissivity data in remote areas where pumping and observation wells for a suitable aquifer pumping test are rare.

16.8 Magnetotellurics

Magnetotelluric methods are a passive means for obtaining information on the conductivity (or resistivity) of subsurface strata, by measuring naturally-occurring, time-varying electric and magnetic fields at the Earth's surface. The magnetotelluric methods were reviewed by Simspon and Bahr (2005) and Berdichevsky and Dmitriev (2008). The underlying concept of magnetotellurics is that natural electromagnetic fields are generated in the Earth's atmosphere by lightning strikes and by the interaction of the solar wind and the ionosphere. The atmospheric field variations induce electric currents in the subsurface.

Magnetotelluric methods involve measurement of the ratio of electric and magnetic fields over frequencies typically within the 10^{-3} Hz to 100 kHz range. The Earth's electromagnetic field has skin effects, which result in higher frequency ranges providing information on shallow depths, and lower frequency ranges providing information from greater depths. At each sampling location, measurements are typically made of the Earth's magnetic field in two horizontal and the vertical directions, and the electric field is measured in two horizontal directions. Measurements of the electric and magnetic fields are made over a range of frequencies. The phase differences between the electric and magnetic components are also recorded. The data are processed using inversion methods to determine variations in resistivity of subsurface strata.

Magnetotelluric methods are attractive tools for groundwater investigations in that they have a greater depth of investigation (up to about 1,000 m; 3,000 ft) compared to electrical resistivity methods and can be used to provide a rapid, cost-effective reconnaissance of an area. Magnetotelluric methods can be used to obtain information on deep structure and stratigraphy, and may also differentiate between aquifers containing fresh and saline waters. The main limitations of magnetotelluric techniques are potential interference from buried conductors, non-uniqueness of data processing, and modest resolution. Magnetotelluric methods have the greatest value as a reconnaissance tool in frontier areas where there are limited geological and hydrogeological data.

16.9 Seismic Reflection

Seismic reflection technology has revolutionized oil and gas exploration, but has had a much more limited application in groundwater investigations. Seismic reflection is an invaluable tool for generating two- and three-dimensional maps of the subsurface structure. The seismic reflection technologies also have the advantage that they can be performed on land, at sea, and in freshwater bodies.

The seismic reflection method is conceptually quite simple. Sound or seismic waves traveling through the earth may be reflected off the boundary between materials with different acoustic impedances, which is defined as the product of the seismic wave velocity and density of the rock. The boundaries off which seismic waves are reflected are referred to as 'seismic reflectors', which often correspond to formation boundaries or the contact between major lithologic changes within a formation. Seismic waves are generated from a controlled source and the two-way travel time is recorded for the waves to reach reflectors and return to a receiver or multiples receivers at the surface. The depth to each reflector can be calculated based on the known seismic velocities of the materials penetrated by the waves and the two-way travel time. The seismic velocities can be obtained from sonic velocity logs run in a nearby borehole or using seismic refraction techniques. Also, when the depths to known reflectors are measured in a nearby well, then the two-way travel time can be used from the seismic reflection profile to estimate the seismic velocities for various units in the profile.

The effort involved in performing seismic reflection surveys greatly varies. The methods used to generate the sound waves include boomers, electrical discharge (sparker), air guns, water guns, and for land-based operations, hammering on a metal plate, explosive charges, and vibrators mounted on trucks. For simple, shallow surveys only several receivers are used, which are referred to as geophones or hydrophones (in marine surveys). Hundreds or thousands of receivers may be employed in three-dimensional seismic surveys. The recorded data require substantial signal and image processing before it can be interpreted. As is the case for other surface geophysical methods, the processing involves an inversion methodology. The calculated depths to reflectors are rough estimates and need to be confirmed against field (well) data.

Although seismic reflection surveys can provide superior information on subsurface structures, it has uncommonly been applied to groundwater investigations mainly because of high costs, particularly for surveys that have good resolution at depth. Marine or freshwater (river) surveys are less expensive, but have the

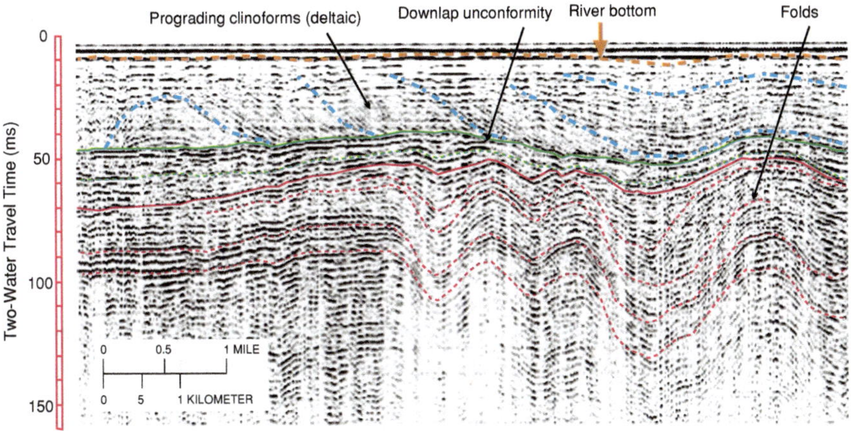

Fig. 16.6 Segment of a seismic reflection profile run on the Caloosahatchee River in southwestern Florida, USA (Cunningham et al. 2001). The profile reveals subsurface depositional and structural features (e.g., prograding clinoforms, folds) for which there is no suggestion at land surface

limitation that surface water bodies are often not located near potential wellfield sites or other areas of interest.

The principal application of seismic reflection surveying for groundwater investigations is for the evaluation of subsurface structures that could impact groundwater flow. Seismic reflection surveys could be performed, for example, to locate suspected faults or delineate the boundaries of local aquifers. The U.S. Geological Survey has performed considerable seismic reflection profiling in Florida, which provides some good examples of applications of the technology. Continuous seismic reflection surveys were run in estuarine and near shore areas to map shallow aquifers (Missimer and Gardner 1976; Cunningham et al. 2001). The surveys identified pronounced folding for which there is no suggestion at land surface, which is essentially flat lying (Fig. 16.6).

The U.S. Geological Survey has successfully performed high-resolution seismic reflection surveys in the freshwater bodies of Florida, using a small boat that can be readily transported and lowered into a canal or lake using a truck-mounted crane. The seismic equipment utilized generally provided usable information only down to approximately 30 m (100 ft) (Kindinger 2003). Seismic reflection surveys have been performed in lakes in the central and northeastern part of the state to identify zones of karstic collapse (e.g., Kindinger et al. 1994, 2000; Spechler 2001) in the limestones, which are suspected as possibly resulting in enhanced vertical flow to the underlying aquifer. The U.S. Geological Survey also performed continuous seismic surveys of Lake Okeechobee as part of a regional aquifer analysis for Comprehensive Everglades Restoration Plan aquifer storage and recovery program.

Recent land-based seismic reflection investigations have been conducted in the coastal plain of the eastern Red Sea in Saudi Arabia to assess inferred faults that impact sediment thickness and the groundwater flow associated with wadi channels (King Abdullah University of Science and Technology geophysical research program). This research has conferred the existence of listeric faults that parallel the Red Sea rift and cause significant variation in continental sediment thickness atop the Pre-Cambrian Shield rocks.

16.10 Airborne Geophysics

Airborne geophysical surveys, performed using fixed-wing aircraft and helicopters, have the advantages of great areal coverage in short time periods and the ability to be performed in inaccessible areas. Helicopters have lower and slower flying platforms and the potential for tighter flight paths than fixed wing aircraft (American Society of Civil Engineers 1998). The rapidity of measurements can result in substantial cost savings over land-based surveys with similar areal coverage. The major disadvantage is a reduced accuracy. Airborne geophysical survey data have the same limitation as surface geophysical data in the non-uniqueness of the interpretation. Airborne geophysical survey data alone are seldom sufficient for hydrogeological interpretations. The surveys instead need to be performed in conjunction with evaluation of well data and land-based geophysical surveys. Ground-truthing is an important element of all surface geophysical data analyses in general.

The applications of airborne geophysics to groundwater investigations (i.e., airborne hydrogeophysics) were reviewed by the National Research Council (1995) and Paine and Minty (2005), which also provided a good bibliography. Airborne surveys cannot substitute for methods that require ground contact, such as seismic reflection and resistivity profiles (Paine and Minty 2005). Common airborne hydrogeophysical methods include electromagnetic induction (both frequency and time domain), gamma-ray spectrometry, magnetics and magnetotellurics.

Airborne hydrogeophysical studies have greatest value for relatively coarse-scale investigations of basin geology and hydrogeology. The airborne hydrogeophysical data may allow for more accurate interpolation and extrapolation of sparse, but more precise, ground and borehole measurements of the same physical properties, provided that they are scalable to the airborne platform (Paine and Minty 2005). Airborne electromagnetic (AEM) surveys have the greatest utility for evaluating groundwater resources at shallow and intermediate depths (<300 m, <1,000 ft) aquifers. Airborne magnetic and magnetotelluric surveys can provide useful information on deeper subsurface geology as part of an initial regional reconnaissance. Airborne magnetic surveys can be used to determine basin geometry and to identify shallow local magnetic anomalies caused by faults, paleochannels, or man-made features, such as wells or pipelines (Paine and Minty 2005).

As is the case for any tool, an evaluation needs to be made concerning the type and quality of information that could be provided by airborne geophysical surveys and whether the potential data obtained are worth the cost to perform the surveys and process the data. Airborne geophysical data can have great value for

- initial reconnaissance of study areas,
- detection of subsurface geologic features, particularly features that might not be detected using methods with a more limited areal coverage,
- providing guidance concerning locations of preferred locations or targets for test wells and surface geophysical surveys, and
- interpolation between wells.

Interference from anthropogenic features is an important limitation of airborne geophysical surveys, which, as a result, are typically performed in undeveloped or rural lands. Airborne geophysical surveys are generally not suitable for densely developed coastal areas.

Electromagnetic surveys using induction methods (continuous wave and transient) can be performed from aircraft because they do not require ground contact. AEM is the most widely performed airborne geophysical method because it can provide information on water salinity and porosity, which are variables of great concern for groundwater resources evaluations. AEM surveys can provide cross-sections and high resolution two- and three-dimensional maps of apparent conductivity (resistivity), which can be further processed to obtain groundwater salinities. However, resistivity values are a function of groundwater salinity, water content (in vadose zone), porosity, and lithology (particularly clay content). Human cultural interference (e.g., roads, powerlines, pipelines, and railroads) can compromise AEM data. AEM surveys, therefore, cannot provide definitive data on local hydrogeological conditions, without some knowledge of local geology and ground-truthing, and calibration against field data. AEM surveying techniques were reviewed by Paterson and Bosschart (1987). There are several applications for airborne geophysical surveys.

16.10.1 Mapping Aquifers in Arid Lands

An important application of AEM in arid lands is mapping the presence of potential groundwater resources. In arid lands, groundwater tends to be conductive because of the presence of significant amounts of dissolved solids and, as a result, AEM may differentiate between dry and saturated sediments. AEM is particularly useful for differentiation between fresh and saline water because of the very high conductivity signature of saline water.

Wynn (2002) documented the combined use of magnetometer and AEM data to map the presence of groundwater resources in an arid region, the San Pedro Valley, of southern Arizona. The magnetometer data provided information on the depth to the crystalline basement and thus the potential storage capacity of basin sediments. An anomalously shallow apparent depth to basement was detected in one area, which was actually a volcanic flow filling a paleochannel. The AEM maps were used to identify more conductive groundwater-bearing sediments that are potential targets for groundwater development.

The Texas Bureau of Economic Geology performed airborne TDEM studies of two 260 km^2 (100.4 mi^2) areas in the Lower Rio Grande Valley (Cameron, Hidalgo and Willacy Counties) of southern Texas (Paine 2000). The goal of the investigation was to assess groundwater resources, particularly depths to the water table and the quality of potential water sources. The studied strata consisted of fluvial and deltaic siliciclastic deposits, in which silts and clays are juxtaposed with sand-rich units that have more abundant groundwater resources. Sinuous features, suggesting the presence of channel or channel complex deposits, were found in the conductivity images. The sinuous features were less conductive than adjacent areas, which suggested that they are coarse deposits that contain relatively fresh water.

The airborne and ground TDEM data gave similar trends, but different absolute values (Paine 2000). The results of the Lower Rio Grande Valley study demonstrated that AEM has great value for identifying favorable targets for groundwater exploration by providing information on depths, lithology, and water quality of aquifers. However, there is still considerable uncertainty concerning the actual values of hydrogeological properties of the target zones or aquifers.

16.10.2 Mapping Coastal the Saline-Water/Fresh-Water Interface

The saline-water interface is a suitable target for AEM because of the great contrast in electrical conductivity between saline and fresh water. Fitterman and Deszcz-Pan (2001, 2002) and Stewart et al. (2002) documented the use of helicopter electromagnetic (HEM) survey to map saline-water intrusion in the Everglades National Park of South Florida, USA. The survey area is 1,036 km^2 (400 mi^2). The Everglades National Park is an ideal area for a helicopter electromagnetic (HEM) investigation of saline-water intrusion because the interface occurs at shallow depths and the strata are flat-lying to subhorizontal and do not contain significant amounts of clay (Fitterman and Deszcz-Pan 2001, 2002). Inasmuch as the area is largely undeveloped, there is also minimal anthropogenic interference. The HEM survey results clearly distinguish variations in the resistivity, and thus salinity, in the water-table aquifer (Fig. 16.7). AEM (HEM) may still be useful in areas with more complex geology because the strong salinity signal tends to overprint geological signals (Fitterman and Deszcz-Pan 2001).

The HEM data were processed to obtain formation resistivity versus depth slices. The formation factor (F) of Archie's (1942) law was determined from dual induction logs and specific conductance data from 23 boreholes drilled and tested in the study area. The water-quality data were also used to obtain a relationship between specific conductance and chloride concentration. Surface-geophysical data were also collected in accessible areas to further ground-truth the HEM data. The major advantage of HEM in the Everglades National Park was that it allowed for detailed mapping of the interface that would not be practically possible using land surface geophysics and wells. The park consists predominantly of very poorly accessible wetlands. The HEM maps resolved the impacts on roads and canals on the position of the saline-water interface.

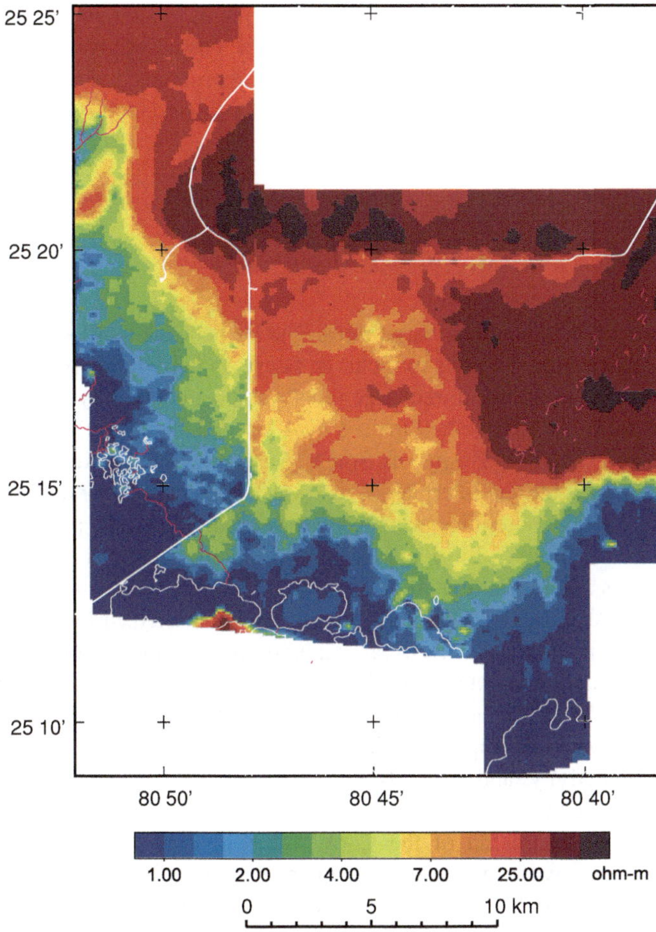

Fig. 16.7 HEM apparent resistivity map for approximate 5–10 m depth in the Everglades National Park, Florida (Stewart et al. 2002). Groundwater salinity varies from freshwater (north) to seawater to the south

16.10.3 Detection of Saline-Water Contamination and Other Contamination Plumes

The contrast between fresh-water and saline-water allows the detection and mapping of various contaminant plumes that have a strong resistivity contrast (e.g., saltwater, oilfield brines, landfill leachate, etc.). Paine and Collins (2003) documented AEM surveys performed in West Texas. A combination of magnetometer and AEM surveys performed using a fixed wing aircraft allowed for the detection of both the location of oil and gas exploration and production wells and local groundwater salinization that was associated with the drilling and operation of the wells. The magnetometer could detect isolated individual wells or clusters of closely spaced wells. In another study area (El Paso County), the AEM and magnetometer data provided information on basin structural features (e.g., faults) and allowed for the mapping of a broad zone of low-conductivity, low-salinity groundwater.

16.10.4 Recharge and Salinity Mapping

AEM has been used to map recharge rates. The basis of the method is to correlate electrical conductivities of soils to recharge rates. An early helicopter-borne AEM survey performed to delineate recharge rates was documented by Cook and Kilty (1992). The study was performed in an 8 by 4 km (5.1 by 2.5 mi) semiarid region in the western Murray Basin of South Australia. The estimated electrical conductivities reflected the interplay of soil water content, salt content, texture and mineralogy. Twenty boreholes were drilled beneath the flight paths and gravimetric water contents and chloride concentrations were measured in the boreholes. Recharge rates were estimated using the chloride mass balance method (Sect. 11.5.1). Areas with low recharge rates are characterized by high dissolved salt concentrations (and thus electrical conductivities). On the contrary, areas that were cleared of native vegetation (Eucalyptus) experience increased recharge and the leaching of salts. Recharge rates are also dependent on soil texture. A theoretical model was developed to describe the relationship between apparent electrical conductivity and groundwater recharge under the agricultural lands within the study area.

An important conclusion of the Cook and Kilty (1992) study is that AEM techniques can provide a means for interpolating estimates of groundwater recharge from field data, rather than providing a means for directly estimating recharge rates. AEM provides only one source of data. Information on topography, geomorphology, soils, vegetation, and groundwater are also needed for understanding recharge processes (Cook and Kilty 1992). A large decrease in prediction accuracy occurs when moving from smaller to larger areas, which was ascribed to greater variation in geology at larger scales.

Edwards and Webb (2003) performed ground-truthing of a previously performed AEM study of the Kamarooka catchment in central Victoria, southwestern Australia, which is badly affected by dryland salinization. The ground-truthing involved comparison of four conductivity depth slices with salinity data from groundwater monitoring wells screened to the same depth intervals. The ground-truthing results indicated that the AEM signal had strong porosity dependence and that the AEM conductivities were not a reliable proxy for groundwater salinity. The strong dependence of AEM data on porosity suggests that it may have a greater value for resolving the physical hydrogeology of the study area.

16.11 Applications of Surface Geophysics to Groundwater Investigations

Surface geophysical techniques are not part of most investigations of groundwater resources. Both ground and airborne geophysical methods have the advantage of allowing for a higher density of measurements and greater areal coverage than is practicably possible using borehole-based techniques. The greater coverage comes at the expense of a lesser accuracy and greater uncertainty. Surface geophysical data are processed using inversion techniques, which do not provide unique solutions. Very impressive colored graphics can now be readily generated, but they not may be accurate. Surface geophysical data thus does not eliminate the need for borehole data, which are still critical for ground-truthing.

Nevertheless, the greater spatial coverage and lesser cost afforded by surface geophysical investigations compared to borehole-based techniques are compelling, particularly for the initial reconnaissance of a project area. Surface geophysical data can also be used for interpolation and extrapolation of limited borehole data. Surface geophysical techniques require specialized professional expertise to collect, process, and interpret the data. Professionals involved in the development of the scope of groundwater investigations should be aware of the various surface geophysics techniques that are available, including their capabilities and limitations. A critical question that needs to be addressed at the start of a project is whether or not surface geophysical techniques can accurately provide the desired information in a cost effective manner. This is a particularly relevant question in arid lands, where only very limited amounts of hydrogeological data are present over vast areas.

References

Abdalla, O. A. E., Ali, M., Al-Higgi, K., Al-Zidi, H., El-Hussain, I., & Al-Hinai, S. (2010). Rate of seawater intrusion estimated by geophysical methods in an arid region: Al Khabourah, Oman. *Hydrogeology Journal, 18*, 1437–1445.

Abdul Nassir, S. S., Loke, M. H., Lee, C.-Y., & Nawawi, M. N. M. (2000). Salt-water intrusion mapping by geoelectrical imaging surveys. *Geophysical Prospecting, 48*, 647–661.

Abraham, J. D., & Lucius, J. E. (2004). *Direct current resistivity profiling to study distribution of water in the unsaturated zone near Amargosa Desert Research Site, Nevada.* U.S. Geological Survey Open-File Report, 2004-1319.

American Society of Civil Engineers. (1998). *Geophysical exploration for engineering and environmental investigations.* Virginia: ASCE Press.

Archie, G. E. (1942). *The electrical resistivity log as an aid in determining some reservoir characteristics* (Vol. 146, pp. 54–67). Transactions American Institute of Mining Metallurgical and Petroleum Engineers.

Baker, G. S., Jordan, T. E., & Pardy, J. (2007). An introduction to ground penetrating radar (GPR). In G. S. Baker & H. M. Jol (Eds.), *Stratigraphic analyses using GRP* (Vol. 432, pp. 1–18). Boulder, CO: Geological Society of America Special Paper.

Benkabbour, B., Toto, E. A., & Fakir, Y. (2004). Using DC resistivity method to characterize the geometry and the salinity of Plioquaternary consolidated coastal aquifer of the Mamora plain, Morocco. *Environmental Geology, 45*, 518–526.

Berdichevsky, M. N., & Dmitriev, V. I. (2008). *Models and methods of magnetotellurics*. Berlin: Springer.

Boucher, M., Faureau, G., Vouillamos, J. M., Nazoumou, Y., & Legchencko, A. (2009). Estimating specific yield and transmissivity with magnetic resonance sounding in an unconfined sandstone aquifer (Niger). *Hydrogeology Journal, 17*, 1805–1815.

Burger, H. R., Sheehan, A. F., & Jones, G. H. (2006). *Introduction to applied geophysics: Exploring the shallow subsurface*. New York: W.W Norton.

Cook, P. G., & Kilty, S. (1992). A helicopter-borne electromagnetic survey to delineate groundwater recharge rates. *Water Resources Research, 28*, 2953–2961.

Cunningham, K. J., Locker, S. D., Hine, A. C., Bukry, D., Barron, J. A., and Guertin, L. A. (2001). *Surface-geophysical characterization of ground-water systems of the Caloosahatchee River Basin, Southern Florida*. U.S. Geological Survey Water-Resources Investigations Report 01-4084.

De Franco, R., Biella, G., Tosi, I., Teatini, P., Lozej, A., Chiozzotto, B., et al. (2009). Monitoring the saltwater intrusion by time lapse electrical resistivity tomography: The Chioggia test site (Venice Lagoon, Italy). *Journal of Applied Geophysics, 69*, 117–130.

Degnan, J. R., & Brayton, M. J. (2010). *Preliminary investigation of paleochannels and groundwater specific conductance using direct-current resistivity and surface-water seismic geophysical surveys at the Standard Chlorine of Delaware, Inc. Superfund Site, Delaware City, Delaware, 2008*. U.S. Geological Survey Open-File Report 2010-1058.

Driscoll, F. G. (1986). *Groundwater and wells* (2nd ed.). St. Paul, Minnesota: Johnson Filtration Systems.

Eastern Research Group. (1993). *Use of airborne, surface, and borehole geophysical techniques at contaminated sites: A reference guide*. U.S. Environmental Protection Agency, Office of Research and Development, Document EPA/625/R-92-007.

Edwards, M. D., & Webb, J. A. (2003). Ground-truthing of a TEMPEST airborne electromagnetic survey in the salinised Kamarooka catchment, near Bendigo in central Australia. In I. D. Roach (Ed.), *Advances in regolith* (pp. 110–114). Australia: CRC LEME.

Fitterman, D. V., & Deszcz-Pan, M. (2001). *Saltwater intrusion in Everglades National Park, Florida measured by airborne electromagnetic surveys*. In First International Conference on Saltwater Intrusion and Coastal Aquifers, Monitoring, Modeling, and Management, Essaouria, Morocco, April 23–25, 2001.

Fitterman, D. V., & Deszcz-Pan, M. (2002). *Helicopter electromagnetic data from Everglades National Park and surrounding areas, Florida: Collected December 9–14, 1994*. U.S. Geological Survey Open-File Report 02-101.

Fitterman, D. V., & Stewart, M. T. (1986). Transient electromagnetic soundings for groundwater. *Geophysics, 51*, 995–1006.

Fitterman, D. V., Menges, C. M., Al Kamali, A. M., & Essa Jamma, F. (1991). Electromagnetic mapping of buried paleochannels in eastern Abu Dhabi Emirate, U.A.E. *Geoexploration, 27*, 111–133.

Fitterman, D. V., Deszcz-Pan, M., & Stoddard, C. E. (1999). *Results of time-domain electromagnetic soundings in Everglades National Park, Florida*. U.S. Geological Survey Open-File Report 99-426.

Fretwell, J. D., & Stewart, M. T. (1981). Resistivity study of a coastal karst terrain, Florida. *Ground Water, 19*, 156–162.

Goldman, M., Rabinovich, B., Rabinovich, M., Gilad, D., Gev, I., & Schirov, M. (1994). Application of the integrated NMR-TDEM method in groundwater exploration in Israel. *Journal of Applied Geophysics, 31*, 27–52.

Huisman, J. A., Hubbard, S. S., Redman, J. D., & Annan, A. P. (2003). Measuring soil water content with ground-penetrating radar: A review. *Vadose Zone Journal, 2*, 476–491.

Jadoon, K. Z., Slob, E. C., Vanclooster, M., Vereecken, H., & Lambot, S. (2008). Uniqueness and stability analysis of hydrophysical inversion for time-lapse ground-penetrating radar estimates of shallow soil hydraulic properties. *Water Resources Research, 44*, W09 421. doi:10.1029/2007WR006 639.

Jol, H. M. (Ed.). (2009). *Ground penetrating radar theory and applications*. Amsterdam: Elsevier.

Kindinger, J. (2003). *Lake Belt study area: High-resolution seismic reflection survey, Miami-Dade County Florida*. U.S. Geological Survey Open-File Report 02-325.

Kindinger, J. L., Davis, J. B., & Flocks, J. G. (1994). High-resolution single-channel seismic reflection surveys of Orange Lake and other selected sites of North Central Florida. U.S. Geological Survey Open-File Report 94-616.

Kindinger, J. L., Davis, J. B., & Flocks, J. G. (2000). Subsurface characterization of selected water bodies in the St Johns River Water Management District, Northeast Florida. U.S. Geological Survey Open-File Report 00-180.

Kobr, M., Mareš, S, & Paillet, F. (2005). Borehole geophysics for hydrogeological studies: Principles and applications. In Y. Rubin & S. S. Hubbard (Eds.), *Hydrogeophysics* (pp. 291–331). Dordrecht: Springer.

Kowalsky, M. B., Finsterle, S., Peterson, J., Hubbard, S., Rubin Y., Majer, E., et al. (2005). Estimation of field-scale soil hydraulic and dielectric parameters through joint inversion of GPR and hydrological data. *Water Resources Research, 41*, W11425. doi:10.1029/2005WR004237.

Lambot, S., Rhebergen, J., Slob, E., Lopera, O., Jadoon, K. Z., & Vereecken, H. (2009). Remote estimation of the hydraulic properties of a sandy soil using full-waveform integrated hydrogeophysical inversion of time-lapse, off-ground GPR data. *Vadose Zone Journal, 8*(3), 743–754.

Legchenko, A., Baltassat, J.-M., Beauce, A., & Bernard, J. (2002). Nuclear magnetic resonance as a geophysical tool for hydrogeologists. *Journal of Applied Geophysics, 50*, 21–46.

Legchenko, A., Baltassat, J.-M., Bobachev, A., Martin, C., Robain, H., & Vouillamoz, J. M. (2004). Magnetic resonance sounding applied to aquifer characterization. *Ground Water, 42*, 363–373.

Lunt, I. A., Hubbard, S. S., & Rubin, Y. (2005). Soil moisture content estimation using ground-penetrating radar reflection data. *Journal of Hydrology, 307*, 254–269.

Mills, T., Hoekstra, P., Blohm, M., & Evans, L. (1988). Time domain electromagnetic soundings for mapping seawater intrusion in Monterey County, California. *Ground Water, 26*, 771–782.

Milson, J. (2003). *Field geophysics* (3rd ed.). Chichester, UK: Wiley.

Minsley, B. J., Ajo-Franklin, J., Mukhopadhyay, A., & Morgan, F. D. (2011). Hydrogeophysical methods for analyzing aquifer storage and recovery systems. *Ground Water, 49*(2), 250–269.

Missimer, T. M., & Gardner, R. A. (1976). *High-resolution seismic reflection profiling for mapping shallow aquifers in Lee County, Florida*. U.S. Geological Survey Water Resources Investigations/NTIS 76-45.

National Research Council. (1995). *Mexico City's water supply, improving the outlook for sustainability*. Washington, DC: National Academy Press.

Paine, J. G. (2000). *Indentifying and assessing ground water in the Lower Rio Grande Valley, Texas, using airborne electromagnetic induction*. Report prepared for the Texas Water Development Board by the Texas Bureau of Economic Geology.

Paine, J. G., & Collins, E. W. (2003). Applying airborne electromagnetic induction in groundwater salinization and resources studies, West Texas. In *Proceedings, Symposium of the Application of Geophysics to Engineering and Environmental Problems* (pp. 722–738). Environmental and Engineering Geophysical Society.

Paine, J. G., & Minty, B. R. S. (2005). Airborne hydrogeophysics. In Y. Rubin & S. S. Hubbard (Eds.), *Hydrogeophysics* (pp. 333–357). Netherlands: Springer.

Paterson, N. R., & Bosschart, R. A. (1987). Airborne geophysical exploration for ground water. *Ground Water, 25*, 41–50.

Reynolds, J. M. (1997). *An introduction to applied and environmental geophysics* (796 pp). Chichester, UK: Wiley.

Rizzo, E., Suski, B., Revil, A., Straface, S., & Troisi, S. (2004). Self-potential signals associated with pumping-tests experiments. *Journal of Geophysical Research, 109*, B10203.

Roy, J., & Lubczynski, M. (2003). The magnetic resonance sounding technique and its use for groundwater investigations. *Hydrogeology Journal, 11*, 455–465.

Samsudin, A. R., Haryono, A., Hamzah, U., & Rafek, A. G. (2008). Salinity mapping of coastal groundwater aquifers using hydrogeochemical and geophysical methods: A case study from North Kelantan, Malaysia. *Environmental Geology, 55*, 1737–1743.

Schirov, M., Legchenko, A., & Creer, G. (1991). A new direct-invasive groundwater detection technology for Australia. *Exploration Geophysics, 22*, 333–338.

Seara, J. L., & Granda, A. (1987). Interpretation of IP time domain/resistivity soundings for delineating sea-water intrusions in some coastal areas of the Northeast of Spain. *Geoexploration, 24*, 153–167.

Sharma, P. (1997). Chlorine-36 in fossil rat urine: An archive of cosmogenic nuclide deposition during the past 40,000 years. *Science, 277*, 538–541.

Sherif, M., El Mahmoudi, A., Garamoon, H., Kacimov, A., Akram, S., Ebraheem, A., et al. (2006). Geoelectrical and hydrochemical studies for delineating seawater intrusion in the outlet of Wadi Ham, UAE. *Environmental Geology, 49*, 536–551.

Simpson, F., & Bahr, K. (2005). *Practical magnetotellurics*. Cambridge, UK: Cambridge University Press.

Skinner, D., & Heinson, G. (2004). A comparison of electrical and electromagnetic methods for the detection of hydraulic pathways in a fractured rock aquifer, Clare Valley, South Australia. *Hydrogeology Journal, 12*(3), 576–590.

Spechler, R. M. (2001). The relation between structure and saltwater intrusion in the Floridan Aquifer System, Northeastern Florida. In E. L. Kuniansky (Ed.), *U.S. Geological Survey Karst Interest Group Proceedings* (pp. 25–29). U.S. Geological Survey Water-Resources Investigations Report 01-4011.

Stewart, M. A., Bhatt, T. N., Fennema, R. J., & Fitterman, D. V. (2002). *The road to Flamingo: An evaluation of flow pattern alterations and salinity intrusion in the Lower Glades, Everglades National Park*. U.S. Geological Survey Open-File Report 02-59.

Stonestrom, D. A., Prudic, D. E., Walvoord, M. A., Abraham, J. D., Stewart-Deaker, A. E., Glancy, P. A., et al. (2007). Focused ground-water recharge in the Amargosa Desert Basin. In D. A. Stonestrom, J. Constantz, T. P. A. Ferré & S. A. Leake (Eds.), *Ground-water recharge in the arid and semiarid southwestern United States* (pp. 107–136). U.S. Geological Survey Professional Paper 1703.

Subsurface Detection Investigations, Inc. (1995). Time domain electromagnetic mapping of salt water in the Floridan Aquifer in Northeast and East-Central Florida, St. Johns River Water Management District, October 1995. Palatka, FL: SJRWMD Special Publication SJ95-SP2.

Sushakov, O. I. (1996). Groundwater NMR in conductive water. *Geophysics, 61*, 998–1006.

Taylor, K., Widmer, M., & Chesley, M. (1992). Use of transient electromagnetics to define local hydrogeology in an arid alluvial environment. *Geophysics, 57*, 343–352.

Telford, W. M., Geldart, L. P., & Sheriff, R. E. (1990). *Applied geophysics*. Cambridge, UK: Cambridge University Press.

Titov, K., Revil, A., Konasovsky, P., Straface, S., & Troisi, S. (2005). Numerical modeling of self-potential signals associated with a pumping test experiment. *Geophysical Journal International, 162*(2), 641–650.

Van Overmeeren, R. A. (1989). Aquifer boundaries explored by geoelectrical measurements in the coastal plain of Yemen: A case of equivalence. *Geophysics, 54*, 38–48.

Van Overmeeren, R. A., & Ritsema, I. L. (1988). Continuous vertical electrical sounding. *First Break, 6*(12), 313–324.

Vanessa, N., Adam, P., & Rosemary, K. (2011). Application of an extended Kalman filter approach to inversion of time-lapse electrical resistivity imaging data for monitoring recharge. *Water Resources Research, 47*, W10525.

Vouillamoz, J. M., Chatenoux, B., Mathieu, F., Baltassat, J. M., & Legchenko, A. (2007). Efficiency of joint use of MRS and VES to characterize coastal aquifer in Myanmar. *Journal of Applied Geophysics, 61*, 142–154.

Voulillamoz, J. M., Faureau, G., Massuel, S., Boucher, M., Nazoumou, Y., & Legchenko, A. (2008). Contribution of magnetic resonance sounding to aquifer characterization and recharge estimate in semiarid Niger. *Journal of Applied Geophysics, 64*, 99–108.

Wilson, S. R., Ingham, M., & McConchie, J. A. (2005). The applicability of earth resistivity methods for saline interface definition. *Journal of Hydrology, 316*, 301–312.

Worthington, P. F. (1977). Geophysical investigation of groundwater resources in the Kalahari Basin. *Geophysics, 42*, 838–849.

Wynn, J. (2002). *Evaluating groundwater in arid lands using airborne magnetic/EM methods: An example in the Southwestern U.S. and Northern Mexico* (pp. 62–64). Tulsa, OK: The Leading Edge.

Yaramanci, U. (2000). Surface nuclear magnetic resonance (MRS)—A new method for exploration of ground water and aquifer properties. *Annali di Geofisica, 43*, 1159–1175.

Yaramanci, U., & Müller-Petke, M. (2009). *Surface nuclear magnetic resonance—A unique tool for hydrogeophysics* (pp. 1240–1247). Tulsa, OK: The Leading Edge.

Young, M. E., de Bruijn, R. G. M., & bin Salim Al-Ismaily, A. (1998). Exploration of an alluvial aquifer in Oman by time-domain electromagnetic soundings. *Hydrogeology Journal, 6*, 383–393.

Young, M. E., Macumber, P. G., Watts, M. D., & Al-Toqy, N. (2004). Electromagnetic detection of deep freshwater lenses in a hyper-arid limestone terrain. *Journal of Applied Geophysics, 57*, 43–61.

Zohdy, A. A. R. (1974). Electrical resistivity. In A. A. R. Zohdy, G. P. Eaton & D. R. Mabey (Eds.), *Application of surface geophysics to ground-water investigations* (pp. 5–66). Techniques of Water-Resources Investigations of the United States Geological Survey, Book 2, Chapter D1.

Zohdy, A. A. R., Eaton, G. P., and Mabey, D. R. (1974). *Application of surface geophysics to ground-water investigations*. Techniques of Water-Resources Investigations of the United States Geological Survey, Book 2, Chapter D1.

Chapter 17
Borehole Geophysical Techniques

17.1 Introduction

In arid lands water resources investigations, it is important to maximize the amount of data that is obtained from the drilling of test or exploratory wells. A key means of obtaining detailed data on the groundwater system being investigated is through borehole geophysical logging. Valuable hydrogeological data can be obtained from both newly drilled and existing wells. Borehole geophysical logging techniques allow for the continuous in situ measurement of the petrophysical properties and composition of the penetrated rock, sediment, and formation fluids. Borehole geophysical logging has long been a critical tool in the oil and gas industry because of the wealth of data that it can cost-effectively provide. Schlumberger (1989) and Asquith and Krygowski (2004) provide detailed reviews of geophysical logging principles and applications with a focus on the oil and gas industry. The applications of borehole geophysical logging to groundwater investigations are reviewed by Driscoll (1986), Keys (1989, 1990, 1997), Collier (1993), Kobr et al. (2005), Maliva et al. (2009), and Maliva and Missimer (2010).

Borehole geophysical logs can provide valuable information for groundwater resources investigations on

- aquifer hydrogeology, particularly the locations of aquifers and confining strata,
- groundwater salinity,
- porosity and permeability,
- location, size, and orientation of fractures and other secondary porosity features,
- aquifer heterogeneity, including the location of preferential flow zones, and
- aquifer mineralogy.

The greatest value can be obtained from borehole geophysical logging when it is performed in conjunction with other aquifer testing methods, such as aquifer performance (pumping) tests, packer tests, and core analyses. For example, aquifer pump testing can provide information on the average hydraulic conductivity and

R. Maliva and T. Missimer, *Arid Lands Water Evaluation and Management*, Environmental Science and Engineering, DOI: 10.1007/978-3-642-29104-3_17,

transmissivity of the test interval. The combination of pump testing and flowmeter logging can provide additional information on aquifer heterogeneity, particularly the presence and hydraulic properties of flow and non-productive zones within an aquifer.

An overview of basic and advanced borehole geophysical techniques as applied to groundwater investigations is provided in this chapter. The emphasis is on the types of information that can be obtained from the various techniques and the limitations of the methods. Borehole geophysical logs can be roughly divided into two suites, basic and advanced, based on their sophistication, information provided, availability, and cost to run. The basic geophysical log suite typically run for groundwater investigations includes some or all of the following logs: caliper, natural gamma ray, resistivity (long-short normal and dual induction), spontaneous potential, temperature, fluid resistivity, flowmeter, and sonic (acoustic). Nuclear borehole logs (neutron and density logs) are less commonly used because of concerns (or local prohibitions) over the use of radioactive materials in freshwater aquifers (i.e., fear that a tool with its radioactive source may be lost in a well).

Some of the basic geophysical logs that are still widely used for groundwater investigations have long been supplanted by more advanced logs in the oil and gas industry. Nevertheless, they still provide useful information and are typically inexpensive to run. Local geophysical loggers or well-drillers with logging equipment can run at least some of the basic geophysical logs. Advanced borehole geophysical logs, such as nuclear magnetic resonance, elemental capture spectroscopy, and imaging logs, can provide fine-scale information on aquifer composition and properties that may be of great value for hydrogeologically complex projects (e.g., managed aquifer recharge systems; Maliva et al. 2009).

Qualitative interpretation of geophysical logs is not complicated and many groundwater professionals can perform an elementary analysis of borehole geophysical logs. However, if the investment is made to run geophysical logs, then it should follow that an effort is made to obtain the maximum value from the logs through quantitative interpretation. Greater technical expertise is necessary for accurate quantitative analysis of borehole geophysical data. Commonly, geophysical logs are run and are given only a cursory examination before they are filed with the geologic logs. This is a lost opportunity to more carefully analyze and assess aquifer properties.

17.2 Quality Assurance and Control

The amount and accuracy of information that can be derived from geophysical logs depends upon the type and quality of the logs and the knowledge and effort-expended by scientists or engineers in processing and interpreting the logs. Log data accuracy is a major area of weakness in the groundwater logging industry. Logging for water resources investigations can be greatly improved by running suites of logs that are compatible with borehole conditions and the petrophysical properties of the aquifer (Collier 1993). Cursory examination of geophysical logs can provide qualitative information on lithologic and water quality variations within a borehole. More detailed analyses can provide quantitative data on the petrophysical properties of the tested strata, provided that the logs are of high-quality. The importance of quality control in borehole geophysical logging cannot be over-emphasized if the data are to be quantitatively interpreted. The authors have encountered numerous geophysical logs in which reported resistivity and sonic transit times are clearly incorrect, and the data are not quantitatively interruptible. However, such logs may still be usable for some qualitative analyses, in which relative, rather than absolute, values are considered.

Two important tasks in the quantitative analysis of geophysical logs are calibration and processing. The accuracy of quantitative analyses of geophysical log data depends upon the accuracy of the unprocessed raw data, and the "correctness" of the algorithms used to process the data. Calibration and standardization are important parts of quantitative geophysical log interpretation, which are discussed by Keys (1989). Standardization is the process of checking the response of the logging tools in the field against known portable standards (Kobr et al. 2005). Sections of the log should be repeated to determine whether or not there is any drift or artifacts in the measurements. Calibration is the process of converting the measured units of log response into units that measure rock characteristics (Kobr et al. 2005). Logging tools can be calibrated under laboratory conditions using specially prepared boreholes.

The raw data recorded during geophysical logging requires processing in order to obtain the petrophysical data of interest. For example, transit time data from a sonic log is processed to obtain porosity values, which is often the primary data of interest. Processing involves applying algorithms to the raw data, many of which include empirical constants. Data processing also involves applying corrections for borehole and geological conditions, such as for borehole diameter, invasion, bed thickness, and adjacent beds (Collier 1993). Geophysical log data should be calibrated against known field data, if available. For example, geophysical log-based interpretations of porosity and pore-water salinity can be checked against core porosity data and water quality data from packer tests. If site-specific data are not available for local groundtruthing of geophysical log interpretations, then often algorithms and associated empirical constants derived from historic logging experience in the investigated lithologies in the project site region (i.e., local default values) can be used satisfactorily. As a general principle, geophysical log data should be viewed as best estimates, for which the confidence of accuracy increasing with the degree to which the data can be checked or calibrated against well-specific or local data.

The accuracy of the depths should also be checked against known depths, such as the base of casing strings. Many logging tools include a natural gamma ray detector, which is very useful for determining if the multiple logs run on a borehole are in sync concerning depth, and for making any necessary depth corrections during processing.

17.3 Caliper Logs

Caliper logs provide a continuous measure of borehole diameter. The caliper log is an integral component of most geophysical logging programs, because the response of some other logs is dependent upon borehole diameter. The most widely used type of caliper tool in groundwater investigations is the mechanical probe, which consists of one or more (typically three or four) metal arms that are maintained in contact with the borehole wall by spring pressure. In a typical caliper log, the arms are connected to a potentiometer, in which resistance is linearly proportional to the extent to which the arms are opened. Changes in resistance are converted to voltage changes, which are sent to the surface either directly or converted to a varying pulse rate (Keys 1989).

Acoustic or sonic caliper tools are also available for special applications, such as logging very large diameter holes. Acoustic caliper logs measure borehole diameter from the return time of high frequency acoustic pulses. The logs may record four traces at 90° spacings or provide 360° profiles of the borehole.

Three-armed averaging probes have arms that move together and measure the average diameter of the borehole. A commonly used, and preferred design, is the X–Y probe, which consists of two orthogonally-oriented pairs of independently linked arms, which provide measurements of borehole diameter in two directions. The X–Y caliper log can thus provide information on asymmetry of the borehole. Probes with four or six independent arms, which can provide greater detail on borehole irregularities, are less commonly used in groundwater investigation.

The caliper log has multiple uses. Variations in borehole diameter are often correlated with variations in the hardness of the rock. Poor-indurated formations may be "washed out" or experience greater erosion during drilling, and thus are evident on caliper logs as intervals of increased borehole diameter. Well-indurated rock on the contrary may have borehole diameters close to the bit size (i.e., close to gauge). Large open fractures or cavernous intervals may be evident on caliper logs by greatly enlarged borehole diameters. A spiky caliper log pattern may be indicative of fractured rock (Fig. 17.1).

Caliper logs should be run before the installation of casing strings. Borehole diameter data can used to calculate theoretical cement and gravel pack volumes, and to confirm that the borehole is clear prior to the installation of the casing. The caliper log can also detect swelling of clays, which could constrict a borehole to the degree that a casing cannot be properly installed and grouted in place.

Many other geophysical logs cannot be interpreted within having detailed borehole diameter data available. The intensity of a gamma ray log is dependent on the distance between the formation and the receiver. The logs produced by every geophysical tool that emits and receives any type of artificial pulse, whether radio waves, electric fields, or radiation, is dependent on the distance between the source, the wall of the borehole, and the collector. Therefore, a caliper log should be run as part of every borehole geophysical log suite.

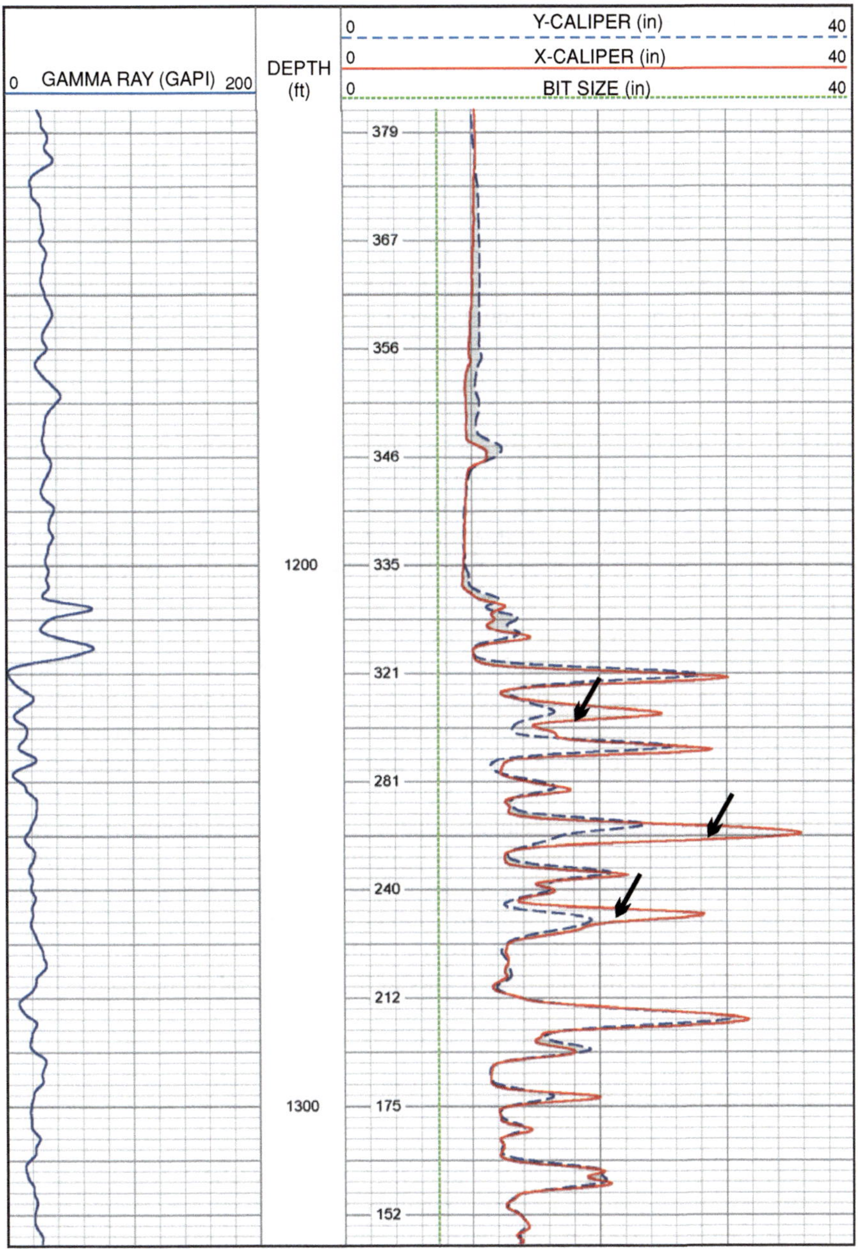

Fig. 17.1 Gamma ray and caliper log. Fractured zones are evident below 1,218 ft, as evidenced by a spiky caliper log pattern. Asymmetrical boreholes are indicated by large differences between x- and y-caliper logs (*arrows*)

17.4 Natural Gamma Ray Log

The natural gamma ray log (also just referred to as the gamma ray log) is one of the most widely used and useful basic geophysical logs in groundwater investigations. The natural ray log records the total gamma radiation detected within a selected energy range (Keys 1989). Gamma ray activity is commonly expressed in terms of American Petroleum Institute (API) gamma ray units. The detected gamma radiation is derived predominantly from the decay of potassium 40 and the uranium and thorium decay series. Rock and sediment types that have relatively high concentrations of potassium, uranium, and thorium, thus tend to have relatively high gamma ray log responses.

In general, clay minerals, and thus shales, tend to have a greater amount of natural radioactivity than clean quartz sands, carbonate rocks, gypsum and anhydrite. Phosphate minerals, and some feldspars and micas, also tend to have high natural radioactivities (Keys 1989). The primary use of gamma ray logs is lithological identification and correlation. Within a mixed siliciclastic aquifer, the gamma ray log can be used to differentiate between clay-rich beds, which tend to be confining units, and relatively clean sands, which tend to be productive units (Fig. 17.2).

Gamma ray logs are particularly useful when combined with a high-quality analysis of well cuttings. As an in situ measurement, the gamma ray log can be used to provide a more refined depth control for the well cuttings samples, which may be important for well design. For example, accurate determination of the depth intervals of more productive sand units in an aquifer is critical for the installation of well screens at optimal depths. Gamma ray log patterns can be used for inter-well correlations, particularly where there is some lithological control.

It is common for most logging tools to include a gamma ray detector, which allows for depth corrections between logs run on the same borehole. For example, if a logged interval has a pronounced gamma ray peak (or other clear response signature) at a given depth, then the depths of all logs run on the well or borehole should be adjusted so that the peak occurs at the same depth in each log.

17.5 Spontaneous Potential

Spontaneous potential (SP) is the oldest and simplest borehole geophysical log. The SP log measures the electrical potential (voltage) generated by the interaction of formation waters, conductive drilling fluids and certain ion-selective rocks, such as shales (Schlumberger 1989). The SP response includes an electrochemical component and in some cases a typically minor electrokinetic (streaming) component caused by the flow of electrolytes (ions) in a permeable medium. An electrokinetic potential may be produced by mud filtrate flowing through the mudcake deposited on borehole walls and into adjoining permeable strata.

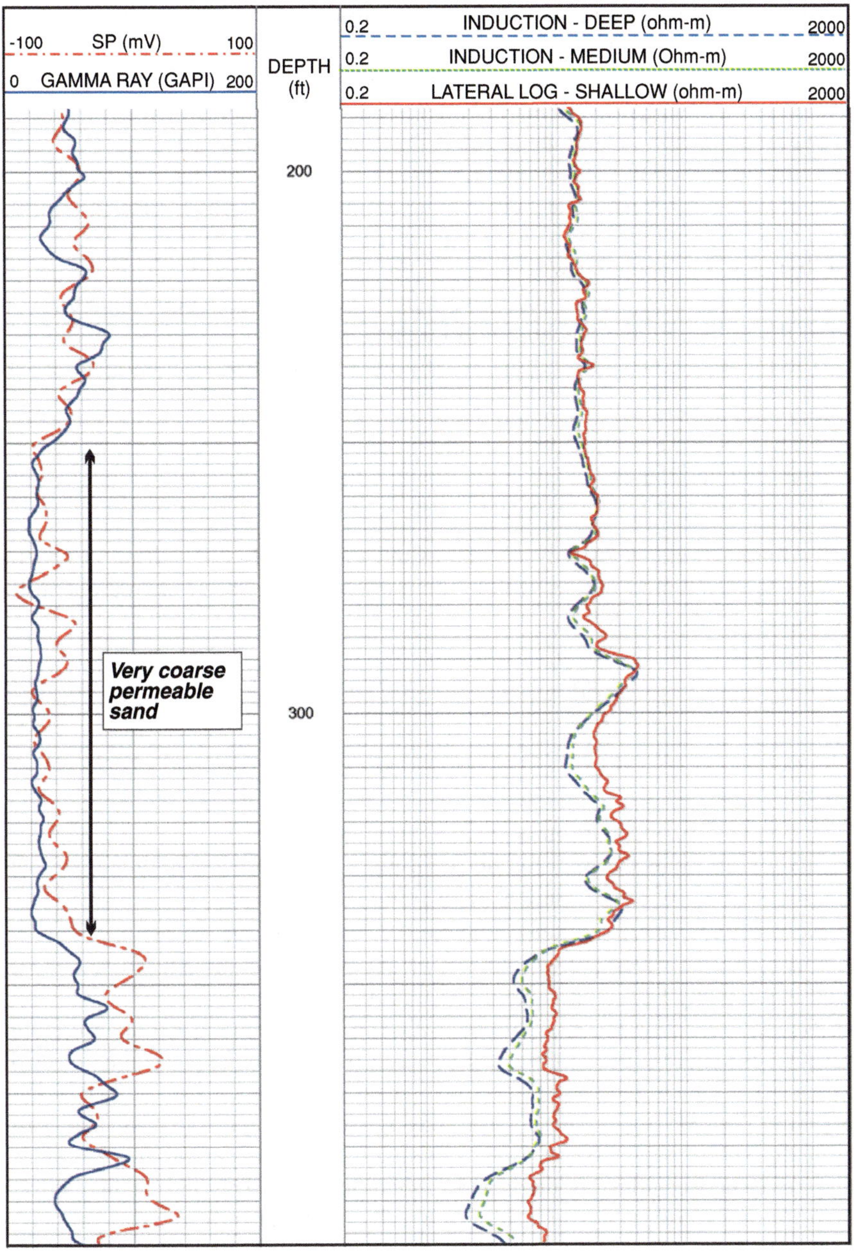

Fig. 17.2 SP, gamma ray, and dual induction logs of a very coarse permeable sand. The sand can be differentiated from adjoining sands by a deflection of the SP curve to the left (indicating a greater salinity than the drilling fluids) and decreased gamma response due to a lower clay content

The primary applications of SP logs are to identify and map strata containing interbedded shale (or clay) beds. The log requires that there is a salinity difference between the drilling fluid and native formation water. The SP log units are millivolts (mV). SP log data are not interpreted in absolute values, but rather in terms of the direction and magnitude of the deflection from a shale baseline. Permeable units are identified by deflections of SP logs from the shale baseline. If the formation water in a permeable unit is more saline (less resistive) than the drilling fluid (e.g., mud filtrate) in the adjoining borehole, then the reflection would be to the left (i.e., the potential will be less) (Fig. 17.2). Conversely, if the formation water is fresher than the drilling fluid, then the deflection at a permeable unit will be to the right (i.e., the potential will be greater).

The main potential applications of SP logs for groundwater investigations are

- locations of permeable bed,
- location of shale or clay beds (confining units),
- inter-well correlation, and
- determination of formation water resistivity.

Formation water resistivity can be estimated from the SP deflection and the resistivity of the mud filtrate (drilling fluid after mud has been filtered out) using the equation (Keys 1989)

$$SP = -(60 + 0.133T_f)\log\left(\frac{R_{mf}}{R_w}\right) \tag{17.1}$$

where,

SP = log deflection in millivolts
T_f = borehole temperature in degrees Fahrenheit
R_{mf} = resistivity of the mud filtrate (ohm-m)
R_w = resistivity of the formation water (ohm-m).

Keys (1989, 1990) cautioned that Eq. 17.1 is based on the assumptions that (1) the formation water is very saline, (2) NaCl is the predominant salt, and (3) the mud is relatively fresh and contains no unusual additives.

The SP log tends to be of limited value in most groundwater investigations because of small salinity differences between borehole fluids and native groundwater. Also, many aquifers do not have shale beds needed to establish a shale baseline. The SP log greatest value may lie in evaluating brackish and saline groundwaters in siliciclastic aquifer systems, where a greater salinity differences may occur between drilling fluids and formation waters. The SP logs may also be used for qualitative analysis of salinity trends and the identification of permeable beds, particularly in conjunction with other logs. In practice, the SP log is very inexpensive to run and is routinely included with resistivity logs.

17.6 Resistivity Logs

Resistivity concepts are discussed in Sect. 16.2 with respect to surface resistivity techniques. A number of different types of resistivity logs have been developed and used in the oil and gas industry and groundwater investigations. A voluminous literature exists on resistivity logging techniques and interpretative methods (e.g., Asquith and Krygowski 2004). In order to quantitatively analyze resistivity logs, calculated apparent resistivity (R_a) values must be converted to true resistivity (R_t) values by correcting for the various extraneous factors that affect resistivity measurements. The extraneous factors include the resistivity of the invaded zone (the zone in which drilling fluids entered the formation), the depth of invasion, resistivity of the drilling fluid (mud), borehole diameter, bed thickness, mudcake presence and thickness, and thickness and resistivity of adjacent beds (Keys 1989; Collier 1993). The methods for correcting for extraneous factors and obtaining true resistivity values are reviewed by Schlumberger (1989) and Asquith and Krygowski (2004).

The normal resistivity and lateral logging tools consist of two current electrodes located near the top and bottom of the tool and intervening potential electrodes. The potential electrodes measure the voltage drop when a constant alternating current is applied to the current electrodes. The volume of investigation increases with increasing potential electrode spacing. Commonly, potential electrodes, having spacings of 40.6 and 162.6 cm (16 and 64-inches) are used, which are referred to as the short and long-normal electrodes of logs. The short-normal log is considered to investigate only the invaded zone, whereas the long-normal log is considered to investigate both the invaded zone and the zone where native formation water is present.

The dual-induction log is based on Faraday's law in which a changing electromagnetic field can generate an electric current. A high-frequency alternating current is passed through transmitter coils in the logging tool (sonde), which generates an alternating magnetic field in the formation. The alternating magnetic field in turn generates current loops in the formation, which flow in paths coaxial to the sonde. The current loops produce their own magnetic field, which induces a current when they cross the receiver coils of the sonde. The recorded signal is proportional to the conductivity of the formation, which is the reciprocal of the formation resistivity. Practical induction logging tools use several arrays of coils, which are designed to achieve a specific targeted focusing and depth of investigation.

A typical dual induction sonde consists of a deep-induction and a medium-induction array on the same sonde, in which the two arrays share the same transmitters, but have different receivers. The dual-induction log is often combined with a shallow laterolog, which is referred to as either the dual induction laterolog (DIL) or dual induction focused log. DIL resistivity logs thus have three tracks, shallow lateral log and medium and deep induction, which record resistivity at different depths of investigation (Fig. 17.2). The greater depth of investigation of the deep induction log and focusing of the current allows for the determination of more accurate values of the true formation resistivity (R_t).

Separation of the three DIL or normal resistivity tracks is a manifestation of the invasion of drilling fluids into the formation, with the shallow logs tending to reflect the composition of drilling fluids and deep logs tending to reflect the composition of the formation waters. Drilling-fluid invasion is thus indicative of the presence of permeable strata. However, if the drilling fluids and formation waters have similar salinities (resistivities) then invasion may not be detectable. It is, therefore, important to record the resistivity of drilling fluids during logging. The depth of invasion is inversely proportional to porosity, with greater depths occurring in low porosity strata (Collier 1993). For a given volume of invasion, the fluid will occupy a greater volume of sediment or rock at lower porosities.

Resistivity log data can be quantitatively analyzed using Archie's equation (Eq. 16.3) to determine porosity or salinity. If the salinity (and thus resistivity) of the formation waters is known, then Archie's equation can be used to estimate formation porosity. Similarly, if the porosity of the formation is known, then the Archie equation can be used to estimate the resistivity of the formation water. The resistivity of water varies with temperature. After correcting log-derived formation water resistivity data for the values at 25°C, specific conductance (SC, μhmos/cm or μS/cm) can be calculated as

$$SC = \frac{10,000}{R_{w,25^{\circ}C}} \tag{17.2}$$

where,

$R_{w,25^{\circ}C}$ = temperature corrected formation water resistivity.

Specific conductance can in turn be converted to total dissolved solids (TDS) or chloride concentration using empirical relationships established from local water quality data.

Thus, a combination of resistivity, temperature, and porosity logs can be processed to provide a salinity versus depth profile. In groundwater investigations, commonly porosity data from a sonic log (Sect. 17.7) and formation resistivity from a deep induction log are used to generate log-derived TDS plots (Fig. 17.3). Where quantitative data on formation porosity are not available, formation water salinity can still be estimated by using porosity values estimated from examination of cuttings or bracketing the likely porosity values.

17.7 Sonic Logs

Sonic or acoustic logs provide information on the physical properties of the tested formation using the velocity, amplitude, and phase relationships of transmitted sound waves. The sonic logging tool consists of one or more transmitters and typically two or more receivers. The most commonly used sonic log is the velocity

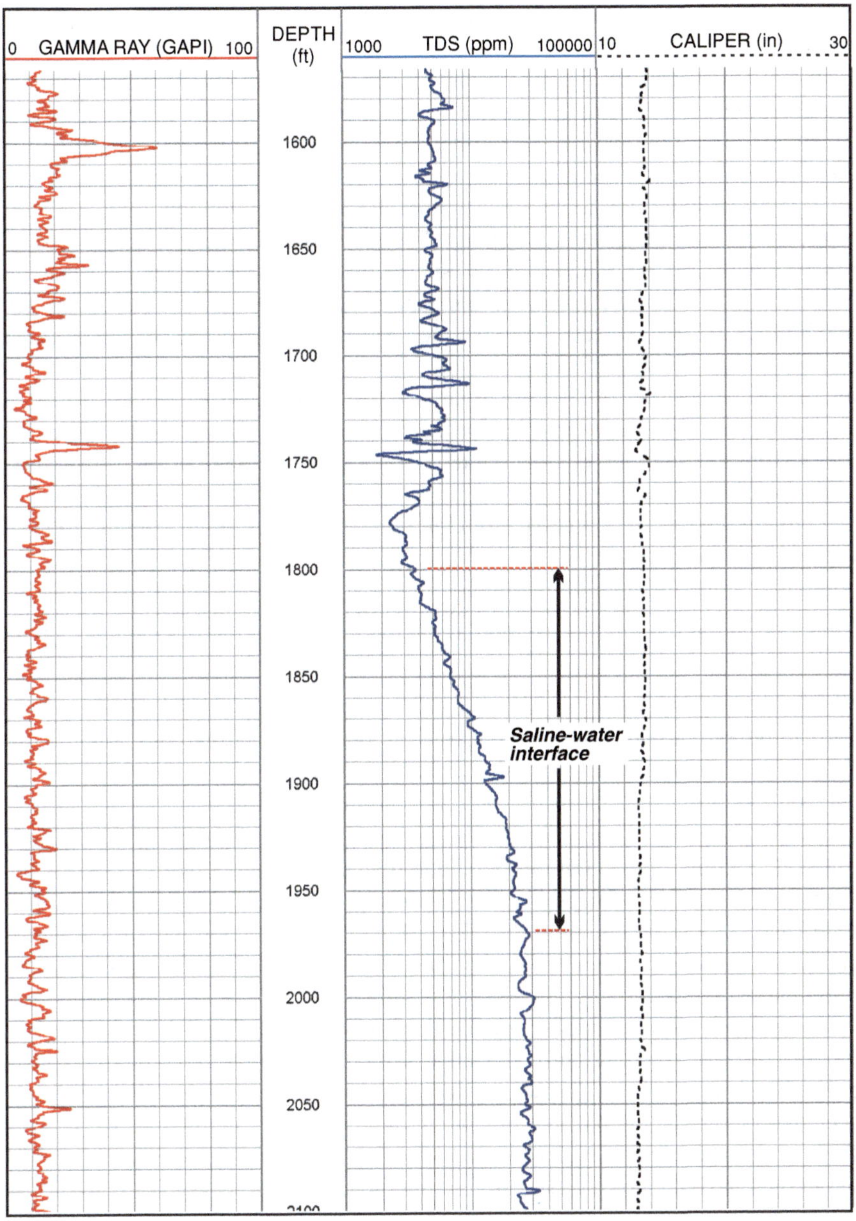

Fig. 17.3 Example of a log-derived total dissolved solids (TDS) plot, which shows a pronounced downhole increase in salinity between about 1,800 and 1,970 ft., approaching seawater values

or transit log, which records the travel time of compressional sound wave pulses from a transmitter to one or more receivers. The velocity of sound waves is a function of both the rock or sediment type (matrix) and its porosity. The travel time of a sound wave from the transmitter to a receiver, includes three components (1) travel time from the transmitter through the borehole fluid to the formation, (2) travel time within the formation, and (3) travel time from the formation through the borehole fluid to the receiver. The travel time within the borehole fluid is dependent upon the diameter of the borehole and the orientation and position of the sonde within the borehole.

In order to determine travel times within the formation, logging tools have two or more receivers at different spacings. The travel time in the formation between the receivers can be calculated by subtracting the travel times between the receivers, which would result in the canceling out of (or compensation for) borehole travel time effects. Transit times within the formation are expressed as interval transit times using the units of either microseconds per meter (μs/m) or microseconds per foot (μs/ft). Borehole effects can be further compensated for by having transmitters above and below the receivers and averaging the calculated transit times. Borehole compensated sonic logs are essential for quantitative analysis of the sonic log data.

Porosity (φ) can be estimated from interval sonic transit times using the Wyllie et al. (1958) formula

$$\varphi = \frac{\Delta t_{\log} - \Delta t_{ma}}{\Delta t_f - \Delta t_{ma}} \tag{17.3}$$

where,

Δt_{log} = interval transit time of the formation (measured; μs/m; μs/ft)
Δt_{ma} = interval transit time of the matrix (μs/m; μs/ft)
Δt_f = interval transit time in the wellbore fluid (μs/m; μs/ft).

Some standardly used interval transit times (Δt_{ma} and Δt_f) are provided in Table 17.1. Sonic logs commonly include tracks for sonic transit time and sonic porosities, which are typically, calculated using a constant matrix interval transit time based on the predominant lithology (Fig. 17.4). This approach can result in significant errors in characterizing formations containing interbedded strata of different lithologies.

Sonic logs often include a variable density log (VDL), which is an intensity modulated-time presentation of the sonic wave train in time from the transmitter pulse. The amplitudes of the wave forms produce a variable density that is presented versus time, shown in color or, more commonly, as gray or black bands (Fig. 17.5). The horizontal scale of the VDL log is in units of time (μs), in which the earliest arrivals are on the left. The frequency of the waves is related to the width of the bands.

The sonic transit time and VDL logs are useful for the detection of fractures and cavities. Unfractured rock usually has a VDL log pattern of continuous parallel bands, whereas fractured intervals may have an offset or a disrupted (chaotic) pattern and exceedingly long sonic transit times (Fig. 17.5).

Porosity values calculated using the Wyllie et al. (1958) formula are primary or matrix (intergranular or intercrystalline) porosity and will be underestimations of

Table 17.1 Interval transit times of different matrices and fluids

Matrix or fluid type	Interval transit time (μs/ft)	Interval transit time (μs/m)
Limestone	47.6–52.6	156.2–172.6
Dolomite	42.0–47.6	137.8–156.2
Sandstone (slightly consolidated)	58.8–66.7	192.9–218.8
Sandstone (consolidated)	52.6	172.6
Shale	62.5–167.0	205.1–547.9
Anhydrite	50	164
Freshwater	200	656
Brine	189	620
Fresh mud	189	620

Sources Keys (1989), Schlumberger (1989), and Asquith and Krygowski (2004)

total porosity if vuggy or fracture porosity is present. The percentage of vuggy or fracture porosity can be calculated by subtracting the sonic porosity from the total porosity obtained by other means (Schlumberger 1989; Asquith and Krygowski 2004). An additional empirical compaction factor should be applied for the calculation of the porosity of unconsolidated sands using sonic log data.

17.8 Nuclear Logging

Nuclear logs include the density and neutron logs, which are used to determine the porosity of formations. Density and neutron logs have the potential to be very useful for groundwater investigations, but are not commonly run because they utilize radioactive sources. The use and transport of radioactive sources is controlled by various governmental jurisdictions, and prohibitions may be in place against the use of radioactive sources in aquifers containing freshwater that are (or may be) used as drinking water sources. The legal liabilities and obligations associated with the use of radioactive sources may result in an unacceptable risk to some loggers, consultants, and well owners. This liability stems from regulatory requirements that obligate the logger, well owner, or consultant to recover the source if it is lost in the borehole.

17.8.1 Density Log

The density log measures the density of electrons in a formation, which is related to the bulk density of the formation. The logging tool consists of a gamma ray source, either Cobalt-60 or Cesium-137, which emits gamma rays into the formation that collide with electrons in the formation. At each collision, the gamma ray loses some of its energy to the electron and continues with a diminished energy

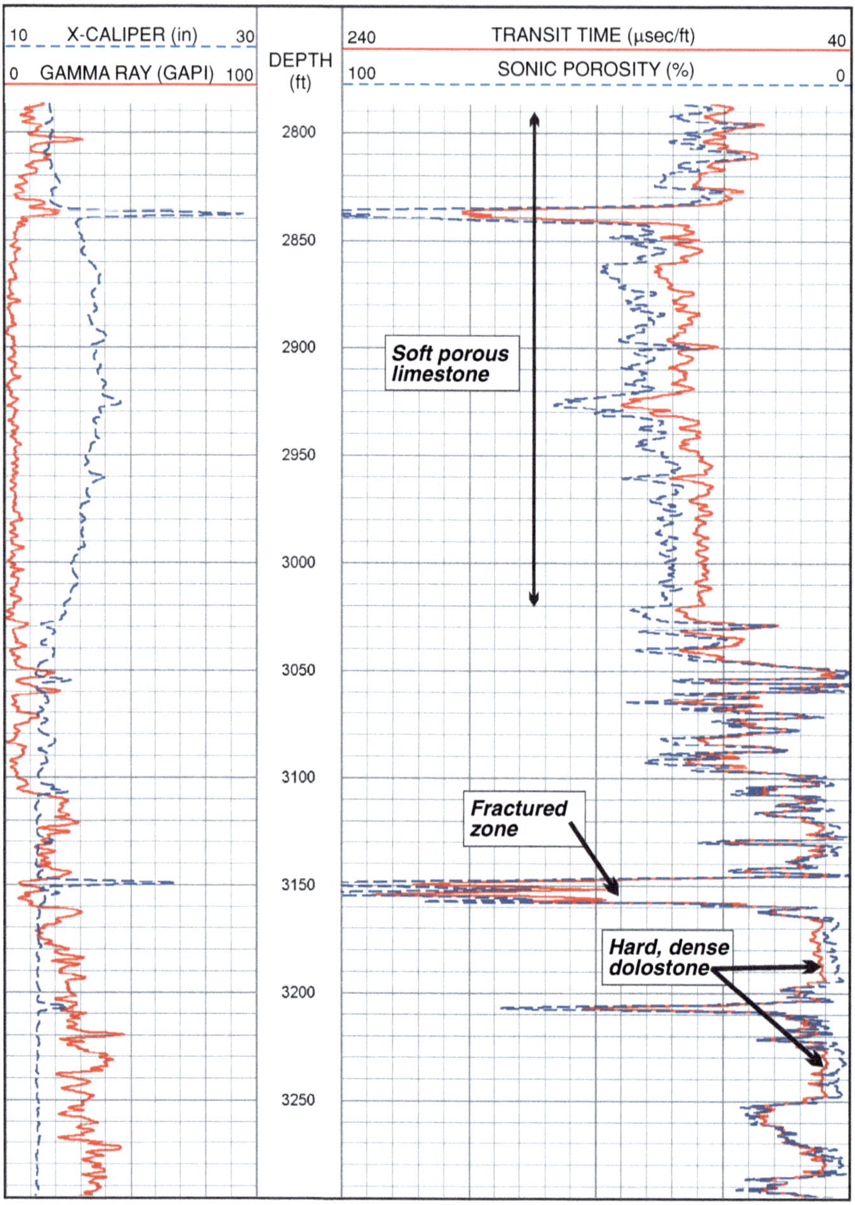

Fig. 17.4 Example of a sonic log that illustrates some basic log responses. Soft porous limestone is indicated by borehole enlargement and sonic porosities of about 35%. Fractured strata are manifested by exceedingly long transit times. Hard dense dolostones are indicted by minimal borehole enlargement and short transit times

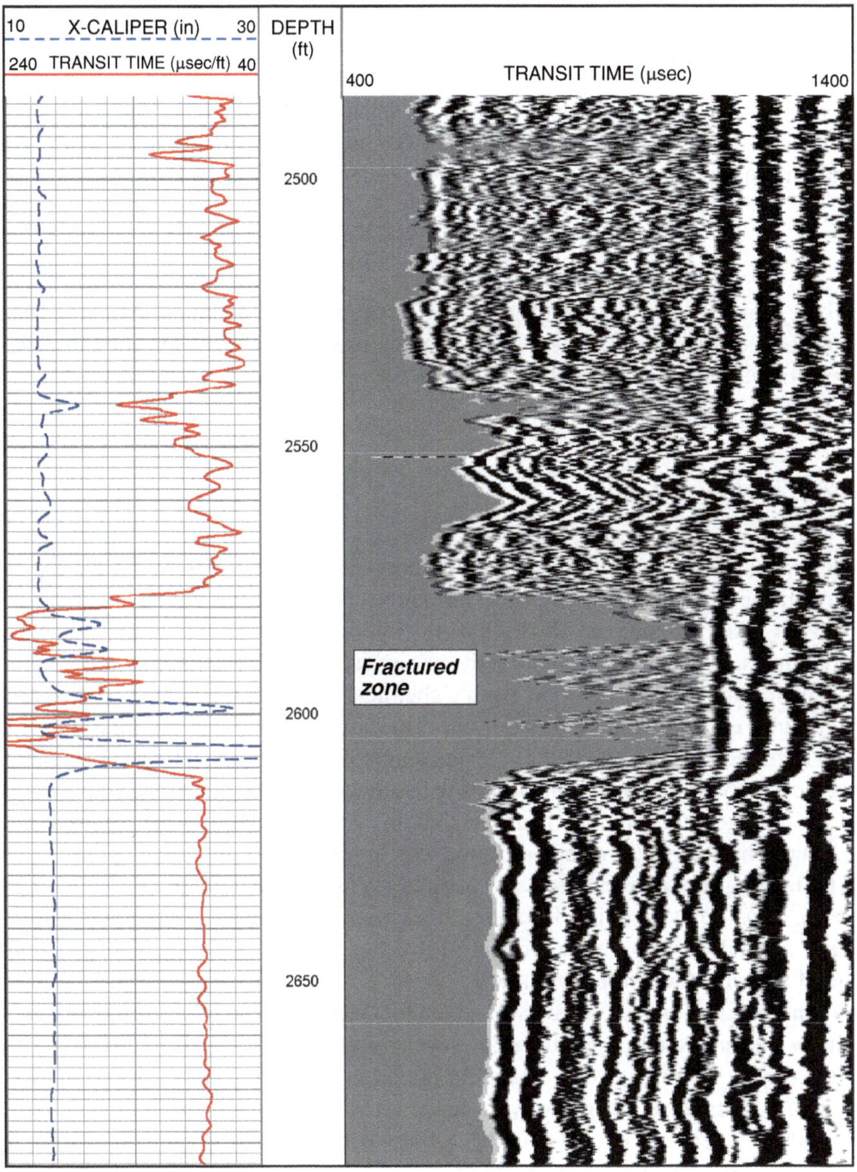

Fig. 17.5 Example of a sonic log and VDL log. Unfractured rock below 2,620 ft is manifested by sharp parallel bands. Fractured rock is indicated by disrupted and chaotic responses

by an interaction known as Compton scattering. The scattered gamma rays that reach a detector at a fixed distance from the source are counted. The density of electrons in a formation, and thus the number of scattered electrons that reach the detector is related to the bulk density (ρ_b) of the formation, as quantified in g/cc.

For water filled sandstones, limestones, and dolostones, the bulk density measured by the density log is practically identical to the true bulk density of the rock. For a few substances, such as gypsum, anhydrite, and coal, a small correction factor needs to be applied to obtain a true bulk density (Schlumberger 1989).

The porosity of a formation (ϕ) is a function of the bulk density of the formation, matrix density (ρ_{ma}), and pore fluid density (ρ_{f}), as follows

$$\phi = \frac{\rho_{ma} - \rho_b}{\rho_{ma} - \rho_f} \tag{17.4}$$

where densities are in units of g/cc. For freshwater to moderately brackish water and fresh drilling muds, a ρ_{f} of 1 g/cc can be used. The matrix densities (ρ_{ma}) can be estimated from the specific gravity of the predominant mineral phase.

17.8.2 Neutron Log

Neutron logs are used to determine the porosity of formations by measuring the amount of hydrogen present in the formation. For clean formations having minimal clay mineral contents, the bulk of the hydrogen in the formation is present in pore fluids, which is water in groundwater investigations. The neutron logging tool generates high-energy neutrons from a chemical source that commonly consists of a mixture of americium and beryllium. The fast nuclei collide with the nuclei of the formation material, each time losing some of their energy. The greatest energy loss occurs when the neutron collides with a nucleus of similar mass, which would be a hydrogen nucleus. The neutrons that are slowed down (thermal neutrons) are captured by the nuclei of other atoms, which then emit high-energy gamma rays. The logging tool either detects the gamma rays or thermal neutrons. The slowing down of neutrons, and thus formation of thermal neutrons and emission of gamma rays, depends largely on the amount of hydrogen in the formation, and thus fluid-filled porosity (Schlumberger 1989).

The neutron tool measures all water in a formation, including water bound in shales and hydrated minerals (e.g., clays). Neutron logs are also affected by the lithology of the matrix rock, and are reported in limestone, dolomite or sandstone porosity units. Neutron porosity is equal to true porosity if the lithology corresponds to neutron porosity units. Otherwise the neutron porosity must be corrected to true porosity (Schlumberger 1989; Asquith and Krygowski 2004). Neutron and density log measurements are affected by aquifer lithology in a different manner. Cross plots of density readings versus neutron porosity values allow for the determination of both aquifer lithology and true porosity (Fig. 17.6).

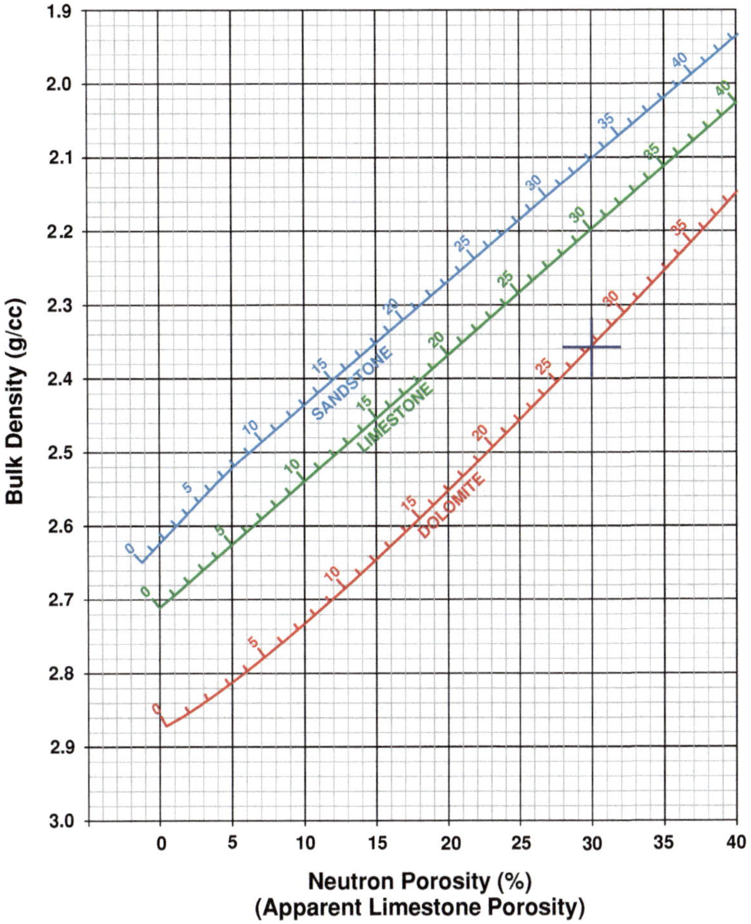

Fig. 17.6 Chart for interpreting neutron porosity and density log data for boreholes filled with freshwater (Schlumberger 1977). For example, a neutron porosity of 30% (in limestone porosity units) and a density of 2.36 g/cc, would indicate the presence of dolomite with a true porosity of 27.5%

17.9 Flowmeter Logs

17.9.1 Introduction

Flowmeter logs measure the vertical velocity of water in a well at the depth at which the tool is positioned. The principle objective of most flowmeter logging programs is the determination of variations in the flow rate (volume/time) with depth, which is related to aquifer heterogeneity. If the transmissivity (product of hydraulic conductivity and thickness) of an aquifer is determined from a pumping test, then flowmeter log data can be used to subdivide the aquifer into zones and apportion the total transmissivity between each zone, thereby, allowing hydraulic conductivity values to be assigned for various units. Flowmeter logging and its applications to hydrogeologic investigations has been discussed by Javandel and Witherspoon (1969), Keys (1989), Molz et al. (1989, 1990, 1994), Kabala (1994), Paillet (1998), Paillet and Crowder (1996) and Pailett and Reese (2000).

Flowmeter logs can be run under either ambient conditions (static logs) or pumping conditions (dynamic logs). Typically, both static and dynamic logs are run. The static logs can detect flow within the well caused by differences in heads between aquifer zones (inter-zone flow). Dynamic logs data must be corrected for the effects of any significant inter-zone flow, which is evaluated through the static log.

Flowmeter readings can be performed while the tool is being raised or lowered (trolling measurements) or at a number of depths in the well while the tool is not being moved (stationary readings). Trolling logs provide a continuous flow profile of the well, whereas the stationary mode provides measurement of the flow in the well at specific depths. Trolling flowmeter log readings depend upon (1) the velocity of water flow in the well at the measuring point and (2) the rate and direction at which the logging tool is being moved. Often a dynamic trolling log is run first on a well and then a series of stationary readings are made to further evaluate identified flow zones.

Three main types of flowmeter logs are commonly run in groundwater investigations: (1) spinner, (2) electromagnetic, and (3) heat-pulse. The impeller or spinner-type flowmeter logs are most commonly used, particularly in flowing wells and for dynamic tests. Each method has it strengths and weaknesses based on the borehole conditions and rate of flow.

17.9.2 Spinner

The spinner or impeller flowmeter is the oldest and still most widely used flowmeter technique in groundwater investigations. The spinner flowmeter method uses an impeller that rotates in response to fluid flow past the tool. The log output is the rotation rate, which is recorded as counts or rotations per second, which is proportional to the fluid flow velocity. Spinner logs can be run under both static and dynamic conditions and in the trolling and stationary modes.

Typically, trolling logs are run under both static and dynamic conditions. Dynamic logs are often run while the logging tool is being lowered through the well and then while the tool is being retrieved. In practice logging upwards tends to provide less useful information (and may be omitted) because the logging tool is being moved in the direction of the water flow, and thus the flow velocity past the tool is low. Spinner flowmeters have poor resolution at low flow velocities and the tools have a tendency to stall. It is imperative that the tool be retrieved at a constant rate and that the pumping rate or natural flow rate also be held constant during dynamic logging. The tool (line) speed should always be recorded when running a trolling flowmeter log.

The location of flow zones can be identified as depth intervals in which the flow rate in a well increases (i.e., more water is entering a well). Flow zones can be readily identified on flowmeter logs as intervals in which there is a pronounced separation of the static and dynamic tracks, provided that there is not a change in borehole diameter (Fig. 17.7).

Spinner flowmeter logs provide information about flow velocity (V) within the tested borehole. The key variable of concern is discharge rate (Q), which is the product of flow velocity and borehole cross-sectional area (A)

$$Q = V * A \tag{17.5}$$

Unless, the flowmeter log is run in a screened interval (in which the diameter is constant), a caliper log should also be run to measure borehole cross-sectional areas. The measured velocities must be normalized to the borehole cross-sectional area to assess flow at all intervals (see Sect. 17.9.5).

The main short-coming of spinner-type flowmeters is the lack of sensitivity to slow velocity flow (Keys 1989). Flow velocity also varies within the borehole, with the greatest velocities occurring in the center of the borehole and lower velocities occurring near the borehole wall as the result of friction. It is, therefore, important for the flowmeter tool to be centralized within the borehole. Large variations in borehole diameter and borehole wall roughness (rugosity) can cause high degrees of turbulence, which can also impact flowmeter log readings.

17.9.3 Electromagnetic Borehole Flowmeter (EBF)

The electromagnetic borehole flowmeter (EBF) is based on Faraday's law of induction. A voltage is induced by a conductor (such as water) moving at right angles through a magnetic field, which is proportional to the velocity of the moving conductor. The water flows through a vertical tube in the center of the probe, which is surrounded by an electromagnet. The EBF has no moving parts. Some applications of the EBF are discussed by Young et al. (1998) and Dinwiddie et al. (1999).

The EBF can be run in both trolling and stationary mode, although most of the reported applications use the stationary mode. The EBF also can be used to measure either the actual flow (discharge) rate or flow velocity in the same manner as a spinner flowmeter. An inflatable packer or skirt can be used to divert the entire well flow through the EBF tool. Inasmuch as the central tube of the EBF tool has a known constant diameter, the EBF can provide a direct measure of flow rate. The EBF thus acts in the same manner as the electromagnetic flowmeter widely used on horizontal pipes. In large diameter wells, the EBF can be operated without a packer. The measured flow through the EBF would be proportional to the total flow through the well.

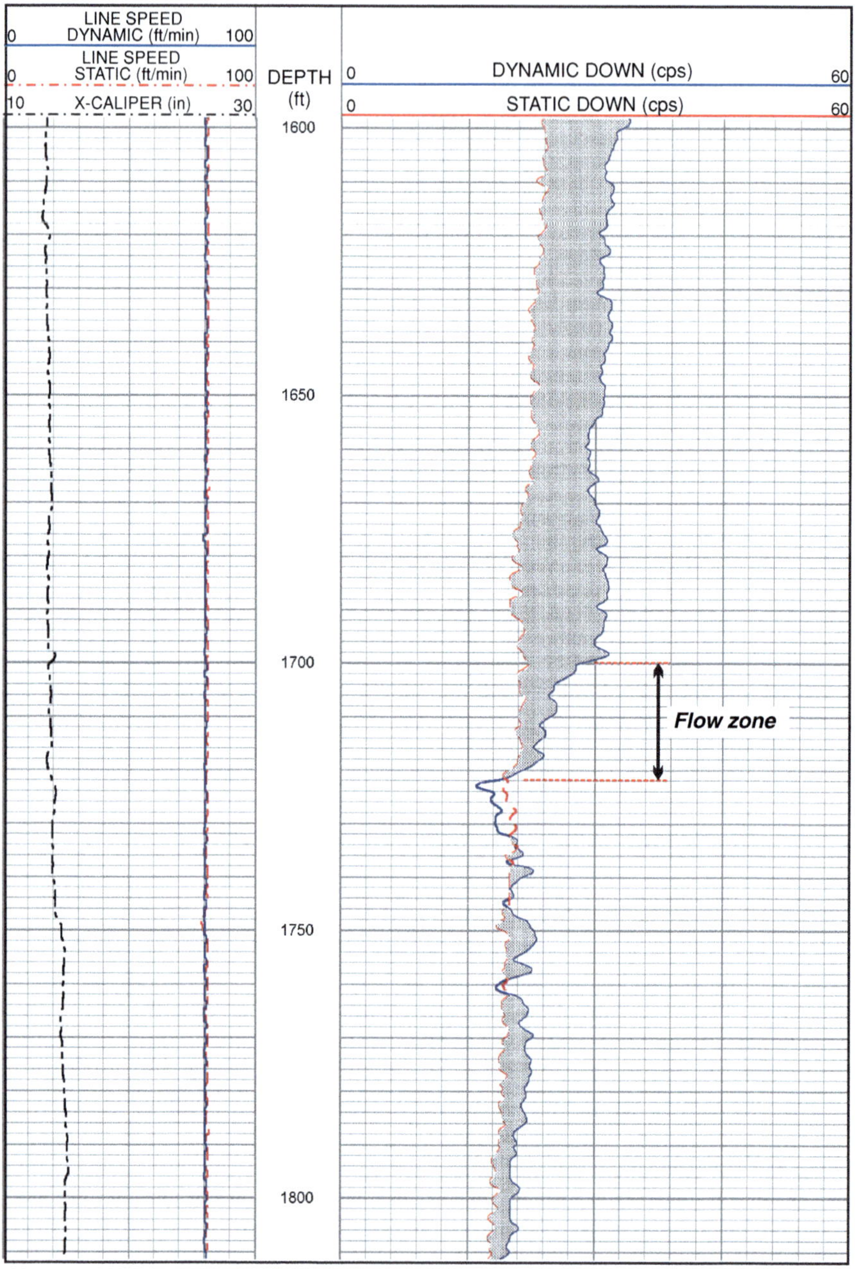

Fig. 17.7 Flowmeter logs run under state and dynamic (pumping) conditions. Logs indicate that most of the water is produced from a 22 ft (6.7 m) thick flow zone

The main advantages of the EBF is that it can operate at lower flow velocities than the spinner log and when used with a packer, it can provide a direct measurement of flow rate without the need to correct for variations in borehole diameter. The sensitivity of the EBF to low flow rates enables it to better detect inter-zone flow within a well.

The limitations of the EBF include that there may be a variable head loss through the EBF, which will be more of a problem at high pumping rates as head loss is a function of the velocity of flow through the tool (Dinwiddie et al. 1999). The head loss will be greatest at the top part of the tested interval, which has the greatest flow rate and velocity. It is therefore important to select a pumping rate that is adequate to stress the zone of interest, but not be too high as to cause excessive and variable head losses within the EBF tool during logging. Bypass flow around the packer may be significant in screened and gravel packed wells (Dinwiddie et al. 1999). Background electromagnetic currents may also affect EBF readings (Young et al. 1998). At the present time, EBF flowmeter logging equipment is not as widely available as the spinner logging equipment.

Calibration of the EBF tool is important. Calibration of the EBF tool can be checked by running the tool in the upper cased part of the well while pumping at a known rate. Where a high degree of accuracy is required, multiple runs should be performed at different pumping rates.

17.9.4 Heat-Pulse Flowmeter

The heat-pulse flowmeter (HPFM), developed by Hess (1982), measures water flow velocity by recording the time of movement of a heated packet of water. The basic design of the logging probe consists of a horizontal wire-grid heating element and heat sensors (thermistors) located above and below the grid. Pulses of electric current are applied to the heating grid, which produces a packet of heated water. The direction and rate of water flow can determined from the time that elapses between the application of the electric pulse and the detection of the warmed water by either the upper or lower thermistor. A flow rate can be determined from the elapsed time and the distance between the heat grid and thermistor. The major advantage of the HPFM is that it is a very sensitive instrument and can measure very low flow rates. It is particularly useful for detecting inter-zone flow in wells. The HPFM is not suitable for measuring flow rates greater than about 8 l/min (2 gpm).

17.9.5 Interpretation of Flowmeter Log Data

For log runs in a screened well or a well in which the cross-sectional area is near uniform, the discharge will be proportional to the measured flow velocity.

A correction may be required for large variations in the thickness of the gravel pack in a screened well. In open boreholes in which there is a significant variation in diameter and cross-sectional area, the flowmeter log data must be corrected using borehole dimensional data obtained from a caliper log. A basic processing technique for evaluating the fraction of the total well discharge (F_d) passing through the borehole at various depths (d) is to normalize the product of flowmeter log velocity readings (v; unprocessed log output) and cross-sectional area at the sample depth (v_d and A_d) with the product within the casing (v_c A_c) as:

$$F_d = \frac{(v_d A_d)}{(v_c A_c)} \tag{17.6}$$

The conversion factor to change the raw flowmeter log readings in rotation rate to velocity need not be considered as it would cancel out. The discharge at any depth (Q_d) in the well can be calculated from the well pumping rate (Q_T) as:

$$Q_d = F_d * Q_T \tag{17.7}$$

The fraction or amount of the total flow entering a well from the aquifer at any depth interval is the difference between the F_d and Q_d values calculated for the top and the bottom of the interval. Plots of the percentage of the well discharge versus depth (flowmeter interpretation log) allows for the location and quantification of flow-zone contributions within the tested intervals.

The average hydraulic conductivity of an aquifer layer (K_i) can be calculated from the flowmeter log data as (Javandel and Witherspoon 1969; Molz et al. 1989, 1990)

$$K_i = \frac{T\left(\frac{\Delta Q_i}{Q_T}\right)}{\Delta Z_i} \tag{17.8}$$

where,

K_i = average hydraulic conductivity of aquifer interval "i" (m/d; ft/d)
T = aquifer transmissivity (m^2/d; ft^2/d)
ΔQ_i = change in flow rate over layer (m^2/d)
Q_T = the total flow rate of well (m^3/d; ft^3/d)
ΔZ_i = the thickness of layer (m; ft).

Equation 17.8 assumes pseudo-steady state conditions (ΔQ_i and Q_T do not change over time), horizontal flow, minimal screen and head losses within the well, and no ambient flow in the well. Pseudo-steady conditions will occur when (Javandel and Witherspoon 1969; Molz et al. 1989, 1990)

$$\frac{r_w^2 * S}{4 * T * t} < 0.01 \tag{17.9}$$

where,

r_w = well bore radius (m; ft),
S = aquifer storage coefficient (dimensionless), and
t = time since the start of pumping (days).

In practice, this condition is met very quickly after the start of pumping and should not be an issue for an aquifer transmissive enough to be considered for use as a water supply. The effects of ambient inter-zone flow (i.e., vertical head gradients) can be compensated for by subtracting two steady–steady flow profiles according to the equation (Paillet 1998; Paillet and Reese 2000):

$$\frac{T_i}{\sum T_i} = \frac{Q_i^a - Q_i^b}{\sum (Q_i^a - Q_i^b)} \tag{17.10}$$

where,

T_i = transmissivity of aquifer interval "i", and Q_i^a and Q_i^b are the outflows from zone "i" under two steady state conditions.

One of the steady state runs could be performed under ambient (static) conditions. Although the basic data collection and analyses involved in the flowmeter method are quite simple, care must be taken to come as close as possible to meeting all assumptions and measuring only the actual flow caused by pumping (Molz et al. 1990). Attention to quality assurance, such as instrument calibration and test repeatability, is also important.

17.10 Temperature and Fluid Resistivity Logs

Perhaps the simplest borehole log is the temperature log, in which the operative element of the probe is a high-resolution thermistor. Fluid resistivity logs measure the resistivity of the fluids within the borehole using electrodes. Similarly, fluid conductivity logs measure the conductivity (reciprocal of resistivity) of borehole fluids. Temperature and fluid resistivity (conductivity) logs are often run using a combined probe in a single pass.

Temperature and fluid resistivity logs may be run under either static or dynamic (pumping conditions) flow conditions. Temperature logs record the temperature of the borehole fluid, not the formation. In wells that have been recently disturbed due to pumping or well drilling activities, the borehole fluid may not be in thermal equilibrium with the formation. Dynamic temperature logs record the weighted average temperature of the water that flowed into a well at or below the probe depth. If most of the water flowing into a well is derived from a deeper flow zone (well below the probe), then a dynamic temperature log will reflect the temperature of the flow zone, not the temperature of rock adjacent to the probe.

Temperature and fluid resistivity logs have several applications in groundwater investigations. Water temperature is needed for the quantitative interpretation of some other geophysical logs and data sets, particularly the determination of salinity (TDS concentrations) from resistivity logs. Information on borehole fluid resistivity is also needed for the interpretation of SP logs. Formation temperature data are needed for geochemical modeling, because equilibrium constants are a function of temperature.

Temperature and fluid resistivity logs are also used for the location of flow zones within a well. Abrupt changes in borehole fluid temperature and fluid resistivity trends in dynamic logs are indicative of the presence of a local flow zone (Fig. 17.8).

Temperature logs are also performed as an element of mechanical integrity testing programs, such as for injection wells systems. The curing of cement is an exothermic reaction. The generated heat of hydration of cement emplaced in the annulus between the casing and formation can be readily detected and measured using a temperature probe run through the casing. If curing proceeds too rapidly, the temperature will "flash" resulting in a spike in the temperature log. Conversely, a significant drop in temperature across a section of casing may indicate the absence of cement in part of the annulus. Temperature logs run on cased intervals after the cement cooled may also provide information on mechanical integrity. Sharp changes in temperature within the casing would suggest the presence of flow zones and thus breaches in the casing.

17.11 Borehole Imaging Logs

Borehole imaging logs provide images of the borehole wall that can be used to identify sedimentary structures, fractures, and other hydrogeological features (e.g., voids), which may have significance for groundwater flow. Imaging techniques are often used to identify secondary porosity features, such as fractures and cavities, which may have a high permeability and, as a result, may be preferential loci for groundwater flow. Borehole imaging logs vary in their image type, such as optical, acoustic, and resistivity. Lovell et al. (1999) provide a good overview of borehole imaging technology and applications, which are also summarized by Hurley (2004).

17.11.1 Borehole Video Survey

Borehole video surveys are performed by slowly lowering a video camera down the well and recording the optical images. Borehole video surveys are commonly run in hydrogeological and water well investigations. Downhole video cameras have become relatively inexpensive and most local geophysical loggers and some well drillers can run a survey at a modest cost. The surveys are used to

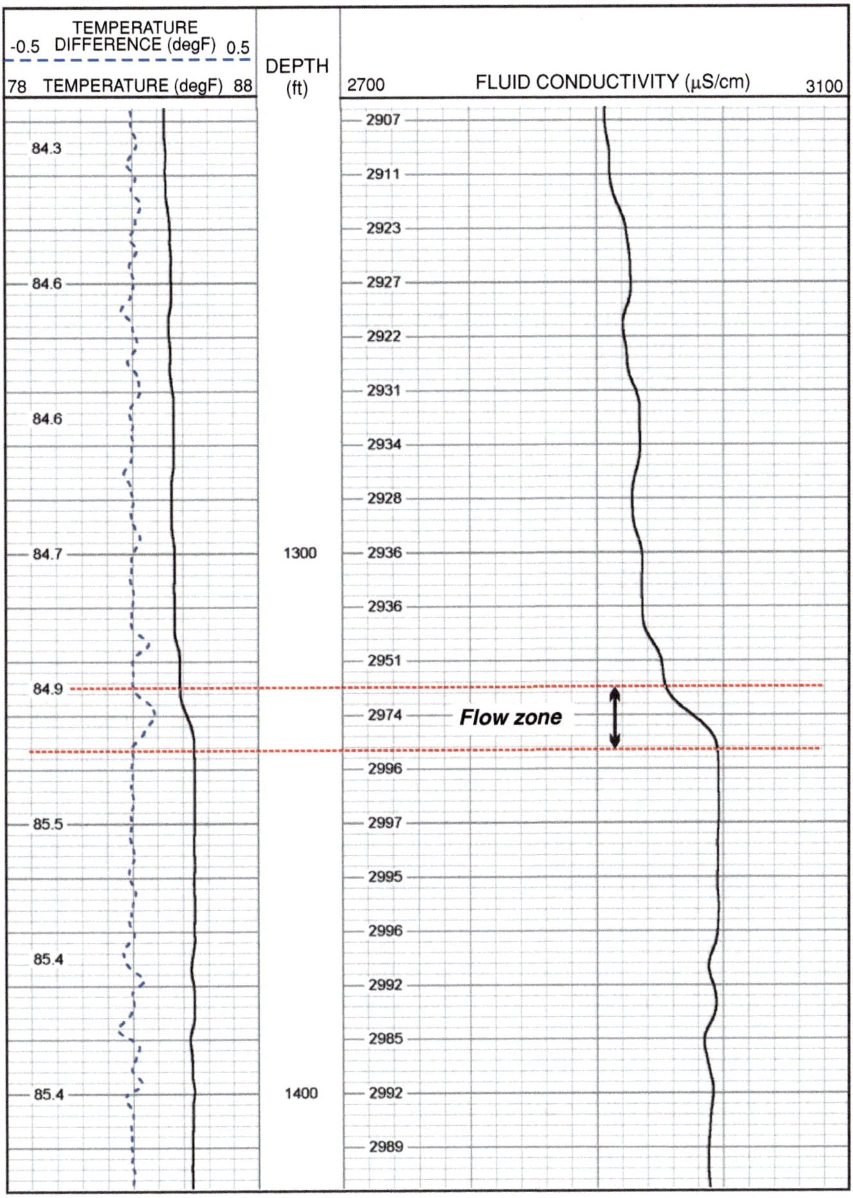

Fig. 17.8 Dynamic temperature and fluid conductivity log. Location of a flow zone is indicated by a relatively rapid change in both temperature and fluid conductivity

- obtain data on local hydrogeology,
- evaluate and document the mechanical integrity of wells,
- evaluate the condition of wells and determine the cause of well problems, such as clogging, and
- evaluate the effectiveness of well rehabilitation activities (before and after surveys).

Borehole video surveys require that the wells be filled with clear (non-turbid) water. It is strongly recommended that a tool be run that has both a downhole and a lateral view camera and records in color. The lateral camera allows for close-up images to be made of the well casing and screen, and borehole wall in open-hole intervals. Most downhole cameras with vertical and lateral image capabilities can move on a swivel to change the field of view which allows very detailed observations to be made of critical well construction features, such as the base of the casing (casing seat). It also allows images to be made of the interior of cavities and close-up views of sedimentary structures and fossils.

17.11.2 Optical Televiewer

The optical televiewer (OPTV) tool generates a continuous oriented, high-resolution 360° image of the borehole wall using an optical imaging system (Cunningham et al. 2004; Williams and Johnson 2004; Roberson and Hubbard 2010). OPTV images can be readily and rapidly interpreted to identify sedimentologic and diagenetic features that are significant for groundwater flow. The images can interpreted to determine the orientation of fractures, faults, and bedding plans, as well as the width of fracture apertures. The digital images can also be processed to provide estimates of vuggy porosity based on grayscale contrasts between vugs and adjoining limestone matrix. OPTV logs require that the boreholes be filled with clear water and may underestimate vuggy porosity if some pores are not darker than the adjoining matrix (Cunningham et al. 2004). A color contrast is required for borehole wall features to be visible.

17.11.3 Acoustic-Televiewer Log

The acoustic-televiewer log (ATV), which is also referred to as the borehole televiewer log (BTV), can provide a continuous high-resolution image of the borehole wall, even in mudded holes. The ATV logging method consists of the emission of ultrasound pulses from a rotating sensor and the recording of the amplitude and travel time of the signals reflected off the borehole wall. A 360° image of the borehole wall is generated, which is typically presented as a flat "unwrapped" diagram (Fig. 17.9). Low amplitude or high transit-time features by convention are shaded a dark color (Hurley 2004). High-amplitude and low

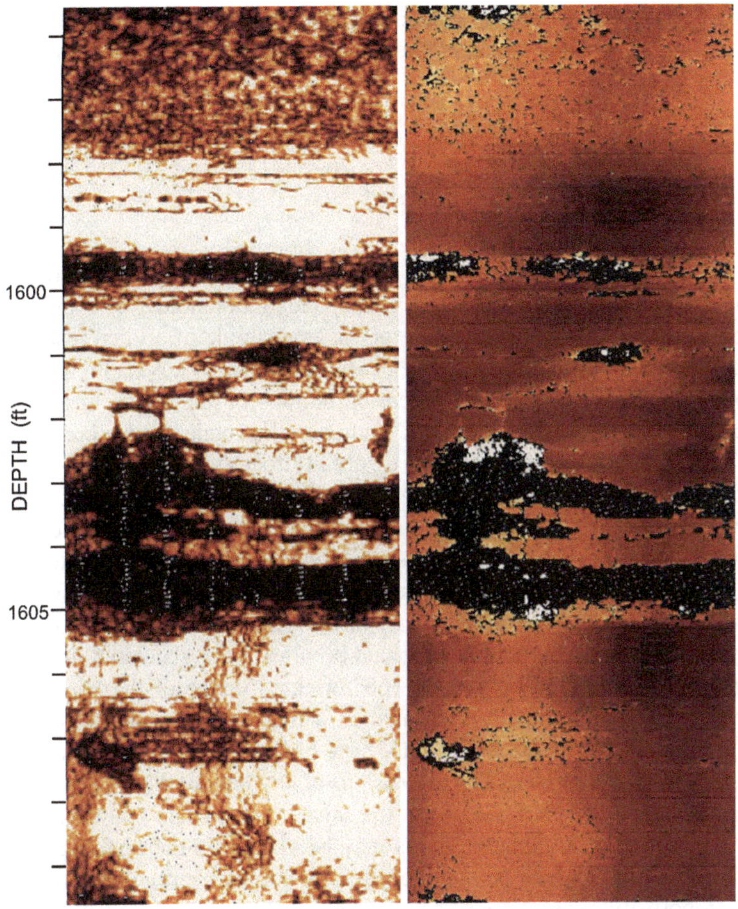

Fig. 17.9 Acoustic televiewer log. Cavities are shaded black

transit-time features are shown as lighter colors (shades of brown, orange, yellow, or white). The ATV can provide qualitative information on porosity, which is related to transit time.

The ATV log is very useful for revealing features such as fractures, fault planes, and bedding. Resolution is on the millimeter scale (fracture of an inch). The ATV log can be used to determine the orientation of fractures and other planar features (e.g., bedding planes). Vertical fractures appear as vertical lines and horizontal fractures appear as horizontal lines. Fractures or beds that dip between vertical and horizontal appear as sinusoidal traces. Angles of dip can be calculated from the amplitude of the sinusoidal traces and borehole diameter.

The ATV is not commonly used in groundwater investigations because of the high cost of the equipment and the need for experienced operators. It has been applied in investigations where information is required on the abundance and extent of fractures and other secondary flow features. In Florida, for example, the ATV log has become a standard tool for confinement analyses of deep injection well systems.

17.11.4 Microresistivity Imaging

Microresistivity imaging is another technique that can be used to obtain high-resolution images of borehole walls in mudded boreholes. The Schlumberger FMI™ Fullbore Formation MicroImager is an example of a microresistivity tool. The basic concept of the FMI is that an applied voltage causes alternating current to flow from each of up to 192 electrode buttons on the sonde through the formation to a receiver electrode located higher on the sonde. As the current emerges from a button on a tool pad, its path is initially focused on a small volume of the formation directly facing the button. The current path then expands rapidly to cover a large volume of formation between the button and upper electrodes. The measured current consists of a high-resolution component, modulated by resistivity variations in the formation directly facing the button, and a low resolution component modulated by the resistivity of the zone between the button and upper electrode. The high-resolution component dominates the generated image because its value varies between buttons. A 360° oriented image of the borehole wall is generated, which is typically presented as a flat "unwrapped" colored diagram (Fig. 17.10), similar to that of the ATV. The FMI log has a vertical and azimuthal resolution of 5 mm (0.2 in). The dimensions of any feature that is 5 mm (0.2 in) or larger can be readily estimated from the image. Finer-scale features, as small as 50 μm, that contain conductive fluids may be visible on the FMI logs. The resolution of the FMI log can be compromised by borehole irregularity (rugosity).

Macroporosity measurements for the FMI log are used to quantify the distribution of porosity in carbonate rocks. The matrix (intergranular and intercrystalline) porosity of a formation can be quantified by the subtracting the FMI macroporosity from total porosity obtained from other logs (e.g., sonic, neutron, density, NMR). In a conductive drilling mud environment, open fractures will be much darker than the surrounding rock due to mud filling the fractures, and their aperture, orientation, and density can be calculated from the image. FMI images allow for the visualization of sedimentary and structural features such as bedding, lamination, brecciation, and slumping. Structural strike and dip can be determined from the orientation of bedding.

17.12 Nuclear Magnetic Resonance Log

Nuclear magnetic resonance (NMR) logging provides a measure of the total fluid-filled porosity and pore-size distribution of a formation from which the bound and

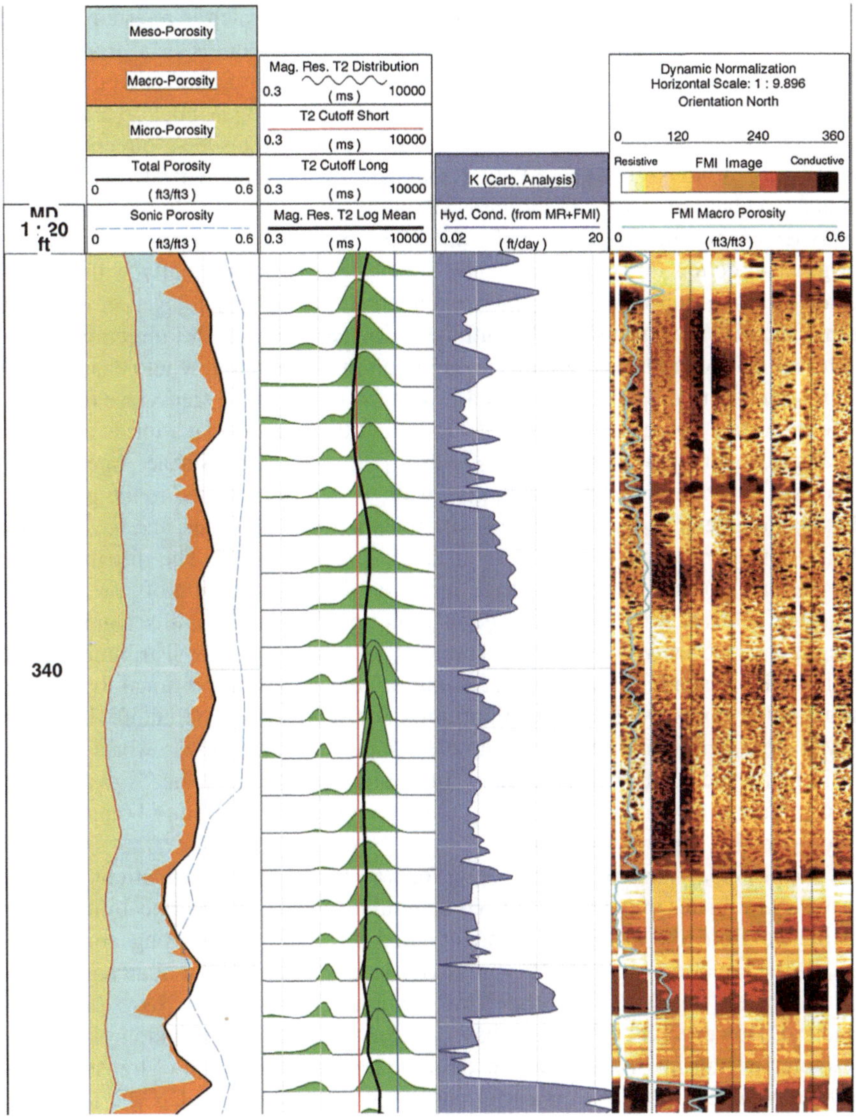

Fig. 17.10 NMR and FMI log from an ASR exploratory well in Florida. Bedding and small vugs caused by the dissolution of fossils are evident in the FMI log (*right track*). The T2 distributions (*second track from left*) and FMI macroporosity are used to determine the pore size distribution (*left*) track and hydraulic conductivity (*third track from left*)

moveable water distribution and permeability (and in turn hydraulic conductivity) are estimated. Current NMR logging techniques are discussed by Kenyon et al. (1995), Coates et al. (1999), Allen et al. (2000), and Freedman (2006). The NMR tool contains a large permanent magnet that aligns the non-lattice bound hydrogen atoms in the formation along the magnetic field vector, which is referred to as the longitudinal direction. The non-lattice bound hydrogen atoms in a groundwater system occur almost entirely in water molecules. A series of magnetic pulses from a radio frequency antenna on the tool causes the hydrogen nuclei (protons) to precess around the direction of the polarization field (transverse direction). The precessing protons create oscillating magnetic fields that generate weak radio signals, which are measured by the NMR tool.

After each radio frequency (RF) pulse, the protons start to relax towards the original direction of polarization. The signal decays exponentially with a characteristic time constant (T_2), which is called the transverse relaxation or decay time. Ideally, the protons will continue to precess around the direction of the external magnetic field until they encounter an interaction that would change their spin orientation to become out of phase with the others in the transverse direction. The important process for the NMR log is that the rate of relaxation is primarily related to interaction of the protons with a pore wall (grain surface interaction). The rate of proton interaction with pore walls is inversely related to the pore size. In small diameter pores, the protons will reach a pore wall faster, and have shorter relaxation times, than would occur in larger diameter pores. The distribution of relaxation (T_2) times thus provides a measure of the distribution of pore sizes.

NMR relaxation data is displayed as a plot of T_2 times in milliseconds versus incremental porosity (Fig. 17.10). The total area under the T_2 distribution curve reflects the relaxation of all the precessing protons and is proportional to the total water-filled porosity. Empirical algorithms and relaxation time cutoffs based on NMR measurements of thousands of core samples from around the world are used to interpret the T_2 distribution. Relaxation time cutoffs divide the T_2 distribution into different pore size bins (e.g., clay-bound water, capillary-bound water, microporosity, and mesoporosity).

The principal value of NMR logging for water resources investigations is that it provides estimates of permeability and hydraulic conductivity. Permeability can be estimated from the NMR total porosity and T_2 distributions using empirically established relationships. The transmissivity values of an interval of an aquifer can be estimated by integrating the NMR hydraulic conductivity over the depth interval. However, the NMR log may under estimate the contribution of discrete fractures to transmissivity in dual-porosity systems. Comparison of transmissivity values obtained from aquifer performance (pumping) tests and NMR logs may provide an estimation of the contribution of fractures and conduits to the transmissivity of the tested interval.

17.13 Elemental Capture Spectroscopy Log

The Elemental Capture Spectroscopy (ECS^{TM}) log provides a means for measuring the elemental composition and mineralogy of a formation. Elemental spectroscopy logs are reviewed by Barson et al. (2005). The ECS logging tool (sonde) uses a standard americium beryllium (AmBe) neutron source (similar to that used for neutron logs; Sect. 17.8.2) and a bismuth germanate (BGO) detector to measure gamma rays released as the result of the collision and interaction of the emitted neutrons with atomic nuclei in the formation. The Reservoir Saturation Tool (RST^{TM}) is similar to the ECS except that it employs a pulsed electronic neutron generator, and thus does not have a radioactive source. The RST log can be run where the use of tools with a radioactive source is prohibited or restricted. The released gamma rays have characteristic energies that depend upon the specific element and type of interaction. The ECS differs from the neutron log in that the sonde measures the full spectrum of gamma rays generated from neutron-element interactions.

The measured gamma ray energy spectrums are processed using an algorithm to determine the contributions from specific elements based on the known detector responses of the important constituents in the investigated strata. The relative elemental spectral yields are then converted to dry-weight elemental concentrations using an oxide closure method. Matrix properties and quantitative dry-weight lithologies are then calculated from the dry-weight elemental concentrations using empirical relationships derived from core chemistry and mineralogical databases. The primary application of the ECS and RST logs is determination of mineralogy of logged formations. The ECS and RST logs can provide measurements of the abundance of common sedimentary rock types and constituents, including calcite, dolomite, total clay (and some clay mineral types), QFM (quartz, feldspar, and mica), siderite and pyrite. The ECS log measurements have sensitivities of 1% dry-weight concentration or less.

17.14 Development of Borehole Geophysical Logging Programs

The fundamental issues for the development of a borehole geophysical logging program for a water resources project are:

- determination of the specific data needs that could potentially be met by geophysical logging,
- identification of the types of logs that are locally available that could meet the data requirements, and
- evaluation of whether the logs can be run and provide interpretable data under the site-specific borehole conditions.

A logging program should then be developed that cost-effectively provides the required data for project. Consideration should also be given to the fact that there may be only one practical opportunity to run some logs (e.g., before the casing is set) and that a valuable opportunity could be lost to obtain site-specific hydrogeological data that might be useful in the future. Often the costs to run some additional logs are very modest, particularly if they can be combined on a logging tool (i.e., simultaneously run). Mobilization and site time are major costs for logging programs.

Caliper and natural gamma ray logs should be run on all wells. Borehole diameter can be important for evaluation of other logs and casing installation. Gamma ray logs are useful for correlation between boreholes and for depth corrections of logs run on the same well.

If groundwater salinity is of concern for an investigation, then resistivity (or induction) logs should be run. A combination of a resistivity, temperature, and porosity (sonic) log would provide sufficient data for quantitative analysis of the salinity versus depth trend. A baseline logging program might consist of the following logs

- Natural gamma ray,
- X–Y caliper,
- Dual induction-laterolog (DIL),
- Sonic, and
- Temperature.

Time series runs of these logs could be used to determine the location of changes in salinity over time. For example, repeat runs of DIL logs over the operational life of a well completed with an open-hole, could be used to determine which aquifer zones are experiencing changes in salinity.

Where information is needed on aquifer heterogeneity (i.e., location of flow and poorly productive zones), a flowmeter logging program would be appropriate. Dynamic logs involve the pumping of a borehole, which requires that the borehole be stable. Flowmeter log programs can thus be run on either open boreholes or screened wells, but not on mud-filled wells. The recommended logs would include:

- Natural gamma ray,
- X–Y caliper,
- Static and dynamic flowmeter (spinner or electromagnetic),
- Static and dynamic temperature, and
- Static and dynamic fluid resistivity (or conductivity).

A time-series of flowmeter logs can be used to determine the locations of aquifers or zones within aquifers that have experienced clogging and thus reduced flow.

A sonic log and imaging log may also be useful for identification of fracture or cavernous zones that may dominate local groundwater flow, which is an important issue for projects or systems in which solute-transport is important. High-degrees of aquifer heterogeneity have been shown to be the principal cause of the failure of some aquifer storage and recovery systems (Maliva and Missimer 2010).

Nuclear magnetic resonance logs can provide data on fine-scale porosity and permeability distribution. NMR logs have the great advantage of being able to be run in mud-filled boreholes and could be used, for example, to locate permeable sands in unconsolidated formations in which to set screens. Sonic, caliper, resistivity, and elemental capture spectroscopy logs could be used to identify clay-rich, low-permeability beds that may impede vertical flow (i.e., act as confining units).

Geophysical logs represent a considerable investment. An important consideration is obtaining the maximum value from the investment. The next step is incorporation of the borehole geophysical log data and other aquifer characterization data into groundwater flow and solute-transport models, which can be performed using workflow software (e.g., Petrel^{TM}). Much too often the full value of borehole geophysical logging is not realized in groundwater resources investigations. Suites of borehole geophysical logs are commonly run, but much of the potential available data are not extracted and incorporated into the groundwater models that are used as the primary groundwater resources evaluation tool.

References

Allen, D., Flaum, C., Ramakrishnan, T. S., Bedford, J., Castelijns, K., Fairhurst, D., et al. (2000). *Trends in NMR logging: Oilfield review, 12*(3), 2–19.

Asquith, C., & Krygowski, S. (2004). *Basic well log analysis* (2nd ed., Vol. 16). American association of petroleum geologists methods in exploration.

Barson, D., Christensen, R., Decoster, E., Grau, J., Herron, M., Herron, S., et al. (2005). Spectroscopy: The key to rapid reliable petrophysical answers. *Oilfield Review, 17*(2), 14–33.

Coates, G. R., Xiao, L., & Prammer, M. G. (1999). NMR logging, principles and applications. Houston: Halliburton Energy Services.

Collier, H. A. (1993). Borehole geophysical techniques for determining the water quality and reservoir parameters of fresh and saline water aquifers in Texas (2 Vols.). Austin, TX: Texas Water Development Board Report 343.

Cunningham, K. J., Carlson, J. L., Wingard, G. L., Robinson, E., & Wacker, M. A. (2004). *Characterization of aquifer heterogeneity using cyclostratigraphy and geophysical methods in the upper part of the karstic Biscayne Aquifer*. Southeastern Florida: U.S. Geological Survey Water-Resources Investigations Report 03-4208.

Dinwiddie, C. L., Foley, N. A., & Molz, F. J. (1999). In-well hydraulics of the electromagnetic flowmeter. *Ground Water, 37*, 305–315.

Driscoll, F. G. (1986). *Groundwater and wells* (2nd ed.). St. Paul, Minnesota: Johnson Filtration Systems.

Freedman, R. (2006). Advances in NMR logging. *Journal of Petroleum Technology, 58*(1), 60–66.

Hess, A. E. (1982). *A heat-pulse flowmeter for measuring low velocities in boreholes*. U.S. Geological Survey Open File Report 82-699.

Hurley, N. F. (2004). Borehole images, In C. Asquith & S. Krygowski (eds.), *Basic well log analysis* (2nd ed.). American Association of Petroleum Geologists Methods in Exploration.

Javandel, I., & Witherspoon, P. A. (1969). A method of analyzing transient fluid flow in multi-layered aquifers. *Water Resources Research, 5*, 856–869.

Kabala, Z. J. (1994). Measuring distribution of hydraulic conductivity and specific storage by the double flowmeter test. *Water Resources Research, 30*, 685–690.

Kenyon, B., Kleinberg, R., Straley, C., Gubelin, G., & Morriss, C. (1995). Nuclear magnetic resonance imaging—technology for the 21st century. *Oilfield Review, 7*(3), 19–33.

Keys, W. S. (1989). *Borehole geophysics applied to ground water investigations*. Dublin: National Water Well Association.

Keys, W. S. (1990). Borehole geophysics applied to ground-water investigations. Techniques of water-resources investigations of the United States Geological Survey, Book 2, Chapter E2.

Keys, W. S. (1997). *A practical guide to borehole geophysics in environmental investigations*. Boca Raton: Lewis Publishers.

Kobr, M., Mareš, S, & Paillet, F. (2005). Borehole geophysics for hydrogeological studies: Principles and applications. In Y. Rubin & S. S. Hubbard (Eds.), *Hydrogeophysics*. (pp. 291–331). Dordrecht: Springer.

Lovell, M. A., Williamson, G., & Harvey, P. K. (1999). *Borehole imaging: Applications and case studies*. Geological Society of London Special Publication no. 159.

Maliva, R. G., & Missimer, T. M. (2010). Aquifer storage and recovery and managed aquifer recharge using wells: Planning, hydrogeology, design, and operation. Houston, Schlumberger Water Services, Methods in Water Resources Evaluation Series No. 2.

Maliva, R. G., Clayton, E. A., & Missimer, T. M. (2009). Application of advanced borehole geophysical logging to managed aquifer recharge investigations. *Hydrogeology Journal, 17*, 1547–1556.

Molz, F. J., Güven, O, & Melville, J. G. (1990). *A new approach and methodologies for characterizing the hydrogeologic properties of aquifers*. U. S. Environmental Protection Agency, Robert S. Kerr Environmental Research Laboratory. Ada, Oklahoma Report EPA/600/2-90/002.

Molz, F. J., Boman, G. K., Young, S. C., & Waldrop, W. R. (1994). Borehole flowmeters: Field applications and data analysis. *Journal of Hydrology, 163*, 347–371.

Molz, F. J., Morin, R. H., Hess, A. E., Melville, J. G., & Güven, O. (1989). The impeller meter for measuring aquifer permeability variations: evaluation and comparison with other tests. *Water Resources Research, 25*, 1677–1683.

Paillet, F. L. (1998). Flow modeling and permeability estimation using borehole flow logs in heterogeneous fractured formations. *Water Resources Research, 34*, 997–1010.

Paillet, F. L., & Crowder, R. E. (1996). *A generalized approach for the interpretation of geophysical well logs in ground-water studies: Ground Water, 34*, 663–898.

Paillet, F. L., & Reese, R. S. (2000). Integrating boreholes logs and aquifer tests in aquifer characterization. *Ground Water, 38*, 713–725.

Roberson, S., & Hubbard, B. (2010). Application of borehole optical televiewing to investigation the 3-D structure of glaciers: Implications for the formation of longitudinal debris ridges, midre Lovénbreen, Svalbard. *Journal of Glaciology, 56*, 143–156.

Schlumberger (1977). *Log interpretation charts*. Houston: Schlumberger.

Schlumberger (1989). *Log interpretation principles/applications*. Schlumberger: Sugarland, Texas.

Williams, J. H., & Johnson, C. D. (2004). Acoustic and optical borehole-wall imaging for fractured-rock aquifer studies. *Journal of Applied Geophysics, 55*(1–2), 151–159.

Wyllie, M. R. J., Gregory, A. R., & Gardner, G. H. F. (1958). An experimental investigation of the factors affecting elastic wave velocities in porous media. *Geophysics, 23*, 459–493.

Young, M. E., de Bruijn, R. G. M., & bin Salim Al-Ismaily, A. (1998) Exploration of an alluvial aquifer in Oman by time-domain electromagnetic soundings: *Hydrogeology Journal, 6*, 383–393.

Chapter 18
Remote Sensing

18.1 Introduction

Remote sensing (RS) can be defined as (Jha et al. 2007, p. 431)

> the observation of targets or processes from a distance (without physical contact), in contrast to in situ measurements wherein measuring devices are in touch or immersed in the observed system and/or process

There are two main types of remote sensing: passive and active. Passive remote sensing involves the measurement of natural emitted energy from the target. Active remote sensing involves the transmission of artificially generated signals to the target and measurement of properties of the return signal. Radar and Light Detection and Ranging (LIDAR) are examples of active remote sensing. Remote sensors may provide information on one or more bands of the electromagnetic spectrum. Remote sensing also includes aircraft and satellite-based geophysical surveys (Sect. 16.9).

The great advantage of remote sensing is that it can provide spatial, spectral, and temporal data and rapid coverage of large as well as inaccessible areas (Vrba and Verhagen 2006). Remote sensing involves observation of the surface of the Earth. Therefore, a link must be established between surface observations and subsurface (e.g., groundwater) phenomena of interest (Jackson 2002).

Remote sensing is not an end in itself, but is instead an early step in a hydrologic workflow. The remote sensing data are typically processed and evaluated in a geographic information system (GIS) framework (Chap. 19). A primary objective in groundwater resources management is to integrate remote sensing data with other data sources (e.g., well testing data and field observations) to develop surface water or groundwater models or fully integrated models.

Remote sensing allows for the collection of a wide variety of data, which involves a considerable range of required sophistication and effort for its processing and interpretation. Aerial photographs of the world can now be readily

R. Maliva and T. Missimer, *Arid Lands Water Evaluation and Management*, Environmental Science and Engineering, DOI: 10.1007/978-3-642-29104-3_18,

obtained on-line from Google Earth, commercial providers, and some government agencies, which can be relatively easily evaluated for some basic information on hydrology and land use. Advanced remote sensing techniques, such as quantitative interpretation of satellite-data, are complex and require specialists on the project team who are intimately familiar with the intricacies of data acquisition, processing, and interpretation.

A huge amount of global remote sensing data is available through the space programs. Remote sensing data are available for arid or semi-arid lands, including regions that are remote and have limited surface access. Satellites are broadly defined into two types, Earth resources satellites and environmental satellites. The former observe the same area of the Earth infrequently, but with relative high spatial resolutions. The latter observe frequently (on order of hours), but with relatively low spatial resolutions, and are used to provide information on weather conditions and large-scale surface phenomenon (Jha et al. 2007). The data are put to a myriad of uses include meteorology, climatology, agriculture, environmental studies, oceanography, and hydrologic investigations.

Remote sensing has become a major academic discipline with numerous dedicated journals (e.g., International Journal of Remote Sensing, Geoscience and Remote Sensing, Journal of Applied Remote Sensing, GIS and Remote Sensing Journal, Journal of Photogrammetry and Remote Sensing). A variety of books published have been published on the subject including textbooks by Siegal and Gillespie (1980), Jensen (2007), Sabins (2007), Schowengerdt (2007), Campbell (2008), and Lillesand et al. (2008). The applications of remote sensing to hydrology and groundwater investigations were reviewed by Salomomson and Rango (1980), Farnsworth et al. (1984), Waters et al. (1990), Engman and Gurney (1991), Rango (1994), Meijerink (1996, 2000), Becker (2006), Brunner et al. (2007), Jha et al. (2007), papers in a Hydrogeology Journal theme issue (2007, v. 15, n. 1), and Jasmin and Mallikarjuna (2011).

Applications of remote sensing to water resources investigations in arid lands include:

- location and mapping of geological features of hydrological significance, such as lineaments in hard rock terrains and buried channel deposits,
- hydrogeological mapping (differentiating between different surface deposits),
- land use mapping, such as the locations and size of irrigated areas from which consumptive water use may be estimated,
- land cover and vegetation mapping with both the type and density of vegetation mapped,
- measurements of parameters important for water budgets, such as precipitation, evapotranspiration, soil moisture, snow cover, and snowmelt runoff,
- mapping of groundwater discharge areas (e.g., sabkhas, playas, streams) and their changes over time,
- mapping of areas of groundwater emergence using temperature anomaly data (differences between groundwater temperature and surface temperatures),
- detailed mapping of topography,

- floodplain delineation,
- detection of changes in the mass of water in storage (GRACE program, Sect. 14.5), and
- measurement of land subsidence (InSAR, Sect. 15.2.4).

A great variety of remote sensing techniques are now available. Project leaders need to be aware of the technologies that are available, the types of data that can be obtained, the limitations of the technologies, and the costs of performing the surveys. An important cost element for remote sensing studies is that of ground-truthing (validation of) the data. Knowledge of local hydrogeological processes occurring in an area is necessary to choose appropriate surface or subsurface indicators.

The objective of this chapter is to provide an overview of some of the main applications of remote sensing techniques to water-resources investigations in semiarid and arid lands. The focus of the overview is on the types of information that remote sensing can provide, the limitations of the techniques, and examples of applications of the technique.

18.2 LIDAR

LIDAR, which is also known as laser altimetry, involves the transmission of pulses of laser light to the ground and measurement of the time of pulse return in order to obtain distances and, in turn, altitudes. Airborne LIDAR is well-suited for generating high-resolution Digital Elevation Models (DEMs). The National Oceanic and Atmosphere Administration (NOAA 2008) prepared an excellent introduction to LIDAR. The major advantage of airborne LIDAR is that it produces a rapid collection of points (more than 70,000 per s) over a large collection area. LIDAR cannot penetrate through vegetation and most of data points will record the top of the canopy rather than land surface. However, even if only a small percentage of the data points reach ground surface through the trees, sufficient data may be available to map land surfaces (NOAA 2008).

LIDAR requires that the altitude and position of the aircraft be known with a very high degree of accuracy, which is achieved using Inertial Navigation Systems (INS) or Inertial Measuring Systems (IMS) and a very high precision GPS. The accuracy of LIDAR is assessed through comparison of LIDAR data with measured altitudes on the ground (ground control points). In practice, independent field measurements are compared with a surface created from the LIDAR points, because LIDAR points, in most cases, will not fall exactly on the spot where the field measurements were made (NOAA 2008).

LIDAR data are commonly stored in a binary form, which at a minimum includes x, y, and z coordinates. DEMs are created using simple (e.g., nearest neighbor) to complex (e.g., kriging) gridding routines, which can create slightly different surface types (NOAA 2008). Contours are among the most commonly used representations for elevation and are commonly stored in vector formats (e.g., shapefile, dxf). Contours derived directly from LIDAR data are accurate, but not "clean" and often require interpolation, simplification, smoothing, or manual editing to achieve the intuitive product most people expect. In the process of cleaning and editing the vectors, the actual elevations can be generalized and lose the accuracy of data from which they were created (NOAA 2008).

18.3 Aerial Photograph Interpretation

Photographs taken from aircraft and satellite platforms are widely used in groundwater investigations to identify physical features of the landscape that might impact surface water and groundwater flow. Relevant surface features include lineaments, landforms, drainage patterns and textures, and vegetation cover. Such surficial information may provide insights to underlying groundwater hydrology. For example, in arid regions the type and density of vegetation can indicate groundwater depth and quality.

Features on aerial photographs are interpreted based upon shape, pattern, size, color (tone), shadow, texture, and association. Many features on aerial photographs are obvious and their interpretation is unequivocal. However, many other features on aerial photographs are subtle and can easily be missed or misinterpreted by neophytes. Experienced workers develop the ability to discern and understand the origin of less obvious features.

Interpretation of aerial photograph is facilitated if they are orthogonally (ortho) rectified. Digital ortho-rectification involves correcting photographs so that the scale is uniform and corresponds to real-world coordinates. Distances and locations can be accurately measured in ortho-rectified photographs in the same manner as they can be determined from maps.

Aerial photographs may be either in true color or false color. True color photographs retain the approximate color visible to the normal human eye. In a false color image, the correspondence between the subject color and image color has been changed. False color images generated in remote sensing studies may represent measured intensities from parts of the electromagnetic spectrum outside of visible range (e.g., infrared, ultraviolet). Colors are assigned to the various signal intensities. For example, infrared light enhances discrimination between healthy vegetation and vacant or poorly vegetated land. Healthy vegetation has a high reflectance of infrared radiation, and is shaded bright red, whereas surface water has a high absorbance of infrared radiation, and is shaded black (Fig. 18.1).

Airborne and satellite photographic systems may collect data in multiple wavelength bands. For example, the Landsat 7 satellite Enhanced Thematic Mapper Plus (ETM+) collects images in seven wavelength bands ranging from blue-green (0.45–0.52 μm) to thermal infrared (10.40–12.50 μm) (NASA 2001; U. S. Geological Survey 2003). The initial Landsat images were in black and white. Composite false-color images are created by assigning a primary color to each of the up to three bands that are being combined (Fig. 18.2). Objects on the

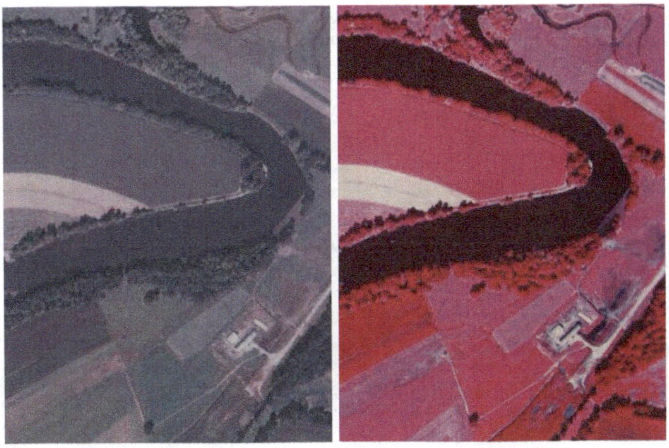

Fig. 18.1 True color (*left*) and false color infrared (*red*) aerial photographs from near Burlington, Vermont (USA). Surface water appears *black* in the infrared photograph and green healthy vegetation appears *red*. *Source* U.S. Geological Survey (2001)

Earth reflect different wavelengths of light. Bands are selected in the creation of composite images in order to best differentiate the objects of interest. For example, the false color image in Fig. 18.2 was created to highlight area of healthy vegetation, which corresponds to well-irrigated farm fields.

18.4 Applications of Remote Sensing to Groundwater Management

18.4.1 Lineament Mapping

18.4.1.1 Introduction to Lineaments

Lineaments are linear features evident at land surface that are an expression of an underlying geological structure. Lineaments may represent fundamental zones of weakness in the lithosphere and are often highly permeable pathways that may persist over long periods of time, especially in tensional fracture environments (Vrba and Verhagen 2006). Lineament analysis is a widely used tool for groundwater exploration in metamorphic and igneous terrains because the greatest amount of water tends to be found near fractures, which may constitute the only significant available porosity and permeability. In semi-arid regions, weathering in hard rock terrains tends to be concentrated along fracture zones. Such localized weathering processes tend to etch lines of weakness, which are evident on aerial photographs and satellite images as lineaments (Carruthers et al. 1991). However, mapped linear features can present a variety of geological phenomena (Sander 2007) and may not necessarily mark the location of underlying major groundwater flow zones (e.g., Carruthers et al. 1991). The

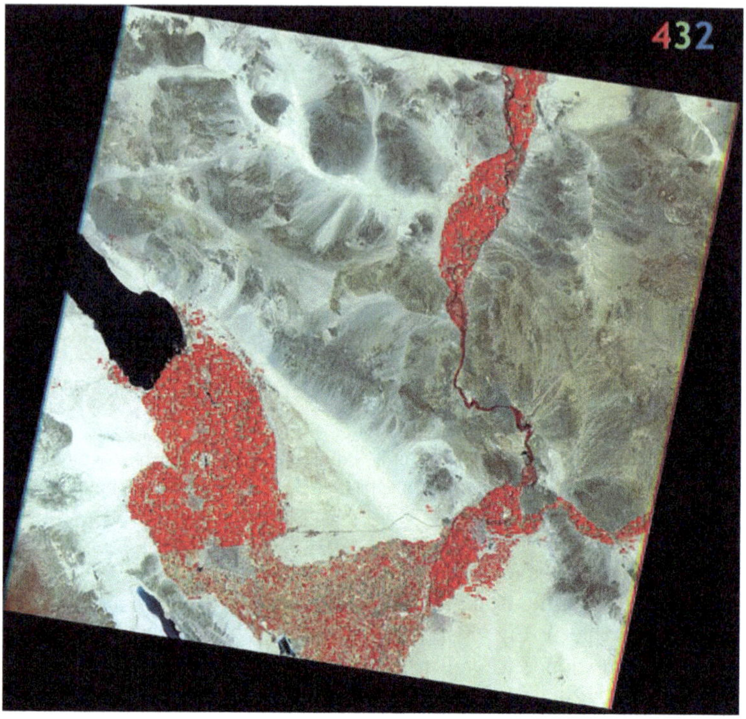

Fig. 18.2 False-color image Landsat 7 ETM+ image of the Salton Sea and Imperial Valley of California (USA). Irrigated areas with healthy vegetation in this desert environment appear *bright red*. The border between the United States and Mexico can be located from the difference in vegetation. *Source* NASA, http://landsat.gsfc.nasa.gov/education/compositor/pdfs/Landsat_7_Compositor.pdf

geologic cause of most lineaments is obscured by soils, regolith, or alluvial covers. Lineaments could also be low permeability features (i.e., dry lineaments). The translation of remote sensing and geophysically-extracted properties into hydrogeological properties is thus often ambiguous (Rhén et al. 2007). Lineaments, if they represent fracture traces, may also mark the direction of hydraulic conductivity tensors that have an influence on regional groundwater flow (e.g., Zeeb et al. 2010).

Important lineaments include (Hobbs 1904)

- crests of ridges or the boundaries of elevated areas or depressions,
- drainage lines,
- coast lines, and
- boundary lines of geologic formations, rock types, or lines of outcrop.

Variations in drainage may be manifested by differences in soils, vegetation, moisture or other surface expressions. Lineaments are two-dimensional manifestations of three dimensional features, which can be detected by a variety of remote sensing techniques including (Sander 2007)

- geophysical data (e.g., aeromagnetic surveys),
- topographic maps,
- DEMs,
- radar data, and
- optical data from satellite images and aerial photographs.

Digital and satellite images have been used to compile maps of lineaments and fracture zones. Multispectral satellite analysis is useful because different bands may reveal surface properties indicative of underlying linear features (Vrba and Verhagen 2006). Greater confidence occurs when lineaments are resolved in multiple bands (Teeuw 1995). Surface altitude data, particularly highly detailed DEMs generated from LIDAR data can be used to detect lineaments.

Brittle deformation zones may also be identified or confirmed using geophysical signatures (Rhén et al. 2007) such as:

- Magnetic lineaments—oxidation of magnetite around fractures reduce the magnetic susceptibility. Low magnetic linear features may thus be a sign of a flow zone.
- Electrical conductivity (resistivity) anomalies—The increased water content or clay minerals that form in fractures may be manifested by anomalies in electrical conductivity. The host non-porous crystalline rock is usually highly resistive.
- Reduced seismic velocities and diffracted or scattered patterns on seismic reflection profiles. Fracturing causes greater degrees of porosity thereby slowing seismic velocities and causing wave scattering.

Lineaments should be analyzed in the context of the geological history and structural environment of the study area (Bisson and Lehr 2004). It is important to determine whether or not a lineament is actually a true surface expression of underlying geological structure consistent with the known tectonic fabric of the region. It is also important to evaluate whether or not the underlying geology is conducive for the existence of flow systems. The local geology should be evaluated for the tendency of the different rock types for fracturing (brittleness) and for fractures to remain open (competency) (Bisson and Lehr 2004).

A multivariate statistical analysis of the factors related to well yield in the fractured bedrock aquifer of New Hampshire (USA) indicates a positive correlation of well yields with proximity to lineaments that are correlated with the primary fracture direction (Moore et al. 2002). Lineaments were related to fracture orientation through either domain analysis (i.e., orientation of fractures in a map grid cell), or discretely by comparing the strike of each lineament in a data set with the strike of nearby steeply dipping planar features or the trend of linear features in a structural geology data set (Fig. 18.3). Well yields were found to be inversely related to lineament density, which was attributed to high densities of observable lineaments tending to occur in the areas of thin overburden. Thin overburden in itself is inversely related to well yield (Moore et al. 2002). Well yields, in general, tend to be positively correlated to lineament densities that are a manifestation of intersecting sets of fractures.

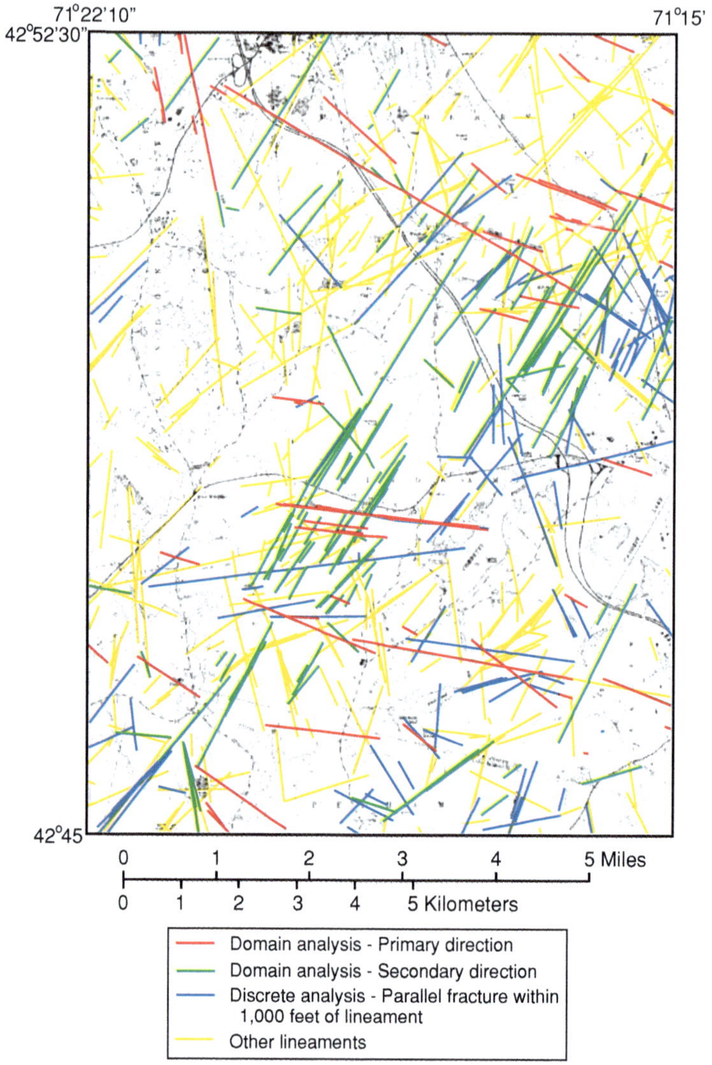

Fig. 18.3 Example of a lineament map prepared by the U.S. Geological for the Windham, New Hampshire, quadrangle (Moore et al. 2002). A key issue is determining which of the numerous lineaments are hydraulically significant

LIDAR data were used to detect lineaments in crystalline bedrock at a nuclear repository site in Sweden (Nyborg et al. 2007; Rhén et al. 2007). Groundwater flow is dominated by deformation zones, which are regions of multiple fractures. Deformation zones are frequently manifested as linear escarpments and distinct troughs because brittle fracturing reduces the weathering resistivity of the rocks. LIDAR was demonstrated to be a useful tool for the identification of lineaments, especially where bedrock is exposed at land surface. Lineaments in the length interval of a few meters (ft) up to at least 50 m (164 ft) where detectable using LIDAR, which compliments data on lineaments longer than 100 m (328 ft) obtained by the combined used of terrain models and geophysics (magnetics; Nyborg et al. 2007). However, LIDAR is more expensive to use than conventional photogrammetrical methods. Remote sensing data are less useful for identification of lineaments where the bedrock is covered with Quaternary sedimentary deposits.

18.4.1.2 Limitations of Lineament Analysis

Lineament mapping has often been criticized for the poor reproducibility between different operators and the apparent random distribution of lines without support of a proper geologic understanding of the area (Sander 2007). Inter-observer and intra-observer (repeat analyses by the same person) reproducibility is low. In a study by Mabee et al. (1994) involving three workers, each performing two temporally separate analyses, individuals were only able to reproduce lineaments at the same geographic location at an average of about 30% of the time. Less than 1% of the lineaments mapped by three observers were coincident. Other studies have reported greater percentages of coincident lineaments and mapped linear features (e.g., Sander et al. 1997). Lineaments considered by individual interpreters to be hydrologically most significant were found to be the most frequently reproduced.

Nevertheless, poor reproducibility will be an on-going limitation of lineament analysis. Automation of the lineament identification process will increase reproducibility and reduce operator bias. It is important to also recognize that the biases of experienced analysts could be beneficial in terms of identifying features of hydrological significance. As emphasized by Sander et al. (1997), professional judgment by experienced personnel is too valuable not to be included in the mapping of features for groundwater development.

It is generally recognized that lineament analysis for groundwater investigations of fractured hardrock aquifers is not credible without field verification (Mabee et al. 1994; Moore et al. 2002). Linear features do not necessarily have underlying structural significance. Linear features of hydrologic significance can be obscured by anthropogenic activities and some linear features may have to be omitted because of possible anthropogenic origins (Teeuw 1995). Best targets for sustainable well sites are located where the fracture zone can interact with storage in overlying unconsolidated material, which means that the fracture zones are locally concealed, and may have to be located through extrapolation of mapped features (Sander 2007). The storage capacity of the unlithified sediments tends to be orders of magnitude greater than the storage within the fractured bedrock. This is a key point because fractured crystalline rock typically has a very low porosity and much of the water storage may occur in overlying sediments (regolith; Sect. 4.2.5).

It is also widely recognized that a fundamental requirement for lineament analysis is that one must start with a good understanding of local geology and hydrogeology (Sander et al. 1997; Sander 2007). Lineament analysis, as is the case of other remote sensing technologies, is not a substitute for traditional field-based hydrogeologic investigation, but is rather an additional complimentary technology. Effective sitting of boreholes requires detailed knowledge of local subsurface hydrogeology, which in most cases cannot be obtained from remote sensing alone (Carruthers et al. 1991). Study area-specific investigations of the relationships between different types of lineaments and well yields are necessary to determine both whether lineament analysis is useful and, if so, how to obtain the most value from the data. For example, the lineament itself may not be the main permeable zone. Sander et al. (1997) documented that the wells completed closest to lineaments had the lowest yields, possibly due to the presence of clay gouge, which often occurs in old, faulted environments. Wells located further away, but within 250 m (820 ft), of lineaments had a relatively higher success rate.

Lineament analysis is used to prioritize areas for further investigation, using field surveys, surface geophysics, and test wells to assess hydrogeologic conditions (e.g., Carruthers et al. 1991). Information on the width and dip of fractures, extent of weathering, thickness of the regolith, and the depth to the top of the saturated zone can potentially be determined from surface geophysical techniques (Carruthers et al. 1991).

Another limitation of lineament analysis is the very large number (1,000s) of lineaments that can be mapped in regional studies, which precludes field checking of each one to determine whether or not they are of hydrologic significance. Filtering techniques are necessary to determine which lineaments warrant further investigation. Mabee et al. (1994) presented a two-step filtering process to reduce the number of lineaments to be further investigated and increase the probability of identifying features of hydrologic significance. The first step is a reproducibility filter in which overlapping linear maps prepared by different workers (or the same worker at different times) are used to identify coincident lineaments, which increases the confidence that the lineaments are real. The second step is a domain overlap analysis in which domains (areas) are identified in the field in which fractures have the same azimuths. Coincident lineaments are then screened to identify "fracture-correlated" lineaments whose orientations coincident with the local fracture (or fault) trend. The two-step screening process reduced the lineament database from 6,500 to 217. Mabee et al. (1994) found that wells located within 30 m (98 ft) of fracture-correlated lineaments were more productive than wells positioned at other locations.

A coincidence test was also used in a study of lineaments and well specific capacity in karstic limestones in Vietnam (Tam et al. 2004) in order to reduce uncertainty. Lineament analyses were performed both automatically using Landsat 7 ETM imagery, and manually by an experienced geologist using black and white aerial photographs with stereoscopic visualization. Tam et al. (2004) emphasized the importance of having professional experience with lineament analysis, and screening the data for non-geological features, such as road and canals. Two parameters were quantified using GIS; lineament length density (L_d) and lineament frequency (L_f). Specific capacity data from 16 wells were used to evaluate the significance of lineaments on well yields. A poor correlation was observed between L_d and specific capacity. The well data were then divided into two groups by cluster analysis of geomorphic and hydrogeological features. A correlation was observed between L_d and specific capacity for wells in recharge areas, but not for the wells in river valleys with relatively flat terrains (discharge areas). The absence of a correlation in discharge areas was suggested to be caused by lineaments being covered with alluvial deposits and thus not manifested by remote sensing data.

Although much higher degrees of technology are now used to identify and map lineaments, it is not clear whether the practical results are much better than in earlier times when lineament analyses were performed primarily manually using aerial photographs. Sander (2007, p. 73) noted with respect to lineament mapping and interpretation for water supply that

> The practical use of the results is sometimes compromised in favor of advanced image processing algorithms and numerical wizardry of mainly academic interest

It is clear that lineament analysis has significant value for groundwater development in hardrock terrains in arid and other regions despite its limitations. Lineament analysis is not an infallible technique for identifying productive rock for well sites, but rather, in some circumstances, can increase the success rate of well drilling (Sander et al. 1997).

18.4.2 Identification of Structural and Sedimentological Features of Hydrogeological Significance

Airborne radar can penetrate through dry sand and collect images of shallow subsurface features in arid regions. Radar can be helpful for identifying shallow groundwater reserves in buried stream channels that are covered with aeolian sands (Vrba and Verhagen 2006). Several studies have demonstrated the usefulness of space shuttle-based radar for identifying paleodrainage features that could have modern water resources importance. McCauley et al. (1982), Abdelsalam et al. (2000), and El-Baz (2001) used different generations of satellite and Shuttle Imaging Radar (SIR) to identify paleodrainage features in the Sahara Desert, which are believed to be relevant to current groundwater resources. The topographic depressions in which most major sand accumulations (sand seas) were deposited are interpreted to have been filled with freshwater in the past, which recharged the underlying aquifer (see Chaps. 3 and 4).

Robinson et al. (2007) utilized remote sensing techniques to evaluate the distribution of paleodrainage features in the East Oweinot region of Southwest Egypt and their effects on water quality within the Nubian Sandstone. Synthetic aperture radar (SAR) was used to map ancient water courses because of the unique ability of radar to penetrate through the surficial sand cover. The depth of near-surface SAR imaging varies according to the moisture content of the sand at the time of imaging and the radar wavelength used for imaging. A Digital Elevation Model (DEM) was produced from Shuttle Radar Topographic Mission (SRTM) elevation data. The DEM was used to establish the slope and confirm the flow direction from the data analyzed from the SAR images. Areas with slopes to 1–5% were appraised as having the best recharge potential from runoff. Areas having very low slopes (0–1%) suggest unfavorable infiltration conditions and the possible ponding and evaporation of surface water in the past.

Robinson et al. (2007) observed that a relationship exists between the spatial organization of fluvial and structural features and the occurrence of relatively low-salinity groundwater. Fluvial and structure controls of recharge left an imprint on the quality of fossil groundwater and contributed to its present-day heterogeneities of occurrence. From a practical perspective, newly identified alluvial fans are targets for new freshwater resources. It was also determined that wells in the vicinity of structural features contain lower salinity water than wells that are not. Previous studies indicated that aquifers formed in anticlines and that the up thrown sides of faults have better water quality.

18.4.3 Identification of Recharge and Discharge Areas

Remote sensing techniques are well suited for the differentiation of land areas into discharge and recharge areas, because each area may have diagnostic vegetation cover patterns, topography, and electrical conductivity. Tweed et al. (2007), for example, provides a good example of the application of remote sensing and GIS to identify groundwater discharge and recharge areas in a basalt aquifer in south-eastern Australia. The normalized difference vegetation index (NDVI) from Landsat data were used to evaluate the amount of green vegetation present in an area. The NDVI is based on the difference in reflectance by healthy green vegetation of visible red and infrared light. In discharge areas, the availability of groundwater allows photosynthesis to be sustained longer into the summer dry season than in recharge areas. Discharge areas would thus have a lower seasonal variability of vegetation photosynthetic activity. Tweed et al. (2007) quantified the variability in terms of the standard deviation of NVDI values (SDVI).

In addition to a lower standard deviation of NVDI values, discharge areas would tend to occur in topographic low locations with accompanying shallow water table positions. Groundwater flow lines will also tend to converge towards discharge areas (Tweed et al. 2007). Recharge areas in the study tend to be associated with volcanic features (identified from Landsat ETM+ images), are topographically high areas, and have a lesser degree of weathering of the basalts. The latter can be identified by relatively high potassium and low thorium concentrations from airborne radiometric (gamma ray spectroscopy) data. Recharge areas may also have relative low salinities and electrical conductivities.

As would be expected, the SDVI indicator gave good results in the drier, north part of the basalt aquifer and had more limited results in the southern region where rainfall is higher (Tweed et al. 2007).

Tweed et al. (2007) noted that their mapping identified the spatial extent of the processes occurring and does not provide quantitative information on the annual net influx and outflux of water. Tweed et al. (2007) also cautioned that areas could temporally vary between recharge and discharge characteristics depending on local climate conditions, such as the rainfall rate.

Münch and Conrad (2007) used remote sensing and GIS to map groundwater dependent ecosystems (GDEs) in the semi-arid Western Cape of South Africa. GDEs were identified using a combination of biomass indicators and physical wetness indicators. The latter considered factors such as the depth to water, lineament density, slope, and flow accumulation, which were evaluated using a weighted overlay function. An important issue is the time of collection of the biomass data. Münch and Conrad (2007) noted that optimal image acquisition date should be at the time when GDE areas exhibit significantly different characteristics from the surrounding land cover, which usually coincides with a period of seasonal climate change and associated vegetation response. Münch and Conrad (2007) used data from shortly after the start the onset of the winter rains in a dry period.

Remote sensing could identify areas of elevated 'wetness' and 'greenness' but there was also a noted marked difference in these indicators for GDEs. Clumps of GDE indicator species were too fragmented with spectral characteristics similar to adjacent land cover to be detected with the resolution of the Landsat images used (Münch and Conrad 2007). GDEs may also be masked by the effects of land use activities, such as land clearance and cultivation (Münch and Conrad 2007). The end product was maps in which GDE probability was classified as low, medium, and high.

Münch and Conrad (2007) emphasized the importance of prior knowledge of the study area, particularly the expected landscape structure and land-cover and land-use characteristics. The study results demonstrated that remote sensing and GIS are a cost-effective method for predicting the location of GDEs. The results were described as being "adequate", and that there are opportunities to improve the accuracy and efficiency of the method, such as through the use of other biomass indicator combinations (Münch and Conrad 2007).

Temperature (infrared thermal analysis) data has been used to identify groundwater discharge areas and saturated soils. Shallow groundwater typically has a much narrower annual range of temperatures than land surface and surface-water bodies. Springs, seeps and other groundwater discharge areas may be detected by the difference in temperature between non-discharge areas. The contrast in temperatures will be greatest at the times of the year when surface temperatures are near the ends of their annual ranges (i.e., winter and summer) (Becker 2006). Saturated soils act as heat sinks in the summer and heat sources in the winter (Becker 2006).

Airborne electromagnetic surveys can detect variations in the electrical conductivity, and thus salinity, of the shallow aquifers. The distribution of salinity is of obvious importance for solute-transport modeling. The distribution of salinity may also provide insights on the local distribution of recharge and discharge (Sect. 16.10.4). Areas that are sites of recharge with relatively fresh water may have lesser electrical conductivities than surrounding areas where evaporative concentration of salts (salinization) is occurring (Brunner et al. 2007).

18.4.4 Vegetation Mapping

Vegetation types and density are strongly dependent on water availability in semiarid and arid regions. As discussed in Sect. 18.4.4, vegetation density can be used to identify areas in which groundwater is close to land surface (e.g., discharge areas) and areas with more abundant groundwater resources (e.g., fracture zones in hard rock terrains).

Elmore et al. (2003) used a combination of remote sensing, groundwater monitoring data, and field vegetation surveys to evaluate changes in vegetation cover and types in the heavily utilized Owens Valley of California. Cloud-free Landsat Thematic Mapper data were analyzed by Spectral Mixture Analysis (SMA) in which spectral properties at each pixel were modeled as linear combinations of the spectral properties of fundamental, basic properties (i.e., end members). The four end members used in the Owens Valley study were light soil, dark soil, vegetation, and shade. The key end member of interest was percent vegetation land cover (%LC).

Data were analyzed for the periods September 1986–1998 and comparisons were made against the initial data and in 2-year pairs. Cluster analysis was used to identify land surfaces exhibiting common change histories. The combined field vegetation survey, depth to water, and %LC data allowed for the evaluation of changes in water-table position on plant communities. A key observation is that phreatophyte communities were more vulnerable to changes in the depth to water, compared to xeric communities. Impacted phreatophyte communities experience a decrease in %LC during droughts, which created an opportunity for the invasion of exotic annual non-phreatophytic vegetation once the drought ended.

18.4.5 Evapotranspiration and Recharge Rate Quantification

In arid and semiarid areas where net surface-water runoff is negligible, long-term recharge rates are the difference between the precipitation and evapotranspiration rates with

$$R = P - ET \tag{18.1}$$

It would clearly be desirable if both P and ET rates could be determined using satellite remote sensing data. Land and satellite based techniques to measure rainfall are discussed in Sects. 7.2.2 and 7.2.3. Evapotranspiration rates can be estimated using satellite spectral information and ground meteorological data using the surface energy budget as (Bastiaanssen et al. 1998, Alberich 2002)

$$R_n = G + H + \lambda E \tag{18.2}$$

where,

R_n = net radiation (watts per m^2)
G = soil heat flux (watts per m^2)
H = sensible heat flux (watts per m^2)
λE = latent heat flux (watts per m^2)
Equation has no English unit conversion

The latent heat flux, which is the energy used for evaporation, is calculated as the residual of the surface energy balance. The fundamental limitation of this method is that errors associated with the P and ET estimates are large relative to the small recharge rates in arid and semiarid regions. However, even if the absolute values of the estimated recharge values are inaccurate, they can still be used for a meaningful zonation of recharge (Alberich 2002; Brunner et al. 2004).

18.4.6 Soil Moisture Measurement

Soil moisture is of obvious importance for agriculture and also has great importance for partitioning rainfall between infiltration and runoff. Soil moisture has historically been determined by point measurements (e.g., gravimetric, TDR, neutron probe analyses), with their inherent spatial limitations. There is much interest in using remote sensing techniques to measure soil moisture over broad areas.

Water has a greater specific heat capacity relative to solids, which results in water being able to store much more heat or cold than the same mass of aquifer rock or sediment. Water also has a substantially greater thermal conductivity than air. Saturated and moist soils will therefore tend to a have greater thermal inertias compared to dry soils. It will thus take moist soils longer to cool at night and longer to heat up during the day than dry soils. A modeling study by Van de Griend et al. (1985) demonstrated that daily maximum and minimum surface temperatures (obtainable using remote sensing) may be used to discriminate several soil characteristics, such as soil moisture and soil types. Soil surface temperatures are influenced by numerous factors and, as a result, diurnal surface temperature variation has yet to be developed into a practical and accurate quantitative tool.

Microwaves are electromagnetic waves with frequencies between 0.3 and 300 gigahertz (GHz). The use of microwave remote sensing to measure soil moisture was reviewed by Jackson (2002), Oldak et al. (2004), and Wagner et al. (2007a, b). Microwave surveys can be performed using either aircraft or satellites. Satellites have the advantage of large area mapping and long-term repetitive coverage, but have a lesser spatial resolution than aircraft surveys. At low microwave frequencies (0.3–30 GHz), the magnitude of the microwave signal is related to the moisture content of the upper several cm (in) of soil.

The basic principle behind soil moisture measurements using microwave radiation is that surface reflectivity and emissivity are a function of the dielectric constant of the soil and the viewing angle. Dielectric constant (also referred to as relative permittivity) is the ratio of the permittivity of a material to the permittivity of a vacuum. Permittivity is a measure of the ability of a material to be polarized by an electric field. The dielectric constant of soils is a composite value of its components; air, soil (mineral grains and organic matter), bound water, and free water. Water has a much greater dielectric constant at low microwave frequencies than the other components, and thus surface reflectivity is sensitive to the soil moisture content.

Microwave remote sensing is performed either in the passive or active mode. Passive mode (e.g., radiometry) involves measurement of natural thermal emissions of the surface (brightness and temperature). The intensity of emitted microwave radiation is related to surface temperature and the emissivity of the surface. When the surface temperature is known, then the surface emissivity and reflectivity can be derived, which can in turn be used to calculate the soil dielectric constant and soil moisture content. Soil roughness, density, texture, and structure effects need to be considered in the soil moisture measurements. Active mode [e.g., synthetic aperture radar (SAR) and scatterometry] consists of the sending and receiving of microwave pulses. The power of the received signal is compared to that of the sent signal to determine the backscattering coefficient, from which soil moisture is estimated.

Microwave sensors operating at very low frequencies (<6 GHz) provide the best soil moisture information. The advantages of the microwave remote sensing are that observations can be made through cloud cover and are independent of solar radiation. Soil moisture measurements are affected by vegetation types and density and surface roughness, for which corrections need to be made to the data set. Current satellite systems are, therefore, more effective for measuring changes in shallow soil moisture rather than absolute values. Change detection techniques are attractive because they provide a simple, albeit indirect, way for accounting for surface roughness effects and heterogeneous land covers (Kerr 2007; Wagner et al. 2007a).

Wagner et al. (2007b) compared several published satellite remotely sensed soil moisture datasets against in situ observations over a test site in Spain (REHMEDUS network). The datasets were:

- Advanced Microwave Scanning Radiometer on board NASA's Aqui satellite (2 datasets), which provides global passive microwave measurements,
- scatterometer data from the European Remote Sensing Satellites ERS-1and ERS-2, and
- METEOSAT imagery and precipitation data using an Energy and Water Balance Monitoring System.

There was much temporal variability in the satellite data, which was explained in part by their measuring of surficial soils (upper 2–8 cm; 0.8–3.1 in), which respond very quickly to atmospheric forcing. The satellite remote testing systems were found to be effective for monitoring soil moisture trends, but not absolute soil-moisture values. The differences in absolute values reflect the choice of auxiliary soil data (e.g., soil type, structure, and hydrologic properties) and not information stemming from the satellites (Wagner et al. 2007a). Retrieval algorithms play an equally important role for the quality of a soil-moisture dataset as the technical specifications and performance of the satellite system (Wagner et al. 2007b).

The next step in microwave remote sensing is processing the data to infer subsurface parameters and features, such as soil texture and drainage rates, soil moisture profiles, and hydraulic conductivity, from surficial soil moisture data (Jackson 2002). A major current limitation of satellite soil moisture measurements is that they provide data on only the surface layer and not soil moisture in the root zone and the lower vadose zone. Assumptions need to be made relating surface soil moisture to root zone moisture (Kerr 2007, Wagner et al. 2007b). An additional constraint is that high-resolution satellite systems have a low temporal measurement frequency. For example, the frequency of the ERS-1 satellite is 35 days (Kerr 2007).

A limitation of the technique in arid and semiarid regions is that soil moisture contents are low and the depth to the water table is often great, which increases the difficulty of using the technique to estimate the main parameters of interest, such as recharge rate. Microwave remote sensing can provide spatial and temporal data, which can complement (but not replace) monitoring and modeling of groundwater recharge (Jackson 2002).

18.4.7 Interferometric Synthetic Aperture Radar (InSAR)

Satellite Interferometric Synthetic Aperture Radar (InSAR) can be used to map local and regional land subsidence caused by groundwater withdrawals (Sect. 15.2.4). Other applications of InSAR for groundwater investigations include (Galloway and Hoffman 2007)

- Improvement in conceptual models. InSAR may help identify faults and other structures that affect groundwater flow by the resulting differential ground displacement. Steep displacements across typically linear alignments on displacement maps can indicate potential fault barriers.
- Mapping the distribution of compressible sediments. The differences may be caused by variations in sediment composition or over-consolidation history.
- Storage coefficient estimation. Aquifer elastic skeletal storage coefficients (S_{ke}) can be calculated from paired time series of head changes and displacement.
- Recognition and mapping of residual compaction from times series of groundwater level and compaction data. Residual compaction may be indicated by on-going compaction while groundwater levels are constant.
- Identification and mapping of areas prone to earth fissuring and surface faults.

18.5 Remote Sensing and Groundwater Modeling

Groundwater modeling requires spatially-distributed input data (and in some models also temporally distributed input data), such as aquifer hydraulic properties, water levels, evapotranspiration, precipitation, and recharge rates. Such data have historically been provided from point measurements, such as individual wells or weather stations. Point data are then applied either uniformly across the model, assigned to fields or zones, and or the values are interpolated and extrapolated by kriging of other mathematical processes. An application of remote sensing to groundwater modeling is that pattern information obtained from remote sensing combined with point information from ground stations allows for improved spatial distributions of the parameters in question compared solely to point source data

(Brunner et al. 2007). For example, relative precipitation intensity may be estimated from remote sensing data such as cloud temperature. A map of absolute precipitation can be obtained by the scaling relative precipitation abundance to the point measurement data (Brunner et al. 2007). Remote sensing data can also be used to define boundary conditions in models, such as streams, drainage basins, wetlands, seepage areas, and recharge areas (Becker 2006).

A major limitation on the application of remote sensing to groundwater models is that it does not provide information on some of the key parameters required for model development, particularly aquifer hydraulic properties. The bulk of the remote sensing data relevant for groundwater modeling are data that can be used for the quantification of the distribution of recharge and discharge (Brunner et al. 2007). Determination of recharge rates from remote sensing data alone is very problematical. Recharge is largely the long-term residual between precipitation and evapotranspiration. While both precipitation and evapotranspiration can be estimated from remote sensing data, the values obtained for both parameters from remote sensing alone have low accuracies. There are thus large errors associated with remote sensing-derived recharge estimates, especially in areas where both quantities are of similar magnitude, which is often the case in arid and semiarid regions (Brunner et al. 2007).

The spatial distribution of recharge can be estimated using RS and GIS from land use and cover and soil types. If the average recharge of a study area can be determined by other means, then the distribution of recharge can be estimated by applying to the average rate weighting factors for the different land use and soil types (Brunner et al. 2007). Field point measurements can be used with remote sensing data to greatly improve the accuracy of the overall dataset and to improve spatial estimates of parameters.

Remote sensing can be used to identify lineaments and other surface structures than may impact subsurface groundwater flow. Analysis of lineaments and magnetic anomalies was used in the Kayne aquifer of Botswana to develop a conceptual model of aquifer compartmentalization (Brunner et al. 2007). Dikes were identified that act as flow boundaries. Revision of the local groundwater model to include the compartmentalization resulted in model improvement that allowed for more accurate prediction of the large drawdowns that occurred in some wells.

18.6 Limitations of Remote Sensing

Remote sensing typically does not measure hydrologic properties directly. Instead, the values of different hydrological parameters are inferred from readings of parts of the electromagnetic spectrum or other geophysical data. Remote sensing using the electromagnetic spectrum commonly detects changes at ground surface or in a shallow layer (<1 m depth; <3.3 ft) of the Earth.

A critical issue associated with remote sensing data in general is that it is not a substitute for in situ data collection, which is still essential to verify the accuracy of remote sensing data and for their interpretation. The field work for ground-truthing remote sensing data is the expensive part in using the data for groundwater modeling (Brunner et al. 2007). Brunner et al. (2007, p. 16), raised the critical point that

> Quantities that are derived from remote-sensing data and which are not checked against ground truth are of low value. They contain spatial pattern information but the weighting with absolute values may be completely erroneous

The most realistic uses of remote-sensing data seem to be the reduction of degrees of freedom and conditioning of both deterministic and stochastic models (Brunner et al. 2007).

Many remote sensing techniques require a high degree of technical sophistication to perform, which has been an impediment to their implementation within developing countries in arid and semiarid lands. Some of the specific technical, educational and institutional limitations to remote sensing and GIS implementation in developing countries (Jha and Chowdary 2007) are

- inadequate groundwater monitoring data,
- poor facilities for distribution of remote sensing data,
- inadequate local digital data and high costs for remote sensing data use and transfer,
- expensive (and thus often unaffordable) remote sensing and GIS software,
- poor knowledge of remote sensing and GIS technologies and their potential benefits,
- lack of adequate infrastructure, training, and support for remote sensing and GIS technologies,
- restrictions on access to remote sensing data for national security reasons, and
- poor or absence of a linkage between researchers, decision makers, and planners, and water users.

In conclusion, the use of remote sensing for water resource development and management in poor, Third World regions must be developed to help meet United Nations Millennium goals. Technical support to conduct the appropriate remote sensing investigation will likely have to be provided by international support organizations through aid and technical assistance programs.

References

Abdelsalam, M. G., Robinson, C., El-Baz, F., & Stern, R. J. (2000). Applications of orbital imaging radar for geologic studies in arid regions: The Saharan testimony. *Photogrammetric Engineering & Remote Sensing, 66*, 717–726.

Alberich, C. (2002). Estimation of the spatial distribution of evapotranspiration using optical remote sensing. In W. Kinzelbach, W. Aeschbach, C. Alberich, I. B. Goni, U. Beyerle, P. Brunner, W.-H. Chiang, J. Rueedi & K. Zoellmann (Eds.), *A survey of methods from groundwater recharge in arid and semiarid regions* (pp. 57–64). Early Warning and Assessment Report Series UNEP/DEWA/RS.02-2, Nairobi, Kenya, United Nations Environment Programme.

Bastiaanssen, W. G. M., Menenti, M., Feddes, R. A., & Holtschlag, A. A. (1998). A remote sensing surface energy balance algorithm for land (SEBAL). *Journal of Hydrology, 212–213*, 198–229.

Becker, M. W. (2006). Potential for satellite remote sensing of ground water. *Ground Water, 44*, 306–318.

Bisson, R. A., & Lehr, J. H. (2004). *Modern groundwater exploration: Discovering new water resources in consolidated rocks using innovative hydrogeologic concepts, exploration, drilling, aquifer testing, and management methods*. Hoboken, NJ: Wiley.

Brunner, P., Bauer, P., Eugster, M., & Kinzelbach, W. (2004). Using remote sensing to regionalize local precipitation recharge rates obtained from the chloride method. *Journal of Hydrology, 294*, 241–250.

Brunner, P., Hendricks Fransen, H.-J., Kgotlhang, L., Bauer-Gotlwein, P., & Kinzelbach, W. (2007). How can remote sensing contribute to groundwater modeling? *Hydrogeology Journal, 15*, 5–18.

Campbell, J. B. (2008). *Introduction to remote sensing* (4th ed.). New York: Guilford Press.

Carruthers, R. M., Greenbaum, D., Peart, R. J., & Herbert, R. (1991). Geophysical investigation of photolineaments in southeast Zimbabwe. *Quarterly Journal of Engineering and Hydrogeology, 24*, 437–451.

El-Baz, F. (2001). Remote sensing of groundwater basins in the eastern Sahara. In *Proceedings, Regional Aquifer Systems in Arid Zones—Managing Non-renewable Resources, Tripoli, Libya, 20–24 November 1999. Paris* (pp. 19–25). UNESCO: IHP-V, Technical Documents in Hydrology No. 42.

Elmore, A. J., Mustard, J. F., & Manning, S. J. (2003). Regional patterns of plant community responses to change in water, Owens Valley, California. *Ecological Applications, 13*, 443–460.

Engman, E. T., & Gurney, R. J. (1991). *Remote sensing in hydrology*. London: Chapman and Hall.

Farnsworth, R. K., Barret, E. C., & Dhanju, M. S. (1984). *Application of remote sensing to hydrology including ground water*. Paris, UNESCO, IHP-II Project A, 1.5.

Galloway, D. L., & Hoffmann, J. (2007). The application of satellite differential SAR interferometry-derived groundwater displacements in hydrogeology. *Hydrogeology Journal, 15*, 133–154.

Hobbs, W. H. (1904). Lineaments of the Atlantic border region. *Geological Association of America Bulletin, 15*, 483–506.

Jackson, T. J. (2002). Remote sensing of soil moisture: Implications for groundwater recharge. *Hydrogeology Journal, 10*, 40–51.

Jasmin, I., & Mallikarjuna, P. (2011). Review: Satellite-based remote sensing and geographic information systems and their application in the assessment of groundwater potential, with particular reference to India. *Hydrogeology Journal, 19*, 729–740.

Jensen, J. R. (2007). *Remote sensing of the environment: An Earth resource perspective*. Upper Saddle River, NJ: Prentice Hall.

Jha, M. K., & Chowdary, V. M. (2007). Challenges of using remote sensing and GIS in developing nations. *Hydrogeology Journal, 15*, 197–200.

Jha, M. K., Chowdhury, A., Chowdary, V. M., & Peiffer, S. (2007). Groundwater management and development by integrated remote sensing and geographic information systems: Prospects and constraints. *Water Resources Management, 21*, 427–467.

Kerr, Y. (2007). Soil moisture from space: Where are we? *Hydrogeology Journal, 15*, 117–120.

Lillesand, T., Kiefer, R. W., & Chipman, J. (2008). *Remote sensing and image interpretation* (6th ed.). Hoboken, NJ: Wiley.

Mabee, S. B., Hardcastle, K. C., & Wise, D. U. (1994). A method of collecting and analyzing lineaments for region-scale fractured-bedrock aquifer systems. *Ground Water, 32*, 884–894.

McCauley, J. F., Schaber, G. C., Breed, C. S., Grolier, M. J., Haynes, C. V., Issawi, B., et'al. (1982). Sub-surface valleys and geoarcheology of the Eastern Sahara revealed by shuttle radar. *Science, 218*, 1004–1020.

Meijerink, A. M. J. (1996). Remote sensing applications to hydrology: Groundwater. *Hydrological Sciences Journal, 41*, 549–561.

Meijerink, A. M. J. (2000). Groundwater. In G. A. Schultz & E. T. Engman (Eds.), *Remote sensing in hydrology and water management* (pp. 305–325). Berlin: Springer.

Moore, R. B., Schwarz, G. E., Clark, S. F. Jr., Walsh, G. J., & Degnan, J.R. (2002). *Factors related to well yield in the fractured-bedrock aquifer of New Hampshire.* U.S. Geological Survey Professional Paper 1660.

Münch, Z., & Conrad, J. (2007). Remote sensing and GIS based determinations of groundwater dependent ecosystems in the Western Cape, South Africa. *Hydrogeology Journal, 15*, 19–28.

NASA (National Aeronautics and Space Administration). (2001). The Landsat Program: http://landsat.gsfc.nasa.gov/. Last Accessed on May 23, 2011.

NOAA (National Oceanic and Atmospheric Administration) (Coastal Services Center). (2008). *Lidar 101: An introduction Lidar technology, data, and applications.* Charleston, SC: NOAA.

Nyborg, M., Berglund, J., & Triumf, C.-A. (2007). Detection of lineaments using airborne laser scanning technology: Laxemar-Simpevarp, Sweden. *Hydrogeology Journal, 15*, 29–32.

Oldak, A., Jackson, T. J., Starks, P., & Elliot, R. (2004). Mapping near-surface soil moisture on regional scale using ERS-2 SAR data. *Remote Sensing, 24*, 4579–4598.

Rango, A. (1994). Application of remote sensing methods to hydrology and water resources. *Hydrological Sciences Journal, 39*, 309–319.

Rhén, I., Thunehed, H., Triumf, C.-A., Follin, S., Hartley, L., Hermansson, J., et'al. (2007). Development of a hydrogeological model description of intrusive rock at different investigation scales: An example from south-eastern Sweden. *Hydrogeology Journal, 15*, 47–69.

Robinson, C. A., Werwer, A., El_Baz, F., El-Shazly, M., Fritch, T., & Kusky, T. (2007). The Nubian Aquifer in Southwest Egypt. *Hydrogeology Journal, 15*, 33–45.

Sabins, F. F. (2007). *Remote sensing: Principles and interpretation* (3rd ed.). New York: Freeman.

Salomonson, V. V., & Rango, A. (1980). Water resources. In B. S. Siegal & A. R. Gillespie (Eds.), *Remote sensing in geology* (pp. 607–633). New York: Wiley.

Sander, P. (2007). Lineaments in groundwater exploration: A review of applications and limitations. *Hydrogeology Journal, 15*, 71–74.

Sander, P., Minor, T. B., & Chesley, M. M. (1997). Ground-water exploration based on lineament analysis and reproducibility tests. *Ground Water, 35*, 888–894.

Schowengerdt, R. A. (2007). *Remote sensing: Models and methods for image processing* (3rd ed.). Burlington, MA: Academic Press.

Siegal, B. S., & Gillespie, A. R. (Eds.). (1980). *Remote sensing in geology.* New York: Wiley.

Tam, V. T., de Smedt, F., Batelaan, O., & Dassargues, A. (2004). Study on the relationship between lineaments and borehole specific capacity in a fractured and karstified limestone area in Vietnam. *Hydrogeology Journal, 12*, 662–673.

Teeuw, R. M. (1995). Groundwater exploration using remote sensing and a low-cost geographical information system. *Hydrogeology Journal, 3*(3), 21–30.

Tweed, S. D., Leblanc, M., Webb, J. A., & Lubczynski, M. W. (2007). Remote sensing and GIS for mapping groundwater recharge and discharge areas in salinity prone catchments, southeastern Australia. *Hydrogeology Journal, 15*, 75–96.

U.S. Geological Survey. (2001). *Understanding color-infrared photographs.* U.S. Geological Survey Fact Sheet 129-01.

U.S. Geological Survey. (2003). *Landsat: A global land-observing program.* Fact Sheet 023-03.

Van de Griend, A. A., Camillo, P. J., & Gurney, R. J. (1985). Discrimination of soil physical parameters, thermal inertia, an soil moistures from diurnal surface temperature fluctuations. *Water Resources Research, 21*, 997–1009.

Vrba, J., & Verhagen, B. T. (2006). *Groundwater for emergency situations: A framework document* (94 pp). International Hydrological Program (IHP) VI, Series on Groundwater No. 12: Paris, UNESCO.

Wagner, W., Blöschl, G., Pampaloni, P., Calvert, J.-C., Bizzarri, B., Wigneron, J.-P., & Kerr, Y. (2007a). Operational readiness of microwave remote sensing of soil moisture for hydrologic applications. *Nordic Hydrology*, *38*, 1–20.

Wagner, W., Naeimi, V., Scipal, K., de Jeu, R., & Martínez-Fernández, J. (2007b). Soil moisture from operational meteorological satellites. *Hydrogeology Journal*, *15*, 121–31.

Waters, P., Greenbaum, P., Smart, L., & Osmaston, H. (1990). Applications of remote sensing to groundwater hydrology. *Remote Sensing Review*, *4*, 223–264.

Zeeb, C., Göckus, D., Bons, P., Al Ajmi, H., Rausch, R., & Blum, P. (2010). Fracture flow modelling based on satellite images of the Wajid Sandstone, Saudi Arabia. *Hydrogeology Journal*, *18*(7), 1699–1712.

Chapter 19
Geographic Information Systems

19.1 Introduction

A geographic information system (GIS) is any system that captures, stores, analyzes and displays data that are linked to location (i.e., are georeferenced). GIS broadly refers to all aspects of managing, manipulating, and using digital georeferenced data. It is no exaggeration to state that GIS has rapidly become an integral tool for groundwater management in arid lands, as well as humid lands. GIS has several main applications for groundwater management (Gogu et al. 2002, and others) including.

- groundwater data management and general hydrogeological analysis,
- hydrogeological map elaboration,
- site evaluation and assessments using overlap and index methods,
- quantitative evaluation of remote sensing and other data to determine values of hydrologic parameters,
- processing and interpretation of remote sensing data,
- hydrogeological database support for process-based groundwater modeling (flow and geochemical), and
- facilitation of the sharing of data.

GIS allows for the analysis of spatial data quickly and easily. Databases can be updated continuously as new data are obtained. The principal value of GIS for water resources management occurs where there are spatial variations in properties of concern and data are available on the property. The most effort associated with GIS is commonly entering the data into system (i.e., data capturing). However, increasingly large amounts of data are already available in GIS usable format. GIS has become indispensable in processing and analyzing hydrogeological data, allowing for the evaluation of huge amounts of geo-referenced data (Vrba and Verhagen 2006). GIS facilitates conjunctive analysis of multi-parameter thematic data, which can assist in decision-making for groundwater management planning (Saraf and Choudhury 1998).

R. Maliva and T. Missimer, *Arid Lands Water Evaluation and Management*, Environmental Science and Engineering, DOI: 10.1007/978-3-642-29104-3_19,

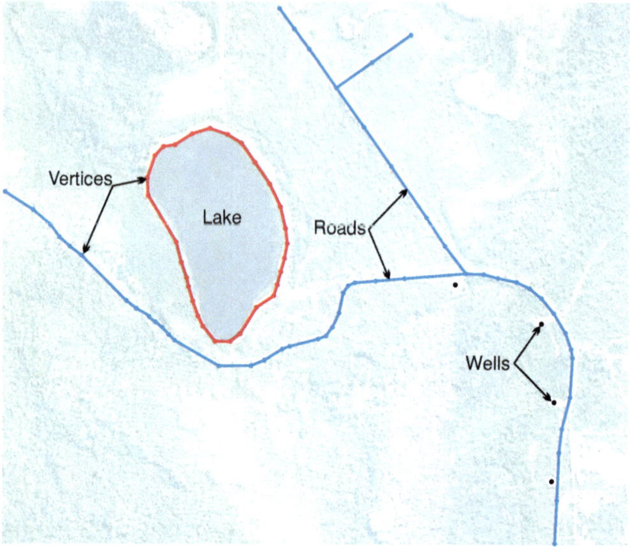

Fig. 19.1 Examples of GIS vectors, which are geometrically composed of one or more connected vertices. Wells are represented as points, roads are represented as polylines, and areal features, such as lakes, as polygons

GIS has become ubiquitous with numerous applications in the physical, natural, and social sciences, and increasingly in everyday life through technologies, such global positioning system (GPS) tools. Virtually all universities now offer courses in GIS. Numerous books have been written on the subject including introductory texts by Clarke (2010), DeMers (2005, 2009), Harvey (2008), and Korte (2001).

19.2 GIS Principles

The underlying feature of GIS is the association of geographic data (coordinates) with non-geographic data (attributes). There are two basic types of data in GIS: vector data and raster data. Vector data are stored as a series of x, y and optionally z coordinates with respect to a coordinate reference system (e.g., latitude and longitude, Universal Transverse Mercator system). A vector can be a point, line, polyline, or polygon (Fig. 19.1). Vectors are made up of one or more interconnected vertices, which have coordinates describing their position in space. A point has a single vertex; lines and polylines are defined by multiple vertices in which the first and last are not the same (i.e., there is no closure). Polygons are defined by multiple (4 or more) vertices in which the first and last are the same (i.e., there is closure). Each vector can be assigned one or more properties.

Examples of polygons are wetlands, land use zones, soil types, vegetation types, surficial geology zones, municipal boundaries, and property ownership. For example, a wetland area would be geometrically described as a polygon defined by a series of vertices. Attributes that could be assigned to the wetland polygon, could include its classification as a wetland, vegetation types, and ET rate. Similarly, a well would be defined as a point and could be assigned attributes such as its production aquifer, depth, water salinity, and aquifer hydraulic parameters obtained from testing of the well.

Raster data consists of a grid of columns and rows of cells (pixels) each of which stores a single value that represents the conditions in the cell. Common types of raster data are digitized georeferenced aerial and satellite photographs and remote sensing data. Digital elevation models (DEMs) are another commonly used type of raster data in which each pixel represents altitude above a datum (e.g., sea level).Vector data have the advantage of being more readily edited and updated and have smaller file sizes. Raster data can be converted into vector data, which is referred to as vectorization.

The basic structure of GIS is vector and raster layers, which can be overlain to generate maps or used for spatial analysis. Each vector layer may contain only one type of geometric data and commonly they are set up to represent one type of feature (theme). For example, a thematic layer may contain only wetlands that are geometrically described as polygons or wells that are represented as points. Each layer is then commonly stored as a shape-file, which consists of a minimum of three subfiles: a main file that stores geometric information (.shp), a shape index file (.shx), and a file that stores attribute data (.dbf).

GIS can be used to generate a wide variety of maps through the overlaying of layers. However, the greatest value of GIS lies in its capabilities for vector spatial analysis. For example, spatial overlay techniques can be used to establish relationships between features. Output vector layers can be generated by combining values from two or more input layers. GIS can also be used to create buffer zones around geometric elements. Another common type of application of GIS is pattern analysis, which may be used to test whether the distribution of a feature is random or not and to identify spatial trends and relationships.

GIS is widely used in groundwater modeling to interpolate limited known point values to areas in which data are not available. For example, water quality and hydraulic data from a relatively small number of wells is interpolated to populate an entire model layer.

19.3 GIS Applications

GIS is used for general applications in water resources management in semiarid and arid lands such as the storage of georeferenced data (e.g., well locations, water use permits) and map generation. GIS is now routinely used for pre- and post-processing of modeling data. GIS has been proven to be a valuable tool for investigations that involve evaluation of multiple types of georeferenced data. Examples are provided in this chapter on the use of GIS in recharge and site evaluation studies. Perhaps the ultimate use of GIS in water-resources management is to establish a centralized data management system in which all data of hydrologic significance in countries or regions are stored and are readily available to multiple users for their diverse needs.

19.3.1 Groundwater Model Development

Numerical groundwater models require as inputs the spatial distribution of the values for a number of parameters in each model layer. A value for each model parameter needs to be assigned to each cell or node that is represented in each model layer. The model parameters include physical parameters (e.g., aquifer thickness, and top and bottom altitudes), hydrogeological parameters (e.g., hydraulic conductivity, storativity, leakance, water levels and heads), and chemical properties (e.g., salinity). GIS is used in groundwater modeling to (Gogu et al. 2002).

- place the model grid into a georeferenced mode,
- create digital databases of the data relevant to the model,
- manipulate and prepare data as input for the various model parameters, and
- display model output (e.g., drawdown and concentration contours).

The most common linkage between GIS and groundwater models is a "loose coupling" whereby the GIS software are used to pre-process and post-process the spatial data, which is transferred to and from the modeling software. Modeling software packages have been developed that include both GIS and modeling software (e.g., Visual MODFLOW™). GIS data storage, modification, and analyses are often performed using more-sophistical, dedicated GIS software (e.g., ArcView), and then imported into the modeling software package, where it is used to populate a numerical groundwater model.

Remote sensing and GIS can be used to determine the areal distribution of hydrologic parameters needed to develop a groundwater model, particularly where the parameters can be correlated to surface features. GIS can be of particular value for large regional models since it provides a means for the efficient management of large data sets, model calibration, and the improved visual display and analysis of model results (Watkins et al. 1996). Digital enhancement and processing of satellite images can provide detailed data on geology and hydrogeology, land use and cover, and hydrologic conditions.

GIS is very commonly used to develop or obtain base maps for groundwater models. GIS is also used to transfer data between the professionals involved in a project. For example, a biologist may provide a shape-file showing the locations of wetlands and other sensitive environments to the groundwater modeler. Alternatively, the modeler may provide a shape-file showing simulated drawdowns to the biologists and other investigation team members.

19.3.2 Recharge Investigations

Determination of rates and locations of recharge is an important use of GIS in semiarid and arid lands. Maps of the estimated recharge obtained using remote sensing can be scaled against actual recharge values obtained from other methods, such as the chloride mass balance method (Chap. 11). For example, Brunner et al. (2004) mapped recharge in the Kalahari Desert of Botswana using GIS and remote sensing data. It was assumed that the long-term recharge rate is the difference between precipitation and evapotranspiration. It was recognized that the estimated absolute values of recharge have a larger error in arid and semiarid lands because of the small difference between ET and P. Precipitation was mapped using the GOES Precipitation Index (GPI) algorithm using data from the Meteorsat satellite and rain gauge reports. The reference evapotranspiration rate was calculated by a surface energy balance approach using multispectral satellite images (NOAS-AVHRR).

The remote sensing data systematically overestimated rainfall and underestimated evapotranspiration, and thus overestimated recharge. Recharge rates calculated using the remote sensing data were correlated with values obtained using the chloride mass balance (CMB) method. The remote sensing calculated recharge values could then be calibrated against absolute recharge values obtained using the CMB method to obtain a map of actual recharge. Brunner et al. (2004) noted that their method is subject to a string of assumptions, such as the limitations on the CMB method (Sect. 11.5.1) and that anomalous site conditions (e.g., subsurface peat fires) could distort the remotely sensed surface energy balance. Nevertheless, the Brunner et al. (2004) method is useful means for obtaining a map of estimated recharge that is needed for some groundwater models.

GIS can be used for the estimation of total recharge in an area, by determining the areal distribution and cumulative areas of different surface coverings and assigning a recharge rate to each. Fayer et al. (1996), for example, used GIS to identify all possible combinations of soil and vegetation types at the Hanford nuclear site in Washington (United States). An appropriate estimate of recharge was assigned to each soil and vegetation type based on field data and simulation results. Total recharge could be estimated from the areal coverage of each soil-vegetation combination.

Harvesting of ephemeral runoff has not been systematically explored in northeastern Jordan because of a poor understanding of the hydrological constraints in the region. Shallow perched aquifers associated with stream beds and recharged by ephemeral run-off are an historically used water source and have potential for providing additional freshwater to the region. GIS was used in the Tulul al Ashagif region of northeastern Jordan to develop an understanding of where shallow aquifers can be found, to estimate the volume of water that might be available, for determination of best locations for sitting recharge facilities, and to define issues related to long-term protection of the water resources (Al-Qudah and Abu-Jaber 2009). GIS analysis using satellite and aerial photographs and topographic maps, was used to find areas with similar characteristics (slope, nature of alluvium, geomorphology, and plant cover) to use for new well locations. Key characteristics that were found to be indicative of the presence of shallow aquifers were altitude, drainage geometry, and sites falling on the linear trend of earlier identified aquifers. Recharge volumes were estimated from rainfall intensity information and known run-off properties in the area.

Groundwater recharge within a groundwater basin can be calculated using the hydrologic budget method from the totals of all inflows and outflows and change in storage. Manghi et al. (2009) presented a GIS method for estimating total groundwater recharge (R_t) from the total groundwater withdrawals (Q_w) and the sum of local changes in storage as

$$R_t = \sum_{i=1}^{i=n} R_i = Q_w \pm \sum_{i=1}^{i=n} (\Delta h_i A_i S_{yi}) \tag{19.1}$$

where,
R_i = recharge in grid cell 'i',
Δh_i = change in head in cell 'i'
A_i = area of cell 'i'
S_{yi} = specific yield of cell 'i'.

The groundwater recharge term includes net surface inflow and outflow. Using the Manghi et al. (2009) method, groundwater recharge can be calculated from groundwater level and extraction data and the specific yield of the aquifer sediments or rock in which water level changes are occurring. The former two data sets may be readily available to water managers.

Recharge rates are sensitive to the specific yield values, which is the parameter for which there is great uncertainty. Manghi et al. (2009) used a method named Well Tool (WET) in which specific yield was estimated from descriptions of the soil materials included in well driller logs. The GIS procedure used was as follows:

(1) Polygon shape-files were created for the basin of interest and bedrock elevation.
(2) Specific yield values were calculated for water level change depths in each well.
(3) Point shape files were created for specific yield data.
(4) Shape files were created for groundwater level measurements for the start and end times of the recharge evaluation period.
(5) The GIS statistical program was used to divide the basin into a grid and assign a groundwater level, specific yield, and bedrock elevation value to each grid cell. Cells containing a bedrock elevation higher than the extrapolated groundwater level are assumed to have zero groundwater volume.
(6) Recharge was calculated as the sum of changes in groundwater volume in the grid cells.

The main advantage of the hydrologic budget method used by Manghi et al. (2009) is that it is based on what is often readily available data. The accuracy of the method depends on the accuracy of the data used. There is considerable uncertainty over the specific yield values, particularly if their determination relies upon lithologic logs prepared by well drillers, which have a huge variation in quality.

Stone et al. (2001) presented a methodology for the evaluation of the spatial distribution of the total recharge in groundwater basins located in mountainous regions that produce runoff, which subsequently infiltrates at lower altitudes. Runoff from mountainous sub-basins infiltrates into down-gradient alluvial sediments. GIS was used to divide the total basin recharge determined by other means, such as by the Maxey and Eakin (1949) method, into sub-basin increments. The Stone et al. (2001) method uses the following data and analysis methods:

(1) The studied basin is divided into sub-basins based on digital elevation model (DEM) surface topography data.
(2) The mean annual precipitation to elevation relationship is determined from weather station data. GIS and DEM are used to define precipitation zones at various elevations.
(3) Average annual precipitation in each basin is calculated from the DEM and precipitation-elevation relationship.
(4) The US Department of Agriculture Soil Conservation Service (SCS) Run-Off Number Method (USDA 1986) is used to determine the percentage of the precipitation within mountainous sub-basins that results in runoff to alluvial basins. A key variable is the amount of water that is retained in each sub-basin in surface depressions or intercepted by vegetation with immediate, water losses caused by evapotranspiration, and local infiltration (i.e., initial abstraction). The initial abstraction is related to soil cover conditions through curve numbers, which are related to hydrologic soil groups (obtained from the literature).
(5) The SCS method is applied to each sub-basin for individual precipitation events. The total annual precipitation in each sub-basin is divided into rainfall events using the average number of precipitation events and the percentage of the total volume of each event as obtained from nearby weather stations. No runoff is assumed to occur if the precipitation event is less than the initial abstraction. Runoff from each sub-basin is determined from the fractions of annual precipitation in excess of the initial abstraction and fraction of the excess precipitation that results in runoff.
(6) Potential recharge in each receiving alluvial basin is determined as the sum of the incident precipitation and the runoff from topographically up-gradient mountainous sub-basins.
(7) The actual recharge in each sub-basin is determined by scaling the potential recharge in each sub-basin to the basin-wide ratio of actual recharge to potential recharge.

The Stone et al. (2001) method requires an estimate of the basin-wide average annual recharge. The value of the method is that it can provide more accurate estimates of where in the groundwater basin recharge is mainly occurring. Such data, in turn, can lead to a more refined groundwater flow model. The main potential source of error in the method is the accuracy of the initial abstraction values and the basin-wide total recharge value.

19.3.3 Site Evaluation Studies

The selection of the preferred locations for production wells, recharge systems, and other infrastructure in general, is often performed using a weighted scoring system. Potential sites are scored for each of the various attributes that effect feasibility and performance. The process typically used involves the following steps

- identification of the criteria that are considered important for the planned activity or system,
- development of a scoring system for each criteria, based upon the degree to which site conditions meet performance and economic goals,
- assignment of a weighting factor for each criteria, which reflects its relative importance for the satisfactory performance of the activity or system,
- assignment of a score for each of the criteria at each of the potential sites, and
- calculation of a cumulative score for each potential site.

Site selection using GIS still involves considerable subjectivity, because decisions are made concerning the criteria needed to establish the assignment of data into classes, whether Boolean or fuzzy logic is used, membership functions, and weighting factors. Scoring systems inherently involve professional knowledge and judgment, which are used to determine the criteria to be considered, the value (score) that is assigned for some non-quantitative criteria, and the weighting factor assigned to each criteria. Although numerical evaluation systems have the advantage that they are seemingly objective, the opportunity exists for personal biases to enter the evaluation process.

Rather than restricting site analyses to discrete sites, GIS can be used for regional site selection analyses using a weighted overlay function. Thematic layers are developed for each criterion. Various remote sensing techniques (Chap. 18) are well suited for the development of some thematic layers. Many remote sensing data are obtained in a digital form (e.g., satellite images). Non-digital data can be digitized and processed to construct thematic layers. For example, topographic maps can be digitized and then processed to create a DEM. Groundwater elevation data can also be processed to generate a potentiometric surface map. The overlaying of the thematic layers in GIS is then used to create a map illustrating the relative suitability of different areas based on the cumulative scores.

An early study by Teeuw (1995) in northern Ghana used Landsat Thematic Mapper data to locate and map lineaments, which were used to determine locations most favorable for successful well construction. The goal was to improve the success rate for well construction (i.e., frequency of wells with high yields). Proximity maps were generated for the distances from villages for which local people would be willing to travel to obtain well water. The proximity maps were merged with maps of (1) lineament coincidence (number of Landsat bands in which a lineament was detected), (2) lineament length, and (3) lineament network (degree of interconnection). The merged maps were used to identify locations that have a greater potential for successful well construction. It was emphasized by Teeuw (1995) that the remote sensing and GIS analysis is the first step in the groundwater exploration effort. The GIS analysis is to be followed by with a site visit (ground truthing), surface geophysics, and then borehole drilling.

There have been numerous studies in which GIS was used for site evaluation using a weighted overlay scoring approach. GIS has been used to evaluate potential sites for artificial recharge systems. For example, Saraf and Choudhury (1998) used remote sensing and GIS to evaluate artificial recharge sites in hard rock terrains in Madhya Pradesh, India. The study area consists mostly of the Deccan Trap basalts. Linear imaging, self-scanning data from the Indian Remote Sensing Satellite program (IRS-LISS II) were used to generate geological, geomorphological, lineament, normalized differential vegetation index (NDVI), and land use maps. Groundwater recharge was estimated by multiplying water level fluctuation (pre- and post-monsoon) by the specific yield of each geological formation. Areas of recharge downstream from surface-water reservoirs are evident by the development of dry season vegetation, which was detected and mapped using remote sensing data.

Remote sensing and GIS were similarly used to evaluate recharge potential in Occidental Lebanon (Shaban et al. 2006). A combination of data from satellite imaging (Landsat 7 ETM), aerial photographs, geological maps, topographic maps, and field observation were used to map five factors that were believed to be significant for groundwater recharge, including: lineaments, drainage, lithology, karst domains, and land cover/land use. For each factor, five descriptive levels were established in order of importance and scores were assigned. A map of recharge potential was generated by applying a weighting multiplier to each factor and overlaying the factor maps to obtain a total score.

GIS and remote sensing were used to delineate zones of varying groundwater use potential in the central Highlands of Eritrea based on integration of thematic maps for (1) lithology, (2) lineaments, (3) geomorphology, and (4) slope (Solomon and Quiel 2006). A geologic map was generated from colored enhanced satellite images and geomorphology was mapped by processing DEM data. Areas considered having very good or good groundwater resources potential tend to be located near large lineaments (Fig. 19.2).

Ghayoumian et al. (2007) similarly performed a GIS site selection analysis for recharge basins in the Gavbandi River Basin in southern Iran. The site selection criteria used were (1) slope, (2) infiltration rate based on soil texture permeability relationships, (3) depth to groundwater, (4) groundwater salinity from electrical conductivity, and (5) land use and geomorphology. Thematic maps were created for each of the criteria, each of which was divided into 4 or 5 classes.

The thematic layers were classified, weighted and integrated in a GIS environment using either Boolean or fuzzy logic (Ghayoumian et al. 2007). In Boolean logic, only satisfactory or unsatisfactory conditions are considered (i.e., areas as assigned a valued of 0 or 1). In fuzzy logic, membership is expressed as a continuous scale from 0 to 1. Each class is assigned a membership function value, which reflects its degree of acceptability (1 being the optimal condition). Themes

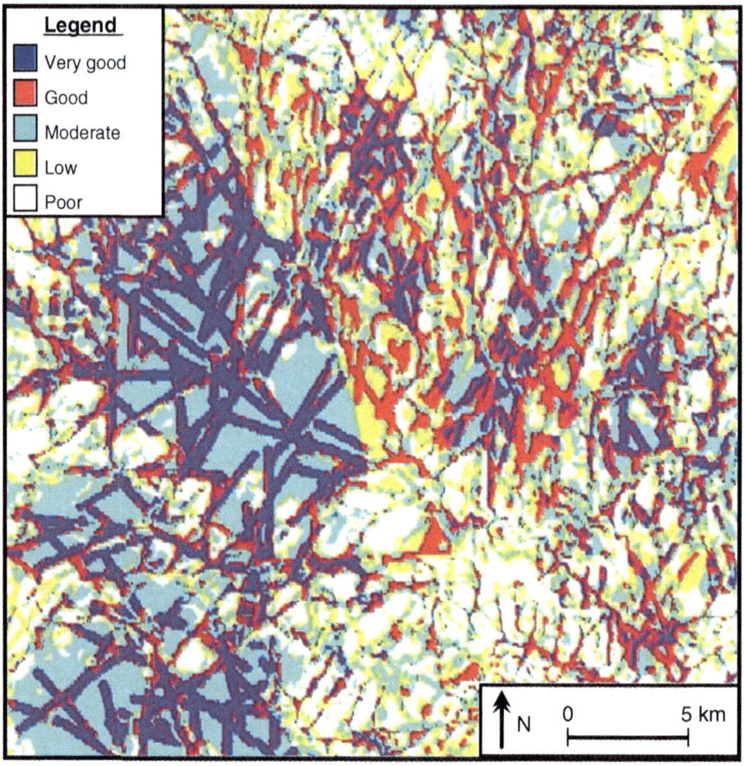

Fig. 19.2 Weighted overlay groundwater resources potential map for an area of the Central Highlands of Eritrea. Highest well yields are related to large lineaments. *Source* Solomon and Quiel (2006)

can be weighted based on their relative importance. The combined membership is a product of all the membership functions.

A weighted-function approach using GIS can also be used to locate areas with the greatest potential of groundwater resource development. Sener et al. (2005) used images obtained from the Landsat 7 Enhanced Thematic Mapper Plus (ETM+) instrument and from conventional sources to create maps of (1) annual rainfall, (2) land use, (3) topographic elevation, (4) slope, (5) drainage density, (6) lithology, and (7) lineament density. The final derived map was the sum of the weighted values of each criterion which allowed identification of areas that have the greatest groundwater development potential.

A weighted overlay approach was used to map ground-water development potential in the West Godavari District, Andhra Pradesh State, India (Kamaraju et al. 1996). An Arc database was generated consisting of 40 layers for 40 data components, out of which 37 were related to groundwater parameters and the remaining three are related to base map details. Higher order databases were next created by grouping the data layers into five thematic layers, including: lithology, geomorphology, structures, recharge conditions, and base map details. The input data on groundwater characteristics was in a descriptive form. Descriptive information was converted into "groundwater favorability index" (GWFI) values, which were rated with numerical values from 0.0 to 10.0. A rating of 10.0 was optimal (100% favorability).

Kamaraju et al. (1996) assigned weighting factors to the GWFI values based on their importance in controlling the occurrence of groundwater. The weighted overlay procedure applied resulted in the demarcation of groundwater units, in which each unit had a combination of groundwater controlling parameters and a corresponding "groundwater potentiality index" (GWPI; i.e., cumulative score). The end product was a map of groundwater development potential that shows 11 distinct classes of groundwater development potential ranging from none to excellent. A groundwater development potential map that was generated was designed to be a tool to aid for decision makers in the planning, exploitation, and management of groundwater resources (Kamaraju et al. 1996). Khan et al. (2006) similarly used remote sensing data and GIS to generate groundwater prospect maps in western Rajasthan, India.

GIS and remote sensing and watershed modeling data (SCS-CN Method in Watershed Modeling System) were used in weighted spatial probability modeling (WSPM) to identify groundwater development potential in the Sinai Desert of Egypt (Elewa and Qaddah 2011). Eight thematic layers were used: (1) rainfall, (2) net groundwater recharge, (3) lithology, (4) lineament density, (5) terrain slope, (6) drainage density, (7) depth to water, and (8) water quality. The WSPM results were reported to correlate well with actual borehole data.

Al Saud (2010) produced a thematic map of potential areas for natural groundwater storage in the Wadi Aurmah Basin of western Saudi Arabia using remote systems and GIS analyses. Six variables were considered; rainfall, lithology, rock fractures, slope, drainage characteristics, and land use and cover. Rock fracture density was identified by lineament analysis using satellite images. High lineament densities were considered to be indicative of high permeability and porosity. Bare, fractured rock was considered to have a high potential for groundwater recharge and human settlements a very low potential. Areas were rated for each variable and a weighting factor applied. Validation of the study was performed by comparing well yields to mapped potential groundwater storage. The results show about a 55% coincidence between well productivity classification and GIS-derived potential groundwater storage. The coincidence was increased to about 64% if the high and very high categories were merged.

19.3.4 Mapping of Contamination Vulnerability

The vulnerability of areas to groundwater contamination depends upon the presence of pollution sources, location of sensitive receptors (e.g., potable water supply wells), and local hydrogeology (e.g., aquifer depth, soil properties). GIS is thus well suited for the mapping of vulnerability to contamination through an overlay process, in which each factor considered is represented as a separate thematic layer.

GIS was used to generate maps of aquifer vulnerability to contamination and contamination risk maps for the semi-arid Metropolitan Zone of the Mexico City Basin (Ramos Leal et al. 2010). Maps of aquifer vulnerability to contamination were generated by overlaying thematic layers for seven physical and hydrogeological parameters that affect the local potential for the downwards transport of contaminants into the underlying aquifer. Pollution source index (PSI) maps were generated based on the density of potential pollution sources. Contamination risk maps were then generated by overlaying the aquifer vulnerability map and PSI map.

19.4 Hydrologic Data Management

The starting point for assuring the sustainable development of any surface water or groundwater system is the establishment of a comprehensive database. GIS has become an integral part of data base systems to assist in organizing, storing, retrieval, and displaying a substantial array of needed information (Alley et al. 1999). Some of data required for effective management of groundwater systems are listed in Table 19.1.

An ideal situation would be to incorporate all historic and current (real-time) data relevant to groundwater resources management into a single GIS-based system, which would be available to all water managers, regulators and authorized users via the internet. The data should be readily importable into analytical software (e.g., groundwater modeling, geochemical modeling, decision support software). A core spatial database would be available for many users for different purposes.

The technical resources now exist to support an ideal GIS groundwater data base system. For example, HydroManager™ water-resources information management system is an example of a customizable system that can provide centralized, secure access to water resources data (Schlumberger Water Services 2009). An important feature of HydroManager and similar systems is a flexible graphical user interface (GUI) that provides user friendly access to the data and analysis capabilities.

The St. Johns River Water Management District (Florida, USA) GIS hydrologic data system is a good example of a user-friendly publically-accessible system. Both current and historic data on precipitation, surface water and groundwater levels, and stream and river discharge are available. A coarse-scale map shows all the monitoring locations for each data type in the District (Fig. 19.3). Zooming into a location brings up an aerial photograph showing the precise location of monitoring stations. Available hydrologic data are obtained by clicking on the monitoring station icon.

Table 19.1 Principal data required for analysis of groundwater systems (after Alley et al. 1999)

Hydrogeological and physical data	Topographic maps showing land surface elevations, landforms, surface water bodies, drainage features, water use areas, and water management structures
	Geological maps of surficial deposits and bedrock
	Hydrogeologic maps showing three-dimensional extent, boundaries, and positions of aquifers and confining units
	Data of subsurface rock types, properties, and chemistries
	Geological and geophysical logs
Hydraulic data	Well performance (specific capacity) and pumping test data
	Aquifer and confining units hydraulic conductivity data
	Aquifer and confining unit storativity data
	Leakance values for confining units
	Saturated thickness maps of unconfined aquifers
	Water level (potentiometric surface) data for aquifers
Hydrologic and water budget data	Precipitation data
	Evaporation data (potential or reference and actual)
	Recharge rate data
	Stream and river flow data (including measurements of gains and losses between gauging stations in the study area
	Current and historic spring flows
	Estimated movement of water between surface water bodies and groundwater
	Current and historic water use data, including locations of wells
	Estimated return flows to aquifers from water uses
	Historic hydrographs
	Current and historic stream flows
Water chemistry and quality, and contamination information	Groundwater water chemistry
	Surface water chemistry
	Locations of known contamination sites and chemical present
	Locations, types, and concentrations of non-point source contamination
	Maps of current and past land uses

Gogu et al. (2002) provided a good example of a more advanced approach for managing hydrogeological data, the HYGES scheme developed for the Walloon region of Belgium. The HYGES scheme then consisted of 17 layers encompassing a very wide variety of physical, hydrological, hydrogeologic, and land use data that were obtained from multiple sources. The hydrogeological data stored in the GIS database can be easily used with different groundwater numerical models,

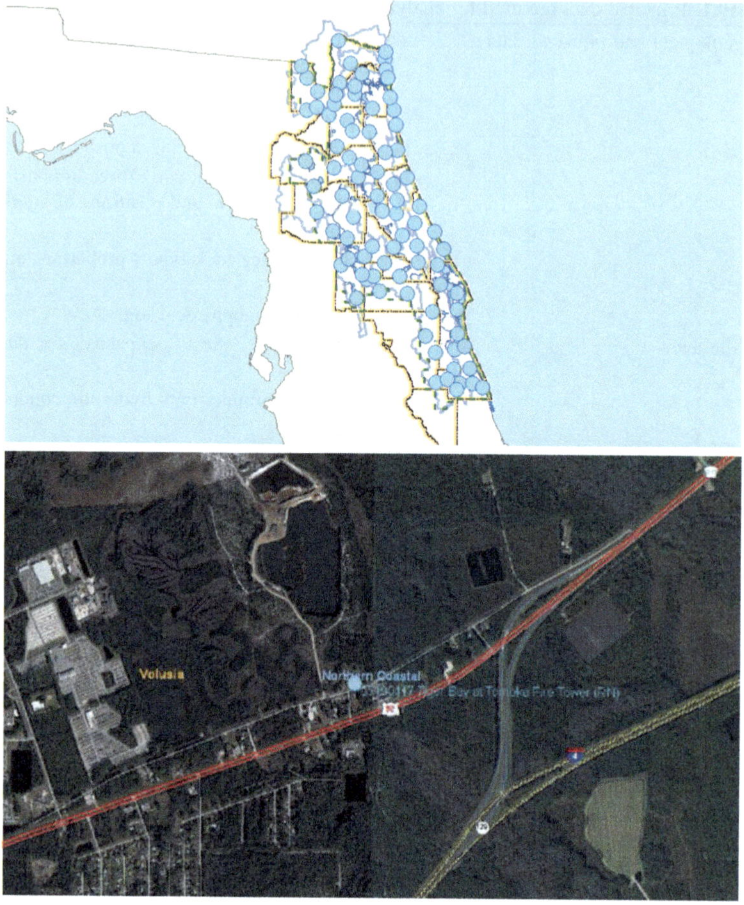

Fig. 19.3 St. Johns River Water Management District (Florida) publically accessible GIS hydrologic data system. Upper map shows all current rainfall monitoring stations. Zooming in on a station brings up an aerial photograph showing the station location. Available data can be obtained by clicking on station icon. http://webapub.sjrwmd.com/agws93/hdsnew/map.html

through linkages between the GIS database and the models. The collected and stored data can be used for predictive modeling in support of groundwater management decisions.

The challenge of GIS system development is encoding all of the data found in a number of different sources into a GIS compatible format. Hydrogeological data are very commonly found in a wide variety of different formats and some are non-digital. The encoding of data is highly manpower intensive. In most regions of the world, the support (political, financial, and human resources) does not exist to create the groundwater management technical framework that is possible using current technology. It is also critical that after a hydrologic data management system is developed, human and financial resources are made available to maintain and continuously update the system with new data as it becomes available.

Nevertheless, progress is being made in a step by step manner, and a major attraction of GIS technology is its expandability. It is readily possible to add additional layers of data, so that the entire system does not need to be developed all at once.

19.5 The Future of GIS in Arid Lands Water Management

As populations increase and the water resources become more depleted in arid regions of the world, the importance of access to detailed and accurate water-resources data will become imperative. GIS-compatible water-resources databases are being developed in a number of arid countries (e.g., Saudi Arabia, Ethiopia) to allow a better level of decision-making to become a reality. GIS-based data storage and retrieval systems will allow assessments to be made on changing conditions, which may improve the accuracy of predictive tools to assess the onset of severe droughts and the localized impacts of global warming. GIS data systems allow the evaluation of changes in hydrologic condition changes over time, assessment of the effectiveness of existing water management schemes, and assist in the development of mitigation measurements in areas that are currently stressed or will be stressed in the near and distant future. Perhaps the incorporation of historic and current hydrologic data into GIS database systems should be an economic priority in all arid lands because only through accurate information analysis can models be developed to allow key resources decisions to be made that will influence the public health and well-being of the people living in these areas and the natural environments upon which they depend.

References

Alley, W. M., Reilly, T. E., & Franke, O. L. (1999). Sustainability of ground-water resources. *U.S. Geological Survey Circular, 1186*, 79.

Al-Qudah, K., & Abu-Jaber, N. (2009). A GIS database for sustainable management of shallow water resources in the Tulul al Ashaqif Region. *NE Jordan: Water Resources Management, 23*, 603–615.

Al Saud, M. (2010). Mapping potential areas for groundwater storage in Wadi Aurnah Basin, western Arabian Peninsula, using remote sensing and geographic information system techniques. *Hydrogeology Journal, 18*, 1481–1495.

Brunner, P., Bauer, P., Eugster, M., & Kinzelbach, W. (2004). Using remote sensing to regionalize local precipitation recharge rates obtained from the chloride method. *Journal of Hydrology, 294*, 241–250.

Clarke, K. C. (2010). *Getting started with geographic information systems* (5th ed.). New Jersey: Upper Saddle River Prentice Hall.
DeMers, W. N. (2009). *GIS for dummies*. Hoboken, NJ: Wiley.
DeMers, W. N. (2005). *Fundamentals of geographic information systems* (3rd ed.). Hoboken, NJ: Wiley.
Elewa, H. M., & Qaddah, A. A. (2011). Groundwater potentiality mapping in the Sinai Peninsula, Egypt, using remote sensing and GIS-watershed-based modeling. *Hydrogeology Journal, 19*, 613–628.
Fayer, M. J., Gee, G. W., Rockhold, M. L., Freshley, M. D., & Walters, T. B. (1996). Estimating recharge rates for groundwater model using GIS. *Journal of Environmental Quality, 25*, 510–518.
Gogu, R. C., Carabin, G., Hallet, V., Peters, V., & Dassargues, A. (2002). GIS-based hydrogeological databases and groundwater modeling. *Hydrogeol Journal, 9*, 555–569.
Ghayoumian, J., Mohseni Saravi, M., Feiznia, S., Nouri, B., & Malekian, A. (2007). Application of GIS techniques to determine areas most suitable for artificial groundwater recharge in a coastal aquifer in Southern Iran. *Journal of Asian Earth Sciences, 30*, 364–374.
Harvey, F. (2008). *A primer of GIS: Fundamental geographic and cartographic concepts*. New York, NY: Guildford Press.
Korte, G. (2001). *The GIS book* (5th ed.). New York, NY: Thomson Learning Albany.
Kamaraju, M. V. V., Bhattacharya, A., Sreenivasa Reddy, G., Chandrasekhar Rao, G., Murthy, G. S., & Malleswara Rao, T. C. (1996). Ground-water potential evaluation of West Godavari District, Andhra Pradesh State. *India-A GIS Approach: Ground Water, 34*, 318–325.
Khan, M. A., Narrain, P., & Moharana, P. C. (2006). Prospecting ground water resources using RS-GIS—a case study from arid western Rajasthan, India. *Journal of the Indian Society of Remote Sensing, 34*, 171–179.
Manghi, F., Mortazavi, B., Crother, C., & Hamdi, M. R. (2009). Estimating regional groundwater recharge using a hydrological budget method. *Water Resources Management, 23*, 2475–2489.
Maxey, G. B., & Eakin, T. E., (1949). Ground water in White River Valley, White Pine, Nye, and Lincoln counties, *Nevada: State of Nevada, Office of State Engineer, Water Resources Bulletin, 8*, 59.
Ramos Leal, J. A., Noyola Medrano, C., & Tapia Silva, F. O. (2009). Aquifer vulnerability and groundwater quality in mega cities: Case of the Mexico Basin. *Environmental Earth Science, 61*, 1309–1320.
Saraf, A. K., & Choudhury, P. R. (1998). Integrated remote sensing and GIS for groundwater exploration and identification of artificial recharge sites. *International Journal of Remote Sensing, 19*, 1825–1841.
Stone, D. S., Moomaw, C. L., & Davis, A. (2001). Estimating recharge distribution by incorporating runoff from mountainous areas in an alluvial basin in the Great Basin Region of the southwestern United States. *Ground Water, 39*, 807–818.
Shaban, A., Khawlie, M., & Abdallah, C. (2006). Use of remote sensing and GIS to determine recharge potential zones. *The Case of Occidental Lebanon: Hydrogeology Journal, 14*, 433–443.
Solomon, S., & Quiel, F. (2006). Groundwater study using remote sensing and geographic information systems (GIS) in the central highlands of Eritrea. *Hydrogeology Journal, 14*, 1029–1041.
Sener, E., Davraz, A., & Ozcelik, M. (2005). An integration of GIS and remote sensing in groundwater investigations: a case study in Burdur, Turkey. *Hydrogeology Journal, 13*, 828–834.
Schlumberger Water Services. (2009). HydroManager water resources information management system. http://www.swstechnology.com/pdfs/technology_sheet/HydroManager.pdf.
Teeuw, R. M. (1995). Groundwater exploration using remote sensing and a low-cost geographical information system. *Hydrogeology Journal, 3*(3), 21–30.
U.S. Department of Agriculture (1986). Technical Release 55, Urban hydrology for small watersheds: U.S. Department of Agriculture, Soil Conservation Service.

Vrba, J., & Verhagen, B. T. (2006). *Groundwater for emergency situations. A framework document: International hydrological program (IHP) VI, Series on groundwater No. 12*. Paris: UNESCO

Watkins, D. W., KcKinney, D. C., Maidment, D. R., & Lin, M. -D. (1996). Use of geographic information systems in ground-water flow modeling. *Journal of Water Resources Planning and Management, 122*, 88–96.

Chapter 20
Groundwater Flow and Solute-Transport Modeling

20.1 Introduction

Numerical modeling has become an integral component of most water-resources investigations. The primary objective of groundwater modeling is the prediction of aquifer responses to current and future groundwater use and other hydrologic changes. An additional benefit of groundwater modeling is that it can provide insights on the properties of hydrologic systems through the calibration process. Two basic types of modeling have wide applications for groundwater resources modeling: flow and solute transport. Groundwater flow models simulate the hydraulic behavior of the aquifer such as the flow of water, changes in volume of water in storage, and changes in aquifer water levels or heads (pressure). Solute-transport models simulate the fate and transport of dissolved constituents in groundwater, which include salts and contaminants. Additional common types of modeling include heat-flow and integrated surface water-groundwater interactions.

Computer modeling has to a large degree supplanted the previously used analytical (equation-based) modeling for the following reasons:

- Computer modeling techniques provide much greater flexibility and capability to simulate complex geologic systems.
- The quality of groundwater modeling codes, particularly their flexibility and ability to simulate natural systems, has progressively improved.
- The historical exponential increase in computation power allows for complex models to be quickly executed on personal computers.
- Pre-processing and post-processing (graphical user interface or GUI) software packages (e.g., Visual Modflow^{TM}) have greatly simplified the construction and running of models, analysis and interpretation of the modeling results, and the preparation of high-quality graphic outputs, such as contour maps and three-dimensional animations.

R. Maliva and T. Missimer, *Arid Lands Water Evaluation and Management*, Environmental Science and Engineering, DOI: 10.1007/978-3-642-29104-3_20,

Groundwater modeling is a major subdiscipline onto itself and there is a voluminous literature of journal articles and technical papers on all aspects of the subject, but few dedicated textbooks on applied numerical groundwater modeling. "Applied Groundwater Modeling, Simulation of Flow and Advective Transport" by Anderson and Woessner (1992) provides a good introduction and overview of groundwater modeling techniques and is still the most widely used text. Hill and Tiedeman (2007) provide an overview of model calibration and Zheng and Bennett (2002) provide a good introduction to contaminant transport modeling.

Arguably, it has become too easy now to perform groundwater modeling (Maliva and Missimer 2010). Groundwater modeling software packages have become so user friendly that they allow groundwater flow and solute-transport models to be developed by groundwater professionals who lack both formal modeling training and a familiarity with the mathematical and hydrogeological assumptions and limitations inherent in the model codes. It has become readily possible to develop and run models and generate impressive graphic outputs, which may be fundamentally wrong. Groundwater models are also increasingly being developed by well-trained modelers in corporate "modeling centers," who lack the critical understanding of local hydrogeology. Groundwater modeling is best performed as team activity, in which an experienced modeler teams with people having local hydrogeological expertise to produce a superior product.

20.2 Groundwater Modeling for Arid and Semiarid Lands

Groundwater modeling procedures for arid and semiarid lands are basically similar to those used for more humid climates. However, arid and semiarid regions pose some greater challenges caused by the great spatial and temporal variation in precipitation, evapotranspiration, surface water flow, and groundwater recharge, which are the primary inputs of the water-balance. Arid lands aquifers are also stressed to a greater degree than aquifers in humid regions (Lloyd 2004), with stress considered as the ratio of annual abstractions to recharge. Accurate estimation of aquifer recharge and discharge is particularly difficult to perform because of the inherent difficulties in making such measurements. In addition, current hydrological conditions that may not be representative of prior wetter conditions during which most recharge occurred. Large, regional aquifers may not be in hydraulic equilibrium with current natural recharge and discharge, much less recent well abstractions.

Hydrologic data are commonly sparse in arid lands with a low density of boreholes both spatially and with depth (Lloyd 2007). Data are particularly sparse for the deep sections of aquifers. Hydraulic data are often obtained from single well tests performed on partially-penetrating wells. Hydraulic data obtained from shallow wells (e.g., storativity, average hydraulic conductivity) are often applied to the entire aquifer thickness, not considering likely changes with depth or the issue of partial penetration corrections (Lloyd 2007). The absence of data on aquifers "pales into insignificance when characteristics of intervening 'non-aquifer' units are considered!" (Lloyd 2007, p. 43). Data are usually not available on the vertical hydraulic conductivity of semi-confining units, which have a major influence on inter-aquifer flows.

Evaluation of large-scale flow systems also poses technical challenges. Multi-aquifer flow systems are often present within groundwater basins, with evaluation of cross-boundary flows between aquifers being extremely difficult, if not impossible, to evaluate, based on the available data. Assessment of regional aquifers needs to consider local faulting, folding, and facies changes, which can have a considerable influence on groundwater flow, particularly by causing discontinuities within aquifers (Lloyd 2007). The sequence stratigraphy of sedimentary aquifers, which greatly affects the vertical and horizontal distribution of hydraulic conductivity, rarely has been characterized (Maliva and Missimer 2010).

Interbasin groundwater flow (intermediate and regional flow system) may exist to varying degrees in both arid and humid regions. However, interbasin flow may have a proportionally greater effect on to basin water budgets in arid lands (and thus needs to be considered in groundwater models) because of low recharge rates. Groundwater models in arid lands can provide a broad understanding of large-scale groundwater flow, but there are often insufficient data to accurately simulate local groundwater flow (Lloyd 2007).

20.3 Limitations of Groundwater Modeling

The most fundamental limitation on groundwater modeling is that it does not result in unique solutions. Even if a model result is consistent with present and past observational data, there is no guarantee that the model will perform at an equal level when used to predict the future (Oreskes et al. 1994), particularly if future input conditions are well outside the range of historic conditions. The results of predictive simulations should therefore be viewed as best estimates. There can be a false perception that as a numerical technique, modeling inherently produces more scientific, objective, and reliable results. Perhaps the most effective use of predictive models is as a tool under continuous refinement in which model predictions are regularly compared to hydrologic monitoring data and the model is updated and recalibrated as necessary.

Although numerical groundwater models can incorporate considerable hydrogeologic detail, they are still simplifications of complex hydrogeological systems in which there will never be sufficient data to fully characterize the aquifer system in question. There will always be some uncertainty associated with modeling results. The pertinent question is whether or not a model output adequately represents real world conditions for the purposes for which the model is to be used. Models do not have to be perfect to be useful from a practical water management perspective.

The quality or accuracy of the output from groundwater models is dependent on the quality of the data incorporated into the model. The computer software adage "garbage in, garbage out" is certainly applicable to groundwater modeling. Groundwater modeling has very limited value in the absence of quality hydrogeologic data and accurate conceptual models. There is a movement towards more stochastic (probability-based) groundwater modeling, particularly in the academic community. However, detailed and high-quality field data are still necessary to provide the geostatistical parameters needed for stochastic modeling. Without accurate field data, the principle of non-uniqueness can render any groundwater modeling results to be inconclusive and of questionable value. Hydrogeological data are critical for constraining model construction and narrowing the envelope of possible solutions, and thus uncertainty.

The uncertainty associated with modeling results can be evaluated in part through sensitivity analyses. Large-scale models developed for groundwater management should not viewed as an end product, but rather as a work in progress, with progressive refinement and improvement anticipated as more data become available. Numerical modeling is an invaluable tool for groundwater management, but it can be abused when it is used to make decisions beyond the capabilities of the models (Missimer et al. 2010). Problems lie not in the uncertainty associated with models, but rather in the failure to understand and consider the uncertainty inherent in models when using them to make decisions.

An additional source of uncertainty, which may be important in some arid and semiarid lands, lay in the input data. For example, the values of key input variables, such as groundwater pumping rates, may have a large uncertainty if there is inadequate monitoring data. A model may thus provide a good representation of aquifer behavior, but predictive simulations using the model may have large errors if, for example, current and future groundwater pumping rates are substantially greater than expected. Discrepancies between observed and measured data may thus be indicative of problems associated with the input and monitoring (observational) data.

20.4 Groundwater Model Development

The development of a groundwater model includes several stages or processes, which may or may not be performed as discrete project phases. For example, an experienced modeler may intuitively know which model code is most appropriate for a given project. A general outline of the model development process, which is similar to the model protocol proposed by Anderson and Woessner (1992), is herein presented.

20.4.1 Statement of Problem and Modeling Objectives

The specific reason for performing modeling and the desired model output should be established. The modeling objectives should also considered, such as the desired accuracy of simulated drawdowns, and whether or not they are realistically obtainable. Consideration should also be given to the economic and technical resources that are available for a project, which will dictate what is possible.

20.4.2 Conceptual Model

Bredehoeft (2005, p. 38) described the role of the conceptual model as

> Every model has as its foundation a conceptual model. The conceptual model is the basic idea, or construct, of how the system or process operates; it forms the basic idea for the model.

The conceptual model may include such factors as the model boundaries and boundary conditions, aquifers included and their interrelationship, recharge and discharge sources and rates, predominant type of groundwater flow (matrix versus conduit), aquifer heterogeneity, and active geochemical processes. The later stage of conceptual model development involves consideration of the number and geometric array of the cells (or elements) in either finite-difference or finite element models.

If the conceptual model does not accurately represent the basic hydrogeologic conditions of the study area, then the results of any modeling based on the conceptual model will be questionable. Numerical modeling is the method by which the appropriateness of the prevailing concept is tested. Conceptual and numerical model development should be viewed as an iterative process in which the conceptual model is continuously reformulated and updated (Bredehoeft 2005). Poeter (2007) claimed that the uncertainty associated with conceptual models is typically larger than measures of uncertainty that can be calculated for any individual model (i.e., uncertainty associated with estimates of parameter values).

The term ‘surprise’ is defined as (Bredehoeft 2005, p. 38)

> The collection of new information that renders one’s original conceptual model invalid.

In other words ‘surprise’ results in a complete paradigm shift. Bredehoeft (2005) proposed that surprises are more frequent (perhaps 20–30% of models) than is commonly realized. Ideally, alternative conceptual-mathematic models should be constructed and tested to determine the best fit and to fully capture the uncertainty associated with a modelling investigation (Poeter 2007). However, this idea is typically not put into practice (Bredehoeft 2005), because of the great amount of effort it would require. Instead, the problems associated with incorrect conceptual models can be reduced by (Bredehoeft 2005)

- collecting as much data as feasible and necessary using all applicable methods and
- remaining open to fact that the conceptual model can change dramatically as more information is collected.

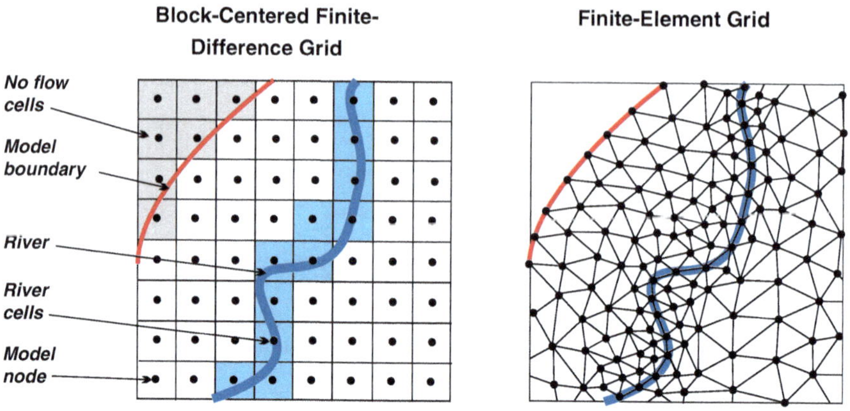

Fig. 20.1 Diagram of finite-difference and finite-element discretization. The finite-element discretization has greater flexibility to model irregularly shaped features. The finite-difference discretization can be used to simulate finer scale features by reducing the grid size

20.4.3 Choice of Model Code

Some of the commonly used groundwater models are discussed in Sect. 20.5. The model code used for an investigation should be amenable to meeting the modeling objectives. Model choice issues include:

- finite difference versus finite-element discretization,
- whether or not surface water-groundwater interaction needs to be simulated,
- whether or not simulation of flow through the unsaturated zone is needed,
- whether or not solute-transport (either density dependent or independent) needs to be simulated,
- model acceptability and defensibility, and
- open-source versus proprietary code.

Two main types of numerical models are used for three-dimensional modeling of groundwater flow in aquifers, which are the finite-difference and finite-element types. The model types differ in the manner of discretization or subdividing the aquifer into a three-dimensional grid. The finite-difference method subdivides the aquifer into regular rectangular grid blocks. In block-centered, finite-difference models, such as MODFLOW, hydraulic heads are calculated at nodes located in the center of the grid blocks (Fig. 20.1). The finite-element method uses an irregular arrangement of node points commonly arranged is triangular shapes (Fig. 20.1). The finite-element grid has greater flexibility to simulate irregularly shaped features, such as rivers and wells, which can be placed on nodes. Most geological and hydrogeological features do not have an orthogonal geometry, so the flexibility of the finite-element grid is advantageous. However, finite-difference codes are mathematically simpler and more stable.

The model codes used for an investigation must meet the technical requirements of the project. The code must also meet the requirements of the system owner, regulatory agencies, and other groups or professionals that may use or review the model. Accessibility of a model to future potential users is an important consideration, where a model is to be used as a long-term water management tool. From an owner's perspective, there is great value in having people other than the model developer be able to review a model and in the future, update it and perform simulations as needed. Therefore, widely used and available model codes should be used where possible.

An important distinction occurs between public-domain, open-source codes, and proprietary software in which the source code is a trade secret. As a general principle, public domain software is preferred over a proprietary software package, unless the latter have features or capabilities necessary for a project that are not readily available in public domain software. The critical advantages of public domain software include the following:

- The source codes for public-domain modeling programs are available for independent review and analysis, which is not possible for proprietary software where the source code is kept secret. Proprietary software is in essence a "black box" in which the users must accept their validity and accuracy largely as a leap of faith.
- Public domain software programs, such as MODFLOW, benefit from refinements and additions that are possible from the pool of users.
- Proprietary software programs may not be admissible in legal settings if the source code is not released.
- Public domain software packages may have greater acceptance, particularly in regulatory settings.
- Public domain software is free, although proprietary pre- and post-processing (graphic user interface) software is often privately developed and commercially distributed.

Proprietary model codes may not be acceptable in some circumstances. In the United States, modeling results are commonly ruled inadmissible in various court settings during litigation unless the model code is also provided. Usually the developer of the proprietary model code will not release the model code because it is a trade secret, which would make simulations run with the code inadmissible.

Many water management and regulatory agencies also prefer or require that modeling results submitted in support of permits must be performed using commonly used public domain codes, so as to allow for their independent review. As a general principle, transparency is desirable in science and open-source, public-domain codes have a clear advantage in this respect. However, some projects may have modeling needs that can be best met with advanced proprietary software. If the capabilities of a proprietary code are needed, then preference should be given to codes that are widely commercially available. Sometimes, it is necessary to use a proprietary code to assess a problem and then re-model the problem using an open-source so that it can be used in courtroom testimony or in water management decisions. This method is fraught with problems because the assumptions between the models may not be similar and the results may not be exactly the same.

20.4.4 Model Development and the Population of the Model Grid

Values for the various physical, hydraulic, and chemical parameters must be assigned to all points or cells in the model grid. A wide variety of different types of data are collected as part of hydrogeological investigations. A challenge exists concerning how to best integrate and incorporate data from diverse sources, such as lithologic logs, borehole geophysical logs, surface geophysics, packer (drill stem) tests, remote sensing, and aquifer testing into groundwater models. Much of the hydrogeological data is obtained from point sources (e.g., wells) and must be interpolated and extrapolated in order to populate model grids. GIS has become a powerful tool for the population of model grids because its capabilities to store and process georeferenced data (Chap. 19).

In the absence of site-specific data, data from regional models may have to be used or best estimates based on aquifer lithologies. A variety of methods have been developed in the oil and gas industry to predict the three-dimensional distribution of rock and sediment types, and thus hydraulic parameters, based on data from a limited number of wells. Sedimentologic (facies) and sequence stratigraphic analyses provide a means for projecting the distribution of sediment and rock types based on the concept that these sediments were deposited in a non-random, predictable manner. Stochastic methods also exist for populating a grid based on limited data.

Storage of geologic and hydrogeologic data in three dimensions allows interpolation of three-dimensional (3D) hydrostratigraphic units, designation of measured or interpolated properties to the units, volumetric calculations, morphology analysis, representation of subsurface structures, parameter flux (i.e., groundwater flow, chemical diffusion) between units, and interpolation of hydrologic properties (National Research Council 2008). Three-dimensional data management and visualization software are currently commercially available (e.g., HydroGeoAnalyst™).

Much more sophisticated workflow software have been developed for the oil and gas industry, which can incorporate the wide variety of data collected from reservoir investigations and development activities into models that simulate complex reservoirs. For example, the Petrel workflow developed for the oil and gas industry offers a set of aquifer characterization, modeling, and simulation programs that can help develop the most efficient and cost-effective solutions to managed aquifer recharge and other water management problems (Herrmann 2006). Petrel allows hydrogeologists, geochemists, geologists, and water managers, to share data and results within the same environment.

The Petrel workflow system allows the integration of the following disciplines:

- geophysical interpretations,
- surface imaging and mapping,

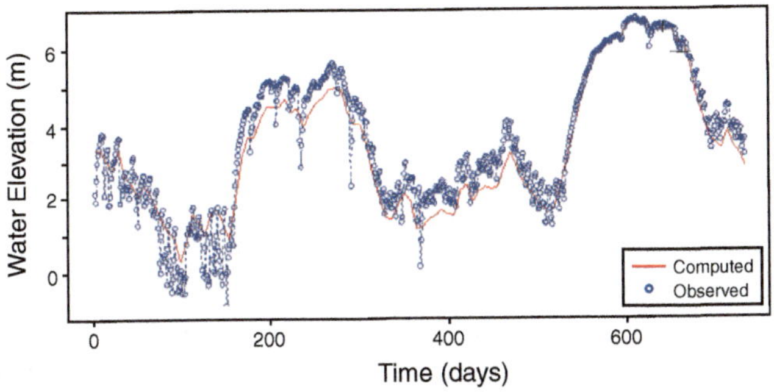

Fig. 20.2 Example of calibration of a groundwater flow model to water elevation data from an observation well (Lee County, Florida). Model captures the seasonal variation in water levels

- geological log interpretation and well correlation,
- complex faulting and fracture modeling,
- facies and geophysical modeling,
- aquifer testing results,
- uncertainty analysis,
- surface and subsurface interaction,
- up-scaling processes, and
- flow and mass transport simulation.

The ultimate data management goal should be to have all of the hydrogeologic data for a region in a single central GIS-based database (Sect. 18.3.5), from which modelers could readily extract the data needed for modeling investigations. The data would also include input data (e.g., pumping and rainfall records) and observational data for model calibration (e.g., monitoring well locations and historic data).

20.4.5 Model Calibration

Model calibration is the process by which the values of model inputs (hydraulic parameters) are adjusted so that the model output matches observed data (Fig. 20.2). The observed data, referred to as calibration targets, may include

- water levels (pressures) in wells (static or in time-series),
- surface-water body elevations and flows (static or in time-series), and
- water quality data (e.g., salinity) measured in wells or recovered water (static or in time-series).

The limitation of non-calibrated simulations is that their accuracy can only be as good as the values assigned to the hydraulic parameters. The results of non-calibrated simulations thus have potential large, unquantifiable uncertainties (errors).

Confidence in a model to provide an adequate representation of the system under consideration generally improves with an increase in the number of calibration targets and their geographic distribution, a longer duration of the calibration period, and a greater range of conditions or stresses during the calibration period. For example, it is preferable that the model calibration period includes both wetter and drier than average periods.

Model calibration is evaluated by analysis of the residuals (errors), which are the difference between simulated and observed values. The main calibration statistics used in groundwater modelling are mean error (ME), mean absolute error (MAE), root mean square error (RSME), and weighted least square residual (WLSR).

Mean error, mean absolute error, and root mean square error are defined with respect to heads as:

$$ME = \frac{1}{n}\sum_{i=1}^{n}(h_m - h_s)_i \tag{20.1}$$

$$MAE = \frac{1}{n}\sum_{i=1}^{n}|h_m - h_s|_i \tag{20.2}$$

$$RMSE = \left[\frac{1}{n}\sum_{i=1}^{n}(h_m - h_s)_i^2\right]^{0.5} \tag{20.3}$$

where,
n = number of calibration targets
h_m = measured head (m, ft)
sh_s = simulated head (m, ft)

Mean error is sensitive to the directions of the error (whether simulated values are greater or less than the measured values) and large errors between target calibrations may cancel out if they are in opposite directions. Calibration statistics commonly provide measures of the average error, as all targets are given an equal weight. The weighted least square residual (WLSR) includes a weighting factor (ω_i), which reflects the expected measurement error (i.e., observation data that may have great errors or variance are given a lower weight) or proportional importance allocated to target "i" and is defined as

$$WLSR = \omega_i(h_m - h_s)_i^2 \tag{20.4}$$

The objective of the calibration procedure is to minimize the residuals. Model calibration has been historically performed manually using a systematic trial and error process. The values for a model parameter in all or a spatial zone of the model grid are adjusted either upwards or downwards in increments to try to obtain a better fit (lower residual) to the calibration targets. An understanding of the hydrogeologic system in question and the response of models to changes in the various parameters allows the process to be performed efficiently. A key component of the calibration process is knowledge regarding the real range of the hydraulic parameters within which calibration should lie. Calibration using unrealistic hydraulic parameters outside of the realistic range of variation may lead to models that are well-calibrated but poorly representative of real conditions. In an area where there are a few high quality aquifer performance tests (APTs) with a corresponding high reliability of the hydraulic data, these points can be held nearly constant in the calibration process and the other data varied to meet the calibration targets.

Calibration of groundwater flow models should also be performed using data other than just aquifer heads (if such data are available). For example, environmental isotope data can provide constraints on boundary conditions (Herczeg and Leaney 2011). Groundwater age data can also be used for model calibration, but the limitations inherent in groundwater age data must be considered (Sanford 2011).

Software is available that automatically performs model calibration. The PEST parameter estimation and optimization software program is increasingly being used for model calibration (Doherty 2005). The PEST program takes control of the model and automatically performs as many model runs as it needs, while adjusting the parameters of the model until the weighted least squares residual is reduced to a minimum value. Much has been written on the relative merits of manual versus automated calibration. Automated calibration (like manual calibration) is the determination of parameter values that produce the best model fit with observations for a given conceptual model (Poeter and Hill 1997). The major advantages are that it expedites that process of adjusting parameters, and thus allows for more calibration runs to be performed in a given time period, and that it may reveal issues overlooked in manual calibration (Poeter and Hill 1997).

A criticism of manual calibration based on a limited number of homogenous parameters is that it results in a great degree of simplification (parsimony). Hunt et al. (2007) presented the advantages of automated calibration through regularized inversion, which can maximize insights gained from field data while also introducing some spatial variability to improve fit.

However, because of the non-uniqueness of model calibration, the best fit solution may not necessarily result in the most accurate predictive model, particularly where the predictive simulation scenario differs significantly from the calibration model conditions. Adjustments in model parameters during model calibration should be hydrogeologically reasonable. If the best fit obtained by automated calibration requires the use of hydrogeologically unreasonable values for parameters, then the results may indicate problems with the conceptual model or perhaps the observational data (e.g., pumping information from wells in the model area) (Poeter and Hill 1997). A key point is that automated calibration, when properly employed, is not a quick and easy method, but still requires a high level of technical sophistication and local hydrogeological understanding by the modeler and team of other experts working on the project.

An important issue in groundwater modeling is that simulated heads are largely a function of hydraulic conductivity and recharge rates (Anderson and Woessner 1992). A decrease in hydraulic conductivity and increase in recharge can both act to increase simulation heads. There is thus no unique solution to the model calibration.

Recharge rates are poorly known, especially in arid and semiarid regions. The range of possible recharge rates in arid and semiarid ranges can be narrowed by independent estimates of the recharge rate obtained from groundwater ages. For example, Zhu (2000) calibrated simulated ^{14}C ages from a numerical flow and transport model against observed ^{14}C age, in addition to a calibration against aquifer heads. Such an approach narrows the envelope of potential recharge rate and hydraulic conductivity values. The age-correction of observed ^{14}C values allowed for the use of a non-reactive transport model, which needed to simulate only advection-dispersion/diffusion and radioactive decay (Zhu 2000).

A sensitivity analysis is a critical component of the model calibration process as it can provide approximate error bars for the model results. Sensitivity analyses are performed by running predictive simulations in which the values of hydraulic parameters are adjusted by fixed amounts (e.g., upwards and downwards by 50%). The goal of the sensitivity analysis is to quantify the effects of the inherent uncertainty in the values of hydraulic parameters on model results. If the results of the sensitivity analysis show only a minor change in predictive simulations results, then a high degree of confidence can be placed on the model. Conversely, if a modest adjustment in the value of a hydraulic parameter that has a high degree of uncertainty causes a substantial change in simulation results, then a large uncertainty (error bars) exists for the simulation results.

20.4.6 Model Verification, Post Audits, and Refinement

The purpose of model verification is to establish greater confidence in a calibrated model by performing additional simulations using new stress conditions and observational data (Anderson and Woessner 1992). For example, a model could be calibrated for a 2-year period, and then a verification simulation could be performed on the data for a third year. If there is a good match between the simulated and observed data for the third year, then greater confidence may be placed in the ability of the model to adequately represent real conditions.

The term verification means the establishment of truth by demonstration. It is not possible to verify or validate any numerical model in the sense that the veracity of a model is established (Oreskes et al. 1994). Instead the burden on a modeler is to "demonstrate the degree of correspondence between the model and the material world it seeks to represent and to delineate the limits of that correspondence" (Oreskes et al. 1994, p. 644).

The model verification process, as the term is widely used in the modeling community, should be an ongoing process. Post audits should be performed in which predictions are compared to what actual happened (i.e., good-quality monitoring data) in order to confirm that the model does indeed provide an acceptable representation of real world conditions and to obtain insights into future possible model enhancements (Anderson and Woessner 1992; Lloyd 2004). If there is a significant discrepancy between predicted and observed data, then the model should be refined and recalibrated. Models should also be revaluated and refined, as necessary, as new hydrogeological data become available, particularly if it does not support the conceptual model of the study area.

20.4.7 Predictive Simulations

The end product of groundwater model development should be a model that is well calibrated against available monitoring data and, as such, can accurately predict future system performance. Predictive simulations can then be performed to assess issues of water management concern, such as changes in aquifer water levels as the result of the continuation of current pumping rates or as a result of changes in future water use (new groundwater development).

20.5 Groundwater Flow Modeling Codes

The choice of software for a groundwater modeling project depends upon:

- model code capabilities versus project needs,
- accessibility of the model code (private versus public domain), and
- acceptability of the model to the system owner, regulatory agencies, courts, and the potential user community.

An additional factor that comes into play during model code selection is provincialism. In the United States, most groundwater modeling is performed using software codes developed by or for the U.S. Geological Survey, U.S. Environmental Protection Agency, or other government agencies. These codes are nearly universally accepted, are usually publically available, and have undergone extensive review and improvements because they are in the public domain. The U.S. Geological Survey MODFLOW code is very widely used for groundwater flow modeling and has great regulatory acceptability. The public domain codes have been packaged into proprietary commercial software products (e.g., Visual MODFLOW). The U.S. government agency codes are less commonly used in Europe and elsewhere in preference for locally developed software. There may also be an element of inertia in the choice of codes as a modeler or project owner may be more familiar with or normally prefer the use one particular code.

Model codes vary in their discretization (finite-difference versus finite-element) and their capabilities concerning simulation of solute transport and heat flow. Solute-transport models are required to simulate the fate and transport of contaminants and the interface between fresh and saline waters. A density-dependent solute transport model is needed for simulations of systems in which there are significant differences in fluid density such as those resulting from differences in salinity or temperature. Reactive solute-transport codes can simulate geochemical processes during groundwater flow, such as mineral dissolution and precipitation.

20.5.1 MODFLOW

The U.S. Geological Survey MODFLOW code (McDonald and Harbaugh 1988) is the industry standard in the United States and is widely used elsewhere. MODFLOW has evolved into a family of codes, which were built upon the basic MODFLOW groundwater flow code. McDonald and Harbaugh (1988) provide an excellent description of the model and its underlying mathematics. Domenico and Schwarz (1998) and Fetter (2001) provide a summary of the MODFLOW family codes, which continues to grow. Codes that have been developed upon or utilize the outputs of MODFLOW include MODPATH (Pollock 1994; particle tracking), MT3DMS (Zheng and Wang 1999, solute transport), SEAWAT (Guo and Langevin 2002, density dependent solute-transport), PHT3D (Prommer et al. 2003, reactive solute transport), and SEAWAT Version 4, which includes heat transport (Langevin et al. 2008)

MODFLOW is a three-dimensional, finite-difference, ground-water flow model. The modular design and public-domain status of MODFLOW has allowed new capabilities to be added to the original model. The great advantage of the modular design is that new features or processes can be added to the model with minor modification of the existing core code. The original code included packages (modules) to simulate basic features and processes, such as rivers, recharge, evapotranspiration, drains, and wells.

The MODFLOW code is widely used in both academia and industry, and has undergone a great deal of review. New or improved modules have been developed to better simulate a diversity of hydrogeological processes. The community using MODFLOW is very large, which has allowed numerous improvements to be made to the model. There are international MODFLOW conferences at which new or expanded uses of the model are discussed.

20.5.2 MicroFEM

MicroFEM is a three-dimensional, finite-element program for steady-state and transient groundwater flow modeling (Hemker and de Boer 2009). The Microsoft Windows version of MicroFEM includes a graphic user interface and packages for grid generation, data input, presentation of results, and model optimization.

20.6 Groundwater Model Codes for Solute Transport

A number of solute-transport codes have been developed that have or can be used for the simulation of contaminant transport, managed aquifer recharge, saline-water intrusion, and other processes of water management concern. Differences between codes include whether finite-difference or finite-element discretization is used, and whether or not the models simulate density-dependent flow and heat transport. Solute-transport codes incorporate groundwater flow codes (or are built upon groundwater flow codes such as MODFLOW) and can therefore be used to model groundwater flow alone. Some of the more commonly used solute transport models are herein summarized.

20.6.1 MT3DMS

MT3DMS (Zheng and Wang 1999) is a modular three-dimensional multispecies transport model for simulation of advection, dispersion and chemical reactions of contaminants in groundwater systems. The MT3DMS code is developed for use with a block-centered, finite-difference flow model such as MODFLOW, and is based on the assumption that changes in the solute concentration do not significantly impact groundwater flow (flow is not density dependent). The MT3DMS code has a modular structure similar to MODFLOW, which allows for the simulation of advection, dispersion/diffusion, source/sink mixing, and chemical reactions. The modular design allows for other transport processes or reactions to be added to the model readily without having to modify the existing code. The combination of MODFLOW and MT3DMS can be used to simulate density-independent solute transport.

20.6.2 SEAWAT

SEAWAT (Guo and Langevin 2002; Langevin et al. 2008) is a coupled version of the MODFLOW and MT3DMS codes designed to simulate three-dimensional, variable-density, saturated groundwater flow. The latest version (Version 4) of SEAWAT added the capability to simulate heat transport. SEAWAT is now widely used in the United States to simulate density-dependent solute-transport in coastal settings. A SEAWAT model can be developed using existing MODFLOW models for the initial flow component. SEAWAT is suitable for the simulation of ASR systems in aquifers containing brackish or saline waters and for the simulation of saline-water intrusion.

20.6.3 SUTRA

SUTRA (Saturated Unsaturated Transport Code; Voss and Provost 2002) was originally a two-dimensional computer program developed by the U.S. Geological Survey that simulates fluid movement and the transport of either energy or dissolved substances in a saturated or unsaturated subsurface environment. The code was later modified for three-dimensional simulations. SUTRA simulations are based a hybridization of the finite-element and integrated finite-difference methods. The discretization consists of quadrilateral elements in two dimensional simulations and hexahedral elements in three-dimensional simulations. The hybrid method preserves the geometric flexibility of the finite-difference method, while taking advantage of the finite-difference efficiency (Voss and Provost 2002). The SUTRA codes can simulate density-dependent flow and are thus suitable for simulating systems involving saline-water intrusion and ASR systems using brackish-water aquifers.

20.6.4 HST3D

HST3D (Heat- and Solute-Transport; (Kipp 1987, 1997) is a three-dimensional code developed by the U.S. Geological Survey for the simulation of groundwater flow and associated heat and solute transport. HST3D is a finite-difference, variable-density code that can be used for the analysis of subsurface waste injection, landfill leaching, saltwater intrusion, freshwater recharge and recovery, radioactive-waste disposal, hot-water geothermal systems, and subsurface-energy storage.

20.6.5 FEFLOW

FEFLOW (Finite Element subsurface FLOW system; Diersch 1998) is a three-dimensional, finite-element simulation package that can be used to model groundwater flow and solute transport in porous media under saturated and unsaturated conditions. FEFLOW can be used in two or three-dimensional modes, and for steady state and transient simulations. FEFLOW can simulate coupled-density dependent flow and is thus capable of simulating saline-water intrusion and ASR systems using brackish-water aquifers. FEFLOW is proprietary software and is commercially available in a package that includes pre- and post-processing software.

20.6.6 ECLIPSE®

ECLIPSE is proprietary reservoir simulation software developed by Schlumberger Limited for the oil and gas industry, and has been the benchmark core in reservoir simulations for over 25 years. ECLIPSE is a finite-volume simulator that allows for the modeling of multi-component (water, oil, and gas) and multi-phase (liquid, gas, gas in solution) systems. ECLIPSE can be used to model individual wells and near well-bore conditions, dual porosity and permeability systems, chemical reactions, and complex reservoir geologies. The simulation capability of ECLIPSE greatly exceeds the needs of most groundwater investigations. The utility of ECLIPSE is greatly enhanced when combined with the Petrel* workflow software, which allows for the integration of geological, geophysical, and hydrogeological data into reservoir or aquifer models (Sect. 15.2). The great capability of ECLIPSE has applications for the simulation of projects involving geologically complex aquifers as found in fractured, dual-porosity systems. ECLIPSE also has the capability of simulating subsurface carbon sequestration systems.

20.6.7 Hydrus

Hydrus is a Microsoft Windows based finite-element for simulating two- and three-dimensional movement of water, heat, and multiple solutes in variably saturated media (Radcliff and Šimůnek 2010). The program may be used to simulate water and solute-transport in unsaturated, partially saturated, or fully saturated media. Hydrus is well suited for the simulation of vadose zone processes and can also simulate seepage face boundaries. A Windows-based graphical user interface is used for grid design and editing, parameter allocation (data inputs), model execution, and visualization and analysis of results. Hydrus is distributed in five different versions (levels) of different sophistication and price, which allows users to acquire the segments of software that are most appropriate for their needs.

20.6.8 PHT3D

The PHT3D code (Prommer et al. 2003) is a general purpose reactive 3D multi-component model that combines the widely established MODFLOW/MT3DMS (Zheng and Wang 1999) with a batch-type geochemical model PHREEQC-2 (Parkhurst and Apello 1999) to compute a wide range of biogeochemical reactions. The PHT3D code has the advantage of being based on widely used, public domain models that are all widely tested and documented.

20.7 Simulation of Compaction and Land Subsidence

Land subsidence caused by aquifer depressurization resulting from groundwater pumping is major problem in some regions (Chap. 15). Compaction can also alter the hydraulic properties of aquifers and confining units. Water released during compaction of confining and semiconfining units may be a significant source of non-renewable groundwater.

The release of water and land subsidence can be simulated using packages developed for the MODFLOW code. The interbedded storage package (ISB1) (Leake and Prudic 1991) simulates storage changes from both elastic and inelastic compaction of compressible fine-grained beds that are within or adjacent to an aquifer (i.e., interbeds). Elastic and inelastic compaction is assumed to be proportional to the change in head. The ISB1 package assumes instantaneous equilibrium of the heads in the interbeds with heads in the surrounding aquifers. This assumption is valid for thin interbeds with very short time constants (Hoffmann et al. 2003a).

The Subsidence and Aquifer-System Compaction (SUB) Package simulates elastic (recoverable) compaction and expansion, and inelastic (permanent) compaction of compressible fine-grained beds (interbeds) within the aquifers (Hoffmann et al. 2003b). The SUB package differs from the ISB1 package in that it can be used to simulate the delayed release of water from storage or uptake of water into storage in the interbeds. The SUB Package supersedes the Interbed Storage Package (IBS1) for MODFLOW, which assumes that water is released from or taken into storage with changes in head in the aquifer within a single model time step. If the time-dependent drainage is turned off, the SUB Package gives results identical to those from IBS1.

The Subsidence and Aquifer-System Compaction package for Water-Table Aquifers (SWT) (unconfined aquifers) simulates vertical compaction in models of regional groundwater flow. The program simulates groundwater storage changes and compaction in discontinuous interbeds or in extensive confining units, accounting for stress-dependent changes in storage properties (Leake and Galloway 2007). Geostatic stress can be treated as a function of the water-table altitude, and compaction is a function of computed changes in effective stress at the bottom of a model layer. Thickness of compressible sediments in an unconfined model layer can vary in proportion to saturated thickness.

An important modeling issue is the treatment of the delayed dissipation of overpressures in thick interbeds. If short-term responses of hydraulic heads to changes in pumping rate are to be simulated, then the presence of delayed drainage in interbeds must be accounted for in the model (Hoffmann et al. 2003a). More water will initially be required to be derived from storage in the aquifer to supply the pumping wells, which would result in larger drawdowns than would occur without delayed drainage from the interbeds. Delayed drainage effects will have a small impact on long-term model runs unless the interbeds are of great thickness.

20.8 Integrated Surface Water-Groundwater Modeling

An objective in water resources management is to develop more holistic models that incorporate the interaction of surface water and groundwater. Major modeling goals are (1) to more accurately simulate aquifer recharge from climatological data (particularly precipitation and temperature), (2) to simulate the effects of groundwater withdrawals on surface water environments and wetlands, and (3) to predict the effects of land and water use changes on surface water flow and groundwater levels. Incorporation of recharge into groundwater flow models has long been a challenge. Often, recharge is treated as the residual in the water balance, i.e., the model is used to determine recharge through the calibration process, rather than recharge being a primary input variable.

The linkage between surface water and groundwater with respect to downward flow (recharge) can be addressed in two main manners. The historically more common method is to not implicitly deal directly with the vadose zone and surface waters in the model, but rather just apply a net recharge as a boundary condition to the upper layer of the model (i.e., water table). An average recharge rate can be estimated using a variety of methods (Chap. 11). Recharge from channel transmission losses can be estimated from rainfall-runoff modeling.

A technically more sophisticated approach is integrated surface water—groundwater modeling, which combines a surface water flow model with a groundwater flow model. Two general approaches have been taken to develop integrated models that simulate surface-water and ground-water interactions (Markstrom et al. 2008). Fully integrated approaches simultaneously solve surface and subsurface flow equations. The alternative is the "coupled regions" approach, which partitions the surface and subsurface systems into separate regions with the governing equations that describe flow in each region integrated (coupled) using iterative methods (Markstrom et al. 2008). For example, separate equations can be used to simulate runoff and infiltration, horizontal and vertical flow in the soil zone, vertical gravity driven flow in the vadose zone, and three-dimensional flow in the saturated zone (Markstrom 2008). For water resources investigations of arid environments, the surface water model could provide recharge rates (e.g., from transmission losses in wadis), which would be input into the groundwater flow model.

Both fully integrated and coupled models are capable of simulating coupled surface-water and ground-water systems. There is no inherent overall advantage of one method over the other. The choice of model codes should depend on project-specific needs, computation efficiency (run times), availability of technical expertise, cost, and whether or not the code is open source. As a general rule, widely used open-source codes are preferred because of their transparency and they benefit from testing and improvements (upgrades) developed by a large pool of users. However, proprietary codes may have specific features needed for a given project, and for this reason, may be the best option.

The U.S. Department of Agriculture SWAT code has been integrated with MODFLOW (Sophocleous et al. 1999; Sophocleous and Perkins 2000; Kim et al. 2008; Chung et al. 2010). Conceptually, recharge rate calculated using SWAT is input into MODFLOW and discharge from MODFLOW is input into SWAT. Recharge elements input from SWAT to MODFLOW are stream transmission losses, percolation beyond the root zone, and pond seepage. Subroutines were developed to represent spatial heterogeneity within subbasins in SWAT and to associate the geographic domain in SWAT with the MODFLOW finite-difference grid.

The GSFLOW code developed by the U.S. Geological Survey (Markstrom et al. 2008) is a good general example of a coupled regions integrated surface-water and groundwater model. GSFLOW combines the USGS Precipitation-Runoff Modeling System (PRMS) surface water model (Leavesley et al. 1983, 2005) with the 2005 version of the USGS three-dimensional modular groundwater flow model MODFLOW (Harbaugh 2005). GSFLOW simulates flow within and between three regions; the top of the canopy to the base of the soil zone (PRMS), streams and lakes (PRMS), and subsurface flow below the soil zone (MODFLOW).

The MIKE SHE/MIKE 11 codes, developed by DHI, is an example of a proprietary fully-integrated surface-water and groundwater modeling system. It includes a full suite of pre- and post-processing tools, plus a flexible mix of advanced and simple solution techniques for each of the hydrologic processes (DHI Software 2007). MIKE SHE/MIKE 11 allows for the simulation of the major processes in the hydrologic cycle and includes process models for evapotranspiration, overland flow, unsaturated flow, groundwater flow, and channel flow and their interactions. Each of these processes can be represented at different levels of spatial distribution and complexity.

Although modeling systems are now available to simulate integrated surface-water and groundwater systems, a pertinent question is the value of such modeling for actual water resources management in arid and semiarid lands, and in other areas. There are an increasing number of ever more sophisticated tools for evaluating water resources. The key issue is determining whether the benefits they provide in terms of improved water management is commensurate with the cost of their deployment.

The fundamental challenge and limitation associated with integrated surface-water and groundwater modeling is the huge amount of data that are required, which is typically greatly beyond that actually available with any certainty for study areas. For example, the large uncertainty inherent in parameters such as actual evapotranspiration rates, stream bed conductance, and unsaturated soil hydraulic conductivities in general, are incorporated into models. Soil zones often have a high degree of heterogeneity with respect to their properties, which is a particular difficult problem for the simulation of complex areas, such a mountain front systems. Climatological data may also be sparse. The large numbers of parameters (i.e., degrees of freedom) compounds the non-uniqueness associated with model simulations in general. That an integrated surface-water and groundwater model is well calibrated to limited observation data (e.g., a downstream gauge) does not imply that it is an accurate tool for predicting future runoff and aquifer recharge. The issue of convergence in an integrated surface-water/groundwater model is critical because of the time difference between surface-water flows and corresponding groundwater flow rates. If the model fails to converge, the predictions can be very inaccurate, particularly for surface-water discharge rates and stages.

Distributed parameter models, in which values have to be assigned for numerous variables contributing to the surface water budget, are particularly problematic because of the invariably huge data deficiency in project areas relative to the specific data requirements. The common practice of using general or regional default values for parameters may introduce large potential errors if these values are not in fact locally appropriate. This is a very important issue in that there is little advantage in using a distributed parameter model with great data requirements if most of the values used are, at best, educated guesses. Use of lumped-parameter models is preferred in areas with large data deficiencies. For example, it may be more accurate to simulate wadi recharge in a groundwater model using an empirical rainfall-runoff-recharge relationship function, than by creating a detailed integrated surface-water/groundwater model for which the distribution of the values of most of the parameters are very poorly constrained.

Integrated surface-water/groundwater models are therefore most appropriately used in studies of semi-arid or arid lands for which an exceptional quantity of hydrologic data are available. It is more appropriate to use a less complex model when first evaluating a given area that has a poor data base. This initial model will allow the identification of data deficiencies and could provide some initial predictions based on the error ranges caused by the paucity of data. As additional data are obtained, the modeling should be upgraded to an integrated model that may then provide a more comprehensive and accurate set of simulations. A key conclusion is that the sophistication of the model code used for water resources management in a given area should evolve with the ability to input the appropriate quantity and quality of hydrologic data.

References

Anderson, M. P., & Woessner, M. W. (1992). *Applied groundwater modeling, simulation of flow and advective transport*. San Diego: Academic Press.

Bredehoeft, J. (2005). The conceptualization model problem—surprise? *Hydrogeology Journal, 13*, 37–46.

Chung, H.-M., Kim, N.-W., Lee, J., & Sophocleous, M. (2010). Assessing distributed groundwater recharge using integrated surface water-groundwater modeling. *Hydrogeology Journal, 18*, 1253–1264.

DHI Software (2007). *MIKE SHE uses manual* (Vol. 1). User Guide.

Diersch H. J. (1998). FEFLOW: Interactive, graphics-based finite-element simulation system for modeling groundwater flow, contaminant mass and heat transport processes. In *Getting started; user's manual; reference manual, version 4.6*. Berlin, Germany; WASY, Institute for Water Resources Planning and System Research Ltd.

Doherty, J. (2005). *PEST: model-independent parameter estimation* (5th ed.). Brisbane, Australia: Watermark Numerical Computing.

Domenico, P. A., & Schwartz, F. W. (1998). *Physical and chemical hydrogeology* (2nd ed.). New York: Wiley.

Fetter, C. W. (2001). *Applied hydrogeology* (4th ed.). New Jersey: Englewood Cliffs Prentice Hall.

Guo, W., & Langevin, C. D. (2002). *User's guide to SEAWAT: A computer program for simulation of three-dimensional variable-density ground-water flow*. U.S. Geological Survey Open-File Report, 01-434.

Harbaugh, A. W. (2005). *MODFLOW-2005, the U.S. Geological survey modular ground-water model–the ground-water flow process*. U.S. Geological Survey Techniques and Methods, 6-A16

Hemker, C. J., & de Boer, R. G. (2009). *MicroFem* (Version 4.10.02).

Herczeg, A. L., & Leaney, F. W. (2011). Review: Environmental tracers in arid-zone hydrology. *Hydrogeology Journal, 19*(1), 17–30.

Herrmann, R. (2006). ASR well field optimization un unconfined aquifers in the Middle East. In *Recharge systems for protecting and enhancing groundwater resources, proceedings of the 5th international symposium on management of aquifer recharge* (pp. 109–114). Berlin, Germany, Paris: UNESCO. 11–16 June 2005.

Hill, M. C., & Tiedeman, C. R. (2007). *Effective groundwater model calibration: With analysis of data, sensitivities, predictions, and uncertainty*. Hoboken, NJ: Wiley-Interscience.

Hoffmann, J., Galloway, D. L., & Zebker, H. A. (2003a). Inverse modeling of interbed storage parameters using land subsidence observations. *Antelope Valley, California: Water Resources Research, 39*(2), 1031. doi:10.1029/2001WR001252,2003.

Hoffmann, J., Leake, S. A., Galloway, D. L., & Wilson, A. M. (2003b). *MODFLOW-2000 ground-water model–user guide to the subsidence and aquifer-system compaction (SUB) package*. U.S. Geological Survey Open-File Report 03-233.

Hunt, R. J., Doherty, J., & Tonkin, M. J. (2007). Are models too simple? *Arguments for Increased Parameterization: Ground Water, 45*, 254–262.

Kim, N. W., Chung, I. M., Won, Y. S., & Arnold, J. G. (2008). Development and application of the integrated SWAT-MODFLOW model. *Journal of Hydrology, 356*, 1–16.

Kipp, K. L., Jr. (1987). HST: *A computer code for simulation of heat and solute transport in three-dimensional ground-water flow systems*. U.S. Geological Survey Water-Resources Investigations Report 86-4095.

Kipp, K. L., Jr. (1997). *Guide to the revised heat and solute transport simulator; HST3D* (Version 2). U.S. Geological Survey Water-Resources Investigations Report 97-4157.

Langevin, C. D., Thorne, D. T., Jr., Dausman, A. M., Sukop, M. C., & Weixing, G. (2008). *SEAWAT version 4: a computer program for simulation of multi-species solute and heat transport*. U.S. Geological Survey Techniques and Methods Book 6, Chapter A22.

Leake, S. A., & Galloway, D. L. (2007). *MODFLOW ground-water model—user guide to the subsidence and aquifer-system compaction package (sub-wt) for water-table aquifers*. U.S. Geological Survey, Techniques and Methods 6-A23.

Leake, S. A., & Prudic, D. E. (1991). *Documentation of a computer program to simulate aquifer-system compaction using the modular finite-difference ground-water flow model*. U.S. Geological Survey Techniques of Water—Resources Investigations, book 6, chapter. A2.

Leavesley, G. H., Lichty, R. W., Troutman, B. M., & Saindon, L. G. (1983). *Precipitation-runoff modeling system: user's manual*. U.S. Geological Survey Water-Resources Investigations Report 83-4238.

Leavesley, G. H., Markstrom, S. L., Viger, R. J., & Hay, L. E. (2005). USGS modular modeling system (MMS)—precipitation-runoff modeling system (PRMS) NMS-PRMS. In V. Singh & D. Frevert (Eds.), *Watershed models* (pp. 159–177). Boca Raton, FL: CRC Press.

Lloyd, J. W. (2004). Do we adequately understand the potential value of our regional aquifers? *Proceedings, International Conference on Water Resources in Arid Environments*.

Lloyd, J. W. (2007). The difficulties of regional groundwater resources assessments in arid regions. *Arabian Journal for Science and Engineering, 32*(1C), 35–47.

Maliva, R. G., & Missimer, T. M. (2010). Aquifer storage and recovery and managed aquifer recharge using wells: Planning, hydrogeology, design, and operation. *Houston, Schlumberger Water Services, Methods in Water Resources Evaluation Series*.

Markstrom, S. L., Niswonger, R. G., Regan, R. S., Prudic, D. E., & Barlow, P. M. (2008). *GSFLOW—couple ground-water and surface-water flow model based in the integration of the Precipitation-Runoff Modeling System (PRMS) and the Modular Ground-Water Flow Model (MODFLOW-2005)*. U.S. Geological Survey Techniques and Methods 6-D1.

McDonald, M. G., & Harbaugh, A. W. (1988). *A modular three-dimensional finite-difference ground water flow model*. U.S. Geological Survey Techniques of Water-Resources Investigation Report 06-A1.

Missimer, T. M., Maliva, R. G., & Guo, W. (2010). Sustainability and the management of water resources in Florida. In F. Bloetcher (Ed.), *Sustainable water resources: A compendium of issues and trends* (pp. 153–161). Denver: American Water Works Association.

National Research Council (2008). *Prospects for managed underground storage of recoverable water*. Washington, DC: National Academy Press.

Oreskes, N., Shrader-Frechette, K., & Belitz, M. (1994). Verification, validation, and confirmation of numerical models in the Earth Sciences. *Science, 263*, 641–646.

Parkhurst, D. L., & Appelo, C. A. J. (1999). *PHREEQC (Version 2)—A computer program for speciation, batch reaction, one-dimensional transport, and inverse geochemical calculations*. U.S. Geological Survey, Water-Resources Investigations Report 99-42549.

Poeter, E. (2007). "All models are wrong, how do we know which are useful?"—looking back at the Darcy lecture tour. *Ground Water, 45*(4), 390–391.

Poeter, E., & Hill, M. C. (1997). Inverse models: A necessary next step in ground-water modeling. *Ground Water, 35*(2), 250–260.

Pollock, D. W. (1994). *User's guide for MODPATH/MODPATH-PLOT Version 2: A particle tracking post-processing package for MODFLOW, the U.S. Geological Survey finite-difference ground-water flow model*. U.S. Geological Survey Open-File Report 94-464.

Prommer, H., Barry, D. A., & Zheng, C. (2003). MODFLOW/MT3DMS based reactive multi-component transport modeling. *Ground Water, 41*, 347–257.

Radcliff, D., & Šimůnek, J. (2010). *Soil physics with HYDRUS: Modeling and applications*. Boca Raton, FL: CRC Press.

Sanford, W. (2011). Calibration of models using groundwater age. *Hydrogeology Journal, 19*, 13–16.

Sophocleous, M. A., Koelliker, J. K., Govindaraju, R. S., Birdie, T., Ramireddygari, S. R., & Perkins, S. P. (1999). Integrated numerical modeling for basin-wide water management: The case of the Rattlesnake Creek basin in south-central Kansas. *Journal of Hydrology, 214*, 179–196.

Sophocleous, M. A., & Perkins, S. P. (2000). Methodology and application of combined watershed and ground-water models in Kansas. *Journal of Hydrology, 236*, 185–201.

Voss, C. I., & Provost, A. M. (2002). *SUTRA, A model for saturated-unsaturated variable-density ground-water flow with solute or energy transport*. U.S. Geological Survey Water-Resources Investigations Report 02-4231.

Zheng, C., & Bennett, G. D. (2002). *Applied contaminant transport modeling* (2nd ed.). New York: Wiley.

Zheng, C., & Wang, P. P. (1999). *MT3DMS: A modular three-dimensional multi-species model for simulation of advection, dispersion and chemical reactions of contaminants in ground water systems: Documentation and user's guide*. U.S. Army Engineer Research and Development Center, Vicksburg, Mississippi, report SERDP-99-1.

Zhu, C., & Murphy, W. M. (2000). On radioactive dating of ground water. *Ground Water, 38*, 802–804.

Part V
Water Management Techniques

Chapter 21
Ancient Water Management

21.1 Introduction

Lessons learned about water supply in ancient civilizations that have survived and flourished in arid and semiarid lands, can place our current water challenges within an historical framework and provide starting points for new (or rediscovery of old) solutions today. The lessons from ancient civilizations that ultimately collapsed can also provide insights into strategies that are ultimately unsustainable. It has been noted that some ancient water management systems were unsustainable and ultimately contributed to, or were the primary cause of, the collapse of the civilizations (e.g., Diamond 2005). However, the concept of sustainability needs to have some fixed time frame in order to be meaningful (Chap. 9). Nothing lasts forever. To tie the concept of sustainability to "forever," causes it to lose its significance in that human civilizations are inherently dynamic. If the water management solutions adopted by an ancient society or culture allowed it to survive or even thrive in a water scarce environment for many centuries, it should not be viewed as a failure because the culture ultimately collapsed. For example, the water management solutions adopted by the Anasazi people in the southwestern United States allowed them to survive for over 700 years in a water scarce environment, which is no mean feat (Sect. 21.6.1).

The water management solutions of ancient civilizations had to meet the needs of much smaller populations compared to present. It is important, therefore, to avoid the romanticism that we can go back to the water management strategies of an older and simpler time to solve current water management problems. Solutions that might work for a population in the thousands would not be expected to be the primary solution to meeting the water demands of rapidly growing urban areas with populations in the millions.

Rainwater harvesting was fundamental to ancient water management in communities living in arid and semiarid lands that did not have access to a perennial river, spring, or other water source. Modern rainwater harvesting techniques are

R. Maliva and T. Missimer, *Arid Lands Water Evaluation and Management*, Environmental Science and Engineering, DOI: 10.1007/978-3-642-29104-3_21,

discussed in Chap. 22. Rainwater harvesting is not a substitute for modern water treatment and supply systems. However, people living in water scarce lands need to consider how to combine modern supply systems with old harvesting methods (AbelKhaleq and Ahmed 2007). In water scare regions “every drop counts”. Indeed, rainwater and stormwater harvesting are increasingly being recognized as practical and cost-effective elements of both water supply and stormwater management in arid and semiarid lands.

There are a limited number of practical water supply options in arid regions, and ancient civilizations had independently discovered and adopted some common solutions. Basic water supply and management elements include techniques to divert, retain, and store ephemeral flows such as local rain and flood waters (Heathcote 1983; Agnew and Anderson 1992). For example, cisterns, which are essentially waterproof, typically underground, storage tanks, were a common water supply tool in many areas of the world, because they are a very effective and practical water storage solution. Ancient cisterns ranged in scale and sophistication from natural or excavated depressions in solid rock, to lined structures sized to serve a single household, to ultimately cathedral-sized structures such as the Basilica Cistern (Yerebatan Sarayı) in Istanbul, Turkey, which has a capacity of almost 85,000 m^3 (2.24 million gal) (Lendering 2008).

A series of examples of the strategies that some ancient civilizations used to secure water supplies in arid and semiarid lands are described. This discussion is by no means comprehensive and the focus of this chapter is on providing examples of the diversity of water supply options employed in areas that lack significant perennial water supplies. The civilizations that developed in relatively water-rich areas such as the flood plains of the Tigris, Euphrates, and Nile Rivers faced different water challenges and their lessons are less widely applicable.

21.2 Ancient Jordan

Jordan is currently one of ten most water poor countries in the world with 90% of country receiving less than 200 mm (7.8 in) of rainfall per year. Despite its arid and semiarid climate, rainwater harvesting allowed several ancient civilizations to flourish in Jordan (AbelKhaleq and Ahmed 2007). Techniques such as deflection dams, cisterns, and reservoirs were planned and constructed in cities by 3000 BC in order to capture water available during the few intense winter rainfall events for later use during the long dry periods between significant rains (Helms 1981; Wåhlin 1997; AbelKhaleq and Ahmed 2007). Cisterns were historically used in Jordan, and throughout the Levant, for collection of water for household use and the watering flocks, but not for the growing of crops (Wåhlin 1997). Household cisterns had small individual capacities, but the combined capacity was enormous and they had the great advantage of be located precisely where people lived (Wåhlin 1997).

21.2.1 Jawa

The lost city of Jawa in the Black Desert of Jordan contains remarkably well preserved ruins of Late Chalcolithic (pre Early Bronze) age. Jawa is especially noteworthy from a water management perspective in that it has the most comprehensive urban water system yet discovered anywhere at such as early date (Helms 1981). The only water that was available in Jawa was that captured from brief winter rains. No readily available groundwater is present in the area and certainly none that would have been accessible during the occupation of Jawa. Jawa thus provides an excellent example of early rainwater harvesting techniques, which were sufficient to allow the city of survive for a while in what is a very hostile environment.

The Jawa water management system was documented in detail by Helms (1981). The key element of the Jawa water system was episodic flows from Wadi Rajil. The explosive kinetic energy of flood flows within Wadi Rajil would disintegrate almost any man-made structure. A permanent total dam thus could not survive. Water was obtained instead by diverting a small fraction of the total flow during storms. Deflection dams were used, which are partial obstructions that allow nearly all the water to pass, while deflecting a small percentage (estimated 3% at Jawa; Helms 1981) into adjoining canals. Storm water that entered the canals at Jawa flowed under gravity to storage areas in which it had to last through the long dry season. Sluice gates were also used to divert water from the gravity channels into some farm fields and individual storage areas. Microcatchments were also utilized to capture local rainfall. Deflection walls were constructed along nearby hillsides to lead water into storage pools or reservoirs.

Dams were used at Jawa to create reservoirs for the storage of captured rainwater. Helms (1981) documented that the reservoir dams at Jawa had remarkably modern design elements. For example, the cores of the dams contained clay and ash for waterproofing and to minimize seepage through the dam. Downstream revetments were also used to allow for the safe drainage of the small amount of seepage, which would otherwise waterlog the base of the dams and allow erosion to occur. A natural lava flow cave was also used as a cistern, which had the benefit of reduced evaporation.

21.2.2 Petra

The city of Petra in southwestern Jordan is one of the archeological wonders of the world. The magnificent buildings carved into the red sandstone cliffs (Fig. 21.1) have become a major global tourist attraction and Petra has been designated both a World Heritage Site and one of the "New Seven Wonders of the World". The city of Petra was the administrative, cultural, and religious center of the Nabataean civilization (Taylor 2002). Nabataeans apparently originated from a tribe from

Fig. 21.1 The Treasury at the ancient Nabataean city of Petra, Jordan. Buildings were carved into the sandstone cliffs. *Source* Weixing Guo

southern Arabia and became an important trading empire because they were located at the intersection of important caravan routes. The Petra site was occupied since prehistoric times, but the earliest structures date from about 300 BC (Guzzo and Schneider 2002). The city of Petra was unknown to the western world until the "lost city" was rediscovered by the Swiss explorer Johann Ludwig Burckhardt in 1812.

Beyond the beauty and historical interest of its buildings and the romantic story of its discovery, Petra is of great interest from a water resources management perspective. The water supply and distribution system of Petra is described in an excellent paper by Ortloff (2005). A critical challenge for the Nabataean people was the construction of a water supply system that provided a constant, year-round supply for water for its growing urban population given its arid climate and variation in seasonal rainfall and spring flow. Petra has been estimated to have had a peak population of 30,000 (Guzzo and Schneider 2002). As discussed by Ortloff (2005), the Nabataeans borrowed and improved upon pre-existing technologies from surrounding societies. The Nabataea water supply solutions were also sustainable as witnessed by the survival of Petra as an urban center for centuries.

The key element to the success of the Petra water supply system is that it took advantage of all potential water resources. The core of Petra lies in a valley surrounded by high mountains into which many canyon streambeds (wadis) drain. The wadis were an important source of water, but can also be the loci for destructive floods. Dams served a dual water supply and flood control function (Fig. 21.2). The Nabataea's simultaneously utilized all possible above-ground and below-ground water supply and storage options, of which storage of rainfall runoff and springs both inside and outside of the city were the main components (Ortloff 2005). The ntegrated water system of Petra included such elements as (Ortloff 2005)

- surface systems to capture run-off from the wadis,
- deep underground cisterns,
- multiple pipeline and channel systems,
- floodwater control through diversion dams and tunnels,
- system redundancy from multiple springs and reservoir sources, and
- pipeline sand particle filtration and removal basins.

Major springs were used for the continuous supply of water to multiple city locations, whereas reservoirs and cisterns provided supplemental water. Covered cisterns minimized evaporative water losses.

A remarkable feature of the Petra water system is that the hydraulic design of its pipeline system is in accordance with modern engineering best practices. The Petra pipeline system designs incorporated features that minimized leakage, maximized the flow rate, minimized clogging, eliminated transient flow instabilities, and included purification by means of settling basins (Ortloff 2005). The water supply and distribution system also required a high degree of bureaucratic oversight regarding the release and storage of water both on a daily and seasonal basis.

Although the technologies used at Petra were not new, the Nabataea system was unique in the much larger scale of water conservation employed and the degree of redundancy in the system (Ortloff 2005). The Nabataeans clearly understood the importance of efficiently using all water resources available in an arid climate with a highly variable water supply. No water could be wasted. The most important lesson from Petra is the importance of water storage in regions that experience intermittent rainfall. Cisterns were previously employed for households. An innovation of Petra was that cisterns were used on a large scale as a major component of the city-wide water supply (Ortloff 2005).

Fig. 21.2 An example of a recent dam in Petra, Jordan, which replaced an earlier Nabataean structure. Dams across ephemeral channels both store freshwater and provide flood control. *Source* Weixing Guo

21.3 Negev Desert

Ancient water management in the Negev Desert of Israel was discussed in detail by Evanari et al. (1971). The most flourishing period for the Negev desert civilization was the Nabataea-Roman-Byzantine era starting at the end of the 3rd century (BC) and lasting 800–900 years. The Nabataea's engaged in runoff farming and learned to control flash floods created by runoff from hill sides. A key issue for run-off agriculture is assuring that runoff is not concentrated in gullies and is instead spread across and absorbed by terraced fields (Evanari et al. 1971). Three basic systems were used for ancient runoff agriculture (Evanari et al. 1971)

- Individual terraced wadis, which is the simplest and most primitive method. A series of low stone walls were constructed across wadis which served an erosion and flood control function. Stone walls also captured soil. Water infiltrates directly into the soil and is also ponded for later infiltration.
- Terraced fields with farmsteads, which consisted of a catchment area on hillside and conduits that collected the runoff and conveyed it to terraced farm fields. The conduits were rock and soil embankments and in some instances also excavated channels. The average ratio of catchment to cultivated area was approximately 20:1. The principle was to concentrate sparse rainfall to farm fields.
- Diversion systems, which were associated with main wadis. Construction of a dam and diversion channel raised the water level above the wadi and conveyed it to terraced farm fields.

Preparation of catchment areas can increase water yields. A smooth, stone-free soil surface was observed to have increased runoff compared to stone-covered surfaces. This was apparently understood by ancient farmers in the Negev and appears to explain mounds of stone found in some catchments.

Drinking water for domestic supply for family and flocks was obtained by taking advantage of surface water stored in natural rock basins eroded in rock exposed at wadi bottoms and cisterns. Cisterns were filled from runoff-collecting conduits after first passing through a small settling basin. Three main types of cisterns were used including (Evanari et al. 1971):

- hillside, which were filled from hillside catchment areas,
- public town, which stored runoff from streets, roofs, and catchment areas outside of town, and
- private, which were built inside of a house and were commonly bottle-shaped with a narrow (1 m; 3 ft) entrance.

Evanari et al. (1971, p. 171) observed that *if one wanders through the ruins of Shiva or Avdat, one gets the impression that the whole planning of the towns was based on water collection and storage*. Both roofs and pavement areas served as catchments for rainwater. Experimental farms in the Negev Desert documented by Evanari et al. (1971) notably demonstrated the feasibility and viability of runoff agriculture even under drought conditions. The remarkable aspect of Nabataean agriculture was how well tuned it was to the realities of the water supply and environment. Evanari et al. (1971, p. 325) future observed that

> The ancient farmer fitted his artificially created agricultural ecosystem into nature and used landscape and topography to his best advantage without damaging his environment. He neither caused erosion nor brought about salination of his agricultural soils.

It was also recognized that ancient farmers also made mistakes and had to pay for them. Arid regions can be very unforgiving.

21.4 Greek and Roman Civilizations

The remarkable sophistication and scale of ancient Greek and Roman water supply and distribution systems can more appropriately be described as modern rather than ancient despite their antiquity. Ancient Greek and Roman urban water supply has received much academic interest and was reviewed by Hansen (1983, 2006); Couch (1990, 1993), Mays (2007a, 2010), Mays et al. (2007), Buxó (2008), and Koutsoyiannis et al. (2008). The Minoan Civilization on Crete, by the 2nd millennium BC, had developed aqueducts, cisterns, wells, fountains, bathrooms, and other sanitary facilities, which suggest lifestyle standards close to those of present (Angelakis and Koutsoyiannis 2003; Mays 2010).

For reasons not well understood, ancient Greeks avoided rivers and lakes in the choice of the locations of major cities and instead chose dry areas, which necessitated the development of advanced hydraulic systems for the collection and transport of water (Koutsoyiannis et al. 2008). The Romans built upon Greek technology with construction of mega-water supply systems, which provided citizens with a more reliable water supply than currently occurs in some modern cities. Ancient Greek and Roman water supplies consist of three main elements:

- groundwater from dug wells,
- importation of water from springs and streams via aqueducts (Fig. 21.3), and
- rainwater harvesting and storage using cisterns.

The scale of Roman aqueduct system and the technical prowess involved in its design and construction is extraordinary especially considering it was built over 1700 years ago. The eleven aqueducts that served the city of Rome had a combined length of 503 km (313 miles; Aicher 1995; Hodge 2002). The estimated capacity of the aqueduct system was 1,127,220 m^3/d (297,780,010 U.S. gallons/d; Hodge 2002).

The Greek tradition emphasized a multiplicity of supply sources, each as a safeguard against failure of the others (Couch 1990). Both ancient Greeks and Romans made extensive use of cisterns throughout the Mediterranean to store rainwater. The systems range from household systems to large-capacity systems that served the community. Cisterns in ancient Greek cities appear to have been

Fig. 21.3 Roman Aqueduct at Segovia, Spain, 1977 (photograph by T. Missimer)

used to store water mainly for non-potable uses, but could be used for drinking water during water shortages (Crouch 1993; Mays 2007a). Both the Romans and Greeks preferred running water for drinking and distrusted stagnant water, but accepted the use of water from cisterns for human consumption (Buxó 2008).

The contribution of the three water supply components in a given city varied over time. For example, the construction of an aqueduct to Athens during the reign of the Roman emperor Hadrian (117–138 A.D.) provided the Athenians with a reliable large supply of freshwater, which resulted in the abandonment of private installations such as wells and cisterns (Koutsoyiannis et al. 2008). Water security was an important consideration in ancient Greece. Aristotle wrote of rainwater fed cisterns as being a secure supply of water in the case of war (Mays 2007a; Koutsoyiannis et al. 2008). Aqueducts were also constructed underground to increase the security of water supply.

21.5 Ancient China

Rainwater harvesting has been practiced in China for over 4,500 years. The earliest techniques involved flood diversion and various strategies (terraces fields and mini-dams) to retain runoff and prevent erosion (Li et al. 2002). Underground clay-lined earthen water-storage cisterns have been used in the Gansu Province since the Ming Dynasty over 600 years ago (Li. et al. 2002).

21.6 Southwestern United States

Three major cultures were present in the southwestern United States during pre-European contact period; the Anasazi, Hohokam, and Mogollon. The fundamental problem in the American Southwest is determining how to obtain enough water to grow crops in an environment in which most of the rainfall is so low and unpredictable that little or no farming is practiced today (Diamond 2005). A series of good years with sufficient rainfall or with a sufficiently shallow water table may result in population growth, resulting in the society becoming increasingly complex and interdependent, and no longer locally self-sufficient. Such a society may not be able to cope with and rebuild itself after a series of bad years (Diamond 2005).

The ancient cultures of the American Southwest thus provide two sets of lessons. The first relates to how these cultures managed to survive and thrive in a water scarce environment. The second lesson relates to why these cultures eventually collapsed, which is still a subject of considerable debate.

Ancient water harvesting traditions in the American Southwest were summarized by Glanzburg (1994). Peoples of the southwest used a variety of water harvesting techniques and conservation methods to grow their food. Check dams and terraces or linear borders were used to catch soil and water. Waffle and grid gardens, in which farm beds were either sunken or surrounded by berms or walls, were used to capture rainwater and hold soil in place. Gravel (lithic) mulch conserved water by reducing evaporation and also acted to reduce wind and water erosion (Sect. 21.9). Floodwater (akchin) farming was practiced in alluvial fans below arroyos (wadis) and small canyons.

21.6.1 Anasazi

The Anasazi are an Ancient Pueblo people who lived in the Four Corners region of the United State (area in the southwestern United States where the states of Arizona, Colorado, New Mexico and Utah meet). The name "Anasazi" has come to mean "ancient people," although the word itself is Navajo, meaning "enemy ancestors" (Grahame and Sisk 2002). The golden age of the Anasazi occurred between 900 and 1130 AD. Chaco Canyon located in northwestern New Mexico was a major Anasazi cultural, administrative, and economic center. The latter part of the Anasazi culture is marked by the construction of multistorey great houses, which is interpreted to indicate that the culture evolved into a relatively sedentary life style (Benson et al. 2007a). The Chaco Canyon ruins, such as Pueblo Bonita (Fig. 21.4), are some of the most spectacular and important archaeological sites in the United States. Chaco Canyon is of great interest from a water management perspective, because of the skills required by the Anasazi to maintain a sizeable population in a harsh environment.

Fig. 21.4 Pueblo Bonita, Chaco Culture National Historic Park, New Mexico, USA. *Source* U.S. National Park Service

Agricultural options in the southwestern United States, where perennial water sources are sparse include (Diamond 2005)

- Dryland agriculture, which involves growing crops in upland areas, when there is enough rain to promote growth in the fields on which the rain fell.
- Farming in areas with a shallow water table into which plant roots can grow. These include canyon bottoms with intermittent and permanent streams.
- Collection of stormwater runoff using ditches and canals to irrigate fields.

Lister and Lister (1981) prepared a classic reference on the history of Chaco Canyon. A variety of rainwater harvesting techniques was employed to take advantage of episodically available freshwater. Chaco Canyon irrigation engineering included such elements as:

- small stone check dams across minor drainages (arroyos),
- large earthen dams,
- water diversion devices that channeled ephemeral surface water flow to agricultural fields and terraces and small reservoirs,
- storage tanks, and
- agricultural terraces and garden plots.

Fig. 21.5 Multiple-story Hohokam abode building. Casa Grande Ruins National Monument, Coolidge, Arizona. *Source* U.S. National Park Service

21.6.2 Hohokam

The Hohokam culture was centered on the Salt River and Gila River valleys in what is now the greater Phoenix, Arizona (USA) metropolitan area. Hohokam is the name given by the present-day Pima Indians and means "the people who vanished." The Hohokam culture or tradition existed from about 300 BC–1450 AD, and was dependent upon local river systems. Within the Salt River Valley they constructed over 450 km (300 miles) of major canals and over 1100 km (700 miles) of distribution canals (Mays 2007b). Ruins of Hohokam adobe buildings (Fig. 21.5) and their canals (Fig. 21.6) are still locally preserved.

The construction and operation of the Hohokam canal system was discussed by Masse (1991), Howard and Huckleberry (1991), Howard (1992), Andrews and Bostwick (2000), and Mays and Gorokhovich (2010). In order to direct water from the river into the canals, archaeologists assume that the Hohokam constructed weirs of log and brush. The canals were engineered in such a way that they were wide at the headwaters and gradually narrowed. In this manner, water levels were kept constant and there was an even flow of water from the beginning to the end. Canal gradient or slope was crucial to the proper engineering functioning of the canals. If the gradient was too steep, then the flow of water would be too fast and erosion of the canal banks would occur. If the grade was too gentle, then water would run slowly, dropping its sediment load, and causing the canal to silt up. Hohokam engineers were thus keenly aware of local topography. Canals narrow from river to their terminus. By reducing the width of channels, the velocity of the water remained relatively constant and between the two critical thresholds for the

Fig. 21.6 Remains of a Hohokam irrigation canal. Park of the Canals, Mesa, Arizona

minimization of erosion and the prevention of silting, both of which would require additional maintenance if they occurred.

The canal system was constructed over several centuries and not all Hohokam canals were in use at any one time. The canal system and all its elaborate parts indicate that the Hohokam had a complex sociopolitical structure. Construction, maintenance, and operation of the canal systems would have required a substantial and well organized effort. Individuals would likely have been required or expected to contribute to system construction and maintenance.

The Hohokam lesson is that long-term survival in arid Sonoran Desert of the American Southwest was possible through the mastery of the limited available surface water resources. The eventual failure of the canal systems appears to have contributed to the destruction of their civilization. A combination of factors may have been involved including flood damage to the canal system, siltation, and soil salinization.

21.7 Qanats

Qanats are a sustainable water supply technology that has been used in arid and semiarid lands for millennia. The technology was developed in ancient times and continues to be used today. A qanat is a near horizontal channel or gallery excavated into an alluvial fan aquifer located at the base of a mountain or foothill.

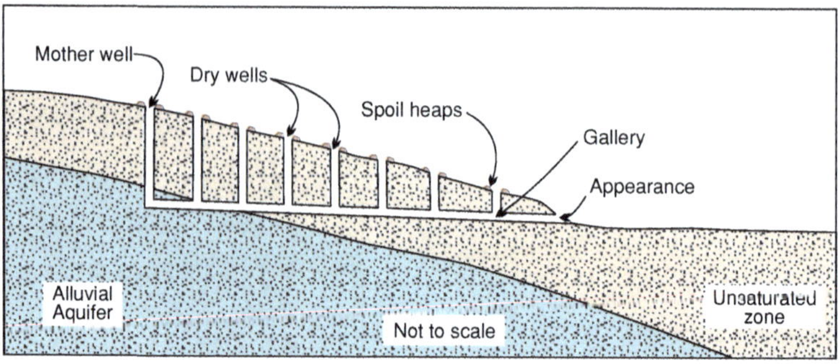

Fig. 21.7 Conceptual diagram of a qanat

Qanats produce water by gravity flow. The distal end of the qanat is located below the water table and acts as a drain. Water infiltrates into the qanat and is conveyed to a nearby village. A diagram of qanat is provided in Fig. 21.7. The main components of a qanat are (Cressey 1958; Wulff 1968; Bybordi 1974; Boustani 2008):

- appearance—place where water comes into view at the surface,
- gallery—gently sloping conveyance canal,
- wet zone—infiltrating canal inside the gallery,
- dry zone—portion of the canal between the wet zone and appearance,
- dry wells—vertical shafts that facilitate soil extraction and provide ventilation, and
- mother well—shaft dug down into the water table that is the furthest water infiltrating shaft(s).

Galleries are usually 50–100 cm (1.6–3.3 ft) wide and 90–150 cm (2.9–4.6 ft) high. The gallery has to be constructed with a slope that is sufficient for easy flow of water and to prevent stagnation, but cannot be too steep as erosion would occur. The longest qanats have a length of about 70 km (34 miles). However, over 80% of qanats in Iran are less than 5 km (3 miles) long (Beaumont 1971). Three different types of qanats have been identified in the United Arab Emirates, which differ in whether or not they flow seasonally or year round (Al Daudi type), and whether their source is local groundwater storage (Al Gheli types) or deep circulation through fractures and fissures (Al Hadouri type) (Alsharahan et al. 2001; Rizk and Alsharhan 2003). Donut-shaped spoil heaps around the tops of the vertical shafts are a characteristic feature in qanat-watered landscapes and mark the routes of Qanats (English 1968, 1998). Water from the qanats is distributed to users through a series of channels with gates (Fig. 21.8).

Qanats are believed to have been first constructed in Persia in about 800 BC (Boustani 2008). There are estimated to be over 30,000 qanats in Iran (Salih 2006: Hussain et al. 2008), which has the greatest density. Qanats have been used in at least 35 additional countries. Most qanats were dug more than 3,000 years ago

Fig. 21.8 Falaj (qanat) distribution system. Al Ain oasis, Abu Dhabi, United Arab Emirates

during the heydays of the Persian Empire. Wherever the Persians went they took the secrets of the qanats, which are called "aflaj", "falaj", or "ain" in Arabia, "fogarra" in North Africa, "karez" in Pakistan and Afghanistan, "galleria" in Spain, and "surangams "in India (Basak et al. 1997; Pearce 2006; Hussain et al. 2008). Qanats have also been constructed in North America (Mexico). The spread of qanats during the pre-modern period can be explained by them being an efficient mix of available capital and technology to supply critically need water (English 1998). The number of operational qanats has decreased dramatically in recent decades. The total discharge from qanats in Iran was estimated to have decreased from 598 billion m^3 [1.6×10^{14} gal] (18,960 m^3/s; 5 million g/s) in the 1960s to 9.9 billion m^3 [2.6×10^{12} gal] (313 m^3/s; 83,000 gal/s) in 1998 (Pearce 2006). Boustani (2008) reported a greater total production of usable water of 750–1,000 m^3/s (198,000–264,000 gal/s). Haeri (2006) attributed approximately 8.6 billion m^3 [2.3×10^{12}] (272 m^3/s; 71,800 gal/s) of water supply to qanats in Iran in 2000. Only about 50 of 150 aflajes in the UAE are still active, mainly due to aquifer drawdowns from large groundwater extractions and low recharge (Murad et al. 2007).

Qanats were traditionally constructed by a group of skilled laborers and need to be maintained on a regular basis. A tremendous effort is involved in the construction of qanat. Beaumont (1971) reported that a 3 km (1.8 miles) qanat near Kerman, Iran, took seventeen years of daily labor to construct. Water yields are reduced when the seepage walls become coated with mineral deposits (Lightfoot 1996). Qanats have been abandoned because they filled with silt or, much more importantly, because pumping of vertical boreholes lowered the water table and dried them out (Beaumont 1971; Lightfoot 1996; Wessels and Hoogeveen 2002; Peace 2006: Remini et al. 2010). Qanat systems are also susceptible to damage from earthquakes. The 2003 earthquake in the Kerman region of Iran was reported to have caused extensive damage to parts of the qanat systems (Salih 2006). Vibrations from vehicles and other activities in towns may also cause collapses of qanats (Remini et al. 2010).

An attraction of qanats is that they are a sustainable source of water. Qanats are self-regulating in that they can produce water only to the limit of natural replenishment. All the water must be replaced by recharge otherwise a qanat will dry up. The flow of water is controlled by the altitude of the water table with respect to the gallery. Qanats cannot cause excessive drawdown because their flow varies directly with the water table altitude (Boustani 2008). Qanats also have minimal evaporative losses, are more resilient to drought than vertical wells, do not use electrical power or fossil fuels, and do not have motors or other equipment that require maintenance (English 1968; Haeri 2006). Historically, some qanats have been a source of power with mills installed along the main gallery to take advantage of the water power. If maintained, qanats have an operational life of centuries whereas the life span of vertical wells is about 20 years (Haeri 2006). Qanats often flow year round so some water may go to waste during times when it is not needed. However, some qanat systems have storage reservoirs and watertight gates that can seal off the entrance (English 1998).

Pearce (2006) noted that although the qanats are an attractive water supply technology, their sociology is wrong in the modern age. Construction and maintenance of qanats is dangerous and time-consuming work. A hereditary class of professional qanat diggers called 'muquannis' build and repair the systems in Iran. The availability of the skilled manpower required for the construction and maintenance of qanats has declined due to reasons such as extremely low wages, poor insurance benefits, and a dissatisfaction with the social ranking (Haeri 2006). Younger generations are literally abandoning the qanats (Wessels and Hoogeveen 2002). Qanats also have the limitation that they do not supply enough water for modern large-scale agriculture (Wessels and Hoogeveen 2002) and the economic incentive to maintain them may be lost if agricultural activities are locally on the decline. Nevertheless, efforts are being made in some areas (e.g., Oman, Syria, Pakistan, parts of China) to renovate some qanats (Wessels and Hoogeveen 2002; Pearce 2006). Qanats are still an important, if not the sole, water source for many communities. However, policies designed to preserve and improve qanats are futile if water tables are allowed to continue to decline (Lightfoot 1996).

Haeri (2006) noted that there are some other social benefits of qanats in that they are a technology that is closely linked to the local community. The local community bears the responsibility for maintaining the local qanat(s) and managing the water so that it is distributed equitably. Traditionally co-owners of qanat systems have the right to the discharge for a specified period of time. In a system in Syria, described by Wessels and Hoogeveen (2002), five main landowning households had rights to irrigate their land every five days. English (1968, 1998) noted a much higher degree of fragmentation of ownership in Iran, where a qanat may have 200–300 owners, with some divided into as many as 10,000 time shares. In one example, the smallest owner was reported to be entitled to only 30 s of water once in every 12 days. Water from a qanat right may be auctioned by its owner for a day or mortgaged on a long-term basis to other irrigators (Ministry of Regional Municipalities, Environment and Water Resources 2005).

Strong social cohesion, clear ownership of the qanat, and an existing system of rights and regulations of water are important factors for the feasibility of renovation and long-term maintenance of qanats (Wessels and Hoogeveen 2002). On the contrary, landowners of the villages where qanats arise may loath to let water that they could capture with electrical or diesel-pumped wells flow away to the lands of another village that owns the qanat (Lightfoot 1996). Conflicts among shareholders and other cultural or tribal factors are an important cause for the abandonment of some qanats (Hussain et al. 2008). A key job is the distributor who has the responsibility to ensure that all water users obtain their entitled lots of water with minimum loss. The distributor must be trusted by all because he can manipulate everyone's lot (Sadr 2001).

The skill for maintaining qanats is also part of the indigenous knowledge of a country and thus has a heritage value (Haeri 2006). With the abandonment of qanats, indigenous knowledge also disappears, which leads to the abandonment of more qanats (Wessels and Hoogeveen 2002). The qanats are also part of a nation's cultural heritage as ancient engineering works. Some qanats, located in holy sites (e.g., Ain Zubaida in the Kingdom of Saudi Arabia), have global cultural significance (Hussain et al. 2008).

Qanats are thus a technology that is well suited for their environment as evidenced by the use of the technology for millennia. The main limitation of qanats is that they are expensive to build and produce limited volumes of water. Construction of new qanats is not economically viable today, especially relative to cost of drilling vertical tube wells which may have greater yields. However, it is difficult to incorporate a potential operational life in the centuries in a life-cycle cost analysis. The vulnerability to impacts from aquifer dewatering from pumping is also a major limitation.

21.8 The Collapse of Ancient Cultures and Climate

The collapse of some ancient civilizations has been attributed to long-term drought, which was either the direct cause or a major destabilizing factor. Long-term climate change effects agricultural production, and thus the food production per capita, which increases the likelihood of armed conflicts, social unrest, famines, and epidemics (Zhang et al. 2007). Ancient societies adapted to and thrived in marginal environments that experienced large interannual climate variability, but some civilizations could not survive unprecedented droughts that continued for decades and centuries (deMenocal 2001). Multidecadal and multicentury-length droughts that started abruptly, were unprecedented in the experiences of the existing societies, and were highly disruptive to their agricultural foundations (Weiss and Bradley 2001).

Binford et al. (1997, p. 246) observed that:

> Human cultures adapt to changing environmental conditions within a range of normal variation. "Normal" is usually defined by recent and short time scales, rather than by long-term variability during which thresholds at environmental extremes can significantly affect cultural adaptability. In commonly defined normal periods, thresholds can be exceeded for short periods without seriously affecting a civilization. However, in the long term, lower frequency variations with larger amplitudes may exceed the limits of human adaptability.

In the case of the Mayan collapse, multiple-year droughts occurred during a much longer period of drier than normal conditions (Peterson and Haug 2005). The resulting sudden collapse was manifested by regional abandonment, replacement of one subsidence base by another, and a reduction in societal complexity (de-Menocal 2001; Weiss and Bradley 2001). The fundamental question is whether the drought was so severe that the Maya succumbed directly to its effects or was it one of a number of debilitating factors that combined in some way to cause its demise (Robichaux 2002).

Erickson (1999) cautioned against extreme neo-environmentalism is which *humans are considered passive pawns at the mercy of drought and floods*. It can be very tempting to adopt a simplistic climate-driven explanation for the collapse of an ancient society, when the real cause is likely much more complex. The "new ecology" approach instead emphasizes that environments are dynamic and there is no "normal" climate around which rainfall or temperature varies or changes. Humans are considered to be active and dynamic agents who not only respond to the challenges of the fluctuating climate conditions, but also create, shape, and transform their environment (Erickson 1999).

Diamond (2005) concluded that societal collapse may occur when people living in fragile and difficult environments are confronted with external environmental changes or human-caused environmental changes beyond their experiences and that they could not have anticipated. Some societies were inherently more vulnerable to climate change than others. The response of a human population to climate change is largely a function of the sensitivity of that population's resources base to that change (Benson et al. 2007b). Diamond (2005) also emphasizes that the decisions that societies make can determine their fate. Societies differ in their flexibility and adaptability to climate change, and some choose what turned out to be wrong strategy for dealing with environmental change. Societal adaptive capacity is further discussed in Chap. 39 in the context of current climate change.

Collapses of some ancient cultures that appear to have had a climatic driver are further discussed. A fundamental question is whether climate change (e.g., prolonged drought) was a strong, irresistible cause of societal collapse or whether the societies that collapsed were already at a "tipping point" and the additional effects of climate change pushed them over the edge. An already weakened society may not have had the adaptive capacity to deal with additional stresses from prolonged drought conditions, whereas a stronger society might have survived.

21.8.1 Akkadian Empire, Mesopotamia

The Akkadian empire was located between the Tigris and Euphrates Rivers in what is today Syria and Iraq. Archeological data from 2300 to 2000 BC document a thriving imperial economy that sustained long-distance trade, construction of monumental buildings, and massive agricultural projects (Weiss et al. 1993; deMenocal 2001). The Akkadian empire suddenly collapsed at about 2200 BC, which is associated with imperial collapse, abandonment of settlements, regional desertion, and large-scale population dislocation (Weiss et al. 1993). Both soil and sediment features in the area suggest the establishment of a marked increase in aridity and intensification of wind circulation. Desertification is interpreted to have significantly reduced soil moisture reserves, which resulted in reduced agricultural productivity (Weiss et al. 1993).

A core from the Gulf of Oman indicates a pronounced increase in aeolian deposition of Mesopotamian origin at about 2075 $\pm$ 150 BP, which documents an abrupt increase in aridity (Kerr 1998; Cullen et al. 2000). During dry periods there is increased deposition of Mesopotamia-derived dust, which characteristically has high concentrations of detrital dolomite, calcite, and quartz, and neodymium and strontium ratios (Cullen et al. 2000). Cullen et al. (2000) noted other published data indicating shifts to much drier conditions commencing between 2050 and 2250 BP.

21.8.2 Classic Maya civilization, Central America

The Mayan Civilization occupied southern Mexico and all or parts of Belize, Guatemala, Honduras and El Salvador. The Mayans are renowned for their extraordinary rich architecture and art, and the development of a written language and astronomical calendar. The Mayan civilization reached its peak during the Classic Maya period, which extends circa 250–900 AD. After reaching its zenith at about 750 AD the civilization imploded from 750 to 900 AD.

The collapse of the Classic Mayan Civilization and the role of drought are discussed by Hodell et al. (1995), Gill (2000), deMenocal (2001), Haug et al. (2003), and Peterson and Haug (2005). The Mayan homeland experiences a pronounced winter dry season, which spurred the development of large-scale water-collection systems. The karstic geology of the region also results in groundwater being located deep below land surface and thus is not easily accessible. Multiple lines of evidence indicate that the 750–900 AD period contain several multiple year droughts that were superimposed on overall drier than normal conditions. The decline occurred from the south to north, which has been suggested as being due to groundwater being more accessible in the north.

The case that a long-term period of drier than normal conditions (i.e., a prolonged drought) was a factor in the decline of the Classic Mayan Civilization is very strong. However, the degrees to which foreign military intrusions, internal conflict, political and economic mismanagement, and other cultural factors were relevant to the collapse and how they interplayed with the occurrence of drought still remains to be worked out (Robichaux 2002). Lucero et al. (2011) proposed that royal power rested in what the Mayan kings provided their subjects, which included water during the annual dry period via massive artificial reservoirs. Once the Mayan kings lost their ability to supply water to the people, it is proposed that their subjects, supporters, and laborers left them and the Classic Mayan civilization collapsed. However, well-adapted Mayan farmers were able to survive and continue through the present.

21.8.3 Tiwanaku Civilization, Bolivian and Peruvian Altiplano, Near Lake Titicaca

The Tiwanaku civilization occupied the Bolivian and Peruvian altiplano, near Lake Titicaca between 300 BC and 1100 AD (deMenocal 2001). The civilization developed a highly productive water-dependent form of raised-field cultivation that stimulated dense human populations, which could not be supported under drier conditions (Binford et al. 1997). The collapse of the Tiwanaku civilization at about 1100 AD corresponds to a start of prolonged drought period, which is indicated by a 12–17 m (39.4–55.8 ft) decline in the water level of Lake Titicaca and a decrease in ice accumulation in the Quelccaya ice cap of Peru (Binford et al. 1997). Raised fields dated between 1100 and 1200 AD were located in the lowest-lying areas or near stream mouths, suggesting that they may have been the last to dry. Presumably the population dispersed and colonized new environmental niches as their intense raised-field agricultural became unviable (Binford et al. 1997).

Erickson (1999) noted that the people inhabiting the Lake Titicaca region had developed a complex indigenous knowledge that includes a sophisticated agricultural technology and elaborate social strategies to mitigate climate fluctuations. Major drought periods cause great hardships to some people, but also created opportunities as enormous areas of the Lake Titicaca bed became exposed, which has deep, organic-rich soils that is highly prized by local farming communities (Erickson 1999). The archeological evidence indicates a long history of building and rebuilding of what were ephemeral settlements.

21.8.4 Mississippian Chiefdoms (Cahokia), United States

The Mississippian (Cahokian) culture developed in the middle Mississippi River valley in what is now southwestern Illinois. The Cahokia area was the site of intense growth between 1050 and 1130 AD including construction of over 120 pyramid mounds (Benson et al. 2007b). Indications of political unrest appeared in the 1100s when walls were first constructed and, at the same time, there is evidence that farming was declining (Benson et al. 2007b). The initial abandonment of Cahokia is evidenced from 1275 to 1300 AD. Tree-ring based climate reconstructions indicate that the 1276–1297 drought, documented by Douglass (1929) in the western United States, also substantially impacted the region occupied by the Cahokians (Benson et al. 2007b). The final phase of abandonment of Cahokia (Sand Prairie Phase), which may have persisted until about 1400, coincides with a period of below average moisture for 46 of 58 years from 1344 to 1401 AD (Cook et al. 2007).

21.8.5 Angkor, Cambodia

The Khmer Empire of Angkor was dependent on the annual monsoon flooding of Cambodia's lowland to support a vast and complex agricultural system. The civilization was thus vulnerable to variability in the strength and intensity of the monsoon (Buckley et al. 2010). Tree-rings based hydroclimatic reconstructions indicate that a multidecadal period of weakened monsoon rains occurred in the middle to late 14th century and that shorter, and at times more severe, droughts occurred in the early 15th century, which time period coincides with the demise of Angkor Empire (Buckley et al. 2010). The drought period corresponds to the time of the transition from the Medieval Climate Optimum (Medieval Warm Period) to the Little Ice Age. Buckley et al. (2010) proposed that weakened summer monsoon rainfalls were punctuated by abrupt and extreme wet episodes, which likely caused flooding that damaged flood control infrastructure.

21.8.6 Neolithic Cultures of the Central Plain of China

Several distinct and roughly contemporaneous Neolithic cultures in the Central Plain of China collapsed around 4200–4000 yr BP (2250–2050 BC). Signs of the collapse include (1) fewer archeological sites found and in a more restricted area, (2) less sophisticated materials used, and (3) disappearance or sharp decline in large-scale architecture (Wenxiang and Tungsheng 2004). The collapse of the Neolithic cultures is marked by a cooling trend and decrease of the monsoon. In the northern areas, temperature declined and aridity increased, whereas in the south, increased precipitation lead to flooding or water-logging conditions (Wenxiang and Tungsheng 2004). Climate change may have led to the rise of dynastic state-level society because warfare intensified as population pressures increased in the more vulnerable agriculture lands (Wenxiang and Tungsheng 2004).

21.8.7 Anasazi, Southwestern United States

The Anasazi people (Sect. 20.6.1) abandoned most of the great houses by about 1150 AD and abandoned the Four Corners region of the southwestern United States by about 1300. Douglass (1929) developed a continuous tree-ring (dendrochronology) record for the Four Corners region and was able to precisely date a period of great aridity in the late 13th century. Harsher growing conditions during drought periods are manifested by thinner annual growth rings. Frequent droughts were identified between 1247 and 1299 AD, with a period of great drought between 1276 and 1299. It is hypothesized that failure of the summer monsoon was associated with great house abandonment and the migration of the Anasazi out of the region. The loss of the summer moisture is suggested to have resulted in a reduced yield of maize, the dietary staple of the Anasazi, forcing them to abandon areas that were marginal for dryland farming and to migrate to better watered areas (Benson et al. 2007a). Benson et al. (2007b) cautioned that drought by itself probably can never totally account for the migration of an entire group that occupied a particular geographic region. Drought may have contributed to social instability and unrest.

The diversion of water into canals combined with the clearing of vegetation (deforestation) resulted in the erosion of deep arroyos to depths below the fields being irrigated (Lister and Lister 1981; Diamond 2005; Mays 2007b). The resulting lowering of the water table is proposed to have reduced agricultural productivity. However, Lister and Lister (1981) suggested that irrigation from tributary arroyos may have allowed for the continuation of the agricultural community even after arroyo cutting lowered the water table.

Chaco Canyon was capable of coping with local short-term periods of dryness, but apparently could not sustain itself over a 60-year period of moisture deficiency. Its social fabric is proposed to have been steadily eroded under unabated pressures of food shortages (Lister and Lister 1981). The question is whether the cause of the collapse of Chaco Canyon and the Anasazi culture was due to human impacts on the environment or because of drought? The answer appears to be both (Diamond 2005). People were living increasingly close to the margin of what the environment could support. The proximate cause was drought, which appears to have pushed them over the edge.

21.9 Conclusions

Ancient societies managed to survive in water scare regions by creatively taking advantage of all available freshwater resources. The basic elements independently adopted by ancient societies in different parts of the world include:

- employment of a wide variety of rainwater harvesting techniques,
- storage of periodically available water, and
- conveyance of water from springs and other perennial water sources to population and agricultural centers.

People living in fragile and difficult environments developed solutions that were brilliantly successful and understandable in the short run, but failed or otherwise created fatal problems in the long run, when people were confronted with external environmental changes or human-caused environmental changes beyond their adaptive capacity (Diamond 2005). Communities living in arid and semiarid lands are inherently vulnerable if they are overly reliant on non-perennial surface-water resources. As populations grow and water demands approach the maximum available supply, there is little room left for error and societies become increasingly vulnerable to any disruptions or reductions in water supplies.

Modern societies potentially have more water management tools available to cope with climate change. Surface-water supplies can be made more reliable and resilient through storage in surface-water reservoirs and underground (managed aquifer recharge; Chap. 23). Modern societies also have the benefits of being able to extract deep groundwater resources (both renewable and non-renewable (Chap. 36); and develop alternative water resources, such as desalination (Chaps. 27 and 28) and wastewater reuse (Chap. 29). Modern societies may also take advantage of the importation of food and fiber (i.e., virtual water trade; Chap. 34) to reduce local agricultural water demands.

It is now clear that the long-term droughts (i.e., megadroughts) that contributed to or were the primary cause of the collapse of the Anasazi and Cahokia cultures, easily exceed the duration of the worst droughts of the 20th century (Cook et al. 2009). A critical issue for evaluating the adaptive capacities of modern societies to future climate change is that historical experiences are not indicative of the range of climatic conditions that may occur in the future. The possibility exists based on climatic modeling results that greenhouse gas forcing may result in megadroughts of the kind experienced by North America in past (Cook et al. 2009). Other regions may also experience droughts of duration and severity unprecedented for the past several centuries. Global climate change and adaptation options are addressed in Chaps. 38 and 39.

Modern developed societies have more tools to deal with climate-induced changes in water supply but also have much greater water demands to meet. Water supply systems can be made more robust by the development of a diversified water supply, which is perhaps the most important lesson from ancient civilizations. Excessive reliance on any one water source can lead to a high degree of vulnerability. For example, the primary water source for the city of Atlanta, Georgia (USA) is Lake Lanier which in 2008 approached record lows due to prolonged drought. Serious questions arose as to what would happen if a metropolitan area with over 5 million people ran out of water. Fortunately the drought broke in the Atlanta area before those questions become a reality, but the vulnerability remains.

Developed countries have the financial resources to develop diversified water supply systems and thus increase their adaptive capacity. The issue is whether there is the foresight and will to make the necessary investments. However, the impetus to invest in measures to prove resilience against droughts tends to wane once the rains return and the past droughts become more distant memories. As discussed in Chap. 39, the greatest vulnerability to adverse climate change occurs in developing countries in arid and semiarid lands, where water supplies are inadequate under current climate regimes and financial resources are inadequate to improve water supplies. Poor subsistence farmers are particularly vulnerable, as their adaptive capacity tends to be very low.

References

AbdelKhaleq, R. A., & Ahmed, I. A. (2007). Rainwater harvesting in ancient civilizations in Jordan: Water science & technology. *Water Supply, 17*(1), 85–93.
Agnew, C., & Anderson, W. (1992). *Water in the arid realm*. London: Routledge.
Aicher, P. J. (1995). *Guide to the aqueducts of ancient Rome* (p. 183). Wauconda: Bolchaz-Garducci.
Alsharhan, A. S., Rizk, Z. A., Nairn, A. E. M., Bakhit, D. W., & Alhajari, S. A. (2001). Hydrogeology of an arid region: The Arabian Gulf and adjoining areas. Amsterdam. Elsevier.
Andrews, J. P., & Bostwick, T. W. (2000). Desert farmers at the river's edge. The Hohokam and Pueblo Grande. Phoenix, Pueblo Grande Museum.
Angelakis, A. N., & Koutsoyiannis, D. (2003). Urban water engineering and management in ancient Greece. In B. A. Stewart & T. Howell (Eds.), *The encyclopedia of water science* (pp. 999–1007). New York: Dekka.
Basak, P., Raghavendra Prasad, P. M., and Sreedharan, K. E. (1997). Western Ghats. In A. Agarwal & S. Narain (Eds.), Dying wisdom: Rise, fall and potential of India's traditional water harvesting systems (pp. 220–223). New Delhi: Centre for Science and Environment.
Beaumont, P. (1971). Qanat systems in Iran. *Bulletin of the International Association of Scientific Hydrology, 16*(1), 39–50.
Benson, L., Petersen, K., & Stein, J. (2007a). Anasazi (Pre-Columbian Native-American) migrations during the middle-12th and late-13th centuries—were they drought induced? *Climate Change, 83*, 187–213.
Benson, L. V., Berry, M. S., Jolie, E. A., Spangler, J. D., Stahle, D. W., & Hattori, E. M. (2007b). Possible droughts of early-11th-, middle-12th-, and late-13th-century droughts on western Native Americans and the Mississippian Cahokians. *Quaternary Science Reviews, 26*, 336–350.
Binford, M. W., Kolata, A. L., Brenner, M., Janusek, J. W., Seddon, M. T., Abbott, M., et al. (1997). Climate variation and the rise and fall of an Andean civilization. *Quaternary Research, 47*, 235–248.
Boustani, F. (2008). Sustainable water utilization in arid region of Iran by qanats. *Proceedings of the World Academy of Science, Engineering and Technology, 33*, 213–216.
Buckley, B. M., Anchukaitis, K. J., Penny, D., Fletcher, R., Cook, E. R., Sano, M., et al. (2010). Climate as a contributing factor in the demise of Angkor, Cambodia. *Proceedings National Academy of Sciences, 107*, 6748–6752.
Buxó, R. (2008). The water management in an ancient Greek-Roman city (I): An example in the north-east of Spain: Options Méditerranéennes, Series A, no. 83, pp. 9–16.
Bybordi, M. (1974). Ghanats of Iran: Drainage of sloping aquifer. *Journal of the Irrigation and Drainage Division, 100*(3), 245–253.
Cook, E. R., Seager, R., Cane, M. A., & Stahle, D. W. (2007). North American drought: Reconstructions, causes and consequences. *Earth Science Reviews, 81*, 93–134.
Cook, E. R., Seager, R., Heim, R. R., Jr., Vose, R. S., Herweijer, C., & Woodhouse, C. (2009). Megadroughts in North America: Placing IPCC projects of hydroclimatic change in a long-term paleaoclimate context. *Journal of Quaternary Science*. doi:10.1002/jqs.1303.
Cressey, G. B. (1958). Qanats karez, and foggaras. *Geographical Review, 48*, 27–44.
Crouch, D. P. (1990). Planning water management for an ancient Greek city. In J. C. Hooghart (Ed.), *Hydrological processes and water management in urban areas* (pp. 271–277). IAHS Publ. No. 198 (Proceedings of the Duisberg Symposium, April 1988).
Crouch, D. P. (1993). *Water management in ancient Greek cities*. New York: Oxford University Press.
Cullen, H. M., deMenocal, P. B., Hemming, S., Hemming, G., Brown, F. H., Guilderson, T., et al. (2000). Climate change and the collapse of the Akkadian empire:Evidence from the deep sea. *Geology, 28*, 379–382.
deMenocal, P. B. (2001). Cultural response to climate change during the late Holocene. *Science, 292*, 667–673.

Diamond, J. (2005). *Collapse, how societies choose to fail or succeed*. London: Penguin Books.
Douglass, A. E. (1929). The secrets of the Southwest solved by talkative tree rings. *The National Geographic Magazine, 56*(6), 737–770.
English, P. W. (1968). The origin and spread of qanats in the old world. *Proceedings of the American Philosophical Society, 112*(3), 170–181.
English, P. W. (1998). Qanats and lifeworlds in Iranian plateau villages, In J. Albert, M. Berhardsson & R. Kerna (Eds.), *Transformation of Middle Eastern natural environments: Legacies and lessons* (pp. 187–205). New Haven: Yale School of Forestry & Environmental Studies Bulletin 103.
Erickson, C. L. (1999). Neo-environmental determinism and agrarian 'collapse' in Andean prehistory. *Antiquity, 73*(281), 634–642.
Evanari, M., Shanan, L., & Tadmor, N. (1971). *The Negev: The challenge of a desert*. Cambridge: Harvard University Press.
Gill, R. B. (2000). *The great Maya Droughts: water, life, and death*. Albuquerque: University of New Mexico Press.
Glanzburg, J. (1994). Water harvesting traditions in the desert southwest. *Permaculture Drylands Journal, 20*, 8–11.
Grahame, J. D., & Sisk, T. D. (2002). Canyons, cultures and environmental change: An introduction to the land-use history of the Colorado Plateau. Last accessed 6 March, 2011: http://www.cpluhna.nau.edu
Guzzo, M., & Schneider, E. E. (2002). *Petra*. Chicago: University of Chicago Press.
Haeri, M. R. (2006). Kariz (Qanat); an eternal friendly system for harvesting groundwater: Adaptation Workshop, November 12 to 13, 2003, New Delhi, India.
Hansen, R. D. (1983). Water and waste water in Imperial Rome. *Water Resources Bulletin, 19*, 263–269.
Hansen, R.D. (2006). Water and wastewater systems in Imperial Rome: www.waterhistory.org
Haug, G. H., Günther, D., Peterson, C., Sigman, D. M., Hughen, K. A., & Aeschlimann, B. (2003). Climate and the collapse of Maya civilization. *Science, 299*, 1731–1735.
Heathcote, R. L. (1983). *The arid lands: their use and abuse*. London: Longman.
Helms, S. W. (1981). *Jawa, lost city of the Black Desert*. Ithaca: Cornell University Press.
Hodell, D. A., Curtis, J. H., & Brenner, M. (1995). Possible role of climate in the collapse of Classic Maya civilization. *Nature, 375*, 391–394.
Hodge, T. (2002). *Roman aqueducts and water supply* (2nd ed.). London: Butterworth.
Howard, J. B. (1992). Hohokam legacy: desert canals. Pueblo Grande Museum Profiles No. 12, p. 5.
Howard, J. B., & Huckleberry, G. (1991). The operation and evolution of an irrigation system: The east papago canal study: soil systems publication in archaeology number 18. Phoenix: Arizona Department of Transportation, Environmental Planning Services.
Hussain, I., Abu-Rizaiza, O. S., Habib, M. A. A., & Ashfaq, M. (2008). Revitalizing a traditional dryland water supply system: The karezes in Afghanistan, Iran, Pakistan and the Kingdom of Saudi Arabia. *Water International, 33*, 333–349.
Kerr, R. A. (1998). Sea-floor dust shoes drought felled Akkadian Empire. *Science, 279*, 325–326.
Koutsoyiannis, D., Zarkadoulas, N., Angelakis, A. N., & Tchobanoglous, G. (2008). Urban water management in ancient Greece: Legacies and lessons. *Journal of Water Resources Planning and Management, 134*, 45–54.
Lendering, J. (2008). Constantinople (Istanbul): Basilica Cistern. Revised August 4, 2008; accessed July 17, 2011: http://www.livius.org/cn-cs/constantinople/constantinople_cistern_basilica.html
Li, X. Y., Zhang, R., Gong, J. D., and Xie, Z. (2002). *Effects of rainwater harvesting on the regional development and environmental conservation in the semiarid Loess Region of Northwest China: Proceedings 12th International Soil Conservation Organization Conference*, Beijing, May 26–31, 2002, pp. 482–486.
Lightfoot, D. R. (1996). Syrian qanat Romani: History, ecology, and abandonment. *Journal of Arid Environments, 33*, 321–336.

Lister, R. H., & Lister, F. C. (1981). *Chaco canyon*. Albuquerque: University of New Mexico Press.
Lucero, L. J., Gunn, J. D., & Scarborough, V. L. (2011). Climate change and Classic Maya water management. *Water, 3*, 479–494.
Masse, B. (1991). Prehistoric irrigation in the Salt River Valley, Arizona. *Science, 214*(23), 408–415.
Mays, L. W. (2007a). Ancient urban water supply systems in arid and semi-arid regions: International Symposium on New Directions in Urban Water Management, 12–14 September 2007, UNESCO, Paris.
Mays, L. W. (2007b). Water sustainability of ancient civilizations in Mesoamerica and the American Southwest: Water Science & technology. *Water Supply, 17*(1), 229–236.
Mays, L. W. (2010). A brief history of water technology during antiquity: Before the Romans. In L. W. Mays (Ed.), *Ancient water technologies* (pp. 1–28): Dordrecht: Springer.
Mays, L. W., & Gorokhovich, Y. (2010). Water technology in the ancient American societies. In Mays, L. W. (Ed.), *Ancient water technologies* (pp. 170–200): Dordrecht: Springer.
Mays, L. W., Koutsoyiannis, D., & Angelakis, A. N. (2007). A brief history or urban water supply in antiquity Water Science and technology. *Water Supply, 7*, 1–12.
Ministry of Regional Municipalities, Environment and Water Resources (2005). Water Resources of Oman: Ministry of Regional Municipalities, Environment and Water Resources, Sultanate of Oman.
Murad, A. A., Al Nuaimi, H., & Al Hammadi, M. (2007). Comprehensive assessment of water resources in the United Arab Emirates (UAE). *Water Resources Management, 21*, 1449–1463.
Ortloff, C. R. (2005). The water supply and distribution system of the Nabataean city of Petra (Jordon), 300 BC–AD 300. *Cambridge Archaeological Journal, 15*, 93–109.
Pearce, F. (2006). *When rivers run dry* (p. 324). Massachusetts: Beacon Press.
Peterson, L. C., & Haug, G. H. (2005). Climate and the collapse of Maya civilization. *American Scientist, 93*, 322–329.
Remini, B., Achour, B., & Albergel, J. (2010). Timimoun's foggara (Algeria): A heritage in danger. *Arabian Journal of Geosciences*. doi:10.1007/s12517-010-0139-9.
Rizk, Z. S., & Alsharhan, A. S. (2003). Water Resources in the United Arab Emirates. In A.S. Alsharhan & W.W. Wood (Eds.), *Water resources perspectives: Evaluation, management and policy* (pp. 245–264). Elsevier: Amsterdam.
Robichaux, H. R. (2002). On the compatibility of epigraphic, geographic, and archaeological data, with a drought-based explanation for the classic Maya collapse. *Ancient Mesoamerica, 13*, 341–345.
Sadr, K. (2001). Water markets and pricing in Iran. In N. I. Faruqui, N. I. Biswas & M. J. Bino (Eds.), *Water management in Islam* (pp. 84–92). Tokyo: United Nations University Press, and Ottawa, International Research Centre.
Salih, A. (2006). *Qanats a unique groundwater management tool in arid regions: The case of Bam region in Iran: Proceedings International Symposium on Groundwater Sustainability*, Alicante, January 24–27, 2006, pp. 79–87.
Taylor, J. (2002). *Petra and the lost kingdom of the Nabataeans* (p. 224). London: Tauris I.B.
Wåhlin, L. (1997).The family cistern: 3,000 years of household water collection in Jordan. In M. Sabour & K. S. Vikør (Eds.), *Ethnic encounter and culture change: Papers from the third nordic conference on middle eastern studies*, Joensuu, Finland, June 19–22, 1996 (pp. 233–249). London: Hurst.
Weiss, H., & Bradley, R. S. (2001). What drives societal collapse? *Science, 292*, 609–610.
Weiss, H., Courty, M.-A., Wetterstrom, W., Guichard, F., Senior, L., Meadow, R., et al. (1993). The genesis and collapse of Third Millennium north Mesopotamian civilization. *Science, 261*, 995–1004.
Wenxiang, W., & Tungsheng, L. (2004). Possible role of the "Holocene Event 3" on the collapse of Neolithic cultures around the Central Plain of China. *Quaternary International, 117*, 153–166.
Wessels, J. & Hoogeveen, R. J. A. (2002). Renovation of qanats in Syria. In Z. Adeel (Ed.), *Sustainable management on marginal drylands, application of indigenous knowledge for*

coastal drylands: *Proceedings UNU-UNESCO-ICARDA International Workshop*, September 21–25, 2002, Alexandria.

Wulff, H. E. (1968). The qanats of Iran. *Scientific American, 218*(4), 94–100.

Zhang, D. D., Brecker, P., Lee, H. F., He, Y. -Q., & Zhang, J. (2007). Global climate change, war, and population decline in recent human history. *Proceedings National Academy of Sciences (PNAS), 104*, 19214–19219.

Chapter 22
Rainwater, Stormwater, and Fog Harvesting

22.1 Introduction

Rainwater harvesting can be succinctly defined as the collection, storage, and reuse of rainwater. The closely related term "water harvesting" refers to the process of concentrating rainfall or runoff from a larger area for use in a smaller target area, typically for agricultural purposes (Oweis et al. 1999). Runoff farming water harvesting (RFWH) refers to the technique wherein collected water is diverted directly to the cropped area (Oweis et al. 1999). RFWM is practiced in areas where rainfall is normally too low to support continuous cropping with a reasonable economic value (Oweis et al. 1999).

Rainwater harvesting dovetails into stormwater management, which has a primary flood control objective. Stormwater harvesting is essentially rainwater harvesting on a larger scale and commonly involves recovery of water from a surface water management system. Rainwater harvesting also dovetails with managed aquifer recharge (Chap. 23), where the primary goal is local aquifer recharge. The distinction between stormwater harvesting and surface water use lies in that the former usually involves the capture of ephemeral flows from individual precipitation events within a small local catchment rather than extracting water from perennial stream flows.

The attraction of rainwater harvesting in arid and semiarid lands is self-evident. Where water is scarce it obviously makes sense to capture any available freshwater for later beneficial use. Rainwater harvesting has been used for millennia as either the primary or a supplemental source of water in arid and semiarid lands that do not have a reliable surface water source (Chap. 21). Recent studies have implicated climate change and drought as a primary agent in human migration, cultural separation, population dislocation, and collapse in prehistoric and early historic societies (Sect. 21.8). The historic response of cultures to changing conditions was to not give up until they had exhausted available options for survival in their homeland (Pandey et al. 2003). Pandey et al. (2003) noted that the cultural response to a fluctuating climate in the archaeological and historic record was

R. Maliva and T. Missimer, *Arid Lands Water Evaluation and Management*, Environmental Science and Engineering, DOI: 10.1007/978-3-642-29104-3_22,

heightened earthwork construction for rainwater harvesting. People tended to respond to a changing climate first by resorting to rainwater harvesting in order to continue to occupy their homelands rather than migrating to other locations. Conversely, as the water supplies become more secure there is the opposite tendency towards abandoning rainwater harvesting techniques.

Rainwater harvesting is also a green technology and there is a rapidly growing popular and technical literature on the subject. Rainwater harvesting methods have been reviewed by National Academy of Science (1974), Waller (1989), and Lancaster (2006, 2008), Kinkade-Levario (2007) and Downey (2010). Water harvesting for agriculture was reviewed by the United Nations Environment Programme (1984), Cricthley and Siegert (1991), and Oweis et al. (1999).

There are two basic types of rainwater harvesting systems, earthworks and storage systems (Lancaster 2006). Earthwork systems are used to harvest and hold water in the soil or recharge the underlying aquifer. Storage systems are used to harvest and store rainwater in tanks, cisterns, or reservoirs. Storage is critical for rainwater harvesting for year-round water supply, particularly in arid lands where rainfall is highly sporadic.

There is great variation in the scale and technical sophistication of rainwater harvesting systems. Rainwater harvesting, for example, may range in scale from a simple barrel used for storage of roof runoff for irrigation use or microcatchments constructed around a tree, to sophisticated systems involving treatment facilities, managed aquifer recharge, and distribution systems to serve multiple users.

Rainwater harvesting is now receiving much greater attention as an element of the water supply in arid and semiarid lands in developing countries. It may be implemented as relatively inexpensive, small-scale methods of water supply that can be adapted to the indigenous needs of rural communities and households (Li 2003). It can have the advantages of a small-scale, simple operation, high adaptability, and low cost, and is, therefore, ideally suited for the socioeconomic and biophysical conditions of semiarid areas. Rainwater harvesting systems can also empower local communities to manage their own resources.

Large-scale implementation of rainwater harvesting can reduce demands on fresh ground and surface-water resources and reduce stormwater flows that may otherwise be a disposal problem. On the other hand, rainwater harvesting under some circumstances could impact other water users by reducing downstream flows. Rainwater harvesting is a highly decentralized technology in much of both the developing and developed world that is performed without significant governmental control and oversight.

Rainwater is perceived to be a pristine water source, but it may in fact contain a variety of chemicals and pathogens. Rainwater can be readily contaminated from microorganisms, chemicals, and metals that accumulated on or are present in the collecting surfaces and storage systems. Harvested water should thus receive some treatment before potable use. Treatment is often not required for non-potable uses, particularly for water collected from an impervious surface. The Australian guidelines for stormwater reuse and recycling for non-potable purposes (NRMMC-EPHC-NHMRC 2009b) address health risks associated with stormwater harvesting. Stormwater collected from channels may require treatment to reduce turbidity before it can be used for irrigation systems because of the potential for clogging of equipment.

Rainwater harvesting systems have limitations that must be understood. The most fundamental constraint is a limited storage capacity. When cisterns are used for storage, the main problem is water balance; storing enough water to meet the needs of a family or community until the next rainfall event. There are also practical limits as to how much rainfall may be harvested and stored at a given site. Cisterns are suitable for storing water for domestic and high efficiency irrigation uses, but not flood irrigation, which has much greater water requirements. MAR can provide large storage volumes for harvested rainwater. The opportunities for rainwater harvesting-MAR integration also vary greatly in scale. For example, roof-water may be diverted to a small gravel-filled infiltration trench for local aquifer recharge. Alternatively, stormwater may after treatment be stored in large-scale ASR systems.

Rainwater harvesting also involves work and active management, which most people would rather avoid if they can have their water supply delivered directly to them via a centralized system. The adoption of personal rainwater harvesting in developed countries is to a large degree a life-style decision (i.e., a “green” practice), or part of a strategy to live off the grid. Rainwater harvesting may also be driven by economic considerations where the cost of freshwater is great. On tropical islands where freshwater is scarce, cisterns are still widely used to store rainwater. For example, building codes in the U.S. Virgin Islands require new homes and businesses to have a cistern or other means of self-supply if they are not connected to the public water system.

Rainwater harvesting systems may lack the local prestige of large-scale water systems, which can undermine support. Wåhlin (1997) noted with respect to the decline of cistern use in Jordan that households apparently no longer feel the responsibility to help themselves to achieve part of their water needs, especially as the turning of a tap is so much more convenient than maintaining a cistern and its collection area. Another example of this attitude is the experience of the El Tofo fog-harvesting system in Chile, which was a technical success, but was abandoned because of local social issues (Dale 2003; Sect. 22.8). It has been noted that local populations may believe that one aspect of being a “developed” area is having water brought to you by the state and that you should never have to think about it (Dale 2003). Agarwal and Narain (1997) observed in India a strong antipathy by some towards traditional water harvesting technology in preference to modern technology (i.e., massive inter-basin water transfers). Obtaining and maintaining public support for rainwater harvesting is critical for its long-term success. Oweis et al. (1999) noted that the chance of success of RFWH systems in developing countries is greater if the farmers and community are involved in the early planning stages onwards. The benefits of the system should be apparent to farmers as early as possible and risks and profit potentials for the investment of labor and other inputs must also be acceptable. The new technology should also be seen as one component of a larger, village-level or regional water management improvement project.

22.2 Rainwater Harvesting Concepts

Rainwater harvesting systems include some or all of the following elements, a catchment area, a means to convey water from the catchment area to the point of use or storage, a storage system, and a point of use. The collected water may be stored in a reservoir, tank (cistern), aquifer, or the soil profile.

The amount of rainfall that may be harvested depends upon the duration and intensity of rainfall events, and the area and properties of the catchment area. The catchment area is the runoff producing surface. The runoff coefficient is the ratio of the runoff volume to the rainfall volume. Runoff coefficients vary with each rainfall event depending upon its intensity and duration, especially where bare soil is used for the catchment (Oweis et al. 1999). For light rainfall events, most of the water is required for wetting, infiltration and filling of surface storage capacity, and runoff is therefore minimal.

Runoff coefficients are also a function of the catchment surface characteristics. Infiltration into pervious soils is discussed in Sect. 5.6.2. High infiltration rates are desirable for earthwork systems in which the goal is to store water in the soil zone or locally increase groundwater recharge. In harvesting systems in which the water is to be conveyed to a storage tank system or farm fields, the goal is to have minimal infiltration into the catchment surface in order to maximize runoff. Runoff inducement is the practice of treating and sealing land surfaces to decrease infiltration and water retention, and increase runoff. Runoff inducement techniques include land clearing, soil smoothing, surface treatments to seal surfaces, application of a rigid surface (e.g., concrete), and application of a flexible surface (e.g., plastic) (Oweis et al. 1999).

Area ratio is the ratio of the catchment area to the crop area, which is the most important variable in the design of RFWH. Area ratios integrate the effects of runoff coefficients, rainfall characteristics, and soil and crop factors (Oweis et al. 1999). The ratio should be high enough to provide sufficient water to meet crop demands most years.

The simplest rainwater harvesting techniques involve the construction of structures or excavations that slow the flow of, or temporarily retain, runoff in order to increase the amount of infiltration. For example, a shallow depression excavated on a slope or within or adjacent to an ephemeral stream channel will retain runoff and increase local infiltration. The local land surface should be contoured to direct water to the basin. Terraces constructed parallel to topographic contours (and thus perpendicular to the slope) have been used for millennia in some regions to capture runoff, control erosion, and create flat farm fields in sloped land. The terraces may be simple earthen berms or may have retaining walls (Fig. 22.1). Ridge and furrows constructed parallel to topographic contours are also effective in reducing runoff and increasing infiltration and soil moisture (Fig. 22.2).

Check dams constructed across ephemeral stream channels (Fig. 22.3) are effective in reducing flow velocity, increasing infiltration, temporarily storing water, and controlling soil erosion. Check dams are small enough so that the bulk of flood flows can readily pass. The dams may be permanent masonry features or

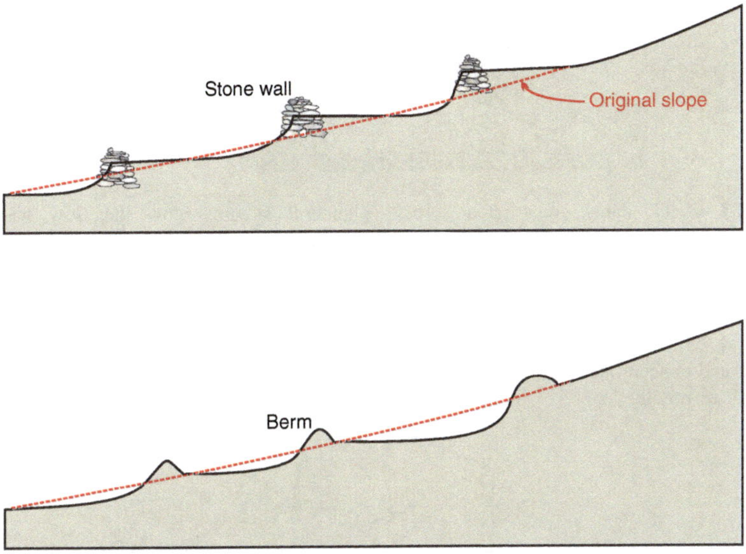

Fig. 22.1 Conceptual diagram of terraces. Walls or berms constructed parallel to the land slope retard the down-slope flow of water, increase infiltration, trap soil and sediment, and create flat areas for farming

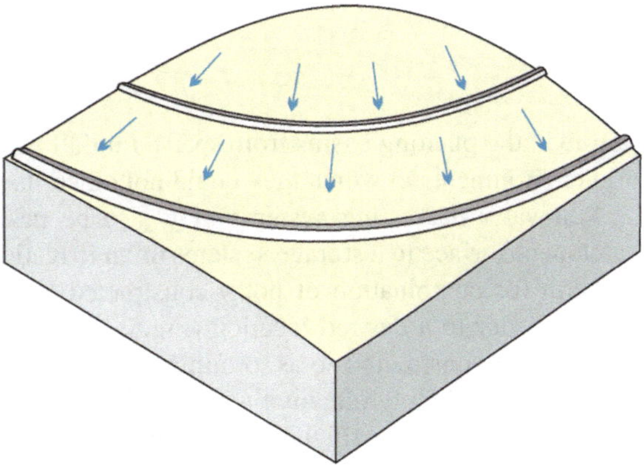

Fig. 22.2 Berms or low ridges constructed parallel to topographic contours on hillsides act to retain runoff and increase infiltration

may be temporary structures constructed of earth, rock, or wood. In the latter case, the dams are expected to be destroyed during major floods and require rebuilding.

Another class of rainwater harvesting systems involves the diversion of runoff using channels, swales, berms, and low walls to a collection area. A simple example of this type of system is the microcatchment, which is a contoured area with berms that is designed to capture and concentrate runoff in a small planting basin, which may contain a single tree (Fig. 22.4). Numerous microcatchments may be constructed on a single hillside. Microcatchments can add enough water to the soil

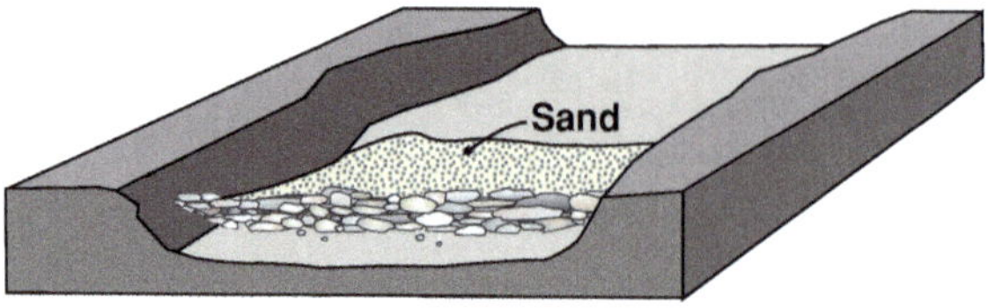

Fig. 22.3 Check dams constructed across ephemeral streams slow the flow water and temporarily retain water allowing for increased infiltration. Water is also retained in sand trapped behind the dam

Fig. 22.4 Microcatchments capture and concentrate down-slope runoff

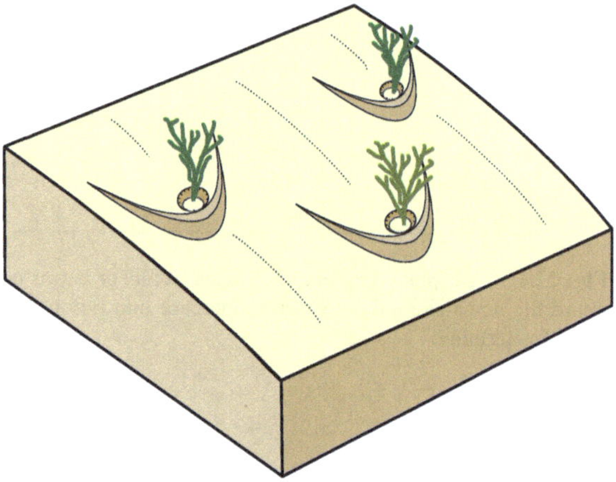

profile in the planting basins from sparse rainfall events to allow growth of trees in arid environments in which they could not otherwise survive (Evanari et al. 1971).

Rainwater harvesting systems may also be designed to divert water from a catchment surface to a storage systems or an irrigation site. For example, a channel or berm (or combination of both) constructed at a slight angle to a hillside can convey water to a desired location (Fig. 22.5). The catchment should be chosen, prepared, or constructed so as to minimize infiltration and maximize run-off. Bare rock is can be highly effective as a catchment surface because of its low permeability (Fig. 22.6). Artificial catchment surfaces are constructed of relatively impervious materials such as concrete, asphalt, or plastic sheeting. Roof-tops (Fig. 22.7) are widely used as collection areas because they are designed to be impervious and are located off the ground (Sect. 22.4). Gutters are used to convey water to either a storage tank (cistern) or an infiltration area.

Paved roads also act as catchment areas. Properly constructed roads are designed with a cross slope to enable surface water to drain from the traffic lanes and shoulders as quickly as possible. Accumulation of water on road surfaces creates a hazardous condition by reducing surface friction and vehicle stability. The water drains directly to swales or commonly via curbs to catch basins and then to infiltration or retention basins (Fig. 22.8).

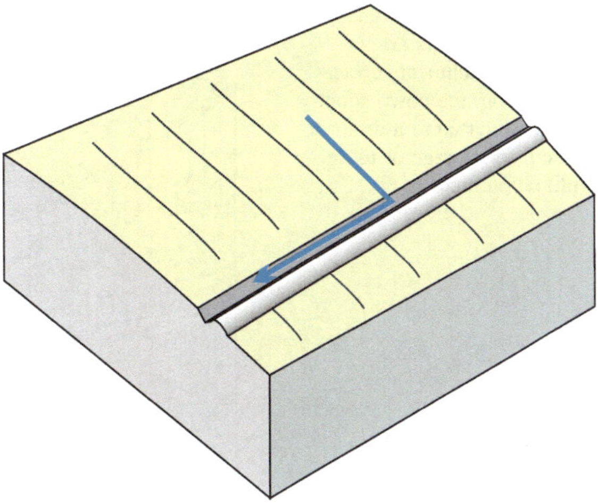

Fig. 22.5 Conceptual diagram of diversion channel and berm that intercept down-slope runoff and conveys it to either a point of use or storage

Fig. 22.6 Rainwater harvesting system in Jordan in which bare rock is used as the catchment surface. The harvested water is discharged to a sandy area near a well, which provides filtration. Photo by Weixing Guo

There is a great range of techniques and potential degrees of implementation of rainwater harvesting, which will depend upon the needs for water, rainfall amount

Fig. 22.7 Sloped or gabled roofs with gutters are effective catchment surfaces. Water from the down spout can be diverted to a drum or cistern for storage or to an infiltration area

and distribution, and available social and economic resources. Systems may range from basic roof rainwater harvesting systems to the conversion of the entire local landscape into an integrated rainwater harvesting system (Lancaster 2006).

22.3 Domestic Scale Rainwater Harvesting Systems

There are a growing number of publications and internet sites devoted to domestic-scale rainwater harvesting. Brad Lancaster (2006, 2008) published two excellent books on the subject. The Texas Water Development Board (2005) published the Texas Manual on Rainwater Harvesting, which is available on-line. A single domestic rainwater harvesting system may meet much of a households water needs, but will

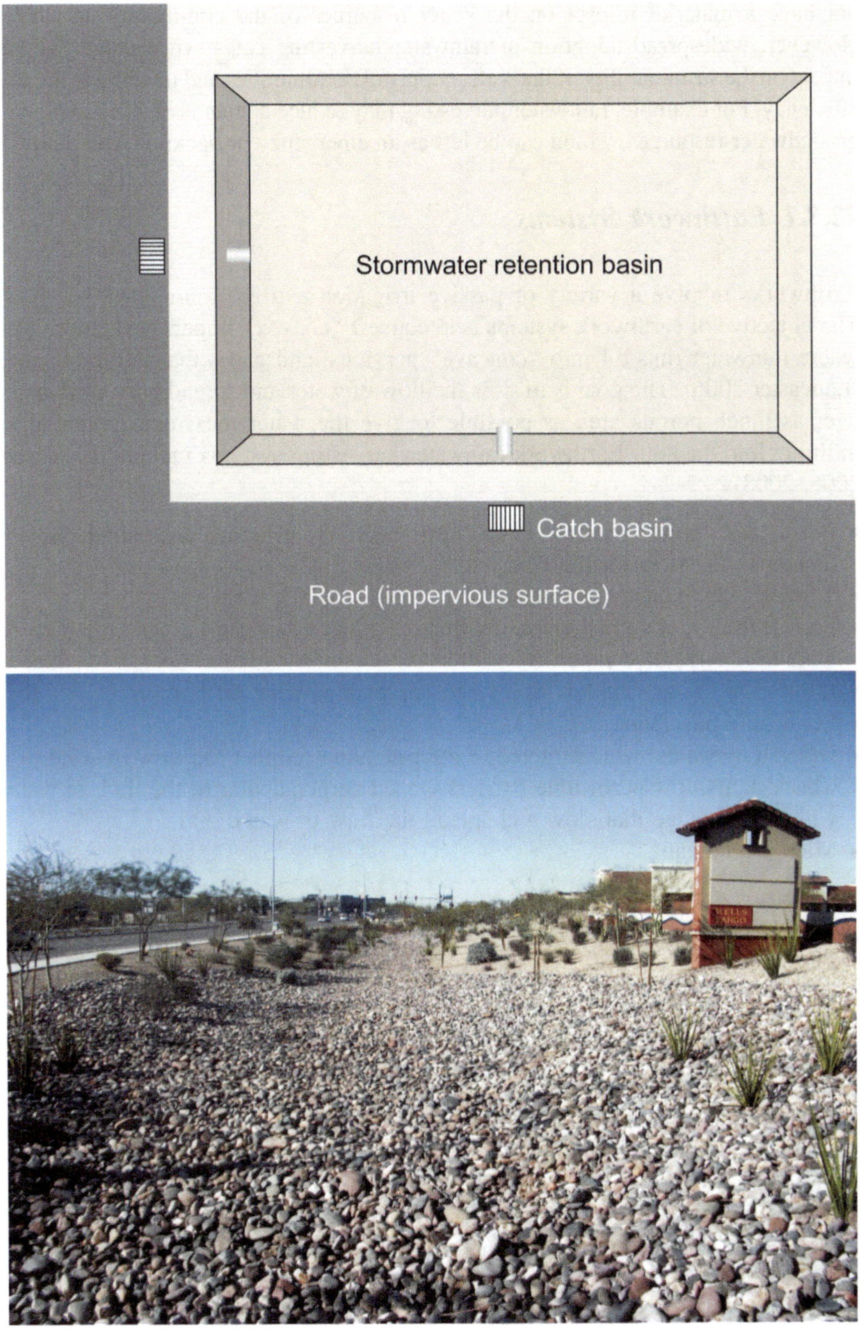

Fig. 22.8 Stormwater management systems may serve an aquifer recharge function. Rainwater runoff from roads is often conveyed to retention basins or swales where it infiltrates into the soil

not have a material impact on the water resources of the community as whole. However, widespread adoption of rainwater harvesting can have a great positive impact on the sustainability of the water supply of communities and increase water use efficiency. For example, rainwater harvesting may reduce demands on over-exploited groundwater resources, which can be left as an emergency or backup water supply.

22.3.1 Earthwork Systems

Earthworks involve a variety of passive irrigation and erosion-control schemes. The objective of earthwork systems is to convert "convex" impervious landscapes where rainwater runs off into "concave" pervious landscapes that infiltrate water (Lancaster 2006). The goal is to slow the flow of water and spread harvested water over as much porous area as possible to give the water maximum potential to infiltrate into the soil. Earthwork rainwater harvesting systems include (Lancaster 2006, 2008):

- Berms and basins; shallow basins surrounded by a berm constructed perpendicular to the slope of the land,
- Terraces; flat benches surrounded by a berms or low walls,
- French drains; trenches or basins filled with porous material (gravel) with or without a perforated pipe,
- Infiltration basins; shallow excavated depressions with flat bottoms,
- Permeable pavement,
- Diversion swales; which intercept, infiltrate, and redirect the flow of water,
- Check dams; low permeable barriers placed perpendicular to the flow of water within a drainage that slow and spread the flow or water, and
- Microcatchments.

22.3.2 Storage Systems

There is a great variety of rainwater harvesting system options that differ in the scale, sophistication, and cost. Residential and small-scale storage-type rainwater harvesting systems have six basic components (Texas Water Development Board 2005):

1. Catchment surface,
2. Gutters and downspouts that channel water from the collection surface (e.g., roof),
3. Leaf screens, first-flush diverters, roof-washers, and other devices that remove debris and dust from rainwater before it enters the storage tank,
4. Storage tanks,
5. Delivery system, and
6. Treatment (purification) system for potable water systems.

22.4 Roof Rainwater Harvesting Systems

The environment and characteristics of the catchment surface affect the quality of the harvested water. The roofs of buildings, as impervious surfaces, are attractive rainwater collecting systems, particularly if they have existing gutters that already capture and divert the rainwater. Roof rainwater harvesting (RRWH) systems can be either single residence systems in which the water has a local, in-site use, or multiple unit systems.

Roofs have the obvious disadvantage in that they are continuously exposed to the elements and thus contamination sources. The ideal roofing material for roof rainwater harvesting (RRWH) is stainless steel or enamel coated steel (Singh et al. 2006), which is seldom used because of its great cost. Commonly used roofing materials, such as asphalt shingles, galvanized steel, concrete or clay tiles, and treated or untreated wood shingles may support the growth of mold, algae, and bacteria, and may leach chemicals and metals (Singh et al. 2006).

Other roof characteristics to avoid include (NRMMC-EPHC-NHMRC 2009b):

- structures above the roof that may corrode and provide a resting place for birds,
- overhanging vegetation that can also provide a resting place for birds and shed leaves onto the roof,
- locations with significant exposure to and atmospheric deposition of pollutants (e.g., industrial areas),
- public and vehicle access
- discharge, overflow, and bleed-off from roof-mounted appliances, and
- use of roof-cleansing biocides and other hazardous chemicals.

The quality of harvested water may be improved by keeping the roof and gutters as clean as possible. A roof-washer should also be employed, which captures and discards the first several liters (gallons) of rainwater during a storm before conveying the rest to the storage facility (Texas Water Development Board 2005; Singh et al. 2006). The "first flush" water typically has the highest contaminant concentrations. Contaminant concentrations also vary depending upon rainfall intensity and the number of dry days preceding a rainfall event (Yaziz et al. 1989; Thomas and Greene 1993). If there is a long time gap between rainfall events, then greater concentrations of contaminants may accumulate on the rooftop collecting surface.

Rooftop collected water often is of poor microbial quality and may contain a variety of different pathogens (e.g., Thomas and Greene 1993; Simmons et al. 2001; NRMMC-EPHC-NHMRC 2009b). At a minimum, rooftop collected rainwater needs to be disinfected prior to potable use because of the presence of microbial pathogens. Potable water systems require treatment. The preferred option in the state of Texas is two in-line cartridge filters (5-μm fiber and 3-μm activated charcoal) followed by ultraviolet disinfection. Other treatment and disinfection options are ozone, membrane filtration, and chlorination (Texas Water Development Board 2005).

Water collected by RRWH systems may either be stored in a cistern or used for groundwater recharge. Cisterns should be opaque and designed with screens to prevent algae growth, mosquito breeding and the entrance of vermin. A wide variety of materials can be used for cistern construction including fiberglass, polypropylene, wood, metal (galvanized steel), concrete, ferrocement, and steel drums (Texas Water Development Board 2005). Systems used for potable water supply should be constructed or lined with a food grade material.

The Texas Water Development Board (2005) recommends testing potable water systems for microbiological contamination before use and periodically thereafter. Testing for total and fecal coliform bacteria is relatively inexpensive and affordable for household systems. However, indicators such as fecal coliform measurements may prove to be inadequate for determining microbial risks associated with consumption of water from rainwater catchment systems because a variety of other pathogens may be present (Lye 2002). The dilemma is that the cost for testing for other pathogens of concern (e.g., *Cryptosporium, Giarda*, and viruses) is great (several hundred to thousands of U.S. dollars) and usually unaffordable for individual household systems. The solution, and widely implemented strategy in practice, is the use of proven treatment technologies and forgo testing of all individual systems (other than perhaps for coliform indicators). For example, if the dual filtration and UV disinfection system is demonstrated to be effective in removing or inactivating *Cryptosporium* and *Giarda cysts*, than testing of individual samples for these cysts would not be necessary. It is also important to recognize that in rural areas of many developing countries very simple and inexpensive technologies are needed to reduce the health risks associated with rainwater harvesting for potable use.

Rooftop collection systems may have special risks in areas where active mosquito control is undertaken using aerosol spraying and airborne control systems. The active ingredients and solvents used in some areas (e.g., Fenthion (Baytex) and diesel fuel) are known or suspected carcinogens, which may not be removed by some household treatment systems. Caution is necessary when using the collected water for potable purposes. Local knowledge is a key to assessing the safety of such systems.

22.5 Stormwater Harvesting

Stormwater captured from artificial or natural channels may be directly used or stored in MAR systems. The Australian guidelines for stormwater reuse (NRMMC-EPHC-NHMRC 2009b) provides an excellent, detailed framework for assessing and managing risks associated with the non-potable stormwater harvesting systems in urban and suburban areas. However, the Australian guidelines are more appropriate for developed countries rather than developing countries with much more limited financial and human resources. The public health and environmental risks associated with non-potable stormwater irrigation often have a very low significance and priority relative to other societal challenges. Where irrigation with poorly or untreated wastewater is uncontrolled, management of irrigation with often much cleaner stormwater will not be a priority issue. Nevertheless, the reduction of any health and environmental risks associated with stormwater harvesting should be a long term goal.

The NRMMC-EPHC-NHMRC (2009b) provides guidance addressing all key aspects of the implementation of stormwater harvesting and reuse systems including:

- evaluation of the potential for human exposure,
- water treatment requirements,
- human access and exposure control,
- commissioning and validation monitoring,
- operation and maintenance including staff training and experience,
- catchment surveillance,
- operational monitoring,
- monitoring of incidents and reporting, and
- system auditing.

It is recognized by the NRMMC-EPHC-NHMRC (2009b) that the monitoring program should be appropriate for the system-specific risks, which depend upon water quality, uses of the waters, and the exposed population. An important recommendation of the NRMMC-EPHC-NHMRC (2009b) is that there should be a 72-h (minimum) buffer time between collection of stormwater and release for irrigation. The buffer allows for the averaging (equalization) of spikes in water quality (e.g., first flush) and provides time to respond to a spill or unexpected water quality event. The storage can be provided by ponds or constructed wetlands.

Stormwater will need to meet water quality standards for its irrigation use (e.g., public access irrigation). Turbidity may need to be reduced through settlement in ponds and wetlands or by the use of filtration systems. High suspended solids concentration can foul irrigation systems, such as by clogging the nozzles of spray irrigation systems. Disinfection requirements are less than those for wastewater because of the lower pathogen concentrations of stormwater. Water quality issues associated with aquifer recharge using stormwater are discussed in Sect. 23.5.

Another key issue for the development and operation of stormwater recycling systems is surveillance of the catchment areas for activities that may add contaminants to the harvested water. An assessment of the catchment should be performed as part of the initial design phase (desktop study) for a project. On-going surveillance should be performed to identify activities that could introduce contaminants (i.e., add risks) beyond those identified during system design.

22.6 Legal Issues

Depending upon where one lives, the rain that falls on one's roof and driveway may not belong to you. For example, surface water is governed in the western United States under the prior-appropriation doctrine, in which senior water rights holders have priority over surface-water flows. The prior appropriation doctrine is often expressed by the aphorism "first in time, first in right". Under law, the states have an obligation to protect all vested water rights. The prior appropriation doctrine has been interpreted in some jurisdictions, such as the states of Colorado and Utah, to prohibit the harvesting of rainwater in most river drainages as it may ultimately result in a reduction in stream flow and thus impact downstream surface water rights. Other states treat surface water rights differently. For example, in Texas rainwater is owned by the property owner until it enters a water course.

The Colorado Senate passed Bill 09-080 in 2009, which allows for limited rainwater collection in rural residences. The water may only be collected from roof tops in areas that are not connected to the domestic water systems. Thus, within urban and suburban areas served by municipal water systems, rainwater harvesting is still illegal. The collected water must be used for ordinary household purposes; fire protection; watering of animals and livestock; and irrigation of not more than 4,047 m^2 (1 acre) of gardens and lawns. A person wanting to capture rooftop precipitation who meets the qualifications in the bill must submit an application and pay a fee to the state engineer. Owners of certain wells are allowed to collect rooftop precipitation under the same use limitations as contained in their well permits.

Although an outright ban on even small-scale rainwater harvesting is extreme, it is important to realize that large-scale adoption of the technologies can have adverse impacts. Widespread interruption of rainfall in upland watersheds potentially may have severe impacts on downstream areas and habitats (Narasimham 2006).

At the other end of the spectrum, the City of Tucson, Arizona passed Ordinance 10597 in 2008 that requires that the site plans for any new business, corporate and commercial structures submitted after June 1, 2010, have to show that half of the estimated yearly landscape water budget will be provided with rainwater harvested on site. The state of Texas has exempted rainwater harvesting system equipment from state sales tax and allows local governments to exempt systems from ad valoreum (property) taxes (Texas Water Development Board 2005).

Rainwater harvesting systems may also fall under the jurisdiction of a variety of governmental agencies that have oversight over building and construction, land development, water use, and environmental protection.

22.7 Large-Scale Rainwater and Stormwater Harvesting

The primary implementation of rainwater and stormwater harvesting in most developed countries has been shallow aquifer recharge as part of stormwater management systems. The runoff from pervious areas is conveyed to a wide variety of systems, such as infiltration or retention basins, swales, soakaways, French drains, and dry wells (Sect. 22.5). The primary function of the systems is stormwater disposal rather than water supply, although increased aquifer recharge is recognized as an important incidental benefit.

Rainwater harvesting has been embraced in some arid and semiarid lands as a critical water supply tool. As discussed in the following sections on India and China, rainwater harvesting is the only remaining economical source of new water supplies, as local aquifers are in a serious overdraft conditions.

22.7.1 Rainwater Harvesting in India

Rainwater harvesting has been continuously practiced in India since at least the third millennium BC. The history of rainwater harvesting was reviewed in great detail in the book *Dying Wisdom* edited by Agarwal and Narain (1997). India has high evapotranspiration rates and many regions face seasonal water scarcity. The people of India have long recognized that human society could not grow without extending the bounties of the monsoon water from the wet months to the dry months (Agarwal and Narain 1997). As documented by Agarwal and Narain (1997), a wide variety of rainwater harvesting technologies was utilized in India mainly to capture the seasonal monsoon rains. Most common were dams and storage reservoirs constructed on ephemeral streams (locally called tanks and a variety of other local names). Various types of subsurface storage vessels (cisterns) with microcatchments are also used, particularly in drier semi-arid regions.

During pre-colonial times, Indian villages were highly resilient, socially and economically cohesive, and self-sufficient communities that necessarily managed their own natural resources, including water (Agarwal and Narain 1997). Community government structures and local rules and regulations were developed to construct and manage the systems. Typically, beneficiaries of the systems were required to provide labor for their maintenance. If the labor could not be personally provided, then a party could provide or pay for a substitute. Sanctions were in place for not contributing to the system management and for the misuse and damage to it (Agarwal and Narain 1997).

The strong local authority to maintain the water harvesting systems was undermined by the centralized government of the British colonial period (Agarwal and Narain 1997). After independence, farmer managed systems were dismissed as being "primitive and inefficient and, therefore, useless" (Singh and Pandem 1997, p. 49). The breakdown in community maintenance structures resulted in the degradation and destruction of tradition water harvesting systems. As a villager quoted by Mohnot and Kolarkar (1997, p. 149) noted "*What was once everybody's business has become nobody's business.*" The degradation of water harvesting systems included deforestation, contamination, and development of catchment areas and reservoirs, the abandonment of canal and other water conveyances, and the dumping of garbage in water storage structures.

Rainwater harvesting was rediscovered in the 1960s in response to declining groundwater availability caused by the rapid expansion of irrigation pumping. The increase in aquifer recharge and rainwater harvesting in India over the past three to four decades is of such an extent that it has been called a "groundwater recharge movement", which has behind it both secular and spiritual proponents (Sakthivadivel 2007). A remarkable feature of the groundwater recharge movement in India is that it is largely a "bottom up" movement initiated by local communities, rather than being a "top-down" movement initiated and mandated by the central government. The movement is a successful example of a community-based effort to manage a common property resource (Sakthivadivel 2007).

The primary recharge method is surface spreading using percolation tanks (ponds) and check dams constructed across or near streams and drainage channels in order to impound runoff and retain it for a longer time to increase the opportunity time for recharge (Sakthivadivel 2007). In India there are more than 1.5 million tanks, ponds, and earthen embankments in 660,000 villages (Pandey et al. 2003). The tanks act both as surface-water storage reservoirs and to recharge underlying aquifers. India receives on the average 1170 mm (46.1 in) of rainfall each year. However, 80% of the rainfall is confined to the monsoon months of June to September and is restricted to 30–60 days during the season. The rainfall also occurs in intense short-duration events of an hour or less, which results in most of the water running off (Radhakrishna 2004). There is thus a huge volume of water that could be potentially retained through the revitalization and expansion of rainwater harvesting. Agarwal and Narain (1997) claimed that rainwater harvesting structures can provide every house in India, even today, with all the water it needs.

In the semiarid hardrock region of western India, runoff captured during a few rainy days of the monsoon season percolates to the water table where it has a residence time that is often sufficient to extend to the next rainy season (Limaye 2002). Percolation tanks feed downstream wells, up to 2–3 km (1.2–1.9 miles) below the tank, with water that in some areas has better quality than the native groundwater. The overall recharge efficiency of tanks is 40–70%, with efficiency decreasing over time because of siltation (Limaye 2002).

Narasimham (2006) reported that rainwater harvesting has been enthusiastically embraced in India, with some urban areas having mandates that every house must have an overhead cistern or other rainwater harvesting device. It was cautioned that in some states, tens of thousands already exist, which suggests that rainwater harvesting may have already attained its limits. The urban roof water harvesting systems may also be more expensive and less reliable than regional water supply schemes (Kumar 2004) (Sect. 22.4).

Traditional rainwater harvesting can have great value for providing water security and could improve the quality of life in rural communities of desert areas of India. Grey and Sharma (2006) described use of cisterns (locally called "taankas") fed by microcatchments to capture monsoonal rains in the Thar Desert of Rajasthan State. In years without local rainfall, the taankas are filled with water from other sources in the period after the monsoon season when water is more accessible and cheaper to acquire. Water from other sources, such as ponds, is also utilized when available. Usually taankas are used only when other supplies are not available (Mohnot and Kolarkar 1997). Taankas are thus not the sole water source, but an important water supply element for families and communities to achieve water security.

Western India areas hit hardest by groundwater depletion, spontaneously created with the support of NGOs a massive well-recharge movement based on the principle "water on your roof stays on your roof; water on your fields stays on your fields; and water on your village stays in your village" (Shah et al. 2000). The recharge program includes modification of some 300,000 wells to divert rainwater into them, and the construction of thousands of ponds, check dams and other rainwater harvesting and recharge structures on the self-help principle to keep rainwater from gushing into the Arabian Sea.

Traditional harvesting technologies require active maintenance. Catchments areas need to be kept clean in order to provide unpolluted water. Silting reduces the infiltration rate and volume of tanks, therefore regular cleaning (desilting) is necessary. In tanks constructed into hardrock areas, vertical boreholes may be blasted to create fractures that may increase recharge. Conflicts have occurred between groundwater users taking surface water for groundwater recharge and downstream surface water users (Sakthivadivel 2007; Limaye 2010). Construction of tanks and other similar reservoirs may also result in the submergence of cultivable lands.

Riverbank filtration (RBF) is also employed with percolation tanks. Drinking water wells are installed in or near the tank bed in order to take advantage of the natural filtration (Sakthivadivel 2007). In some areas there is a problem with uncontrolled RBF whereby farmers install wells near surface reservoirs owned or maintained by others.

22.7.2 Rainwater Harvesting in the Loess Plateau, China

Large parts of China face, especially the semiarid loess plateau in the northwestern part of the county, serious water supply problems, caused by a combination of minimal water resources, high population pressure, and ecosystem vulnerability, which are constraining the sustainability of agriculture and economic development (Li et al. 2002). Rainfall in the region is governed by the prevailing monsoon climate in which approximately 60% of the rain falls between July and September. Not only is total rainfall often inadequate, it also has a high inter-seasonal and annual variation (Li et al. 2000).

Gansu provincial government recognized that rainwater harvesting could be a solution to the water supply problem in rural communities. Rainwater harvesting agriculture (RHA) was developed by scientists as an integrated synergistic system for sustained water management. RHA consists of three main components; (1) rainwater harvesting (2) water-saving irrigation, and (3) highly effective crop production (Li et al. 2000). Rainwater harvesting is the core component of RHA.

The provincial government launched the 1-2-1 rainwater harvesting program in 1995 that assisted each rural household in building one (1) concrete catchment surface with an area of about 100 m^2 (1076 ft^2), two (2) concrete storage tanks and an irrigation system for one (1) mu (1/15 ha) [0.16 acres] of cropland for the production of high market value crops. In this program, the government provided the building materials and technical services, and the farmers contributed the labor (Li et al. 2000). Through 2000, the program has helped farmers construct 2.18 million storage tanks, which supply 1.97 million people in Gansu province (Li et al. 2002). Within the whole of China rainwater-harvesting has solved the drinking water problem for about 23.80 million rural residents as of 2000 (Li et al. 2002). The storage tank (i.e., well or cistern) can have a variety of shapes and may also have a traditional earthen construction.

Other types of catchments used include rooftops, courtyards, paved and unpaved roads, and cleared and smoothed hillsides (Li 2003). Various types of barrier materials (e.g., concrete, soil cement, asphalt-fabric membranes and plastic film), mechanical treatments (e.g., smooth and compacting), and chemical treatments have been used to waterproof the surfaces and increase precipitation runoff (Li et al. 2000; Li 2003).

Because of the limited water from the rainwater harvesting system, supplemental irrigation must be used efficiently (Li 2003). Therefore, farmers have been adjusting agricultural production to crops, such as vegetables and fruits, that offer higher economic returns under supplemental irrigation conditions. They are adopting high-efficient irrigation techniques (e.g., drip and micro-sprinkler; Li 2003). Experimental studies indicate the drip irrigation under plastic (DIUP) had the greatest irrigation efficiency (Huang et al. 2002).

In addition to meeting drinking water needs and improving crop yields and incomes, rainwater harvesting has also significantly contributed to ecological and environmental conservation (Li et al. 2002; Li 2003). Rainwater harvesting contributed to soil and water conservation by capturing and storing runoff and reducing erosion. Increased productivity in the irrigation fields also reduced the pressure to convert steeply sloped and highly erodible forestlands and grasslands into cropland (Li et al. 2000, 2002).

The China experience demonstrates that rainwater harvesting can be successfully implemented on a large-scale. RHA is well suited for northwestern China in that the small scale of the individual systems makes it easy to construct and operate at the small farm level, which is the basic unit of agricultural production in China today (Li et al. 2000). Li et al. (2000) pointed out that success and spread of RHA will depend upon:

- whether research scientists and extension agents can provide long-term training and technical services to local farmers, increasing their ability to operate and maintain the systems,
- the development of a wide variety of rainwater harvesting technologies that are ecologically sound and economically viable for farmers, and
- the design and implementation of policy instruments and strategies to facilitate the spread of RHA.

22.8 Fog Harvesting

22.8.1 Introduction

Fog is essentially a large cloud that is in contact with land surface. Fog normally forms when the relative humidity is close to 100% and atmospheric cooling occurs. The amount of water that air can hold is directly related to temperature; cooling conditions can result in the condensation of water vapor into the fine water droplets that constitute fog and clouds. It has long been recognized that fog can be a natural source of water at land surface. Fog naturally collects on the surfaces of plants and other objects, and drips to the ground. In some arid coastal regions (e.g., northern Chile, southern Peru, the Arabian Gulf coast of the United Arab Emirates, and many others), fog may be the main source of freshwater available to local ecosystems.

The earliest reference to fog collection is the fountain or rain trees, which were noted on Hierro Island in the Canary Islands since their discovery at the beginning of the 15th century (Gioda et al. 1995). The Garoé tree, also known as the fountain tree, weeping tree, and magic rain tree, has been a legend since its discovery by Jean de Béthencourt between 1402 and 1405. The tree had what was considered to be the amazing ability to catch fog droplets on its leaves and to convert these droplets back into water, which was available to plants and man. The fountain tree was likely a laurel species, *Ocotea foetens* that was located in a favorable position for fog collection (Gioda et al. 1995). Fountain trees are still used in part of Hierro Island to supply sheep and goat herds with water.

There have been numerous studies since the first recording of fog collection of its importance to some local ecosystems and on the artificial harvesting of fog for human purposes. Research has accelerated starting in the 1980s to assess the potential for harvesting fog as a water source in arid and semiarid lands. Fog harvesting has the advantage of being a renewable source of water that does not depend on precipitation, groundwater, surface water, and water from the oceans (Schemenauer and Cereceda 1997).

Fog harvesting involves the collection of the droplets of water that constitute fog. Collection of fog is a simple impaction process. An object (collector) is placed in the path of the droplets. Some droplets strike and adhere to the surface of the collector, which is typically a plastic mesh, while others flow around the collector. The collected water drips from the mesh and is collected by gravity in a trough (Schemenauer and Cereceda 1997). The water collected from each collector is then stored in one or more tanks (Fig. 22.9). The mesh design allows the wind-carried fog droplets to pass into and through the collector. If the collector were a solid barrier, then the fog-bearing wind will pass around the collector (Schemenauer and Cereceda 1994).

The collected water is typically of potable quality. However, after a dry period, the initially collected water may have a relatively high salinity (i.e., levels not suitable for potable water use), because of windblown salt deposition on the

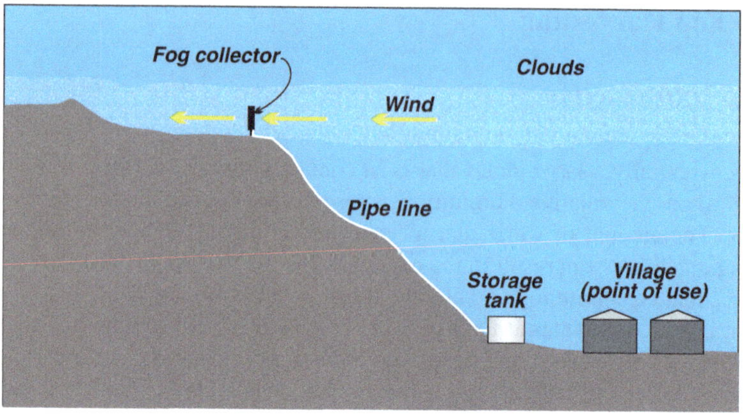

Fig. 22.9 Conceptual diagram of a fog harvesting system. Collectors are located at topographic high points were wind blown clouds reach land surface as fog. The collected water is piped to a storage tank and its point of use

collector (Olivier 2002). The collector may need to be either washed down with freshwater or some mechanism would have to be introduced to allow the first water collected during a wet period to be discarded (Olivier 2002). Factors that affect the system capacity include the number of collectors, their size and efficiency, fog frequency, fog liquid water content, and wind speed (Schemenauer and Cereceda 1997). Advection fogs, which are clouds that formed over sea and are pushed onto mountains, often have the advantage of a relatively high wind speed and uniform wind direction, thereby producing a higher flux rate of water.

A most notable early investigation of fog harvesting is the Comanchaca project, that was performed at the El Tofo site in Northern Chile from 1987 to 1989. Fifty 48 m^2 (517 ft^2) collectors produced an average of 7,200 L/d (1,901 gpd) during three drought years (Cereceda et al. 1992; Schemenauer and Cereceda 1997). The system was expanded to 75 collectors with a capacity of 11,000 L/d (2,941 gpd). Commercially available polypropylene mesh shade screen was used as collector material. The El Tofo site has become the prototype for community scale fog harvesting systems. However, ten years after its installation, the El Tofo system was in a total state of disrepair, despite its success at bringing a steady supply of clean water to the community of Chungungo (Dale 2003). The El Tofo system was regarded by local officials as being an unreliable, irregular, and insufficient source of water, and the officials instead lobbied for pipe line from a river 20 km (12.4 mi) away. Alternative water projects often suffer from a lack of local prestige and communities may have an ambition to move to what they perceive as a higher technological plateau (Dale 2003).

The success of the El Tofo system lead to a tripling of the local population and the original number of collectors could not reliably meet the increased demand (Dale 2003). The option to install additional collectors was not pursued. Another problem is that the villagers were not involved in any decisions involving the design and development of their water supply system, and were subsequently unprepared to commit to the long-term operation and maintenance of the system. The El Tofo project, originally developed as a research project, had multiple stakeholders, but no single authority looking over the whole system (Dale 2003). A key lesson of the El Tofo experience is that social issues can be as important as technical issues in the long-term success of fog-harvesting and other alternative water supply projects (Dale 2003). To be successful, all projects must involve the people who will benefit from the water provided (United Nations Environment Programme 1998; Schemenauer et al. 2004a, b). Irrespective of its outcome, the El Tofo project was unquestionably a great success in that it demonstrated the feasibility of fog-harvesting to provide water to a community and it spawned additional projects elsewhere in Chile and in other countries.

Costs depend upon system size, local labor and materials, and the ease of site access. Commercially available shade cloth was used for the Comanchaca project collectors (Schemenauer and Cereceda 1997). Additional costs include pipelines, storage tanks, and a small chlorination facility. Schemenauer and Cereceda (1997) estimate that the cost of water delivered to the village was approximately US $2 per m^3 ($3.78 per 1,000 gal).

Schemenauer and Cereceda (1994) proposed a standard fog collector (SFC) device to measure the deposition of fog on a vertical surface. The SFC device allows for the comparison of fog collection potential between different locations. The SFC consists of one meter (3.28 ft) square double layer of commercially available 35% shade cover polypropylene mesh mounted on a vertical frame. The SFC is set a uniform height of 2 m (6.56 ft) above land surface, with the mesh perpendicular to the prevailing wind direction. The water is collected by gravity using a plastic trough under the mesh. Modifications to the basic SFC including mounting it on a swivel to keep the collector normal to the wind and automating the measurement of collection rates, such as by using a tipping bucket (Schemenauer and Cereceda 1994). The SFC will also record precipitation onto the collector. A conventional rain gauge would allow for the differentiation between collected fog and precipitation (Schemenauer and Cereceda 1994).

Juvik and Nullet (1995) noted that the SFC has several short comings, particularly (1) a fixed orientation (2) no separation (quantification) of vertical and horizontal precipitation, and (3) a variable catchment surface caused by the density of the shade cloth depending on the manufacturer and installation. Juvik and Nullet (1995) proposed using a cylindrical louvered collector as an alternative to the SFC. From an academic perspective, the Juvik and Nullet collector is superior for accurate quantification of fog collection. Inasmuch as the SFC design is essentially a small-scale version of an operational fog collector, it has the practical advantage of providing data on the likely actual performance of operational systems.

Fog harvesting will be obviously restricted to areas that experience frequent fog events. In many areas of the world that experience water scarcity and frequent fogs, studies have already been performed as to the feasibility of fog harvesting. As fog is clouds in contact with the land, fogs often occur at relatively high altitudes (where the land surface rises to cloud level) and in coastal areas where is a high humidity from the sea.

22.8.2 Saudi Arabia

Fog harvesting was investigated in the Asir Highlands, located in the southwestern part of the Kingdom of Saudi Arabia, by Al-hassan (2009). Settlements in the Asir Highlands are located above an altitude of 2,000 m (6,560 ft). Three SFC were installed and monitored from April 2006 to April 2007. The average annual daily collection rate was 6.2 L/m^2 (0.15 gal/ft^2) at the two highest altitude sites (3,133 and 3,177 m above sea level) [10,276 and 10,421 ft] and 3.3 L/m^2 (0.08 gal/ft^2) at the lower altitude (2,400 m) [7,872 ft] site. The greatest daily water collection rate was 24 L/m^2 (0.6 gal/ft^2).

Monthly water production data indicate that the winter months (December, January, and February) are the most productive, averaging 11.5 L/m^2/day [0.3 gal/ft^2/day] at the Al-Sahab Park site (the most productive site). Production drops to an average of 2.25 L/m^2/day [0.06 gal/ft^2/day] during June and July. The collected water was very pure with low concentrations of dissolved solids and was found to be suitable for potable consumption under World Health Organization and Saudi Arabian standards (Al-hassan 2009).

22.8.3 Western South Africa

Olivier (2002, 2004) documented feasibility studies and the operational testing of a fog harvesting system in western South Africa. Western South Africa experiences both arid conditions and poor-quality groundwater, so there is a clear interest in developing alternative potable water supplies. Initial feasibility studies indicated that fog harvesting is feasible in western South Africa and the mission station of Lepelfontein, which has a population of about 200 people. A fog harvesting system was constructed with an area of 70 m^2 (753 ft^2). The collector was of standard design, constructed of 30% shade cloth mounted on a rigid frame orientated normal to the prevailing wind direction. Over the period September 1991 through August 2001, the system produced an average of 323 L/d (85.3 gal/day), which is equivalent to 4.6 L/m^2/day (0.11 gal/ft^2/day). Approximately 88% of the collected water was fog and 12% rainfall, as measured at an on-site automated weather station. The collected water was suitable for human consumption and was a vast improvement over the water quality previously locally available. The Lepelfontein experimental system is considered to be a success that could be expanded to other water scare communities in Western South Africa.

22.8.4 Yemen

The Yemen Arab Republic experiences a dry winter in which local communities experience water shortages. Fog locally develops during the winter dry season and this fog is a potential water supply. A feasibility study, involving the installation of ten SFC at sites in the Sadaa Governorate in the northern part of the country, was performed from January 14, 2004 through March 31, 2004 (Osses et al. 2004). The best sites produced an average of 4.5 L/m^2/d (0.1 gal/ft^2/day). The best producing sites were located at the highest altitudes. Osses et al. (2004) noted the fog harvesting could be combined with rainwater harvesting using cisterns. The rainy season occurs from April and August, so the end of the main fog-harvesting period is close in time to the start of the rainy season. Cisterns are usually empty during the fog period and could be used to store harvested water.

A two phase fog-harvesting investigation was carried out in the Hajja Governorate (Schemenauer et al. 2004a). The first phase consisted of the installation of 26 SFCs during the 2003 winter in order to locate the site(s) with the greatest potential for fog-collection (greatest yields). Reported yields ranged from 0.02 to 4.54 L/m^2/d (0.0005 to 0.1 gal/ft^2/day). The second phase consisted of the construction of an operational system in the Mabijan district. The operational system was constructed with 25 large fog collectors (LFCs) each with a 40 m^2 (431 ft^2) mesh collecting surface, for a total areas of 1,000 m^2 (10,764 ft^2). The average yield during the winter of 2004 was 180 L (47.5 gal) per day per collector (4.5 L/m^2/d) [0.1 gal/ft^2/day]. Operational problems that arose during operational testing were related to anchoring the collectors and damage to collectors from extremely strong winds, which may have been caused by increased wind loads from the use of a 45–50% shade coefficient mesh (Schemenauer 2004a, b).

22.8.5 Valencia, Spain

The Valencia Region of western Spain is a semiarid region with an average annual precipitation of 300–500 mm (11.8–19.7 in). Estrela et al. (2008) investigated fog harvesting at four stations in the mountains near the coast. Cylindrical string fog collectors were used, which were 26 cm (10.2 in) in diameter, 46 cm (18.1 in) in height, and strung with 5 rows of 0.8 mm (0.03 in) nylon line. The collector contained 1000 vertical rows of closely spaced strings and had an effective collection surface if 0.12 m^2 (1.3 ft^2) (diameter times height). Cylindrical collectors were used instead of SFCs in order to avoid the effects of variation in wind direction on collector efficiency.

Corrections were made to the collected water volumes for the effects of rainfall. The vertical rainfall was measured using a standard rain gauge. The collected water may also contain some horizontal rain, which was quantified as the product of vertical rainfall and an empirical function of wind velocity. The function of horizontal to vertical rainfall was determined from rainfall and collector data collected on fog-free days (Estrela et al. 2008).

Estrela et al. (2008) reported data for one year of operation of the collectors. The two collectors at the southern site (Mt. Monduver and Mt. Montoó) had the best performance with average annual fog collection rates of 7.3 and 7.0 L/m^2/d (0.18 to 0.17 gal/ft^2/day) respectively. The two northern sites (Mt. Bartolo and Mt.

Peñaroya) had collection rates of 2.0 and 2.9 L/m^2/d (0.05 and 0.07 gal/ft^2/day), respectively. Except for the summer, fog episodes that delivered sizeable volumes of water were coupled to rainfall.

22.8.6 Southern Peru

Southern Peru and northern Chile is one of the driest regions in the world, but receives considerable amounts of advective fog of marine origin. Semenzati (1996) documented the initial work on a European Commission (EC) project in coastal Peru. Fog collection was investigated as a tool for the recuperation of large areas of coastal hills of Peru, which is a fragile ecosystem that has been undergoing rapid degradation and desertification due mainly to overgrazing. SFCs were installed at ten locations. The main fog season is usually from May to September. The greatest fog collection rates occurred in September (during the 1995 study period) and averaged 13.9 L/m^2/d (0.34 gal/ft^2/day). The greatest maximum daily rate reported was 107.35 L/m^2/d (2.63 gal/ft^2/day) Fog harvesting systems are currently operational in the village of Bellavista near Lima and in some Peruvian coastal communities (Fields 2009). Some of the water obtained from the collectors is used to irrigate freshly planted trees with the anticipation that once they grow large enough, they will act as their own fog collectors and become self-sustaining and recharge the groundwater (Fields 2009).

22.8.7 Other Areas

Studies have demonstrated fog-harvesting to be a viable water source in Oman (Abdul-Wahab and Lea 2008), the Canary Islands (Marzol 2002), and the Dominican Republic (Schemenauer et al. 2004b).

22.8.7.1 Lessons Learned

A very important feature for the successful implementation of fog harvesting is the establishment of a routine quality control program, which includes the following elements (United Nations Environment Programme 1998):

- inspection of cable tension and fasteners,
- inspection of horizontal mesh tension,
- maintenance of mesh nets, including repair of tears and cleaning off of dust and algal growth,
- maintenance of collection drains, pipelines, and storage tanks, and
- monitoring of dissolved chlorine if water is used for potable water supply.

22.8.8 Conclusions

Fog harvesting has been demonstrated to be a feasible method of obtaining water in areas that experience frequent fogs. Collection systems have modest yields per square meter (square foot) of collector, but because the collectors are relatively inexpensive to build, large collectors can be economically constructed. Nevertheless, fog harvesting is still an expensive method to obtain water relative to most conventional freshwater sources, but may be significantly less expensive than other alternative water supplies (e.g., seawater desalination).

Fog harvesting has some definite limitations. First is that system success depends on the existence of a favorable combination of geographical and meteorological conditions. Water production will depend on the geographic and temporal availability of fog. The occurrence of fog has a definite seasonal variation, which results in a pronounced seasonality variation in water production. Average monthly production may vary by a factor of five or more between the high and low production months (e.g., Al-hassan 2009). Where the fog season is short, larger collection and storage systems may be needed to collect and store larger volumes of water when conditions are favorable in order to meet year-round water demands.

The technology is very sensitive to changes in local climatic conditions that affect the water content and frequency of fogs. A backup water supply is therefore recommended for use during periods of unfavorable climatic conditions (United Nations Environment Programme 1998).

22.9 Lithic Mulch Agriculture

Lithic mulch agriculture (LMA) is an agricultural strategy in which stones (gravel, rock, pebbles, and ash) are used as mulch for improving crop yields. This method has been used for over 1,000 years (Lightfoot 1996). Materials are raked or piled into mounds, ridges, terraces, or extensive surface layers. Although not strictly a rainwater harvesting technology, LMA is a small-scale technology that functions to optimize the use of the available water resources. LMA was often used in conjunction with rainwater harvesting. LMA increases soil moisture, controls and reduces surface erosion, increases average soil temperature and moderates diurnal extremes (Lamb and Chapman 1943; Lightfoot 1994, 1996). Soil moisture is increased by reducing the evaporation rate and reducing soil crust formation, which increases infiltration capacity.

Stones act as solar collectors, absorbing solar energy and heat during the day, and radiating much of the absorbed heat into the soil throughout the day and night. Change in reflectivity alters the heat balance and microclimate of the soil and air near the gravel. Such variation in soil materials may have profound consequences by increasing the incidence and amount of dew, which can have important consequences for vegetation and animals (Walton 1969). However, some studies dispute the role of stone piles in dew collection (e.g., Evanari et al. 1971).

Lightfoot (1994, 1996) reviewed the history of LMA, which has been very widely practiced (e.g., Middle East, Mediterranean region, Atacama Desert of-Peru, Central China, southwestern United States, New Zealand, Canary Islands). He had difficulty in evaluating the history of LMA in that piles of stones could have multiple potential uses. For example, piles of stones could be LMA structures or alternatively may have been convenient disposal points for stones from field clearing, with the crops grown between piles (Lightfoot 1996). Dual-functions can also not be discounted.

LMA has been practiced primarily in regions that experience a deficiency of soil moisture during the growing season, which suggests that it has been primarily used for evaporation control. LMA failed to spread geographically beyond a few locations where it was independently developed, was rarely used for any significant length of time, and generally contributed little to the overall food supply (Lightfoot 1994, 1996). Excavation and spreading of lithic mulch involved a considerable investment in time and labor, which appears from the historic record to have been employed as occasional drought-evasive strategy, and as means to allow for agriculture on marginal lands, and in cooler upland areas with a short growing season (Lightfoot 1994, 1996; Lightfoot and Eddy1995). For example, detailed cultural-ecological studies of LMA demonstrates a connection between drought and adoption of the technique amongst the ancient Sinagua people of New Mexico, USA (Lightfoot 1994). Experimentation with and the spread of LMA would be abbreviated when drought conditions ceased and less labor intensive strategies produced adequate crop yields or other traditional food-stress coping mechanisms proved sufficient to survive a drought (Lightfoot 1994).

There are some negative impacts of LMA. The technique can result in a reduction in the recycling of essential nutrients and the organic content of the soil, which may become a detriment to LMA over a long period of time. Also, it inhibits the application of fertilizer (Lightfoot 1994, 1996). LMA could be used today in arid and semiarid regions, but its impact on large-scale food production would be minimal compared to irrigated agriculture (Lightfoot 1994, 1996).

22.10 Rainwater Harvesting Economics

Rainwater harvesting should be evaluated in the same manner as other water supply options, with full consideration given to its capacity, reliability, and costs, particularly relative to other options. Rainwater harvesting includes a wide variety of technologies and system scales. Household rainwater harvesting, particularly in developed countries, may be more of a life style rather economic matter. For example, a 208 L (55-gallon) barrel of harvested rainwater in the United States is worth about 15 to 25 US cents using common marginal utility water rates, which hardly justifies its cost and effort. Even though less expensive water supply options may be available, rainwater harvesting is increasingly being employed because it is considered to be more environmentally sound and part of a low-impact "green" life style. Economics is just one of many factors involve in general personal decision-making.

Kumar (2004) provided a value evaluation of the viability of roof rainwater harvesting in India. As Kumar (2004) and others have reported, India has a great and long tradition of water harvesting, and in some regions it is part of the local culture and deeply rooted in the socio-economic fabric. With respect to roof water harvesting systems (RWHS) the key technical issues are:

- How much water can be harvested from different types of housing stocks?
- At what scale can the technology be adapted in rural and urban environments?
- Is the technology economically feasible?

The technical viability of RWHS depends on the variation in rainfall, which is greatest in drier climates. In simpler terms, will there be adequate rainfall to store enough water to meet demands. Another important issue is the spacing of rainfall events. If most of the rainfall occurs during intense storms that occur close together, than very large storage capacities will be needed to store enough water for dry periods.

Kumar (2004) performed a supply-side analysis for the city of Ahmedabad, which has a semi-arid, tropical climate (average rainfall is 688.2 mm [27.1 in]). The results of the evaluation indicate that RWHS would not be adequate for meeting even basic survival needs. The greatest benefits of RWHS would be realized by higher and middle income groups with single-story homes with relatively large roof areas and storage spaces. The least benefits would be achieved in urban multi-story buildings and slums in which most of the structures do not have adequate size roofs. Kumar (2004) concluded that RWHS cannot ensure domestic water security and can only supplement other water supply systems. It was also noted that the cost of RWHS may be greater than regional water supply schemes and may be comparable to desalination in certain situations. In urban areas, RWHS lack a significant economy scale, particularly compared with regional water supply schemes in areas with concentrated populations. RWHS can be economically viable as a supplementary source of water in rural areas with dispersed populations and in hilly areas with high rainfalls (Kumar (2004).

Kumar (2004) determined that government and non-government organization (NGO) programs that push RWHS can actually do more harm than good to low income people who would receive less benefits from the technology, and that RWHS could divert funds from more economical expansions of water utilities that could provide greater benefits to the poor.

Srinivasan et al. (2010) compared the economics of desalination, efficiency improvement, and rainwater harvesting in Chennai, southern India, as elements in the development of a sustainable water supply. Rainwater harvesting was considered only as method for aquifer recharge, rather than storage with cisterns. By increasing groundwater recharge, rainwater harvesting allows more groundwater to be available for consumers using private wells. Rainwater harvesting alone would result in improved groundwater availability, but would generate no new water that could be used in the piped system. A combination policy was proposed consisting of steep tariff increases, efficiency improvements (reducing pipe leakage loses), and rainwater harvesting, which meet the criteria for efficiency, reliability, equity, financial responsibility, and revenue generation.

References

Abdul-Wahab, S. A., & Lea, V. (2008). Reviewing fog water collection worldwide and in Oman. *International Journal of Environmental Studies, 65*, 485–498.

Agarwal, A., & Narain, S. (1997). *Dying wisdom. Rise, fall and potential of India's traditional water harvesting systems* (p. 404). New Delhi: India Centre for Science and Environment.

Al-hassan, G. A. (2009). Fog water collection evaluation in Asir region-Saudi Arabia. *Water Resources Management, 23*, 2805–2813.

Cereceda, P., Schemenaur, R. S. & Suit, M. (1992). An alternative water supply for Chilean coast desert villages. *Water Resources Development, 8*, 53–59

Critchley, W., & Siegert, K. (1991). Manual for the design and construction of water harvesting schemes for plant production. Rome: Food and Agriculture Organization of the United Nations.

Dale, S. (2003). Collecting fog on El Tofo: International Development Research Centre. Accessed March 28, 2010: http://www.idrc.as/en/ev-30617-201-1-DO_TOPIC.html

Downey, N. (2010). *Harvest the Rain, how to enrich your life by seeing every storm as a resource* (p. 228). Santa Fe: Sunstone Press.

Estrela, M. J., Valiente, J. A., Correll, D., & Millán, M. M. (2008). Fog collection in the western Mediterranean basin (Valencia region, Spain). *Atmospheric Research, 87*, 324–337.

Evanari, M., Shanan, L., & Tadmor, N. (1971). *The Negev. The challenge of a desert* (p. 345). Cambridge: Harvard University Press.

Fields, H. (2009). Fog collectors bring water to parched villages: National Geographic, July 2009: http://news.nationalgeographic.com/news/2009/07/090709-fog-catchers-peru-water-missions.html.

Giodia, A., Hernández, Z., Gonzáles, E., & Espejo, R. (1995). Fountain trees in the Canary Islands: Legend and reality. *Advances in Horticultural Sciences, 9*, 112–118.

Grey, N. O., & Sharma, O. P. (2006). Traditional rainwater harvesting technologies: Key to drinking water security for desert communities in arid regions of India. In *Recharge Systems for Protecting and Enhancing Groundwater Resources. Proceedings of the 5th International Symposium on Management of Aquifer Recharge*, Berlin, Germany, 11–16 June 2005 (pp. 196–203). Paris: UNESCO.

Huang, Z, B., Shan, L., Wu, P. T., Gao, J. E., Yang, X. M., & Ben-hur, M. (2002, May 26–31). *Artificial rainwater harvesting system and the using for agriculture on Loess Plateau of China: Proceedings 12th International Soil Conservation Organization Conference* (pp. 463–468), Beijing.

Juvik, J. O., & Nullet, D. (1995). Comments on "a proposed standard fog collector for use in high-elevation regions". *Journal of Applied Meteorology, 34*, 2108–2110.

Kinkade-Levario, H. (2007). *Design for water: Rainwater harvesting, stormwater catchment, and alternate water reuse* (p. 235). Gabriola Island: New Society Publishers.

Kumar, M. D. (2004). Roof water harvesting for domestic water security: Who gains and who losses? *Water International, 29*(1), 43–53.

Lamb, J., Jr., & Chapman, J. E. (1943). Effect of surface stones on erosion, evaporation, soil temperature, and soil moisture. *Agronomy Journal, 35*, 567–578.

Lancaster, B. (2006). *Rainwater harvesting for drylands and beyond: Guiding principles to welcome rain into your life and landscape* (Vol. 1, p. 183). Tucson: Rainsource Press.
Lancaster, B. (2008). *Rainwater harvesting for drylands and beyond: Water-harvesting earthworks* (Vol. 2, p. 419). Tucson: Rainsource Press.
Li, X. Y. (2003). Rainwater harvesting for agricultural production in the semiarid loess region of China. *Food, Agriculture & Environment, 1*, 282–285.
Li, F., Cook, S., Geballe, G. T., & Burch, W. R., Jr. (2000). Rainwater harvesting agriculture: an integrated system for water management on rainfed land in China's semiarid areas. *Ambio, 29*, 477–483.
Li, X. Y., Zhang, R., Gong, J. D., and Xie, Z. (2002). *Effects of rainwater harvesting on the regional development and environmental conservation in the semiarid Loess Region of Northwest China: Proceedings 12*th *International Soil Conservation Organization Conference* (pp. 482–486). May 26–31, 2002, Beijing.
Lightfoot, D. R. (1994). Morphology and ecology of lithic-mulch agriculture. *The Geographical Review, 84*, 172–183.
Lightfoot, D. R. (1996). Syrian qanat Romani: History, ecology, and abandonment. *Journal of Arid Environments, 33*, 321–336.
Lightfoot, D. R., & Eddy, F. W. (1995). The construction and configuration of Anasazi pebble-mulch gardens in the Northern Rio Grande. *American Antiquity, 60*(3), 459–470.
Limaye, S. D. (2002). Aquifer recharge management for sustainable water supply in semi-arid hard rock region in western India. In P. Dillon (Ed.), *Management of aquifer recharge for sustainability* (pp. 309–313). Lisse: Swets & Zeitlinger.
Limaye, S. D. (2010). Review: Groundwater development and management in the Deccan Traps (basalts) of western India. *Hydrogeology Journal, 18*, 543–558.
Lye, D. J. (2002). Health risks associated with consumption of untreated water from household roof catchment systems. *Journal of the American Water Resources Association, 38*, 1301–1306.
Marzol, J. (2002). Fog water collection in a rural park in the Canary Islands (Spain). *Atmospheric Research, 64*, 239–250.
Mohnot, S. M., and Kolarkar, A. S. (1997). Thar Desert. In A. Agarwal & S. Narain (Eds.), *Dying wisdom. Rise, fall and potential of India's traditional water harvesting systems* (pp. 104–149). New Delhi: India Centre for Science and Environment.
Narasimham, T. M. (2006). Ground water in hard-rock areas of peninsular India: Challenges of utilization. *Ground Water, 44*, 130–133.
National Academy of Sciences. (1974). *More water for arid lands promising technologies and research opportunities* (p. 154). Washington, DC: National Academy of Sciences.
NRMMC-EPHC–NHMRC (Natural Resource Management Ministerial Council, Environment Protection and Heritage Council and National Health and Medical Research Council) (2009b). Australian Guidelines for Water Recycling: Managing Health and Environmental Risks. Phase 2: Stormwater Harvesting and Reuse (July 2009).
Olivier, J. (2002). Fog-water harvesting along the West Coast of South Africa: A feasibility study. *Water SA, 28*, 349–360.
Olivier, J. (2004). Fig harvesting: An alternative source of water supply on the west coast of South Africa. *GeoJournal, 61*, 203–214.
Osses, P., Schemenauer, R. S., and Leibbrand, N. (2004, October 11–15) Fog collection evaluation in the Sadaa Governorate, Yemen. In H Rautenbach & J. Oliviera (Eds.), *Proceedings of the Third International Conference on Fog, Fog Collection and Dew* (p. 4), Cape Town.
Oweis, T., Hachum, A., and Kijne, J. (1999). *Water harvesting and supplementary irrigation for improved water use efficiency in dry areas. SWIM 7* (p. 41). Colombo: International Water Management Institute.
Pandey, D. N., Guota, A. K., & Anderson, D. M. (2003). Rainwater harvesting as an adaptation to climate change. *Current Science, 85*(1), 46–59.
Radhakrishna, B. P. (2004). Man-made drought and the looming water crisis. *Current Science, 87*(1), 20–22.

Sakthivadivel, R. (2007). The groundwater recharge movement in India. In M. Giordano & K. G. Villholth (Eds.), *The agricultural groundwater revolution: Opportunities and threats to development* (pp. 195–210). Wallingford: CAB International.
Schemenauer, R. S. & Cereceda, P. (1994). A proposed standard fog collector for use in high-elevation regions. *Journal of Applied Meteorology, 33*, 1313–1322.
Schemenauer, R. S., and Cereceda, P. (1997, October 6 to 9). *Fog-water collection in arid coastal locations: Proceedings IV Reunion National Sobre Sistemas de Captación de lluvia* (pp. 279–290) (1997), Torreón.
Schemenauer, R. S., Osses, P. & Leibbrand, M. (2004a, October 11–15). Fog collection evaluation and operational projects in the Hajja Governorate, Yemen. In H. Rautenbach & J. Oliviera (Eds.), *Proceedings of the Third International Conference on Fog, Fog Collection and Dew* (p 4). Cape Town.
Schemenauer, R.S., Osses, P., Lara, F., Zywina, C., & Cereceda, P. (2004b, October 11–15). Fog collection in the Dominican Republic. In H. Rautenbach & J. Oliviera (Eds.), *Proceedings of the Third International Conference on Fog, Fog Collection and Dew* (p. 4), Cape Town.
Semenzato, R. (1996, June 17–18) Fog: A water resource for the development of arid regions. In *Proceedings, Fifth Joint Conference on Agriculture, Food, and the Environment* (p. 8), Padova.
Shah, T., Molden, D., Sakthivadivel, R., & Seckler, D. (2000). *The global groundwater situation: Overview of opportunities and challenges* (p. 19). Colombo: Sri Lanka International Water Management Institute.
Simmons, G., Hope, V., Lewis, G., Whitmore, J., & Gao, W. (2001). Contamination of potable roof-collected rainwater in Auckland, New Zealand. *Water Research, 35*, 1518–1524.
Singh, A., & Pande, U. C. (1997). Western Himalaya. In A. Agarwal & S. Narain (Eds.), *Dying wisdom. Rise, fall and potential of India's traditional water harvesting systems* (pp. 36–49). New Delhi: Centre for Science and Environment.
Singh, R. P., Pathak, B., Parthvi, R., & Parthvi, P. (2006, June 11–16). Roof-top rain water harvesting to recharge groundwater reservoir of a group of houses. In *Recharge Systems for Protecting and Enhancing Groundwater Resources, Proceedings of the* 5th *International Symposium on Management of Aquifer Recharge, Berlin, Germany* (pp. 223–228). Paris: UNESCO.
Srinivason, V., Gorelick, S. M., and Guolder, L. (2010). Sustainable urban water supply in south India: Desalination, efficiency improvement, or rainwater harvesting? *Water Resources Research, 46*(W10504), 15. doi:10.1029/2009WR008698.
Texas Water Development Board (2005). *Texas Manual on Rainwater Harvesting* (3rd ed.) (p. 58). Austin: Texas Water Development Board.
Thomas, P. R., & Greene, G. R. (1993). Rainwater quality from different roof catchments. *Water Science and Technology, 28*(3/5), 291–299.
United Nations Environment Programme. (1984). *Rain and stormwater harvesting in rural areas* (p. 238). Dublin: Tycooly International Publishing.
United Nations Environment Programme (1998). Fog Harvesting (Section 1.3). In *Sourcebook of alternative technologies for freshwater augmentation in Latin America and the Caribbean.* Paris: United Nationals Environmental Programme, Division of Technology, Industry, and Economics.
Wåhlin, L. (1997, June 19–22). The family cistern: 3,000 years of household water collection in Jordan. In M. Sabour & K.S. Vikør (Eds.), *Ethnic encounter and culture change: Papers from the Third Nordic Conference on Middle Eastern Studies, Joensuu, Finland* (pp. 233–249). London: Hurst.
Waller, D. H. (1989). Rain water—an alternative source in developing and developed countries. *Water International, 14*, 27–36.
Walton, K. (1969). *The arid zone* (p. 175). Chicago: Aldine Publishing Co.
Yaziz, M. I., Gunting, H., Sapari, N., & Ghazali, A. W. (1989). Variations in rainwater quality from roof catchments. *Water Research, 23*(6), 761–765.

Chapter 23
Managed Aquifer Recharge

23.1 Introduction

Water scarcity can be caused primarily by a shortage in the total amount of water available, or to temporal imbalances between supply and demand. In many areas of the world, periodic water shortages can be alleviated by storing water available during rainy periods (or other times of excess supply) for later use during dry periods. The storage and recovery cycle could be seasonal or inter-year with water stored in years with higher than normal rainfall for use during subsequent dry years. Even in arid lands, where the average annual total supply of water is inadequate to meet all demands, seasonal or inter-year storage of water may still be of value by capturing water that would otherwise not be available for beneficial use. Mutiso (2003) observed that in most arid and semiarid lands, the quest for water does not depend on the absolute amount of precipitation, but on the fraction retained.

Dillon (2005, p. 313) defined the term "management of aquifer recharge" or MAR as describing the "intentional banking and treatment of waters in aquifers." The term "management of aquifer recharge" has been superseded by "managed aquifer recharge". The term MAR was introduced as an alternative to "artificial recharge", which has the connotation that the use of the water was in some way unnatural (Dillon 2005). As the construction of large dams is becoming an unviable option in most areas due to a lack of suitable sites, great costs, and environmental concerns, storage of water underground has become an increasingly attractive option. Bouwer (2000, p. 220) observed that

> considering that more than 98% of world's fresh liquid water supplies already occurs underground, there is plenty of room for more

A primary purpose of MAR is to augment the supply of available groundwater. MAR systems may also be constructed and operated with a primary goal of improving the quality of recharged waters. The augmentation of groundwater supply may be direct by increasing the amount of water in storage, or indirect, by ameliorating the impacts of groundwater use, which may be constraining the

R. Maliva and T. Missimer, *Arid Lands Water Evaluation and Management*, Environmental Science and Engineering, DOI: 10.1007/978-3-642-29104-3_23,

utilization of available resources. An example of the latter is the strategic recharge of freshwater along coastal areas to control saline-water intrusion. Storage of water in MAR systems is also recognized to be an adaptive tool to help overcome the adverse effects of climate change (Sukhija 2008) (Chap. 39).

MAR includes a wide variety of processes by which water is intentionally added into an aquifer or induced to flow into and through an aquifer for treatment purposes (Table 23.1). MAR, as defined by Dillon (2005), includes two main end-member types of technologies; (1) methods that are used primarily to provide temporal storage of water, and (2) methods that are used primarily for water treatment. The term "aquifer recharge and recovery" (ARR) refers to the artificial recharge of water into an aquifer and its later recovery for subsequent use with a primary goal of water treatment. ARR of reclaimed water can be a very cost-effective wastewater treatment technology, which can have considerable value in developing countries where the financial and technical resources to operate more sophisticated wastewater treatment plants may not be available. ARR includes technologies, such as aquifer storage transfer and recovery, in which water is injected into an aquifer for subsequent recovery after treatment, and soil-aquifer treatment, in which wastewater is recharged in a controlled manner by surface spreading and later recovered from the aquifer.

Huisman and Olsthoorn (1983) differentiated between indirect and direct artificial recharge. Indirect recharge (also referred to as induced recharge) occurs by locating groundwater abstraction close to areas of rejected recharge or natural discharge (e.g., surface water bodies). Bank filtration is a prime example of indirect recharge and is separately discussed in Chap. 24. Direct recharge involves the conveyance of water into a suitable aquifer to recharge it by either surface or subsurface methods. MAR systems involving stormwater may have a dual stormwater management (flood control) and aquifer recharge function.

23.2 Types of MAR Systems

MAR systems are categorized herein into three main groups that vary in their primary recharge process, including

- systems that involve subsurface injection using wells,
- systems that involve surface or near surface application or spreading of water, and
- systems that result in an enhancement of natural recharge processes.

23.2.1 Injection or Recharge Wells

Injection using wells is the preferred method of aquifer recharge where low permeability strata are present between land surface and the target aquifer that cannot be practically disrupted or otherwise bypassed by excavation or trenching.

Table 23.1 Managed aquifer recharge techniques

Technique	Description
Aquifer storage and recovery	Injection of water into an aquifer and its later recovery using wells
Aquifer recharge using wells	Injection of water into an aquifer with the goal of increasing aquifer water levels
Aquifer storage transfer and recovery	Injection of water into an aquifer and its recovery using different, nearby wells with the goal of using flow through the aquifer as a treatment process
Salinity barrier system	Injection or recharge of water by surface spreading to create a hydraulic mound to prevent saline-water intrusion
Soakaways	Pits, chambers, or trenches that drain stormwater directly to the ground
Infiltration ponds, recharge ponds, rapid infiltration basins	Constructed basins in which water is diverted with the goal of recharging the underlying aquifer
Surface flooding and ditch and furrow systems	Systems that recharge water through sheet flow of water or shallow ditch and furrow systems
Infiltration galleries, pits, and trenches	Shallowly excavated structures used for infiltration where surface infiltration rates are low. Often filled with sand or gravel
In-channel infiltration systems	Check dams and levees constructed in channels to back-up, spread, and retain water to increased wetted area and duration of flooding
Percolation tanks	Basins created in ephemeral streams to capture part of the monsoon flows in order to recharge the underlying aquifer. Term is used in India
Soil-aquifer treatment	Infiltration of wastewater into shallow basins with the goal of improving its quality by vadose and saturated zone processes. As originally defined, the flow of water is controlled and restricted to a limit area of the aquifer
Sand dams	Low dams constructed on ephemeral streams in order to capture sand and create an artificial aquifer used to store storm flows
Wadi dams	Dams constructed across ephemeral streams (wadis) in order to increase recharge and manage flooding
Recharge releases	Slow release of water from surface reservoirs on ephemeral streams in order to enhance recharge of downstream underlying
Bank filtration	Extraction of water using wells (or other means) near a surface water body in order to take advantage of filtration provide by bed and aquifer sediments
Dune filtration	Infiltration of water into sand dunes and its recovery with the goal of improving its quality

Source Oaksford (1985), Bouwer (1999, 2003), Dillon (2005), Gale (2005), and Balke and Zhu (2008)

Injection wells must also be used where the aquifer to be recharged is confined. Wells may also be the preferred option in situations where the water table is located a great distance below land surface. Wells are also the preferred recharge mechanism where surface infiltration facilities are subject to high evaporation loss rates. Aquifer recharge using wells can be performed either under gravity (i.e., water is allowed to flow into a well which has a water level below land surface) or by injection under pressure. Managed aquifer recharge using wells, including specifically aquifer storage and recovery (ASR), was reviewed by Huisman and Olsthoorn (1983), Pyne (2005), and Maliva and Missimer (2010).

Huisman and Olsthoorn (1983, p. 237) noted that

> Without any doubt, the most important drawback to the use of injection wells is the danger of clogging, primarily caused by an entrance rate into the aquifer which is one to two orders of magnitude higher than that with spreading ditches.

Injection wells force a large volume of water through a relatively small surface area, the borehole wall. The suspended load of the injected water is either filtered out at the borehole wall or passes through the borehole wall surface into the adjoining aquifer, which greatly increases the rate of mechanical clogging compared to the much larger surface area of spreading basins. Clogging of injection wells may also occur as the result of gas (air) binding, biological growth, and chemical precipitation (Sniegocki 1963; Huisman and Olsthoorn 1983). Source water quality is, therefore, critical for the successful performance of injection wells. Injection wells may seriously clog in days if the injected water is of poor quality, not properly pretreated (e.g., disinfected), or is not geochemically compatible with the aquifer rock or native groundwater.

Water may also be injected either directly into an aquifer or into the vadose zone above the water-table aquifer. The latter, which are referred to as vadose recharge wells, are used where the water table is located a great distance below land surface. The main advantages of vadose recharge wells is that they are less expensive to construct than deeper phreatic injection wells and provide the opportunity for water quality improvement through vadose zone processes. However, vadose recharge wells are more prone to clogging and are difficult to rehabilitate and, as a result, it can be difficult to maintain their injection rates.

A fundamental issue with injection well systems is optimizing and maintaining injection capacity. Injection capacity is usually quantified in terms of specific injectivity, which is defined as the injection rate divided by pressure increase. Injection well performance depends upon the following primary factors:

- source (injected) water quality,
- pretreatment,
- injection well design and operation, and
- well rehabilitation program.

All of the above factors were reviewed in detail by Maliva and Missimer (2010), who provided an extensive bibliography. A basic design and operational issue is that injection wells should not necessarily be designed in the same manner as purely extractive production wells. Well efficiency needs to be maximized in order to reduce the clogging rate. Wells also need to be designed to allow for ready well rehabilitation, which includes periodic back-flushing and less frequent, more major rehabilitation. For example, wellheads need to be readily accessible to a drilling rig or hoist truck for well rehabilitation activities.

The regulatory requirements of injection wells may also be stricter than those for surface spreading. For example, in the United States, injection wells are regulated under the U.S. Environmental Protection Underground Injection Control rules. Injection directly into an aquifer is perceived as posing a greater risk to groundwater quality than its application by surface spreading because of the contaminant attenuation processes that occur during infiltration and percolation to the water table.

Three main applications of injection wells for MAR are:

- aquifer, storage, and recovery, in which water is stored during times of excess supply for later use during periods of high demand or low supply,
- aquifer recharge, which is performed to increase the volume of water stored in an aquifer, and
- salinity barrier systems in which water is injected to create a barrier against saline-water intrusion (Chap. 25).

23.2.1.1 Aquifer Storage and Recovery

Aquifer storage and recovery was defined by Pyne (1995) as

> The storage of water in a suitable aquifer through a well during times when water is available, and the recovery of the water from the same well during times when it is needed.

Pyne's definition of ASR includes three main components: (1) water is stored underground, (2) the water is emplaced underground using wells, and (3) the water is recovered using the same well as was used for emplacement. A simplified conceptual diagram of a single well ASR system used to store freshwater in a brackish aquifer is provided in Fig. 23.1. The injected water flows radially outward from the injection well, displacing and partially mixing with the ambient native groundwater in the storage aquifer. A transition or mixing (buffer) zone develops between the injected and ambient water. During recovery, the stored water, mixing zone, and ambient water are drawn back towards to the ASR well.

Injection and recovery using the same well has obvious cost benefits compared to the construction of separate dedicated injection and recovery wells. Well pumps used for recovery can also be used for well rehabilitation (periodic back-flushing). However, there are circumstances in which the use of dedicated injection and recovery wells may have overriding operational advantages. If the injected water is migrating laterally or vertically due to buoyancy, a greater amount of the stored water may be recovered by having a dedicated recovery well located down gradient from the injection well.

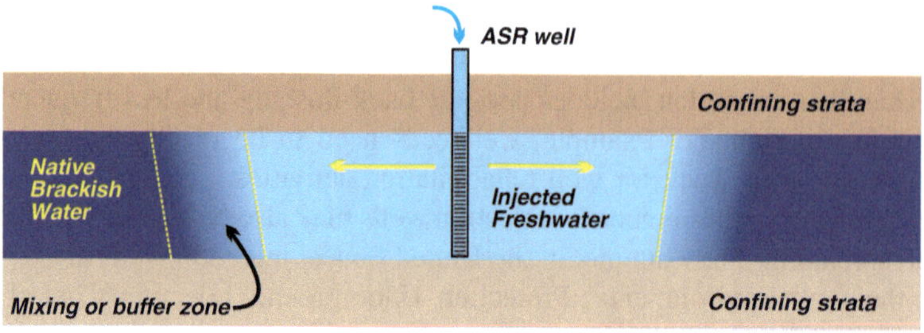

Fig. 23.1 Conceptual diagram of a chemically bounded (brackish-water) ASR system. Injection of freshwater laterally displaces the native brackish groundwater. The stored freshwater and native brackish water are separated by a buffer or mixing zone. During recovery (not shown) native groundwater and the mixing zone are drawn towards the ASR well

The essential, defining feature of ASR is that it involves the injection, storage, and recovery of the *same* water, allowing for some mixing with ambient groundwater. In other words, ASR involves the local storage of water within an aquifer. Pyne's (1995) definition of ASR was therefore modified by Maliva and Missimer (2010, p. 5) as follows:

> ASR is the storage of water in a suitable aquifer through a well during times when water is available, and the recovery of the same or similar quality water using a well during times when it is needed.

ASR is a storage technology and requires the periodic availability of excess water that can be stored for later use. Water stored in existing ASR systems includes potable water, treated wastewater, treated surface water, stormwater, and desalinated water. ASR has numerous advantages, which were discussed by Pyne (1995) and Maliva and Missimer (2010). Of particular value is that ASR systems have much lower costs and land requirements than surface storage options, avoid the evaporative losses of surface reservoirs, and have a lesser susceptibility to the contamination of the stored water. The main disadvantages or limitations of ASR systems include that unfavorable hydrogeologic conditions can result in low recoverability of stored water and adverse changes in water quality may occur due to fluid-rock interactions (e.g., metals leaching). However, in properly performing systems, water losses and deterioration of stored water quality are usually less than would occur in surface reservoirs. Underground storage of "impaired" waters (e.g., treated wastewater, stormwater, and surface water) can result in an improvement in water quality through natural attenuation process (Chap. 31).

Maliva and Missimer (2008, 2010) noted that there are several distinct types of ASR systems that differ in how they achieve the useful storage of water. Injection is considered to result in useful storage if it results in an additional supply of water at a useable quality that would not otherwise have been available. Chemically bounded ASR systems, such as conceptually illustrated in Fig. 23.1, achieve useful storage by displacing water of poorer quality. For example, freshwater injected into an aquifer containing brackish water may later be recovered at a salinity acceptable for its intended use. Injection results in a new freshwater resource at the time of recovery, that would otherwise not have been available had the injection of freshwater not been performed.

The performance of chemically bounded ASR systems is typically measured in terms of recovery efficiency (RE), which is defined as the percentage of the volume of injected water (V_{inj}) that is recovered (V_{rec}) at a quality suitable for its intended use:

$$RE(\%) = 100\left(V_{rec}/V_{inj}\right) \qquad (23.1)$$

Recovery efficiency can be calculated over the entire operational history of a system (system recovery efficiency; SRE) or over an individual operational cycle (operational recovery efficiency; ORE) (Sheng et al. 2007). ORE tends to improve over time (operational history) as the native groundwater is flushed from the vicinity of ASR wells by repeated operational cycles.

Unless there is only a small difference between the quality (e.g., salinity) of the injected water and native groundwater, recovery efficiencies will usually be less than 100%. A reasonable long-term ORE target for an ASR system using a brackish-water aquifer as a storage zone is 70–80%. An important lesson of the historic ASR experience is the danger of setting overly optimistic expectations for ASR system performance. Setting unrealistic targets, such as a 100% recovery efficiency target, can result in systems that are performing well, as allowed by local hydrogeological conditions, being improperly perceived as failures (Maliva and Missimer 2010).

A second major type of ASR system is physical storage systems in which injection increases the total volume of water present in an aquifer, as manifested by an increase in water levels or pressure (heads; Fig. 23.2). Physical storage ASR systems typically involve the injection of freshwater into freshwater aquifers. A critical aspect of physical ASR systems is that the increase in heads must persist until the time of recovery otherwise no net storage has occurred. The aquifer in essence acts as enormous storage tank. If water is pumped into a tank and it leaks out before it is needed, then the pumping of the water into the tank results in no net benefits. The performance of physical storage ASR systems must be evaluated by water level criteria, whereas the performances of chemically-bounded ASR systems are evaluated using water quality criteria.

A difficulty with using water levels to quantify ASR (and MAR in general) induced changes in storage volume is that other processes (such as recharge and pumping) will also affect the volume of stored water and aquifer water levels. The proper baseline to evaluate physical storage ASR systems is not necessarily water levels at the time of injection, but rather the water levels that would have occurred at the time of recovery in the absence of injection. Groundwater modeling is, therefore, typically required to quantitatively determine the waters levels that would have occurred at the time of recovery in the absence of recharge and pumping.

Over-drafted unconfined aquifers (i.e., aquifers in which water levels have declined because of excessive pumping) are good candidates for physical storage ASR systems because over-draft results in available storage space. A key technical

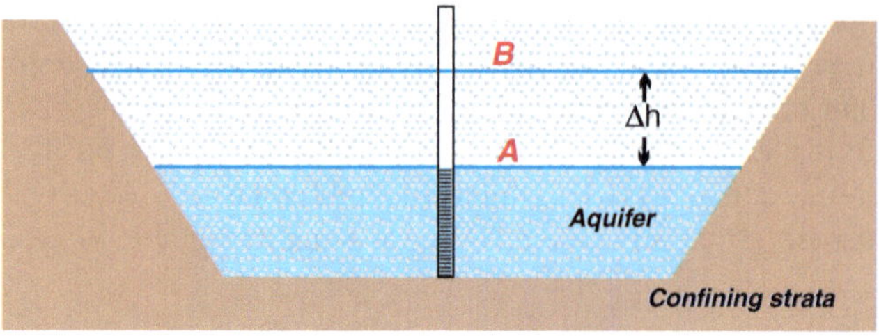

Fig. 23.2 Conceptual diagram of a physical storage ASR system. Injection of water into the aquifer results in an increase in water levels or heads from elevation "*A*" to "*B*". The amount of stored water is the product of the increase in head (Δh) and the area and storativity of the aquifer

issue is selecting an aquifer in which there is sufficient lateral and underlying confinement so that most of the stored water is retained (i.e., does not leak out) until the time of recovery. Intermontane basins bounded by crystalline bedrock are good candidates for physical-storage ASR systems. The best example of a large-scale physical storage system is the Las Vegas, Nevada (U.S.A.) system in which treated surface water from the Colorado River is stored in an over-drafted basin fill aquifer (Maliva and Missimer 2010).

A third type of ASR system common in parts of the United States is regulatory storage systems in which injection of water confers the right to later pump additional groundwater, which would not otherwise be allowed. Depending upon the jurisdiction, the system owner may either obtain a 100% credit for injected water or a partial credit. The rationale for regulatory storage ASR systems is that the injection and recovery of water does not adversely impact long-term aquifer water levels (i.e., contribute to over exploitation). Regulatory storage ASR systems may still have adverse water resources and environmental impacts by increasing groundwater withdrawals during dry periods. Environmental impact assessments need to consider the impacts of both injection and recovery. Regulatory storage systems, which are also referred to as groundwater banking systems, are further discussed in Sect. 23.10.

ASR systems vary in the duration of storage. The majority of ASR systems that have been constructed to date are designed and operated to provide seasonal (intra-year) storage. In areas with a pronounced seasonality of rainfall (e.g., India monsoon), water is stored during wet periods and later recovered during following dry or high demand periods. Water may also be stored to meet peaks in demand, due to temporal spikes in population (e.g., tourism seasons or holiday periods). It is often far less expensive to construct an ASR system to meet short-term peak demands than to construct additional desalination capacity, which may be needed for only a short time each year. ASR systems may also be designed to provide longer-term, strategic water storage to meet demands during infrequent droughts or as a result of a catastrophic disruption of primary water sources. Strategic storage ASR systems are being developed in the Arabian Gulf area (e.g., Abu Dhabi) as a strategic reserve of potable water that could be tapped in event of a disruption of the desalination plants, which are the primary potable water source in the region.

Maliva and Missimer (2010) provide a comprehensive overview of the planning, design, construction, and operation of ASR systems, including a review of global ASR practices and experiences. Perhaps the most important lesson is that ASR system performance is highly dependent upon site-specific hydrogeology. Aquifer characterization, therefore, is critical for evaluating whether or not ASR will likely be successful at a site. Recovery efficiencies of ASR systems will often be less than 100% as inevitably some water may be lost to mixing and leakage. However, a system with realistic recovery efficiency of 70% can still provide additional water during dry or high demand periods at a fraction of the cost of other supplemental supply options. Economics of an ASR system are more important than the recovery efficiency and should be used to ascertain the system viability.

23.2.1.2 Aquifer Recharge Using Wells

ASR is a subcomponent of MAR with the specific requirement that *local* storage of water occurs using wells. Water may also be injected into an aquifer with the primary goal of increasing aquifer water levels or pressures. Injection wells are often used where local hydrogeological or physical conditions are unfavorable for recharge by surface spreading (land application). Surface spreading (Sect. 23.2.2) may not be feasible where impervious strata are present above the water table, the target aquifer is confined, or adequate land is not available, cost-prohibitive, or contaminated. Recharge wells associated with saline-water intrusion control are discussed in Chap. 25.

The goal of the recharge may be either to obtain an overall increase in aquifer storage or pressures, or to address local hydrogeologic concerns, such as saline-water intrusion and land subsidence. Implicit in the concept of the aquifer recharge is that its intended purpose is to allow for continued or expanded use of aquifers. Water could be later produced from the aquifer using either dedicated production wells or, perhaps, the same well used for injection (ASR). If local storage of water and water quality are not issues, then it makes no difference whether water is produced from dedicated production wells or from the same wells used for injection. From an operational perspective, the former option may be more practical. Water could be injected near the recharge water source and production could occur near the point-of-use. The aquifer can be used, in essence, to convey water between the storage and use areas, typically with the injection being up-gradient from the withdrawal location.

In unconfined aquifers, time lags occur between the artificial recharge and down-gradient hydraulic effects. Barber et al. (2009) document the results of a modeling investigation in which the time lag was found to be advantageous for meeting water management goals. A major water management priority in Western Washington state and Idaho is the protection and restoration of late summer minimum in-stream flows in the Spokane and Little Spokane Rivers. Summer in-stream flows have been adversely impacted by increasing groundwater withdrawals from the Spokane Valley—Rathdrum Prairie (SVRP) aquifer. The SVRP aquifer is a regional unconfined aquifer that is composed of glacial sediments with a relatively high hydraulic conductivity and is hydraulically connected to the rivers.

Excess freshwater is available during the winter from the Lake Pend Oreille reservoir. Barber et al. (2009) used a calibrated regional groundwater flow model to simulate the effects of artificial recharge using infiltration basins or injection wells on summer stream flows. The modeling results indicate that, depending on the artificial recharge location, as much as 30% of the winter artificial recharge could be lagged to improve summer flows. The augmented summer flow could potentially allow for increased summer groundwater withdrawals (important for the region's economic growth), while still meeting regional river flow requirements.

Another potential application of aquifer recharge is the complete alteration of a brackish-water or saline water aquifer into a freshwater system. There are a number of locations around the world that contain confined aquifers with naturally-occurring high salinity water (commonly trapped connate waters). Where saline-water aquifers are located adjacent to a seasonal source of fresh surface water, particularly unallocated floodwater, it would be possible to construct a large number of injection wells to store freshwater harvested from the surface-water body. In this case the objective would be to create a strategic storage of water that would be used only if absolutely necessary. It would involve a long term plan, perhaps on the order of 50–100 years and assumes that the economics are viable to create the long term water bank or conversion of a saline-water bearing aquifer into a freshwater system. This would be particularly viable in locations where the known freshwater resources will be insufficient in the future and the only other option may be expensive desalination treatment of seawater or perhaps in inland locations not close to any viable seawater source. Such a system has been discussed in South Florida for the storage of excess freshwater, which is currently discharged into estuarine water bodies, causing seasonal damage to marine ecosystems.

A critical issue for aquifer recharge and groundwater banking systems (Sect. 23.9) is, therefore, that it is insufficient to evaluate the systems solely on the basis of basin-wide water budget criteria. Detailed groundwater modeling is needed to evaluate the temporal and spatial impacts of system operation. Aquifer recharge wells have the same design and operational issues as ASR wells, which were discussed in detailed by Maliva and Missimer (2010). Recharge wells need to be designed and constructed to most efficiently emplace the target volume of water. Management of clogging is an important operational challenge.

23.2.1.3 Aquifer Storage Transfer and Recovery

The use of separate injection and recovery wells may be desirable as a means of improving stored water quality by providing additional residence time and to take better advantage of the filtration and other treatment processes provided by the aquifer (Fig. 23.3). Rinck-Pfeiffer et al. (2006) referred to the use of separate injection and recovery wells for the purpose of chemical and microbial contaminant attenuation as "aquifer storage transfer and recovery" (ASTR). The essential,

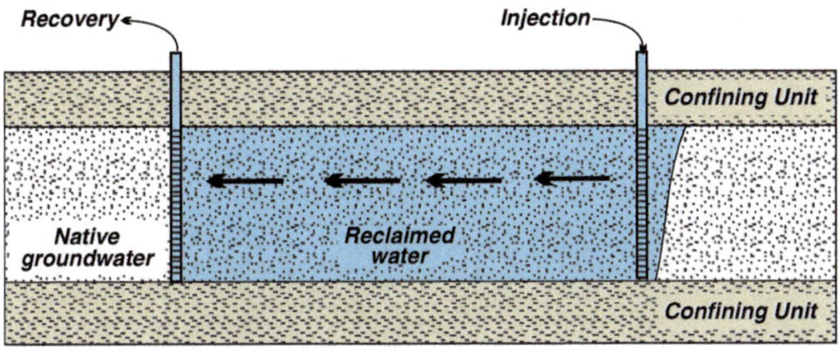

Fig. 23.3 Conceptual diagram of an ASTR system in a confined aquifer. Water quality is improved through the natural attenuation of contaminants by physical, chemical, and biological processes as the water flows through the aquifer from the injection well to the recovery well

defining feature of ASTR is the intentional use of injected water flow through an aquifer as a treatment method.

An excellent operational example of an ASTR system is the El Paso Water Utilities (Texas, USA) Hueco Bolson Recharge Project, which involves injection of highly-treated reclaimed water from the Fred Hervey Water Reclamation Plant into the upper Hueco Bolson Aquifer and its later recovery for potable use (National Research Council 1994; Sheng 2005). The recharge and recovery wells have a minimum spacing of 782 m (2,566 ft) in order to ensure an adequate aquifer residence time (two year minimum) for complete inactivation of viruses in the recovered water.

A key technical issue for ASTR systems is accurately determining the travel times between injection and recovery wells, which may be more rapid than expected because of aquifer heterogeneity (e.g., presence of flow zones with very high hydraulic conductivity). Aquifer characterization, modeling, and tracer testing can be used to assess aquifer travel times. Differences in water chemistry between the injected water and native groundwater can be used as effective natural tracers. Conservative ions (e.g., chloride, bromide, fluoride) and salinity-related parameters (specific conductance, and total dissolved solids) can be particularly useful. If there is a significant difference in salinity between injected water and the native groundwater, then a time-series of resistivity logs run on the recovery wells may provide insights on the presence and location of preferred flow zones between the injection and recovery wells. It is important to implement a well-conceived monitoring plan at the start of a project in order to capture the valuable data that can be provided by the initial breakthrough of injected water in the recovery wells.

23.2.1.4 Vadose-Zone Recharge Wells

Vadose wells are essentially injection wells that are completed above the water table (Fig. 23.4). They have the general advantage of injection wells in that they have a much smaller surface footprint than surface spreading systems and may have relatively high recharge rates, which depends upon the transmissivity of the

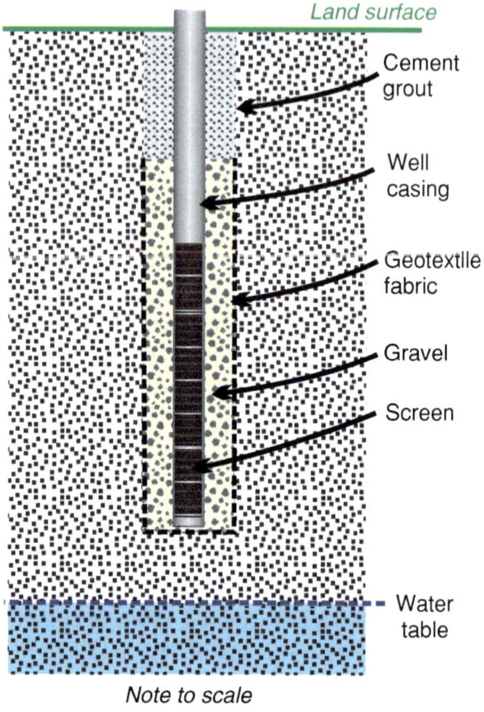

Fig. 23.4 Conceptual diagram of a vadose recharge well

injection zone and well design. There are two main types of vadose wells that have an aquifer recharge function. By far the most numerous are "dry" wells that are used for stormwater disposal, which are further discussed in Sect. 23.4. These wells have a relatively simple construction and typically involve gravity drainage (i.e., water is not pumped under pressure). Although they are installed primarily for stormwater disposal, injection into the vadose zone also results in the recharge of the local unconfined aquifer.

Less common are vadose recharge wells that are installed as part of aquifer recharge projects. Vadose wells have similar construction as ASR and other injection wells (Maliva and Missimer 2010). Inasmuch as the wells are completed above the water table, cascading and associated air entrainment is a major design and operational issue. Injection is normally performed using a smaller-diameter injection tube (with or without an orifice plate) or using a downhole flow control valve. The major operational issue associated with vadose recharge wells is management of clogging. The wells are suitable only for high-quality water with low suspended solids. Inasmuch as the wells are dry, maintaining injection capacity (specific injectivity) by periodic backflushing is not an option.

The City of Scottsdale Water Campus operates a noteworthy example of a large-scale vadose recharge system. Dry wells were chosen initially over deep injection wells because they were much less expensive to construct and could be replaced more economically should it become necessary due to clogging or diminished infiltration rates (Marsh et al. 1995). The injected water is reclaimed water that has been treated by microfiltration and reverse-osmosis, and so is of very high quality. Recharge started in 1997 and the initial system consisted of 27 wells with total depths of approximately 55 m (180 feet). The system was subsequently expanded and as of 2010, consists of 27 recharge standard (RS) and 28 recharge emergency (RE) wells. The West Campus recharge system is designed for an eventual total of 80 wells. The RS wells are operated on a daily basis and the RE wells are designed to temporarily accommodate wet weather flows (Gastélum et al. 2009). The RS wells have a screened completion and injection is performed using either an injection tube with a fixed orifice plate or a downhole flow control valve.

The original 27 injection wells experienced an overall decrease in specific injectivity between 1999 and 2007, ranging from −12 to −89% with a system average of −48%. The RS wellfield capacity decreased from 123,500 m^3/d (32.63 Mgd) in 1999 to 65,300 m^3/d (17.26 Mgd) in 2007 (Gastélum et al. 2009). The reduction in specific injectivity is attributed to physical clogging from suspended solids and chemical clogging (calcium carbonate scale) (Lluria 2009; Gastélum et al. 2009).

23.2.2 Surface Spreading

23.2.2.1 Introduction

Surface spreading is the simplest, oldest, and mostly widely used method for managed aquifer recharge (Todd 1980; Asano and Cotruvo 2004). Stormwater, river water, treated wastewater and other waters may be either applied to the land surface or locally impounded in infiltration basins with the objectives of water disposal or increasing (or initiating) local ground water recharge. Water infiltrates through the bottom and sides of the basin and percolates downwards to the water table (Fig. 23.5).

With respect to systems used for wastewater disposal, a basic distinction amongst surface spreading systems is between slow-rate and high-rate systems. Slow-rate systems rely primarily upon crops and other vegetation for the removal of nutrients and other constituents. The rate of application is thus tied to the rate of plant uptake. Slow-rate systems have the disadvantage of a large land requirement. In arid-regions slow-rate land application systems also have the disadvantage of the build-up of salt in the plant root zone caused by evapotranspiration (Bouwer 1974). High-rate land application systems have infiltration rates much greater than the rate of evapotranspiration so that there is minimal evaporative concentration of salts in the recharge water and greater recharge rates. High-rate systems may have lesser rates of removal of nutrients and contaminants through vadose zone processes.

Surface spreading systems vary depending upon whether the infiltration area is permanently, occasionally, or never flooded (i.e., has standing water). The infiltrated volume is the product of infiltration rate, area, and time. The former is a function of

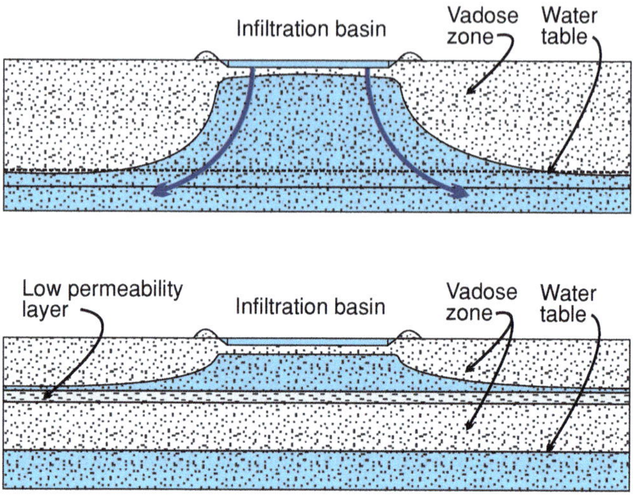

Fig. 23.5 Conceptual cross-sectional diagrams of an infiltration basin. Systems are designed so that water infiltrates through the bottom and sides of the basin and flows downward to the water table where it recharges the water-table aquifer (*top panel*). Recharge may cause the water table to locally rise to land surface (not shown). The presence of an intervening low permeability layer can result in a perched aquifer condition with little water actually reaching the water table (*bottom panel*)

the vertical hydraulic conductivity of the underlying strata and hydraulic gradient (standing water elevation). Surface spreading methods were reviewed by Huisman and Olsthoorn (1983), Oaksford (1985) and Roscoe Moss Company (1990).

The most common type of surface spreading system used in arid and semiarid environments is infiltration basins, which are essentially bermed and/or excavated impoundments constructed atop pervious sediments. Intermittent wetting and drying is necessary in order to maintain infiltration rates (manage clogging; Sect. 23.2.2.3). The alternation of wetting and drying also changes redox conditions, which is important for the removal of nutrients in wastewater systems (Chap. 31).

For systems designed for stormwater, wetting and drying cycles occur naturally. Where infiltration basins are used continuously, such as is the case for wastewater systems, infiltration systems are divided into cells with the system operated so that some cells are receiving water while others are drying (Fig. 23.6).

Infiltration rates decrease over time as a result of clogging at or just below the bottom of the basins. The total infiltrated volume is the integral of the infiltration rate versus time curve (Fig. 23.7). In order to minimize system size, locations of surface spreading systems should be selected that have high infiltration and percolation rates (vertical hydraulic conductivities). However, high vertical hydraulic conductivities could be caused by secondary porosity or other types of sedimentologic heterogeneity, which can allow water to bypass natural contaminant attenuation processes. An understanding of the vadose zone hydrogeological process active in a project area is critical for both site selection and system design.

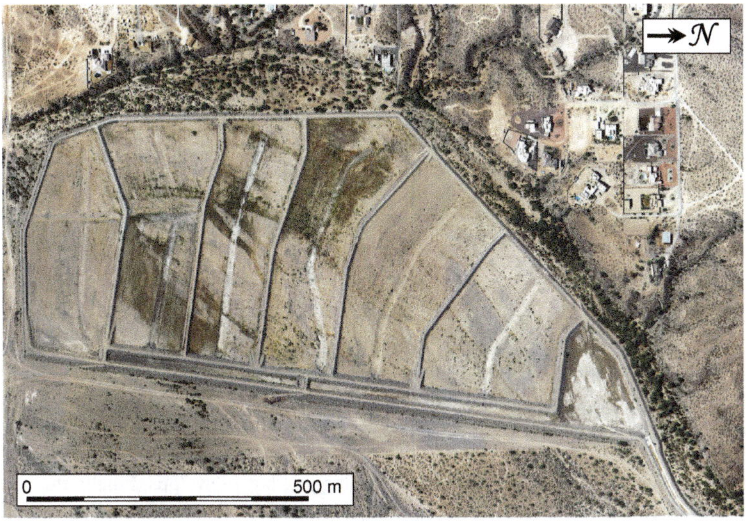

Fig. 23.6 Aerial photograph of the Agua Fria Recharge Project near Peoria, Arizona, in which the infiltration basin is divided into cells. *Source* U.S. Geological Survey

Another important design consideration is the location of the recharge system with respect to aquifer users and sensitive environments, and the goals of the recharge program. In other words where will the recharged water go? If the water is impaired, then an important consideration is avoiding transport of contaminants to nearby potable water supply wells. Another important consideration is avoiding locations where existing soil or shallow groundwater contamination may be mobilized by infiltration. Infiltration systems should also not cause unintentional flooding in nearby low-lying areas.

In the case of over-drafted aquifers, the system goal is commonly to increase the total amount of water in storage or at least to arrest or decrease the decline in water levels. In this case, the location of the system in the groundwater basin is not critical and systems can be located near the source of recharge water. The aquifer can act as a net conveyor of water in the sense that recharge occurs at one location to offset pumping at another location.

Recharge systems can also be designed for water quality improvement purposes. For example, soil-aquifer treatment systems (Sect. 23.2.2.6) are a specialized type of surface spreading system that is specifically designed to cause an attenuation of the concentration of nutrients, organic compounds, and contaminants in wastewater.

23.2.2.2 Design Issues

The fundamental design issue for infiltration basins and other types of surface spreading systems is to maximize the infiltration rate and to retain the water so as to increase the time available for recharge. Infiltration rates are most commonly

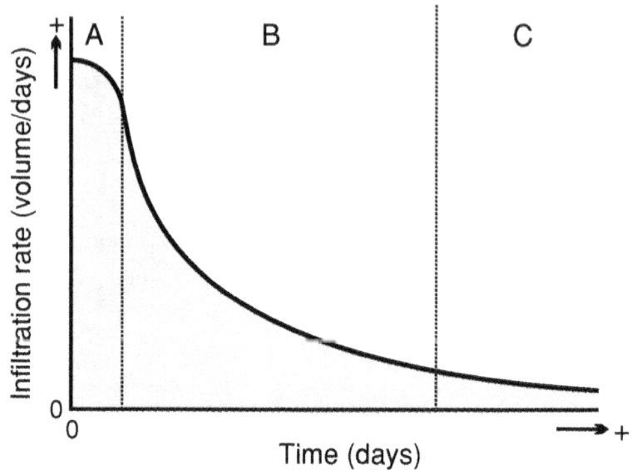

Fig. 23.7 Conceptual infiltration rate versus time curve for a basin in which water level is kept constant. The curve can be divided into three parts (Behnke 1969). Initially there is little change in infiltration rate as not enough material has been deposited to form a limiting layer (*A*). *B* Reflects the rapid decrease in infiltration as the clogging layer develops. Finally the infiltration rate stabilizes as the rate of flow through the layer is so slow that little new material is being added. The area under the curve (*shaded gray*) represents the total infiltrated volume

and conveniently expressed in units of length over time (e.g., cm/d or inches/d). Volumetric rates can be obtained by multiplying infiltration rates by basin areas. Retention of surface water in basins also increases the time available for evaporative losses, so maximization of the infiltration rate is of primary concern in hot arid regions with high potential evaporation rates. Another commonly used parameter is hydraulic loading rate, which is the long-term infiltration rate including drying periods.

Hydrogeological characterization is critical for selecting the most appropriate system design for a given location. The characterization includes testing of infiltration rates and the evaluation of strata between the recharge system surface and the water table. The presence of intervening confining strata can result in a perched aquifer condition rather that recharge to the water table (Fig. 23.5). The main issues that should be considered in determining locations for infiltration basins are (Bouwer 1978; Oaksford 1985)

- Soils should be sufficiently permeable to yield acceptable infiltration rates.
- The vadose zone should not contain low permeability layers that could result in perched groundwater conditions.
- The regional water table should be sufficiently deep to keep the groundwater mound below the base of the infiltration basins, but not so deep so that large quantities of water are needed to wet the vadose zone before recoverable water reaches the water table.
- The receiving aquifer should be unconfined and sufficiently transmissive to allow lateral movement of recharge water and prevent groundwater mounds that reach the basin bottom or nearby land surface.

A fundamental element required in investigations of potential infiltration basin sites is the performance of a series of infiltration tests (Sect. 10.5.1) in order to determine the hydraulic properties of surficial soils. If the results of double-ring infiltrometer testing are favorable, then the next step in the development of a large-capacity system should be to perform a longer duration test on a greater area. For example, testing may be performed on a temporary small-scale infiltration basin or pit constructed in the proposed infiltration basin site. Test borings (including cores) and wells are also needed to determine the hydraulic properties of the vadose zone strata and underlying aquifers, including the presence of any low permeability strata that could impede vertical flow. Testing requirements will vary depending upon the size of the system and target loading rates. Clearly, the hydrogeology of large systems that are to be a primary disposal (recharge) point for a wastewater treatment plant need to be thoroughly evaluated in order to ensure long-term operational reliability. Aquifer characterization requirements are also greater if a system is to be relied upon to perform a treatment role.

Analytical equations are available for evaluating mounding beneath various geometries of surface spreading systems. Analytical approaches are discussed by Bouwer (1978) and Huisman and Olsthoorn (1983). However, analytical approaches have become superseded by numerical computer modeling, which allows for greater flexibility. Numerical models that include modules for variable saturated flow (e.g., MODFLOW) can be used as a design tool to evaluate different system design options and have the advantage that they can much more readily incorporate hydrogeological heterogeneity, provided of course that such data are available from a detailed aquifer characterization.

Recharge basin shape is often dictated by land availability and other logistical issues. Large recharge basin systems typically contain several basins, which allow for part of the system to be dried for management of clogging and other maintenance activities while the remaining basins are receiving water. Where the recharge water flow is substantially below the system capacity, sending the flow to only one or two basins can reduce evaporative losses and increase net recharge. Upstream basins may be used as sediment-collection basins or clarifiers to reduce suspended solids concentrations of the recharge water before it enters the main basins.

Basin shape affects the hydrologic response of the underlying surficial aquifer to recharge. Rastogi and Pandey (2002) simulated groundwater mound development below recharge basins of different shapes. For a given total area and recharge rate, the groundwater mound height decreased as the basin perimeter increased. Circular basins resulted in the greatest mounding, rectangular basins the least. Narrow rectangular basins are thus preferred where mounding needs to be decreased in order to increase system capacity. More equant shapes are preferred when mound build up is desired, such as for salinity barrier systems (Rastogi and Pandey 2002). The perimeter to volume ratio becomes more important based on the initial position of the water table with the highest water table position requiring the greatest perimeter because the groundwater flow must be primarily horizontal away from the mound rather than vertical. Once the underlying aquifer is fully saturated beneath the central part of the basin, essentially no vertical flow exits the basin in that area and all outflow is horizontal.

Recharge basin systems need to have a sufficient area to accommodate the design recharge flows. The required basin area (A) can be calculated as the injected volume (Q) divided by the infiltration rate (I). However, calculation of basin area must also consider the times that basins are off-line for drainage and drying, and the decreases in infiltration rate over time due to clogging. In arid regions, the total application time is often less than 50% of the total time due to drainage and drying time (Asano et al. 2007). The decline in infiltration rate over time is system specific and cannot be accurately predicted in advance of system operations, or at least without pilot testing. The need for operational flexibility thus should to be incorporated into system design.

Infiltration tests performed in the Sultanate of Oman by Haimerl et al. (2002) illustrate some key relationships concerning infiltration and recharge by surface spreading. Tests were performed using an infiltration basin in which 11 tensiometers were installed at different depths. Groundwater levels were measured in an adjacent borehole. The arrival of the wetting front can be identified by the initiation of a decrease in water tension (suction). As would be expected, infiltration rates were greatest in dry soils, but actual recharge was greater in pre-wetted soil. The high matric potential for dry soil draws water into the soils, but the water tends to be retained. Pre-wetted soil requires a lesser infiltration volume until percolation to the water table (i.e., groundwater recharge) can occur. An unexpected result of the Haimerl et al. (2002) test was that during steady percolation, the water tension was greater than the theoretical value assuming the hydrostatic pressure, which suggests that the soil was not fully saturated.

A critical relationship is that surface-spreading systems with small water loading (volume/area) rates may have large infiltration rates, but little or no recharge as the water is retained in soil and subsequently the water lost to evapotranspiration. Smaller wetted areas and longer application times can allow for greater recharge rates (Haimerl et al. 2002). From a design perspective, small recharge basins may result in greater recharge rates than a larger basin, for the same volume of applied water. Alternatively, a large basin could be divided into subbasins.

In areas with a deep water table, injection wells may be a more efficient method for groundwater recharge. However, groundwater recharge in areas with deep (>100 m; 328 ft) water tables may still be possible. Izbicki et al. (2008) documented testing of an infiltration pond in the western Mojave Desert (California) in which the water table was located 113–121 m (371–397 ft) below land surface. Migration of the wetting front was monitored using a series of tensiometers installed in a test borehole. It initially took three years for the infiltrated water to each the water table. The time was reduced to about a year in subsequent tests due to the increased hydraulic conductivity with increased water saturation from previous testing and reduced retention of water in the unsaturated zone. An important hydrogeological and operational issue is the presence of clay layers (paleosols) that limit the downward movement of water and enhance the spreading of water away from the source of recharge. The water at the wetting front also had a high salinity caused by the mobilization of salts, but the salinity quickly decreased to the value of the infiltrating water.

23.2.2.3 Migration of Infiltrated Water

In situations where indirect potable reuse is a possibility (or reality), determining the travel pathways, distance, and residence time from recharge sites to production wells is a major technical and regulatory issue. For existing, operational systems, travels times and flow pathways may be inferred from tracer and age data. Artificial tracers may be employed, but commonly recharged water has sufficient differences in composition from native groundwater so that its presence can be readily identified and often quantified.

Groundwater modeling is also a valuable tool for determining the fate and transport of recharged reclaimed water. Different levels of sophistication may be used in model development. Flow and particle tracking programs (e.g., MODFLOW and MODPATH) can provide information on flow direction and travel times. Non-reactive transport models (e.g., MT3DMS) can be used to evaluate the mixing of waters and the origin of produced water. Even more sophisticated is reactive solute-transport modeling, in which contaminant and pathogen attenuation processes during groundwater flow are also simulated. However, the increase in modeling sophistication increases the data requirements and thus uncertainty within the produced model.

Surface geophysical methods can also provide value information on the movement of infiltrated water in the vadose zone. For example, a time series of geophysical surveys (4-D survey) starting prior to the initiation of recharge may be used to map changes in water content in the vadose zone. Microgravity surveys can detect changes in the water content in the vadose zone from the increase in mass (density) of saturated or partially saturated sediments. Resistivity-based methods can be used to map changes in water content from the associated decrease in resistivity (increase in conductivity) with increasing water content.

The Orange County Water Department (OCWD) (California, USA) recharge system was simulated by Thompson et al. (1999). Age ($^{3}H/^{3}He$) data indicate that the age of water varies with depth. The Orange County groundwater basin consists of irregularly alternating sands, gravels, and conglomerates, and semi-permeable silts, siltstones and clays, which results in complicated flow paths. Groundwater flow modeling demonstrated the importance of aquifer heterogeneity and the need for high-resolution three-dimensional models to realistically represent flow pathways and travel times in this system. Geostatistical methods (transition probability/Markov indicator technique) were used to create a hydrogeological model that honored borehole information. The modeling and geochemical data indicated that groundwater entering production wells spanned a large range of ages (decades or more). Water in some production wells appears to be derived from multiple sources. The modeling results also indicated that flow pathways are very sensitive to hydraulic conductivity values.

23.2.2.4 Clogging

Clogging is the major issue affecting long-term of performance of surface spreading systems. Clogging can result from the deposition and concentration of suspended sediments or the formation of an organic matter-rich (schmutzdecke) layer. Earlier studies of the clogging of surface spreading basins provided some basic insights into the processes involved. Experimental studies by Behnke (1969) demonstrated that the infiltration rate versus time curves have three segments (Fig. 23.7). The initial segment shows little change in infiltration rate with time as not enough material has been deposited to form a flow-limiting surface layer. The second segment exhibits a rapid decrease in infiltration rate with time as the clogging layer is horizontally continuous and progressively increases in thickness. The last segment shows little change over time as the rate of flow through the clogging layer is so slow that little new material is being added to the layer.

Clogging is essentially a surface sealing process as evidenced by tension developing a short distance below the surface (Behnke 1969). Behnke (1969) proposed that the clogging mechanism is composed of two processes. Initially, clogging occurs by gravitation settling, which produces a graded layer as coarser sediments settle first. Infiltration rates decline as finer material is deposited and the layer increases in thickness. Interstitial straining occurs as the uppermost pores become small enough to strain out the remaining particles and the infiltration rate is markedly reduced.

An important issue is the depth of clogging below the system surface. The accumulation of suspended materials near land surface is desirable because it can be more readily removed. The transport and deposition of suspended sediments deeper below land surface is much more problematic. It is very difficult, if practicably possible at all, to remove fine sediments that have accumulated in pore spaces well below land surface. This deeper clogging can seriously affect the long-term performance of surface spreading systems.

Goss et al. (1973) performed infiltration tests using suspensions of sediments in which cesium 237 was added as a radioactive tracer. Cesium is readily sorbed onto clay minerals. The cesium tagged sediments traveled deep below land surface >45.7 cm (>18 inches) in the test in which large pores were exposed at land surface. In the test where the large pores were destroyed by cultivation, most (>90%) of sediment was filtered out in the upper 2.5 cm (one-inch) of sediment, which can be readily removed by scrapping or other means. Very coarse sediments or macropore recharge can thus adversely impact the long-term performance of infiltration basins even though they can result in very high initial infiltration rates.

Clogging rate is clearly is a function of the suspended solids concentration of the water being recharged, which varies greatly between systems. Systems in Europe commonly recharge treated water of very high quality, and thus clogging rates are low. In contrast in the western United States, untreated surface water or reclaimed water treated to different degrees is most commonly recharged by surface spreading and, as a result, management of clogging is a major operational concern. Maintaining or increasing infiltration rates is of particular great importance in urban systems (e.g., Orange County, California) where the cost of additional land for system expansion is prohibitive.

Phillips et al. (2007) documented the results of column testing to assess the controls of percolation rate over time. The tests were performed using actual foulant and sediment samples from the Orange County Water District (OCWD; California, USA) recharge basins. The change in percolation rates was found to follow a log-decay expression with respect to accumulated foulant as follows:

$$Q_L = Q_0 e^{-rL} \tag{23.2}$$

where,

Q_0 = initial percolation rate (m/d) (ft/d)
Q_L = percolation rate at end load L (m/d) (ft/d)
L = total foulant accumulated per unit area (mg/m^2) (lb/ft^2)
r = sediment/foulant interaction coefficient (m^2/mg) (ft^2/lb).

The sediment/foulant coefficient was found to be a function of both foulant composition and total suspended solids (TSS) concentration. For a given total foulant accumulation, a greater reduction in percolation rate occurred when the load was applied more rapidly because of a greater TSS concentration. Reduction in TSS concentration was, therefore, identified as means of improving system performance (Phillips et al. 2007).

Pavelic et al. (2006) investigated the effect of ponding depth on infiltration rates. An increase in ponding depth would be expected to result in an increase in infiltration rate because of a greater hydraulic gradient across the clogging layer and a greater wetted area. Greater ponding depths can have negative impacts, such as allowing greater time for algal growth to occur. The column tests did indeed show an increase in infiltration rate with ponding depth in the short term, but the increase did not persist into the long term. A reduction in infiltration rates at greater ponding depths was suggested to be caused by compaction of the clogging layer (filter cake compression) and greater penetration depths of particulates into the pore spaces of the soil matrix or both processes.

Infiltration rates decrease over time because of processes such as:

- silt and clay accumulation,
- organic layer formation, and
- clay swelling and dispersion.

Maintaining or at least reducing the rate of decline of infiltration rates is both a major design and operational consideration. For example, off-channel recharge systems may include desilting (sedimentation) basins that allow fine sediments to settle out before they enter the main recharge basins. Maintenance activities include:

- periodic drying of basins to crack or chip surface layers,
- harrowing and other techniques to physical disrupt surficial layers and mix them within the underlying soil,

- removal of accumulated fine sediments by scrapping, and
- removal of vegetation to reduce evapotranspiration losses.

Periodic drying of the basins is also effective in preventing or reducing the build-up of algae and other aquatic plant growth. However, periodic drying does not result in the restoration of infiltration rates to their original values. Over time, infiltration rates tend to progressively decrease and basins tend to clog more rapidly (e.g., Shuh 1990), which necessitates more aggressive treatments to restore system capacity.

Harrowing or plowing methods are effective in increasing short-term infiltration rates, but the progressive increase in the concentration of fine sediment and organic matter in the mixed layer can lead to a long-term decline in infiltration rates, which necessitates removal of the layer. A spring tooth harrow was found to be the preferred technique in the Central Arizona Project recharge basins because it tends to break up surficial soils in an upwards motion as opposed to churning it deeper (Gorey and Dent 2007).

Maintaining infiltration rates in large-scale surface spreading systems is a critical issue for operators. Special machinery has been developed or adapted for the scarifying and scraping of infiltration basins, which are usually operated in dried basins (Huisman and Olsthoorn 1983). A more innovative approach was taken by the OCWD that involved the development of several generations of remote-operated subaqueous Basin Cleaning Vehicles that could remove clogging layers while a basin remains in service (Hutchinson 2007). A self-contained, self-propelled beach cleaner was also found to be effective in selectively removing the clogging layer (Hutchinson 2007).

Vegetation control can be accomplished by the use of herbicides or physically removing larger shrubs and trees manually or using heavy equipment. Allowing goats to graze in the recharge areas has been found to be an effective, economical, and environmental friendly-vegetation control method (Gorey and Dent 2007).

A key operation issue is reducing the system down time for drying and minimizing the costs associated with basin maintenance. The OCWD used a percolation rate decay model to determine the optimal basin cleaning, which considered factors such as (Hutchinson 2007):

- lost recharge from down time when a basin is out of service,
- costs associated with basin cleaning activities, and
- costs associated with draining deep basins.

Infiltration basins may also be designed with the reduction of clogging as a goal. Vertical or angled surfaces are less prone to clogging than horizontal surfaces because suspended materials more readily settle out of suspension on the latter. For example, Peyton (2002) described a ridge and furrow system for basin floors that was tested in California. The concept is that the slope of ridges remains relatively free of fine sediment, and therefore tends to retain their permeability. The fine sediments in the recharge water instead accumulate in the furrows. The loss of permeability caused by deposition in the furrows is significantly offset by the sustainability of recharge on the ridges. The main maintenance activity is inducing a mild wave action to re-suspend and wash-off fine sediment deposits in the ridge slopes, which can be achieved by lowering the water level to the height of the ridges. Furrow systems composed of closely spaced sloped, flat-bottom ditches were also described by Schiff (1957).

Determination of the optimal maintenance program necessarily involves site-specific adaptive management. Nevertheless, experiences from other facilities can provide insights into the potential effectiveness of various maintenance options. Unfortunately, there have been few comparative studies of treatment options. Mousavi and Rezai (1999) documented the effectiveness of five treatment options: (1) no treatment, (2) removal of deposited sediments, and (3) removal of deposited sediments plus 5 cm (2 in), (4) 10 cm (4 in), and (5) 15 cm (6 in) of the underlying soil. Removal of deposited sediment and 15 cm (6 in) of the underlying soil was the most effective method, restoring infiltration capacity to 68.3% of the original value. The Mousavi and Rezai (1999) study results indicate the importance of infiltration of clay and silt-sized material into underlying soil in reducing infiltration rate. Scrapping or scratching the deposited sediments was found to not be very efficient in restoring infiltration rates.

23.2.2.5 Monitoring

Monitoring requirements for infiltration basins includes operational monitoring (system performance parameters), groundwater levels, and groundwater quality. The key operational parameter for an infiltration basin is the infiltration rate, which can be quantified in terms of the volumetric rate for the entire basin (volume/time) or in terms of a height of water (length/time) obtained by dividing the volumetric rate by the basin area. Monitoring of basin infiltration rate requires data on the flow into a basin and the change in water levels. If the infiltration rate is low, water loses caused by evapotranspiration also need to be considered.

A great variation exists in the sophistication of monitoring programs. At the low end, systems can be operated with essentially no monitoring, other than perhaps periodic inspection to determine if the basins are being over topped. Water is diverted to channels or basins with no measurement of rates and volumes. Alternatively, large-capacity modern recharge systems can be fully automated with monitoring data sent to a central off-site location from which pumps and motorized valves are operated using a programmable logic controller (PLC) system (Gorey and Dent 2007).

Groundwater level monitoring provides information on the aquifer response to infiltration and thus the recharge rate. Groundwater monitoring is also important to evaluate potential impacts to off-site properties. The rise in the water table at infiltration basin sites could cause surface ponding in low-lying down-gradient areas. Hydraulic gradient data from a network of monitoring wells installed around an infiltration basin are needed for the evaluation of the rate of groundwater flow and thus contaminant transport.

Groundwater quality monitoring is implemented to provide data on the impact of infiltration basins on groundwater quality, which can be either beneficial or detrimental. In areas in which the shallow groundwater is of relatively poor quality (for example, due to high-salinities in coastal areas), recharge of freshwater may improve groundwater quality. Conversely, infiltration basins used for the recharge of wastewater may result in an increase in the concentrations of nutrients, pathogens, metals, and chemical contaminants. The number of wells, test parameters, and frequency of sampling should depend upon the quality of recharge water, site-specific impact concerns, and regulatory requirements.

23.2.2.6 Soil Aquifer Treatment

Soil-aquifer treatment (SAT) is a high-rate land application system that involves the application of partially-treated domestic wastewater to the soil surface to provide natural treatment as the water infiltrates into the soil and flows through underlying aquifers. The wastewater quality improvements from SAT are addressed in Chap. 31. Properly designed and managed SAT systems provide a clear, essentially odorless renovated water that can be used for unrestricted irrigation, primary contact recreation, and other purposes (Bouwer 1985). Post-treatment requirements will depend upon the quality of the recovered water and its intended use. Disinfection (chlorination) may only be required for non-potable uses. Some organic compounds may pass through the SAT systems and additional treatment (e.g., granulated activated carbon and reverse osmosis) may thus be required if the water is to be used for potable water supply.

SAT, as originally defined, differs from groundwater (aquifer) recharge using sewage effluent in that the recharged water is to be recovered and its extent in the aquifer controlled (Bouwer 1985, 1989, 1991; Idelovitch et al. 2003). Where the receiving aquifer contains freshwater, an integral part of the design and operation of the SAT system is controlling the flow of recharged water in the aquifer so that it can be collected instead of migrating away and eventually entering wells used for potable water supply (Bouwer 1989, 1991). The same natural treatment processes are active in other types of surface spreading systems, whether or not the recharged water is locally contained or controlled. The term SAT has recently been more generally applied to surface application systems that use vadose zone transport as a natural treatment process for groundwater recharge. SAT systems may be operated to also provide seasonal and multiannual storage of reclaimed water (e.g., Kanarek and Michail 1996).

Bouwer (1974, 1989, 1991) presented several conceptual design types of SAT systems. The infiltration area creates a local hydraulic mound and the recovery areas (whether natural or engineered) result in a cone of depression that captures the recharged water (Fig. 23.8). The key design consideration is configuring the infiltration and recovery locations and rates so that the movement of the plume of effluent is controlled. Groundwater modeling is thus an essential tool for system design.

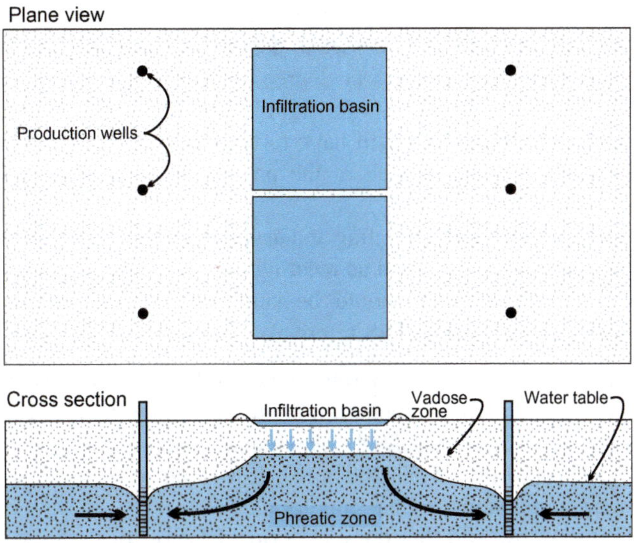

Fig. 23.8 Conceptual diagram of a SAT system with central infiltration basins surrounded by production wells

The effluent is usually applied within engineered infiltration basins. The vadose (unsaturated) zone is used as a natural filter to remove suspended solids, biodegradable organic matter, and pathogenic microorganisms. Significant reductions in nutrients and heavy metals may also occur by sorption and a variety of biologically mediated reactions (Bouwer 1989, 1991; Pescod 1992; Fox et al. 2001a, b). Additional filtration and removal of contaminants occurs as the water travels through the aquifer.

SAT systems do not result in a reduction in salinity (total dissolved solids). Salinity may actually increase somewhat as the result of evaporation, leaching of salts in the soil, and the atmospheric deposition of salt as dust and aerosols. In situations where the wastewater flows have a relatively high salinity; SAT alone may not produce irrigation quality water.

The primary advantage of SAT systems is that they are technically simple to construct and operate and have relatively low construction and operational costs. A disadvantage of SAT systems is that they have relatively large land requirements, especially for large capacity systems. System performance is also dependent on local hydrogeological conditions, which may not be favorable. The basic design considerations for SAT systems are (Bouwer 1974, 1989, 1991; Pescod 1992)

- Achieving a sufficiently high infiltration (hydraulic loading) rate so that the target effluent flows can be accommodated in an economically-sized infiltration basin system.

- Soil and aquifer sediments should allow for sufficient filtration and retention time for natural attenuation processes to achieve water quality goals. Very high loading rates may result in a lesser degree of treatment.
- If the applied sewage effluent is of a significantly lesser quality than the native groundwater, then the system will have to be designed and operated to prevent migration of the effluent outside of the part of the aquifer used for the SAT system (containment).
- Development of the optimal flooding and drying schedules must be accomplished to achieve treatment goals, such as meeting nutrient concentration targets.
- Pretreatment of wastewater should be sufficient to allow for an acceptable (manageable) clogging rate (TSS removal).

Pretreatment is additionally important in controlling the redox state of the SAT system. High organic carbon concentrations result in high total oxygen demands, and in turn the removal of dissolved oxygen and creation of anaerobic conditions in the saturated zone (Fox et al. 2001a, b). Low total oxygen demands are necessary in order to maintain aerobic conditions.

Bouwer (1985) observed that the most critical factor for the successful operation of an SAT system is unquestionably having an adequate basin area to handle the design flow and to allow for a system-specific optimal wetting and drying schedule. System design must, therefore, consider the inevitable reductions in infiltration and hydraulic loading rates due to clogging. Greater travel time and distance to production wells, in general, result in a better quality of the recovered water, at least up to a certain limit (Bouwer 1974). Most of the improvement in water quality occurs in upper 1–2 m (3.3–6.6 ft) of the infiltrated soil. There are thus diminishing returns concerning water quality improvement with increasing retention time and travel distance.

The recommended pretreatment for municipal wastewater for SAT includes primary and secondary treatment or a stabilization pond. Tertiary treatment and disinfection are also commonly performed in the United States (Asano and Cotruvo 2004), where SAT may serve more of a polishing function. Alternatively, the SAT process is used in Spain and France as a tertiary treatment processes to prepare the water for unrestricted irrigation (Asano and Cotruvo 2004). Additional filtration may be required to lower the TSS concentration, which relates to bacterial and viral control.

SAT systems should be constructed in areas with granular soils that have sufficient permeability to give high infiltration rates, but yet be fine enough to provide good filtration. The best soils are in the fine sand, loamy sand, and sandy loam range (Bouwer 1985, 1989, 1991; Pescod 1992). Ideally, the fine sands should be underlain by coarser materials. A critical issue is that the vadose zone and shallow aquifer underlying should not contain beds of impermeable material such as clays that would impede the vertical flow of water. The infiltration basins should also be located and designed so that the basin floors are at least one meter (3 feet) above the water table at all times (Pescod 1992). Sites with shallow water tables are unsuitable for SAT because they have an inadequately thick vadose zone.

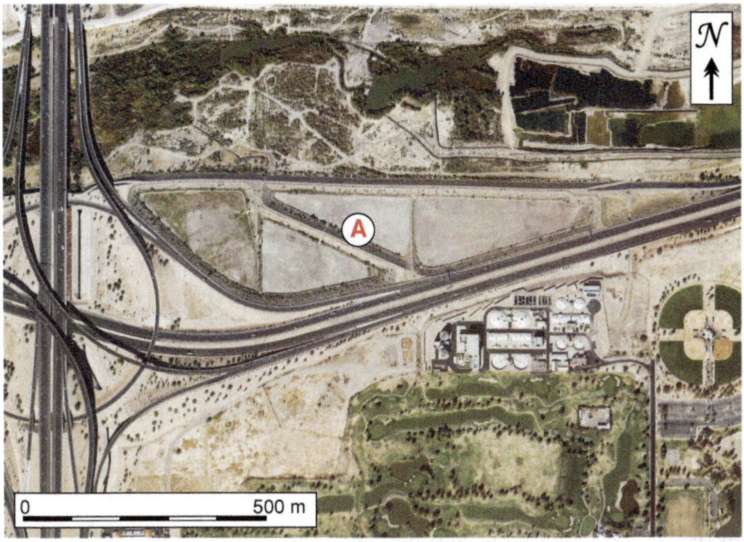

Fig. 23.9 Aerial photograph of the Mesa Northwest Water Reclamation Plant and soil-aquifer treatment system recharge basins (*A*). *Source* U.S. Geological Survey

Operationally, the filtering out of suspended material at and just below the soil surface will result in the formation of a clogging layer that will progressively reduce the vertical hydraulic conductivity and infiltration rate. Management of the clogging layer is a critical operational concern for SAT systems. Periodic drying of basins is therefore necessary to dry and crack the clogging later in order to restore infiltration capacity. The drying process results in the formation of desiccation cracks in the clogging layers and curling of the flakes, which restores system performance. SAT systems are thus designed with multiple basins, which allow some basins to dry while one or more basins are being flooded (Fig. 23.9). The clogging layer must be periodically removed. Complete removal of the clogging layer is preferred to mixing it with the underlying soil using, for example, a disk harrow (Pescod 1992). The later will improve system performance in the short-term, but the accumulation of fine material in the upper soil will eventually require the upper (mixed) soil to be replaced.

The effluent sent to the SAT system should have some treatment to reduce the concentration of suspended solids. High suspended solids concentrations will result in more rapid clogging, which would necessitate larger basin areas and operational efforts to maintain system performance. Biological activity in the infiltration basins should also be controlled. Excessive algae growth could cause accelerated clogging through both the accumulation of organic matter and induced calcium carbonate precipitation by the removal of carbon dioxide (Bouwer 1991; Asano and Cotruvo 2004). Bacterial biofilms that can form on the floor of infiltration basins can become essentially impervious after only a minor thickness of organic material has accumulated. Weed and insect control is also a concern for SAT systems.

SAT systems can also remove nitrogen compounds by the bacterially-mediated alternation of nitrification under aerobic conditions, and denitrification under anaerobic conditions. Nitrate in the recharge water can be removed by denitrification under anaerobic conditions whereby it is converted to nitrogen gas. Where the nitrogen is in the form of ammonium, it can be removed by first converting it to nitrate under aerobic conditions (nitrification) and then converting it to nitrogen by denitrification with a transition to anaerobic conditions.

The form and concentration of nitrogen compounds in water passing through SAT systems will depend upon the hydraulic loading rates and flooding and drying period schedule for the infiltration basins (Bouwer 1974, 1985, 1991; Pescod 1992). System specific flooding and drying schedules need to consider factors such as (Bouwer 1991):

- ammonium and carbon concentrations in the sewage effluent,
- infiltration rate,
- cation exchange capacity of the soil,
- exchangeable ammonium percentage in the soil,
- depth of oxygen penetration into the soil during drying periods, and
- soil and effluent temperature.

These noted factors can determine whether aerobic or anaerobic processes predominate within the SAT system. Frequent, short-duration (several days long) flooding and drying periods promotes aerobic processes, which favor the nitrification of ammonium to nitrate. Longer (one month or greater) flooding and drying periods lead to anaerobic conditions. Rapid infiltration systems that are continuously flooded maintain anaerobic conditions, which are not conducive for nitrogen removal (Benhan-Blair & Associates and Engineering Enterprises 1979). Intermediate (1–2 week) flooding and drying periods can result in a succession of aerobic and anaerobic conditions in the upper part of the soil profile, which stimulates nitrification and then denitrification, respectively (Pescod 1992). Optimization of the operation of SAT systems is an exercise in adaptive management. The optimal site-specific schedule for flooding and drying, and the cleaning of the basins, must be evaluated based on on-site experimentation and monitoring (Pescod 1992).

A design and operational consideration for SAT systems is that contaminant removal includes both biotic (renewable) and abiotic (exhaustible) removal mechanisms (Quanrud et al. 1996). Soils have a finite sorptive capacity, which may eventually become exhausted. If sorption is the primary removal mechanism, the breakthrough of contaminants could eventually occur. Biological removal processes may continue indefinitely and are thus considered to be sustainable. Some organic compounds that are absorbed onto soil undergo biodegradation over time and thus do not result in the long-term loss of the sorptive capacity of the soil (Idelovitch et al. 2003). Even where non-sustainable processes (adsorption and chemical precipitation) are important components of the treatment process, these processes can last for a very long term (tens or hundreds of years; Idelovitch et al. 2003). The Dan Region Project (Israel) SAT system still had excellent and stable removal of trace elements and phosphorous after 25 years of operation (Idelovitch et al. 2003). Long-term system performance will depend on local hydrogeological conditions, and system design and operation.

23.2.2.7 Recharge (Infiltration) Trenches, Galleries, and Shafts

Recharge trenches and galleries are excavations that are commonly filled with high-permeability coarse-grained sediment (usually gravel). Water is applied through screened (slotted or perforated) pipe. Trenches and galleries have essential the same construction. Galleries differ from trenches in that they are wider and a single excavation contains multiple screened pipes. The gravel used to fill trenches serves mainly to support the trench and prevent collapse of the trench wells, and to provide water storage. Geotextile fabric liners may be used to separate the fill and native aquifer material. Where the surficial sediments are unconsolidated, temporary sheet piling may be required in order to prevent collapse of the walls during excavation and construction.

The major advantage of recharge trenches and galleries is that their vertical walls provide a larger area for infiltration compared to a basin of the same area. The vertical walls are less prone to clogging due to suspended sediment settling. Recharge trenches and galleries can be covered with turf, pavement, or otherwise made unobtrusive. Trenches and galleries also have the advantages that there is less potential for unsupervised or uncontrolled contact with the recharged water and there is no potential for the mosquito breeding associated with ponds (Bekele et al. 2009). However, a limitation of recharge trenches is that they are prone to clogging, similar to recharge wells and infiltration basins.

Recharge trenches and galleries have a wide variety of uses. They are very commonly used for small-scale local stormwater disposal systems. The drain fields of household septic systems are another very widely employed example of infiltration galleries. Galleries may also be used for water supply. Missimer (2009) discusses the applications of beach galleries for raw water supply for membrane treatment facilities.

A variety of design options are available for recharge trenches and galleries, which vary in their length, width, and depth, the number, orientation, and design of screens, and the type of fill material used. The depth of the trench depends on the depth to the water table, system capacity, and local hydrogeological conditions. Small, low-capacity, systems may be simply designed using data from infiltration (percolation) tests and projected flows. Optimization of the design of larger-scale systems used for aquifer recharge typically requires groundwater modeling, as is the case for surface spreading systems. Various design options may be modeled in order to determine the design that can most efficiently achieve system target capacities and minimize or avoid potential adverse impacts.

A conceptual diagram of a typical gravel-filled infiltration trench is provided in Fig. 23.10. Infiltration trenches in the Ruhr region of Germany were described by Hantke and Schlegel (1995), which have a basic design of a 1 m (3.3 ft) width and depths of up to 6 m (19.7 ft). The trenches are filled with coarse sand and covered with either, wood, corrugated steel, or concrete slabs. Clogging was not a significant

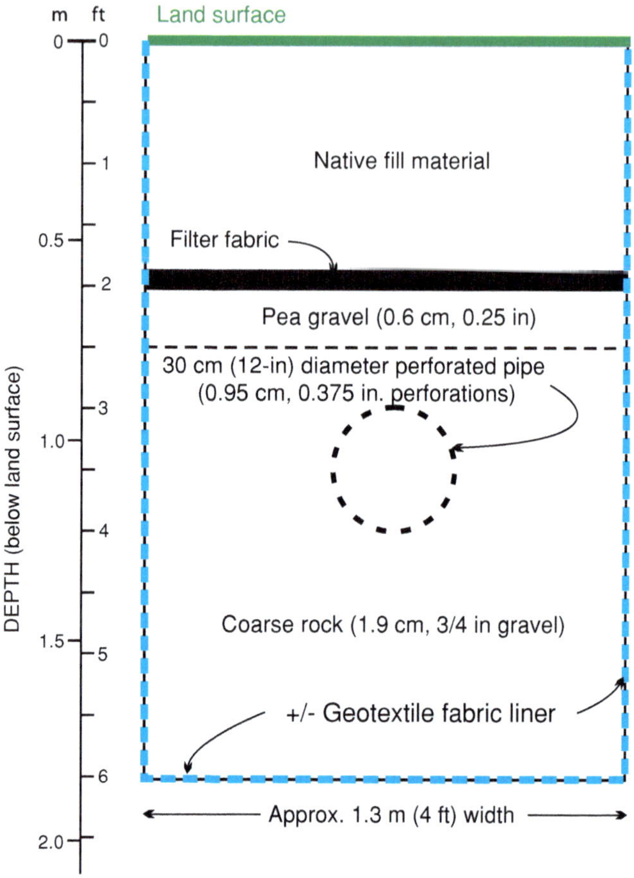

Fig. 23.10 Example of a basic infiltration trench design

problem when high quality water (e.g., river Rhine water purified to drinking water standards) was recharged.

Design options other than a gravel fill are available and have been implemented. For example, the Atlantis® Infiltration Tank System (developed by Atlantic Water Management) uses modular polypropylene crates (Matrix™ tank modules) as a substitute for gravel. The Atlantis® system has the benefits of rapid construction, light weight, strength, and high storage capacity (90% void space). The Rainstore® system (developed by Invisible Structures, Inc.) consists of stackable thin-walled cylindrical columns constructed of either high-impact polypropylene (HIPP) or high-density polyethylene (HDPE) plastic. The system can be used for water storage (detention) through the use of an impermeable liner or for recharge (retention) through the use of a permeable geotextile liner.

Field testing was performed in Western Australia of secondary-treated wastewater disposal using infiltration galleries constructed with either a gravel-fill or the

Atlantis® system (Bekele et al. 2009). The gallery constructed using the Atlantis® was found to be less prone to clogging. The cause of clogging of the gravel-filled trench was uncertain, but may be related to the presence of plant roots. Although the Atlantis® system was more expensive to construct, its superior performance in terms of less clogging and reduced maintenance costs far outweighed its greater construction cost (Bekele et al. 2009).

Recharge shafts are similar to recharge trenches, with the exception that they are circular and have a smaller areal extent. Chadha (2003) documented recharge shafts (dry wells) in India, which are vertical shafts 2–3 m (3.3–6.6 ft) in diameter up to 6 m (19.7 ft) deep that are excavated into relative high-permeability granular strata. Shafts are filled with sand and gravel pack to act as an inverted filter so as to provide silt-free water for recharge.

The monitoring requirements for recharge trenches and galleries are system specific. Small-scale systems, such as used for local stormwater disposal and septic drain fields usually involve no monitoring other than observation of system performance. For large capacity systems, flow rates should be recorded and monitoring wells installed both within and adjacent to trenches and galleries in order to allow for evaluation of increases in the water level differential that would be indicative of significant clogging. As the trench or gallery walls clog, water levels within the trench during recharge should become progressively greater than those in the adjoining aquifer. Operation of a systems may be automated though the use a float switch and programmable logic device, which is programmed to adjust flow rates in order to maintain a target water level in the trench or gallery (Bekele et al. 2009).

23.3 MAR by Enhancement of Natural Recharge Processes

Aquifer recharge may be increased by a variety of processes that increase infiltration. The general concept is that recharge from ephemeral flows may be increased by employment of measures that slow the rate of water flow, retain water, and increase the rate of infiltration. MAR by natural recharge enhancement processes is in essence a larger-scale version of rainwater harvesting using earthworks (Sect. 22.3.1). Commonly natural recharge enhancement processes involve modification of ephemeral stream channels.

23.3.1 In-Channel Recharge Methods

In-channel recharge methods include discharge to natural channels with no improvements (e.g., discharge of stormwater and treated wastewater into wadis) and various modifications to the stream channel in order to increase the infiltration capacity. In-channel system modifications may involve the following:

- construction of dams (permanent or inflatable) to retain stormwater,
- conditioning the bed by scraping off fine surface sediments or crusts, or the replacement of surficial sediments with more permeable coarser-grained sediments,
- leveling or widening of the channel, and
- construction of permanent or temporary check dams and levees to slow the flow.

The key design considerations are maximizing the volume of infiltration by increasing the area exposed to water, conditioning the channel bed to increase the infiltration rate (volume per time per unit), retaining water (infiltration duration), and increasing water depth and thus hydraulic gradient.

The Orange County Water Department (OCWD; California, USA) in-channel recharge system in the Santa Ana River is a textbook example of the use of temporary sand levees to create a serpentine flow path (Fig. 23.11). The levees increase the length of the flow path, reduce the flow velocity, and increase the wetted area, and, as result, increase the infiltration rate. During major floods the temporary levees are washed out and thus cease to be an impediment to the flow of flood waters. Inflatable rubber dams are also used to divert water from the channel into off-channel recharge basins (Markus et al. 1995).

23.3.2 Wadi Dams

Wadi flows, by definition, are intermittent as the result of infrequent, and often short and intense, nature of rainfall events in arid and semiarid lands. Due to the ephemeral nature of wadi floods, floodwater retention or recharge dams have been constructed in arid regions, and are especially common in the Middle East (e.g., Saudi Arabia, Oman, United Arab Emirates). The dams serve two primary purposes: flood control and aquifer recharge. Flash floods in wadis can be extremely intense and result in severe damage to physical structures and loss of life. Wadi dams are thus a critical flood control element. Wadi dams also act to retain water for recharge that would otherwise flow away and not be available for beneficial use. The reservoirs act as both recharge sites and siltation basins.

Permanent wadi dams can have a variety of construction designs, similar to dams constructed on perennial rivers, including rock and earth-fill embankment dams, masonry dams, gravity dams and arch-gravity dams (U.S. Department of the Interior 1987). Dams must be constructed in accordance with standard engineering design principles in order to protect downstream communities. More recently constructed large wadi dams tend to be of the concrete gravity design. Smaller dams may be of less robust earthen construction. For example, percolation ponds and tanks are very widely used in India with purpose of harvesting surface water runoff during the monsoon period. Percolation tanks are commonly simple earthen dams, several 100 m (feet) long and several meters (feet) high that are constructed across ephemeral stream channels (Sukhija 2008).

Fig. 23.11 Santa Anna River, Orange County, California in-channal and off-channel recharge facilities. A series of low berms in the main river channel (*A*) create a more tortuous and slower flow. Off-channel recharge basins that parallel the main channel (*B*) are present further to the north. *Source* U.S. Geological Survey

Wadi dams act to increase recharge by:

- increasing the time over which infiltration occurs,
- increasing the water depth and thus vertical hydraulic gradient,
- allowing for the settling of fines, which may increase downstream recharge, and
- increasing the area of potential recharge (wetted area).

The reservoir bed may lose much of its infiltration capacity from the deposition of silt and clay. The reservoirs actually act instead as large siltation basins. Cleaner water discharged downstream of the dam more readily infiltrates into the wadi alluvial sediments (where downstream flow can occur based on the dam design). Depending upon the system, much of the recharge may, therefore, actually occur downstream of the dam (e.g., Zeelie 2002; Ministry of Regional Municipalities, Environment and Water Resources 2005). The effects of wadi dam recharge systems can be dramatic. For example, Zeelie (2002) reported that the wadi dam and infiltration basin in the Omaruru Delta of Namibia increased the sustainable yield of the alluvial aquifer during the study period from 2.8 to 5.9 million m^3/d (739–1,558 million gallons). However, it was noted that future sustainable yields could be reduced by a reduction in the dam storage capacity (e.g., due to siltation) and that Omaruru River can experience multiple year periods during which there is no runoff stored.

Wadi dam systems can consist of either of a single large gravity dam or a cascade of smaller check dams. The advantages of check dams include (Alderwish 2010):

- regulate flow by reducing velocity,
- delays (extends) the duration of flow in the wadi channel and hence increases infiltration opportunity,
- removes suspended materials, providing clearer water downstream, which infiltrates more readily, and
- check dams may be less expensive and safer than larger gravity dams.

Recharge estimates for new and existing dam sites can be obtained using water-balance methods based on monitoring data, Darcian flow-net approaches, and groundwater modeling (Alderwish 2010). The key variables affecting the recharge rate are the vertical hydraulic conductivity of the reservoir bed sediments, reservoir area, the depth to the water table beneath the channel, and the retention time. Some of the water that infiltrates into the wadi channel bed sediments is lost to subsequent evapotranspiration and never reaches the water table to become recharge.

Water budget estimates for two recharge reservoirs in central Saudi Arabia indicate that a high percentage (82 and 94.5%) of water in reservoirs is taken into the soil (Al-Turbak and Al-Muttair 1989). Al-Turbak (1991) investigated groundwater recharge at a reservoir in central Saudi Arabia (Al-Amalih). Water budget estimates indicate that 93.3% of water stored during the 1985–1986 season infiltrated into the soil. The groundwater outflow was estimated using Darcy's law from the hydraulic conductivity estimated from a pumping test, the hydraulic gradient, and the cross-section area of flow determined from a geophysical survey and water table altitude data. The study results suggest that surface reservoirs are very effective in infiltrating water, but that a significant fraction of the water remains far away from where it is needed. A critical issue is that where there is a great depth to the water table, a large amount of water that infiltrates through the bottom of a reservoir may not recharge the underlying aquifer.

The efficiency of infiltration in wadi reservoirs is greatly dependent on the fine-sediment load of the streams. In high-gradient outwash wadis, the sediment load is extreme and a single flood event can create a mud layer over the entire bottom of the reservoir and the overlying water can help compact this layer. With age the wadi reservoir becomes much less effective for recharge if the bottom is not cleared of sediment. Low recharge rates (as opposed to infiltration rates), in the range of 10–25%, have been recorded in numerous Middle East wadi reservoirs. The estimated recharge in percolation tanks in India using the CMB method is 20–60%, with the greatest rates occurring where tanks are located on sediment or rock with a high hydraulic conductivity (Sukhija 2008).

Because of relatively high land surface slopes in upstream areas of wadis, recharged water will flow downstream in alluvial aquifers rather than result primarily in local storage. Recharge results in an increase in down-gradient water levels and an improvement in water quality. For example, analysis of water quality data using geostatistical methods in a GIS environment has demonstrated the improvement in water quality (a decrease in total dissolved solids concentration) that occurs downstream of artificial recharge dams in Oman (Bajjali 2005). The recharge water acts to push saline-water seawards.

Recharge dams can provide significant water resources benefits to local communities. For example, Ibrahim (2009) documented how recharge dams across intermittent streams in Gadarif City, Sudan provided important benefits to the community. The first dam was constructed in 1995 for flood control purposes, but it was noted that water stored behind the dam had infiltrated into the soil and raised the water table by a few meters (more than 6 ft). A second dam was constructed in 1998 on another stream in the city to further increase aquifer recharge. The recharge dams resulted in the renewed availability of water in some previously dry wells and allowed the villagers to drill new boreholes and wells to obtain water for drinking and farming. Recharge from the dams resulted in 2,200 m^3/d (580,800 gpd) of new groundwater resources, which is approximately 12% of the total demand.

Wadi recharge dams have the same operations and maintenance issues as other surface spreading recharge systems, particularly declines in recharge rates caused by the accumulation of silts and organic matter. Management of recharge dams in Saudi Arabia was investigated by Al-Muttair et al. (1994). Management options investigated in a comparative study were:

- release of water to downstream beds,
- release to downstream artificial recharge basins,
- silt removal, and
- scratching the reservoir beds using a plow with chisels.

Infiltration rates were measured from the decline in water levels corrected for evapotranspiration, as measured using Class A evaporation pans at each of the studied dam sites. All four management methods provided benefits. Wadi channel recharge was determined to be a cost-effective option due to the naturally high infiltration rates of the coarse wadi sediments. Scratching was effective in the basin with a modest silt accumulation in which the chisels penetrated through the entire thickness of the silt layer. Discharge to recharge basins showed the least benefits, which was attributed to site-specific conditions at a study site. In this case, the reservoir already had a high infiltration rate and the infiltration rate of the basin happened to be relatively low.

Optimizing of aquifer recharge in wadis is a research area that merits further investigation. More data are needed on the both the infiltration rates and recharge rates for existing wadi dam systems. Additional research is particularly needed on the effectiveness of different options for increasing recharge rates. Storage of water in the reservoirs before release allows for more time for settling of suspended solids, which is beneficial for downstream infiltration rates, but is also allows for more time for evaporative losses. More wadi dams need to be designed and constructed that have the ability to discharge downstream via gate and controlled spillway systems. Economic aspects need to be taken into consideration, in addition to recharge rate and water quality. In light of the remote location of some

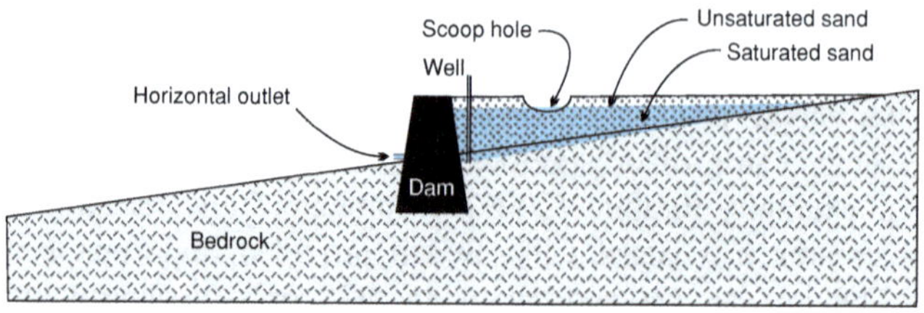

Fig. 23.12 Conceptual diagram of a sand dam. Water stored in sand deposited upstream of the dam may be recovered using a horizontal outlet near the base of the dam, a vertical well, or a scoop hole

wadi dams and high sporadic nature of runoff events, automated or passive systems are preferred over options that require considerable on-site action.

23.3.3 Sand Dams

Sand dams, which are also referred to as trap dams, sand-storage dams, and barrier dams, are impermeable structures constructed across ephemeral river valleys to trap sand and gravel and thus create an artificial aquifer (Fig. 23.12; Baurne 1984; Nilsson 1988; Van Haveren 2004; Aerts et al. 2007). Sand dams are different from subsurface dams (also referred to as groundwater dams), which are impervious barriers constructed in existing sediment deposits in order to arrest flow within a natural aquifer (Fig. 23.13). Sediment deposition occurs primarily during heavy rainfall events. The reduction in flow velocities in the reservoir behind the dam results in the deposition of the bedload and the infiltration of water. The distinction between sand dams and groundwater dams is not clear cut as sand dams may serve a dual function by intercepting local groundwater flow (Nilsson 1988).

Both sand sands and groundwater dams are increasingly being looked at as an element of small-scale water supply in developing countries (Nilsson 1988). There are practical limits on the quantity of water that can be stored up-gradient of both types of dams, which limits their potential contribution to large urban areas. However, the volume of water that can be stored in sand and groundwater dam impoundments can significantly contribute to the domestic and irrigation water supply of villages.

As reviewed by Nilsson (1988), Agarwal and Narain (1997) and Van Haveren (2004), sand dams have a very long history for creation of domestic, irrigation, and stock water supplies. The principal advantage of sand dams is that they have greatly reduced evaporative losses compared to surface reservoirs. The reduced evaporation, particularly during prolonged dry periods, results in sand dam created aquifers being able to provide a comparable or greater volume of water than a comparable sized surface reservoir, depending on local conditions. The infiltration of water through the trapped sediments can also result in an improvement in water quality through the filtering out of pathogenic microorganisms. Additional benefits

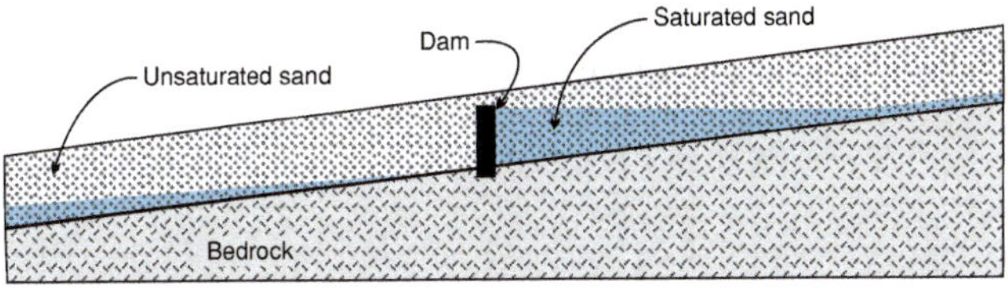

Fig. 23.13 Conceptual diagram of a groundwater dam. An impermeable barrier is constructed across an aquifer, perpendicular to the groundwater flow direction. Groundwater accumulates on the upstream side of the dam

include that a series (cascade) of dams can increase regional aquifer recharge and water levels, moderate stream flow (i.e., provide flood control benefits), and allow for sand harvesting and rehabilitation of gullies. Sand dams also tend to occupy low value land, are low maintenance, and are a low sophistication-level technology that is inexpensive enough to be implemented by local communities with available resources (Mutiso 2003).

Sand dams function both as a barrier and as a spillway (Van Haveren 2004). During flash floods, the sand and gravel aquifer is recharged, while the bulk of the flow passes over the dam. The stored water is later captured by digging scoop holes, construction of vertical wells, or use of a horizontal outlet (delivery) pipe installed through the dam. Scoop holes are the least expensive option, but are vulnerable to contamination. Outlet pipes have the advantage that the water flows under hydrostatic pressure, but they also have the disadvantages that they may weaken the dam structure, maintenance is complicated, and it is a more expensive option (RAIN 2008).

The dams are constructed above ground and are commonly up to 4–6 m (13.1–19.7 ft) high. The dams may be constructed of either reinforced concrete, stone masonry, compacted earth with a concrete spillway, or stone gabion or blocks either sealed with a clay layer or have a clay core (Nilsson 1988). Ideally, locally available materials should be used to the degree practical. A key design issue is to fill the reservoir with coarse sediments, avoiding silt and clay deposition, which could impede vertical flow and recharge of the sand reservoir. Larger dams should be constructed in stages to ensure that only sand and gravel are retained behind the dam. High dams would tend to also retain fine sands and silts, which would reduce the effective storage capacity. Finer grain sediments also result in significantly greater evaporative losses (Hellwig 1973). Baurne (1984) recommended that dams should be constructed with a closable v-shaped or rectangular notch that can be opened so as to prevent clay and silt from settling during low flow periods.

The amount of evaporation depends upon capillarity, which is inversely related to grain size (Hellwig 1973). Coarse sands and gravel thus tend to have lesser evaporative losses. Coarse deposits also are more permeable and thus have greater infiltration rates, which allow the artificial aquifers to more rapidly fill during flood events.

The amount of extractable water (V_e) depends upon the volume of sediment behind a dam, its specific yield (S_y), and any net inflows and outflows from the aquifer from baseflow and leakage. Extractable water value (V_e) can be estimated as follows (after Aerts et al. 2007):

$$V_e = \left(X * W * D * S_y\right) + B - L_v - L_{out} + L_{in} \quad (23.3)$$

where,

X = length of river segment (m) (ft)
W = average width of river segment (m) (ft)
D = average thickness of riverbed sediments (m) (ft)
B = baseflow into river sediments from riverbanks (m^3) (ft^3)
L_v = vertical leakage out bottom of aquifer (m^3) (ft^3),
L_{out} = horizontal leakage out of the aquifer downstream of dam (m^3) (ft^3)
L_{in} = horizontal leakage into the aquifer from upstream of dams (m^3) (ft^3).

The dams act to obstruct the downstream groundwater flow through the river bed. The volume of water that is available for abstraction is considerably greater than the volume present in the artificial aquifer because of the additional water stored in the riverbanks and baseflow (RAIN 2008).

Sand dams are a relatively inexpensive community based project (Van Haveren 2004). The importance of community involvement throughout the development, operation, and maintenance of a project is strongly stressed (RAIN 2008). Community acceptance is critical for the initiation of a sand dam project and in the selection of dam sites. As is the case for water storage projects in general, some arrangement is necessary for the equitable use of the stored water. Commitment and organization of the community are very important for the success of sand dams projects (Lasage et al. 2008).

Mutiso (2003), Aerts et al. (2007) and Lasage et al. (2008) discussed sand dams in the Kitui District of Kenya, which has the greatest density of dams in the world. Sand dams are a community-scale water management strategy in the District. The construction and maintenance of the dams is dependent on inputs and commitments of the local community in conjunction with technical and financial support from a NGO (Sahelian Solution Foundation, SASOL). Socio-economic indicators and hydrologic data indicate that the sand dams are a successful local adaptation to deal with droughts. By providing a local supply of water, less time is needed to fetch water, which has benefits such as significantly increasing school attendance and making more time available for income producing activities (RAIN 2008).Currently, approximately 500 dams capture a small fraction of the annual steam flow and, therefore, do not have a significant impact on downstream water resources. However, if global climate change increases the variability in rainfall, then during some dry years the water captured by the dams could cause downstream water shortages. More frequent water shortages would also occur if the number of dams were significantly increased.

The Rainwater Harvesting Implementation Network (RAIN 2008) prepared a very useful guide to sand dam implementation. It must be stressed that dam construction projects should include an experienced engineer in the project team.

Dams must be designed to withstand peak flows, which must be discharged safely without causing erosion of the river bank, and the dam must be anchored in the bedrock and stream banks to prevent failure during floods and underflow losses from the created aquifer (Nilsson 1988). A critical part of sand dam projects is determining suitable locations. The selection of locations of suitable catchments and riverbeds needs to consider the following factors (RAIN 2008):

- proximity to potential water users,
- river width (should be no more than 25 m; 82 ft),
- river slope gradient (0.3–4%, preferably 2–4%),
- availability of coarse sediments in the catchment,
- maximum flood height,
- capability of the riverbed to store water; dam should be built on bed rock or an impermeable layer, and
- riverbank height; both banks should be high enough so that the river will never overflow its banks during maximum flood events.

If the river slope gradient is too low, then coarse grained material cannot normally be transported and predominantly fine-grained materials will be deposited in the reservoir. Where the slope is steep, the storage volume for a given dam height is less or a higher dam must be constructed to achieve storage goals. If the bedrock below a dam is relatively permeable (for example, from the presence of fractures), then efforts should be made to seal the fractures, such as by pressure grouting or pouring very thin mortar into the fracture network (Nilsson 1988).

Sand dams, if well constructed, should require little if no maintenance. However, the community organization should be capable and willing to perform whatever maintenance tasks are necessary. Potential maintenance requirements include (RAIN 2008):

- repairing cracks and weak points in the dam and abutments,
- cleaning the well or outlet, and
- removing silt from the top of the aquifer.

23.4 MAR Systems Involving Subsurface Engineering

Most MAR systems store water by taking advantage of aquifers in their natural condition. Storage capability of aquifers can be greatly enhanced by various subsurface construction techniques, particularly the installation of horizontal flow barriers. A groundwater dam can increase the volume of groundwater locally in storage by obstructing the downstream flow of groundwater within alluvial aquifers. A physical storage ASR system can be created by surrounding a volume of porous and permeable strata with a hydraulic flow barrier.

23.4.1 Groundwater Dams

Groundwater dams are not strictly an MAR technology because they do not involve actual additional recharge of water. Groundwater dams are of water resources value as a water conservation technology because they facilitate the collection of shallow groundwater that might otherwise not be available for beneficial use. However, the impacts to downstream areas must be considered during the planning stage of groundwater dam schemes (Nilsson 1988). Construction of a groundwater dam could deprive downstream shallow aquifer users of water upon which they depend.

Well-defined narrow valleys underlain and bordered by impervious strata are preferred locations for groundwater dams. Natural impermeable features such a rock bars that form natural underground dams (Nilsson 1988) and valley or channel constrictions may be preferred dam locations. As is the case with surface water dams, avoidance of vertical and lateral leakage (flow around the dam) is a key design issue. There is a wide variety of construction options for groundwater dams because they are subject to much lesser stress that surface dams. Construction materials that may be used include (Hanson and Nilsson 1986; Nilsson 1988; Onder and Yilmaz 2005):

- local clay, which may be protected by plastic sheeting,
- poured bentonite with a variable amount of cement to produce rigidity as needed,
- concrete,
- stone masonry,
- plastic or tar felt sheet mounted on a wooden frame,
- sheets of steel, corrugated iron or PVC, either installed in a trench of driven in (i.e., sheet piling), and
- injection of bentonite or cement grout.

Typically, the crest of the dam is kept some distance below land surface to avoid water logging and erosion by bed load transported sediments. Groundwater dams are essentially linear features that take advantage of existing lateral confinement. The groundwater dam concept can be taken a step further by constructing four ‘walls’ to create a subsurface storage ‘tank’.

An important technical issue associated with groundwater dams is constructing them so that they actually arrest the flow of water and result in higher groundwater levels. A thorough hydrogeologic analysis is therefore necessary to evaluate the magnitude of the horizontal flow, the storage properties of the sediments, and the vertical and lateral confinement (Nilsson 1988). Monitoring wells (piezometers) should be installed upstream and downstream of the dam site to determine the existing hydraulic gradient and for the later evaluation of dam performance (Nilsson 1988).

Siting criteria for groundwater dams in hard-rock terrains were reviewed by Ramasesha et al. (2002), which include

- The valley should be well-defined with ideally a narrow outlet (bottle necked), which reduces the necessary size of the dam and thus cost.
- The land surface and groundwater hydraulic gradient should not be too steep (<1%) since high gradients reduce the storage volume behind a dam of a given height.
- The unconfined aquifer should not be too deep (≤10 m; 32.8 ft) and have a well-defined impermeable base.
- The thickness of the unconfined alluvial aquifer should be adequate (>5 m; >16.4 ft) so that the quantity of groundwater stored is commensurate with the effort and investment.
- The water table should be located well below the riverbed in order to provide storage space and avoid up-stream water logging.
- The dam should not be located upstream of an area of heavy groundwater use as it will result in a reduction in downstream groundwater flow.

A great advantage of groundwater dams is there is no recurring expenditures for maintenance and the life of the structures is expected to be longer than conventional artificial recharge structures (Ramasesha et al. 2002). As a subsurface construction, groundwater dams have a minimal surface footprint and do not interfere with local land use.

Groundwater dams have been used for over a century. An early example in Los Angeles, California, USA, was documented by Slichter (1902) and Kresic (2009). Groundwater dam implementation has been largely a small-scale technology implemented to provide water to local communities and farms (i.e., subsistence level water supply). Foster (2002) provided an economic evaluation of groundwater dam implementation in Brazil. About 500 dams were installed in the Pernambuco state in the 1990s. Approximately half of the 151 investigated dams were in active use. Inactivity of the other dams was caused by either construction problems making their use impossible or the availability of other sources of water. Economic analysis indicates that the costs of the dams could be recovered within a few years of operation. Foster (2002) noted that the human factor is essential for the success of groundwater dams. If there is no cooperative effort with the construction of the dams, and subsequent ownership by the community, then effective operation and adequate maintenance are unlikely to follow.

Groundwater dams constructed to date are predominantly small-scale structures (<10 m; <32.8 ft in depth). Large -scale groundwater dams could be used to store large volumes of water for strategic purposes or to meet seasonal needs. A groundwater dam (20–70 m; 66–230 ft) high was proposed for the Holy City of Makkah, in the Kingdom of Saudi Arabia (Al-Ghamdi 2009; Khairy et al. 2010). The primary water supply to the city is a desalination plant located on the Red Sea, which provides approximately 90% of the potable supply. A need for a strategic reserve of water was recognized in the case of a planned or accidental shut-down of the plant or pipeline. The threat would be aggregated if it occurred immediately before or during the pilgrimage season. The groundwater dam would be constructed across Wadi Naman and the recommended construction material is a plastic (high-bentonite) concrete installed using diaphragm wall techniques.

The main advantages of a subsurface dam for strategic water supply are (Al-Ghamdi 2009; Khairy et al. 2010):

- It does not result in the submergence of land and therefore does not damage the environment or cause socio-economic problems.
- It avoids significant evaporative losses that occur in surface reservoirs in hot, arid regions.
- It has low maintenance.
- There is no threat to downstream communities if the dam should ever fail.
- Groundwater quality is often good and there is a low susceptibility to contamination.

More advanced engineering is required for the design of groundwater dams of great depths. Numerical modeling is necessary to analyze vertical and horizontal stresses and the stability of cut-off wall design options (e.g., Khairy et al. 2010). For large-volume groundwater dam systems designed for strategic (i.e., long-term) water storage, water management measures should be instituted to restrict use of the stored water and prevent contamination (Al-Ghamdi 2009). The water management rules and regulations should control land use and activities in the immediately vicinity and upstream of the underground reservoir.

23.4.2 Artificial Aquifers

Sand dams take advantage of natural sedimentological processes to create a storage aquifer. An artificial aquifer can be created by excavating a basin, lining it with plastic or other impervious material (e.g., clay) and then filling the lined basin with clean porous sand or gravel (Helweg and Smith 1978). An overlying gravel mulch layer can further reduce evaporation losses. Excavation can be minimized by constructing the artificial aquifer within an existing gully or depression. Artificial aquifers were proposed as an appropriate technology for supplying water to developing countries in semi-arid regions as it can take advantage of relatively inexpensive labor and low equipment, materials, and energy costs (Helweg and Smith 1978). The artificial aquifers have the same advantages of conventional sand dams in that they take advantage of sand filtration to improve water quality and avoid most of the evaporative losses of surface reservoirs.

23.4.3 Physical Storage ASR Using Horizontal Flow Barriers

Physical storage ASR systems utilize aquifers, in essence, as a sediment or rock-filled storage tank. A fundamental requirement for physical storage ASR systems

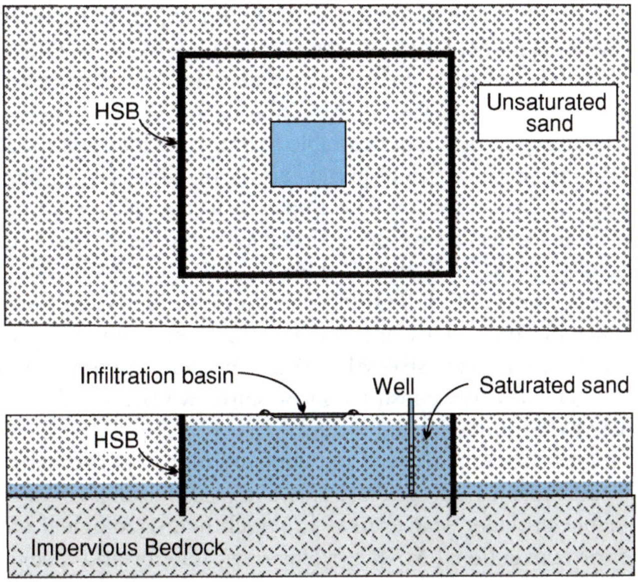

Fig. 23.14 Conceptual diagram of a physical storage ASR system constructed using horizontal salinity barriers (HSB). The HSBs are set in an impervious bedrock in order to prevent downward leakage losses. Recharge could be performed using either wells or infiltration basins

is that the aquifer be well confined at both its bottom and sides in order to minimize the loss of stored water to leakage. Where lateral confinement is not naturally present, horizontal flow barriers (HFBs) may be used to create "sides to the tank."

The basic concept is to install the HFBs to surround a volume of aquifer that is to be used as a storage reservoir (Fig. 23.14). The HFBs should be installed into the top of an underlying confining unit (e.g., shale or clay unit), which would retard downward flow of the stored water. HFBs could include slurry walls and grout curtains. Slurry walls are installed in near surface unconsolidated sediments (e.g., sand units), commonly using a trenching device that can simultaneously remove the sediments and emplace a slurry material (e.g., bentonite clay) into the excavated trench. Grout curtains are installed in rock or deeply buried sediments by injecting cement grout into the formation through closely spaced borings at high pressures. A variety of methods have been developed over the past two centuries for emplacing grout curtains including jet grouting, compaction grouting, permeation grouting, hydrofracture grouting, and consolidation grouting (Glossop 1960, 1961; Houlsby 1990; Weaver and Bruce 2007). The properties of the slurry walls, particularly their permeability and ability to maintain large head differentials, can be varied depending on the composition of the grout used, spacing of bore holes, and emplacement procedures.

HFBs are a widely and long-used technology for controlling groundwater flow. Grout curtains are a particularly important technology for the stabilization of dam foundations, in which much greater water pressures are involved than would normally occur in ASR systems. Construction of an ASR system using HFBs is technically feasible provided that favorable hydrogeological conditions are locally present, which include a sufficient porous and permeable storage unit and underlying confinement. The water could be emplaced using wells or surface spreading.

The Lake Belt In-Ground Reservoir Technology Project, which is part of the Comprehensive Everglades Restoration Plan (CERP) for South Florida, includes the use of approximately 30 m (100 ft) deep slurry walls to form a seepage barrier around an area of mine lakes and porous limestone with the goal of storing stormwater (USACOE and SFWMD 2002). Earlier proposed projects in the Middle East would involve construction of slurry walls around an area of sand dunes in order to store desalinated water.

A limitation of the technology is its relatively high cost compared to conventional ASR and other MAR systems. However, ASR using HFBs may be economically viable in water scarce areas in which hydrogeological conditions are unfavorable for conventional MAR systems (e.g., interior arid lands).

23.5 Stormwater Recharge

Urbanization commonly results in a reduction in permeable surface area and thus a decrease in recharge and an increase in runoff. MAR using stormwater serves three main functions; (1) recharge of the surficial (water-table) aquifer, (2) reduction of flows to combined stormwater and sewer systems, and (3) prevention or reduction of flooding. Stormwater picks up contaminants as it flows off roofs and land surfaces to recharge areas. Contaminants such as heavy metals, oil and grease, petroleum products, pesticides, industrial chemicals, pathogenic microorganisms, salt, and nutrients may accumulate on surfaces exposed to rainwater and can be mobilized in stormwater. Chemical pollutants include toxic organic chemicals such as volatile organic aromatic compounds, polynuclear aromatic hydrocarbons, phenols, and phthalates. MAR using stormwater thus poses a threat for groundwater contamination and may require some pretreatment before storage.

There has been considerable research on the groundwater pollution risks of stormwater recharge. Pitt (1996) and Pitt et al. (1999) provide an excellent overview of the groundwater contamination potential from urban stormwater infiltration practices. The NRMMC-EPHC-NHMRC (2009b) provides a very good review of the risks associated with stormwater harvesting and reuse. The contaminants present in stormwater and their concentrations will vary depending upon the activities conducted and chemical use in the collection area.

Rainfall itself may contain significant concentrations of contaminants. For example, Bucheli et al. (1998a) documented high concentrations of pesticides and their degradation process in rainfall in Switzerland. During the main pesticide application time, their concentration in rainwater may exceed the Swiss and European drinking water standards of 500 ng/L for some specific contaminants. The highest concentrations occur in the first few mm (0.1 inch) of rainfall (i.e., first flush), as the chemicals are washed out of the atmosphere. The high concentrations of the first flush are dampened out in the groundwater system by dilution, filtration, sorption, precipitation and other chemical and biological processes. Rainfall runoff may also pick up chemicals from roofing materials. Bucheli et al. (1998b), for example, reported concentrations of the biocide (R, S) mecoprop, which originates from a roof protection agent, in roof runoff at concentrations that exceeded the Swiss and European drinking water standards by up to a factor of 1×10^4. Runoff from roofing may also contain elevated concentrations of metals, such as copper and zinc.

A wide variety of systems are used for stormwater management, which can be divided into systems that primarily serve as a pretreatment function, those that discharge water onto land surface, and those that discharge below land surface. Pretreatment systems, which are commonly considered best management practices, include:

- retention ponds,
- detention ponds,
- catch basins and other sediment sumps,
- grass filter strips, and
- grass swales (vegetated open channels).

Surface application systems to recharge groundwater include infiltration basins (percolation ponds) and pervious pavements. Subsurface application systems include dry wells and soakaways, infiltration trenches, and French drains.

Stormwater management systems often serve dual functions. For example, retention ponds can serve a pretreatment and storage function as well as the site of groundwater recharge. Grass swales may also provide pretreatment, function as a stormwater conveyance, and allow for groundwater recharge.

Metals tend to have low solubilities and high removal rates caused by sorption, precipitation, and sedimentation processes. Metals and organic carbon compounds tend to accumulate in a sludge layer that forms at or near the base of the application system or in soil layers (e.g., Appleyard 1993; Pitt 1996; Mikkelsen et al. 1997; Barraud et al. 1999; Göbel et al. 2008). The sludge layer can become a source of pollution and eventually reduce system performance, so therefore, it should be periodically removed. Periodic street-cleaning can also substantially reduce the contaminant load entering stormwater management systems and subsequently, the groundwater system (Pitt 1996).

The sorption of contaminants reduces soil quality while impacts to groundwater are reduced. Most infiltration systems are ineffective in retaining soluble organic compounds (e.g., COD, TOC, TKN, some pesticides) present in the stormwater. Chemicals vary in their tendency to be sorbed onto soils and aquifer materials. Depending upon their physical and chemical properties, they will undergo degradation by chemical, biological, and physical (e.g., volatilization) processes.

The rate and degree of degradation is based both on the properties of the contaminant and the chemical and biological processes occurring in the retention, detention, and shallow groundwater receiving areas. Salts are typically not removed during stormwater treatment. The concentration of salts may actually increase if they are leached out of the vadose zone by infiltrating stormwater.

A wide variety of pathogenic microorganisms may be present in stormwater. Stormwater often has high concentrations of the indicator parameter fecal coliform bacteria. Specific bacteria reported in stormwater include *E. coli*, *Salmonella*, *Shigella*, and *Pseudomonas*. The concentration of pathogenic bacteria is usually below the infective dose level and they tend to be removed during stormwater treatment and recharge (Pitt 1996). Pitt (1996) indicated that enteroviruses pose a greater potential health risk because of their low infective doses, smaller size, and greater resistance to natural attenuation process. Pathogens are removed during infiltration and percolation through the vadose zone by straining, sorption, and inactivation. Drying of soil in most cases (duration of drying is a key issue) will kill both bacteria and viruses.

Impacts to groundwater from stormwater infiltration systems depend on the type of infiltration device, type (and thus quality) of runoff, and local hydrogeological conditions (Göbel et al. 2008). There is a general inverse relationship between infiltration rates and metals removal. Coarser grained sandy sediments with a high permeability allow for higher infiltration rates, but are less effective for the attenuation of metals concentrations and other contaminants because there is a lesser opportunity of sorption and degradation processes (Fischer et al. 2003). Shallow water tables, causing a thin vadose zone, also result in a lesser opportunity for attenuation processes to remove contaminants.

The effectiveness of infiltration systems for contaminant attenuation depends largely on the physical, chemical, and biological processes active in the vadose zone. An important issue involving the effectiveness of infiltration systems is bypass flow in which stormwater rapidly passes through the vadose zone, and thus, escapes attenuation processes. More rapid migration, and lesser contaminant attenuation, may occur if flow pathways (i.e., macropore flow) exist that allow the infiltrating water to bypass the filtration provided by shallow soils and sediments.

Infiltration systems can have mixed impacts on water quality, depending on the concentrations of contaminants in the runoff and the receiving ambient groundwater. For example, in a study of urban retention basins in Karachi, Pakistan, high infiltration volumes in the post-monsoon season resulted in lower than background concentrations of some metals due to dilution, but high concentrations of some metals (e.g., copper and cadmium) that are present at elevated concentrations in the runoff (Zubair et al. 2010).

The choice of the infiltration system type depends on stormwater flows and water quality, local hydrogeological conditions, logistical issues (e.g., land availability), and economic considerations. Surface infiltration devices that have a substantial thickness of underlying soils above the water table are preferable to using subsurface infiltration devices unless the runoff is known to be relatively free of pathogens (Pitt 1996; Pitt et al. 1999). Pretreatment involving sedimentation is

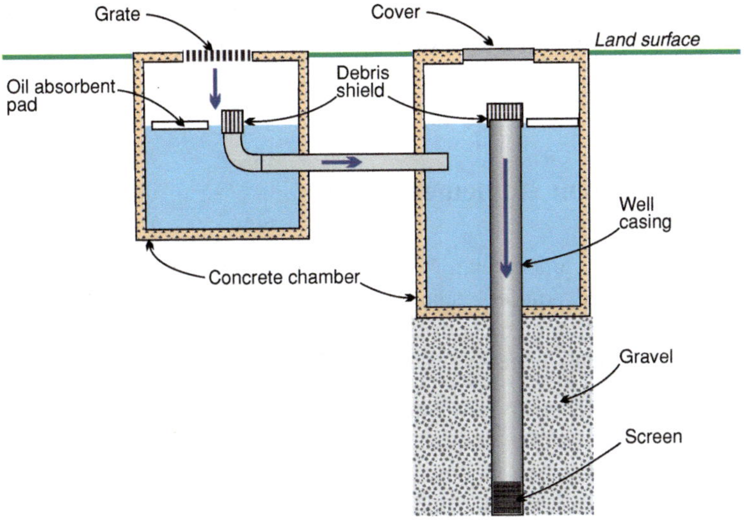

Fig. 23.15 Diagram of a catch basin with a settling system. Oil and grease can be removed using absorbent pads that require periodic replacement. *Source* USEPA (2009)

also recommended, if practical, which can consist of grass filters, sediment sumps, and wet detention ponds.

Catch basin systems can be designed to provide some pretreatment prior to discharge to drainage wells or other disposal or recharge sites. For example, catch basins may designed with dual chambers that allow for settling of suspended materials, separation of floating materials, and removal of oil and grease (Fig. 23.15). Oil absorbent pads placed in catch basins can reduce the potential for petroleum products to enter the groundwater system.

Numerous studies have documented that the first flush water in a storm tends to contain substantially higher solute concentrations (Schiff and Tiefenthaler 2011; and references therein). Diversion of the first flush from infiltration systems is also recommended, if local groundwater contamination is an important concern. However, diversion of the more contaminant-rich first flush water to surface water outfalls may also have significant environmental consequences, perhaps greater than impacts the local groundwater system.

Seasonal flushing may also occur in which the runoff from the initial storms of the season have greater concentrations or loads than storms later in the season (Li and Barnett 2008; Schiff and Tiefenthaler 2011). Seasonal flushing is of particular importance in arid regions because of extended dry periods during which contaminants may accumulate on land surfaces (Schiff and Tiefenthaler 2011). Watersheds can have a tremendous storage capacity for suspended solids and potentially other contaminants of concern. For example, in the Santa Anna River of Southern California, U.S.A., half of the discharge of total suspended solids mass occur in less than one-third of the volume of the water for the entire year

(Schiff and Tiefenthaler 2011). Seasonal flushing of suspended solids and contaminants can be eliminated by extended detention, such as in reservoirs (Schiff and Tiefenthaler 2011).

23.6 MAR System Implementation

The driver for the implementation of MAR is the recognition of the need or benefits of increased aquifer recharge or underground storage of water, or an alternative or supplement to conventional wastewater treatment. Dillon (2009) proposed five elements for every successful MAR project:

- suitable demand for recovered water,
- adequate source of water for recharge (in terms of both quantity and quality),
- suitable aquifer in which to store and recover water,
- sufficient land to harvest and treat water, and
- capability to effectively manage a project.

To these five elements, the following can be added:

- adequate financial resources to test, construct, and operate the system, and
- regulatory framework that is favorable in terms of allowing the technology, providing measured oversight, and protection of the systems.

MAR projects are typically implemented in a phased manner. Maliva and Missimer (2010) presented a general strategy for MAR implementation that focused on ASR systems, which consists of the following phases:

Phase I	Desktop Feasibility Evaluation
Phase II	Field Investigations and Testing of Potential System Sites
Phase III	Design
Phase IV	Pilot System Construction and Testing
Phase V	Project Review and Adaptive Management
Phase VI	System Expansion

The completion of each phase is Go/No Go decision point, in which the results of the phase are evaluated and the feasibility of the project is reassessed. Project feasibility also includes an evaluation of potential risks and mitigation options (Sect. 23.7). MAR should be approached with goal of avoiding the financial commitment to construct a full-scale system until there is a high degree of certainty that the project will meet performance expectations set at the start of the project.

The Australian Managed Aquifer Recharge Guidelines (NRMMC-EPHC-NHMRC 2009a) provide an excellent step-by-step plan for the implementation of MAR. However, it must be recognized that MAR implementation strategies will depend upon national and local technical and financial resources. The level of governmental regulatory oversight of MAR in the United States, Australia, and

Europe is not a realistic model for most of the rest of the world. For example, widespread rainwater harvesting and MAR has been implemented in India in a highly decentralized manner and in most instances by farmers or local communities without direct government control or involvement (Sect. 22.7.1).

23.6.1 Phase I: Desktop Feasibility Assessment

The ASR desktop feasibility evaluation consists of a review of available information on water supply, demands and quality, hydrogeology, potential system sites and land availability, potential risks to public health, water resources and the environment, utility infrastructure, and regulatory requirements. It is referred to as "desktop" study in that it does not involve significant field work, such as well drilling, other than site inspections. A key element of the desktop feasibility assessment is the determination of which MAR options are likely feasible and identification of the MAR option that will best meet water storage or treatment requirements. However, it is recognized that for some projects, the MAR system type is known from the start. Another important element of desktop feasibility assessments is the identification of hydrogeological data deficiencies and the development of a field testing program that will obtain the information needed to confirm system feasibility and for system design.

23.6.2 Phase II: Field Investigations and Testing of Potential System Sites

Field investigation programs are designed to obtain site-specific hydrogeological data on the aquifer that is to be used as a storage or treatment zone. Aquifer characterization techniques are summarized in Section IV and are described in detailed with respect to ASR and MAR projects using wells by Maliva and Missimer (2010). Depending upon the project type and local hydrogeology, field investigations may involve test (exploratory) wells, test borings or cores, infiltration tests, and source water and native groundwater testing. Surface geophysical techniques may also be part of field investigations (Chap. 16). Field investigation programs should also consider data requirements for groundwater model development, if modeling is an integral component of system MAR design.

The testing performed as part of exploratory well programs is system specific and should be based on project-specific data requirements. In some locations, sufficient data are available from existing wells in the project site vicinity so that the approximate depths, aquifer transmissivity (hydraulic conductivity) and other hydraulic properties, well yield, mineralogy, and water quality can be reasonably ascertained from a review of the technical literature and well records. It is also important to recognize that the scope of work of field investigations must be commensurate with system size, cost, and complexity. Clearly, much more data are required to design a multiple well ASR or salinity barrier system, than to install a series of small check dams in a wadi channel. Excessive field investigations add to system costs and thus impact system economic viability.

As a potential cost savings, the exploratory wells should be designed and located so that they could serve a useful function for the final system. For ASR systems and other MAR systems using wells, exploratory wells are often used as one of the long-term monitoring wells for the system and perhaps as an ASR well.

23.6.3 MAR System Design

The successful design of MAR systems depends greatly on local conditions. It is therefore not possible to provide generic standards, guidelines and blueprints (Bouwer 1999). The design of MAR systems requires knowledge of system types, their processes, controls and limitations, and a thorough assessment of local water quality, aquifer hydraulics, and hydrogeological conditions. Bouwer (2002, p. 140) succinctly noted that

> Design and management of artificial recharge systems involves geological, geochemical, hydrological, biological, and engineering aspects. Because soils and underground formations are inherently heterogeneous, planning, design, and construction of groundwater recharge schemes must be piecemeal, first testing for fatal flaws and general feasibility and then proceeding with pilot and small-scale systems until the complete system can be designed and constructed.

The American Society of Civil Engineers (2001) "Standard Guidelines for the Artificial Recharge of Groundwater" provides a general overview to the planning, design, and operation of some types of MAR systems. Maliva and Missimer (2010) provide an overview of the design of ASR systems and other MAR systems that use wells.

The selection of system type depends upon site-specific hydrogeological conditions, land availability, existing infrastructure, and economics. The selection of system type and design also depends upon whether the objective of the system is storage, treatment, or both. All of these factors need to be considered in order to obtain the optimal solution for a given location. A basic design constraint is the vertical hydraulic conductivity of surface and subsurface strata, which controls infiltration and recharge rates (Schiff 1957). If shallow strata have a lower hydraulic conductivity than deeper strata, then the preferred system may be one that penetrates the restricting layer, such as infiltration trenches and injection wells. If infiltration is impeded by a low-permeability surficial layer, such as hard pan, then greater infiltration rates may be achieved by disrupting or removing the layer (e.g., ripping, chiseling, or trenching).

There are relatively few comparative studies of MAR methods for individual sites. Legg and Sagstad (2002) documented a study of spreading basins, recharge trenches, and vadose recharge wells for the recharge of treated wastewater in Glendale, Arizona. Spreading basins were determined to be the most efficient method for aquifer recharge, but the recharge trenches and wells had the advantage of a much lower land requirement. The susceptibility of the trenches and wells to clogging was also a critical issue. At other sites, where the near surface sediments have low permeabilities, spreading basins may turn out to be a poor option for recharge.

23.6.4 Phase IV: Pilot System Construction

Large MAR systems should be designed and constructed in a phased manner, which initially includes a pilot test. For a multiple well ASR system, pilot testing normally involves the construction and operational testing of a single well ASR system. For a large infiltration basin or SAT system, pilot testing might include the operation of a single cell (basin), which may be temporarily subdivided to allow alternating wetting and drying cycles. The primary objectives of pilot testing are to determine if the system will operate as designed and to obtain additional data that could be used to optimize the design of the full-scale system.

Pilot testing is also important for evaluating water chemistry issues associated with MAR systems including the occurrence of adverse geochemical reactions (e.g., arsenic leaching) and beneficial processes (e.g., attenuation of contaminant concentrations). Operational testing should, to the extent possible, mimic actual anticipated long-term operational conditions. For example, the water recharged during operational testing should be the essentially the same water as will be recharged in the completed systems. Also, the duration of the recharge, storage, and recovery cycles should be similar to the anticipated mode of system operation.

23.6.5 Phase V: Project Review and Adaptive Management

Adaptive management is a structured iterative decision-making process in the face of uncertainty. Uncertainty is reduced and management improved by system monitoring and adapting to new and different information. Adaptive management is often simply referred to as "learning by doing." Adaptive management approaches are employed when there is uncertainty over the performance and impacts of a proposed activity. For example, an adaptive management approach is being taken in the multi-billion (U.S.) dollar Comprehensive Everglades Restoration Plan (CERP) because it is not possible to accurately predict the performance and interaction of each the numerous elements of the project. The key feature of adaptive management plans is that they encourage and incorporate flexibility in the design and operation of systems to allow for adaptation to unexpected results.

Adaptive management should be considered during system design. For example, MAR systems should have a robust design that can accommodate some unexpected conditions. There should also be design and operational flexibility, which allows for opportunities to adjust the design or operation of the system to adapt to information obtained from monitoring or project experience. Adaptive management also involves leaning (continuous on-going system evaluation) in which, for example, monitoring data are compared to initial predictive modeling results and the model is recalibrated or refined as necessary based on new information.

The performance of MAR systems should be periodically and objectively reviewed. Operational procedures may need to be adjusted in order to improve system performance. For example, injection wells typically experience a loss of performance over time. An adaptive management approach should be taken in order to identify operational changes and well rehabilitation strategies that allow for the wells to maintain performance standards at the lowest costs. An adaptive management approach may also be fruitful for wadi recharge and infiltration basin systems, in which different operational procedures are tested in order to find the operational protocol that most cost-effectively results in the greatest recharge rates.

23.6.6 Phase VI: System Expansion

Most MAR systems are expandable in that system capacity may be increased by the installation of additional wells or infiltration basins. The operational data and experiences from pilot testing and subsequent project phases should be used to improve the design for further system expansion. A groundwater model calibrated against operational data can be a valuable tool for optimizing the design of a MAR system expansion. The initial system design and location should also consider the potential for future system expansion, even if the need may not occur until well into the future. For example, potential project sites should ideally have room for additional wells or infiltration basins, which might be needed at some time in the future. Pretreatment systems should also be readily expandable.

23.7 Managed Aquifer Recharge System Risks and Risk Assessment

An essential element of assessing the feasibility of MAR systems is objectively evaluating the uncertainty and potential system risks, and identifying data deficiencies and uncertainty. The risks or hazards associated with MAR projects are summarized in Table 23.2. The risks associated with MAR projects determine whether or not a project is technically and economically feasible and meets regulatory requirements. Risk assessment is therefore a critical element of all MAR investigations. The sophistication of risk assessments required for a MAR system

Table 23.2 Risks of hazards associated with MAR projects

Risk type	Examples
Operational	1. System not meeting performance objectives (a) Low recovery efficiency (b) Does not meet target hydraulic response (increase in water levels or heads) 2. Excessive well or aquifer clogging 3. Loss of well mechanical integrity
Recharge water quality	1. Recharge water does not meet applicable water quality standards and impairs aquifer or adversely impacts other aquifer users (a) Microbiological (parameters) (b) Salinity (c) Nutrients (d) Chemicals (e) Radionuclides 2. Anticipated water quality improvements (natural aquifer treatment) not realized; recovered water does not meet water quality standards 3. Excessive migration of recharged water
Fluid-rock interaction	1. Deterioration of recharge water quality (e.g., arsenic and metals leaching)
Hydraulic impacts	1. Pressure-induced fracturing 2. Land subsidence 3. Aquifer dissolution (loss of integrity caused by low pH or other problem)
Environmental impacts	1. Impacts on groundwater dependent ecosystems (e.g., reduction is stream or spring flow and wetland hydroperiods during recovery) 2. Water logging
Greenhouse gases	1. Excess energy consumption relative to other options

Sources NRMMC-EPHC-NHMRC (2009b), Maliva and Missimer (2010)

varies greatly depending upon system type and system-specific circumstances. Systems storing water for potable water use clearly require greater scrutiny that non-potable systems. Risk assessments should also take advantage of the experiences of similar systems, especially if they located in the same geographic area and have similar hydrogeological conditions.

The Australian MAR guidelines (NRMMC-EPHC-NHMRC 2009a) provide an excellent framework for assessing the risks associated with MAR projects. The guidelines are not mandatory and have no formal legal status, but are designed to provide guidance for individual governments in the development of their MAR policies and regulations. The application of the guidelines to a trial ASTR project is discussed by Page et al. (2010a). The phasing of the Australian MAR guidelines coincides with the normal MAR project phasing of a desktop study, followed by field testing, and then construction and testing of a pilot system (Sect. 23.5). The Australian MAR guidelines assess 12 hazards common to MAR projects in a phased manner. The risk assessment procedures include four steps, herein described, with increasing costs of acquiring information as confidence in the viability of the project increases.

Two types of risk assessments are performed under the NRMMC-EPHC-NHMRC (2009a) MAR guidelines. First is a "maximal" risk assessment, which identifies inherent risks in the absence of preventative measures. The second type of risk assessment evaluates residual risks after the consideration of potential preventative measures. For example, a maximal risk assessment of a reclaimed water ASR system would likely identify pathogens in the recovered water as a significant risk element. The residual risk assessment might consider risks remaining after natural attenuation during the planned storage period and post-treatment of the recovered water such as disinfection.

The recommended project phasing and risk assessment procedures under the NRMMC-EPHC-NHMRC (2009a) MAR guidelines are as follows:

Stage 1—Desktop Study. The stage 1 desktop study determines if a project is likely to be viable and the likely difficulty in its implementation. Stage 1 involves consideration of both technical and regulatory issues and identifies data that will need to be obtained in Stage 2.
Stage 2—Investigations and Assessment. Stage 2 involves collection of additional information needed to assess risks such as aquifer residence time, source (recharge) and native groundwater chemistry and quality, and characterization of reactive aquifer minerals. A 'maximal risk assessment' is undertaken to estimate the risks in the absence of any controls or preventive measures. A pre-commissioning residual risk assessment is performed to evaluate potential preventative measurements and operational procedures that could be implemented to ensure acceptable low residual risks to human health and the environment. The residual risk assessment can be repeated until the residual risk is acceptable, for example, by the addition of extra preventive measures.
Stage 3—Commission Trials. Stage 3 involves the testing of a pilot system. The MAR system is trialed to validate the effectiveness of preventive measures and operational controls, and to assess the suitability of the recovered water for its intended uses. Stage 3 includes monitoring for the validation of processes that could not be measured prior to the start of recharge. The commission trials include an operational risk assessment, which has the purpose to assess whether on-going operation of the project has acceptably low human health and environmental risks. The aim is to identify unforeseen risks and required preventive measures.
Stage 4—Operation. Stage 4 involves the actual operation of the MAR system with a management plan and regular operational monitoring. Validation monitoring is performed to assess whether the quality of the recovered water is acceptable and to verify that the environmental value of the affected aquifer is protected.

The application of Australian MAR guidelines to the Parafield ASTR system in South Australia, for example, identified in Stage 2 that the risk for turbidity, color, iron, and fecal indicator organisms in recovered water would be high for a potable water use and that additional treatment of the recovered water would be required (Page et al. 2010b). The Australian MAR guidelines provide a technically sound and flexible approach for assessing risks with universal applicability. However, it is important to recognize that degree of effort and technical sophistication applied in the risk assessment tasks will necessarily vary between locations depending upon local financial and technical resources.

23.7.1 Microbiological Processes

Evaluation and management of the microbiological risks associated with MAR systems involving wastewater (reclaimed water) requires consideration of:

- the pathogens present in the recharged wastewater and their concentrations,
- the attenuation rate of pathogens in the storage aquifer between the times of recharge and recovery, which is a function of residence time and pathogen decay (inactivation) rate,
- the degree to which natural attenuation processes are being relied upon for pathogen removal,
- the uses of the reclaimed water and water quality requirements, particularly health-based water quality criteria (Chap. 31), and
- post-recovery treatment processes.

MAR projects that rely upon aquifer treatment as a barrier against microbial pathogen entry into the drinking water supply will always require a greater degree of scrutiny and validation of the removal of a range of specific target microbial pathogens in the recharged water (NRMMC-EPHC-NHMRC 2009a). Pathogens are also a concern where there is the possibility of unplanned consumption of the stored water. On the contrary, systems that store water for an irrigation use that is already of a suitable quality for irrigation and are not relying upon aquifer treatment for pathogen reduction, pose much less risk. MAR systems that store very highly treated water for later potable use may also require minimal pathogen monitoring if it can be demonstrated that the water treatment processes produce water that does not pose an unacceptable pathogen risk.

The greatest risks occur where the recharge water has a high initial concentration of pathogens (e.g., untreated wastewater) and the residence time is low (Page et al. 2010a; Toze et al. 2010). The least risk occurs where the recharge water undergoes a high degree of treatment and long residence times occur. The simplest approach for addressing the health risks associated with pathogenic microorganisms involves relating the recharge retention time to the inactivation (removal) rate of each pathogen within the aquifer. Inactivation is the only factor that should be used to assess aquifer treatment, and other processes such as retardation should only be considered to be an added benefit (NRMMC-EPHC-NHMRC 2009a).

Representative groups of pathogens need to be used for pathogen risk assessment because it is not practically and economically possible to test for all pathogens. For example, three representative pathogens, rotavirus, *Cryptosporidium*, and *Campylobacter*, were considered in a comparative study of subsurface pathogen treatment by Page et al. (2010a). Conservative pathogens (i.e., species with slow inactivation rates) should be chosen. For example, in Australia enteric viruses and protozoa are most resilient to inactivation and, therefore, are recommended to be included in survivor studies (NRMMC-EPHC-NHMRC 2009a).

Pathogen numbers in recharged water, aquifer residence time, and pathogen decay rates are the most important variables in predicting the risk of wastewater MAR systems (Toze et al. 2010). Recharge water sampling and analyses can constrain the likely values of pathogens. The actual values for the aquifer residence time and pathogen decay rates are often poorly known on a site-specific basis. Pathogen decay rates are a function of numerous variables, including physiochemical parameters and the indigenous microorganism community (Sect. 30.3), and thus need to be determined on a site-specific basis.

NRMMC-EPHC-NHMRC (2009a) cautioned that comparison of pathogen numbers between recovered and recharged water is rarely adequate for an appropriate risk assessment. Direct testing of pathogen decay rates was strongly recommended. In situ diffusion chambers have been used to determine $\log_{10}$ reduction in a pathogen numbers for MAR research sites (e.g., Page et al. 2010a, Toze et al. 2010), but in practice are too involved to be a routine tool for most MAR projects.

In the absence of site-specific pathogen decay data, conservatively long values obtained from the scientific literature could be used, particularly if data are available from geochemically similar systems. Thus, for each pathogen identified in the recharge water, the minimum retention time needed to achieve a specified water quality goal can be estimated. For example, if pathogen "A" has an estimated $\log_{10}$ removal rate of 6–10 days (based on published data), the World Health Organization (2006) target of 6–7 $\log_{10}$ pathogen removal could be achieved by a 70 day retention time. An additional safety factor could be added to provide additional assurance.

Aquifer characterization and testing (e.g., tracer tests) and groundwater (solute-transport) modeling can be used to evaluate the travel time from the recharge point to any nearby aquifer user or the recovery wells for an ASTR system. Travel (retention) time and distance are also critical for evaluating the potential for impacts to sensitive receptors. The NRMMC-EPHC-NHMRC (2009a) provides some analytical equations for estimating retention times. In practice, numerical modeling is now widely performed to evaluate aquifer hydraulic responses for MAR systems and to estimate travel times.

A critical issue for evaluating travel times is aquifer heterogeneity. The presence of zones of greater hydraulic conductivity in an aquifer can result in flow rates orders of magnitude greater than would occur in homogenous aquifers (Maliva and Missimer 2010). Standard geophysical logs (e.g., flow meter) and advanced borehole geophysical logs (e.g., nuclear magnetic resonance) can provide data on the aquifer heterogeneity, which can be used to develop more accurate solute-transport models (Maliva et al. 2009). Tracer tests can provide more direct data on travel times within an aquifer.

Although field testing can reduce uncertainty in the value of some key parameters, some uncertainty as to the actual values still remains. Residence times and decay rates and water pathogen concentrations are often incorporated into risk assessments as probability distribution functions (PDFs) and evaluated using statistical methods such as Monte Carlo simulations.

Toze et al. (2010) documented a static QMRA performed for a wastewater MAR research project in Western Australia, Because this was a research project, the scope of work was well beyond that practical for most projects, especially those not involving indirect potable reuse. Realistically, system-type specific goals need to be established for regulatory jurisdictions, which can be either pathogen concentrations at the point of recovery or number of $\log_{10}$ removals. The risks associated with a given MAR systems can be assessed using system specific aquifer residence times and pathogen decay rates (or conservative 'worse-case' rates from the literature). Determination of pathogen decay rates, likely exposure frequencies and volumes, and dose–response functions need to be determined in a government or university research environment.

23.7.2 Chemical Contaminants

The health impacts of compounds of emerging concern (CECs) with respect to wastewater reuse and potable water supplies have received explosive interest over the past two decades. It has also been documented that some, but not all, CECs are attenuated in groundwater environments. CECs are a risk concern for MAR systems in the same sense as microbial pathogens. The fate and transport of CECs must be assessed under site-specific aquifer conditions where natural attenuation is being relied upon as part of a multiple barrier approach. The fate and transport of CECs are discussed in Sect. 30.3.5.

23.8 Economics of MAR Projects

The economics of MAR projects need to be considered in the same manner as other water supply and storage options. The basic economics of ASR and other MAR systems were discussed by Maliva and Missimer (2010). MAR systems should be economically evaluated in the same manner as other engineering projects, which typically involve some form of net present value, present worth, or life cycle cost analysis. The cost of MAR systems includes both the capital (construction) costs and operation and maintenance costs. Amortized costs for MAR systems should be compared to the value of the additional water provided and the costs of other supply or storage options.

Economic analysis of ASR and other artificial recharge projects is complicated by the difficulty of applying an economic value to water. Todd (1965) in a pioneering paper noted with respect to the economics of groundwater artificial recharge that

> in assessing the benefits of artificial recharge, consideration must be given to the importance of water to the total economy, to the value of water for various uses, as well as to the direct and intangible benefits that may accrue.

Intangible benefits are varied and difficult to quantify. For example, the value of a strategic water supply to meet potable demands in an extreme emergency is inestimable (Todd 1965). Aquifer recharge and ASR systems used to maintain irrigation water supplies have a direct value to the farmers who use the water, and also provide economic benefits to local businesses dependent upon the farmers as customers and add social and cultural benefits to communities dependent upon agriculture.

Analysis of the costs and benefits of a MAR system in arid and semiarid environments is complicated by often very large temporal variations in the supply of water suitable for recharge. An important decision is the size or capacity of the MAR system. For example, in systems designed to capture and store ephemeral stream flows, the decision needs to be made as to how much of the flow should be captured for recharge. A conflict exists between the goals of (1) recharging as much of the available water as possible, which necessitates construction of very large systems, and (2) maximizing the efficiency (use rate) of the system, which would involve construction of relatively small capacity systems (Murillo et al. 2002). It is inefficient to construct a very large capacity system in which much of the capacity will very seldom be needed unless it is for emergency use when there would otherwise be no water supply.

Optimization of system capacity involves analysis of historic time-series data of available water (e.g., stream flow) versus system capacity. This data can be used to estimate the number of days per year on average that a system would operate at its design capacity (i.e., use factor) and the percentage of available water that is captured (Murillo et al. 2002). The most hydrologically efficient system in terms of use of capacity may not be the best choice when economics are considered. For example, if water has a high value and land is inexpensive, construction of a very large capacity spreading basin systems may be the optimal choice.

Economic analyses of MAR systems are simplest when the volume and value of additional water can be accurately quantified, which would likely be the case for a utility-operated ASR system in which the recovered water is sold. The volume of recovered water would be metered. The value of the recovered water could be the revenue generated from its sale or perhaps the cost savings versus other supplemental supply options.

Cost-benefit analyses are more complicated for aquifer recharge systems, especially those that involve enhancement of natural recharge processes. For small-scale systems in which the work is self-performed by the local community, it can be difficult to assign a true cost for the labor. The total volume of recharge and its economic value may have a high uncertainty. The additional value provided by systems, such as the flood control benefits of wadi dams, also need to be considered.

Kalantari et al. (2010) provide an example of a cost-benefit analysis of check dam and recharge basins for aquifer recharge in southwest Iran. The costs for recharge schemes depend on the

- degree of source water treatment and siltation,
- construction (materials and labor),
- land value,
- maintenance, and
- type of recharge system.

Consideration was given to the percentage of recharged water that is recoverable and the irrigation return flows to an unconfined aquifer system. The primary benefits of the aquifer recharge are that as more water becomes available for irrigated farming, there is an increase in agricultural production and income to the farmers. The economic benefits were quantified as the increase in income from each unit of recharged water after subtracting costs (net value of additional production, not gross revenues). The estimated cost-benefit ratio was determined to be 1:1.32 with a payback period of three years. There was an additional unquantifiable benefit of mitigation of flood impacts.

23.9 Unmanaged Aquifer Recharge

Unmanaged aquifer recharge, which is also referred to as "culturally modified recharge" (Stephens 1996), is a locally important component of water budgets in some areas. The NRMMC, EPHC & NHMRC (2009a) differentiated between unintentional recharge and unmanaged recharge. The former includes unintended recharge, such as from pipe leakage. The latter includes intentional activities that have a primary disposal function, such as septic system leach fields.

Unmanaged aquifer recharge (UMAR) includes such processes such as (Lerner 1986, 1990):

- irrigation in excess of crop consumption and ET losses,
- on-site sewage disposal systems—septic systems,
- pipe leakage, and
- canal seepage.

Leakage of water mains and wastewater force mains would be expected to contribute more water to recharge because they are pressurized. Sewers and storm drains are typically not pressurized (except for force mains), but usually receive less maintenance than water mains (Lerner 1986).

Changes in vegetation can also result in an unmanaged increase in recharge. Removal or replacement of deep-rooted native vegetation with shallow-rooted crops and grasses can result in a reduction in ET and an associated increase in recharge. For example, Peck and Williamson (1987) documented how the clearing of native forest vegetation and its replacement with pasture or crops in Western Australia resulted in a substantial rise in the water table and the mobilization of salts in the soils. In areas that were cleared for agriculture, the potentiometric surface moved upward at more than 2.6 m/yr (8.5 ft/yr) as averaged over several years. The potentiometric surface rise was equivalent to increased recharge estimated as

6–12% of rainfall depending on the value of the specific yield used for the aquifer. Reforestation has been demonstrated to result in decreases in groundwater levels relative to areas still under pasture (Bell et al. 1990).

UMAR can become part of the local hydrology and impact local water levels and ecosystems, particularly if it is locally a long-term phenomenon. The All-American Canal in the southern California is a good example of the transboundary issues that may arise when unmanaged seepage from an unlined canal becomes in essence the local norm over time, even to the extent that local ecosystems have become dependent on it. As further discussed in Sect. 37.6.12, international conflicts can arise between the owner of the water being 'lost" to seepage and those who have become dependent on it. In the All-American Canal example, groundwater users and some sensitive environments in Mexico had become dependent on canal leakage in the United States, which had been occurring for many decades.

In Central Mexico, 60 m^3/s (2,119 ft^3/s) of untreated wastewater is transmitted through unlined canals from Mexico City 80 km (49.7 mi) north to the Tula Valley, where it is used to irrigate 90,000 hectares (347 mi^2) of farm land referred to as "El Mezquital". Water quality improves significantly during transit and the farmers benefit from its nutrient content. It is estimated that 25 m^3/s (883 ft^3/s) of the wastewater flow artificially recharges aquifers in the Tula Valley through a combination of canal seepage and irrigation return flows from very high irrigation rates (1.5–2.2 m/yr; 4.9–7.2 ft/yr) (Jimenez and Chávez 2004). The consequences of the artificial recharge have been a rising water table (and associated flooding and salinization of some farm fields), increasing river flow, and the appearance of several springs (Jimenez and Chávez 2004).

The Tula Valley aquifers were investigated as a potential drinking water source for Mexico City. Recovery of some of the artificially recharged water would allow for the recovery of some flooded and saline lands (Jimenez and Chávez 2004). Agricultural use and the canal seepage process were found to be very effective in removing pollutants, but the recovered water may have elevated salinity and nutrients (e.g., nitrate). Biological analyses of the water discharged from a newly formed spring indicate that the water is of very high quality. The results of the study indicate that it is feasible to use the UMAR water for drinking water after treatment by membrane processes.

Sidi Abid and Sfax City area of southeastern Tunisia, on-site disposal of wastewater and application of reuse water in excess of crop demands resulted in a local rise in water levels. The rising shallow groundwater levels are associated with increased nutrient concentrations and, in the case of on-site disposal, pathogens in the shallow aquifer (Bouri et al. 2008; Chamtouri et al. 2008).

In many urban areas in semiarid and arid regions, rising water tables from UMAR are creating problems with local flooding and impacts to underground structures (e.g., basements, buried utilities, and building foundations). Recharge losses from leaking water mains was investigated by Lerner (1986) for the city of Lima, Peru, and Hong Kong. Recharge from leakage was evaluated from piezometer and groundwater chemistry data, water balances and groundwater modeling, and conventional leak detection tests (minimum night flow tests).

Lerner (1986) estimated that 40% of the average potable water supply at the time of study was becoming recharge through leakage. A groundwater model of the city would not calibrate unless the recharge from pipe leakage was included. Leakage was also determined to be the controlling factor on groundwater heads and slope stability in Hong Kong, but the actual recharge rate was not determined.

Rising groundwater levels are a problem in some cities in the Middle East. Some residential areas of Kuwait City have experienced flooding of basements and stagnant water occurs on land surface from a rise in water levels from residential irrigation, septic systems, and leakage from water mains. Drainage of this water is a potential source of water depending on its quality (Al-Otaibi and Mukhopadhyay 2005). Parts of Riyadh and Jeddah in the Kingdom of Saudi Arabia are also experiencing rising water levels, necessitating drainage for protection of building foundations, basements, and underground structures (Alhamid et al. 2007; Al-Sefry and Sen 2006).

Management of UMAR requires an identification of its sources and quantities, which usually involves an evaluation of utility operational (e.g., system water losses) and water chemistry data, and groundwater modeling. The Vázquez-Suñe et al. (2010) study of UMAR in Barcelona, Spain, provides a good introduction to potential geochemical tracers and a useful template for similar studies. Studies of UMAR should start with a preliminary conceptual model that identifies potential sources of recharge, assesses potential groundwater geochemical conditions, and includes reasonable estimates concerning which tracers might display a broad range of concentrations to identify the potential recharge sources. Mixing models are then used to determine the contributions of potential recharge waters to the water present in studied wells.

The basic challenges associated with mixing models used to ascertain the relative contribution of these recharge sources are (Vázquez-Suñe et al. 2010):

- identification of conservative (non-reactive) species that have markedly different concentrations in each recharge source,
- identification of at least n − 1 species, where n = number of potential recharge sources, and
- accurate determination of the concentrations of the tracer species in each source end-member, which may vary in both space and time.

For example, the chemistry of water leaking from sewage lines may vary seasonally and between different locations in the collection network. A single set of values for the concentration of each tracer species may not be adequately representative of "sewage line leakage" recharge water. Mixing ratios are very sensitive to end-member concentrations. Tracers used in study of urban recharge in Barcelona (Vázquez-Suñe et al. 2010) were Cl, SO_4^{-2}, ^{34}S, B, F, Br, EDTA, Zn, ^{18}O, D, total N, and residual alkalinity.

Management of UMAR in urban environments may include a program to reduce leakage from water and wastewater mains and, in some instances, groundwater pumping to lower water levels in shallow aquifers. Depending upon its quality, the recovered water may be suitable for some non-potable uses.

23.10 Groundwater Banking

Groundwater banking is a form of MAR in which an aquifer is used to store water. The term "groundwater banking" has been applied to both the physical storage of water in an aquifer and to the accounting system associated with the MAR. Groundwater banking is performed in a regulatory context in which the recharge of a given volume of water confers the later right to withdraw water. The aquifer is essentially managed in a similar manner as a financial bank, in which the system owner or regulator tracks and authorizes credits and debits. Groundwater banking is also referred to as regulatory storage ASR (Maliva and Missimer 2008, 2010).

Groundwater banking systems may allow for the more effective conjunctive use of surface water and groundwater. Water banking systems also provide a flexible framework within which temporary water transfers can efficiently occur (Frederick 1995). The systems facilitate a spot market in which sellers and buyers of water are pooled. Credits obtained for the recharge of a given volume of water may be sold to another water user who has a greater need and may be willing and able to pay a higher price for the water.

The basic requirement for a groundwater banking system is that use of the storage aquifer is tightly regulated and not open to new allocations. There would be no point to a water banking system if new users could freely withdraw additional water from the aquifer without contributing to the water stored in the system. A key issue is that for a water bank to be successful, the bank's depositors must have assurance that water deposited in the bank will be available to them for withdrawal at a later time (Sandoval-Solis et al. 2011).

Water is banked typically either through direct recharge of water into the aquifer or by "in-lieu recharge". In-lieu recharge is credit given for allocated groundwater that is not withdrawn. With respect to the water budget of an aquifer, the net effects of recharging a given volume of water or not withdrawing the same volume of water that would otherwise be used are similar. Deposits may also occur by other activities that result in a net recharge to the aquifer, such as canal lining (and thus leakage reduction) and changes in agricultural irrigation practices (Contor 2009).

The simplest type of groundwater banking system would occur in a groundwater basin that is well confined at its sides and base, and thus there is minimal leakage out of the system. The aquifer would literally behave in a similar manner as a large tank. Operation of the groundwater bank per se would not have a significant impact on the aquifer water budget so long as withdrawals do not exceed deposits and the water levels did not rise above the extinction depth for evapotranspiration to occur. Recharge would increase the volume of water in storage and subsequent withdrawals would abstract the stored water, theoretically returning water levels to pre-recharge, baseline levels. Where some of the recharged water is lost by leakage or other means, a discount could be applied to withdrawals (Contor 2009, 2010). Less than 100% of the recharged water may be allowed to be recovered to account for system water losses. Groundwater losses could be subtracted from the aquifer bank accounts proportionally to the amount of storage in each account (Sandoval-Solis et al. 2011).

Groundwater banking systems could be operated by and for the benefit of a single user, such as a governmental water management agency or utility. Alternatively, groundwater banking systems could have multiple participants in which credits are issued for water recharge, which could be used for later withdrawals. Systems could also be operated entirely or partially on a cash basis in which landowners with excess water would receive a cash payment for recharged water and users of the stored water would purchase a specified amount of water (Purkey et al. 1998). Some of the collected funds would be used to pay for the operation of the system.

Groundwater banking systems have technical and operational challenges. Purkey et al. (1998) discuss the technical and regulatory issues associated with groundwater banking in the state of California, USA. While many of the regulatory issues are state specific, the general concepts are broadly applicable. A basic requirement for any water bank scheme is that some mechanism must be in place to prevent the stored water from being abstracted by other aquifer users, particularly those who are not participating in the system.

A prerequisite for groundwater banking systems is a thorough analysis of basin hydrogeology and the potential hydrologic responses to both recharge and recovery. For example, in some areas of California the lowering of the water table has resulted in a new equilibrium between pumping and recharge, as additional recharge from surface water bodies has been induced. Raising the water table through aquifer recharge, may decrease the amount of induced recharge resulting in a lesser net increase in aquifer storage (Purkey et al. 1998).

Another important issue for groundwater banking is avoiding third party impacts, which can result from changes in the location and timing of recharge and withdrawals. For recovery, a major issue is avoiding injury to other groundwater users. Clear rules and limits on the recovery of water must be defined and the groundwater levels should not be allowed to drop below levels that would occur in the absence of a conjunctive use program (Sandoval-Solis et al. 2011). The effects of surface water withdrawals and groundwater-surface water interactions on the environment are another key issue. In California, maintaining minimum environmental flows and temperatures required for the spawning of anadromous fish is a critical environmental issue. Increased drawdowns during recovery periods should not impact sensitive environments and stream flows, upon which fish populations depend.

A critical issue for groundwater banking is the overall aquifer use pattern. A fundamental requirement is that recharge (either direct or in lieu) results in an increase in the volume of water in storage in the aquifer as manifested by an increase in water levels or pressures (Maliva and Missimer 2010). In the case where an aquifer is in overdraft and groundwater levels are stable or declining, some or all of the water recharged in a groundwater banking system will be extracted by other users. Consequently, progressively more credits will be accumulated by system participants, but the volume of water stored in the aquifer is not correspondingly increasing. The end result is that much of banked water may not be recoverable.

In lieu recharge can result in excessive credits being accumulated if they are based on a baseline set too high. Current hydrologic conditions within a groundwater basin reflect actual historic water use rates, which may be significantly less than permitted rates. In some instances, groundwater allocations for irrigation may be inflated and exceed actual normal irrigation water use rates. Depending upon weather conditions and farm operations, an agricultural user may not need to use all of its allocation in most years. Credits may, therefore, accumulate for unused allocations that did not represent an actual decrease in water use from an historic water use baseline.

The Las Posas Basin of Southern California illustrates some the challenges, which are summarized by Maliva and Missimer (2010) that may arise from groundwater banking. Fox Canyon Groundwater Management Agency (FCGMA), the regulatory agency that has jurisdiction over the project, noted that in the historic accumulation of credits within the FCGMA has been steadily increasing, approaching 678×10^6 m^3 (550,000 AF, 179,200 Mg) in 2006 (Fox Canyon Groundwater Management Agency 2007). The estimated total net credits balance in the East, West, and South Las Posas Basin at the end of calendar year 2006 was 143×10^6 m^3 (116,002 AF, 37,805 Mg) compared to an annual extraction of 33.6×10^6 m^3 (27,234 AF, 8,875 Mg,). The accumulated credits are over four times the annual extraction rate. The volume of water represented by the credits that are accumulating through the operation of the ASR system and in-lieu recharge greatly exceeds the amount of water that could be extracted during a short-time period (e.g., major drought). The Fox Canyon Groundwater Management Agency (2007) noted that

> Should there be an extended period with limited groundwater recharge by either natural or anthropogenic sources, a significant number of credits could be used in a short period of time, ultimately overstressing, and possible permanently damaging the resources. Thus, although the credit system represents a low-cost groundwater-use option that can assist individual operators during extended dry periods, it also represents a threat to the regional groundwater resource since, under the current Ordinance, it lacks limits that would mitigate cumulative regional overuse during these same periods.

It was noted that even a 5% use of the total amount of credits currently available would result in a net 24% increase in annual extraction, which could result in persistent depressions in groundwater altitudes, land subsidence, and seawater intrusion (Fox Canyon Groundwater Management Agency 2007). Other water banking systems in California, such as systems in Kern County, experienced large drawdowns and associated impacts from large recovery volumes during a drought, which spurred multiple lawsuits (Barringer 2011).

Additional considerations are the timing and location of withdrawals. Groundwater abstractions result in *local* aquifer drawdowns, which can have adverse local impacts such as reductions in stream and spring flows and wetland water levels, and the lowering of water levels in wells. Local adverse impacts can occur even though the system is neutral in terms of the overall aquifer water budget. Impact assessments need to consider the temporal and spatial effects of both recharge and proposed withdrawals.

Groundwater banking systems are further complicated when aquifers are in contact with one or more surface-water bodies. Changes in groundwater levels can impact the direction and rate of flow between surface-water bodies and underlying aquifers. For gaining streams, the increase in groundwater levels caused by recharge may increase the flow into the stream, resulting in a lesser amount of net stored water. For loosing streams, recharge may reduce the natural leakage rate. Recharge will thus cause loosing streams to lose less and gaining streams to gain more water (Contor 2009).

Thus, where the storage aquifer is connected to surface waters, mass balance assessment alone cannot account for impacts to surface-water users and the environment, because of differences between the timing and location of recharge and withdrawals. It is, therefore, necessary, to include surface water-aquifer response functions in the management of these systems (Contor 2009, 2010). For example, groundwater modeling can be employed to assess the impacts of proposed recharge and subsequent withdrawals on surface-water bodies and the aquifer. The modeling would need to consider both the location and timing of recharge and withdrawals. Depending upon local circumstances, a substantial discount may have to be applied to avoid adverse impacts if the withdrawals are to occur during dry periods when surface-water bodies are most vulnerable. From an economic perspective, the application of aquifer response functions turns water into a homogenous quantity as the withdrawal point and time are accounted for and no longer makes a difference with respect to surface water impacts (Contor 2010).

In summary, groundwater banking has significant potential as an element of integrated water resources management, because it creates incentives for conservation and facilitates both the conjunctive use of groundwater and surface water, and water reallocation (trading). Groundwater banking needs to be implemented within a sound hydrogeological framework, in which the impacts of recharge and recovery are clearly understood. The main technical issue is ensuring that banked water represents a true net addition to the volume of water stored in an aquifer and unacceptable impacts (both to system participants and third parties) do not occur during recovery. A discount will typically need to be applied to stored water in order to account for non-recoverable water. However, if the discount is too high (even if technically justifiable) the water resources and economic values of the system to the participants may be lost.

References

Aerts, J., Lasage, R., Beets, W., de Moel, H., Mutiso, G., Mutiso, S., et al. (2007). Robustness of sand storage dams under climate change. *Vadose Zone Journal, 6*, 572–580.

Agarwal, A., & Narain, S. (1997). *Dying wisdom. Rise, fall and potential of India's traditional water harvesting systems*. New Delhi: India Centre for Science and Environment.

Alderwish, A. M. (2010). Induced recharge at new dam sites—Sana's Basin, Yemen. *Arabian Journal of Geosciences, 3*, 283–293.

Al-Ghamdi, A. S. (2009). Development of strategic water reserve for the Holy City of Makkah, Saudi Arabia. *Water Science and Technology: Water Supply, 9*(5), 533–540.

Alhamid, A. A., Alfayzi, S. A., & Hamadto, M. A. (2007). A sustainable water resources management plan for Wadi Hanifa in Saudi Arabia. *Journal King Saud University Engineering Sciences, 19*(2), 209–222.

Al-Muttair, F. F., Sendil, U., & Al-Turbak, A. S. (1994). Management of recharge dams in Saudi Arabia. *Journal of Water Resources Planning and Management, 120*, 749–763.

Al-Otaibi, M., & Mukhopadhyay, A. (2005). Options for managing water resources in Kuwait. *Arabian Journal for Science and Engineering, 30*, 55–68.

Al Sefry, S. A., & Şen, Z. (2006). Groundwater rise problem and risk evaluation in major cities of arid lands—Jeddah case in Kingdom of Saudi Arabia. *Water Resources Management, 20*, 91–108.

Al-Turbak, A. S., (1991). Effectiveness of recharge from surface reservoir to an underlying unconfined aquifer. In G. Schiller, R. Lemmela & M. Spreafico (Eds.), *Hydrology of natural and man-made lakes* (*Proceedings of the Vienna Symposium*, August, 1991) (pp. 191–196). International Associations of Hydrological Sciences (IAHS) Publication No. 206.

Al-Turbak, A. S., & Al-Muttair, F. F. (1989). Evaluation of dams as a recharge method. *Water Resources Development, 5*(2), 119–124.

American Society of Civil Engineers. (2001). *Standard guidelines for artificial recharge of groundwater (EWRI/ASCE 34-01)*. Reston, VA: American Society of Civil Engineers.

Appleyard, S. J. (1993). Impact of stormwater infiltration basins on groundwater quality Perth metropolitan region, Western Australia. *Environmental Geology, 21*(4), 227–236.

Asano, T., Burton, F. L., Leverenz, H. L., Tsuchihashi, R., & Tchobanoglous, G. (2007). *Wastewater reuse issues, technologies and applications*. New York: McGraw-Hill.

Asano, T., & Cotruvo, J. A. (2004). Groundwater recharge with reclaimed wastewater; health and regulatory considerations. *Water Research, 38*, 1941–1951.

Bajjali, W. (2005). Model the effect of four artificial recharge dams on the quality of groundwater using geostatistical methods in GIS environment, Oman. *Journal of Spatial Hydrology, 5*(2), 1–15.

Balke, K.-D., & Zhu, Y. (2008). Natural water purification and water management by artificial groundwater recharge. *Journal of Zhejiang University Science B, 9*(3), 221–226.

Barber, M. E., Hossan, A., Covert, J. J., & Gregory, G. J. (2009). Augmentation of seasonal low stream flow by artificial recharge in the Spokane Valley—Rathdrum Prairie aquifer of Idaho and Washington, USA. *Hydrogeology Journal, 17*, 1459–1470.

Barraud, S., Gautier, A., Bardin, J. P., & Riou, V. (1999). The impact of intentional stormwater infiltration on soil and groundwater. *Water Science and Technology, 39*, 185–192.

Barringer, F. (2011) *Storing water for a dry day leads to suits*. New York Times, July 26, 2011.

Baurne, G. (1984). Trap dams: artificial subsurface storage of water. *Water International, 9*, 2–9.

Behnke, J. J. (1969). Clogging of surface spreading operations for artificial ground-water recharge. *Water Resources Research, 4*, 870–876.

Bekele, E., Toze, S., Patterson, B., Devine, B., Higginson, S., & Fegg, W., et al. (2009). Chapter 1. *Design and operation of infiltration galleries and water quality changes*. CSIRO: Water for a Healthy Country National Research Flagship.

Bell, R. W., Schofield, N. J., Loh, I. C., & Bari, M. A. (1990). Groundwater response to reforestation in the Darling Range of Western Australia. *Journal of Hydrology, 115*, 297–317.

Benhan-Blair & Affiliates and Engineering Enterprises. (1979). *Long-term effects of land application of domestic wastewater: Milton, Wisconsin, rapid infiltration site* (128 pp). Robert S. Kerr Environmental Research Laboratory, U.S. Environment Protection Agency, Report EPA-600/2-79-145.

Bouri, S., Abida, S., & Knanfir, H. (2008). Impacts of wastewater irrigation in arid and semi arid regions; case of Sidi Abid. *Environmental Geology, 53*, 1421–1432.

Bouwer, H. (1974). Design and operation of land treatment systems for minimum contamination of ground water. *Ground Water, 12*(3), 140–147.

Bouwer, H. (1978). *Groundwater hydrology*. New York: McGraw-Hill.

Bouwer, H. (1985). Renovation of wastewater with rapid-infiltration land treatment systems. In T. Asano (Ed.), *Artificial recharge of groundwater* (pp. 249–282). Stoneham, MA: Butterworth.

Bouwer, H. (1989). Ground water recharge with sewage effluent. In A. I. Johnson & D. J. Finlayson (Eds.), *Artificial recharge of ground water* (pp. 170–185), *Proceedings of the International Symposium,* Anahiem, CA, August 23–27.

Bouwer, H. (1991). Groundwater recharge with sewage effluent. *Water Science and Technology, 23*, 2099–2108.

Bouwer, H. (1999). Artificial recharge of groundwater: Systems, design, and management. In L. W. Mays (Ed.), *Hydraulic design handbook* (pp. 24.1–24.44). New York: McGraw-Hill.

Bouwer, H. (2000). Integrated water management: Emerging issues and challenges. *Agricultural Water Management, 45*, 217–228.

Bouwer, H. (2002). Artificial recharge of groundwater: Hydrogeology and engineering. *Hydrogeology Journal, 10*, 121–142.

Bouwer, H. (2003). Integrated water management for the 21st century: Problems and solutions. *Food, Agriculture and Environment, 1*(1), 118–127.

Bucheli, T. D., Müller, S. R., Heberle, S., & Schwarzenbach, R. P. (1998a). Occurrence and behavior of pesticides in rainwater, roof runoff, and artificial infiltration. *Environmental Science and Technology, 32*, 3457–3464.

Bucheli, T. D., Müller, S. R., Heberle, S., & Schwarzenbach, R. P. (1998b). Bituminous roof sealing membranes as major sources of the herbicide (R, S)-mecoprop in roof runoff waters; potential contamination of groundwater and surface waters. *Environmental Science and Technology, 32*, 3465 –3471.

Chadha, D. K. (2003). State of art of artificial recharge applied on village level schemes in India. In A. Tuinhof & J. P. Heederick (Eds.), *Management of aquifer recharge and subsurface storage* (pp. 19–24). National Committee—International Association of Hydrogeologists (NNC-IAH) Publication No. 4.

Chamtouri, I., Abida, H., Khanfir, H., & Bouri, S. (2008). Impacts of at-site wastewater disposal systems on the groundwater aquifer in arid regions; case of Sfax City, southern Tunisia. *Environmental Geology, 55*, 1123–1133.

Contor, B. A. (2009). Groundwater banking in aquifers that interact with surface water: aquifer response functions and double-entry accounting. *Journal American Water Resources Association, 45*(6), 1465–1474.

Contor, B. A. (2010). Status of ground water banking in Idaho. *Journal of Contemporary Water Research and Education, 144*, 29–36.

Dillon, P. (2005). Future management of aquifer recharge. *Hydrogeology Journal, 13*, 313–316.

Dillon, P. (2009). Water recycling via managed aquifer recharge in Australia. *Boletin Geológico y Minero, 120*(1), 121–130.

Fischer, D., Charles, E. G., & Baehr, A. L. (2003). Effects of stormwater infiltration on quality of groundwater beneath retention and deformation basins. *Journal of Environmental Engineering, 129*(5), 464–471.

Foster, S. (2002). Subsurface dams to augment groundwater storage in basement terrain for human subsistence—Brazilian experience. GW MATE and World Bank Case Profile Collection No. 5.

Fox Canyon Groundwater Management Agency. (2007). Annual report for calendar year 2006. Fox Canyon Groundwater Management District, Venture, CA.

Fox, P., Houston, S., & Westerhoff, P. (2001a). *Soil aquifer treatment for sustainable water supply* (203 pp). Denver: Colorado American Water Works Association Research Foundation.

Fox, P., Houston, S., Westerhoff, P., Drewes, J. E., Nellor, M., Yanko, W., et al. (2001b). *Investigations of soil-aquifer treatment for sustainable water reuse.* Tempe, AZ: National Center for Sustainable Water Supply.

Frederick, K. D. (1995). Adapting to climate impacts on the supply and demand of water. *Climatic Change, 37*, 141–156.

Gale, I. (2005). *Strategies for managed aquifer recharge (MAR) in semi-arid regions* (30 pp). Paris: UNESCO.

Gastélum, J., Lluria, M., & Small, G. G. (2009). Vadose zone recharge wells: Ten years later at the City of Scottsdale's water campus facility. *Arizona Hydrological Society 2009 Annual Water Symposium Proceedings*.
Glossop, R. (1960). The invention and development of injection processes, Part 1: 1802–1850. *Géotechnique, 10*(3), 91–100.
Glossop, R. (1961). The invention and development of injection processes, Part II: 1850–1960. *Géotechnique, 11*(4), 255–279.
Göbel, P., Zimmermann, J., Klinger, C., Stubbe, H., & Coldewey, W. G. (2008). Recommended urban storm water infiltration devices for different types of run-off under varying hydrogeological conditions. *Journal of Soils and Sediments, 8*, 231–238.
Gorey, T., & Dent, P. (2007). Maintenance techniques utilized by Central Arizona Project to optimize recharge. In P. Fox (Ed.), *Management of aquifer recharge for sustainability: Phoenix* (pp. 404–411). Arizona: Acacia Publishing.
Goss, D. W., Smith, S. J., Stewart, B. A., & Jones, O. R. (1973). Fate of suspended sediment during basin recharge. *Water Resources Research, 9*, 668–675.
Haimerl, G., Zunic, F., & Strobl, T. (2002). An infiltration test to evaluate the efficiency of groundwater recharge dams in arid countries. In V. P. Singh, M. Al-Rashed & M. M. Sherif (Eds.), *Surface water hydrology* (pp. 313–322). Lisse: Swets & Zeitlinger.
Hanson, G., & Nilsson, A. (1986). Ground-water dams for rural-water supplies in developing countries. *Ground Water, 24*(4), 497–506.
Hantke, H. H., & Schlegel, W. (1995) The use of seepage trenches for artificial ground water recharge. In A. I. Johnson & R. D. G. Pyne (Eds.), *Artificial recharge of ground water II* (pp. 177–187), *Proceedings of the Second International Symposium on Artificial Recharge in Ground Water*. American Society of Civil Engineers, New York.
Hellwig, D. H. R. (1973). Evaporation of water from sand, 4: The influence of the depth of the water-table and the particle size distribution of the sand. *Journal of Hydrology, 18*, 317–327.
Helweg, O. J., & Smith, G. (1978). Appropriate technology for artificial aquifers. *Ground Water, 16*(3), 144–148.
Houlsby, A. C. (1990). *Construction and design of cement grouting: A guide to grouting in rock foundations*. New York: Wiley.
Huisman, L., & Olsthoorn, T. N. (1983). *Artificial groundwater recharge*. Boston: Pitman Advanced Publishing.
Hutchinson, A. S. (2007). Challenges in optimizing a large-scale managed aquifer recharge operation in an urbanized area. In P. Fox (Ed.), *Management of aquifer recharge for sustainability: Phoenix* (pp. 346–354). Arizona: Acacia Publishing.
Ibrahim, M. B. (2009). Rainwater harvesting for urban areas: a success story from Gadarif City in central Sudan. *Water Resources Management, 23*, 2727–2736.
Idelovitch, E., Icekson-Tal, N., Avraham, O., & Michail, M. (2003). The long-term performance of soil aquifer treatment (SAT) for effluent reuse. *Water Science and Technology Water Supply, 3*(4), 239–246.
Izbicki, J. A., Flint, A. L., & Stamos, C. L. (2008). Artificial recharge through a thick, heterogeneous unsaturated zone. *Ground Water, 46*, 475–488.
Jimenez, B., & Chávez, A. (2004). Quality assessment of an aquifer recharged with wastewater for its potential use as drinking water source: "El Mezquital Valley" case. *Water Science and Technology, 50*, 269–276.
Kalantari, N., Rangzan, K., Thigale, S. S., & Rahimi, M. H. (2010). Site selection and cost-benefit analysis for artificial recharge in the Baghmalek Plain Khuzestan Province, southwest Iran. *Hydrogeology Journal, 18*, 761–773.
Kanarek, A., & Michail, M. (1996). Groundwater recharge with municipal effluent: Dan Region Reclamation Project, Israel. *Water Science and Technology, 34*(11), 227–233.
Khairy, A. T., Al-Ghamdi, A. S., & Gutub, S. A. (2010). Analysis and design of a deep subsurface dam. *International Journal of Civil and Environmental Engineering IJCEE-IJENS, 10*(3), 32–41.

Kresic, N. (2009). *Groundwater resources: Sustainability, management, and restoration*. New York: McGraw-Hill.

Lasage, R., Aerts, J., Mutiso, G.-C. M., & de Vries, A. (2008). Potential for community based adaptation to drought: Sand dams in Kitui, Kenya. *Physics and Chemistry of the Earth, 33*, 67–73.

Legg, C., & Sagstad, S. (2002). Optimization and use of various recharge techniques for reclaimed wastewater at a sensitive site in Glendale, Arizona. In P. Dillon (Ed.), *Management of aquifer recharge for sustainability* (pp. 333–338). Lisse: Swets & Zeitlinger.

Lerner, D. N. (1986). Leaking pipes recharge ground water. *Ground Water, 24*, 654–662.

Lerner, D. N. (1990). Recharge due to urbanization. In D. N. Lerner, A. S. Issar, & I. Simmers (Eds.), *Groundwater recharge, a guide to understanding and estimating natural recharge: International Associations of Hydrogeologists, Contributions to Hydrogeology* (Vol. 8, pp. 210–214). UK: Kenilworth.

Li, M. H., & Barnett, M. E. (2008). Relationship between antecedent dry period and highway pollutant: Conceptual models of buildup and removal processes. *Water Environment Research, 80*, 1556–1564.

Lluria, M. R. (2009). Successful application of managed aquifer recharge in the improvement of water resources management of semi-arid regions: Examples from Arizona and the southwestern U.S.A. *Boletin Geológico y Minero, 120*, 111–120.

Maliva, R. G., Clayton, E. A., & Missimer, T. M. (2009). Application of advanced borehole geophysical logging to managed aquifer recharge investigations. *Hydrogeology Journal, 17*, 1547–1556.

Maliva, R. G., & Missimer, T. M. (2008). ASR, useful storage, and the myth of residual pressure. *Ground Water, 46*, 171.

Maliva, R. G., & Missimer, T. M. (2010). Aquifer storage and recovery and managed aquifer recharge using wells: Planning, hydrogeology, design, and operation. *Methods in water resources evaluation series no 2*. Houston: Schlumberger Water Services.

Markus, M. R., Thompson, C. A., & Ulukaya, M. (1995). Enhanced artificial recharge utilizing inflatable rubber dams. In A. I. Johnson & R. D. G. Pyne (Eds.), *Artificial recharge of ground water II* (pp. 120–128), *Proceedings at the Second International Symposium on Artificial Recharge in Ground Water*. New York: American Society of Civil Engineers.

Marsh, F. L., Dueker, L. L., & Small, G. G. (1995). Recharge well technology at the Water Campus Project, Scottsdale, Arizona. In A. I. Johnson & R. D. G. Pyne (Eds.), *Artificial recharge of ground water II* (pp. 220–230), *Proceedings at the Second International Symposium on Artificial Recharge in Ground Water*. New York: American Society of Civil Engineers.

Mikkelsen, P. S., Häflinger, M., Ochs, M., Jacobsen, P., Tjell, T. C., & Boller, M. (1997). Pollution of soil and groundwater from infiltration of highly contaminated stormwater—a case study. *Water Science and Technology, 36*(8–9), 325–330.

Ministry of Regional Municipalities, Environment and Water Resources. (2005). *Water resources of Oman: Ministry of Regional Municipalities*. Sultanate of Oman: Environment and Water Resources.

Missimer, T. M. (2009). Water supply development, aquifer storage, and concentrate disposal for membrane water treatment facilities. *Methods in water resources evaluation series no. 1*. Houston, Texas: Schlumberger Water Services.

Mousavi, S.-F., & Rezai, V. (1999). Evaluation of scraping treatments to restore infiltration capacity of three artificial recharge projects in central Iran. *Hydrogeology Journal, 7*, 490–500.

Murillo, J. M., López-Geta, J. A., de la Orden-Gómez, J. A., & Gavilán, C. J. (2002). Method for dimensioning artificial recharge plants where water supply is irregular. In P. Dillon (Ed.), *Management of aquifer recharge for sustainability* (pp. 405–408). Lisse: Swets & Zeitlinger.

Mutiso, S. (2003). The significance of subsurface water storage in Kenya. In A. Tuinhof & J. P. Heederick (Eds.), *Management of aquifer recharge and subsurface storage* (pp. 25–31).

Netherlands National Committee—International Association of Hydrogeologists (NNC-IAH) Publication No. 4.
National Research Council. (1994). *Ground water recharge using waters of impaired quality*. Washington, DC: National Academy Press.
Nilsson, A. (1988). *Groundwater dams for small-scale water supply*. Burton-on-Dunsmore: ITDG Publishing.
NRMMC-EPHC-NHMRC (Natural Resource Management Ministerial Council, Environment Protection and Heritage Council, National Health and Medical Research Council). (2009a). Australian Guidelines for water recycling: Managing Health and Environmental Risks (Phase 2): Managed Aquifer Recharge, July 2009.
NRMMC-EPHC-NHMRC (Natural Resource Management Ministerial Council, Environment Protection and Heritage Council and National Health and Medical Research Council). (2009b). Australian Guidelines for water recycling: Managing Health and Environmental Risks (Phase 2): Stormwater Harvesting and Reuse, July 2009.
Oaksford, E. T. (1985). Artificial recharge: Methods, hydraulics, and monitoring. In T. Asano (Ed.), *Artificial recharge of groundwater* (pp. 69–127). Stoneham, MA: Butterworth.
Onder, H., & Yilmaz, M. (2005). Underground dams: European. *Water, 11*(12), 35–45.
Page, D., Dillon, P., Toze, S., Bixio, D., Genthe, B., Jiménez Cisneros, B. E., et al. (2010a). Valuing the subsurface pathogen treatment barriers in water recycling via aquifers for drinking supplies. *Water Research, 44*, 1841–1852.
Page, D., Dillon, P., Vanderzalm, J., Toze, S., Sidhu, J., Barry, K., et al. (2010b). Risk assessment for aquifer storage transfer and recovery with urban stormwater for producing water for a potable quality. *Journal of Environmental Quality, 39*, 11.
Pavelic, P., Mucha, M., Dillon, P., & Barry, K. (2006). Laboratory column study on the effect of ponding depth of infiltration rates during SAT. In *Recharge systems for protecting and enhancing groundwater resources. Proceedings of the 5th International Symposium on Management of Aquifer Recharge* (pp. 624–629), Berlin, Germany, 11–16 June 2005. UNESCO, Paris.
Peck, A. J., & Williamson, D. R. (1987). Effects of forest clearing on groundwater. *Journal of Hydrology, 94*, 47–65.
Pescod, M. E. (1992). *Wastewater treatment and use in agriculture, FAO Irrigation and Drainage Paper 47*. Rome: Food and Agriculture Organization of the United Nations.
Peyton, D. (2002). Modified recharge basin floor to control sediment and maximize infiltration efficiency. In P. Dillon (Ed.), *Management of aquifer recharge for sustainability* (pp. 215–220). Lisse: Swets & Zeitlinger.
Phillips, D. W., Jr, Lyon, S., & Hutchinson, A. S. (2007). Development of a percolation decay model to guide future optimization of surface water recharge basins. In P. Dillon (Ed.), *Management of aquifer recharge for sustainability* (pp. 433–446). Lisse: Swets & Zeitlinger.
Pitt, R. (1996). *Groundwater contamination from stormwater infiltration*. Chelsea, MI: Ann Arbor Press.
Pitt, R., Clark, S., & Field, R. (1999). Groundwater contamination potential from stormwater infiltration practices. *Urban Water, 1*, 217–236.
Purkey, D. R., Thomas, G. A., Fullerton, D. K., Moench, M., & Axelrad, L. (1998). *Feasibility study of maximal program of groundwater banking (December 1998)*. San Francisco: Natural Heritage Institute.
Pyne, R. D. G. (1995). *Groundwater recharge and wells*. Boca Raton, FL: CRC Press.
Pyne, R. D. G. (2005). *Aquifer storage recovery, a guide to groundwater recharge through wells* (606 pp). Gainesville, FL: ASR Systems.
Quanrud, D. M., Arnold, R. G., Wilson, L. H., & Conklin, M. H. (1996). Effects of soil type on water quality improvement during soil aquifer treatment. *Water Science and Technology, 33*(10–11), 419–431.
RAIN. (2008). *A practical guide to sand dam implementation*. Amsterdam: Rainwater Harvesting Implementation Network.

Ramasesha, C. S., Nandakumaran, P., & Suresh, S. (2002). Efficacy of sub-surface dykes as groundwater conservation structures in hard rock terrain of Tamil Nadu, India. In P. Dillon (Ed.), *Management of aquifer recharge for sustainability* (pp. 399–404). Lisse: Swets & Zeitlinger.

Rastogi, A. K., & Pandey, S. N. (2002). Influences of artificial recharge shapes on the mound underneath in phreatic aquifers. In P. Dillon (Ed.), *Management of aquifer recharge for sustainability* (pp. 295–297). Lisse: Swets & Zeitlinger.

Rinck-Pfeiffer, S., Pitman, C., & Dillon, P. (2006). Stormwater ASR in practice and ASTR under investigation in Salisbury, South Australia. In *Recharge systems for protecting and enhancing groundwater resources. Proceedings of the 5th International Symposium on Management of Aquifer Recharge* (pp. 151–159) Berlin, Germany, 11–16 June 2005. UNESCO, Paris.

Roscoe Moss Company. (1990). *Handbook of ground water development*. New York: Wiley.

Sandoval-Solis, S., McKinney, D. C., Teasley, R. L., & Patino-Gomez, C. (2011). Groundwater banking in the Rio Grande Basin. *Journal of Water Resource Planning and Management, 137*(1), 62–71.

Schiff, L. (1957). The Darcy Law in the selection of water-spreading systems for ground-water recharge, *Symposium Darcy* (pp. 99–110), Dijon 20–26 Septembre 1956, IAHS Publication No. 41.

Schiff, K. C., & Tiefenthaler, L. L. (2011). Seasonal flushing of pollutant concentrations and loads in urban stormwater. *Journal American Water Resources Association, 47*(1), 136–142.

Sheng, Z. (2005). An aquifer storage and recovery system with reclaimed water to preserve native groundwater resources in El Paso, Texas. *Journal of Environmental Management, 75*, 367–377.

Sheng, Z., Arthur, J. A., Restrepo, J., & Allen-King, R. (2007). On recovery efficiency of managed underground storage of recoverable water systems. In P. Fox (Ed.), *Management of aquifer recharge for sustainability. Proceedings of the 6th International Symposium on Managed Aquifer Recharge of Groundwater* (pp. 84–95). Acacia Publishing: Phoenix.

Shuh, W. M. (1990). Seasonal variation of clogging of an artificial recharge basin in a northern climate. *Journal of Hydrology, 121*(1–4), 193–215.

Slichter, C. S. (1902). The motions of underground waters. U.S. Geological Survey Water-Supply and Irrigations Papers 67, Washington, DC.

Sniegocki, R. T. (1963). Problems in artificial recharge through wells in the Grand Prairie Region. U.S. Geological Survey Water-Supply Paper 1615-F, Arkansas.

Stephens, D. B. (1996). *Vadose zone hydrology* (347 pp). Boca Raton, FL: CRC Press.

Sukhija, B. S., (2008). Adaptation to climate change: strategies for sustaining groundwater resources during droughts. In W. Dragoni & B. S. Sukhija (Eds.), *Climate change and groundwater* (pp. 169–181). Geological Society of London Special Publication 288.

Thompson, A. F. B., Carle, S. F., Rosenberg, N. D., & Maxwell, R. M. (1999). Analysis of groundwater migration from artificial recharge in a large urban aquifer: A simulation perspective. *Water Resources Research, 35*, 2981–2998.

Todd, D. K. (1965). Economics of ground water recharge. *Journal of the Hydraulic Division, Proceeding of the American Society of Civil Engineers, 91*, 249–270.

Todd, D. K. (1980). *Groundwater hydrology* (2nd ed.). New York: Wiley.

Toze, S., Bekele, E., Page, D., Sidhu, J., & Schackleton, M. (2010). Use of static quantitative microbial risk assessment to determine pathogen risks in an unconfined aquifer used for managed aquifer recharge. *Water Research, 44*, 1038–1049.

U.S. Department of the Interior (Bureau of Reclamation). (1987). The design of small dams: A water resources technical publication. Washington, DC: U.S. Government Printing Office.

USACOE & SFWMD. (2002). Project management plan. Lake belt in-ground reservoir technology project, U.S.A. Army Corps of Engineers and South Florida Water Management District, April 2002.

Van Haveren, B. P. (2004). Dependable water supplies from valley alluvium in arid regions. *Environmental Monitoring and Assessment, 99*, 259–266.

Vázquez-Suñe, E., Carrera, J., Tubau, I., Sánchez-Vila, X., & Soler, A. (2010). An approach to identify urban groundwater recharge. *Hydrology and Earth System Sciences, 14*, 2085–2097.

Weaver, K. D., & Bruce, D. A. (2007). *Dam foundation grouting: Reston* (504 pp). Virginia: American Society of Civil Engineers.

World Health Organization. (2006). *WHO guidelines for the safe use of wastewater, excreta, and greywater* (Vol. 1). Geneva: WHO Policy and regulatory aspects.

Zeelie, S. (2002). OMDEL dam and recharge ponds to enhance recharge in the Namib Desert. In P. Dillon (Ed.), *Management of aquifer recharge for sustainability* (pp. 387–392). Lisse: Swets & Zeitlinger.

Zubair, A., Hussain, A., Farooq, M. A., & Abbasi, H. N. (2010). Impact of storm water on groundwater quality below retention/detention basins. *Environmental Monitoring and Assessment, 162*, 427–437.

Chapter 24
Riverbank Filtration

24.1 Introduction

Riverbank filtration (RBF), which is also referred to as bank filtration, is an old technology for treating surface water. Rather than obtaining water directly from a river or other surface water body and then treating it, surface water is drawn indirectly using wells located on land near the surface water body. RBF systems take advantage of the natural filtration that occurs as water passes through the riverbed sediments and underlying aquifer sediment or rock (Fig. 24.1). The pumping of a well lowers the pressure (head) in the aquifer and river bed sediments, which creates a downwards hydraulic gradient from the surface water body into the aquifer (or in some instances increases a natural downward gradient). The downward flow of water into the underlying aquifer is referred to as "induced infiltration" and "induced recharge."

Most RBF systems are, as the name implies, located near rivers, but the technology can be applied at other surface water bodies. For example, RBF is increasingly being used as an alternative intake for seawater desalination facilities (Missimer 2009). RBF-based alternative intakes can produce better quality water that reduces pretreatment requirements (and thus operational costs) and can also avoid many of the environmental impacts associated with conventional intakes such as the impingement and entrainment of marine or freshwater organisms.

The major attraction of RBF is that it can provide substantial cost savings over conventional surface water treatment systems that include open intakes and various filtration, coagulation, and settlement steps. It also has the great advantage in developing countries in that it is a simple technology that does not require a high degree of local technical sophistication and is often substantially less expensive than conventional surface water treatment options.

A variety of different well technologies can be used in RBF systems including:

- vertical wells,
- horizontal wells installed in trenches,

R. Maliva and T. Missimer, *Arid Lands Water Evaluation and Management*, Environmental Science and Engineering, DOI: 10.1007/978-3-642-29104-3_24,

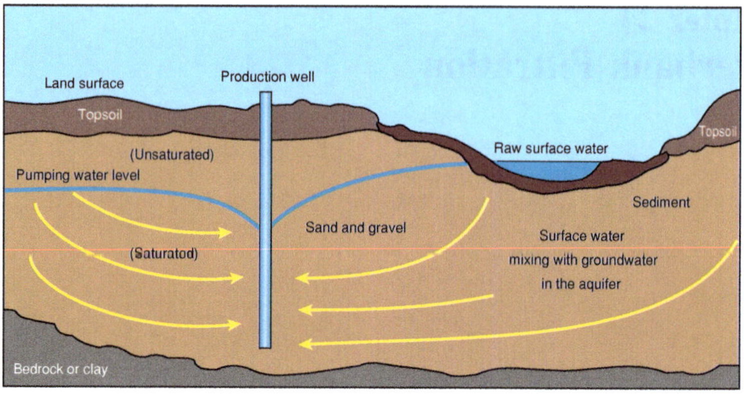

Fig. 24.1 Generalized diagram on a simple river bank filtration system (from Ray et al. 2002a)

- horizontally or obliquely drilled wells,
- collector wells (Ranney collectors),
- galleries constructed along shores, and
- horizontal seepage wells.

The earliest RBF system is reported to have been constructed in Glasgow, Scotland in 1810 (Huisman and Olsthoorn 1983; Ray et al. 2002b). RBF has been used along the lower Rhine River in Germany for over a century (Sontheimer 1980). The Düsseldorf, Germany, system dates back to 1870 (Ray et al. 2002b). The technology likely goes back for centuries, but was not recognized as such (Kuehn and Mueller 2000).

The primary advantages of RBF systems are that they can support well production through recharge of the aquifer and they can result in an improvement in water quality. RBF systems were initially developed to provide water that was directly suitable for human consumption without further treatment. In the lower Rhine River of Germany, increasing pollution of the river in the 1950s resulted in deterioration in the quality of water produced by RBF systems. The water quality deterioration includes increases in iron and manganese concentrations, break-through of odor and taste-causing substances, and chemical pollutants (Sontheimer 1980). Today nearly all water utilities along the lower Rhine River treat the produced water using granulated activated carbon (GAC), often combined with ozonation and filtration (Sontheimer 1980). The water utilities thus now use RBF as a pretreatment step, prior to more advanced treatment (Ray et al. 2002a).

At a minimum, RBF acts as a pretreatment step in drinking water production and, in some instances, can serve as the final treatment step before disinfection (Ray et al. 2002b). RBF systems cost-effectively reduce the concentrations of particulates (suspended solids, turbidity), pathogenic and non-pathogenic microorganisms, dissolved organic carbon, and many (but not all) organic and inorganic compounds. RBF systems thus provide both the intake and initial treatment of surface waters. An additional benefit of RBF technology is that underground passage may compensate for peaks of contaminant concentrations and shock loads (Kuehn and Mueller 2000). Peaks in contaminant concentrations, such as from a major shock event (e.g., spill), are smoothed out by the variable travel times from the surface water body to the well or other type of RBF intake. The greatest vertical head gradients, and thus infiltration rates, also occur during peak flow periods when there may be the greatest dilution of contaminants. RBF also provides a time buffer between a surface-water contamination event and the contaminated water reaching a water treatment plant, usually at a lower concentration.

24.2 RBF System Design Issues

The key design issue for RBF systems is obtaining the optimal balance between system yield, water quality improvement, system reliability, and cost. RBF systems behave essentially as slow sand filters. If the grain size is too coarse, then the rate of flow will be high but the degree of filtration will be too low. The selection of the type of RBF system should be based on site-specific hydrogeological and surface water hydrologic conditions. There is no one universal best design. The choice of design should consider system performance, construction (capital) costs, and operational and maintenance costs, including the need for backup capacity.

Horizontal systems (horizontal collections, infiltration galleries, and horizontally drilled wells) may be the preferred option where the production aquifer or aquifer zone is relatively thin, and, as a result, vertical wells could have only short screened intervals (Hunt et al. 2002). Vertical wells may be less expensive to construct and are also much easier to rehabilitate than horizontal wells. For large capacity systems, horizontal collectors or galleries may be preferred over numerous vertical wells because of the large number of wells that would be required. For any new projects, the advantages and disadvantages and costs of the various options need to be carefully considered.

The key hydrogeological issues for RBF system design are the rate of induced infiltration, effectiveness of filtration and contaminant attenuation, and clogging management, which depend upon the following variables (Huisman and Olsthoorn 1983; Kuehn and Mueller 2000; Gollnitz 2002; Hoehn 2002; Hunt et al. 2002; Schubert 2006a, Gollnitz et al. 2004; Caldwell 2006; Hubbs 2010):

- river stage altitude,
- water viscosity (i.e., temperature dependence of hydraulic conductivity),
- stream bed thickness,
- stream bed hydraulic conductivity,
- wetted streambed area,
- stream flow velocity,
- local soil and water chemistry,
- stream and aquifer physical properties (e.g., grain and pore throat sizes),

- connectivity of streambed and aquifer (vertical leakance),
- vertical and horizontal separation of collection points from surface water bodies, and
- retention time.

An important design and operational consideration is the fraction or percentage of the produced water that is derived from local induced infiltration. Wells installed near a surface-water body will also produce local groundwater. Some of the improvement in quality between the RBF and surface water is caused by dilution with groundwater. Depending upon the circumstances, groundwater production in RBF systems may or may not be a significant issue. Groundwater production in RBF systems could contribute to local environmental impacts such as the dehydration of wetlands. Water produced by bank filtration may be of better or lesser quality than the native groundwater, so the fraction of each in the produced water may be an important issue. Most RBF systems use alluvial sand and gravel aquifers with hydraulic conductivities greater than 1×10^{-4} m/s (2.6 ft/day) (Grischek et al. 2002). If the hydraulic conductivity of the stream bed sediments is too low or the connection to the groundwater system too tight, then the induced infiltration rate will be low and a greater proportion of the produced water will be native groundwater.

A biologically active layer, analogous to the biologically active 'schmutzedecke' layer of slow sand filters, can form at the riverbed–aquifer interface. The fine-grained, organic-rich layer is an important element of the filtration process as indicated by substantial reductions in the concentrations of organic compounds within the first few centimeters of the flow path (Ray et al. 2002a, b). Concerns have been expressed that stream bed scour during high flow-velocity periods could cause the loss of the fine-grained layer and jeopardize the filtration process (Gollnitz 2002, Gollnitz et al. 2004). In slow sand filters, a 'ripening' period of usually around 2 days is required until the schmutzedecke layer is reestablished and peak filtering performance occurs (Ray et al. 2002b). Studies of an RBF system near Cincinnati, Ohio (USA) indicate that streambed scour did not occur to a significant degree during high stage events. The biologically active layer was reported to take only 2–3 days to be reestablished after river flow stabilizes at a system on the Elbe River at Torgau, Germany (Baveye et al. 2002). Gollnitz et al. (2004) noted that even if the surface layers were removed, the underlying aquifer should be still capable of removing particulate material and most of the larger active bacteria and oocysts.

If the biologically active layer becomes too thick, it could adversely impact system performance by reducing the infiltration rate. Periodic scouring of the surface layer may be a necessary to maintain system performance, which naturally occurs in many rivers. In locations with stagnant water, the surface layers may have to be manually removed or disrupted in order to maintain system yields.

The optimal distance between wells and the surface water body depends on (Huisman and Olsthoorn 1983; Grischek et al. 2002):

- expected vertical and horizontal leakage rate of the river into the aquifer,
- preferred flow path length,

- preferred retention time, and
- acceptable groundwater drawdown.

Considering all the variables that can affect the performance of an RBF system, it is clear that performance will greatly vary between systems. Prospective RBF systems must be independently evaluated on a case by case basis. Huisman and Olsthoorn (1983) provide analytical equations for the hydraulics of basic RBF systems. However, the current state of the art is the use of numerical (computer) models, which allow for the consideration of more complex system designs and hydrogeology. However, the accuracy of numerical models depends upon the quality of the data used in their construction and population. A critical variable is the hydraulic conductivity of the riverbed material, which is difficult to measure and changes over time due to clogging.

24.3 Evaluation of RBF System Performance: Pathogen Removal

RBF system performance, in terms of water quality issues, has been evaluated on two levels. RBF systems can be evaluated based on the quality of the produced water itself, which is dependent on the quality of the source water and the efficacy of the RBF treatment. If the produced water is of a quality suitable for its intended use, then the RBF system can be considered to be properly functioning. A relatively low level of treatment (i.e., contaminant removal) would be required if the source water already has a high quality. Alternatively, RBF systems can be evaluated solely on their efficacy at contaminant removal (e.g., percent reduction in concentrations). For RBF and other natural filtration systems, the objective is not necessarily the observation of no surface-water indicators, but rather adequate removal so as to minimize the risk to public health (Gollnitz et al. 2003).

Within the United States, water produced from RBF systems will commonly be considered to be groundwater under the direct influence of surface water (UDI), as determined by the detection of surface water organisms and debris in a microscopic particle analysis (MPA) test. Groundwater that is determined to be UDI must meet surface-water treatment requirements. Under current US Environmental Protection Agency (USEPA) rules (Long-Term Enhanced Surface Water Treatment Rule) surface-water treatment facilities must meet specified treatment goals, such as a 3-log (99.9%) removal of *Cryptosporidium* oocycts. In order for RBF systems to receive treatment credits, they must demonstrate a quantitative treatment capability with respect to microorganisms, such as *Cryptosporidium* and *Giardia*, rather than just their absence (Gollnitz et al. 2005).

Water quality improvement in RBF systems is reviewed in a series of papers edited by Ray (2002), Ray et al. (2002c), and Ray and Shamrukh (2011). The removal of pathogenic microorganisms was discussed in detail by Schijven et al. (2002). Pathogens are removed by a variety of processes including straining, sedimentation within pores, colloidal filtration, attachment (sorption), and inactivation. Horizontal and vertical setbacks and groundwater travel times are key variables that affect pathogen removal (Schijven et al. 2002). Longer travel times provide more time for natural attenuation processes to operate. The European philosophy is to design for a minimum average retention time of 3 weeks to allow for the removal of microbial pathogens, balance shock loads of contaminants, and improve the biodegradation of more resistant organic compounds (Schubert 2006b).

Removal processes are most efficient when the groundwater flow velocity is low and the aquifer is composed of sand or other granular materials. The efficiency of straining depends on the ratio of the diameter of the granular media (and thus pore and pore throat size) and the diameter of suspended particles. Straining is insignificant when the ratio is greater than 20 and no particle penetration occurs when the ratio is less than 10 (Schijven et al. 2002). Most pathogen removal occurs during the first several days and meters (feet) of transport in the riverbed sediments with the schmutzedecke layer being a critical barrier.

Direct quantification of *Cryptosporidium* and *Giardia* removal rates is hampered by their usually low concentrations in natural waters relative to the method detection limits. RBF systems are not amenable to microbial challenge (pathogen spiking) studies because of the difficulties and great costs of seeding large quantities of pathogens or surrogates into surface-water bodies (Gollnitz et al. 2004). The difficulty is even more acute in flowing rivers in which the pathogens or surrogates would have to be continuously supplied. Studies have been performed using natural surrogates in the river water, such as coliform bacteria, enterococci, total aerobic endospores, algae, diatoms, turbidity, and particle counts. All of the these potential surrogates are present in greater quantities than *Cryptosporidium* and *Giardia* oocysts and are, therefore, more conducive to quantitative analysis of their removal rates.

Microorganism removal rates were examined in the Charles M. Bolton groundwater system at the Greater Cincinnati Water Works (Gollnitz et al. 2003, 2004). The goal of the investigation was to demonstrate whether or not RBF can provide surrogate reduction at a level equal to or better than that achieved by conventional surface-water treatment. Surrogate reduction was studied along flow paths to wells that were considered to have either relatively high and low vulnerabilities to contamination based on their distance from the water source. Approximately 39 and 11% of the surface water samples from the Great Miami River tested positive for *Giardia* and *Cryptosporidium,* respectively, but neither pathogen was detected in any of 285 groundwater samples collected from the wells. All of surrogates showed a greater than 3.5 log removal rate, with the lowest rates of 3.6 and 3.7 logs occurring for particle counts, which may have been influenced by clay and silt within the sediment matrix. Total algae and endospores had a 5 log or greater removal rate. The testing results indicated that RBF was highly effective and better than conventional surface water treatment plants. The majority of the total surrogate removal occurs in streambed sediments rather than in the aquifer.

Gollnitz et al. (2005) also documented the results of surrogate testing at the Central Wyoming Regional Water System (CWRWS), which is located on the

North Platte River near Caspar, Wyoming (USA). *Giardia* was present in virtually all river samples and *Cryptosporidium* was periodically detected, but neither was ever detected in groundwater samples collected from the RBF system. The surrogate testing results indicate that the aquifer provides a minimum 2-log removal of surrogates during the summer months when water production and induced infiltration are greatest.

Data from the Louisville, Kentucky, RBF system also shows a pronounced improvement in water quality (Wang 2002). The Louisville RBF system is highly effective in reducing turbidity, biodegradable organic matter, and microorganisms, such as total coliform bacteria (3.8 log removal), HPC bacteria (2.0 log removal) and total aerobic spores (3.0 log removals). Significant ($\approx$50%) removal of natural organic matter and disinfection byproduct precursors also occurs. An important observation is that much of the water quality improvement, particularly for particulate matter (turbidity and microorganisms), occurs during the first meter (3.3 ft) of travel (Wang 2002).

RBF has great potential as the primary treatment technology in developing countries. Dash et al. (2008), for example, documented the performance of a RBF system at Nainital Lake in the state of Uttarakhand in northern India. Water is obtained from tube wells installed adjacent to the lake, whose water is non-potable water because of high concentrations of organic matter and coliform bacteria. The proportion of lake water in the produced water was estimated using oxygen isotope data to be about 80% in the monsoon season and 25 to 40% during the non-monsoon period. The water recovered from the tube wells did not contain coliform bacteria and was found to be of better quality than that produced by the rapid sand filtration used at the water treatment plant. The superior performance of the RBF system lead to the closure of the water treatment plant and the installation of additional tube wells.

Dash et al. (2010) documented the performance of a RBF system located on banks of the River Ganga in western Uttarakhand state, Haridwar district, of northern India. The studied production well data indicated that RBF provided 2.5 and 3.5 log removals of total coliform and fecal coliform bacteria, respectively, during non-monsoonal periods, and provided 4.4 and 4.7 log removals of total coliform and fecal coliform bacteria, respectively, during monsoonal periods when the concentration of coliform bacteria in the river are greater. The Dash et al. (2010) study found that RBF followed by final disinfection with chlorine provides clean, safe drinking water.

24.4 Evaluation of RBF System Performance: Chemical Contaminant Removal

The primary factor that will influence the composition of water produced by RBF systems is the composition of the infiltrated water and native groundwater, and the proportion of each in the pumped water. The composition of the produced water may be furthered modified by fluid-rock (or sediment) interactions that occur along the flow path from the river to the production well, which includes cation exchange, adsorption, redox reactions, and mineral precipitation and dissolution. Some processes documented as operating in RBF systems in The Netherlands include (Stuyfzand, et al. 2006):

- adsorption and degradation of dissolved organic compounds and metals,
- redox reactions that result in a decrease in dissolved oxygen and increase in total inorganic carbon and sulfate, and
- carbonate dissolution, which results in an increase in calcium and bicarbonate concentration.

Published RBF studies for systems located along the lower river Rhine focused more on organic and inorganic compound removal. In general, RBF systems are effective in reducing the dissolved organic carbon (DOC) concentration of Rhine River water by about 50% (Kuehn and Mueller 2000). Similar reductions in total organic carbon (TOC) concentrations were also documented at three RBF sites in the United States (Hoppe-Jones et al. 2010). The majority of TOC reduction occurred very rapidly during the initial phase of infiltration. Non-aromatic TOC is preferentially removed.

The amount of reduction in organic carbon (TOC and DOC) is related to the dissolved oxygen (DO) concentration of the infiltrating water. During summer months, the Rhine River has lower DO concentrations, which limits the biological degradation of organic carbon and results in anaerobic conditions (Eckert et al. 2006). The anaerobic period may be beneficial in allowing for additional attenuation of some micropollutants (Eckert et al. 2006). RBF systems are also effective in reducing the concentrations of phosphate and iron, and some heavy metals. Chromium and arsenic concentrations were reduced by in the Duisberg, Germany, RBF system by over 90%, whereas other metals (silver, selenium, beryllium, tin) had removal rates of 20% or less (Sontheimer 1980).

Riverbank filtration is also effective in the removal of some compounds of emerging concern (CECs) (Heberer 2002; Heberer et al. 2004; Verstraeten et al. 2002; Zuehlke et al. 2004; Massmann et al. 2008; Hoppe-Jones et al. 2010). CECs are discussed in Sects. 31.2.5 and 31.3.5. CECs present in wastewater influent are removed to varying degrees by wastewater treatment processes. However, some CECs are present in treated effluent that is discharged to surface water, where their concentrations are reduced by dilution and removal by physical and biological processes. CEC concentrations in river water also depend upon the proportion of wastewater in the river flow. Water use patterns in the watershed may also impact CEC concentrations. For example, Hoppe-Jones et al. (2010) documented that CEC compounds found in European rivers were not detected in the Ohio River and Cedar River RBF sites in the United States. It was suggested that their absence may be due to a greater dilution of contaminants in wastewater in the United States resulting from a greater per capita water use.

CECs undergo further attenuation and dilution in the groundwater environment during RBF. Reduction in concentration of the compounds known to be attenuated during subsurface flow (as opposed to the more refractory compounds) is most rapid during the initial infiltration period, during which time the majority of the total organic carbon reduction also occurs (Hoppe-Jones et al. 2010). The rate of attenuation of CECs is reduced at colder temperatures (<10°C). However, some of the more refractory CECs (e.g., carbamazepine and primidone) are not significantly attenuated during RBF and have been detected at very low concentrations in groundwater samples from production wells. Additional treatment of recovered water would be required if the detected CECs are determined to represent a significant health risk at the concentrations detected in produced groundwater.

24.5 Clogging

Clogging of the sediment–water interface is an important design and operational issue for RBF systems. The downward flow of water across the interface can result in clogging and a resulting reduction in vertical hydraulic conductivity, by the deposition of suspended fine solids, chemical precipitation, and biological activities (e.g., biofilm formation). Clogging of riverbed sediments decreases the amount of water produced by induced infiltration (infiltration rate) and, as result, the percentage of groundwater in the produced water increases. The main factors that affect the reduction in vertical hydraulic conductivity are (Huisman and Olsthoorn 1983; Schubert 2002a, b, 2006a; Hubbs 2006):

- variations in suspended solids concentration of the river water,
- infiltration rate, which depends upon the hydraulic gradient from the river to the aquifer which, in turn, depends upon the river stage and drawdown (pumping rate), and
- self-cleaning mechanism by bed load transport.

Schubert (2006a, b) emphasized that clogging is unavoidable and needs to be considered during system design. Management of clogging is viewed by Schubert (2006a) as a "perpetual search for a balance correlated to the fluctuating river-aquifer interactions". Clogging rates may be reduced for a given system capacity by decreasing the induced infiltration rate, which requires either spreading out the wellfield (increasing area) or moving the wells further away from the river, which may increase the contribution of groundwater produced. Hubbs (2006) proposed that riverbed clogging could be caused or increased by the development of local unsaturated conditions just below the sediment–water interface. Unsaturated conditions would result in an increase in effective stress (as the hydrostatic pressure is reduced), which would be expected to increase the rate of compaction of sediments in the clogging layer and thus reduction in porosity and permeability.

Current conditions in the source water body also play a critical role in clogging. The current velocity needs to be great enough to exert a high enough shear force to result in bedload transport (i.e., rolling, skipping, and sliding of sand grains) and to keep very fine sediments in suspension (Schubert 2002a, b). Areas experiencing either erosion or the deposition of very fine sediments are prone to clogging. Areas of high flow velocity in which the streambed is paved with coarse (gravel and cobble-sized) material may have a relatively high susceptibility to clogging because bed material is immobile. Clogging may occur by deep filtration, wherein suspended material accumulates in pore spaces below the sediment–water interface. The accumulated clogging materials are not readily removed by erosion because the sediment–water interface is armored by the coarse materials. Hence, sandy streambeds may allow for greater long-term induced-infiltration rates than initially more permeable gravel beds because the clogging layer at the sediment–water interface can be more readily removed by scouring (Stuyfzand et al. 2006).

Hubbs (2006) discussed methods for estimating shear stress at the sediment–water interface. However, an important starting point for a RBF investigation should be a field sedimentological study of proposed sites. Basic field observations can provide important insights as to scour and sediment deposition and transport patterns.

Where RBF-type systems are used as alternative intakes systems for seawater desalination facilities, wave action can be effective in causing self-cleaning of the sediment–water interface (e.g., Maliva and Missimer 2010). In addition, shallow marine sediments are typically intensely bioturbated. Benthic organisms are highly effective in churning up shallow sediments, which would also be expected to disrupt any clogging layers that form at the sediment–water interface. Biological activity appears to also be an important cleaning mechanism to restore permeability in lake-bank filtration systems (Schubert 2006a, b; Dash et al. 2008).

24.6 Modeling of RBF Systems

Numerical groundwater flow modeling of RBF systems, including alternative seawater intake systems, can be used to perform the following functions:

- optimization of system performance; evaluation of various design and operational options for meeting target system capacity,
- evaluation of the source of the produced water (balance between bank filtered water and groundwater),
- evaluation of travel times from surface water body to production wells,
- assessment of the potential environmental impacts from wellfield operation, and
- delineation of wellfield protection areas.

Solute transport and reactive transport modeling has the potential for simulating water quality changes for different design and operational options (e.g., pumping rates and schedule). A key issue for developing an appropriate modeling strategy is to determine the degree to which solute transport is a concern, i.e., where the produced water originates and its composition. If solute transport is an important concern, then aquifer heterogeneity is a critical issue. The model must have greater spatial resolution and there is a much greater data requirement. Aquifer heterogeneity can impact the performance of RBF systems in the following ways:

- High hydraulic conductivity flow zones can result in more rapid flow to production wells and the bypass of filtration.
- Low hydraulic conductivity confining strata can separate the production zone from the surface water body and greatly reduce the amount of water obtained by filtration of surface waters.
- Wells completed in relatively high hydraulic conductivity flow zones may have enhanced well yields. The flow zones could act in a similar manner as a horizontal gallery installed below the surface water body.

The temperature effects on water viscosity and thus, hydraulic conductivity should also be assessed in areas that experience large seasonal temperature fluctuations. The viscosity of water increases with decreasing temperature, which can result in significant reductions in well yields and specific capacity. Temperature variation has had a significant documented effect on the yield from RBF systems (Caldwell 2006).

Non-calibrated models can provide useful initial insights into the behavior of RBF systems. However, a fundamental technical issue is the connection between the surface water body and underlying aquifer, for which there is typically great uncertainty. The hydraulic properties of the surface water bed sediments or rock can vary greatly depending upon the development of the schmutzedecke layer. Aquifer testing is critical, which can provide targets for the development of a calibrated flow and solute-transport model. The test should be of sufficient duration to detect the flow from the surface-water body. The model calibrated against pumping test results (transient calibration) can then be used as a predictive tool to evaluate various design options (Schafer 2006).

However it must be stressed that data from pumping tests and initial system operation may not be indicative of long term system performance because of the clogging at the sediment–water interface (Schubert 2006b; Hubbs 2006). A balance is eventually reached between degenerative clogging processes and regenerative scouring processes, assuming that biological and chemical clogging processes are not an issue.

24.7 Modeling of Contribution of Bank Filtered Water

The water produced from wells constructed near a surface water body will consist of bank filtered (BF) and native groundwater (GW) fractions. The withdrawals of native groundwater can result in adverse environmental impacts such as the lowering of water levels in wetlands. The lowering of the water table may also present a risk of land subsidence from the compaction of clays and peats (Stuyfzand et al. 2006). In some systems, native groundwater may be of poorer quality than river water. For example, native groundwater near rivers may have high natural dissolved iron concentrations or anthropogenic contaminants associated with industrial activities along the river (Ray and Prommer 2006; Wett 2006).

The BF/GW ratio can be determined from field data if there is a consistent difference in the concentration of a conservative chemical parameter between the river water and native groundwater. The ratio will vary depending on the sampling point. For example, in the Louisville RBF system, the laterals of the horizontal collector closest to Ohio River had a 10% dilution with groundwater after about one year, whereas the landward lateral had a 70% dilution (Wang 2002).

Groundwater flow modeling can be used to determine the BF/GW ratio. For example, Shankar et al. (2009) developed a three-dimensional, finite-element groundwater flow model of the Grind RBF system, located near Düsseldorf, Rhine Valley, Germany. The BF/GW ratio was determined by the flow across control lines located along the Rhine River (BF flux) and across a peninsula upon which the wellfield is located (GW flux). The model results indicated that average BF/GW ratio is 75/25, and that the ratio varies depending on factors such as river stage. Grischek et al. (2002) performed a theoretical MODFLOW model, which confirms the expected relationship that the least mixing occurs when the RBF wells are located either on an island or the inside of a meander.

The BF/GW ratio can also be determined by solute transport modeling using either actual water quality data or by arbitrarily assigning the two waters different concentrations. Non-reactive solute transport codes (e.g., MT3D) can be used to simulate the mixing of waters from different sources provided that there is a distinct chemical difference between the waters. Aquifer testing or actual operational data can be used to calibrate the model. For example, nitrate was found to be a suitable tracer for modeling of a system in Austria in which the local groundwater concentration was much greater than that in the River Enns (Wett 2006).

Alternatively, mixing can be evaluated using MT3D by assigning the surface water a concentration of 100 and the native groundwater a concentration of zero. The modeled concentration in the recovered water would be the percentage of bank filtered water. Modeling of RBF systems can be taken a step further through reactive, multi-component transport models such as PHT3D, which allows for the simulation of redox reactions and contaminant attenuation processes (e.g., Ray and Prommer 2006). Again, it must be emphasized that the fundamental constraint of modeling solute-transport in RBF systems is not the modeling process itself, but rather the abundance and quality of hydrogeological data to populate the models.

24.8 Limitations and Opportunities of RBF

RBF is in principal an efficient system for the removal of microbial contaminants, but its efficiency can be diminished by short flow paths, high heterogeneity, high hydraulic gradients and accompanying high flow velocities (Schijven et al. 2002). An important benefit of RBF processes is that they are always working to minimize contaminant concentrations. The liability is that a temporary RBF failure to completely remove pathogenic microorganisms could result in a difficult to recognize short period of moderate contaminant concentrations (Schijven et al. 2002).

The performance of RBF systems is dependent on local hydraulic and hydrogeological conditions, which may not be favorable for the desired system performance. The potential public health risk is greatest when the RBF water receives no other treatment other than disinfection. Although RBF systems are effective in removing or reducing the concentrations of some chemical contaminants of concern (including many compounds of emerging concern), some compounds and elements are resistant to removal.

A fundamental limitation to the application of RBF in arid and semiarid lands is that it requires a surface-water source. Significant perennial freshwater resources are not available in most arid and semiarid lands, which are, by definition, water scare. The greatest potential application of "classic" RBF is in the flood plains of major rivers that pass through arid lands such as the Nile, Ganges, and Tigress-Euphrates Rivers.

The limitations of RBF do not negate the great advantages of the technology. In developing countries in arid and semiarid regions, where hydrological and hydrogeological conditions allow, RBF may produce water that is of substantially better quality and safer than existing water supplies. RBF may be a viable water treatment option where the technical and financial resources for advanced water treatment facilities are not available (Ray 2011). Even if the water produced by an RBF system does not meet potable standards and will receive additional treatment, RBF may be a less expensive option than a conventional surface water intake and filtration system.

Opportunities also exist to use RBF as an element of wastewater reuse. Where wastewater is discharged to lakes and ephemeral stream channels, recovery of the water for reuse using nearby wells may result in a significant improvement in water quality and reduction in health risks. The use RBF to recover wastewater for reuse is transitional to ARR and SAT.

References

Baveye, P., Berger, P., Schijven, J., & Grischek, T. (2002). Research needs to improve the understanding of riverbank filtration for pathogenic microorganism removal. In C. Ray, G. Melin & R. B. Linksy (Eds.), *Riverbank filtration. Improving source-water quality* (pp. 311–319). Dordrecht: Kluwer Academic Press.

Caldwell, T. G. (2006). Presentation of data for factors significant to yield from several riverbank filtration systems in the U.S. and Europe. In S. A. Hubbs (Ed.), *Riverbank filtration hydrology* (pp. 299–344). Dordrecht: Springer.

Dash, R. R., Mehrotra, I., Kumar, P., & Grischek, T. (2008). Lake bank filtration at Naintal, India: Water quality evaluation. *Hydrogeology Journal, 16*(6), 1089–1099.

Dash, R. R., Bhanu Prakash, E. V. P., Kumar, P., Mehrotra, I., Sandhu, C., & Grischek, T. (2010). Riverbank filtration in Haridwar, India: Removal or turbidity, organics, and bacteria. *Hydrogeology Journal, 18*(4), 973–938.

Eckert, P., Rohns, H. P., & Irmscher, R. (2006) Dynamic processes during bank filtration and their impact on raw water quality. In UNESCO (Ed.) *Recharge systems for protecting and enhancing groundwater resources, Proceedings of the 5th International Symposium on Management of Aquifer Recharge* (pp. 17–22). Berlin, 11–16 June 2005, Paris.

Gollnitz, W. D. (2002) Infiltration rate variability and research needs. In C. Ray, G. Melin & R. B. Linksy (Eds.), *Riverbank filtration. Improving source-water quality* (pp. 281–290). Dordrecht: Kluwer Academic Press.

Gollnitz, W. D., Clancy, J. L., Whitteberry, B. L, and Vogt, J. A. (2003). RBF as a microbial treatment process. *Journal American Water Works Association, 95*(12), 56–66.

Gollnitz, W.D., Whitteberry, B.L, and Vogt, J.A., 2004, Riverbank filtration: Induced filtration and groundwater quality. *Journal American Water Works Association, 96*(12), 98–110.

Gollnitz, W. D., Clancy, J. L., McEwen, J. B., & Garner, S. C. (2005). Riverbank filtration for IESWTR compliance. *Journal American Water Works Association,* 97(12), 64–76.

Grischek, T., Schoenheinz, D., & Ray, C. (2002). Siting and design issues for riverbank filtration schemes. In C. Ray, G. Melin & R. B. Linksy (Eds.), *Riverbank filtration. Improving source-water quality* (pp. 291–302). Dordrecht: Kluwer Academic Press.

Heberer, T. (2002) Tracking persistent pharmaceutical residues from municipal sewage to drinking water. *Journal of Hydrology,* 266(3–4), 175–189.

Heberer, T., Mechlinski, A., Franck, B., Knappe, A., Massmann, G., Pekdeger, A., et al. (2004). Field studies on the fate and transport of pharmaceutical residues in bank filtration. *Ground Water Monitoring and Remediation, 24*, 70–77.

Hoehn, E. (2002). Hydrogeological issues of riverbank filtration—a review. In C. Ray (Ed.), *Riverbank filtration: Understanding contaminant biogeochemistry and pathogen removal* (NATO Series VI: Earth and Environmental Sciences) (pp.17–41). Dordrecht: Kluwer.

Hoppe-Jones, C., Oldham, G., & Drewes, J. E. (2010). Attenuation of total organic carbon and unregulated trace organic chemicals in U.S. riverbank filtration systems. *Water Research, 44*, 4643–4659.

Hubbs, S. A. (2006). Evaluating streambed forces impact the capacity of riverbed filtration systems. In S. A. Hubbs (Ed.), *Riverbank filtration hydrology*. Dordrecht: Springer.

Hubbs, S. A. (2010) Laboratory-simulated RBF particle removal processes. *Journal American Water Works Association,* 102(12), 57–66.

Huisman, L., & Olsthoorn, T. N. (1983). *Artificial groundwater recharge*. Boston: Pitman Advanced Publishing.

Hunt, H., Schubert, J., and Ray, C. (2002). Conceptual design of riverbank filtration systems. In C. Ray, G. Melin & R. B. Linksy (Eds.), *Riverbank filtration. Improving source-water quality* (pp. 19–27). Dordrecht: Kluwer Academic Press.

Kuehn, W., & Mueller, U. (2000). Riverbank filtration: an overview. *Journal American Water Works Association, 92*(12), 60–69.

Maliva, R. G., & Missimer, T. M. (2010). Self-cleaning beach-gallery design for seawater desalination plants. *Desalination and Water Treatment, 13*, 88–95.

Massmann, G., Dünnbier, U., Heberer, T., & Taute, T. (2008). Behaviour and redox sensitivity of pharmaceutical residues during bank filtration—Investigation of residues of phenazone-type analgesics. *Chemosphere, 71*(8), 1476–1485.

Missimer, T. M. (2009). Water supply development, aquifer storage, and concentrate disposal for membrane water treatment facilities, Methods in Water Resources Evaluation Series No. 1. Schlumberger Water Services: Houston.

Ray, C. (2002). Riverbank filtration: Understanding contaminant biogeochemistry and pathogen removal (NATO Series VI: Earth and Environmental Sciences). Dordrecht: Kluwer.

Ray, C. (2011) Riverbank filtration concepts and applicability to desert environments. In C. Ray & M. Shamrukh (Eds.), *Riverbank filtration for water security in desert countries* (pp. 1–4). Dordrecht: Springer.

Ray, C., & Prommer, H. (2006). Clogging-induced flow and chemical transport simulation in riverbank filtration systems. In S.A. Hubbs (Ed.), *Riverbank filtration hydrology* (pp. 155–177) Dordrecht: Springer.

Ray, C., and Shamrukh, M. (2011). Riverbank filtration for water security in desert countries. Dordrecht: Springer.

Ray, C., Grischek, T., Schubert, J., Wang, J. Z., & Speth, T. F. (2002a). A perspective or riverbank filtration. *Journal American Water Works Association, 94*(4), 149–160.

Ray, C., Schubert, J., Linsky, R.B., and Melin, G. (2002b). Introduction. In C. Ray, G. Melin & R. B. Linksy (Eds.), *Riverbank filtration. Improving source-water quality* (pp. 1–18). Dordrecht: Kluwer Academic Press.

Ray, C., Melin, G., & Linksy, R. B. (2002c). *Riverbank filtration. Improving source-water quality*. Dordrecht: Kluwer Academic Press.

Schafer, D. C. (2006). Use of aquifer testing and groundwater modeling to evaluate aquifer/river hydraulics at Louisville Water Company, Louisville, Kentucky, USA. In S. A. Hubbs (Ed.), *Riverbank filtration hydrology* (pp. 179–198). Dordrecht: Springer.

Schijven, J., Berger, P., & Miettinen, I. (2002). Removal of pathogens, surrogates, indicators, and toxins using riverbed filtration. In C. Ray, G. Melin & R. B. Linksy (Eds.), *Riverbank filtration. Improving source-water quality* (pp. 73–116). Dordrecht: Kluwer Academic Press.

Schubert, J. (2002a). German experience with riverbank filtration systems. In C. Ray, G. Melin & R.B. Linksy (Eds.), *Riverbank filtration. Improving source-water quality* (pp. 35–48). Dordrecht: Kluwer Academic Press.

Schubert, J. (2002b). Hydraulic aspects of riverbank filtration—field studies. *Journal of Hydrology*, *266*, 145–161

Schubert, J. (2006a) Experience with riverbed clogging along the Rhine River. In S.A. Hubbs (Ed.), *Riverbank filtration hydrology* (pp. 221–242). Dordrecht: Springer.

Schubert, J. (2006b). Significance of hydrologic aspects on RBF performance. In S.A. Hubbs (Ed.), *Riverbank filtration hydrology* (pp. 1–20). Dordrecht: Springer.

Shankar, V., Eckert, P., Ojha, C., & König, C. M. (2009). Transient three-dimensional modeling of riverbank filtration at Grind well field, Germany. *Hydrogeology Journal, 17*, 321–326.

Sontheimer, H. (1980). Experience with riverbank filtration along the Rhine. *Journal American Water Works Association,* 72(7) 386–390.

Stuyfzand, P. J., Juhàsz-Holterman, M. H. A., & de Lange, W. J. (2006). Riverbank filtration in the Netherlands: Well fields, clogging, and geochemical reactions. In S. A. Hubbs (Ed.), *Riverbank filtration hydrology*. Dordrecht: Springer.

Verstraeten, I. M., Heberer, T. & Scheytt, T. (2002). Occurrence, characteristics, and transport and fate of pesticides, pharmaceutically active compounds, and industrial and personal care products at bank filtration sites. In C. Ray, G. Melin & R. B. Linksy (Eds.), *Riverbank filtration. Improving source-water quality* (pp. 175–277). Dordrecht: Kluwer Academic Press.

Wang, J. (2002). Bank filtration case study at Louisville, Kentucky. In C. Ray, G. Melin, & R. B. Linksy (Eds.), *Riverbank filtration. Improving source-water quality* (pp. 117–145). Dordrecht: Kluwer Academic Press.

Wett, B. (2006). Managing clogging of a RBF-system at the River Enns, Austria. In S. A. Hubbs (Ed.), *Riverbank filtration hydrology* (pp. 259–280). Dordrecht: Springer.

Zuehlke, S., Duennbier, U., Heberer, T., & Fritz, B. (2004). Analysis of endocrine disrupting steroids: Investigations of their release into the environment and their behavior during bank filtration. *Ground Water Monitoring and Remediation,* 24, 78–85.

Chapter 25
Coastal Water Management Issues

25.1 Introduction

Coastal areas in arid and semiarid lands face additional groundwater management challenges from salinization compared to more inland areas. Saline–water intrusion, in particular, is a major threat to coastal aquifers. Saline–water intrusion occurs in areas where the rate of groundwater extraction significantly exceeds the rate recharge. Over extraction disturbs that natural water balance and induces migration of saline-water into parts of aquifers that naturally contain fresh groundwater. Saline–water intrusion is an ever-growing problem in many coastal areas because of a combination involving:

- population growth and the concomitant increase in water demands, which is often concentrated in coastal urban areas,
- paucity of alternative economical water sources, especially for agriculture, and
- land development activities that provide conduits for intrusion (e.g., canal construction) and reductions in recharge (e.g., increases in impervious area).

Arid and semiarid lands areas are particularly vulnerable to saline–water intrusion because of their low recharge rates. The distribution of rainfall is often also highly variable resulting in drought periods in which even modest groundwater withdrawals can result in the development of a landwards hydraulic gradient that can induce saline–water intrusion. Protecting precious freshwater resources in arid and semiarid lands from saline–water intrusion is thus a critical water management concern.

Wadi Ham in the United Arab Emirates (UAE) is a good example of a coastal aquifer that is vulnerable to saline–water intrusion (Sherif et al. 2006; Sherif and Kacimov 2007). Wadi Ham drains into the Gulf of Oman along the Fujairah and Kalbha coast of the UAE. The main aquifer consists of seawards thickening alluvial wedge of gravel (Quaternary Alluvium Aquifer) that is underlain by ophiolite deposits (Semail Ophiolites). Groundwater is exploited intensively for irrigation in the coastal plain and for domestic water supply in this area.

R. Maliva and T. Missimer, *Arid Lands Water Evaluation and Management*, Environmental Science and Engineering, DOI: 10.1007/978-3-642-29104-3_25,

Groundwater modeling results and monitoring data indicate that a seaward flow of groundwater occurs throughout the aquifer during wet periods (Sherif and Kacimov 2007). During dry periods, the flow is reversed in the southeastern part of the aquifer, towards a wellfield. Severe saline–water intrusion is reported to occur during dry periods (Sherif and Kacimov 2007), which is caused by intensive pumping exacerbated by drought conditions. Groundwater modeling results demonstrated that managed aquifer recharge (MAR) can be effective in locally improving groundwater quality and pushing the saline–water interface seawards.

Saline–water intrusion in coastal settings has been intensely studied and a voluminous literature has been produced on the subject over the past several decades. Bear et al. (1999) provides a detailed overview of seawater intrusion concepts, methods and practices. There have also been numerous conferences and conference sessions dedicated to saline–water intrusion and related coastal groundwater issues. Numerous case study papers have been published and there are few major areas that currently are experiencing significant saline–water intrusion that have not already been investigated to some degree.

There has been great research interest in methods to:

- locate and characterize the freshwater/saline–water interface,
- develop analytical and numerical models to simulate the behavior and movement of the interface,
- monitor the changes in the interface position over time, and
- prevent or reverse saline–water intrusion.

The tools needed to achieve these goals are all currently available, but will be certainly continue to be progressively refined in the future. The key issue for applied hydrogeology is optimization, i.e., determining the most efficient and cost-effective means to locate, simulate, and monitor the saline–water interface, and to mitigate saline–water intrusion.

25.2 Causes of Saline–Water Intrusion

Anthropogenic saline–water intrusion is caused by groundwater pumping (or other activities) altering natural groundwater flow patterns so that saline water is induced to move towards freshwater resources. Under natural, pre-development conditions, the seaward flow of fresh groundwater controls the landward migration of the saline–water interface (Fig. 25.1). The interface between fresh and saline groundwater is typically roughly wedge-shaped because of density stratification, with the greatest landward extent of saline water occurring at the base of the aquifer, which is commonly referred to as the "toe" of the wedge. Fresh and saline groundwaters are separated by a mixing zone, the width or thickness of which varies between aquifers and specific site conditions. Unless there is a perturbation of the system, such as a change in freshwater heads or sea level, equilibrium is reached and the position of the interface becomes stable over time. However, some

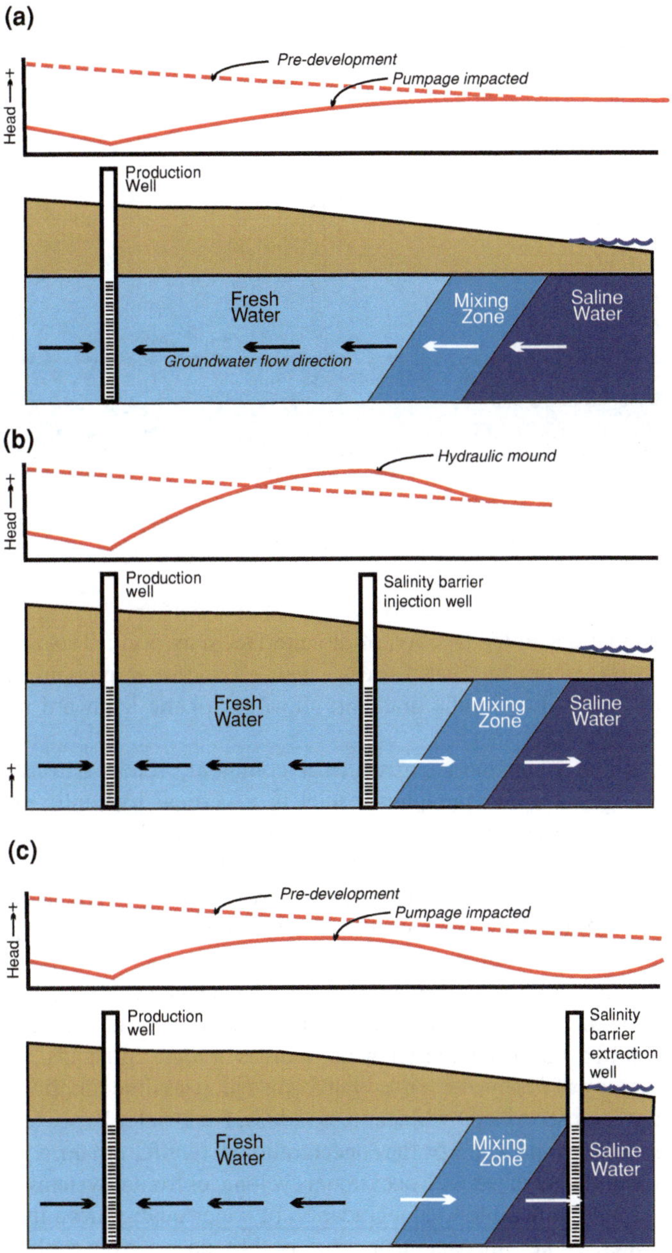

Fig. 25.1 Schematic diagram of a horizontal salinity barrier. **a** Production well pumping creates a landward hydraulic gradient, which induces the landwards migration of saline water. **b** Injection landwards of the mixing zone would create a hydraulic mound, reversing the direction of groundwater flow seawards of the barrier. **c** A seaward hydraulic gradient can also be induced by pumping seawards of the freshwater/saline-water interface.

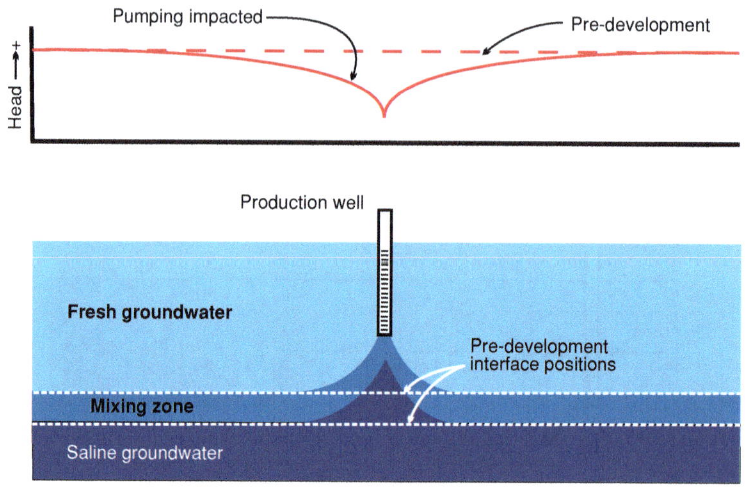

Fig. 25.2 Conceptual diagram of vertical saline–water intrusion (up-coning). The lowering of heads at a pumped well results in a vertical hydraulic gradient that allows saline-water to migrate upwards

seasonal fluctuations in the position of the interface may occur. Horizontal saline–water intrusion occurs in coastal areas when groundwater pumping reduces or reverses the seaward hydraulic gradient, allowing for the landward migration of saline–water interface.

Vertical saline–water intrusion, which is commonly referred to as up-coning, occurs when groundwater pumping results in a vertical hydraulic gradient and induces the upward migration of saline water into wells (Fig. 25.2). Vertical saline–water intrusion can potentially occur wherever an aquifer used for groundwater production is underlain by more saline water. Vertical saline–water intrusion tends to be more common in coastal settings because of the presence of saline waters at shallow depths.

Saline–water intrusion is a critical issue for the management of freshwater lenses that are present on islands and in some inland areas. Freshwater lenses are a volumetrically limited, but still important water sources in some arid lands. For example, the hydrogeology of freshwater lenses that formed atop a saline limestone aquifer in coastal Oman was described by Macumber (2003). Freshwater lenses form in broad shallow depressions and wadis by the concentration of runoff from infrequent periods of intense rainfall associated with occasional cyclonic or frontal systems. Freshwater lenses are highly vulnerable to a degradation of water quality from the up-coning of saline water. A key management issue is thus not to over-exploit any lens. Macumber (2003) recommended a management strategy of distributing groundwater withdrawals amongst numerous lenses in order to avoid over-exploitation of any one lens. Commonly, galleries, horizontal wells, or shallow trenches are used to develop these aquifers by skimming water off the top of the aquifer (e.g., the Bahamas, Western Australia).

An often overlooked issue is that increases in groundwater salinity (i.e., salinization) in coastal areas may be due entirely or in part to processes other than vertical and horizontal saline–water intrusion. Salinization of coastal aquifers may have multiple causes. For example, in Bahrain, four possible causes of salinization were identified (Zubari 1999):

- horizontal seawater invasion,
- up-coning of brackish water from underlying aquifers,
- migration of sabkha water, and
- infiltration of irrigation return flows (or salination of soils).

The lowering of the potentiometric surface in a coastal aquifer can induce a horizontal landward hydraulic gradient, an upward hydraulic gradient from underlying strata, and increase the downward hydraulic gradient from overlying strata, which can potentially induce the movement of saline waters into a freshwater aquifer from three different directions. Management of salinization, thus, requires knowledge of the specific cause(s) of the increase in salinity. Sea spray, saline irrigation return flow infiltration, evaporation of irrigation water from soils, and dissolution of evaporite minerals can also be significant salinity sources (Custodio 2004).

Incorrect identification of salinity sources can lead to ineffective management (Zubari 1999; Custodio 2004; Milnes and Renard 2004; Trabelsi et al. 2007). The source of salinity can be determined using environmental tracer data, such as major and minor element concentrations and stable isotopes. A non-seawater source may be indicated by ratios of sodium to chloride significantly different from that of seawater. The geochemical data can be evaluated and interpreted using multivariate statistical methods (e.g., Trabelsi et al. 2007), mixing curves (e.g., Buszka et al. 1994), and inverse modeling techniques (Campbell et al. 1997a, b).

Saline irrigation return flow can result in salinity increases through solute recycling into the aquifer. Part of the irrigation water leaves the soil-groundwater system by evapotranspiration, whereas the solutes in the irrigation water are left behind and eventually enter the groundwater. Commonly, commercial fertilizers contain some salt which adds to the overall salinity of the irrigation return flows. The salinity of the shallow groundwater may progressively increase over time, the rate of salinity increase being related to the aquifer turnover time, which is in turn related to the aquifer extraction rate and water volume (Milnes and Renard 2004). Saline irrigation return flow may be superimposed on the horizontal seawater intrusion process and may be recognized by a correlation of salinity parameters (e.g., chloride) with agriculture-related parameters (e.g., nitrate, pesticides) (Milnes and Renard 2004). Where saline irrigation return flow is significant, the solute mass return flow from irrigation needs to be simulated in numerical models as a coupled boundary condition in which the flux is a function of well pumping.

25.3 Location and Characterization of Saline–Water Interface

It is now well understood that the interface between fresh groundwater and saline groundwater is often not a sharp interface. The fresh and saline groundwaters are separated by a transition (i.e., mixing) zone of varying thickness. The nature of the interface is of importance for understanding and simulating saline–water intrusion. For example, the depth of the interface may be calculated using the well known Ghyben-Herzberg relation:

$$z = \frac{\rho_f}{(\rho_s - \rho_f)} h \tag{25.1}$$

where,

z = depth of the presumed sharp saline–water interface below sea level (m or ft)
h = height of the freshwater zone above sea level (m or ft)
ρ_s = density of seawater (g/cm^3; lbs/ft^3)
ρ_f = density of freshwater (g/cm^3; lbs/ft^3)

However, the Ghyben-Herzberg relationship is frequently not valid because the underlying assumptions of a sharp interface and the absence of freshwater flow to the sea (i.e., hydrostatic conditions) are not met. The assumption of a sharp interface and applicability of the Ghyben-Herzberg relation was referred to as a 'hydromyth' by Custodio (2004). The complexity of the interface between the fresh and saline groundwater requires consideration of geological and hydrogeological structure, and present and past hydrodynamic conditions (Custodio 2004). Aquifer heterogeneity (vertical and horizontal) can profoundly impact the shape, position, and rate of movement of the saline–water interface.

Theoretical numerical modeling and Monte Carlo simulations demonstrated that when heterogeneity (i.e., variance of hydraulic conductivity) increases, it is not sufficient to use an equivalent hydraulic conductivity to accurately represent the position of the saline-water wedge, the width of the mixing zone, or the saltwater flux (Kerrou and Renard 2010). The position of the wedge is controlled by the variance in hydraulic conductivity, which in turn controls the horizontal effective hydraulic conductivity and longitudinal and transverse macro-dispersivities (Kerrou and Renard 2010). As would be expected, the modeled penetration length of the toe of the saline-water wedge increases with increasing heterogeneity (effective horizontal conductivity). The presence of a flow zone (i.e., strata with a relatively high hydraulic conductivity) near the base of any aquifer would allow saline-water to migrate inland more rapidly and to a greater extent than would occur under more homogeneous conditions.

In reality, it is not possible to accurately predict the location and shape of the saline–water/fresh water interface by theoretical methods alone. It is instead necessary to determine its location in the field. The position of the saline–water interface can be mapped in the field using well and surface geophysical methods. Most production wells are open to relatively thick intervals and thus provide limited data on the depth of the saline–water interface. The depth of the interface can be determined either through the collection of discrete water samples at differing depths during or after well drilling or through borehole geophysics. Off bottom (single) packer or straddle packer tests are the preferred methods for the collection of depth-specific water samples during drilling as they can isolate a selected depth interval for sample collection. Dual tube and cable tool drilling methods can also provide acceptable quality water samples as the strata above the drill bit are cased off during drilling.

Borehole geophysical logs are also commonly used to obtain salinity versus depth profiles. A combination of a resistivity log and porosity log (e.g., sonic) are routinely processed to obtain a profile of formation water resistivity versus depth, which can be further processed to obtain a salinity versus depth profile (Sect. 17.6). In situations where the lithology is uniform and an accurate estimate of porosity is available, an estimated salinity versus depth profile can be obtained using only a resistivity log.

Surface resistivity geophysical methods are very well suited for mapping the vertical and geographic position of the salinity–water interface because the interface is typically marked by a sharp change in resistivity. The application of VES and TDEM for mapping the salinity–water interface is further discussed in Sect. 16.5.1. For the most accurate interpretations, all three data sources should be utilized. Borehole water quality and geophysical data allow for the calibration and ground-truthing of the surface geophysical data.

25.4 Monitoring of Saline–Water Intrusion

The position of the saline–water interface may change over time in response to groundwater pumping and changes in recharge. Monitoring of the position of the interface is thus a critical part of groundwater management in coastal areas. If saline–water intrusion is occurring, i.e., the saline–water interface is migrating landwards, mitigation actions may be necessary to protect the aquifer.

Saline–water intrusion can be monitored through a series of wells located perpendicular to the coast, and thus the saline–water interface. The preferred monitoring system is a series of well clusters with each well in the cluster open to different depth, which allows the detection of the movement of different parts of the wedge-shaped interface. The wells could be periodically sampled for salinity-related parameters such as specific conductance, chloride, or total dissolved solids. Another option is to equip the wells with a conductivity probe and data logging system.

The preferred monitoring well design is to have the wells screened or open to a small depth interval (2–5 m; 6.6–16.4 ft) rather than the entire aquifer thickness. The flow of water into wells open to the entire aquifer thickness represents an average value that is weighted by the transmissivity of each bed. If most of the flow enters a well from a flow zone located near the top of the aquifer, it is possible that considerable movement of the interface may occur without detection. The movement of the interface might not be detected until the top of the saline-water wedge reaches the well.

Installation of multiple monitoring wells can be quite expensive, particularly for a deep aquifer. Dual- or tri-zone monitoring wells open to different depth intervals may be less expensive than separate wells. However, multiple-zone monitoring wells are more complicated to properly construct so that the monitoring zones are hydraulically separated. Maintaining the mechanical integrity of the wells is also a concern.

Multilevel monitoring systems are an alternative to multiple separate wells or multiple-zone monitoring wells. Multilevel monitoring systems, such as the Westbay™ System, consist of multiple packers installed in a single borehole that isolate selected zones of an aquifer (Fig. 25.3). The Westbay System allows for the collection of both pressure data and water samples from each discrete monitoring a zone. Westbay Systems were installed for the Orange County Water District (OCWD, California, USA) as part of its extensive groundwater resource management program (Black et al. 1988). The OCWD has had 57 Westbay System monitoring wells installed that provide for testing, monitoring, and sampling of approximately 550 zones. The OCWD estimated a savings of 55–65% from the installation of Westbay Systems compared with alternative technologies. A Westbay System installed at the City of Daytona Beach, Florida, for saline–water intrusion monitoring, provided evidence for hitherto unrecognized hydraulically separated zones in the Upper Floridan Aquifer as demonstrated by differences in heads.

For wells with an open-hole construction, running a time series of resistivity (dual induction) logs may reveal changes in the vertical position of the saline–water interface. Changes in salinity near the borehole may be caused by inter-zone flow in the well. The well should, therefore, be pumped prior to logging to ensure that native groundwater is present near the borehole throughout the well.

Running a time-series of surface resistivity surveys can also be used to map changes in the position of the saline–water interface over time (Sect. 16.5.1). For example, Abdalla et al. (2010) documented an approximately 600 m (about 2,000 ft) seaward migration of a saline-water in Oman using TDEM surveys conducted in 2002 and 2007. The time-series of surveys should ideally be performed on the same transects. Either the electrode network can be reinstalled for each survey or a series of electrodes can be permanently left in place. The latter option is less labor intensive, but may be problematic on a long-term basis as electrodes may deteriorate over time (e.g., undergo oxidation), local ground conditions may change, or they may be destroyed by surface activity.

From a practical perspective, where saline–water intrusion poses a serious threat to groundwater resources, a multiple-element approach should be taken in order to determine the position of the saline–water interface (or interfaces) and to detect changes in its position. There is no adequate substitute to a monitoring

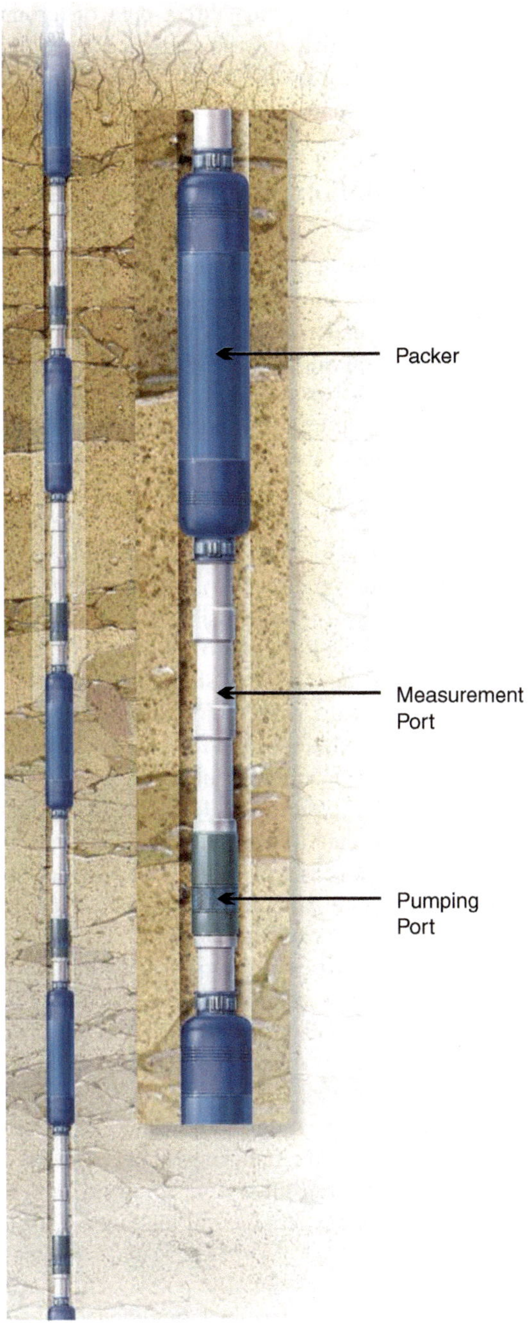

Fig. 25.3 Westbay System diagram (© Schlumberger)

well network from which direct data on groundwater salinity can be obtained. Monitoring well data may be augmented by borehole and surface geophysical data. The latter may be invaluable for extrapolating between monitoring wells.

25.5 Simulation of Saline–Water Intrusion

Density-dependent solute-transport codes are available that can be used to simulate the saline–water interface and intrusion of saline water. Density dependent models are critical because heads (water levels in wells) are dependent on salinity. The U.S. Geological Survey SEAWAT code (Guo and Langevin 2002; Langevin et al. 2007), which is based in part on the widely used MODFLOW code, is a density-dependent solute-transport code that is well suited for simulation of groundwater flow and solute transport in coastal settings. General groundwater modeling procedures are discussed in Chap. 20.

A key issue in the simulation of saline–water intrusion is the accuracy and resolution of the models. Current modeling capability allows for the evaluation of whether or not the current transition zone is in equilibrium with present-day sea level. However, it is not possible to accurately forecast salinity changes at a given well (Sanford and Pope 2010). Finer spatial resolution, which requires spatial time discretization and the use of short duration time steps, increases computation time. However, even if infinite computer capacity was available, that would not guarantee that a simulation could forecast salinity increases at a particular well because there would be insufficient data to estimate hydraulic parameters in the needed spatial detail (Sanford and Pope 2010). Modeling capability is limited by the availability of data to populate the model rather than the modeling process itself.

Accurate modeling of solute transport, whether for simulating the response of coastal aquifers to pumping or the operation of managed aquifer recharge systems, requires much more data than is needed for the simulation of bulk groundwater flow. Heterogeneity with respect to hydraulic conductivity and porosity needs to be evaluated and incorporated into models. Detailed aquifer characterization, therefore, is critical in order to obtain the data needed for model development. Aquifer characterization techniques are summarized in Chap. 10 and with respect to ASR and other MAR systems using wells by Maliva and Missimer (2010). Aquifer characterization programs typically require aquifer performance testing (using pumping and monitoring wells) in order to determine aquifer bulk hydraulic properties (e.g., transmissivity, storativity, and leakance, if necessary) and packer testing or borehole geophysical logging (ideally both) in order to quantify aquifer heterogeneity.

Well data are essentially point sources of information, which need to be extrapolated in order to populate a model. Sequence stratigraphy is a very widely used tool in the oil and gas industry, but has seldom been used in groundwater investigations. Sequence stratigraphic applications to managed aquifer recharge projects were reviewed by Maliva and Missimer (2010). The power of sequence stratigraphic analysis is that it provides an element of predictability to the vertical and horizontal sedimentology (and thus hydrogeology) of aquifer sediments and rock. Once the basic sequence stratigraphic framework of an area has been determined, sequence stratigraphic principles can be used to make more informed interpolations and extrapolations of available well and geophysical data.

A sequence stratigraphic approach was evaluated for simulation of seawater intrusion in the Dominguez Gap area of Los Angeles, California (Nishikawa et al. 2009). Sequence stratigraphy was used to overcome some of the limitations of the traditional approach taken in the development of groundwater models, which are based on the correlation of lithostratigraphic-based units. The value of sequence stratigraphy is that it can account for lateral variations in lithology and the continuity and interconnection of coarse-grained facies that are the main transmissive intervals (Nishikawa et al. 2009). Folding and faulting, and their effects on downward migration of seawater where also incorporated into the Dominguez Gap model.

The development of aquifer characterization programs and workflows to analyze and process the data and then incorporate it into groundwater flow and solute-transport models is an area where much work is still needed to optimize the management of coastal water resources. There is considerable potential for the application of oilfield technologies to water resources problems. For example, advanced borehole geophysical logs, such as nuclear magnetic resonance, can provide fine-scale data on aquifer heterogeneity (Sect. 17.12) and allow more detailed input to solute transport models.

25.6 Coastal Regions and Climate Change

Coastal regions are highly vulnerable to the effects of climate change, particularly projected rises in sea level. The average rate of sea level rise measured by tide gauges from 1961 to 2003 was 1.8 mm ± 0.5 mm/year (0.07 ± 0.02 in/year). Sea level is projected to rise by the end of century (2090–2099) by 0.18–0.59 m (0.6–1.9 ft) relative to the baseline period of 1980–1999 (Kundzewicz et al. 2007; Committee of Environment and Water Resources 2008).

The key local issue is change in relative sea level (i.e., sea level relative to local land surfaces) rather than global sea level. Relative sea level may also change because of either natural changes in land surface altitude (e.g., isostatic rebound, or tectonic uplift or subsidence) or anthropogenic causes (e.g., subsidence due to groundwater pumping or hydrocarbon production). The most significant adverse impact of sea level rise is the loss of land, coastal habitats, and infrastructure, and associated ecological, cultural, and subsistence values (Klein and Nicholis 1999).

Local coastal communities and nations vary in their vulnerability to and ability to adapt to climate change induced rises in sea level. Analysis of coastal vulnerability starts with some notion of the natural system susceptibility to the biogeophysical effects of sea-level rise and its natural ability to cope with the effects (Klein and Nicholis 1999). Clearly, low-lying coastal areas and islands are much more susceptible to sea level rise than higher altitude coastal areas. Societies differ greatly in their adaptive capacity to environmental change in general (Chap. 39). It is important to understand the dynamic interaction that takes place between natural and socioeconomic systems.

The vulnerabilities of coastal communities to sea level rise include (Bloetscher et al. 2010):

- higher water-table elevations, resulting in a lower capacity to store rainwater and thus more runoff is produced,
- runoff rates exceeding stormwater drainage system capacity may occur,
- increased risk of groundwater contamination from seawater inundation and tropical storm surge,
- inundation of land on which septic systems provide on-site treatment and disposal,
- increased saline water infiltration into wastewater systems, which may impair treatment and reuse systems, and
- increased saline–water intrusion and associated contamination of freshwater resources.

Rising sea level will result in both the inundation of low-lying areas and an increase in saline-water heads. The change in the position of the saline-interface will depend upon the magnitude of the sea level rise and local hydrogeological conditions. The response to sea-level rise will also depend upon whether the coastal aquifer is close to the flux-controlled system or head-controlled system end members (Werner and Simmons 2009). In a flux-controlled system, the groundwater discharge to the sea is not changed; groundwater levels rise and the hydraulic gradient is maintained. In a head-controlled system, inland freshwater heads remain the same despite the sea level rise, because of various surface controls such as drains and rivers. Analytical model results indicate that minimal inland movement will occur in flux-controlled systems, whereas large changes may occur in head-controlled systems (Werner and Simmons 2009). The migration of the toe of the saline-water wedge is greater with decreasing recharge rates, increasing hydraulic conductivity, and increasing depth of the aquifer below sea level (Werner and Simmons 2009). The potential impacts of sea level rise can be evaluated by numerical modeling on a gross scale (e.g., Sherif and Singh 1999; Bobba 2002).

25.7 Mitigation of Saline–Water Intrusion

There are four basic measures available to control lateral saline–water intrusion, including:

- reduce pumping in order to reduce or eliminate over extraction,
- relocate pumping inland,

- aquifer recharge (positive salinity barriers), and
- pumping of brackish or saline water, seawards of the saline–water interface (extractive barriers).

Reduction and relocation of pumping are strategies aimed at restoring the hydraulic equilibrium at the saline–water interface. The most common historic reaction to saline–water intrusion is to relocate wells further inland. From a technical perspective, reduction and relocation of pumping are obvious mitigation options. However, from a practical perspective they are often very difficult to implement in water scarce regions in which alternative water sources are not available. Continuation of water use practices that are unsustainable in the long run may be politically expedient, as opposed to the immediate discord that could arise from forcing local farmers to stop irrigating. Commonly, the relocation of wells inland is coupled with an increase in production, thereby causing saline–water intrusion to occur to an even greater degree and exacerbate the overall water supply problem.

Aquifer recharge can be implemented over a broad upland area in order to restore the water balance (or at least reduce the overdraft) or strategically to manage the movement of the saline–water interface. Salinity barriers are systems designed to control or reverse the movement of the saline–water interface. The key feature of salinity barriers is the restoration or creation of a seaward hydraulic gradient within the fresh water side of the saline–water interface. The gradient can be created by managed aquifer recharge landward of the interface or by the pumping of brackish or saline-water seawards of the interface (Fig. 25.1).

Salinity barrier options are (Banks and Richter 1953; Bruington 1969; Todd 1974, 1980; Oude Essink 2001):

- recharge barriers (injection and surface application),
- extractive barrier,
- combination recharge and extractive barrier, and
- physical barrier.

Creating a salinity barrier by MAR can be a critical component of IWRM. Reclaimed water can be used for recharge in the salinity barrier, which serves the additional goal of providing a means of putting the water to an environmentally sound beneficial use and avoiding adverse environmental impacts associated with wastewater disposal (see Chap. 29).

25.7.1 Positive Salinity Barriers Systems

Positive salinity barrier systems involve creation of a hydraulic mound landward of the saline–water interface. Positive salinity barrier systems were reviewed by Maliva and Missimer (2010). Prime examples are the Talbert Gap barrier in Orange County, California and the West Coast Basin Barrier, Dominguez Gap

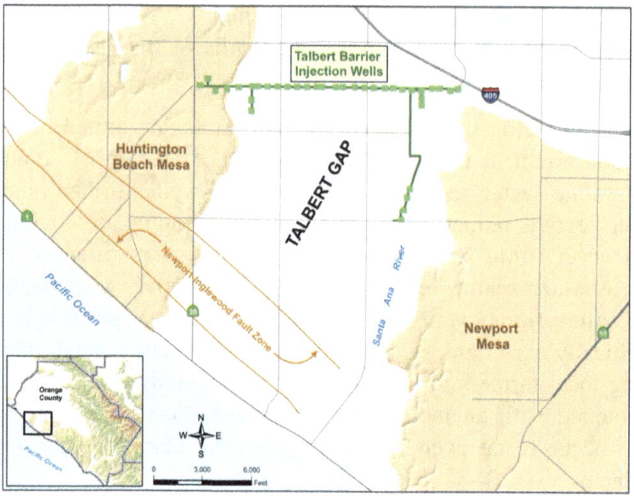

Fig. 25.4 Map showing the location of the Talbert Gap salinity barrier system which consists of an alignment of injection wells between the Huntington Beach Mesa and the Newport Mesa (courtesy of the Orange County Water District)

Barrier, and Alamitos Gap systems in Los Angeles, California. All of these salinity barrier systems prevent migration of saline water from the Pacific Ocean into landward groundwater basins that are intensely used for water supply.

The Talbert Gap salinity barrier operated by the Orange County Water District (OCWD) is perhaps the best known operational large-scale salinity barrier system. The Newport-Inglewood fault zone in coastal Orange County forms the southwestern boundary for all but the shallow aquifers and appears to be an effective barrier against saline–water intrusion. Saline water is able to migrate inland into the Orange County Groundwater Basin through gaps in the Newport-Inglewood fault zone where shallow aquifers connect the basin to the Pacific Ocean. The Talbert Gap is approximately 4 km (2.5 miles) wide and is located between the cities of Newport Beach and Huntington Beach. Geologically, the Talbert Gap is located between the Huntington Beach Mesa and the Newport Mesa (Fig. 25.4).

The salinity-barrier system originally consisted of 23 multiple-zone (nested) injection wells with a total of 81 injection points. There are four injection zones (shallow aquifers), which are referred to, in descending order, as the Talbert aquifer, and alpha, beta, and lambda aquifers (Fig. 25.5). Each multi-zone injection well is open to 2–4 of the zones. The injection wells are constructed at approximately 180 m (600 ft) intervals along a city street (Ellis Avenue). The injected water consists of highly treated reclaimed water produced at the state-of-the-art Groundwater Replenishment System (GWR), which has been described by Markus (2009). The GWR treatment process consists of microfiltration (MF) pretreatment followed by reverse osmosis (RO) and ultraviolet light and hydrogen peroxide treatment to break down remaining organic compounds through an

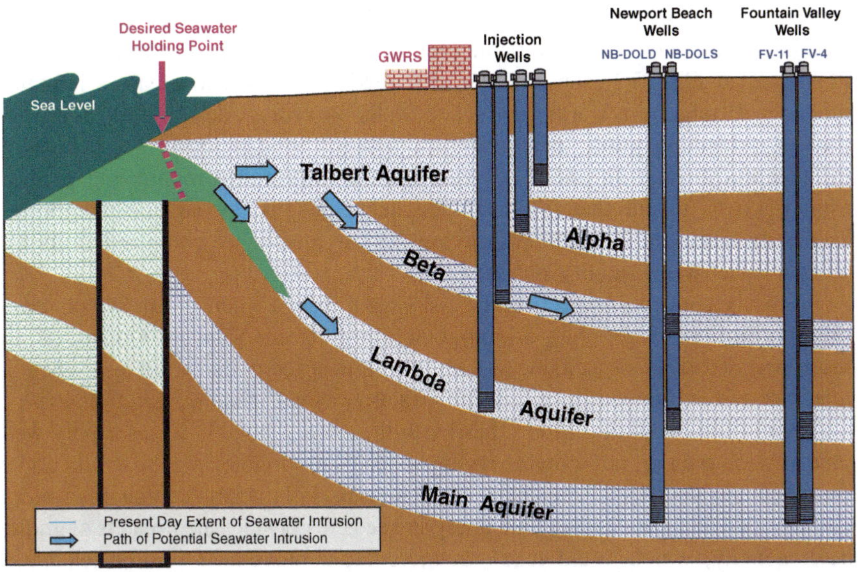

Fig. 25.5 Schematic north–south cross section of the Talbert Gap salinity barrier. Highly treated wastewater is injected into four separate aquifers (courtesy of Orange County Water District)

accelerated decomposition (oxidation) process. The result of the treatment process is ultra-pure water. Such a high level of treatment is required because some of the recharged water will eventually enter the potable supply (i.e., indirect potable reuse will occur).

A salinity-barrier system has been constructed at the city of Salalah in coastal Oman, which has been described by Shammas and Jacks (2007) and Shammas (2008). Household sewage was traditionally disposed of using septic tanks, soakaways, and cesspit systems, which appeared to be the main source of bacteriological contamination of the groundwater. The Salalah Central Sewage Treatment Plant (STP) was constructed to provide centralized sewage treatment service to the community. The wastewater receives tertiary treatment and disinfection. The goals of the salinity barrier system are to (Shammas 2008):

- act as a barrier to saline–water intrusion,
- increase the potentiometric surface of the coastal aquifer, and
- provide a source of water for crop irrigation.

The first operational phase of the Salalah salinity barrier system consists of 40 injection wells and 40 observation wells. The wells are located 1.5–2.0 km (0.9–1.3 miles) from the shoreline and are located approximately 300 m (1,000 ft) apart with a total depth of 40 m (120 ft). The salinity-barrier system is located near the area where the aquifer is used for agricultural water supply. The potable water wellfield is located further inland and is, thus, further separated from the salinity barrier.

There are many other coastal areas in which positive salinity barriers using reclaimed water could be a valuable tool for managing saline–water intrusion. As illustrated in Fig. 25.1, some of the recharged water used to create a hydraulic mound will migrate towards production wells. A critical design and operational issue is protecting public health. Where indirect potable reuse is a possibility, the recharged water should be of very high quality and a monitoring program established to ensure that the potable water supply is not adversely impacted, especially if natural attenuation processes in the aquifer are being relied upon in order to meet water quality standards.

From a technical perspective the risks are far less when groundwater near a salinity barrier system has a non-potable use (e.g., Salalah, Oman). If the reclaimed water is of suitable quality for unrestricted irrigation use, then its injection and subsequent recovery for irrigation should pose no additional risks.

Travel times from the salinity barrier to the nearest potable water supply wells (and any intervening non-potable wells) are critical variables for the evaluation of the potential risks of a salinity barrier system, especially when the recharge water is not of potable quality. Travel time analysis needs to consider the impacts of aquifer heterogeneity, which could result in more rapid movement of injected water. A detailed, accurate aquifer characterization is the critical foundation for evaluating the fate and transport of any contaminants present in the recharged water. The risk assessment also needs to consider the degree of treatment that the produced groundwater receives and how effectively that treatment process removes potentially harmful pathogens. In order to increase travel times from the salinity barrier to production wells, it may be cost-effective to move, abandon, and replace some production wells located nearest to the coast and the proposed salinity barrier. There are numerous design, operation, and regulatory issues that need to be considered to optimize the water-resource benefits of a salinity barrier system.

The implementation of salinity barrier systems is similar to that of MAR systems in general and is also discussed by Maliva and Missimer (2010). An important additional task is accurately locating the position of the saline–water interface, which in many areas is not well established.

25.7.2 Extractive Salinity Barriers

Mitigation of saline–water intrusion by brackish or saline-water pumping has received much less attention than positive (recharge) salinity barriers, although they may be practical solutions or an element of mitigation plans. Theoretical discussions and the results of modeling studies of extractive salinity barriers are provided by Coe (1972), van Dam (1999), Sherif and Hamza (2001), Kacimov et al. (2009), Pool and Carrera (2010). Key technical issues are the number and location of the production wells and their completion zone. The optimal design may involve extraction of water from near the base of the aquifer, at or near the toe of the saline–water interface (van Dan 1991).

The concept of a purely extractive barrier to control saline–water intrusion has been considered for many years. An experiment was conducted in Oxnard County, California in the late 1960s to access the feasibility of a strictly extractive barrier (California Department of Water Resources 1970). Todd (1980) described the dynamics of extractive saline–water intrusion systems.

Key operational issues are the disposal of the produced water and the minimization of the extraction of freshwater that will be drawn towards the production (extraction) wells. One possibility is use of the produced water as feedwater for desalination. Avital et al. (2010) described the Ashdod Nir-Am Eastern Interceptor under development in the Southern Coastal Aquifer of Israel. The final planned system will consist of a row of 34 production wells and 2 brackish water desalination plants. The wellfield is located along the eastern margin of the aquifer and will prevent saline water from migrating into the freshwater in the aquifer to the west. Modeling results indicate that the system will eliminate 15,000–17,000 metric tons (16,500–18,700 tons) of salt per year from flowing to the west. The interceptor system serves a dual purpose of protecting existing freshwater resources and being a source of raw water for desalination. It is recognized that the interceptor system is not the sole solution to the rehabilitation and restoration of the Southern Coastal Aquifer. Water use will also have to be reduced to achieve a balance with the recharge rate.

As a strategy to minimize the loss of freshwater, Pool and Carrera (2010) theoretically investigated a double pumping barrier system design. The negative barrier would consist of two wells (or rows of wells) configured with a seawater well located at the saline–water interface and a freshwater well located further inland. The freshwater well would capture freshwater than would otherwise be captured by the seawater well. The modeling results indicate that the double pumping barrier system is feasible. However, achieving long-term performance goals would require a well-balanced design that would be determined through groundwater modeling.

25.7.3 Injection and Extraction Barrier Systems

The commonly proposed design for a combined injection and extraction barrier system has the extractive wells located seawards of the saline–water interface and the freshwater injection wells located landwards of the interface (Todd 1974; Tsanis and Song 2001; Pool and Carrera 2010). In theory, a combination of a positive salinity barrier and extractive salinity barrier could be more effective than either barrier operated alone.

An alternative design proposed by Sheahan (1977) has an extraction well located landwards of the injection well. Both are located inland of the saline–water interface. The proposed salinity barrier system (for Santa Clara, California) would use reclaimed water for injection. The rationale for the more inland freshwater extraction well is that it would prevent any possible degradation of fresh groundwater supplies. The extraction wells would capture injected wastewater and prevent its inland migration.

25.7.4 Physical Barrier Systems

Physical barriers are impervious or semi-impervious subsurface barriers, for which there is a variety of construction material options, such as bentonite clay, concrete, grout, or sheet piles (Todd 1974; Oude Essink 2001; Allow 2011; Luyun et al. 2011). The concept and construction are essentially the same as groundwater dams (Sect. 23.3.6). A physical barrier constructed landward of the saline–water interface could prevent the inland movement of the water at the barrier. The cost of physical barriers depends upon their depth, length, and construction materials and method. Physical barriers would have a relatively high construction cost, but there would be minimal subsequent operational and maintenance costs.

Physical barrier systems have a very low operational flexibility, so it is imperative that they be correctly located with respect to the saline–water interface. A barrier that is effective against inland migration of saline water will also be effective in preventing the seaward flushing of any saline water trapped on the landward side of the barrier. Physical barrier systems should be designed so that saline-water cannot migrate inland around the barrier into areas of concern.

A physical barrier was recently installed on a site located on New Providence Island, Bahamas (Missimer Groundwater Science 2005). The barrier was constructed to limit the impacts of the construction of a marina on the position of the saline–water interface and to prevent the drainage of the local freshwater lens. Solute transport modeling, performed using the finite-element model FEFLOW, showed that the barrier had to be wrapped around most of the landward perimeter of the marina basin and that a collection pipe system had to be located on the upgradient side of the barrier to collect the ponded freshwater for recovery and transmittal to an inland storage basin.

25.7.5 Optimization

The key issue for salinity barrier systems is the optimization of system design in terms of the number and locations of wells, the depths of their injection (or extraction) zones, and the injection (extraction) rates per well. Density-dependent solute-transport modeling is an important tool for evaluating salinity barrier options. An important issue is developing a model based on accurate aquifer characterization, which must adequately incorporate aquifer heterogeneity. For positive salinity barriers, it is critical to have an accurate understanding of where injected water will go in an aquifer and how it will impact the position of the saline–water interface. Similarly, for extractive salinity barriers, it is critical to know where in an aquifer pumped water is derived. Once a well-calibrated model is developed for the groundwater system at the barrier site, simulations can be performed to evaluate different designs and operational options.

The results of theoretical modeling using the SEAWAT code and a flow tank experiment by Luyun et al. (2011) indicate that most effective location for salinity barrier injection wells is near (seaward) to the toe of the saline-water wedge. However, Luyun et al. (2011) noted that their results are valid only for isotropic and homogenous aquifers.

Several published studies discuss the applications of groundwater modeling in the evaluation and design of salinity barrier systems. Tsanis and Song (2001) simulated remedial options for saline–water intrusion control in the Upper Florida Aquifer in South Carolina using the SUTRA code. Bray and Yeh (2008) developed a calibrated simulation model for the Alamitos Barrier Project in Los Angeles, California. The predictive simulations were performed to evaluate the optimal scheduling of injection, and well locations.

Allow (2011) simulated both a subsurface physical barrier and injection well salinity barrier in the Damsarkho (Latakia), Syria, coastal plain. The SEAWAT code was used. The modeling results indicate that both salinity barrier options would be effective in the prevention of seawater intrusion. Ru et al. (2001) demonstrated using numerical modeling the likely effectiveness of a subsurface dam on Okinawa, Japan, in preventing saltwater intrusion. Bloetscher et al. (2005) reported the results of solute transport modeling of a proposed reclaimed water salinity barrier system in Hollywood, Florida. The modeled migration of the injected reclaimed water was then incorporated into a risk analysis.

An important lesson is that salinity barrier systems are conceptually simple, but local hydrogeology in coastal areas is often complex. Overly simplified approaches to salinity barrier systems can thus lead to ineffective results. The technical challenge lies in their optimization, which involves determining the most cost-effective construction and operational protocols that achieve saline–water intrusion management goals, while protecting public health and the environment. Optimization requires thorough aquifer characterization, high-quality groundwater solute-transport modeling, and also an adaptive management approach, which allows for improvements in system performance to be based upon operational experiences. Adaptive management of salinity barrier systems requires that they be designed and constructed to allow for operational flexibility.

Testing of a small-scale pilot system is critical for the development of large-scale systems. Pilot testing results can provide a strong indication of whether or not the planned salinity barrier system will likely perform as anticipated. Pilot testing results will also provide critical data required to refine and calibrate the groundwater flow and solute-transport model that is needed for final system design. Additionally, monitoring data from the pilot testing can provide information on geochemical processes operating in the system (e.g., effectiveness of natural contaminant attenuation processes) and operational issues, such as the susceptibility of injection wells to clogging and associated losses in capacity.

References

Abdalla, O. A. E., Ali, M., Al-Higgi, K., Al-Zidi, H., El-Hussain, I., & Al-Hinai, S. (2010). Rate of seawater intrusion estimated by geophysical methods in an arid region: Al Khabourah, Oman. *Hydrogeology Journal, 18*, 1437–1445.

Allow, K. A. (2011). The use of injection wells and a subsurface barrier in the prevention of seawater intrusion: A modeling approach. *Arab Journal of Geociences*. doi:10.1007/s12517-001-0304-9.

Avital, A., Lumlesky, S., Moss, D., Gev, I., Negev, I., & Oron, H. (2010, October 3 to 7). *Desalination as a key component for the southern Israeli coastal aquifer rehabilitation and restoration plan: Proceedings, EuroMed 2010, Desalination for Clean Water and Energy, Tel Aviv*.

Banks, H. O., & Richter, R. C. (1953). Sea-water intrusion into ground-water basins bordering the California coast and inland bays: Transactions. *Transactions, American Geophysical Union, 34*(4), 575–582.

Bear, J., Cheng, A. H.-D., Sorek, S., Ouazar, D., & Herrera, I. (1999). *Seawater intrusion in coastal aquifers—concepts, methods and practices*. Dordrecht: Kluwer.

Black, W. H., Goodrich, J. A., & Patton, F. D. (1988). Groundwater monitoring for resource management: In *Proceedings ASCE International Symposium on Artificial Recharge of Groundwater, Anaheim, California, August 21–29, 1998*.

Bloetscher, F., Englehardt, J. D., Chin, D. A., Rose, J. B., Tchobanoglous, G., Amy, V. P., & Gokgoz, S. (2005). Comparative assessment of municipal wastewater disposal methods in Southeast Florida. *Water Environment Research, 77*(5), 489–490.

Bloetscher, F., Meerfoff, D. E., Heimlich, B. W., Brown, A. R., Batler, D., & Loucraft, M. (2010). Improving resilience against the effects of climate change. *Journal American Water Works Association, 102*(11), 36–46.

Bobba, A. G. (2002). Numerical modeling of salt-water intrusion due to human activities and sea-level change in the Godavari Delta, India. *Hydrological Sciences-Journal-des Sciences Hydrologiques, 47*, S67–S80.

Bray, B. S., & Yeh, W. W.-G. (2008). Improving seawater barrier operation with simulation optimization in southern California. *Journal of Water Resources Planning and Management, 134*, 171–180.

Bruington, A. E. (1969). Control of sea-water intrusion in a ground-water aquifer. *Ground Water, 7*(3), 9–14.

Buszka, P. M., Brock, R. D., and Hooper, R. P. (1994). *Hydrogeology and selected water-quality aspects of the Hueco Bolson Aquifer at the Hueco Bolson Recharge Project area, El Paso, Texas*. U.S. Geological Survey Water-Resources Investigations Report 91-4092.

California Department of Water Resources. (1970). *Oxnard Basin experimental extraction-type barrier*. Sacramento: California Department of Water Resources Bulletin 147–6.

Campbell, B. G., Mirecki, J. E., Conlon, K. J., & Petkewich, M. D. (1997a). Evaluation of aquifer storage recovery in the Santee Limestone/Black Mingo Aquifer near Charleston, South Carolina, 1993–1995. In D. R. Kendall (Ed.), *Conjunctive use of water resources: Aquifer storage and recovery: Proceedings AWRA Symposium* (pp. 231–239). California: Long Beach.

Campbell, B. G., Conlon, K .J., Mirecki, J. E., and Petkweich, M. D. (1997b), *Evaluation of aquifer storage recovery in the Santee Limestone/Black Mingo Aquifer, near Charleston, South Carolina, 1993–1995*. U.S. Geological Survey Water-Resources Investigations Report 96-4283.

Coe, J. J. (1972). Seawater extraction barrier. *Journal of the Irrigation and Drainage Division, ASCE, 98*(3), 387–403.

Committee on Environment and Natural Resources (2008). *Scientific assessment of the effects of global climate change on the United States: A report on the committee on environment and natural resources*. Washington: U.S. National Science and Technology Council.

Custodio, E. (2004). Myths about seawater intrusion in coastal aquifers. In L. Araguás, E. Custodio & M. Manzano (Eds.), *Groundwater and saline intrusion. Selected papers from the 18th Salt Water Intrusion Meeting, 18 SWIM* (pp. 599–608), Cartagena.

Guo, W., & Langevin, C. D. (2002). *User's Guide to SEAWAT: A computer program for simulation of three-dimensional variable-density ground-water flow*. U.S. Geological Survey Open-File Report 01-434.

Kacimov, A. R., Sherif, M. N., Perret, J. S., & Al-Mushikhi, A. (2009). Control of sea-water intrusion by salt-water pumping: Coast of Oman. *Hydrogeology Journal, 17*, 541–558.

Kerrou, J., & Renard, P. (2010). A numerical analysis of dimensionality and heterogeneity effects on advective dispersive seawater intrusion processes. *Hydrogeology Journal, 18*, 55–72.

Klein, R. J. T., & Nicholis, R. J. (1999). Assessment of coastal vulnerability to climate change. *Ambio, 28*(2), 182–187.

Kundzewicz, Z. W., Mata, L. J., Arnell, N. W., Döll, P., Jimenez, B., Miller, K., Oki, T., Sen, Z., and Shiklomanov, I (2007). Freshwater resources and their management. In M. L. Parry, O. F. Canziani, J. P Palutikof, van der Linden P .J & C. E. Hansen (Eds.), *Climate change 2007: Impacts, adaptation and vulnerability. contribution of working group ii to the fourth assessment report of the intergovernmental panel on climate change* (pp. 173–310). Cambridge: Cambridge University Press.

Langevin, C. D., Thorne, D. T., Jr., Dausman, A. M., Sukop, M. C., & Weixing, G. (2007). SEAWAT (Version 4): A computer program for simulation of multi-species solute and heat transport. U.S. Geological Survey Techniques and Methods Book 6, Chapter A22.

Luyun, R., Jr., Momii, K., & Nakagawa, K. (2011). Effects of recharge wells and flow barriers on seawater intrusion. *Ground Water, 49*(2), 239–249.

Macumber, P. G. (2003). Lenses, plumes and wedges in the Sultanate of Oman: A challenge for groundwater management. In A. S. Alsharhan & W. W. Wood (Eds.), *Water resources perspectives: Evaluation, management and policy* (pp. 349–370). Amsterdam: Elsevier.

Maliva, R. G., & Missimer, T. M. (2010). *Aquifer storage and recovery and managed aquifer recharge using wells: Planning, hydrogeology, design, and operation*. Houston: Schlumberger Water Services, Methods in Water Resources Evaluation Series No. 2.

Markus, M. R. (January 15, 2009). Groundwater replenishment & water reuse (pp. 1–9). The Water Report, Issue 59.

Milnes, E., & Renard, P. (2004). The problem of salt recycling and seawater intrusion in coastal irrigated plains: An example from the Kiti aquifer (southern Cypress). *Journal of Hydrology, 288*, 327–343.

Missimer Groundwater Science. (2005). *Hydrogeology of the Albany House Development site and surrounding lands, New Providence Island, Bahamas*. Nassau: Unpubished consultants report to New Providence Development Corporation, Ltd.

Nishikawa, T., Siade, A. J., Reichard, E. G., Ponti, D. J., Canales, A. G., & Johnson, T. A. (2009). Stratigraphic controls on seawater intrusion and implications for groundwater management, Dominiguez Gap area of Los Angeles, California, USA. *Hydrogeology Journal, 17*, 1699–1725.

Oude Essink, G. H. P. (2001) Improving fresh groundwater supply—problems and solutions. *Ocean and Coastal Management, 44*(5–6), 429–449.

Pool, M., & Carrera, J. (2010). Dynamics of negative hydraulic barriers to prevent seawater intrusion. *Hydrogeology Journal, 18*, 95–105.

Ru, Y., Jinno, K., Hosokawa, T., and Nakagawa, K., (2001, April 23–25) Study of effect of subsurface dam in coastal seawater intrusion. In *First International Conference of Saltwater Intrusion and Coastal Aquifers—Monitoring, Modeling, and Management*. Essaouira, Morocco.

Sanford, W. E., & Pope, J. P. (2010). Current challenges using models to forecast seawater intrusion: lessons from the Eastern Shore of Virginia, USA. *Hydrogeology Journal, 18*, 73–93.

Shammas, M. I., & Jacks, G. (2007). Seawater intrusion in the Salalah plain aquifer, Oman. *Environmental Geology, 53*, 575–587.

Shammas, M. I. (2008). The effectiveness of artificial recharge in combating seawater intrusion in Salalah coastal aquifer, Oman. *Environmental Geology, 55*, 191–204.

Sheahan, N. T. (1977). Injection/extraction well system—A unique seawater intrusion barrier. *Ground Water, 15*(1), 32–50.

Sherif, M., El Mahmoudi, A., Garamoon, H., Kacimov, A., Akram, S., Ebraheem, A., et al. (2006). Geoelectrical and hydrochemical studies for delineating seawater intrusion in the outlet of Wadi Ham, UAE. *Environmental Geology, 49*, 536–551.

Sherif, M. M., & Hamza, K. I. (2001). Mitigation of seawater intrusion by pumping brackish water. *Transport in Porous Media, 43*, 29–44.

Sherif, M., & Kacimov, A. (2007). Seawater intrusion in the coastal aquifer of Wadi Ham, UAE. In W. Sanford, C. Langevin, M. Polemio & P. Povinec (Eds.), *A new focus on groundwater-seawater interactions* (pp. 315–325). International Association of Hydrological Sciences (IAHS) Publication 312.

Sherif, M. M., & Singh, V. P. (1999). Effect of climate change on sea water intrusion in coastal aquifers. *Hydrological Processes, 13*(8), 1277–1287.

Todd, D. K. (1974). Salt-water intrusion and its control. *Journal American Water Works Association, 63*(3), 180–187.

Todd, D. K. (1980). Groundwater hydrology (2nd ed.). New York: John Wiley & Sons.

Trabelsi, R., Zairi, M., & Ben Dhia, H. (2007) Groundwater salinization of the Sfax superficial aquifer, Tunisia. *Hydrogeology Journal, 15*, 1341–1355.

Tsanis, I. K., & Song, L.-F. (2001). Remediation of sea water intrusion: A case study. *Ground Water Monitoring and Remediation, 21*(3), 152–161.

Van Dam, J. C. (1999). Exploitation, restoration, and management. In J. Bear, A. Cheng, S, Sorek, D. Ouazar, & A. Herrera (Eds.), *Seawater intrusion in coastal aquifers: Concepts, methods, and practices* (pp. 73–125). Netherlands: Kluwer.

Werner, A. D. & Simmons, C. T. (2009). Impact of sea-level rise on sea water intrusion in coastal aquifers. *Ground Water, 47*(2), 197–204.

Zubari, W. K. (1999). The Dammam aquifer in Bahrain—hydrochemical characterization and alternatives for management of groundwater quality. *Hydrogeology Journal, 7*(2), 197–208.

Chapter 26
Domestic and Agricultural Water Conservation

26.1 Introduction

Several related terms are germane to discussions of water conservation. Dictionary definitions of the word "conservation" refer to the prevention of wasteful or harmful use of a resource (e.g., McKean 2004). The term "efficiency" means the state or quality of being efficient, which is in turn defined as achieving the maximum productivity with minimum wasted effort or expense (McKean 2004). The terms 'water conservation' and 'water saving' are not synonymous (Pereira et al. 2002). Water conservation, as strictly defined, refers to practices and policies aimed at conserving or preserving water resources (both quantity and quality). Water saving refers to activities aimed at limiting and controlling water demand, including avoiding waste and misuse of water. In practice, the terms water conservation and water saving have been used interchangeably. Water conservation and savings are also referred to as "demand-side" water management, whereas development of new water resources is part of "supply-side" management. The essential aim of demand management is to ensure that a given supply of water is distributed to accord more closely with its optimal use pattern (Winpenny 1994).

Efficiency can be simply defined as the ratio of output to input:

$$Efficiency = \frac{Output}{Input} \tag{26.1}$$

Water use efficiency (WUE) is broadly defined as the output per unit volume of water. Output can be variously defined in economic terms (e.g., the economic value of goods produced), yield terms (e.g., the mass or volume of crops produced), and for agriculture, in terms of transpired water (i.e., water actually used by plants). Oweis et al. (1999) recommended use of the term "productivity," where the numerator is a measure of either the quantity of the product or its economic value. Input is normally expressed in terms of a standard volume of water (e.g., cubic meter or cubic kilometer; gallons or acre-feet). For example, WUE or productivity is sometimes

R. Maliva and T. Missimer, *Arid Lands Water Evaluation and Management*,
Environmental Science and Engineering, DOI: 10.1007/978-3-642-29104-3_26,

expressed in terms of value of goods generated in U.S. dollars per cubic meter of water (or per cubic foot, acre-foot or 1000 gallons). Efficient water use achieves the maximum economic value or yield from the use of water with minimal waste.

The goal of water conservation, which ties into the concept of sustainability (Chap. 9), can be achieved in part through water savings, which in turn can be realized through increased water use efficiency. Water use efficiency relates to both the use of water for a given activity (e.g., growing a specific crop) and the allocation of water between sectors, which is addressed under water policy (Chap. 32). Water savings may be achieved by adopting agricultural practices that, for example, increase the number of kilograms (tons) of wheat grown per cubic meter (cubic foot or acre-foot) of water, or by reallocating water to activities that generate more wealth per unit volume of water used.

Postel (1992, p. 23) observed that

> Doing more with less is the first and easiest step along the path toward water security. By using water more efficiently, we in effect create a new source of supply.

Postel (1992) in a landmark book referred to this new supply of water that could be made available by more efficient water use, particularly for irrigation, as the "last oasis." There are numerous publications on water conservation and efficiency in all aspects of society. Water conservation and saving methods are comprehensively reviewed by Vickers (2001) and Pereira et al. (2002), and specific subjects have been addressed in numerous journal articles. Several motivations have been proposed for conservation, including economic efficiency, equity (fairness), and moral principle (Peterson 2003). Economic efficiency means that resource use should yield the greatest net benefit to society.

Market-based incentives use the price of water to promote a more efficient use of water (Winpenny 1994). The price of water could be increased or pollution charges implemented in order to decrease water use and pollutant discharges. Alternatively, a water market could be established which relies upon market forces to increase the efficiency of water use. Water saved through more efficient water use, and as a result made available for other uses, is often much less expensive than the marginal costs of developing new alternative water supplies. Where water use is driven by market considerations, the owner of a resource will naturally want to obtain the maximum value from the resource and avoid waste. As related by Peterson (2003) with respect to the use of the High Plains aquifer of the mid-continent of the United States (semi-arid area), groundwater use often lacks economic incentives for efficient use because the private costs to pump water are less than the social costs of withdrawing water. Social costs include impacts to other aquifer users from lowered water levels and environmental impacts. The groundwater owner typically only pays for the cost to extract the water, and not the costs of the impacts associated with water use. As discussed in Chap. 34, legal recognition of water rights as something that are owned and can be transferred provides a strong economic incentive for conservation and increased efficiency in the use of water.

Equity or fairness relates to the distribution of water within and across generations of users. Current groundwater users can impact other aquifer users and the ability of future generations to use the aquifer. The equity principle requires that current water use be sustainable. The final motivation for conservation is a moral principle. Conservation of natural resources is believed by some to reflect a moral duty, which can have either a secular or religious foundation (Paterson et al. 2003). For example, in Islam there are explicit admonitions against the waste of water (Sect. 33.2).

Opportunities exist for water savings through increased efficiency in all sectors of water use (domestic, agricultural, and industrial). The need for increased water use efficiency is obvious in water scarce regions. However, wasteful water use is still widespread in many arid and semiarid lands, because consumers lack direct incentives to save water. Where water is relatively inexpensive relative to incomes or where water is heavily subsidized, there is little economic incentive for users to adopt or invest in water saving practices and technologies. On the contrary, where water is very expensive or difficult to obtain, water use is inherently highly efficient. As an extreme example, where people have to manually transport water to their homes from a distant source, water is a precious resource that is managed very carefully. On the contrary, where water is inexpensive and can be obtained in essentially unlimited quantities by just turning on a tap, there is very little economic incentive to curtail wasteful activities.

Water conservation in two main sectors are discussed herein; domestic use in urban and suburban areas and irrigated agriculture. Domestic water use is a small fraction of total water use, but is important in that it directly affects people lives. Domestic water use is essentially an economic issue because virtually any area can be provided with all of the water that it needs (or wants) provided that adequate financial resources are available. Coastal areas potentially have an unlimited supply of water through desalination. However, financial resources are not unlimited and providing adequate water to meet the needs of rapidly growing urban areas is an increasing challenge.

Irrigated agriculture is by far the largest water use and this is the area in which the greatest increase in water use efficiency is needed in order to meet the food requirements of a rapid growing global population. There is a great need for "more crop per drop" (Pearce 2006). Agricultural water use also has strong economic drivers. Small farmers in developing countries often do not have the financial resources to adopt more water efficient irrigation technologies. Where water for irrigation is available at no or a very low (often subsidized) cost, there is no economic incentive to invest in improved irrigation efficiency. The capital and operational costs to implement more efficient irrigation technologies are greater than any cost savings from decreased water use. It is increasingly being accepted that recognition of water as an economic good is necessary to increase the efficiency of water use (Sect. 34.2).

Evaluations of water conservation and efficiency also need to consider the recycling of water. Water that is "wasted" through inefficient water use is not necessarily lost for beneficial use. The degree of basin-scale efficiency achieved by conservation depends upon what happens to water after use (Seckler 1996). Water lost through conveyance leakage and over-irrigation may recharge underlying aquifers and be recoverable for future irrigation use. Excessive domestic water use (e.g., long showers) may increase wastewater flows and thus, increase the supply of reclaimed water, which may be put to a beneficial irrigation use. Similarly, decreasing domestic use through increased efficiency may result in no net water savings if the resulting decrease in wastewater flow decreases the amount of water available to "downstream" users (Seckler 1996). Analyses of water efficiency need to differentiate between "paper" or "dry" water savings and "real" or "wet" water savings (Keller and Keller 1995; Seckler 1996).

Although the water is recycled, its quality may deteriorate, which impacts future uses. There are also economic costs associated with the recycling of water, such as those of water and wastewater treatment, which also need to be considered.

26.2 Domestic Water Conservation and Saving

26.2.1 Domestic Water Savings Opportunities

There is great interest in increasing the efficiency of both water and energy use in new and existing homes. Amy Vicker's (2001) "Handbook of Water Use and Conservation" is perhaps the most comprehensive single reference on the subject. However, a quick search of the internet will reveal numerous publications and sites providing water saving tips and advice. Water utilities very commonly provide educational information on water conservation to their customers. Local governments are also a good source of information on water conservation.

The main domestic water uses in developed countries are:

- outdoors lawn and landscape irrigation,
- clothes washing,
- sanitary—toilet flushing,
- sanitary—showering, baths, washing, shaving, and brushing teeth,
- potable use, food preparation, and dish washing, and
- leaks.

There are multiple opportunities for water savings for each of the listed domestic water uses. For example, outdoors water use is commonly the largest household use in the United States. Water savings are possible through the avoidance of over-irrigation, reduction in irrigated areas, and xeriscaping (i.e., use of native and adaptive plants that require very little water to thrive). In some residential communities in the Sonoran Desert of Arizona, colored gravel is used instead of turf lawns (Fig. 26.1).

Appliances are available with high water and energy efficiencies, which can result in substantial water and lower operational costs. The United States Environmental Protection Agency (USEPA) WaterSense® Program encourages water conservation through the labeling of plumbing fixtures that meet both efficiency and performance standards (http://www.epa.gov/WaterSense/). Products and services that have earned the WaterSense® label have been certified to be at least 20% more efficient without sacrificing performance.

Fig. 26.1 Xeriscape landscaping in Sun City, Arizona (U.S.A.). Colored pebbles are used instead of turf grass

Water savings are also possible from changes in personal behavior. An obvious example is to take shorter showers. Water savings are also possible through the running of dish washers and clothes washers only with full loads. Savings can also be achieved by not allowing faucets to run while brushing teeth or shaving. There is a myriad of small changes in personal behavior that can cumulatively result in substantial water savings, especially if they become widely adopted.

From a water management perspective, the issue is not whether there are technologies and behavioral changes that are available to reduce water use, but rather how to motivate water consumers to adopt them and conserve water. A small percentage of most populations have a very strong conservation ethic and will be inclined to be as water and energy efficient as possible, even if it results in decreased performance. However, it is recognized that a key principle for the successful implementation of changes towards high water-efficient fixtures and appliances is that they will take hold in the marketplace only if they result in equal or improved homeowner satisfaction (Selover 2010). Performance is the key issue. For example, if water efficient showerheads lack sufficient spray power, people will take longer showers, negating any water efficiency gains, or they will replace them with better performing but less efficient models.

Selover (2010) summarized the Environments for Living ® Certified Green Program, which combines indoor water quality, water conservation, and appliance efficiency to make buildings more efficient and increase homeowner satisfaction by creating homes that perform better than houses constructed using conventional methods. Elements include:

- high efficiency toilets,
- low-flow faucet aerators,
- low-flow showerheads,
- water efficient clothes washers, and
- design of homes and plumbing systems.

House designs are often not conducive to water efficiency. For example, a plumbing design issue is the time required (and thus water wasted) to get hot water at various taps in the house from a central water heater. A lot of water is wasted letting showers and faucets run until hot water from a distant water heater arrives. More water efficient designs might include a more centrally located water heater or using separate heaters at some points of use. However, water inefficient designs may have beneficial elements. For example, locating a water heater in a garage, further from points of water use, is preferred from a safety perspective in the event of a failure.

The Leadership in Energy and Environmental Design (LEED) building certification system developed by the U.S. Green Building Council (USGBC) includes requirements for water efficiency, as well as for energy savings, indoor air quality, carbon dioxide emissions reduction, and stewardship of resources and minimization of environmental impacts. Buildings are awarded LEED points on a 100-point scale based on numerous criteria, with credits weighted to reflect their potential environmental impacts. With respect to water use efficiency, points are awarded for water efficient landscaping (50% reduced irrigation or, better yet, no irrigation) and for reduced indoors water use through the utilization of efficient plumbing fixtures. The highest certification is platinum, which for new construction and major renovation requires a score of 80 or greater. The LEED certification system has become internationally recognized as the standard for "green" building construction and there is increasing desire by private developers and governments that their buildings be LEED certified. Information on the LEED certification program is available on the USGBC website (www.usgbc.org).

26.2.2 Domestic Water Savings Implementation

Water conservation measures have been implemented through legal mandates, market forces, and development of a conservation ethic. As a general principle, development of a water conservation ethic is more effective than coercion through legal mandates. However, legal mandates can serve an education function. Religious teachings can also be instrumental in the development of a water conservation ethic (Chap. 33). Market forces act to reduce wasteful water use when water consumers bear the full costs of water.

Implementation of water demand management can be stressful where cheap water is taken for granted. Demand management can be implemented by simultaneous and full reform of all relevant policies and institutions (Big Bang approach) or by gradual adjustments to accustom the public to the change (Winpenny 1994).

26.2.2.1 Legal Mandates

Water savings can be forced upon a society through legal means. Laws or ordinances can be passed that impose penalties for the improper or excessive use of water. For example, in regions suffering from drought and dwindling water supplies, restrictions on outdoor water use are often implemented, because this is a major category of non-critical water use. The restrictions may include elements such as prohibitions against vehicle washings (in non-recycling systems) and the limitation of lawn and landscape irrigation to a few hours on selected days each week or even an outright ban if the shortage is very severe.

Conservation ordinances and statutes may fail to be effective because there is often little or no enforcement (Pape 2010). The implementation of water restrictions may be more effective as an education measure. The implementation of restrictions on water use may convince people that a water shortage does indeed exist and that there are means that they can employ to reduce water use. It has been the authors observations that most people will obey water restrictions because they believe it is the right thing to do and they are by nature law-abiding, rather than because they fear sanctions for non-compliance. However, the absence of enforcement can weaken the effects of the ordinances. Compliance decreases if people see some of their neighbors flaunting the ordinances without sanctions. On the other hand, in some communities residents are active in the enforcement of water use restrictions by reporting violations, such as lawn watering outside of permitted times.

Compliance with legal mandates during water shortage periods is also dependent on the consistency of the agency having oversight regarding water use. When a real crisis in water use occurs, people must believe that it is not just a political issue, but a true water shortage. The authors have observed water management agencies trying to enforce extreme conservation measures during periods of flooding or normal rainfall to "condition" the general populations to impending new measures or to exert a degree of power over local utilities. When the public does not perceive the oversight water agency is credible, then legal mandates are ignored. This is a critical issue throughout water management. Water management can only be successful if the stakeholders have confidence in the competency, objectivity, and integrity of those entrusted to make water management decisions.

An effective means of the use of legal mandates for reducing water use is through green plumbing and building codes. Jurisdictions may implement ordinances that require the use of water efficient fixtures, appliances, and plumbing designs. A major advantage of the green building codes is that pre-established code enforcement structures (e.g., local building code and enforcement departments) may exist and building departments are already inspecting installations where permits are required (Pape 2010). Codes will apply to new construction and remodeling that requires a permit, but usually not for the retrofitting of existing buildings. Pape (2010) advocated that implementing water conservation strategies at the time of construction and remodeling is the most cost-effective means to achieve sustainable water efficiency for both the water utility and its customers.

26.2.2.2 Consumer Education

Consumer education is a key element in developing a water conservation ethic as well as enabling market forces to bring about increases in water efficiency. Very often water consumers are not aware of wasteful practices or conditions, such as leaks. Water conservation education can include a wide variety of tools such as:

- flyers included in utility bills,
- mass market advertisements (television, radio, newspapers ads, billboards),
- presentations to clubs and organizations,
- presentations to schools,
- tours of water treatment facilities,
- presentations (booths) at public gatherings,
- xeriscape demonstration gardens, and
- in-home visits and water audits.

Keen et al. (2010) documented the results of an in-home visit program tested in the City of Phoenix, Arizona. The objective of the program was to provide personalized water conservation tips and technologies. High-volume users were offered a variety of tips and guides for conservation techniques, evapotranspiration controllers for use with automatic sprinkler systems, and devices that disperse a liquid to form a transparent monomolecular layer on a swimming pool surface to reduce evaporation. Follow up visits were performed and changes in metered water use were examined to determine the effectiveness of the program. The participants used 18% less water the year after the visits than the year before, which was due in part to wetter conditions and a tendency for high-water users to use less water in the subsequent year (i.e., regression towards the mean). The overall reduction was about 4% above the reductions that occurred during similar climatic conditions in non-visited households.

The Phoenix study demonstrated that households have a varying responsiveness to water conservation even after a very personalized high-contact education program. The education program was most effective in reducing watering of lawns and landscaping that were previously overwatered and in prompting the repair of leaks (both indoors and outdoors) (Keen et al. 2010). The pilot program was viewed as a success and the City of Phoenix hired water auditors to focus on detecting leaks and reducing outdoors water use (Keen et al. 2010).

26.2.2.3 Market Incentives

Water demand is elastic beyond the level required to meet basic human needs. Increases in the price of water will affect overall consumer behavior by reducing non-essential (discretionary) water use. Conservation rate structures provide water users with pricing information that conveys (at least in part) the marginal cost of developing new water resources. The effectiveness of conservation rate structures depend upon the magnitude of the price signal and degree of elasticity of water demand (coefficient of elasticity), which will vary between users. The response of some water users to an increase in price will be highly elastic in that it will result in a large reduction in water use. Other users may not respond at all (i.e., their coefficient of elasticity is equal to zero). Even after rate increases to encourage conservation, water rates often still represent only a small fraction of the total expenditures of households in developed countries.

The logic behind conservation rate structures is that they encourage efficient water use and discourage waste by ensuring that customer bills communicate the full cost of providing water services, including the costs of new supplies (Chesnutt and Beecher 1998). The principle economic value is that conservation rates, and other conservation elements, can result in a postponement or deferral of the construction of additional treatment and source capacity (Chesnutt and Beecher 1998).

Conservation rates can be implemented to reduce year-round water use or can be focused on reducing water use during peak demand or low supply periods. Peak demand is a critical economic issue for water suppliers as treatment infrastructure has to be able to meets peaks in demand, which may occur over only a relatively short time of the year. During the rest of the year, water infrastructure constructed to meets peaks in demand is underutilized.

In the absence of government subsidies, water utilities finance their operations through the sale of water. Utilities need a stable and predictable long-term revenue stream that covers current operating and maintenance costs, debt service, replacement of aging infrastructure, and system expansion to meet increases in demands. Conservation rates or surcharges can benefit utilities in two manners (Mann and Clark 1993). If future water use is highly insensitive to the imposition of surcharges, then customers will provide funding for projects that expand the water supply. Alternatively, if future water demands are sensitive to the surcharges, then the utility and its customers benefit by avoiding the costs associated with constructing new water treatment and distribution infrastructure.

Conservation rates include the following options (Cuthbert and Lemoine 1996; American Water Works Association 2000):

- uniform rate; fixed rate per unit volume of water,
- flat seasonal rate with a higher rate charged during peak demand periods,
- inverted block rates, and
- excess-use rates.

Uniform rates involve the changing of the same rate for each unit volume of water throughout the year. It is a conservation measure in the sense that water bills are proportional to water use. Uniform rates introduce conservation incentives that are not present with a decreasing-block rate (rates decrease with increasing use) and fixed fee (unmetered) billing.

Flat seasonal rates incorporate two or more uniform rates for different periods during the year. A higher rate is charged during peak demand periods in order to selectively reduce demand during these periods. A lower rate is charged during low demand (or high supply) periods, in which there is a lesser need to reduce demand.

Inverted-block rate billing structures incorporate increasing rates with higher use. Blocks are specified volumes of water used during a billing period, which are typically derived from a water budget that defines an efficient level of water use for each customer. The first block of water, which includes non-discretionary uses, is billed at the lowest rate. Subsequently, blocks of water, which consist mostly of discretionary use, are billed at higher rates. Inverted-block rate billing is an attractive strategy because it benefits low-income users with minimum water use. The rate schedule can be constant throughout the year, or higher block rates can be charged during peak demand periods.

Excess-use rate involves an extra charge for water use above an average value during peak demand periods. It is similar to the inverted-block rate structure in that it focuses on reducing nondiscretionary use. Excess charges can be increased based on the amount of use above the baseline rate.

The idea behind conservation-oriented pricing is to charge customers for the full cost of water service and, over the long term, bring supply and demand into balance. It has been argued that margin-cost pricing sends better price signals in terms of achieving efficiency in supply and that a true conservation rate structure should be set to equal the marginal cost of new supplies (Chesnutt and Beecher 1998). However, full margin-cost pricing is generally not practical because large cost differentials often exist between existing and new water supplies. Setting the rate for all water at the marginal cost for new supplies, would result in a large, politically unacceptable windfall for utilities if most of the water is still produced from lower-cost existing sources. It is possible to overshoot with water conservation rates, which is undesirable in that it deprives consumers of the use of water they want, imposes an unnecessary financial burden, and may reduce the revenue stream to a utility to below that necessary to finance its operations (Chesnutt and Beecher 1998). Conservation rates need to be adjusted over time (gradually implemented) in order to achieve the required balance between water supply and demand and revenues.

Rate structures attempt to balance revenue generation, interpersonal equity, and conservation incentives. Block rates should be set to equalize the marginal price elasticity of both low and high income users to encourage all to conserve water, which entails a steep rate progression (Agthe and Brillings 1987). High income households tend to have low price elasticity because water represents a smaller proportion of their total expenditures than is the case for lower income groups. Initial block rates may be set sufficiently low as a "lifeline" rate structure to subsidize low income customers for the amount of water required for sanitary and other basic needs (Agthe and Brillings 1987).

A number of studies have been performed to evaluate the effectiveness of conservation rate structures. The empirical data indicate that conservation rates result in a significant reduction in water use, particularly when rates are implemented in conjunction with an active conservation program (Cuthbert and Lemoine 1996; Jordan and Albani 1999). However, it can be difficult to differentiate between the effectiveness of the conservation rate structures and other elements of conservation plans. Conservation rates may be less effective for utilities with consolidated billing (combined water, wastewater, electric, and/or gas) in which the water price signal is less clear.

The response of conservation rates (elasticity) may be greater in the long run that short run (Carver and Boland 1980). Most consumers are probably unaware of the quantities of water used for various tasks and it may take some time to obtain an understanding of the effects of rate changes on their water bills and the benefits of conservation. Homeowners will also tend to gradually invest in more water efficient fixtures and landscaping (Carver and Boland 1980; Chesnutt and Beecher 1998). Appliances have a relatively long life-cycle and consumers will usually not prematurely replace a functional major appliance with a more efficient model unless the replacement costs will be recouped by operational cost savings. Conversely, after the initial impact of a rate increase, consumers may come to accept the increase as normal and revert to previous behavior.

The results of utility surveys suggest that conservation rates are more effective in reducing non-residential than residential water use and in reducing demand during peak periods (Jordan and Albani 1999). However, an early study in Washington DC had the opposite results with residential use decreasing, but little change in commercial and government use (McGarry and Brusnigham 1979).

Public acceptance of conservation rates can be improved by (Cuthbert and Lemoine 1996)

- gradual implementation (i.e., phasing in of higher rates),
- setting rates that are based on the cost service, rather than solely and arbitrarily setting penalty rates to curtail use, and
- public involvement and education.

Conservation rate structures and other conservation elements introduce an element of instability and uncertainty in the utility revenue stream (Cuthbert and Lemoine 1996; Jordan and Albani 1999, American Water Works Association 2000; Borisova and Rawls 2010). Inverted-block rates result in a greater percentage of total revenues being obtained from high-volumes water users which have more variability. Instability can be addressed through a revenue stabilization fund (American Water Works Association 2000). Public relations can also be a challenge as some consumers will oppose any rate increase. Concerns have also been expressed that inverted-block rates penalize large households, but a study in Washington DC, discovered no correlation between large families and increased water bills (McGarry and Brusnigham 1979).

26.2.2.4 Conservation and Utilities Operations

Water conservation is clearly important as there is little justification for the inefficient use of any scare resource. Conservation measures can have adverse impacts on water utilities if they are implemented suddenly and without planning. Water sales provide the cash flows for utilities that cover the variable costs of production and the substantial fixed costs of utility infrastructure (Beecher 2010), which includes debt service. Conservation measures that reduce water sales may result in rising infrastructure costs having to be recovered from a shrinking sales base (Beecher 2010). Rising infrastructure costs stem from the continuous need to maintain existing infrastructure that has a finite operational life. The sudden implementation of water conservation requirements can result in at least the short-term probability of stranded and unused investments. The reward to consumers for water conservation may be higher rates, although not as high as they would otherwise be (Beecher 2010).

Water conservation on a long term basis is beneficial to utilities and consumers as it can reduce or avoid additional capital and operating costs. The marginal cost of developing new water sources is typically substantially greater than current average water costs. Ideally, conservation will offset increases in demands from increases in population within a utilities service area.

Conservation measures should be implemented in a planned manner consistent with the realities of water utility operations. Most water utilities operate under tight financial conditions in which there is limited, if any, excess cash flow that can accommodate a sudden drop in revenues. For example, the sudden mandatory implementation of residential water restrictions in South Florida during a local drought caused a sudden drop in utility revenues, which for some utilities necessitated staff reductions and deferral of maintenance.

26.2.2.5 Non-Revenue Water

Potable water supply is a capital intensive activity. Water utilities must fund their operations through revenues generated from the sale of water, which may be subsidized to some degree through general revenues. In developing countries water system expansions may be funded by loans and foreign aid. A critical issue for water utilities, particularly in urban areas in developing countries, is non-revenue water use (NRW). NRW is the difference between water put into the distribution system and the amount billed to customers and includes the following elements (Kingdom et al. 2006):

- physical losses, mainly from leakage in distribution network and theft,
- commercial losses (also referred to as apparent losses), which consists of unbilled or underbilled customer use, and
- unbilled authorized uses (e.g., firefighting flows, and utility uses).

NRW can seriously affect water utilities through lost revenues and increased operational costs. Water physically lost if saved could be used to increase the supply of water available to customers. Where water is unbilled or under-billed, there is little, if any, economic incentive to use water frugally.

The water supply and financial impacts of NRW on utilities, particularly in developing countries is staggering, as documented by Kingdom et al. (2006) and includes:

- worldwide cost of $14 billion (US), with 1/3 borne in developing countries,
- an estimated 45 MCM (11,880 million gallons) are lost daily in developing countries through leakage in the distribution network, which is enough to serve nearly 200 million people,

- 30 MCM (7,920 million gallons) is delivered each day to customers but is not billed because of pilferage, employee corruption, and poor metering,
- estimated NRW in the developing world is probably more than 40–50%, and around 35% in developed countries, and
- reduction of NRW in developing countries by half would provide 8 BCM/yr (2,112 billion gallons) of already treated water and save $2.6 billion (US) per year.

Such large losses have a particularly great impact on developing countries that are starving for additional revenues for maintenance of existing services and expansion of services (Kingdom et al. 2006; Mutikanga et al. 2009). The main reasons why utilities struggle with such high NRW is that they operate under weak governance and economic constraints. Some specific issues hindering the reduction of NRW are (Kingdom et al. 2006):

- NRW reduction is inherently complex,
- lack of full understanding of the magnitude, sources, and costs of NRW,
- lack of technical capacity,
- missing management focus and lack of individual (management and staff) incentives, and political focus,
- NRW reduction is less attractive than building new facilities with associated ribbon-cutting ceremonies,
- opposition from those benefiting from commercial losses by receiving 'free' or discounted water, and
- lack of flexibility in implementing key elements of NRW reduction.

Reduction in NRW represents a cost-effective, viable option for improving the financial performance of utilities and financing of new infrastructure development. Reducing commercial losses is almost always cost-effective. Reducing physical losses through leakage can be expensive, requiring significant technical knowledge, and must be carried out extensively to bring results (Kingdom et al. 2006). Cost-benefit analyses should, therefore, be performed to determine if a leakage reduction program makes economic sense (Kingdom et al. 2006). The estimated cost of reducing physical leakage is $215–$500 US per m^3/d ($815–$1,895 per 1000 gallons/day) in the developing world with greater costs in developed countries due to higher labor costs (Kingdom et al. 2006). These costs need to evaluated and compared with the costs to produce additional water. There may be other considerations, such as leakage of water adversely affecting below-ground infrastructure by flooding or corrosion.

The World Bank study (Kingdom et al. 2006) proposed that private sector assistance could be part of the solution to the NRW problem in both developing and developed countries. The private sector could provide economic and technical resources that underfunded utilities lack. However, it was recognized that the traditional technical assistance contracts in which the contractor provides services, but does not guarantee results, have often failed to meet objectives. The contractor is compensated for input not results. Kingdom et al. (2006) recommend performance based service contracting in which the payment is linked to actual results achieved in NRW reduction. Such contracts would involve both a fixed fee element and payment for meeting contractually enforced performance measures, such as a specified reduction in NRW. A critical issue in performance-based contracts is to develop an effective incentive framework that encourages the private sector to deliver results in the most cost-effective manner and allocates risks appropriately between parties (Kingdom et al. 2006). NRW reduction strategies should also implemented that will have long-term benefits that can be maintained.

26.3 Irrigation and Agricultural Water Conservation

26.3.1 Agricultural Irrigation Primer

Irrigation can be defined as the provision of artificial water supplies to guarantee an increase in crop production in areas where there is a water deficiency (Heathcote 1983). Irrigation can result in a tremendous increase in productivity versus rain-fed (dry land) agriculture and can allow a greater variety of crops to be grown. Irrigation is essential for economically viable agriculture in arid and semiarid regions (Pescod 1992). There is little question that irrigation is vital for meeting current and future food demands.

The following basic conditions should be met in order to make irrigated agriculture a success on a farm level (Pescod 1992):

- the required amount of water should be applied,
- irrigation water should be of an acceptable quality,
- water application should be properly scheduled,
- appropriate irrigation methods should be used,
- salt accumulation in the root zone as the result of plant ET should be prevented by means of leaching,
- rise of the water table should be controlled by means of appropriate drainage, and
- plant nutrients should be managed in an optimal manner.

There are five basic types of irrigation that are widely used for agriculture (Pescod 1992):

1. Flood irrigation—water is applied over the entire field to infiltrate into the soil.
2. Furrow irrigation—water is applied to furrows between ridges upon which crops are grown.
3. Sprinkler—water is applied from above (sprinkler systems included solid-set, travelling, spray gun, and center-pivot).
4. Sub-irrigation—water is applied below the root zone in such a manner that it wets the root zone by capillary rise.
5. Localized irrigation—A variety of methods are included, such as micro-irrigation, drip, bubblers, and micro-sprinklers (micro-jet) that apply water in a very targeted manner.

Fig. 26.2 Drip irrigation at a commercial aloe vera plantation in Curacao

Flood irrigation is typically the least efficient irrigation method, whereas localized irrigation techniques are employed because of their high efficiencies. Localized irrigation systems, such as drip systems (Fig. 26.2) are designed to apply water to the soil only in the vicinity of plants and consequently, have the highest efficiency.

The amount of water required for irrigation is dependent on the plant ET requirements, which must be adjusted for effective rainfall, leaching requirements, application losses (irrigation inefficiency), and other factors (Pescod 1992). Irrigation water application should be matched to plant requirements. Water should be applied to crops before the soil moisture potential reaches a level for which the ET rate is likely to be reduced below its potential rate (Pescod1992). The timing and amount of irrigation should be tied to maintaining soil moisture at levels sufficient to allow plants to fully meet their water requirements for optimal growth.

Irrigation application in excess of plant requirements results in some deep percolation beyond the plant root zones (i.e., return flow), which can be an important local source of groundwater replenishment in irrigated arid lands (Bouwer 1978).

Depending upon local circumstances, irrigation can result in either a net input or output to local groundwater resources. In the case where an external water source is used, such as imported surface or reclaimed water, irrigation can cause the water table to rise as it results in a net addition to water budget. If a local shallow aquifer is used for water supply, irrigation can result in a lowering of aquifer water levels, due to the net loss of water caused by evapotranspiration and drainage.

Agricultural water use and irrigation is a complex subject which cannot be given justice in a single chapter or even a single book. Nevertheless, it is important to present an overview of irrigation concepts and issues because of its obvious importance as the major water use in arid and semiarid regions.

26.3.2 Soil Salinization

Agricultural water use and water budgets need to consider the potential for soil salinization. Dissolved solids (e.g., salts, nutrients, and chemicals) within the irrigation water will accumulate in the soil and shallow groundwater unless there is a corresponding outflow of the dissolved solids out of the groundwater basin. Salinization of soils is a widespread problem that plagues agricultural areas in arid and semiarid lands. Elevated salt concentrations in the root zone of plants can reduce productivity, and eventually reach a level at which the land is no longer arable and must be taken out of production. Some salinization processes occur concomitantly with precipitation of insoluble minerals, thereby causing the formation of hardpan or calcrete, which can also result in the degradation and abandonment of farmlands.

The primary cause of soil salinization is evaporative concentration of salts in the capillary (root) zones of crops. The direct evaporation of water from the capillary zone and transpiration by plants removes essentially pure water, while leaving salts behind. The primary factors controlling soil salinization is the position of water table (or perched aquifer level), soil permeability, and the salinity of irrigation water. Soil salinization is of particular concern in arid and semiarid lands because of their high evapotranspiration rates and low precipitation rates. ET concentrates salts and the low precipitation rates result in a lesser degree of flushing of accumulated salts out of the soil zone. Irrigation practices are also critical for preventing and managing soil salinization. Salts contained in the irrigated water will accumulate in the soil zone unless they are flushed out by the periodic application of additional water.

Elevated sodium concentration in soils can also cause defloculation of clays and loss of permeability. Sodic soils refer to soils in which the presence of introduced sodium could adversely impact soil structure and properties. Sodic soils have excess sodium on the exchangeable complexes of soils, which is commonly quantified using the soil exchange sodium percentage (ESP):

$$\mathrm{ESP} = 100 * \mathrm{Na}/(\mathrm{Ca} + \mathrm{Mg} + \mathrm{Na} + \mathrm{K} + \mathrm{Al}) \qquad (26.2)$$

The concentration units are equivalents (milliequivalents) of exchangeable cations per 100 g of soil. Soils are considered sodic if they have a high ESP. Soils with an ESP of greater than 6% are considered to be sodic in Australia, whereas a cutoff of 15% is used in the United States (Halliwell et al. 2001). However, soils may be still be adversely impacted by the addition of sodium at lesser ESPs.

Water with high sodium to calcium ratios can cause cation exchange with clay minerals, in which calcium ions are replaced by sodium ions. Hydrated sodium ions have greater volumes resulting in swelling of clay colloids and clay dispersion, with an associated reduction in porosity and permeability. Clay dispersion has an adverse impact on soil structures by breaking down soil aggregates. The impacted soil has a reduced infiltration rate and becomes hard and compact when dried.

A commonly used parameter in the soil sciences is the sodium adsorption ratio (SAR)

$$SAR = \frac{Na^{+}}{\sqrt{\frac{1}{2}[Ca^{2+} + Mg^{2+}]}} \tag{26.3}$$

where Na^+, Ca^{2+}, and Mg^{2+} are concentrations expressed in milliequivalents per liter (meq/L). Halliwell et al. (2001) described two types of SAR, practical and effective. Practical SAR is calculated from total concentrations and includes Ca and Mg whose activity has been reduced through organic complexing. Effective SAR is calculated from the active (i.e., uncomplexed) concentrations of Ca and Mg, which may be significantly less than the total concentrations.

The effects of sodium adsorption on infiltration rates are summarized by Ayers and Wescott (1985). There is no single threshold for an acceptable SAR value, although values above 10 may be problematic, especially in fresh water. SAR hazards vary with crop sensitivity, soil type, and salinity. At higher salinities, higher SAR values can be accommodated without causing infiltration problems. However, the introduction of fresher water (e.g., rainfall, irrigation water) can increase the risk of infiltration problems. If water use for irrigation has a high SAR, then the soil may be treated by adding amendments containing calcium (or magnesium), such as gypsum.

The strategies available for the managing of soil salinization depend upon its specific local cause. The primary management methods are leaching and drainage, i.e., lowering of the water table. Some deep percolation is necessary in order to prevent salt build-up in the plant root zone, which can adversely impact plant growth. The salt balance can be evaluated using the mass-balance equation (Bouwer 1978):

$$C_i D_i = C_d D_d \tag{26.4}$$

where,

C_i = TDS concentration of the irrigation water (mg/L)
D_i = amount of irrigation water applied (mm)
C_d = TDS concentration of deep-percolation water (mg/L)
D_d = amount of deep percolation water leaving the root zone (mm).

The leaching requirements will depend upon the salt tolerance of the crops and the salinity of the irrigation water, with the amount of required leaching increasing with increasing plant sensitivity and irrigation water salinity. Where groundwater levels are high, drains need to be installed to lower the water table in order to prevent water logging and salinization of soils. Drainage water is a source of pollution from salts, nutrients, and agricultural chemicals. When wastewater effluent is used for irrigation, the drainage water could contain elevated concentrations of industrial chemicals, CECs, and other containments present in the effluent (Bouwer 2000). Depending upon its quality, drainage water may be recycled for irrigation use. Where the drainage water has elevated salinities, it may still be recycled if it can be diluted with fresher water.

An important implication of soil salinization is that it may practically limit irrigation efficiency because of the need for periodic excess water applications required to leach out salts. However, highly efficient irrigation systems result in less water being applied to the soil. The resulting lower salt load applied to the soil reduces the amount of leaching that is required.

26.3.3 Agricultural Irrigation Efficiency

Increasing the efficiency of water use in agriculture is a critical issue for water resources management in arid and semiarid lands because it is the largest use of water and in many regions the most inefficient use. Irrigation entered its heyday in the 1950s through a combination of cheap energy and improved pumping technologies, and has become the cornerstone of global food security by increasing crop yields (Postel 1992, 1993). In many large surface water networks, less than half of the diverted water actually benefits crops (Postel 1993). For example, Deng et al. (2004) noted that in the dry North China Plain an estimated half of the irrigation water is lost during transfer to the farmer's fields with additional substantial losses from inefficient irrigation practices occurring once the water reaches the fields. Most of the water loss is from unlined canals or the percolation of unused water through the soil. Water lost due to irrigation inefficiency is not necessarily lost for overall use as it may recharge an underlying unconfined aquifer, but it may suffer degradation in quality (Postel 1992, 1993). Excessive irrigation (and thus deep percolation) is undesirable as it results in additional costs, wastes water, and can result in leaching of fertilizer and natural nutrients from soils into the underlying aquifers, commonly resulting in contamination.

Water productivity needs to be considered on local, regional and global scales. On a global scale, food and feed crop demand will nearly double in the coming 50 years, driven by population growth and dietary change (Comprehensive Assessment of Water Management in Agriculture 2007). Without further improvements in water productivity and major shifts in production patterns, the amount of water consumed by agricultural evapotranspiration will correspondingly increase by 70–90%. However, much of the potential gain in physical water productivity has already occurred in some high productivity areas, such as in parts of China and Egypt, and developed countries (Comprehensive Assessment of Water Management in Agriculture 2007). The challenge of irrigation water supply is expected to be exacerbated in some arid and semiarid lands by climate change, which may increase temperatures and decrease rainfall.

Fedoroff et al. (2010) observed that meeting future water supplies will involve a radical rethinking of agriculture. Both land productivity (i.e., crop production per unit area of arable land) and water productivity (i.e., crop production per unit volume of water) will have to be substantially improved. New agricultural practices will be required, including integration of nutrient flows, improved dryland agriculture, and the use of saline waters. It was also recognized that meeting future food demands will require improvements in conventional and molecular breeding, as well as genetic modifications to increase productivity and adapt our existing food and fiber crops to changing climatic conditions. Developing countries are especially vulnerable to climate change because they depend heavily on agriculture, they tend to be warm already, they lack infrastructure to respond well to increased climatic variability, and they lack capital to invest in innovative adaptations (Lybbert and Summer 2010). Innovation and diffusion of technology to small farmers in developing countries is critical for their being able to better cope with current conditions and adapt to future climate change (Lybbert and Summer 2010).

An important option is to improve rain-fed (dryland) agriculture through better water management practices (Comprehensive Assessment of Water Management in Agriculture 2007). Better soil and land management practices can increase water productivity. Adding an irrigation component is often an important element of upgrading dryland agriculture to mitigate risks inherent in its being dependent on rainfall. Supplemental irrigation and water harvesting have been shown to allow for substantial improvements in water productivity in the dry farming areas of West Asia and North Africa (Oweis et al. 1999; Oweis and Hachum 2006). Improving agricultural productivity in areas that are dependent on rainfall has the greatest potential to reduce poverty and hunger, especially for Sub-Saharan Africa and large parts of Asia. Current yields in many rain-fed areas are low and improving farming practices could double or quadruple yields (Comprehensive Assessment of Water Management in Agriculture 2007). For countries with a GDP dependent largely on agriculture, increasing agricultural productivity is the greatest opportunity for reducing poverty, and irrigation can act as a springboard for economic development (Comprehensive Assessment of Water Management in Agriculture 2007).

More efficient water use and improved environmental management can allow the same resources to serve a greater proportion of the population. However, developing countries are often unable to support the high equipment, installation, and operation and maintenance costs of more efficient technology (Cassardo and Jones 2011). Increased water use efficiency (WUE) in some circumstances may not be economically efficient. The additional economic value obtained from the implementation of more efficient irrigation practices may be less than costs to implement the practices or there may be a greater opportunity cost in that the money invested to improve agricultural WUE could result in greater returns if invested elsewhere in the economy.

Efficiency is generally defined as the ratio of outputs to inputs. A variety of definitions have been proposed for WUE, which are reviewed by Hansen et al. (1980), Sinclair et al. (1984), Bos (1985), Keller and Keller (1995), and Howell (2001), among numerous publications on the subject. Agricultural WUE can be most simply defined as the ratio of crop yield to water use:

$$WUE = \frac{Crop\ yield}{Water\ use} \tag{26.5}$$

This definition considers total water use, not irrigated water use. Total water use includes soil moisture from precipitation, which for dryland agriculture is the source of all of the water used. Indeed, in areas blessed with abundant rainfall, water is not a limiting factor on agricultural production and WUE is not a concern.

Irrigation water efficiency (WUE_i) is defined as:

$$WUE_i = \frac{Y_i - Y_d}{I_i} \tag{26.6}$$

and

$$WUE_i = \frac{ET_c - P_e}{V_d} \tag{26.7}$$

where,

Y_i = yield for irrigation level "i"
Y_d = yield for dryland or rainfed irrigation of the same area
I_i = amount of irrigation water applied for irrigation level "i"
ET_c = crop transpiration (m^3 or m^3/m^2)
P_e = effective precipitation (m^3 or m^3/m^2)
V_d = volume of water delivered (m^3 or m^3/m^2).

Various units of crop yield may be used for Eq. (26.6) (e.g., tons of wheat). The complication of applying equation is 26.6 is determining the value of Y_d, which perhaps could be estimated from other non-irrigated farms in the same geographic area as the farm under consideration (Howell 2001). Equation (26.6) is most easily applied in arid and semiarid lands where the crop under consideration could not be grown without irrigation and thus, Y_d is equal to zero. Supplemental irrigation (i.e., application of a limited amount of water when rainfall is insufficient) can be highly efficient as a relatively small amount of water can substantially increase production compared to non-irrigated areas (Oweis et al. 1999; Oweis and Hachum 2006). Equation (26.7) defines irrigation WUE in terms of crop transpiration (instead of crop yield), since it is the parameter associated with plant growth and yield (Hansen et al. 1980; Wallace and Batchelor 1997). However, from a practical applied water management perspective, the key variable is the yield of food or fiber.

WUE can also be estimated as

$$WUE = \frac{Yield}{(P_e + I + SW)} \tag{26.8}$$

where,
I = irrigation water applied (m^3/m^2)
SW = soil water depletion from root zone during the growing season

Equation (26.8) and other more involved definitions or equations may be scientifically more rigorous, but suffer from the limitation that the values of parameters are not usually accurately known, particularly on a normal operational level (as opposed to a research site). These definitions also do not consider the fate of return flows that are available to other users.

Irrigation involves a chain of sequential processes from the water source (reservoir or well) to crop yield. The WUE defined by Eq. (26.5) can be considered a total efficiency, if water use is defined as extraction from the water source. Hsiao et al. (2007) addressed how total efficiency is related to the efficiency of each step in the water supply and irrigation chain. Total efficiency (E_{Total}) is mathematically related to the efficiency of each step (E_i) as

$$E_{Total} = E_1 \times E_2 \times E_3 \ldots \quad \prod_i E_i \tag{26.9}$$

Where, efficiencies are expressed in fractions (dimensionless) of the water available or used during each step. The multiplicative nature of total efficiency can be understood by considering a situation where half of the water from a reservoir is available after conveyance to the farm ($E_1 = 0.5$) and half of the remaining water is available after on-site storage ($E_2 = 0.5$). The total efficiency of both steps is 0.25. Therefore, the output of the conveyance and storage system is 25% of the input.

An important conclusion of Hsiao et al. (2007) is that because of the multiplicative nature of the efficiency chain, small improvements in efficiency at each link (step) in the chain can result in very large improvements in total efficiency. It is, therefore, more effective to make modest improvements in several or more steps than to concentrate on major improvements in only one or two steps. Efforts to improve WUE efficiency should start with an assessment of the actual and attainable efficiencies for the given situation and then consideration of what steps in the chain can improvements be made at a reasonable cost (Hsiao et al. 2007).

Wallace (2000) roughly estimated that only about 13–18% of the initial water resources in both irrigated and dryland agriculture is used as transpiration by crops, which is important as transpiration is the only water use associated with actual plant growth and yield. This low efficiency indicates that there is plenty of room for improvement (Wallace 2000). Water use efficiency can be improved by (Wallace and Batchelor 1997; Wallace 2000):

- using more of the water resource as transpiration, and
- fixing more carbon per unit of water transpired.

There are four main types of water losses associated with irrigated agriculture:

- losses during conveyance and storage,
- evaporation from soil,
- runoff from farm fields, and
- drainage from the plant root zone.

A variety of techniques are well known for reducing all of the listed water losses (Sect. 26.3.4). Increasing the amount of carbon fixed per unit of water transpiration (i.e., increasing transpiration efficiency) may be achieved through the choice of crops and crop varieties, breeding and genetic manipulation, and manipulation of crop microenvironments (Wallace and Batchelor 1997).

A key point that needs to be emphasized is that where water is scarce, the focus should be on water use efficiency rather than land use efficiency (i.e., crop yields per unit area of land), although water use efficiency and land use efficiency are not necessarily incompatible (Deng et al. 2004). Increasing agricultural productivity extends beyond just crop yield per unit area of land or volume of water. The choice of crops is a critical issue affecting the economies of farms. Reduction in post-harvest losses is one of the best opportunities for effectively improving crop productivity (Lybbert and Summer 2010). Post-harvest losses may reach 50% is some circumstances. Eliminating such losses is equivalent to a 100% increase in crop production.

The effects of changes in agricultural irrigation practices need to be considered in the context of the total water balance (Foster and Perry 2010). Keller and Keller (1995, p. 7) succinctly observed that *one user's inefficiency can be the next users' supply of water*. Excess irrigation water (i.e., the irrigation inefficiency) may be reused either for shallow aquifer recharge or by downstream irrigators from drainage canals. For example, at a farm level, water use in the Nile River valley of Egypt is inefficient, but at national level, Egyptian water utilization in agriculture through reuse is over 70% (Keller and Keller 1995; Allan 2001). Such a high level of efficiency is rarely achieved elsewhere. Allan (2001) noted that hydrologic conditions in Egypt are favorable for reuse of irrigation water and that such conditions may not be present in other areas.

Keller and Keller (1995) defined effective water use efficiency (WUE_e) as follows:

$$WUE_e = \frac{ET_c}{V_e} \tag{26.10}$$

where,

V_e = effective use of water (inflows − outflows), which considers recycling and water quality changes

Salgot and Torrens (2008) and Foster and Perry (2010) cautioned that reduction of water use in agriculture is not as clear a practice as it may seem. Excess water in agricultural fields contributes to the leaching of salts from the soil zone (Salgot and Torrens 2008). If insufficient water is applied, salts may accumulate in the soil and cause a loss of its productivity. The accumulation of salts in the soil will be particularly acute in arid regions with high evapotranspiration rates and relatively high initial salinities.

Paradoxically, measures that allow farmers to grow higher value crops per unit of water pumped have the implication of making groundwater use more profitable and encouraging farmers to increase irrigated areas (Foster and Perry 2010). Adoption of more efficient irrigation practices has in some areas increased the use of groundwater (Keller and Keller 1995; Howell 2001; Foster and Perry 2010; Sophocleous 2010). Improvement in WUE can also adversely impact downstream users who may be reliant on return flows for their water supply. Decisions intended to raise water use efficiency that are based on classical farm-scale efficiency calculations often do not result in real water savings (Keller and Keller 1995). Improvement of WUE should be evaluated on the integrated catchment level, in which consideration is given to the impacts of changes in water use in different parts of a catchment on the overall catchment efficiency and the economic, social, and institutional issues associated with the implementation of improved water use efficiency techniques (Wallace and Batchelor 1997).

26.3.4 Irrigation Water Use Efficiency Techniques

Postel (1992, 1993) observed that most of the world's farmers still irrigate the way their ancestors did 5,000 years ago by flooding or channeling water by gravity across crop land. The most common method is furrow irrigation in which the water flows through shallow furrows that separate crops. This method is inefficient as the water is typically not distributed evenly across an irrigated fields and excess water needs to be applied to ensure that the entire field receives sufficient water.

Wallace and Batchelor (1997) identified four main improvement categories for increasing irrigation efficiency at a field level: agronomic, technical, managerial, and institutional. Agronomic options include measures to improve crop yields such as improved crop husbandry, introduction of higher-yielding varieties, and maximizing cropped area to times of low potential evapotranspiration and high rainfall. Technical options include the adoption of more efficient irrigation technologies. Managerial options include adoption of demand-based irrigation schedules using soil-moisture sensors and improved maintenance of equipment. Institutional options include creating incentives and disincentives that encourage efficient water use and discourage waste, and improved training and education. Elements from all four categories should be employed in order to achieve the greatest improvements in WUE (Wallace and Batchelor 1997).

Postel (1992, 1993) reviewed some of the technical and managerial modifications to conventional irrigation methods and alternative irrigation methods that can increase the efficiency of irrigation, which include:

- surge irrigation, in which water is applied to alternative furrows at pre-established time intervals,

- low-energy precision application (LEPA) sprinkler systems, which reduce the energy required and water losses of standard overhead sprinklers,
- laser leveling of farm fields, which increases the efficiency of furrow irrigation,
- recycling of water,
- soil moisture monitoring—irrigation schedule times and amounts are based on the actual crop water needs,
- microirrigation such as drip irrigation—water is applied using porous or perforated piping at or just below land surface,
- improved overall system management to discourage excess irrigation applications by some users so that all users receive adequate water,
- conjunctive use of groundwater and surface water—pumping groundwater in water logged areas as opposed to applying more surface water, and
- development of salt-tolerant, drought-resistant, and water-efficient plant breeds.

There are thus numerous tools available to increase water productivity. The optimal choice of techniques for a given area will depend upon local conditions. Improving water management in agriculture requires learning by doing and a flexible, adaptive approach (Comprehensive Assessment of Water Management in Agriculture 2007). It is also recognized that agriculture should be viewed as an integrated multiple use system and as an agroecosytem. The goal is to increase the productivity of water use while preserving or enhancing ecosystems and reducing poverty (Comprehensive Assessment of Water Management in Agriculture 2007).

Jin et al. (1999) proposed the concept of temporal and spatial management of soil water (TSMSW) as a means to increase water use efficiency. The basic premise of TSMSW is to harmonize as much as possible crop water demand and soil water availability. Jin et al. (1999) reported that in Heilonggang region of North China, 80% of the rainfall occurs from June through September. It is therefore desirable to preferentially plant crops, such as summer maize and cotton, which have main growth periods coinciding with the rainy season, when soil moisture alone can provide all or most of the plant water needs. The area planted for crops that grow during the winter dry season, such as winter wheat, should be reduced because irrigation is required or otherwise crop yields will markedly decrease. The main elements of TSMSW are (Jin et al. 1999):

- adjustment of crop structures and rotations to fit changes in soil moisture,
- increasing soil water resources,
- reduction of soil water evaporation (such as by the use of mulches and plastic sheets), and
- management of soil water to meet temporal and spatial crop water demands.

26.3.5 Saline-Water Irrigation

Freshwater resources may be conserved by the irrigation of agricultural crops with saline-water, which is often locally abundant, especially in coastal areas. There has been growing interest over the past century in the potential use of salt tolerant plants (halophytes) for agricultural, fuel, and ornamental uses in water scarce regions. The use of ocean water for the irrigation of crops, if feasible, represents a fantastic opportunity (O'Leary 1988). Halophytic crops could be grown in coastal areas with inadequate freshwater resources or in agricultural areas impacted by soil salinization. Halophytic crops in the latter case would allow areas impacted by salinization to remain in production.

There are two potential options for the development of halophytic crops. Existing crops could be breed, selected, or genetically modified to become salt tolerant. Alternative, existing halophytic plants could be selected and bred for superior performance (O'Leary 1988). Modification of existing plants for salt tolerance poses a much greater technical challenge. Compensatory processes must occur in the tissues of plants to maintain osmotic balance in the absence of high internal salt concentrations, which consumes energy and substrates that might have otherwise gone into growth. The increase in salt tolerance in grain crops, for example, has come at the expense of decreased yield (O'Leary 1988).

It would certainly be easier to breed and select halophytes for increased yields and food quality than to make a halophyte out of a non-halophyte (O'Leary 1988). The main use of halophytes to date has been for animal feed (fodder). The genus *Atriplex* has been used since the beginning of century for fodder in widespread areas of the world. Productivities have been achieved for some cultivated halophytes irrigated with seawater that are comparable to those of conventional crops irrigated with freshwater (Aronson et al. 1988; O'Leary 1988).

The nutritional and digestibility of halophyte fodder is an important concern. Palatability is also an issue related to the salt content of fodder. The amount of halophyte that an animal will voluntarily eat is a function of its salt content. The high salt content of halophyte forage can also have a significant effect on the amount of water consumed by animals feeding on the material (O'Leary 1988). Solutions to the salt content issue include blending the halophyte fodder with other components in the feed mix or developing a mechanism for washing the salt out of the fodder (O'Leary 1988).

Other potential uses of halophytes are as seed crops, ornamental vegetation, and fuel. Halophytes have received relatively little attention as seed crops because of the small size of their seeds (O'Leary 1988). Ornamental vegetation is perhaps the easiest application to implement. Seashore *Paspalum* grass is a salt-tolerant turf grass that is being used for golf courses in some water challenged areas. Seashore *Paspalum* can be irrigated with brackish or saline water where the soils are sandy and not subject to salinization.

Irrigation of halophytes also requires proper soil water management, particularly drainage (O'Leary 1988). A major issue for irrigation with saline water is the need to maintain salinity levels in the root zone at an acceptable level, which will vary between crops. A key technical issue is the need to deliver water in known and consistent patterns because poor uniformity will result in spatial differences in salinity within the root zone (Jaradat 2005). Boron accumulation in the soil is another potential long-term problem (Ayars et al. 1993, 2006). When high salinity water is used for irrigation, the technology and expertise required are a level higher than what is required for conventional irrigation design and operation, and the impacts of poor system design and management are much greater (Jaradat 2005). Specific management practices may require the blending of saline water with good quality irrigation water or to use the saline water intermittently, especially during the growth stages when the crop is most tolerant to salinity (Rhoades et al. 1992; Stenhouse and Kijne 2006).

The use of saline water for irrigation can result in an increase in the salinity of the underlying aquifers, which may preclude their use. Where halophytes are grown, such as for landscape purposes, salinity increases in the underlying aquifers can be prevented by deficit irrigation, in which the application of water is less than the reference ET rate and water does not pass the root zone, which is, again, tricky because the salts cannot be permitted to accumulate in the root zone (Glenn et al. 2009).

Although the use of saline-water for the irrigation of salt tolerant plants offers great opportunities, much has to be done to achieve the potential benefits. In practice, farmers have adopted few of these species for commercial cultivation, which probably reflects the failure of these nonconventional crops to compete economically with alternative crops or some other limitations to their commercialization, such as lack of familiarity of farmers with the plant species, absence of well-established production packages, and lack of established markets and post-harvest processing mechanisms (Stenhouse and Kijne 2006).

26.3.6 Irrigation Water Use Efficiency Incentives

Irrigation has an important economic aspect in that water has been, and continues to be, heavily subsidized in both developed and developing countries. The economics of water and subsidies is addressed in Chap. 34. It is clear that farmers realistically cannot afford to pay the true cost of the vast majority of water projects, and this is especially true in developing countries where farm incomes are very low or farming is at a subsistence level. Stiles (1996) noted with respect to developing countries, investment in more efficient irrigation and improvement in irrigation water distribution are capital intensive and almost always require some degree of public investment regardless of the eventual beneficiary. In Africa and other developing areas, the true costs of irrigation water development are rarely passed on to farmers, partly at least because so many irrigation projects are funded by donors. Maintaining social stability is given precedence over the pure economic yield of the provided water. Water pricing schemes have frequently floundered on political opposition, compounded by the difficulties in measuring water deliveries and collecting fees from large number of small users (Comprehensive Assessment of Water Management in Agriculture 2007).

However, the cost of water affects water use. Farm-level choices are influenced by farm-level resource constraints (El-Sadek 2010). Farmers with a limited supply of land, but a relatively abundant water supply will choose crops that maximize returns to land, rather than to water, which leads to inefficient water use. Where inexpensive water is available, there is little impetus for increasing the efficiency of its use.

The Comprehensive Assessment of Water Management in Agriculture (2007) noted that the incentives for producers (i.e., irrigators) often differ from those of the broader society (less water for agriculture, more for cities and the environment) and rather than trying to charge farmers more for their water use, the parts of society benefiting from re-allocations may need to compensate farmers for less water use. It may thus be more effective to give farmer's a positive incentive to conserve water than to penalize them by trying to charge them more (or for the first time) for water, for which they may be unwilling or unable to pay.

References

Rhoades, J. D.; Kandiah, A.; Mashali, A. M. (1992). *The use of saline waters for crop production.* FAO Irrigation and Drainage Paper 48. Rome: Food and Agriculture Organization of the United Nations.

Agthe, R. E., & Brillings, R. B. (1987). Equity, price elasticity and household income under increasing block rates for water. *American Journal of Economics, 46*(3), 273–286.

Allan, J. A. (2001). *The Middle East water question. Hydropolitics and the global economy.* London: I.B. Tauris.

American Water Works Association. (2000). *Principles of Water Rates, Fees, and Charges (AWWA Manual M1).* Denver: CO American Water Works Association.

Aronson, J. A., Pasternak, D., & Danon, A. (1988). Introduction and first evaluation of 120 halophytes under seawater irrigation. In E. E. Whitehead, C. F. Hutchinson, B. N. Timmermann & R. G. Varady (Eds.), *Arid lands. Today and tomorrow. Proceedings of an International Research and Development conference* (pp. 737–746). Boulder: Westview Press.

Ayars, J. E., Hutmacherm, R. B., Schoneman, R. A., Vail, S. S., & Pflaum, T. (1993). Long term use of saline water for irrigation. *Irrigation Science, 14*, 27–34.

Ayars, J. E., Christen, E. W., Soppe, R. W., & Meyer, W. S. (2006). The resource potential of in situ shallow ground water use in irrigated agriculture: A review. *Irrigation Science, 24*, 147–160.

Ayers, R. S., & Wescott, D. W. (1985). *Water Quality for Agriculture.* FAO Irrigation and Drainage Paper 29. Rome: Food and Agriculture Organization of the United Nations.

Beecher, J. A. (2010). The conservation conundrum: How declining demand affects water utilities. *Journal American Water Works Association, 102*(2), 78–80.

Borisova, T., & Rawls, C. (2010). Conservation water rates for residential customers: A practical overview: Florida. *Water Resources Journal, 62*(8), 16–22.

Bos, M. G. (1985). Summary of ICID definitions of irrigation efficiency. *ICID Bulletin, 34*, 28–31.

Bouwer, H. (1978). *Groundwater hydrology*. New York: McGraw-Hill.

Bouwer, H. (2000). Integrated water management: Emerging issues and challenges. *Agricultural Water Management, 45*, 217–228.

Carver, P. H., & Boland, J. J. (1980). Short- and long-run effects of price on municipal water use. *Water Resources Research,* 16(4) 609–616.

Cassardo, C., & Jones, A. J. A. (2011). Managing water in changing world. *Water, 3*, 618–628.
Chesnutt, T. W., & Beecher, J. A. (1998). Conservation rates in the real world. *Journal American Water Works Association, 9*(2), 60–70.
Comprehensive Assessment of Water Management in Agriculture. (2007). *Water for food, water for life: A comprehensive assessment of water management in agriculture* (p. 645). London: Earthscan, and Colombo: International Water Management Institute.
Cuthbert, R. W., & Lemoine, P. R. (1996). Conservation-oriented water rates. *Journal American Water Works Association, 88*(11), pp. 68–78.
Deng, X. -P., Shan, L., Zhang, H., & Turner, N. C. (2004, 26-Sep to 1-Oct). *Improving agricultural water use efficiency in arid and semiarid areas of China, New directions for a diverse Planet. Proceedings of the 4th International Crop Science Congress* (p. 14). Brisbane.
El-Sadek, A. (2010). Virtual water trade as a solution for water scarcity in Egypt. *Water Resources Management, 24*, 2437–2448.
Fedoroff, N. V., Battisti, D. S., Beachy, R. N., Cooper, P. J. M., Fischhoff, D. A., Hodges, C. N., et al. (2010). Radically rethinking agriculture for the 21st century. *Science, 327*, 833–834.
Foster, S. S. D., & Perry, C. J. (2010). Improving groundwater resource accounting in irrigated areas: A prerequisite for promoting sustainable use. *Hydrogeology Journal, 18(2),* 291–294.
Glenn, E. P., McKeon, C., Gerhart, V., Nagler, P.L., Jordan, F., & Artiola, J. (2009). Deficit irrigation of a landscape halophyte for the reuse of saline waste water in a desert city. *Landscape and Urban Planning, 89*(3–4), 57–64.
Halliwell, D. J., Barlow, K. M., & Nach, D. M. (2001). A review of the effects of wastewater sodium on soil physical properties and their implications for irrigation systems. *Australian Journal of Soil Research, 39*(6), 1259–1267.
Hansen,U. E., Stringham, G. E., & Israelsen, W. O. (1980). *Irrigation principles and practices* (4th ed.). New YorK: Wiley.
Heathcote, R. L. (1983). *The arid lands: Their use and abuse* (p. 323). London: Longman.
Howell, T. A. (2001). Enhancing water use efficiency in irrigated agriculture. *Agronomy Journal, 93*(2), 281–289.
Hsiao, T. C., Steduto, P., & Fereres, E. (2007). A systematic and quantitative approach to improve water use efficiency in agriculture. *Irrigation Science, 25*, 209–231.
Jaradat, A. A. (2005). Saline agriculture in the Arabian Peninsula: Management of marginal lands and saline water resources. *Journal of Food, Agriculture & Environment, 3*(2), 302–306.
Jin, M., Zhang, R., Sun, L. & Gao, Y. (1999). Temporal and spatial soil water management: a case study in the Heilonggang region, PR China. *Agricultural Water Management*, 42, 173–187.
Jordan, J. L., & Albani, R. (1999). Using conservation rate structures. *Journal American Water Works Association, 91*(8), 66–73.
Keen, A. H., Keen, D., Francis, G. E., & Wolf, A. (2010). High-contact, hands-on outreach program changes customers' water use behavior. *Journal American Water Works Association, 102*(2), 38–45.
Keller, A. A., & Keller, J. (1995). *Effective efficiency: A water use efficiency concept for allocating freshwater resources.* Arlington: Center for Economic Policy Studies, Winrock International.
Kingdom, B., Liemberger, R., & Marin, P. (2006). *The challenge of reducing non-renewable water (NRW) in developing countries. How the private sector can help: A look at performance-based service contracting.* Washington, DC: The World Bank Group, Water Supply and Sanitation Board Discussion Papers Series, Paper No. 8., December 2006.
Lybbert, T. & Summer, D. (2010). Agricultural technologies for climate change mitigation and adaptation in developing countries: Policy options for innovation and technology diffusion, Issues Brief No. 6 (p. 30). Geneva: International Centre for Trade and Sustainable Development, and Washington, DC: International Food & Agricultural Trade Policy Council.
Mann, P. C. &d Clark, D. M. (1993). Marginal-cost pricing: its role in conservation. *Journal American Water Works Association, 85*(8), 71–78.
McGarry, R. S., & Brusnigham, J. M. (1979). Increasing water and service rate schedules: a tool for conservation. *Journal American Water Works Association, 71*(9), 74–479.

McKean, E. (2004). *The new oxford american dictionary* (2nd ed.). New York: Oxford University Press.

Mutikanga, H. E., Sharma, S., & Vairamoorthy, K. (2009). Water loss management in developing countries: Challenges and prospects. *Journal American Water Works Association, 101*(12), 57–68.

O'Leary, J. W. (1988). Saline environments and halophytic crops. In E. E. Whitehead, C. F. Hutchinson, B. N. Timmermann & R. G.Varady (Eds.), *Arid lands. Today and tomorrow. Proceedings of an International Research and Development Conference* (pp. 773–790). Boulder: Westview Press.

Oweis, T., & Hachum, A. (2006). Water harvesting and supplemental irrigation for improved water productivity of dry farming systems in West Africa and North Africa. *Agricultural Water Management, 80*, 57–73.

Oweis, T., Hachum, A. & Kijne, J. (1999). *Water harvesting and supplementary irrigation for improved water use efficiency in dry areas. SWIM 7* (p. 41). Colombo: International Water Management Institute.

Pape, T. E. (2010). New codes are making it easier to be green. *Journal American Water Works Association, 102*(2), 57–63.

Pearce, F. (2006). *When rivers run dry*. Boston: Beacon Press.

Pereira, L. S., Cordery, I., & Iaconides, I. (2002). *Coping with water scarcity* (p. 269). UNESCO, International Hydrological Programme, IHP-VI, Technical Documents in Hydrology No. 58.

Pescod, M. E. (1992). *Wastewater treatment and use in agriculture*. FAO Irrigation and Drainage Paper 47. Rome: Food and Agriculture Organization of the United Nations.

Peterson, J.M. (2003). Conserving the Ogallala Aquifer: Efficiency, equity, and moral motives. *Choices*, First Quarter 2003, 15–18.

Postel, S. (1992). *Last oasis: Facing water scarcity*. New York: WW Norton.

Postel, S. (1993). Water and agriculture. In P. H. Gleick (Ed.), *Water in crises. A guide to the world's fresh water resources* (pp. 56–66). Oxford: Oxford University Press.

Salgot, M., & Torrens, A. (2008). Impacts of climate change on water resources: The future of groundwater recharge with reclaimed water in the South of Europe. In: W. Dragoni & B. S. Sukhija (Eds.), *Climate change and groundwater* (pp. 145–168). London: Geological Society of London Special Publications No. 288.

Seckler, D. (1996). *The new era of water resources management: From dry to wet water savings. Issues in Agriculture 8*. Washington, DC: Consultative Group on International Agricultural Research.

Selover, G. (2010). Ensuring a more energy-efficient future by changing the way homes are built. *Journal American Water Works Association,*102(2), 46–51.

Sinclair, T. R., Tanner, C. B., & Bennett, J. M. (1984). Water use efficiency in crop production. *BioScience, 34*, 36–40.

Sophocleous, M. (2010). Review: Groundwater management practices, challenges, and innovations in the high plains aquifer, USA—lessons and recommended actions. *Hydrogeology Journal, 18*, 559–575.

Stenhouse, J., & Kijne, J. W. (2006). *Prospects for productive use of saline water in West Asia and North Africa*. Comprehensive Assessment Research Report 11. Colombo: Comprehensive Assessment Secretariat.

Stiles, G. (1996). Demand-side management, conservation, and efficiency in the use of Africa's water resources. In E. Racked, E. Rathgeber, & D. B. Brooks, (Eds.), *Water management in Africa and the Middle East* (pp. 8–27). Ottawa: International Development Research Centre.

Vickers, A. (2001). *Handbook of water use and conservation*. Amherst: WaterPlow Press. p. 464.

Wallace, J. S. (2000). Increasing agricultural water use efficiency to meet future food production: *Agriculture, Ecosystems and Environment, 82*, 105–109.

Wallace, J. S., & Batchelor, C. B. (1997). Managing water resources for crop production. *Philosophical Transactions Royal Society London B, 352*, 937–947.

Winpenny, J. (1994). Managing water as an economic resource. London: Rutledge.

Part VI
Desalination

Chapter 27
Desalination: Desalination in Arid Lands

27.1 Introduction

Modern technology currently allows the treatment of virtually any type of water to potable standards. The critical issue is the cost, primarily the economics of the treatment processes and conveyance of the treated water, as related to the value of the use for the desalinated water. The National Research Council (2008, p. 12) reported that

> The potential for desalination to meet water demands in the United States is constrained not by the source water resources or the capabilities of current technology, but by a variety of financial, social, and environmental factors.

Therefore, even in well-developed economies, such as in the United States, the use of desalinated water may be limited because of its relative cost, especially in comparison to other available supplies. However, the key issue question should be, when there is no other additional source of water supply other than desalination, what is the real economic value of the water?

In arid and semiarid lands, desalination is one available option to meet the potable water requirements of growing populations, particularly of those countries that border the oceans and seas. The oceans are essentially an unlimited supply of water that can be used for desalination purposes and can be also used for the disposal of the high-salinity wastewater generated by the process, which is termed concentrate. Although desalination does provide some or nearly all of the potable water supply solution for some countries (mostly in the Middle East region), desalted water comes at a high price, not only in terms of cost, but also in terms of energy consumption and, in some situations, environmental impacts.

Conventional desalination is energy intensive and has a large carbon footprint when fossil fuels are used to provide the thermal and electrical energy required to desalt water. The water-power nexus is obvious in that water and power are intertwined in terms of solutions to meet both demands. There is considerable ongoing research in assessing how to reduce the energy consumption of desalination

R. Maliva and T. Missimer, *Arid Lands Water Evaluation and Management*, Environmental Science and Engineering, DOI: 10.1007/978-3-642-29104-3_27,

and make the process more efficient. Also, research is being conducted on the use of alternative energy sources to provide the power necessary to run the desalination plants. Renewable energy sources are being used to provide energy for desalination in Australia (wind power) (Sect. 27.5) and a number of large-scale solar-power facilities are being investigated in the Middle East. A small solar-powered facility was first put on-line in 1981 (Boesch 1982). Renewable energy sources allow some desalination plants to operate "off the grid" and reduce their overall carbon footprint. However, systems powered by renewable energy sources must still be attached to the grid or have alternative on-site power generating or energy storage capacity if they are to operate during periods when the renewable source is insufficient to meet power requirements, such as at night (solar) or when the wind is not blowing.

Analyses of alternative power costs for desalination facilities indicate that in many cases systems renewable energy sources are more expensive to construct and operate based on current global energy market conditions. However, renewable energy sources become more economically competitive as the cost of conventional energy sources increases. Consideration is currently being given to the use of nuclear energy to provide the power for the desalination of seawater. Desalination facilities co-located with nuclear plants near the sea could also share the intakes and outfalls, which would provide process water for the desalination plant and cooling water for the nuclear plant (Zarg 2003; Megahed 2001, 2008; Khamis 2009; Sect. 27.5.4). However, the nuclear option may not meet public acceptance at the present time based on the combined earthquake, tsunami wave, and nuclear power plant disaster that struck northern Japan in March, 2011.

Desalination facilities that have a sufficient capacity to meet large demands are divided into two general classes, thermal and membrane. The current global distribution of desalination technologies used to produce drinking water is shown in Fig. 27.1. Thermal desalination methods still constitute the largest percentage of the global capacity at 53%. Membrane technologies provide 47% of the global capacity and are very rapidly gaining market share because of their lower energy usage (Global Water Intelligence). Most large thermal desalination plants are co-located with power generation facilities to allow the use of steam for both power generation and thermal desalination.

There are several other technologies that can be used for smaller-scale desalting or are in the research stage with potential future commercial use. Crystallization processes, including freezing and the gas hydrate process, have been assessed for many years and have not shown much promise for commercial or extensive use at large scales. Humidification processes that have been or are currently being evaluated include the dew evaporation process, seawater greenhouse, membrane distillation, mechanically-intensified evaporation, and atmosphere water vapor processes. Several of these processes have small-scale applications that are economic. Solar stills are being used extensively in agricultural applications, particularly in greenhouses. Some other processes that are being evaluated at a research level are forward osmosis, ion exchange, flow through capacitors, liquid–liquid extraction, centrifugal reverse osmosis, and rotary vapor compression. Miller (2003) provides a review of alternative desalination processes.

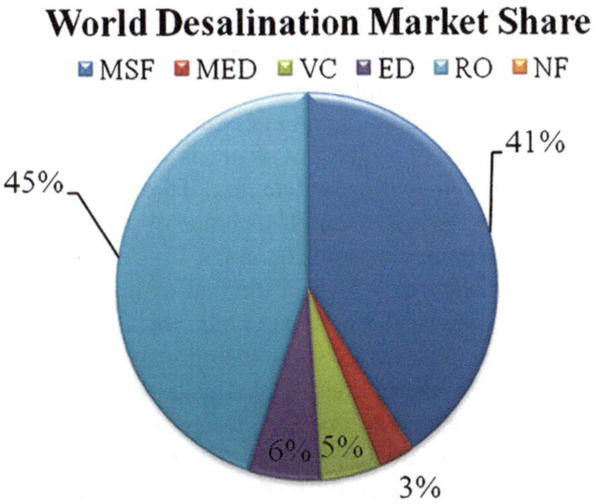

Fig. 27.1 Global distribution of desalination technologies currently used (Global Water Intelligence)

A fundamental concept of desalination is that a certain amount of energy is required to produce freshwater from a saline water source. The amount of energy required increases with the salinity of the water source. The minimum unit of power required to convert saline water into freshwater water is termed the thermodynamic limit. For a 1% saltwater solution at 25°C, the thermodynamic limit is about 0.78 kWh/m^3 of energy (Thu et al. 2010). In nature, the sun evaporates seawater to form water vapor that condenses and becomes fresh water that returns to the Earth's surface as rainfall. The natural process requires about 475 kWh/m^3 of energy to make seawater into fresh water using the energy of the sun (Dr. Kim Choon Ng, personal communication). The natural process is quite inefficient, but it costs nothing to humanity. Desalination processes cannot produce freshwater from seawater below the thermodynamic limit consumption of energy, but they can treat water with a much greater efficiency compared to natural solar heating and condensation. The key to successful desalination is to make it as economically viable as possible by maximizing efficiency.

There are five fundamental aspects of desalination that must be considered in the feasibility, design, operation, and economics of water production using desalination technology. These elements are

- Raw water development (intakes),
- Pretreatment of the raw water,
- Desalination process,
- Post-treatment of desalinated water, and
- Concentrate management (residuals).

All five of these issues affect the viability and economics of desalination and have potential impacts to human health and the environment if handled improperly. Development of the raw water supply is a critical aspect of desalination and it is very important to feed the desalination process with the highest quality water possible.

This is primarily an economic and environmental issue. High-quality water lowers pretreatment costs and the source of raw water determines the environmental impacts. For example, surface intakes for seawater desalination facilities produce impingement and entrainment issues and are vulnerable to contamination.

Pretreatment of the raw water is required in order to protect the desalination process from fouling and failure. The degree of pretreatment depends upon the quality of the raw water. The choice of which desalination process to use is based on local conditions and economics. Co-development of a desalination facility with a power plant may favor a thermal or hybrid process, while the reverse osmosis process may be the optimal choice for a stand-along facility. Whichever process used must be capable of producing water of the quality desired.

Post-treatment of desalinated water is required to make it safe for potable use. Thermal treatment produces water with very low concentrations of dissolved solids, which are required to maintain human health. Failure to post-treat the desalinated water could lead to health problems as the result of the flushing of minerals and salts from the body with reduced replacement (Cotruvo 2006; World Health Organization 2005, 2006b, 2009).

The residuals created during the desalination process, including water rejected during the treatment process and cleaning and treatment chemicals that are part of the waste stream termed "concentrate." Concentrate from desalination facilities must be disposed of in an environmentally acceptable manner or there could be serious environmental damage, particularly to the nearshore coastal zone where surface discharges are made. All of these factors must be considered when evaluating and designing desalination systems to supply water in arid lands. Detailed reviews of desalination have been presented by Buros et al. (1980); Buros (2000); Watson et al. (2003); National Research Council (2008), and the World Health Organization (2007).

From a water management perspective in arid lands, desalinated seawater is truly "new water" and adds to the water in the natural system. Desalinated brackish groundwater is also new water, but is a resource that can be depleted if not managed wisely. A discussion of the relevant desalination technologies currently used to provide water in many arid lands throughout the world follows. The discussion includes existing and new technologies, and their relative energy usage and cost. Possible uses of alternative energy sources for desalination are also discussed, because power production and desalination is a nexus being fully linked.

27.2 Thermal Desalination Processes

Thermal desalination processes have been utilized in some form starting with the boiling of seawater to make freshwater on ships beginning in about 200 A.D. Sir Richard Hawkins reported from the South Seas in 1662 that he had supplied his men with fresh water by using shipboard distillation (personal communication from John Tonner). The oldest distillation desalination unit in the United States was installed at Fort Pulaski on Cockspur Island, Georgia, in 1862. Seawater desalination was

Fig. 27.2 Photograph of the Seawater Distillation facility located on Malta dated 1881 (courtesy of John Tonner)

initiated in Malta in 1866 (Riolo 2001). A photograph of a building which housed an early seawater distillation facility in Malta is provided in Fig. 27.2. Another distillation unit was operated at Fort Jefferson in the Dry Tortugas, Florida, during the eighteenth century. This was a Lillie unit, which is a process that was patented in 1888 (personal communication with John Tonner).

Most of the thermal desalination facilities in operation use a form of distillation, which is based on the principle that water vapor evaporated from saline water has a very low salinity. The condensed water vapor is collected to provide very high-quality water that is nearly devoid of dissolved solids. There are three primary types of commercial, thermal desalination processes that are currently being used, which are multi-stage flash distillation, multiple-effect distillation, and vapor compression (Buros et al. 1980; Buros 2000; National Research Council 2008). A new and very promising technology, adsorption desalination, is also included in this discussion because of its great potential of reducing overall desalination costs, particularly in arid lands.

27.2.1 Multi-Stage Flash Distillation

Multi-stage flash distillation (MSF) involves the distillation of seawater in a series of stages through the successive lowering of saturation pressures and temperature. The underlying principle of MSF is that seawater will boil at progressively lower temperatures as the pressure in a confined container is deceased. This is achieved by the use of a constrictor or depressurization nozzle that lowers the pressure within the confined container of the subsequent stage. In the MSF process, water is evaporated and condensed on the surfaces of the cooler tubes, which are the feed seawater pipes, allowing the latent heat of evaporation to be recovered for reuse by pre-heating the seawater. The first commercial MSF plant was a 4-stage plant with a low efficiency, which was designed and constructed in Kuwait in 1957 (Al-Wazzan and Al-Modaf 2001).

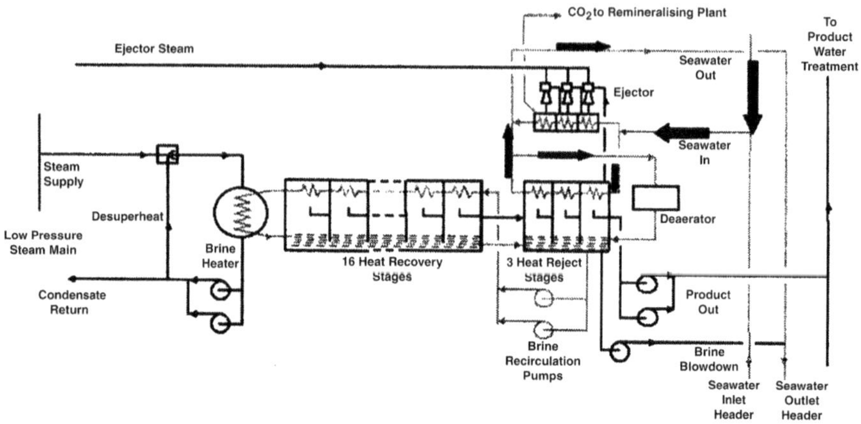

Fig. 27.3 Multi-stage flash distillation desalination process diagram (modified from Awerbach 2007)

The conceptual process of MSF is not complex (Fig. 27.3). A simplified overview of the process is that seawater is heated using heat exchangers with energy in the form of steam obtained from a co-located power plant. The heated seawater passes through a series of chambers, referred to as stages, each having progressively lower pressures. In each stage, the water boils or "flashes" and the vapors are collected using condensers that are cooled by the inflowing seawater. Awerbach (2007) provided a very good overview of the MSF process. More detailed descriptions of the MSF process, operating details, and historical improvements in the process are described by (Hamed 2000; Sommariva et al. 2001; Borsani and Ferro 2006; Khawaji et al. 2008; Ettouney 2009).

There are several critical details of the MSF process that affect system performance. Seawater is pumped into the heat rejection part of the plant where it exchanges energy with the lowest temperature vapors (Awerbach 2007). Most of the cooling water recycles back to the sea, but some of it is used as makeup water. The makeup water is treated to remove dissolved gases and a scale inhibitor is added to reduce calcium sulfate scaling of the exchange tubes and other parts of the process. The initial seawater feed mixes with the recycled stream and is pressurized and heated to the maximum top brine temperature (TBT). Maximum operating temperature for the TBT is achieved in a brine heater that utilizes stream. This thermal energy is most commonly obtained from low pressure steam (2.5–3.5 atmospheres) originating from the end stages of an electric plant generating turbine. The heated brine then flows into a chamber that is maintained at a pressure slightly less than the saturation pressure of the water (Awerbach 2007). At this location, a certain fraction of the water flashes to steam, which is stripped of suspended brine droplets as it passes through a mist eliminator. The steam condenses on the surfaces of the heat transfer tubing and is conducted into collection trays as hot product water.

The hot brine passes into the next chamber, which is at a lower pressure, where some additional water flashes to stream at a lower temperature and the same process of brine stripping, condensation, and collection is repeated. Another key feature is that the distillate from the first stage also passes into the distillate tray of the second stage to assist in the heating processes and simultaneously to cool the stream (Awerbach 2007). The process is repeated in each stage until the brine is discharged as blowdown water and the treated water leaves the plant as product.

An MSF plant operates in a series of vessel evaporators or stages that each contains a progressively lower internal pressure that reduces the temperature required to create stream ("flashing") in each stage. MSF systems use a bulk liquid boiling process to avoid mineral scaling on the heat transfer tubes. These systems commonly operate in as many as 20 heat recovery stages and 4 heat rejection stages. Once the salinity of the heated water reaches a critical threshold or concentration, it must be discharged to prevent precipitation of calcium sulfate (gypsum) and calcium carbonate, which collectively can cause scaling of the system.

A key element of the design of MSF plants is to optimize the use of heat energy, thus increasing the energy efficiency of the plant. In order to increase plant efficiency, it is common to recycle of fraction of the heated blowdown brine by combining it with the feedwater. The latent heat of condensation is removed by the recirculating stream that flows through the interior of the tubes used to condense the vapor in each stage (Awerbach 2007). This allows the circulating brine to be preheated to the maximum operating temperature of the process and recovers the energy of the condensing vapor. This is the heat recovery part of an MSF plant.

Additional removal of waste heat occurs at the cool end of the plant, where a separate set of tubes is installed. Feedwater is used as the coolant in the last stage of heat rejection. A major portion of this feedwater is discharged back to the sea, while some portion of it is added to the makeup water stream after pretreatment and air removal.

Awerbach (2007) suggests that MSF desalination technology has many advantages, including that plant energy consumption is not dependent on the feedwater quality in terms of salinity compared to most other desalination techniques and MSF plants produce nearly distilled water with a total dissolved solids of 25 mg/L or less. Based on the high quality and low dissolved solids concentration of the distillate, the product water can be blended with seawater reverse osmosis water to form a hydrid plant producing potable water at high efficiency (see Sect. 27.4.1). The distillate may be blended with brackish groundwater to increase the potable water production and provide needed salts.

Most large-scale MSF plants are co-located with steam or gas turbine power generation plants to maximize efficient energy use (Miller 2003). While high pressure stream is used to generate power by passing through turbines, low pressure stream is used to heat water for the desalination process. The efficiency of an MSF facility is typically expressed as the gained output, which is defined as the mass of water product per mass of the heating steam (Spiegler and El-Sayed 2001). The specific electricity consumption to produce freshwater using the MSF process is about 4 kWh/ton of distillate. However, since the process also requires thermal energy (steam), the total energy required by the MSF process to treat a cubic meter of seawater to freshwater is about 57.14 kWh/m^3 or more. There are a variety of different methods used to calculate these numbers and some methods produce slightly different results. These estimates were provided by Dr. Kim Choon Ng (King Abdullah University of Science and Technology and National University of Singapore). An example of a large MSF plant in an arid region is the facility at Jubail, Saudi Arabia.

27.2.2 Multiple Effect Distillation

Multiple effect distillation (MED) is the oldest of the commercial thermal water treatment technologies. The first MED seawater plants, which consisted of six facilities, each with a capacity of 75 m^3/day of product water, were installed in Egypt in 1912 (El-Dessouky 2007). Numerous MED treatment plants were constructed in the 1950s, but many of the plants suffered problems with extensive mineral scaling on the heat transfer tubes and the technology was, therefore, sparingly used in favor of MSF until a design solution was found (Miller 2003).

In the MED process, feedwater flows over a heat transfer surface in the first chamber (effect) that is heated by primary stream (Awerbach 2007). This results in the evaporation of a small percentage of the feedwater. In most modern MED plants, the primary steam and downstream vapors flow into a horizontal tube, where condensation occurs resulting in the production of treated water and the release of its latent heat of condensation. Feed seawater, with normal or concentrated salinity, is sprayed onto the exterior of the tube, producing additional vapor upon receiving the latent heat from the inner surface. Each chamber or effect in the series contains a vacuum at a progressive lower pressure. Brine evaporation occurs in each effect, which allows for additional evaporation at a progressively lower temperature. Efficiency is brought to the system by channeling the vapor from the one effect into the next, where it delivers latent heat to assist in evaporation of water from the brine flowing on the opposite wall of the tube. The process is repeated through each effect with the combined condensed water vapor constituting the product (Awerbach 2007).

Older MED units used submerged tubes and scaling occurred because of the high heating of the concentrated brine. In modern MED systems, the seawater feed is preheated by a fraction of the vapor from the effects. The system is fed forward and does not scale because the most concentrated brine is exposed to the lowest temperature (Awerbach 2007). Within each of the chambers or effects, the feedwater is heated to a boiling temperature in the plenum (with lower temperature in each effect moving downstream because of reduced pressure), which is the location where the heated feedwater is diverted into the tubes to undergo additional evaporation. Awerbach (2007) suggests that the heat transfer surface must be uniformly wetted to avoid dry spots that encourage the deposition of scale.

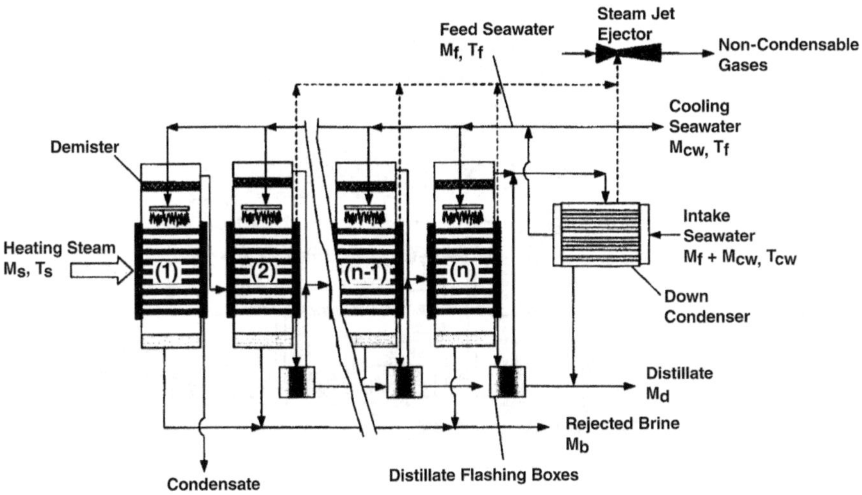

Fig. 27.4 Multiple-effect distillation process diagram (modified from El-Dessouky et al. 2000)

Because of the reduction in temperature in each effect moving downstream in the plant, each effect functions as a condenser for the upstream effect.

Efficiency and capacity in MED plants are achieved by increasing the number of effects and the heat transfer area or by increasing the operating temperature (Awerbach 2007). Most MED systems have between 10 and 16 evaporation effects (chambers) and have an operating temperature of about 70°C. Systems operating at lower temperatures have lower rates of scaling.

The parallel feed multiple-effect evaporation process is shown in Fig. 27.4. This configuration is considered to be perhaps the most efficient MED process design. El-Dessouky et al. (2000, p. 1687–1688) describes the general process referring to the figure as follows:

> The effects are numbered from 1 to n from the left to right (the direction of the heat flow). Each effect constitutes a heat transfer area, vapor space, mist eliminator and other accessories. In the parallel feed system, the vapor flows from left to right, in the direction of falling pressure, while the feed seawater flows in a perpendicular direction. As for the parallel/cross flow system, the brine stream leaving the first stage flows to the second, where it flashes and mixes with the feed seawater. Either system contains a number of evaporators, a train of flashing boxes, a down condenser, and a venting system. The parallel and the parallel/cross flow systems contain (n−1) flashing boxes for the distillate product. In the parallel/cross flow system, brine flashing takes place inside effects 2 to n. The two configurations utilize the horizontal falling tubes, which are characterized by their ability to handle seawater scaling. This is because of the wetting rates and efficient water distribution over the heat transfer surfaces by large spray nozzles. Thus, dry-patch formation or water mal-distribution is eliminated. This configuration offers the additional advantages of positive venting and disengagement of vapor products and/or non-condensable gases, high heat transfer coefficients, and monitoring of scaling or fouling materials……

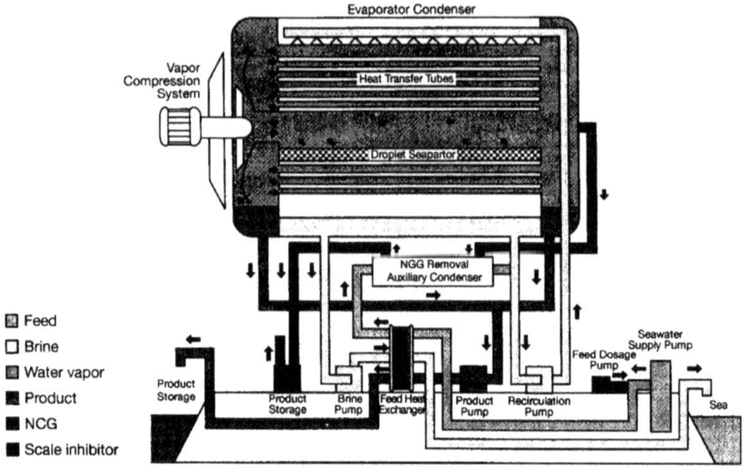

Fig. 27.5 Vapor-compression desalination process diagram (modified from Awerbuch 2007)

Energy consumption used by an MED plant to treat seawater to freshwater is considerably less compared to an MSF plant. The specific electricity consumption for the production of freshwater from seawater is about 1.8 kWh/ton and the total primary energy required about is 43.21 kWh/m^3 or more.

The capacity of MED desalination has been rising rapidly because of its lower energy consumption and it is being coupled with seawater reverse osmosis desalination in new hybrid systems. More detailed descriptions on the MED process, innovations in MED designs, and the future of MED are given in El-Dessouky et al. (2000); Sommariva et al. (2001); Khawaji et al. (2008), and Ettouney (2009). An example of a large, operating MED desalination plant is the Fujaurah II facility in Qidfa, UAE with a capacity of 454,600 m^3/day (121.4 MGD).

27.2.3 Vapor Compression Distillation

The vapor compression distillation (VCD) process produces freshwater from seawater by developing heat from the compression of vapor. It works by reducing the boiling point of seawater by pressure reduction, similar to MSF and MED. There are two different methods used to generate the heat needed to produce vapor, which are mechanical compression and the steam jet.

In the VCD process, feed seawater is pumped into a preheater where it obtains thermal energy from concentrate and product water effluents (Awerbach 2007). The preheated seawater is then sprayed over the heat exchange tubing which is at a higher temperature (Fig. 27.5). Part of the sprayed seawater evaporates and the vapor is compressed using a mechanical or thermal compression unit. The vapor is then pumped inside of the heat exchange bundles. Condensation of the compressed water vapor occurs inside the tubing, releasing energy that is transferred to the seawater sprayed over the exterior of the tubing, thereby, causing evaporation. The condensed distillate and hot seawater are pumped through the preheater to exchange thermal energy with the seawater feed. The condensed distillate is then sent to post-treatment after discharging from the system. A certain percentage of the concentrated seawater is discharged back to the sea, while part is blended with the feed seawater and returned to the process. The VCD process can use electrical energy to function with some units using steam in the thermocompression unit to increase the energy of the water vapor.

Low temperature VCD is a reliable and generally efficient process that can operate at below 70°C, which reduces the scaling potential. However, VCD units are generally used for small seawater desalination facilities (Khawaji et al. 2008). The total primary energy use for VCD conversion of seawater to freshwater is 31.71 kWh/m^3.

27.2.4 Adsorption Desalination (New Technology Not Yet Fully Commercial)

An intriguing new desalination technology is adsorption desalination that employs the sorption affinity of a vapor onto the surfaces of a mesoporous adsorbent. This novel approach to seawater (and brackish water desalination) was patented by Ng et al. (2006). Compared to other more established thermal desalination technologies, adsorption desalination systems have considerably lower energy consumption because it can be operated at a low activation process temperature where the heat source supply is free, either solar, waste heat, or a geothermal source. Thu et al. (2010) report that the thermodynamic limit for the specific energy consumption to desalt a saline solution at 1% concentration of salt at 25°C is about 0.78 kWh/m^3. This limit is the minimum energy consumption that is required to produce potable water at a given solution concentration regardless of the desalination method employed. The adsorption desalination (AD) method uses either waste or solar heat and it achieves a specific electricity consumption of 1.38 kWh/m^3, which is only about twice that of the thermodynamic limit (Wang and Ng 2005; Wang et al. 2007). This is the lowest specific energy consumption ever reported for any desalination technology.

Adsorption desalination technology has three fundamental components which are the evaporator, the condenser, and the adsorption and desorption beds. The adsorption and desorption beds contain high porosity, silica gel that is packed in between the fins of the tube-finned heat exchangers. The silica gel is held in the heat exchanger by stainless steel mesh. A conceptual diagram of an AD desalination plant is provided in Fig. 27.6 along with a close-up of the adsorption/desorption beds. A photograph of the solar-powered AD research plant at the King Abdullah University of Science and Technology is shown in Fig. 27.7.

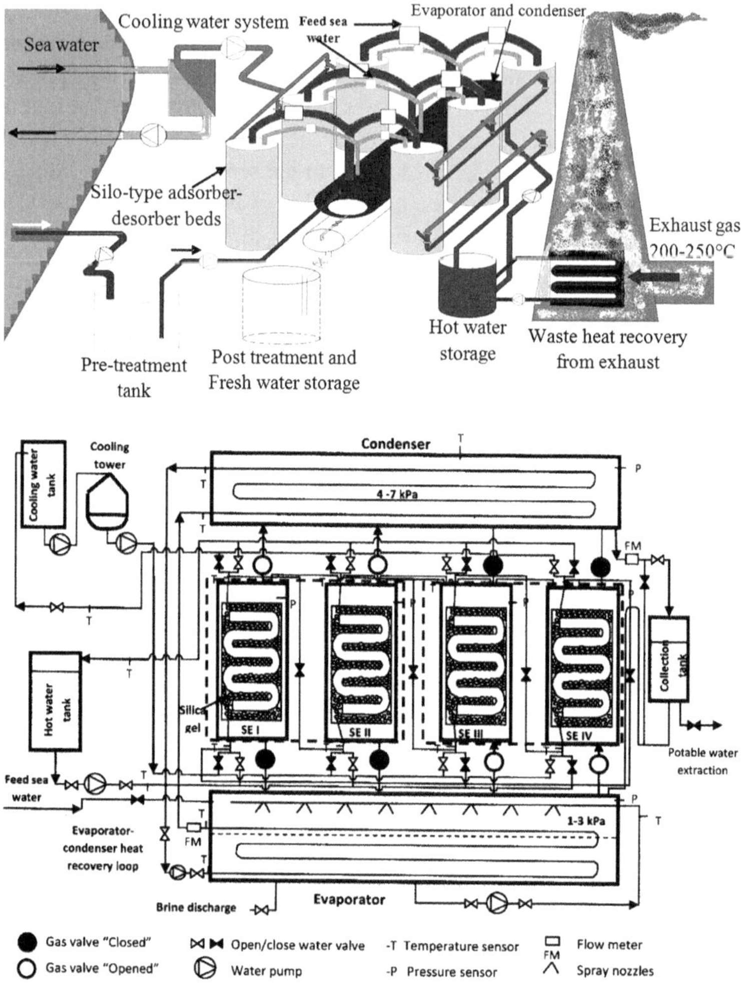

Fig. 27.6 Adsorption desalination process diagram with close-up of the adsorption/desorption beds (from Thu et al. 2010)

The AD cycle requires that seawater first be de-aerated prior to its feed into an evaporator. De-aeration removes any volatiles that are present. The evaporator, normally at saturation temperature and pressure, enables desalting to occur by a boiling process in which vapor is generated at the warmer tube surfaces. The tubes are supplied with an external coolant, emanating either from a chilled-water recirculation loop of an air-conditioning unit or a cooling tower. Water vapor from the evaporator is adsorbed onto the surfaces or pores of the adsorbent with the adsorber bed, such as silica gel. Concomitantly, heat is supplied to the desorber bed to drive out the water vapor from the pores in the silica gel. Desorbed vapor

Fig. 27.7 Photograph of the research adsorption desalination plant at the King Abdullah University of Science and Technology, Thuwal, Saudi Arabia

moves to the cooler condenser tube surfaces when the connecting valves are open. Cooling tower water (such as seawater or brackish water cooling tower) is supplied to the condenser for heat rejection, a feature that is required of all thermodynamic machines. Hence, it is observed that with only one low temperature heat source input, two useful effects are obtained from the AD cycle, namely the cooling effect at the evaporator and the production of potable water. This batch-operated cycle has a half-life cycle interval of 5–10 min depending on the temperature of the heat source, type of heat recovery and the number of pairs of adsorber and desorber beds in the plant.

The total primary energy, comprises the summation of the thermal and electrical input normalized respectively by the boiler and grid efficiencies, required by the AD process to convert seawater to freshwater is 39.8 kWh/m^3, which is comparable to other thermal conversion methods and higher than reverse osmosis. Since AD uses either primary waste heat, solar heating, or geothermal energy, which are deemed to be free of charge, the cost of seawater conversion to freshwater is merely the cost of electricity for pumping the heating and cooling fluids in the cycle. A life-cycle cost is estimated to be \$0.30/m^3. Hence, the AD technology has not only an edge in cost reduction of seawater to freshwater conversion; it is also environmentally friendly because the carbon dioxide emission of the cycle is much lower when compared with other desalination methods. An estimate of the primary energy usage indicates that the AD cycle emits about 0.65 kg of CO_2 per cubic meter of product water whilst RO, MED and MSF are 6, 9, and 11 times higher respectively.

The key issue with regard to AD is the ability to scale up the technology to a capacity that would allow commercial use. It is likely that this technology will be initially coupled with other thermal desalination technologies to produce new hybrid systems that will produce desalinated water at a greater efficiency.

27.3 Membrane Desalination Processes

Use of membranes to desalinate water was invented in the late 1950s (Reid and Breton 1959) and has matured into a mainstream desalination process over the past 50 years. There are currently four membrane processes used to treat water of varying salinities and a fifth membrane technology that uses membranes in combination with electrical separation (electrodialysis and electrodialysis reversal). Of the four primary membrane processes (reverse osmosis, nanofiltration, ultrafiltration, and microfiltration), only the reverse osmosis process can desalinate seawater and brackish water. Nanofiltration is a very useful process, but removes primarily hardness, large cations and anions, and large organic molecules. The ultrafiltration and microfiltration processes are used to remove microorganisms and large organic molecules, which limits their use to fresh feedwater sources. The electrodialysis reversal process is primarily used to treat brackish water, especially where the water chemistry causes low efficiency of conversion by standard low pressure reverse osmosis (i.e., high sulfate concentration in the feedwater).

Reverse osmosis treatment of seawater and brackish water to freshwater is a very important process used to provide potable water to many arid countries in the Middle East and water-short semi-arid countries in other parts of the world. Of the various commercial desalination technologies currently used, reverse osmosis is the least energy intensive (see Sect. 27.8). The use of the RO for seawater desalination has been increasing relative to thermal desalination for the past 2 decades. In regions where the feedwater quality is difficult to treat in terms of high dissolved solids concentrations, such as the Arabian Gulf with a salinity range of 42–55 parts per thousand, reverse osmosis desalination is being collocated with a thermal process to allow mixing of the treated waters to obtain an overall lower cost of water production. These facilities are known as hybrids (see Sect. 27.4).

27.3.1 Seawater Reverse Osmosis

When fluids of differing salinity are placed adjacent to a semi-permeable membrane, flow occurs from the low salinity fluid to the high salinity fluid. This natural process is termed osmotic flow or osmosis. Membrane desalination occurs when the natural osmotic flow is reversed by pressurizing the high salinity side of the membrane (Fig. 27.8). The reverse osmosis (RO) process desalts seawater or brackish water by creating sufficient pressure to overcome the natural osmotic pressure of the concentrated solution and forcing freshwater to pass from the seawater (feed) side of the membrane to the freshwater side (discharge). Therefore, the amount of pressure required for desalination to occur is dependent on the salinity of the raw water being treated. The osmotic pressure (P_{osm}) can be calculated for any concentration of dissolved solids in solution using the van Hoff equation

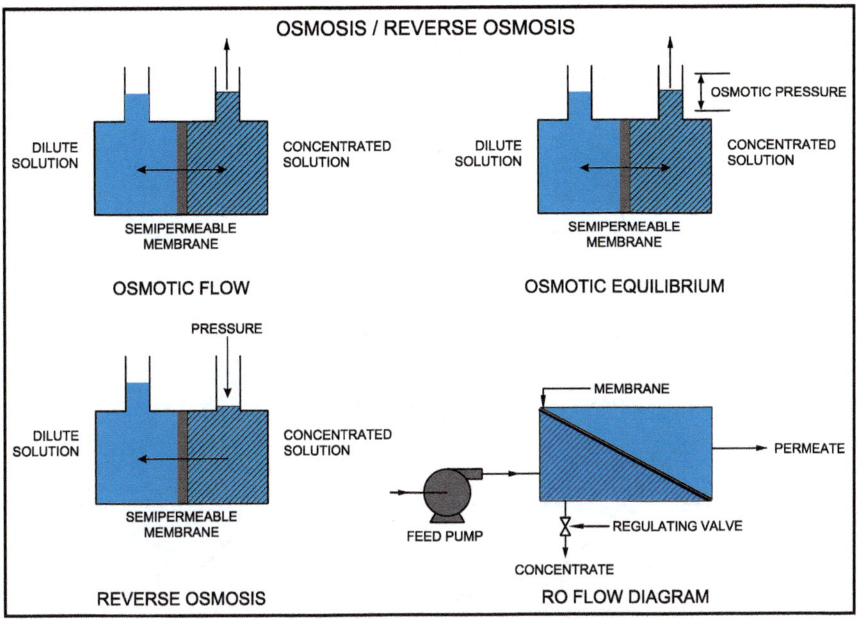

Fig. 27.8 Diagram showing the processes of osmosis and the general reverse osmosis treatment process (from Missimer 2009)

$$P_{osm} = R(T + 273) \sum m_i \qquad (27.1)$$

where,

R = universal gas constant (0.082 L·bar/mol °K)

T = temperature (°C)

m_i = molar concentration constituent "i" (mol/L) in solution (ions and uncharged species)

For example, the osmotic pressure of normal seawater with a total dissolved solids concentration of about 35,000 mg/l is about 27 bars (2,700 kPa or 385 psig). However, the optimal operational pressure for a membrane plant must take into consideration not only the required osmotic pressure to pass water through the membrane, but also the friction loses and required flows in the membrane, vessel, and connecting lines. A typical operating pressure for a seawater membrane system ranges from 55.2 to 70.3 bars (800–1000 psig).

Seawater RO systems can be designed as either single-pass or two-pass systems depending on the chemistry of the raw water and the specific rejection criteria set of the facility. Two-pass systems are quite common, especially to meet the boron rejection requirements (Fig. 27.9; Redondo et al. 2003; Xu et al. 2010). A majority of seawater RO plants use the spiral-wound membrane configuration (Fig. 27.10) with a few facilities using hollow-fiber membranes (e.g., Jeddah, Saudi Arabia).

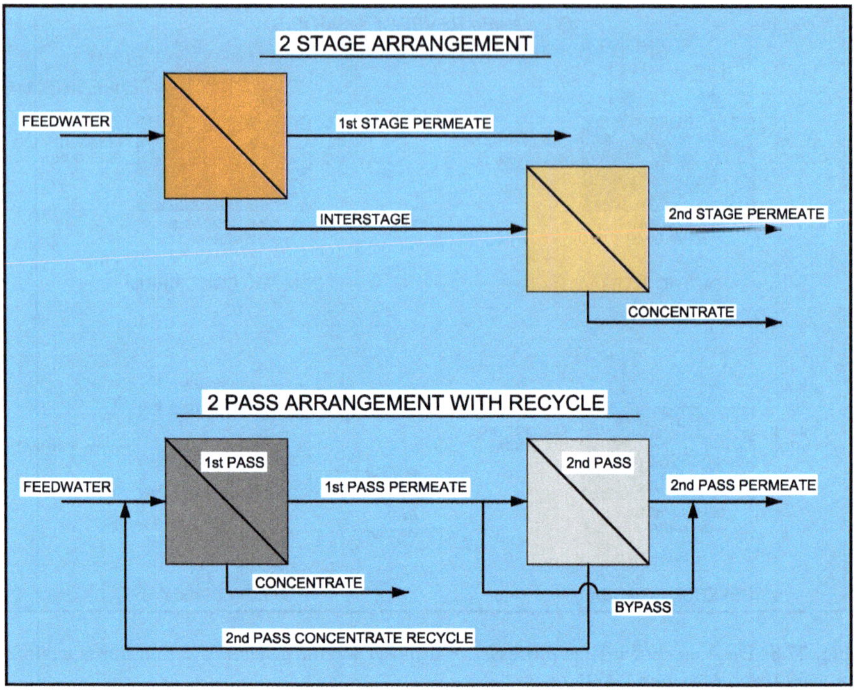

Fig. 27.9 Diagram showing the stages and passes for a sweater RO system (from Missimer 2009)

RO is rapidly becoming the most popular process used to desalinate seawater, because primarily of the lower energy consumption and cost for potable water production. The total primary energy required to desalinate seawater by RO is 14.29 kWh/m^3 compared to the thermal processes that range from 31.71 to 57.14 kWh/m^3 (calculated by Dr. Kim Choon Ng, King Abdullah University of Science and Technology and National University of Singapore)). However, the cost of seawater RO is significantly affected by the quality of raw water and the persistent problem of membrane biofouling. Biofouling reduces the life-expectancy of the membranes and adds operating cost.

A significant part of the capital and operating costs for seawater RO involves the construction and operation of the pretreatment systems. Raw seawater contains large biological organisms, debris, silt and clay, microscopic organisms (algae and bacteria), and dissolved organic compounds. Most of this material must be removed from the feedwater before it enters the RO treatment process, otherwise the membranes can become fouled. Membrane fouling causes reduction in treatment capacity as less water can pass through a membrane at a given operational pressure. As flux is increased to compensate for the biofouled membrane, the quality of the produced water can deteriorate. Therefore, the pretreatment process

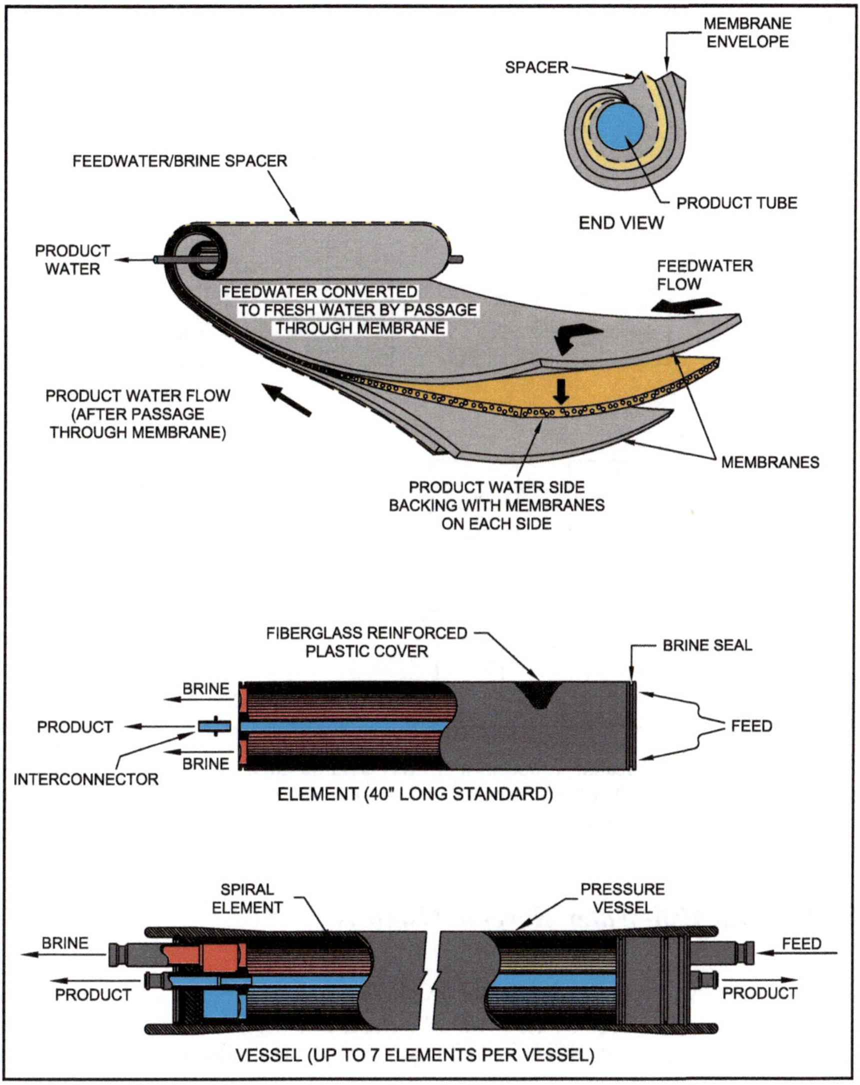

Fig. 27.10 The spiral-wound membrane configuration and pressure vessel (from Missimer 2009)

trains can be complex and expensive to operate. A series of four different pretreatment process trains are shown in Fig. 27.11. One of the first three processes is being used in most seawater RO plants today. The great costs of pretreatment can be reduced by improving raw water quality by using subsurface intakes instead of open-ocean intakes (Missimer 2009; Missimer et al. 2010; Sect. 27.7.3).

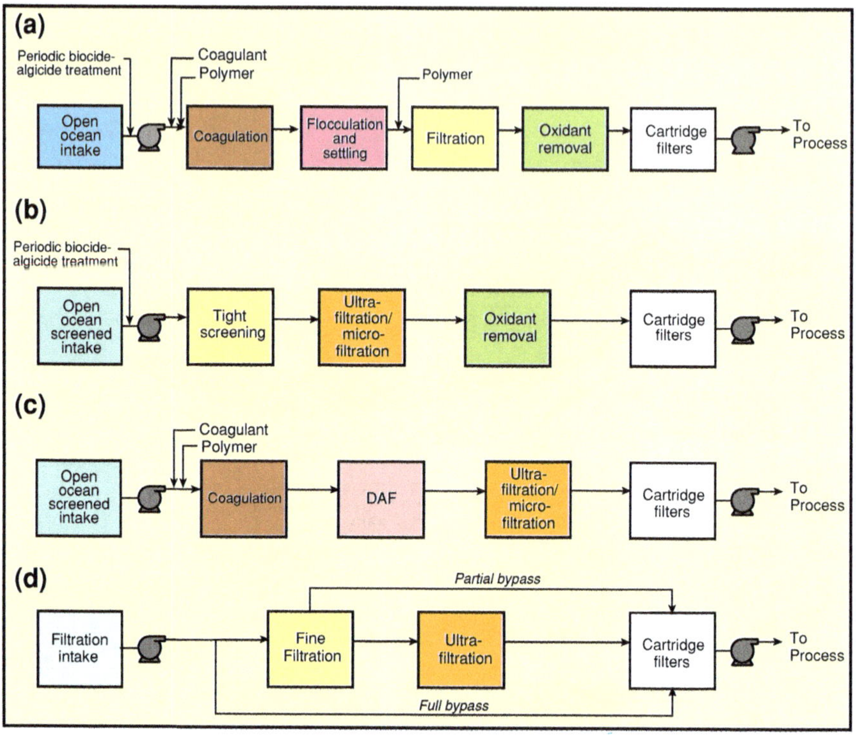

Fig. 27.11 Various pre-treatment process trains for based on the quality of the raw feedwater (from Missimer et al. 2010). **a** Conventional pretreatment **b** alternative pretreatment **c** alternative pretreatmnt **d** alternative subsurface intake

27.3.2 *Brackish-Water Reverse Osmosis*

The seawater RO market is dominated by the Middle East region where there is a considerable need to create "new water". However, the use of RO desalination technology to treat brackish water has been dominated by the United States. RO desalination of brackish groundwater has been implemented in Florida, the Carolinas, the mid-continent (Kansas and Colorado), Texas, and the arid regions of the Southwest (Arizona and New Mexico). Interest in brackish groundwater desalination is growing rapidly as many areas are at the limits of sustainable freshwater resource utilization, but have abundant brackish-water resources.

The primary water supply for brackish water RO is groundwater with a range of salinities from 1,000 to 12,000 mg/L. Cost and energy consumption to treat brackish water by RO are considerably less than for seawater. Therefore, numerous investigations are being conducted to assess brackish groundwater resources in all parts of the world. Many arid regions contain abundant supplies of brackish groundwater that are either in the initial stages of development, such as the lower part of the Hueco

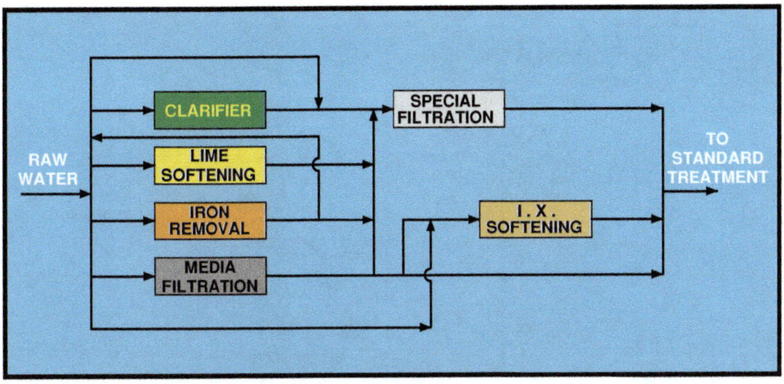

Fig. 27.12 A schematic diagram of brackish-water RO pretreatment (from Missimer 2009)

Bolson Aquifer near El Paso, Texas (Sayre and Livingston 1945; Groschen 1994; Sheng and Devere 2005), or have yet to be fully investigated. It is estimated that the state of Texas has approximately 3,330 km^3 (2.7 billion acre feet) of brackish groundwater resources (Texas Water Development Board 2007).

Development of brackish-water aquifers as supplies to RO facilities requires considerable scientific evaluation, because the aquifers tend to yield water with a progressively greater salinity over time (Missimer 2009; Maliva et al. 2011). Raw water chemistry needs to be predicted over the operational life of the desalination system, which requires sophisticated density-dependent solute-transport modeling. Brackish groundwater, in certain cases, may constitute a non-renewable resource, such as in Bahrain (Zubari 2002).

Treatment of brackish water using the RO process is generally less complex than seawater treatment. The process train includes either standard or special pre-treatment, the membrane process, and post-treatment (Fig. 27.12). Since the primary source of feedwater is groundwater, the pre-treatment process commonly involves only pH modification or addition of an anti-scalant (or both). Additional pre-treatment process can be required if the feedwater contains high concentrations of iron, manganese, or silica. Brackish-water RO systems most commonly achieve conversion efficiencies ranging from 70 to 85% depending upon the salinity and overall water chemistry. Feedwaters that have high sulfate to chloride ratios may suffer lower conversion rates and alternative processes, such as electrodialysis reversal, may be more effectively used (Ian Watson, personal communication).

27.3.3 Electrodialysis Reversal

Another process used to desalinate brackish water is electrodialysis reversal (ER), which is a modern version of the electrodialysis process. The basic principal of this process is that ions in a solution will be attracted to electrically-changed anodes

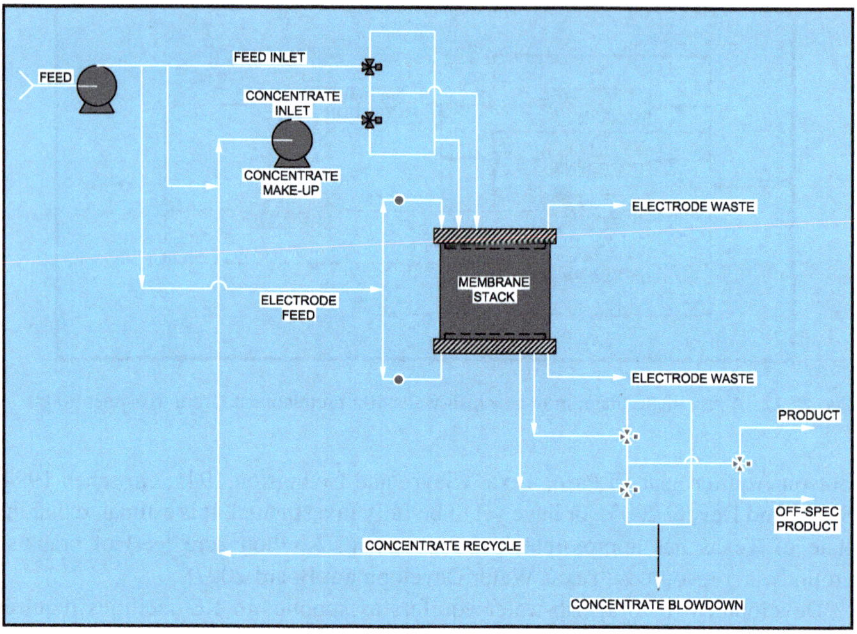

Fig. 27.13 A schematic diagram of a typical electrodialysis reversal process (from Missimer 2009)

(+) and cathodes (−) within a stacked membrane system. The poles are reversed to reduce scaling. A schematic of the ER process is given in Fig. 27.13.

The ER process has been used in many types of industrial applications and, with the new improvements in the process efficiency, it has become more economical to use in larger-scale potable water applications. ER can produce very high treatment conversion efficiencies up to 94%, such as is achieved in the Suffolk, Virginia facility (Missimer 2009). The disadvantages of using this process include the need to remove iron and manganese (when present in the feedwater) in a pretreatment stage, a larger floor area requirement compared to RO, and that it achieves virtually no organics removal. The advantages are a potentially high rate of water conversion, lower operating pressure, and the ability to treat unusual feedwater qualities, particularly those with high sulfate to chloride ratios, which have a high potential for $CaSO_4$ scaling chemistry in the RO treatment process. ER has a role for arid lands desalination, particularly in small systems with unique water quality issues.

27.3.4 Forward Osmosis

Osmosis is the natural process involving the potential fluid flow from high salinity to low salinity across a membrane. It occurs in many natural systems including the human body. For many years, desalination technology has focused on the use of membranes to overcome the natural osmotic pressure of a fluid by applying pressure to a high salinity solution (i.e., seawater or brackish water) and forcing it through a membrane to leave the salt behind and recover the freshwater. This reverse osmosis technology is highly developed and is a mainstream desalination process.

Forward osmosis is a process that is being used in many applications today and is being investigated for use in desalination (Cath et al. 2006). All applications of forward osmosis involve using a draw solution of higher salinity on one side of a membrane and a fresher fluid on the opposite side of the membrane. The osmotic gradient pulls the fresher fluid into the higher salinity fluid. To achieve desalination, the draw solution must be treated (such as by heating) to separate the freshwater.

Recent bench-scale research has shown that a draw solution formed by mixing of ammonium carbonate and ammonium hydroxide in specific proportions can be used to desalt seawater (McCutcheon et al. 2005, 2006; McGinnis and Elimelech 2007). It was shown that a driving force of 238 bar could be generated with a salt solution of 0.05 M NaCl and 127 bar for a salt solution of 2 M NaCL, which is similar to concentrate generated from conventional desalination (McCutcheon et al. 2005). Another draw solution using ammonia and carbon dioxide can desalt seawater with salt rejection at 95% and fluxes as high as 25 $L/m^2/h$ using an FO CTA membrane (McCutcheon et al. 2006).

Forward osmosis desalination must be considered to be an emerging technology that has promise to produce cost-effective water treatment, perhaps ultimately at lesser expense for some applications compared to existing thermal and membrane methods. Cath et al. (2006) suggest that two significant limitations currently constrain use of the technology. These constraints are the lack of high-performance membranes and the necessity to develop a draw solution that would allow for easy separation of the product water. High recovery percentages will require that the draw solution has the properties to create an osmotic gradient sufficient to efficiently move water through the membrane. It is likely that forward osmosis will be used in concert with some other desalination processes in the future.

27.4 Hybrid Facilities

Over the past few decades, there has been a realization that the combined use of several desalination processes coupled with power generation can significantly improve the efficiency of potable water production. These combined systems are termed hybrids or hybrid systems. Perhaps the simplest and first of the hybrid systems implemented was the combination of power and thermal distillation.

Combining water treatment and power generation is a logical and economic concept based on the realization that it takes a considerable amount of power to generate the heat necessary to desalt seawater using any thermal process. An obvious synergy is to use waste heat from power generation facilities to heat the feedwater for thermal desalination facilities. Also, while treated water can be stored for later use, electricity in direct form cannot. The integration of the power and water cycle produces efficient use and recycling of energy. Awerbach (2007) suggested that the key design issues for a water/power hybrid system are:

- Seasonal demands for electricity and water,
- Power-to-water ratio,
- Minimization of fuel consumption and increasing power plant efficiency, and
- Minimization of the environmental impact of carbon dioxide including potential consideration of a CO_2 tax credit.

The reason that the first hybrid systems, including co-located and operated electric power generation and desalination facilities, were developed involves the simple principal of recovering waste heat from the power generation process and recycling it into any of the thermal desalination processes reduces cost. This recycling and heat flow process for each of the primary thermal desalination processes is described in Sects. 27.2.1, 27.2.2, and 27.2.3. A large amount of heat is wasted during the power generation process, particular for cooling within the power plant, which is put to beneficial use in desalination.

Co-located desalination and power plants may have an additional synergy in that surface water intakes and outfalls can be shared. The much greater cooling water flows of power plants provides for substantial dilution of the desalination concentrate.

27.4.1 Combined Power Generation, Multistage Flash Distillation, and Reverse Osmosis

A relatively new hybrid system concept involves combining power generation with both multi-stage flash distillation and reverse osmosis treatment. This combination not only saves energy and operating costs, but helps solve another problem involving the production of potable quality water from high salinity feedwater, such as those in the Arabian Gulf and Red Sea. Normal seawater has a dissolved solids concentration of about 35,000 mg/L, but the Arabian Gulf locally has salinities in the range of 40,000 mg/L to over 55,000 mg/L (e.g., Beltagy 1983). The Red Sea has an average salinity of about 40,000 mg/L. The rejection rate of a standard two-pass seawater RO system using these high salinity waters produces product water with a dissolved chloride and total dissolved solids concentration above the drinking water standards. Therefore, combining the product water from MSF, which is nearly pure water (at about 25 mg/L TDS), and the product water from RO treatment, produces a blend that meets the drinking water standards for TDS and chloride. The water blend is then post-treated to add hardness for health purposes.

This hybrid combination produces a number of significant advantages over stand-alone systems.

According to Awerbach (2007, p. 406–407), these advantages are:

- A common, considerably smaller seawater intake can be used.
- Product waters from the RO and MSF plants can blended to obtain a suitable water quality,

- A single pass RO process can be used,
- Blending distilled water with membrane product reduces the strict requirements for boron removal by RO,
- The useful RO membrane life can be extended,
- The feedwater temperature to the RO plant is optimized and controlled by using cooling water from the heat-reject section of the MSF/MED or power plant condenser,
- The low-pressure steam from the MSF plant is used to de-aerate or use de-aerated brine as a feedwater to the RO plant to minimize corrosion and reduce residual chlorine,
- Some components of the seawater pretreatment process can be integrated,
- One post-treatment system is used for the product water from both plants, and
- The brine discharged-reject from the RO plant is combined with the brine recycle in the MSF.

There are several different configurations of the combined power and treatment processes that may be used to obtain a more efficient scheme to produce potable water. Awerbach (2007) termed these methods as the "classic" scheme and the alternative to the classic scheme. Some of the largest desalination facilities in the world are hybrid power-MSF-RO facilities, such as those located at Jubail, Saudi Arabia and at Fujairah, UAE (Fig. 27.14).

27.4.2 Combined Power Generation, Multistage Flash Distillation, and Nanofiltration (and Reverse Osmosis)

The combined use of power generation, MSF, and nanofiltration (NF) is becoming an important hybrid innovation. It is likely that reverse osmosis water treatment will be added to this hybrid to make efficiencies even greater.

NF is a process that primarily removes hardness and large organic compounds from water. The addition of the NF process to treat the feedwater to a thermal desalination plant allows the facility to run at a higher temperature and with greater salinities in the recycled brine without causing scaling (Hamed 2006). Using NF to pretreat marine RO feedwater will also have operational benefits, particularly a potential reduction in the biofouling rate. Therefore, the power generation-MSF-NF and the power generation-MSF-RO-NF desalination hybrids can both have a significant impact on treatment efficiency and cost.

27.4.3 Combined Power Generation, Multiple-Effect Distillation, and Reverse Osmosis

The power generation-MED-RO hybrid operates in a very similar manner to the MSF-RO hybrid system. There are some differences in the routing of the heated

Fig. 27.14 Photograph and diagram of the hybrid power plant-MSF-RO plant at Fujairah, UAE (from Tom Pankratz)

water based on the general differences in the operations of the two types of thermal processes. It is also possible to treat the overall system feedwater using NF to again increase efficiency by allowing a higher operating temperature for the MED process.

There are a few of these small hybrid systems in operation within the United Arab Emirates. The new Fujairah II-UAE IWPP facility will incorporate power production with 454,600 m^3/day (100 MIGD) of MED and 136,400 m^3/day (30 MIGD) of RO.

27.4.4 Combined Multiple Effect Distillation and Adsorption Desalination

Perhaps the most interesting possible new hybrid system would be a combination of power generation with MED and AD desalination technologies (Ng et al. 2011). The waste heat from the power generation plant and the recycled heat within the MED units can be combined to create, perhaps the most energy-efficient desalination system developed to date. The combined technology could bring the cost of seawater desalination down to about \$0.26/m^3, which is considerable less than any seawater desalination system operating in the world today. In addition, owing to the manner in which the cycle is designed, it minimizes the effects of scaling and corrosion significantly as the salinity of the seawater feed increases in the direction of decreasing stage temperatures. The AD component could also be used simultaneously to produce cold water from the second cycle. This added feature could be used onsite to run the HVAC system required to maintain various electrical control units and indoor facility rooms cool during the hot, arid months. This combination would lower the overall combined operating costs for the site by an even greater amount. The AD technology can operate using any heat source, including alternative sources, particularly solar and geothermal sources.

27.5 Alternative Energy Use in Desalination

All desalination technologies using thermal or membrane processes are energy intensive. Most of the energy consumed for producing freshwater from seawater or brackish water originates in electric power plants using hydrocarbon fuels for generation. Therefore, most desalination plants have a very high carbon footprint. The overall largest contribution to anthropogenic carbon dioxide emitted to the atmosphere originates in electric power production processes (Intergovernmental Panel on Climate Change 2007) and utilities use a large portion of the generated power. Use of alternative energy sources for desalination is rapidly becoming a significant option for reducing environmental impacts. Currently, the percentage of desalination facilities using renewable energy sources is only about 1% of the total global capacity (Delyannis 2003). Alternative-energy seawater desalination processes were reviewed by Manwell and McGowan (1994) and Kalogirou (2005).

There is currently considerable interest and research on the combination of desalination with renewable energy sources (RESs). Both solar and wind energy can be used to power desalination systems. RES-powered desalination have been demonstrated to be a viable option for some sites that are "off the grid" and thus, lack ready access to electrical power and fossil fuel supplies. However, under current energy prices, RES-powered desalination is typically more expensive that desalination powered using conventional energy sources, but the cost differential is decreasing as RES systems are becoming more efficient and conventional energy costs increase.

The cost of desalination systems, in terms of the unit cost of water (C_W), is the sum of the amortized capital cost ($C_{capital}$), energy costs (C_{energy}) and operational and maintenance costs ($C_{O\&M}$) dividing by the water production (V_w). Unit costs are commonly expressed as US dollars or Euros per cubic meter or 1,000 gallons.

$$C_w = \left(C_{capital} + C_{energy} + C_{O\&M}\right)/V_w \tag{27.2}$$

In most circumstances, desalination using renewable energy sources currently cannot economically compete with systems operating with conventional energy sources under current energy prices (Suri et al. 1989; García-Rodríguez and Gómez-Comacho 1999; Agha et al. 2006). RES based systems become more economically viable in situations where conventional energy sources are not readily available or are expensive. Trieb (2007) estimated that solar power currently has an equivalent cost of about $50/barrel of fuel oil and that future improvements in technology and scaling caused by the operation of larger plants could reduce the cost by up to 50%. At the current cost of crude oil of roughly US $100/barrel, RES desalination technologies have become more cost-competitive with other conventional desalination technologies.

There are two basic types of renewable energy-powered desalination systems; (1) systems in which the renewable energy supply and desalination are inexorably integrated and (2) electrically powered systems in which renewable energy is used primarily as a cost savings to grid power or other sources of electrical power. An example of the latter is a wind-powered RO system in an area connected to the electrical grid (e.g., Kwinana Desalination Plant, located near Perth, Western Australia). Reverse-osmosis desalination and wind-powered electrical generation are separate processes that are linked primarily by economic considerations. Wind-powered RO makes economic sense if the cost of the wind generated electricity is less than the cost of electrical power provided by conventional generation facilities. The cost of electrical power should ideally consider externalities, such as the impacts of greenhouse gas emissions (carbon footprint). The same economics restraints apply for RO systems powered by electricity generated from solar energy. Integrated systems, on the contrary, use renewable energy (particularly solar) to heat water as part as the thermal distillation process.

The future importance of RES powered desalination for water supply in semiarid and arid regions will be great, because these regions have an abundance of solar energy. Trieb (2007) and others have pointed out that the use of concentrated solar thermal power (CSP) to power seawater desalination is a rather obvious approach to solving water scarcity problems in the MENA region, which has outstanding potential for solar power. Each square kilometer of land in the MENA region receives each year an amount of solar energy that is equivalent to 1.5 million barrels of crude oil (Trieb 2007). Deficits in sustainable water resources in the MENA region are being covered by seawater desalination and non-renewable groundwater resources. There are no other sustainable and affordable alternatives to CSP powered seawater desalination in the region (Trieb 2007), other than perhaps nuclear energy-powered facilities. The key issue to meet projected increases in future water demands is the speed of market introduction of the concept and implementation of additional water use efficiency measures (Trieb 2007). Desalination systems can either have a dedicated RES system or can be provided electrical power off of a RES-powered grid.

Another alternative power source for desalination is nuclear energy. There is a considerable literature base on the use of nuclear power for desalination and the possible co-location of nuclear power and desalination facilities. There is a considerable compatibility between the waste heat generated during cooling of the reactors and the operation of any one of the thermal desalination processes. Although nuclear energy facilities may not be currently popular, the potential hybrid use of nuclear generation and thermal desalination cannot be ignored, particularly in light of the low carbon footprint of nuclear power.

27.5.1 *Solar-Powered, Direct and Indirect Solar Desalination*

Most arid land areas have a high average temperature and low levels of cloud cover, thereby, making them ideal for the development and use of solar energy systems. Direct desalination of seawater and brackish water has been accomplished on a small scale for decades using the general concept of solar stills and solar ponds. Indirect use of solar-powered electrical generation coupled with desalination processes is also quite feasible in arid areas. For example, the average thermal radiation at a latitude of 22° N in Saudi Arabia is about 22.5 $MJ/m^2/day$ based on between 9 and 13 h of sunshine per day. The corresponding thermal energy rating is 1,250 $kWh/m^2/year$ compared to the humid country of Singapore, located on the equator, which produces 965 $kWh/m^2/year$ (personal communication, Dr. Kim Choon Ng). In either case, the quantity of solar energy available for use is an extremely high number.

Solar-powered desalination options have been reviewed by Glueckstern (1995); Abu-Jabal et al. (2001); Kalogirou (2005); Agha et al. (2006), Rizzuti et al. (2007), and Tiwari and Tiwari (2008). Solar desalination systems can be either direct or indirect systems. Direct systems are single systems that use solar energy to distill seawater, brackish water, or contaminated freshwater. Indirect systems involve two or more subsystems, particularly a subsystem that collects solar energy and a subsystem that performs the actual desalination. In indirect systems, a traditionally used energy source is replaced by solar energy or another renewable energy source.

Four main options exist for the use of solar energy in desalination of seawater and brackish water:

- Direct distillation (solar stills),
- Thermal energy collection systems that are a source of hot water for distillation systems,
- Thermal energy collection systems that generate electrical energy to power desalination systems, and
- Photovoltaic systems that use solar energy to generate electrical energy used to power a desalination system.

A basic issue for solar energy systems is that they produce energy only during daylight hours. Also, solar energy (insolation) seasonally varies, and production can be reduced by cloud cover. Solar-powered desalination systems either operate only during daylight hours or energy storage is required to allow the system to operate continuously. Photovoltaic systems could be provided back-up electrical power by either connecting to the electrical grid or the provision of a convention fuel-powered generator.

27.5.1.1 Direct Distillation Using Solar Stills

Various forms of solar-still type desalination have been used for centuries. Malik et al. (1985) reported that the first use of solar desalination was documented by

Mouchot in 1869. He suggested that Arab alchemists used polished Damascus mirrors for solar distillation in the fifteenth century. The ancient Greeks and Romans also used fire alembics to produce a distillate (Kalogirou 2005). A number of solar distillation methods were used in the late 1500s in Europe and in 1774 the French chemist Lavoisier developed a solar still that used large glass lenses to concentrate sunlight onto flasks to cause evaporation and then collected the condensate. The first American patent on solar distillation was filed by Wheeler and Evans in 1870. A solar distillation plant was built in Las Salinas, Chile in 1872 by Carlos Wilson. This plant produced 22.7 m^3/d (6,000 gal/d) of fresh water and was operated for 36 years (Kalogirou 2005). A number of new concepts in solar distillation were developed in the twentieth century with a large number of patents filed.

The Office of Saline Water (OSW) was founded in the United States in 1952 to investigate the science and technology of saline-water conversion. A series of five solar still demonstration plants, funded by OSW, were constructed, including one at Daytona Beach, Florida (Talbert et al. 1970). Kalogirou (2005) documented the construction operation of solar distillation plants on four Greek islands constructed between 1965 and 1970. These plants had capacities between 2,044 and 8,640 m^3/day (540,000 and 2,280,000 gal/d). A number of additional solar distillation plants have been constructed and used on other islands, such as Santo Porto and Madeira, Portugal and in India. Agricultural irrigation from solar stills is used in Jeddah, Saudi Arabia at a medium to small scale.

The great advantage of soil stills is that they have low costs, low maintenance, and are energy independent. A major disadvantage of solar stills is that they have low yields. The production rate of traditionally designed solar stills is proportional to the system area. The cost of water produced depends predominantly on the capital investment to construct the systems, since there is minimal or no energy requirements. Unfortunately, there is little, if any, economy of scale with solar still systems. However, solar stills may be the most economic option for low capacity (<200 m^3/d, 53,000 gal/d) desalination systems (Kalogirou 2005). Solar stills can also be a survival strategy for small-scale, emergency water supplies, but are not a viable option for large-scale water supply.

There is on-going research concerning how to increase the output of solar distillation systems so that they can become a viable source of household or small community water supply. Schwarzer et al. (2010), for example, described a modular system consisting of a thermal collector and a multiple-stage desalination tower that can produce yields of approximately 35 L/day (9.2 gal/d) per unit. Although such technology is simple and low maintenance, the critical issue is capital cost.

27.5.1.2 Indirect Desalination Using Solar Ponds

Thermal systems use solar energy to either directly heat water or thermal oil coupled with a heat-exchanger. A variety of configurations are possible. Solar ponds and collectors have been used for thermal energy collection, and then coupled with a heater exchanger to allow operation of MED and MSF systems for desalination of feedwater. MED has the advantage for coupling with solar-powered distillation because it operates at lower temperatures, has lesser scale formation, and has stable operation between virtually zero and 100% output (Kalogirou 2005; Agha et al. 2006). MSF systems have operational flexibility for varying steam supplies. The new AD technology has a much greater flexibility for use with solar collectors and would operate at the lowest energy consumption of all of the thermal desalination processes.

Salt-gradient solar ponds are a simple technology for capturing of solar energy. Solar energy is transferred to water as heat. As the density of water decreases with temperature, the heated water will tend to rise to the surface of water bodies. A key design is the suppression of the natural convection, which would result in a loss of the captured solar energy to the environment. The density of water increases with salinity. Surface-water bodies thus tend to be become density stratified with salinity increasing with depth. A key feature of salt-gradient solar ponds is that the increase in density caused by increasing salinity is greater than the decrease in density caused by increasing temperature. The result is that heat is not lost from the bottom of the ponds through upwards convection and the water progressively increases in temperature with depth, in some instances reaching values over 80°C. The heated water can be used for a variety of purposes requiring hot water, such as direct heating, electrical generation, and desalination.

The advantages of solar ponds include

- It is a "green" technology and negligible external energy is required.
- It is technologically simple.
- It provides heat storage; heat for desalination can be obtained day and night.
- Systems are readily scalable and large capacities are possible.
- Generated salt could be sold.

Solar ponds require a source of saline water to replenish the system and work best in tropical areas with abundant sunlight. They could be coupled with the desalination process by placement of the concentrate into the ponds to produce larger amounts of energy and as replacement water. The systems should be designed so that the ponds do not contaminate underlying aquifers. An impervious liner is typically used to prevent contamination and the loss of heated brines. Solar-powered desalination has been studied for many years at El Paso, Texas, where a solar pond system was operated for 16 years. Lu et al. (2001) reported that distillate was produced at a test rate of 1.63–5.0 L/min (619–1,900 gal/d) using a multi-effect, multistage distillation unit.

27.5.1.3 Indirect Desalination Using Solar Collectors

Solar thermal collectors are devices that collect heat by absorbing sunlight. There is great interest in solar thermal collectors and it is not possible to describe all the designs developed. Technology varies greatly from industrial and utility-scale facilities employing high technology to "backyard" systems that can be constructed by an individual for a modest cost.

There are two main types of solar thermal collectors, non-concentrating and concentrating. Non-concentrating collectors use a dark-colored surface to absorb sunlight. A commonly used and relatively inexpensive type of non-concentrating collector is the flat plate collector (Fig. 27.15). A heat transport fluid can be water flowing through tubes in the stationary absorber that provides heat. The heated fluid can be used directly or through a heat exchanger. Roof-top, flat plate-type solar thermal collectors are widely in parts of the United States for heating swimming pools. An alternative non-concentrating collector design is evacuated-tube collectors, which consist of parallel rows of round or oval-shaped glass tubes connected to a header pipe. Each tube has the air removed from it to eliminate heat loss through convection and radiation, which allows for greater efficiencies and higher temperatures to be achieved. Evacuated tube collects are substantially more expensive than flat plate collectors. However, their rounded shape reduces the accumulation of dust, which reduces maintenance costs in arid regions. Solar collectors are used in concert with AD technology to produce the most energy efficient desalination system currently known (see Sect. 27.2.4).

Concentrating solar collectors use mirrors or reflectors to focus sunlight onto collection tubes or pipes containing a coolant. A common design is parabolic dish and tube reflectors in which sunlight is concentrated at a point or horizontal tube. Other designs include the solar towers, where mirrors are used to concentrate sunlight at a point and funnel it through a lens system. Concentrating collectors can heat fluids to very high temperatures. The fluid, which can be a thermal oil or liquid sodium, is sent to a boiler where it is used to generate steam. The heated fluid acts as both a transfer and storage medium.

A variety of different combinations of thermal collectors and desalination systems have been shown to be possible, and have been tested. For example, a system in Abu Dhabi, UAE used a combination of evacuated tube collectors and MED (El-Nashar and Ishii 1985; El-Nasher and Samad 1998; El Nashar 2009). A demonstration AD plant at the King Abdullah University of Science and Technology is being used for direct desalination of seawater and to produce cooling fluids. It utilizes a flat plate collector system mounted on the roof (Fig. 27.15). A parabolic trough collector (PTC) was used in a test system at a solar energy research center in Spain (Platforma Solar de Almería) to heat thermal oil, which was used to generate steam for a MED plant (García-Rodríguez and Gómez-Comacho 1999). A combination scheme using solar collectors and the MED was suggested for the Middle East region by Al-Karaghouli et al. (2009) (Fig. 27.16).

An important research focus on solar energy, in general, and solar-powered desalination is improving system efficiency and thus total costs. For example, the coupling of a heat pump to the MED resulted in a significant reduction in cost of water in the Platforma Solar de Almería system (García-Rodríguez and Gómez-Comacho 1999). The AD research work is currently focused on combining AD and solar collection is perhaps the most economical solution to large-scale desalination.

Fig. 27.15 Photograph of a solar flat-plate collector system at the King Abdullah University of Science and Technology, Saudi Arabia. The hot water generated on these plates feeds a pilot adsorption desalination system

27.5.2 Wind-Powered Desalination

There has been a very large global increase in the production of electricity using wind-powered turbines over the past two decades. The theoretical maximum aerodynamic efficiency of converting wind to mechanical power is 59% (Kalogirou 2005). There have been substantial improvements and innovations in wind power electric generation turbine technology. Ackermann and Söder (2002) have reviewed wind power and energy production technology.

Development of new, large-scale desalination facilities in technologically developed arid or semi-arid lands, such as Spain and Australia, has run into opposition from environmental advocacy groups and the general population, because of their high energy consumption and associated carbon footprint. Concerns over the development of conventional power plants to produce the energy required to desalinate seawater has delayed or curtailed development of new desalination facilities. A compromise solution to this opposition was reached in Western Australia, allowing the construction of the Kwinana Desalination Plant (previously discussed). A wind-power power facility was developed to offset the energy demand of the new desalination plant, thereby reducing its carbon footprint. The reverse-osmosis desalination plant and the wind-powered electrical generation are separate facilities and linked only via the electric power grid. Base-load power for operation of the RO plant during times when the wind turbines are offline is met by conventional power facilities.

Wind-powered RO makes economic sense if the cost of the wind generated electricity is less than the cost of electrical power provided by conventional

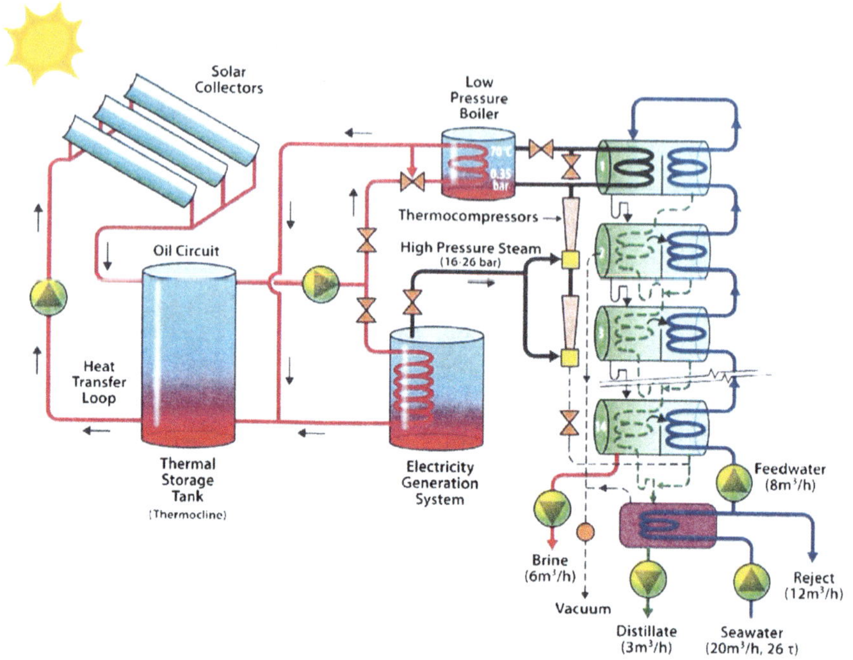

Fig. 27.16 Schematic diagram of a combined solar collection and MED system suggested for the Middle East region by Al-Karaghouli et al. (2009)

generation facilities. The cost of electrical power should ideally consider externalities, such as the impacts of greenhouse gas emissions (carbon footprint). While wind generation facilities do not always directly power the seawater RO plant, it adds renewable energy into the grid to offset the energy consumption of the desalination process. Direct wind-powered desalination may also be viable economic water supply options for areas off the main power grid. Direct wind powering of desalination plants at a demonstration scale has been documented by Petersen et al. (1979) and a few small operating sites by Harrison et al. (1996).

27.5.3 Geothermal Powered Desalination

In arid regions that have high natural heat fluxes, the use of geothermal energy to directly or indirectly power desalination facilities could be quite effective. There are several types of potential geothermal heat exchange systems that could be used with desalination including:

- Produced hot water from wells,
- Closed-loop, shallow exchange systems using feedwater for heat exchange,

- Thermal springs heat exchange systems,
- Dry deep well closed-loop heat exchange systems using a specific type of oil or other fluid pumped to a surface heat exchange system, and
- Deep well closed-loop system with circulating feedwater.

Bourouni et al. (1999a, b) demonstrated that naturally warm (70–90°C) brackish groundwater in Tunisia could be desalted using an evaporator and condenser system utilizing heat exchangers made of plastic. The cost of distilled water production was estimated to be 10–15% lower than other thermal treatment systems. A review of desalination using geothermal energy was presented by Goosen et al. (2010). They found that a large number of geothermal power plants have been developed in many areas in the world, including some in arid areas such as Ethiopia, semi-arid Greek islands, and a semi-arid region of China. Goosen et al. (2010, p. 1425) noted that

> The combination of a renewable energy source, such as wind, solar and geothermal, with desalination systems holds immense promise for improving potable water supplies in arid regions.

Karytsas et al. (2004) reported that a geothermal energy project on the Greek Island of Milos will utilize geothermal fluids with a temperature range of 300–323°C at depths between 800 and 1,400 m below sea level. A project was developed that will combine a MED/thermal vapor compression desalination plant with a capacity of 80 m^3/h (500,000 gal/d) and a power generating unit with a capacity of 470 kWe. Parts of Western Saudi Arabia have generally high heat flows and very hot temperatures can be expected at shallow depths in the crust, especially in the vicinity of Medina, where volcanic activity has occurred in the historic past. This region may be excellent for a geothermal power and desalination project.

An economic study on the use of a geothermal brine source in Israel with temperatures of 110 and 130°C showed that a low price for desalination could be achieved (Ophir 1982). This analysis considered only the conventional thermal desalination technologies available at that time. An analysis using the new adsorption desalination technology coupled with a geothermal heat source would show even lower potential desalination costs.

27.5.4 Nuclear-Powered Desalination

Nuclear desalination is a realistic and potentially cost-competitive means of providing fresh water in arid and semi-arid regions. The term "nuclear desalination" was defined by the International Atomic Energy Agency in 1996 as

> ...the production of potable water from seawater in a facility in which a nuclear reactor is used as the source of energy for the desalination process. Electrical and/or thermal energy may be used in the desalination process. The facility may be dedicated solely to the

> production of potable water, or may be used for the generation of electricity and the production of potable water, in which only a portion of the total energy output of the reactor is used for water production.
>
> In either case, the notion of nuclear desalination is taken to mean an integrated facility in which both the reactor and the desalination system are located on a common site and energy is produced on-site for use in a desalination system. It also involves at least some degree of common or shared facilities, services staff, operating strategies, outage planning, and possible control facilities and seawater intake and outfall structures. Non-nuclear desalination is understood to be the production of potable water in a facility in which a fossil-fuelled plant and/or the electrical grid is used as the source of energy for the desalination process.

It is clear from the definition that there are two possible configurations for nuclear desalination, which are: (1) direct heating of feedwater for desalination accomplished by a thermal process, such as MSF, and (2) indirect water production by use of nuclear-generated electricity to power any of the desalination processes. Ragheb (2010) suggested that from a thermodynamic perspective either a single-purpose or a dual-purpose combination of nuclear desalination could be considered, but a dual-purpose system does not necessarily yield a greater amount of thermodynamic efficiency. He found that a dual-purpose facility should be designed for the relative needs for water and electricity, because the water and energy production processes are thermodynamically competitive rather than complementary. The method used by Ragheb (2010) considers only the external energy losses and ignores the internal losses, which may capture only about 50% of the true energy losses (personal communication with Dr. Kim Choon Ng). However, the lesson learned is that some synergies between nuclear power generation and use of the waste heat in thermal desalination occur, but the design and processes employed must be carefully balanced and optimized to meet the collective requirements of electricity and water production.

There is experience in the use of nuclear desalination in Kazakhstan, India, Japan, Pakistan, and China (Megahed 2001; World Nuclear Association, 2011). Also, a number of feasibility studies have been performed on nuclear desalination in various arid regions, particularly the northern African coastal countries and the Middle East (Oak Ridge National Laboratory 1971a, b, c; Hedayat et al. 1977; Al-Mutaz 2001; Megahed 2001, 2008). The largest operational nuclear desalination facility was in Aktau, Kazakhstan, where a liquid metal-cooled fast reactor (BN-350) was used for multiple purposes, including supplying local industry and the population with electricity and powering an 80,000 m^3/day (21.1×10^6 gal/d) capacity desalination plant using the MED and MSF processes. The Aktau nuclear desalination system operated for a 26-year period from 1973 to 1999 (Muralev et al. 1997). The process used for this nuclear desalination facility is shown in Fig. 27.17.

A series of desalination plants have been used for many years at various Japanese nuclear power facilities. These plants have capacities from 1,000 to 3,000 m^3/day (264,000 to 793,000 gal/d) with the water being used primarily for facility process water (Goto 1997). In India, a hybrid nuclear desalination demonstration facility is being operated at the Madras Atomic Power Station at Kalpakkam.

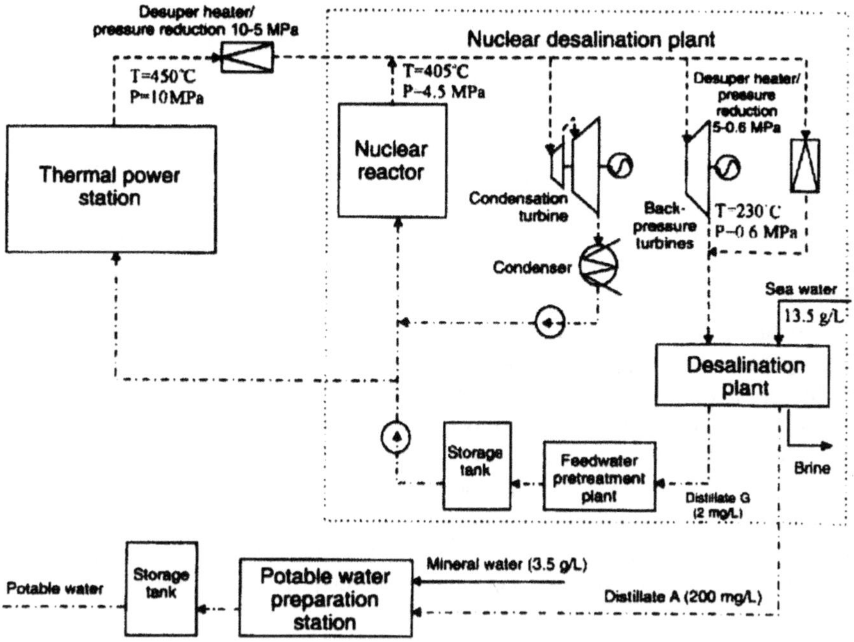

Fig. 27.17 A schematic diagram for the nuclear desalination facility operated at Aktar, Kazakhstan (from Muralev et al. 1997)

It contains a reverse osmosis unit with an 1,800 m^3/day (476,000 gal/d) capacity and an MSF unit with a 4,500 m^3/day (1.19×10^6 gal/d) (World Nuclear Association 2011). Another facility was installed at Kudankulam in 2009 to supply 10,200 m^3/day (2.69×10^6 gal/d) of process water to the new plant. Additional facilities have been commissioned in Pakistan (125 MW nuclear plant with 4,800 m^3/day, 1.27×10^6 gal/d MED plant) and in China at the Guangdong Nuclear Power Plant (10,080 m^3/day, 2.66×10^6 gal/d desalination facility) (World Nuclear Association 2011). Most of the existing or proposal new nuclear desalination facilities use some type of hybrid desalination system involving power plant generated heat and electricity with MSF, MED, RO, and various combinations of these processes (Al-Mutaz 2003; Faibish and Ettouney 2003).

Nuclear desalination remains an option to power desalination because it is an alternative energy source that has low carbon dioxide emissions. However, Megahed (2001) points out that there are a number of critical issues that require assessment before nuclear-powered desalination facilities become a fully viable option for large facilities. A significant issue is safety, which is a reactor issue rather than a desalination issue. In systems using stream from the reactor to drive a thermal process, additional monitoring for radiation would be required. The reactor safety issue is no longer a minor issue based on the recent nuclear disaster in Japan caused by an earthquake and a subsequent tsunami wave. However, newer generation nuclear power facilities have additional redundancies or fail-safe features, which greatly increase safety. Reliability of the combined facility is another issue because if the nuclear reactor must be shut down for any problem, whether it is for a safety or maintenance reason, there must be backup power or a process to allow water production to continue. Nuclear desalination must be demonstrated to be economically viable in comparison to other technologies and hybrids.

Although the combined process does produce electricity and water at competitive costs, the permitting, public approval, construction, and monitoring costs for operation of the nuclear reactor remain high and are somewhat unpredictable. Financing to design and build nuclear desalination facilities may be problematical due to the issue of uncertainty in the project delivery costs caused by the permitting and public approval processes. The issues of reactor security and non-proliferation also add to the issue of uncertainty. The final underlying issue is public acceptance. Nuclear reactor operation creates fear within many sectors of the public and strong opposition among other groups. Based on the current global political climate, nuclear desalination will continue to be investigated, but few large facilities are likely to be developed until many of the critical policy issues are resolved or until conventional power sources become unacceptable or limited due to resource exhaustion or environmental concerns (e.g., global warming).

27.6 Future Innovations and Energy and Cost Reductions

Desalination is a mature and economically viable industry that provides a "new water" source that is critical to future development and economically viability of arid regions. Although desalination processes have been improved considerably over the past 30 years, they still use a large amount of energy and are costly to build and operate. There are still many improvements that need to be made to reduce the energy consumption and cost. Great strides have been made in efficiency improvement, reduction in energy use, and systems cost reductions caused by development of new membranes, creation of thermal/membrane hybrid systems, and the entry of the private sector into competitive bidding to supply water. The challenge for the future will be to continue to make additional improvements in the existing components, but also to develop completely new technologies and improve the quality of the feedwater, particularly in seawater systems.

27.6.1 Membrane Pretreatment of Distillation Treatment Plant Intake Water (and RO in Hybrid Systems)

Thermal desalination facilities have been co-located with power plants to recycle waste heat and to use common intakes and outfall for cooling and feedwater. Additional efficiencies have been achieved by adding the RO process to the MSF or MED processes co-located with a power plant. Other options exist for improving efficiencies such as using nanofiltration to pretreat the raw water entering both the thermal and membrane desalination processes. Nanofiltration can reduce the hardness of the feedwater to allow the thermal plants to operate at higher temperatures without causing scaling in the condensing tubes. In hybrid systems nanofiltration can simultaneously remove particulate and some organic compounds that cause fouling of RO membranes.

The use of nanofiltration in these hybrid facilities has already been tested on a pilot basis and is in the design of new proposed plants. However, as this technology advances, it will be increasingly incorporated into retrofits of existing facilities as it can provide improved efficiency and corresponding reduced cost.

27.6.2 Use of Waste Heat, Solar Energy, and Geothermal Energy with Adsorption Desalination

Adsorption desalination is a new technology that is described in Sect. 27.2.4. Because the process can operate efficiently at relatively low temperatures, the use of waste heat, solar generated heat, or low-grade geothermal heat can produce very low costs for desalination of seawater, brackish water, or even hypersaline waters. The potential for AD technology is only at the beginning stage and when coupled with renewable energy sources, it has the potential to solve various water-supply problems at a large range of scales from individual buildings to regional water supply facilities of large capacity. This technology has an additional advantage that it not only produces distilled water, but the process lowers the heat of the feedwater during the adsorption cycle, therefore creating a cold water stream. The cold water stream can be used for air conditioning. Thus, an AD plant can be seasonally adjusted to favor either desalination or cooling, which makes the technology particularly attractive to locations that have large seasonal variations in rainfall and average temperature.

27.6.3 Use of Subsurface Intakes for Seawater RO Plants

Most large-capacity seawater desalination facilities utilize open-ocean intake systems to provide feedwater. Open-ocean intakes have some associated environmental impacts, such as impingement and entrainment of marine life, which must be evaluated during the planning and design of the overall desalination facility. These environmental impacts can significantly raise the cost of facility permitting and construction thus raising the treated water delivery cost. Also, the raw seawater passing through intakes contains particulate inorganic and organic matter that requires removal before the feedwater can enter the process train of membrane treatment facilities. Therefore, a complex and costly pretreatment system must be designed, constructed, and operated to protect the primary membrane process and to reduce the rate of biofouling. Improvements to the design and operation of intakes can improve considerably the quality of the feedwater and reduce pretreatment costs. There are several methods to improve the intake structures to reduce cost. Conventional velocity cap or open-channel intakes can be improved by replacement with passive screens. Also, open-ocean intakes can be replaced by subsurface intakes that can vastly improve raw water quality.

The simple use of passive screen intakes that use a wedgewire screen (i.e., Johnson barrel screens) at the end of the intake pipe can reduce the influx of large marine debris. When coupled with air-burst cleaning technology, passive screens can operate in a cost-effective manner with low maintenance. Passive screens also significantly reduce impingement and entrainment, which reduces environmental impacts and the amount of marine debris that must be removed from travelling screens and other parts of the pretreatment process. Passive screens, however, will not removal microscopic organisms that can foul pretreatment or membranes, such as the dinoflagellates in a red tide event.

A viable alternative to open-ocean intakes is the use of subsurface intake systems. These systems remove microscopic organisms and suspended solids as well as some organic compounds. There are several types of these systems, including (Missimer 2009; Missimer et al. 2010):

- Wells, conventional vertical type,
- Wells, horizontal configuration,
- Horizontal collector wells (Ranney wells),
- Beach galleries, and
- Seabed filtration systems,

Subsurface intakes provide primary filtration and are biologically active, and thus can remove some organic compounds as the seawater passes though the groundwater system or the engineered filter. All of the subsurface intake designs provide a lower energy, more "natural" approach to pretreatment of seawater. There are existing subsurface intake systems that provide feedwater to seawater RO facilities with large capacities. For example, the well intake at Sur, Oman has a capacity of 65,000 m^3/day (17.2×10^6 gal/d; David et al. 2009). There are several subsurface intakes being designed for very large seawater RO facilities, such as a project being considered in Mexico that would have a capacity of 1.3 million m^3/ day (343×10^6 gal/d).

The type of subsurface alternative intake that is the best solution for a given project will depend upon local geologic conditions at the shoreline or offshore. In some cases, the use of conventional wells is not practical because of the large number that would be required to obtain the desired capacity. Beach galleries or a seabed filters could instead be economic design solutions. Beach galleries are an engineered filter constructed beneath the beach between the high and low tide levels. They function in a similar manner as rapid sand filters with the wave action at the shoreline continuously cleaning the filter. Seabed filters or galleries are engineered filters that are constructed in the seabed offshore from the beach (Fig. 27.18). Seabed filters may be the preferred design in areas that have no sandy beaches or beaches with low wave activity. Seabed filters are constructed in sandy marine bottom sediments and operate similar to slow sand filters. Seabed filters are cleaned to a degree by the wave orbital motion and marine currents above the filter, but may require periodic removal of the upper 5–10 cm (2–4 in) of sand. Both beach galleries and seabed filter systems can be used for high-capacity desalination facilities. Detailed descriptions and design criteria for these systems are contained in Missimer (2009).

The use of subsurface intakes can reduce the cost of desalination in several ways. Subsurface intakes reduce pretreatment requirements and environmental impacts, which in turn reduces permitting costs and improves public perception. The capital cost for design and construction of subsurface intakes commonly is offset by the reduction in capital costs for construction of the pretreatment processes. However, the real savings is in reduction of operating costs, which could amount to 15–25% depending upon the water quality conditions at a given site (Missimer et al. 2010).

27.6.4 Use of Operational Aquifer Storage and Recovery to Improve Efficiency

Desalination facilities, like all water treatment facilities, must be designed and constructed to meet the peak day demand within the distribution system. In potable water supply systems that have a high peak day to average day ratio, a significant part of the desalination plant is not operated during most of the year. For example, if the peak day to average day ratio is 2–1, then up to 50% of the treatment capacity of the plant is not operated for up to 95% of the year depending upon the temporal distribution of demand during the year. This creates inefficiencies and adds operating costs to the system caused by the extra capital cost required to design and build system capacity to meet peak day demands, added costs to maintain the unused capacity, and some extra costs, such as the lost use of membranes during a large part of their operating life of about 5 years.

Aquifer storage and recovery (ASR) is the use of the groundwater system to store water when there is an excess supply and the later recovery of the water when it is needed to either meet peaks in demand or for emergency supply to be tapped disruptions in the primary water supply system (Sect. 23.2.1.1; Maliva and Missimer 2010). Coupling of an ASR system with a desalination system can allow the desalination facilities to be designed and constructed closer to the average day demand. Some excess capacity is needed to charge the ASR system with enough water to supplement the treated water stream during peak demand periods. The coupling of desalination with ASR adds efficiency and provides cost savings. It may be far less expensive to construct and operate an ASR system to meet peaks

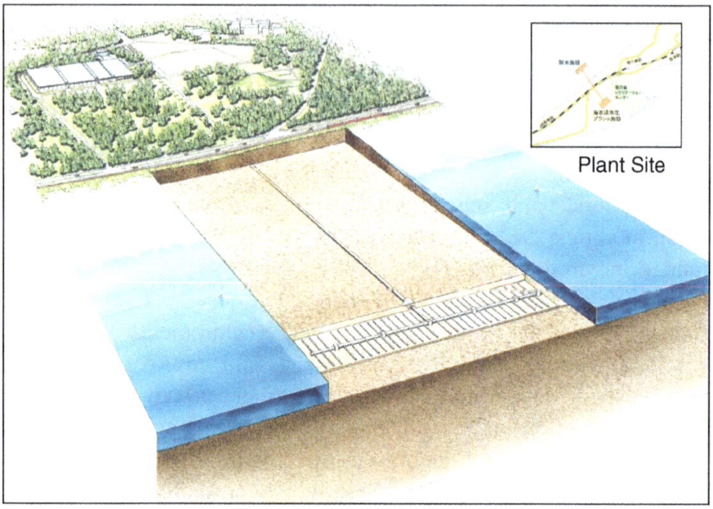

Fig. 27.18 Seabed filtration system at Fukuoka, Japan with a capacity of 83,270 m3/day (from Pankratz 2006)

in demand than to construct additional desalination capacity that is seldom used. A number of ASR systems are used in conjunction with water systems operating desalination facilities, such as the Manatee Road system in Collier County, Florida (Missimer et al. 1992).

27.6.5 Combination of Power Plant, Distillation Plant, Reverse Osmosis Plants and Aquifer Storage and Recovery on a Single Site to Improve Operational Efficiency

Perhaps the greatest operational efficiency for desalination, particularly for the large capacity Middle East systems and other arid lands applications, would be to construct a fully integrated "power/desalination campus" containing a power plant, thermal desalination plant, membrane desalination plant, a nanofiltration pretreatment facility or a subsurface intake, and an ASR system (Fig. 27.19). This fully integrated water treatment and storage approach could provide a balance between electrical daily and annual cycles (base load and peak demands), daily and annual water demands, and the need for operational and strategic long-term storage of treated water. Commonly, the electrical and water supply demand patterns are not congruent and perhaps in conflict with each other. The additional of the ASR system to provide storage increases the overall system efficiency by becoming electrical storage by proxy, since treated water can be stored while electricity cannot. This campus concept could reduce both electrical and desalination costs by vastly improving operational efficiency. Perhaps cost could be reduced even farther by adding AD into the system to maximize the use of waste heat.

27.7 Current Comparative Energy Use and Desalination Economics

Water supply in arid lands areas includes a variety of potential options depending upon the resources available and their relative development and maintenance costs. The overriding water management principle is that the key potable water source must be sustainable within the planning horizon or until an alternative water source can be developed based on the local or regional economic framework.

Desalination of seawater is a sustainable means of providing a water supply within the realm of resource management. The sea is an inexhaustible water resource in terms of providing a source of raw water supply. However, the issue of sustainability must also consider the issue of economic sustainability. Desalination is a viable means of supplying potable water, but is requires that the end user must be able and willing to pay for the water in order to retire the capital construction debt and to operate and maintain the facility. Therefore, in developed countries or regions, the higher cost of potable water is likely not a strain on the economy in general. However, in poor countries, particularly in arid regions, the ability to pay high costs for any water supply may not be viable without government or international subsidies. Desalted water can be used for potable supply and can stimulate economic development, which can provide economic returns commensurate with the costs of the overall system. However, the use of desalinated water for low economic return activities, such as agricultural irrigation or some large-scale industrial uses, is not economically viable. Based on the general realities of economics, it is important to clearly understand the energy requirements and true cost of desalination as it stands today and how this may change in the future. Desalination is a key component of integrated water resources management within arid regions, especially when coupled with other key components including water reuse, conservation, surface-water capture and storage, and surface (dams) and subsurface storage (ASR).

27.7.1 Energy Use of Various Desalination Technologies

A large number of methods have been developed to assess the comparative energy consumption to produce desalinated water. Since the energy consumption rate is highest for the conversion of seawater to fresh water, this will be the focus of the discussion. Brackish-water desalination systems have lower energy consumptions based on the salinity and over water quality of the feedwater. Comparisons between the various technologies are based on the concept of the total primary energy required to convert a unit of seawater to freshwater, in this case 1 m^3 or other English units, such as 1,000 gallons.

In a perfect world, the lowest amount of total primary energy required to desalinate normal seawater is based on the thermodynamic limit. For seawater with a total dissolved solids concentration of 35,000 mg/L, the thermodynamic limit for desalination is 0.78 kWh/m^3. A comparison of the total primary and

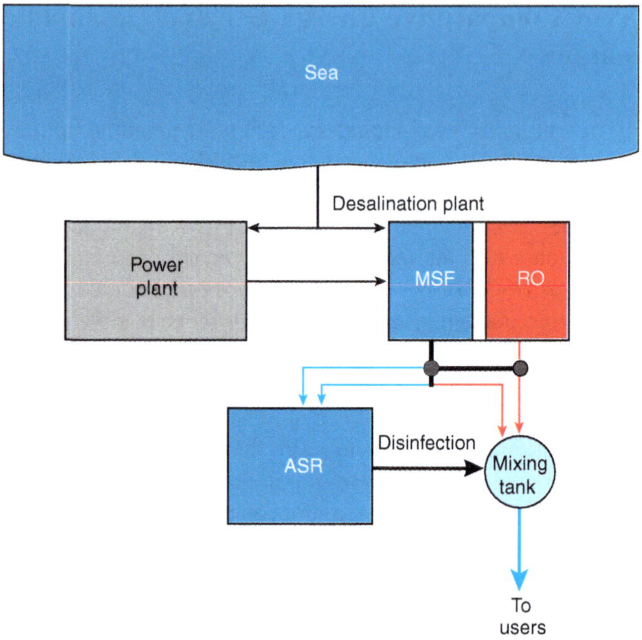

Fig. 27.19 Configuration of the hybrid "desalination campus" concept combined power generation with MSF or MED treatment, RO treatment, NF pretreatment, and ASR. The concept offers the most efficient combined power-desalination operating system (modified from Maliva and Missimer 2010)

electric energy required to desalt seawater is given is Table 27.1. This information is based on analyses published by Spiegler and El-Sayed (2001); Miller (2003); Blank et al. (2007); Thu et al. (2010) and personal communication with Dr. Kim Choon Ng (National University of Singapore and King Abdullah University of Science and Technology), who recalculated all of the values. If seawater is desalinated solely by a solar process, similar to a solar still, the total primary energy required is very high at 475 kWh/m^3, because of the very low efficiency of the conversion. The total primary energy used for various thermal and membrane technologies in descending order of energy use are solar, MSF, MED, VC, AD, RO, and combined AD and MED using waste heat.

There are a number of other uses of energy within the operation of a desalination plant other than the actual process. The pumps and pretreatment processes can also use a considerable amount of energy as well as the high service pumps that convey the treated water into the distribution system. These uses are given in Table 27.1 as electrical energy costs base on the analysis of Blank et al. (2007).

It is important to understand the differences in energy use between processes to analyze potential use of each technology in a given region. The option of co-locating a power plant with a thermal process can be compared to the energy use of a stand-alone facility. A strict energy analysis should be made during the planning process to view these technologies.

Table 27.1 Comparisons of desalination technology energy consumption (seawater desalination)

Technology	Total primary energy (kWh/m^3)	Electrical energy (kWh/m^3)
Solar	475.01	0.63
Multistage flash distillation	57.14	2.50
Multiple effect distillation	43.21	2.00
Adsorption desalination	39.80	1.38
Vapor compression	31.71	11.10
Reverse osmosis	14.29	5.00
Adsorption desalination + Multiple effect distillation (using waste heat)	16.14	1.94

Source Dr. Kim Choon Ng

Table 27.2 Estimated cost of seawater desalination for each technology

Technology	Cost range in \$/m^3	Average cost \$/m^3
Solar[1]	\$0.05	\$0.05
Multistage flash distillation[2]	\$0.70–5.36	\$1.32
Multiple effect distillation[2]	\$0.27–1.49	\$0.92
Vapor compression[2]	\$0.46–1.21	\$0.82
Reverse osmosis[2]	\$0.45–1.62	\$0.96[3]
Adsorption desalination[1]	\$0.30	\$0.30
Adsorption desalination + Multiple effect distillation (using waste heat)[1]	\$0.26	\$0.26

[1] Thu et al. (2010) and Kim Choon Ng (personal communication)
[2] Miller (2003)
[3] Includes data from smaller plants and those with large pretreatment difficulties, most probable average number for large plants is about \$0.78/m^3

27.7.2 Comparative Costs of Current Desalination Technologies

Actual cost of providing desalinated water has been assessed and debated in a large number of publications over the past 20 years. Summaries of these analyses are given in Miller (2003) and (Blank et al. 2007). The true cost of desalination is the sum of the cost to provide the total primary energy required to operate the process, the amortized capital costs, cost of replacement parts on a life-cycle assessment, chemical costs, and maintenance and operating costs including labor. The cost to the consumer is higher because there is a conveyance cost associated with pumping and piping the water, administrative overhead (i.e., billing), and profit or contingency costs. A comparative cost of the conversion of seawater to freshwater using different processes is given in Table 27.2. The numbers in the table are based on a uniform rate of \$5/million btu of natural gas heat generation (personal communication with Dr. Kim Choon Ng).

It is clear that the least costly process is the use of natural solar distillation which has a cost of only \$0.05/m^3. There is considerable variation in the cost for

MED based on whether it is powered by solely electrical heating or uses waste heat. The RO costs are commonly understated because of the continuing problem of biofouling and the need to equip or retrofit facilities with very expensive pre-treatment processes, like dissolved air flotation (DAF) systems in areas subject to red tide development or where oil is present in the feedwater. It should be noted that the AD uses waste heat in this compilation and powering the AD technology with solar and some electricity to run the pumps would increase this cost slightly. The least costly process would be combined AD and MED using waste heat.

When attempting to obtain a complete cost for desalination prior to adding conveyance and administrative overhead costs, it is necessary to add local fuel costs and local electric energy costs. This information can be obtained from the IPCC reports produced annually by the United Nations. The financial cost can be obtained with a capital recovery factor as

$$CRF(n, i) = \frac{i(1+i)^n}{(1+i)^n - 1} \tag{27.3}$$

where,

CRF (n,i)	= amortized yearly capital cost
I	= interest rate
N	= life-span of equipment

Based on some recent competitive bids for long term water supply contracts for seawater desalination facilities normalized to a feedwater salinity of 35,000 mg/L, thermal-membrane hybrid systems ranged from \$0.65 to \$1.10/m^3 and membrane facilities ranged from \$0.65 to \$0.90/m^3. The price is based on plant capacity, financing, local fuel costs, operating and maintenance costs, administrative costs, and profit. It should be noted that many tenders contain special subsidies and financing terms that make direct comparisons of bid prices very difficult to analyze.

The new technologies, such as AD technology, that use alternative energy sources may allow for lower total costs than are presently possible using current commercial desalination technologies. It is believed that the cost of seawater desalination will go below \$0.40/m^3 in the next 10 years. Desalination is a vibrant area of research and an ever more vital technology in arid lands, so it is difficult to foresee where the next technological breakthrough will occur.

References

Abu-Jabal, M. S., Kamiya, I., & Narasaki, Y., (2001). Proving test for a solar-powered desalination system in Gaza-Palestine. *Desalination,137*(1–3), 1–6.

Ackermann, T., & Söder, L. (2002). An overview of wind energy—status 2002. *Renewable and Sustainable Energy Reviews, 6*(1–2), 67–127.

Agha, K. R., Abdel-Wahab, M., & El-Mansouri, K., (2006, December 11–14). Potential for solar desalination in the arid states of North Africa and the Middle East, In *Proceedings Scientific*

Research Outlook and Technology, Development in the Arab World (SRO4) Conference. Damascus, Syria.

Al-Karaghouli, A., Renne, D., & Kazmerski, L. L. (2009). Solar and wind opportunities for water desalination in the Arab regions. *Renewable and Sustainable Energy Reviews, 13*(9), 2397–2407.

Al-Mutaz, I. S. (2001). Potential of nuclear desalination in the Arabian Gulf countries. *Desalination, 135*, 187–194.

Al-Mutaz, I. S. (2003). Hybrid RO MSF: A practical option for nuclear desalination. *International Journal of Nuclear Desalination, 1*(1), 1–10.

Al-Wazzan, Y., & Al-Modaf, F. (2001). Seawater desalination in Kuwait using multistage flash evaporation technology-historical overview. *Desalination, 134*, 257–267.

Awerbach, L. (2007). Hybrid systems and technology. In M. Wilf (Ed.), *The guidebook to membrane desalination technology: Reverse osmosis, nanofiltration, and hybrid systems process, design, application, and economics* (pp. 395–453). L'Aquila, Italy: Balaban Desalination Publishers.

Beltagy, A. I. (1983). Some oceanographic measurements in the Gulf waters around Qatar peninsula. *Qatar University Science Bulletin, 3*, 329–341.

Blank, J. E., Tusel, & Nisan, S. (2007). The real cost of desalted water and how to reduce it further. *Desalination, 205*, 298–311.

Boesch, W. M. (1982). World's first solar-powered reverse osmosis desalination plant. *Desalination, 41*, 233–237.

Borsani, R., & Ferro, A, M. (2006, March 5–6). *MSF innovation-Beyond large size units: Proceedings of the International Desalination Forum on Innovation and Integration*, Dubai, U. A. E.

Bourouni, K., Martin, R., & Tadist, L. (1999a). Analysis of heat transfer and evaporation in geothermal desalination units. *Desalination, 122*, 301–313.

Bourouni, K., Martin, R., Tadist, L., & Chaibi, M. T. (1999b). Heat transfer and evaporation in geothermal desalination units. *Applied Energy, 64*, 129–147.

Buros, O. K. (2000). *The ABCs of desalting* (2nd ed). Topsfield, MA: International Desalination Association.

Buros, O. K., Cox, R. B., Nusbaum, I., El-Nashar, A. M., & Bakish, R. (1980). *The USAID desalination manual*. Englewood Cliffs, NJ: IDEA Publications.

Cath, T. Y., Childress, A. E., & Elimelech, M. (2006). Forward osmosis: Principles, applications, and recent developments. *Journal of Membrane Science, 281*, 70–87.

Cotruvo, J. (2006, April). *Health aspects of calcium and magnesium in drinking water: Proceedings of the International Symposium on Health Aspects of Calcium and Magnesium in Drinking Water*, Baltimore, MD, USA.

David, B., Pinot, J., & Morillon, M. (2009). Beach wells for large-scale reverse osmosis plant: The Sur case study: *Proceeding of the International Desalination Association World Congress on Desalination and Water Reuse*. Dubai, UAE, DB09-16.

Delyannis, E. (2003). Historic background of desalination and renewable energies. *Solar Power, 75*(5), 357–366.

El-Dessouky, H. T. (2007). Sea water desalination: Importance, need, methods, and historical developments. *Journal of Pakistan Materials Society, 1*(1), 34–35.

El-Dessouky, H. T., Ettouney, H. M., & Mandani, F. (2000). Performance of parallel feed multiple effect evaporation system for seawater desalination. *Applied Thermal Engineering, 20*, 1679–1706.

El-Nashar, A. M. (2009). *Multiple effect distillation of seawater using solar energy: Hauppauge*. New York, NY: Nova Science Publishers.

El-Nashar, A. M., & Ishii, K. (1985). Abu Dhabi solar distillation plat. *Desalination, 52*, 217–234.

El-Nasher, A, M., & Samad, M. (1998). The solar desalination plant in Abu Dhabi: 13 years of performance and operational history. *Renewable Energy, 14*(1–4), 263–274.

Ettouney, H. (2009). Conventional thermal processes. In A. Ciollina., G. Micale., & L. Rizzuti. (Eds.), *Seawater desalination and renewable energy processes (Green Energy and Technology)*. Berlin: Springer.

Faibish, R. S., & Ettouney, H. (2003). MSF nuclear desalination. *Desalination, 157*, 277–287.

García-Rodríguez, L., & Gómez-Comacho, C. (1999). Thermo-electric analysis of a solar multi-effect distillation plant installed at the Platforma Solar de Almería (Spain). *Desalination, 122*, 205–214.

Glueckstern, P. (1995). Potential uses of solar energy for seawater desalination. *Desalination, 101*, 11–20.

Goosen, M., Mahmoudi, H., & Ghaffour, N. (2010). Water desalination using geothermal energy. *Energies, 3*, 1413–1442.

Goto, T. (1997). *Operating experience gained with nuclear desalination plants by Japanese electric power companies: Proceedings Series, Nuceal Desalination of Seawater*. IAEA, Vienna.

Groschen, G. E. (1994). *Simulation of groundwater flow and the movement of saline water in the Hueco bolson aquifer, El Paso, Texas, and adjacent areas*. U. S. Geological Survey Open-File Report 92-171.

Hamed, O. A. (2000). Thermal performance of multistage flash distillation plants in Saudi Arabia. *Desalination, 128*, 281–292.

Hamed, O. A. (2006). Overview of hybrid desalination systems-current status and future prospects. *Desalination, 186*, 207–214.

Harrison, D. G., Ho, G. E. & Mathew, K. (1996). Desalination using renewable energy in Australia. *Renewable Energy*, 8(1–4), pp. 509–513.

Hedayat, S. E., Sary, A. A., Nawar, S. M., Attia, M. A., El-Osery, I. M., Sabri, Z. A., et al. (1977). Design of a small single-purpose nuclear desalination plant compatible with the local resources in the Middle East. *Desalination, 20*, 257–266.

Intergovernmental Panel on Climate Change (IPCC) (2007). Fourth assessment report of the Intergovernmental panel on climate change (S. Solomon., D. Qui., M. Manning., Z. Chen., M. Marquis., K. B. Averyl., M. Tignor., & H. L. Millers., Eds.). Cambridge, UK: Cambridge University Press.

Kalogirou, S. A. (2005). Seawater desalination using renewable energy sources. *Progress in Energy and Combustion Science, 31*, 242–281.

Karytsas, C., Mendrinosa, D., & Radoglou, G. (2004). The current geothermal exploration and development of the geothermal field of Milos Island in Greece. *GHC Bulletin, 25*, 17–24.

Khamis, I. (2009). A global view on nuclear desalination. *International Journal of Nuclear Desalination, 3*(4), 311–328.

Khawaji, A. D., Kutubkhanah, I. K., & Wie, J.-M. (2008). Advances in seawater desalination technology. *Desalination, 221*, 47–69.

Lu, H., Walton, J. C., & Swift, A. H. P. (2001). Desalination coupled with salinity-gradient solar ponds. *Desalination, 136*, 13–23.

Malik, M. A. S., Tiwari, G. N., Kumar, A., & Sodha, M. S. (1985). Solar desalination. Oxford: Pergamon Press.

Maliva, R. G., & Missimer, T. M. (2010). *Aquifer storage and recovery and managed aquifer recharge using wells: Planning, hydrogeology, design, and operation*. Houston, Schlumberger Water Services, Methods in Water Resources Evaluation Series No. 2.

Maliva, R. G., Missimer, T. M., & Fontaine, R. (2011). Injection well options for the sustainable disposal of desalination concentrate. *Proceedings, International Desalination Association World Congress on Desalination and Water Reuse*. Perth, Western Australia, IDAWC/PER11-008.

Manwell, J. F., & McGowan, J. G. (1994). Recent renewable energy driven desalination system research and development in North America. *Desalination, 94*, 229–241.

McCutcheon, J. R., McGinnis, R. L., & Elimelech, M. (2005). A novel ammonia-carbon dioxide forward (direct) osmosis desalination process. *Desalination, 174*, 1–11.

McCutcheon, J. R., McGinnis, R. L., & Elimelech, M. (2006). Desalination by a novel ammonia-carbon dioxide forward osmosis process: influence of draw and feed solution concentrations on process performance: Journal of. *Membrane Technology, 278*, 144–123.

McGinnis, R. L., & Elimelech, M. (2007). Energy requirements of ammonia–carbon dioxide forward osmosis desalination. *Desalination, 207*, 370–382.

Megahed, M. M. (2001). Nuclear desalination: history and prospects. *Desalination, 135*, 169–185.

Megahed, M. M. (2008). Feasibility of nuclear power and desalination on El-Dabaa site (Egypt). *Desalination, 246*, 238–256.

Miller, J. E. (2003). *Review of water resources and desalination technologies* (p. 54). Sandia National Laboratories, Report 2003-0800.

Missimer, T. M. (2009). *Water supply development, aquifer storage, and concentrate disposal for membrane water treatment facilities:* Houston, Texas, Schlumberger Water Services, Methods in Water Resources Evaluation Series No. 1.

Missimer, T. M., Maliva, R. G., Thompson, M., Manahan, W. S., & Goodboy, K. P. (2010). Reduction of seawater reverse osmosis treatment costs by improvement of raw water quality: Innovative intake designs. *The International Desalination & Water Reuse Quarterly, 20*(3), 12–22.

Missimer, T. M., Walker, C. W., & Bloetcher, F. (1992). Use of aquifer storage and recovery technology to improve membrane water treatment plant efficiency, Collier County, Florida. *Desalination, 87*, 269–280.

Muralev, E. D., Nazarenko, P. I., Poplavskij, V. M. & Kuznetsov, I. A. (1997). Experience gained in the operation and maintenance of the nuclear desalination plant in Aktau, Kazakhstan. *Proceedings Series, Nuclear Desalination of Seawater, IAEA, Vienna* (pp. 355–366).

National Research Council (2008). *Prospects for managed underground storage of recoverable water*. Washington, D.C: National Academy Press.

Ng, K. C., Wang, L. Z., Gao, L. Z., Chakraborty, A., Saha, B. B., Koyama, S., Akisawa, A., & Kashiwagi, T. (2006). *Apparatus and method for desalination*. WO Patent number 121414.

Ng, K. C., Kyaw, T., Amy, G., Chunggaze, M., & Al-Ghasham, T. Y. (2011). *An advanced cycle for low-temperature driven desalination*. United States Provisional Patent No. 61/450, 165.

Oak Ridge National Laboratory (1971a). *Middle East study: Application of large water-producing center-the study area* (Vol. 1). ORNL-4481, USA.

Oak Ridge National Laboratory (1971b). *Middle East study: Application of large water-producing energy centers-United Arab Republic* (Vol. 2). ORNL-4481, USA.

Oak Ridge National Laboratory (1971c). *Middle East study: Application of large water-producing energy centers-Israel* (Vol. 3). ORNL-4481, USA.

Ophir, A. (1982). Desalination plant using low grade geothermal heat. *Desalination, 40*, 125–132.

Pankratz, T. (2006). Seawater desalination technology overview: St. Simons Island, Georgia. Georgia Joint Comprehensive Desalination Study Committee, 22–23 August 2006.

Petersen, G., Fries, S., Mohn, J., Meuller, A. (1979). Wind and solar powered reverse osmosis desalination units. Description of two demonstration projects. *Desalination*, 31, 501–509.

Ragheb, M. (2010, March 21–24). *Single and dual purpose nuclear desalination: Proceedings of the 1st International Nuclear and Reliable Energy Conference (INREC10)*. Amman, Jordan.

Redondo, J., Busch, M., & Dewitte, J.-P. (2003). Boron removal from seawater using FILMTEC^{TM} high rejection SWRO membranes. *Desalination, 156*, 229–238.

Reid, C., & Breton, E. (1959). Water and Ion flow across Cellulosic Membranes. *Journal of Applied Polymer Science, 1*, 133–143.

Riolo, A. (2001). Maltese experience in the application of desalination technology. *Desalination, 136*, 155–124.

Rizzuti, L., Ettouney, H., & Cipollina, A. (2007). A review of modern technologies and researches on desalination coupled to renewable energies. In *Proceedings of the NATO Advanced Research Workshop on Solar Desalination for the 21st Century, Hammamet, Tunisia*. New York: Springer. 23–25 February 2006.

Sayre, A. N., & Livingston, P. (1945). *Groundwater resources of the El Paso area*. Texas: U. S. Geological Survey Water-Supply Paper 919.

Schwarzer, K., Vieira da Silva, E., Hoffschmidt, B., & Schwarzer, T. (2010). A new solar desalination system with heat recovery for decentralised drinking water production. *Desalination, 251*, 204–221.

Sheng, Z., & Devere, J. (2005). Understanding and managing the stressed Mexico-USA transboundary Hueco bolson aquifer in the El Paso de Norte region as a complex system. *Hydrogeology Journal, 13*, 813–825.

Sommariva, C., Hogg, H., & Callister, K. (2001). Forty-year design life: The next target material selection and operating conditions in thermal desalination plants. *Desalination, 136*, 169–176.

Spiegler, K. S. & El-Sayed, Y. M. (2001). The energetics of desalination processes. *Desalination*, 134, 109–128.

Suri, R. K., Al Marafie, A. M. R., Al-Homoud, A. A., & Maheshwari, G. P. (1989). Cost-effectivness of solar water production. *Desalination, 71*, 165–175.

Talbert, S. G., Eibling, J. A., & Lof, G. O. G. (1970). *Manual on solar distillation of saline water: R&D. Progress Report No. 546*. United States Department of Interior, Battelle Memorial Institute, Columbus Laboratories.

Texas Water Development Board (2007). *Water for Texas 2007* (Vol. 1). Texas Water Development Board Document GP-8-1.

Thu, K., Chakraborty, A., Saha, B. B., Chun, W. G., & Ng, K. C. (2010). Life-cycle cost analysis of an adsorption cycles for desalination. *Desalination and Water Treatment, 20*, 1–20.

Tiwari, G. N., & Tiwari, A. K. (2008). *Solar distillation practice for water desalination systems: Raipur*. India: Anshan publishers.

Trieb, F. (2007, June 20–24). Concentrating solar power for seawater desalination: *Proceedings of the Middle East North Africa Renewable Energy Conference (MENAREC 4)*. Damascus, Syria.

Wang, X. L., Chakraborty, A., Ng, K. C., & Saha, B. B. (2007). How heat and mass recovery strategies impact the performance of adsorption desalination plant: Theory and experiments. *Heat Transfer Engineering, 28*, 1.

Wang, X., & Ng, K. C. (2005). Experimental investigation of an adsorption desalination plant using low-temperature waste heat. *Applied Thermal Engineering, 25*, 2780–2789.

Watson, I.E., Morin, O.J., Jr. & Henthorne, L. (2003). *Desalting handbook for planner's* (3rd ed.). U.S. Bureau of Reclamation, Desalination and Water Purification Research and Development Program Report No. 72.

World Health Organization. (2005). *Nutrients in drinking-water, Sanitation and the human environment: Geneva*. Switzerland: WHO.

World Health Organization. (2006). *Expert committee meeting on health effects of calcium and magnesium in drinking water*. Geneva, Switzerland: WHO Document Production Services.

World Health Organization. (2007). *Desalination for safe water supply, guidance for the health and environmental aspects applicable to desalination*. Geneva: Public Health and the Environment World Health Organization.

World Health Organization (2009). Calcium and magnesium in drinking water-Public health significance. (J. Cotruvo., & J. Bartam., Eds). Geneva, Switzerland: WHO.

World Nuclear Association (2011). *Nuclear desalination*. www.world-nuclear.org/info/in71.html.

Xu, J., Gao, X., Zou, L., & Gao, C. (2010). High performance boron removal from seawater by two-pass SWRO system with difference membranes. *Water Science and Technology: Water Supply, 10*(3), 327–336.

Zarg, M. -C. (2003). Development policy of nuclear power combined with desalination in China. *Nuclear Science and Techniques, 14*(2), 144–148.

Zubari, W. K. (2002). Use of stochastic modeling in the sustainable management of non-renewable brackish groundwater in Bahrain. In M. Sherif., V. P. Sing., & M. F. Al-Rashed, Eds, *Proceedings of the international conference on water resources management in arid regions (WaRMAR)* (Vol. 2, pp. 269–287). Lisse, Netherlands: Groundwater hydrology.

Chapter 28
Environmental Issues in Desalination

28.1 Introduction

Desalination is an energy intensive technology, especially when applied to seawater. Therefore, desalination has both high costs and a very large carbon footprint when conventional energy sources are used to power the various desalination processes. A variety of renewable energy sources are available that can be applied to reduce the carbon footprint of desalination. The global environmental impact of desalination can be reduced in terms of carbon dioxide emissions, but at a cost. Seawater desalination facilities will still have some environmental impacts on local areas related to intakes and outfalls of concentrate. Brackish-water desalination of groundwater can create a potential resource depletion problem and also faces the challenge of finding an economical and environmentally sound means of concentrate disposal, particularly for inland facilities.

The environmental impacts of desalination merit discussion and may be relevant on scales ranging from global to regional to local include (Pontius et al. 1996; Höpner and Windelberg 1997; Einav et al. 2002; Höpner and Lattemann 2002; von Medeazza 2005; Younos 2005; Sadhawani et al. 2005; World Health Organization 2007; National Research Council 2008a, b; Laspidou et al. 2010; Cotruvo et al. 2010):

- increased demand for electricity and thus greenhouse gas emissions if electricity is produced using fossil fuels,
- impacts of the discharge of hypersaline brines on marine benthic communities,
- discharge of chemicals used for pretreatment and membrane cleaning,
- discharge of metals from corrosion (Cu, Fe, Ni, Mo, Cr),
- aesthetic issues, noise visual impacts,
- impacts to aquifers from leaks from seawater and brine pipes,
- damage during construction of intakes and outfalls,
- major ion toxicity of concentrate (more from brackish water) caused by excesses or deficiencies of major ions relative to the native water (Mickley 2000), and
- thermal pollution effects from thermal distillation facilities.

R. Maliva and T. Missimer, *Arid Lands Water Evaluation and Management*, Environmental Science and Engineering, DOI: 10.1007/978-3-642-29104-3_28,

Desalination impacts may be significant, but can be managed to allow facility development and operation without causing unacceptable harm to natural systems or populations. The key management issues involve understanding of the processes, assessing the potential impacts, and creatively designing solutions to minimize or eliminate the impacts.

28.2 Energy Consumption Issues

A key goal over the past two decades, and an on-going research subject, is reduction in the amount of energy required for the desalination process. Reducing energy consumption decreases both costs and environmental impacts. There have been significant advancements in achieving this goal including:

- improvements in thermal desalination methods to increase the quantity of recycled latent heat,
- co-location of electric power generating facilities with thermal desalination plants to capture wasted heat and for co-location of common components (intakes and outfalls),
- creation of hybrid desalination systems, such as MSF-UF-RO, MED-UF-RO, and others,
- improvement of energy recovery systems on seawater and brackish water RO systems,
- improvements in membrane materials to improve recovery efficiency,
- improvements in intake designs to improve feedwater quality and reduce pretreatment system costs and energy consumption, and
- development of new low-energy desalination technologies, such as adsorption desalination (AD).

Within arid regions there are many opportunities to use renewable (RES) or alternative energy systems to provide the power required by desalination facilities. Currently, desalination using alternative energy usually costs considerably more than conventional powered systems that use predominantly hydrocarbon fuels to generate heat or electricity or both. The main exception is desalination facilities located in areas in which conventional energy sources are not available. When atmospheric environmental impacts are given a specific cost, such as the use of carbon credits, then the comparative real cost of RES will come down and will be quite competitive with conventional-powered systems. Alterative renewable energy sources will also become more cost competitive, if the costs of conventional energy source (hydrocarbons and coal) continue to rise in response to increasing global demands. Additionally, environmental impacts of new more energy efficient desalination technologies, such as AD, have the potential to significantly reduce the carbon footprint of desalination facilities and reduce global environmental impacts.

28.3 Intake Issues

Environmental impacts of conventional seawater intakes have become contentious issues over the past decade. The primary issues are the local and perhaps regional impacts to the marine environment caused by the entrainment and impingement of marine biota within the intake systems. Surface-water intake issues have been a major consideration in the power industry for the past 50 years, but since seawater desalination has not been used significantly in the United States and the European Union until relatively recent years, the entrainment and impingement issues for conventional open-ocean intakes were not considered to be problematical (or at least given much attention).

In the United States beginning in the 1970s, power plant intakes, used primarily to supply cooling water, were required to comply with the US Environmental Protection Agency (EPA) rules adopted under Section 316(b) of the Clean Water Act. This act requires that intakes must use the "Best Available Technology" (BAT) to minimize adverse environmental impacts. The specific environmental impact of greatest concern is the death of marine organisms caused by impingement and entrainment within the intake structure and pretreatment systems. These rules apply to both power and desalination plant intake and pre-treatment systems (Missimer 2009).

Impingement is the trapping of marine organisms on the filter screens of marine intakes as the result of the high inflow water velocity. Depending upon the specific marine species, its size, age, and the water conditions, some of the organisms will die while others will survive and can be returned to the environment. However, the estimated impingement survival rate for all marine biota taken as a group is less than 15%.

Entrainment of organisms occurs when they pass through an intake screen and enter additional filtration processes. There is no viable method to keep these organisms hydrated or recycle them into the sea. Therefore, the mortality rate for entrained organisms is considered to be 100% (Missimer 2009).

The degree of impingement and entrainment are dependent on the design of the intake and the corresponding velocity of the feedwater inflow. If the velocity is maintained at a low rate, then many species can swim away from the influence of the intake. Depending on the type of intake (i.e. canal end, velocity cap, etc.), even large fish can enter the intake and become trapped. Passive intakes with a slow velocity through the intake screen slots tend to have low impingement rates and entrain only fish larvae or eggs. Water temperature can also have an impact on impingement and entrainment, as the ability of specific species to swim away from the intake can diminish with decreasing temperature (cold effect).

Recent performance standards have been adopted by the USEPA under Section 316(b) of the Clean Water Act Phase II rule. This rule requires an impingement reduction of 80–95% from uncontrolled levels and a corresponding entrainment reduction of 60–90%. Impingement and entrainment studies are currently required in the facility permitting process by various agencies in the United States, the European

Union and other political jurisdictions. Some impacts must be mitigated by restocking fish species to compensate for losses. However, this type of environmental mitigation is commonly opposed by many environmental activist groups.

There are a number of methods that can be employed to reduce or eliminate impingement or entrainment in desalination intakes (Missimer 2009). Properly designed passive screen intakes have a very low impingement rate and will only pass very small marine organisms into the pretreatment system. Subsurface intakes, such as wells, beach galleries, and seabed filters, can completely eliminate the impingement/entrainment problem. Use of subsurface intakes is limited to regions that have a compatible geology that would support the design and construction of a given intake type. Therefore, there will always be some need for surface intake systems, which will need to be carefully designed in order to meet requirements to reduce environmental impacts associated with impingement and entrainment.

28.4 Concentrate Disposal Issues

Concentrate disposal is perhaps the most significant of all desalination technology environmental issues with regard to the siting and operation of desalination facilities (Watson et al. 2003). The National Research Council 2008a, b (p. 133) stated

> The effects of desalination concentrate discharge to oceans, seas, and estuaries, appears to vary widely. The impacts depend on the site-specific environment, the organisms examined, the amount of dilution of the concentrate, and the use of diffuser technology.

Although concentrate management can be a major environmental issue, requiring considerable investigation and permitting activities, a study by the California Department of Water Resources (2003) concluded that *"with proper design and location of outfalls, concentrate disposal may not be a major impediment to desalination."* The issue of concentrate disposal is and will continue to be a significant issue in desalination facility design and implementation (Andrews et al. 1991; Conlon 1988, 1989, 1990; Conlon and Smith 1991; Hashim and Hajjaj 2005; Höpner and Lattemann 2002; Höpner and Windelberg 1997; Lattemann and Höpner 2008; MEDRC 2002; Mackey and Seacord 2008; Malomosr et al. 2004; Mickley et al. 1993; Mickley 1996, 2000, 2001, 2002, 2003, 2007; Morton et al. 1996; Thompson and Brodeur 1991).

Depending upon the desalination system design, the volume of concentrate requiring disposal, and the local geologic conditions, and the region of the world where the facility is located, there are a variety of possible concentrate disposal options. The available concentrate disposal methods include (Missimer and Pankratz 2009):

- Surface-water discharge (primarily to tidal waters),
- Deep well injection (subsurface),
- Subsurface infiltration galleries (tidal waters),

- Discharge to a municipal sewer system,
- Evaporation ponds (can be coupled with solar evaporation), and
- Thermal evaporation process (zero-liquid discharge).

28.4.1 Physical and Chemical Properties of Concentrate

There is a wide range in the physical and chemical properties of concentrate, depending on the source of the feedwater (i.e. seawater or brackish water), the chemicals added to the water, and the composition of the chemicals used for process cleaning and back-washing of filters in the pretreatment system. The general properties of concentrate originating in seawater desalination facilities, both thermal and membrane plants, have been described by Mickley (1993), MEDRC (2002), Lattemann and Höpner (2003), World Health Organization (2007), Mickley (2007), and Lattemann and Höpner (2008). A description of the typical physical and chemical characteristics of concentrate from seawater desalination facilities is given in Table 28.1. The water chemistry of brackish-water concentrate is quite different compared to seawater concentrate (Table 28.2). The range of concentrations of various constituents is very large based on the more variable chemistry of the brackish groundwater developed as a source of the feedwater compared to seawater.

In seawater desalination systems, the physical and chemical parameters of most concern are usually the salinity of the discharge water, the temperature of the water (thermal), the concentrations of chlorine (used to prevent biofouling), copper (a corrosion product), and the iron compounds used to clean and back-flush the pretreatment filters. The concentrate from brackish-water systems typically contains an imbalance in the ratios of major ions, hydrogen sulfide, radium 226 and 228, gross alpha activity, and sometimes high iron concentrations, all of which can be problematical.

The physical and chemical properties of concentrate control what disposal options are viable and the post-treatment of the concentrate that may be required to prevent environmental damage and meet regulatory standards. Perhaps the most sensitive environmental issues occur when concentrate is discharged to tidal waters within estuaries or nearshore marine systems.

28.4.2 Surface Discharge to Tidal Waters

The vast majority of desalination facilities dispose of concentrate by discharge to tidal surface waters (Mickley 2003; Lattemann and Höpner 2008). The interaction of concentrate with sensitive marine and estuarine environments has the potential to cause significant adverse impacts. The potential impacts of seawater desalination systems are different than those of brackish-water systems.

Table 28.1 Physical and chemical properties of concentrate from seawater desalination facilities

	RO	MSF
Physical properties		
Salinity	Up to 65,000–85,000 mg/l	About 50,000 mg/l
Temperature	Ambient seawater temperature Specific cases from +5 to +10°C	+5 to +25°C above ambient
Plume density	Negatively buoyant	Positively, neutrally or negatively buoyant depending on the process, mixing with cooling water from co-located power plants and ambient density stratification
Dissolved oxygen	If wells used, typically below ambient seawater DO because of the low DO content of the source water. If open intake used, approximately the same as the ambient seawater DO concentration.	Could be below ambient seawater DO because of physical deaeration and use oxygen scavengers
Biofouling control additives and by-products		
Chlorine	If chlorine or other oxidants are used to control biofouling, these are typically neutralized before the water enters the membranes to prevent membrane damage	Approximately 10–25% of source water feed dosage, if not neutralized
Halogenated organics	Typically low content below harmful levels	Varying composition and concentrations, typically trihalomethanes
Removal of suspended solids		
Coagulants	May be present if source water is (e.g., iron-III-chloride) conditioned and the filter backwash is not treated, May cause effluent coloration if not equalized prior to discharge	Not present (treatment not required)
Coagulant aids	May be present if source water is (e.g., polyacrylamide) conditioned and the filter back-wash water is not treated	Not present (treatment not required)
Scale control additives		
Antiscalants	Typically low content below toxic levels	Typically low content below toxic levels
Acid (H_2SO_4)	Not present (reacts with seawater to cause harmless compounds, i.e. water and sulfates; the acidity is consumed by the naturally alkaline seawater, so that the discharge pH is typically similar or slightly lower than that of ambient seawater)	Not present (reacts with seawater to cause harmless compounds, i.e. water and sulfates; the acidity is consumed by the naturally alkaline seawater, so that the discharge pH is typically similar or slightly lower than that of ambient seawater)

(continued)

Table 28.1 (continued)

	RO	MSF
Foam control additives		
Antifoaming agents	Not present (treatment not required	Not present (treatment not required (e.g., polyglycol)
Contaminants due to corrosion		
Metals	May contain elevated levels of iron, chromium, nickel, molybdenum if low-quality stainless steel is used	May contain elevated copper and nickel concentrations if inappropriate materials are used for the heat exchangers. Copper concentration may be elevated if located with power plant, usually not a problem
Cleaning chemicals		
Cleaning chemicals	Alkaline (pH 11–12) or acidic (pH 2–3) solutions with additives such as: detergents (e.g., docecysulfate), complexing agents (e.g., EDTA), oxidants (e.g., sodium perborate), biocides (e.g., formaldhyde)	Acidic (pH 2) solution containing corrosion inhibitors such as benzotriazole derivates

Modified from Lattemann and Höpner 2008

Table 28.2 Typical concentrate physical and chemical properties of brackish groundwater reverse osmosis and electrodialysis reversal treatment facilities

Physical Properties	
Salinity	8,000–30,000 mg/l
Temperature	18–30°C
Dissolved oxygen	0–1 ppm
Hydrogen sulfide	0–10 ppm
Chemical Properties	
Ion ratios	Unbalanced (not in natural ratios)
Heavy metals	Enriched, but normally low
Iron	0–6 mg/l
Radium 226,228	0–20 pCi/l
Gross alpha	0–30 pCi/l
Scale Control	
Polyphosphates	Low concentrations
Polyacralytes	Low concentrations
Acid (H_2SO_4)	Low concentration, neutralized in most cases
Cleaning chemicals	
Alkaline or acidic	Typically not significant, only periodic discharge

Modified from Lattemann and Höpner 2008

Environmental impacts of seawater desalination systems can be classified into three general categories, including thermal discharge issues, salinity issues, and chemical or solids discharges. Each of these discharge issues has a different set of parameters that must be evaluated for their specific impacts. Obtaining regulatory approval for concentrate disposal can be a major cost and time consuming item for desalination plants. Pontius et al. (1996) reviewed the regulatory issues that can impact concentrate disposal in the United States.

Thermal pollution has been an issue associated with power plant cooling streams for decades. Commonly, power and desalination facilities are being developed together to gain efficiencies, especially with use of common intakes and discharge lines. Based on a survey of facilities, including stand-alone and coupled desalination and power facilities, the temperature differences between the concentrate (or combined concentrate/power plant cooling stream) generally range between 5 and 15°C (Mickley et al. 1993; Lattemann and Höpner 2008). The World Health Organization (2007) suggested a range from 5–25°C. The low differentials are mostly for stand-alone plants and the higher values are for power plant/desalination plant combinations with a thermal process used for desalination. The largest thermal differentials occur in combined power/desalination facilities, which tend to dilute the salinity of the concentrate discharge and make the composite discharge neutrally buoyant or buoyant (Lattemann and Höpner 2008).

Thermal discharges may have significant impacts on the marine vegetation and fisheries, particularly in cool or cold marine waters (Naylor 1965; Bramer 1990, 1995; Abbaspour et al. 2006). Naylor (1965) summarized the various environmental impacts that can be caused by thermal discharges, which include:

- heat death,
- change in organism's metabolism, growth rate, and form,
- changes in reproduction,
- behavior change (i.e. congregation of marine mammals causing conflicts with other human activities, such as boat collisions),
- elimination of species or change in diversity, and
- species changes (infauna relocation).

Thermal discharges may also affect the population densities of marine crustaceans and bivalves (Bramer 1990, 1995). Changes in marine or estuarine vegetation can also occur. While not all of these changes are harmful, thermal discharges in warm climate regions can increase the heat death rate of organisms that live near the limit of thermal tolerance level in the natural system. Thermal effects from concentrate discharge alone would likely be insufficient to create a major thermal impact compared to the much greater cooling stream from a power plant.

All desalination facilities produce a concentrate that is higher in salinity than the source of the feedwater. Therefore, concentrate discharged into the marine environment has a significantly greater salinity, in most cases, than the ambient seawater at the discharge location. Concentrate generated from seawater desalination facilities contain elevated salinities plus chemicals used for pretreatment and membrane cleaning, and metals released by corrosion. The impacts of concentrate disposal can occur on two scales. Concentrate discharge can cause localized impacts at the point of discharge and basin-wide impacts caused by long-term discharges by a large numbers of plants in bodies of water with limited seawater exchange or restricted circulation (Höpner and Lattemann 2002).

Local salinity increases can adversely impact benthic organisms. Concentrate from seawater desalination is denser that seawater and thus tends to sink. Most marine organisms live in an osmotic balance with their environment and an increase in the concentration of salts may result in dehydration of cells, decreases of the turgor pressure, and ultimately, death of the organism (Einav et al. 2002). There is considerable variation in the sensitivity of various marine species to changes in salinity. Mobile species can and do swim away from adverse salinity conditions, but sessile or benthic infauna do not have the ability to move and therefore, can expected to be affected to a greater degree. Local impacts to marine benthic communities can be reduced through site selection and outfall location and design (site field impact analysis). Since the impacts of concentrate salinity are significant, it is important to reduce the concentrate salinity as quickly as possible upon discharge into the marine environment. The degree of mixing of concentrate with seawater is caused by a variety of factors including wave orbital motion, tides, bathymetry, currents, and water depth. Mixing can also be enhanced by the use of diffusers. Therefore, site selection and the outfall design are import in minimizing marine impacts.

Höpner and Lattemann (2002) examined the impacts of brine discharge from seawater desalination plants in the Red Sea. The northern Red Sea, particularly the Gulf of Aqaba, has a relatively limited exchange of seawater and is vertically density-stratified. Brine discharge was found to have a negligible effect on overall salinity because salinity is controlled by the high evaporation rate in the arid climate of the region. It was recognized that salinity increases could have primarily a local impact on benthic organisms, which may already be at the upper limit of their environmental tolerance, similar to the conclusions of Einav et al. (2002). Höpner and Lattemann (2002) concluded that the conflict between the Red Sea as an acceptor and the desalination industry as a source of brine is far below a critical level.

Discharge of biocides, metals, chemicals, and particulate discharges into marine waters is also an environmental issue. Chlorine, which is used to prevent biofouling of the intake, pretreatment system, and sometimes the membranes, is of greatest concern because it is a biocide and reacts with dissolved organic carbon to produce a variety of halogenated compounds, some which are carcinogenic or otherwise harmful (Höpner and Latteman 2002). Copper is released as the result of corrosion of copper-nickel alloys used in power plant condensers and in heat exchangers of MSF or MED plants. Releases of copper are considered a concern near the point of discharge where it could accumulate in sediments. Antiscalant concentrations appear to be present at concentrations too low to create a significant environmental risk. The discharge of ferric chloride that is used in the coagulation process within pretreatment systems for the RO process is usually not problematical (Lattemann and Höpner 2008). However, during the backflushing of filters, a red particulate discharge can occur that is quite stark and creates public disapproval, but has little environmental impact.

Concentrate from brackish-water facilities treating groundwater can actually present greater challenges to elimination of environmental impacts and in obtaining surface discharge permits. Commonly, the concentrate is devoid of oxygen and contains high concentrations of hydrogen sulfide. When present, the anoxic water can have very severe impacts on the biota within both the water column and in the sediments. In virtually all cases, the hydrogen sulfide must be removed from the concentrate prior to discharge and the waste stream must be aerated. Another water quality problem of limited, but perceived significance is the occurrence of radionuclides in relatively small concentrations. Radium 226 and 228, as well as gross alpha activity is naturally-occurring in some groundwater sources. When present in the concentrate, it is very difficult to obtain a permit to discharge in the United States and in the European Union. In the United States the US Environmental Protection Agency, under the National Pollution Discharge Elimination System (NPDES) permitting program, can grant a "mixing zone" for contaminants with concentrations above the discharge standards. However, even trace amounts of radioactive substances discharged to the environment can raise severe public interest group reactions, typically delaying permit issuance for years or blocking the process entirely. While in most cases, the risk analysis performed for these radionuclide discharges shows that they are not environmentally harmful. However permit conditions can be added that require some type of treatment, which is not economically feasible. This requires that another concentrate disposal technique must be used if the project is to move forward.

Environmental impacts of desalination in providing essential freshwater need to be kept in perspective. Impacts from desalination are very commonly outranked by other development and industrial activities (Höpner and Lattemann 2002). Lattemann and Höpner (2008) conclude that environmental impacts from concentrate discharges into the marine environment can be mitigated by proper location of the outfall, post-treatment of the concentrate (i.e. hydrogen sulfide removal and aeration), and proper design of the concentrate diffusor system to enhance natural mixing. Location of a concentrate outfall in a water body that already shows poor water quality characteristics will likely lead to significant environmental damage, so the location is important. They suggest that a standard Environmental Impact Assessment for concentrate impacts should be developed to avoid unnecessary environmental impacts. A key point made by Einav et al. (2002) is that a balanced environmental evaluation should be performed that considers (1) societal benefits of desalination (2) its impacts on sensitive environments (3) costs of minimizing the impacts of desalination, and (4) economic and environmental costs of other alternative water supply options.

In conclusion, there are a few key methods required to reduce the environmental impacts of concentrate disposal by surface-water discharge, which include:

- Rapid mixing of the concentrate with ambient marine waters after discharge,
- Minimizing temperature differences between the discharged concentrate and local marine waters,
- Minimizing the discharge of toxic cleaning chemicals (separation of the waste streams), and
- Use of due diligence in the site selection for the outfall.

28.4.3 Subsurface Disposal Methods

Disposal of concentrate by injection into the subsurface has been successfully used for many years as documented in Muniz and Skehan (1990), Walker and Missimer (1996), Maliva and Walker (1998), Missimer (2009), and Maliva et al. (2011). There are a number of different types of subsurface disposal systems available for concentrate disposal, including:

- Low-pressure injection into deep, well-confined aquifers with high hydraulic conductivities,
- High-pressure injection into deep, well-confined aquifers with low hydraulic conductivities,
- Low-pressure injection into shallow, semi-confined or unconfined aquifers containing seawater,
- Low-pressure injection into horizontal wells constructed beneath the seabed,
- Moderate-pressure injection into oil and gas salinity disposal wells (Class II well under UIC), and
- Low-pressure disposal into beach galleries.

Perhaps the first wells used for concentrate disposal were shallow injection wells constructed on carbonate islands adjacent to seawater production wells. This is a common disposal method throughout the Bahamas, Grand Cayman Island (Missimer and Winters 2003), Providenciales (Turks and Caicos Islands), the Maldive Islands, and many other locations. Some of these wells are located on semi-arid islands. These wells are typically drilled relatively close to existing production wells, but cased to greater depths, such as in the Bahamas, where they have a minimum of 150 m (492 ft) of casing. The production wells have a maximum depth of about 85 m (279 ft). In many locations, the vertical separation between the feedwater production zone and the injection zone is only 20–30 m (66–98 ft). This is acceptable where the hydraulic conductivity of the production and injection aquifer is very high because the higher density of the concentrate (nearly double that of seawater) causes the injectate to move downwards by gravity. Recirculation of the concentrate from injection wells back into production has not been documented at any site. However, care must be taken in the design and separation of the production and injection zones. Some of these wells have a very high injection capacity and could provide a viable concentrate disposal option for large capacity seawater desalination plants.

Shallow, semi-confined aquifers have been used in a number of locations in the past. The partial confinement prevents upward movement of the injected concentrate and can aid in the permitting of the wells. While shallow injection disposal of concentrate is a quite environmentally low impact disposal method in coastal systems located near tidal water, the practice has been eliminated in parts of the United States in areas where there may be a deep aquifer than contains water with a dissolved solids concentration of less than 10,000 mg/l. Despite a high degree of confinement and no possibility of aquifer contamination because of an upwards directed hydraulic gradient, current regulations make it very difficult to use this technique. Typically, these injection wells are suitable for relatively small capacity desalination systems.

Horizontal wells drilled beneath the seabed are a very viable possibility for concentrate disposal by low-pressure injection. The key to success is favorable hydrogeologic conditions within the injection zone. In regions where there is a thick carbonate aquifer of Tertiary age, there is a high probability of developing a success horizontal injection system. Such a system could possibly accept a high capacity injection rate and would be a feasible disposal method for small to large desalination facilities.

Beach gallery disposal systems have been proposed for some seawater desalination facilities in coastal California. While this possibility should be viable for large capacity systems, the water will discharge through the top of the gallery and will not likely travel horizontally through the shallow sediments. If located offshore, the impacts of the dense water moving along or ponding by density on the bottom would likely adversely impact sessile organisms or infauna within the bottom sediments, such as mollusks. These impacts could be severe and would likely be unacceptable. An alternative would be to design the galleries within the intertidal zone (surf zone), where breaking waves would rapidly mix the concentrate with ambient quality seawater. This mixing would essentially eliminate impacts to marine life. The only other impacts would occur during construction when a dewatered excavation would be necessary to install the discharge pipes and gallery structure. The facility would be underground and would not be visible to people visiting the beach or swimming in the surf.

Perhaps the largest capacity injection wells used for concentrate disposal are deep, confined-aquifer wells penetrating karstic limestone or fractured dolomite (Walker and Missimer 1996). Deep injection wells are commonly used for the disposal of concentrate of brackish-water RO plants in South Florida (Fig. 28.1), where an injection zone with an extremely high transmissivity is present (the so-called "boulder zone" of the Lower Floridan Aquifer) that is capable of accepting very large injected water flows. Injection rates in these wells are greater than 3,800 m^3/day (1×10^6 gal/d) with some injection wells used for municipal wastewater disposal in Florida in southeastern Florida having capacities of up to 83,000 m^3/day (21.9×10^6 gal/d (Walker and Missimer 1996; Maliva et al. 2011). An example of a deep high-capacity injection well system

Fig. 28.1 Two deep injection wells under construction in Hialeah, Florida (USA). Each of the wells is capable of disposing of the 22,000 m^3/d (5.8 Mgd) plant design concentrate flow

used for concentrate disposal in an arid region is the system constructed at the Kay Bailey Hutchison Desalination Plant, located in El Paso, Texas, which has a capacity to dispose of 11,360 m^3/day (3×10^6 gal/d) of brackish-water RO concentrate. Potential high-capacity injection zones are located in some areas of the Middle East, such as within the Paleocene Umm er Radhuma Formations along the eastern Arabian Gulf or in some of the deeper, Cretaceous carbonate formations.

Use of petroleum production wells in depleted fields or as part of secondary recovery systems has been suggested for concentrate disposal. Either dedicated injection wells or depleted petroleum wells have been used by the oil and gas industry for decades for disposal of produced water (i.e. water that accompanies oil during production). The Texas Water Development Board and Texas Bureau of Economic Geology investigated the potential use of depleted petroleum wells for disposal of concentrate at inland locations in Texas (Nicot and Chowdhury 2005; Mace et al. 2006; Burnett 2007). These wells could be economically used for small to medium-capacity desalination systems that are located within semiarid and arid lands in which surface outfalls are not feasible.

Deep, high-pressure injection of brines into confined, low-hydraulic conductivity aquifers is being used in locations such as Paradox Valley, Colorado (Ake et al. 2002; Mahrer et al. 2005) and is widely used for the disposal of produced brines in the oil and gas industry. High injection pressures, often greater than 7,000 kPa (1,000 psi), are required to overcome the low hydraulic conductivity of the injection zone (Maliva et al. 2011). Some of these injection wells can dispose of large volumes of concentrate. However, high-pressure injection is commonly limited to low volume injection and may be subject to rapid clogging (depending upon the quality of the injected water), which in some instances is not reversible. Great care must be exercised in the design and operation of these wells and a detailed assessment of the geologic and geochemical conditions must be made. There have been cases where the injection of brines has induced seismicity (minor earthquakes) as demonstrated at Paradox Valley, Colorado (Ake et al. 2002; Mahrer et al. 2005). This particular issue can be controlled to a large degree by reducing the injection rate and pressure.

Subsurface disposal of concentrate generally has minimal environmental impacts, but requires careful design and technical evaluation. It is a particularly attractive option in interior locations where there are favorable geologic conditions and no other economical and environmentally sound disposal options. Typically, the injection rate is limited, so very large-capacity seawater desalination systems may not be able to use the subsurface injection disposal option, except for the possible use of beach galleries. Some brackish-water RO facilities are being designed with an additional process to reduce the volume of concentrate by running the plant discharge through a seawater membrane or thermal process. This secondary concentration process reduces the water volume requiring disposal, but also increases the fluid density, thereby requiring greater pressure for injection and raising the potential for geochemical clogging of the injection zone. There are some systems in which the water quality or density of injected fluids can actually control the desalination plant process design, in order to keep the concentrate denser that the injection zone fluid and thus eliminating the possibility of buoyancy-driven upward migration of the injectate (Missimer et al. 2002).

28.4.4 Evaporation Ponds

Discharge of concentrate using evaporation ponds is a potential disposal option in arid lands areas with high potential evaporation rates. The ponds must be lined to protect underlying groundwater from becoming contaminated (Mohamed et al. 2005). A very interesting approach is the discharge of concentrate into solar ponds, thereby both disposing of a waste stream and generating heat, and in turn, possibly electric power that could power the desalination process (Ahmed et al. 2001; Lu et al. 2001; Lu et al. 2004). In this hybrid process, the heat is harvested from the pond to help run an MED desalination process, electricity is generated using heat exchange, the concentrate is disposed of in the pond, and the salt is harvested as a product. This option is particularly attractive for desalination systems located within inland areas which have hot arid climates. There is likely a limit on the capacity of the desalination facility that could use the system because the size of the pond required and associated volume of salt that would require disposal could become problematical. This operation merges with the concept of zero liquid discharge and must be considered to have a very low environmental impact.

Another possible means of concentrate disposal in both coastal and interior basin areas of arid lands is discharge to naturally saline environments. Sabkhas (coastal and inland types) and salinas are two types of naturally-occurring evaporation basins. Coastal sabkhas are naturally-occurring salt flat areas in which there is no vegetation and salt occurs at land surface. Coastal sabkhas are present in many parts of the Middle East region. Inland sabkhas or salinas occur in closed basins in which water gathers becoming evaporatively concentrated and forming vast high salinity lakes or salt flats, such as the Great Salt Lake in Utah (USA), the Dead Sea, and the salt flats of Death Valley, California. Where these features occur, concentrate could be discharged into the environment to evaporate and add salt to an environment already containing abundant natural salt. There would be minimal impacts and it would be a low energy disposal solution. In coastal areas, some damming could be required to prevent discharge of hypersaline brine to tidal waters.

28.4.5 Zero Discharge/Crystallization

Considerable attention has been given to the concept of zero liquid discharge (ZLD) for desalination facilities (Bond and Veerapaneni 2008, 2009; Mickley 2009; Oren et al. 2010; US Bureau of Reclamation 2002). The concept of desalting seawater and reducing the concentrate to a dry crystal state either requires a large land area or is extremely energy-intensive (and thus expensive) and has a large carbon footprint. ZLD thus tends to be the concentrate disposal option of last resort. Perhaps the best opportunities for development of cost-effective zero discharge facilities would be to either discharge to a natural environment that is unaffected by the saline water (e.g. sabkhas) or to find some offsetting economic value for the dry residue. Discharge to solar ponds is a possible solution, but for limited capacity plants.

Concentrate volume reduction techniques include thermal evaporators (brine concentrators) and secondary RO desalination. The former can produce high purity water. Techniques to reduce the concentrate to a solid include crystallizers, spray dryers, and evaporation ponds. Bond and Veerapaneni (2007) proposed the following treatment series for brackish groundwater ZLD desalination:

- primary RO desalination,
- concentrate treatment (e.g., fluidized bed crystallization or granular media filitration,

- secondary RO desalination, and
- brine concentration (thermal desalination) and evaporation.

The concentrate treatment system used in the Bond and Veerapaneni (2007) pilot studies was fluidized bed crystallization with addition of sodium aluminate to remove silica. The pilot testing results were favorable in demonstrating the technical feasibility of concentration volume reduction using secondary RO desalination. However, learning how to operate the fluidized bed crystallization system was challenging. The fluidized bed crystallization process depends on numerous factors including degree of supersaturation, pellet diameter, specific pellet area, bed porosity, superficial velocity, fixed bed height, temperature, and pellet density. System design also depends on source water chemistry and treatment goals.

There is on-going research interest in developing techniques that can reduce the costs and energy requirements of ZLD. For example, Bond et al. (2011) documented the successful pilot testing a ZLD system that utilizes electrodialysis metathesis (EDM) to reduce the volume of concentrate produced by RO desalination of brackish water. The concentrated brine is then solidified using a crystallizer. ZLD using EDM appears to be less expensive than ZLD using thermal processes for desalination systems that treat water with total dissolved solids (TDS) of less than 5,000 mg/l. Thermal processes are less expensive for systems treating water with a TDS concentration of greater than 10,000 mg/l.

Laspidou et al. (2010) investigated the use of existing saltworks for ZLD. The major cost item was transporting brine from existing or potential desalination plant locations to saltworks facilities. Cost to operate a zero-discharge desalination plant is partially offset by the profit realized by selling the extra salt and cost savings from not having to construct a marine outfall. Cost to transport brine by truck is prohibitively expensive. To make process viable, there is a need to substantially reduce concentrate volume perhaps to slurry that could be pumped through a pipeline or to dryness.

Recovery of high quality salt for sale has been thoroughly investigated and can only be viable if the desalination plant is located in an area in which there is an industrial demand for the salt or extracted chemicals. Jeppesen et al. (2009) investigated the recovery of various substances from the concentrate stream and concluded that it may be financially feasible to recover, sodium, magnesium, potassium, rubidium, phosphorus, cesium, and germanium. As many of these metals become more valuable, the cost of extraction may become sufficiently profitable to offset the cost of desalination with ZLD. Even if the revenues generated from the sale of industrial compounds from concentrate cannot recoup the additional treatment costs, the technology may still be economically viable if it reduces concentrate disposal costs that would otherwise be incurred.

It is not feasible to use the ZLD/crystallization methods for the disposal of concentrate from large seawater desalination facilities. The sheer volume of salt produced in a Middle East large-capacity desalination facility would produce enough bulk salt in a few days or weeks to meet all of the industrial needs of the region for a year. Smaller-capacity desalination systems used to treat brackish water may produce a concentrate that could be reduced to dry salt, which could then be disposed of in a landfill (Oren et al. 2010).

28.5 Desalination Environmental Impacts with Future Innovations

Research on desalination is dynamic and new technologies and methods are being developed to achieve the following:

- improve treatment efficiency (energy recovery, aquifer storage, and recovery, etc.),
- reduce the net energy consumption during treatment,
- increase the viability of renewable energy use,
- reduce environmental impacts of intakes, and
- reduce environmental impacts of concentrate disposal.

The innovations discussed at the end of Chap. 27 will not only impact the energy use and cost of desalination, but they will correspondingly decrease the environmental impacts of desalination. Any reductions in the use of fossil fuels to power desalination, including greater energy efficiency, the use of renewable energy sources, and higher conversion rate of feedwater to product, will collectively reduce global impacts of the technology.

Concentrate disposal is still the leading cause of environmental impacts of desalination. The absence of an economical and environmentally sound means of concentrate disposal has been and continues to be a fatal flaw for some proposed desalination projects. For example, in some inland areas of the southern United States, brackish groundwater resources are present, but the main concentrate disposal options (surface water outfalls, blending with wastewater, and deep-well injection) are not locally viable.

Localized environmental impacts can be minimized by using proper protocols in the siting and design of desalination facilities. Sites near viable, environmentally-acceptable concentrate disposal locations can offset some additional conveyance costs for the treated water. Post-treatment of the concentrate must be achieved to make it acceptable for discharge into tidal waters. Additional research will be required to economically achieve ZLD scenarios for large-scale desalination facilities.

It can be concluded that desalination does have some negative environmental impacts, but they are far less significant compared other activities of man, particularly power generation using fossil fuels and industrial emissions of a variety of pollutants. The environmental impacts need to be considered in the context of it providing a vital resource, potable water. The environmental impacts of desalination should be viewed in terms of the adverse impacts associated with water scarcity and the environmental impacts associated with other water development projects. For example, even dams have major environmental impacts that are comparatively large compared to desalination. There is currently much research focused on making desalination more environmentally friendly and an economic option available for use by all populations requiring a new water source.

References

Abbaspour, M., Javid, A. H., Moghimi, P., & Kayhan, K. (2006, September 4–6). Modeling of thermal pollution in the northern coastal area of the Persian Gulf and its economic and environmental assessment. In *Proceedings, 8th Conference on Modelling, Monitoring and Management of Water Pollution, Water Pollution* (pp. 445–453), WP06, Bologna, Italy.

Ahmed, M., Arakel, A., Hoey, D., & Coleman, M. (2001). Integrated power, water and salt generation: A discussion paper. *Desalination, 134*(1), 37–45.

Ake, J., Mahrer, K., O'Donnell, D., & Block, L. (2002). Deep-injection and closely monitored induced seismicity at Paradise Valley, Colorado. *Seismological Society of America Bulletin, 95*, 664–683.

Andrews, L. S., Moore, R. D., & Rose, C. T. (1991). Permitting the discharge of reverse osmosis concentrate to surface water. In *Proceedings, American Water Works Association Seminar on Membrane Technologies in the Water Industry* (pp. 458–501).

Bond, R., Batchelor, B., Davis, T., & Klayman, B. (2011). Zero liquid discharge desalination of brackish water with an innovative form of electrodialysis: Electrodialysis metathesis. *Florida Water Resources Journal, 63*(7), 36–44.

Bond, R., & Veerapaneni, S. (2007). *Zero liquid discharge for inland desalination*. Denver: AWWA Research Foundation.

Bond, P., & Veerapaneni, S. (2008). Zeroing in on ZLD technologies for inland desalination. *Journal of the American Water Works Association, 100*(9), 76–89.

Bond, P., & Veerapaneni, S. (2009). New strategies for managing desalination concentrate with zero discharge from two WRF research projects. In *Proceedings of World Environmental and Water Resources Congress 2009: Great Rivers* (Vol. 342, pp. 5478–5487).

Bramer, R. N. (1990). Power station thermal effluents and marine crustaceans. *Journal of Thermal Biology, 15*(1), 91–96.

Bramer, R. N. (1995). The influence of rising background temperature on the effects of marine thermal effluents. *Journal of Thermal Biology, 20*(1–2), 105–110.

Burnett, D. B. (2007). Desalination of brackish water and disposal into waterflood injection wells. *Final Report Submitted to The Stripper Well Consortium*. College Station, Texas: Texas A&M University.

Conlon, W. J. (1988). Disposal of concentrate from membrane process plant. In *Proceedings, American Water Works Association Seminar 54 on Membrane Process Principles and Practice, 62nd Annual Conference and Exposition*.

Conlon, W. J. (1989). Disposal of concentrate. *Waterworld News, 5*(1), 18–19.

Conlon, W. J. (1990). Historical development of concentrate regulations. *Desalination, 78*, 11–16.

Conlon, W. J., & Smith, J. E. (1991). Brine waste diposal from membrane process. In *Proceedings of the American Water Works Association/WPCF Joint Residuals Management Conference*. Durham, North Carolina, USA.

Cotruvo, J., Voutchkov, N., Fawell, J., Payment, P., Cunliffe, D., & Lattemann, S. (Eds.), (2010). *Desalination Technology: Health and Environmental Impacts*. Boca Raton, Florida: CRC Press.

Einav, R., Harussi, K., & Perry, D. (2002). The footprint of the desalination processes on the environment. *Desalination, 152*, 141–154.

Hashim, A., & Hajjaj, M. (2005). Impact of desalination plants fluid effluents on the integrity of seawater, with the Arabian Gulf in perspective. *Desalination, 182*, 373–393.

Höpner, T., & Lattemann, S. (2002). Chemical impacts from seawater desalination plants—A case study of the Northern Red Sea. *Desalination, 152*, 133–140.

Höpner, T., & Windelberg, J. (1997). Elements of environmental impact studies on coastal desalination plants. *Desalination, 108*, 11–18.

Jeppersen, T., Shu, L., Keir, G., & Jegatheesan, V. (2009). Metal recovery from reverse osmosis concentrate. *Journal of Cleaner Production, 17*, 703–707.

Laspidou, C., Hadjibiros, K., & Gialis, S. (2010). Minimizing the environmental impact of sea brine disposal by coupling desalination plant with solar saltworks: A case study for Greece. *Water, 2*, 75–84. doi:10.3990/w20100075.

Lattemann, S., & Höpner, T. (2003). Seawater desalination: Impacts of brine and chemical discharge on the marine environment. LAquila, Italy: Balaban Desalination Publications.

Lattemann, S., & Höpner, T. (2008). Environmental impact and impact assessment of seawater desalination. *Desalination, 220*, 1–15.

Lu, H., Walton, J. C., & Swift, A. H. P. (2001). Desalination coupled with salinity-gradient solar ponds. *Desalination, 136*, 13–23.

Lu, H., Swift, A. H. P., Hein, H. D., Jr., & Walton, J. C. (2004). Advancements in salinity gradient solar pond technology based on sixteen years of operational experience. *Journal of Solar Energy Engineering, 126*, 759–767.

Mace, R. E., Nicot, J.-P., Chowdhury, A. H., Dutton, A. R., & Kalaswad, S. (2006). Please pass the salt: Using oil fields for the disposal of concentrate from desalination plants. *Texas Water Development Board Report 366.*

Mackey, E. D., & Seacord, D. (2008). *Regional solutions for concentrate management.* Alexandria, Virginia: WateReuse Foundation.

Mahrer, K., Ake, J., Block, L., O'Donnell, D., & Bundy, J. (2005). Injecting brine and inducing seismicity at the world's deepest injection well, Paradox Valley, Southwest Colorado. In C.-F. Tsang & J. A. Apps (Eds.), *Underground injection science and technology, Developments in Water Science 52* (pp. 361–375). Amsterdam: Elsevier.

Maliva, R. G., Coulibaly, K., Guo, W., & Missimer, T. M. (2011). Confined aquifer loading: Implications for groundwater management. *Ground Water, 49*(3), 302–304.

Maliva, R. G., & Walker, C. W. (1998). Hydrogeology of deep-well disposal of liquid wastes in Southwestern Florida, U.S.A. *Hydrogeology Journal, 6*, 538–548.

Malomosr, P., Lozier, J., Mickley, M., Reis, R., Russel, J., Schaefer, J., et al. (2004). Current perspectives on residuals management for desalting membranes. *Journal of the American Water Works Association, 96*(12), 73–87.

MEDRC. (2002). Assessment of the composition of desalination plant disposal brines: Project NO-98-AS-026, Middle East Desalination Research Center (MEDRC), Oman.

Mickley, M. (1996). Environmental considerations for disposal of desalination concentrates. *The International Desalination and Water Reuse Quarterly, 5*(4), 56–61.

Mickley, M. (2000). *Major ion toxicity in membrane concentrate.* American Water Works Association.

Mickley, M. (2001). Membrane concentrate disposal: Practices and regulation. *Bureau of Reclamation, Report.* Denver, Colorado, US.

Mickley, M. (2002). *Municipal membrane plants: Growth trends, and concentrate/backwash disposal practices and issues, Proceedings.* Tampa, Florida: American Membrane Technology Association Biennial Conference and Exposition.

Mickley, M. (2003). Desalination concentrate management and issues in the United States. In *Proceedings of the International Desalination Association World Congress on Desalination and Water Reuse.* Paradise Island, Bahamas, Paper BAH03-155.

Mickley, M. (2007). RO concentrate management. In M. Wilf (Ed.), the guidebook to membrane desalination technology: Reverse osmosis, nanofiltration, and hybrid systems process, design, application, and economic (pp. 375–394). L'Aquila, Italy: Balaban Desalination Publishers.

Mickley, M. (2009). Desalination and water purification research and development and program no. 155: Treatment of concentrate. *U.S. Bureau of Reclamation, Contract No. DIBR 04-FC-81-1050, Final Report.*

Mickley, M., Hamilton, R., Gallegos, L., & Truesdall, J. (1993). Membrane concentrate disposal. *American Water Works Association, Report*. Denver, Colorado: American Water Resources Association Research Foundation.

Missimer, T. M. (2009). *Water supply development, aquifer storage, and concentrate disposal for membrane water treatment facilities: Methods in Water Resources Evaluation Series No. 1*. Houston: Schlumberger Water Services.

Missimer, T. M., Guo, W., Martin, W. K., & Thompson, D. M. (2002). Concentrate chemistry as a control of deep well disposal and effects on membrane treatment plant design. Tampa Bay Brackish-water R. O. plant design. *Proceedings, American Membrane Technology Biennial Conference and Exposition*.

Missimer, T. M., & Pankratz, T. (2009). Concentrate water disposal. In T. M. Missimer (Ed.), *Water supply development, aquifer storage, and concentrate disposal for membrane water treatment facilities* (pp. 253–280). Houston, Texas: Schlumberger Water Services, Methods in Water Resources Evaluation Series No. 1.

Missimer, T. M., & Winters, H. (2003). Reduction of biofouling at a seawater RO plant in the Cayman Islands. In *Proceedings of the International Desalination Association World Conference on Desalination and Water Reuse*, BAH03-190.

Mohamed, A. M. O., Maraqa, M., & Al Handhaly, J. (2005). Impact of land disposal of reject brine from desalination plants on soil and groundwater. *Desalination, 182*(1–3), 411–433.

Morton, A. J., Callister, I. K., & Wade, N. M. (1996). Environmental impacts of seawater distillation and reverse osmosis processes. *Desalination, 108*, 1–10.

Muniz, A., & Skehan, S. T. (1990). Disposal of concentrate from brackish water desalting plants by use of deep injection wells. *Desalination, 78*, 41–47.

National Research Council (2008a). *Prospects for managed underground storage of recoverable water*. Washington, DC: National Academy Press.

National Research Council (2008b). *Desalination—A national perspective*. Washington, DC: National Academies Press.

Naylor, E. (1965). Effects of heated effluents upon marine and estuarine organisms. *Advanced Maine Biology, 3*, 63–103.

Nicot, J.-P., & Chowdhury, A. H. (2005). Disposal of brackish water concentrate into depleted oil and gas fields: A Texas study. *Desalination, 181*, 61–74.

Oren, Y., Korngold, E., Daltrophe, N., Messalem, R., Volkman, Y., Aronov, L., et al. (2010). Pilot studies on high recovery BWRO-EDR for near zero liquid discharge approach. *Desalination, 261*, 321–330.

Pontius, F. W., Kawczynski, E., & Koorse, S. J. (1996). Regulations governing membrane concentrate disposal. *Journal of the American Water Works Association, 88*(5), 44–52.

Sadhawani, J. J., Veza, J. M., & Santana, C. (2005). Case studies on environmental impact of seawater desalination. *Desalination, 185*, 1–8.

Thompson, M. A., & Brodeur, T. P. (1991). The key to membrane applications-concentrate disposal. In *Proceedings, International Desalination Association World Conference on Desalination and Water Reuse*.

U.S. Bureau of Reclamation (2002). Zero discharge waste brine management for desalination plants. *U.S. Bureau of Reclamation, Report no. DRDT-REPT*.

von Medeazza, G. L. M. (2005). "Direct" and socially-induced environmental impacts of desalination. *Desalination, 185*(1–3), 57–70.

Walker, C. W., & Missimer, T. M. (1996). Deep well injection of concentrates, The Collier County, Florida experience. In *Proceedings, Biennial Conference and Exposition of the American Desalting Association, The Water Puzzle, How Does Desalting Fit In*? (pp. 122–138).

Watson, I. E., Morin, O. J., Jr., & Henthorne, L. (2003). Desalting handbook for planner's (3rd ed.). *US Bureau of Reclamation, Desalination and Water Purification Research and Development Program Report No. 72*.

World Health Organization (2007). *Desalination for safe water supply, guidance for the health and environmental aspects applicable to desalination*. Geneva: Public Health and the Environment World Health Organization.

Younos, T. (2005). Environmental issues of desalination. *Journal of Contemporary Water Research and Education, 132*, 11–18.

Part VII
Wastewater Reuse in Arid Lands

Chapter 29
Wastewater Reuse

29.1 Introduction

Wastewater reuse has a long history throughout the world. Indications of reuse of wastewater go back in time at least 5,000 years to the Minoan civilization of ancient Greece (Asano and Levine 1996; Vignesnaran and Sundaravadivel 2004). The major historic and current use of wastewater is irrigation, both agricultural and non-agricultural (e.g., landscape, parks, golf courses). Irrigation is a preferred use of reclaimed water (treated sewage effluent), because it does not require water of the highest quality and wastewater contains nutrients needed for crops. Wastewater may also be recycled for industrial uses and, less commonly, to augment the potable water supply. An extensive literature has developed on all aspects of wastewater reuse. Asano et al. (2007) provides an excellent overview of wastewater reuse issues, technologies, and applications. The health risks associated with wastewater reuse and managed aquifer recharge (MAR) are discussed in Chap. 30. The use of natural contaminant attenuation processes in wastewater MAR in order to improve water quality is addressed in Chap. 31.

Wastewater reuse has five main benefits

(1) It can provide needed additional water in water scarce regions.
(2) It can provide needed plant nutrients and thus reduce fertilization costs.
(3) Reuse of wastewater may prevent or reduce the adverse impacts associated with wastewater disposal.
(4) It is a reliable source of "new" water.
(5) It is the only source of additional water that increases as population increases.

Wastewater resources necessarily must be an important component of integrated water resources management in most arid and semiarid countries (Chap. 35). In countries where current freshwater reserves are or will be in the near future at their sustainable limits, recycled wastewater is the only significant low-cost alternative water source for agricultural, industrial, and urban non-potable

R. Maliva and T. Missimer, *Arid Lands Water Evaluation and Management*, Environmental Science and Engineering, DOI: 10.1007/978-3-642-29104-3_29,

purposes (Lazarova et al. 2000; Miller 2006). In many parts of the world, using water once simply is no longer an option (Levine and Asano 2004).

Pescod (1992) noted that whenever good quality water is scarce, water of marginal quality will have to be considered for use in agriculture. Marginal water is defined as water that possesses characteristics that have the potential to cause problems when it is used for its intended purposes. Municipal wastewater may be of marginal quality, because of associated health hazards and impacts to crops. With respect to the use of wastewater for irrigation, a marginal quality requires more complex management practices (Pescod 1992).

One of the important characteristics of reclaimed water is that it is a very reliable source of water. The production of reclaimed water is relatively constant throughout the year, and is almost constant between years (Dillon 2000; Friedler 2001). The reliability of reclaimed water is a major advantage for water users, because it allows for greater confidence in agricultural investments (Friedler 2001). Reclaimed water typically also has a relatively constant quality. However, the quality of reclaimed water may not be ideal for agricultural irrigation and some other uses. Reclaimed water in arid and semiarid lands and coastal areas often has elevated salinities caused by poor quality of the freshwater supply, evaporative concentration, and seepage of saline waters into collection systems.

Wastewater disposal can exacerbate water scarcity by polluting available freshwater resources. In water scarce regions, wastewater may be the least expensive source of new water, especially if fresh ground and surface waters have reached or are close their sustainable limits. However, wastewater reuse may pose a public health risk based on the presence of pathogenic microorganisms and, to a lesser degree, chemical contaminants of variable toxicity. Sewage effluent should be treated so that it meets the quality requirements for its intended use. The reliability of treatment systems must also be tied to the uses of the reclaimed water. Large-scale use of sewage effluent for irrigation often requires that it be treated for unrestricted irrigation so that farmers can grow what they want and the water can be used for other purposes that may involve public contact, such as landscape irrigation (Bouwer 1991). However, the reality is that in many periurban areas of developing countries, wastewater is being used for agricultural irrigation with little or no treatment.

A politically and technically challenging question is the specific degree of treatment necessary for different types of wastewater recycling. There is a natural and understandable tendency towards adopting the "precautionary principle", founded on the concept of "better safe than sorry", and thus over-treatment of the wastewater occurs. However, Law (2003) pointed out that the precautionary principle is a term described by some as a reason for doing nothing. It is important to recognize that over-treatment can unnecessarily increase project costs to level at which a water recycling project becomes economically unviable and the water resources benefits are lost. Over-treatment also results in an inefficient use of what are often limited financial resources. Nevertheless, over-treatment may still be necessary to obtain public and regulatory support for a project, but its opportunity costs should be recognized.

Proven wastewater treatment and purification processes currently exist to produce water of virtually any quality desired (Asano and Levine 1996; Mujeriego and Asano 1999). Wastewater can be treated to such a degree that it poses essentially no public health threat, even if the water were to be directly consumed. For example, the wastewater treatment system for the Groundwater Replenish System in Orange County, California, includes microfiltration (MF) followed by reverse osmosis (RO) and ultraviolet light and hydrogen peroxide treatment, which produces a water of a quality far beyond that produced by many potable water systems (Markus 2009). Technical risks no longer represent the major concern for development of water reclamation and reuse projects. Other issues, such as financing, failure of management, and social acceptance, have become more critical (Bixio et al. 2005). However, such high treatment levels come at a great cost and are not necessary or economically viable for most wastewater reuse applications. Affordability and technical resources to operate wastewater treatment systems are important considerations, especially in developing countries. An important water resources challenge in water scarce regions is obtaining a balance between cost effectively obtaining the maximum water resources benefits from wastewater reuse, while at the same time providing a level of acceptable protection to public health and the environment.

Within some developing countries, Scott et al. (2010) noted that

> Investments in technologically complex treatment processes and policies have failed due to ill-planned, badly operated and badly managed facilities, under-resourced institutions, limited human-resources capacity and severe financial challenges.

In some developing countries, expensive needed wastewater treatment facilities sit unused, because the human, technical, and economic resources are unavailable for their operation and maintenance. Regrettably, international donors tend to be more willing to support new construction projects than to later finance their operation.

Potential or hypothetical health risk considerations have limited expansion of the use of reclaimed municipal wastewater for groundwater recharge when such recharge may ultimately constitute a significant fraction of the groundwater that is use for potable water supplies (Asano and Cotruvo 2004). However, accidental or unplanned groundwater recharge with wastewater already widely occurs through irrigation and land treatment, conveyance system leaks, and disposal of municipal and industrial wastewater via percolation ponds and infiltration (Asano and Cotruvo 2004). Asano and Cotruvo (2004) further noted that a properly planned and managed water reuse project can produce higher quality finished water than unplanned reuse as it is currently commonly practiced. They also noted that (p. 1944) that:

> The irony is that water derived from 'natural' but obviously imperfect sources, often receives only basic treatment (filtration and disinfection). The final product might not be as high quality as the reclaimed wastewater that has been subject to much more rigorous treatment, water quality control, and management.

An additional consideration is that it may be far less expensive to prevent a contaminant from entering the wastewater stream than later removing the contaminant prior to reuse. A source control program to limit potentially harmful constituents from entering the wastewater collection system should be an integral part of wastewater reuse projects. For example, discharges of industrial wastes to sanitary sewer systems should be prevented or at least closely controlled, if the wastewater is to be reused in a manner that may result in public contact. Improvements in the general health of the population living in collection areas will reduce pathogen concentrations in the raw wastewater. Reuse of wastewater needs to be integrated into comprehensive land, water, and wastewater management plans, rather than being an afterthought, so that the reclaimed water supply and quality meets user requirements (Bahri 1999).

Discussions of planned wastewater reuse should also consider that in many areas large amounts of unplanned reuse may occur, through aquifer recharge from on-site sewage systems (septic systems, cesspools, pit latrines), leaking sewage collection and transmission pipes, excess irrigation using wastewater, and wastewater discharged to dry or flowing rivers and streams. Common wastewater handling and reuse practices already incidentally result in high rates of infiltration to underlying aquifers in many urban areas, which represents a water resources benefit, a potential health hazard (Foster and Chilton 2004), and a potential subsurface infrastructure problem (e.g., flooding of basements and transportation corridors). In rapidly growing urban centers in developing countries, most wastewater is discharged to convenient water courses with minimal treatment. The potential range of pollutants includes pathogenic microorganisms, excessive nutrients and dissolved solids, dissolved organic carbon (DOC), toxic heavy metals, and contaminants of emerging concern (CECs).

Foster and Chilton (2004) observed that there is a need to confront the reality of current wastewater management and disposal practices and identify areas where cost-effective interventions and incremental investments can reduce risks to groundwater users rather than constructing conventional wastewater treatment plants that may be of questionable operational sustainability. The different economic realities in affluent developing countries and less wealthy developing countries unavoidably result in different approaches to wastewater disposal and reuse, irrespective of any egalitarian sentiments. It is also important to recognize that decentralized (e.g., household) reuse of wastewater is already widespread in many developing countries with minimal oversight and attention to health and safety issues. There is a need to implement inexpensive, and culturally and socially acceptable measures that can improve the safety of decentralized wastewater, especially in rural areas where centralized treatment systems will not be constructed in the foreseeable future.

Table 29.1 Wastewater terminology (after Levine and Asano 2004; World Health Organization 2006)

Term	Meaning
Wastewater	Liquid wastes discharged from domestic and commercial premises to individual or municipal disposal or treated systems. The term often refers to untreated (raw) liquid wastes.
Municipal wastewater	Wastewater that is produced mainly by household and non-industrial commercial activities.
Sewage	Mixture of human excreta and water used to flush the excreta, and water used for domestic purposes.
Graywater (greywater)	Household wastes from kitchen, bath, or laundry, which general does not contain excreta (toilet wastes)
Effluent	Liquid that flows out of a process such as wastewater treatment.
Excreta	Human waste (feces and urine)
Treated sewage effluent (TSE)	Treated water produced by a sewage treatment facility
Reclaimed water Recycled water	Treated water produced by a sewage treatment facility (generally synonymous with treated sewage effluent) that meets water criteria intended for reuse.
Reuse water	Treated wastewater that can be used for beneficial purposes.
Sludge	Mixture of solids and waste that is the byproduct of the wastewater treatment process.
Water recycling	Alternative term for water reclamation and reuse, which is hoped to result in a better public reception as it does not include the word 'waste'.
Biosolids	Treated sludge.

29.2 Wastewater Terminology

A variety of terms are used in the technical literature with respect to wastewater, which differ with respect to the source of the water and its degree of treatment. Wastewater terminology is summarized in Table 29.1. Terminology has evolved over time. What were originally referred to as "sewage treatment" plants later became more politely referred to as "wastewater treatment" plants or facilities, and more recently as "water reclamation" facilities. The term "treated sewage effluent" (TSE) refers to the treated water produced by water reclamation facilities. The terms "reclaimed water" and "reuse water" refer to treated sewage effluent that is of a quality suitable to reuse. However, treated sewage effluent and reclaimed water are commonly used interchangeably.

29.3 Wastewater Treatment Processes

Wastewater treatment systems can be divided in two main categories: electromechanical and "natural" systems (Jiménez et al. 2010). Electromechanical systems require large amounts of electrical energy to operate, high levels of

Table 29.2 Basic wastewater treatment process

Natural treatment processes	Waste stabilization ponds
	Wastewater storage and treatment reservoirs
	Septic systems
	Constructed wetlands
	Soil aquifer treatment
	High-rate aerobic ponds.
Preliminary and primary treatment	Coarse screening and grit removal
	Grit chambers or channels
	Sedimentation/clarification
	Skimming of floating materials
Secondary treatment	Aerobic microbiological removal of biodegradable organics
	–Activated sludge process
	–Trickling filters
	–Rotating biological contactors
	–Stabilization ponds
	–Membrane bioreactors
	–Surface aerated basins
	Secondary sedimentation
Tertiary treatment—filtration	Depth filtration (sand, dual-media)
	Surface filtration
	Membrane filtration
	Dissolved air floatation
Tertiary—advanced wastewater treatment	Coagulation-sedimentation
	Nitrification & denitrification
	Phosphorous removal
	Granulated activated carbon
	Reverse osmosis
Disinfection	Chlorine
	Chloramines
	Ozone
	Ultraviolet

financial investment, and skilled manpower for their successful operation and maintenance. Natural systems, on the contrary, have much lower electrical energy requirements, are relatively inexpensive to construct and operate, require less skilled manpower, and are thus more suitable for developing countries.

Wastewater treatment processes are categorized into preliminary, primary, secondary, and tertiary or advanced treatment processes. Some of the common treatment processes are listed in Table 29.2. Prescod (1992), Treweek (1985), National Research Council (1994) and Mujeriego and Asano (1999) provide concise summaries of the wastewater treatment processes, and more detailed descriptions can be found in wastewater engineering text and reference books (e.g., Asano et al. 2007; Tchobanoglous et al. 2003). Natural treatment systems suitable for development countries were reviewed by Mara (2003) and Jiménez et al. (2010).

Preliminary treatment involves the removal of coarse solids and other large materials using screens, grit removal systems, and in some systems, the comminution of large objects. Primary treatment basically involves the removal of settleable solids by sedimentation and the removal of floating materials by skimming. Solids removal is typically performed using sedimentation tanks or clarifiers. The collected sludge is treated by a variety of processes including aerobic or anaerobic digestion, stabilization in ponds, and land application. Many sludge disposal systems dry the sludge for sale as a soil enhancer or pellitize it using a drying apparatus for sale or disposal.

Secondary treatment includes the additional step of the removal of residual soluble, colloidal, and suspended biodegradable solids by aerobic biological processes and secondary sedimentation or clarification. Commonly used technologies include activated sludge, trickling filters, and rotating biological contactors. Secondary treatment processes vary primarily in how oxygen is supplied and the rate at which microorganisms metabolize the organic matter. Secondary treatment also results in some reductions in nutrient and pathogen concentrations.

Tertiary treatment involves the addition of one or more treatment steps beyond secondary treatment to further remove suspended solids and dissolved substances to meet water quality targets. Typically, tertiary or advanced treatment is applied to meet reuse water quality requirements. Additional filtration is often performed to achieve a water quality that is suitable for reliable disinfection and to eliminate suspended solids on which bacteria and viruses can attach.

The term "advanced wastewater treatment" or AWT is used when tertiary treatment includes nutrient (nitrogen and phosphorous) removal. The final treatment step is disinfection, which is performed to kill or deactivate any remaining pathogenic organisms present in the wastewater. Tertiary treatment and disinfection can include a wide variety of physical, biological, and chemical processes, and can result in production of a very high-quality, safe water suitable for a wide variety of reuse applications.

Advanced membrane treatment processes, which may include microfiltration, nanofiltration or reverse osmosis, are effective in reducing DOC and the concentrations of some trace organic compounds. Tests using reclaimed water compounds (EDTA, NTA, and APECs) revealed that some organic compounds may pass through membranes and are detected in the permeate (Drewes et al. 2003). Membrane treatment processes are generally less effective in the removal of low molecular weight ($\leq$500 Dalton) organic compounds.

Key issues in determining the appropriate advanced treatment technology to use are the operational reliability of the each unit or process and the overall capability of the complete treatment system to provide reclaimed water that meets water quality criteria for the planned uses (Mujeriego and Asano 1999). Asano et al. (2007) emphasized that reclaimed water treatment processes must perform reliably if the public is to have confidence that the practice of water reuse is acceptable, especially if indirect potable reuse may occur. Monitoring of performance is thus an important part of reuse.

Wastewater treatment system reliability depends upon the (Asano et al. 2007)

- ability to treat influent with temporal variation in wastewater flows and quality,
- degree of instrumentation and automation,
- mode of operation (continuous, seasonal, or intermittent),
- availability of standby (back-up) processes and equipment to operate during upsets or surges, and
- robustness of system design.

The technologies must also be affordable and within the technical capabilities of local wastewater treatment plant personal to operate and maintain them. Hence, the treatment solutions employed in affluent developed countries are often not practical in developing countries.

Natural biological treatment processes are also widely used, particularly outside of the developed countries. Wastewater stabilization ponds (WSPs) are a type of natural biological treatment system that utilizes trains of ponds to treat wastewater. WSPs are a safe and effective treatment method that is especially effective and widely used in warm climates. WSP systems are reviewed by Ramadan and Ponce (2010). WSPs are a series of shallow holding basins or ponds that are used for secondary treatment of wastewater. The most common design consists of two parallel trains of ponds that may include, in order:

- anaerobic ponds,
- facultative ponds, and
- maturation ponds (or constructed wetlands).

The anaerobic and facultative ponds are designed primarily for biochemical oxygen demand (BOD) removal and the maturation ponds are designed for pathogen removal, although both processes occur in all three types of ponds. Anaerobic ponds are deep (commonly 3–5 m; 10–17 feet) in order to exclude oxygen and encourage the growth of anaerobic bacteria that breakdown organic matter in the effluent. They function in essence as large open septic tanks and also allow for the settling of solids as sludge.

Facultative ponds are shallower ponds (1–2 m; 3–7 feet deep) that are designed for BOD removal through aerobic processes. Organic matter is metabolized by heterotrophic bacteria with dissolved oxygen provided by microalgae rather than by aeration equipment. The algae also remove nutrients. Maturation ponds are very shallow (<1 m deep; 3 feet deep) ponds that are designed to provide tertiary treatment. Constructed wetlands may also be used for maturation (i.e., polishing of the wastewater).

The WSPs should be constructed in areas where the surficial sediments have a low permeability or the ponds should be lined. A steady effluent flow has been shown to encourage the rapid and continuous growth of bacteria involved in the biological breakdown of the effluent. It is also important to have a slow enough flow through the system in order to provide sufficient retention time to allow for the biodegradation of organic matter and pathogen die-off. The systems should also not be overloaded with BOD.

WSPs can be a highly effective wastewater treatment method that can achieve >90% BOD removal, 70–90% nitrogen removal, 99.999% fecal coliform reduction, and 100% helminth removal (Ramadan and Ponce 2010). They have the advantage of design, construction, and operational simplicity, relative low costs and skilled labor requirements, and a low energy requirement. WSPs have large land requirements relative to some other wastewater treatment options and have to be located and operated to minimize odor nuisance.

The levels and methods of the sewage treatment are dictated by uses of the reclaimed water, human health concerns, and environmental priorities. A major factor in determining the level of treatment required for wastewater is whether indirect potable reuse (Sect. 29.5.2) is expected or may potentially occur. Reclaimed-water MAR systems that store water for irrigation use in aquifers in which indirect potable reuse will not occur require lesser degrees of treatment. Secondary treatment plus filtration and disinfection should be sufficient. The filtration is useful for further removal of suspended solids and large microorganisms (e.g., *Giardia cysts* and *Cryptosporidium* oocysts). Meeting of disinfection-byproduct (DBP) standards may also be an important regulatory issue for reclaimed water MAR systems, since the relatively high dissolved carbon concentration in reclaimed water is favorable for their formation both before and after recharge.

Wastewater treatment should also not be divorced from economic considerations. The treatment systems must be affordable for the nation, city, or community to construct and operate, and must not exceed the local level of technical expertise. With respect to the reuse for agricultural irrigation, Pescod (1992) noted that the most appropriate wastewater treatment method is that which will produce an effluent meeting the target microbiological and chemical qualities at low costs and with minimal operational and maintenance requirements. It is also recognized that the level of treatment will necessarily vary between regions and countries based on costs and consideration of the difficulties of reliably operating complex systems. It is better to design reuse systems to safely accept a low-grade effluent rather than to rely on advanced treatment processes to produce a reclaimed effluent that must continuously meet stringent water quality standards (Pescod 1992).

Rural areas and small towns and cities in developing countries usually lack centralized sewerage systems, which precludes centralized reuse systems. Wastewater disposal is usually decentralized and largely uncontrolled with common disposal methods including septic systems, pit latrines, or cesspits. Graywater reuse offers considerable promise as a source of water for household-scale (or multiple household-scale) restricted irrigation because it is typically much less dangerous than "blackwater" containing fecal matter. The technical, social, economic, and policy issues associated with graywater use in the Middle East is reviewed by McIlwaine and Redwood (2010). Some treatment of the graywater is necessary to reduce pathogen concentrations, potential soil damage, and environmental impacts such as odors, mosquitoes, and rodents. Advantages of graywater reuse systems include the reduced use of expensive or limited freshwater for irrigation, an increase in crop productivity, and savings from reduced cesspit cleaning. Reuse of graywater can also adversely impact centralized wastewater treatment systems by reducing flows and dilution.

29.4 Wastewater Reuse for Agricultural Irrigation

The use of wastewater for agricultural irrigation was reviewed in the Food and Agriculture Organization Irrigation and Drainage Paper 47 (Pescod 1992) and in much greater detail in the World Health Organization (2006) Guidelines for the Safe Use of Wastewater, Excreta and Greywater. Use of treated wastewater for agricultural irrigation is attractive because of the large water requirements for irrigated agriculture and its lesser quality requirements than for potable use. The suitability of a reclaimed water supply for local irrigation depends on three broad issues, (1) its compatibility with site soil characteristics, crops, and irrigation methods, (2) health and safety issues associated with its wastewater origin, and (3) economic considerations.

Reclaimed water must meet the same water quality criteria as other irrigation water sources, particularly (e.g., Pescod 1992; Asano et al. 2007)

- concentration of soluble salts,
- relative proportion of sodium to other cations (sodicity),
- concentration of boron and other elements that are toxic to plants,
- nutrient concentrations, and
- suspended solids concentrations.

Unacceptable salinity and sodium concentrations can result in

- soil salinization and associated reduction in hydraulic conductivity,
- reduction in crop yields, which depends upon crop sensitivity, and
- contamination of underlying aquifers.

Salinity and sodium concentrations are not reduced during conventional wastewater treatment. Technologies that can reduce salinity, such as reverse osmosis, are by far too expensive to be economically viable for irrigation of most crops and turf. Instead, salinity can be managed through methods such as blending with fresh water or occasional irrigation with low salinity water to flush out salts (Toze 2004).

Discussion of the wastewater use for irrigation needs to start with recognition that there is a large difference in wastewater irrigation practices between most economically developed and developing countries. The reality is that in developing countries, irrigation with wastewater is already an increasingly common practice around most cities (approximately 4 of 5) in the developing world, with most of it being untreated (Bos et al. 2010). In many of the areas where wastewater irrigation is already practiced, effective treatment will not be available for years to come (Bos et al. 2010). Within developing countries, most use of wastewater used for irrigation is unplanned, i.e., outside of the effective planning and control of the government. In many areas there is a laissez-faire attitude towards wastewater reuse as authorities face more important challenges of urbanization and economic development (Qadir et al. 2007). The extent of irrigation with untreated wastewater is often unknown or ignored by governments because of fears of raising attention to the issue, which could have economic repercussions for the trade of agricultural products (Qadir et al. 2007).

There are clear societal gains from the existing practice of irrigation with untreated wastewater in developing countries, particularly the economic value of the crops grown and the improvements in the livelihood of farmers and communities dependent upon farmers (Qadir et al. 2007). A consequence of uncontrolled reuse is that irrigation of vegetable crops with untreated or inadequately treated wastewater is a major source of enteric disease in some countries (Asano et al. 2007). However, it has also been made clear that the approach of banning the largely informal practice of irrigation with untreated wastewater will not work (Scott et al. 2010). The issue is not whether or not to irrigate with wastewater, but how to reduce the risks associated with the practice.

In more economically developed areas, at least some regulations and water quality standards are in place concerning the use of wastewater for agricultural irrigation. A logistical challenge is that the main production centers for treated municipal wastewater are towns and cities, which are becoming more and more distant from agricultural areas (Salgot and Torrens 2008). Agricultural reuse systems thus often require long transmission mains and pump stations, and involve great costs.

The primary health concern associated with the use of wastewater for irrigation is the transmission of pathogenic microorganisms to humans, which can occur where raw or poorly treated wastewater is used to irrigate salad crops or other vegetables that are eaten uncooked (Pescod 1992). The World Health Organization (2006) guidelines provide a sound basic policy framework for the use of wastewater for agriculture. The guidelines also address aquaculture, which is generally not currently practiced to a significant degree in arid and semiarid lands. Conservation of nutrients may be an important consideration for wastewater systems in which the effluent is used for agricultural irrigation, whereas nutrient removal is an objective where the effluent is discharged to surface waters (Jiménez et al. 2010).

Agricultural reuse should also be considered part of the wastewater treatment process. Although reuse has an economic value in terms of crop production and water conservation, it also results in the removal of nutrients that might otherwise have entered shallow aquifers. Another key point that needs to be emphasized again is that wastewater reuse reduces or eliminates environmental impacts associated with wastewater disposal. Economic analyses of wastewater reuse systems need to consider both the benefits associated with the use of reclaimed water and any savings resulting from reduced disposal requirements and impacts.

A key water management goal is to progressively optimize wastewater reuse, by maximizing its water resources and economic benefits while reducing associated public and environmental health risks. The World Health Organization (2006, p. viii) stated that:

> The use of wastewater, excreta, and greywater in agriculture can help communities to grow more food and make use of precious water and nutrient resources. However, it should be done safely to maximize public health gains and environmental benefits.

The World Health Organization (2006, p. 6) also noted that:

> The Guidelines should be considered in the context of national environmental, social, and economic conditions.

Pescod (1992) similarly pointed out that the guidelines need to be interpreted carefully and modified in light of local epidemiological, sociocultural, and environmental factors. The World Health Organization (2006, p. vii), also noted that:

> Overly strict standards may not be sustainable and, paradoxically, may lead to reduced health protection, because they may be viewed as unachievable under local circumstances and, thus, ignored.

And (p. 6)

> The Guidelines set target values designed in such a way as to allow progressive implementation and, therefore, to be achieved over time in a systematic, orderly and incremental way, depending on current realities and the existing resources of each individual country or region. The greatest threats to health should be prioritized and addressed first.

The World Health Organization (2006) thus recognizes the basic reality that more affluent nations can afford higher levels of wastewater treatment than developing nations. Less wealthy nations should focus on implementation of wastewater treatment practices for reuse as resources become available, focusing first on the steps that could provide the greatest public health benefits. It was also advocated that (World Health Organization 2006, p. 35):

> the adverse impacts of wastewater use in agriculture should be carefully weighed against the benefits to health and the environment associated with these practices. Yet this is not a matter of simple trade-offs. Whenever wastewater use in agriculture contributes significantly to food security and nutritional status, the point is to identify associated hazards, define the risks they represent to vulnerable groups and design measures aimed at reducing these risks.

Malnutrition in some circumstances may pose a greater threat to public health than the use of wastewater for agricultural irrigation. However, the World Health Organization advocates that efforts still should be made to reduce the health risks.

The health risks associated with the reuse of reclaimed water are discussed in Chap. 30. The contaminants of greatest concern in wastewater are pathogenic microorganisms and macroorganisms (Pescod 1992; Toze 2004, 2006). Most studied health risks related to the use of wastewater, excreta, and graywater, are those associated with excreta-related infectious diseases (World Health Organization 2006). Organic chemicals usually exist in wastewater at very low concentrations and ingestion over prolonged periods would be necessary to produce detrimental effects on human health.

The potential health impacts associated with infectious diseases extends to anyone who may come in contact with the wastewater or products of the land treated with it. Three main groups of people that are at risk from infectious associated diseases from the use of wastewater for irrigation are (Pescod 1992; World Health Organization 2006):

- farm or pond workers and crop handlers, and their families,
- local communities in close proximity to wastewater use areas and people who otherwise may have contact with fields, ponds, wastewater, excreta, graywater or products contaminated by them, and
- product consumers and food preparers.

Both local consumers and consumers of exported food can be impacted by contaminated food. The latter can adversely impact trade if it results in a loss of confidence in the safety of the produce from a country or region. There have been several events of food poisoning in recent years from contaminated produce that resulted in multiple fatalities.

A variety of different strategies are available for controlling the negative health impacts associated with the use of wastewater for agricultural irrigation, as follows (World Health Organization 2006):

- treatment of wastewater to prevent contaminants from entering the environment, or reducing their concentration,
- crop and produce restrictions—wastewater is used to only irrigate plants that are not eaten directly by people or that are always processed or cooked before being eaten,
- wastewater application techniques and holding periods aimed at reducing contamination of produce and to allow for sufficient time for pathogen die off in the environment prior to harvest,
- exposure control methods to protect workers (e.g., personal protection equipment),
- produce washing and cooking to reduce exposure for produce consumers, and
- improvements in public health and hygiene to reduce pathogen input into wastewater and graywater.

Simple biological treatment methods such as storage of wastewater in ponds or reservoirs prior to application can reduce health risks (Postel 1992). Clearly a much higher degree of treatment and water quality are needed for the irrigation of crops that are likely to be eaten uncooked or where direct contact of humans with the wastewater will occur (Pescod 1992; Postel 1992). Where such a high degree of treatment of wastewater is not possible, health goals could be met by prohibiting the use of wastewater for crops that are eaten raw. The health risk associated with the use of wastewater can be reduced by employing drip irrigation in which the water is applied to or below the soil surface and not on the parts of a plant that are eaten or touched (Postel 1992).

More than one method is typically used in combination to provide increased public health protection (Sect. 30.9). The methods chosen are also dependent upon local resources. The World Health Organization (2006) recognizes that in many developing countries, requiring wastewater treatment to a degree that would ensure the water is free of pathogens (e.g., double membrane filtration) is not an economically feasible option. The very high costs and operational complexity of wastewater treatment processes that can remove all pathogens to safe levels will generally preclude its application in developing countries. The World Health Organization (2006) instead advocates that non-treatment approaches need to be considered to prevent transmission of pathogens or exposure to hazardous chemicals. Modification of behavior (e.g., modifying irrigation practices, produce washing) is far less expensive than constructing complex treatment infrastructure and can result in considerable risk mitigation (Scott et al. 2010).

Behavioral modifications are subject to breakdown on an individual level. For example, produce washing can reduce health risks from pathogens, but it is impossible to ensure that all produce consumers will properly wash their produce before consumption. Nevertheless, behavioral modifications may reduce societal health impacts from wastewater reuse. Returning to produce washing, if most people wash their produce, then the incidence of disease from pathogens in wastewater would be reduced. That some people in a community will invariably chose not to adopt behavioral modifications that could improve their health, does not negate the overall benefits achieved by those who do adopt healthier practices.

It is recognized in the 2006 WHO Guidelines that health-based targets, are to viewed as 'targets' and not absolute limits, and that different countries and communities differ in their ability at a given time to meet the standards. A common metaphor is that countries are on different "rungs" on the "sanitation ladder". Local standards and actual implementation should progressively develop as a country moves up the sanitation ladder (Bos et al. 2010).

The World Health Organization (2006) recommends that the regulation and management of the use of wastewater and graywater should include the following essential functions:

- identification of hazards,
- generating data on health risks and the effectiveness of possible health protection measures to manage them,
- establishing health-based targets to manage health risks,
- implementing health-based targets to manage health risks, and
- system assessment and monitoring.

Assessment of the risks associated with wastewater reuse is addressed in Chap. 30.

29.5 Potable Reuse

Potable reuse is the introduction of wastewater into the drinking water supply. Potable reuse may be either direct or indirect and may occur in either a planned or unplanned manner. Potable reuse is of obvious greater public health concern than agricultural reuse because it is potentially a more direct pathway for human exposure to pathogens and other contaminants in wastewater. Wastewater can be treated to whatever quality desired and reclaimed water can be treated to a potable quality. It has been demonstrated that technologies now exist that can produce reclaimed water of a quality equivalent to or better than that of many potable water sources (Asano and Levine 1996). Nevertheless, potable reuse projects often elicit strong public opposition, even where it can be demonstrated that the projects will pose no significant health risks. Emotional reaction often supersedes rational evaluation of the facts. As Dr. Lucas Van Vuuren noted "Water should not be judged by its history, but by its quality."

29.5.1 Direct Potable Reuse

Direct potable reuse is defined by Asano et al. (2007) as the introduction of highly treated wastewater either directly into the potable water distribution system downstream of the water treatment plant or into the raw water supply immediately upstream of the water treatment plant. To date, direct potable reuse, sometimes distastefully described as "toilet to tap" has seldom been seriously considered because of public perception and opinion issues.

Direct potable reuse was implemented in Chanute, Kansas, as an emergency measure during an extreme drought during 1956–1957. An epidemiological survey showed fewer cases of stomach and intestinal illnesses during the period of reclaimed use than in the following winter when Chanute resumed using river water, which received wastewater from upstream communities (Asano et al. 2007). The City of Denver, Colorado, conducted the Direct Potable Water Reuse Demonstration Project from 1985 to 1992. The project included whole animal testing of the highly treated water. Based on an assessment of the results of an extensive testing program, it was concluded that the reclaimed water met all health standards and was of equal and better quality than the city drinking water. Nevertheless, direct potable reuse was not implemented in Denver.

The world's first direct potable reuse system was commissioned in Windhoek, Namibia, in 1968. The history of direct potable reuse in Windhoek was reviewed by Haarhoff and Van der Merwe (1996), Law (2003), du Pisani (2006) and Asano et al. (2007). Direct potable reuse was implemented because of the lack of alternative economically viable water supply options. A cornerstone of the reclamation system is the separation of industrial wastes from domestic effluent, with the former being sent to a different wastewater treatment plant. The current facility, the New Goreangab Water Reclamation Plant, went into operation in August 2002. The current wastewater treatment system includes the following series of processes: coagulation/flocculation, dissolved air floatation, dual-media filtration, ozonation, granular activated carbon (GAC) filtration, membrane filtration (ultrafiltration), and chlorination (Law 2003; du Pisani 2006).

Although it is certainly possible to treat wastewater to a level for which it poses no significant health risk, there is still great difficulty in obtaining public acceptance of direct potable reuse. Du Pisani (2006) emphasized that to obtain public confidence, quality monitoring and control are of the utmost importance, and that the most difficult challenge for emulating the Windhoek experience would be breaking down the fear-barrier to direct potable reuse. Du Pisani (2006) also observed that direct potable reuse may only be realistically considered in cases where no viable alternatives exist.

29.5.2 Indirect Potable Reuse

Indirect potable reuse was defined by the National Research Council (1998) as "the abstraction, treatment, and distribution of water for drinking from a natural source-water that is fed in part by the discharge of wastewater effluent." Indirect potable reuse includes storage in a natural environment, which in the case of an MAR system, would be an aquifer. Depending upon the local regulatory environment, public sentiments, and system-specific water quality issues, high levels of treatment may be required for systems considered to involve indirect potable use. For example, the wastewater treatment process for the Talbert Seawater Intrusion Barrier in Orange County, California (Water Factory 21 and its replacement, the Groundwater Replenish System), is based on the assumption that indirect potable reuse will occur, and includes multiple treatment processes that are utilized to produce a very high purity water that meets strict drinking water-quality standards (Markus 2009). The infiltration basin wastewater recharge system at Orange County is also operated under the same assumption and an additional mitigative measure is a long time travel to the nearest potable water supply well.

Indirect potable reuse incorporates an environmental buffer between the wastewater and water supply. The environmental buffer can be either surface water or groundwater. Environmental buffers allow for the dilution and degradation of many remaining contaminants by physical and chemical processes. The environmental buffer also provides a time buffer in the event of a treatment system failure.

Public health issues related to indirect potable reuse were reviewed in detail by the National Research Council (1998) and Rodriguez et al. (2009). Indirect potable reuse may be either planned or unplanned. Planned indirect potable reuse is the purposeful augmentation of the water supply with reclaimed water derived from treated municipal wastewater (National Research Council 1998). Planned indirect potable reuse involving underground injection of wastewater or surface spreading has had limited implementation to date, but there is increasing worldwide interest. Unplanned indirect has long been a common reality, particularly with respect to the use of surface waters for potable supply (i.e., the Mississippi River in the United States).

Planned indirect potable reuse systems incorporate a multiple barrier approach that may include some or all of the following elements (Asano et al. 2007):

- source controls,
- robust and redundant conventional wastewater treatment,

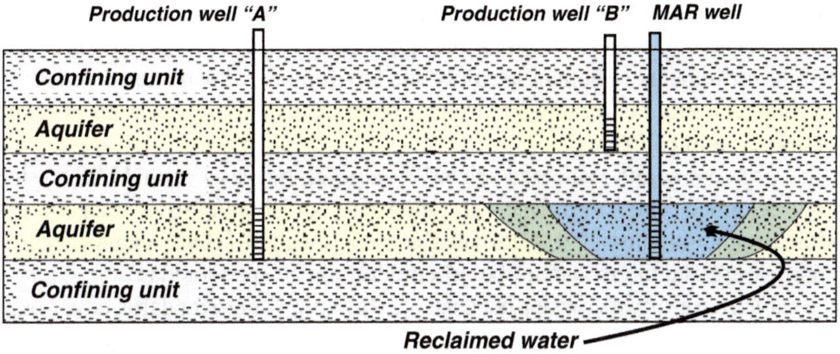

Fig. 29.1 Indirect potable reuse may be avoided by the horizontal separation of recharge of reclaimed water from production wells **A** or through the use of separate aquifers for reclaimed water recharge and water production **B**

- robust and redundant advanced wastewater treatment,
- environmental buffer that allows for additional natural treatment and dilution,
- water treatment, and
- a robust monitoring program.

An important question raised by Asano et al. (2007) is whether to apply advanced treatment technologies (if necessary) to the wastewater or potable water treatment processes. For example, should reverse osmosis be applied to the secondary treated wastewater or to the water recovered for potable supply? Inasmuch as not all of the reclaimed water may enter the potable supply, applying advanced treatment to the water supply may be the more cost-effective option.

Unplanned, indirect potable reuse is the unintentional addition of wastewater to a water supply that is subsequently used as a water source (National Research Council 1998). Unplanned, indirect potable reuse is common in areas in which surface waters are used for potable water supply. Communities using rivers for water supply often locate the intakes for their water treatment system upstream of the community and the wastewater outfall downstream. The water supply for downstream communities thus contains wastewater from all the communities located upstream of their intakes. The wastewater is diluted and naturally treated to varying degrees by the river freshwater flows and biological processes.

In the context of MAR, unplanned indirect potable reuse would involve the unintentional migration of recharged water into a potable water supply well. The potential for unplanned indirect potable reuse can be greatly reduced by a combination of hydraulic and geographic separation of reclaimed-water MAR systems from potable water-supply wells (Fig. 29.1). MAR systems that recharge water for a non-potable use, such as irrigation, may be located in different areas as potable water-supply wells or may use different aquifers.

Indirect potable reuse may still be a regulatory issue for MAR systems in some jurisdictions even though from a technical perspective there is no realistic means for it to occur (no pathway). Regulatory requirements related to the health impacts of indirect potable water reuse may be applied to projects for which there is no potential for impacts to human health or the environment. Such over-conservatism may come at a great cost and could render projects that have great water resources benefits economically unviable.

The primary health concerns related to aquifer recharge with MAR systems using reclaimed and stormwater are microbiological and, to a lesser degree, chemical contaminants. The National Research Council (1998) emphasized that drinking-water standards cannot be relied upon as the sole standard for safety for the potable reuse of wastewater because the drinking water standards cover only a limited number of contaminants and they were intended for conventional, relatively uncontaminated sources of freshwater. Evaluations and the design of wastewater reuse and MAR systems should include an evaluation of what contaminants might be present in the reclaimed water and the fate and transport of those contaminants in the aquifer.

29.5.3 Wastewater MAR

Wastewater MAR provides both water quality and storage benefits. The main applications of MAR using wastewater include

- Additional "natural" treatment. Polishing of wastewater quality occurs through various SAT, ASTR, and ARR applications (Chap. 23).
- Aquifer recharge. Recharge of wastewater may increase water levels in overdrafted aquifers, or at least reduce the rate of decline of water levels.
- Salinity barriers. Strategic recharge in coastal areas may prevent or even reverse saline-water intrusion.
- Temporal storage. Storage of reclaimed water is used to meet later needs for non-potable uses.

The quality of wastewater is improved by the removal of microbial and chemical contaminants by natural attenuation processes (Chap. 31). Aquifers can accommodate large-scale storage of treated wastewater and may also function to transmit the reclaimed water to its point of use. Treated wastewater may be conveyed in an aquifer in ASTR or ARR systems from the recharge site to downstream users (Fig. 29.2). Natural attenuation of contaminants along the flow path would reduce health risks associated with residual contaminants in the treated wastewater. MAR technologies are reviewed in Chap. 23. ASR and MAR using wells for the recharge of wastewater is discussed in detail by Maliva and Missimer (2010). MAR of wastewater differs from MAR using other types of water primarily in the real and perceived health concerns and greater susceptibility of wells and infiltration basins to clogging because of the presence of relatively high concentrations of suspended solids and nutrients.

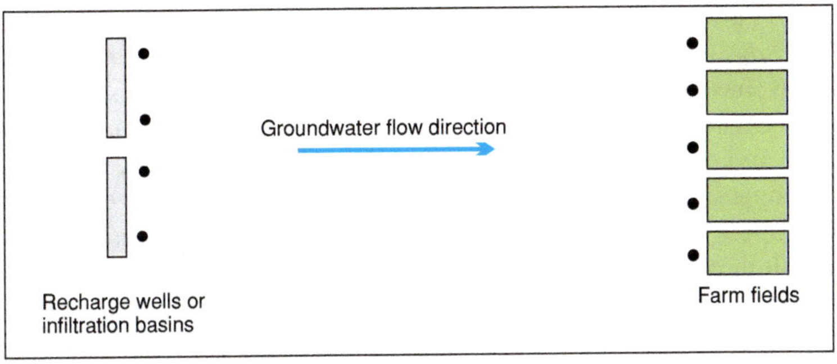

Fig. 29.2 Conceptual diagram of an ASTR or ARR system. Wastewater is recharged upstream of the point of use, which in this example is farm fields. Recharged wastewater flows down-gradient to the farm fields with an associated reduction in contaminant concentrations due to natural attenuation processes at the recharge site and within the aquifer. Alternatively, flow can be induced by a hydraulic gradient created by the recharge and groundwater pumping

Depending upon prior wastewater treatment, when wastewater is to be recharged into an MAR system, it may require some pretreatment. The pretreatment requirements for wastewater to be used for aquifer recharge depend upon the following factors (Asano and Cotruvo 2004):

- quality of the wastewater (presence and likely concentrations of pathogens, dissolved solids, heavy metals, and organic compounds of concern),
- treatment processes that are technically available and economically feasible for removing contaminants of concern in the wastewater before recharge,
- water quality and uses of the receiving aquifer,
- effects of infiltration and percolation, and groundwater passage on water quality,
- spatial and temporal separation of the recharge location from potentially impacted sensitive receptors (i.e., maintaining sufficient travel times for natural attenuation processes to occur),
- degree of groundwater dilution,
- type, capability, and reliability of the water treatment processes to remove contaminants that reach an extraction point,
- extent and type of human exposure and known or potential health risks,
- regulatory requirements, and
- public perception and cultural issues.

The last two factors may not be in tune with the actual health and environmental risks associated with MAR projects involving reclaimed water. For example, fresh and brackish water aquifers (i.e., aquifers containing <10,000 mg/l of total dissolved solids) are considered to be Underground Sources of Drinking Water (USDWs) under U.S. Environmental Protection Agency rules, and are required to be protected as such, even though under many circumstances there is little likelihood of actual potable use (e.g., 10,000 mg/l water cannot be directed consumed without desalination). The lack of specific criteria and guidelines governing aquifer recharge of groundwater is currently hampering the implementation of large-scale groundwater recharge projects in some areas (Asano and Cotruvo 2004). Wastewater recharge may be governed by just the nebulous general mandate that projects should not cause significant environmental harm, which is open to individual interpretation, leading to high costs, project delays and cancelation, and, in some situations, litigation.

Public opposition may be aroused for wastewater MAR recharge projects under the perception that what are perceived to be pristine aquifers will become impaired. It therefore behooves professionals involved in MAR projects to become involved to the extent possible in the rule-making and public education process. MAR using reclaimed water may require a substantial public outreach and educational effort, especially for the first project in an area.

As is the case for wastewater reuse in general, it is important to recognize that the treatment standards for wastewater used for MAR necessarily must consider local socioeconomic conditions. Even if drinking water augmentation is not explicitly foreseen in a MAR project, drinking water quality standards are still commonly applied to the recharged or recovered water in many applications in developed countries (Wintgens et al. 2008). High-end, state-of-the-art systems, such as the Orange County Groundwater Replenishment System (Markus 2009), are not a feasible template for the undeveloped world (nor in most cases in the developed world in economic difficult times). Instead, natural attenuation processes (Chap. 31) along with conventional wastewater and water treatment may be relied upon to improve water quality. The overall reduction in health risks associated with each link in the wastewater treatment, MAR, water treatment, and water use chain need to be considered rather than focusing entirely on any one link.

There is now abundant evidence that MAR can result in substantial improvements in water quality (Chap. 31) and in some instances produce a water of higher-quality than is otherwise available. MAR with treated wastewater or impaired surface water by surface application in areas with thick vadose zones may provide water that is of superior quality to existing groundwater. For example, a laboratory study of MAR in Addis Ababa, Ethiopia, using treated wastewater from the Kaliti Wastewater Treatment Plant and polluted water from the Big Akaki River, indicated that MAR through thick vadose soils could produce water nearly free of total coliform bacteria and significantly reduce concentrations of ammonia, nitrite, COD, and BOD (Abiye et al. 2009). MAR using treated wastewater would both recharge the aquifer, which is experiencing declining water levels from rapidly increasing groundwater pumping, and put the treated wastewater to beneficial use. The recharged water would have comparable or even better water quality compared to that in the existing nearby Akaki Wellfield.

29.6 Economic Issues

Wastewater reuse has two fundamental components (1) actual treatment of the wastewater, and (2) distribution of wastewater to potential customers. The former is a cost that may have to be borne even without reuse, because wastewater will still require treatment before it can be disposed of in an environmentally sound manner. The level of treatment is technically tied to the quality of the water that is being reused and the quality requirements for the intended use of the water, which includes consideration of human and environmental safety. The United Nations Economic and Social Council succinctly noted in the 1958 that "no higher quality water, unless there is a surplus of it, should be used for a purpose that can tolerate a lower grade" (Okum 2002). Secondary or tertiary treated wastewater may still require additional treatment to meet water quality requirements for some uses, such as unrestricted irrigation and the recharge of aquifers used for potable water supply.

The cost of construction of reclaimed water distribution systems vary greatly depending upon local circumstances, and may be cost prohibitive. For example, in older, sprawling metropolitan areas, the local irrigated areas may be capable of accepting only a small percentage of the daily wastewater flows. The installation of dual-piping systems for local residential use in existing communities in many circumstances cannot be justified on purely economic terms. The cost of installing new underground utilities in developed areas is often prohibitive. Dual-piping systems are usually economically viable in new communities with significant open space, in which reuse water lines can be installed at the same time as other utilities. Nevertheless, in water scarce areas, the substantial costs of retrofitting existing communities with a dual-piping system may still be economically attractive relative to the costs of developing new water supplies (e.g., desalination) and constructing new or upgrading existing wastewater treatment facilities (Okum 2002).

The economics of wastewater reuse is a subject that is receiving increasing considerable attention, and has been reviewed by Khouri et al. (1994), Khouzam (2003), Bixio et al. (2005), Hochstrat et al. (2007) and Winpenny et al. (2010). A fundamental issue is that wastewater reuse has great value from an integrated water resources management perspective and can have great societal benefits, but yet many water scare countries are far from optimizing the use of this valuable water resource. The authors have observed that there are many potential wastewater MAR projects that would have considerable water resources values in water scarce regions, but it is difficult to make a business case for a specific government or private entity to invest in the projects under the prevailing economic environment. For example, where reclaimed water is distributed to farmers for free, there is no direct cash flow to pay for system construction and operation.

Wastewater reuse can be economically justified on a societal basis when (Khouri et al. 1994; Winpenny et al. 2010)

- There are high disposal costs or environmental impacts. Reuse may provide cost savings relative to costs of upgrading wastewater treatment system to meet environmental standards.
- Irrigation water demands exceed supply and no other new sources of water are available. Wastewater reuse may be economical even if treatment costs are borne by irrigators.

- Change from irrigation with fresh groundwater or surface water to treated wastewater results in addition freshwater being available for higher value urban and industrial uses.
- Benefits from nutrients in wastewater, which could result in increased crop yields and reduce fertilization costs. Some subsistence farmers cannot afford fertilizer.

A combination of the above factors may be sufficient, even though each benefit alone is not sufficient. It is unlikely that schemes could be justified with reference only to agriculture. However, the major benefit in most cases is likely to be the value of freshwater made available for high-value urban and industrial uses (Winpenny et al. 2010).

Bixio et al. (2005) presented the results of an international survey to identify and evaluate the status of the implementation and operational practices of municipal water reclamation projects. Globally, most municipal wastewater reclamation projects have benefited from financial incentives or government subsidies, especially to cover part of the initial investment. The water supply benefits of water reclamation projects are insufficient to carry the investment costs of improving the effluent. Users seldom pay the full costs of reclaimed water. However, the economic analysis of projects seldom consider the externalities, such as the marginal costs of new sustainable water sources in areas where existing sources are developed at or beyond sustainable levels. For example, subsidized reclaimed water supplies to non-potable users may make economic sense if it eliminates the need to expand much more costly desalination facilities to meet potable demands.

The main economic costs and benefits of wastewater reuse are summarized in Table 29.3. Cost analyses of wastewater reuse systems seldom consider the financial, social and environmental burden of effluent disposal to the environment (Bixio et al. 2005). Wastewater generated in urban areas needs to be properly disposed of anyways, and use of reclaimed water for agriculture (or irrigation uses) may transfer some of the costs to the rural sector (Friedler 2001). The revenues generated from the sale of reclaimed water commonly cannot finance the systems, but the overall economic benefits may be sufficient to justify the societal investment.

The development of reuse systems has been driven as an alternative to upgrading wastewater treatment facilities to meet environment regulations or goals for nutrient removal before discharge to sensitive surface waters (Okum 2002). It may be significantly less expensive to use reclaimed water with relative high nutrient concentrations for turf grass irrigation (where the nutrients are beneficial) than constructing advanced water treatment facilities so that the water can be safely discharged to the surface water bodies. The former option clearly also provides a water supply benefit.

Wastewater reuse projects should be evaluated using similar criteria as other utility projects. Winpenny et al. (2010) described a three-fold approach:

- Cost benefit analysis—project benefits should exceed costs

Table 29.3 Economic costs and benefits of wastewater reuse systems

Costs	Benefits
1. Construction and operational costs for new or upgraded wastewater treatment facilities to meet reuse water quality and reliability requirements.	1. Economic value obtained from the use of the new water source (e.g., expansion of irrigated agriculture).
2. Infrastructure to convey reclaimed water to users.	2. Value of freshwater saved and made available for other uses.
3. Storage facilities to balance supply and demand.	3. Avoidance of costs for wastewater disposal.
	4. Nutrients provided reduce fertilization costs

- Cost effectiveness analysis—compares alternative ways of delivering given benefits. Cost effectiveness analyses are difficult when different options provide uneven results and are not strictly comparable
- Financial feasibility analysis—estimation of financial impacts of a project for public capital and recurring budgets.

Reclaimed water systems also allow for an increase in agricultural development and production in arid regions (Friedler 2001). Distribution of costs and benefits between different stakeholders is crucial to the feasibility of reuse projects. (Winpenny et al. 2010). Farmers will likely economically benefit from wastewater reuse if they don't have to bear the costs of new treatment and distribution infrastructure. They may lose financially if they have to bear these costs. In principle, farmers should pay if they receive significant net benefits, but economic incentives should be used to encourage farmers to join recycling programs (Winpenny et al. 2010). The feasibility of charges to farmers depend on the options they have, which depend upon the costs of currently available water. Where water is needed for new development, for which there are no alternative water supplies, then full-cost recovery may be feasible (Winpenny et al. 2010). On the contrary, where farmers currently receive water for free or at a highly subsidized cost, then there is no economic incentive to invest in reuse systems and the imposition of any charges may be resisted.

Pescod (1992) proposed that farmers should be charged only for any incremental costs associated with additional treatment or distribution required for the use of effluent in agriculture. Farmers should not be expected to pay for the collection and treatment of urban wastewater, which is a societal responsibility for environmental protection objectives. The costs of additional treatment required for agricultural reuse might be less than those required for environmental protection. The costs for upgrading systems for wastewater reuse may also be less than the often unquantified environmental costs associated with wastewater disposal. A coherent national policy for wastewater use in agriculture is essential, which includes a defined division of responsibility (including economic) and adoption of realistic standards and regulations to safeguard public health and the environment (Pescod 1992).

29.7 Public Acceptance of Wastewater Reuse

A key issue noted by many is that public acceptance of reuse projects is vital for the future of wastewater reclamation, recycling, and reuse. Poor public perception may jeopardize projects involving use of reclaimed wastewater (e.g., Asano and Levine 1996; Lazarova et al. 2000). Successful implementation of wastewater reuse technologies will require an understanding of the social environment in which they are to be applied (Lazarova et al. 2000), including cultural and religious values, teachings, and doctrines (Chap. 33). The end users must be willing to accept the water. Highly treated wastewater may still be viewed as being just wastewater even though it may be of superior quality to currently used freshwater sources (Bixio et al. 2005).

Wastewater reuse may become a particularly contentious issue when potable reuse is involved. Direct potable reuse, often disparaged as "toilet-to-tap," in general, has a very low public acceptance. Unplanned indirect potable reuse has long been common place. The often cited example is potable water supplies taken from rivers in which downstream users consume water that includes wastewater from upstream users. Daughton (2004, p. 128) noted that:

> the public often will not accept reliance on existing, proven drinking water standards for planned recharge whereas they have long done so for unplanned reuse.

There is often a visceral revulsion to the concept of intentionally adding wastewater to the water supply in any form, which has no relationship to actual health risks and water quality. The "Law of Contagion" may apply, which suggests that once water has been in contact with contaminants it can be psychologically very difficult for people to accept that it has been purified (Khan and Gerrard 2006).

Perception is often more important than reality, particularly because the decision to pursue or not pursue wastewater reuse is often decided in the political rather scientific realm. Unplanned indirect potable reuse is often taken for granted as the normal state of affairs or is just an unpleasant, unavoidable reality that most people choose not to give any thought. The intentional addition of treated wastewater into the water supply may elicit stronger responses because it is viewed as a choice and is thus avoidable. The results of a survey in Australia revealed that the main perceived risk associated with potable reuse is to the safety of children and animals (Hurlimann 2007).

A key issue in the public perception of indirect potable reuse is the hydrologic distance between the water's origin as waste and its use for personal activities, as well as the number of natural or artificial barriers along the way (Bruvold 1985; Daughton 2004; Marks 2004; Khan and Gerrard 2006; Hurlimann 2007). The passage of water in a natural environment, whether it be a river or aquifer, may reduce its taint as being wastewater and thus make reuse more acceptable (Bouwer 2000; Al-Otaibi and Mukhopadhyay 2005; Dillon et al. 2006a, b). Key points noted by Daughton (2004, p. 132) are:

It is the spatial and temporal proximity of the reuse that is of concern and

the more remote the hydrologic connection, the easier it is for the public to accept water reuse

Water reuse applications that consistently have had the least amount of community support have been those that involve recycling municipal wastewater into drinking water supplies (Alhumoud et al. 2003; Marks 2004; Khan and Gerrard 2006). Projects that reuse treated wastewater for landscape and golf course irrigation usually face little public opposition. For example, results of a survey of public attitude towards wastewater reuse in Kuwait indicates (95.7%) opposition to the use of highly treated wastewater for human uses (showers and bathing, and cloth washing), but wide acceptance (89.6%) for its use for other purposes (Alhumoud et al. 2003).

A common theme in many studies of the public (and potential users) reactions to wastewater reuse, is a lack of information about wastewater quality, health risks, and, for farmers, impacts on crops and soils (e.g., Bruvold 1985; Bahri and Brissaud 1996; Alhumoud et al. 2003; Daughton 2004; Marks 2004; Khan and Gerrard 2006). The lack of knowledge leads to negative perceptions of wastewater reuse. As knowledge of wastewater treatment process increases, such as through visits to water reclamation facilities, there is a boost in acceptance of reuse (Bruvold 1985; Marks 2004). It has been recommended that types of reuse that are perceived to have a low risk be introduced first to obtain public approval prior to considering higher contact uses (Bruvold 1985; Marks 2004). Bruvold (1985) also found that acceptance of wastewater reuse was correlated with the beliefs that a water shortage existed in a respondent's area and that the reuse project would bring economic benefits to the community.

Khan and Gerrard (2006) emphasized the importance of diligently working to obtain stakeholder support for projects as early as possible in the project development (e.g., initial planning phase). Stakeholders include all parties (individual and organizations) that have an interest in a project and include the general public, landowners, potential customers, business interests, and government agencies. Stakeholders need to be convinced that a project will provide benefits and have their fears allayed about possible health and environmental risks. It is important to avoid situations where stakeholders believe that a project is being undertaken in secret and being forced on them, and that their concerns are being ignored. Instead, stakeholder support is more likely when they are involved at the early stage of a project, kept informed through open communication, their opinions and concerns are sought and genuinely considered, and they have some level of control over the process (i.e., they are involved in the decision making process).

Stakeholders must also have confidence and trust in organizations that will be implementing the reuse project and the regulatory agencies that will be overseeing the systems (Marks 2004; Khan and Gerrard 2006; Hurlimann 2007). Trust has been found to be associated with believing that the source of information is expert, unbiased, disinterested, and not sensationalizing (Hurlimann 2007). Trust in organizations and reuse could be lost through failed projects. A local failure could result in a decrease in support for reuse on a national level (Dillon 2000). Trust needs to be backed up with sound science and an understanding of the actual risks posed by water reuse. The more people that understand the actual risks posed by water recycling from pathogens, the greater the confidence the public will have in its acceptance (Gerba and Rose 2003).

Secrecy and delays in passing on information should be avoided (Khan and Gerrard 2006). Failure to engage the stakeholders early in a project may give rise to rumors, increase levels of concern, and cause stakeholders to question the motives and intentions of the project owners and managers. The information vacuum created by the lack of communication may be quickly filled with disinformation, which may be very difficult to later dispel. Despite best efforts to have open and honest communications with stakeholders, the reality is that some stakeholders will be inexorably opposed to reuse projects because they will bear disproportionate impacts or other philosophical reasons. For example, projects that expand the water supply are opposed by some environmental advocacy groups because they enable local population growth.

Wastewater can be treated to meet drinking water standards, but that may not been enough to assuage the public that indirect potable reuse is safe. It has often been correctly noted that there are many harmful chemicals for which there is no drinking water standard and routine testing. From a practical perspective it is not possible to sample either treated wastewater or potable water for all chemicals. Even if such testing were performed, for most chemicals there is little or no data on their health risks to humans and the concentrations that are detected (National Research Council 1994, 1998). It is impossible to ever certify that treated water is absolutely safe because of the extraordinary difficulty of verifying that it is free from any previously unrecognized contaminants (Daughton 2004).

As a general philosophic matter, it is not possible to prove a negative, such as that recycled water will not pose any hazard. Opponents of reuse projects often use this philosophical tactic to argue that the safety of a reuse project has not been proven and, therefore, a project should not be approved or pursued. Even if wastewater were tested for a 1,000 chemicals, the argument might still be made that there is still the possibility of the presence of harmful chemicals having concentrations that were not measured. Public education may be needed to place wastewater reuse in a proper context relative to other risks.

The trace concentrations of endocrine disrupting compounds (EDCs) and other compounds of emerging concern (CECs) in wastewater, surface water, and some drinking waters, have received great public and scientific attention in recent years (Sect. 30.5). However, very little is known about the actual risks they pose to humans. Any adverse impacts from CECs would stem from long-term chronic exposure rather than a one-time consumption of water. The long-term latency of disease onset (if such an onset does indeed occur) can result in the disease escaping detection even by exhaustive epidemiological studies, because of the transient nature of most communities and the problems associated with statistical power related to wide natural variations (Daughton 2004). The difficult in assessing the actual impacts of CECs raises the question (Daughton 2004, p. 134):

> what are the ramifications of revealing trace levels of substances for which little or no toxicological information exists

It is a large jump from impacts to small fish that spend their entire lives immersed in the water, which have been documented in some studies, and assumptions concerning potential human health impacts to minor exposure.

The reporting of the CEC presence in drinking water has had an element of sensationalism, certainly leading to influx of research money to study the issue. However, their presence has commonly not been put into proper perspective relative to their actual human health risks, and much less their risk relative to other activities. The available data on the concentrations of CECs in treated wastewater and their tendency to undergo natural attenuation indicate that human health risks associated with CECs in reuse projects is much more a matter of perception rather than reality. However, this is not to say that the CEC issue should be ignored. The introduction of CECs into the environment through wastewater discharge has real ecological consequences that need to be addressed (Toze 2006).

Forgoing wastewater reuse, because of disproportionate concerns over unproven and unlikely health risks, could have much greater impacts on human health and longevity caused by water scarcity and its ramifications, such as inadequate food supply. It would be folly to accept substantial real health risks associated with water scarcity that could be ameliorated by wastewater reuse, because of the exaggerated perceived risks over the presence of CECs, most (but not all) of which are attenuated by natural process.

From an arid lands water management perspective, the scarcity of water dictates that wastewater reuse must be implemented to some degree because it is too valuable a resource to waste. The challenge is to find a means to maximize its beneficial uses, while protecting public health and the environment, and being in tune with local values and culture.

29.8 Industrial Wastewater Reuse

Freshwater using for cooling, particularly for thermoelectrical power generation, represents the single largest opportunity for industrial water reuse because of both the large water demands and that the water quality requirements for industrial cooling are not high (Vignesnaran and Sundaravadivel 2004). Electrical power generation facilities are often located in urban areas where domestic wastewater is generated in large quantities (e.g., a nuclear power plant located in the Miami area of southern Florida will use domestic wastewater currently being discharged to tide because of its relatively high TDS concentrations). A major problem may be biofilm growth due to the presence of microorganisms and elevated nutrient and dissolved oxygen concentrations in the wastewater.

29.9 Environmental Concerns Over Reuse

Salts and chemicals present in reclaimed water can potentially accumulate in the environment as the result of reuse. If the reclaimed water is more saline than native shallow groundwater, then its use through irrigation can result in a salinity increase, especially in arid and semiarid environments in which there is a high degree of evapotranspiration. Similarly, refractory organic compounds that are not removed in the wastewater treatment process or undergo natural attenuation may also accumulate in the soil, surface waters, and groundwater. Irrigation in general can result in increases in salinity through the concentration of salts in the root zones.

Zoller (2006, p. 683) proposed that:

> Water resuse/recycling and reclamation, wherever, should be applied only if the long-term pollution consequences of the local/regional surface and groundwater can be avoided. Any delay in taking the necessary preventive measures will aggravate the situation—until it is reversible—rather than solve the water shortage problem.

Zoller (2006) identified inorganic salts, which are not removed by wastewater treatment, and biodegradation resistant nonionic surfactants (APEOs), which are only partially removed, as being primary concerns for causing soil, surface water, and groundwater pollution during reuse and reclamation. Zoller (2006, p. 687) also noted that:

> Sustainable development in the semiarid zones requires combined desalination—biological treatment of wastewater before their reuse, recycling, and/or reclamation.

Wastewater reuse may poses a potential risk to shallow aquifers that are used for potable supply. For example, the unconfined Biscayne Aquifer is the primary water source for southeastern Florida, including the Miami and Fort Lauderdale metropolitan area. The concern has been raised that pharmaceutical compounds and other CECs in the treated wastewater may impact the water supply through urban (golf course and landscaping) reuse (Bloetscher et al. 2005a, 2005b; Bloetscher 2009). However, there is no evidence that would suggest that the public health in southeastern Florida has in any way been impacted by wastewater reuse practices.

Whereas in many developing countries wastewater is discharged directly to river channels without any treatment, postponement of wastewater reuse until desalination facilities are added is not realistic. Furthermore, advanced treatment (desalination) of irrigation water will seldom, if ever, be economically viable for large-scale irrigation, especially in developing countries. However, the basic concern is valid in that wastewater reuse may not be sustainable if it leads an unacceptable degradation of the soil and water resources. On-going research is needed to accurately assess and quantify the potential environmental and public health risks associated with wastewater reuse and the development of mitigation measures, as necessary. The environmental concerns for wastewater reuse need to be weighed against the risks and costs of other water supply options and the real health risks associated with water scarcity.

References

Abiye, T. A., Sulieman, H., & Ayalew, M. (2009). Use of treated wastewater for managed aquifer recharge in highly populated urban centers: A case study in Addis Ababa. Ethiopia. *Environmental Geology, 58*(1), 55–59.

Alhumoud, J. M., Behbehani, H. S., & Abdullah, T. H. (2003). Wastewater reuse practices in Kuwait. *The Environmentalist, 23*, 117–126.

Al-Otaibi, M., & Mukhopadhyay, A. (2005). Options for managing water resources in Kuwait. *Arabian Journal for Science and Engineering, 30*, 55–68.

Asano, T., Burton, F. L., Leverenz, H. L., Tsuchihashi, R., & Tchobanoglous, G. (2007). *Wastewater reuse, issues, technologies and applications*. New York: McGraw-Hill.

Asano, T., & Cotruvo, J. A. (2004). Groundwater recharge with reclaimed wastewater; health and regulatory considerations. *Water Research, 38*, 1941–1951.

Asano, T., & Levine, A. D. (1996). Wastewater reclamation, recycling and reuse: Past, present, and future. *Water Science and Technology, 30*(10–11), 1–14.

Bahri, A. (1999). Agricultural reuse of wastewater and global water management. *Water Science and Technology, 40*(4–5), 339–346.

Bahri, A., & Brissaud, F. (1996). Wastewater reuse in Tunisia: assessing a national policy. *Water Science and Technology, 33*(10–11), 87–94.

Bixio, D., De heyder, B., Cikurel, H., Muston, M., Miska, V., Joksimovic, D., et al. (2005). Municipal wastewater reclamation: Where do we stand? An overview of treatment technology and management practice. *Water Science & Technology: Water Supply, 5*(1), 77–85.

Bloetscher, F. (2009, March). Should Florida revisit comparative assessment of municipal wastewater disposal methods in the southeastern part of the state. *Florida Water Resources Journal, 59*(3), 18–32.

Bloetscher, F., Englehardt, J. D., Chin, D. A., Rose, J. B., Tchobanoglous, G., Amy, V. P., et al. (2005a). Comparative assessment of municipal wastewater disposal methods in Southeast Florida. *Water Environment Research, 77*(5), 489–490.

Bloetscher, F., Muniz, A., & Witt, G. M. (2005b). *Groundwater injection: Modeling, risks and regulations*. New York: McGraw-Hill.

Bos, R., Carr, R., & Keraita, B. (2010). Assessing and mitigating wastewater-related health risks in low-income countries. In P. Dreschel., C. A. Scott., L. Raschid-Sally., M. Redwood., & A. Bahri., (Eds.), *Wastewater irrigation and health. Assessing and mitigating risk in low income countries* (pp. 29–47). Ottawa: International Development Research Centre.

Bouwer, H. (1991). Groundwater recharge with sewage effluent. *Water Science and Technology, 23*, 2099–2108.

Bouwer, H. (2000). Integrated water management: Emerging issues and challenges. *Agricultural Water Management, 45*, 217–228.

Bruvold, W. H. (1985). Obtaining public support for reuse water. *Journal American Water Works Association, 77*(7), 72–77.

Daughton, C. G. (2004). Ground water recharge and chemical contaminants: challenges in communicating the connections and collisions of two disparate worlds. *Ground Water Monitoring and Remediation, 24*(2), 127–138.

Dillon, P. (2000). Water reuse in Australia: Current status, projections and research. In P. J. Dillon (Ed.), *AWA water recycling Forum* (pp. 99–104). Adelaide: Proceedings of the First Symposium.

Dillon, P., Pavelic, P., Toze, S., Rinck-Pfeiffer, S., Martin, R., Knapton, A., et al. (2006a). Role of aquifer storage in water reuse. *Desalination, 188*, 123–134.

Dillon, P., Toze, S., Pavelic, P., Vanderzalm, J., Barry, K., Ying, G. -L., Kookana, R., Skjemstad, J., Nicholson, B., Miller, R., Correll, R., Prommer, H., Greskowiak, J., & Stuyfzand, P., (2006b). *Water quality improvements during aquifer storage and recovery at ten sites, in Recharge systems for protecting and enhancing groundwater resources, Proceedings of the*

5th International Symposium on Management of Aquifer Recharge (pp. 85–94) Berlin, Germany, UNESCO, Paris. 11–16 June 2005.
Drewes, J. E., Heberer, T., Rauch, T., & Reddersen, K. (2003). Fate of pharmaceuticals during groundwater recharge. *Ground Water Monitoring and Remediation, 23*, 64–72.
Du Pisani, P. L. (2006). Direct reclamation of potable water at Windhoek's Goreangab Reclamation Plant. *Desalination, 188*, 79–88.
Foster, S. S. D., & Chilton, P. J. (2004). Downstream of downtown: Urban wastewater as groundwater recharge. *Hydrogeology Journal, 12*, 115–120.
Friedler, E. (2001). Water reuse—an integral part of water resources management. *Israel as a Case Study: Water Policy, 3*, 29–39.
Gerba, C. P., & Rose, J. B. (2003). International guidelines for water recycling: Microbiological considerations. *Water Science and Technology. Water Supply, 3*(4), 311 316.
Haarhoff, J., & Van der Merwe, B. (1996). Twenty-five years of wastewater reclamation in Windhoek, Namibia. *Water Science and Technology, 33*, 25–35.
Hochstrat, R., Joksimovic, D., Wintgens, T., Melin, T., & Savic, D. (2007). Economic considerations and decision-support tool for wastewater reuse scheme planning. *Water Science and Technology, 56*(5), 175–182.
Hurlimann, A. C. (2007). Is recycled water risky? *An Urban Australian Community's Perspective: Environmentalist, 27*, 83–94.
Jiménez, B., Mara, D., Carr, R., & Brissaud, F. (2010). Wastewater treatment for pathogen removal and nutrient conservation: suitable systems for use in developing countries. In P. Dreschel., C. A. Scott., L. Raschid-Sally., M. Redwood., & A. Bahri., (Eds.), *Wastewater irrigation and health. Assessing and mitigating risk in low income countries* (pp. 149–169). Ottawa: International Development Research Centre.
Khan, S. J., & Gerrard, L. E. (2006). Stakeholder communications for successful water reuse operations. *Desalination, 187*, 191–202.
Khouri, N., Kalbermatten, J. M., & Bartone, C. R. (1994). *Reuse of wastewater in agriculture: A guide for planners*. Washington, DC: UNDEP—World Bank Water and Sanitation Program.
Khouzam, R. F. (2003, December 16–18). *Economic aspects of wastewater reuse study case: The Arab Nation: Paper presented to the Economic Research Forum for Arab Countries, Turkey and Iran (ERF) 10th Annual Conference.*
Law, I. B. (2003, May). Advancing reuse—from Windhoek to Singapore and Beyond. *Water, 30*, 31–36.
Lazarova, V., Cirelli, G., Jeffrey, P., Salgot, M., Icekson, N., & Brissaud, F. (2000). Enhancement of integrated water management and reuse in Europe and the Middle East. *Water Science and Technology, 42*(1–2), 193–202.
Levine, A. D., & Asano, T. (2004). Recovering sustainable water from wastewater. *Environmental Science and Technology, 38*, 201A–208A.
Maliva, R. G., & Missimer, T. M. (2010). *Aquifer storage and recovery and managed aquifer recharge using wells: Planning, hydrogeology, design, and operation: Houston, Schlumberger Water Services, Methods in Water Resources Evaluation Series No. 2.*
Mara, D. (2003). *Domestic wastewater treatment in developing countries*. London: Earthscan
Marks, J. S. (2004, August). Advancing community acceptance of reclaimed water. *Water, 31*(5), 46–51.
Markus, M. R. (2009, January 15). Groundwater replenishment & water reuse. *The Water Report, 59*, 1–9.
McIlwaine, S., & Redwood, M. (Eds.) (2010). *Greywater use in the Middle East. Technical, social, economic and policy issues*. Burton on Dunsmore: Practical Action Publishing, Ottawa, International Development Research Centre.
Miller, G. W. (2006). Integrated concepts in water reuse: Managing global water needs. *Desalination, 187*, 66–75.
Mujeriego, R., & Asano, T. (1999). The role of advanced treatment in wastewater reclamation and reuse. *Water Science and Technology, 40*(4–5), 1–9.

National Research Council. (1994). *Ground water recharge using waters of impaired quality.* Washington, DC.: National Academy Press.

National Research Council. (1998). *Issues in potable reuse: The viability of augmenting drinking water supplies with reclaimed water.* Washington, DC: National Academy Press.

Okum, D. A. (2002). Water reuse introduces the need to integrate both water supply and wastewater management at local and regulatory levels. *Water Science and Technology, 46*(6–7), 273–280.

Pescod, M. E. (1992). *Wastewater treatment and use in agriculture, FAO Irrigation and Drainage Paper 47.* Rome: Food and Agriculture Organization of the United Nations.

Postel, S. (1992). *Last oasis: Facing water scarcity.* New York: WW Norton.

Qadir, M., Wichelns, D., Raschid-Sally, L., Minhas, P. S., Drechsel, P., Bahri, A., & McCornick, P. (2007). Agricultural use of marginal-quality water—opportunities and challenges. In *Comprehensive assessment of water management in agriculture, Water for food, water for life* (pp. 425–457). London, Earthscan, and Colombo: International Water Management Institute.

Ramadan, H., & Ponce, V. M. (2010). Design and performance of waste stabilization ponds, Version 081218. Retrieved September 9, 2010 from http://stabilizationponds.sdsu.edu/.

Rodriguez, C., Van Buynder, P., Lugg, R., Blair, P., Devine, B., Cook, A., et al. (2009). Indirect potable reuse: a sustainable water supply alternative. *International Journal of Environmental Research and Public Health, 6*, 1174–1209.

Salgot, M., & Torrens, A. (2008) Impacts of climate change on water resources: the future of groundwater recharge with reclaimed water in the South of Europe. In W. Dragoni.,& B. S. Sukhija. (Eds.), *Climate change and groundwater: Geological Society of London Special Publications No. 288* (pp. 145–168).

Scott, C.A., Dreschel, P., Raschid-Sally, L., Bahri, A., Mara, D., Redwood, M., & Jiménez, B. (2010). Wastewater irrigation and health: challenges and outlook for mitigating risks in low-income countries. In P. Dreschel., C. A. Scott., L. Raschid-Sally., M. Redwood., & A. Bahri. (Eds.), *Wastewater irrigation and health. Assessing and mitigating risk in low income countries* (pp. 381–394). Ottawa: International Development Research Centre.

Tchobanoglous, G., Burton, F. L., & Stensel, H. D. (2003). *Wastewater engineering, treatment, disposal and reuse* (4th ed.). New York: McGraw Hill.

Toze, S. (2004, September—October 26). *Reuse of effluent water—benefits and risks, in New directions for a Diverse Planet, Proceedings of the 4th International Crop Science Congress* (p. 11). Brisbane, Australia.

Toze, S. (2006). Water reuse and health risks—real vs perceived. *Desalination, 187*, 41–51.

Treweek, G. P. (1985). Pretreatment processes for groundwater recharge. In T. Asano (Ed.), *Artificial recharge of groundwater* (pp. 204–218). Stoneham, Massachusetts: Butterworth Publishers.

Vignesnaran, S., & Sundaravadivel, M. (2004). Recycle and Reuse of domestic wastewater. In S. Vigneswaran., (Ed.), *Wastewater recycle, reuse and reclamation, in Encyclopedia of life support systems (EOLSS).* Oxford, UK: Eolss Publishers. (http://www.eolss.net).

Winpenny, J., Heinz, I., & Koo-Oshima, S. (2010). *The wealth of water: The economics of wastewater use in agriculture.* Rome: Food and Agriculture Organization of the United Nations.

World Health Organization. (2006). *WHO guidelines for the safe use of wastewater, excreta, and greywater, Policy and regulatory aspects* (Vol. 1). Geneva: WHO.

Zoller, U. (2006). Water reuse/recycling and reclamation in semiarid zones: The Israeli case of salination and "hard" surfactants pollution of aquifers. *Journal of Environmental Engineering, 132*(6), 683–688.

Wintgens, T., Salehi, F., Hochstrat, R., & Melin, T. (2008). Emerging contaminants and treatment options in water recycling for indirect potable reuse. *Water Science and Technology, 57*(1), 99–107.

Chapter 30
Wastewater Reuse Health and Safety Issues

30.1 Introduction

The health hazards associated with wastewater are a function of concentration of contaminants in the wastewater, the duration and level of exposure, and the dose–response function for each hazard. An important issue when evaluating the potential health impacts (on humans, as well as on the general environment) of wastewater reuse is that the risk associated with contaminant exposure depends upon both the duration and intensity of exposure. A fundamental distinction is between acute and chronic exposure. An acute exposure is a one-time exposure to a substance or microorganism that is sufficient to cause a serious health impact. A one-time exposure to pathogenic microorganisms, such as consuming water that contains even small numbers of active *Cryptosporidium* cysts or *Giardia* oocysts or enteric viruses, may be sufficient to cause serious illness.

Chronic exposure, on the contrary, involves repeated, continuous exposure to a hazardous substance over an extended period, in which any single exposure is insufficient to cause serious harm. Most of the chemicals found in raw and treated wastewater would require long-term chronic exposure to induce illnesses. For example, some disinfection byproducts (e.g., trihalomethanes; THMs) have been linked to an increased risk of cancer, but the effect comes from long-term (multi-decadal) consumption of water containing THMs, rather than from a one-time exposure.

30.2 Risks Associate with Wastewater Reuse

The Australian Guidelines for Water Recycling (NRMMC-EPHC-AHMC 2006) provide an excellent overview of risk assessments for reclaimed and graywater recycling (reuse) systems. A basic distinction is made between hazards and risk. Hazards are biological, chemical, physical, or radiological agents that have the potential to cause harm to people, animals, crops, or the environment. Risk is the

R. Maliva and T. Missimer, *Arid Lands Water Evaluation and Management*, Environmental Science and Engineering, DOI: 10.1007/978-3-642-29104-3_30,

likelihood of identified hazards causing harm, and is the product of likelihood of exposure to the hazard and the actual occurrence of harm from the exposure (NRMMC-EPHC-AHMC 2006). Maximal (unmitigated) risk is that occurring in the absence of preventive measures. Residual risk is that occurring after existing and proposed preventive measures.

Risk assessments can be defined broadly as the processes of estimating the probability of occurrence of an event and the probable magnitude of adverse effects on safety, health, and ecology over a specified time period (Asano et al. 2007). Risk assessment involves four main components:

1. Hazard identification, which is the recognition of microorganisms and chemicals that increase the incidence of a health condition.
2. Exposure assessment, which is the evaluation of exposure scenarios and the probability (and frequency) of exposure of an individual to a given chemical or pathogen dose over a specified time period.
3. Dose–response assessment, which quantifies the risk of disease or infection of an individual from a given chemical or pathogen dose.
4. Risk characterization, which combines the exposure and dose–response assessments to estimate the incidence of a given adverse impact on a population.

The risk assessment process can also include management tasks (NRMMC-EPHC-AHMC 2006) such as

- Identification of preventive measures to control identified hazards and to establish monitoring programs to ensure that preventive measures are operating efficiently.
- Verification that the management system consistently provides recycled water of a sufficient quality that is fit for its intended use.

The most significant hazards with respect to humans are health impacts associated with exposure to pathogenic microorganisms (Sect. 30.3). On the contrary, the most significant environmental hazards tend to be chemical and physical hazards. The main chemical hazards associated with wastewater are boron, cadmium, chlorine disinfection residuals, nitrogen, phosphorous, salinity, chloride, and sodium (NRMMC-EPHC-AHMC 2006). Some of these hazards are present in non-recycled waters, and thus not unique to wastewater reuse.

Environmental risk assessment involves the identification and assessment of risks associated with water recycling at all environmental endpoints (i.e., locations where an impact could be identified; NRMMC-EPHC-AHMC 2006). Environmental impacts include those affecting the broad categories of air, soil, fauna and flora, groundwater and surface water. Environmental risk assessment involves the same basic procedures as occur in human health risk assessments (NRMMC-EPHC-AHMC 2006) and include the following elements:

- identification of the hazard,
- estimation of the likelihood that a hazard will occur (i.e., endpoint will be exposed to a hazard in sufficient quantities to cause a detrimental effect),

- estimation of the consequences (impacts) of hazardous events that might occur, and
- characterization of the overall risk of individual and combined hazards.

30.2.1 Risk Management and Reduction

Risks associated with wastewater reuse or other recycling of impaired waters can be reduced through preventive actions. The preferred strategy when human health risks are involved is a multiple barrier approach in which a series of preventive measures are employed so that protection is provided should any one barrier fail. Reduction in the performance of one barrier does not result in a total loss of water quality management. Preventive measures may include the following (NRMMC-EPHC-AHMC 2006):

- source control (reducing the amount of a hazard entering the wastewater system),
- wastewater treatment,
- restrictions on recycled water use, and
- restrictions of site access (exposure control).

30.2.2 Monitoring

Monitoring is an integral element of risk assessment and management. The NRMMC-EPHC-AHMC (2006) recommends four types of monitoring for water recycling projects:

- baseline monitoring (i.e., where are we now; what are current conditions?),
- validation monitoring (i.e., will it work?),
- operational monitoring (i.e., is it working now?), and
- verification monitoring (i.e., did it work?).

Baseline monitoring defines the properties of the receiving environment and provides a point of reference to test for environmental impacts as part of the verification monitoring. Validation monitoring determines whether preventive measures are capable of adequately controlling recycled water quality. Validation monitoring is performed before the start of actual operation testing. An example of validation monitoring is the testing of wastewater treatment equipment before commencing the distribution and use of the recycled water.

Operational monitoring is the routine monitoring of control parameters to determine whether the physical and chemical properties of the recycled water are within the target range for its intended use. Operational monitoring should be performed at critical control points, which are defined as an activity, procedure or process where control can be applied, operational parameters can be monitored, and corrective actions can be implemented (NRMMC-EPHC-AHMC 2006). For example, a critical control point could be the downstream end of the wastewater treatment process where turbidity and chlorine residual are continuously monitored. A corrective action could be an adjustment of the chlorine dose or diverting the wastewater flow from the reuse system until the water meets standards (NRMMC-EPHC-AHMC 2006).

Verification monitoring assesses the overall performance of the system and includes an evaluation of the ultimate quality of the recycled water produced and the quality of the receiving environment. It could also involve an assessment of human health issues, for example, whether there is an increased incidence of illness amongst farm workers.

Monitoring programs need to be system specific with the monitored parameters and frequency of sampling dependent upon the specific identified hazards and level of risk. Economics must also be considered in the development of monitoring programs as excessive monitoring requirements could render a valuable project unviable. "Right-sizing" of monitoring program design is a perpetual challenge in wastewater reuse and MAR projects. Regulatory agencies understandably tend to take a conservative approach and would rather err on the side of too much monitoring, particularly if human contact is likely. A sensible approach is to have a relatively high degree of monitoring (more frequent analyses and a greater number of parameters) at the start of projects, and then gradually decrease the monitoring program based on operational experiences. In other words, an adaptive management approach should be taken in which the monitoring program is adjusted as more is learned about the operation of the system. Commonly, some parameters are not detected at concentrations of concern in wastewater or have a low variation in concentration over time, and may be either dropped from the monitoring program or analyzed less frequently.

30.3 Pathogenic Microorganisms

Microorganisms associated with waterborne disease are primarily enteric pathogens (i.e., they originate within the intestines of humans or other animals), which have a fecal-oral or fecal-dermal route of infection (either human-to-human or animal-to-human) and survive in water (National Research Council 1998; Bos et al. 2010). A variety of protozoa, helminthes, trematodes, bacteria, and viruses have been identified as infectious agents in untreated municipal wastewater. However, in more than half of all reported outbreaks of waterborne disease, the disease-causing agent was never discovered, which suggests the presence of unrecognized pathogens (National Research Council 1998). Wastewater irrigation schemes may also increase the health risks from vector-borne diseases (e.g., malaria; World Health Organization 2006). The infectious agents present in wastewater depend upon the sources of the wastewater, the general health of the contributing population, the existence of "disease carriers" in the population, and the ability of the various infectious agents to survive in environments outside of their host (National Research Council 1994).

Risks from microbial contamination depend not only on the dose of microorganisms, but also on the host's immune status. Limited data are available on the pathogen doses that are necessary to cause infection for most microorganisms. There is also a limited understanding of the relationship between infection and the various forms of illness (Macler and Merkle 2002). Sensitive populations, including children, the elderly, and people with compromised immune systems, stand a greater risk of severe outcomes (National Research Council 1998).

It is important to also recognize that the more commonly used treatment processes may not completely remove pathogens (Toze 2006). Multiple barriers are ideally employed in wastewater reuse projects that involve human exposure, and especially when indirect potable reuse is known to occur or unplanned indirect use is a possibility. Health effects associated with the reuse of wastewater for irrigation in arid lands tend to be less because viruses and bacteria in aerosols are rapidly inactivated by the combination of warm temperatures, low humidity and bright sunlight (Alhumoud et al. 2003).

Pathogens with a long persistence in the environment, low minimum infection doses, that elicit little or no human immunity, and have long latency periods (e.g., helminthes), have a higher probability of causing infections than others (Bos et al. 2010). Routine monitoring for all pathogens that could be present in wastewater is not feasible, would be highly expensive, and cannot be performed in real time (Asano and Cotruvo 2004). From a public health and process control perspective, the most critical group of pathogenic microorganisms is enteric viruses because of the possibility of infection from exposure to low doses and the lack of routine, cost-effective methods for virus detection and quantification (Mujeriego and Asano 1999). It is not practical or affordable to test for all enteric viruses of concern on a routine basis.

Indicator organisms are usually used for pathogen monitoring. Indicator organisms may provide some information on water quality, but the absence of indicator organisms does not guarantee that the water is free of pathogens. The criteria for an ideal indicator of drinking water pollution are (Standridge 2008; Asano et al. 2007):

- occurrence is proportional to that of the pollutant,
- never present in non-contaminated (safe) water,
- always present in contaminated water,
- do no multiply in the environment (including water distribution system),
- die off in the environment more slowly than other pathogens,
- can be easily detected in the laboratory, and
- is safe to work with.

In addition, sample collection and analytical procedures should not be onerous and exceedingly expensive. To date, no ideal indicator organism has been identified (Asano et al. 2007).

Fecal bacteria are commonly used as indicators of the presence of fecal contamination, particularly enteric pathogens. However, the presence of fecal contaminants in a water sample does not necessarily correlate with presence of enteric pathogens, as only a small percentage of a population is infected at any given time and is thus excreting pathogens (Payment and Locas 2011). The number of viruses and other pathogens in water cannot be predicted by statistical methods on the basis of bacterial indicators. Fecal microbial indicators are not an absolute indicator of the presence of pathogens, but are rather an indicator of a probability of their co-occurrence (Payment and Locas 2011).

Total coliform bacteria have historically been used as an indicator of fecal contamination because of the simplicity and low cost of performing the analysis. Total coliforms have classically been described as nonspore-forming, gram-negative rods capable of hydrolyzing lactose to acid and gas end products within 48 h at 35°C (Standridge 2008). Total coliform bacteria do not necessarily have a fecal origin. There is now ample evidence that total coliform and "fecal" coliform bacteria are poor indicators of the presence of fecal contamination, as reviewed by Standridge (2008). Total coliform bacteria are ubiquitous in natural waters and multiply in drinking water environments. The strongest arguments that have been used in favor of the commonly used total and fecal coliform tests are that they are inexpensive and conservative, and not "less protective of public health" (Standridge 2008). The conservatism of the tests comes at the expense of false positive detections of contamination. It is the authors' experience that false positive detections of total coliform bacteria is especially problematic in MAR systems and wells in general.

A strong case has been made that *Escherichia coli (E. coli)* is a much better indicator of fecal contamination of drinking water (Edberg et al. 2000; Standridge 2008). Newly developed enzymatic methods for the detection of *E. coli* are simple, rapid, specific, and sensitive (Standridge 2008). However, *E. coli* testing has its limitations. Only certain variants of *E. coli* can cause serious illness, while many variants are harmless. Under some circumstances, *E. coli* can occur and multiply in environments without fecal contamination. Another limitation of *E. coli* as an indicator is that it does not indicate the presence of non-fecal pathogens, such as *Legionella* and *Mycobacteria*. Nevertheless, *E. coli* and thermo-tolerant coliforms have grained international acceptance (Australia, European Union, World Health Organization) as fecal contamination indicators of choice in preference to the total and fecal coliform tests. The USEPA to date has been a notable exception (Standridge 2008).

30.4 Chemical Contaminants

The National Research Council (1998) recognized three categories of chemical contaminants that are present in reclaimed water:

1. Inorganic chemicals and natural organic matter that are naturally present in the water supply.
2. Chemicals created by industrial, commercial, and other human activities in the wastewater service area.

3. Chemicals added or generated during water and wastewater treatment and distribution processes.

To this list can be added soluble microbial products (SMPs) which are organic compounds formed during the wastewater treatment process by the decomposition of organic matters (e.g., Drewes and Fox 1999).

Anthropogenic organic compounds (category 2) are of greatest concern because of the large number of compounds that might be present in wastewater, the inability to analyze for all of them, and the lack of toxicity information for many of the compounds. Virtually any chemical that is used in the service area of a wastewater treatment facility could potentially occur in the wastewater flow. It may be possible to flag potential compounds of concern in wastewater by documenting the major chemical inputs into wastewater, including both industrial and household sources. Source control is an important element of wastewater reuse systems, in which potential sources of contaminants in the wastewater collection area are identified and mitigation actions implemented to prevent the contaminants from entering the wastewater flow and thus reclaimed water flow.

However, even if all organic chemicals in a wastewater water sample could be identified, there would be no basis for assigning risk to most of the identified compounds (National Research Council 1998). Most petroleum-based chemicals and solvents are at least partially removed in the sewage treatment process or are naturally attenuated in the groundwater environment. Category 2 compounds also include compounds of emerging concern, which are a wide range of organic compounds present in wastewater in trace quantities (Sect. 30.5).

30.5 Compounds of Emerging Concern

Compounds of emerging concern (CECs) are chemical compounds that are currently not regulated but which may have potentially deleterious human health or ecotoxicological effects. CECs are also referred to as "emerging contaminants", "emerging pollutants", "emerging pollutants of concern", and "emerging substances of concern". CECs have received a great deal of attention over the past decade because they have been detected in what were considered to be clean surface waters and in the drinking water supplies for some major cities. A voluminous literature has very quickly accumulated, with thousands of papers published on the presence and concentrations of CECs in treated and untreated wastewater and surface waters, their ecotoxiological effects, and their fate and transport in MAR and reuse systems. CECs have also entered into general news media and, as a result, are of increasing public concern. Fundamental questions still remain as whether or not CECs at the concentrations actually or potentially found in potable water supplies pose a real risk to human health. There is a very clear need for less hyperbole in favor of a sound and balanced risk-based approach to the issue of CECs in the environment.

Virtually all compounds that are used in society have pathways of discharge into the environment, and, therefore, entry into the water supply (Ongerth and Khan 2004). CECs include both the chemicals themselves and their metabolic breakdown products (i.e., metabolites). CECs are a wide variety of organic compounds that include:

- pharmaceuticals (prescription and non-prescription drugs and their breakdown products),
- antibiotics,
- synthetic and natural hormones,
- personal care products,
- detergent metabolites,
- antimicrobial agents (disinfectants),
- brominated and chlorinated flame retardants,
- perfluorooctane surfactants,
- fragrance and flavoring compounds,
- insect repellants (e.g., DEET),
- x-*ray contrast agents (*e.g., Iopromide),
- plasticizers, and
- caffeine.

CECs are now being identified in groundwater, surface water, and reclaimed water because analytical technologies have improved so that these compounds can now be detected at very low concentrations, often on the order of nanograms (10^{-9} g)/L (Sedlak et al. 2000). The presence of CECs in the environment is not a new phenomenon, but rather their presence could not be detected earlier because of their very low concentrations. However, issues still remain concerning the reproducible analytical quantification of many CECs because of their very low concentrations and that relatively few laboratories that can perform these low concentration analyses. Increasingly lower detection limits are creating greater challenges for assessing, communicating, and ameliorating diminishing risks. An "inflating universe" of potential stressors will challenge the feasibility of sustaining regulatory and compliance monitoring on a chemical-by-chemical basis (Daughton 2009).

Daughton and Ternes (1999) provided an overview of emerging contaminants and their environmental occurrence. A key conclusion was that pharmaceutical and personal care products (PPCPs) have the potential for direct release into the environment worldwide, anywhere humans live or visit. Many pharmaceutical compounds after ingestion by humans are discharged (excreted), either partially or wholly unmetabolized, and are removed by wastewater treatment facilities to varying degrees. Pharmaceuticals also enter the wastewater stream when unused pills and solutions are disposed of in toilets, and through the land application of biosolids. Veterinary chemicals may enter the environment through manure applications and their use in aquaculture. The origin, fate, analytical methods, risks, and treatment technologies of PCPPs are reviewed in detail by Ternes and Joss (2006) and Richardson (2006).

The primary route by which active pharmaceutical ingredients enter the environment is thought to be through feces and urine. Daughton and Ruhoy (2009) reviewed secondary routes for PPCPs to enter the environment and recognized that they may also be important for certain PPCPs. Secondary routes include (1) disposal of leftover, unused medications, and (2) bathing, washing, laundering, and dermal transfers of medications and metabolites excreted in sweat or dermally applied. Dermally applied medications include topical medications and drugs administered by transdermal systems.

Other geographically limited activities may be the source of very high concentrations of PPCPs. For example, Fick et al. (2009) documented that bulk pharmaceutical production in the Patanacheru industrial region near Hyderabad, India, has resulted in severe contamination of surface, ground, and drinking water. The source of the problem is contaminated sewage effluent and likely unauthorized dumping of industrial wastes. The most urgent aspect of the drug contamination is that high levels of broad-spectrum antibiotics are likely to promote the development of highly antibiotic resistant microorganisms and possible transfer of resistance factors to human pathogens (Fick et al. 2009).

Drug residuals are found at concentrations typically orders of magnitude (2 or more) below those at which an effective therapeutic dose would result from ingesting the water (Ongerth and Khan 2004). The major human health concern with emerging contaminants is not acute effects from one time exposure, but rather chronic effects from long-term exposure to low doses (Drewes et al. 2003). However, there are a few documented cases of significant allergic reactions to non-therapeutic occurrences of antibiotics in drinking water. The full extent and implication of the presence of emerging contaminants in groundwater (and surface water) are largely unknown. Potential concerns with the presence of these compounds in the environment include abnormal physiological processes, reproductive impairment, increased incidents of cancer, development of antibiotic resistant bacteria, and the potential for increased toxicity of chemical mixtures (additive or synergistic effects; Koplin et al. 2002).

Although there is strong evidence for CECs having endocrine disruption and other effects on aquatic organisms, very little is known on their toxicological effects on humans at the concentrations detected in drinking water. There is a large difference in exposure between small aquatic organisms that live their entire lives in impacted waters and humans and other large terrestrial mammals which have exposure through drinking the effected water. Groundwater quality standards (maximum contaminant levels) have not yet been promulgated for individual emerging contaminants.

The primary source of emerging contaminants is untreated and treated wastewater. There is great variation in the effectiveness of standard wastewater processes in removing the different CECs that may be present in the influent (e.g., Carballa et al. 2004, 2005a, b; Clara et al. 2004, 2005; Joss et al. 2005). The efficiency of the removal of the different CECs varies depending upon the treatment processes and plant layout (e.g., sludge retention time) and the adsorptive behavior of the activated sludge (Kreuzinger et al. 2004). Aerobic treatment (activated sludge process) is effective in partially removing most CECs, with the efficacy of removal improved by increasing the sludge retention time to at least 15 days (Fono and McDonald 2008). Nevertheless, some CECs, such as the antiepileptic (anti-convulsant) drugs carbamazepine and primidone, are little removed during the wastewater treatment process and are stable in the groundwater environment (Clara et al. 2004; Kreuzinger et al. 2004; Guo and Krasner 2009).

Pharmaceuticals have been documented in biosolids concentrated in the wastewater treatment process and concerns have been raised over the potential impacts of biosolid-associated pharmaceuticals on terrestrial systems, including groundwater and surface water. The potential for transport of pharmaceutical compounds from soils into both groundwater and surface water depends on their persistence in the soil environment. The available data demonstrate that degradation of pharmaceuticals associated with biosolids is a very complex issue and depends upon the pharmaceutical compound, other chemicals present in the soil and biosolids, the source and nature of the biosolids, and soil properties and bioactivity (Monteiro and Boxall 2009).

The very large number of CECs practically precludes monitoring for the presence of each parameter in treated wastewater and in the source and finished waters of drinking water systems. Even if such sampling could be cost-effectively performed, the question would still remain as to the significance of detected concentrations. Selection of targeted analytes for environmental monitoring should be developed initially by hazard-based and ultimately risk-based approaches (Brooks et al. 2009). Brooks et al. (2009) and Dickenson et al. (2009) recommend that the selection of targeted analytes or surrogates should be based on the following questions:

- Does a target analyte such as carbamazepine, provide an indicator for other contaminants?
- Does the compound represent a chemical class either known or suspected to present a hazard or risk?
- Does a target analyte have a representative susceptibility to removal by employed treatment processes (e.g., advanced oxidation)?

The selection of target analytes should also consider CEC use and potential releases to the environment within the watershed or basin, and experiences in similar settings elsewhere. In other words, monitoring programs should not be developed in a vacuum.

The difficulties of evaluation of the ecological and human health risks associated with personal care products (PPCPs) is compounded by metabolites, which form biologically and abiologically during wastewater treatment, are excreted by humans and animals, and form by degradation processes in the environment. Degradation does not always result in detoxification and some metabolites may have a greater toxicity than the parent compounds (Celiz et al. 2009). Many metabolites have either not yet been detected or data is lacking on their chronic toxicity at their detected concentrations. Identification of metabolites that form under natural conditions is non-trivial. However, advanced high-sensitivity instrumentation has become commercially available, which will likely lead to an increase in the number of metabolites detected in the environment (Celiz et al. 2009).

Although thousands of technical papers have been published on CECs, there is not a clear picture as to the seriousness of the problem or the best course of action. The rapidly inflating literature of the subject increases the possibility that an ever greater share of the publications will not receive adequate examination or even be read (Daughton 2009). Although it is widely accepted that the presence of CECs in the environment and water supply is undesirable, there is much debate and uncertainty as to what is the appropriate response. One clear issue is the importance of placing CECs into a proper context relative to other environmental risks. Daughton (2009, p. 2491) noted that

> we must also determine where PPCPs fall within the growing list of overarching environmental issues in a world of diminishing resources and continuing emerging new concerns.

and (p. 2494):

> Their place in a larger universe of chemical stressors is essential to appreciate so that the public can develop a more accurate and meaningful perspective of chemical exposure in general and that optimally informed regulatory decisions can be formulated.

Stanford et al. (2010) presented the results of a comparative study of the concentrations of some steroidal hormones and endocrine disruptors (EDCs), and net estrogenic activity in drinking water in the United States and in 10 common foods and beverages. The tested foods and beverages included apple juice, dairy- and soy-based infant formulas, milk, green tea, and vegetable juice. The results indicate that the exposure to natural estrogens and other suspected EDCs from drinking water pales in comparison to exposure through other dietary routes (Stanford et al. 2010). The presence of estrogenic activity in food and beverage items is largely attributed to compounds with potential beneficial health effects (phytoestrogens) rather than those indicative of risk from consumption. Airborne exposure to detected compounds was reported to range from 30 to 36,000 times of that from drinking water. Thus, compared to air exposure, water consumption may represent only a small fraction of pharmaceutical, personal care product, and EDC exposure (Stanford et al. 2010). Stanford et al. (2010) observed that the perceived risks of CECs in drinking water must be carefully balanced with the known environmental and monetary costs of energy-intensive treatment processes and the known risks of dietary and airborne exposure to these and other contaminants.

Ternes and Joss (2006) noted that although the presence of CECs in drinking water are unlikely to cause health effects, their presence indicates that part of the raw water that enters a waterworks stems from municipal wastewater. CECs are thus a useful indicator of the presence of wastewater, which may flag the potential for the presence of pathogens.

Upgrading of wastewater and water treatment facilities to remove all CECs is generally not economically feasible. Advanced water treatment involving reverse osmosis, advanced oxidation processes (e.g., ultraviolet treatment and/or hydrogen peroxide), and granular activated carbon is effective in reducing the concentrations of CECs that pass through conventional secondary and tertiary wastewater treatment (Snyder et al. 2007; Benotti and Synder 2009), but such treatment comes at a high cost. Advanced treatment processes are also energy intensive and thus may have a high "carbon footprint" (Benotti and Synder 2009).

Reactive approaches using end-of-chain approaches (i.e., wastewater treatment) are not as efficient or economical as proactive solutions that optimize source reduction or pollution prevention (Daughton 2009). Proactive solutions for the reduction of CEC input into wastewater are multifaceted involving consumers, physicians, and insurers, all of which can act to minimize excessive use and discharges of PPCPs to the environment (Daughton 2009). Confusion and debate still surrounds what constitutes the best approach for disposal of unwanted, left over medications. Key issues are the safety of the process, effectiveness of removal from the environment, and the prevention of some drugs being removed from the trash and abused by drug abusers (Daughton and Ruhoy 2009). Collection programs can be complicated by the presence of controlled substances, which in some jurisdictions can be transferred by the prescription holder only to law enforcement or other government agents (Daughton and Ruhoy 2009).

It is also important to recognize that there are costs associated with overreaction to CEC contamination. Abandonment or failure to introduce wastewater reuse because of exaggerated concerns over CECs could ultimately have more severe environmental and human health consequences than the CECs might possibly ever cause. In developing countries, the health impacts associated with an inadequate food and water supply likely greatly outweigh possible health effects associated with CECs in wastewater reuse. There is little doubt that the explosive growth of research on CECs will continue. It is hoped that a result of the research will be the placement of the risks associated with CECs in a proper perspective.

30.6 Epidemiological Studies and Risk Assessments

Assessment of the health hazards associated with reuse of wastewater can be evaluated through either epidemiological studies or risk assessments. Both types of studies are complex investigations that are typically not performed for individual wastewater reuse projects. Epidemiological studies and risk assessments require specialized professionals and are typically performed on a research, policy or regulatory level, rather than on a project level. These studies are more commonly used to establish treatment and water quality standards for various types of reuse projects. Wastewater treatment and quality standards established by epidemiological studies and risk assessments are then applied by regulatory agencies to individual reuse projects.

Although most water development and management professionals will never be involved in the performance of epidemiological studies or risk assessments, it is important that they understand the basic underlying concepts. Improper assumptions used in risk assessments can result in soil and water quality standards that bear little relationship to actual risks.

Epidemiological studies compare the prevalence of a disease or infection in an exposed group compared to an unexposed control group. Epidemiological studies require a large amount of accurate data on the health history of the exposed and control groups in order to determine if there is a statistically significant difference in their health. Such data may be very difficult to obtain in developing countries, particularly if the response is not severe, memorable, or normally recorded. For example, it is much more difficult to obtain data on the long-term incidence of a mild diarrheal disease from wastewater pathogens than for more severe diseases that result in hospitalization or death. The latter are more likely to be officially recorded.

Epidemiological studies are further complicated when applied to chemical contaminants in which there is a long latency period. For example, chemically-induced cancers may not occur until years or decades after exposure. People exposed to carcinogenic chemicals may change jobs or move out of the area of exposure. It can be extraordinarily difficult, if not impossible, to locate workers who may have been exposed to a chemical at a farm decades earlier, particularly in developing countries.

Quantitative Microbial Risk Assessment (QRA or QMRA) procedures are reviewed in detail by Haas et al. (1999). Application of the QMRA procedures to wastewater and storm water reuse is presented by Mara et al. (2007). QMRA, and similar risk assessment procedures used for chemical contaminants, are based on evaluations of the likelihood of exposure (frequency and extent) to a contaminant and the dose–response for the contaminant. The dose–response assessment may be either static or dynamic (Asano et al. 2007). Static assessment considers a dose response function solely as the result of a single exposure event. Dynamic assessments incorporate complexity associated with infection/exposure disease processes including indirect exposure (and multiple routes of exposure), and varying susceptibility to infection or illness. Dynamic modeling more closely resembles what occurs in nature, but has much greater data requirements, which introduces additional uncertainty.

The most serious obstacle to the application of risk assessment to wastewater reuse is the very limited availability of dose–response data for most constituents of concern in wastewater (Asano et al. 2007). Both exposure scenarios and dose–response model equations include various parameters, the values of which are often poorly known. Consider, for example, the potential exposure of farm workers to reclaimed water used for irrigation. The exposure evaluation must consider the concentrations of pathogens in the reclaimed water, potential mechanisms of exposure (e.g., consumption of water and soil, inhalation of aerosols), and the likely frequency and magnitude of exposure.

Dose–response relationships may not be linear. For example, a threshold dose value may have to be exceeded for infection to occur. The response to a given pathogen dose will also vary between individuals. There will thus inherently be a great uncertainty in the quantitative evaluation of the risks that farm workers face. Techniques are available for dealing with the uncertainty in the values of the risk assessment parameters. One option commonly used for risk assessments is Monte Carlo simulations, which involves the random selection of parameters within defined linear ranges for the various parameters.

The results of the epidemiological studies and risk assessments are a calculated risk of infection or disease for an individual exposed over a year or other time period. The calculated risk of infection or disease is then evaluated relative to a "tolerable" risk. With respect to the use of wastewater for irrigation, the World Health Organization (World Health Organization 2006) recommends that the tolerable impact from treated water consumption and water reuse in agriculture is $\leq 1 \times 10^{-6}$ disability-adjusted life year (DALY) per person per year.

The DALY considers both the frequency of adverse health impacts *and* their severity. A DALY is equal to one year of healthy life lost and includes both time lost as the result of premature death and time spent disabled by disease. Health impacts are weighted in terms of severity within a range of 0 for no impacts to 1 for death (NRMMC-EPHC-AHMC 2006). DALYs can be calculated as

$$\begin{aligned} \text{DALYs} = {} & \text{YLL (years of life lost)} \\ & + \text{YLD (years lived with a disability or illness)} \end{aligned} \tag{30.1}$$

For example, a mild diarrhea with a severity of 0.1 and duration of 7 days results in a DALY of 0.002 per case (NRMMC-EPHC-AHMC 2006). DALYs per person per year are obtained by multiplying the frequency of disease (number of cases per year) by DALYs per case.

Acceptable or tolerable risk values proposed or implemented by regulatory agencies inherently consider the severity of health impacts. For example, Mara et al. (2007) used a risk of rotavirus infection of 1×10^{-2} per person per year (pppy) as the reference level for an acceptable risk. The relatively high acceptable rate of infection is based on the resulting gastroenteritis tending not to have severe impacts (particularly in adults), provided that dehydration is prevented.

A key point is that neither QMRA nor epidemiology is an exact science and any comparison of the two methods has to take this into account (Mara et al. 2007). In order for there to be a satisfactory agreement between the two methods, the values for the parameters used in the QMRA and Monte Carlo simulations should be close to the actual values likely to occur in the field (Mara et al. 2007).

The threshold for what constitutes an acceptable or tolerable risk depends upon local circumstances and is not static (Asano et al. 2007). The acceptability of a given risk for a give activity depends upon:

- costs (human, social, and economic) associated with the risk,
- costs of implemented measures to reduce the risk,
- societal resources available that could be mobilized to reduce the risk, and
- other risks that could be reduce with available resources.

Public acceptance of risk in general is often based largely on emotion rather than objective analysis. For example, minute risks associated with drinking water quality may be considered much less acceptable than much greater risks routinely assumed, for example, by driving an automobile or smoking. People, in general, tend to be more accepting of risks that they feel they understand and have some control over, rather than risks that they do not fully understand and are imposed upon them by others.

The WHO guidelines have been adopted by some countries, but many countries have developed their own, often stricter standards. Asano et al. (2007) proposed that the WHO guidelines can be considered as an interim measure in the socio-economic realities of many countries, until they have the capacity to produce higher-quality reclaimed water. Each country can and should establish national risk criteria and procedures that suit its epidemiological, social, and economic needs. Risk assessment is the standard means to quantify the possible health consequences of human exposure in particular circumstances, but it is recognized that risk assessments are far from perfect (Asano and Cotruvo 2004). There is inherently great uncertainty in the quantification of small long-term risks from exposure to chemicals at low concentrations. Asano and Cotruvo (2003, p. 175) noted with respect to risk assessments that:

> All such computations and "conclusions" are limited in their reliability and credibility by the quality of the exposure and toxicological data, the mathematical expressions used, and the lack of scientific understanding of the mechanisms of carcinogenesis operating at low environmental doses in genetically diverse humans, compared to the high doses to which test animals are exposed. In addition, little is known of the effects of interactions of low doses of chemicals.

Nevertheless, there is still a necessity to establish some health-based targets for compounds that are known or highly suspected to adversely impact human health even though it is understood that there is considerable uncertainty in the appropriateness of specific target values. The tendency has been to establish conservatively low values, under the implicit philosophy 'better safe than sorry'. The dilemma is that an overly restrictive water quality standard may result in the allocation of societal resources for unnecessarily treatment processes, while the resources could provide greater public health benefits if allocated in other, more productive areas.

Health-based targets can be developed to either (World Health Organization 2006):

1. Establish a defined level of health protection, which could be the absence of a specific disease related to exposure to wastewater or a quantitative level of acceptable health impacts (e.g., $\leq 10^{-6}$ DALY, per person per year).
2. Establish health protection measures to achieve the defined level of protection. The health protection measures (HPMs) could be operational or treatment procedures, a target exposure level (e.g., maximum concentration level, MCL), or target minimum pathogenic removal rate.

30.7 Microbiological Standards and Guidelines

With respect to the use of wastewater for irrigation, the World Health Organization (World Health Organization 2006) recommendations are a useful starting point for consideration of microbiological treatment standards. The WHO recommended health-based target of $\leq 10^{-6}$ DALY per person per year for viral, bacterial, and protozoa pathogens and ≤ 1 per liter concentration of helminth eggs can be achieved by a 6 to 7 $\log_{10}$ pathogen removal. A one-log removal is the time for a 90% reduction in concentration. A 6 $\log_{10}$ removal is equal to a 99.9999% reduction in pathogen concentration.

Log_{10} removals (or just log removal) can be calculated as follows:

$$\log_{10} removals = -\log\left(\frac{c_f}{c_i}\right) \tag{30.2}$$

where,

C_i = initial number of organisms,
C_f = final number of organisms at day t.

A one $\log_{10}$ removal time (or just log removal time; τ) is defined as:

$$\tau = t/\mathrm{Log}_{10}(C_0/C_t), \tag{30.3}$$

where,

t = time (days),
C_0 = number of organisms at day 0,
C_t = number of organisms at day t.

The target 6 to 7 $\log_{10}$ pathogen removal can be achieved by a combination of wastewater treatment processes and other health protection measures. The World Health Organization (2006) presented an example where a 7 $\log_{10}$ pathogen removal could be achieved by wastewater treatment processes that provide a 4 $\log_{10}$ pathogen removal, a 2 $\log_{10}$ removal due to pathogen die-off between the last irrigation and consumption, and a 1 $\log_{10}$ removal due to normal household washing of salad crops and vegetables prior to consumption. The World Health Organization advocated that the treatment alone approach to meeting health-based standards, such as required in California, USA, is overly strict. The World Health Organization recommends a fecal coliform concentration of $\leq 10^3$ to $\leq 10^4$ CFU per 100 mL for unrestricted irrigation, whereas the California standard is ≤ 2.2 CFU per 100 mL of total coliforms.

QMRA and Monte Carlo risk assessments performed by Mara et al. (2007) for several wastewater reuse scenarios indicate that the WHO guidelines are appropriate to overly conservative. Mara et al. (2007) also expressed the common opinion that the California wastewater quality standard of ≤ 2.2 CFU per 100 mL is difficult to justify epidemiologically and is unlikely to be cost effective in protecting public health. It is important, therefore, that countries adopt wastewater reuse regulations and guidelines based on local water supply, public health, and economic conditions, rather than adopt standards of other countries, which may be locally inappropriate.

Gerba and Rose (2003) proposed that fecal coliform should be used as a treatment performance measurement and not as an indicator of virus or parasite performance and risk. Fecal coliform bacteria can grow (reproduce) in warm tropical waters and there is not an established ratio of fecal coliforms or other indicators to parasites and viruses in raw water (Gerba and Rose 2003). Treatment processes also differ in their effectiveness in removing indicator bacteria versus other pathogens. International guidelines based solely on bacteria indicators and treatment requirements may not necessarily reflect the risks posed by the use of recycled waters for different purposes (Gerba and Rose 2003). Gerba and Rose (2003) recommended a multiple indicator approach including viruses. However, as previously noted, routine monitoring for all pathogens that could be present in wastewater is not feasible, would be highly expensive, and cannot be performed in real time (Asano and Cotruvo 2004).

There is clearly a need for a more effective pathogen indicator approach (and associated water quality standards or guidelines) that is both affordable and has a fast laboratory turnaround time. A consistent approach to pathogen indicators is also desirable, so that data and experiences between different projects and regions can be more readily correlated. However, one is reminded of the American essayist Ralph Waldo Emerson's famous quote

> A foolish consistency is the hobgoblin of little minds

The quest for consistency often leads to inflexibility, as it becomes easier to insist on use of a standard monitoring protocol (i.e., set of parameters and frequency) and water quality standards, rather than developing system-specific plans for health protection based on local conditions.

30.8 Implementing Health-Based Targets and Monitoring

A regulatory framework is needed for actually meeting wastewater use objectives, such as the protection of public health, prevention of environmental damage, conservation of water and nutrients, preservation of livelihoods, and meeting produce quality standards for domestic consumption and export (World Health Organization 2006). Creation of a regulatory framework to effectively manage the treatment and use of wastewater may be a great challenge in nations and regions where environmental and public health institutions are not well developed. Different government institutions may have responsibilities that may be in conflict. For example, different agencies may have jurisdiction over water use, wastewater treatment facilities, and environmental protection.

A very important issue raised by the World Health Organization (2006) is that it is easier to make regulations than to enforce them. The overly restrictive water quality standards were impossible to achieve in many situations and were, therefore, ignored rendering the early WHO Guidelines useless (Bos et al. 2010).

Establishing health-based targets can be a relatively simple matter, particularly if existing targets (e.g., World Health Organization, United States Environmental Protection Agency) are adopted. Within developed countries, existing governmental institutions typically exist with the authority and resources to regulate large urban municipal wastewater treatment facilities and the use of reclaimed water. Regulating wastewater use for irrigation is a much more difficult task, if at all possible, in scattered villages in developing countries. In the latter case, education may be more effective that regulatory enforcement.

The final element in the regulation and management of the use wastewater and graywater is monitoring of system performance (Sect. 30.2.2). Monitoring serves three main purposes (World Health Organization 2006):

1. Validation—determination after construction that a system and its individual components are capable of meeting specific performance targets.
2. Operational monitoring—routine monitoring to determine if the treatment system is operating within design parameters and that health protection measures are working as designed.
3. Verification—shows that the end product meets treatment targets and ultimately health-based targets.

The parameters and sampling frequency for wastewater reuse systems is typically determined by the regulatory agency that has jurisdiction over the project. Parameters included in the monitoring program are usually those for which there is a promulgated water quality standard. There is currently no widely accepted generic monitoring program. Ideally, monitoring programs should be system specific and developed based on actual treated sewage effluent quality and end uses (i.e., system-specific risks). Monitoring programs should also be adjusted based on earlier results.

Salgot et al. (2006) proposed a monitoring program that incorporates water quality criteria (both chemical and microbial) based upon specific final uses of the water. The more important parameters are sampled more frequently. Salgot et al. (2006) also considered cost in the development of the monitoring program. Expensive indicator compounds of micropollutants or heavy metals are recognized as needing rare monitoring (every year or 5 years).

Monitoring can represent a significant fraction of the total system operational costs. There is often a natural desire to have an expanded monitoring program in order to try to provide additional assurances that a reuse program does not pose an unacceptable health risk. However, excessive monitoring results in additional costs and may provide no public or environmental health benefits. The challenge is for system owners and regulators to work together to develop a "right sized" monitoring program that provides necessary data on reclaimed water quality while not resulting in an excessive financial burden on the system owner and operator. Too often a "one-size-fits-all" approach is taken, in which a standard monitoring program is required for every project irrespective of project-site realities. Such an approach has the superficial benefits of apparent fairness and simplicity, but may result in considerable wastes of effort and money.

Degree of compliance of wastewater reuse systems over long-periods of time is one of the current weak points in the actual practice of reuse and is especially a concern in developing countries with a poorly developed regulatory system (Salgot et al. 2003). Reliability, with respect to wastewater treatment and reuse, is defined as the probability of successful performance of the unit and facility in satisfying specific operational conditions and producing effluent that meets minimum effluent water quality standards (Salgot et al. 2003). Reliability depends upon system design, operation, and maintenance, and requires frequent sampling to confirm compliance with minimum effluent quality standards. Systems have different potentials for malfunctions and requirements for successful long-term reliability, which need to be considered for the system design. A key treatment facility operational issue is how plant upsets or malfunctions are handled to avoid discharges of inadequately treated effluent to the reclaimed water stream.

Reliability requirements also vary depending on the sensitivity of the intended use. Systems intended for direct or indirect potable reuse require higher degrees of reliability and more frequent sampling (continuous where possible) to ensure that the reclaimed water consistently meets water quality standards. Fail-safe procedures should be incorporated into the system design to ensure that water not meeting the standards cannot enter the water supply.

Additionally, controls need to be in place to prevent unintended, inappropriate uses of the reclaimed water after it leaves the wastewater treatment facility. For example, different colored pipe can be used to differentiate between reclaimed water and potable water lines. Reclaimed water pipe in the United States, Canada, Europe, Australia and much of the rest of the world is colored purple (lavender, lilac) in order to avoid cross-connections. However, despite such an obvious indicator, cross-connections of reclaimed water to residential potable water lines have still occurred in some rare instances (e.g., several homes in Cape Coral, Florida), which highlights the need for vigilance. Warning signs are usually required where reclaimed water is used for public access area irrigation to discourage consumption.

References

Alhumoud, J. M., Behbehani, H. S., & Abdullah, T. H. (2003). Wastewater reuse practices in Kuwait. *The Environmentalist,23*, 117–126.

Asana, T., & Cotruvo, J. A. (2003). Groundwater recharge with recycled municipal wastewater: health and regulatory considerations. In R. Aertgeerts, R. & A. Angleakis (Eds.), *Health risks in aquifer recharge using reclaimed water - State of the art report.* (pp. 171–197) Geneva: World Health Organization.

Asano, T., Burton, F. L., Leverenz, H. L., Tsuchihashi, R., & Tchobanoglous, G. (2007). *Wastewater reuse. Issues, technologies and applications* (p. 1570). New York: McGraw-Hill.

Asano, T., & Cotruvo, J. A. (2004). Groundwater recharge with reclaimed wastewater; health and regulatory considerations. *Water Research,38*, 1941–1951.

Benotti, M. J., & Synder, S. A. (2009). Pharmaceuticals and endocrine disrupting compounds: Implications for ground water replenishment with recycled water. *Ground Water,47*(4), 499–502.

Bos, R., Carr, R., & Keraita, B. (2010). Assessing and mitigating wastewater-related health risks in low-income countries. In P. Dreschel, C. A. Scott, L. Raschid-Sally, M. Redwood & A. Bahri (Eds.), *Wastewater irrigation and health. Assessing and mitigating risk in low income countries* (pp. 29–47). Ottawa: International Development Research Centre.

Brooks, B. W., Huggett, D. B., & Boxall, A. B. A. (2009). Pharmaceuticals and personal care products: Research needs for the next decade. *Environmental Toxicology and Chemistry,28*, 2469–2472.

Carballa, M., Omil, F., Lema, J. M., Llompart, M., Garcia, C., Rodriguez, I., et al. (2004). Behaviour of pharmaceuticals, cosmetics and hormones in a sewage treatment plant. *Water Research,38*, 2918–2926.

Carballa, M., Omil, F., & Lema, J. M. (2005a). Removal of cosmetic ingredients and pharmaceutical in sewage primary treatment. *Water Research,39*, 4790–4796.

Carballa, M., Omil, F., Lema, J. M., Llompart, M., Garcia, C., Rodriguez, I., et al. (2005b). Behaviour of pharmaceuticals and personal care products in a sewage treatment plant of northwest Spain. *Water Science and Technology,52*(8), 29–35.

Celiz, M. D., Tso, J., & Aga, D. S. (2009). Pharmaceutical metabolites in the environment: Analytical challenges and ecological risk. *Environmental Toxicology and Chemistry,28*, 2473–2484.

Clara, M., Strenn, B., & Kreuzinger, N. (2004). Carbamazepine as a possible anthropogenic marker in the aquatic environment: Investigations on the behaviour of carbamazepine in wastewater treatment and groundwater infiltration. *Water Research,38*, 947–954.

Clara, M., Strenn, B., Gans, O., Martinez, E., Kreuzinger, N., & Kroiss, H. (2005). Removal of selected pharmaceuticals, fragrances and endocrine disrupting compounds in a membrane bioreactor and conventional wastewater treatment plants. *Water Research,39*, 4797–4807.

Daughton, C. G. (2009). Chemicals from the practice of healthcare: Challenges and unknowns posed by residues in the environment. *Environmental Toxicology and Chemistry,28*, 2490–2494.

Daughton, C. G., & Ruhoy, I. S. (2009). Environmental footprint of pharmaceuticals: The significance of factors beyond direct excretion to sewers. *Environmental Toxicology and Chemistry,28*, 2495–2521.

Daughton, C., & Ternes, T. (1999). Pharmaceuticals and personal care products in the environment: Agents to subtle change? *Environmental Health Perspectives, 107*(suppl. 6), 907–938.

Dickenson, E. R. V., Drewes, J. E., Sedlack, D. L., Wert, E. C. & Snyder, S. A. (2009). Applying surrogates and indicators to assess removal efficiency of trace organic chemicals during chemical oxidation of wastewaters. *Environmental Science & Technology, 43*(16), 6242–6247.

Drewes, J. E., & Fox, P. (1999). Fate of natural organic matter (NOM) during groundwater recharge using reclaimed water. *Water Science and Technology,40*(9), 241–248.

Drewes, J. E., Heberer, T., Rauch, T., & Reddersen, K. (2003). Fate of pharmaceuticals during groundwater recharge. *Ground Water Monitoring and Remediation,23*, 64–72.

Edberg, S. C., Rice, E. W., Karlin, R. J., & Allen, M. J. (2000). *Escherichia coli*: The best biological drinking water indicator for public health protection. *Journal of Applied Microbiology,88*, 106s–116s.

Fick, J., Söderstrom, H., Lindberg, R. H., Phan, C., Tysklind, M., & Larson, D. G. J. (2009). Contamination of surface, ground, and drilling water from pharmaceutical production. *Environmental Toxicology and Chemistry,28*, 2522–2527.

Fono, L. J., & McDonald, S. (2008). Emerging compounds: a concern for water and wastewater utilities. *Journal of the American Water Works Association,100*(11), 5–57.

Gerba, C. P., & Rose, J. B. (2003). International guidelines for water recycling: Microbiological considerations. *Water Science and Technology: Water Supply,3*(4), 311–316.

Guo, Y. C., & Krasner, S. W. (2009). Occurrence of primidone, carbamazepine, caffeine, and precursors for *N*-nitrosodimethylamine in drinking water sources impacted by wastewater. *Journal of the American Water Works Association,45*(1), 58–67.

Haas, C. N., Rose, J. B., & Gerba, C. P. (1999). *Quantitative microbial risk assessment* (p. 449). New York: John Wiley & Sons.

Joss, A., Keller, E., Alder, A. C., Göbel, A., McArdell, C. S., Trenes, T. & Siegrist, H. (2005). Removal of pharmaceuticals and fragrances in biological wastewater treatment. *Water Research, 39*(14), 3139–3152.

Koplin, D. W., Furlong, E. T., Meyer, M. T., Thurman, M., Zaugg, S. D., Barber, L. B., et al. (2002). Pharmaceuticals, hormones, and other organic wastewater contaminants in U.S. streams, 1999–2000: A national reconnaissance. *Environmental Science and Technology,36*, 1202–1211.

Kreuzinger, N., Clara, M., Strenn, B., & Vogel, B. (2004). Investigation on the behaviour of selected pharmaceuticals in the groundwater after infiltration of treated wastewater. *Water Science and Technology,50*(2), 221–228.

Macler, B., & Merkle, J. C. (2002). Current knowledge of groundwater microbial pathogens and their control. *Hydrogeology Journal,8*, 29–40.

Mara, D. D., Sleigh, P. A., Blumenthal, U. J., & Carr, R. M. (2007). Health risks in wastewater irrigation: Comparing estimates from quantitative microbial risk analyses and epidemiological studies. *Journal of Water and Health,5*, 39–50.

Monteiro, S. C., & Boxall, A. B. A. (2009). Factors affecting the degradation of pharmaceuticals in agricultural soils. *Environmental Toxicology and Chemistry,28*, 2546–2554.

Mujeriego, R., & Asano, T. (1999). The role of advanced treatment in wastewater reclamation and reuse. *Water Science and Technology,40*(4–5), 1–9.

National Research Council. (1994). *Ground water recharge using waters of impaired quality* (p. 283). Washington, DC: National Academy Press.

National Research Council. (1998). *Issues in potable reuse: The viability of augmenting drinking water supplies with reclaimed water* (p. 263). Washington, DC: National Academy Press.

NRMMC-EPHC-AHMC (Natural Resource Management Ministerial Council, Environment Protection and Heritage Council, Australian Health Ministers Conference) (2006). National Guideline for Water Recycling: Managing Health and Environmental Risks (Phase 1), November 2006.

Ongerth, J. E., & Khan, S. (2004). Drug residuals: How xenobiotics can affect water supply sources. *Journal of the American Water Works Association,96*(5), 94–101.

Payment, P., & Locas, A. (2011). Pathogens in water: Value and limitations of correlation with microbial indicators. *Ground Water,49*(1), 4–11.

Richardson, S. D. (2006). Environmental mass spectrometry: Emerging contaminants and current issues. *Analytical Chemistry,78*, 4021–4046.

Salgot, M., Huertas, E., Weber, S., Dott, W., & Hollender, J. (2006). Wastewater reuse and risk: Definition of key objectives. *Desalination,187*, 29–40.

Salgot, M., Vergés, C., & Angelakis, A. N. (2003). Risk assessment in wastewater recycling and reuse. *Water Science and Technology: Water Supply,3*(4), 301–309.

Sedlak, D. L., Gray, J. L., & Pinkston, K. E. (2000). Understanding microcontaminants in recycled water. *Environmental Science and Technology,35*, 508A–515A.

Snyder, S. A., Adham, S., Redding, A. M., Cannon, F. S., DeCarolis, J., Oppenheimer, J., et al. (2007). Role of membranes and activated carbon in the removal of endocrine disruptors and pharmaceuticals. *Desalination,202*, 156–181.

Standridge, J. (2008). *E. coli* as public health indicator of drinking water quality. *Journal American Water Works Association,100*(2), 65–75.

Stanford, B. D., Snyder, S. A., Trenholm, R. A., Holandy, J. C., & Vaderford, B. J. (2010). Estrogenic activity of US drinking waters: A relative exposure comparison. *Journal American Water Works Association,120*(11), 55–65.

Ternes, T. A., & Joss, A. (2006). *Human pharmaceuticals, hormones and fragrances—The challenge of micropollutants in urban water management*. London: IWA Publishing.
Toze, S. (2006). Water reuse and health risks—Real vs perceived. *Desalination, 187*, 41–51.
World Health Organization (2006). *WHO guidelines for the safe use of wastewater, excreta, and greywater* (Vol. 1). Geneva: Policy and regulatory aspects.
World Nuclear Association (2011). Nuclear desalination: www.world-nuclear.org/info/in71.html, 6 pp.

Chapter 31
Managed Aquifer Recharge and Natural Water Treatment Processes in Wastewater Reuse

31.1 Introduction

Both the vadose and phreatic zones are active chemical, physical, and biological environments. The concentrations of pathogens and chemical contaminants tend to be reduced as water infiltrates into the soil, percolates to the water table, and flows through aquifers. The contaminant attenuation processes include filtration, sorption, precipitation, sedimentation, and various biological and chemical degradation processes. Water flowing through the soil (vadose zone) and within aquifers in some instances may experience deterioration in water quality. Trace metals and arsenic have leached into water stored in some managed aquifer recharge (MAR) systems, which was reviewed by Maliva and Missimer (2010).

It has been recognized for centuries that the flow of impaired water through the soil zone and an aquifer can be used as a method of water treatment. The initial applications were mainly concerned with the removal of pathogens from surface waters. Natural water treatment processes include several related technologies that use natural filtration and contaminant attenuation processes to improve the quality of water. The water may be either surface water (impaired to varying degrees) or wastewater. MAR processes specifically designed for water treatment include:

- riverbank (bank) filtration (RBF, Chap. 24),
- soil-aquifer treatment (SAT) and other aquifer recharge and recovery (ARR) systems that incorporate the vadose zone as part of the water treatment process,
- aquifer storage transfer and recovery (ASTR) systems in which recharge is performed using wells.

Other types of MAR systems with a primary storage function may also provide water treatment benefits. There is growing operational and experimental data demonstrating that the concentrations of some contaminants, such as disinfection byproducts, nutrients, organic compounds, and pathogenic microorganisms, are decreased during storage by natural inorganic and microbiological process

R. Maliva and T. Missimer, *Arid Lands Water Evaluation and Management*,
Environmental Science and Engineering, DOI: 10.1007/978-3-642-29104-3_31,

(Harpaz 1971; Dillon et al. 1999, 2006b; Dillon and Toze 2005; Pyne 2002; McQuarrie and Carlson 2003, Pavelic et al. 2005b, 2006b; Vanderzalm and Le Gal La Salle 2005; Wall et al. 2006, 2007; Khan and Rorije 2002; Stuyfzand 2007; Maliva and Missimer 2010). The placement of reclaimed or surface water into storage in natural groundwater environments increases the recycling time and thereby allows more time for biodegradation of contaminants that degrade more slowly (Dillon et al. 2006b). The intentional use of natural attenuation process to improve water quality has been referred to as natural aquifer treatment (NAT).

The occurrence of natural aquifer treatment processes has implications for both the regulation and operation of ASR and other MAR systems. If the concentration of a water-quality parameter will quickly be reduced during storage, then the requirement that the recharged water meet water quality standards at the point of recharge (e.g., ASR wellhead) is unnecessarily restrictive. A more appropriate approach would be to set the water-quality compliance point at monitoring wells located at the perimeter of a zone of discharge (ZOD) for the MAR system (Pyne 2002). For systems designed primary for water treatment, the recovered water must either meet standards for its intended use or receive post-treatment.

MAR projects that rely upon natural aquifer treatment as a barrier against microbial pathogens before recovery for drinking water will always require a greater degree of scrutiny and validation of the removal of a range of specifically targeted microbial pathogens in the aquifer (NRMMC-EPHC-NHMRC 2009a, b). Pathogens are also a concern where there is the possibility of unplanned consumption of the stored water. On the contrary, systems that store water that is already of a suitable quality for irrigation for later non-potable use, and are not relying upon aquifer treatment for pathogen reduction, pose much less risk.

The great attraction of natural treatment processes is that they usually have a low cost. Once the systems are designed and constructed, they are relatively simple and inexpensive to operate. Natural contaminant attenuation processes also continuously operate and do not require uninterrupted human intervention, which is an important advantage in areas where, for example, the electrical power supply is not reliable. Natural treatment processes are thus well suited for developing countries. However, these systems typically require a fairly sophisticated understanding of local hydrogeology in order to properly design systems that will efficiently meet treatment and water quality goals. Using riverbank filtration (Chap. 24) as an example, it is a technically easy matter to install a series of wells or a gallery along a river; the technical challenge is in properly locating and designing a wellfield or gallery system that most efficiently (cost effectively) provides the needed volume of water at the desired quality.

Natural treatment processes vary depending upon whether or not they incorporate vadose (soil) zone treatment. The vadose zone is important from a geochemical perspective because it is an oxic environment, whereas groundwater often becomes chemically reducing at relatively shallow depths within an aquifer (below the water table). The changing redox conditions are important for contaminant attenuation because some chemicals are preferentially removed in one or the other redox environment, or require the transition from oxic to anoxic condition for their removal.

31.2 Water Quality Improvements During SAT

31.2.1 Introduction

There has been considerable recent research on the effectiveness of SAT, and land supplication systems in general, in removing contaminants. The research and operational data all point to SAT as being an effective method for reducing the dissolved organic matter and organic pollutants. SAT does not result in the reduction of the salinity of wastewater and, in some situations, may result in an increase in salinity through evaporative concentration and leaching of salts, and other processes. Further demonstration of the effectiveness of SAT in removing CECs and other contaminants that pass through the conventional wastewater treatment process would increase the value of the technology as a water management tool. A key issue is whether indirect potable reuse through SAT or groundwater recharge through surface spreading would pose a public health risk. A comparative study of SAT and advanced membrane treatment processes (nanofiltration/reverse osmosis) indicates that both processes have similar effectiveness in reducing dissolved organic carbon (DOC) and the concentrations of some CECs (Drewes et al. 2003).

31.2.2 Nutrient Removal During SAT

SAT systems can be operated to remove dissolved nutrients from wastewater by a combination of nitrification and denitrification. Nitrification is the biologically mediated process by which ammonia (NH_3) and ammonium (NH_4^+) are converted to nitrite (NO_2^-) and then nitrate (NO_3^-) under oxic conditions. The basic equations for the nitrification processes are (Stumm and Morgan 1996):

$$NH_3 + 1.5O_2 \rightleftharpoons NO_2^- + H_2O + H^+ \qquad (31.1)$$

$$NH_4^+ + 1.5O_2 \rightleftharpoons NO_2^- + H_2O + 2H^+ \qquad (31.2)$$

$$NO_2^- + 0.5O_2 \rightleftharpoons NO_3^- \qquad (31.3)$$

Denitrification is the bacterially mediated conversion of nitrate to nitrogen gas, which occurs under anoxic conditions, as follows

$$NO_3^- + 1.25\{CH_2O\} + H^+ \rightleftharpoons 0.5N_2(g) + 1.75H_2O + 1.25CO_2 \qquad (31.4)$$

Denitrification requires a source of organic carbon, which in SAT systems is normally provided by the wastewater. SAT systems have been demonstrated to be highly effective in reducing the dissolved nitrogen concentration of wastewater provided that the flooding and drying cycles are properly managed based on site-specific conditions (Sect. 23.2.2.5). The drying period increases the dissolved oxygen of the system, which allows for the nitrification of ammonia and ammonium. Anoxic conditions that become established during subsequent flooding periods allows for the denitrification of nitrate produced by nitrification (and nitrate and nitrite originally present in the wastewater) to nitrogen gas. SAT is also effective in reducing the concentration of phosphorous through chemical precipitation (calcium phosphate) and adsorption.

31.2.3 Pathogen Removal During SAT

SAT has been documented to be highly effective in removing pathogens (e.g., Wilson et al. 1995). Pathogen removal occurs as a result of a variety of physical and biological processes as water passes through the vadose zone and groundwater environments. Pathogen attenuation has not received as much attention as chemical attenuation in SAT systems because of reclaimed water used was already of high quality (commonly tertiary treated with disinfection).

Fox et al. (2001a, b) reported that field data from the Montebello Forebay (California, USA) SAT site suggested that 30 m (100 ft) of subsurface travel is sufficient for a 7-log removal of bacteriophage, which was used a virus tracer. Testing of infiltration galleries using secondary treated wastewater in Australia indicate a 3 log reduction in microorganisms (Bekele et al. 2009).

In general, passage through the vadose zone is insufficient for complete pathogen removal. Pathogens are frequently detected in shallow monitoring wells installed near recharge basins. Pathogen concentrations decrease with increasing distance from the basins. Reduction in pathogen concentrations to non-detectable levels occurs by filtration, sorption, and inactivation processes during groundwater flow. The fate of pathogenic microorganisms in groundwater environments is further examined in Sect. 31.3.1.

31.2.4 Organic Carbon Removal During SAT

It was recognized in the earliest SAT research (e.g., Bouwer 1974; Bouwer et al. 1980), that SAT is effective in reducing the concentration of DOC in wastewater. Wastewater effluent organic matter (EfOM) includes both natural organic matter (NOM) that was already present in the drinking water and soluble microbial products (SMPs) derived from the biological wastewater treatment process (Amy et al. 1987; Drewes and Fox 1999). The NOM is dominated by humic and fulvic compounds.

Field and laboratory studies by Drewes and Fox (1999) demonstrated that short-term SAT results in the preferential removal of ultra-hydrophilic compounds, which consist primarily of amino acids, proteins and polysaccharides, all of which are potentially biodegradable. Sustained biodegradation of more poorly degradable organic compounds occurred during long-term SAT. Long-term SAT also resulted in a change in the structure of the DOC with a loss of aromatic and carboxylic character and an increase in aliphatic compounds. Further study of the fate of EfOM and trace organic compounds at operational SAT sites located Mesa and Tucson, Arizona, indicated that 50–70% of the dissolved organic carbon (DOC) was removed after accounting for dilution (Fox et al. 2001a, b; Drewes et al. 2003; Amy and Drewes 2007). The character of the bulk organics present after long-term SAT resembles the character of NOM present in drinking water. A "naturalization" of wastewater organic matter occurs, which results in a character of the organic matter becoming very similar to the NOM present in surface water and groundwater not affected by recycled water (Drewes 2009). Dissolved organic nitrogen (DON) was almost completely removed and polysaccharides and proteins in the EfOM were completely removed. Humic substances showed a lesser, but still significant reduction in concentration.

Fox et al. (2005) analyzed soil samples from year-long bench-top studies and from an operational SAT system (Mesa Northwest Water Reclamation Plant, Arizona, USA) to determine whether or not organic carbon is accumulating in the soil. This is an important question as it addresses whether or not reductions in DOC measured in SAT systems are caused by sorption or sustainable biological processes. Organic carbon accumulated mostly in the upper 2 cm (0.8 in.) of soil and organic carbon did not accumulate below 8 cm (3.1 in.). The total organic carbon accumulation in the surficial soil was <20% of the total organic carbon load that was applied. The results of the Fox et al. (2005) investigation indicate that organic carbon accumulation in SAT systems is sustainable because it occurs by biodegradation processes rather than by sorption.

Organic carbon removal in an operational SAT system in Tucson, Arizona, showed the typical pattern of rapid DOC removal in the upper 3 m (10 ft) of the soil (Quanrud et al. 2003). Comparison of recharge basins that had been in operation for 10 years with new (less than 2 year old) basins, demonstrated that DOC removal efficiency does not decrease over time and that the systems are sustainable. SAT systems were found to be most effective in the removal of hydrophilic biodegradable organic compounds. THM formation potential was reduced by an average of 91% across the vadose zone. Drewes (2009) also observed that in general the majority of EfOM removal occurs in the initial phase of infiltration.

Bench-top experiments were performed at the University of Arizona to examine the effects of soil type on the removal of nonpurgable dissolved organic carbon (NPDOC) and on the absorbance of ultraviolet light at 254 nm (UV_{254}) (Quanrud et al. 1996). NPDOC consists of virtually all the DOC in treated wastewater as the volatile compounds are largely removed during the wastewater treatment process. UV_{254} is used as measure of the DOC compounds that are disinfection byproduct precursors. The experiment results indicated that soil types had only a modest effect on the removal of NPDOC and no effect on UV_{254}. There is no significant correlation between infiltration and the removal efficiency of either NPDOC or UV_{254}.

31.2.5 CEC Removal During SAT

As discussed in Sect. 30.5, some CECs are not completely removed during conventional wastewater treatment process. The effectiveness of SAT in the removal of CECs has been investigated through both laboratory experiments and field evaluations. Some of these studies are herein discussed. It is clear that SAT is effective in reducing the concentrations of many CECs that survive conventional secondary and tertiary treatment processes. However, some compounds are not completely removed during passage through the vadose zone, but are later attenuated in the groundwater environment.

Cordy et al. (2004) performed laboratory experiments on the effectiveness of SAT in removing CECs and pathogens. The experiment consisted basically of passing an actual secondary treated wastewater sample through a 2.4 m (7.9 ft) long sediment column and comparing the difference in composition between the influent and drainage water. Only eight CEC compounds were common to and persisted in the initial and final influent samples and the discharge sample, which were: carbamazepine, sulfamethoxazole, benzophenone, 5-methyl-1H-benzotriazole, N,N-diethyltoluamide (DEET), tributylphosphate, tri(2-chloroethyl) phosphate, and cholesterol.

Cordy et al. (2004) concluded that under recharge conditions similar to those in arid and semiarid climates, some CECs and pathogens may persist in treated effluent after SAT and reach the groundwater. However, the Cordy et al. (2004) study did not address additional attenuation that will occur to some degree in the groundwater environment.

CECs include endocrine disrupting compounds (EDCs), which are a wide variety of chemicals that act like hormones in the endocrine system and disrupt the physiologic function of endogenous hormones. EDCs include both synthetic and natural hormones. The mechanism of removal of steroids in SAT was investigated in bench-top experiments by Mansell et al. (2004). The studied steroids were 17β-estradiol, estriol, and testosterone. All three compounds were found to have a high tendency to be absorbed onto porous media. Testosterone showed the greatest retardation, followed by estriol and 17β-estradiol. All compounds were also subject to removal by biodegradation regardless of the type of organic carbon present in the sample. The biodegradation occurred under both aerobic and anoxic conditions. In field studies at two sites, estriol and testosterone were not detected in lysimeter samples taken 1.5 m (4.9 ft) below land surface (Mansell and Drewes 2004). Substantially reduced concentrations of 17β-estradiol were detected at 1.5 m (4.9 ft) below surface in the vadose zone, but the hormones were not detected in down-gradient monitoring wells.

The Mansell et al. (2004) and Mansell and Drewes (2004) studies demonstrated that SAT is efficient in removing hormones present in reclaimed water used to augment groundwater supplies. Physical sorption was the primary method of removal and was most efficient in soils containing high concentrations of silt, clay, and organic matter. Microbial degradation resulted in further removal of the hormones. Photolytic decay by ultraviolet light was minimally effective in reducing concentrations.

Field data from SAT systems in Mesa and Tucson, Arizona, also documented the removal of 17β-estradiol, estriol, and testosterone to below detection limits (Amy and Drewes 2007). The only two pharmaceutically active compounds (PhACs) that were detected after SAT were the antiepileptic drugs carbamazepine and primidone. Another CEC that was detected after SAT (in the most down-gradient monitoring well) was the chlorinated flame retardant TCPP [tris-(chloroisopropyl)-phosphate]. However, the concentration of TCPP was still significantly reduced.

A field study of the fate of alkylphenol polyethoxylates (APEOs) metabolites was performed at the Mesa Northwest Water Reclamation Plant by Montogomery-Brown et al. (2003). APEOs are surfactants and with their degradation metabolites are among the most frequently detected aquatic contaminants in the environment (Montogomery-Brown et al. 2003). SAT was found to be effective in the removal of most APEOs metabolites. The alternating flooding and drying cycles of SAT system (and thus alternating aerobic and anaerobic conditions) appear to enhance the overall APEO removal efficiencies (Montogomery-Brown et al. 2003). Drewes et al. (2003) documented that in the long-term SAT reduced the concentrations of EDTA, APECs (alkylphenol polyethoxycarboxylates) and NTA (nitrilotriacetic acid) by 89, 99, and 100% respectively.

However, some CECs are persistent in groundwater environments and thus water treated by SAT. Column experiments conducted by Rauch et al. (2006) demonstrated soil organic matter content and composition affect organic micro-pollutant removal. Different organic matter substrates promoted the establishment of different soil microbial community compositions and concentrations, which affected the removal of different CECs. The study also demonstrated that microbial adaptation to CECs took place, as removal rates increased with increased exposure of the columns to trace pollutants. Bioavailable organic carbon seemed to promote the removal of certain trace compounds (e.g., gemfibrozil), whereas oligotrophic recharge conditions enhance the removal of other CECs, likely by stimulating a slow-growing diverse microbial community (Rauch et al. 2006).

The impact of the low levels of CECs remaining in reclaimed water after SAT must be put into perspective. Amy and Drewes (2007, p. 25) noted that

> The health effects of such low (ng/l) levels are still being debated but it is noteworthy that I70 values (lifetime intake via drinking water based on 70 years and 2 L/day) are far less than daily therapeutic doses for almost all PhACs.

31.3 Water Quality Improvements in the Groundwater Environment

The attenuation of contaminants is caused by a variety of processes that may be active in aquifers including filtration, sorption, and biodegradation processes. The passage of water through different redox environments (oxic and anoxic) maximizes the opportunity for contaminant attenuation (Dillon et al. 2006b; Stuyfzand 2007).

Often water injected into MAR or the induced to recharge shallow aquifers in RBF systems have high dissolved oxygen concentrations. Reducing conditions may later become established in the aquifer, especially if the recharged water has a high organic carbon concentration.

Natural aquifer treatment is the primary intended purpose of ASTR and RBF systems. Natural aquifer treatment has been used to improve water quality for over two centuries using RBF and there is now considerable data on its effectiveness for removing pathogenic organisms, organic carbon, and emerging contaminants (Chap. 24). Improvements in stored water quality may be a secondary benefit of MAR systems that have a primary storage function. An additional issue is that natural aquifer treatment may reduce the stigma associated with reuse of reclaimed water. Public confidence in water recycling projects is increased when the reclaimed water is put back into the natural system, such as rivers and aquifers, before being recovered for reuse (Al-Otaibi and Mukhopadhyay 2005; Dillon et al. 2006a, b).

31.3.1 Fate and Transport of Pathogenic Microorganisms in Groundwater Environments

The fate and transport of microorganisms in subsurface environments has long been of interest because of the health concerns over their entering water-supply wells. The concentration of pathogenic microorganisms is reduced in groundwater environment by physical removal (e.g., filtration, sedimentation, adhesion), inactivation (dying off), and dilution.

Size, shape, surface charge, and hydrophobicity influence the transport, retardation, and adhesion to surfaces of particles by microorganisms. Solute characteristics, including ionic strength, pH, temperature and concentrations of DOC, surfactants, and nutrients, have also been shown to influence particle and microorganism transport and adhesion (National Research Council 2008). The filtering efficiency of any aquifer is dependent upon aquifer characteristics, particularly pore throat size distribution, and the between types of organisms based on their size, shape, and surface characteristics.

Schijven et al. (2000) and Medema and Stuyfzand (2002) proposed that at a study site in The Netherlands, attachment onto iron hydroxides that formed by the oxidation of iron pyrites may be an important removal mechanism for microorganisms. The injection of oxygenated water results in the preferential formation of iron hydroxides near injection wells. At the pH of the injection zone water, the iron hydroxides have positive surface charges and are capable of sorbing negatively-charged microorganisms. At greater distances from the injection well, anoxic conditions prevail and iron hydroxides do not form, and as a result, attachment or sorption of microorganisms may be very low or negligible.

Inactivation is ultimately the primary means for the removal of pathogenic microorganisms in groundwater environments. Physical processes that prevent or retard the movement of microorganisms act to increase the retention time of microorganisms and thus time available for the inactivation. Greater retention times can also be achieved by increasing the distance from the recharge point to the recovery point.

John and Rose (2005) and Toze (2005) summarized the current state of knowledge for the inactivation of many microorganisms of concern in groundwater. Microbial survival is influenced by a variety of factors including the type of organism, temperature, dissolved oxygen concentration, redox state, water chemistry, water source, and the indigenous groundwater organisms. The data included in the John and Rose (2005) study gave median inactivation rates for the studied viruses of 0.02–0.09 $\log_{10}$/day, which correspond to $\log_{10}$ removal times of 11–50 days. The median inactivation rates for the studied bacteria were 0.05–0.2 $\log_{10}$/day. The lowest median $\log_{10}$/day inactivation rate (hepatitis A) of 0.02 $\log_{10}$/day corresponds to a one log removal time of 50 days, and thus, a 7 log/year removal rate. Enteric microorganisms enter an environment already inhabited by native groundwater microorganisms, which lead to greater inactivation rates (John and Rose 2005). *Escherichia coli* bacteria appear to be the more negatively impacted by native groundwater organisms than viruses.

Gordon et al. (2002), Gordon and Toze (2003), Toze and Hanna (2002), Sidhu et al. (2006), Wall et al. (2006, 2007) performed experimental studies on pathogenic microorganism inactivation and indicator organisms that might be present in reclaimed water used for artificial recharge. The Gordon et al. (2002) study included poliovirus and cocksackievirus. Toze and Hanna (2002) studied additional microbial pathogens, including *Salmonella typhimurium*, *Aeromonas hydrophila*, *E. coli*, and coliphage MS2. Both studies showed that the presence of indigenous groundwater microorganisms was the most important factor that influences the decay rate of the studied microorganisms. Other variables, such as nutrient concentrations and dissolved oxygen, were thought to be secondary influences on viral survival through their effect on the activity of native groundwater organisms and increased production of compounds instrumental in viral inactivation. Different enteric pathogens behave differently under similar redox conditions (Sidhu et al. 2006).

Wall et al. (2006, 2007) investigated the specific effects of indigenous groundwater organisms on enterovirus survival. Only 27% of the indigenous bacteria strains studied showed an ability to reduce mean viral copy numbers of the enterovirus *Poliovirus* type 1 and *Cocksackievirus* B3 over time compared to sterile controls. Viral decay rates were observed to be faster in the groundwater microcosm as whole than occurred with individual groundwater isolates. The results suggest that there may be a number of compounds, such as protease enzymes, produced by a small number of the active indigenous groundwater microorganisms that work in conjunction to produce the consistently faster decay times observed in the presence of the entire groundwater microcosm. Antiviral activity by groundwater microorganisms is virus specific and in some instances may not require the direct presence of groundwater microorganism cells (i.e., antiviral activity can be caused by the production of virucidal compounds that can act at some distance).

Little research has been completed on the effects of pressure and temperature on the inactivation of pathogenic organisms in MAR systems. Some wastewater storage or partially treated surface water occurs in deep aquifers that occur more than 300 m (1,000 ft) below surface (e.g., the Everglades Restoration ASR system in Florida). Many of these deep aquifers have elevated pressures and higher temperatures compared to the native environment of the pathogens and these physical factors may also impact the rate of deactivation.

31.3.2 Microbial Risk Evaluation and Management

It must be emphasized that pathogens are the primary health concern associated with wastewater reuse and MAR because a one-time exposure may be sufficient to cause serious illness. Natural contaminant attenuation processes provided by MAR have the greatest value in reducing the health risks associated with wastewater reuse through the elimination or reduction in the concentrations of pathogens. Evaluating and mitigating the risks associated with pathogens is thus critical for MAR systems that store or treat reclaimed water. Health and risk analysis methods are covered in Chap. 30.

Evaluation and management of the microbiological risks associated with MAR requires consideration of:

- pathogens present in the recharged wastewater and their concentrations,
- attenuation processes active in the storage aquifer between the times of recharge and recovery (attenuation rates are a function of residence time and pathogen decay rates),
- the degree to which natural attenuation processes are being relied upon for pathogen removal,
- the uses of the reclaimed water and water quality requirements, particularly health-based water quality criteria, and
- post-recovery treatment processes.

The greatest risk occurs where the recharge water has high concentration of pathogens (e.g., untreated wastewater) and the residence time is low (Page et al. 2010b; Toze et al. 2010). The least risk occurs where the recharge water undergoes a high degree of treatment and long residence times occur. Therefore, the simplest approach for addressing the health risks associated with pathogenic microorganisms involves relating the recharge retention time to the inactivation (removal) rate of each pathogen within the aquifer. Inactivation is the only factor that should be used to assess aquifer treatment, and other processes such as retardation should only be considered an added benefit (NRMMC-EPHC-NHMRC 2009a, b).

Representative groups of pathogens need to be used to make a pathogen risk assessment, because it is not practically and economically possible to test for all pathogens. For example, three representative pathogens, rotavirus, *Cryptosporidium*, and *Campylobacter*, were considered in a comparative study of subsurface pathogen treatment by Page et al. (2010b). Conservative pathogens (i.e., specifies with slow inactivation rates) should be chosen. For example, in Australia enteric viruses and protozoa are the most resilient and it was, therefore, proposed that they should be included in survivor studies (NRMMC-EPHC-NHMRC 2009a, b).

Pathogen numbers in recharge water, aquifer residence time and pathogenic decay rates are the most important variables in predicting the risk of wastewater MAR systems (Toze et al. 2010). Recharge water sampling and analyses can constrain the likely values of pathogens. The actual values for the aquifer residence time and pathogen decay rates are often poorly known on a site-specific basis. Pathogen decay rates are a function of numerous variables, including physiochemical parameters and indigenous microorganism community, and thus need to be determined on a site-specific basis.

NRMMC-EPHC-NHMRC (2009a, b) cautioned that comparison of the numbers of pathogens in recovered and recharged water is rarely adequate for an appropriate risk assessment. Direct testing of pathogen decay rates was strongly recommended. In situ diffusion chambers have been used to determine $\log_{10}$ reductions in a pathogen numbers for MAR research sites (e.g., Page et al. 2010b; Toze et al. 2010), but are too involved to be a routine tool for most MAR projects.

In the absence of site-specific pathogen decay data, conservatively long $\log_{10}$ removal rates obtained from the scientific literature could be used, particularly if data are available from geochemically similar systems. Thus, for each pathogen identified in the recharge water, the minimum retention time is needed to achieve a specified water quality goal. For example, if pathogen "A" has an estimated $\log_{10}$ removal rate of 6–10 days (based on published data), the World Health Organization (2006) target of 6–7 $\log_{10}$ pathogen removal could be achieved by a 70 day retention time. Toze (2006) cautioned that current understanding of the influence of environmental factors on pathogen survival is incomplete and when combined with the complexity of the environment receiving wastewater, it is still difficult to accurately predict the stability of various pathogens in different environments. Additional safety factors are, therefore, prudent to provide additional assurance.

Aquifer characterization and testing (e.g., tracer tests) and groundwater (solute-transport) modeling can be used to evaluate the travel time from the recharge point to any nearby aquifer user or the recovery wells for an ASTR system. Travel (retention) time and distance are also critical for evaluating the potential for impacts to sensitive receptors. The NRMMC-EPHC-NHMRC (2009a, b) provided some analytical equations for the estimation of retention times. In practice, numerical modeling is now widely performed to evaluate aquifer hydraulic responses for MAR systems and to estimate travel times.

A critical issue for evaluating travel times is aquifer heterogeneity. The presence of zones of greater hydraulic conductivity in an aquifer can result in flow rates orders of magnitude greater than would occur in homogenous aquifers (Maliva and Missimer 2010). Standard geophysical logs (e.g., flow meter) and advanced borehole geophysical logs (e.g., nuclear magnetic resonance) can provide data on the aquifer heterogeneity that can be used to develop more accurate solute-transport models (Maliva et al. 2009; Chap. 17). Tracer tests can provide more direct data on the travel times within an aquifer.

Although field testing can reduce uncertainty in the values of some key parameters, uncertainty concerning the actual values still remains. Residence times, decay rates, and water pathogen concentrations are often incorporated into risk assessments as probability distribution functions (PDFs) and evaluated using statistical methods such as Monte Carlo simulations.

Toze et al. (2010) documented a static QMRA (Sect. 29.9) performed for a wastewater MAR research project in western Australia, but as a research project, the scope of work was well beyond that practically possible for most projects, especially those not involving indirect potable reuse. Realistically, system-type specific goals need to be established for regulatory jurisdictions, which can be either pathogen concentrations at the point of recovery or number of $\log_{10}$ removals (or ideally both). The risks associated with a given MAR systems can be assessed using system specific aquifer residence times and pathogen decay rates (or conservative 'worse-case' rates from the literature). Determination of pathogen decay rates, likely exposure frequencies and volumes, and dose-response functions need to be determined in a government or university research environment.

In conclusion, the population of pathogenic organisms present in recharged water will tend to decline during storage in MAR systems. Wide variation in reported decay or inactivation rates between different organisms, as well as between different studies of the same organism, indicates that the degree of movement and survival of pathogens is likely highly site-specific (Toze 2005). Field data indicate that the transport of microorganisms in aquifers tends to be greatly retarded. Very high attenuation rates occur between injection wells and recharge basins, and nearby monitoring wells. The potential health risk associated with reclaimed water MAR systems is, thus, very low provided that at least a modest physical separation is present between the MAR systems wells and nearby potable-water production wells. More detailed assessment of groundwater conditions and their effects on introduced microbial pathogens may be required for each new MAR scheme being investigated (Toze 2005), if the system is being relied on to provide treatment.

31.3.3 Organic Carbon Removal

Organic matter present in injected water is in essence food for microorganisms, whose metabolic processes can result in local changes in the chemistry of injected waters. The first reaction that will occur is the oxidation of organic matter (expressed as CH_2O) using dissolved oxygen, which is basic aerobic respiration in which oxygen is consumed and organic matter is converted to carbon dioxide and water.

$$CH_2O + O_2 \rightarrow CO_2 + H_2O \tag{31.5}$$

Organic matter oxidation using dissolved oxygen is the most energetically favorable redox reaction and will proceed until the dissolved oxygen supply is exhausted. The next major redox reaction in MAR systems is denitrification, which is commonly catalyzed by the bacteria *Pseudomonas dentrificans* (Eq. 31.4). Sulfate reduction is the next important reaction in the redox series, particularly where relatively organic-rich water is stored in a brackish-water or saline-water aquifer. The mixing zone between the injected and stored water may have a combination of high organic concentration and high sulfate concentration, which is highly conducive for sulfate reduction. Sulfate reduction is commonly catalyzed by the bacteria *Desulfuvibrio desulfuricans,* and results in the release of hydrogen sulfide (H_2S) and bicarbonate ions (HCO_3^-).

$$SO_4^{2-} + CH_2O \rightarrow H_2S + 2HCO_3^- \quad (31.6)$$

In brackish and high-salinity groundwater, the concentration of sulfate typically is high relative to the organic matter concentration so that sulfate reduction is the terminal electron acceptor. The supply of available organic matter is exhausted long before the sulfate supply is exhausted. Methanogenesis is the next step in the redox sequence, and can be considered to be the bacterial reduction of carbon dioxide to form methane (CH_4). Using acetic acid (CH_3COOH) as the organic compound, the overall reaction is as follows (Stumm and Morgan 1996):

$$CH_3COOH \rightarrow CH_4 + CO_2 \quad (31.7)$$

Oxidizing and reducing conditions may be either temporally or spatially separated within MAR systems. Organic matter plays a critical role for microbially-driven redox reactions in MAR systems. Sharp spatial variations in redox state appear to be especially important in ASR systems that are used to store reclaimed water or surface water with a relative high organic content. Field and modeling results indicate that redox reactions tend to be concentrated near the immediate vicinity of the ASR injection and recovery wells (Greskowiak et al. 2005a, b, 2006). Aerobic, sulfate-reducing, and methanogenic conditions, with their associated mineralogical reactions, may thus co-exist over short distances. For example, Vanderzalm et al. (2002) documented that a reactive zone forms adjacent to the ASR well at the Bolivar system in South Australia. The reactive zone is characterized by high microbial activity relative to the relatively large flux of nutrients through the aquifer immediately surrounding the ASR wells. Sulfate-reducing conditions, for example, were favorable only in the immediate vicinity of the ASR wells and not detected in an observation well located 4 m (13 ft) away.

The concentration of the organic compounds in injected water tends to decrease over time and with distance from an ASR well. The composition of the organic carbon also changes with distance from ASR wells. The fate and residence time of dissolved organic carbon (DOC) in an aquifer is related to its decomposability by the microbial biomass, its absorption by the solid matrix, and to its flocculation through interaction with cations (Skjemstad et al. 2002, 2005). Skjemstad et al. (2002, 2005) studied the changes in the concentration and composition of DOC in the Bolivar ASR system in South Australia. The composition of the DOC in the injectate and well waters at the Bolivar site are typical of natural organic matter (NOM) present in other surface and groundwater. Samples of the injectate and water samples collected at monitoring wells located 4, 50, and 75 m (13, 164, and 246 ft) from the injection well were analyzed. Filtration of particulate organic matter occurs near the ASR well. Microbial assimilation of the particulate organic matter releases lower molecular weight molecules as DOC.

Skjemstad et al. (2005) documented at the Bolivar ASR site that there was a net loss of DOC during passage through the aquifer, with the high molecular weight and more acidic materials appearing to be preferentially lost. The hydrophilic fraction of the DOC appears to have been preferentially sorbed by the mineral matrix, whereas the hydrophobic fraction is transported more readily through the aquifer. Sorption appears to be a major mechanism for the decreasing DOC concentration. The observed, continued increase in DOC content after breakthrough in a monitoring well suggests a limited sorption capacity of the aquifer material had been gradually overcome. The potential for disinfection byproduct formation is expected to decline with the decrease in DOC. However, the potential for disinfection byproduct formation per unit weight of DOC may be slightly higher towards the later stages of recovery because of the slower rate of decrease in the decline in aromatic organic carbon, which has a relative high potential for disinfection byproduct formation (Skjemstad et al. 2005).

31.3.4 Disinfection Byproducts

The fate of disinfection byproducts (DBPs) in water injected in ASR systems and other MAR systems was summarized by Maliva and Missimer (2010). Disinfection byproducts form by reaction of disinfectants with organic matter present in water. Chlorine is the main chemical disinfectant currently in use for water and wastewater treatment. The main DBPs formed during chlorine-based disinfection are trihalomethanes (THMs) and haloacetic acids (HAAs) (Table 31.1). THMs may also form after the recharge of water through the reaction of residual chlorine with DOC present in either in the recharged or aquifer water.

Long-term chronic exposure to DBPs is believed to result in an increased risk of some cancers. DBPs are of greatest concern for potable water systems, particularly where the recovered water will be disinfected again. DBPs, which are usually detected in parts per billion, are not a health issue associated with non-potable water uses. However, DBPs are still a regulatory concern for reclaimed water ASR systems in the United States because primary drinking water standards (maximum contaminant levels; MCLs) have been established for THMs and HAAs, which are incorporated into groundwater standards.

The concentration of THMs and HAAs in ASR systems may decrease over time by three main processes: (1) dilution with native groundwater in the storage zone, (2) sorption onto aquifer solids, particularly organic matter, and (3) biotransformation

Table 31.1 Main trihalomethanes and haloacetic acids

Compound	Formula
Trihalomethanes (THMs)	
Trichloromethane (chloroform)	$CHCl_3$
Bromodichloromethane	$CHBrCl_2$
Dibromochloromethane	$CHBr_2Cl$
Tribromomethane (bromoform)	$CHBr_3$
Haloacetic Acids (HAAs)	
Monochloroacetic acid	$CH_2ClCOOH$
Dichloroacetic acid	$CHCl_2COOH$
Trichloroacetic acid	CCl_3COOH
Monobromoacetic acid	$CH_2BrCOOH$
Dibromoacetic acid	$CHBr_2COOH$

(also referred to as biodegradation). Abiological transformations (e.g., hydrolysis) of THMs occur too slowly to be a significant process in ASR systems (Nicholson et al. 2002; Pavelic et al. 2005a, b; Nicholson and Ying 2005). Attenuation by adsorption processes will depend upon the organic content of the aquifer and the hydrophobicity of the DBP. In aquifers with low organic contents, adsorption of THMs and HAAs will be minimal (Nicholson et al. 2002; Nicholson and Ying 2005).

Experimental and field studies have shown that the biotransformation rate of THMs and other low molecular weight haloaliphatic compounds is dependent upon the redox environment (Bouwer and Wright 1988). In general, as the environment (electron acceptor condition) becomes more oxidizing, more compounds tend to persist. THMs are resistant to biodegradation under oxic conditions (Singer et al. 1993; Landmeyer et al. 2000; Nicholson et al. 2002; Fram et al. 2003), whereas HAAs undergo biodegradation under both oxic and anoxic conditions (Singer et al. 1993; Landmeyer et al. 2000; Nicholson et al. 2002).

The stability of the carbon–halogen bond influences reaction rates and stability. The ease of reductive dehalogenation in the same redox environment generally follows the order of I > Br > Cl > F. Brominated compounds should therefore undergo reductive dehalogenation more readily than chlorinated compounds (Bouwer and Wright 1988). Experimental studies indicate that highly brominated THMs ($CHBr_3$ and $CHBr_2Cl$) undergo the fastest degradation rates and with chloroform degradation being the slowest. The significant biodegradation of chloroform is initiated either under sulfate-reducing (Gupta et al. 1996a, b) or methanogenic (Bouwer and Wright 1988) conditions. As the geochemical environment becomes progressively more reducing, from denitrifying to sulfate-reducing to methanogenic, the rate of biodegradation of THMs increases (Bouwer and Wright 1988). The preponderance of the experimental and field data indicate that concentrations of HAAs and the brominated THMs are attenuated by biological processes in chemically-reducing groundwater environments, with the most rapid attenuation occurring in methanogenic environments. Chloroform ($CHCl_3$) is the most refractory THM and appears to require highly reducing conditions for rapid biological degradation to occur. Dilution appears to be the main mechanism for decreases in chloroform concentrations under aerobic conditions.

DBPs are not of significant concern to most ASR and MAR systems with storage or recharge zones that contain anaerobic (chemically reducing) groundwater. The concentrations of both HAAs and THMs decrease greatly during a normal several-month or longer storage period. Management of THMs will be a significant operational concern in ASR and MAR systems that inject into zones containing aerobic groundwater. If free chlorine is present in the injected water, then total THM concentrations may increase early in the storage phase as the chlorine reacts with dissolved organic carbon present in the injected water or natural groundwater. The combination of THM formation and low attenuation rates can result in high THM concentrations in the recovered water.

31.3.5 CECs in Groundwater Environment

An issue of growing technical and regulatory concern for MAR systems, particularly systems storing reclaimed and surface water, is the fate and transport of emerging contaminants in aquifers. There is currently a paucity of information on the fate and transport of pharmaceutical residues and related compounds in the aquatic environment and during groundwater recharge to enable an understanding of the efficacy of natural attenuation processes (Heberer 2006). Because of the highly variable chemistries, there is greater variability in the attenuation rates of emerging contaminants in soil and groundwater environments.

Field and experimental studies have demonstrated that the concentrations of some compounds rapidly decrease in soil and groundwater environments through biodegradation and adsorption, whereas other compounds are much more resistant (e.g., Khan and Rorije 2002; Drewes et al. 2002, 2003; Heberer et al. 2004; Snyder et al. 2004; Ying et al. 2004; Zuehlke et al. 2004; Stuyfzand 2007; Ternes et al. 2007; Drewes 2009; Hoppe-Jones et al. 2010). In general, hydrophilic compounds tend to be more persistent, as hydrophobic compounds tend to be adsorbed onto solids. Aquifer redox conditions also strongly influence the rate of attenuation of some emerging contaminants. Persistent compounds often have structural features that prevent enzymatic attack (National Research Council 2008). Absorption of organic compounds is strongly related to the organic content of the media (Khan and Rorije 2002).

Most of the published studies on the behavior of emerging contaminants in groundwater environments are focused on bank filtration and land application systems. Drewes et al. (2002, 2003) documented that caffeine, analgesic/anti-inflammatory drugs (e.g., diclofenac, ibuprofen, ketoprofen, naproxen, and fenoprofen) and blood lipid regulators (e.g., gemfibrozil) were quickly reduced to concentrations near or below detection limits during groundwater recharge at facilities in the Western United States. Antiepileptic drugs (e.g., carbamazepine, primidone) were not removed during recharge under either oxic or anoxic conditions. Other refractory CECs include clofibric acid (a blood-lipid regulator), some antibiotics (e.g., sulfamethoxazole), X-ray contrast media, and chlorinated flame retardants (TCEP, TCPP) (Drewes 2009). Snyder et al. (2004) similarly documented that the variable removal of CECs during artificial recharge by land application in an arid environment (Las Vegas, Nevada area). The CECs were not detected in down-gradient monitoring wells, which suggests continued biodegradation in the saturated zone.

Heberer et al. (2004) studied the fate and transport of CECs at two bank filtration sites in Berlin, Germany. CEC compounds differed in whether their concentrations were reduced only by dilution, or were partially or completely removed. Antibiotics, estrogens, bezafibrate, and indomethacine were completely removed. The following compounds were persistant or underwent only partial removal: carbamazepine, primidone, clofibric acid (metabolite of blood lipid regulating drugs), AMDOPH (drug metabolite), propyphenazone (drug metabolite), bentazone (pesticide residue), mecoprop (pesticide residue), (chloroethyl)-phosphate (TCEP; flame retardant), and tris-(chloroisopropyl)-phosphate (TCPP; flame retardant). AMDOPH and propphenazone residues were believed to have originated mainly from a spill at a former upstream production facility, rather than from treated wastewater. With the exception of AMDOPH, the concentrations of the CECs detected in production wells were less than 500 ng/l.

Zuehlke et al. (2004) studied the fate and transport of estrogenic steroids during bank filtration in Berlin, Germany. The studied estrogens were: 17β-estradiol (E2), a natural estrogenic steroid, 17α-ethylestradiol (EE2), a synthetic steroid, and Estrone (E1). The removal rates during the wastewater treatment process ranged from 76% (EE2) to 94% (E2). Only E1 was detected in the surface-water source body for the studied bank filtration system. E1 was only detected in a shallow monitoring well located very close to the bank. Thus, even a short travel distance was sufficient to remove E1.

Ying et al. (2003) conducted laboratory experiments of sorption and degradation of five endocrine disrupting compounds (EDCs) that have been detected in wastewater. The studied EDCs were the estrogenic steroids E2 and EE2, bisphenol A (BPA), an industrial chemical, 4-tert-octylohenol (4-t-OP) (a surfactant degradation product), and 4-n-nonylphenol (4-n-NP) (a surfactant degradation product). The results of degradation experiments under aerobic conditions, gave calculated half-lives of 2-days for E2, 7 days for 4-n-NP, and 81 days for EE2 (Ying et al. 2003). The concentrations of BPA and 4-t-OP were unchanged. Under anaerobic conditions, the concentrations of the five EDCs remained almost unchanged over 70 days. Only E2 underwent very slow degradation with an estimated half-life of 107 days.

Ying et al. (2003) also observed that sorption behavior could have different impacts on the concentration of EDCs in ASR systems depending upon whether the system uses a single well or separate wells for recharge and recovery. The concentration of EDCs by sorption near a dual-use well could result in a spike in concentration at the start of recovery. In aquifers with high concentrations of organic matter (which is favorable for the sorption of organic compounds), sorption could progressively reduce the concentration of the EDCs between the separate injection and recovery wells. Ying et al. (2004) reported that E2 and EE2 had a moderate affinity for sorption on the aquifer sediment from the Bolivar ASR site.

In summary, field and experimental studies have demonstrated that the concentrations of some compounds rapidly decrease in groundwater environments through biodegradation and adsorption, whereas other compounds are much more resistant (e.g., Khan and Rorije 2002; Drewes et al. 2003; Heberer et al. 2004, Snyder et al. 2004; Ying et al. 2004; Zuehlke et al. 2004; Stuyfzand 2007; Hoppe-Jones et al. 2010). In general hydrophilic compounds tend to be more persistent, as hydrophobic compounds tend to be adsorbed onto solids. Dilution may be an important process for the reduction in the concentrations of CECs to below detection limits where the wastewater flux does not dominate the local water budget. MAR systems may provide a cost effective means for the removal of the less resilient CEC compounds from the stored water (Khan and Rorije 2002; Stuyfzand 2007).

References

Al-Otaibi, M., & Mukhopadhyay, A. (2005). Options for managing water resources in Kuwait. *Arabian Journal for Science and Engineering, 30*, 55–68.

Amy, G., Bryant, C., & Belyani, M. (1987). Molecular weight distribution of soluble organic matter in various secondary and tertiary treated effluents. *Water Science and Technology, 19*(3–4), 529–538.

Amy, G., & Drewes, J. (2007). Soil aquifer treatment (SAT) as a natural and sustainable wastewater reclamation/reuse technology: Fate of wastewater effluent organic matter (EfOM) and trace organic compounds. *Environmental Monitoring and Assessment, 129*, 19–26.

Bekele, E., Toze, S., Patterson, B., Devine, B., Higginson, S., Fegg, W et al. (2009). *Design and operation of infiltration galleries and water quality changes*. Collingwood: CSIRO: Water for a Healthy Country National Research Flagship.

Bouwer, E. J., & Wright, J. P. (1988). Transformations of trace aliphatics in anoxic biofilm columns. *Journal of Contaminant Hydrology, 2*, 155–169.

Bouwer, H. (1974). Design and operation of land treatment systems for minimum contamination of ground water. *Ground Water, 12*(3), 140–147.

Bouwer, H., Rice, R. C., Lance, L. C., & Gilbert, R. G. (1980). Rapid-infiltration research at Flushing Meadows project. *Journal Water Pollution Control Federation, 56*, 2457–2470.

Cordy, G. E., Duran, N. L., Bouwer, H., Rice, R. C., Furlong, E. T., Zaugg, S. D., et al. (2004). Do pharmaceuticals, pathogens, and other organic wastewater compounds persist when wastewater is used for recharge? *Ground Water Remediation & Monitoring, 24*(2), 58–69.

Dillon, P., Pavelic, P., Toze, S., Ragusa, S., Wright, M., Peter, P., et al. (1999). Storing recycled water in an aquifer, benefits and risks. *Australian Water and Wastewater Association Journal, 26*, 21–29.

Dillon, P., Pavelic, P., Toze, S., Rinck-Pfeiffer, S., Martin, R., Knapton, A., et al. (2006a). Role of aquifer storage in water reuse. *Desalination, 188*, 123–134.

Dillon, P., & Toze, S. (2005). *Water quality improvements during aquifer storage and recovery. Volume 1: Water quality improvement processes*. AWWA Research Foundation Report 91056F.

Dillon, P., Toze, S., Pavelic, P., Vanderzalm, J., Barry, K., Ying, G.-L., et al. (2006b). Water quality improvements during aquifer storage and recovery at ten sites, in recharge systems for protecting and enhancing groundwater resources. *In Proceedings of the 5th International Symposium on Management of Aquifer Recharge* (pp. 85–94). Berlin, Germany, 11–16 June 2005. Paris: UNESCO.

Drewes, J. E. (2009). Ground water replenishment with recycled water- water quality improvements during managed aquifer recharge. *Ground Water, 47*(4), 502–505.

Drewes, J. E., & Fox, P. (1999). Fate of natural organic matter (NOM) during groundwater recharge using reclaimed water. *Water Science and Technology, 40*(9), 241–248.

Drewes, J. E., Heberer, T., & Reddersen, K. (2002). Fate of pharmaceuticals during indirect potable reuse. *Water Science and Technology, 46*, 73–80.

Drewes, J. E., Heberer, T., Rauch, T., & Reddersen, K. (2003). Fate of pharmaceuticals during groundwater recharge. *Ground Water Monitoring and Remediation, 23*, 64–72.

Fox, P., Houston, S., & Westerhoff, P. (2001a). *Soil aquifer treatment for sustainable water supply* (p. 203). Denver: American Water Works Association Research Foundation.

Fox, P., Houston, S., Westerhoff, P., Drewes, J. E., Nellor, M., Yanko, W., et al. (2001b). Investigations of soil-aquifer treatment for sustainable water reuse. Tempe: National Center for Sustainable Water Supply.

Fox, P., Aboshamp, W., & Alsamadi, B. (2005). Analysis of soils to demonstrate sustained organic carbon removal during soil aquifer treatment. *Journal of Environmental Quality, 84*, 156–163.

Fram, M. S., Bergamaschi, B., Goodwin, K. D., Fujii, R., & Clark, J. F. (2003). Process affecting the trihalomethane concentrations associated with the third injection, storage, and recovery test at Lancaster, Antelope Valley, California, March 1998 through April 1999. *U.S. Geological Survey Water-Resources Investigations Report 03-4062.*

Gordon, C., & Toze, S. (2003). Influence of groundwater characteristics on the survival of enteric viruses. *Journal of Applied Microbiology, 95*, 536–544.

Gordon, C., Wall, K., Toze, S., & O'Hara, G. (2002). Influence of conditions on the survival of enteric viruses and indicator organisms in groundwater. In P. J. Dillon (Ed.), *Management of aquifer recharge for sustainability* (pp. 133–138). Lisse: A.A. Balkema.

Greskowiak, J., Prommer, H., Vanderzalm, J., Pavelic, P., & Dillon, P. (2006). Quantifying biogeochemical changes during ASR of reclaimed water at Bolivar, South Australia. In *Recharge systems for protecting and enhancing groundwater resources, Proceedings of the 5th international symposium on management of aquifer recharge, Berlin, Germany, 11–16 June 2005* (pp. 360–365). Paris: UNESCO.

Greskowiak, J., Prommer, H., Vanderzalm, J., Le Gal La Salle, C., Pavelic, P., & Dillon, P. (2005a). PHT3D modeling of water quality changes during ASR at Bolivar. In P. Dillon (Ed.), *Water quality improvements during aquifer storage and recovery. Volume 1: Water quality improvement processes* (pp. 245–277). AWWA Research Foundation Report 91056F.

Greskowiak, J., Prommer, H., Vanderzalm, J., Pavelic, P., & Dillon, P. (2005b). Modeling of carbon cycles and biogeochemical changes during injection and recovery of reclaimed water at Bolivar, South Australia. *Water Resources Research, 41*, W10418.

Gupta, M., Gupta, A., Suidan, M. T., & Sayles, G. D. (1996a). Biotransformation rates of chloroform under aerobic conditions-II, sulfate reduction. *Water Research, 30*, 1387–1394.

Gupta, M., Sharma, D., Suidan, M. T., & Sayles, G. D. (1996b). Biotransformation rates of chloroform under aerobic conditions-I. methanogenesis. *Water Research, 30*, 1377–1385

Harpaz, Y. (1971). Artificial ground-water recharge by means of wells in Israel. *American Society of Civil Engineers, Journal of the Hydraulics Division, 97*, 1947–1964.

Heberer, T. (2006). Occurrence, transport, attenuation and removal of pharmaceutical residues in the aquatic environment and their relevance to drinking water supply in urban areas, in recharge systems for protecting and enhancing groundwater resources, *In Proceedings of the 5thInternational Symposium on Management of Aquifer Recharge* (pp. 529–534). Berlin, Germany, 11–16 June 2005. Paris: UNESCO.

Heberer, T., Mechlinski, A., Franck, B., Knappe, A., Massmann, G., Pekdeger, A., et al. (2004). Field studies on the fate and transport of pharmaceutical residues in bank filtration. *Ground Water Monitoring and Remediation, 24*, 70–77.

Hoppe-Jones, C., Oldham, G., & Drewes, J. E. (2010). Attenuation of total organic carbon and unregulated trace organic chemicals in US riverbank filtration systems. *Water Research, 44*(15), 4643–4659.

John, D. E., & Rose, J. B. (2005). Review of factors affecting microbial survival in groundwater. *Environmental Science & Technology, 39*, 7345–7356.

Khan, S. J., & Rorije, (2002). Pharmaceutically active compounds in aquifer storage and recovery. In P. J. Dillon (Ed.), *Management of aquifer recharge for sustainability* (pp. 169–174). Lisse: A.A. Balkema.

Landmeyer, J. E., Bradley, P. N., & Thomas, J. M. (2000). Biodegradation of disinfection byproducts as a potential removal process during aquifer storage recovery. *Journal of the American Water Resources Association, 26*, 861–867.

Maliva, R. G., Clayton, E. A., & Missimer, T. M. (2009). Application of advanced borehole geophysical logging to managed aquifer recharge investigations. *Hydrogeology Journal, 17*, 1547–1556.

Maliva, R. G., and Missimer, T. M. (2010). Aquifer storage and recovery and managed aquifer recharge using wells: Planning, hydrogeology, design, and operation. *Methods in Water Resources Evaluation Series No. 2*. Houston: Schlumberger Water Services.

Mansell, J., & Drewes, J. E. (2004). Fate of steroidal hormones during soil-aquifer treatment. *Ground Water Monitoring & Remediation, 24*(2), 94–101.

Mansell, J., Drewes, J. E., & Rauch, T. (2004). Removal mechanisms for endocrine disrupting compounds (steriods) during soil aquifer treatment. *Water Science and Technology, 50*(2), 229–237.

McQuarrie, J. P., & Carlson, K. (2003). Secondary benefits of aquifer storage and recovery: Disinfection by-product control. *Journal of Environmental Engineering, 129*, 412–418.

Medema, G. J. & Stuyfzand, P. J. (2002). Removal of micro-organisms upon basin recharge, deep well injection and river bank filtration in the Netherlands. In. P. J. Dillon (Ed.), *Management of aquifer recharge for sustainability* (pp. 125–131). Lisse: A.A. Balkema.

Montogomery-Brown, J., Drewes, J. E., Fox, P., & Reinhard, M. (2003). Behaviour of alkyphenol polyethoxylate metabolies during soil aquifer treatment. *Water Research, 37*, 3672–3681.

National Research Council (2008). *Prospects for managed underground storage of recoverable water*. Washington, DC: National Academy Press.

Nicholson, B. C., Dillion, P. J., & Pavelis, P. (2002). Fate of disinfection by-products during aquifer storage and recover. In P. J. Dillon (Ed.), *Management of aquifer recharge for sustainability* (pp. 155–160). Lisse: A.A. Balkema.

Nicholson, B., & Ying, G.-G. (2005). Attenuation of disinfection by-products during aquifer storage and recovery. In P. Dillon & S. Toze (Eds.,). *Water quality improvements during aquifer storage and recovery Volume 1: Water quality improvement processes* (pp. 39–64). AWWA Research Foundation Report 91056F.

NRMMC-EPHC-NHMRC (2009a). *Australian guidelines for water recycling: Managing health and environmental risks (Phase 2) managed aquifer recharge*, July 2009. Australia: Natural Resource Management Ministerial Council, Environment Protection and Heritage Council, National Health and Medical Research Council.

NRMMC-EPHC-NHMRC (2009b). *Australian guidelines for water recycling: Managing health and environmental risks. Phase 2: Stormwater harvesting and reuse*, July 2009. Australia: Natural Resource Management Ministerial Council, Environment Protection and Heritage Council and National Health and Medical Research Council.

Page, D., Dillon, P., Toze, S., Bixio, D., Genthe, B., Jiménez Cisneros, B. E., et al. (2010). Valuing the subsurface pathogen treatment barriers in water recycling via aquifers for drinking supplies. *Water Research, 44*, 1841–1852.

Pavelic, P., Nicholson, B. C., Dillon, P. J., & Barry, K. E. (2005a). Fate of disinfection by-products in groundwater aquifer storage and recovery with reclaimed water. *Journal of Contaminant Hydrology, 77*, 351–373.

Pavelic, P., Dillon, P., & Nicholson, B. (2005b). Analysis of the fate of DBPs at eight ASR sites. In P. Dillon & S. Toze (Eds.), *Water quality improvements during aquifer storage and recovery. Volume 1: water quality improvement processes* (pp. 39–64). AWWA Research Foundation Report 91056F.

Pavelic, P., Mucha, M., Dillon, P., & Barry, K. (2006). Laboratory column study on the effect of ponding depth of infiltration rates during SAT. In *Recharge systems for protecting and enhancing groundwater resources. Proceedings of the 5th international symposium on management of aquifer recharge, Berlin, Germany, 11–16 June 2005* (pp. 624–629), Paris: UNESCO.

Pyne, R. D. G. (2002) Aquifer storage recovery wells: The path ahead. *Florida Water Resources Journal*, 19–27.

Quanrud, D. M., Arnold, R. G., Wilson, L. H., & Conklin, M. H. (1996). Effects of soil type on water quality improvement during soil aquifer treatment. *Water Science and Technology, 33*(10–11), 419–431.

Quanrud, D. M., Hafer, J., Karpiscak, M. M., Zhang, J., Lansey, K. E., & Arnold, R. G. (2003). Fate of organics during soil-aquifer treatment: Sustainability of removals in the field. *Water Research, 37*, 3401–3411.

Rauch, T., Munoz, J., Drewes, J. E., Amy, G. & Choi, H. (2006). Fate of pharmaceutical during soil infiltration leading to artificial recharge. In *Recharge systems for protecting and enhancing groundwater resources, Proceedings of the 5th international symposium on management of aquifer recharge, Berlin, Germany, 11–16 June 2005* (pp. 562–567). Paris: UNESCO.

Schijven, J. F., Medema, G., Vogelaar, A. J., & Hassanizadeh, S. M. (2000). Removal of microorganisms by deep well injection. *Journal of Contaminant Hydrology, 44*, 301–327.

Sidhu, J., Hanna, J., & Toze, S. (2006). Influence of groundwater redox conditions on decay of enteric viruses and Cryptosporidium. In *Recharge systems for protecting and enhancing groundwater resources. Proceedings of the 5th international symposium on management of aquifer recharge, Berlin, Germany, 11–16 June 2005* (pp. 511–517). Paris: UNESCO.

Singer, P. C., Pyne, R. D. G., Mallikarjum, A. V. S., Miller, C. T., & Majonnier, C. (1993). Examining the impact of aquifer storage and recovery on DBPs. *Journal American Water Works Association, 85*(11), 85–94.

Skjemstad, J. O., Hayes, M. H. B., & Swift, R. S. (2002). Changes in natural organic matter during aquifer storage. In P. J. Dillon (Ed.), *Management of aquifer recharge for sustainability* (pp. 149–154). Lisse: A.A. Balkema.

Skjemstad, J., Swift, R., & Hayes, M. (2005). Evaluation of changes in NOM during ASR at the Bolivar site. In P. Dillon, & S. Toze (Eds.). *Water quality improvements during aquifer storage and recovery. Volume 1: Water quality improvement processes* (pp. 99–110). AWWA Research Foundation Report 91056F.

Snyder, S. A., Leising, J., Westerkoff, P., Youn, Y., Mash, H., & Vanderford, B. (2004). Biological and physical attenuation of endocrine disruptors and pharmaceuticals: Implications for water reuse. *Ground Water Monitoring and Remediation, 24*, 108–118.

Stumm, W., & Morgan, J. J. (1996). *Aquatic chemistry* (3rd ed.). New York: Wiley-Interscience.

Stuyfzand, P. J. (2007). Behavior of pharmaceuticals and other emerging pollutants in various artificial recharge systems in The Netherlands. In P. Fox (Ed.), *Management of aquifer recharge for sustainability: Proceedings of the 6th International Symposium on Managed Aquifer Recharge of Groundwater* (pp. 231–245). Phoenix: Acacia Publishing.

Ternes, T. A., Bonerz, M., Herrmann, N., Teiser, B., & Andersen, H. R. (2007). Irrigation of treated wastewater in Braunschweig, Germany: An option to remove pharmaceuticals and musk fragrances. *Chemosphere, 66*(5), 894–904.

Toze, S. (2005). Pathogen survival in groundwater: A review of the literature. In P. Dillon, & S. Toze (Eds.), *Water quality improvements during aquifer storage and recovery. Volume 1: Water quality improvement processes* (pp. 123–140). AWWA Research Foundation Report 91056F.

Toze, S. (2006). Water reuse and health risks—real vs perceived. *Desalination,187*, 41–51.

Toze, S., & Hanna, J. (2002). The survival potential of enteric microbial pathogens in a reclaimed water ASR project. In. P. J. Dillon (Ed.), *Management of aquifer recharge for sustainability* (pp. 139–146). Lisse: A. Balkema.

Toze, S., Bekele, E., Page, D., Sidhu, J., & Schackleton, M. (2010). Use of static quantitative microbial risk assessment to determine pathogen risks in an unconfined aquifer used for managed aquifer recharge. *Water Research,44*, 1038–1049.

Vanderzalm, J. L., Le Gal La Salle, C., Hutson, J. L. & Dillon, P. J. (2002). Water quality changes during aquifer storage and recovery at Bolivar, South Australia. In P. J. Dillon (Ed.), *Management of aquifer recharge for sustainability* (pp. 82–88). Lisse: A.A. Balkema.

Vanderzalm, J., & Le Gal La Salle, C. (2005). Impacts of redox cycling on inorganic chemistry during ASR. In P. Dillon & S. Toze (Eds.), *Water quality improvements during aquifer storage and recovery. Volume 1: Water quality improvement processes* (pp. 25–38). AWWA Research Foundation Report 91056F.

Wall, K., Toze, S., & O'Hara, G. (2006). Interactions of indigenous bacteria with enteric viruses during water quality improvement by aquifer storage and recovery (ASR). In *Recharge systems for protecting and enhancing groundwater resources. Proceedings of the 5th international symposium on management of aquifer recharge, Berlin, Germany, 11–16 June 2005* (pp. 518–524). Paris: UNESCO.

Wall, K., Toze, S., & O'Hara, G. (2007). Indigenous groundwater microorganism processes that influence the decay of enteric viruses. In P. Fox (Ed.), *Management of aquifer recharge for sustainability: Proceedings of the 6th international symposium on managed aquifer recharge of groundwater* (pp. 246–256). Phoenix: Acacia Publishing.

Wilson, L. G., Amy, G. L., Gerba, C. P., Gordon, H., Johnson, B., & Miller, J. (1995). Water quality changes during soil aquifer treatment of tertiary effluent. *Water Environment Research,67*, 371–376.

World Health Organization (2006). *WHO guidelines for the safe use of wastewater, excreta, and greywater. Volume 1, Policy and regulatory aspects*. Geneva: WHO.

Ying, G.-G., Kookana, R. S., & Dillon, P. (2003). Sorption and degradation of selected endocrine disrupting chemicals in aquifer material. *Water Research,37*, 3785–3791.

Ying, G.-G., Kookana, R. S., & Dillon, P. (2004). Attenuation of two estrogen compounds in aquifer materials supplemented with sewage effluent. *Ground Water Monitoring and Remediation,24*, 102–107.

Zuehlke, S., Heberer, T., Duennbier, U., & Fritz, B. (2004). Analysis of endocrine disrupting steroids: Investigation of their release into the environment and their behavior during bank filtration. *Ground Water Monitoring and Remediation,24*(2), 78–85.

Part VIII
Water Policy and Management

Chapter 32
Water Policy and Governance

32.1 Introduction

The bulk of this book is dedicated to the technical issues and methods used to assess, manage, and develop water resources in arid and semiarid lands, with a strong focus on groundwater resources. Sound science (i.e., understanding of hydrogeologic systems in questions and their response to water use) should provide a framework upon which water management decisions are made. However, technical issues are only one part of the story. Where water is scare, decisions must be made as to how available water resources are allocated between competing demands and the extent to which societal resources are expended to create new sources of water, such as through desalination. The human element of the water supply and demand equation may be far more complicated than the technical element.

The word "management" is defined as the "process of dealing with or controlling things or people" (McKean 2005). Water management can be broadly defined as the process of controlling the supply, allocation, use, and quality of water. Water management includes both water governance and policy. Water governance is basically the manner in which water is governed, which includes the legal, political, social, and administrative framework and institutions that are in place for water management. The term "policy" is defined as "a course or principle of action adopted or proposed by a government, party, business, or individual" (McKean 2005).

Water management, governance, and policy are interrelated. Water governance is ultimately a policy decision. Within an existing water governance framework, policy decisions are made concerning how water is managed. Water management inherently has an action element, which involves the implementation of policies. For example, a policy decision may be made concerning how water should be allocated between different sectors, but unless the decision is brought into actual practice, it cannot be assumed that water use is actually being managed. Where water use is not controlled in some manner there is no water management.

R. Maliva and T. Missimer, *Arid Lands Water Evaluation and Management*, Environmental Science and Engineering, DOI: 10.1007/978-3-642-29104-3_32,

The definition of management is neutral in the sense that the management of water in a country or region will fall at some point in the spectrum between highly effective and ineffective. Effectiveness is considered the degree to which water management meets socio-cultural goals of sustainability, equity, and economic efficiency. Equity refers to the fairness of the distribution of water between people in society, including between generations.

An obvious problem immediately arises as to the specific water management goals of a society composed of diverse individuals whose values diverge. For example, people differ in their opinions concerning the importance of economic growth versus environmental protection, and thus have different personal criteria for what constitutes effective water management. Water management is as much about managing people (water and land uses) as about managing water (surface and groundwater resources) (Kresic 2009). Effective water management necessarily involves meaningful stakeholder input and ideally consensus building, to the extent that is possible, as an objective.

Water governance and policy involves addressing the fundamental questions concerning the management of water such as:

- How and by whom are water management decisions made?
- What are the priorities in the allocation of water resources?
- How is water allocated between different users, economic sections, and the environment?
- How is water infrastructure financed and how are users charged?
- How are water conservation measures implemented and financed?

The answer to the questions of how and by whom water management decisions are made in a given jurisdiction may appear to be obvious, but often are not. It has been the authors experience in the United States, that water policy decisions are frequently made by regulatory agency staff who have considerable decision-making latitude within an indefinite water governance and legal framework. For example, there is great latitude in how a legal mandate to protect wetlands in general, is translated into water policy and, in turn, water management on the scale of evaluating whether or not to issue a permit or authorization for a given water use project. The issue of who makes water policy decisions may have serious consequences if the policies reflect the personal biases of the decision makers, which are not in tune with societal values. Returning to the wetlands example, a decision-maker with a pro-business bias might shift policy towards lesser wetland protection and greater water use, whereas another decision maker who is a strong environmental protection advocate may shift policy towards excessive wetland protection at the expense of economic development.

Development of policies without mechanisms for their implementation is a pointless exercise, unless the policy development will realistically serve as a step in the development of an active and functional water management institutional framework. It is also important to recognize that countries vary greatly in the economic and human resources available for water management. The will to effectively manage water resources may exist, but not the means to do so.

Strategic decisions concerning allocation of water amongst disparate interests is greatly complicated by the fact the water so vital to life in general and to many economic activities. A fundamental goal of water policy should be that long-term water use will be sustainable, although the meaning and implications of sustainability are debatable (Chap. 9), particularly with respect to non-renewable water resources (Chap. 36).

Toope et al. (2003) observed that it is unfortunate that the word "management" is used rather than the word "allocation," as the later clarifies the nature of the process in which water uses and policies are engaged. Allocation is politically charged and is implicit in all management regimes (Toope et al. 2003), particularly in arid and semiarid lands where there is little, if any, water that is not already explicitly or implicitly allocated. In simple terms, a fundamental aspect of water management is deciding who gets what water and how much, which can be highly divisive when it involves reallocation from historical use patterns.

Development and implementation of water policies in a region does not occur in vacuum. Water use laws, policies, and traditions may be long entrenched in a region and difficult to change, even though in some cases they were developed during periods of relatively low water demand. Historical water management regimes may no longer be appropriate for conditions of water scarcity. For example, state water law doctrines (prior appropriation) in the arid and semiarid western United States were developed during the pioneer period when the population and water use were very low, and the primary objective was encouraging economic development.

Political and social constraints determine what is feasible in managing and allocating water (Toope et al. 2003). Management of groundwater resources is particularly challenging because of information deficits, which create considerable uncertainty over the impacts of groundwater pumping (Schlager 2006). Groundwater problems and conflicts often develop only after groundwater users are firmly entrenched, and subsequent implementation of institutional arrangements to control water use may be a difficult and expensive undertaking (Schlager 2006).

An additional element of water policy is the manner in which water policy is implemented. A continuum exists between the state using coercive or police powers to adopt and enforce regulations and less intrusive programs that rely on incentives and education to encourage voluntary action (Jacobs and Holway 2004). Regulatory actions are often taken after less intrusive polices are found to be ineffective. However, the regulatory approach tends to lead to a confrontational political environment that creates winners and losers (Jacobs and Holway 2004).

Water policy must also be concordant with prevailing societal values to be successful. It is important to always recognize that water policy decisions are inherently value-laden and are thus not amenable to resolution by pure economic or technical analysis. It is normal for cultural identities, religious beliefs, and long established water use practices to have more weight in water policy discourses than science (Toope et al. 2003). Cultural and religious issues in water use, policy, and management are discussed in Chap. 33. It is very difficult to place a direct economic value on some aspects of water policy, such as environmental protection and the adverse health and social impacts to humans resulting from an inadequate water supply. Nevertheless, economics must clearly play a role in water policy because societies do not have unlimited financial resources and effective water management is a critical element of national prosperity in water scarce regions.

It is probably accurate to state that there is widespread agreement on the general goals and principles of water management. For example, the Dublin Statement on Water and Sustainable Development commended to the world leaders in 1992, at the U.N. Conference on Environment and Development (UNCED) in Rio de Janeiro incorporates the following guiding principles:

Principle No. 1	Fresh water is a finite and vulnerable resource, essential to sustain life, development and the environment.
Principle No. 2	Water development and management should be based on a participatory approach, involving users, planners and policy-makers at all levels.
Principle No. 3	Women play a central part in the provision, management and safeguarding of water.
Principle No. 4	Water has an economic value in all its competing uses and should be recognized as an economic good.

International water law also now recognizes access to water as a human right (Sect. 32.4). The fundamental challenge lies not in reaching agreement on principles of water management, but rather in turning goals and principles into effective action.

No chapter, or even entire book, can provide all the solutions to water management in arid and semiarid lands. Any meaningful discussion of water management in arid and semiarid lands necessarily must address water policy in additional to the technical issues. A discussion is provided herein of some of the basic issues germane to water use in water scarce regions. These issues need to be considered in the development water policies and, more importantly, in the implementation of water management. It is not the intent to provide the answers to difficult water management questions, which inherently would incorporate the author's personal values, but rather to present the basic issues and their implications.

32.2 Tragedy of the Commons

The influential essay "The Tragedy of the Commons" written by Garrett Hardin (1968), is a useful starting point for discussions of water policy and management. The Tragedy of the Commons was written with respect to population growth but its thesis applies to any shared, limited (common-pool) resource, including water resources in arid lands (e.g., Glennon 2002). Hardin used cattle grazing on a common pasture as a metaphor for people sharing a common finite resource. The commons is open to all herdsmen, but has a finite carrying capacity. Too many animals would result in over-grazing and the ruin of the commons. As a rational being, each herdsman seeks to maximize his gain, which would occur by increasing the size of his herd. By adding an additional animal, a herdsman receives the full benefit from the additional animal, while the cost of the additional animal in terms of damage to the resource is borne by all users of the commons. It is, therefore, the rational conclusion of each herdsman that the sensible course of action is to increase his herd. The tragedy as explained by Hardin is that each herdsman becomes locked into a system that compels him to increase his herd without limits, which ultimately leads to the ruin of the shared resource upon which all depend.

The key lesson of "The Tragedy of the Commons" is that people acting in their self-interest can destroy a shared resource even though it is ultimately in no one's interest for this to happen. With respect to water resources, individual water users can maximize their gain by pumping more water to irrigate a greater area or using the water for other purposes, while the impacts or costs of the additional pumping will be shared by all aquifer users. Groundwater users have an incentive to pump all that they can so long as the marginal value product of the water is greater than or equal to the marginal pumping costs (Feitelson 2006). Ultimately, aquifer users acting in their own self-interest can lead to the depletion or otherwise impairment of the aquifer upon which they all depend.

Garrett Hardin's solution to the tragedy of the commons is "coercion," which is mutually agreed upon by the majority of the people affected. It was recognized that appeal to conscience would not be effective. With respect to water resources, mutually agreed upon coercion has been implemented in many areas in the form of water management or groundwater conservation districts. Another solution to the tragedy of the commons is privatization, which gives users a proprietary interest to protect what has become their property. Privatization is means of restricting access to a resource. Tradable permit systems have also been implemented in which the government sets a quota for total harvest of the resource (e.g., fish harvest, water use) and tradable permits are issued for a defined share of the harvest.

Since the publication of "Tragedy of the Commons" there has been great interest in the governance of common-pool resources. Numerous technical papers and dedicated books have been published on the subject. Ostrom et al. (1994, 2002) provide a good overview of the governance on common-pool resources. It is clear that use of common pool resources is far more complex than suggested in the "Tragedy of the Commons." Dietz et al. (2002, p. 1) observed that

> Things are not as simple as they seem in the prototypical model. Human motivation is complex, the rules governing real commons do not always permit free access to everyone, and the resource systems themselves have dynamics that influences their response to human use

and (p. 5) that

> People sometimes do, however, move beyond individual self-interest. Communication, trust, the anticipation of future actions, and the ability to build agreements and rules sometimes control behavior well enough to prevent tragedy.

Rose (2002, p. 234) similarly observed that

> quite ordinary people have the psychological, social, and moral wherewithal to arrive at cooperative arrangements on matters of common interest

Richerson et al. (2002, p. 426) also noted that

> both field and experimental evidence show that people cooperate in ways that are hard to reconcile with the behavior of selfish actors

Indeed, many workers have noted that Hardin's example of the medieval common fields was not tragic at all, but rather is an example of a set of sustainable community-based agricultural practices that lasted for centuries, if not millennia (Rose 2002).

Studies of common-pool resources illustrate examples of both where resources users create institutional arrangements and management regimes that allow for the long-term equitable distribution of the resource, particularly where resources were not successfully protected in the past (Agrawal 2002; and references therein). Users of a common-pool resource have a strong incentive to derive means for protecting the resource against overuse, particularly if they plan on using the resource for a long period of time. Commitment to the longevity of the resource is an important issue as to whether or not an individual user will cooperate and make sacrifices to protect it. Kopelman et al. (2002) noted a consistent finding of the gaming literature is that cooperation drops off as the end of the interaction is near. Thus, as it becomes evident that a common-pool resource will be soon depleted, users may tend to harvest as much as they can, while they can. Another variable is whether or not a user has an "exit option," which is earnings opportunities outside of the commons. Users of a common poor resource may strategize to (1) stay and conserve, (2) stay and degrade, or (3) degrade and leave (Bardham and Dayton-Johnston 2002). Exit options weaken prospects for cooperation.

Governance of common-pool resources must address exclusion (who may have access to a resource and who may not) and subtractability, which means that each 'unit' harvested from a common-pool resource is not available to other users. Both exclusion and subtractability must be addressed for sustainable management of common-pool resources. Appropriation addresses the sharing of the resource amongst users. Provision relates to the development, management, and enhancement of the productive capacity of common-pool resources. Provision of a commons resource must also consider the "free-rider" problem. A user may choose not (or cannot be forced) to contribute to the development and maintenance of a resource, while still receiving the benefits of its provision. With respect to groundwater recharge projects, for example, there is little incentive to invest in a project if others can pump the recharged water for free, i.e., there is no exclusion of new users (Schlager 2007).

A basic question is whether a common-pool resource is best governed by the shared users as common property, as private property, or by governmental bodies, or by some combination of these end members. It is clear that there is no one universal answer and that the optimal governance of each common-pool resource necessarily must depend on resource- and location-specific conditions (Dietz et al. 2002). With respect to government ownership or control of common-pool resources in general, Dietz et al. (2002) noted that nationalization of resources typically leads to:

- rejection of any existing indigenous institutions,
- poor monitoring of resource boundaries and harvesting practices because the government did not have the capability (e.g., human and financial resources) to monitor the resources to which they asserted ownership, and
- de facto open access conditions and a race to use the resources.

Schlager (2007) noted that with respect to management of local resources, people often realize that negotiation is better than having control imposed by central and distant authorities, and regulations would result in bureaucratic control and, therefore, corruption. Case studies of common-pool resources suggest that whenever government does not intervene, users are more likely to manage sustainably (Agrawal 2002). In the absence of external control over a resource, local users have a great incentive to devise, implement, and adhere to their own management solutions. Sanction systems may promote desired behavior, but may also undermine intrinsic motivations for cooperation and other generally helpful factors for community life such as interpersonal trust (Kopelman et al. 2002).

However, communities of groundwater users will likely need active assistance of high levels of government in order to adequately address provision problems affecting many communities of groundwater users. Successful commons management on any scale requires a system of legitimate institutions in which legitimacy is derived from the acceptance by the resource users (Richerson et al. 2002). The legitimacy of government authorities is determined in large part by their commitment to fair allocation and decision-making procedures (Kopelman et al. 2002). Provision problems, such as declining aquifer water levels, saline-water intrusion, and water quality deterioration, are processes that require basin-scale investigations and actions, and are very difficult to resolve. For example, basic groundwater provision problems such as aquifer overdraft have not been resolved in a numbers of basins in the Western United States, despite a high level of technical sophistication and existing governmental regulatory framework that generally has a high-degree of legitimacy among water users.

Self-governance of common-pool water resources was found more likely to be successful if there is a sense among resource users that governance attempts will make a difference and that they will enjoy the benefits of the governance, and that there is trust amongst the shared resource users (Schlager 2007). Self-efficacy, the belief that one's actions can make a difference, is an important motivator for cooperation. Studies of common pool resources, in general, indicate that successful management is more likely if there is a small group size, the resource has well-defined boundaries, members belong to a group, and it is easy to monitor and enforce compliance with agreements (Agrawal 2002, and references therein).

Cause attribution and framing also impact the degree to which people will cooperate towards preserving and cooperatively managing common-pool resources (Kopleman et al. 2002). People need to see themselves as causal agents in order to contribute at higher levels to the solution of environmental or resource problems. Whether a problem is framed as an ethical decision or business decision will also impact behavior. For example, weak economic sanctions and a low probability of detection (for example, against over-pumping groundwater), may result in an increase in "cheating" because the problem becomes viewed as a business decision (Kopelman et al. 2002). The economic benefits from "cheating" may exceed the potential costs of getting caught, especially if there is a low probability that the cheating will be detected and the penalties are low.

Governance of groundwater is more difficult than surface water because:

- it is more accessible,
- it is less visible and more difficult to detect (intake pipes verses hidden wells),
- aquifers have poorly defined boundaries (at least as perceived by local users),
- there are potentially many users, and
- there are lower upfront production and transaction costs (i.e., there is little exclusion; it is easy for new users to tap into the aquifer by just installing a well).

Uncertainty as to the size of common-pool resources in general, such as the volume of water available in an aquifer and the impacts of its use, tends to also reduce cooperation (Kopelman et al. 2002).

Tracking appropriation and provision problems involving groundwater irrigation is not easy because of (Schlager 2007):

- information problems (e.g., aquifer water balance),
- limitations on well construction and pumping are perceived as threatening livelihoods,
- monitoring of the 'invisible' resource is costly and difficult, and
- groundwater users are often not confronted with appropriation and provision problems until long after they invested in well systems and have become accustomed to the benefits of irrigation.

32.3 Water as an Undervalued Asset

The value to society of a safe and reliable water supply has long been recognized. Ancient civilizations have gone to extraordinary lengths to secure water sources for their cities. In light of the importance of a potable water supply, the amount of money that consumers in developed countries typically pay for their water is very low compared to the other expenses in life. For example, at the time the book was written, the water rate in Lee County, Florida (USA), was $2.84 per 1,000 gal (3,785 L) for the first 6,000 gal (22,710 L) plus a $10.6 monthly service and administrative charge. The median household income in Lee County was $45,645 in 2009 (U.S. Census Bureau 2011). For the vast majority of residents in developed countries (such as the people of Lee County), the cost of a reliable, high-quality water supply piped to their homes is an insignificant part (often less than 1%) of their total household budget. Nevertheless, most utility managers can attest that even a small increase in rates required for operation and maintenance, or upgrade a water supply system will elicit some strong public objections. The availability of abundant clean water on demand is taken for granted and is rarely even considered unless there is a disruption in the supply.

Conversely, in developing countries without adequate water treatment and distribution systems, the costs of purchasing or otherwise obtaining water may be a major part of the household budget. There is also an enormous human cost associated with inadequate water supply in some developing countries caused by health impacts and the daily labor involved in obtaining water from often distance sources. Experience by water users, in regions with safe, in-house water supplies, shows that the cost to individuals of not having safe delivery systems is far higher than the real cost of construction and operation of an adequate system (Pereira et al. 2002).

Government support typically allows water utilities throughout the world to sell irrigation water for far less than the cost of supplying it (Postel 1992; Revenga 2001; Johnson et al. 2001). In economic terms, water demand is elastic. The demand for water is related to its cost to consumers (domestic, industrial, and agricultural). As the price of water increases, less water will be used. Low prices and subsidies encourage inefficient use of water and discourage the adaptation of water-saving technologies (Revenga 2001; Johnson et al. 2001). Water users vary in their elasticity. Irrigated agriculture has a greater elasticity to high prices than urban water users in most parts of the world (Ward and King 1998). Cities are willing to pay much more for water than farmers lose in production by taking water out of agriculture. Indoors domestic water consumption appears to largely inelastic, whereas outdoor use is more elastic. Observed elasticities may reflect on water charges; where the cost of water is an insignificant part of the household budget, changes in price may not have a very dramatic effect on water use (Winpenny 1994).

The underpricing of water has serious implications for the operation and development of water supply and distribution infrastructure as it reduces the available financial resources. Racked (1996, p. 7) notes that

> because of underpricing, water utilities may be unable to recover their own costs, which leads to a vicious cycle of inadequate maintenance, poor services, unwillingness to pay even nominal charges, and then still lower revenue, and so on.

The underpricing of water sends the economic signal that water is not valuable. There is growing recognition that water needs to be treated as an economic good and that there is a need for greater economic efficiency in the use of water. The economics of water is further discussed in Chap. 34.

32.4 Water as a Human Right

There has been a strong push over the past two decades to have access to water be considered a human right. The efforts culminated on July 28, 2010, when the United Nations General Assembly voted to expand the Universal Declaration of Human Rights to include the right of water and sanitation as a basic human right. The vote differs from previous statements or resolutions in that for the first time states have the legal binding responsibility to ensure the realization of this right, along with other basic human rights.

Arguments for explicitly acknowledging a human right to water include that it (Gleick 2007):

- encourages the international community and individual governments to renew their efforts to meet the basic water needs of their populations,
- creates specific national and international legal obligations and responsibilities,
- maintains a spotlight of attention on the deplorable state of water management in many parts of the world,
- focuses attention on the need to more widely address international watershed disputes and to resolve conflicts over shared waters, and
- can help specific priorities for water policy, particularly that meeting basic human needs should take precedence over other water management and investment decisions.

Langford (2005, p. 281) observed that the "*history of human rights shows that once people feel and experience something as a human right, it becomes a difficult force to restrain.*"

The declaration of a human right to water establishes a welfare or positive right because it creates governmental obligations to provide citizens with sufficient water resources. A liberty right, on the contrary, ensures noninterference with a right, which in the case of water would be a government denying people access to water (Hardberger 2005). Liberty rights are relatively easy to implement and have minimal costs. Welfare rights, on the contrary, involve a significant deployment of financial and human resources, and strong political will, and are thus much more difficult to implement.

Biswas (2007, p. 7) well summarized the issue as

> It is fair to note that at present no sane individual opposes the concept that all humans should have access to clean water. Similarly, not a single country opposes this concept either. The main issue hinges around not whether water is a human right, but how to ensure that all humans have access to clean water and proper wastewater management within the social, economic, physical and political conditions and constraints within which they live.

And

> Declarations are easy to make. The problem invariably lies as to how the declarations can be translated into reality to improve the quality of life of the poor people in developing countries.

There has been much written concerning the concept and implications of water as a human right. A "Manual on the Right to Water and Sanitation" (COHRE, AAAS, SDC and UN-HABITAT 2007), has been published on its implementation. There is also much disagreement and uncertainty as to what the declaration of water as a human right actually means in terms of obligations, duties, and implications. Declaring water and sanitation as a human right does not put pipes in the ground and build water and wastewater treatment plants.

Segerfeldt (2005, p. 112) noted that to declare water as a human right in documents by governments or bodies such as the United Nations does not increase water distribution to the needy; "*we cannot drink rights on paper, only water.*" Many important questions exist concerning:

- how much water people have a right to and of what quality,
- how is it to be delivered, who pays for it,
- who is responsible for implementation of the right and liable for its non-achievement, and
- what are the obligations of water users, and the timetable for implementation.

There is considerable latitude in the development of potential standards for the amount of water required to be supplied, and its availability (reliability of supply), accessibility, costs, and quality. These standards could be set very loosely causing difficulty in true evaluation of whether or not they are being met.

An extreme interpretation advocated by some activist non-government organizations (NGOs) is that if access to water is a basic human right, then it must be free (or provided at highly subsidized prices) and that private-sector participation is precluded (Biswas 2007; Institute for Human Rights & Business 2009). Human rights are believed by some to transcend economics and that private companies should not make a profit in providing goods. However, Maxwell (2009, p. 28) also noted that

> Despite what some observers may claim, clean water certainly isn't – and can never be – free

Numerous practical questions also exist about the specific obligation of countries and water providers. For example, could a water utility be held guilty of a human rights violation for discontinuing service to users who chose not to pay for provided water or for its failure to provide service to all users? Does a private water utility have an obligation to provide water to the most marginalized communities, who cannot or will not pay on principle? (Institute for Human Rights and Business 2009) Does the human right to water apply to users that waste water?

Clearly, countries and regions will differ in their ability to supply water depending upon the degree of water scarcity and local financial and human resources. On a broad scale, establishment of the right to water and sanitation as a basic human right, pressures governments to give a higher priority to providing services to all people. It is generally recognized that consideration must be given to the resources available to each country to carry out this obligation. Perhaps the minimum requirement for a country is due diligence, which provides an adjustable criterion that depends upon a particular government capabilities and resources (Hardberger 2005).

32.5 Cultural Conservatism

Development and implementation of water policies often have to either accommodate or overcome deep cultural conservatism with respect to water. Water policies "must give close attention to cultural factors especially where the traditional values and 'meaning' attributed to water reinforces those interests of elements of the agricultural communities, elites, and bureaucracies who together influence water allocation and management" (Allan 2001, p. 175). Traditional water management practices in some instances may be obsolete under the realities of growing populations, urbanization, and increasing water scarcity. However, the rediscovery of traditional water management techniques, such as rainwater harvesting (Chap. 22), may be an important element of the development of sustainable local water resources. Religious teachings may be concordant with sound water management principles and form the foundation of a water conservation ethic (Chap. 33).

Keough et al. (2010, p. 125) raise an important question as to whether water management is a technical or social-cultural issue? Specially, "*is water management a matter of social engineering to ensure the smooth operation of a given technological solution, or is the search for an appropriate technological response to a problem embedded in a larger social-cultural process?*" It may ultimately be more difficult to educate a community to accept a technical solution that is contrary to their socio-cultural practices and norms, than to start by accepting the social-cultural constraints of the community and then find a harmonious technical solution.

32.6 Government Policy and Water Law

Groundwater and surface waters can be managed in four basic manners, which should be viewed as end members:

- Unregulated use—water resources are considered a common-pool resource available to all with no restrictions on access.
- Community or stakeholder management—water users formally or informally collectively manage a water resource and control who has access and how the resource is used.
- Market system—water is considered to be an economic good that is privately owned and can be used or traded as the owner sees fit.
- State control—water is owned by the state and its use is subject to government authorization (permitting). The state determines the allocation of water between users and sectors, which is controlled through either limits on water use or the construction of wells.

Unregulated use may be viable where water supplies are plentiful relative to demands, but clearly is not a desirable and sustainable strategy where water is scarce relative to demands. However, unregulated water use may be the *de facto* policy where local government resources are inadequate to control water use. A tragedy of the commons may result from unregulated use of a scarce resource, but the possibility that users may decide to implement voluntary rules for the management of the resource should not be discounted.

It is often argued that groundwater should be legally held in the public domain as a condition for sustainable or acceptable groundwater management. Llamas (2004) pointed out that this assumption is far from evident. Llamas (2004) proposed instead that groundwater management should be in the hands of the stakeholders of the aquifer under supervision of a water authority. Holding water in the public domain arguably may create a "tragedy of the commons" situation (Sect. 32.2), where there is little incentive for users to conserve the resource. The logic of stakeholder management (i.e., a bottom-up management approach) is that it has a greater potential to obtain a consensus and achieve cooperation rather the conflict.

Community or stakeholder management can be successful for relatively small groups of culturally coherent users (stakeholders) managing local resources. It is not a viable option for country-scale water management involving multiple user groups with different interests, and often low levels of inter-group trust. However, it is recognized that local communities and stakeholders should still be involved in the water management decision-making process.

In practice, country-scale water resources management will necessary involve state control or a market system. Actual water policy and law may involve a combination of the latter two factors. Water markets require government intervention to define and protect water rights. The use of water must be regulated in some manner so that ownership is possible and has value. Under some water laws, water is considered to be owned by the state, but permits for its use are tradable.

Water law and policy profoundly effect how efficiently water is used in both the physical and economic sense. As discussed in Sect. 32.2, unregulated use of common pool resources can result in overexploitation and damage to the resource (i.e., tragedy of the commons). Market-based systems (i.e., treating water as an economic good) can lead to greater economic efficiency in the use of water, but may have significant limitations related to transaction costs, market competiveness, and social and environmental cost issues (Chap. 34). Water laws and practices in most countries are biased towards the individual right to withdraw water for private gain and against the general public interest of leaving some water instream for fisheries, recreation, and the integrity of ecosystems (Postel 1992). State control has the limitation that water use decisions may become dominated by political considerations, which may or may not be in accordance with societal values, equity, or overall best interests. Water may be managed in a wise and fair manner or in a foolish and biased manner.

Government water use regulations and water law in some instances may actually encourage wasteful use of water through "use it or lose it" policies. Under such policies, the owner or recipient of a right to use water may retain that right only through continuous use. "Use it or lose it" policies have been formally or informally adopted in water use law and regulations with the intent that surface water and groundwater not be tied up in unused allocations. There is a societal interest in that water resources should be put to beneficial use (reasonable and beneficial doctrine in Western United States water law) and not be hoarded by landowners or other water users who have no current need for the water. However, if allocations are reduced or taken completely taken away if they are not used, then a perverse incentive is established to always fully use water allocations, even for activities that yield little or no economic value. Conservation of water through the adoption of more efficient irrigation practices is of little direct value, if it results in a concomitant decrease in water supply or allocation as the user can no longer demonstrate a need for the original allocation. The 'use it or lose it' policy applies to both individual water users and the relationship between exogenous users. Upstream users have an incentive to establish an historical use of water resources and little incentive to conserve water if it means that the downstream water users may obtain a right to the water. A key lesson is that any water use policy can have both desirable and undesirable results.

32.7 Water Governance Options

Groundwater management in the United States provides some good examples of the various principles developed for the governing of groundwater use. The use of the United States as an example is by no means meant to be an endorsement. Both surface water and groundwater management in the United States are far from optimal. Water governance in the United States also reflects American cultural values, which at the time of development of water law were more concerned with economic growth and property rights rather than equity and environmental protection. It is clear that there is no one universal best method for governing water use and that each of the water governance systems has both advantages and disadvantages. Also, sound governance systems may be poorly implemented or may be corrupted by biased or incompetent management.

Management of groundwater in the United States is primarily the responsibility of the states, with the Federal government having jurisdiction over water quality through environmental protection laws. Five main groundwater doctrines have been adopted in the United States; (1) absolute ownership, (2) reasonable use, (3) prior appropriation, (4) correlative rights, and (5) restatement of torts. The five doctrines are end-members, and some states have adopted combinations of the doctrines. There is considerable variation in the specifics of state groundwater laws.

Groundwater law varies depending upon whether groundwater is considered to be owned by the state or by the landowner. States and jurisdictions also vary on how they address the impacts of groundwater withdrawals on surface waters, and what actually legally constitutes groundwater. A distinction is often made, particularly in the western states, between shallow groundwater that is closely connected to a stream or river, and deeper groundwater with a lesser connection to surface-water bodies. Water use law is also evolving in response to the overdraft of some aquifers. Doctrines that were adopted during an earlier time when the groundwater resources were viewed as being essentially unlimited, have become unviable in an age of rapidly declining aquifer water levels.

32.7.1 Absolute Ownership

The doctrine of absolute ownership, which is referred to as English Rule, holds that the landowner owns the groundwater underlying his or her property. Under the rule of capture, the landowner can withdraw as much water as desired without any liability for impacts to adjacent and more distant landowners. Water belongs to the person with biggest pump. The advantage of the absolute ownership doctrine is its simplicity. Absolute ownership has the disadvantage in that it offers no encouragement for conservation and provides for no long-term security. The water supply of an individual can be compromised by another party installing wells near the property boundary or through cumulative overdraft of the aquifer. Absolute ownership states include Texas (Senate Bill 2 is activity changing this law by implementing groundwater basin rules), Louisiana, Maine, Massachusetts, Indiana, Rhode Island, and Connecticut.

The doctrine of absolute ownership becomes obsolete as available groundwater resources become more limited. For example, surface water in Texas is considered water of the state and its use has historically been tightly controlled primarily under the doctrine of prior appropriation. Texas groundwater law is evolving with establishment of groundwater management areas and groundwater conservation districts. Groundwater conservation districts are now the preferred method managing water and have authority to regulate the spacing of water wells, the production from water wells, or both.

32.7.2 Prior Appropriation Doctrine

The prior appropriation doctrine, which is referred to as 'Western Water Law', is often expressed as 'first in line, first in right.' Prior appropriation is used in the Western United States to govern surface water and has been adopted by a majority of the western states to also regulate groundwater (Bryner and Purcell 2003). Under the prior appropriation doctrine, each water user has a water right that has an annual quantity and an appropriation date. The water right may control pumping rates, well spacing, and well construction. Water rights are issued until all the available water is allocated, with the general requirement that the proposed water use be reasonable and beneficial. The senior appropriator (i.e., holder of the water right with the oldest appropriation date) acquires the right to use a certain volume of water each year, with the remainder of the flow being available to the more junior appropriators in order of their water right dates. No credit is given for unused water each year. During drought periods, or other circumstances, in which water use must be reduced, the pumping by the more junior users must be curtailed first. States that have adopted the prior appropriation doctrine for groundwater are Alaska, Colorado, Idaho, Kansas, Montana, Nevada, New Jersey, New Mexico, North Dakota, Oregon, South Dakota, Utah, Washington, and Wyoming (Wright 1990).

The prior appropriation doctrine has the advantage in that it provides a means for controlling groundwater use and it provides some security to water right holders. The main disadvantage is that it is biased greatly towards senior users. Groundwater and surface-water rights must also be reconciled. Typically, surface-water rights are senior to groundwater rights, and priority is given to minimizing the impacts of groundwater withdrawals on surface-water flow, and thus, surface water rights. For example, in Colorado, a distinction is made between tributary and non-tributary water. When withdrawal of tributary water exceeds a statutory limit for impacts on surface water, its continued use requires a plan for mitigating effects on surface-water systems.

Prior appropriation, as conceived and administered, was not designed to conserve water, but rather was designed to encourage the diversion and use of water (Schlager 2006). The prior appropriation doctrine effectively created incentives for private owners to invest in delivering water to where it was most productive (Anderson and Snyder 1997). The system provides a definite, certain legal right to use water, which is a prerequisite for capital investment in expensive water infrastructure. The right to use water is a real property in the sense that it can be sold, bequeathed, or otherwise transferred so long as it is approved by the local water authorities (Office of Technology Assessment 1983).

Under the prior appropriation doctrine, a water right holder that does not use its full allocation faces loses the right to the unused portion. No consideration is given as to what is the best use of the water and a disincentive to conserve water exists because unused water rights may be lost.

32.7.3 Reasonable Use Doctrine

The reasonable use doctrine, which is referred to as the "American Rule," allows the landowner beneficial use of groundwater as long as the use does not harm other aquifer users, the aquifer system, or the environment. Groundwater use is permitted by a state or a local regulatory agency. The groundwater may be considered to be either owned or held in trust by the state or landowner. In Florida, the water is held by the state in a trust for the common use of the people. The application of reasonable use doctrine may inherently include recognition of priority, in which the impacts of proposed new water uses on existing users is considered in the permitting process. The reasonable use doctrine has been adopted by the majority of eastern states, and also by Arizona and Nebraska (Wright 1990).

Florida is a good example of a reasonable use doctrine state. Water use is regulated by five water management districts, which manage the consumptive use of water, aquifer recharge, well construction, and surface-water management under Chapter 373 of Florida Statutes. The water management districts also administer flood protection programs, perform technical investigations into water resources, and acquire and manage lands for water management purposes. Groundwater use above a minimum rate (i.e., above domestic use rate) requires a consumptive use permit (CUP) or water use permit (WUP).

Groundwater law in Florida is based on Professor Frank Maloney's (1972) Model Water Code as codified in Chapter 373 of Florida Statues. Groundwater use permitting is based on three criteria or prongs. The proposed groundwater use should:

- be reasonable and beneficial,
- not harm existing users, and
- be in the public interest.

Permit applicants must demonstrate that a proposed new use will not have a significant adverse impact on existing permitted and legal users, the aquifer (e.g., cause saline-water intrusion), and the environment. The beneficial use requirement also includes the implementation of conservation measures to minimize the waste of water. A landowner does not inherently have a right to groundwater and may be denied a permit if the requested water use causes unacceptable adverse impacts or causes cumulative groundwater drawdowns to reach an unacceptable threshold.

An individual consumptive use permit (CUP) or water use permit (WUP) has a:

- specific allocation,
- identified source (e.g., aquifer and well locations),
- particular use, and
- defined duration.

The CUP or WUP water allocation is not 'owned' by the applicant, but can be transferred if the new property owner is continuing the same use. Permits have a finite duration and there is not an inherent right to a renewal. In practice, renewals are seldom denied although allocations may be changed.

32.7.4 Correlative Rights Doctrine

Under the correlative rights doctrine, groundwater rights are proportional to the landowner's share of the land overlying the aquifer. The water is considered to be owned by the landowner, but its use is controlled by the state. The use of the groundwater must be reasonable and beneficial, and is correlated with the rights of other groundwater users. In the event of shortages, caused by droughts or excessive aquifer drawdowns, all water users share in a pro-rated reduction in water use related to the land owned; no distinction is made between senior and junior water users or to the use of the water. The correlative use doctrine has been adopted by Hawaii, Iowa, Minnesota, and Vermont. California has adopted elements of both the correlative rights and reasonable use doctrines (Wright 1990).

32.7.5 Restatements of Torts

The restatement of torts doctrine combines elements of the absolute ownership and reasonable use doctrines. A landowner is entitled to withdraw groundwater for a beneficial purpose and is not liable for interference with the water uses of others unless: (1) the withdrawal causes unreasonable harm by lowering water levels or pressures, (2) it exceeds the owners reasonable share of the annual supply of groundwater, or (3) it causes a direct and substantial adverse effect on a water course or lake and causes harm to a person entitled to use its water. The restatement of torts doctrine has been adopted by Michigan, Ohio, and Wisconsin (Wright 1990).

32.8 Closing Thoughts

The greatest challenge in water management in water scarce regions is often inertia when current practices are unsustainable. The course of least resistance is to continue on the existing water management path, even though it may no longer be the best option for current conditions. To use a colloquial expression, it is often easier to just "kick the can down the road" rather than making difficult, politically unpopular changes. There is an inherent conservatism with respect to water management. Changing water governance may be politically very difficult, if not impossible, because many people may have vested interests in maintaining the status quo. Those who have existing rights to water will naturally resist any changes that would result in a reallocation of their water (unless perhaps they are financially well compensated). Those used to free or very cheap water may be strongly opposed to paying more.

It usually takes a crisis to overcome inertia. It is not until most of the stakeholders, whether they be farmers and domestic water users, truly understand and accept the magnitude of a water crisis, will there be the political will to accept major policy changes. Education is thus a critical element of successful water management. It is also imperative that stakeholders participate in the decision-making process and have confidence that the decisions of water managers are based on sound science, are unbiased, and in the best interest of society as a whole.

References

Agrawal, A. (2002). Common resources and institutional sustainability. In E. Ostrom, T. Dietz, N. Dolšak, P. C. Stern, S. Stonich, & E. U. Weber (Eds.), *The drama of the commons* (pp. 41–85). Washington: National Academy Press.

Allan, J. A. (2001). *The middle east water question*. London: Hydropolitics and the global economy.

Anderson, T. L., & Snyder, P. (1997). *Water markets: Priming the invisible pump*. Washington: Cato Institute.

Bardham, P., & Dayton-Johnson, J. (2002). Unequal distribution: Heterogeneity and commons management in large-scale multivariate research. In E. Ostrom, T. Dietz, N. Dolšak, P. C. Stern, S. Stonich, & E. U. Weber (Eds.), *The drama of the commons* (pp. 87–112). Washington: National Academy Press.

Biswas, A. K. (2007). Water as a human right in the MENA region: Challenges and opportunities. In A. K. Biswas, E. Rached, & C. Tortajada (Eds.), *Water as a human right for the Middle East and North Africa* (pp. 1–17). London: Routledge.

Bryner, G., & Purcell, E. (2003). *Groundwater law sourcebook of the western United States.* Boulder, CO: Natural Resources Law Center, University of Colorado School of Law.

COHRE, AAAS, SDC., & UN-HABITAT. (2007). *Manual on the right to water and sanitation.* Geneva: Centre on Housing Rights and Evictions.

Dietz, T., Dolšak, N., Olstrom, E., & Stern, P. C. (2002). The drama of the commons. In E. Ostrom, T. Dietz, N. Dolšak, P. C. Stern, S. Stonich, & E. U. Weber (Eds.), *The drama of the commons* (pp. 3–35). Washington: National Academy Press.

Feitelson, E. (2006). Impediments to the management of shared aquifers: A political economy perspective. *Hydrogeology Journal, 14*(3), 319–329.

Gleick, P. H. (2007). *The human right to water* (5 pp.). Oakland: Pacific Institute.

Glennon, R. J. (2002). *Water follies: Groundwater pumping and the fate of America's fresh waters*. Washington, DC: Island Press.

Hardberger, A. (2005). Life, liberty and the pursuit of water: Evaluating water as a human right and the duties and obligations it creates. *Northwestern Journal of International Human Rights, 4*(2), 331–362.

Hardin, G. (1968). The tragedy of the commons. *Science, 162*, 1243–1248.

Institute for Human Rights and Business. (2009). *Business, human rights and the right to water: Challenges, dilemmas, and opportunities: Roundtable consultative report (Jan. 2009 Draft).* London: Institute for Human Rights & Business.

Jacobs, K. L., & Holway, J. M. (2004). Managing for sustainability in an arid climate Lessons learned from 20 years of groundwater management in Arizona, USA. *Hydrogeology Journal 12*(1), 52–65.

Johnson, N., Revenga, C., & Echeverria, J. (2001). Managing water for people and nature. *Science, 292*, 1071–1072.

Keough, N., Smirat, S., & Benjamin, S. (2010). Lessons from a participatory approach to household greywater use in Jordan. In S, McIlwaine., & M, Redwood (Eds.), *Greywater use in the Middle East. Technical, social, economic and policy issues* (pp. 113–138). Burton on Dunsmore, Practical Action Publishing, and Ottawa, International Development Research Centre.

Kopelman, S., Weber, J. M., & Messick, D. M. (2002). Factors influencing cooperation in commons dilemma: Areview of experimental psychological research. In E. Ostrom, T. Dietz, N. Dolšak, P. C. Stern, S. Stonich, & E. U. Weber (Eds.), *The drama of the commons* (pp. 113–156). Washington: National Academy Press.

Kresic, N. (2009). Groundwater resources: *Sustainability, management, and restoration.* New York: McGraw-Hill.

Langford, M. (2005). The United Nations concept of water as a human right: A new paradigm for old problems? *Water Resources Development 21*(2), 273–282.

Llamas, M. R. (2004). Water and ethics. Use of groundwater: UNESCO International Hydrogeological Programme. 33.

Maloney, F. E. (1972). *A model water code: Text and commentary.* Gainesville: University of Florida Press (University Water Resources Center Publication No. 8).

Maxwell, S. (2009). Moving beyond the "privatization wars"—a call for balance. *Journal American Water Works Association, 10*(1)*–12*. 26–28.

McKean, E. (2005). *The new Oxford American Dictionary* (2nd ed.). New York: Oxford University Press.

Office of Technology Assessment. (Oct, 1983). Institutions affecting western agricultural water use. In *Water-related technologies for sustainable agriculture in the U.S. arid/semiarid lands* (pp. 109–145). U.S. Congress Office of Technology Assessment, OTA-F-212.

Ostrom, E., Gardner, R., & Walker, J. (1994). *Rules, games, & common-pool resources.* Ann Arbor: University of Michigan Press.

Ostrom, E., Dietz, T., Dolšak, N., Stern, P. C., Stonich, S., & Weber, E. U. (2002). *The drama of the commons*. Washington: National Academy Press.
Pereira, L. S., Cordery, I., & Iaconides, I. (2002). Coping with water scarcity: UNESCO, International Hydrological Programme, IHP-VI, Technical Documents in Hydrology No. 58.
Postel, S. (1992). *Last oasis: Facing water scarcity*. New York, WW Norton, 239.
Racked, E. (1996). Introduction, In E, Racked., E, Rathgeber., & D. B, Brooks (Eds.). *Water management in Africa and the Middle East* (pp. 4–8). Ottawa: Canada International Development Research Centre.
Revenga, C. (2001). Will there be enough water? Earth Trends, Water Resources Institute, 5.
Richerson, P. J., Boyd, R., & Paciotti, B. (2002). An evolutionary theory of commons management. In E. Ostrom, T. Dietz, N. Dolšak, P. C. Stern, S. Stonich, & E. U. Weber (Eds.), *The Drama of the commons* (pp. 403–442). Washington: National Academy Press.
Rose, C. M. (2002). Common property, regulatory property, and environmental protection: Comparing community-based management to tradeable environmental allowances. In E. Ostrom, T. Dietz, N. Dolšak, P. C. Stern, S. Stonich, & E. U. Weber (Eds.), *The drama of the commons* (pp. 233–257). Washington: National Academy Press.
Schlager, E. (2006). Challenges of governing groundwater in U.S. Western States. *Hydrogeology Journal, 14*(3), 350–360.
Schlager, E. (2007). Community management of groundwater. In M. Giordano & K. G. Villholth (Eds.), *The agricultural groundwater revolution: Opportunities and threats to development:* (pp. 131–152). Wallingford: CAB International.
Segerfeldt, F. (2005). Water for sale. *How business and the market can reduce the world's water crisis*. Washington: Cato Institute.
Toope, S. J., Rainwater, K., & Allan, T. (2003), Managing and allocation water resources: Adopting the integrated water resources management approach. In A. S, Alsharhan., & W. W, Wood (Eds.), *Water resources perspectives: Evaluation, management and policy* (pp. 1–8). Amsterdam: Elsevier.
U.S. Census Bureau. (2011). State and County Quick Facts. Lee County. Retrieved Aug 7, 2011, from http://quickfacts.census.gov/qfd/states/12/12071.html.
Ward, F. A., & King, J. P. (1998). Reducing institutional barriers to water conservation. *Water Policy, 1*, 411–420.
Winpenny, J. (1994). *Managing water as an economic resource*. London: Rutledge.
Wright, K. E. (Ed.). (1990). *Water rights of the fifty states and territories*. Denver: American Water Works Association.

Chapter 33
Religious and Cultural Influences on Water Management

33.1 Introduction

> Resource management in the arid lands has been influenced not only by the potentials of the arid ecosystem and skills of the peoples of the arid land, but also by the cultural and institutional frameworks within which the resource management has taken place.

(Heathcote 1983, p. 253).

Water resources management involves the consideration or balancing of technical, hydrologic, and socioeconomic issues. Successful water resource management policies must, therefore, consider local cultural and religious traditions and values. Water management policies need to fit the cultural norms of the affected people (Amery 2001a). It was observed by the World Bank that poor people trust religious organizations more than any other organizations except their own social institutions (Palmer and Finlay 2003). However, scientists and engineers are typically trained to view problems from a purely technical perspective. Faruqui (2001a, p. 9) observed that

> There seems to be a belief that comprehensive in-depth studies that look at all aspects of a particular religion or culture are becoming unimportant in a globalizing, urbanizing world, in which religion or beliefs are perceived to be less important as the world proceeds towards some kind of common, material-based culture.

Pereira et al. (2002, p. 35) similarly noted that

> In dealing with cultural issues great care is needed, particularly to understand how the current situation developed historically. To advocate change in some practices may be offensive to the local population, no matter what the logic from a water resource point of view.

The separation of church and state, and thus water policy, is taken for granted in much of the secular states of North America, Europe, Asia, and Australia. However, Faruqui (2001a) noted that the concept of secularism, or the separation

R. Maliva and T. Missimer, *Arid Lands Water Evaluation and Management*, Environmental Science and Engineering, DOI: 10.1007/978-3-642-29104-3_33,

of church or mosque and state, does not occur within Islam as a religion. Clearly, water policy within Muslim countries must be consistent with Islamic teachings. For example, Western agency prescriptions concerning water management in rural Arabia have tended to ignore rural social priorities and the practicalities of implementing such changes in the context of Islamic principles and existing political systems, which adversely impacted their acceptance (Young 2002).

There is a growing school of thought or realization that religious teachings are compatible with and can promote conservation. Religious stories, images, and values can provide powerful ways to capture the attention of legislators, enforcement personnel, and the public at large, and supplement secular ethics in order to develop a more robust environmental and water conservation ethic (Fisher-Ogden and Saxer 2006). For example, water management strategies that are grounded in the teachings of Islam (as opposed to western notions) are likely to have greater appeal to the people of the Islamic world (Amery 2001a). The same can be said of other religions. Palmer and Finlay (2003) related an example of how unsustainable fishing practices in Tanzania (dynamite fishing) were curtailed because local religious leaders taught the communities that such wasteful and destructive practices were contrary to their Islamic faith. Previously attempted secular educational campaigns and government threats were ineffective.

It is critical for successful water management to recognize that water policy decisions are inherently culturally value-laden and policies that are contrary to local religious and cultural values or teachings are likely to face stiff resistance and may ultimately fail. Opposition to water policies that are contrary to local values will be fiercest when they are imposed or proposed by entities external to the culture. In simple terms, people in general do not like to be told to change their ways by outsiders, however virtuous their intentions may be.

Cultural values may arise directly from religious teachings or values or may have a secular origin. However, secular cultural values may still have an ultimate origin in the religious background of the predominant community. It is important to also recognize that religious teachings are often in tune with effective water management. The belief that water and the natural environment are gifts from God should encourage conservation and environmental stewardship.

33.2 Water Management in Islam

An overview of Islamic water principles is provided by Faruqui (2001b) and the papers included in the volume "Water Management in Islam" (Faruqui et al. 2001c). Islam is fundamental to both individual behavior and governmental laws and regulations. The fundamental tenants of law at its highest level in the Middle Eastern Gulf states is that all law is based upon and must conform with Islamic law (shari'ah), derived from the Quran and reports of statements and actions of the prophet Mohamed (the hadiths) (Alsharhan et al. 2001).

Water has great importance in Islam as a purifying and cleansing agent. Muslims must be ritually pure before approaching God in prayer, which is achieved through ablution (Abrams 2001). Islam is unique amongst the major religions in the significance given to water and the number of specific statements and teachings related to water management. As Islam arose and developed in a desert area it, therefore, has much to say about the use, ownership, and transfer of water (Caponera 2001). Faruqui (2001b, p. 15) noted that:

> Water is of profound importance in Islam. It is considered a blessing from God that gives and sustains life, and purifies humankind and the earth.

The fundamental mission of Islam is to ensure the spiritual salvation of each individual Muslim because only the individual is held capable of salvation (Naff and Dellapenna 2002). The primary concern of shari'ah is the moral behavior and spiritual health of the individual (Naff and Dellapenna 2002). Islamic water law, therefore, tends to focus on individual behavior. Water is considered to be a vital community resource of which everyone has a right to a fair share, and a Muslim may not hoard excess water (Faruqui 2001b). Water should not be wasted. The Quran clearly and strongly discourages the waste of water and other resources (Faruqui 2001b; Shah et al. 2001). Specifically, the Quran (6:141; 7:31) states (Shah et al. 2001)

> But waste not by excess: for Allah loveth not the wasters

A hadith holds that excess use of water is forbidden, even where water is abundant (Naff and Dellapenna 2002). The Islamic water management principle may be succinctly summarized as advocating wise use (Amery 2001a). Humans are stewards (*khulafa*) of this life-giving resource (Naff and Dellapenna 2002).

The first priority for water use in Islam is to sustain human life. The second priority is for domestic animals, and the third is for agricultural irrigation. Industrial and recreational uses have a lesser priority. Humans are recognized to be stewards of the environment, and the environment has very strong and specific water rights (Amery 2001b; Faruqui 2001b). Humans, as trustees or stewards of water and natural resources, have the role and responsibility to ensure that the resources are used in a reasonable, equitable, and sustainable manner (Al-Jayyousi 2001).

In Islam, water is considered a gift from God, so no individual literally owns it. Under Islamic law, water is a common entitlement of all people. Government has the responsibility to secure and manage water resources for the benefit of the whole community (Abderrahman 2000). Water resources are considered to be public property, and a permit or concession is required for any use of water (Caponera 2001). However, most Islamic scholars have concluded that water stored in private containers, distribution systems and, reservoirs or privately treated is considered to be a private good (e.g., Faruqui 2001b; Kadouri et al. 2001; Sadr 2001). The investment in treatment, distribution, and storage grants private ownership or a priority right to the water so obtained, but does not give claim to the river, reservoir, or aquifer from which the water comes (Sadr 2001). A water utility that constructs and operates a water supply, treatment, and distribution system thus has the right to recover costs (Faruqui 2001b). The owner of a private good has the right to use, trade, sell, or donate the good. Prices are to be fixed by the market. However, Islam forbids speculation and manipulation of the market to raise prices and increase profits (Kadouri et al. 2001).

In as much as conservation of water is an explicit element of Islam, it has been emphasized in several papers the importance of using Islamic communication channels to increase awareness of Islamic concepts with respect to water conservation (Atallah et al. 2001; Shah et al., 2001). The mosque is the best forum for addressing the general public at all levels on matters concerning issues of general life such as water conservation (Atallah et al. 2001). It is recommended that Imams, as community leaders, should be included in water resources planning and management activities (Atallah et al. 2001).

Faruqui (2001b, p. 32) observed that

> There is no contradiction between what Islam says about water management and emerging international convention on the issue, as reflected by recent accords such as the Dublin Principles or the UN Water Convention. In fact, the Islamic water management principles are not unique. Some of the same principles could be derived by studying other faiths, their holy books, and the lives of their prophets.

Farooq and Ansari (1983) provide an Islamic perspective on wastewater reuse. There are two distinct uses of water in Islam: (1) ordinary mundane purposes such as drinking, cooking, and washing of dishes and clothes, and (2) religious purposes for obtaining the state of ritual purity. In the Islamic tradition, water is divided into three categories based on its purity and the purposes for which it may be used: (1) *tahūr*, (2) *tāhir*, and (3) *mutanajjis* (Farooq and Ansari 1983). The purest water is *tahūr*, which may be used for both mundane and religious purposes. *Tāhir* water is suitable for mundane purposes, but has been significantly changed by the mixture of something extraneous that makes it unsuitable for ritual washing (ablution). *Tahūr* water becomes *tāhir* after being used for ritual washing and cannot be used again for religious purposes. *Mutanajjis* water has become defiled as the result of pollution and is unfit for mundane and religious purposes, but may still be used for irrigation.

The reuse of wastewater effluent appears to be perfectly legitimate from the Islamic viewpoint because the effluent may regain its original pure character by (1) treatment, (2) dilution with a larger quantity of *tahūr* water, or by (3) the passage of time (Farooq and Anasari 1983). The Council of Leading Islamic Scholars in Saudi Arabia issued a fatwa in 1978, stating that reclaimed water can be used for ablution and drinking if it is sufficiently and appropriately treated to ensure good health. The Scholars Council of the Kingdom of Saudi Arabia stated that (Water Authority of Jordan 2006)

> According to the report set by the experts in this regards, a large body of water would be pure from any impurity if such impurity is removed, if more water added to it, or if such impurity is eliminated by the passing of time, the sun, the wind or any other cause that would remove its impurity. Impure water could be purified by the modern filtering

> techniques that are the best and most efficient methods for purification, in which many materials will be added to remove impurities and certified by the water treatment experts. Therefore, this Council believes that such water will be totally pure and it may be used for ritual purification and drinking as long as there are no negative consequences on health. If drinking is to be avoided, it is merely for reasons of public health and safety, not due to any ramifications of Islamic Law. The Council recommended avoiding using this treated water for drinking purposes to avoid health problems and also in consideration of the negative public sentiment about this water.

The fatwa showed the dynamic nature and wisdom of Islamic laws in solving the challenges to the Muslim community (Abderrahman 2000). It was recognized that wastewater reuse is a public health issue and that potable reuse should be avoided if possible because of public aversion to the concept. However, there is still the local perception that treated wastewater is unclean from a religious perspective. For example, in the West Bank of Palestine, there is a strong belief that wastewater reuse is against Islam, and the majority of people are unaware of or do not seem to understand the religious opinion on the subject (Al-Kharouf et al. 2008).

33.3 Water Management in Christianity and Judaism

Water is an important element in Christianity and Judaism as a ritual agent of cleansing and purification, as well as being a symbol of life. However, both the Old Testament and New Testament of the Bible have little to say about the mundane use of water, even though both Christianity and Judaism developed in a water scarce region. Christianity and Judaism are highly fractured in the sense that the religions are split into numerous denominations, and there is no single leader or spokesman. Furthermore, religious beliefs are often tightly interwoven with social and political values. People who hold conservative religious values tend also to be politically and socially conservative. For example, religious conservatives in the United States tend to be more skeptical of the occurrence of global warming than the general population even though there is no clear religious basis for taking a position on either side of the issue. People belonging to more liberal denominations tend to be politically liberal and identify themselves as environmentalists. However, these broad generalizations do not always hold and some conservative Christians are strong supporters of environmental conservation.

The Judeo-Christian outlook on conservation ranges between two poles; that nature exists only for humanity's sake, and humans are charged to be stewards of nature with the need to conserve it (Fisher-Ogden and Saxer 2006). There is conservative fundamentalist Christian hostility to environmentalism that typically finds its endorsement in the book of Genesis. This "dominionist" environmental philosophy focuses on the Bible verse Genesis 1:28 (Maltby 2008):

> And God blessed them, and God said unto them, Be fruitful, and multiply, and replenish the earth, and subdue it: and have dominion over the fish of the sea, and over the fowl of the air, and over every living thing that moveth upon the earth.

This is interpreted as giving man the right to exploit the resources of the Earth for his purposes as he sees fit. Environmentalism has been equated to secularism. This belief leads to an anthropocentric view that nature has no reason to exist other than to save and serve man (White 1967). Lynn White (1967), in his highly influential essay, attributed the current ecologic crises to the anthropocentrism of Christianity. Hundreds of papers have subsequently been written supporting and refuting White's thesis, and many of his assumptions and conclusions have been called into question (Minter and Manning 2005).

A parallel view is that the second-coming of Jesus will occur soon and that true believers will be carried off to heaven (which is referred to as the Rapture), which would render the natural resources of the Earth superfluous and thus the need to conserve them. Maltby (2008) described this philosophy as "*Why care about ecological crises when the true believers will be rescued by the Rapture?*" A similar line of thought would apply to the use of nonrenewable water resources.

There is a growing school of thought within Christianity that the Bible directs man to be steward of the Earth as God's creation and thus, imparts a responsibility to protect the environment (e.g., Berry 1993; Johnson 2000). The Christian stewardship ethic interprets the Genesis commandments as giving man a divine charge to be good stewards and to take care of and protect God's creation (Kearns 1996, 1997). Van Dyke (2006) observed that "*acts of stewardship are acts of moral significance because they are acts that fulfill moral obligations toward the intrinsic value of what God has created.*"

The ethic of conservation is thus not based on the perceived usefulness of the natural environment and nonhuman life to humans or aesthetic considerations, but rather is based on the recognition that nonhuman life was created for God's purposes and is "good" (Van Dyke 2006). The rebellion of humans against God takes many forms including the abuse of the rest of creation (WCC et al. 2003).

Although there are divisions in the Judeo–Christian world concerning environmental protection and conservation, religious values and teachings tend not to enter into modern water management. The lack of a significant impact is related to the absence of explicit doctrines related to water use and the secular nature of predominantly Christian states.

33.4 Buddhism

In Budhism, symbolism and rituals are pointless because followers seek spiritual enlightenment from seeing the reality of unreality (Abrams 2001). Buddhists believe that all things are interconnected and that humans do not have an autonomous existence. Fossey (2003, p. 77) reported a belief in Buddhism is that "*the health of the whole is inseparably linked with the health of the parts, and the health of the parts is inseparably linked with the whole.*"

Buddhism holds that all actions have consequences (karma) and that one's actions will ultimately affect oneself. Respect for life and the natural world is essential, and it is important to live in peace and harmony with nature. The challenge of Buddhist nations is to find a workable balance of economic development, religious tenets, and enforceable environmental standards (Fisher-Ogden and Saxer 2006).

33.5 Hinduism

Hinduism asks its followers to see God in every object in the universe and holds that all life is sacred. Hindus believe (Rao 2003) that *"There is no life that is inferior. All lives enjoy the same importance in the universe and all play their fixed roles. They are to function together and no link in the chain is to be lost."*

The Hindu religion holds that all water is sacred, and holy places are usually located on the banks of rivers, which are viewed as being particularly sacred. The Ganges River is the most sacred of rivers. Polluting rivers is a big sin (Rao 2003). Water is used for cleansing and purification. In the Ganges, the pure are made even more pure, and the impure have their pollution removed, if only temporarily (Abrams 2001). Sacred texts state that the waters of the sacred river Ganges must be kept free of pollution (Fisher-Ogden and Saxer 2006). Paradoxically, the Ganges River is surprisingly one of the most polluted rivers in the world (West 2007).

33.6 Conclusions

West (2007) observed that, it is common, especially in developed countries, for religious influences to be ignored in water policy decisions. The developed countries water ethics are based mainly on secular economic and aesthetic considerations. Although religious teachings may encourage conservation and discourage waste, people vary greatly in their devotion and the degree to which religion influences daily behavior, particularly with respect to mundane activities. Water ethics in developing countries are often based on basic survival considerations such as need for water to support domestic uses and local food production (West 2007). For example, cleaning up the Ganges River has great religious significance, but among many people, a lower priority is given to pollution control and remediation compared to food production and other basic survival concerns. The relationship between religion and actual actions of adherents with respect to water use, pollution, and conservation is a subject that warrants further study.

References

Abderrahman, W. A. (2000). Water demand management and Islamic water management principles: A case study. *Water Resources Development, 16*(4), 465–473.

Abrams, P. (2001). Water in religion. *The water page*, http://www.africanwarer.org/religion.html (May 5, 2011).

Al-Jayyousi, O. (2001). Islamic water management and the Dublin Statement. In N. I. Faruqui, A. K. Biswas, & M. J. Bino (Eds.), *Water management in Islam* (pp. 37–41). Tokyo: United Nations University Press and Ottawa, International Research Centre.

Al-Kharouf, S., Al-Khatib, I., & Saheen, H. Q. (2008). Appraisal of social and cultural factors affecting wastewater reuse in the West Bank: *International Journal of Environment and Pollution, 33*(1), 3–14.

Alsharhan, A. S., Rizk, Z. A., Nairn, A. E. M., Bakhit, D. W., & Alhajari, S. A. (2001). Hydrogeology of an arid egion. *The Arabian Gulf and adjoining areas*. Amsterdam: Elsevier.

Amery, H. A. (2001a). Islamic water management. *Water International, 26*, 481–489.

Amery, H. A. (2001b). Islam and the environment. In N. I. Faruqui, A. K. Biswas, & M. J. Bino (Eds.), *Water management in Islam* (pp. 41–48). Tokyo: United Nations University Press and Ottawa, International Research Centre.

Atallah, S., Ali Khan, M. Z., & Malkani, M. (2001). Water conservation through public awareness based on Islamic teachings in the Eastern Mediterranean. In N. I. Faruqui, A. K. Biswas, & M. J. Bino (Eds.), *Water management in Islam* (pp. 49–57). Tokyo: United Nations University Press and Ottawam International Research Centre.

Berry, W. (1993). Christianity and the survival of creation: *Cross Currents, 43*(2), p. 15.

Caponera, D. A. (2001). Ownership and transfer of water and land in Islam. In N. I. Faruqui, A. K. Biswas, & M. J. Bino (Eds.), *Water management in Islam* (pp. 78–84). Tokyo: United Nations University Press and Ottawa, International Research Centre.

Farooq, S., & Ansari, Z.I. (1983). Wastewater reuse in Muslim Countries: An Islamic perspective: *Environmental Management, 7* (2), 119–123.

Faruqui, N. I. (2001a). Introduction. In N. I. Faruqui, A. K. Biswas, & M. J. Bino (Eds.), *Water management in Islam* (pp. 8–12). Tokyo: United Nations University Press and Ottawa, International Research Centre.

Faruqui, N. I. (2001b). Islam and water management: Overview and principles. In N. I. Faruqui, A. K. Biswas, & M. J. Bino (Eds.), *Water management in Islam* (pp. 15–37). Tokyo: United Nations University Press and Ottawa, International Research Centre.

Faruqui, N. I., Biswas, A. K., & Bino, M. J. (Eds.). (2001). *Water management in Islam*. Tokyo: United Nations University Press and Ottawa, International Research Centre Water Management in Islam.

Fisher-Ogden, D., & Saxer, S. R. (2006). World religions and clean water laws. *Duke Environment Law & Policy Forum, 17*, 63–117.

Fossey, K. (2003). Buddhism. In M. Palmer & V. Finlay (Eds.), *Faith in conservation: New approaches to religions and the environment* (pp. 77–82). Washington: The World Bank.

Heathcote, R. L. (1983). The arid lands: Their use and abuse. London: Longman.

Johnson, W. T. (2000). The Bible on conservation: *A 21st century prescription: Electronic Green Journal, 1*(12), retrieved from http://scholarship.org/uc/item/2z33g35r.

Kadouri, M. T., Djebbar, Y., & Nehdi, M. (2001). Water rights and water trade: an Islamic perspective. In N. I. Faruqui, A. K. Biswas, & M. J. Bino (Eds.), *Water management in Islam* (pp. 72–78). Tokyo: United Nations University Press and Ottawa, International Research Centre.

Kearns, K. L. (1996). Saving creation: Christian environmentalism in the United States: *Sociology of Religion, 57*(1), 55–70.

Kearns, K. L. (1997). Noah's Ark goes to Washington: *A profile of evangelical environmentalism: Social Compass, 44*(3), 349–366.

Maltby, P. (2008). Fundamentalist dominion, postmodern ecology: *Ethics & Environment, 13*(2), 119–141.
Minter, B. A., & Manning, R. E. (2005). An appraisal of the critique of anthropocentrism and three lesser known themes in Lynn White's *The historical roots of our ecologic crisis, 18*(2), *Organization & Environment*, 163–176.
Naff, T., & Dellapenna, J. (2002), Can there be confluence? A comparative consideration of western and Islamic fresh water law. *Water Policy, 4*, 465–489.
Palmer, M., & Finlay, V. (2003). *Faith in conservation: New approaches to religions and the environment*. Washington: The World Bank.
Pereira, L. S., Cordery, I., & Iaconides, I. (2002). Coping with water scarcity: UNESCO, International Hydrological Programme, IHP-VI, *Technical Documents in Hydrology No. 58.*
Rao, S. (2003). Hinduism. In M. Palmer & V. Finlay (Eds.), *Faith in conservation: New approaches to religions and the environment* (pp. 91–96). Washington: The World Bank.
Sadr, K. (2001). Water markets and pricing in Iran. In N. I. Faruqui, A. K. Biswas, & M. J. Bino (Eds.), *Water management in Islam* (pp. 84–92). Tokyo: United Nations University Press and Ottawa, International Research Centre.
Shah, S. M. S., Baig, M. A., Khan, A. A., & Gabriel, H. F. (2001). Water conservation through community institutions in Pakistan: mosques and religious schools. In N. I. Faruqui, A. K. Biswas, & M. J. Bino (Eds.), *Water management in Islam* (pp. 57–62). Tokyo: United Nations University Press and Ottawa, International Research Centre.
Van Dyke, F. (2006). Cultural transformation and conservation: Growth, influence, and challenges for the Judeo-Christian Stewardship Environmental Ethic: *Perspectives on Science and Christian Faith, 58*(1), 48–63.
Water Authority of Jordan. (2006). Religious Grounds, Water Authority of Jordan, Environment & Water Reuse Unit http://www.waj.gov.jo/English/reuse/religious_grounds.htm. (Accessed 7/27/2009).
WCC, EPC, & VFCES (Ecumenical patriarchate of constantinople, world council of churches, Vatican Franciscan center of environmental studies). (2003). Christianity. In M. Palmer & V. Finlay (Eds.), *Faith in Conservation: New approaches to religions and the environment* (pp. 83–86). Washington: The World Bank.
West, C. A. (2007). For body, soul, or wealth: The distinction, evolution, and policy implications of a water ethic. *Stanford Environmental Law Journal, 26*, 201–232.
White, L. Jr. (1967). The historical roots of our ecological crisis. *Science, 155*(3767), 1203–1207.
Young, M. E. (2002). Institutional development for sustainable groundwater management—an Arabian perspective, In Hiscock, K. M., Rivett, M. O., & Davison, R. M., (Eds.), Sustainable groundwater development. London: Geological Society of London Special Publications, (vol. 193, pp. 63–74).

Chapter 34
Economics

34.1 Introduction

Water is a critical element of global economic activity and its supply and distribution involves enormous ongoing capital investment and operations and maintenance costs. The use of water is also subject to market forces. Water has a value, which varies depending on its scarcity, its production, treatment, and transportation costs, and the value of crops and other goods produced through its use. Water also has aesthetic, recreational, and environmental values. The demand for water, beyond that needed for basic human survival, is elastic in that cost influences the amount used (Fig. 34.1). There is some correlation between the rate of water use and the price paid by users (if any). Any meaningful discussion of water management, therefore, cannot ignore economic issues associated with the development and use of water resources.

Water has a dual identity as both a social and economic good. The basic characteristics of water preclude a true free market in water (Sect. 34.6). Allocation of water can be based on either economic efficiency or equity (i.e., fairness of allocation across economically disparate groups) (La Moigne et al. 1997). Although economic efficiency is clearly important, few would argue that maximization of wealth should be the sole objective of nations. Actions that increase societal economic efficiency with respect to water are not necessarily fair and could have the net result in making the poor poorer and the rich richer (Khouzam 2003). Reallocation of water can increase overall societal welfare, but there still may be losers in the process. In practice, countries usually strive to achieve some balance between economic efficiency and equity. The latter is necessary for social stability. The challenge is achieving social goals with respect to water, while also maximizing the economic benefits from the use of water.

Sandra Postel made two important observations concerning global water supply,

R. Maliva and T. Missimer, *Arid Lands Water Evaluation and Management*, Environmental Science and Engineering, DOI: 10.1007/978-3-642-29104-3_34,

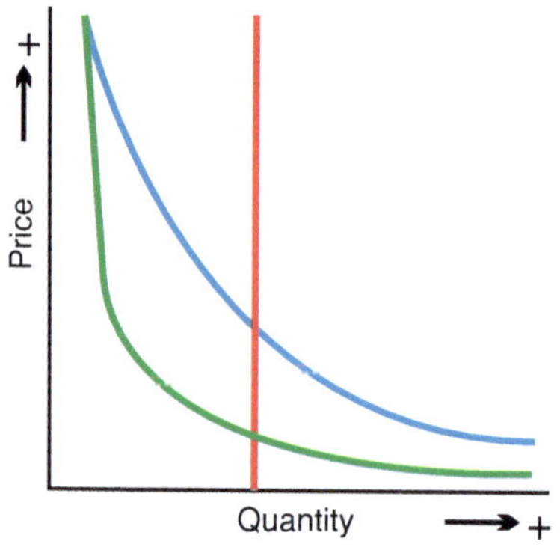

Fig. 34.1 Example of an elastic demand curve (*blue*) in which the quantity of an item consumed decreases with increasing price. Where demand is perfectly inelastic (*red curve*) the quantity of an item consumed is independent of price. On a household level, the demand curve for water tends to resemble the green curve in that the demand for water to meet basic potable needs is inelastic, but more elastic after basic needs are met

> water is consistently undervalued, and as a result is chronically overused (Postel 1992, p. 156).

> many of the water shortages cropping up around the world stem from the widespread failure to value water at anything close to its true worth (Postel 1992, p. 166).

An axiom often attributed to Benjamin Franklin notes than "when the well's run dry, we know the true value of water." In fact, assigning a value to water is far from a straightforward issue. In a free market system, the value of water is related to the willingness to pay for agricultural, domestic, and industrial users. However, the amount of water that users are willing to pay for depends upon the value of the use (e.g., high-value versus low-value crops), how the water is used (e.g., primary supply versus drought protection), and local and user-specific circumstances (Young 1996; National Research Council 1997; Winpenny et al. 2010). Similarly, the value of water can be evaluated by the willingness to accept compensation for giving it up. Willingness to pay or accept compensation is a monetary measure of the intensity of individual preferences (Young 1996).

When markets are absent or do not operate efficiently, economic evaluation of resource allocation decisions requires some means of estimating resource value be found (Young 1996). The value of water used for irrigation (and other purposes) can be quantified in terms of the marginal productivity of water, which the extra value of output that can be obtained from additional applications of water (Young 1996; Winpenny et al. 2010). With respect to irrigation, the value of water is the change in income with and without an irrigation project. This type of residual method is most accurate where water constitutes a significant fraction of the value of the output (Young 1996).

The economic value of water is not a fixed, inherent attribute of a good or service, but rather depends upon time, circumstances, and individual preferences (National Research Council 1997). Reliability is an important consideration.

Domestic, industrial, and agricultural water users all place a higher value on reliable water supplies than on supplies with high risk availability (Young 1996). Even more challenging is assigning a value for the environmental services performed by water. Methods for evaluating the value of water were reviewed in detail by Gibbons (1986); Young (1996); National Research Council (1997); and Winpenny et al. (2010).

A fundamental water management issue is the allocation of water between different sectors in society (e.g., household, agricultural, and industrial). Efficient allocation of water between sectors to gain high returns and high levels of employment is fundamental to economic and political stability (Allan 1998b, 2001). Allocations (and how they are determined) always reflect political considerations together with economic realities (Lithwick 2000). Allocations based on economic considerations tend to promote efficiency in both the production and consumption of water, whereas other modes of allocation do not (Lithwick 2000). However, non-economic, sociopolitical considerations are still important.

Where water is scarce, it needs to be used efficiently in order to obtain the maximum benefits. Water use efficiency includes two basic approaches or components (Allan 2001):

- technical or productive efficiency, which is concerned with obtaining increased returns for water use for a given activity, and
- economic or allocative efficiency, which is concerned with allocating water resources to activities that provide the greatest economic value.

A critical issue is that increases in technical efficiency achieve very modest returns compared to economic efficiency (Allan 2001). Water is efficiently allocated when the marginal value of water (i.e., incremental increase in value that could be obtained by using a little more water) is he same across all classes of water users of water (National Research Council 1997). However, re-allocation is a political act and creates disadvantages to some and benefits to others.

An individual needs approximately 1,000 times more water to raise the food he or she consumes than the volume of water used for drinking. Food production requires most of a community's water. Water for food production is, therefore, the strategic water use in any economy (Allan 1998a, b, 2001). In many arid countries, agriculture constitutes a very large percentage of total water use, but generates a small fraction of the total income of the country (e.g., Allan 1998a, b; Lithwick 2000; Wheida and Verhoeven 2007). The economic value of industrial water uses greatly exceeds the value obtained from the same volume of water used in agriculture. Since agriculture is the major user of water and has a low economic efficiency, it is an obvious target for reallocation to municipal and industrial users. However, increasing water and food demands go hand in hand with population growth, and decreasing irrigation could result in decreasing domestic food production, unless there is a concomitant increase in water use efficiency.

Reallocation of water from agriculture may have great sociopolitical implications. Societies find it difficult to admit that agriculturally marginal regions may not be worth farming (Toope et al. 2003). If a country is going to move out of agriculture, it is going to have to move into something else. A country must therefore determine what products or services (if any) it can provide and have a competitive and comparative advantage (Brooks 2000). The social and political costs of greater unemployment of agricultural workers who cannot find jobs elsewhere in the country may outweigh economic benefits from the reallocation of water. Economic analyses concerning water allocation, thus, need to consider less tangible and difficult price issues, such as social, national security, environmental preservation, recreation, and aesthetic values.

Although it is recognized that socioeconomic objectives should be considered in economic analyses of water use, objectivity is important in estimating the value of socioeconomic goals (Rogers et al. 1998). For example, rural irrigated agriculture has an economic value in alleviating poverty, but that value needs to be objectively quantified when determining the optimal allocation of scarce water resources.

Underpricing of water has serious consequences for the expansion, operation, and maintenance of water utilities. Construction and the operation and maintenance of new water infrastructure are major expenses, and it needs to be recognized that their benefits must justify the associated direct and indirect costs. The National Research Council (1995, p. 64) observed that

> the question of whether consumers are willing to finance a new water supply system is an example of linking cost to demand and is, in fact, the central question guiding policy toward infrastructure improvements in developing countries today: Is the cost justified by demand?

Srinivason et al. (2010) noted that in developing countries, water systems reach a "low-level equilibrium" as water users invest in coping mechanisms. Water utilities may find themselves in a scenario where income from water sales is inadequate to maintain and expand the supply and distribution systems, but the necessary increase in rates (tariffs) for full cost recovery is very difficult or causes default on debt necessitating governmental subsidies. Some supply options, such as desalination, may have a cost than exceeds the economic benefits to the customers. Increased tariffs for piped water become particularly difficult after water users have invested in wells and other coping mechanisms. Once consumers have invested in private wells, they have access to a cheap backstop source of water, and their willingness to pay for expensive piped water system improvements will likely be lower (Srinivason et al. 2010). However, as incomes increase, consumers may choose the higher-cost and higher-quality option of a piped water supply (Srinivason et al. 2010).

Another economic obstruction to the creation of a reliable and high quality centralized water system to cities that are lacking them is indirect impacts. For example, some large developing countries contain major population centers not served by centralized water systems. Water is provided by a private cottage industry of water suppliers that truck water to individual homes, apartments, and businesses. In certain cases, this private water supply industry and the ancillary support businesses (e.g., truckers, tank suppliers, well owners outside of the city, and water treatment specialist) constitute a substantial economic base providing jobs and income to large numbers of people. The centralization of supply is very expensive to the end users especially in major metropolitan areas where disruption of streets and other underground infrastructure occurs during construction. The secondary economic impacts include displacement of the private water supply industry with downstream impacts to individual families and reduction of overall community income.

34.2 Water as an Economic Good

The basic premise behind the recognition of water as an economic good or resource with a monetary value is that the price people are willing to pay for it serves as a guide to its rational management (Fisher 1995). Market forces will allocate water to uses that bring the highest economic returns and create user groups capable and willing to pay a greater price. Treating water as an economic resource means less waste, confining its use to where it is really valuable, and preferring reallocation to new supply schemes where these are costly in economic or environmental terms (Winpenny 1994). Winpenny (1994, p. 112) concluded that

> Promoting water markets and raising tariff decentralises the task of matching demand with supply, and mobilises every party behind solving the problem. Nothing less will suffice.

The concept that water is an economic good in the same sense as other commodities is controversial, particularly amongst farmers and agriculture-dependent economies where access to free or heavily subsidized water is taken for granted (Allan 2001). Part of the resistance is due to fact that many farmers could not survive if they had to pay a fair market price for their irrigation water supply. Artificially low prices send the signal that water has little value. Indeed, inefficient water use is simply the user's rational response to low prices. Users can afford to be wasteful only when water is cheap (Anderson and Snyder 1997).

The issue of treating water as an economic resource in which users should pay the full cost extends into domestic supply. The National Research Council (1995, p. 67) noted that:

> Many people believe that since water is essential to life, authorities are obliged to bring it to the population at little or no cost.

That water should not be sold or not even have an economic value in some instances may have religious, cultural, or traditional bases. Under Islamic law, shari'ah, flowing water and water in channels cannot be sold (Sect. 33.2). Tony Allan noted that

> If nature does not deliver water free to users then users tend to assume that other agencies, such as governments, should emulate nature. However, this inclination to consider all water to be similar to rainfall, that is free, has very powerful consequences with respect to the perception of the value of the water compared with the real costs of engineering the water to points where it can be used. (Allan 2001, p. 113).

Water was recognized as an economic good in one of the four Dublin principles (formulated in the 1992 International Conference on Water and the Environment) and is an integral component of Integrated Water Resource Management (IWRM) (Global Water Partnership 2000). In order to extract the maximum benefits from water, it is recognized in IWRM that it must be perceived and treated as a resource that has value rather than being viewed as a 'free good'. Clean, safe water cannot be produced and delivered without expense and someone has to pay for it (Segerfeldt 2005). Pereira et al. (2002) noted that resolutions to provide running water to the majority of households world-wide have failed, primarily because there has been no firm political will to charge realistic prices for the supply of water. The problem of inferior, inadequate water supply systems in most areas of the world is not one of water scarcity, but one of political determination to charge a sufficient price to cover the real total cost of the supply system (Pereira et al. 2002). Recognizing that water has an economic value does not equate to its being treated as a pure economic commodity, which risks a shift of public perception away from a sense of water as a common good, and from a shared duty and responsibility (Rahaman and Varsi 2005).

There are two schools of thought concerning the specific meaning of the concept that water should be treated as economic good (Van der Zaag and Savenije 2006; Kresic 2009). One school is a pure market approach in that water should be priced through free market with its economic value rising spontaneously from the actions of willing buyers and sellers. The second school of thought is that water use decisions should be based on integrated analyses of all the costs and benefits of alternative options, which does not necessarily involve actual direct financial transactions. Treating water as an economic good is considered to be about recovering the costs of providing water and making informed choices about the use, conservation, and allocation of water (Van der Zaag and Savenije 2006). Pricing should give the signal that water is in indeed a scarce good that should not be wasted (Van der Zaag and Savenije 2006).

Lest the idea that the resistance to water as an economic good be considered a phenomenon of developing countries, irrigation water in the Western United States has historically been heavily subsidized. Agricultural users and their representatives, who otherwise strongly believe in free market economic principles, pushed forward many water supply projects that provided heavily subsidized water. False economic analyses were widely used to give the perception that projects made economic sense, when in fact they could never be economically justified because the farmers could never afford the true cost of the delivered water (Reisner 1999).

The economic cost of water includes three main components (Allan 2001):

- Delivery cost (e.g., costs to operate surface water irrigation systems and to pump groundwater).
- Economic cost, which is the value of water itself as an economic asset that could be put to varied uses. It includes the opportunity cost (scarcity rent) of water use elsewhere in the economy and the value added by the use of water.

- Intangible value as an environmental resource, which can also have tangible values involving fisheries and other water-related resources.

Opportunity costs are the costs borne by society resulting from depriving other sectors of the use of the water. Opportunity costs with respect to water use can be quantified as the difference in the net economic value of the goods that could be produced by different uses of a given volume of water. There is an opportunity cost associated with the use of water for agriculture if reallocation of water could have allowed for greater economic returns elsewhere in the economy. In practice, opportunity costs are difficult to accurately quantify, because water is not the sole or primary limiting factor and cost for many activities. For example, semiconductor manufacturing is a much more economically efficient use of water than agriculture, but whether or not a country has a viable semiconductor industry is invariably due to a deficiency of economic and technical resources, rather than water resources.

Rarely has the inherent economic cost or value of water been fully incorporated into management decisions. Instead, economic analyses of water use typically focus primarily, if not solely, on delivery costs. For example, economic analysis of wellfield projects have historically focused only on the capital costs to install the wells and associated infrastructure, and operational and maintenance costs, not the value of the water itself. It is only when groundwater is recognized and regulated as a finite resource, and is no longer "free for the taking" if one installs a deep enough well, is the value of water recognized.

The market for water can never really be free as is there is seldom, if ever, a truly competitive market, which requires a large number of buyers and sellers. Certain water uses are also often considered to have a social value in addition to the private value placed on it by its users (Fisher and Huber-Lee 2010). There are commonly no substitutes for water. Economic models have been developed that can allocate available water to maximize the total benefits from water, while also incorporating social and economic values (e.g., Fisher 1995, 2002). For example, subsidized water could still be provided to the poor as a social value. The absence of water trade denies potential buyers of water that could be used in a beneficial and profitable manner and also imposes an opportunity cost on the owner of the water rights that is equal to the amount of money that could have been earned through selling the water. Water trading on both an intra-national and international scale can result in win–win scenarios where both parties gain (Fisher 2002).

From an economic perspective, consumers of water should actually pay the marginal cost of water (i.e., the cost to obtain additional supplies) rather than the average cost (Winpenny 1994; Brook 1996). The optimum allocation system from a purely economic perspective is that which equates the marginal cost of supply with the margin benefits (shadow price) of the water in use (Winpenny 1994). The benefit from consuming the last unit of supply is equal to the cost of providing it.

34.3 Self-Sufficiency and Food Security

A conflict exists to varying degrees in many countries between the use of water for irrigation versus non-agricultural uses, particularly industry and urban domestic uses. The water management challenge requires achieving a balance between (1) economic efficiency, (2) food security concerns, and (3) sociocultural values. There is one school of thought that water allocation decisions should be based entirely on economic considerations. Adam Smith (1776) in his influential book "An Inquiry into the Nature and Causes of the Wealth of Nations" (Book IV Chapter II) noted that

> It is the maxim of every prudent master of a family, never to attempt to make at home what it will cost him more to make than to buy…What is prudence in the conduct of every private family, can scarce be folly in that of a great kingdom. If a foreign country can supply us with a commodity cheaper than we ourselves can make it, better buy it of them with some part of the produce of our own industry, employed in a way in which we have some advantage.

The bulk of water use in arid regions is still for irrigation. It has been recognized that the use of water for irrigation is relatively inefficient in an economic sense. For every agricultural worker that is supported by irrigation, sixty to seventy industrial workers could be supported by use of the same water (Heathcote 1983). Hence, the case is made that in water scarce regions, the economically correct decision is to reallocate water from agriculture to other sectors that have a greater economic productivity and instead, purchase food from countries with more abundant water resources.

Putting opportunity costs aside, arid region agricultural projects may provide no pure economic value. Fadlelmawla (2009) noted with respect to farm projects in Kuwait, that production far from satisfies local needs for a few kinds of vegetables, let along providing any degree of food security. The farms also depend on expatriate manpower. Agriculture additionally costs the government substantially more through direct and indirect subsidies and contributes to the depletion of the countries water resources.

However, food security is a sensitive political issue that may supersede economic considerations. People have a deep intuitive aversion to being dependent on other economies for their water and food (Allan 2003a). Allan (2003a, p. 8) observed that

> awareness of a dependence on water and staple food coming from outside their own sovereign territories can be very destabilizing.

Food supply represents a fundamental national security issue. It has been known from ancient sieges that it is possible to starve an opponent into submission. Food can be a more potent weapon than guns and bombs to achieve political objectives. It was suggested by some lawmakers in the United States in the 1970s, that the United States' "grain weapon" could be just as effective as the Arab "oil weapon." Although the use of such trade weapons in a globalized economy would ultimately be self-defeating, the possibility of such ill-advised political actions can never be excluded.

The withholding of food could also occur under non-aggressive situations. For example, if global food production were to undergo a sudden decline due to unexpected climatic conditions, there would be intense political pressure for normally food-exporting countries to stop or reduce exports so as to meet the internal demands. What would a nation dependent on food imports do under such circumstances? Food is a high bulk, perishable commodity, which precludes significant long-term stock-piling particularly for populations in the millions to hundreds of millions. However, in an increasing number of nations in arid and semiarid regions, it is understood that self-sufficiency of the food supply is or will be not achievable, especially on a long-term basis, because irrigation water demands exceed available freshwater resources. Although it is true that total food self-sufficiency is not practicably possibly in some countries and regions, this does not necessarily negate the value of some local food production.

34.4 Virtual Water

The term virtual water is a metaphor that was created by Professor Tony Allan of King's College London and the School of Oriental and African Studies in 1993. It refers to the amount of water use involved in the growing and manufacture of products that are traded. Virtual water has also been referred to as the embedded water, and as Professor Allan (2003a) noted, the basic concept had been previously discussed by others. The term "virtual" refers to the fact the most of the water used to produce a product is not actually present in the product.

A critical implication for the virtual water concept is that water used, for example, to grow a crop (or used in the production of other traded items), is in essence exported by the producer and imported by the consumer. For example, the production of one metric tonne (1,000 kg) of grain requires roughly 1,000 tonnes (cubic meters) of water (Allan 1998a, b, 2003a). The trade of a tonne of grain thus involves the trade of 1,000 tonnes of virtual water. Food imports, particularly of cereal grains, are important indicators of water deficits (Allan 2001). Different foods and products have different virtual water contents. For example, it requires about 16 times as much water to produce a kilogram of meat as to produce a kilogram of wheat (Allan 1998a, b). Valuing virtual water is a problem as the value of water varies with circumstances. In some humid regions, water in the soil profile is essentially free (Allan 2003a).

From a national or regional perspective, water deficits can be addressed through the importation of virtual water. For example, the Middle East/North Africa (MENA) region is extremely water deficient in the sense that existing water resources are inadequate to meet the needs of rapidly growing and urbanizing populations (Allan 1998b). Total water self-sufficiency requires that a nation have sufficient local water to provide drinking water, water for industry and services, and water for food and other essential agricultural production (Allan 2003a). The MENA region ran out of water in the 1970s in the sense that local water resources became inadequate to locally produce the food supply of the region (Allan 1998b, 2001).

Important of virtual water in the form of food allows for global economic processes to ameliorate serious local water shortages (Allan 1998a, b, 2001, 2003a). As a high volume and low unit value commodity, it is by far a less expensive option to import 1,000 tonnes of virtual water in the form of a tonne of grain, than to import or produce by desalination 1,000 tonnes of water to locally grow 1 tonne of grain. Water embedded in food commodities can be mobilized very quickly to address periodic water shortages (Allan 2003a).

The ideal situation is for regions with excess water supply to be virtual water exporters and regions with water deficits to be virtual water importers. Water-scare countries might wish to import products that require a lot of water in their production and export products or services that require less water (Hoekstra and Hung 2002). However, the opposite often occurs where water intensive crops, such as fruits and vegetables and fiber crops, are grown in arid and semi-arid lands. Global trade in virtual water lies at the heart of some of most intractable hydrologic crises on the planet (Pearce 2006).

Scare water resources are used to grow crops for export, which results in the export of virtual water from water scare regions, further depleting the local water resources (e.g., Saudi Arabian wheat exports of the past).

The reliance of virtual water importation is a solution to local water scarcity, but it has national and economic security implications in the same manner as many industrialized countries have become highly dependent on imported energy. Not all factors involved in the evaluation of virtual water strategy are easily quantifiable, such as food security, environmental externalities of large water infrastructure, and political stability (Yang and Zehnder 2007). Food security through trade also leaves countries increasingly vulnerable to changes in agricultural commodity price variations.

Although there appears to be sufficient 'virtual' water to meet regional demands, Allan (2001a, p. 56) cautioned with respect to the MENA region that.

> it behooves the governments of the MENA region to evaluate the position with as much vigilance as those who manage another vulnerable economy, that of Japan, study global energy resources and the future availability of key strategic minerals.

Economic diversity and strength will enable even very seriously water deficient economies to adopt conflict-reducing environmental and economic policy solutions (Allan 2001a). A corollary is that virtual water trade needs the backing of amiable political relationships and that hostilities impede the development of virtual water trade (Yang and Zehnder 2007). With respect to virtual water trade in Egypt, El-Sadek (2010) reported that local scholars did not favor reliance on virtual water because of the perception that food-exporting countries are mostly western countries. Since the relationship between Egypt and western countries is dominated by skepticism and fear of domination, the dependence on virtual water importation could lead to political instability and domination by western countries. El-Sadel (2010) also noted that before adopting the virtual water policy option, Egypt needs to be assured that it can have fair and secure trade with water abundant nations. Such attitudes are understandable and certainly not unique to Egypt; there is general apprehension amongst non-western countries about becoming dependent upon western countries for something as basic as their food supply.

The World Water Council (2004) noted that

> At the global level, virtual water trade has geopolitical implications: it induces dependencies between countries; it is influenced by and has implications on the world food prices as well as on the global trade negotiations and agreements on tariffs and trade. Indeed the issue of virtual water is related to that of globalisation, which raises a concern among the general public. This can be understood from the fact that increasing global trade implies increased interdependence of nations. This can be regarded either as a stimulant for co-operation or as a reason for potential conflict.

Hoekstra and Hung (2002, 2005) quantified the volume of virtual water trade flow between nations in the form of crops in the period 1995–1999 and put the virtual water trade balances of nations in the context of national water needs and availability. The basic approach is to multiple crop trade flows (ton/yr) by associated virtual water content (m^3/yr, gal/yr). Countries with the largest virtual water exports were the United States, Canada, Thailand, Argentina, and India. The largest virtual importers were Sri Lanka, Japan, the Netherlands, Republic of Korea, and China.

The Hoekstra and Hung (2002, 2005) procedure starts with the determination of the crop specific water demand (SWD[n,c]) for crop "c" in location "n" in m^3/ton

$$SWD[n,c] = \frac{CWR[n,c]}{CY[n,c]} \tag{34.1}$$

where,

CWR[n,c] = crop water requirements (m^3/ha, gal/acre)
CY[n,c] = crop yield (ton/ha, tons/acre)

The crop water requirement is the cumulative crop evapotranspiration over the growing season, which Hoekstra and Hung (2002) estimated using the standard procedure of multiplying reference crop ET rate by a crop coefficient.

The virtual water trade (VWT; m^3/yr, gal/yr) flow associated with a crop is calculated by multiplying the crop trade (CT; ton/yr, tons/yr) by the specific water demand.

$$\mathrm{VWT}[\mathrm{n,c}] = \mathrm{CT}[\mathrm{n,c}] * \mathrm{SWD}[\mathrm{n,c}] \tag{34.2}$$

Using this approach the net virtual water import (NVWI) flow for a nation can be calculated as the sum of the virtual water trade flows of all imported crops minus the flows for all exported crops. A negative total indicates that there is a net export of virtual water from the country. The water footprint of country is the sum of its total domestic water use (WU, m^3/yr) and NVWI.

The water dependency index (WD) is a measure of a country reliance on virtual imports to meets its water needs

$$WD = \frac{NVWI}{WU + NVWI} * 100 \quad (34.3)$$

WD is equal to 0 if a country is a net exporter of virtual water. The water self-sufficiency index (WSS) of a nation is a measure of the degree to which a country meets its water needs through domestic sources. The WS and WD indices are related as follows:

$$\text{WSS} = 100 - \text{WD} \quad (34.4)$$

The Hoekstra and Hung (2002) comparison of water dependency versus water scarcity shows that some countries departed from the expected correlation between degree of water dependency and water scarcity, as defined by the ratio of water use to water availability (internal renewable water resources). Some countries have a low water scarcity and high virtual water dependency (e.g., Norway and Switzerland) while other countries have relatively scarce water resources but have a high degree of water self-sufficiency (e.g., Qatar and Egypt).

Yang et al. (2006) examined the water use efficiency embodied in the virtual water trade. A key point raised by Yang et al. (2006) is that globally, the total volume of food imported is approximately equal to the global volume of food exported. However, this equilibrium does not apply to the global virtual water trade because the water used for producing a given amount of food may differ between the importing and exporting sides of the trade. Globally, a water savings results when virtual water is exported by countries with water productivity that is higher than the importing countries. For example, if a country in which 100 m^3 of water is needed to produce a given volume of wheat, instead imports that wheat from a country in which only 70 m^3 is used, then there is a net savings of 30 m^3 of water. The global trade in wheat and maize has resulted in a 41 and 59%, respectively, reduction in global water use (Yang et al. 2006). The importation of virtual water allows the importing country to save domestic water for higher value uses.

The type of water used for irrigation and its opportunity costs also need to be considered when evaluating virtual water trade (Yang et al. 2006). Virtual water is often divided between ‘green water’ and ‘blue water’. Green water refers to the volume of rainwater that is lost to evapotranspiration during the growing of the crop, but is also used to refer to soil moisture (i.e., water source for rain-fed or dryland agriculture). Blue water refers to water obtained from surface water bodies and aquifers that is lost to evapotranspiration. The blue water is equal to the irrigation withdrawals minus any aquifer returns. The term ‘gray water’ is used to refer to the volume of water that becomes polluted during the production of a crop or product (Water Footprint Network 2009).

The opportunity costs of blue water are much greater than that of green water, as blue water can more readily be put to other uses. For example, groundwater can be put to a variety of uses, but there are very few alternatives for the use of soil moisture. Trading of green virtual water is thus more economically efficient than trading blue virtual water from an opportunity costs perspective (Yang et al. 2006; Yang and Zehnder 2007). For the major virtual water exporting countries (United

States, Canada, Australia, France, Argentina, Thailand, and Brazil), the exported water is overwhelming green.

Importing food effectively reduces the domestic demand for water. However, strictly speaking virtual water import typically does not generate a real savings in water in water scare countries in the sense that there as a reallocation of blue water from agricultural to non-agricultural sectors. Instead, virtual water import has allowed people to be better nourished (Yang and Zehnder 2007). More food becomes available than is possible to produce locally with existing water resources.

Virtual water trade can have some adverse impacts. Some net virtual water importers are not water scarce, and the importation of cheap food depresses local crop prices to the detriment of local farmers (Yang et al. 2006). There are also environmental and water resources impacts associated with virtual water exportation. Increased production may come at the expense of natural environments as more land is brought into production. The heavy use of fertilizers and pesticides to achieve high productivities has resulted in groundwater and surface water contamination (Yang et al. 2006). The lack of financial resources deprives poor countries of the opportunity to purchase food and participate in the global virtual water trade (Yang et al. 2006). The need to purchase food also drains limited financial resources that might be better put to economic development activities.

Allan (2003b) also noted that virtual water may be the second best solution rather than optimal solution. The optimal solution would involve measures to increase returns of water by technical and allocation techniques. However, under some of the global climate change scenarios, there may be no alternative to certain regions other than virtual water importation (Chaps. 38 and 39). As population grows and climate change occurs, a full global reevaluation of the economics of virtual water flows will be required to meet demands.

34.5 Water Ownership and Market-Driven Solutions

Who owns water and who has the right to use it is becoming an increasingly contentious issue as demand for water increases. Water management policies that were developed during times when water demands were modest relative to supplies may become obsolete under conditions of water scarcity. However, water management laws, regulations, and policies may develop deep cultural and social roots. Great inertia often exists against changes in long established water use laws and policies. The entities that currently possess rights to water have a very strong vested interest in maintaining the status quo. Clearly, the owner of a right to water, which may have been obtained through the purchase of land at market value, would strongly object to reductions or restrictions on the use of the water, which would diminish the value of the property. Alternatively, it is viewed as perverse by many that the rain that falls on the roof of the house that one owns may be legally belong to a downstream surface-water right holder, which is the current situation in some states in the Western United States.

Much has been written over the past two decades on the merits of free water markets versus government regulation of water, with the expressed opinions typically reflecting the philosophical beliefs of the owner rather than objective empirical data. Water market advocates propound that market systems promote efficient use of water through pricing mechanisms and results in water being allocated to where it results in the greatest net economic returns. Government intervention, it is argued, results in distortions in the market and inefficient water use. It has also been argued that governmental regulation of water results in its politicization. For example, Anderson and Snyder (1997, p. 21) noted that

> It will not do to assume that because they are in the public sector they are working toward the public interest. Votes, campaign coffers, influence and budgets are all factors that affect decisions in the political sector. Even where politicians want to do what is good for the constituents or for the country, they still face the problem of knowing what the people want.

Lack of ownership of water and other impediments to the transfer of water 'rights', discourage market-driven conservation and optimization of uses. It has been proposed by many that a freer market for water would increase the overall efficiency of water use. The increased efficiency would be both physical (hydrological) and economic. In the economic sphere, water may be transferred from low-value uses to high-value uses if individuals hold legal rights that can be sold to others (Anderson and Snyder 1997; Winpenny 1994; Revenga 2001; Johnson et al. 2001; Clayton 2009; Glennon 2009; Alder 2009). The right to a volume of water in a marginally productive agricultural area may be worth more than the value of agricultural operations. There is thus an opportunity cost associated with water use, in that more money might be obtained from selling the water. Postel (1993) noted that almost everywhere the value of water for crop production is less than its value for other uses. For example, in California, agribusinesses have been using 80% of the state's limited freshwater, while producing 3% of its economic output (Solomon 2010). The issue is not whether water is used efficiently in irrigation but whether irrigation is an efficient use of water (Brooks 1997).

Transfers of water rights thus increase the overall economic efficiency of water use. In southwestern United States, municipalities are purchasing water rights and agricultural land with associated water rights to secure supplies to meet growing urban and suburban needs. If the price of water is greater than the valued earned from irrigation, then a farmer could sell or lease the water right and convert his fields to non-irrigated cropland, adopt more efficient irrigation practices, or take the fields out of production altogether. The purchased farms are referred to as 'water ranches'. Winpenny (1994) observed that farmers who decide to sell their water instead of using it on low value crops are doing themselves, as well as society, a favor.

Water markets have also been argued to increase the flexibility of water systems, which will likely be an increasingly important issue in response to climate change (Adler 2009). Water pricing can be effective in reducing demands during drought periods. Market forces could also temporarily reallocate water to its greatest value uses during periods of increased water scarcity.

In addition to increasing economic efficiency, market-drive transfers of water could also increase the hydrologic efficiency of the use of water. As the economic value of water increases, a greater incentive will exist to be more efficient in the use of water and to avoid waste. Farmers may adjust to using less water by becoming more efficient. If a farmer can profit from the sale of water, he will be motivated to invest in conservation practices that will save water and make water available for other remunerative uses. When farmers sell or lease water rights, they may fallow the least productive fields on the farm, shift the crop mix, or change the irrigation system (Glennon 2009). Ownership of water also creates incentives for investments that improve the efficiency of water use and increases the availability of water through exploration and development activities (Donoso 2006). "Salvage legislation" would allow farmers who conserve water to acquire valuable transferable water rights (Adler 2009).

The principal objections to an entirely free water market system stem largely from the recognition that water is also a social good and that water trading can have significant third party effects. There is a common belief that water is a public good that is owned by all and should, therefore, not be subject to economic forces. Similarly, there is the concern that if water is treated purely as a commodity, then the wealthy could be able to monopolize the resource at the expense of the poor. Turning water into an economic commodity also results in a windfall to water rights holders who may have obtained the right for no or just a nominal cost. Additionally, treating water as an economic good could exclude non-market considerations, such as environmental protection and cultural issues. However, all of these concerns currently occur to some degree in jurisdictions in which the transfer of water is restricted.

The importance of historic water use to local culture should not be discounted. For example, Young (2002) noted that in Oman, rural village and aflaj (qanat) supplied oases lie at the heart of Omani culture and tradition. It is, therefore, essential socially and politically to support and maintain this heritage, which is a concern that is typically ignored in estimating the economic value of water.

The question is not whether market solutions used to manage scarce water resources is a good or bad idea, but rather that its results will depend upon how it is implemented. Solomon (2010, p. 380) observed that

> Whether the commodification of water ultimately leads to efficiency gains that ease water scarcity or results instead in an unregulated regime of water pricing and allocation that condemns the water poor to choose between desiccated, unhealthy lives and desperate remedies, depends on the terms by which societies choose to inject market forces into the traditional, public realm of water.

34.6 Water Markets

Once water is recognized to be an economic resource, the next step is to obtain the maximum or at least a net increase in the economic value of the use of water, which can be defined as the net benefits (total benefits minus total costs) obtained from given volume of water. Water is reallocated in some manner from low value uses to higher value uses. The reallocation can occur through market or administrative (command and control) mechanisms. When considering the benefits and drawbacks of water markets, it is important to go beyond philosophical beliefs, either for or against, and consider real-life issues. Water markets have the capability to increase the economic efficiency of water use, but it is clear that the properties of water preclude a true free market. In many aspects, water markets fail to meet the basic conditions necessary to yield accurate incentives and foster efficient resource use (Frederick 1995; Livingston 1995).

The strongest argument for water markets is that economic incentives can increase the efficiency in water use, and prompt the development of new sources of water and alternatives to water use. Morriss (2006, p. 1010) enthusiastically noted that:

> once these markets develop, we will unleash the enormously creative power of entrepreneurs on the problems of inadequate water, inefficient uses of water, and poor quality water.

It must be emphasized that a requisite for any free market is that there be numerous buyers and sellers. Buyers must also have choices. When there are usually no practical alternatives to water supply, then there is no choice, and hence a free market is not possible (Van der Zaag and Sevinije 2006). In the case of water, one also cannot have a free market if a disproportionate amount of the resource is the hands of one or several users.

In a free market, the costs and benefits of water trades must be internalized to the buyer and seller, which is usually not the case for water. Third party impacts are a significant concern (Sect. 34.8). Livingston (1995, p. 205) observed that

> without adequate institutional arrangements, efficient utilization is inhibited because all costs imposed by water users are not necessarily born by the users themselves. That is, private costs and benefits diverge from social costs and benefits, which yield serious distortions in allocation.

In practice, a mix of market and administrative mechanisms are necessary for the reallocation of water to obtain maximum benefits, and that the position on the spectrum between market and administrative control will depend upon country specific conditions (Livingston 1995). Local water use traditions and regulations need to be considered in the development of water marketing systems.

A precondition for water markets is the existence of enforceable property rights to the water (Winpenny 1994). For water markets to flourish, the efficiency gains from transferring water to higher value uses needs to be large enough to offset the transaction costs (Winpenny 1994). A water market also requires a means for the physical transfer of water to occur. The bulkiness and low value per volume of water often makes it uneconomical to transport it from water-rich to water-scare regions.

A distinction is usually made in the regulation of water between the actual ownership of water and the right to use water. Surface- and groundwater are often considered to be property of the state and, therefore, non-transferable to private parties. However, the right to use water may become private property that is tradable. Donoso (2006), based on Chilean experience, proposed that the main conditions to establish a water market system based on water rights are:

- relative water scarcity,
- secure water use rights,
- clearly defined water use rights,
- well regulated water use rights markets,
- adequate inventory of water resources, and
- efficient conflict resolution system.

Water scarcity is an obvious prerequisite for a water market. If new sources of water are still readily available, then there is usually little economic incentive to trade existing rights. A fundamental requirement for water markets (or any other market) is that water rights be legally established and protected. Potential buyers of water rights must have confidence that purchased water rights would provide them actual water. A basic model for water trading is a "cap-and-trade" mechanism whereby a government agency determines the maximum allowable yield of the resource (e.g., total annual groundwater withdrawal from a basin). Allocations (i.e., rights or permits) for the use of shares of the annual yield are considered private property and are tradable.

The initial allocation of water rights is often the most controversial part of water marketing schemes (Tietenberg 2002). Initial allocations of commons in general may be made be based on (Tietenberg 2002):

- random access (lotteries),
- first come, first serve (historic use, priority, grandfathering, prior appropriation),
- administrative rules based on eligibility criteria, and
- auctions.

Historic use or priority are the most commonly used methods for the allocation of water and are a logical bases for the establishment of tradable water rights, because users may have already invested in the development of infrastructure for the use of water and have become dependent on the resource. Anderson and Snyder (1997) argue forcefully that water rights established under the prior appropriation doctrine of the Western United States can serve as the foundation for water markets.

The issue of fairness and equity arises when initial tradable water rights are established. A right to use scarce water can be very valuable. Hence, granting of tradable water rights allows for some users to acquire a substantial amount of wealth for what may have been obtained for a very modest investment (i.e., just filing the required paperwork). In some cases, excessive water rights were obtained by landowners based on inflated estimates of their actual needs. Water rights may have been granted at no direct cost, other than administrative costs. Later users are forced to purchase water at market prices, which the sellers may have obtained essentially for free. However, even where tradable water rights are granted for free, they can still result in an increase in the economic efficiency of water use as they are later tradable to higher value uses (Tietenberg 2002, and references therein).

Effective water markets also require effective monitoring and enforcement against overuse and unpermitted use. Where water has a high economic value temptations exist for water users to cheat and use more water than they are entitled. Unauthorized use by parties that do not hold permits poses a threat to the both the resource and the water market. Water rights have little value if others who do not have a permit can abstract water with impunity. Thus, the paradox exists that in order to have a "free" market, considerable government intervention may still be necessary to protect water rights and to prevent hoarding. The impacts of water trading on third parties must also be considered (Sect. 34.8).

An additional factor that may impact water markets is climate change. The quantity of water, and therefore the value of water rights based on historic use and runoff patterns, becomes less certain as hydrologic conditions change (Frederick 1995).

34.7 Chilean Water Market Experience

The Chilean model of water rights and water resources management has become a paradigmatic example of the free-market approach to water law and economics (Bauer 2004). Water marketing in Chile and the Western United States were compared by Mentor (2001). The 1981 Water Law in Chile declared that water is public property to which the state can grant private right of use. Chilean water law recognizes various types of rights to use water including consumptive, non-consumptive, permanent, contingent, continuous, discontinuous, and alternative, and combinations of these rights (Donoso 2006).

Water rights are free, but an auction is used when there are competing claims on the same water. Water rights are separated from land ownership and once granted water rights are fully protected as private property rights. The goals of the 1981 Water Law were to increase investment in irrigation infrastructure and to increase the productive value of water uses by relying upon market forces to stimulate efficiency and to stimulate the development of high-value agricultural uses (Bauer 2004). In contrast to Western United States water law, a water rights holder in Chile does not have an obligation to use the right. Thus, the Chilean water laws in practice encourage hoarding of water, but do not encourage the use of water for its own sake merely to avoid relinquishment of the water right through nonuse (Mentor 2001). Water rights could be obtained by parties with no intention of using the water in the foreseeable future, if at all, which excludes the water from beneficial use by others. As noted by Bauer (2004, p. 55):

> by defining private water rights in a manner that includes no duties or obligations to public interest, the code favored private speculation, hoarding, and monopoly of water rights, and had undermined the incentives for using water rights in economically productive activities

While it was widely recognized that hoarding of water was economically inefficient, proposals to re-appropriate unused water rights have met with significant opposition because it was considered by some to be an unconstitutional infringement of property rights (Bauer 2004; Donoso 2006). Options considered include a tax on all water rights (offset by a corresponding decrease in the land tax) and a fee on unused water rights. The Chilean Congress accepted a proposal in 2005 to tax unused water rights to create an economic disincentive against hoarding (Donoso 2006).

Performance of the Chilean water code has been variable with a key conclusion that water rights transactions are uncommon in most parts of Chile, and that the great majority of transactions took place within the agricultural sector rather than involving inter-sector reallocation (Bauer 2004). Water transactions are more prevalent in parts of the country where water is scarce and thus has greater economic value (Bauer 2004; Donoso 2006). As noted by Bauer (2004), some observers have suggested that the lack of water rights transactions showed that the existing allocation of water resources was already efficient. Water rights owners rarely sell used or surplus water rights, but instead tend to hold onto them for protection against occasional drought years or in anticipation that they will increase in value over time (Bauer 2004).

Problems with the water law independent of market allocation include (Bauer 2004; Donoso 2006):

- unavoidable transaction costs related to modification of distribution infrastructure,
- externalities; negative effects on third parties, particularly the loss of water to downstream farmers dependent on spillover water, and the environment, and
- uncertainties over the availability of water, such as inter-year variability in discharge.

Problems related to market allocation of water include:

- lack of timely and adequate information on water rights,
- uncertainty over resource availability (e.g., are the purchased water rights fully available),
- unavoidable transaction costs (e.g., investigations, legal, administrative),
- speculation and hoarding; obtainment of water rights by parties who will not use the resources, which creates an inefficient use of water from a societal perspective, and
- absence of a clear and effective mechanism for resolving disputes.

Bauer (2004) argues that Chilean water law is not compatible with IWRM because it fails to address elements such as social equity, coordination of multiple uses, river basin management, resolution of conflicts, environmental protection, and non-consumptive water rights. Specifically, Bauer (2004, p. 133) observed that:

> Because the Chilean approach to managing water as an economic good puts all the emphasis on water as private good and tradable commodity, it is very difficult to recognize or enforce the other aspects of water as a public good.

A major challenge of the Chilean system is that by establishing water rights as private property and having strong constitutional and judicial protection of property rights, the water management system inherently has a low degree of flexibility.

34.8 Third Party Effects of Water Trading

Spillover or third party effects may result from water rights transfers, which need to be considered when developing water policy. Third parties include a range of economic, environmental, and social interests impacted by a transfer, but have a non-proprietary stake in the process. The Committee on Western Water Management of the United States National Research Council conducted a study of water transfer issues in the Western United States, focusing on third party effects (National Research Council 1992). The study is still germane to water transfers in general although the specific regulatory discussions are applicable only to the Western United States.

Third parties include (National Research Council 1992):

- other water rights holders and users (e.g., loss of return flows),
- agriculture (including businesses and farmers in area of water origin),
- the environment,
- urban interests,
- ethnic communities,
- non-agricultural rural communities, and
- taxpayers.

A key observation of the National Research Council (1992) is that the primary parties in a voluntary water transfer (i.e., the buyer and seller), negotiate their own best interests and exercise control over whether or not the transfer will occur. Public policies must be concerned with the interests of third parties who stand to be affected by the transfer, but are not represented in the negotiations.

Considerable concerns have been raised about the economic impacts of water transfers. Transfers typically result in the transfer of water from rural to urban communities (Henderson and Akers 2008). The discontinuation of agricultural activities results in a loss of farm jobs and farm incomes, less agricultural business activity, and less household spending in the community, which effects non-farm related businesses (Office of Technology Assessment 1983; National Research Council 1992; Henderson and Akers 2008; Clayton 2009; Glennon 2009). The local tax base is affected by a decrease in economic activity and the transfer of land to non-taxpaying municipal ownership.

The discontinuation of farming operations can thus have a ripple effect throughout the local community. The indirect financial contribution of agriculture to the local economy may be equal to or several times greater than the value of direct crop sales. Where the transfer of water occurs locally, the change in the local economic base may be neutral or net beneficial, but there may still be people left behind, such a farm workers left out as the economy shifts away from agriculture (Office of Technology Assessment 1983).

The effects of economic dislocations from the transfer of water may be a highly contentious issue, especially if the water is transferred out of a community. Adverse impacts may be experienced at the point of origin, whereas benefits are accrued at the transfer location. Water may be the basis of present and future economic vitality and environmental amenities (National Research Council 1992). Local communities can lose the economic benefit associated with the water in its present (use) and possible future uses.

There is also a scale factor associated with water transfers in the sense that transfers can have a negative impact on local communities, but result is a greater net benefit to the state or country as a whole (National Research Council 1992). The question of the importance of rural community preservation is a social, political and economic issue for which there is no universal answer. There is no consensus in the value of rural agricultural communities (National Research Council 1992). Changes in the use of water are necessary and desirable in a dynamic society (National Research Council 1992). Community preservation concerns need to be weighed against overall societal benefits. One option to mitigate impacts of water transfers on local communities is that mechanisms be emplaced to compensate communities for transfer-related losses of tax base and that third party effects should be internalized as a cost of the transfer (National Research Council 1992).

There is also the basic social desirability issue of loss of agricultural production needed to feed a growing global population. Although there is little question that water use for agriculture generates less wealth than the use of the same volume of water to manufacture semiconductors or other industrial uses, there is a societal benefit in having affordable food. A common theme envisioned for transfer of water is a regulated market with governmental responsibility to protect third parties from potential harm caused by water transfers and to ensure that the transfers do not harm the environment (Glennon 2009). The National Research Council (1992) noted that the quest is for a balance point that allows for broader participation in decision making while not inhibiting desirable transfers.

Clayton (2006) proposed the following six elements in market-driven water transfer system, particularly where inter-basin transfer is involved:

1. incorporation of public review,
2. simplification of water transfer protocols and legal systems to ease trading, reduce costs, and increase flexibility,
3. incorporating expiration dates for traded water,
4. the use of water banks, particularly for water trades across state lines,
5. requiring that a useable portion of the water remain associated with its previous use, and
6. including externalities (e.g., environmental protection, aesthetic and recreational values) in the pricing of water.

Inter-sector transfers of water can be performed through a free market system or may be implemented as an element of governmental water management policies. In this case national priorities are set concerning which sectors will be given priority. For example, within Israel, the policy is that domestic and urban uses have the first priority for water, followed by industrial uses and finally agriculture (Faruqui 2001b). Agricultural needs are increasingly being met by urban wastewater reuse.

Intersector transfers must also give consideration to cultural values. Preservation of rural agricultural communities may be given a higher priority based on cultural issues than can be justified based on their contribution to the gross national product. Small towns in agricultural land areas may be especially harmed by the transfer of water to urban areas. Some political systems may give disproportionate power to farming areas. For example, the Senate of the United States has two members from each state irrespective of their populations. Low population farm states have a disproportionate influence in the Senate relative to their population, than the more densely populated more urban states. Religious values and teachings may also be a consideration for intersector transfers of water. In theory the relocation of water from agriculture to domestic use should be easier to support in Muslim countries than in non-Muslim countries because of the primacy in Islam of water use for domestic purposes (i.e., the right to quench thirst and the right and requirement to be clean; Faruqui 2001b).

Brooks (1997) claimed with respect to the Middle East that the combination of modern market approaches with their emphasis on efficiency and traditional institutional approaches, with their emphasis on equity, will prove superior in almost every way to any of the mega-projects for water supply under consideration in the region.

34.9 Subsidies

Very few would argue that water should be treated entirely in a free market manner in which those who cannot afford to pay for water should be denied it. Some subsidization is necessary. It also recognized that subsidies and other types of government support must be geared to protecting those at the bottom of the income scale and that "*no serious industry observer or market advocate would argue with this assertion*" (Maxwell 2009, p. 28).

The subsidization of water for irrigation and urban use is ultimately based on political and social grounds. Subsidization occurs when users do not pay the full cost to construct and operate the water supply and irrigation systems. When they receive energy below cost, groundwater pumping is encouraged. In many areas, irrigation systems are built, maintained, and operated by public agencies, with irrigators charged next to nothing for these services (Postel 1993). This reduces incentives for conservation and increased water efficiency and allows water to be used for low-value crops, which would not be economically viable if the farmers had to pay the full cost of the water. For some crops, the value added by irrigation is less than the average cost to supply the water (Brooks 1996). However, Faurés et al. (2007) observed with respect to subsidized investments in irrigation, that the systems need to be reviewed within the context of rural development, rather than simply agricultural development. Increases in farmer's incomes fuel the rest of the rural economy.

A critical point is that subsidies are implemented when society (or at least the part of the society that decides on the issue) places a greater value on a particular use of water than its market value. Otherwise, there would be no need for a subsidy. The major problem is that when subsidies are used to artificially depress the price of water (or any other good), then that water will be consumed in quantities that are greater than economically efficient quantities. In elementary economics, a commodity that is supplied too cheaply will sooner or later need to be rationed by more or less arbitrary means. Those who are fortunate to be provided water will use it excessively and wastefully (Winpenny 1994). There is little economic incentive to conserve water. However, effective water pricing, particularly for irrigation water is a highly sensitive issue, particularly in low income countries, where agriculture still dominates the economy and most farmers have limited incomes. Maintaining social stability is given a greater value than the market value of water.

Stile (1996) noted that with respect to developing countries, investment in more efficient irrigation and improvement in irrigation water distribution are capital intensive and almost always require some degree of public investment regardless of the eventual beneficiary. In Africa and other developing areas, the true costs of irrigation development are rarely passed on to farmer, partly at least because so many irrigation projects are funded by donors. Farmers may not be able to pay the full cost of water because they may not be able to recover fair value for the crops due to price setting on food, which is implemented with the goal of protecting consumers. In this case, there may be both a subsidy for farmers in providing the water at below market value, and an indirect "tax" on farmers for keeping food prices artificially low, below market value.

Subsidization can be justified for domestic water supply. It is a generally accepted development goal that water to meet the basic needs of individuals and communities should be made available at prices consistent with the economic capacity of respective communities (Allan 2001). The initial development of the basic water infrastructure in much of Europe (and other developed areas) involved massive government subsidies (i.e., investments that were not paid for directly through water rates) (Rahaman and Varis 2005). It is absurd to believe that the construction or substantial improvement of a comparable water infrastructure can be implemented in developing countries without some form of subsidies.

An important point raised by the Global Water Partnership (2000) is that subsidies should be transparent and it should be recognized that all subsidies have to be paid by someone. Subsidies have very often been hidden from public view, because they would otherwise not be politically acceptable. For example, many water projects in the Western United States represent massive subsidies to farmers because the water provided by these government-financed projects has been sold to them a fraction of its true cost and funded via congressional funding through the U. S. Bureau of Reclamation (Reisner 1986). The true economics of these water projects was commonly obscured to make them politically acceptable.

34.10 Privatization of Public Water Treatment and Distribution Infrastructure

An important distinction is made between ownership of water and ownership and management of municipal water supply and distribution infrastructure. A question that is receiving more interest in recent years is whether water and wastewater utilities should be publically or privately owned and operated. Both private and public water utilities operate in the United States and, from the experience of the authors' there is little difference between the quality of services provided by both. Yet in many areas of the world, privatization of water is a highly controversial issue.

The Committee on Privatization of Water Services in the United States of the National Research Council (CPWSUS 2002, p. 57) observed that

> Private organizations do not necessarily operate efficiently, nor are they inherently more efficient than public organizations. By the same token, public organizations do not always deliver on their promises. The point is that neither public nor private organizations automatically entail "effectiveness." Well-run and poorly run organizations exist in both sectors.

There are many permutations involving private ownership and operation of water and wastewater utilities. Privatization of water utility services can take a wide variety of forms from providing specific services (e.g., billing or some maintenance activities) to outright ownership of the water supply and distribution infrastructure. There are four basic types of utility privatization (CPWSUS 2002):

- outsourcing of specific support services, such as laboratory services, billing, meter reading, some maintenance activities, and vehicle maintenance,
- outsourcing of the full operation and maintenance of the water utility,
- design, build, and operate contracts for new, expanded, or up-dated facilities, and
- outright sale of utility assets to a private company.

A common type of privatization is concessionary agreements in which a private company rents the infrastructure from the public sector and undertakes to maintain the system and finance any improvements in return for being allowed to collect payments from users (Segerfeldt 2005). The performance requirements of concessionary contracts may include requirements to supply water to the poor, maintain infrastructure, and meet water quality and environmental objectives. Concessionary contracts can also provide public oversight of water rates. Privatization does not imply that the private utility will have a free hand to set water rates. Private companies typically fall under the regulatory oversight of local public utility commissions or other government agencies, which have the authority to review and approve or reject proposed rate increases. This process can be contentious and lead to contractual disputes.

The arguments presented in support or against privatization of water utility services often reflect strong philosophical biases of authors with selected anecdotal evidence presented in support of positions. Maxwell (2009, p. 26) noted:

the participants in this debate usually talk past one another and fail to move the discussion forward; indeed, such forums often serve as little more than soapboxes for airing the same old arguments between the advocates and opponents of privatization.

Opponents of privatization commonly object to privatization as a matter of principle against corporations profiting from need of the people for water, which is considered to be a basic human right (Segerfeldt 2005). It has been suggested that private companies have little incentive to maintain infrastructure, protect the environment, provide service to the poor as well as the rich, and to monitor for water quality (Glennon 2009). The alternative thesis that has been proposed is that private water corporations are more likely than government bureaucracies to handle water with care. Profit motives are suggested to give strong incentives to conserve water, provide services to customers, and reduce waste (Segerfeldt 2005). Segerfeldt (2005) argued that the strongest benefit of private involvement is the profit motivation to reach as many users as possible. However, it is important to also recognize that it is often unprofitable to serve some areas because of low population densities or inabilities to pay. In a free market, the natural tendency is to harvest the 'low-hanging fruit' and decline to provide unprofitable services. It is possible that private contractors might achieve cost savings (i.e., economic efficiency) by cutting staff, not making necessary investments in operations and maintenance, and reducing necessary long-term investments (CPWSUS 2002).

Privatization is not the same as competition. Once a long-term contract is signed, competitive pressures to operate efficiently and meet public expectations are gone (CPWSUS 2002). Competition is limited to the period when competitive bids are being accepted.

The strongest argument in favor of privatization is that can provide needed resources in areas where public systems cannot provide adequate water supplies or cannot do so in an economical manner. There is no practical reason to consider privatization in areas where the public water system has the financial and technical resources, ability, and desire to cost-effectively provide safe water to the entire local population. Privatization is considered where public systems cannot provide what is considered to be adequate water service or does not have the financial resources to meet current and future system needs. The major advantage of privatization is that it can provide much needed access to capital, managerial competence, and technical expertise and experience that are not otherwise available (Segerfeldt 2005). Privatization or long-term Design-Build-Own-Operate (DBOO) or Design-Build-Own-Operate-Transfer (DBOT) contracts allow private companies to bring innovation into the operating world and can lower the cost of providing potable water to municipalities (e.g., large desalination facilities in the Middle East or North Africa). The ownership transfer option DBOT can be used both to obtain industrial innovations in design and as a financing mechanism to allow the expansion of utility systems where public funds are inadequate. Also, financial guarantees can be obtained from third parties to protect successful bidders from default losses, particularly in developing countries (e.g., World Bank, certain NGO's).

Privatization results in a reduction in local control, which can have both beneficial and negative impacts on the local communities. Private companies may be less susceptible to political pressures. The primary objective of private utilities is to provide a reasonable return for the owners or shareholders. Economic efficiency may be sacrificed in the public sector in order to achieve other community goals such a local utility jobs and supporting local suppliers (CPWSUS 2002). Privatization can also result in a loss of local capabilities and expertise should a decision be made in the future to transfer the entire utility or functions back to the public sector (CPWSUS 2002).

Privatization often results in an increase in the cost of water as private companies must recoup their costs and make a profit (unless innovation in treatment processes reduces cost). Public systems are commonly subsidized to some extent and revenues collected from water sales do not cover system capital, operations, and maintenance costs. In the absence of government subsidies, costs must rise to meet expenses. However, it is important to recognize that people without water main service are paying more for water today than they would if connected to a distribution network. The higher costs are direct costs from purchasing water from vendors, and indirect costs associated with the time and effort required to secure water and the health impacts of poor quality water (Kadouri et al. 2001; Segerfeldt 2005). A realistic water price would allow for reinvestment into water supply and distributions systems to serve the poor that have no service. However, a common thread in almost all developing countries is the urban middle and upper class areas are more likely to be served and also have more political clout than the poorer members of the society. Those who are fortunate to receive subsidized service are often unwilling to pay greater rates to cover the costs of expanding the systems into poorer areas.

The success of the regulating monopolies, such as private water utilities, depends upon the strength of political institutions responsible for their oversight. Privatization may be less successful in developing countries because ineffective or corrupt governments lack the capacity or will to adequately regulate the private sector (Glennon 2009). Privatization is not a panacea for financial woes and decayed infrastructure (Glennon 2009).

References

Adler, J. H. (2009). *Warming up to water markets: Regulation*. Washington, DC: Cato Institute.

Allan, J. A. (1998a). Virtual water: A strategic resource. Global solutions to regional deficits. *Ground Water, 36*, 545–546.

Allan, J. A. (1998b). "Virtual Water:" As essential element in stabilizing the political economies of the Middle East. In J. Albert, M. Berhardsson & R. Kerna (Eds.), *Transformation of Middle Eastern natural environments: Legacies and lessons* (pp. 141–149). Yale School of Forestry and Environmental Studies Bulletin 103.

Allan, J. A. (2001). *The middle east water question hydropolitics and the global economy*. London: I.B. Tauris.

Allan, J. A. (2003a). Virtual water—the water, food, and trade nexus. Useful concept or misleading metaphor? *Water International, 28*(1), 4–11.

Allan, J. A. (2003b). Integrated water resources management is more of a political than a technical challenge. In A. S. Alsharhan & W. W. Wood (Eds.), *Water resources perspectives: Evaluation, management and policy* (pp. 1–8). Amsterdam: Elsevier.

Anderson, T. L., & Snyder, P. (1997). *Water markets: Priming the invisible pump*. Washington, DC: Cato Institute.

Bauer, C. J. (2004). *Siren song: Chilean water law as a model for international reform*. Washington, DC: Resources of the Future Press.

Brooks, D. B. (1996). Between the Great Rivers: Water in the heart of the Middle East. In E. Racked, E. Rathgeber & D. B. Brooks (Eds.), *Water management in Africa and the Middle East* (pp. 44–57). Ottawa: International Development Research Centre.

Brooks, D. B. (1997). Between the Great Rivers: Water in the heart of the Middle East. *Water Resources Development, 13*, 291–309.

Brooks, D. B. (2000). Keynote address: Access to water in the Eastern Mediterranean. In D. B. Brooks & O. Mehmet (Eds.), *Water balances in the Eastern Mediterranean* (pp. 1–12). Ottawa: International Development Research Centre.

Clayton, J. A. (2009). Market-driven solutions to economic, environmental, and social issues related to water management in the western USA. *Water, 1*, 19–30.

CPWSUS (Committee of Privatization of Water Services in the United States, National Research Council). (2002). *Privatization of water services in the United States: An assessment of issues and experience*. Washington, DC: National Academy Press.

Donoso, G. (2006). Water markets: Case study of Chile's 1981 Water Code. *Ciencia e investigación Agraria, 33*, 157–171.

El-Sadek, A. (2010). Virtual water trade as a solution for water scarcity in Egypt. *Water Resources Management, 24*, 2437–2448.

Fadlelmawla, A. (2009). Towards sustainable water policy in Kuwait: Reforms of the current practices and the required investments, institutional and legislative measures. *Water Resources Management, 23*, 1969–1987.

Faruqui, N. I. (2001). Islam and water management: Overview and principles. In N. I. Faruqui, A. K. Biswas & M. J. Bino (Eds.), *Water management in Islam* (pp. 15–23). Ottawa: United Nations University Press, International Research Centre.

Faurès, J.-M., Svendsen, M., & Turral, H. (2007). *Reinventing irrigation, in water for food, water for life: A comprehensive assessment of water management in agriculture* (pp. 353–394). London: International Water Management Institute.

Fisher, F. M. (1995). The economics of water dispute resolution, project evaluation and management: An application to the Middle East. *Water Resources Development, 11*(4), 377–390.

Fisher, F. M. (2002). Water value, water management, and water conflict: A systematic approach. *GAIA, 11*(3), 187–190.

Fisher, F. M., & Huber-Lee, A. (2010). The value of water: Optimizing models for sustainable management, infrastructure planning, and conflict resolution: EuroMed 2010, Desalination for Clean Water and Energy Proceedings, Tel Aviv, October 3–7, 2010.

Frederick, K. D. (1995). Adapting to climate impacts on the supply and demand of water. *Climatic Change, 37*, 141–156.

Gibbons, D. C. (1986). *The Economic Value of Water*. Washington, DC: Resources for the Future.

Glennon, R. (2009). *Unquenchable. America's water crisis and what to do about it*. Washington, DC: Island Press.

Global Water Partnership. (2000) *Integrated water resources management*. Global Water Partnership TAC Background Paper No. 4.

Heathcote, R. L. (1983). *The arid lands: Their use and abuse*. London: Longman.

Henderson, J., & Akers, M. (2008). Can markets improve water allocation in rural America? *Federal Reserve Bank of Kansas Economic Review, Fourth Quarter, 2008*, 97–117.

Hoekstra, A. Y., & Hung, P. Q. (2002). *Virtual water trade. A quantification of virtual water flows between nations in relation to international crop trade*: Value of Water Research Report 11, IHE Delft.
Hoekstra, A. Y., & Hung, P. Q. (2005). Globalisation of water resources: International virtual water flows in relation to crop trade. *Global Environmental Change, 15*, 45–56.
Johnson, N., Revenga, C., & Echeverria, J. (2001). Managing water for people and nature. *Science, 292*, 1071–1072.
Kadouri, M. T., Djebbar, Y., & Nehdi, M. (2001) Water rights and water trade: An Islamic perspective. In N. I. Faruqui, A. K. Biswas & M. J. Bino (Eds.), Water management in Islam (pp. 72–78). United Nations University Press: Ottawa.
Khouzam, R. F. (2003). *Economic aspects of wastewater reuse study case: The Arab Nation*. Paper presented to the Economic Research Forum for Arab Countries, Turkey and Iran (ERF) 10th Annual Conference, 16–18 December 2003.
Kresic, N. (2009). *Groundwater resources: Sustainability, management, and restoration*. New York: McGraw-Hill.
Le Moigne, G., Dinar, A., & Giltner, S. (1997). *Principles and examples for the allocation of scarce water resources among economic sectors*: Options Méditerranéenesss, Sér. A., no. 31, pp. 87–102.
Lithwick, H. (2000). Evaluating water balance in Israel. In D. B. Brooks & O. Mehmet (Eds.), *Water balances in the Eastern Mediterranean* (pp. 29–58). Ottawa: International Development Research Centre.
Livingston, M. L. (1995). Designing water institutions: Market failures and institutional responses. *Water Resources Development, 9*, 203–220.
Maxwell, S. (2009). Moving beyond the "privatization wars"—a call for balance. *Journal American Water Works Association, 101*(12), 26–28.
Mentor, J., Jr. (2001). Trading water, trading places: Water marketing in Chile and the western United States. In *Groundwater and water resources management: The changing value of water*. AWRA/IWLRI: University of Dundee International Specialty Conference.
Morriss, A. P. (2006). Real people, real resources, and real choices: The case for market valuation of water. *Texas Tech Law Review, 38*, 973–1010.
National Research Council. (1992). *Water transfers in the west, efficiency, equity, and the environment*. Washington, DC: National Academy Press.
National Research Council. (1995). *Mexico City's water supply, improving the outlook for sustainability*. Washington, DC: National Academy Press.
National Research Council. (1997). *Valuing ground water, economic concepts and approaches* (p. 189). Washington, DC: National Academy Press.
Office of Technology Assessment. (1983). Institutions affecting western agricultural water use. In *Water-related technologies for sustainable agriculture in the U.S. arid/semiarid lands*. (pp. 109–145). U.S. Congress Office of Technology Assessment OTA-F-212.
Pearce, F. (2006). *When rivers run dry*. Boston: Beacon Press.
Pereira, L. S., Cordery, I., & Iaconides, I. (2002). *Coping with water scarcity: UNESCO, International Hydrological Programme*. IHP-VI, Technical Documents in Hydrology No. 58.
Postel, S. (1992). *Last oasis: Facing water scarcity*. New York: WW Norton.
Postel, S. (1993). Water and agriculture. In P. H. Gleick (Ed.), *Water in crises. A guide to the world's fresh water resources* (pp. 56–66). Oxford: Oxford University Press.
Rahaman, M. M., & Varis, O. (2005). Integrated water resources management: Evolution, prospects and future challenges, Sustainability, Source, Practice and Policy, *1*(1), 15–21.
Reisner, M. (1999). *Cadillac desert: The American West and its disappearing water* (2nd ed.). New York: Penguin.
Revenga, C. (2001). *Will there be enough water? Earth Trends*. Water Resources Institute.
Rogers, P., Bhatia, R., & Huber, A. (1998). *Water as a social and economic good. How to put the principle in practice*. TAC Background Papers No. 2. Global Water Partnership: Stockholm, Sweden.

Segerfeldt, F. (2005). *Water for sale. How business and the market can reduce the world's water crisis*. Washington, DC: Cato Institute.

Smith, A. (1776). An inquiry into the nature and causes of the wealth of nations. London: W. Strahan & T. Cadell.

Solomon, S. (2010). *Water. The epic struggle for wealth, power, and civilization*. New York: Harper Collins.

Srinivason, V., Gorelick, S. M., & Guolder, L. (2010). Sustainable urban water supply in south India: Desalination, efficiency improvement, or rainwater harvesting? *Water Resources Research, 46*, W10504. doi:10.1029/2009WR008698.

Stiles, G. (1996). Demand-side management, conservation, and efficiency in the use of Africa's water resources. In E. Racked, E. Rathgeber & D. B. Brooks (Eds.), *Water management in Africa and the Middle East* (pp. 8–27). Ottawa: International Development Research Centre.

Tietemberg, T. (2002). The tradable permits approach to protecting commons. In E. Ostrom, T. Dietz, N. Dolšak, P. C. Stern, S. Stonich, & E. U. Weber (Eds.), *The drama of the commons* (pp. 197–232). Washington, DC: National Academy Press.

Toope, S. J., Rainwater, K., & Allan, T. (2003). Managing and allocation water resources: Adopting the integrated water resources management approach. In A. S. Alsharhan & W. W. Wood (Eds.), *Water resources perspectives: Evaluation, management and policy* (pp. 1–8). Amsterdam: Elsevier.

Van der Zaag, P., & Savenije, H. H. G. (2006). *Water as an economic good: The value ofpricing and the failure or markets* (Value of Water Research Report Series No. 19). Delft: UNESCO-IHE.

Water Footprint Network. (2009). Water footprint, introduction and glossary, world footprint network. http://www.waterfootprint.org.

Wheida, E., & Verhoeven, R. (2007). An alternative solution of the water shortage problem in Libya. *Water Resources Management, 21*, 961–982.

Winpenny, J. (1994). *Managing water as an economic resource*. London: Rutledge.

Winpenny, J., Heinz, I., & Koo-Oshima, S. (2010). *The wealth of water: The economics of wastewater use in agriculture*. Rome: Food and Agriculture Organization of the United Nations.

World Water Council. (2004). E-Conference synthesis: Virtual water trade—conscious choices.

Yang, H., Wang, L., Abbaspour, K. C., & Zehnder, A. J. B. (2006). Virtual water trade: An assessment of water use efficiency in the international food trade. *Hydrology and Earth System Sciences, 10*, 443–454.

Yang, H., & Zehnder, A. (2007). *"Virtual water": An unfolding concept in integrated water resources management. Water Resources Research, 41*, W12301. doi: 10.1029/2007WR006048.

Young, R. A. (1996). *Measuring economic benefits for water investments and policies*. Washington, DC, World Bank Technical Paper No. 338.

Young, M. E. (2002). Institutional development for sustainable groundwater management—an Arabian perspective. In K. M. Hiscock, M. O. Rivett & R. M. Davison (Eds.), *Sustainable groundwater development* (pp. 63–74). Geological Society of London Special Publications 193.

Chapter 35
Integrated Water Resources Management

35.1 Introduction

Since the 1980s, Integrated Water Resources Management (IWRM) has become the most often espoused paradigm for the management of water resources. The most widely used definition of IWRM is that of the Global Water Partnership (2000):

> IRWM is a process which promotes the coordinated development and management of water, land, and related resources, in order to maximize the resultant economic and social welfare in an equitable manner without compromising the sustainability of vital ecosystems.

IWRM and related concepts such as"total water management" and "holistic water management" share the overriding goals of improving overall water managements, balancing of interests and interdependencies in water decisions, and increasing shareholder participation in the decision making process (Griggs 1999, 2008; Thomas and Durham 2003). Integrated urban water management (IUWM) is a component of IWRM that is concerned with the specific problems of urban areas. Conjunctive use of surface water and groundwater is an integral element of IWRM that involves consideration of all available water resources and developing strategies that obtain the maximum value from the combined resources while minimizing the undesirable physical, environmental and economical effects associated with their use.

IWRM evolved in the 1980s, when it became increasingly realized that the historical supply-driven water management approach, which focused on continuously seeking to maximize the water supplies, could not be sustained (Al Radif 1999). IWRM is also a response to the historically fragmented management of water, in which the various components of water management have been regulated (to the extent that they had been regulated at all) in isolation from the other aspects. Groundwater, surface water, wastewater, and land use are still commonly managed separately, even though they are closely inter-related. The overriding

R. Maliva and T. Missimer, *Arid Lands Water Evaluation and Management*, Environmental Science and Engineering, DOI: 10.1007/978-3-642-29104-3_35,

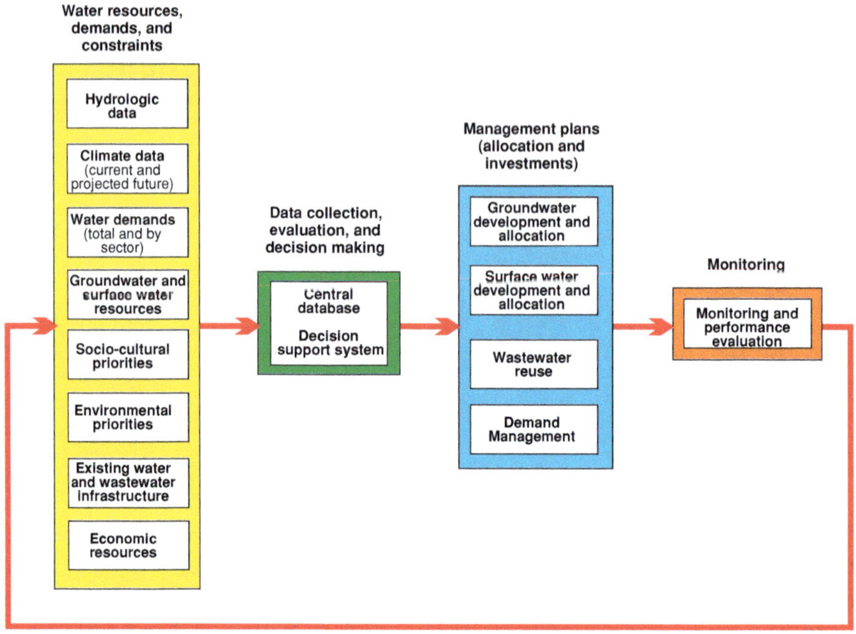

Fig. 35.1 Basic integrated water resources management workflow

theme of IWRM is the more formal recognition of the hydrologic cycle and the need for a more holistic approach to water management. Water resources should be managed with the objectives of optimizing the use of all potential water resources, managing water use to achieve the maximum economic and societal benefits, and minimizing adverse impacts of human activities on water resources and water use on natural environments. A basic workflow for IWRM is presented in Fig. 35.1. Data on all types of available water resources, water demands (total and by sector), social and environmental goals, existing infrastructure, and economic resources are collected and analyzed in order to develop water management plans that optimize the use of all water resources.

35.2 Meaning and Implementation of IWRM

Pertinent questions are what does IWRM specifically mean and how does one achieve it? The Global Water Partnership (2000) noted that

> Whereas certain principles underlying IWRM may be commonly applicable independent of context and stage of economic and social development, there is no universal blueprint as to how such principles can be brought into practice.

Bauer (2004, p. 9) cautioned that

> it is evident that integrated water resources management is an ideal concept rather than a set of specific guidelines and practices. Like "sustainable development", it is a phrase that can mean all things to all people; because it seems to mean everything, it may end up meaning nothing.

Biswas (2004, p. 249) similarly observed that

> A comprehensive and objective assessment of the recent writings of the individuals and institutions that are vigorously promoting integrated water resources management indicates that not only no one has a clear idea as to what exactly this concept means in operational terms, but also their view of it in terms of what it actually means and involves vary very widely.

IWRM as presented by the Global Water Partnership (2000) includes the integration of a variety of social, political, economic, and engineering/construction elements. IRWM cannot be implemented in vacuum. All of the elements already exist in societies with varying degrees of effectiveness. Water management is also just one aspect of a society, which also needs to manage a variety of other sectors (ideally in an integrated manner) such as energy, food supply, and transportation.

Perhaps IWRM should be considered an ideal towards which societies approach on a step-by-step manner, but will likely never fully reach, at least over a short to mid-range (decadal) time. Medema et al. (2008) observed that

> IWRM is not an end state to be achieved, it is a continuous process of balancing and making trade-offs between different goals and views in an informed way.

However, Biswas (2004, p. 249) cautioned against this approach, noting that:

> Some have argued that integrated water resources management is a journey, and not a destination, and the concept provides only a road map for the journey. The problem, however, with such reasoning is that in the area of water management, we are long on road maps, but short on drivers!

Many areas of the world have water problems that need solutions now, and there is neither the time nor the resources to develop and implement an all-encompassing IWRM plan, assuming that it is practicably possible. Devising IWRM principles and strategies is an interesting intellectual exercise, but the important matter is actually implementing plans that lead to more effective management of water resources. Medema et al. (2008) noted with respect to management frameworks in general that the community developing the framework (mostly academics) is unlikely to be the community that is actually implementing the framework. Ideas may 'appear good on paper', but may not be workable in the real world. Furthermore, the desire to develop a better long-term plan can become an excuse to not do anything now.

Mitchell (2008) raised a subtle but important issue in that there is a difference between a "comprehensive" approach and "integrated" approach. A comprehensive approach involves the consideration of all variables and relationships. However, an integrated approach involves examination of only the key variables and relationships that account for most of the variation in the system under consideration. Key variables may be identified based on existing data and knowledge, particularly local knowledge. An integrated approach may be more streamlined than a comprehensive approach.

It has been recognized by many that there is a gap between theoretically agreed IWRM policies and actual implementation, which may stem in part from the ambiguity of the concept (e.g., Rahaman and Varis 2005; Jeffrey and Gearey 2006; Biswas 2008; Medema et al. 2008). When there are basic issues that need to be resolved, and many parties share responsibility, it is not clear who has the lead role, who should pay, and how such decisions are made (Griggs 1999). Stakeholder participation in the decision-making process is inherently desirable, but how are decisions to be made when some of the parties involved have essentially irreconcilable economic interests?

Implementation of IWRM needs to be considered in the context of how real-world water management is performed. Those responsible for water management face an existing complex of inter-related water, wastewater, social, economic, political, and environmental elements, that are often in conflict with each other. Actual water management decisions are typically made for a 1–5 year planning and budget period. Longer-term water management plans are often developed by water utilities and water management agencies, but usually become out-dated for projects anticipated to be constructed 10 years or more into the future and thus have little relevance.

Decisions have to made as to what concrete steps a country, community, or water utility can implement in the short-term (i.e., now) that will materially improve water management. The actions will depend upon the resources that can be mobilized and political realities. Reorganizing a society with respect to its management of water will not occur in the short term. However, initiating a water reuse, conservation, or aquifer recharge programs is achievable within a constricted timeframe.

Allan (2003b) noted that IWRM is an intensely political process because water users have interests and do not want them to be diminished by such intervention. A critical point is that there will be resistance to change, particularly by those who benefit by or are accustomed to the status quo. Allan (2003b) also raised the important issue of there being a great schism in the acceptance of IWRM between developed countries (the North) and developing countries (the South). Knowledge of sustainability, IWRM, and water use efficiency is not readily assimilated by poor billions in the developing world, particularly when it contradicts deeply held beliefs and expectations about water (Allan 2003b). Water users and communities must know about, understand, and want new water management systems before they can be introduced and operated effectively.

From a practical perspective, the primary value of IWRM may be that it encourages a wider perspective and holistic view of water resources. For example, water management plans should consider all potential water supplies and all demands, and strive for an optimal strategy for the allocation of water resources that meets long-term societal needs, while maintaining natural environments. An integral component of applied IWRM is the consideration of a broad spectrum of options and applying systematic evaluation tools to the selection process (Chesnutt et al. 2008).

A variety of strategies and technologies should be considered for increasing water supplies and making more efficient existing use of water resources. Elements of IWRM that countries or communities may adopt include:

- wastewater reuse,
- aquifer recharge,
- stormwater management and flood control,
- storage (surface and underground),
- demand management (conservation and reallocation),
- conjunctive use of surface water and groundwater,
- development of alternative water supplies,
- water importation/exportation, and
- virtual water trade.

These elements are already being implemented throughout the world, but not necessarily under the banner of IWRM. It is the experience and observations of the authors that global water management is moving towards more integrated water management out of necessity. In large parts of the world there are simply no more readily available additional freshwater resources and existing resources are under threat. Therefore, societies are being forced to make better use of existing resources or, if financial resources permit, develop new sources of water such as desalination. The changes are occurring largely within the existing social and political frameworks, with the rate of progress being controlled largely by economic resources and institutional inertia.

A risk of creating excessive, unrealistic expectations concerning what can be achieved through IWRM is that it can lead to disillusionment or frustration and abandonment of the entire concept. However, adopting an integrated approach may be successful for smaller-scale, better defined problems, particularly when there is consensus that such an approach is needed and a clear, specific objective has been developed.

Few would argue against that the concept that water should be managed in a more holistic and comprehensive manner and that all stakeholders should be able to participate in its management to some degree. The core of IRWM is in fact just an elaboration of what should be considered good water management and sound decision-making. Although many criticisms of IWRM concerning its ambiguity, lack of objective evaluation criteria, and difficulty of implementation (e.g., Biswas 2004, 2008) are valid, the question becomes what is the alternative? Clearly, the alternative of a myopic approach to water management is not preferable.

Perhaps the greatest value of IWRM may be that it may encourage some societies to adopt more effective water management schemes by viewing water issues in a broader context and considering more options. With respect to integrated urban water management, Mays (2009) noted that it is usually impossible to have a completely integrated urban water management system, but considerable potential exists for improving urban water management through a better 'integrated' consideration of the various aspects and options.

Despite the flaws in concept and the timeframe constraints inherent within IRWM, there are documented cases of IRWM being successfully applied for long-term water management. Perhaps the most successful urban application is that occurring in Singapore. This is a small country which is essentially one large, urban area located on an island. The government of Singapore, under the leadership of the Public Utilities Board (PUD) has developed and implemented a system linking water supply (freshwater in country and freshwater imported from Malaysia), new water (desalination), stormwater management, wastewater reuse, and integration with the power supply. The program has very specific goals and objectives with timeframes to complete projects. The overall plan is widely supported by the public and business sectors, despite the substantial costs involved in the implementation of the plan.

35.3 Conjunctive Use

Conjunctive use is an integral element of IWRM and can be defined as the coordinated use of available water resources with the objective of optimizing overall water management. Conjunctive use involves maximizing water availability (both total supply and reliability) and minimizing capital and operational costs. The main water resources that may be available in arid and semiarid lands are groundwater, surface water, treated wastewater, rainwater, and desalinated water. The latter is typically reserved for domestic and commercial uses because of cost and energy consumption.

Conjunctive use typically revolves around making the best use of groundwater resources. Groundwater use offers both potential opportunities and perils for social and economic prosperity in both the developed and developing world. Increased groundwater irrigation has great potential for reducing rural poverty in some areas of the world, such as South Asia and Africa. The advantages of groundwater use are compelling (Shah et al. 2000):

- it is produced at the point of use and requires little transport,
- it offers individual farmer's irrigation "on demand",
- its use entails significant incremental cost of lift, so farmers tend to be economical in its use,
- it is less capital intensive to develop than surface water systems,
- it is usually self financing,
- it is typically a more reliable (less "flashy") supply than surface water.

However, three main problems dominate groundwater use (Shah et al. 2000):

- depletion caused by overdraft (excessive pumping),

- salinization and waterlogging caused mostly by inadequate drainage (lack of flushing with fresh water), and
- pollution, caused by agricultural, industrial, and human activities.

The greatest value of groundwater lies in stabilization of the water supply (Tsur 1990; Purkey et al. 1998; Reichard et al. 2010). Groundwater can be used to mitigate undesirable fluctuations in surface-water supply and improve the timing of water application within the growing season. Purkey et al. (1998, p. 52) observed that:

> In any situation where surface water supplies are variable, the presence of groundwater resources that can be tapped "as needed" for municipal, agricultural or other uses, carries a stabilization value well beyond that associated with increases in water supply alone.

Oweis et al. (1999) and Oweis and Hachum (2006) document how supplemental irrigation, which is the conjunctive use of rain and irrigation, can result in a substantial improvement of yields over rain-fed agriculture alone. Supplemental irrigation can be highly water efficient since a relatively small amount of water applied during periods of low rainfall can result in a substantial increase in productivity.

Surface freshwater flows, if not used or stored, are lost to tide, evaporation, or other sinks, and are thus not available for future beneficial use. Surface water, therefore, should be the first source of water used and groundwater use curtailed in order to allow aquifers to recharge and groundwater levels to recover (or at least decline no further). Reclaimed water should also be used as much as possible for non-potable purposes, in order to reduce stress on groundwater resources. Fresh groundwater should serve the role of a strategic reserve for coping with water scarcity. Under drought conditions, available groundwater reserves may be relied upon to meet a temporary lack of surface-water resources. Groundwater can also be used as an emergency supply in the event of other disruptions in supply, either natural or man-made. For example, Reichard et al. (2010) discussed how groundwater could mitigate the impacts of a major earthquake that could disrupt the surface-water importation system to Southern California. Managed aquifer recharge (MAR) can enhance the ability of groundwater to play a stabilization role by allowing for the capture of excess surface water and reclaimed water, and increasing the available supply of groundwater.

Reichard et al. (2010) performed an economic analysis of the impacts of increased groundwater use during a water emergency. The evaluated impacts of increased pumping are increased lift costs, land subsidence, and saline-water intrusion. Managed aquifer recharge provides economic benefits by reducing the impacts of increased groundwater use. There is great uncertainty in assessing the physical impacts and associated costs of a period of increased groundwater use. Reichard et al. (2010) provided a framework for quantifying the incremental costs and economic benefits of the use of groundwater as a backup water supply. Various simulation and optimization models including interactive decision support systems have been developed to evaluate proposed conjunctive use options (Mariño 2001).

Groundwater is ideal for use as a water supply for stabilizing variations in the supply of rainfall and surface water. However, there are risks associated with developing supplemental groundwater supplies. In water scarce regions, the temptation will exist to utilize any readily available water source. The primary danger with drought relief wells is that after a drought they may continue to be used and effectively increase demands on the source aquifer and water use in general. The end result is to make the system more vulnerable to the next drought (Loucks and Gladwell 1999).

35.4 Optimization

IWRM does not lead to perfect solutions in the sense that all water demands are met and all stakeholders are happy with the results. The goal instead is to approach as much as possible an optimal solution. Optimization is defined as the process or methodology of making something as fully perfect, functional, or effective as possible. The optimal water management solution is the best or most favorable possible under existing circumstances. In practice, optimization invariably involves tradeoffs between competing and conflicting objectives on multiple levels.

A tradeoff is defined as an exchange in which one benefit is given up for another considered more desirable. Tradeoffs are a fairly simple matter when only one party is involved, and are a frequent, normal part of life. Economic decisions are commonly tradeoffs, when one decides that a benefit of a purchase is more desirable than the benefits of saving the money or making an alternative purchase. Tradeoffs become far more complicated when multiple parties are involved and some will benefit and others will be harmed by a change.

For example, a fundamental conflict in many regions is between the uses of water for the maximization of economic returns versus the reservation of water for environmental goals. There is no perfect solution to that tradeoff. Social considerations (e.g., maintaining small farms and equity) are often also in conflict with economic efficiency. Reallocation of water from agriculture may harm farmers, while benefiting urban dwellers and industrial users. The tradeoff may be net beneficial to society as a whole, but there will still be winners and losers. With respect to wastewater reuse, competing and conflicting objectives include economic benefits, environmental objectives, and public health objectives (Al-Juadi et al. 2010).

Optimization can be viewed as the process of making the set of tradeoffs that most closely achieve overarching societal objectives and priorities. The great challenge of implementation of IWRM on a large (country or region-wide) scale, is the large number of elements that need to be considered and thus the number of tradeoffs for which an optimal solution needs to be found. Furthermore each tradeoff may have a chain of consequential events that also need to be considered. For example, reallocation of water from agriculture will directly impact the effected farmers, but can have a range of consequential effects including reduced incomes for businesses that serve farmers, increases in migration of farm workers to urban areas, reduced food production, and social and political instability in farming communities.

35.5 Multiple Criteria Decision Analysis

IWRM with respect to water supply will involve a set of supply options and a set of priorities and objectives. The priorities and objectives include economic, social, and environmental factors or elements. Multiple criteria decision analysis (MCDA) techniques are available to assist decision-makers to select between numerous and sometimes conflicting objectives. A wide variety of MCDA methods are available, many of which were reviewed by Belton and Stewart (2002). The first step in MCDA is identification of policy objectives or elements, and the feasible range of actions for each element (Stewart and Scott 1995). Scenarios are generated by combinations of feasible options for each element. MCDAs differ in the mathematical algorithms utilized, means by which options for each element or objective is scored, the method of application of weights, and the manner in which stakeholder preferences are considered.

Weighted scoring systems are widely used for MCDA evaluation of water and wastewater infrastructure options and have applications for IWRM decisions. The weighted sum is perhaps the simplest and most understandable MCDA. The performance score for scenario "i" (Z_i) is calculated as

$$Z_i = \sum_{j=1}^{n} w_j z_{ij} \tag{35.1}$$

where,
W_j = weight factor for criteria "j"
Z_{ij} = performance value "j" in scenario "i".

The similar weighted product methods use multiplication rather than addition as the main mathematical operation. Values are assigned with the optimal performance receiving the greatest scores. Decreasing scores are assigned with increasing departures from the optimum value. Assignment of performance values inherently involves considerable subjectivity, particularly where performance cannot be readily quantified. For example, it can be very difficult to objectively assign performance values to a preservation of an ecosystem function criterion. Even where a criterion is quantitative, subjectivity may still enter the scoring. For example, a scenario that results in a 50% reduction in stream flow would not necessarily warrant a performance value of 0.5 relative to an optimal no impact score of 1.0, if the impacts of the reduction are severe.

Al-Juadi et al. (2010) used the weighted goal programming method to evaluate wastewater reuse options in Gaza, Palestine. The weighted goal programming method evaluated competing objectives using a single composite objective function in which the users assigns weights to the deviation of each objective from the target level. Deviations are weighted according to the relative importance attached to each goal by the decision maker. For example, if economic benefit is the primary goal, then that goal would be given a greater weight factor. Different decision-makers may have different preferences and, thus, assign different weights to each element. The key issue is gaining an agreement on the weighting of all factors used in the analysis, if MCDA is used in a final decision-making mode.

MCDA techniques can be further applied for group decision support for policy evaluations. Interested and affected parties may differ greatly in their views and aspirations with respect to a given project (Stewart and Scott 1995). For example, local farmers may give highest priority (i.e., greatest weight) to irrigation water supply, whereas environmental advocacy groups may give irrigation supply a low priority relative to maintaining instream flows for environmental conservation. Group evaluation of scenarios can be performed by MCDA based on criteria that are relevant to their specific interests (Stewart and Scott 1995). A weighting scoring system could be used in which groups assign their own weighting factor for each criteria. The preferred scenario may be the solution that receives the greatest aggregate score from all groups. The chosen scenario would likely not be considered ideal by any group, but would instead be the most acceptable compromise. Stewart and Scott (1995) emphasize that the MCDA should be an iterative process involving the refinement of potential scenarios. The process will only work if the facilitator is seen by all parties as being a dispassionate and honest broker, and that all parties are honestly seeking the best common good (Stewart and Scott 1995). The parties involved must be willing to accept a scenario that is not ideal from their perspective. A single optimal solution that meets all objectives is replaced with a set of tradeoffs (Simonovic 1996). Commonly, it is useful to use decision evaluation software with a graphical interface to allow groups to observe the impacts of tradeoffs on changes in weighting factors on the outcome produced by the decision, such as economic impacts. The STELLA programs or similar software can be effectively used for this purpose and is a very good tool to use in public meetings wherein numerous potential solutions are reviewed and ranked.

35.6 Decision Support Systems

Decision support systems (DSSs) are computer programs that use analytical methods and data management systems to formulate and evaluate multiple solutions to problems. The great attraction of DSSs is that they allow for the examination of numerous considerations involved in attempting to design and manage increasingly sustainable water resources systems (Loucks and Gladwell 1999).

DSS can be used to evaluate alternative sets of assumptions or water management alternatives, such as different operating policies, costs, demand strategies, different water supply options, and hydrologic assumptions (e.g., normal climate variation and climate change). Water management options or scenarios may be evaluated in terms of their impacts on water resources, environmental impacts, and social and economic benefits and costs. DSSs are quickly becoming standard visual and multimedia tools, which are now irreplaceable in transmitting technical knowledge to decision makers, policy makers, and the public (Kresic 2009).

The basic components of DSSs are a graphic user interface (GUI), databases, and simulation and optimization models. With respect to water resources, simulation models can evaluate water resources behavior and may include groundwater and surface water flow models of differing degrees of sophistication. Socio-economic management systems address human demand for water, and how water is stored, allocated, and delivered.

Models are composed of objects and processes (Jordan 2006). Objects are defined by their properties and behaviors and include various water supply components (e.g., reservoirs, conveyances, treatment systems, and groundwater) and demand components. Objects are linked by processes that quantitatively describe how objects communicate (interact) with other objects. Objects are related to other objects by inputs and outputs, such as water flow into and out of a reservoir (Jordan 2006). Analytical solutions or look-up tables are used to describe the relationship between objects, such as the relationship between well pumping and groundwater levels, or the costs of pumping water and supply and demand components.

The Water Evaluation and Planning Version 21 IWRM model (WEAP21) is good, example, of a sophisticated water management DSS (Yates et al. 2005; Stockholm Environment Institute 2011). WEAP 21 can be used to evaluate water management scenarios with regard to factors such as supply sufficiency, costs, average costs of delivered water, hydropower production, and meeting in-stream flow requirements (Yates et al. 2005). Subroutines can be used to solve water management allocation problems with the objective to maximize satisfaction of demand, subject to user defined supply priorities, demand site preferences, mass balances, and other constraints. The basic model procedure is to first develop a reference ("business as usual") scenario and then develop one or more "what if" policy scenarios with alternative assumptions about future developments. The alternative assumptions can be based on policies, costs, technological development, and other factors that may affect demand, pollution, supply, and hydrology (e.g., climate change).

DSSs vary greatly in their sophistication in terms of their scales and issues considered and modeling procedures. Water management DSSs were reviewed by Chemonics International (2004). Some DSSs (e.g., WEAP21) can be linked to groundwater flow models such as MODFLOW. The WEAP code was also modified to produce a tool that can address long-term strategic planning for water utilities (Huber-Lee et al. 2006; Purkey and Huber-Lee 2006). Generic dynamic system simulation software (e.g., STELLA) may also be employed. The capability of DSSs may be increased through the incorporation of artificial intelligence, such as expert systems, neural networks, and fuzzy programming (Simonovic 1996).

The main challenges associated with DSSs are gathering the correct information, formulating proper questions, and interpreting the output (Purkey and Huber-Lee 2006). As is the generally the case with any sort of predictive modeling, the accuracy of the simulations will depend upon the quality of the data used in the simulation. An important question is whether the effort (and cost) to develop DSS models is commensurate with the benefits. Does a DSS provide new insights or is it used to validate what is already known or could be determined using simpler methods? There is always the danger with computer models with impressive graphics is that their results may have a greater acceptance (particularly by lay people) than is warranted by the accuracy of the raw data used and underlying assumptions.

35.7 IWRM and Water Management Planning

IWRM principles need to be incorporated into the water management planning process. Planning, in general, involves the following seven main elements:

(1) establishment of goals and objectives (setting of priorities),
(2) identification and analysis of problems,
(3) identification of potential alternatives,
(4) evaluation of alternatives, based on effectiveness, costs, and ancillary impacts (adverse and beneficial),
(5) decision making,
(6) development of implementation strategy, and
(7) design, construction, operation, and adaptive management.

There are opportunities to incorporate IWRM principles throughout the planning process. Setting of priorities involves input from user groups and other stakeholders. Identification and analysis of water resources management problems should be performed in the context of the wider range of societal problems. For example, inadequate urban water supply may be related to wider social-economic conditions (e.g., a high degree of poverty). The decision-making process in water management requires a holistic approach in which all potential water sources are considered as well as social, economic and environmental factors.

It has been the observation of the authors, that IWRM principles are in fact being incorporated into water management on both a water utility and governmental water management district level. The adoption of IWRM principles is not an explicit philosophical change, but rather is a pragmatic approach borne out of necessity. Utilities are looking for the least expensive solutions to meet water demands in their service areas, while avoiding environmental harm (or at least meeting regulations), and are well aware of the political ramifications of not meeting their stakeholder (i.e., customers) expectations. That IWRM has not progressed further in actual water management reflects the slow, typically incremental nature of change in water management. Water managers typically inherit an existing water supply and distribution infrastructure, which represents a great investment and out of necessity it will continue to be used in the future. Abrupt, major changes in course are simply not possible. Water systems can be become more integrated as systems expand or existing facilities reach the end of their useful life and are replaced.

35.8 IWRM in Arid and Semiarid Lands

The fundamental reality in arid and semiarid lands is that fresh water supplies will be obtained from some combination of:

- surface water,
- renewable groundwater,
- non-renewable groundwater,
- reuse of wastewater,
- new water resources (desalination), and
- rainwater harvesting.

In most arid regions, renewable groundwater use is already at, or more commonly exceeds, sustainable levels. Most arid regions either do not have significant surface-water resources or they are already fully utilized. Some regions are blessed with large quantities of non-renewable groundwater resources, which by definition are a finite resource that will some day run out if intensely used. The reality is that additional water supplies will have to be obtained from some combination of the reuse of wastewater, new water resources, and rainwater harvesting.

Managed aquifer recharge will also be an increasing used technique to optimize the management of existing water resources. Aquifer storage and recovery (ASR) or other MAR systems may be used to store excess surface water, stormwater, or treated wastewater that would otherwise go to waste (Chap. 23). MAR systems designed for wastewater treatment could also be used to increase the supply of water that is of suitable quality for reuse.

The fundamental IWRM challenge in arid and semiarid lands is finding the optimal blend of the available water sources in order to meet societal water supply requirements, while also meeting economic, social, and environmental objectives. The demand side of the equation must also be addressed. There is little question that increased water efficiency is needed to obtain maximum value from a scarce resource. Increasing the water use efficiency for individual uses (e.g., agricultural irrigation) is a relatively non-controversial means for increasing overall societal water use efficiency. Taking a holistic approach to water conservation, including consideration of economic and social issues, may lead to a local solution that creates the proper incentives to invest in water efficiency technology.

The far more challenging issue is increasing water use efficiency through inter-sector reallocation of water. This IWRM exercise involves consideration of all the water resources that are locally available, the various demands on that water, and social, cultural, economic, and political ramifications of reallocation. MCDA methods are well suited for evaluation of water reallocation from an economic perspective in which costs and benefits can be more readily quantified. However, evaluation of the human and environmental impacts is much more difficult to quantify (e.g., human, social, and environmental impact of large dam projects, such as the Chinese Three Gorges Dam).

The reality is that the actual implementation of IWRM in arid and semiarid lands will be a slow and gradual process because of the conservatism of the countries (and in fact most countries throughout the world). It is folly to believe that rapid change is possible in such a fundamental areas as water management in which there are well-entrenched historic use patterns and interests. However, the fact that people involved in water management in arid and semiarid lands are increasingly aware and talking about IWRM is very positive. There is a growing convergence about most of the concepts underlying IWRM (Savenije and Van de Zaag 2008). IWRM will ultimately not be implemented in a revolutionary manner, but rather water management will evolve in the IWRM direction, as a more holistic approach is taken in decision-making processes.

References

Al Radif, A. (1999). Integrated water resources management (IWRM): An approach to face the challenges of the next century and to avert future crises. *Desalination, 124*, 145–153.

Al-Juadi, A. E., Kaluarachichi, J. J., & Kim, U. (2010). Multi-criteria decision analysis of treated wastewater use for agriculture in water deficit regions. *Journal American Water Resources Association, 46*(2), 395–411.

Allan, J. A. (2003). Integrated water resources management is more of a political than a technical challenge. In A. S. Alsharhan & W. W. Wood (Eds.), *Water resources perspectives: Evaluation, management and policy* (pp. 1–8). Amsterdam: Elsevier.

Bauer, C. J. (2004). *Siren Song: Chilean water law as a model for international reform.* Washington, DC: Resources of the Future Press.

Belton, V., & Stewart, T. J. (2002). *Multiple criteria decision analysis: An integrated approach.* Norwell, MA: Kluwer Academic Publishers.

Biswas, A. K. (2004). Integrated water resources management: A reassessment. *Water International, 29*(4), 248–256.

Biswas, A. K. (2008). Integrated water resources management: Is it working? *Water Resources Development, 24*(1), 5–22.

Chemonics International Inc. (2004). *International survey of decision support systems for integrated water management.* Bucharest: Report prepared for U.S. Agency for International Development/Romania.

Chesnutt, T. W., Fiske, G., Pekelney, D. M., & Beccher, J. (2008). Water efficiency programs for integrated water management. *Journal American Water Works Association, 100*(5), 132–141.

Global Water Partnership. (2000). Integrated water resources management. *Global Water Partnership TAC Background Paper No. 4.*

Griggs, N. S. (1999). Integrated water resources management: who should lead, who should pay? *Journal American Water Resources Association, 35*(3), 527–534.

Griggs, N. S. (2008). *Total water management: Practices for a sustainable future.* Denver: American Water Works Association.

Huber-Lee, A., Schwartz, C., Sieber, J., Goldstein, J., Purkey, D., Young, C., Soderstrom et al., E., Henderson, J., & Raucher, R. (2006). *Decision support systems for sustainable water supply planning: Denver.* Colorado: AWWA Research Foundation.

Jeffrey, P., & Gearey, M. (2006). Integrated water resources management: Lost on the road from ambition to realisation? *Water Science and Technology, 53*(1), 1–8.

Jordan, D. (2006). Understanding decision support systems: A tool for analyzing complex systems. *Southwest Hydrology, 5*(4), 16–34.

Kresic, N. (2009). *Groundwater resources: Sustainability, management, and restoration*. New York: McGraw-Hill.

Loucks, D. P., & Gladwell, J. S. (1999). *Sustainability Criteria for Water Resources Systems*. Cambridge, UK: Cambridge University Press, International Hydrology Series.

Mariño, M. A. (2001, July). Conjunctive management of surface water and groundwater, In Regional Management of Water Resources, *Proceedings of a symposium held during the 6th IAHS Scientific Assembly in Maastricht vol. 268* (pp. 165–173). The Netherlands: IAHS Publication.

Mays, L. W. (Ed.). (2009). *Integrated urban water management: Arid and semi-arid regions*. Paris, Leiden: UNESCO, Taylor & Francis.

Medema, W., McIntosh, B. S., & Jeffery, P. J. (2008). From premise to practice: A critical assessment of integrated water resources management and adaptive management approaches in the water sector. *Ecology and Society, 13*(2), 29. Retrieved from http://www.ecologyandsociety.org/vol12/iss2/aert29/.

Mitchell, B. (2008). Resource and environmental management: Connecting the academy with practice. *The Canadian Geographer, 52*(2), 131–145.

Oweis, T., Hachum, A., & Kijne, J. (1999). *Water harvesting and supplementary irrigation for improved water use efficiency in dry areas. SWIM 7*. Colombo, Sri Lanka: International Water Management Institute.

Oweis, T., & Hachum, A. (2006). Water harvesting and supplemental irrigation for improved water productivity of dry farming systems in West Africa and North Africa. *Agricultural Water Management, 80* 57–73.

Purkey, D. R., Thomas, G. A., Fullerton, D. K., Moench, M., & Axelrad, L. (1998, December). *Feasibility study of maximal program of groundwater banking*. San Francisco: Natural Heritage Institute.

Purkey, D. R., & Huber-Lee, A. (2006). A DSS for long-term water utility planning. *Southwest Hydrology, 5*(4), 18–31.

Rahaman, M. M., & Varis, O. (2005). Integrated water resources management: Evolution, prospects and future challenges: Sustainability. *Source Practice and Policy, 1*(1), 15–21.

Reichard, E. G., Li, Z., & Hermans, C. (2010). Emergency use of groundwater as a backup supplu. *Quantifying hydraulic impacts and economic benefits: Water Resources Research, 46*. W 09524, doi:10.1029/2009WR008208.

Savenije, H. H. G., & Van der Zaag, P. (2008). Integrated water resources management: Concepts and issues. *Physics and Chemistry or the Earth, 33*, 290–297.

Shah, T., Molden, D., Sakthivadivel, R., & Seckler, D. (2000). *The global groundwater situation: Overview of opportunities and challenges*. Colombo, Sri Lanka: International Water Management Institute.

Simonovic, S. P. (1996). Decision support systems for sustainable management of water resources: 1. General principles. *Water International, 21*, 223–232.

Stewart, T. J., & Scott, L. (1995). A scenario-based framework for multicriteria decision analysis in water resources planning. *Water Resources Research, 31*(11), 2835–2843.

Stockholm Environmental Institute. (2011). *WEAP, User Guide* Water Evaluation and Planning System.

Thomas, J.-S., & Durham, B. (2003). Integrated water resources management: Looking at the whole picture. *Desalination, 156*, 21–28.

Tsur, Y. (1990). The stabilization role of groundwater when surface water supplies are uncertain: The Implications for groundwater development. *Water Resources Research, 26*(5), 811–818.

Yates, D., Sieber, J., Purkey, D., & Huber-Lee, A. (2005). WEAP21: A demand-, priority-, and preference-driven water planning model. *Water International, 30*(4), 487–500.

Chapter 36
Non-Renewable Groundwater Resources

36.1 Non-Renewable and Fossil Groundwater Definitions

Perhaps no other water policy issue has greater philosophic disagreement than the exploitation of non-renewable groundwater. Non-renewable groundwater was defined by Margat et al. (2006) as a

> groundwater resource available for extraction, of necessity over a finite period, from the reserves of an aquifer which has a very low current rate of average annual renewal but a large storage capacity.

This definition is satisfactory because it recognizes that aquifers are seldom absolutely non-renewable in the sense that no recharge is occurring. The key issue is that recharge is insignificant within the framework of the current water budget of the aquifer and that water is being produced primarily (if not entirely) from storage. It is also important in the characterization of non-renewable groundwater to recognize that there may be large temporal variation in recharge. A groundwater basin may receive minor periodic, localized recharge events (e.g., during rare wet periods), which would prevent the aquifer from being categorized as non-renewable in a stricter definition of the term (Houston and Hart, 2004). The concept of renewability, thus, necessarily involves a time frame. However, from a practical water management perspective renewability should be considered relative to actual or potential use rates and extractable storage volumes.

Non-renewable water is often considered to be synonymous with fossil water or paleowater. Margat et al. (2006) made the distinction that fossil water, by definition, is of great age (millennia). An aquifer may contain large volumes of fossil groundwater at depth, which if extracted would be replaced by more modern recharge. Aquifers, therefore, may contain both modern recharged water and paleowater that may be differentiated using environmental tracers (Edmunds 1999). Old groundwater samples may be obtained down-gradient in aquifers with long flow paths (Bourdon 1977; Houston and Hart 2004). Recharge may be

R. Maliva and T. Missimer, *Arid Lands Water Evaluation and Management*, Environmental Science and Engineering, DOI: 10.1007/978-3-642-29104-3_36,

occurring but it takes a very long time for water to travel from the recharge area to the discharge point. Younger water may be present near the top of aquifers if there is some recent vertical recharge.

Therefore, fossil water is not necessarily non-renewable. Non-renewable groundwater is by definition a finite resource. Once it has been extracted it is gone forever, at least on a human time scale. The exploitation of a non-renewable groundwater is commonly referred to as "groundwater mining", as groundwater use is analogous to depletion of mineral deposits by mining. The use of non-renewable groundwater resources is also analogous to oil and gas production.

36.2 Utilization of Non-Renewable Groundwater

Utilization of non-renewable groundwater can be classified into two categories, planned and unplanned schemes (Foster et al. 2006; Margat et al. 2006). In planned schemes, mining of the aquifer is contemplated from the onset, usually for a specific development project in an arid region. Unplanned schemes involve the incidental depletion of an aquifer as the result of intense, often uncontrolled groundwater abstraction.

Many arid and semiarid regions are underlain by aquifers containing huge quantities of fresh or brackish water. The volume of water that is required to directly sustain life is very small compared to the total amount used for other economic activities and could be provided by desalination or other methods. The key water resources management question is the whether or not (or the degree to which) non-renewable water should be used for agricultural and industrial purposes.

From an economic perspective, agriculture has a relatively low economic value so the cost of pumping water is an important consideration. Progressively increasing groundwater withdrawal results in a lowering of aquifer water levels, which can undermine the economic and social vitality of traditional groundwater dependent communities (Margat et al. 2006). Irrigation will be terminated once water levels drop to a depth that it is no longer economically viable to pump it. Extraction of non-renewable water for potable use may continue longer because it has a higher economic value compared to irrigation and most industrial water uses. Economics will thus control water use and there will ultimately be a self-limitation in the use of non-renewable groundwater.

The immediate choices that societies face with respect to the use of non-renewable groundwater are to either (Khouri 2001):

- maximize groundwater production to enhance socioeconomic development, with the hope of implementing a replacement solution once the water runs out,
- adopt a plan of exploitation that prolongs the period of production, or
- not use or greatly limit the use of the resource.

There are various philosophical positions on this question. The use of any non-renewable resource is inherently not sustainable using a strict definition of sustainability. Therefore one position is that non-renewable water should not be developed at all because all development activities should be sustainable. Alternatively, groundwater perpetually left in the ground is not providing any benefit to local communities and nations. If there is an obligation to preserve non-renewable groundwater for future generations, then future generations would also be obliged to preserve the water for generations yet further in the future, with the end result being that a valuable resource would never be used (Loucks and Gladwell 1999).

Llamas (2004) described the notion that groundwater mining is always over-exploitation and goes against ecological and ethical principles as a "hydromyth." It has been advocated that groundwater mining should be evaluated using the same criteria as other non-renewable resources, such as oil, coal, and iron (Issar and Nativ 1988). It is customary to develop supplies of natural resources by mining them until they are depleted, and then moving on to another location. Mining groundwater is much the same as mining any other mineral commodity, but it is perceived differently (Shomaker 2007). The difference in perception is due largely to water being essential to our lives and necessarily must be obtained from relatively nearby sources (Shomaker 2007).

Groundwater mining can often be beneficial to society if such developments are well designed and controlled. Llamas (2005) raised the issue that

> Fossil groundwater has no intrinsic value if left in the ground except as a potential resource for future generations, but are such future generations going to need it more than present ones?

Postel (1992, p. 176) raised the contrary point that non-renewable water

> will be valued far more highly by future generations faced with feeding a much larger world population.

A fundamental water resources question is whether the policy of sustainable development can be applied to regions that are going through the process of drying up. Is it acceptable to allow local populations to suffer, if the suffering could be alleviated through the use of non-renewable groundwater (Issar 2008)? Issar (2008) proposed the concept of "progressive development" as a reform of the rigid concept of 'sustainable development' that is suitable for situations experiencing an increase in aridity. The use of groundwater is seen as an integral component of a comprehensive plan of regional development (Issar and Nativ 1988; Issar 2008). Maxey (1968, p. 19) noted over 40 years ago with respect to non-renewable water that

> development can be logically and effectively planned so that mining of groundwater is highly beneficial and constitutes a rational step in orderly optimum water resources development.

The critical issue for the management of non-renewable water is developing a realistic plan for the use of resources that fully considers the volume of recoverable water and extraction rates (i.e., resource life), alternative water supplies, collateral impacts from groundwater extraction, the economic efficiency of use of the resources, and social, political, and ecological issues. Groundwater resources in general need to be evaluated in terms of their total economic value (National Research Council 1997). An important concern is obtaining the maximum returns on the resource, although it is recognized that some of the economic parameters are very difficult to assess (Younger 1992; Llamas 2004). In addition, the relevance of the conflict between groundwater development and natural conservation and its solution will differ between countries and will also change over time (Llamas 2004). For example, environmental preservation may be given a low priority in a country facing famine conditions, while affluent countries may be much more sensitive to the ecological consequences of groundwater development.

Llamas (2001, 2004) proposed that groundwater mining may be a reasonable action if various conditions are met:

- the amount of groundwater reserves can be estimated with reasonable accuracy.
- the rate of reserve depletion can be guaranteed for a long period.
- the environmental impacts of such groundwater withdrawals are properly assessed and are considered clearly less significant than the socio-economic benefits from groundwater mining.
- solutions are envisaged for the time when the groundwater is fully depleted.

Al-Eryani et al. (2006) proposed that the use of non-renewable groundwater should focus on social sustainability which would involve maintaining or improving the well-being of communities involved as meeting present and future social, economic, cultural, and environmental needs, and maintaining inter-generational equity in terms of the benefits derived from the resource. Al-Eryani et al. (2006, p. 26) proposed that:

> The overall goal should be to use groundwater in a manner that maximizes long-term economic and social development of the community and decreases, over time, the frequency and severity of threats to society leaving the people better prepared to cope with socio-economic stresses associated with increasing water scarcity as aquifer storage is depleted.

Issar (2008) similarly observed that developing societies benefiting from increased water availability through the use of non-renewable groundwater resources will need to undergo rapid technological advancement with regard to the development of new water resources and the development of more efficient water conservation and re-use practices. Utilization of non-renewable groundwater resources is viewed as a transitional step in societal development.

The fundamental question that arises concerning the use of non-renewable groundwater (Issar 2008) is

> what happens in the future when this resource is exhausted?

An appropriate "exit strategy" needs to be identified, developed, and implemented before the time the resource is substantially depleted. Development of an exit strategy normally implies that the society in question will advance economically, socially, and technically so as to enable future generations to develop substitute water sources at tolerable capital and operational costs (Al-Eryani et al. 2006). Gleeson et al. (2010) also suggest that sustainability goals for water quality and quantity should be set on a multigenerational time horizon of 50–100 years. The issue of inter-generational equity is addressed by the improvement in well-being of the people and the creation of opportunity for younger generations.

A more pessimistic view is that exploitation of non-renewable groundwater resources could allow a country to temporarily continue to rely upon inefficient agricultural operations and retard social, economic, and technical development. Reliance on non-renewable resources could, thus, potentially be detrimental to a society.

It is the intent of this chapter to present some of the basic philosophical issues related to the utilization of non-renewable groundwater, not to advocate any one position on the issue. A pertinent question is the means by which the decisions are made concerning whether or not and how (to what degree) to exploit non-renewable groundwater. Al-Eryani et al. (2006) indicted that a just and effective decision-making process is needed, which requires the participation and social cooperation of stakeholders and the upholding and promoting the right of communities to participate meaningfully in decisions concerning groundwater that affects their livelihoods. A regulatory framework is also needed to enforce decisions made concerning the use of non-renewable groundwater resources.

The critical issue for managing non-renewable groundwater resources is to start by firmly recognizing that one is dealing with a limited resource. The volume of recoverable water needs to first be determined. Informed decisions can be then be made as to the exploitation strategy, such as annual withdrawal rates and thus the life of the resource. Development strategies should focus on obtaining the maximum short and long-term value from the resource.

36.3 Assessment of Non-Renewable Resources

If non-renewable groundwater is to be managed effectively, it is critical that aquifer characterization be performed that is sufficient to assess water availability for the planned abstraction period and the impacts of abstractions on the resource, environment, and other aquifer users (Foster et al. 2006).

The methods for assessing groundwater resources in general and in arid and semiarid lands are discussed in Sect. 36.4 (Chaps. 10–20). The assessment methods for non-renewable groundwater resources are similar to those used for renewable resources. The major goal of aquifer assessment programs is to obtain sufficient data to allow for development of a model of a groundwater basin. The model is then use to simulate the aquifer response under current and potential future pumping scenarios. The key point is that effective management of non-renewable resources requires information on both potential total and extractable groundwater reserves and the impacts of various use scenarios. The most critical impacts are the aquifer depletion rate and associated aquifer drawdowns and potential large-scale impacts, such as land subsidence.

The major tasks required in the characterization of non-renewable water resources are (after Puri et al. 2006):

- quantification of storage reserves, which requires information on the three-dimensional variation of specific yield,
- assessment of contemporary recharge rates, which by definition of non-renewable groundwater resources are likely to be small,
- appraisal of depletion trends and the risk (or actual occurrence) of water quality deterioration (especially salinity increases),
- prediction of potential ecological impacts, which typically are not significant for deep confined aquifers with the possible exception of ecosystems dependent on artesian spring flows, and
- potential assessment for other adverse impacts from water use, such as land subsidence.

In asmuch as aquifers containing non-renewable aquifers produce water mainly out of storage, evaluation of the total amount of water in storage is a critical concern. If the aquifer area and thickness are known, then the total water reserves can be estimated from storativity and specific yield values. However, the geometry of non-renewable aquifers may not coincide with basin boundaries. The distribution of non-renewable stored water may reflect relict drainage networks (Edmunds 2003).

A distinction needs to be made between total storage reserves and extractable groundwater reserves. The latter involves consideration of the technical and economic limits of aquifer exploitation (Puri et al. 2006). Variation in groundwater quality is also an important factor in evaluating extractable groundwater reserves, because salinity may increase to unacceptable levels with prolonged pumping and reduce the potential viable yield of the aquifer.

The exploitation of non-renewable groundwater resources is usually evaluated on one level by the ratio of the total extractable or recoverable reserves and the proposed annual pumping rate. As a hypothetical example, if an aquifer has estimated reserves of 10,000 km^3 (2.6 billion gal.) and the proposed annual extraction rate is 5 km^3 (1.3 MG), then the conclusion may be made that the proposed withdrawals would be acceptable as it involves only 0.05% of the total reserves. Similarly, it may be concluded that withdrawal rates could be sustained for 2,000 years. Additional technical factors need to be considered for the meaningful evaluation of the use of non-renewable groundwater, including:

- In confined aquifers, the initial water produced from storage will be due to compression of the aquifer solid matrix and expansion of the water. Inasmuch as the confined storativity is a small fraction of specific yield, initial drawdowns from pumping will be relatively large. The rate of drawdown will decrease once water levels drop below the top of the aquifer and water is produced by dewatering of the aquifer. A fundamental point that is often overlooked is the difference in timing between the causation of pressure changes and the actual flow of water. Within confined aquifers, pressure changes (such as from pumping) are induced very quickly, but the actual flow of water is very slow within the aquifer (Bourdon 1977).

- Accurate determination of specific yield values is difficult, particularly in a confined aquifer that will convert to unconfined during pumping. Long-term pumping tests or data are required to measure delayed yield in unconfined systems, but cannot be used to do so in confined aquifers that will convert to unconfined systems over time. The use of cores and associated porosity measurement and borehole geophysics may be needed to perform these estimates. Gravimetric data may provide useful information on specific yields (Chap. 14).
- In some circumstances, a large fraction of the water produced from an aquifer may be derived initially from the dewatering of compressible aquitards internal to the aquifer rather than from the aquifer itself.
- Present day regional hydraulic gradients may be the result of on-going decay of a paleogradient established during early pluvial periods when major recharge occurred.

Environmental tracers, such as stable and radioactive isotopes (^{2}H, ^{3}H, ^{18}O, ^{14}C, ^{36}Cl), noble gases, and anthropogenic compounds (e.g., CFCs) can be used to date groundwater and to estimate recharge rates (Chap. 12). Groundwater modeling can allow for evaluation of conceptual models of the behavior of non-renewable groundwater resources, particularly the storage properties and depletion rates. Puri et al. (2006) noted the importance of carrying out repeated model audits and updates (recalibration) as new data become available. New data will include monitoring of pumping rates, water levels, and water quality and any other additional hydrogeologic data (e.g., land subsidence, climate information).

The large regional aquifers of the Middle East and North Africa are not in steady state with current (recent pre-development) recharge and discharge. Aquifer water levels are slowly declining in places from highs reached during earlier pluvial periods (Bourdon 1977). For artificial abstractions on a time scale of 10's of years, the background, long-term unsteady-state conditions may be regarded as quasi-steady initial conditions (Thorweihe and Heinl 1999). Thus, when calculating new pumping-induced drawdowns, the "natural" decline of water levels may be neglected (Thorweihe and Heinl 1999).

Houston and Hart (2004) similarly document the presence of fossil gradients in aquifers in the Andes of Chile. They concluded that

> it is incorrect to label groundwater resources as either fossil or actively recharged, as there is, in reality, a continuum between the two, and all such aquifers require site-specific evaluation and new, more flexible guidelines for development.

Water quality considerations may also limit non-renewable groundwater use. Intensive pumping may induce saline-water intrusion. It has recently been discovered that some non-renewable groundwater in Middle East has radium concentrations an order of magnitude or more greater than international groundwater standards and thus, poses a risk to long-term utilization of the resource

(Vengosh et al. 2009). The elevated radium concentrations are believed to be caused by alpha recoil from aquifer solids combined with relatively low adsorption in clay-poor lithologies. Radium can be removed by treatment processes such as ion exchange, reverse osmosis, and lime softening (Vengosh et al. 2009), which may still allow safe use of the water for potable supply. However, the radium in non-renewable water may limit its use for some agriculture and aquaculture activities, especially if radium is accumulating in the soils or in living organic matter (Vangosh et al. 2009).

Monitoring data are also critical for the management of non-renewable groundwater resources. The value of detailed monitoring of groundwater abstraction and use, and the aquifer response (ground-water levels and quality) to such abstractions, cannot be overemphasized (Foster et al. 2006). Monitoring programs should include systematic groundwater sampling (levels and water quality) and commence as early as possible in the aquifer development cycle (Puri et al. 2006). Monitoring data provide a means to compare actual aquifer system response to pumping to simulated responses. Actual data on long-term aquifer depletion is a key element in the aquifer management decision-making process (Puri et al. 2006).

36.4 Legal and Institutional Controls of Non-Renewable Groundwater Use

The legal framework for the management of groundwater resources in general varies between countries and, in some countries, between regions and states within the country (Chap. 32). Many countries still lack coherent policies and strategies for the management of groundwater resources and the institutional and economic resources to adequately administer groundwater management legislation (Nanni et al. 2006). Management plans may also have to be implemented in countries in which existing knowledge of groundwater resources is limited. It is often much easier to establish rules, policies, and regulations than to actually enforce them. Nanni et al. (2006) advocates aquifer management organizations (AMOs) as a means of allowing stakeholders to participate in the development and implementation of groundwater management plans. All stakeholders need to appreciate that they are dealing with a finite resource and that some controls are necessary to prolong the useful life of the resource. A "bottom up" approach in groundwater management has a greater potential for obtaining acceptance by aquifer users, then a 'top down' approach in which regulations are imposed by what are perceived to be distant authorities.

Non-renewable groundwater resources are storage dominated and, therefore, should not be addressed in the same manner as the flow of rivers or renewable groundwater resources (Nanni et al. 2006). Management of the latter usually focuses on sustainable or safe yields, which are not applicable in the hydrologic sense to non-renewable groundwater. Management plans should therefore focus on extractable resources and management goals including the allowable decline in water levels over a given period of time. The plans must consider the impacts of the declining aquifer water levels, which will include decreasing well yields and increasing pumping costs, and possibly also deteriorating water quality.

It is essential to limit the duration of water use rights to a finite period so as to facilitate the periodic review process and adjustment of rights or allocations as necessary (Nanni et al. 2006). There are a variety of methods available to control groundwater extraction rates and water level declines in non-renewable aquifers including (Abderrahmam 2006; Nanni et al. 2006):

- restrictions on crops allowed (banning high water use crops),
- restrictions on irrigated acreage,
- permit system to limit and control withdrawal rates,
- well registration and restrictions on new well construction,
- licensing of well drillers and restrictions on well construction including depths and pump capacity,
- mandatory equipping of wells with meters and charges for water use,
- metering of wells to quantify use,
- control leakage and water losses from supply networks, and
- implement artificial recharge.

36.5 Transboundary Non-Renewable Groundwater Resources

Management of non-renewable resources has an additional challenge when the aquifer in question extends across a national or other politically significant boundary. Groundwater flow does not respect national boundaries. Pumping by one nation can affect other nations that share the aquifer. International law has sparsely taken into account transboundary aquifer systems whether renewable or non-renewable (Nanni et al. 2006; Chaps. 32 and 37).

The worst case scenario would be accelerated groundwater pumping by all parties under the belief by each party that the water they do not extract would be taken by their neighbors. Similarly, a rush to development could occur in order to use as much water as possible now so as to establish an 'historic' existing use of the water, and thus, obtain leverage in any future negotiations. The rush to development would be an extreme example of the tragedy of the commons (Sect. 32.2). Such a scenario has a greater likelihood of occurring when the countries involved are hostile towards each other.

There is currently no international law that explicitly addresses management issues specific to transboundary non-renewable groundwater. The United Nations Law of Transboundary Aquifers adopts the principle of equitable and reasonable utilization (Chap. 37). However, if reasonable utilization means sustainable utilization, then the principle of equitable and reasonable utilization does not apply to non-renewable aquifers as these aquifers cannot be sustainably used on a continuous, long-term basis without eventually exhausting the resource (Davids 2005). In the absence of any widely accepted international law concerning the use of non-renewable groundwater, the most practical solution to the management of transboundary non-renewable groundwater resources is bilateral or multilateral cooperation and agreements between the nations that share a resource, which is adopted in the Law of Transboundary Aquifers.

It is in the long-term interest of all parties that share a non-renewable resource, to share information and jointly manage the resource so as to obtain the maximum value. For example, a Joint Authority was established between Egypt and Libya, and later joined by Chad and Sudan, to study and manage the Nubian Sandstone Aquifer System (Sect. 36.6.3). Similar cooperation is occurring among the nations that share the North Western Sahara Aquifer System (Sect. 36.6.2).

An option proposed for the management of transboundary non-renewable groundwater resources is to treat them in a similar manner as oil and gas in which the costs and benefits (i.e., share of producible water) is allocated equally, in proportion to the resource located in each state, or based on some other agreed upon compromise (Eckstein and Eckstein 2005). Various options are available for the equitable management of transboundary non-renewable groundwater resources, provided the parties involved genuinely seek an equitable resolution.

36.6 Examples of Non-Renewable Groundwater Management

Several examples of the management of non-renewable groundwater resources are presented herein. The examples were not selected as examples of either effective or ineffective management, but rather to illustrate the range of issues and responses that countries or groups of neighboring countries face. The one common denominator is that there is an appreciation that exploitation of the resources will eventually have to be phased out as the aquifers become depleted. Exit strategies are less clear, or are least a subject that has not been addressed in the technical literature. If use of non-renewable groundwater is the foundation of local agriculture, then the future may hold greatly reduced agriculture activity with obvious implications for national food security ambitions.

36.6.1 Saudi Arabia

Saudi Arabia has limited renewable freshwater resources, but large volumes of non-renewable groundwater, which are utilized along with alternative water sources (desalination) to meet domestic, industrial, and agricultural demands. The total estimated groundwater reserves in Saudi Arabia down to a depth of 300 m

(984 ft) is approximately 2,185 billion m^3 (Bm^3; 5.8×10^{14} gallons), compared to an average annual recharge rate of 2,763 million m^3 (Mm^3; 7.3×10^{11} gallons) (Abderrahmam 2006). The total domestic and industrial demand in 2000 was approximately 2,030 Mm^3 (5.4×10^{11} gallons) about half of which was supplied using desalinated water. The agricultural use was about 19,074 Mm^3 (5×10^{12} gallons). Domestic and industrial water demands are projected to more than double between 2000 and 2020, due to an anticipated approximate doubling of the total and urban population (Abderrahmam 2006).

Intensive use of groundwater in the Kingdom of Saudi Arabia started after the increase in oil revenues in 1973. The use of non-renewable groundwater resources is viewed in the Kingdom as a strategy to support socio-economic development. Agricultural development is considered an essential tool for achieving a social balance between urban and rural areas and for alleviating low income levels in rural communities (Abderrahmam 2002). Socio-economic development of rural areas is considered an important national goal, and a way of maintaining rural population and preventing further migration to urban areas. Maintaining rural populations also has a national security dimension in that they give support to security and defense authorities in remote areas. Agricultural development is supported in the Kingdom of Saudi Arabia by interest free loans, price supports, and financial assistance for the costs of machinery and wells (Abderrahman 2005, 2006). A decision was made that carefully planned, intensive use of non-renewable groundwater was expected to have acceptable impacts and support national development. It was concluded that intensive use of non-renewable groundwater resources for socioeconomic development for a limited number of years should not always be rejected (Abderrahmam 2002).

Cultivated areas increased by three times between 1974 and 2000 and the Kingdom of Saudi Arabia became a wheat exporter. Wheat production in 1992 was 4.22 million metric tons (9.3 million tons), which greatly exceeded the predicted national demand of 1.22 million metric tons (2.7 million tons). This overproduction hindered diversification of agricultural production and resulted in unnecessary consumption of large volumes of groundwater. In 1993, the government reduced the area of wheat cultivation eligible for price support to 25% of its previous size. The reduction was meant to encourage crop diversification and reduce irrigation water consumption (Abderrahman 2001). In 2008, the Saudi Arabia government announced it is abandoning its 30-year program to grow wheat in order to save water (Karam 2008).

Groundwater use in Saudi Arabia has been placed under the jurisdiction of the Ministry of Water and Energy. The non-renewable groundwater management goals currently are (Abderrahmam 2006):

- recovery of aquifers or an orderly use of aquifer reserves,
- minimizing water quality deterioration,
- maximization of groundwater productivity, and
- promoting social transition to a less water-dependent economy.
- measures that have been taken to improve groundwater management include (Abderrahmann 2005):

- demand management to reduce irrigation water consumption, such as reductions to wheat support and irrigated areas,
- pipe and canal leakage control measures, and
- augmentation of water resources from non-conventional sources as desalination, wastewater reuse, and aquifer recharge.

36.6.2 North Western Sahara Aquifer System

The North Western Sahara Aquifer System (NWSAS) contains two principle aquifers that cover an area of more than 1 million km^2 (386,000 square miles); the Continental Intercalary (CI, deeper aquifer) and Complex Terminal (CT) aquifers. The CI consists mainly of Upper Jurassic and Lower Cretaceous sandstones and the CT consists primarily of Upper Cretaceous limestones. The NWSAS is shared by Algeria, Tunisia, and Libya and has a total recharge rate of approximately 1 Bm^3/year (264 billion gallons/year). The annual exploitation reaches 2.2 Bm^3/year (581 billion gallons/year), with about 1.3 Bm^3/year (343 billion gallons) in Algeria, 0.55 Bm^3/year (145 billion gallons) in Tunisia, and 0.33 Bm^3/year (87 billion gallons) in Libya (Mamou et al. 2006). Annual exploitation is thus more than double the recharge rate and it is projected to increase in the future. Initial signs of aquifer deterioration are evident including (Mamou et al. 2006):

- increased potential for conflict between countries,
- water salinization,
- disappearance of artesian flows,
- outlets (discharge points) drying up, and
- excessive drawdowns in pumped wells

Studies of the NWSAS are underway under a project known as the Systeme Aquifere du Sahara Septentrional (SASS), which is under the supervision of the Observatoire due Sahara ET du Sahel (OSS). The key elements of the NWSAS studies are the implementation of a shared database and the development of a groundwater model that is capable of representing aquifer behavior under the proposed development schemes and acting as a management tool for the aquifer to meet the common interests of the countries that share it (Mamou et al. 2006; Salem 2007). Predictive simulations have been performed of various management scenarios. The objective of the modeling is to identify the most vulnerable areas of the NWSAS, minimize harmful effects from groundwater extraction, and identify sites favorable for future exploitation. The Chotts region of coastal Algeria and Tunisia was identified as being particularly vulnerable because of its high population density and susceptibility to saline-water intrusion into the CT aquifer from vertical recharge near the coastline. The modeling results suggest that appreciable increase in production is possible, but at the cost of dispersing additional pumping to more distant wellfields. The importance of confirming the modeling results was stressed by Mamou et al. (2006).

The NWSAS studies are an excellent example of cooperation between countries by sharing data to improve knowledge of the hydrology of an aquifer system upon which that they all rely (Mamou et al. 2006). The second phase of the SASS project will emphasize the development of a consultation mechanism among the NWSAS basin countries. It is recommended that a permanent institution be established in which the department responsible for management of the aquifer system will be placed.

36.6.3 Nubian Sandstone Aquifer System

The Nubian Sandstone Aquifer System (NSAS) underlies 2.2 million km^2 (849,000 mi^2) in North Africa and is shared primarily by Egypt, Libya, Sudan, and Chad (Bakhabakhi 2006). The NSAS contains two major units, the Nubian Sandstone Aquifer (NSA) and the Post-Nubian Aquifer System (PNAS). The NSA is largely unconfined except in the north, where it is overlain by PNAS. The PNAS is present in the northern halves of Egypt and Libya. The NSA consists mainly of Jurassic and Lower Cretaceous sandstones and clay-rich units. An extension of the Nubian Sandstone also underlies the Sinai Peninsula and Negev Desert of Israel, where it is also being increasingly exploited for water supply. Structural geological features and sedimentologic facies changes control the thickness and transmissivity of the Nubian sandstone and thus groundwater flow.

No significant recent recharge is detectable in the NSAS. Groundwater in many basins of the NSAS in northern Africa and Middle East have low oxygen and hydrogen isotope ratios ($\delta^{18}O$ and δD) relative to modern precipitation in these regions, which has been interpreted as reflecting paleo-recharge during periods of high humidity during the Late Pleistocene (Abd El Samie and Sadek 2001; Vengosh et al. 2007, references therein). The isotopic data also indicate the possibility of an older major recharge event (Sturchio et al. 2004; Patterson et al. 2005; Vengosh et al. 2007). A large-scale modeling investigation of the NSAS indicates that the aquifer system has been in a transient state for the last 3000 years and that the main episodes of recharge occurred at 20,000–25,000, and 5,000–10,000 years before present (Gossel et al. 2004). Natural discharge of the NSA occurs in the lowlands or depressions in the western desert of Egypt and the Libyan Desert, and associated with deep-seated fault complexes that define the River Nile and Gulf of Suez (Sultan et al. 2007).

There is much locally much heterogeneity in the NSAS, both horizontally and vertically, with respect to hydrochemical properties and age. There is a general consensus that negligible recharge of the NSAS occurs in North Africa and Arabia under present arid conditions. A relationship exists, where the NSA is unconfined, between the spatial organization of fluvial and structural features and salinity, which reflect past recharge patterns (Robinson et al. 2007). Isotopic data indicate that some recent recharge of the NSA occurs in the Sinai Peninsula, where the formation outcrops and is exposed in the cores of folded structures (Abd El Samie and Sadek 2001). Wide scatter in the isotopic and hydrogeochemical data indicate that considerable variation in recharge conditions occurs (e.g., rate, altitude, temperature, air masses, location, overland flow). The reported recharge rate is approximately 4.9 Mm^3/yr (1,264 Mgal/yr) versus an estimated 200,000 Mm^3 (528 billion gallons) of water held in storage under the Sinai Peninsula and Negev Deserts (Isar 1985; Abd El Sami and Sadek 2001).

It has been recognized for some time that the NSA groundwater basins in North Africa and Arabia are not in an equilibrium condition and that the disequilibrium conditions extend back in time long before the start of intensive development for irrigation. The aquifers were found to have sloping potentiometric surfaces even through modern recharge was thought to be negligible (Bourdon 1977). Residual heads from previous pluvial periods were suggested to be the most important mechanism contributing to the maintenance of present-day flow in the fossil (non-renewable) groundwater (Bourdon 1977). The non-equilibrium conditions in the NSA of Egypt are also evidenced by the estimated natural inflow into the aquifer being lower than the total natural outfall (Ebraheem et al. 2002; and references therein). The system was already in a deplenishing process (with declining water levels) in Egypt before irrigation started in 1960 (Heinl and Brinkmann 1989; Thorweihe and Heinl 1999; Ebraheem et al. 2002).

The total volume of freshwater in storage in the NSA is approximately 370,000 Bm^3 (9.8×10^{16} gallons). The northern part of the NSA contains saline water. The PNAS contains approximately 73,000 Bm^3 (1.9×10^{16} gallons) of freshwater (Bakhabakhi 2006). Pressure on the NSAS has increased rapidly over the past three decades because of population growth, food demands, and economic growth. In the past 40 years approximately 40 Bm^3 (1.1×10^{13} gallons) has been extracted in Libya and Egypt, which contain the greatest freshwater reserves and have the greatest use rates (Bakhabakhi 2006).

The estimated extractable freshwater reserves in the NSAS are about 14,500 Bm^3 (3.8×10^{15} gallons), assuming a maximum allowable decline in water levels of 100 m (328 ft) (Bakhabakhi 2006). The extraction rate from the NSAS was estimated to be 2.17 Bm^3/yr (5.7×10^{11} gallons) (Bakhabakhi 2006). A comparison between extraction rates and estimated extractable freshwater reserves suggest that the NSAS has a long productive life.

Despite the long life expectancy, mining of the aquifer will eventually lead to a rise in extraction costs as water levels decline and water quality deteriorates (Bakhabakhi 2006). The effects of groundwater extraction will also vary geographically with adverse impacts occurring much earlier in high groundwater use areas. The rate of groundwater extraction will also increase significantly over time. Groundwater modeling of the NSAS in Egypt demonstrated that groundwater reserves are currently being mined and that groundwater levels will continuously decline over time if pumping continues at current or greater rates (Ebraheem et al. 2002, 2003). Long-term aquifer management plans necessarily are based on a threshold of maximum acceptable drawdowns and a time horizon. Water levels of more than 100 m (328 ft) below land surface are considered to be uneconomical for irrigation (Ebraheem et al. 2002, 2003). Assessment of potential groundwater development projects in southwestern Egypt were performed using a 100 year time horizon (Ebraheem et al. 2003). This a particularly acute problem in Egypt as the withdrawals from the Nile River will be reduced by international agreements and by natural fluctuations in flow caused by global warming.

A commission for the coordination between Libya and Egypt of the management of the NSAS was created in October 1989, at a summit meeting in Tobruk (Salem 2007). Both Sudan and Chad later joined the Commission and became full members. The mandate of the 'Joint Authority for the Study and Development of the Nubian Sandstone Aquifer System' is to collaborate and development co-operative activities for the development of the aquifer system (Abu-Zeid and Abdel-Meguid 2006). The objectives and functions of the Commission are (Salem 2007, p. 111):

- collection of data, information and study results from participating countries,
- performance of complementary studies to determine the present state of the aquifer,
- preparation of plans for the development of the resources and proposing and implementing joint polices for the exploitation and use of water resources at national and regional levels,
- management of the aquifer on a sound scientific basis,
- cooperation in the field of training and capacity building,
- involvement of the public by calling for rational use of the NSAS,
- study of the environmental impacts of water development, and
- organization of scientific workshops and dissemination of aquifer related information and strengthening of ties with regional and international organizations of common interest.

The Nubian Sandstone Aquifer System Project was initiated in 1998, with the Center for Environment and Development in the Arab Region and Europe (CEDARE) as an executing agency. A primary aim of the project is the development of a regional mathematical model that is capable of representing the aquifer conditions and simulating its future behavior in response to planned development schemes (Abu-Zeid and Abdel-Meguid 2006; Salem 2007). The Nubian Aquifer Regional Information System (NARIS) was developed, which is a system that allows for the easy exchange and flow of data and information between the concerned countries.

36.6.4 Great Man-Made River Project

The Great Man-Made River Project (GMRP) in Libya is the most ambitious project yet developed to extract non-renewable groundwater resources for social and economic development. Libya has a scarcity of renewable water resources. The traditionally used groundwater resources in the northern coastal belt, where most of the population resides, have become increasingly at risk due to saline-water intrusion. The coastal aquifers have been contributing more the 2,400 Mm^3 (6.3×10^{11} gallons) compared to an average annual recharge of less than 650 Mm^3 (1.7×10^{11} gallons) (Salem 2007). The GMRP was initiated in 1984 with the goal of transporting water from aquifers in the southern part of the country to farms located in coastal belt. The GMRP goal is to minimize water deficits in the affected coastal zones. Seventy percent of the GMRP water is allocated to agricultural development. The Great Man-Made River Authority (2008) reports that:

> The agricultural development plan aims to realise the highest possible rate of self-sufficiency in grain and fodder crops, achieve food security, increase the capital investment and job opportunities in the agricultural sector, produce raw materials for the food processing industry, encourage and support agricultural settlement and enhance foreign investment opportunities in agriculture.

The GMRP produces and transports groundwater from four groundwater basins (Kufra, Sirt, Murzuk, and Hamadah), which are part of the NWSAS and NSAS. The groundwater basins contain an estimated 10,000–12,000 Bm^3 ($2.6–3.2 \times 10^{15}$ gallons) of economically extractable water (Great Man-Made River Authority 2008). Phase III of the project has been completed, which increased the system capacity to 3.68 Mm^3/d (972 million gallons/day) or 1.343 Bm^3/yr (355 billion gallons/year]. The final system capacity will exceed 6 million m^3/d [1.584 billion gallons/day] or 2.1 Bm^3/yr (554 billion gallons/year). Although the extracted volumes are enormous, they constitute only small fraction of the estimated groundwater reserves ($\leq$ 0.06% per year). The scale of the project is extraordinary. At planned build-out, the GMRP will consist of 1,300 wells and almost 4,000 km (2,485 miles) of pre-stressed concrete pipe, primarily 4 m (13.1 ft) in diameter (Great Man-Made River Authority 2008).

Studies determined that it was more viable to convey water in the desert to existing population centers rather than to relocate people to the source of water in the desert. At the time the project was initiated over 20 years ago, the long-distance water transfer of the GMRP was competitive with seawater desalination (Alghariani 2004). However, the costs of seawater desalination have since dropped dramatically. The results of economic analyses of the GMWRP versus seawater desalination will vary depending upon whether the value of the groundwater is considered (i.e., its depletion cost) and the additional costs of piping water from the desalination plant to the farm field are included (Alghariani 2004). Externalities of seawater desalination, such as its carbon footprint and environmental impacts from concentrate disposal also need to be considered in economic analyses (Alghariani 2004).

The country faced a difficult choice between two development options (Bakhbakhi and Salem 2001):

- large-scale extraction of groundwater for maximum benefit of the present generation, or
- limited extraction that ensures sustainable development and conservation of the resource base.

It is recognized that the GMRP is by no means an everlasting solution to Libya's water supply, but rather a vital transitory stage to a more sustainable water supply, which will involve alternative sources (e.g., desalination and wastewater reuse) and increased efficiency of water use (Bakhbakhi and Salem 2001).

The GRMP has been transformative in improving the quality of life in Libya's coastal cities. Questions remain concerning how long the water will last and whether the irrigated agricultural operations will provide the quantity and quality of crops anticipated (Watkins 2006).

36.6.5 Great Artesian Basin

The Great Artesian Basin (GAB) underlies approximately 1/5 of the land area of Australia, encompassing the large arid and semiarid region west of the Great Dividing Range. The GAB is the largest and most important groundwater resource in Australia and has enabled human settlement and establishment of an important pastoral industry in an otherwise dry and often desolate environment (Great Artesian Basin Consultative Council 1998). The primary uses of the non-desert area are low intensity grazing of beef cattle sheep for wool production. More intensive land uses including wheat farming, cotton and fruit crops, and heavy grazing are generally restricted to the wetter more eastern and southeastern parts of the basin (Great Artesian Basin Consultative Council 1998). The area containing the GAB also contains petroleum and mineral reserves.

The GAB regional aquifer is a multi-layered aquifer system of Middle Triassic to Late Cretaceous age that consists mainly of quartzose sandstone aquifers and confining beds composed of siltstone, mudstone and marine argillaceous sediments. The main aquifer is primarily of Jurassic age (Habermehl 1985, 2001). The GAB aquifer consists of four main fining upwards cycles, with the aquifer sandstones at the base of each cycle (Great Artesian Basin Consultative Council 1998). The basin is greater than 3,000 m (9,840 ft) thick in its deepest area.

The GAB aquifer was discovered in the late 1870s and the first free-flowing artesian wells were drilled in 1878 (Great Artesian Basin Consultative Council 1998). The aquifer contains good quality water (500–1500 mg/L TDS), but has high temperatures, up to 100°C near the center of the basin. Cl^{36} and C^{14} dating indicate that the water is about 5000 years old near the recharge area and up to two million years old or greater in downgradient areas (Great Artesian Basin Consultative Council 1998). Transmissivity ranges from 10–2000 m^2/d (108–21,520 ft^2/day) (Great Artesian Basin Consultative Council 1998).

The GAB aquifer system contains approximately 8.7×10^6 gigaliters (GL, 3.3×10^{13} gallons) of water in storage and an average recharge rate of 1,000 GL/yr (2.64×10^{11} gallons/yr) (Great Artesian Basin Consultative Council 1998; Herczeg 2008). Hence the groundwater of the GAB aquifer system is not strictly non-renewable. Diminishing flows and pressures were a concern from the very early stages of development of the GAB aquifer system, and the first legislation aimed at control of use of the aquifer system was passed in the 1910s (Great Artesian Basin Consultative Council 1998). Water levels in the aquifer locally dropped 100 m (328 ft) [or more], with the greatest rate of change occurring from 1900 to 1920 (Great Artesian Basin Consultative Council 1998; Herczeg 2008).

Prior to the 1880s there are were about 3,000 flowing springs that ringed the GAB and provided the only access to the water (Ponder 2002). The drilling of 1,000s of boreholes led to the extinction of about a third of the springs and reduced flows for the latter. Springs have rich endemic flora and fauna and have considerable significance as part of the cultural heritage of Aboriginal people, whose occupancy of some springs extends back 30,000 years (Great Artesian Basin Consultative Council 1998).

Key issues concerning the GAB (Great Artesian Basin Consultative Council 1998) include:

- continued declining artesian pressure in parts of the basin from excess extraction,
- reduced natural discharges are causing detrimental impacts on water dependent ecosystems,
- waste of water (estimated 2/3 of extracted water), particularly from free-flowing bore drains,
- increased competition for access to artesian groundwater,
- basin water use values tend to be managed in isolation from other values such as environmental protection, and
- inability or unwillingness of some water users to finance the upgrade of groundwater infrastructure.

Initially, wells were frequently sited adjacent to ephemeral watercourses. Uncontrolled well discharges flowed along the natural drainage lines until the flow was exhausted by seepage and evaporative losses. Up to 90% of water in drains goes to waste (Great Artesian Basin Consultative Council 1998). In many remote regions, channel equipment (liners) was not readily available or not economically viable for the less wealthy pastoralists. Installation of piping was also not economically viable early on and there was a total absence of any incentive to economize on the use of water (Great Artesian Basin Consultative Council 1998; Herczeg 2008).

The “Cap and Pipe the Bores Program” (note that “bores” in Australia refers to wells) was implemented for the Great Artesian Basin to reduce some the waste of water. The saved water is split 70/30 between the aquifer and community, which serves to both protect the environment and aquifer, and make additional water available to economically important sectors such as tourism and mining (Costa 2009). The share of the water saved allocated to the community is auctioned off. It was estimated that 47,000 megaliters (ML) (1.24×10^{16} gallons) will have been saved through the Cap and Pipe the Bores Program through the end of 2009.

36.6.6 Northern Chile

Northern Chile is one of the most arid regions of the world. Parts of the Atacama Desert have not had rainfall since record keeping has begun and the average rainfall is less than 1 mm/yr. Groundwater is the most important water source because the very low precipitation and high evaporation rates results in only minor surface water flow. The Region de Antofagasta in northern Chile is also host to unique mineral deposits (copper, gold, lithium). Severe conflicts have developed in the region between water for mining and urban purposes and protection of traditional irrigation agriculture and natural ecosystems (Grosjean and Veit 2005). Traditional agricultural irrigation has high water demands and very low economic value (i.e., contribution to the Chile GDP) compared to mining, but is part of a several thousand year old cultural heritage (Grosjean and Veit 2005).

The Pacific Andean Slope and Atlantic Andean Slope Hydrogeological Provinces groundwater resources occur mainly in alluvial fans, volcaniclastic materials, and lacustrine deposits (Reboucas 1999). Aquifers in northern Chile are discretized because of structural discontinuities and may be differentiated on the basis of the source of contained waters (Houston and Hart, 2004). Many groundwater basins occur in grabens formed by normal faulting. Evaporation is the principle discharge mechanism under current natural conditions. Natural recharge occurs mainly from river floods from mountains and in some areas snow melt. Periodic major floods may result in substantial recharge of alluvial fan deposits (Houston 2002). Dry periods were interrupted by low frequency storms and (sub) decadal-scale humid spells associated with discrete spells of groundwater recharge (Grosjean and Veit 2005).

Higher altitude basins may have some current recharge, while basins at lower levels may contain predominantly non-renewable water. There are concerns that global climate change is adversely impacting water resources, particularly those associated with snow melt discharge. Some predictions indicate that the Pacific Decadal Oscillation will cause droughts to have durations on the order of decades rather than the recent historic 1–2 year periods (Neary and Garcia-Chvesich 2008).

Isotopic and geochemistry data from two hydrologically closed groundwater basins in Northern Chile, Pampa del Tamarugal and Salar de Atacama, indicate that a large portion of the groundwater resources should be considered non-renewable (Fritz et al. 1981; Aravena 1995). Oxygen and hydrogen isotopic data indicate that the groundwater basins contain multiple aquifers that are recharged at different elevations. Stable isotope and tritium data from the Romeral Basin indicated that wells may tap deeper aquifers that are recharged at higher altitudes than shallower aquifers (Squeo et al. 2006). All rivers that drain into the basins disappear before reaching the lower basin, which contains central salt flats (salares), except during high runoff events in very wet years. Most of the water that reaches the lower basin is lost to evaporation. The main loci of recharge may also vary depending on climatic conditions. In the Romerol Basin, recharge occurs mainly in the higher parts of the basin during dry periods, whereas during wet periods more recharge occurs in the lower and middle parts of the basin (Squeo et al. 2006). In wet periods, especially during El Niño Southern Oscillation (ENSO) events the water table recovers very quickly and the stable isotope data showed that local precipitation plays an important role in groundwater recharge.

Magaritz et al. (1990) proposed that deep circulation systems occur within the Andes that contribute to basinal aquifers. The prime evidence is that the younger water, based on isotope and temperature data, is found in the center of the study area (Pampa del Tamarugal), and older water is present in the alluvial fans to the east at the mouth of canyons. The groundwater in the center of the basin is also fresher. Groundwater is posited to flow from the recharge areas in the higher areas of the Andes through fractured and/or volcanic units and faults. Montgomery et al. (2003) documented the occurrence of inter-basin groundwater flow as evidenced by hydraulic gradient data. Groundwater movement is occurring between basins that are closed with respect to surface water. Surface water divides thus do not correspond to groundwater basin boundaries. Montgomery et al. (2003) estimated that the underflow from the high Salar de Michincha basin to the topographically lower Salar de Coposa basins was 200–240 L/s (3,170–3,804 gpm), which is the same order of magnitude as estimated recharge and discharge rates. Houston and Hart (2004) noted that deep circulation is open to question and Montgomery et al. (2003) indicated that inter-basin underflow does not occur in all basins.

Groundwater modeling by Houston and Hart (2004) demonstrated that the observed groundwater gradients in small basins can only be explained by head decay following a period of recharge under wetter climate conditions, apparently hundreds of years ago. Modeled decay times were less than 1,000 years. Current water levels cannot be explained by current recharge conditions. The groundwater of the relatively small Andean basins cannot be of great antiquity or the basins would have already reached equilibrium conditions with zero discharge and horizontal water tables (Houston and Hart 2004). The declining water levels in some basins may thus be in part natural decay of heads established during earlier pluvial periods. The groundwater basins may, therefore, contain non-renewable water under current conditions, but may be renewable under a longer time period that encompasses periodic pluvial periods.

Despite the very arid conditions, there are increasing demands on the existing water resources. One of the major challenges facing water resources managers in Northern Chile is allocating the sparse resource between communities and mining companies without disturbing the delicate equilibrium that prevails in the local ecosystems (Aravena 1995; Madaleno and Gurovich 2007). Gold and copper mining are much more important to the economy of the country than subsistence farming. Indigenous populations and their livestock are being forced out of the region for water scarcity reasons (Madaleno and Gurovich 2007). The northern Chile experience well illustrates the difficult decisions water managers face in trying to balance the many competing demands on limited resources in water scarce regions.

Presently almost all of the surface water resources in Chile are allocated under the Chilean Water Code; the rights given to waters users are in perpetuity regardless of the benefits. The 1981 Water Law declared that water is a public property and that the state can grant private rights of use, which can be traded in the market (Mentor 2001). The Direccion General de Aguas (DCA) grants requests for new rights free of charge whenever water is physically and legally available.

Many aquifers in northern Chile have reached the limits of exploitation. Under the 2000 modification of Water Code there is now a clear distinction between a water right and a 'right to use water' (Suzuki 2007). In order to obtain a right to use water, new extractions of groundwater require approval of an 'Early Action Plan' to avoid irreversible environmental damage in sensitive areas protected by the state (Rodríguez 2006). The objectives of an Early Action Plan are to forecast, follow-up, evaluate, and verify the expected effects or impacts at the moment of granting the groundwater use rights. Groundwater modeling is an integral element of the process. An Early Action Plan must include actions that should be taken away in the case that monitoring data indicate defined limits are approached or reached (Suzuki 2007).

The development of the Monturaqui-Negrillar-Tilipozo (MNT) aquifer in northern Chile, as documented by Anderson et al. (2002), provides insights into some of the technical and regulatory issues involved in groundwater management in Chile. The aquifer consists of Tertiary sediment infill of a north–south trending graben. A very thorough investigation of the aquifer was performed which included 50 km (31 miles) of TDEM profiles, and over 80 boreholes were drilled. The MNT aquifer has an extremely large volume of water in storage, estimated to be as much as 10^{10} m^3 (2.64×10^{12} gallons), but a very modest estimate through flow of 250–900 L s^{-1} (2.16–7.78×10^4 m^3 d^{-1}) [5.7×10^6–2.1×10^7 gallons]. Recharge occurs at very low rates through surface inflows at the edges of the aquifer system. Stable isotope data suggest the recharge is from rain and snow fall at greater than 4,000 m (13,120 ft) above sea level.

The Tilipozo wetlands were the only indentified groundwater dependent natural feature and its protection is the main criterion for sustainability of groundwater use (Anderson et al. 2002). The maximum tolerable water level reduction in the Tilipozo wetlands was determined to be 0.25 m (0.82 ft), which corresponds to a 6% reduction in the recharge rate. Modeling results indicate that only about 5% of the total water in storage in the aquifer can be abstracted if the sustainability criteria are to be fulfilled (Anderson et al. 2002).

References

Abd El Samie, S. G., & Sadek, M. A. (2001). Groundwater recharge and flow in the Lower Cretaceous Nubian Sandstone aquifer in the Sinai Peninsula using isotopic techniques and hydrochemistry. *Hydrogeology Journal, 9*, 378–389.

Abderrahmam, W. A. (2001). Water demand management in Saudi Arabia. In: N. I. Faruqui. A. K. Biswas., & M. J. Bino (Eds.), *Water management in Islam* (pp. 62–68). Tokyo, Ottawa: United Nations University Press, International Research Centre.

Abderrahmam, W. A. (2002). Should intensive use of non-renewable groundwater resources always be rejected? In R. Llamas & E. Custodio (Eds.), *Intensive Use of Groundwater: Challenges and Opportunities* (pp. 191–203). Lisse: A.A. Balkema.

Abderrahman, W. A. (2005). Groundwater management for sustainable development of urban and rural areas in extremely arid regions: a case study. *Water Resources Management, 21*, 403–412.

Abderrahmam, W. A. (2006). Saudi Arabia aquifers. In S. Foster & D. P. Loucks (Eds.), *Non-renewable groundwater resources. A guidebook on socially-sustainable management for water-policy makers* (pp. 63–67). Paris: UNESCO.

Abu-Zeid, K., & Abdel-Meguid, A. (2006). *Pioneering action in managing the transboundary Nubian Sandstone groundwater aquifer*. Truckee, CA: Technical Report, Centre for Environment and Development for the Arab Region and Europe (CEDARE).

Al-Eryani, M., Appelgren, B., & Fosters, S. (2006). Social and economic dimensions of non-renewable resources. In S. Foster & D. P. Loucks (Eds.), *Nonrenewable groundwater resources: A guidebook on socially sustainable management for water-policy makers*. Paris: United Nations Educational.

Alghariani, S. A., (2004) *Water transfer versus desalination in North Africa: Sustainability and cost comparison, and comments by The Center of Data, Studies & Researches, Great Man-Made River Authority, Libyan Arab Jamahiriya*. London, UK: School of Oriental and Africasn Studies (SOAS)/King's College London Occasional Paper 49.

Anderson, M., Low, R., & Foot, S. (2002) Sustainable groundwater development in arid, high Andean basins, In: K. M, Hiscock., M. O, Rivett & R. M.Davison (Eds.), *Sustainable groundwater development* (pp. 133–144). London, UK: Geological Society, Special Publications 193.

Aravena, R. (1995). Isotope hydrology and geochemistry of northern Chile groundwaters. *Bulletin Institute Francois Etudes Andines, 24*, 495–503.

Bakhabakhi, M. (2006). Nubian Aquifer System. In S. Foster & D. P. Loucks (Eds.), *Non-renewable groundwater resources. A guidebook on socially-sustainable management for water-policy makers* (pp. 75–81). Paris: UNESCO.

Bakhbakhi, M.,& Salem, O. (2001). Why the Great Man-made River Project. In *Proceedings, Regional Aquifer Systems in Arid Zones: Managing Non-renewable Resources, Tripoli, Libya* November 1999, (pp. 1–16). IHP-V, Technical Documents in Hydrology No. 42. Paris: UNESCO.

Bourdon, D. J. (1977). Flow of fossil groundwater. *Quarterly Journal of Engineering Geology and Hydrogeology, 10*, 97–124.

Costa, P. (2009, May 26). *Water auction to provide stimulus for Great Artesian Basin*. Minister Phillip Costa. *Weehawken*, NJ: MP press.

Davids, J. (2005). Is it reasonable to use the Nubian Sandstone Aquifer System unsustainable under International Law? CEMLP Annual Review. *The Centre for Energy, Petroleum, and Mineral Law and Policy*. Dundee: University of Dundee.

Ebraheem, A. M., Riad, S., Wycisk, P., & Seif El-Nasr, A. M. (2002). Simulation of impact of present and future groundwater extraction from the non-replenished Nubian Sandstone Aquifer in southwest Egypt. *Environmental geology 43* (pp. 188–196).

Ebraheem, A. M., Goramoon, H. K., Riad, S., Wycisk, P., & Seif El Nasr, A. M. (2003). Numerical modeling of groundwater resource management option in East Oweinat area. *SW Egypt: Economic Geology, 44*, 433–444.

Eckstein, Y., & Eckstein, G. E. (2005). Transboundary aquifers: Conceptual models for development of international law. *Ground Water 43* (679–690).

Edmunds, W. M. (1999, November). Integrated geochemical and isotopic evaluation of regional aquifer systems in arid regions. In *Proceedings, Regional Aquifer Systems in Arid Zones: Managing Non-renewable Resources, Tripoli, Libya* (pp. 107–118). IHP-VI, Technical Documents in Hydrology No. 42. Paris: UNESCO.

Edmunds, W. M. (2003). Renewable and non-renewable groundwater in semi-arid and arid regions. In A. S Alsharhan & W. W Wood (Eds.), *Water resources perspectives: Evaluation, management and policy* (pp. 265–280). Amesterdam: Elsevier.

Foster, S., Nanni, M., Kemper, K., Garduño, H., & Tuinhof, A. (2006) Utilization of non-renewable groundwater, a socially-sustainable approach to resource management. *The World Bank Sustainable Groundwater Management Concept & Tool Brief Note Series* Note 11, p. 6.

Fritz, P., Silva, C., Suzuki, O., & Salati, E. (1981). Isotope hydrology of groundwater in the Pampa del Tamarugal, Chile. *Journal of Hydrology, 53*, 161–184.

Gleeson, T., VanderSteen, J., Sophocleous, M. A., Taniguchi, M., Alley, W. M., Allen, D. M., et al. (2010). Groundwater sustainability strategies: *Nature Geoscience, 3*, 378–379.

Gossel, W., Ebraheem, A. M., & Wycisk, P. (2004). A very large scale GIS-based groundwater flow model for the Nubian sandstone aquifer in Eastern Sahara (Egypt, northern Sudan and eastern Libya). *Hydrogeology Journal, 12*, 698–713.

Great Artesian Basin Consultative Council. (1998). *Great Artesian Basin Resource Study Summary*. Canberra: Great Artesian Basin Consultative Council.

Great Man-Made River Authority. (2008). GMRA. Retrieved November 25, 2008, from http://www.gmmra.org/en/.

Grosjean, M., & Veit, H. (2005). Water resources in the arid mountains of the Atacama Desert (Northern Chile): Past climate changes and modern conflicts. In U. M Huber., H. K. M. Bugmann, & M. A. Reasoner (Eds.), *Global change and mountain regions. An overview of current knowledge* (pp. 93–104). Dordrecht: Springer.

Habermehl, M. A. (1985). Groundwater in Australia. In *Hydrogeology in the service of man, memoires of the 18th Congress of International Associations of Hydrogeologists* (pp. 31–52). UK: Cambridge.

Habermehl, M. A. (2001). Hydrogeology of the Great Artesian Basin, Australia. In *Proceedings, Regional Aquifer Systems in Arid Zones: Managing Non-renewable Resources, Tripoli, Libya* (pp. 123–142). 20–24 November, 1999. Paris: UNESCO.

Heinl, M., & Brinkmann, P. J. (1989). A groundwater model of the Nubian aquifer system: *Hydrological Sciences Journal, 34*, 425–447.

Herczeg, A. L. (2008). *Background report on the Great Artesian Basin*. Australia: A Report to the Australian Government from the CSIRO Murray-Darling Basis Sustainable Yields Project; CSIRO.

Houston, J., & Hart, D. (2004). Theoretical head decay in closed basin aquifers: an insight into fossil groundwater and recharge events in the Andes of northern Chile. *Quarterly Journal of Engineering Geology and Hydrogeology, 37*, 131–139.

Houston, J. (2002). Groundwater recharge through an alluvial fan in the Atacama Desert, northern Chile: mechanisms, magnitudes and causes. *Journal of Hydrology, 16*, 3019–3035.

Issar, A. S., & Nativ, R. (1988). Water beneath the deserts. *Keys to the past a resource for the present Episodes, 11*(4), 256–262.

Issar, A. (1985 July). *Fossil water Under the Sinai-Negev* (pp. 104–111). Peninsula: Scientific American.

Issar, A. S. (2008). Progressive development in arid environments: adapting the concept of sustainable development to a changing world. *Hydrogeology Journal, 16*, 1229–1231.

Karam, S. (2008, January 8). Saudi scraps wheat growing to save water. Middle East.

Khouri, J. (2001). Impacts of intensive development on regional aquifer systems in arid zones. In *Proceedings, Regional Aquifer Systems in Arid Zones—Managing Non-renewable Resources, Tripoli, Libya* (pp. 339–365), 20–24 November 1999. Paris: UNESCO.

Llamas, M. R. (2001). Considerations on ethical issues in relation to groundwater development and/or mining. In *Proceedings, Regional Aquifer Systems in Arid Zones: Managing Non-renewable Resources, Tripoli, Libya* (pp. 475–487), 20–24 November 1999. Paris: UNESCO.

Llamas, M. R. (2004). Water and ethics. *Use of groundwater*. Paris: UNESCO International Hydrogeological Programme.

Llamas, M. R. (2005). and Martínez-Santos. *Intensive groundwater use: a silent revolution that cannot be ignored: Water Science & Technology, 51*(8), 167–174.

Loucks, D. P., & Gladwell, J. S. (1999). *Sustainability Criteria for Water Resources Systems: Cambridge*. UK: Cambridge University Press, International Hydrology Series.

Madaleno, I. M., & Gurovich, . (2007). Conflicting water usages in northern Chile. *Boletín de al Asociación de Geógrafos Españoles, 45*, 439–443.

Magaritz, M., Aravena, R., Peña, H., & Grilli, A. (1990). Source of ground water in the deserts of Northern Chile: Evidence of deep circulation of ground water from the Andes. *Ground Water, 28*, 513–517.

Mamou, A., Besbes, M., Abdous, B., Latrech, D. J., & Fezani, C. (2006). North Western Sahara Aquifer System. In S. Foster & D. P Loucks (Eds.), Non-renewable groundwater resources. A guidebook on socially-sustainable management for water-policy makers, IHP-Vim Series on *Groundwater* 10, (pp. 68–74). Paris: UNESCO.

Margat, J, Foster, S., & Droubi, A. (2006). Concept and importance of non-renewable resources. In S. Foster & D. P Loucks (Eds.), Non-renewable groundwater resources. *A guidebook on socially-sustainable management for water-policy makers* (pp. 13–24). Paris: UNESCO.

Maxey, G. B. (1968). Hydrogeology of desert basins. *Hydrogeology Journal, 6*(5), 10–22.

Mentor, J., Jr. (2001). Trading water, trading places: Water marketing in Chile and the western United States. In *Groundwater and water resources management: The changing value of water AWRA/IWLRI*. Dundee: University of Dundee International Specialty Conference.

Montgomery, E. L., Rosko, M. J., Castro, S. O., Keller, B. R., & Bevacqua, R. S. (2003). Interbasin underflow between closed Altiplano basins in Chile. *Ground Water, 41*, 523–531.

Nanni, M., Burchi, S., Mechlem, K., & Stephan, R. M. (2006). Legal and Institutional Considerations. In S. Foster & D. P Loucks (Eds.), Non-renewable groundwater resources. *A guidebook on socially-sustainable management for water-policy makers* (pp. 49–58). Paris: UNESCO.

National Research Council. (1997). *Valuing Ground Water, Economic Concepts and Approaches*. Washington: National Academy Press.

Neary, D. G., & Garcia-Chevesich, P. (2008). Climate change impacts on municipal, mining, and agricultural water supplies in Chile. *Hydrology and Water Resources in Arizona and the Southwest, 38*, 53–57.

Patterson, L. J., Sturchio, N. C., Kennedy, B. M., van Soest, M. C., Sultan, M., Lu, Z.-T., Lehmann, B., Purtschert, R., El Alfy, Z., El Kaliouby, B., Dawood, Y., & Abdallah, A. (2005). Cosmogenic, radiogenic, and stable isotopic constraints on groundwater residence time in Nubian Aquifer, Western Desert of Egypt. *Geochemistry, Geophysics, Geosystems 6* Q01005. doi:10.1029/2004GC00077.

Ponder, W. F. (2002). Desert Springs of the Australian Great Artesian Basin. In *Conference Proceedings. Spring-Fed Wetlands*. Las Vegas: Important Scientific and Cultural Resources of the Intermountain Region. Retrieved from (http://www.wetlands.dri.edu).

Postel, S. (1992). Last oasis. *Facing water scarcity* (pp. 239). New York: WW Norton.

Puri, S., Margat, J., Yurtsever, Y., & Wallin, B. (2006). Aquifer characterization techniques. In S. Foster & D. P Loucks (Eds.), *Non-renewable groundwater resources. A guidebook on socially- sustainable management for water-policy makers* (pp. 35–47).Paris: UNESCO.

Reboucas, A. C. (1999). Groundwater resources in South America. *Episodes, 22*, 232–237.

Robinson, C. A., Werwer, A., El Baz, F., El-Shazly, M., Fritch, T., & Kusky, T. (2007). The Nubian Aquifer in Southwest Egypt. *Hydrogeology Journal, 15*, 33–45.

Rodríguez, J. M. (2006). The Monturaqui—Negrillas—Tilopozo aquifer of Chile. In S. Foster & Loucks, D. P (Eds.), *Non-renewable groundwater resources. A guidebook on socially-sustainable management for water-policy makers, IHP-Vim Series on Groundwater No. 10* (pp. 89–92). Paris: UNESCO.

Salem, O. (2007). Management of shared groundwater basins in Libya. *African Water Journal, 1*(1), 106–177.

Shomaker, J. (2007). What shall we do with all this ground water? *Natural Resources Journal, 47*, 781–791.

Squeo, F. A., Aravena, R., Aguirre, E., Pollastri, A., Jorquera, B., & Ehleringer, J. R. (2006). Groundwater dynamics in a coastal aquifer in north-central Chile: implications for groundwater recharge in an arid ecosystem. *Journal of Arid Environments, 67*, 240–254.

Sturchio, N. C., Du, X., Purtschert, R., Lehmann, B. E., Sultan, M., Patterson, L. J., et al. (2004). One million year old groundwater in the Sahara revealed by krypton-81 and chlorine-36. *Geophysics Research Letters, 31*, 05503. doi: 10.1029/2003GL01923.

Sultan, M., Yan, E., Sturchio, N., Wagdy, A., Abdel Gelil, K., Becker, R., et al. (2007). Natural discharge: a key to sustainable utilization of fossil groundwater. *Journal of Hydrology, 335*, 25–36.

Suzuki, O. (2007). *The future of Northern Chile and water resources: Port of Entry* Retrieved March 23, 2010 from http://www.portofentry.com/site/root/resources/analysis/4406.html.

Thorweihe, U., & Heinl, M. (1999). Groundwater resources of the Nubian Aquifer System. In *Proceedings Regional Aquifer Systems in Arid Zones—Managing Non-Renewable Resources Tripoli, Libya* (pp. 239–251). 20–24 November, 1999. IHP-VI Tcehnical Documents in Hydrology. Paris: UNESCO.

Vengosh, A., Hening, S., Ganor, J., Mayer, B., Weyhenmeyer, C. E., Bullen, T. D., et al. (2007). New isotopic evidence for the origin of groundwater from the Nubian Sandstone Aquifer in the Negev, Israel. *Applied Geochemistry, 22*, 1052–1073.

Vengosh, A., Hirschfield, D., Vinson, D., Dwyer, G., Raanan, H., Rimawi, O., et al. (2009). High naturally occurring radioactivity in fossil groundwater from the Middle East. *Environmental Science and Technology, 43*, 1769–1775.

Watkins, J. (2006, March 18). *Libya's thirst for 'fossil water'*. London: BBC World Service.

Younger, R. A. (1992). Managing aquifer over-exploitation and policies. In I. Simmers, F. Villarroya & L. F Rebollo (Eds.) *Selected papers on overexploitation* (pp. 199–222). Hannover: Heise.

Chapter 37
Transboundary Water Management

37.1 Introduction

Water has been attributed to be the cause of conflicts or used as a military or political tool throughout history. Gleick (2003) documented 112 instances of water conflict through August 2003 and increased the documented conflicts to 167 through 2006 (Gleick 2006). Examples on the use of water as a military tool include destruction of dams and levees, intentional flooding, diversion of water, and poisoning of wells. Wolf (2000), however, noted that there is a lack of evidence for water actually being the primary cause of a war, as opposed to being a secondary issue, although there is disagreement on this matter. Many believe that a primary cause of the Six-Day War between Israel and its neighbors was triggered by the action of Syria to divert the Jordan River. Ariel Sharon, a military commander and prime minister, wrote

> The Six-Day War really began on the day Israel decided to act against the diversion of the Jordan....While the border disputes were of great significance, the matter of water diversion was a stark issue of life and death (Pearce 2006, p. 168).

The role of water in the Six-Day War was reviewed by Shemesh (2004) and Seliktar (2005). Actual military actions occurred over water in this area in 1951, 1953, and 1964 (Gleick 2006).

An important distinction is made between water as tool, target, or victim of armed conflict versus water being the primary cause of a conflict. Wolf (2000) purported that the last and only war over water of size actually occurred 4,500 years ago between the Sumerian city-states of Lagash and Umma. Smaller, but direct conflicts have been fought over water on a small scale between Bedouins and western armies and between American Indians and earlier settlers of the American West (e.g., the famous Frederick Remington painting, The Battle for the Water Hole).

The historical record reveals that shared water does lead to tension, and even some localized violence, but usually not to war. Perhaps the reason most water

R. Maliva and T. Missimer, *Arid Lands Water Evaluation and Management*, Environmental Science and Engineering, DOI: 10.1007/978-3-642-29104-3_37,

conflicts do not rise to the status of war is that it is a common practice for the more powerful government to exact an unequal quantity of water, therefore, leaving the less powerful government unable to react by military conflict. Today, most of these disputes are settled in international court, not on a battlefield. Flashpoints associated with water generally induce the parties to enter negotiations and occasionally into creative and resilient working arrangements (Wolf 2007). A greater threat than violence associated with transboundary water management is inaction. It may take decades to resolve some conflicts, all the while, water quantity and quality degrade to where the health of dependent populations are damaged or destroyed (Wolf 2007). The gradual decline in water quantity and quality can affect the internal stability of a nation or region, which may in turn impact the international arena (Wolf 2007).

Water management challenges increase greatly when a resource is shared between different political entities. Intra-national conflicts, such as between states, can be resolved using existing national courts or legislative or political institutions, although not necessarily to the satisfaction of all parties involved. Water has often been the cause of intra-national violence, generally between tribes, water use sectors and provinces (Wolf 2000). Water scarcity can lead to instability and internal conflict.

The difficulty of resolving conflicts over water is compounded when sovereign states are involved and the appropriation of water resources is considered vital to multiple national interests. No international institutions exist that can force a solution to a transboundary water conflict if all the parties involved are not willing to abide by a third party decision on the conflict. It has been said that:

> In war, truth is the first casualty (Aeschylus Greek tragic dramatist, 525–456 BC)

Regrettably, water resources claims on contested shared resources can become an exercise in "political hydrology" in which science is abused to justify political objectives. As Allan (2001, p. 76) noted

> Integral to the politics of natural resources is the construction of knowledge to reinforce the position of those with power.

Although many areas will experience increasing water scarcity in the future, gloomy scenarios of wars over water are by no means certain to come to pass. Fisher and Huber-Lee (2010, p. 3) observed that the "*idea that the next war will be about water is a myth—provided that the disputants will think about the matter rationally*," because of the low actual value of water and the high cost of war. No matter how important water is, it cannot be worth more than the cost of replacing it (Fisher 1995). Fisher (2002) cited the late economist Gideon Fishelson, who made the astute observation that water as a scarce resource has a value and that the cost of seawater desalination places an upper bound on the value of water to any country that has access to a seacoast. Fisher (2002, p. 190) additionally noted that

> Every country with a seacoast can have as much water as it is wants if it chooses to spend money to do so. Hence so far as water is concerned, every country with a seacoast can be self-sufficient in its food supply if it is willing to incur the cost of acquiring the necessary water. Disputes over water among countries are merely disputes over costs, not over life and death.

The critical reality is that the value of the water that is in dispute (e.g., in the Israel-Palestinian dispute), is trivial relative to the great cost of waging war (Fisher 1995, 2002; Fisher et al. 2005). Fisher (1995, p. 388) also noted that when one focuses on the value and price of water, the dialogue changes from one of "*repeated statements of irreconcilable historical claims to a dispassionate discussions of facts and assumptions.*" An important distinction is between the ownership and use of water. Fisher and Huber-Lee (2010, p. 4) observed that

> Water ownership is thus a property right entitling the owner to the economic value of the water. Hence a dispute over water ownership can be translated into a dispute over the right to monetary compensation for the water involved, taking into account social and environmental values.

Once water is understood to be an economic good that can be traded, possibilities arise for win–win scenarios. For example, a trade in water rights can provide one party with needed water, while the other party receives money that could be put to higher values uses.

Nevertheless, water is perceived as a vital national interest and unilateral actions that significantly reduce the water supply of a nation can be viewed as an act of aggression irrespective of objective economic analysis. Conflicts are often started over issues involving national pride, self-righteousness, and greed. Wars seldom make sense from an economic perspective when the full costs are considered. Countries often take actions that are ultimately against their national interest, in order to pursue policies that are "a matter of principle." One cannot, therefore, discount water issues as triggering a future conflict especially if there is an underlying animosity between the nations involved. McCaffrey (1993) noted with respect to water treaties that

> Where treaty regimes are lacking it is often because of tensions between the states concerned, tensions that are sometimes due to precisely to the scarcity of water resources in the region.

The principle impediment to transboundary water agreements, particularly over shared aquifers, can be in the domestic arena (Feitelson 2006). Groundwater users may be small, but cohesive and well organized interest groups that have disproportionate political influence. Limitations on groundwater use are of obviously much greater importance to groundwater pumpers than to the general electorate. Any attempt to impose limitations on domestic abstractions, may thus face stiff opposition from well entrenched interest groups composed mainly of existing users and those that serve them (e.g., well drillers) (Feitelson 2006). The domestic politics are wrong for what is perceived (or is in fact) as taking water away from domestic farms in order to give it to foreign water users. Agreements over the shared aquifers may be more readily reached and ratified, if water is a secondary issue or subsumed in a more comprehensive set of negotiations (Feitelson 2006). The greatest threat of the global water crisis to human security comes from the fact that millions of people lack access to sufficient quantities of water of acceptable quality for their well-being and the result is societal instability (Wolf 2007).

37.2 International (Transboundary) River Water Law

The four main legal theories concerning international river basins are summarized by Agnew and Anderson (1992) and Benvenisti (1996):

(1) Theory of absolute territorial sovereignty. States in possession of the upper waters of a river do not have a general obligation to refrain from diverting its water and denying downstream states its full flow.
(2) Theory of absolute territorial integrity. A state is forbidden to stop or divert the flow of a river or otherwise use such water so as to prevent a downstream state from making proper use of the flow. A state may cause no harm to other states.
(3) Theory of community or integrated drainage basin development. Riparian states should join each other to make full utilization of the waters of a river.
(4) Theory of limited territorial sovereignty or equitable utilization. States may make use of international waters to the extent that it does not interfere with reasonable use by other riparians.

Theory of absolute territorial sovereignty is also referred to as the "Harmon Doctrine" after the 1896 legal opinion of the United States Attorney General Judson Harmon. The Harmon Doctrine states the rules, principles and precedents of international law impose no liability or obligation on the United States. According to the Harmon Doctrine, nations have complete freedom of action with respect to any international water course within its territory. McCaffrey (1993) noted how the proposed application of the Harmon Doctrine has been selective and self-serving and varies depending upon the position of a country (upstream or downstream) with respect to its neighbors. For example, the United States had advocated that Harmon Doctrine applies with respect to Rio Grande flows from the United States into Mexico, but rejected the doctrine and espoused the right of prior appropriations with respect to the flow of the Columbia River into the United States from Canada.

International law lacks such features as compulsory jurisdiction and centralized enforcement that are characteristic of domestic legal systems. Nevertheless, states tend to observe international law because it is their self-interest to do so, as they may be unwilling or unable to bear the ultimate costs of violations (McCaffery 1993). With respect to shared water resources, the rules of international law are derived from either treaties or international customs (McCaffery 1993). Most treaties, either bilateral or multilateral, reflect the principle of equitable utilization. Equitable utilization is an attractive concept, but cannot be precisely defined and does not provide a set of guidelines to structure negotiations. There is no accepted definition of 'equity' in international law (Hussein and Al-Jayyousi 2001). The definition of equitable is typically determined separately for each water resource issue assessed based on resource-specific conditions and circumstances.

With respect to transboundary surface-water resources and groundwater, upstream riparians or groundwater using states, tend to underestimate the volume of water in the natural system. Upstream riparians want to establish a low figure for the volume of the 'natural' resource so that any negotiated agreement results in downstream users receiving a share of the flow that is smaller than actually available (Allan 2001). Conversely, downstream users tend to overestimate the natural flow in order to obtain as high an entitlement as possible.

Water users may also increase current use in order to establish a greater claim on water resources. For example, groundwater of the West Bank is a strategic issue for Israel, Jordan, and Palestine. During heavy rainfalls during and after 1992, groundwater levels recovered to a level of a decade earlier, which prompted a period of increased Israeli pumping. Because of awareness of a potential peace process and negotiated division of water resources, Israeli behaved as normal downstream user and began to increase withdrawals to increase its level of 'current use' (Allan 2001).

General (customary) international law with respect to shared water resources has widely relied upon the work of recognized experts. Bodies that have proposed resolutions or rules with respect to shared international waters include the Institut de Droit International (Institute of International Law), International Law Association (ILA), and the International Law Commission of the United Nations (McCaffery 1993). The ILA produced the Helsinki Rules on the Use of Waters of International Rivers. There is no consensus on the issue of equitable use versus no harm requirements. An unresolved issue is the right of later users of water to obtain an equitable share of water, if to do so would harm existing water users (McCaffrey 1993).

37.2.1 Helsinki Rules and Berlin Rules

The Helsinki Rules on the Uses of Waters of International Rivers (International Law Association 1967) are early guidelines on the regulation of transboundary rivers. The Helsinki Rules were adopted in August 1966, but have no legal standing. The Helsinki Rules adopted the equitable utilization principle. Article IV states "Each basin State is entitled, within its territory, to a reasonable and equitable share in the beneficial uses of the waters of an international drainage basin."

Article V lists the relevant factors, which are to be considered when determining what are reasonable and equitable shares, which include, but are not limited to:

- the geography of the basin, including in particular the extent of the drainage area in the territory of each basin state,
- the hydrology of the basin, including in particular the contribution of water by each basin state,
- the climate affecting the basin,
- the past utilization of the waters of the basin, including in particular existing utilization,
- the economic and social needs of each basin state,
- the population dependent on the waters of the basin in each basin state,
- the comparative costs of alternative means of satisfying the economic and social needs of each basin state,

- the availability of other resources,
- the avoidance of unnecessary waste in the utilization of waters of the basin,
- the practicability of compensation to one or more of the co-basin states as a means of adjusting conflicts among uses, and
- the degree to which the needs of a basin state may be satisfied, without causing substantial injury to a co-basin state.

It is recommended in the Helsinki Rules that questions or disputes be referred to a joint agency, and if that fails to result in a resolution, the states should jointly request the mediation of a third state, qualified international organization or person, or finally an arbitration tribunal or the International Court of Justice.

The Helsinki Rules were superseded by the Berlin Rules, which were adopted by the ILA on August 21, 2004. The Berlin Rules still embrace the equitable and reasonable utilization principal. The Berlin Rules integrate traditional rules regarding transboundary waters with rules derived from the customary international environmental and international human rights laws that apply to all waters, national as well as international (International Law Association 2004). The Berlin Rules are not legal obligations but, in the judgment of the ILA, are emerging as rules of customary international law. In other words, some of the Berlin Rules express the progressive development of the relevant international law (International Law Association 2004).

37.2.2 1997 Convention on the Non-navigation Uses of International Watercourses

The Convention on the Non-Navigation Uses of International Watercourses (United Nations 2005) was adopted by the General Assembly of the United Nations on May 21, 1997, but has not been ratified. The Convention contains several key articles with respect to transboundary river water law. Article 5 requires that states utilize international water resources in their respective territories in an equitable and reasonable manner and requires that all relevant information and circumstances be taken into account. Article 5 is essentially the Helsinki Rules requirements.

Article 7 includes the obligation to not cause significant harm to other water course users. If significant harm does occur, then a state has an obligation, in consultation with the effected state, to eliminate or mitigate such harm and, where appropriate, to discuss the question of compensation. The Convention on the Non-Navigation Uses of International Watercourses goes beyond the Helsinki Rules by including the no harm requirement, which is a controversial provision as its preferentially protects existing uses in downstream states and could restrict some water uses (particularly new uses) in upstream states that are necessary for economic development.

Article 9 requires regular exchange of hydrological, meteorological, hydrogeological, ecological, and water quality data. Article 12 states that watercourse states have an obligation to notify other states concerning planned measures with possible adverse effects. Conflicts are to be resolved through third parties, mediation, or by submitting the dispute to arbitration of the International Court of Justice.

37.3 Conflict Resolution

> The assertion of water rights is easy; gaining recognition for them is difficult; attaining the goals announced in the political rhetoric of water rights is impossible (Allan 2001, p. 175).

> Riparians do what is in their own interest to the limits that their political antennae tell them is possible (Allan 2001, p. 292).

Water resources can be "resolved" at least temporarily by the threat or actual use of force. The party with the greatest military strength (local or region hegemon) can unilaterally dictate regional water policy. For example, the construction of a dam by an upstream user can be prevented by a downstream user if the downstream user has sufficient military supremacy that could be used to prevent construction of or destroy the dam (e.g., Israeli destruction of the East Ghor Canal built by Jordan to divert the Yarmonk in 1969) (Samson and Charrier 1997). An actual example is that Israel as a result of its military supremacy has gained control over much of the water resources within the Jordan Basin and could dictate groundwater use restrictions in the occupied territories (Allan 2001). The economically and militarily strongest nation is in a position to act in its own interests irrespective of impacts on its neighbors, who have little recourse.

At the international level, the advantages of low levels of regulation to the hegemon, whether a super-power or major regional power, causes progress toward adoption of an international regime for water to continue to be very slow (Allan 2001). In the absence of an effective and binding international legal framework for resolving conflicts with respect to water, conflicts over water are most frequently resolved through negotiated bilateral or multilateral agreements. However, the historical experience has been that progress in resolving transboundary water conflict is usually exceedingly slow. Equitable and reasonable utilization is a sound basis for resolving transboundary water issues, but a party may not be interested in an "equitable" arrangement if it results in a decrease in its available water. A state that currently enjoys an inequitably large share of a transboundary water resource has a vested interested in opposing equitable solutions. Ultimately conflicts may be resolved when the parties involved come to realize that the costs associated with the absence of an agreement exceed the costs associated with an agreement.

37.4 Transboundary Aquifers

Groundwater flow systems do not recognize international boundaries. The effects of groundwater pumping and land use activities may extend across international boundaries. The most obvious scenario is where the drawdown from groundwater pumping in one country extends across an international boundary. Contamination from land uses may also migrate across international boundaries. Transboundary impacts may be more subtle. Land use changes may cause changes in aquifer recharge rates that affect down gradient aquifer water levels.

Management of transboundary groundwater necessarily includes two main elements. First is the development of an understanding of aquifer dynamics, particularly hydraulic properties, storage volume, recharge and discharge locations and rates, location of sensitive environments, and current utilization and its impacts. The second element is development of a management agreement that is acceptable to the states that share the aquifer.

Eckstein and Eckstein (2003, 2005) proposed six models or scenarios in which groundwater resources can have transboundary consequences. The six models summarized below encompass the vast majority of transboundary aquifers:

Model A: Unconfined aquifer that is hydraulically linked with a river and both of which flow along an international boundary. The river may act as a hydraulic barrier.

Model B: Unconfined aquifer intersected by an international boundary and linked hydraulically with a river that is also intersected by the international boundary. One country is thus upstream and up-gradient of the other, with the associated implications for pollution transport.

Model C: Unconfined aquifer intersected by an international boundary and hydraulically linked to a river that flows entirely on one state.

Model D: Unconfined aquifer completely in one state, but is hydraulically linked to a river that flows across the international boundary. The upstream state can impact river water quality and in turn groundwater quality in the downstream state.

Model E: Confined aquifer intersected by an international boundary and not connected with a surface water body. The aquifer may be recharged in only one state.

Model F: Confined aquifer intersected by an international boundary that is largely disconnected from the hydrologic cycle (i.e., contains non-renewable water).

Management of transboundary aquifers is especially challenging in arid and semiarid regions were aquifers are being overexploited and the extractions in one nation can materially affect the groundwater resources of its neighbors. Non-renewable aquifers are a particular challenge as nations are drawing upon an essentially finite reserve of water (Chap. 36). Non-renewable water that one nation withdraws near the border is not available to the other nation.

Nations may have different visions as to the use of the aquifer, and these visions may be conflict. For example, one nation may view current intense groundwater use as an important element in its social and economic development, while its neighbors may be inclined towards conserving the resource or may simply not be in a position now to intensely use the water for its own purposes. A tragedy of the commons scenario (Sect. 32.2) may exist with respect to transboundary aquifers in which a state obtains the full benefit of groundwater pumping whereas the impacts are shared by all neighboring states.

Historically, water in transboundary aquifers is appropriated by those who have the largest pumps (Sheng and Devere 2005). The situation can become even more acute where it is recognized that the water is non-renewable and a rush exists for each state to pump as much as possible, because they fear that if they do not utilize the resource, the other state will. Another possibility is a 'fugitive resource scenario', when water moves from one state to another in a consistent, single direction (Barberis 1991; Jarvis et al. 2005). Upstream users may receive all the benefits from ground water use, while the costs are shared, or perhaps borne entirely by downstream users. Reaching agreement on transboundary water management is complicated by the difficulty of one nation monitoring another's compliance with an agreement. The nations may not have the resources to accurately monitor water use or trust each other to comply with agreements.

Brooks (1996, p. 57) noted with respect to the Middle East and North Africa

> Discussions about international waters, including those in the Middle East, typically conclude with a call for basin-wide or aquifer-wide commissions to manage them as a unit. In my view, such schemes are visionary or, at best, premature. There is simply too little trust among these nations to consider joint management. It has taken the United States and Canada many years to establish joint procedures for management of the St. Lawrence River and almost as long for the Netherlands and Belgium to learn how to manage the aquifer that underlies their border.

In practice, transboundary aquifers are managed either by Hathaway (2011):

- de facto groundwater allocation occurring through a progression of events, including unilateral action by a dominant entity or a lack of action,
- cooperative means using a variety of social and political frameworks (e.g., negotiated agreements), or
- groundwater allocation effected through existing surface water agreements in which a surface water/groundwater connection exists and is recognized.

Hathaway (2011) concludes that identification of appropriate governance mechanisms and selecting the necessary tradeoffs under shortage conditions will constitute the greatest challenge in achieving effective transboundary groundwater management.

37.5 Transboundary Groundwater Law

As reviewed by Eckstein and Eckstein (2005), transboundary aquifer management, to the limited extent that it has been addressed, has historically been arranged by bilateral agreements or treaties between parties, rather than being settled though non-existent international law. Eckstein and Ekstein (2003, p. 205) noted that:

> With ground water consumption reaching and even exceeding sustainable withdrawals in many parts of the world, and in order to avoid future disputes and maximize beneficial use of this shared but finite resource, there is a need to clarify the rights and obligations that states enjoy vis-à-vis transboundary and international ground water resources.

The development of international law with respect to groundwater was reviewed by Hayton and Utton (1989), Matsumoto (2002), Eckstein and Eckstein (2003, 2005), Hall (2004), and Jarvis et al. (2005). With respect to international law, transboundary aquifers are usually addressed in a cursory, ill-defined manner with a lack of consensus regarding applicable international agreements and law (Jarvis et al. 2005).

37.5.1 Bellagio Draft Treaty

The Bellagio Draft Treaty for transboundary groundwater was developed by a multidisciplinary group of specialists over an eight-year period to provide a mechanism by which international aquifers could be managed by mutual agreement rather than continuing to be subjected to unilateral taking (Hayton and Utton 1989). The final revision of draft was completed in 1989, and is presented and discussed by Hayton and Utton (1989). In order to minimize intrusion into the sovereign sensitivities of independent countries, control is to be asserted only in zones considered to be critical, and actual enforcement is left to the administrative agencies of each country. Oversight and facilitating responsibility is lodged in an international agency ("the Commission"), whose discretion is limited.

The intent of the Bellagio Draft Treaty was to provide a technically and legally adequate and fairly comprehensive set of provisions, explained and supported by commentary, addressed to the matter in hand (Hayton and Utton 1989). The created Commission has a research, advisory, planning, and oversight role, rather than an enforcement role. The Commission is also charged with the creation and maintenance of a comprehensive and unified database pertaining to transboundary groundwater system of concern. In the event that a dispute cannot be resolved through the Commission, the governments are to enter into formal negotiations. If they cannot come to an agreement, the parties are to refer the matter to mediation, conciliation, arbitration, the International Court of Justice, or any other means of peaceful settlement.

37.5.2 Internationally Shared Aquifer Resource Management Programme

The International Association of Hydrogeologists (IAH) and the UNESCO International Hydrological Programme (IHP) established the Internationally Shared Aquifer Resource Management (ISARM) program in 2000. The ISARM program was summarized by Puri and Aureli (2005). The goal of the ISARM Program is to promote cooperative study and management of transboundary aquifers throughout the world. The program is working to develop scientific, legal, socioeconomic, and institutional guidelines and recommendations to aid sharing nations in the management of their transboundary aquifers.

The focus of the program is the development of a multidisciplinary toolkit or guidance package that might consolidate the best practices, guidance and information on transboundary resource management. It is recognized that transboundary aquifers are complex and that no one uniform approach is likely to apply to all aquifers. A key component of the ISARM program is the performance of regional case studies in different parts of the world. The case studies are to be selected so that that each makes a contribution to the overall understanding of the management of transboundary aquifers (Puri and Aureli 2005). The focus of the ISARM program is not to provide an external solution to transboundary aquifer management, but rather to provide expertise to nations to find their own solutions.

37.5.3 1997 Convention on the Non-Navigation Uses of International Watercourses

The 1997 Convention on the Non-Navigation Uses of International Watercourses (Watercourse Convention) is a milestone in the development of international law related to groundwater resources in that a watercourse was defined as system of surface waters and groundwater that constitute a unitary whole (Eckstein and Eckstein 2003). Under the Watercourse Convention, the much better developed international norms and principles applied to surface-water resources are extended to apply to groundwaters that fall under the definition of an "international water course". The most notable principles are reasonable and equitable utilization, no substantial harm, cooperation, and good faith negotiations (Eckstein and Eckstein 2003). However, not all types of transboundary aquifers (e.g., non-renewable groundwater resources) fall under the rubric of the Watercourse Convention (Eckstein and Eckstein 2003).

37.5.4 Law of Transboundary Aquifers

The United Nations General Assembly on December 11, 2008, adopted resolution A/RES/63/124, "The Law of Transboundary Aquifers." The resolution encourages states to make appropriate bilateral or regional arrangements for the proper management of their transboundary aquifers, taking into account the principles of the draft articles. It is the intent that states work together to resolve their groundwater management issues. It was also recommended that a convention be elaborated based on the principles of the draft articles of The Law of Transboundary Aquifers. Some of the main principles of the draft articles include

- Each aquifer state has sovereignty over the portion of a transboundary aquifer or aquifer system located within its territory. It shall exercise its sovereignty in accordance with international law and the present articles.
- Aquifer states shall utilize transboundary aquifers or aquifer systems according to the principle of equitable and reasonable utilization.
- Aquifer states shall establish individually or jointly a comprehensive utilization plan, taking into account present and future needs, and alternative water sources available for the aquifer states.
- Aquifer states shall not utilize a recharging transboundary aquifer or aquifer system at a level that would prevent continuance of its effective functioning.
- Aquifer states shall, in utilizing transboundary aquifers or aquifer systems in their territories, take all appropriate measures to prevent the causing of significant harm to other aquifer states or other states in whose territory a discharge zone is located.
- Aquifer states shall, on a regular basis, exchange readily available data and information on the condition of their transboundary aquifers or aquifer systems, in particular of a geological, hydrogeological, hydrological, meteorological and ecological nature and related to the hydrochemistry of the aquifers or aquifer systems, as well as related forecasts.

The Law of Transboundary Aquifers also lists relevant factors *that should be taken into account in order to evaluate equitable and reasonable utilization, which include*:

(a) the populations dependent on the aquifer or aquifer system in each aquifer state,
(b) the social, economic and other needs, present and future, of the aquifer states concerned,
(c) the natural characteristics of the aquifer or aquifer system,
(d) the contribution to the formation and recharge of the aquifer or aquifer system,
(e) the existing and potential utilization of the aquifer or aquifer system,
(f) the actual and potential effects of the utilization of the aquifer or aquifer system in one aquifer state on other aquifer states concerned,
(g) the availability of alternatives to a particular existing and planned utilization of the aquifer or aquifer system,
(h) the development, protection and conservation of the aquifer or aquifer system and the costs of measures to be taken to that effect, and
(i) the role of the aquifer or aquifer system in the related ecosystem.

Although the principles of the Law of Transboundary Aquifers are sound, the fundamental challenge is how to apply them to a real world situation, such as an aquifer in overdraft that is shared by states with serious water needs and perhaps different historical aquifer uses. The basic challenge is the conflict between equitable use and no harm to existing historic users.

37.6 Transboundary Groundwater Management: United States and Mexico

Every transboundary aquifer has aquifer-specific water management issues. The transboundary aquifers shared by the United States and Mexico illustrate some of the types of water management issues that may arise between friendly countries. The challenges are obviously much greater when there is a high degree of animosity and distrust between the parties sharing an aquifer.

Difficulty in developing an effective management strategy for the transboundary aquifer shared by the two countries is exacerbated when there are different legal systems in the two countries. Groundwater management in Mexico is centralized, with water being considered a national resource subject to central allocation, while water law in the United States is under the purview of each state (Hall 2004; Mumme 2005). The border-states within the United Sates have different legal traditions for the management of groundwater and are jealous of their state sovereignty. Mumme (2005) also recognized that economic conditions are an impediment to effective management of transboundary aquifers. Water users, particularly the predominant agricultural users, have little economic incentive to economize their water use because their cost is so low. Groundwater users essentially pay for only the cost of pumping the water, not for the water itself.

The 1944 Water Treaty between the United States and Mexico established the International Boundary and Water Commission (IBWC) to resolve transboundary water disputes. The IBWC has historically focused almost entirely on surface water issues. Minute 242 of the IBWC struck in August 1983, was the first bilateral agreement the two countries to limit groundwater pumping in one specific area (Mumme 2005). Minute 242 also commits the two countries to consultation on new development in shared aquifers along the entire border. However, development of the Minute 242 commitment on groundwater lacked the level of political support to move forward in the policy area (Mumme 2005). In the United States, concerns were expressed that it infringed upon state's rights to regulate groundwater. The agreement was also vague concerning what actually constitutes the obligation to consult (Mumme 2005). Difficulties also lie in the decentralization of water regulation in the United States and the political difficulty of regulating private use of groundwater supplies (Mumme 2005). The political difficulties are exacerbated when the reductions in private groundwater use in the United States, which is viewed as a property right (primarily in western states), would be perceived as being for the benefit of Mexican users.

37.6.1 Hueco Bolson Aquifer

The El Paso area is an arid environment with a limited water supply. The average annual rainfall in the City of El Paso is only about 22 cm (8.7 in) and the natural recharge is thus low. The Hueco Bolson aquifer is a regional aquifer that is shared by two countries (US and Mexico) and three states (Texas, New Mexico, and Chihuahua). Most of the population is centered in the cities of El Paso, Texas, and Ciudad Juarez, Chichuahua. The area is located within the Basin and Range Province, in which aquifers occur in sediment-filled, normal block-faulted basins. The Hueco Bolson is a roughly north–south oriented trough approximately 322 km (200 miles) long and 40 km (25 miles) wide and consists of alluvial deposits with a maximum thickness of over 2,740 m (9,000 ft), although freshwater is present down to a maximum depth of about 430 m (1,400 ft) (Ashworth 1990; National Research Council 1994; Sheng et al. 2001; Sheng and Devere 2005). The Hueco Bolson aquifer is the primary or sole water source for the cities of El Paso and Cuidad Juarez, Fort Bliss Military Reservation in Texas, and some smaller communities (Sheng et al. 2001).

Recharge occurs through mountain-front recharge, seepage from the Rio Grande and canals, irrigation return flows, deep well recharge, and inter-basin flow (Ashworth 1990; Hibbs 1999; Sheng and Devere 2005; Hutchinson 2006; Druhan et al. 2008). Environmental isotope data indicate that mountain-front recharge and inter-basin from the Tularosa Aquifer to the northern are two distinct recharge sources in the northern Hueco Bolson aquifer (Druhan et al. 2008). Modeling results indicate that the drawdowns within the Hueco Bolson have increased interbasin groundwater flow (Hutchinson 2006; Druhan et al. 2008).

Long-term pumping rates have significantly exceeded both natural and artificial recharge. Because of the low recharge rates, the aquifer has been in overdraft since the 1900s (Ashworth 1990). Total groundwater withdrawals in 2001 were approximately 312 Mm^3 (82.4 billion gallons) versus 9.6 mm^3 (2.5 billion gallons) of natural and artificial recharge (Sheng and Devere 2005). However, there is considerable uncertainty in the mountain-recharge rate based on the assumptions inherent in the methods of estimation (Druhan et al. 2008). Pre-development inter-basin flow into the northern Hueco Bolson aquifer is estimated to be on the order of 4.6–7.4 Mm^3 (1.2–1.95 billion gallons), with Hueco Bolson drawdowns causing an addition 22.2 Mm^3 (5.8 billion gallons) of inflow (Hutchinson 2006; Druhan et al. 2008). Irrespective of uncertainties in recharge and inflow estimates, it is clear that the current groundwater withdrawals are at least an order of magnitude greater than the total water inputs into the basin.

The Hueco Bolson Aquifer is, therefore, susceptible to overdraft, and water levels have been in decline since the initial development of the aquifer in the early 1900s. The history of groundwater management in El Paso was documented by Hutchinson (2006). Heavy water usage has resulted in drawdowns in the area of Ciudad Juarez and El Paso of more than 60 m (200 ft) and increases in the salinity of the aquifer water (Heywood and Yager 2003). Brackish-water intrusion occurs from several different processes including leakage from interbeds and confining strata, lateral encroachment, upconing of saline water, and irrigation drain return flows (Hibbs 1999; Hutchinson 2006; Sheng and Devere 2005).

Several options have been identified to better manage or extend the life of the Hueco Bolson aquifer (Sheng et al. 2001)

- Artificial recharge (aquifer storage and recovery).
- Use of brackish water for blending or desalination both as an alternative water source and as means of controlling movement of brackish water towards fresh water. There are an estimated at 24.7 Bm^3 (6.5 trillion gallons) of brackish groundwater in El Paso County alone (Sheng and Devere 2005).
- Increase use of surface water by the conversion of agricultural surface water use to municipal use through linings of canals, retirement of agricultural lands, and improvement of irrigation efficiency. Groundwater should be used more strategically as a resource for coping with drought periods.
- Optimization of pumping so as the minimize salinization.

The underlying brackish groundwater is being used as feedwater for the world's largest inland desalination facility. The El Paso Water Utilities (EPWU) Kay Bailey Hutchison Desalination Plant has a design capacity of 104,000 m^3/d (27.5 Mgd).

Artificial recharge using reclaimed water is recognized to be one component in the long-term management of the Hueco Bolson Aquifer. The recharge would not stop the overdraft, but would contribute towards extending the lifetime of the aquifer. The EPWU Hueco Bolson Recharge Project involves the injection of highly-treated wastewater from the Fred Hervey Water Reclamation Plant into the upper Hueco Bolson Aquifer. The reclamation plant has a design capacity of 37,850 m^3/d (10 Mgd) and produces water of quality that is comparable to that of the water currently present in the aquifer and meets all USEPA and Texas Department of Health primary and secondary drinking water standards (National Research Council 1994).

One of the purposes of the Hueco Bolson Recharge Project is the local restoration of groundwater levels (Sheng 2005). From a hydrologic perspective, groundwater levels are restored locally so long as injection continues. Once injection stops, however, aquifer heads will recede to background levels. The major hydrologic benefit of the ASR system is that it contributes towards a more sustainable water use. The recharge of recovered water reduces the net extraction of the City of El Paso. However, it is recognized that restriction of pumping on the Texas side of the border alone will not stop the depletion of the remaining freshwater storage in the aquifer or prevent further brackish water intrusion (Sheng and Devere 2005).

The greatest difficulty facing management of the Hueco Bolson aquifer is getting federal, state, and local agencies involved in water management on both sides of the border to cooperate with each other (Sheng and Devere 2005). There is no legal framework to bring all the parties together to address the depletion of the freshwater resources of the Hueco Bolson aquifer, which will profoundly affect both sides of the border. Different states have different natural resources and economic capabilities to develop alternative water supplies. In addition the "white map syndrome" is a fact of life in the region, in which nothing is considered to exist on the other side of the border (Chafez 2000). Chafez (2000) also noted that unless all the actors in the region really work together to jointly develop alternatives for future water supply, the region is on a collision path.

37.6.2 All-American Canal Lining

The All-American Canal is an aqueduct that conveys Colorado River water to the Imperial Valley of Southern California. The name of the canal is derived from its being constructed entirely on the United States side of the United States-Mexico border. The canal, which was opened in 1942, was constructed unlined and seepage recharge became an important component of the Mexicali-Imperial Valley aquifer water budget. The seepage loss also represents a loss of water to southern California water users. It was recognized that additional water would be available to southern California water users if the canal were lined to reduce the seepage losses. The San Diego County Water Authority agreed to pay part of the $285 million cost of a project to line the canal, in exchange for the estimated 95 Mm^3 (77,700 acre-feet) of water saved from the lining. Plans to line the canal developed into an international controversy that raised some basic issues concerning trans-national groundwater management and law, which were discussed by Huber (2008) and Kibel (2009) and are herein summarized.

The U.S. Bureau of Reclamation obtained final approval in 2005 to line a portion of the All-American Canal with concrete. Users of the Mexicali-Imperial Valley aquifer had become accustomed to the seepage recharge. The reduction in recharge associated the canal lining would lower aquifer water levels, which would have detrimental impacts to farmers, cities (particularly city of Mexicali), and hydrologically connected wetlands. Of particular concern was the Andrade Mesa wetland area in Northern Mexico, which is an important feeding location for migratory birds.

Mexican and American non-profit groups challenged the approval of the lining project in United States Federal District Court. The plaintiffs argued that although the canal lining would occur on United States soil, the environmental, social, and economic impacts to the Mexican side of the canal would revert back to the United States. For example, drying up of the Andrade Mesa wetlands would impact birds that spend part of their lives in the United States. The position of the Bureau of Reclamation was that the seepage water belonged to the United States as part of its Colorado River allocation under the 1944 Waters Treaty and that its status remained unaffected by any conservation measures the United States should take. The litigation prompted the United States Congress to adopt legislation in December 2006 that exempted the project from compliance with United States environmental laws. The lining of 23 miles of canal was completed in 2009.

Fundamental questions raised include:

- Does the principle of equitable utilization (i.e., not to cause significant harm to other basin states) apply because Mexican farmers and cities had developed a dependency on the recharge over the course of decades?
- Was the recharge water abandoned over the course of decades, and thus available for the beneficial use of others, or was there intent all along to recapture the water at some time in future?

- What are the obligations of states to assess cross-boundary environmental impacts (transnational environmental impact assessments)?
- What are the obligations to minimize detrimental impacts to recharge and discharge processes under the United Nations International Law Commission Draft Articles on the Law of Transboundary Aquifers?

A unilateral decision was made by the United States, and Kibel (2009, p. 207) concluded:

> In sum, the experience to date with the All-American Canal reveals that while the United States and Mexico may share a common aquifer, there is presently little common ground as to the appropriate legal framework for managing this groundwater in an equitable and ecologically sound manner.

Hathaway (2011) observed that neither groundwater impacts associated with the lining of the All American Canal or science were brought into the decision-making process in a meaningful way.

References

Allan, J. A. (2001). *The Middle East water question: Hydropolitics and the global economy*. London: I.B. Tauris.

Anderson, M. P., & Woessner, M. W. (1992). *Applied groundwater modeling, simulation of flow and advective transport*. San Diego: Academic Press.

Ashworth, J. B. (1990). *Evaluation of ground-water resources in El Paso*. Texas: Texas Water Development Board Report 324.

Barberis, J. (1991). The development of international law of transboundary ground water. *Natural Resources Journal, 31*, 167–186.

Benvenisti, E. (1996). Collective action in the utilization of shared freshwater: The challenges of international water resources law. *American Journal of International Law, 90*, 384–415.

Brooks, D. B. (1996). Between the Great Rivers: Water in the heart of the Middle East. In E. Racked, E. Rathgeber & D. B. Brooks (Eds.), *Water management in Africa and the Middle East: Ottawa* (pp. 44–57). Canada: International Development Research Centre.

Chavez, O. E. (2000). Mining of internationally shared aquifers: The El Paso-Juarez Case. *Natural Resources Journal, 40*, 237–260.

Druhan, J. L., Hogan, J. F., Eastoe, C. J., Hibbs, B. J., & Hutchinson, W. R. (2008). Hydrogeologic controls on groundwater recharge and salinization: A geochemical analysis of the northern Hueco Bolson aquifer Texas, USA. *Hydrogeology Journal, 16*, 281–296.

Eckstein, G., & Eckstein, Y. (2003). A hydrogeological approach to transboundary ground water resources and international law. *American University International Law Review, 19*, 201–258.

Eckstein, Y., & Eckstein, G. E. (2005). Transboundary aquifers: Conceptual models for development of international law. *Ground Water, 43*, 679–690.

Feitelson, E. (2006). Impediments to the management of shared aquifers: A political economy perspective. *Hydrogeology Journal, 14*(3), 319–329.

Fisher, F. M. (2002). Water value, water management, and water conflict: A systematic approach. *GAIA, 11*(3), 187–190.

Fisher, F. M., & Huber-Lee, A. (2010). The value of water: Optimizing models for sustainable management, infrastructure planning, and conflict resolution. EuroMed (2010) *Desalination for Clean Water and Energy Proceedings*, Tel Aviv, October 3–7, 2010.

Fisher, F. M., Huber-Lee, A., Amir, I., Arlosoroff, S., Eckstein, Z., Haddadin,& M. J., et al. (2005). *Liquid assets: An economic approach for water management and conflict resolution in the Middle East and beyond.* Washington, D.C.: Resources for the Future.

Fisher, F. M. (1995). The economics of water dispute resolution, project evaluation and management: An application to the Middle East. *Water Resources Development, 11*(4), 377–390.

Gleick, P. H. (2003). Water conflict chronology (updated August 18, 2003), unpublished spreadsheet.

Gleick, P. H. (2006). Environment and security: Water conflict chronology version 2006–2007. In P. Gleick (Ed.), *The world's water, 2006–2007: The biennial report on freshwater resources* (pp. 189–218). Washington: Island Press.

Hall, R. E. (2004). Transboundary groundwater management: Opportunities under international law for groundwater management in the United States-Mexico border region. *Arizona Journal of International and Comparative Law, 21*, 873–911.

Hathaway, D. L. (2011). Transboundary groundwater policy: Developing approaches in the western and southwestern United States. *Journal American Water Resources Association, 47*(1), 103–113.

Hayton, R. D., & Utton, A. E. (1989). Transboundary groundwaters: The Bellagio Draft Treaty. *Natural Resources Journal, 29*, 663–720.

Heywood, C. E., & Yager, R. M. (2003). *Simulated ground-water flow in the Hueco Bolson, an alluvial aquifer-basin aquifer system near El Paso*, Texas: U.S.G.S. Water-Resources Investigations report 02-4108.

Hibbs, B. J. (1999). Hydrogeologic and water quality issues along the El Paso/Juarez Corridor; an international case study. *Environmental and Engineering Geoscience, 5*, 27–39.

Huber, S. S. (2008). The legalities of lining the All-American Canal: CEMLP Annual Review (26 pp.). The Centre for Energy, Petroleum, and Mineral Law and Policy, University of Dundee.

Hussein, I., & Al-Jayyousi, O. (2001). Management of shared water: A comparison of international and Islamic law. In N. I. Faruqui, A. K. Biswas & M. J. Bino (Eds.), *Water management in Islam* (pp. 100–105). International Research Centre: Tokyo, United Nations University Press, and Ottawa.

Hutchinson, W. R. (2006). *Groundwater Management in El Paso, Texas.* Boca Raton, FL: Dissertation.com.

International Law Association Berlin Conference. (2004). *Water Resources Law.* London: International Law Association.

International Law Association. (1967). *The Helsinki Rules on the uses of waters of international rivers.* London: International Law Association.

Jarvis, T., Giordano, M., Puri, S., Matsumoto, K., & Wolf, A. (2005). International borders, ground water flow, and hydroschizophrenia. *Ground Water, 43*, 764–770.

Kibel, P. S. (2009). A line drawn in water: Aquifers beneath the Mexico-United States Border. *Water Law Review, 12*, 191–207.

Matsumoto, K. (2002). Transboundary ground water and international law: Past practices and current implications. Unpublished research paper, Department of Geosciences, Oregon State University.

McCaffrey, S. C. (1993). Water, Politics and International Law. In P. H. Gleick (Ed.), *Water in crisis, a guide to the world's freshwater resources* (pp. 92–104). New York: Oxford University Press.

Mumme, S. P. (2005). Advancing binational cooperation in transboundary aquifer management on the U.S.-Mexico Border. *Colorado Journal of International Environmental Law and Policy, 16*, 77–94.

National Research Council. (1994). *Ground water recharge using waters of impaired quality.* Washington, DC: National Academy Press.

Pearce, F. (2006). *When rivers run dry.* Boston, MA: Beacon Press.

Puri, S., & Aureli, A. (2005). Transboundary aquifers: A global program to access, evaluate, and develop policy. *Ground Water, 43*, 661–668.

Samson, P., & Charrier, B. (1997). International freshwater conflict: Issues and prevention strategies. Green Cross International. http://www.greencrossinternational.net/GreenCrossPrograms/waterres/gcwater/report.html.

Seliktar, O. (2005). Turning water into fire: The Jordan River as the hidden factor in the Six Day War. *The Middle East Review of International Affairs*, *9*(2), Article–4.

Shemesh, M. (2004). Prelude to the Six-Day War: The Arab-Israel struggle over water resources. *Israel Studies, 9*(3), 1–45.

Sheng, Z., & Devere, J. (2005). Understanding and managing the stressed Mexico-USA transboundary Hueco bolson aquifer in the El Paso de Norte region as a complex system. *Hydrogeology Journal, 13*, 813–825.

Sheng, Z., Mace, R. E., & Fahy, M. P. (2001). The Hueco Bolson: An aquifer at the crossroads. In R. E. Mace, W. F. Mullican III & E. S. Angle (Eds.), *Aquifers of West Texas* (pp. 66–75). Texas Water Development Board Report 356.

Sheng, Z. (2005). An aquifer storage and recovery system with reclaimed water to preserve native groundwater resources in El Paso, Texas. *Journal of Environmental Management, 75*, 367–377.

United Nations. (2005). Convention on the Law of the Non-navigational Uses of International Watercources 1997. Official Records of the General Assembly, Fifty-first Session, Supplement No. 49 (A/51/49).

Wolf, A. T. (2000). Trends in transboundary water resources: Lessons for cooperative projects in the Middle East. In D. B. Brooks & O. Mehmet (Eds.), *Water balances in the Eastern Mediterranean* (pp. 137–156). Ottawa: International Development Research Centre.

Wolf, A. T. (2007). Shared waters: Conflict and cooperation. *Annual Review of Environment and Resourcs* 2007, 32.3.1.–3.29, doi: 10.1146/annurev.energy.32.041006.101434.

Part IX
Global Climate Change

Chapter 38
Overview and Impacts on Arid Land Water Resources

38.1 Climate Variability and Change: An Overview

38.1.1 Natural Climate Changes: Cyclic, Abrupt and General Variation on a Large Scale

Variations in climatic conditions control the hydrologic cycle on the Earth at many scales, both spatial and temporal. Natural factors have produced cycles of global climate change ranging from extremely warm conditions, during which no polar ice existed on the planet (referred to as "Greenhouse Earth"), to the "Snowball Earth" (also called "Icehouse Earth") when a large part of the entire planet was frozen. Natural climate changes were created by variations in the solar radiation reaching the Earth, some of which were controlled by changes in the rotation of the Earth, caused by eccentricity of the Earth's orbit around the Sun, the obliquity of the ecliptic (i.e., angle of the Earth's axial tilt with respect to the plane of the Earth's orbit), and precession (i.e., direction of the Earth's axis of rotation relative to the fixed stars) (precession) (Milankovitch 1938, 1941). These natural climatic variations, commonly known as the Milankovitch cycles, operate on the temporal scales of 100,000, 41,000, and 23,000 years, respectively, or close to these general ranges (Berger 1978), (Fig. 38.1). Imbrie et al. (1984) suggested that there are three eccentricity cycles with periods of 95,000, 123,000, and 413,000 years and the precessional cycle can be subdivided into 19,000 and 23,000 year cycles.

Natural climatic variations caused the ice ages and the interglacial periods between them. The major and minor glacial and interglacial events occurred over thousands or tens of thousands of years, with some rather abrupt transitions from cold to warm and warm to cold climates. However, the changes in climate were generally slow within the overall warming (interglacial) or cooling (glacial) periods. The glacial cycles coincided with natural variations in atmospheric carbon dioxide concentrations as shown in Fig. 38.2.

R. Maliva and T. Missimer, *Arid Lands Water Evaluation and Management*,
Environmental Science and Engineering, DOI: 10.1007/978-3-642-29104-3_38,

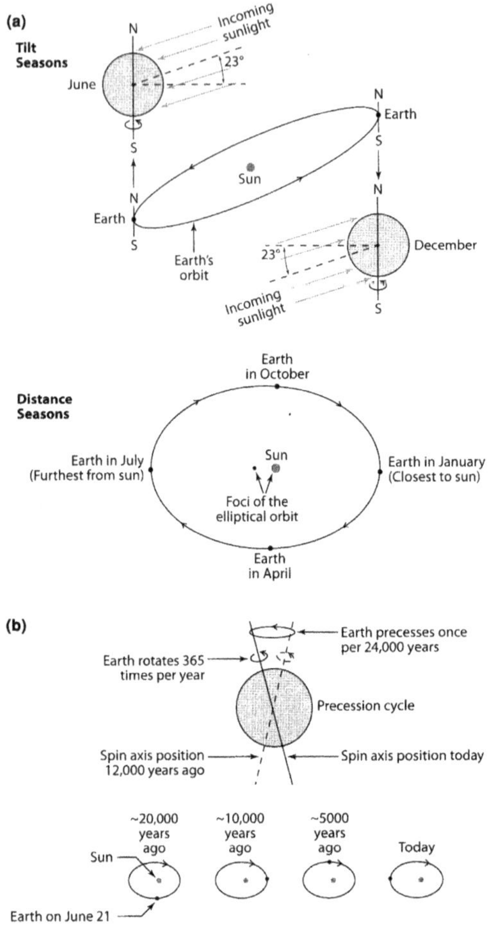

Fig. 38.1 The relationship between natural global climate cycles and the characteristics of the Earth's orbit around the Sun, the obliquity of the Earth, and the precession of the orbit known as the Milankovitch cycles (from Broecker 2010)

Superimposed within the natural global climate cycles are other rather abrupt and severe events, which are defined as large changes that occur within less than 30 years (Clark et al. 2002) or as transitions in climate that are much more rapid than the orbital cycles (Rahmstorf 2003; Alley et al. 2003). These externally-forced events can be rather dramatic, such as the Dansgard-Oeshger events, which caused a rapid cooling in the Greenland area of 8–16°C in only a few decades (Masson-Delmotte et al. 2004; Broecker 2010). The most recent of these events was the Younger Dryas which occurred between about 12,700 and 11,300 years ago with an estimated duration of 1,350 years. It marked a major cooling event from the warm Bolling Allerod period to the very cool Younger Dryas period. These events are numerous in the recent (Pleistocene and Holocene) geologic record and have durations on the millennial timeframe (Broecker 2010). For example, a series of rapid climate changes known as the Heinrich events (Hemming 2004) resulted in periods of enhanced iceberg discharges into the Atlantic Ocean. These events had durations of

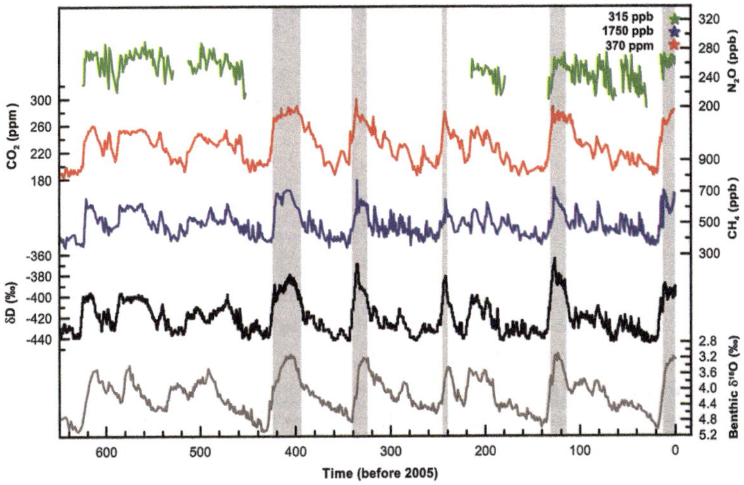

Fig. 38.2 Variations in greenhouse gas concentrations, deuterium, and oxygen isotopes in the Earth's atmosphere for the past 650,000 years showing the last five interglacial warm cycles (from Jansen et al. 2007)

hundreds to thousands of years. Another natural rapid change occurred about 8,200 years ago and lasted about 80 years. These abrupt cold events produced hemispheric effects that were out of phase and more centered in the North Atlantic, suggesting that there was not a general global lowering of mean temperature (Blunier et al. 1998; Landais et al. 2006).

The Heinrich events are believed to be caused by ice sheet instabilities while the Dansgard-Oeshger events, particularly the Younger Dryas and the 8.2 k events, were likely caused by large influxes of freshwater into the North Atlantic, thereby disrupting the ocean conveyor system that circulates warm water from the south and moderates temperature in the northern Atlantic (Alley and Clark 1999; Broecker 2010). Abrupt natural climate changes are of considerable concern today and tend to be underestimated in current climatic models (Alley et al. 2003). There is still the unanswered question concerning the ability of mankind to trigger an abrupt event based on the discharge of greenhouse gases (Broecker 1999). A sudden reduction in the North Atlantic region of 8–16°C would have catastrophic consequences with regard to the northern hemispheric population and the world as a whole (Broecker 1999).

38.1.2 Natural Climatic Variations Affecting Global Human Populations

With the evolution of mankind and the increasing influence of man on the environment of the Earth, there is now great importance given to assessment of both natural variations and man-caused (anthropogenic) impacts to the climate. Variations in the climate cause changes in the hydrologic system and, in turn, impact the available water supply and associated links to the food supply. Past natural climate variations have had impacts on human history and are expected to have even greater impacts in the future.

Global and regional climate changes have been the norm rather than the exception throughout Earth's history. Natural climate change has had profound impacts on human history, with the critical link being food supply. For example, the drying of huge areas of North Africa, Arabia, Afghanistan, and Rajasthan in the third Millenium BC appears to have had a major effect on the development of urban civilization in the region. As Lamb (1995) observed, if pasture and stocks of wild game were failing in the drying regions, then the obvious advantages of cultivation in more or less irrigated river valleys (such as those of the Trigis and Euphrates) would be obvious to those faced with abandonment of an age-old way of life. Climate change was thus a driver for the transition from a nomadic life style to agricultural and incipient urban life styles. Transitions to more arid climatic conditions were likely the cause, or at least a contributing factor, to the decline of some ancient civilizations (Chap. 21).

The Medieval Warm Period (MWP) (also referred to as the Medieval Climate Optimum and Medieval Climate Anomaly) from AD 950–1250 in Europe and North America allowed for the growth of population, technological advances, an agricultural revolution, and cultural achievements such as the building of great cathedrals in Europe (Lamb 1995). It also corresponds to the period of Viking exploration of Greenland and Newfoundland, which was facilitated by the retreat of sea ice and fewer severe storms (Lamb 1995). The MWP was followed in Europe by the Little Ice Age (LIA), which was manifested by colder winters and wetter springs and autumns. The LIA resulted in the advance of glaciers, and harvest failures and famines, which caused the desertion of villages in marginal (higher elevation and more northern) areas (Lamb 1995).

The most fundamental factor affecting long-term climate change is the radiation balance between incoming radiation originating from the sun and outgoing radiation from the surface of the Earth. Local and regional climate changes are driven by atmospheric circulation patterns. Historic climatic data reveal that major regional climatic changes have occurred rapidly (with 50 years or less) in the past, which were induced by natural causes. Changes in global wind circulation, and thus wind and ocean heat transport, may accelerate the switch to warmer or colder climate, particularly if it occurs at a critical stage of a long upwards or downwards trend in the energy budget (Lamb 1995). Climate changes documented in the southwestern United States developed in a matter of years and, in the case of the most severe past mega-droughts, persisted for decades (Cook et al. 2008).

38.1.3 Climate Change Induced by Man's Activities

> Humanity is in the process of conducting a great global experiment. If unpleasant effects are encountered they cannot be quickly reversed (Abelson 1977, p. 941).

Natural climatic cycles, whether long term or abrupt, are rooted in the complex natural processes occurring on the Earth. These complex processes produce both temporal and spatial natural variability in climate. It is important to separate natural climatic variability from anthropogenic climatic change caused by the discharge of greenhouse gases. The United Nations Educational, Scientific and Cultural Organization (2006, p. 18) suggested:

> There is frequent confusion between climate change and climate variability. Climate change is associated with global warming and is a long-term change with its origins in natural factors and, as is now accepted, human activities. Climate variability, on the other hand, has always been part of the Earth's climate system, although it has so far received surprising little attention from the water sector. It affects water resources by way of floods, droughts, waterborne disease, and so on. It is not just the extremes of climate variability that are of concern to the water sector: the increasing and extreme variability in the hydrologic cycle and climate systems, together with the dynamic processes that lie behind it, impact on countries water resources and can make it difficult to meet Millennium Development Goals (Lenton 2004).

The amount of research and the resulting publications on global climate changes has been truly explosive since the 1990s. The evidence is now indisputable that global climate has been warming since about 1900 and that the increase in greenhouse gases, including carbon dioxide, methane, and nitrous oxide have been a major cause of the increase (Solomon et al. 2007), (Fig. 38.3). Greenhouse gases affect the atmospheric radiation balance. Greenhouse gases are largely transparent to incoming solar radiation, which is mainly in the visible light and nearby wavelengths. About half of the solar radiation is absorbed by and warms the Earth's surface and radiates outwards as longer-wavelength, infrared heat. Greenhouse gases absorb some of the heat that is radiated upwards by the Earth's surface and re-radiates it in all directions, including downwards, back towards the earth surface. The re-radiated thermal energy acts to heat the Earth's surface. An increase in concentration of greenhouse gases thus should increase overall atmospheric and surface temperatures.

The description of the greenhouse effect is highly simplistic as numerous factors affect the relationship between changes in the concentration of greenhouse gases and the Earth's temperature on planet-wide, regional, and local scales. The Earth's climate has clearly been warming since the middle twentieth century. Uncertainty still exists as to the overall contribution of natural (non-anthropogenic) climate variation to the observed warming, especially since the Earth is in an interglacial, natural warming period. There is still considerable variation in the magnitude and distribution of climate changes predicted by the now numerous different atmospheric general circulation models (GCMs). Nevertheless, there is a very strong consensus in the scientific community that global warming will

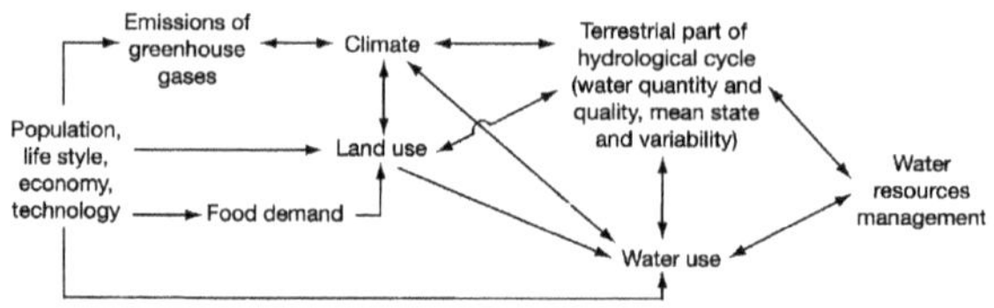

Fig. 38.3 Impact of human activities on freshwater resources and their management, with climate change being only one of the various pressures (from Kundzewicz et al. 2007 as modified from Oki 2005)

continue, but the magnitude of the increase and its effects are still uncertain and will depend upon actions taken to curb the increase in the concentrations of greenhouse gases in the atmosphere. Global climate change is expected to result in an increased variability (more extreme conditions) around a gradually increasing mean temperature.

38.2 Effects of Global Climate Change on Hydrologic Systems

38.2.1 General Effects

There is a complete integration between climate as a whole and the hydrologic cycle. Global climate change will alter the natural system and there are a series of projected changes that will likely occur. Predicted changes in the water cycle include (Trenberth et al. 2007; Bates et al. 2008; Karl et al. 2009):

- changes in precipitation patterns and intensity,
- changes in the incidence and geographic extent of drought,
- widespread melting of "permanent" snow and ice with reductions in the volume of glaciers,
- reductions in annual winter snow accumulation,
- increases in atmospheric water vapor,
- increases in evaporation,
- increases in water and soil temperatures,
- reductions in lake and river ice cover, and
- changes in soil moisture and runoff.

The rates of these changes are critical within the world of water management. Some rate changes have already been detected, while others are postulated to occur based on model predictions.

There is considerable uncertainty concerning the effects of global warming on precipitation rates, especially at regional or subregional scales. Global climate models (GCMs) are used to make future predictions on the various changes in the hydrologic cycle based on real data inputs from weather stations, historic records, paleoclimate records from ice cores, and various assumptions on the future input of greenhouse gases. GCMs predict that the total amount of water in the atmosphere will increase at a rate of about 7% per degree C increase in temperature (Trenberth et al. 2007, p. 254), as warm air can hold more water at a given relative humidity. The predicted increase in precipitation is only 1–3% per degree C of surface warming (Wentz et al. 2007). However, satellite observational data for the 1987–2006 period do not support the muted response predicted by the GCMs (Wentz et al. 2007). In any event, over the long-term total precipitation must equal total evapotranspiration based on atmospheric equilibrium and mass balance.

Climate models predict that in general precipitation in coming decades will be concentrated into more intense events, with longer periods of little precipitation in between (Bates et al. 2008). The increase in the number of consecutive dry days is projected to be most significant in the North and Central America, the Caribbean, northeastern and southwestern South America, southern Europe and the Mediterranean, southern Africa and Western Australia (Bates et al. 2008).

Changes predicted in the amount, intensity, frequency and type of precipitation include (Bates et al. 2008; Karl et al. 2009):

- pronounced increase in precipitation in eastern North America, southern South America, northern Europe,
- pronounced decrease in precipitation in the Mediterranean region and most of Africa and southern Asia,
- higher latitudes are projected to receive more precipitation,
- the dry belt that lies just outside the tropics is expected to expand polewards,
- increases in tropical precipitation during the rainy season, and
- more precipitation is expected to fall as rain rather than snow.

Huntington (2006) reviewed the data on the frequency of extreme weather events, such as hurricanes, typhoons, floods and droughts. No evidence was found for increased frequency of flooding, tropical storm frequency and intensity, and the duration of the storm season. The Intergovernmental Panel on Climate Change (IPCC) concluded that it is more likely that future tropical cyclones will become more intense, with greater peak wind speeds and heavier precipitation associated with ongoing increases in tropic sea surface temperatures (Kundzewicz et al. 2007; Committee on Environment and Natural Resources 2008; Bates et al. 2008). Huntington (2006) also cautioned that the lack of detected increase in frequency and intensity of tropical storms during the twentieth century should not be taken as evidence that further warming will not lead to such changes in the futures.

Global climate change is expected to have serious impacts to regions in which the water supply is dominated by either snow fall or glaciers. A key distinction is snowfall that seasonally melts is a renewable resource, whereas much the ice in glaciers is essentially a non-renewable resource. The impact of a warming climate on water availability in snow- and glacier-dominated regions was reviewed by Barnett et al. (2005). The winter snow pack in snow-dominated hydrologic regimes, such as the Western United States, provides an important storage function. The precipitation that falls as snow is stored at high altitudes and is more slowly released in the spring and summer as the snow melts. In a warmer world, less winter precipitation will occur as snow and the accumulated snow will melt earlier in the spring. The peak river run-off will shift to the early spring away from the summer and fall when it is needed most. Floods may become more common and intense during the spring (Gleick 1993). The shift in the timing of run-off without adequate storage will lead to regional water shortages that may impact agricultural and municipal water supplies, hydroelectric production, and environmental flows.

Glacier-dominated hydrologic regimes, such as western South America and the Himalaya-Hindu Kush region, face even more serious water supply challenges (Barnett et al. 2005). As essentially fossil water, there is no replacement to the glacial water supply once the glaciers melt. In the short term (next several decades), glacier-dominated regions may see an increase in water, because of the accelerated melting of the glaciers and associated run-off. Any increase in water supply is expected to be followed by an abrupt decrease in water supply as the glaciers disappear. Some regions dependent on glacial water are expected, based on current projections, to experience heavily depleted dry season water resources once the glaciers have disappeared (Barnett et al. 2005; Bates et al. 2008).

38.2.2 Effects on Freshwater Systems and Uncertainty

Freshwater resources are among the systems that are particularly vulnerable to climate change (Bates et al. 2008; Kundzewicz et al. 2008). The effects of global climate change will be superimposed upon existing stresses to water resources caused by growing populations and economic development within a quite complex interconnection of numerous factors (Fig. 38.3). The consequence of increasing global temperatures will be an overall acceleration of the hydrologic cycle with increased rates of both evaporation and precipitation. In agreement with the global assessment, the U.S. Global Change Research Program (Karl et al. 2009, p. 41) concluded that:

> Warmer climate will result in increased evaporation of water from land and sea and allows for more moisture to be held in the atmosphere, and hence increased overall precipitation. The regional and seasonal distribution of precipitation will change and more precipitation will come in heavier rains rather than light events. A warmer world is expected to produce both wetter and drier conditions.

Historical climatic data does appear to support the acceleration of the hydrologic cycles and the actual climatic observations are mixed (Table 38.1). However, an analysis of extreme climatic data for the second half of the twentieth century supports that the world is becoming both warmer (Fig. 38.4) and wetter (Fig. 38.5) (Frich et al. 2002; Trenberth et al. 2007), but evidence for observed interannual variability based on a rigid statistical analysis is still quite sparse (Trenberth et al. 2007). The Frich et al. (2002) study is biased towards North America, Europe,

Table 38.1 Observed climatic trends of various components of the global hydrologic system (modified from Trenberth et al. 2007 and Lemke et al. 2007)

Hydrologic Feature	Observed climate-related trends
Precipitation	Increasing over land north of 30 °N over the period 1901–2005
	Decreasing over land between 10 °S and 30 °N after the 1970s
	Increasing intensity of precipitation
Cryoshpere	
Snow cove	Decreasing in most regions, especially in spring
Glaciers	Decreasing almost everywhere
Permafrost	Thawing between 0.02 m/yr (Alaska) and 0.4 m/yr (Tibetan Plateau)
Surface waters	
Streamflow	Increasing in Eurasian Arctic, significant increases or decreases in some river basins
	Earlier spring peak flows and increased winter base flows in Northern America and Eurasia
Evapotranspiration	Increased actual evapotranspiration in some areas
Lakes	Warming, significant increases and decreases of some lake levels, and reduction in ice cover
Groundwater	No evidence for ubiquitous climate-related trend
Floods and droughts	
Floods	No evidence for climate-related trend
Droughts	Intensified droughts in some drier areas since the 1970s
Water quality	No evidence for climate-related trend
Erosion and sediment transport	No evidence for climate-related trend
Irrigation water demand	No evidence for climate-related trend

northern and eastern Asia, and South Africa because of the availability of data. No data were presented for most of North Africa, the Middle East, and South Asia. The available data indicate that global precipitation increased by about 2% between 1990 and 1998 and that there has been considerable regional variation in the changes in average precipitation (Huntington 2006). The global compilation presented by Trenberth et al. (2007) shows a marked increase in the slope of mean global temperature over the past 25 years. However, an examination of ice core data shows that such variations are common within the climatic record found in ice cores over the past several thousand years. Local and regional precipitation distribution is governed primarily by atmospheric circulation patterns, the availability of moisture, and surface terrain effects. The first two are influenced by temperature.

The attribution of changes in global precipitation pattern to climate change is uncertain, because precipitation is strongly influenced by large-scale patterns of natural variability (Bates et al. 2008; Cook et al. 2008). For example, with respect to the American Southwest, Cook et al. (2008, p. 168) cautioned that

> it would be premature to state that the recent drought heralds a period of anthropogenic drying as opposed to the continuation of natural decadal and multidecadel variations

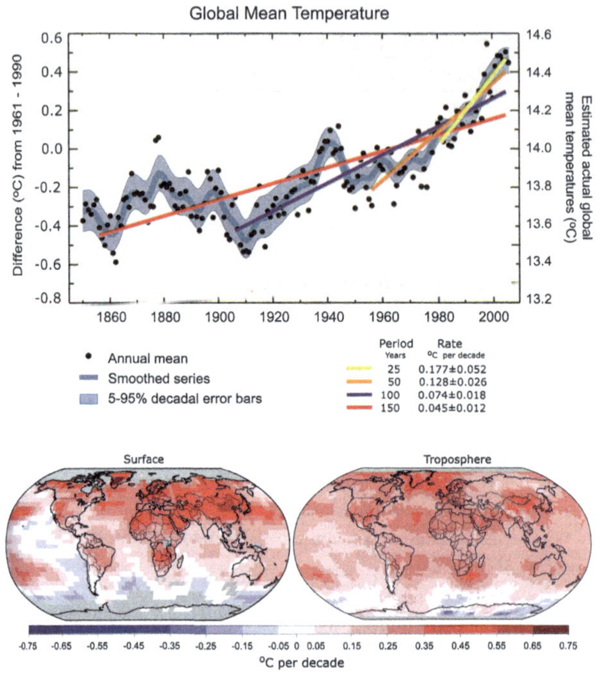

Fig. 38.4 Atmospheric concentrations of carbon dioxide, methane, nitrous oxide, and the associated radiative forcing (from Jansen et al. 2007)

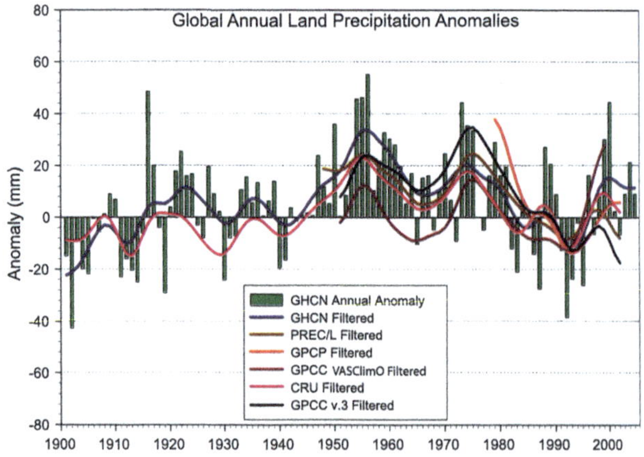

Fig. 38.5 Historical annual global mean observed temperatures with simple fits to the data. The slope of the mean fit to the data shows that the slope of the temperature increase has steepened in the last 25 years. (from Trenberth et al. 2007)

Huntington (2006) similarly cautioned that there are substantial uncertainties regarding the trends in the hydrological and climatic variables caused by the variable quality and regional coverage of the data and that regional analyses are variable and sometimes contradictory. The specific influence of climate change is only one of the many interactive factors that produces climate change as suggested by Oki (2005); (Fig. 38.1). Again, the paleoclimate data indicate that very large hydrological changes have naturally taken place over the past millennium that rival in magnitude the expected changes in the next several decades and centuries (Cook et al. 2008). A current period of local increased aridity could represent either a drought that is soon to end, the start of a natural long-term mega-drought, or the emergence of an anthropogenic drying trend (Cook et al. 2008).

The paleoclimate data offers a sobering picture of just how severe droughts can be under natural climate conditions (Cook et al. 2008). Recent climate data are insufficient to capture the frequency and occurrence of mega-droughts that clearly exceed anything in the historic records in many regions (Cook et al. 2008). For example, tree-ring data extending back to AD 1,000 documented mega-droughts in the American West during the Medieval Climate Anomaly (MCA) that persisted for several decades or longer, far beyond any droughts in the historic record (Herweijer et al. 2007). The MCA droughts dwarfed the modern droughts by their longevity rather than by their severity. The climate was just as variable as today, but the variation was around a drier mean rainfall (Herweijer et al. 2007). Herweijer et al. (2007) raised the ominous question concerning how would modern American Western society react to a mega-drought comparable to that the MCA? It has been also suggested that the climatic forcing responsible for the mega-drought could be induced by anthropogenic global warming (Cane et al. 1997; Herweijer et al. 2007).

From a practical, water management perspective, the question of the anthropogenic contribution to global climate change is not relevant. It is a fact that the global temperature is rising. The specific contribution of the natural interglacial cycle, anthropogenic warming caused by greenhouse gas emissions, man-induced changes in the surface albedo, warming induced by cities, and other natural factors are interacting to produce climate change. The key issue is how to adapt to climate changes within a rational framework of time and space. Abrupt changes in climate can occur faster than the time scales needed for human and natural systems to adapt (Cook et al. 2008). These changes cause environmental changes on the Earth's surface, in the oceans, and they are a causative agent of the failure of civilizations, along with a series of other human factors (Diamond 2005).

If climate change, for whatever causes, is impacting local water resources, then adaptive strategies need to be developed to meet societal and environmental water needs. However, the linkage between water management and greenhouse gases should not be ignored. The major concern is that current water management practices may not be robust enough to cope with the impacts of climate change. Indeed, in many locations, water management practices cannot satisfactorily cope with current climate variability (Bates et al. 2008). In arid lands, the consequences of climate change and variability and the potential reaction time to make water management changes is particularly acute because of the high level of stress already placed on the scarce resources.

38.2.3 Modeling of Global Climate Change with Emphasis on Semi-Arid and Arid Lands Precipitation

Climate modeling is a means to simulate the function of natural systems for the purpose of making predictions on how they will respond to both natural and man-caused changes. Climatic models provide guides concerning how the natural system functions and how it will respond to changes in the short- and long-term time scales. As a rule, models are only as good as the data that are incorporated into them and the validity of underlying assumptions. They do not provide absolute answers, but assist scientists in assessing what data are needed to monitor changes in climate accurately and the degree to which different atmospheric parameters affect warming and cooling. The interactive sun-atmosphere-ocean-land mass system is extraordinarily complex and all of the parameters that must be incorporated into models are not known to the degree desired. Therefore, the mathematical architecture and calibration of atmospheric models or Global Climate Models (GCMs) has a major effect on their accuracy in terms of prediction. It is necessary to assess the sensitivity of models to various input parameters to ascertain which parameters cause the highest degree of uncertainty, so that input measurements can be improved. While GCMs have been vastly improved over the past decade, they still cannot be used to actually predict past, large-scale climate changes, such as ice ages and sudden climate changes. These models, however, are useful in assessing the impacts of greenhouse gas forcing on climate change and are currently the best tools available to make such assessments.

There is considerable uncertainty concerning the effects of global warming on precipitation rates, especially on the regional or subregional scales. GCMs are used to make future predictions on the various changes in the hydrologic cycle based on real data inputs from weather stations, historic records, paleoclimate records from ice cores, and various assumptions on the future input of greenhouse gases. Projected likely (high to very high confidence) future changes in water resources (summarized from Fourth Assessment Report for Working Group II of IPCC 2007; (Bates et al. 2008; Kundzewicz et al. 2007, 2008; Committee of Environment and Water Resources 2008; U.S. Global Change Research Program 2009) include:

- Many presently water stressed arid and semiarid lands are likely to suffer decreasing water availability as both river flows and groundwater recharge decline. Areas expected to experience decreased water availability (run-off) are the Mediterranean region, southern Africa, southwestern South America, Central America, southwestern North America, and southern Australia.
- Precipitation in regions that are expected to become drier will become less frequent but more intense with longer dry periods between extreme events.
- Extreme rain extreme events will become more common. There will be greater risks of droughts and floods.

An example of a model scenario for the semi-arid area of southern Europe and the arid area of northern Africa was developed by subtracting the measured

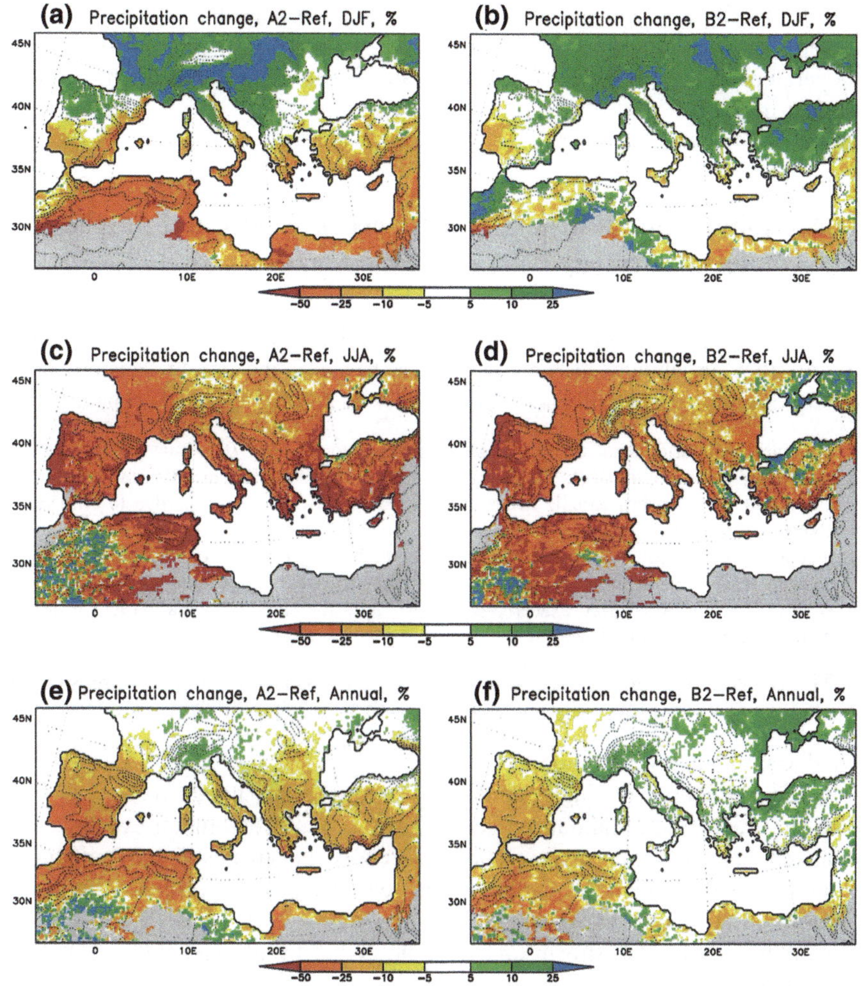

Fig. 38.6 Modeled changes in precipitation from the period 1961–1990 compared to 2071–2100 in southern Europe and North Africa (from Gao and Giorgi 2008). The winter month model runs are for winter at the top, the middle are the summer, and the bottom are the annual (Fig. 38.6)

precipitation for the period 1961–1990 from the projected precipitation for the period 2017–2100 and comparing the results based on six conceivable scenarios (Gao and Giorgi 2008) (Fig. 38.6). This region was chosen as an example, because it has been projected to be one of the most impacted areas in the world and the modulated climate change is referred to as G06. Two different greenhouse gas scenarios were modeled; the IPCC (2000) (Nakicenovic and Swart 2000) A2 scenario, with high ranges of CO_2 concentration reaching 850 ppm by 2,100, and the B2 scenario, with low ranges of CO_2 concentration reaching about 570 ppm. The region was modeled with a fine grid spacing of 20 km^2 (7.7 mi^2) The A2 model runs show an annual reduction in precipitation of up to 50% while the B2 scenario shows a little over 25% reduction. Gao and Giorgi (2008) also modeled temperature change, potential ET change, soil water change, and the impacts on drought indices.

38.3 Global Climate Change: Impacts to Arid Land Surface-Water Hydrology

Alterations of the hydrologic cycle will likely be amplified in dry lands areas because they are already susceptible to stress based on the lack of moisture and a variety of anthropogenic impacts. Kundzewicz et al. (2007, p. 175) conclude that

> Semi-arid and arid areas are particularly exposed to impacts of climate change on freshwater (high confidence). Many of these areas (e.g., Mediterranean basis, western USA, and north-eastern Brazil) will suffer decreases in water resources due to climate change (very high confidence). Efforts to offset declining surface water availability due to increasing precipitation variability will be hampered by the fact that groundwater recharge will decrease considerably in some already water-stressed regions (high confidence), where vulnerability is often exacerbated by the rapid increase in population and water demand (very high confidence).

There are several effects of climate change on the surface-water resources of the Earth as a whole and more specifically on arid lands. These affects are

- reduced annual average flows,
- increased intensity of single rainfall and corresponding flood events,
- general reduction in rainfall producing smaller number of runoff events,
- altered annual stream hydrographs, especially in streams fed by mountain snowfall with high flows earlier in the season and more extended low lows during the typically dry season,
- possible cessation of stream and river flows fed by glaciers (full ice-melt in some regions), and
- rising sea levels will flow estuarine tidal areas forcing seawater further inland and affecting surface-water resource availability.

On a global basis, a warmer climate will result in greater rainfall because of increased rates of ET. However, runoff and stream flows will be redistributed on a regional basis. Milly et al. (2005) made a global assessment of the effects of climate change on runoff starting by comparing historical records for the periods 1900–1970 and 1971–1998. The expected changes for the period 2041–2060 were assessed using an ensemble of 12 different models. The twenty first century data show that runoff has changed regionally, ranging from increases of 30% to decreases of 20%. The model for the period 2041–2060 projects changes of 40% higher to at least 30% lower (Fig. 38.7). The key predictions for arid lands are significant reductions in runoff in western North America, Mexico, southwestern South America, North Africa and the southern Mediterranean area, southern

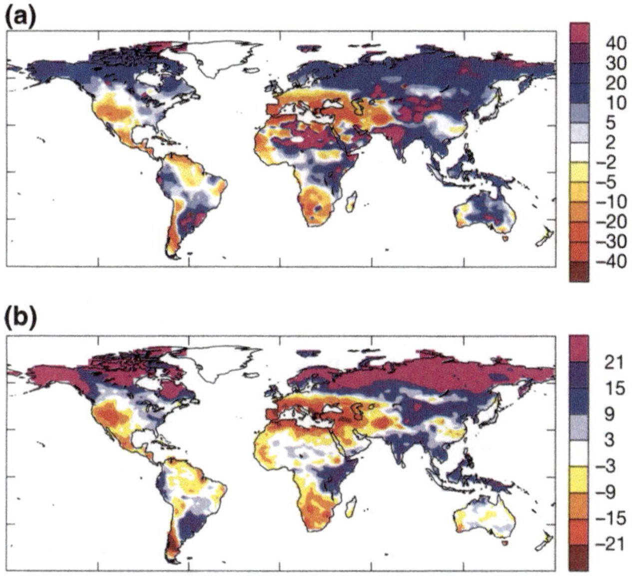

Fig. 38.7 Relative changes in runoff in the twenty first century (from Milly et al. 2005). **a** Ensemble (arithmetic mean) of relative (percentage) change. **b** Number of pairs of runs showing a positive change minus number showing a negative change

Africa, the Middle East in general (some increases in the Arabian peninsula), and western Australia. The reduction in many of these regions is based on the overall lessening of annual rainfall. Additionally, the flows of several of the streams feeding arid lands with glacier meltwater will lessen or disappear. Milly et al. (2002) also warned that there will be significant increases in flooding in all regions based on the extremity of events.

Global climate change will increase the variability in precipitation and corresponding runoff, and will likely have a significant effect on the frequency and duration of drought conditions, particularly in arid areas. As warming occurred during the twenty first century, the Palmer Drought Severity index shows that droughts have become more frequent (Trenberth et al. 2007; Dai et al. 2004). This is illustrated in Fig. 38.8. The most affected arid regions include western South America, the Saharan area of northern Africa, southern Europe, Iran, and both eastern and western Australia. Care must be taken in the interpretation of gaging records because many rivers and streams have been dammed with the reservoir storage reducing peak seasonal flows (Ye et al. 2003).

The historical changes and the anticipated future changes in stream flow characteristics raises a number of serious questions regarding the current methods being used to manage and control surface water flows. Historic water stages and discharge records have been used globally to manage surface water systems and to make projections for management of water supplies and for flood control.

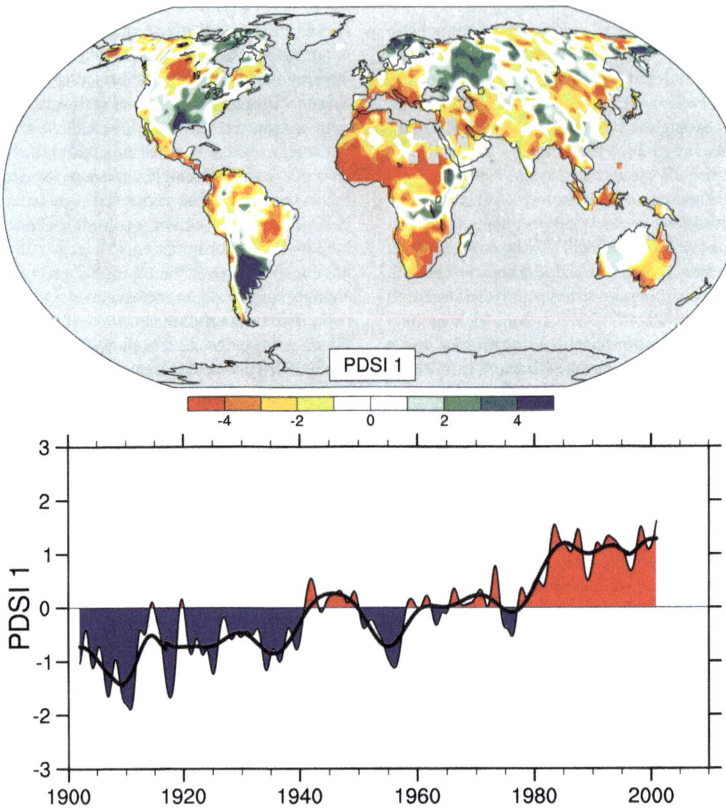

Fig. 38.8 Changes in the Palmer Drought Index between 1900 and 2000 (from Dai et al. 2004 as illustrated in Trenberth et al. 2007)

The projections being made under various global climate change scenarios raise the question concerning the usefulness of using historic records, when more extremes are going to be the norm in the future. The possibility of having several 100 year storms in a single year based on historical stage records will become a real possibility. Therefore, the methods of surface-water management may be forced to change in the near future (see Sect. 39.2).

38.4 Global Climate Change: Impacts to Groundwater Systems

The effects of global climate change on groundwater resources have considerably uncertainties and the impacts are sure to vary between regions. Changes in rainfall patterns can affect groundwater recharge and thus the amount of water ultimately

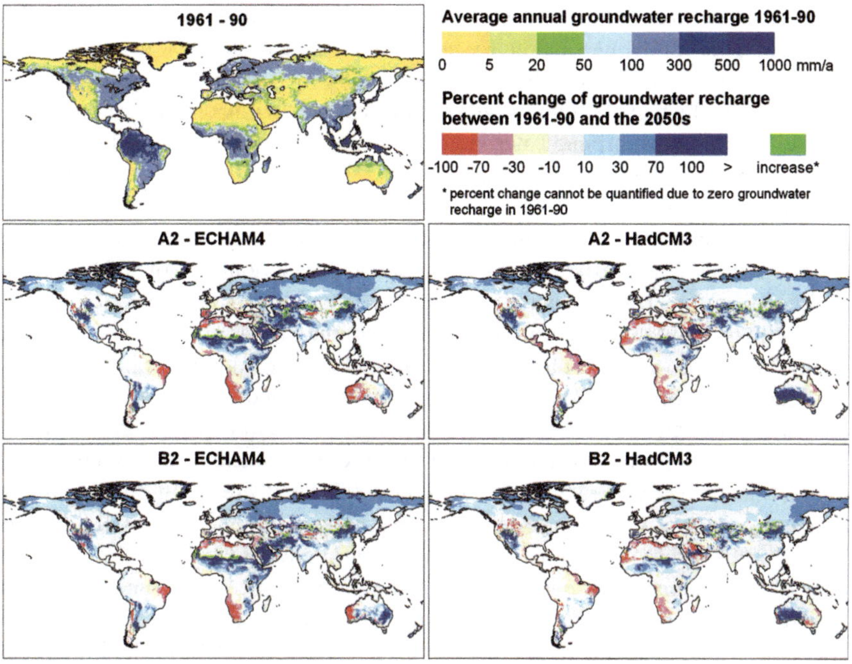

Fig. 38.9 Simulated changes in diffuse groundwater recharge based on global climate change modeling (from Doll and Florke 2005 as illustrated in Kundzewicz et al. 2007)

available for use. Changes in temperature and rainfall are also expected to affect the demand for groundwater. However, groundwater will also be an important element of adaptation to climate change because of the great volume of storage provided by aquifers. Groundwater can provide a valuable temporal buffer against variations in water supply, especially with the expected changes in the patterns of streamflow. Depleted aquifers can have large available capacity for the storage of excess surface water and reclaimed water using various managed aquifer recharge techniques (Chaps. 23 and 39).

Global models tend to predict that groundwater recharge rates, based on the greater amount of water in the atmosphere, will increase, but not at the same rates as stream runoff (Doll and Florke 2005). Kundzewicz et al. (2007) suggest that modeling using the ECHAM4 interpretation of the A2 scenario shows a 9% increase for the combination of total runoff, including groundwater recharge and fast surface and subsurface runoff for the compared periods of 1961–1990 to the 2050s. True groundwater recharge only increases by 2% for the same comparison. The simulated changes between 1961–1990 and 2041–2070 are shown in Fig. 38.9. Careful examination of these simulations for arid regions shows considerable variation with the most negatively affected areas being the northernmost and southern most parts of Africa and parts of southwestern South America. It is interesting to note that parts of Saudi Arabia are shown to have nearly a 100%

increase in recharge rates and parts of the Western United States show a 30% increase. This modeling does not consider the impacts of saltwater intrusion in coastal areas that is induced by higher sea levels.

Predicting the fate of groundwater resources in a changing climate is complicated by the fact that most groundwater systems have already been, and will continue to be, altered by human activities that are not related to climate change and many hydrological and biological process involve non-linear responses to atmospheric conditions associated with climate change (Green et al. 2007). Groundwater resources will be increasingly stressed by global population growth and be affected by impacts to other water sources, particularly rainfall and surface water. Climate change may indirectly affect groundwater resources if surface water supplies are locally reduced, which would prompt additional groundwater pumping.

Changes in precipitation and temperature will affect groundwater resources primarily through their effect on the recharge rate. Groundwater recharge is a sensitive function of local climate factors (micro-climates) and geology, topography, and land use (Dragoni and Sukhija 2008). A decrease in precipitation or temperature-induced increase in the evapotranspiration rate will decrease groundwater recharge and thus ultimately the groundwater supply. Climatological data clearly indicate that over the past half century some areas, such as southern Europe, have experienced both increasing temperature and decreasing precipitation (Ducci and Transfaglia 2008; Polemio and Casarano 2008). Monitoring data also indicate a lowering of aquifer water levels or heads and spring discharges, but it is difficult to make a direct link to changes in recharge because groundwater exploitation has also increased (e.g., Ducci and Transfaglia 2008).

Owing to a lack of data on historic recharge rates and groundwater levels, and the slow reaction of groundwater systems to changing recharge conditions, climate-related changes in groundwater recharge has not been clearly observed (Alley 2001; Bates et al. 2008). An important consideration involving evaluation of groundwater resources in arid regions is that steady state conditions may not prevail and aquifers, particularly deeper aquifers containing old water, may have long-duration transient responses to stresses (Seiler et al 2008). The implications of transient responses to, for example a decrease in aquifer recharge, may not be evident at the present time. Even if climatic conditions were stabilized now, the consequences of the present trends of climate change will continue in subsurface well into the future because of the delayed responses of all large subsurface reservoirs (Seiler et al. 2008).

Climate change will affect recharge both through the temperature and precipitation effects, particular in coupled riverine-fluvial systems where losing streams are the primary source of annual recharge. In some arid and semiarid mountainous regions, snowmelt is the major source of recharge water. As temperature rises more of the precipitation with fall as rain rather than snow. Snowmelt on the contrary provides a slower and steadier source of water, of which a greater proportion may recharge underlying aquifers. Rising temperatures may also result in more rapid snowmelt.

The tendency for greater variability in rainfall can have both positive and negative impacts on groundwater recharge. Recharge may decrease in humid regions because the infiltration capacity of soils will be exceeded more frequently, and will result in a greater proportion of the rainfall running off (Bates et al. 2008). On the contrary, more intense rainfall events may increase recharge in arid regions, as only high intensity rainfall events are able to infiltrate fast enough for water to escape loss to evapotranspiration. Alluvial aquifers are recharged mainly by inundation during floods (Bates et al. 2008; Kundzewicz and Döll 2009). The more intense rainfall events will also allow greater recharge of fractured bedrock and basalt-flow aquifers, thereby allowing greater groundwater discharge to ephemeral streams (Al-Sefry et al. 2004).

In addition to the direct effect of climate change on hydrological systems, climate change is expected to result in indirect impacts on groundwater resources. How humans respond to climate change will affect the ultimate impacts to water resources. The relationship between physical hydrology and socioeconomic responses can be highly complex. Holman (2006) discussed the indirect socio-economic impacts resulting from the impacts of climate change on land use patterns such as urbanization, flooding, and crop production. The direct impacts of climate change are generally more important than those of socioeconomic scenarios. However, locally the impacts of socioeconomic scenarios can be highly significant, especially where they lead to major land use changes (Holman 2006). For example, warmer temperatures and a longer growing season can necessitate increased irrigation, which could result in a local increase in groundwater recharge from the return flow (Toews and Allen 2009). If the local increase in ET due to increased temperatures is greater than the increase in precipitation, then the result could be a decrease in soil moisture, which may result in decreased agricultural production and increased irrigation requirements (Gleick 1993).

38.5 Regional Variability in Surface Water and Groundwater Impacts

Modeling of the responses of groundwater in arid and semiarid regions to climate change is complicated by recharge tending to be highly episodic. The main issue is not changes in annual rainfall, but rather changes in the abundance, intensity, and timing of recharge-producing rainfall events. While the general global models (Sect. 38.4) are useful for identifying regions that will have the most significant changes, local models will have to be developed to assess specific regional water management strategies.

Development of regional and subregional climate models to assess changes in surface-water and groundwater resources is even more complex than the development of standard surface water, groundwater, or integrated models (see Chap. 20). The overprint of the atmospheric forcing models in conjunction with the local water resources models adds the requirement for more data, which are commonly not available. Some assumptions need to be made concerning the future climate scenarios for a region or subregion, based on a coarse-grid, large-scale climate model, and then a finer-scale local model is developed based on the water-resources data available. Projections of water resource management scenarios can then be made with the full understanding that there will be large error ranges and perhaps some unpredictable extreme scenarios, such as more intense droughts or floods.

From a data input perspective, the development of surface water response models is somewhat easier to achieve compared to complex groundwater models that are developed to predict recharge changes and water availability in the long term. The changes in rainfall predictions based on global and regional climate models can be reasonably translated into regional and subregional settings by reducing the grid spacing and adding more local climate data stations locations (where such data are available). Translations of the climate forcing model scenarios into fine-grid groundwater models is more difficult because fewer data are generally available to populate the model layers and the response times are orders of magnitude slower compared to correlative surface water responses.

An example of a regional model is that developed by Ng et al. (2010) to assess the effects of climate change on recharge in the semiarid Southern High Plains on Texas and New Mexico. Current recharge occurs mainly by diffuse vertical flow below agricultural (cotton) fields. An ensemble forecasting technique was employed that generates many equally likely outcomes that can be statistically analyzed. This technique allows full, reasonable ranges of outcomes to be assessed to develop management strategies to deal with the extremes. The inputs were data-conditioned realizations of soil and vegetation properties and meteorological time series realizations generated with a stochastic weather generator for a range of CGM predictions.

Possible precipitation changes spanned a range of −25 – +20%, whereas modeled recharge change varied from −75 to +35%, which is similar to that projected in other semiarid areas, such as southern Europe (Gao and Giorgi 2008). The predicted changes in groundwater recharge have a substantially greater magnitude than the causative climatic changes. The differential response was related to the role of episodic recharge events. A major variable is the timing of intense rainfall events with respect to annual variations in potential evaporation (ET_p). The maximum recharge occurs during high intensity rainfall events during the winter and spring when ET_p rates are lowest (Ng et al. 2010). The driest scenario predicted a much wetter August during which there is no increase in percolation because of dry antecedent soil conditions and high crop actual ET. The Ng et al. (2010) study illustrates the great uncertainty that exists over the impacts of climate change on groundwater resources.

38.6 Impacts of Climate Change on Agriculture and Food Supply Security

The largest projected impacts of climate change on the hydrologic cycle and, consequently, the availability of water resources are expected to become quite apparent in the second half of the twenty first century. The acceleration of impacts as warming becomes greater under several of the greenhouse gas forcing scenarios between 2050 and 2100 is of considerable concern from a global perspective because some experts predict a corresponding peak in global population in about 2075 at near 9.22 billion (United Nations 2004). Recently, the United Nations has questioned their 2004 projection and has raised the projected population by 2010 to 10.1 billion, with continued growth based on a correction for projected fertility in Africa. In either case, the global requirement for water solely for irrigation to meet food security issues would be critical, even without global climate change.

In arid lands, typically the need for supplemental irrigation to grow food crops is greater compared to wetter areas. Therefore, the impact of warmer climate will tend to increase the demand for irrigation water to maintain existing production rates. Also, the increase in the intensity of floods is also a great concern in arid areas, where commonly crops are raised in the floodplains of both perennial and ephemeral streams. Flood damage to crops could become a greater risk in arid lands based on the increased intensity of the floods (Ragab and Prudhomme 2002).

Jager and Ferguson (1991) summarized the expected impacts of climate change on global agriculture in temperate and tropical areas as follows:

Temperate regions

- The growing season will increase, benefiting the crop yield.
- Increased CO^2 concentration in the atmosphere may benefit crops.
- An increase in plant disease and pest incidence could occur.
- The demand for irrigation water will increase.

Tropical regions

- The growing season will be reduced, adversely affecting crops.
- Higher temperature and high humidity could cause higher sterility in rice.
- Increased CO_2 may benefit dry matter production but not grain yield.
- An increase in disease and pest incidence could occur with new pests emerging.
- The uncertainties of the monsoon will make crops more vulnerable to water deficit.
- Cloudiness or low radiation at ground level will adverse influence crop yields.

While the projected impacts of climate change are generalized for all climates, many of the changes will be amplified in arid and semiarid lands. In all cases it is clear that food production will require more water under a warmer scenario, local sources of water supply will be stressed by extended drought periods and changes in stream flow patterns (higher peak flows and reduced low flows). Other impacts include greater upstream penetration of tidal, saline water, and perennially arable land area may be reduced by the increased flooding. The issue of agricultural production and food supply becomes even more complex when the realization occurs that climate change will likely increase the quantity of water required to meet current production needs of a seven billion global population without consideration of the future needs of perhaps ten billion people.

Perhaps another key question is: what impact will global climate change have on the amount of arable land available for agricultural development? The amount of arable land is a function of the availability of water, soil conditions, the ability to withstand flooding, and the duration of the growing season. Climate change will cause some reductions in arable area based on sea level rises, which will flood some low-lying coastal farm lands with tidal water. A similar reduction will occur in agricultural lands adjoining inland estuarine areas. Increased reliance on groundwater for irrigation may reduce arable land area by resource depletion or soil salinization. Again, high intensity flooding of stream flood plains may impact the viability of vast agricultural land areas depending upon the frequency and intensity of flood events. However, increased atmospheric moisture and rainfall in many temperate areas may increase the amount of arable land and lengthen the growing season. Currently, the balance between reduction and increases in arable land area under various climate change scenarios can only be described as quite uncertain.

The uncertainty of crop production on a global scale and regional changes to either increased or decreased crop production will likely force a global assessment of food trading. The concept of virtual water trading via food sales and purchases between countries and regions will be a critical part of global and local water management in the future, particularly in arid and semiarid areas where the water availability is not adequate for the production of the necessary volume of food (Hoekstra and Hung 2002; Chapagain and Hoekstra 2008).

References

Abelson, P. H. (1977). Energy and climate. *Science, 197*, 4307–4941.

Alley, W. M. (2001). Ground water and climate. *Ground Water, 39*, 161.

Alley, R. B., & Clark, P. U. (1999). The deglaciation of the northern hemisphere: A global perspective. *Annual Reviews of Earth and Planetary Science, 27*, 149–182.

Alley, R. B., Marotzke, J., Nordhaus, W. D., Overpeck, J. T., Peteet, D. M., Pielke, R. A., Jr, et al. (2003). Abrupt climate change. *Science, 299*, 2005–2010.

Al-Sefry, S. A., Sen, Z. Z., Al-Ghamdi, S. A., Al-Ashi, W. A., & Al-Bardi, W. A. (2004). Strategic ground water storage of Wadi Fatimah, Makkah region. *Technical Report SGS-TR-2003-2*. Jeddah: Saudi Geological Survey.

Barnett, T. P., Adam, J. C., & Lettenmaier, D. P. (2005). Potential impacts of a warming climate on water availability in snow-dominated regions. *Nature, 438*, 303–309.

Bates, B., Kundzewicz, Z. W., Wu, S., & Palutikof, J. (Eds.). (2008). Climate change and water. *Intergovernmental Panel on Climate Change Technical Paper VI*, Geneva.

Berger, A. L. (1978). Long-term variations of caloric isolation resulting from the Earth's orbit elements. *Quaternary Research, 9*, 139–167.

Blunier, T., Chappellaz, J., Schwander, J., Dallenbach, A., Stauffer, B., Stocker, T. F., et al. (1998). Asynchrony of Antarctic and Greenland climate change during the last glacial period. *Nature, 394*, 739–743.

Broecker, W. (1999). What if the conveyor were to shut down? Reflections on a possible outcome of the great global experiment. *Geology Today, 9*(1), 1–7.

Broecker, W. (2010). *The great ocean conveyor: Discovering the trigger for abrupt climate change*. Princeton: Princeton University Press.

Cane, M. A., Clement, A. C., Kaplan, A., Kushnir, Y., Murtugudde, R., Pozdnyakov, D. et al. (1997). 20th century sea surface temperature trends. *Science, 275*, 957–960.

Chapagain, A. K., & Hoekstra, A. Y. (2008). The global component of freshwater demand and supply: An assessment of virtual water flows between nations as a result of trade in agricultural and industrial products. *Water International, 33*(1), 19–32.

Clark, P. U., Pisias, N. G., Stocker, T. F., & Weaver, A. J. (2002). The role of the thermohaline circulation in abrupt climate change. *Nature, 415*, 863–869.

Committee on Environment and Natural Resources. (2008, May). Scientific assessment of the effects of global climate change on the United States. *A Report on the Committee on Environment and Natural Resource*. National Science and Technology Council.

Cook, E. R., Bartlein, P. J., Diffenbaugh, N., Seager, R., Shuman, B. N., Webb, R. S. (2008). Hydrological variability and change. In *Abrupt Climate Change: A Report by the U.S. Climate Change Science Program and the Subcommittee on Global Change Research*, Reston, VA, pp. 143–257.

Dai, A., Trenberth, K. E., & Qian, T. (2004). A global data set of Palmer Drought Severity Index for 1870–2002: Relationship with soil moisture and effects of surface warming. *Journal of Hydrometeorology, 5*, 1117–1130.

Diamond, J. (2005). *Collapse, how societies choose to fail or succeed*. London: Penguin Books.

Doll, P., & Florke, M. (2005). Global-scale estimation of diffuse groundwater recharge. *Frankfurt Hydrology Paper 03*. Institute of Physical Geography, Frankfurt University.

Dragoni, W., & Sukhija, B. S. (2008). Climate change and groundwater: A short review. In W. Dragoni & B. S. Sukhija (Eds.), *Climate change and groundwater* (pp. 1–12). Geological Society of London Special Publications No. 288.

Ducci, D., & Transfaglia, G. (2008) Effects of climate change on groundwater resources in Campania (southern Italy). In W. Dragoni & B. S. Sukhija (Eds.), *Climate change and groundwater* (pp. 25–38). Geological Society of London Special Publications No. 288.

Frich, P., Alexander, L. V., Della-Marta, P., Gleason, B., Haylock, M., Klein Jack, A. M. G. et al. (2002). Observed coherent changes in climatic extremes during the second half of the twentieth century. *Climate Research, 19*, 193–212.

Gao, X., & Giorgi, F. (2008). Increased aridity in the Mediterranean region under greenhouse gas forcing estimated from high resolution simulations with a regional climate model. *Global and Planetary Change, 62*, 195–209.

Gleick, P. H. (1993). Water in the 21st century. In P. H. Gleick (Ed.), *Water in crisis, a guide to the world's freshwater resources* (pp. 105–113). New York: Oxford University Press.

Green, T. R., Taniguchi, M., & Kooi, H. (2007). Potential impacts of climate change and human activity on subsurface water resources. *Vadose Zone Journal, 6*, 531–532.

Hemming, S. R. (2004). Heinrich events: Massive late Pleistocene detritus layers of the North Atlantic and their global imprint. *Reviews of Geophysics, 42*, RG1005. doi:10.1029/2003RG000128.

Herweijer, C., Seager, R., Cook, E. R., & Emile-Gray, J. (2007). North American drought of the last Millennium from a gridded network of tree-ring data. *Journal of Climate, 20*, 1353–1376.

Hoekstra, A. Y., & Hung, P. Q. (2002). Virtual water trade. A quantification of virtual water flows between nations in relation to international crop trade. *Value of Water Research Report 11*. IHE Delft.

Holman, J. P. (2006). Climate change impacts of groundwater recharge—uncertainty, shortcomings, and the way forward? *Hydrogeology Journal, 14*, 637–647.

Huntington, T. G. (2006). Evidence for intensification of the global water cycle review and synthesis. *Journal of Hydrology, 319*, 83–95.

Imbrie, J., Hays, J. D., Martinson, D. G., McIntyre, A. C., Mix, A. C., Morley, J. J., et al. (1984). The orbital theory of Pleistocene climate: support from a revised chronology of the marine $\delta^{18}O$ record. In A. Berger, J. Imbrie, J. D. Hays , G. Kukla & B. Saltzman (Eds.), *Milankovitch and climate: Understanding the response to astronomical forcing, NATO science series c* (pp. 269–305). Dordrecht: Kluwer Academic Publishers.

Intergovernmental Panel on Climate Change (IPCC). (2000). Emissions Scenarios. In N. Nakicenovic & R. Swart (Eds.) *Special Report of the Intergovernmental Panel on Climate Change*. Cambridge: Cambridge University Press.

Intergovernmental Panel on Climate Change (IPCC). (2007). In S. Solomon, D. Qui, M. Manning, Z. Chen, M. Marquis, K. B. Averyl, M. Tignor, & H. L. Millers. (Eds.), *Fourth Assessment Report of the Intergovernmental Panel on Climate Change*. Cambridge: Cambridge University Press.

Jager, J., & Ferguson, H. L. (Eds.). (1991). Climate change: Science impacts and policy. *Proceedings of the Second World Climate Conference*. Cambridge: Cambridge University Press.

Jansen, E., Overpeck, J., Briffa, K. R., Duplessy, J.-C., Joos, F., Masson-Delmotte, V., et al. (2007). Paleoclimate. In S. D. Solomon, D. Qin, M. Manning, Z. Chen, M. Marquis, K. B. Avery, M. Tignor & H. L. Miller (Eds.), *Climate change 2007: The physical science basis, Contribution of Working Group I to the fourth assessment report of the Intergovernmental Panel on Climate Change* (pp. 431–497). Cambridge: Cambridge University Press.

Karl, T. R., Melillo, J. M., & Peterson, T. C. (Eds.). (2009). *Global climate change impacts in the United States*. Cambridge: Cambridge University Press.

Kundzewicz, Z. W., & Döll, P. (2009). Will groundwater ease freshwater stress under climate change? *Hydrological Sciences-Journal-de Sciences Hydrologiques, 54*(4), 665–675.

Kundzewicz, Z. W., Mata, L. J., Arnell, N. W., Döll, P., Jimenez, B., Miller, K. et al. (2007). Freshwater resources and their management. In M. L. Parry, O. F. Canziani, J. P. Palutikof, P. J. van der Linden & C. E. Hansen (Eds.), *Climate Change (2007) Impacts, Adaptation and Vulnerability. Contribution of Working Group II to the Fourth Assessment Report of the Intergovernmental Panel on Climate Change* (pp. 173–310). Cambridge: Cambridge University Press.

Kundzewicz, Z. W., Mata, L. J., Arnell, N. W., Döll, P., Jimenez, B., Miller, K., et al. (2008). The implications of projected climate change on freshwater resources and their management. *Hydrological Sciences, 53*(1), 3–10.

Lamb, H. H. (1995). *Climate, history and the modern world* (2nd ed.). New York: Routledge.

Landais, A., Masson-Delmotte, V., Jouzel, J., Raynaud, D., Johnsen, S., Buber, C., et al. (2006). The glacial inception as recorded in the North GRIP Greenland ice core: Timing structure and associated abrupt temperature changes. *Climate Dynamics, 26*(2–3), 273–284.

Lemke, P., Ren, J., Alley, R., Allison, I., Carrasco, J., Flato, G., Fujii, Y., Kaser, G., Mote, P., Thomas, R., & Zhang, T. (2007). Observations: Change in snow, ice and frozen ground. In S. Solomon et al. (Eds.), *Climate change 2007: The physical science basis. Contribution of Working Group I to the Fourth Assessment Report of the Intergovernmental Panel on Climate Change* (pp. 337–384). Cambridge: Cambridge University.

Lenton, R. (2004). Water and climate variability: Development impacts and coping strategies. *Water Science and Technology, 49*(7), 17–24.

Masson-Delmotte, V., Jouzel, J., Landis A., Stievenard, M., Johnsen, S.J., White, J. W. C., Werner, M., Sveinbjornsdottir, A., & Fuhrer, K. (2004). GRIP deuterium excess reveals rapid and orbital-scale changes in Greenland moisture origin. *Science, 309*(5731), 118–121.

Milankovitch, M. (1938). Astronomische mittel zur erforschung der erdgesschichtlichen climate. *Handbuch der Geophysik, 9*, 593–698.

Milankovitch, M. (1941). *Kanron der Erdbestrahlang und seine Andwendung auf das Eiszeeiten problem*. Serbian Akademie Beographie Special Publication 132.

Milly, P. C. D., Wetherald, R. T., Dunne, K. A., & Delworth, T. L. (2002). Increasing risk of great floods in a changing climate. *Nature, 415*(31), 514–515.

Milly, P. C. D., Dunne, K. A., & Vecchia, A. V. (2005). Global pattern of trends in streamflow and water availability in a changing climate. *Nature, 438*(17), 347–350.

Nakicenovic, N., & Swart, R. (Eds.). (2000). *Special report on emissions scenarios: A special report of working group III of the Intergovernmental Panel on Climate Change*. Cambridge: Cambridge University Press.

Ng, G.-H., McLaughlin, D., Entekhabi, D., & Scanlon, B. R. (2010). Probablistic analysis of the effects of climate change on groundwater recharge. *Water Resources Research, 46*, W07502, 18 p. doi:10.1029/2009WR007904.

Oki, T. (2005). The hydrologic cycles and global circulation. In M. G. Anderson (Ed.), *Encylopedia of hydrological sciences*. Chichester: Wiley.

Polemio, M., & Casarano, D. (2008) Climate change, drought and groundwater availability in southern Italy. In W. Dragoni & B. S. Sukhija (Eds.), *Climate change and groundwater* (pp. 39–51). Geological Society of London Special Publications No. 288.

Ragab, R., & Prudhomme, C. (2002). Climate change and water resources management in arid and semi-arid regions: Prospective and challenges for the 21st century. *Biosystems Engineering, 81*(1),3–34.

Rahmstorf, S. (2003). Timing of abrupt climate change: A precise clock. *Geophysical Research Letters, 30*(10), 1510. doi:10.1029/2003GL017115

Seiler, K.-P., Gu, W.-Z., & Stichler, W. (2008). Transient response of groundwater systems to climate changes. In W. Dragoni & B. S. Sukhija (Eds.), *Climate change and groundwater* (pp. 111–119). Geological Society of London Special Publications No. 288.

Solomon, S., Qin, D., Manning, M., Alley, R. B., Berntsen, T., Bindoff, N. L. et al. (2007). Technical summary. In S. Solomon, D. Qin, M. Manning, Z. Chen, M. Marquis, K. B. Averyt, M. Tignor, & H. L. Miller (Eds.), *Climate Change (2007) the Physical Science Basis. Contribution of Working Groupto the Fourth Assessment Report of the Intergovernmental Panel Onclimate Change*. Cambridge: Cambridge University Press.

Toews, M. W., & Allen, D. M. (2009). Simulated response of groundwater to predicted recharge in a semi-arid region using a scenario of modeled climate change. *Environmental Research Letters, 4*,19, 035003 doi:10.1088/1748-9326/4/3/035003.

Trenberth, K. E., Jones, P. D., Ambenje, P., Bojariu, R., Easterling, D., Klein Tank, A. et al. (2007). Observations: Surface and atmospheric climate change. In S. D. Solomon, D. Qin, M. Manning, Z. Chen, M. Marquis, K. B. Avery, M. Tignor & H. L. Miller (Eds.), *Climate Change (2007) The Physical Science Basis. Contribution Of Working Group I to the Fourth Assessment Report Of the Intergovernmental Panel on Climate Change* (pp. 235–336). Cambridge: Cambridge University.

United Nations. (2004). *World population to 2300*. New York: United Nations.

United Nations Educational, Scientific and Cultural Organization (UNESCO). (2006). Water: A shared responsibility. *United Nations World Development Report 2*, Paris.

U.S. Global Change Research Program (USGCRP). (2009). Global climate change impacts in the United States. *A report of the USGCRP*. Washington: USGCRP.

Wentz, F. J., Ricciardulli, L., Hilburn, K., & Mears, C. (2007). How much more rain will global warming bring? *Science, 317*(5835), 233–235.

Ye, X. A., Yang, D. Q., & Kane, D. L. (2003). Changes in the Lena River streamflow hydrology: Human impact verses natural variation. *Water Resources Research, 39*, 1200. doi:10.1029/2003WR001991.

Chapter 39
Adaptation and Mitigation of Global Climate Change in Arid Lands

39.1 Adaptive Strategies in Response to Climate Change

Humans have throughout history adapted to climate change, and will out of necessity adapt to future climate change. However, compared to climatic changes in the past, the changes that are presently taking place are occurring in a world where many vast areas are densely populated and have a high water demands (Dragoni and Sukhija 2008). Even without climate change, many areas of the world are now facing, now or will face in the future, severe water crises because of population growth. The vulnerability of a region, country, or community to climate change is a function of its exposure, sensitivity, and adaptive capacity to climate stimuli. It is also clear that with respect to climate change, there will be winners and losers and the losers may wind up paying a very high price.

There is an exponentially growing literature on adaptation to climate change, which has become an academic discipline in its own right. However, much of the literature on adaptation to climate change is long on theory and short on practical guidance. Smit and Wandel (2006, p. 285) observed that

> Research that focuses on the implementation processes for adaptation is still not common; at least, it is not common under the label of "adaptation" research, and certainly not in the climate change field.

Adaptations can be either proactive (i.e., anticipatory) or reactive, as well as autochthonous or planned. As the focus of this book is on water management, the pertinent question is what actions can be taken to specifically prepare for and respond to adverse impacts water supplies caused by climate change. However, it is useful to first consider how societies adapt to changing climatic conditions in general. Adger et al. (2005, p. 78) defined adaptation as:

> adjustment in ecological, social, and economic systems in response to observed or expected changes in climatic stimuli and their effects and impacts in order to alleviate adverse impacts of change or take advantage of new opportunities.

R. Maliva and T. Missimer, *Arid Lands Water Evaluation and Management*, Environmental Science and Engineering, DOI: 10.1007/978-3-642-29104-3_39,

The three cornerstones of adaptation are (Adger et al. 2005)

- reduction in the sensitivity of the system to climate change,
- altering the exposure of the system to climate change, and
- increasing the resilience of the system to cope with change.

Fankhauser et al. (1999) noted that successful adaptation to a large extent depends on three elements:

- timely recognition of the need to adapt,
- an incentive to adapt, and
- the ability to adapt.

The incentive to adapt might be taken for granted, but potential responses to climate change have economic consequences that could favor, at least in the short term, maladaptive actions. For example, it may be in a farmer's short-term economic self-interest to continue more profitable, high-rate irrigation, which is a maladaptation to increasing local water scarcity. Continued overuse of a "commons" resource, such as water, often leads to the destruction of the resource, especially in the absence of the recognition of the necessity to change the management model (Hardin 1968).

Adaptation to climate change can occur on different societal levels and have different drivers. Some adaptation would automatically occur without specifically designed policy changes. Farmers may adjust cropping patterns, planting, and harvesting times, and other practices as evidence of persistent changes in temperature and precipitation accumulates (Frederick 1995). Other adaptations, involving large-scale water management will require governmental action. Intergovernmental or individual government cooperation is very difficult to obtain as evidenced by the lack of a negotiated agreement to control the discharge of greenhouse gases.

Response sensitivities of human societies and their adaptive capacity are two factors that determine the vulnerability and degree to which societies succumb to the adverse effects of climate change (Sukhija 2008). Adger et al. (2005) also noted that adaptation can involve both building adaptive capacity and implementation of specific adaptation decisions (i.e., transforming capacity into action). Of course, the actions must be the correct ones as adaptations may amplify the impacts of climate change if they are ineffectual and unsustainable. Adger et al. (2005) gave the example of increased use of fossil-fueled powered air-conditioning as an unsustainable response to climate change.

There has been much discussion as to what is mean by adaptive capacity, how it is acquired, and differences in adaptive capacity between societies. Jeffrey and Gearey (2006, p. 2) noted that

> Crudely, adaptive capacity details the way in which individuals, organisations and economic sectors will need to adjust to uncertain or ill-defined change. Adaptivity is the ability to cope with changing circumstances.

Adaptive capacity is essentially synonymous with social resilience. Adger (2003) discussed the relationship between adaptive capacity and social capital, which is a term described "*as relations of trust, reciprocity, and exchange; the evolution of common rules; and the role of networks.*" Social capital was also recognized to have both public and private elements, both of which are based on trust, reputation, and reciprocal action. The main thesis of Adger (2003) is that social capital is central to adaptive capacity as it provides the opportunity to act collectively. A number of workers have pointed out that adaptive capacity depends on local circumstances and all levels of societies (both state and private). Adger (2003, p. 400) further observed that:

> In the context of climate change, many potential risks necessarily involve intervention and planning by the state, yet adaptation strategies are equally dependent on the ability of individuals to act collectively in the face of risks.

Co-management, in which resource stakeholders work together with government agencies, has been suggested as a more effective avenue for building community resilience (Tompkins and Adger 2004). However, inclusive institutions and the sharing of responsibility for natural resources go against the dominant hierarchical institutional forms of most governments throughout the world (Tompkins and Adger 2004).

Yohe and Tol (2002) proposed that the determinants of adaptive capacity include a variety of system, sector, and location specific characteristics, including:

- the range of available technological options for adaptation,
- availability of resources and their distribution across the population,
- the structure of critical institutions, including the allocation of decision-making authority, and the decision criteria that would be employed,
- the stock of human capital including education and personal security,
- the stock of social capital including the definition of property rights,
- the system's access to risk spreading processes,
- the ability of decision-makers to manage information, the processes by which decision-makers determine which information is credible, and the credibility of the decision-makers, and
- the public's perceived attribution of the source of stress and the significance of exposure to its local manifestations.

Smit and Wandel (2006) cautioned that the broad factors or determinants that influence sensitivities and constrain the abilities of communities to deal with hazards or stressful conditions are too general to guide in practical adaptation programs. Vulnerability to climate change will vary considerably on a local and regional scale, primarily because of differing degrees of impacts from changes and differences in the adaptive capacity of local populations and their management skills and resource availability (Sukhija 2008). The latter is a critical issue. Adaptive capacity depends on physical, technical, and economic resource availability in addition to social and political factors. Societies greatly differ in the resources and tools that they have available to adapt and respond to climatic changes. Clearly, the United States and European Union countries have greater resources to deal with climate change than the Sahel countries. However, the generally very low regard for and distrust of decision-makers in some countries (e.g., United States in 2010–2012) can result in a very low adaptive capacity for change in general despite a great resource base. Perhaps the will to adapt in countries of diverse interests, such as the United States, evolves solely from impending tragedy or a state of crisis.

Research on specific adaptations often involves generating a list of possible adaptation activities, and then evaluating them according to common principles or criteria, such as benefit-cost, cost effectiveness, and multiple criteria procedures (Smit and Wandel 2006). Variables that are considered in the analyses include benefits, costs, implementability, effectiveness, efficiency, and equity. With respect to water supply, a community could increase its adaptive capacity by taking measures that increase its ability to cope with droughts of increasing length and severity. Once a suite of options is identified, the preferred option(s) is determined through benefit-cost analysis or other means.

At the present time, relatively little meaningful action is being taken to increase adaptive capacity with respect to climate change because it is not a sufficiently popular issue and there is no political consensus to drive collective action, particularly in the western and developing countries where such action would involve significant sacrifices. Adger et al. (2005) argue that adaptation will require large-scale investment that is likely to be triggered through extreme events that raise the awareness of climate change within the policy making sphere and hence, give legitimacy to global action. Smit and Wandel (2006) observed that it is extremely unlikely for any type of adaptive action to be taken in light of climate change alone, and instead, that adaptive actions for climate change will tend to be integrated into resource management, disaster preparedness, and sustainable development programs.

Once adaptive responses are implemented, the question then becomes whether or not they are successful. Adger et al. (2005) proposed that adaptations need to be evaluated using the criteria of economic efficiency, effectiveness, equity, and legitimacy, and that the relative importance of these criteria varies between regions, countries, sectors, and local circumstances. Effectiveness is difficult to quantify particularly where it depends upon the future unknown state of the world, which includes climate, economic, and social conditions. In simpler terms, it is difficult to determine whether an action was effective if there is no benchmark concerning how conditions would have been different if the action had not been taken. Effectiveness and equity also address who specifically benefits from actions and who may be harmed (i.e., downstream impacts).

A number of workers have also noted that many societies are ill-adapted to current climate conditions, much less, more challenging future conditions. The development of adaptation strategies to reduce vulnerability to future climate change should start by identifying and quantifying current vulnerability to climate extremes (such as droughts, floods, and sea level changes) and developing adaptive measures that reduce current vulnerabilities (Lasage et al. 2008; Bates et al. 2008). However, Bates et al. (2008) cautioned that

> Continued investment in adaptation in response to historical experience alone, rather than projected future conditions that will include both variability and change, is likely to increase the vulnerability of many sectors to climate change.

Given the prevailing uncertainty concerning the specific local impacts of climate change, the best strategy is to increase the flexibility of systems to function under a wide range of climatic conditions, as well as their robustness to withstand more severe climate shocks (Fankhauser et al. 1999). Such measures to increase flexibility and robustness could be worthwhile in their own right, independent of climate change considerations (Fankhauser et al. 1999).

39.2 Use of Stationarity or Non-Stationarity in Future Water Resources Planning and Management with Consideration of Global Warming

An important water management consideration with respect to climate change is that the historic probability density functions for climatic parameters may become obsolete. Milly et al. (2008) described this situation as the death of "stationarity", wherein natural systems fluctuate within an unchanging envelope of variability. For example, the 100 year flood or drought may have a significantly different magnitude in the future. Vogel et al. (2011) developed a "flood magnification factor" to model non-stationarity flood events in the United States and from that the analyses found a factor of two to five times historical flood discharges and may appropriate. Milly et al. (2008) proposed that hydrologic variables could instead be modeled stochastically to describe the likely evolution of the probability density functions and provide estimates of uncertainty.

The issue of non-stationarity in water resources management is not new and does not only apply to future global warming scenarios. Hirsch (2011) points out that non-stationarity can arise from a variety of man-caused conditions, such as urbanization of a watershed and the resulting changes in flood-frequency, groundwater drawdowns resulting in diminished base flow, man-made reservoirs that reduce peak flows and increase low flows, and others. In arid lands streams that have perennial flow, the depletion of baseflow by groundwater development can create a severe non-stationarity, which may increase for extended time periods as a result of lengthy lag times. Climate change will greatly affect the timing of stream discharge in areas affected by snowfall and melting because of the earlier melt times and the possible increase in the peak discharge, with an accompanying reduction in low flow discharge rate and duration (Hirsch 2011).

However, the variation between the current GCM model predictions and low spatial resolution cast great doubts over the ability to predict local climate change, particularly with respect to the direction and magnitude of changes in water

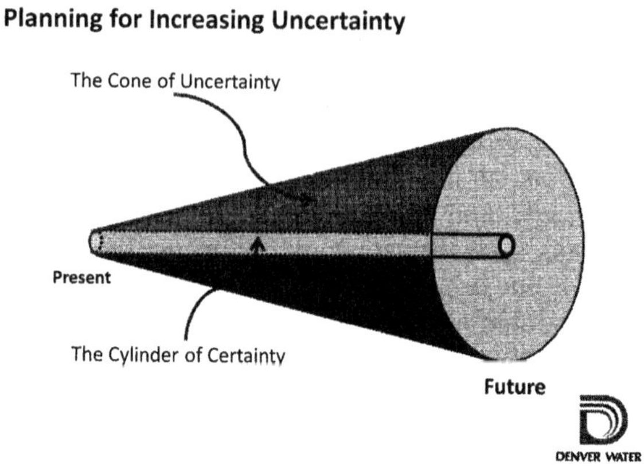

Fig. 39.1 The water planning cylinder of future uncertainty in time (from Waage and Kaatz 2011)

availability and flooding. Barsugli et al. (2009) note that the model suggested temperature changes are more accurate than projected precipitation and streamflow projections. This results in large errors in trying to assess effects on small stream basins (Barsugli et al. (2009). In arid lands, the projections used to assess flash flood frequency and discharges rises to the highest level of uncertainty. In conclusion, the current state of the art in climate modeling is insufficient for predicting future local water resources changes, but is of value for identifying possible future scenarios that need to be considered for long-term planning. This issue in no way suggests that continued monitoring and improvements in monitoring networks to assess both surface-water and groundwater flows should be abandoned. Field data are needed now more than ever to calibrate models that are essential for predicting the response of hydrologic systems to future changes in inputs (e.g., precipitation).

A more practical strategy for dealing with non-stationarity is expanding the envelope of variability considered in water management decisions (Fig. 39.1). Means et al. (2010) suggest that four basic steps should be used in planning for utilities to adapt for climate change:

- understand climate science and climate model projections (particularly uncertainties),
- assess water system vulnerabilities to potential climate change,
- incorporate climate change into water utility planning, and
- implement adaption strategies.

Rather than relying upon predictions from past climatological records and climatological models, water systems should be designed to accommodate greater climatic variability in both directions (e.g., locally wetter or drier conditions). Water-resource management should be flexible so as to be able to cope with more frequent and severe floods and droughts, and periods of surplus and shortage (McClurg 1998; Bouwer 2000, 2003). In other words, water management plans should incorporate both robustness and adaptive management (Bates et al. 2008; Brekke et al. 2009; Waage and Kaatz 2011). Robust water management plans perform well over a wide variety of possible future scenarios (Waage and Kaatz 2011), although their performance may not be optimal for some given scenarios. Systems are designed so as to minimize the possibility of an adverse surprise rather than to perform optimally under anticipated future conditions, which may not come to pass (Fiering and Kindler 1987).

39.3 Adaptive Water Management Strategies

Adaptive management strategies essentially involve reevaluation of decisions and the making of adjustments as new information becomes available. Systems are designed with flexibility so that changes in direction are possible in the future. The vulnerability of a nation to climate change depends as much, if not more, on its wealth than on the degree of climate change it would experience (Goklany 1995). Goklany (1995) emphasized that sustainable economic growth would make adaptation and mitigation measures more affordable and reduce the vulnerability of societies to global climate change. On a broad scale, strategies for enhancing adaptation include:

- technological changes to ensure greater productivity and efficiency,
- sustainable economic growth, and
- free, unsubsidized trade.

With respect to water, adaptations to climate change can include both supply-side and demand-side measures. Supply-side measures include activities that increase supply, such as (Bates et al. 2008):

- development of additional groundwater resources,
- increased storage (reservoirs and aquifer storage and recovery),
- desalination,
- increase amount of water reuse,
- rainwater harvesting, and
- water transfers.

Supply-side solutions also include improved water management (integrated water resources management; Chap. 35), which can result in more efficient and flexible water systems that can better accommodate climatic change and variability (Frederick 1995). Demand-side measures include activities that decrease water use and more efficiently manage resources. For example, consumptive users can adapt to an altered hydrologic regime (decrease in supply) by (Tarlock 2000):

- conservation,
- reallocation of existing uses,
- temporarily forgoing a use, and
- permanently forgoing a use.

Bates et al. (2008) pointed out that demand-side options may lack practical effectiveness because they rely on the cumulative actions of individuals, who tend to act in their own best interest and not in the interest of the "commons" as defined by Hardin (1968).

Adaptation to climate change becomes even more complicated with respect to the management of transboundary water resources. Tarlock (2000) explored the relationship between international water treaties and the projected impacts of global climate change on major river basins. The purpose of international allocation treaties is to generally allow the construction of upstream and downstream dams. The treaties usually assume a fixed, perpetual flow regime with no provisions for changed circumstances. The parties involved in such international treaties are likely to insist that the status quo be maintained irrespective of decreased supply (Tarlock 2000).

Tarlock (2000) proposed that adaptation to projected adverse impacts of climate change requires the presence of a reasonably well-developed property rights regime in the effected basin. A property rights regime is necessary as it can facilitate the necessary reallocation of water in a way that allows users to share risks and shift water uses fairly and efficiently among competing consumptive and non-consumptive uses. As emphasized by Turton (1999) the capability of societies to reallocate water between sectors is a critical element in their adaptive capacity to water scarcity. Tarlock (2000) also suggests that adaptation in transboundary river basins also applies the same to transboundary aquifers, which have been managed within a tenuous framework of untested international law (Chap. 37).

39.4 Increase in Water Supply by Development of New Water Resources: Supply Side Approach

Global climate change will place additional stresses on water supplies that are already in crisis in many regions of the world. All water management systems will require complex integration to balance water supply and demand within the context of the available financial resources. Whereas countries that are developed and have adequate financial resources to plan and mitigate for changes in the hydrologic system can potentially cope with changes in their water supplies, developing countries and those with limited fiscal resources will be unable to deal with the expected coming changes from a financial viewpoint. Unfortunately, many water-stressed regions also happen to be those that have large population growth rates and projected corresponding increases in future water demand.

An obvious adaptation to water shortages is to develop new supplies, which can be either true sources of new water (e.g., desalination) or capturing existing water flows that are not being fully put to beneficial use, such as by rainwater harvesting and managed aquifer recharge. The alternative, end-member, adaptation strategy to decreased water supplies is to reduce demands, which can be achieved by either discontinuing or reducing some water intensive activities or adopting measures that increase the water use efficiency of individual activities.

39.4.1 Development of Additional Groundwater Resources

New groundwater resources can be developed in certain regions, but in many locations groundwater resources are already under stress or will become depleted as in the case of some non-renewable water resources of the Middle East. Many groundwater sources are also connected to surface-water systems that provide natural recharge, so these sources will be affected in a similar manner to surface-water resources to reduced rainfall and runoff

Groundwater resources will become increasingly important in a changing climate scenario because they (Sukhija 2008):

- provide a large storage capacity that can allow for the storage of runoff that would otherwise be lost,
- are less vulnerable to droughts,
- are less vulnerable to contamination, and
- do not require large-scale engineering structures for their exploitation and augmentation.

There are several different strategies that may be applied to the use of groundwater under global climate change scenarios. First, new groundwater resources can be developed only under sustainable scenarios. Second, groundwater can be used on a temporary basis during extreme droughts when surface-water sources are unavailable. The groundwater sources used should have sufficient capacity to meet demands and not be dependent upon short-term recharge from drought-vulnerable surface waters. The groundwater source used should also be recharged between drought events.

Adaption to a more variable climate must also involve preparation for emergency situations, such as droughts. Emergency situations also include extreme meteorological and geological events, such as earthquakes, volcanos, tsunamis, landslides and hurricanes and typhoons. Vrba and Verhagen (2006) and Sukhija (2008) proposed that deep confined aquifers could be used as an emergency source of water to cope with drought periods. Deep confined aquifers (renewable or non-renewable), if available in a region, are the most suitable source of safe and usually good quality water in an area impacted by droughts. An alternative emergency groundwater source is unconventional aquifer systems such as fractured or fault zone aquifers with deep circulating groundwater (Vrba and Verhagen 2006). Aquifers containing old groundwater are ideal for emergency supplies because they are less vulnerable to contamination. Groundwater use during emergency situations may be a temporary, non-sustainable exploitation until normal supplies are restored (Vrba and Verhagen 2006). The primary problem is controlling water use so that aquifers do not become over exploited during non-emergency periods. Development of new groundwater sources and the use of groundwater on a temporary basis require resource management at the highest level because once the source is depleted or contaminated there may be no other low-cost water supply option.

SKM (2009) suggested in a report to the World Bank, "Adaption options for climate change on groundwater resources", as cited by Dillon et al. (2009), that groundwater management must include

- management of groundwater recharge (MAR),
- protection of groundwater quality,
- management of groundwater discharge, and
- management of groundwater storage.

Dillon et al. (2009, p. 2) pointed out that

> In many cases, adaptations to reduce the vulnerability of groundwater dependent systems to climate pressures are the same as those required to address non-climate pressures, such as over allocation and over use of groundwater.

Effective use of groundwater to help mitigate for climate changes will be dependent on a number of other strategies needed to maintain the groundwater source, including managed aquifer recharge, increased storage, and water reuse.

39.4.2 Increased Storage (Reservoirs and Aquifer Storage and Recovery)

Anticipated changes in the flow characteristics of streams (Sect. 38.3), such as earlier peak discharges of higher magnitude, extended periods of lower discharge, and lower base flows, will require both physical and policy changes in the management of watersheds. The primary issue, in many cases, will not be that there will be a decrease in overall supply of water available, but rather that the timing of discharge will change and the flows will be more concentrated. A key mitigation measure is the creation of more storage. The storage component must be viewed as a conjunctive plan involving both surface water and groundwater storage.

The general strategy to create storage is to build more dams, create larger reservoirs, and develop aquifer storage and recovery systems (or a variety of other managed aquifer recharge systems) linked to the reservoirs. In recent decades the number of dams and reservoirs being decommissioned has likely outnumbered the new ones being constructed. Dam and reservoir systems have large environmental and economic impacts that have been commonly ignored in the past because of the low-cost power generated by hydroelectric facilities tied to dam/reservoir projects, flood control benefits, and the ability to convey irrigation water via canal systems.

Many observers believe that the era of large-scale dam and reservoir construction is over (Pearce 2006). However, in the coming years, as flood events become more intense, the wise development of carefully planned dams and reservoirs will be required to mitigate for global climate change. These facilities will likely require new design concepts not commonly used in the past, as wells as new management techniques to avoid unplanned downstream environmental and economic impacts. For example, all reservoirs have a limited life-expectancy due to

Fig. 39.2 Dam across Wadi Abha in Saudi Arabia. The dam captures ephemeral flows that are used to supply the city of Abha (photo by Samir Al-Mashharawi)

siltation, which is the infilling of the reservoir with sediments carried by the inflowing streams. Larger magnitude flood events in the upper basins may increase the sediment load into the new reservoirs, thereby decreasing their life-expectancy based on the current design model for dams. New designs need to be developed to allow pass-through of sediment by creating dams with under-flushing or other sediment diversion methods, yet undefined.

Large river basins with perennial flow can no longer be managed based on solely an individual basin concept without consideration given to the surrounding region, other basins, or the global water issue (Hoekstra 2011). There will need to be more frequent interbasin transfers of water occurring in the future either by design or accident (e.g., groundwater effects) (Sect. 39.4.6).

Dams will need to be constructed on ephemeral streams to control intense floods in the future and to capture and store water for aquifer recharge. Construction of numerous dams in wadis has occurred in many locations in the Middle East (Fig. 39.2). The conceptual design of these dams may require some modification with the addition of managed aquifer recharge systems to allow more of the captured and detained water to recharge the underlying and bounding alluvial aquifer system before the detained water evaporates. The major limiting factor for recharge in wadi reservoirs is the sealing of the connection between reservoir bottom and the underlying aquifer with fine-grained sediments carried in the flood waters. The amount of recharge can be increased if the suspended solids concentration of the stored water is reduced. Some wadi dam reservoirs are operated as siltation basins, with the clearer upper water layer (which forms as suspended sediments settle) skimmed off the top at the spillway and discharged downstream of the dam. Alternatively, recharge could be performed using wells.

Within the realm of global climate change, the definition of "winners and losers" depends on the scale of the affected area (O'Brien et al. 2004). It is recognized that the storage provided by various managed aquifer recharge schemes will necessarily have to be a major component for adapting to both current local water scarcity and as well as to increased variability in water supply due to climate change. The effectiveness of this storage for mitigation against climate change will be based on the issue of scale (Dragoni and Sukhija 2008; Dillon et al. 2009). Large geographic areas with extremely high water consumption rates may not be able to store a sufficient volume of water to make a material positive impact on water management.

Another concept that will require investigation is the use of a combination of dams and off-stream reservoirs to capture and store flood flows. Such systems could be designed as diversions or could be equipped with massive pumps to fill the reservoirs. In arid land areas, wadi floods entering cities could be diverted or captured in the upper part of the basins before flash floods cause downstream damage and loss of life. Back-pumping schemes could be developed to convey the flood water to reservoirs higher in the basins and use gravity systems for managed aquifer recharge and irrigation systems.

It is clear that storage will be a key mitigation measure in adapting to global climate change. There is a need to develop new concepts to allow efficient capture, diversion, and storage of stormwater flows in the future. While the cost of these schemes will be high, they may be far less expensive than desalination or other more energy-intensive methods to add new water supplies, particularly for non-potable uses.

39.4.3 Increase Amount of Water Reuse

Globally, a small fraction of municipal wastewater is treated and reused for beneficial purposes (Chap. 29). Only one city of size in the world currently treats municipal wastewater to full potable standards and returns it into a utility system for direct potable use (Windhoeck, Namibia). Under any of the global climate change scenarios or even without change, increasing the reuse of water is just common sense and is quite cost effective in comparison with the alternatives.

Effective reuse of municipal wastewater is dependent upon the water being effectively treated to remove harmful pathogens and chemicals that would adversely affect the end user of the water. If the water will have a potable use, then advanced treatment processes may be required, which are expensive and energy intensive. Less expensive means of using MAR to give wastewater a "polishing treatment" to remove trace organic compounds and viruses, are being developed and researched to allow implementation of large-scale systems in arid lands (Missimer et al. 2012). The reuse of municipal wastewater will be a major mitigation measure in water management in a climate change period.

Reuse within other categories of water users will also be important to stabilize water supplies in times of great imbalances in supply and demand. Reuse of water has been effectively accomplished within industry for decades. The cost of developing and treating new water for industrial purposes in a manufacturing facility can be greater than treating and recycling process water within the facility. Recycling also reduces or eliminates the environmental impacts associated industrial wastewater disposal. Industrial wastewater recycling will have to be increased in developing countries, not only in industrialized locations. Opportunities also abound in the better management of agriculture return flows, to allow for both greater recycling and protection of shallow groundwater quality.

The timing between production of municipal wastewater and the demand may not be synchronous. Therefore, storage of the treated wastewater may be necessary. The use of aquifer storage and recovery for wastewater storage would help balance the supply and demand cycle and could assure that a greater percentage of wastewater is used (Dragoni and Sukhija 2008; Maliva and Missimer 2010).

39.4.4 Desalination

Desalination of seawater is truly a means to develop "new water" because the reservoir of seawater is a renewable and essentially an infinite quantity of water (Chap. 27). The primary constraints on the use of desalination as a major water-supply source are energy consumption and cost. While wealthy areas of the world can develop large cost-effective desalination systems using a variety of technologies, desalination is often not an economically viable option in many developing countries, unless there are external subsidies. The largest water user, agriculture, normally cannot be sustained using desalinated water. As the cost of desalination goes down and renewable energy sources are increasingly used to generate the required electrical energy, it will become a more viable water supply source to mitigate climate change caused imbalances of supply and demand. Most countries that border the coastlines or have coastal access could develop seawater desalination systems. Such systems could also be used as backup systems when extended droughts become problematical.

Brackish-water desalination is also a means to provide water supplies in some arid areas. Brackish groundwater is abundant is some inland and coastal areas and is less expensive to treat than seawater. Brackish-water desalination will be part of the solution to mitigate for climate-change induced supply variations, but in some areas may have limited applications based on the availability of water supply and the sustainability of the source. A critical issue for any large-scale implementation of brackish groundwater desalination is a thorough aquifer evaluation including solute-transport modeling to predict water quality over the planned duration of the project.

The use of desalination based on the current energy consumption requirements creates another problem. If hydrocarbon fuels are used to produce the power necessary to desalt water, then emissions of greenhouse gases will be increased, exacerbating global climate change. Continued improvements will be needed to reduce the energy consumption required to desalt water. Desalination facilities are also expected to be increasingly powered by alternative energy schemes, if for no other reason than alternative energy costs are decreasing relative to hydrocarbon fuels (Chap. 28). The need to reduce energy requirements is one reason why the development of new technologies, such as adsorption desalination, is so important to water supply development in the coming years. Additional innovations in desalination technology are certainly expected, as this is a very active academic and industrial research field.

39.4.5 Rainwater Harvesting

Rainwater harvesting is a viable means of supplying water for small users of water (Chap. 24). Widespread adoption of rainwater harvesting by numerous small users could capture and store enormous quantities of water. As atmospheric moisture increases, rainwater harvesting may have greater application, especially using schemes like fog water harvesting in dry coastal areas. Capturing roof runoff and other stormwater catchment schemes could be used on a larger scale to help mitigate climate change induced supply fluctuations. The areas which have an actual gain in rainfall will benefit the most from rainfall harvesting methods.

39.4.6 Water Transfers and Diversions

Major upper-basin diversions of water may become more common-place to meet water supply demands of downstream water users that have different timing and supply requirements. There are several types of diversions that could be used to mitigate for climate change. Diversions and transfers can be used to:

- to transmit water into a more productive watershed,
- to change the timing of water delivery based on crop need or human consumption cycling,
- to avoid massive flooding of populated downstream floodplain areas,
- to avoid the loss of water to non-production evaporation,
- to divert sediment to downstream flows and perhaps to reestablish reservoir storage, and
- to maintain boat traffic on rivers to maintain commerce.

Interbasin water transfers could be made for irrigation of productive crop lands that may not coincide with the basin having more favorable resources for agricultural production. For example, if one river basin has a narrow floodplain and little productive arable land is present, flow could be diverted into a bounding basin having better soils and potentially greater crop production potential. This type of transfer, and all of the others, would be predicated on the acceptability of impacts of the diversion on the environment and people residing in the basin facing the loss of some water flow. Perhaps the basin having water diverted from it may have greater, unneeded flows based on climate changes. Nevertheless, projects involving large-scale transfer of water typically elicit very strong opposition from the donor side of the project, perhaps forcing the development of an acceptable economic compensation system.

Timing of river flow is another issue that could justify diversion. For example, certain crops require irrigation water during specific growth periods and not during the entire crop cycle. During planting, sugarcane requires that water levels in the irrigation ditches be lowered so that the soils can be ploughed and planting can occur. Water levels are raised in the irrigation ditches during dry periods to provide irrigation and lowered during harvest times to allow field access. Similar cycles are required in many other crops, such as rice production. Coordination between river basins could be developed to allow diversions to match the water availability with demand. This coordination may require a change in the crop type raised within one or both of the basins to allow a more productive crop scheme to be developed.

All of the global climate models suggest that flooding will be a large problem in the future with greater "spiking" of river flows. This flooding within a basin could create devastating results to both the populations that live within the floodplain and to crops grown on the floodplain. Diversions from one basin to another could be used to lower flood levels to manageable stages allowing populations to avoid massive impacts. Other alternatives may be to physically remove people from floodplain areas or to build massive dam and levee systems in the absence of knowledge concerning the projected intensity of flooding. Diversions could also be used to purposely produce controlled floods that would replenish soils within the floodplains to aid in agricultural productivity in one or more of the basins.

Within arid or semi-arid lands, many rivers and streams terminate in fully closed basins, thereby creating hypersaline lakes, seas, salinas, playas, or pans (e.g., Dead Sea, Great Salt Lake, Salar de Atacama). The freshwater discharging into these basins is lost for productive use. In some arid lands ephemeral stream basins, there will be an increase in the frequency and magnitude of rainfall-induced flood events. The diversion of this water to productive uses would actually create a new source of water supply. This could be achieved by the design and construction of dams and reservoirs with connections to other basins via pipelines or tunnels. This type of diversion does not mean that all water should be diverted from downstream saline lakes or seas. The documented lowering of the Dead Sea is an example of what should be avoided in terms of management. However, the proposed Red-Dead Project would be perhaps the largest diversion project in history with seawater pumped from the Red Sea to land surface in Jordan, which would then would be allowed to flow by gravity down into the Dead Sea Basin where part of the water would be used as a source for a desalination plant and part to generate power as the water cascades into the deep Dead Sea Basin. The project will produce potable water, will generate power, water allow environmentally friendly disposal of the concentrate (into the Dead Sea), and will help stabilize or raise the stage of the Dead Sea. This is an example of how diversions should be used to allow greater quantities of water to be developed.

One of the large problems with the construction and operation of dams and reservoirs is how to deal with sediment accumulation within the reservoirs and the erosion problems caused downstream by the trapping of sediment behind dams. It may be possible that in new dams a diversion system could be developed to capture and transmit sediment in flood water downstream or into adjacent basins lacking sediments. The large-scale problem of reservoir capacity loss by sediment infilling requires that a new look be taken on how to release this sediment from the respective reservoirs and perhaps simultaneously aid the environment. The basic problem is the enormous volumes of sediment trapped behind dams preclude economical dredging using currently available technology. The cost of dredging enough sediment to materially improve reservoir capacity is usually prohibitive. New ideas will be required to assess and resolve these problems because of the acute need for additional storage caused by climate change.

Periodic diversion of river basin flows may be needed to maintain river stages in certain regions to allow river boat and ship traffic to be maintained. There are regions that are solely dependent on rivers as pathways for transmitting and receiving goods. Careful design to allow controlled diversions to aid in navigation is another potential mitigation measure that may be required to respond to climate change.

Surface-water diversions may be vulnerable to climate-induced changes in source water flows. For example, the Colorado River is the source of water for Central Arizona Project (Fig. 39.3), which supplies the cities of Phoenix and Tucson, Arizona (U.S.A.). Climate change-induced reductions in the flow of the Colorado River would adversely impact the water supply of Arizona and other states that have become dependent on the river.

39.5 Demand Reduction

Global water demand will continue to increase into the future based on the current models of water use, which consider population growth, requirements for food production, and economic development. Perhaps of greatest concern is continued population growth. Some believe that a population peak will be reached, as suggested by the United Nations (2004), and then a slight decline will occur to a stable but generally older population (Pearce 2010). However, the recent United Nations adjustment in the population projections based on higher than previously believed fertility rates in Africa is of great concern, especially within the context of global climate change.

The greatest control on growth of water demand would be a strict control on population growth. This concept is laden with social and religious overtones and will likely not be successful on a global scale. The great population control experiment in China has had mixed results, but has reduced the population growth

Fig. 39.3 Central Arizona Project canal near Lake Pleasant, Arizona (U.S.A.)

rate with some unexpected societal results. Without population growth control, there will need to be some very serious changes in how water is used, especially to produce food. A worst case scenario will be nations or regions suffering simultaneous prolonged, severe famine caused by a combination of population growth, decreasing water supply induced by global climate change, and drought conditions that are part of the normal climatic cycle. The reality is that many countries and regions are on a trajectory of increasing maladaptation to climate change as their water scarcity under the current climate regime is increasing. A key issue for countries that will facing increasing water scarcity in the future is how to much more effectively manage water demand by reducing water use in all fundamental use areas including human consumption, agriculture, power generation, and assorted industrial uses even with population growth. Even natural system water uses may have to be evaluated within the context of global water management.

39.5.1 Conservation

There are many approaches to water conservation involving utility types of issues, such as water loss management from distribution systems and in-house water use reductions (water-conserving toilets and showers), industrial water reuse schemes, and a variety of agricultural water-saving measures (Chap. 26). Water can be conserved and demand reduced using these methods within the realm of what can be termed "soft use", wherein known excess quantities of water are being used with minimal benefit. Care must be taken when reducing soft use because in a way it is a buffer during emergencies that can be immediately taken advantage of without causing harm to the water users. Water conservation methods cannot be applied to "hard uses" or those that are a necessity for life and critical economical activities. When water cutbacks are applied to hard uses, there are resulting economic impacts, and important activities may have to be ceased until adequate water is available. Water used for drinking and sanitary purposes is considered to be a hard use as well as water required to meet minimal irrigation needs for growing of food crops.

In a warmer climate and with a growing population, the trend will be for more people to live in cities, requiring water to be delivered in utility systems. Therefore, the management of water loss in utility pipelines is a key factor in conservation. Kingdom et al. (2006) estimated that 32×10^9 m^3 (8,454 billion gallons) globally leaks from utility lines each year. Water theft and non-beneficial diversions cause an additional loss of 16×10^9 m^3/year (4,227 billion gallons/year). Utility system losses can be managed by use of metering and distribution system pressure monitoring to detect leaks, leading to repairs. Leak detection and repair is a particularly difficult problem to address in older utility systems in large metropolitan areas, such as Damascus, Rome, or even New York. Developing countries are giving close attention to system losses as new treatment and distribution systems are be designed and constructed to meet increased demands (Mutikanga et al. 2009).

39.5.2 Reallocation of Existing Uses (Permanent and Temporary)

A key issue in the management of water within the context of global climate change is flexibility. It is probable that large-scale reallocation of water uses may be required on a permanent or temporary basis to respond to the anticipated changes in basin hydrology. Most river basin systems currently are managed to meet a variety of demands including potable use, flood control, agricultural uses (irrigation and flood control), and environmental flows necessary to maintain the biological productivity of the stream, fisheries, and the downstream estuarine areas. Some type of reallocation scheme will be necessary to respond to extreme changes in basin hydrology, particularly droughts in arid regions.

Temporary reallocation of water and land uses is not new. In southern California, a semi-arid region, the large municipal water users have contracted with farmers to use water allocated from the Colorado River. In particularly dry periods, the utility can reallocate water from the farmers to add flow to the aqueducts feeding the municipal water system. For this temporary reallocation, the local government compensates the farmers for not planting crops or for crop damage during the times in which the water is diverted. Long-term contracts have been negotiated that have benefits for both water users.

Another reallocation scheme involves cooperation between ranchers and water managers in wet basins that can expect to have greater flood frequency at higher than historical stages. In this case, an agreement has been reached between the South Florida Water Management District and some ranchers within the Caloosahatchee River Basin (Southwest Florida, U.S.A.) to allow the temporary storage of flood water on ranch lands during extreme high water periods. This allows the District to avoid construction of very expensive new reservoirs and compensates the ranchers for the flood control service and the temporary removal of livestock from the land. Similar schemes may be applied in wadi areas of the Middle East, wherein grazing and small food plantations in wadis could be moved or could be allowed to flood during small events with some compensation paid to the farmers to allow the service.

Permanent reallocations will be required, especially in river basins that experience severe impacts such as the permanent loss of seasonal meltwater from mountain glaciers. Agricultural water use in many of these basins may become non-sustainable and any rights to the remaining water should be reallocated to higher uses. This issue will be contentious and will need to be negotiated in an equitable manner. Unfortunately, under most global warming scenarios there will be clear winners and losers in terms of water resources. The key is to begin the reallocation process before the resources are no longer available, thereby reducing conflicts.

39.5.3 Agriculture Water Use Changes and Water Trading

The current paradigm of agricultural water use and production is not sustainable within the context of global climate change (or even without it) based on future population increases with associated food demands. One can project the agricultural water use required to meet the demands of a 9.2 billion population by the year 2075 using current agricultural irrigation practices with fresh water. This projection for all uses including potable, industrial, and agricultural would amount to between 15 and 20% of the total rainfall on the Earth. A large part of the projected use, perhaps 65% or more, is for agricultural production to meet food needs.

Water conservation methods should be applied to all large-scale farming operations. For example, open-ditch irrigation is extremely inefficient on a farm level with high evaporation and channel losses. Drip and microjet irrigation methods need to be applied where practical for most non-cereal food crops, especially vegetables and fruits. In semi-arid areas that will become wetter, dryland farming may become more viable in the future. However, adoption of high-efficiency irrigation will not occur in areas where water is inexpensive (or free) and thus a business case cannot be made by farmers for the investment.

New methods for raising food crops on a large scale must be initiated to meet food security demands over the next 50 years. This will require the expansion of some current practices and the development of totally new concepts. Raising large quantities of high-value food crops in greenhouses will be an increasingly important concept. These facilities will need to be large and contain a fully controlled environment with water collected, treated, and recycled internally. Large-scale aquaculture and mariculture facilities exist today in which fish and shrimp are raised within a fully controlled environment. Such facilities need to be expanded to all forms of food types that can be economically raised within this environment. Also, genetic alterations to food plants need to be developed that will allow these crops to be grown using seawater or brackish water for irrigation within greenhouse areas. Research on this issue is being conducted at the Desert Agriculture Center at the King Abdullah University of Science and Technology in Saudi Arabia, and at other institutions.

Another method of growing some food is urban agriculture, which is becoming more popular in many cities. Rooftop areas are used to grow crops, such as tomatoes and other vegetables, with a very low costs and high productivity. Water is gathered from the rooftops and recycled within the systems. Greenhouses could also be developed on rooftop areas and increase the overall footprint of land that can be used to raise food crops. These systems have little impact on water supplies.

With increasing water scarcity, it will be necessary to abandon agriculture in some areas or transition to very low water use activities, because the necessary water may be unavailable or the growing costs will become too high. In this case, the concept of water trading via the export and important of food will become very important. The "virtual water" market will have to be increased to meet global food demands (Hoekstra and Hung 2002; Chapagain and Hoekstra 2008). The trading of food between regions and nations will become a necessity so that more water-rich regions can be the primary food suppliers, while water-poor areas are the receivers. While the concept is simple and direct, the political and national security issues created by food trading is complex and greater global cooperation will be required to meet the demands within a global climate change condition.

39.5.4 Temporarily Forgoing a Use

Projected variability in climate, particularly the frequency and duration of droughts, may cause temporary changes in new water use projects or schemes. For example, there may not be sufficient water available to develop new farmlands in a particular basin, because of drought-related reductions in water flows. Such a project would have to be delayed to a future time. Also, there may be a necessity to reassess crop rotation schemes, wherein during drought periods a low water use crop is planted or certain fields are left fallow, while in wet times a higher water use crop is planted.

Decisions on water use will have to be coordinated within local and region frameworks with a greater degree of cooperation between users and political jurisdictions than currently exists. Temporary changes can be negotiated with greater success than permanent changes that tend to have protracted, serious economic impacts.

39.5.5 Permanently Forgoing a Use

All future global climate change scenarios show that some areas will be winners and others will be losers within the realm of water supply. The scale of the areas within these categories plays an important role in the decision-making process, particular on the feasibility of developing new water uses and continuing old ones (O'Brien et al. 2004). Because of the increased variability in yield of water supply sources in the future, careful attention will have to be given to new water users with high, fixed demands and users that have little flexibility in their water needs. If a new large water user cannot provide its own sustainable water supply (e.g., seawater desalination or purchase of existing water rights) or show how it could reduce its needs during an extended drought period, then it would not be a viable operation. The key to managing water supplies within the global climate change context is flexibility and fixed uses will have to be limited below a threshold based on projected drought flows, particularly in arid and semi-arid lands.

The concept of "highest and best use" within an economic framework will be required to assess what is in the overall best interest of the basin residents and region. Water needs to be managed as an "economic good," although there is considerable latitude in how economic considerations are incorporated into water management. Reallocation decisions may have severe economic impacts to various water users and it will likely be necessary to develop economic buyout schemes and perhaps a water rights trading methodology to manage complex watersheds. Water rights have value and in a market economy, the trade of these rights may reduce the economic impacts to individuals (or compensate individuals for adverse impacts) and provide an equitable management of water.

39.6 Reduction and Control of Greenhouse Gas Emissions in the Future

Based on the past and present efforts at obtaining a global agreement on greenhouse gas emissions, there appears to be no global political will to move forward with policies significant enough to reverse the current trends. While agreements such as the Kyoto Protocols appeared to provide a framework of agreement, it was not signed by the United States and "third world" countries were exempted (e.g., China).

Therefore, in terms of controlling global greenhouse gas emissions, it was ineffective at curtailing future changes to the degree necessary to make meaningful improvements.

Unfortunately, it appears that significant traction in the political landscape to allow significant changes in policy to be made will likely only occur once global warming creates a large-scale crisis that is obvious and severe enough to reach the attention of the world leaders and general populations. Therefore, water managers must plan and develop the infrastructure based on the worst-case assumption that little change will be made in the general scenarios generated by the Intergovernmental Panel for Climate Change (i.e., greenhouse gases will continue to rise with associated climatic impacts). The issue of global climate change will be a key factor in water management centuries into the future. Many societies will have to increase their capacity to adapt to changing climatic conditions far beyond their historic experiences.

References

Adger, W. N. (2003). Social capital, collective action, and adaptation to climate change. *Economic Geography, 79*, 387–404.

Adger, W. N., Arnell, N. W., & Thompkins, E. L. (2005). Successful adaptation to climate change across scales. *Global Environmental Change, 15*, 77–86.

Barsugli, J. C., Anderson, C., Smith, J. B., & Vogel, J. M. (2009). Options for improving climate modeling to assist water utility planning for climate change. *Water Utility Climate Alliance*. (http://www.wucaonline.org/assests/pdf/actions_whitepaper_120909.pdf).

Bates, B., Kundzewicz, Z. W., Wu, S., & Palutikof, J., (Eds.). (2008). Climate Change and Water. *Intergovernmental Panel on Climate Change Technical Paper VI,* Geneva.

Bouwer, H. (2000). Integrated water management: Emerging issues and challenges. *Agricultural Water Management, 45*, 217–228.

Bouwer, H. (2003). Integreated water management for the 21st century: Problems and solutions. *Food, Agriculture & Environment, 1*(1), 118–127.

Brekke, L. D., Kiang, J. E., Olsen, J. R., Pulwarty, R. S., Raff, D. A., Turnipseed, D. P., et al. (2009). Climate change and water resources management—A federal perspective. *U.S. Geological Circular 1331*.

Chapagain, A. K., & Hoekstra, A. Y. (2008). The global component of freshwater demand and supply: An assessment of virtual water flows between nations as a result of trade in agricultural and industrial products. *Water International, 33*(1), 19–32.

Dillon, P., Gale, I., Contreras, S., Pavelic, P., Evans, R., & Ward, J. (2009). Managing aquifer recharge and discharge to sustain irrigation livelihoods under water scarcity and climate change. In G. Blöschl, N. van de Giesen, D. Muralidharan, L. Ren, F. Seyler, U. Sharma, & J. Vrba (Eds.), *Improving integrated surface and groundwater resources management in a vulnerable and changing world* (pp. 1–12). International Association of Hydrological Sciences (IAHS) Publication, 330.

Dragoni, W., & Sukhija, B. S. (2008). Climate change and groundwater: A short review. In W. Dragoni & B. S. Sukhija (Eds.), *Climate change and groundwater* (pp. 1–12). Geological Society of London Special Publications No. 288.

Fankhauser, S., Smith, J. B., & Tol, R. S. J. (1999). Weathering climate change: Some simple rules to guide adaptation decisions. *Ecological Economics, 30*, 67–78.

Fiering, M., & Kindler, J. (1987). Surprise in water resources design. *International Journal of Water Resources Development, 2*(4), 1–10.
Frederick, K. D. (1995). Adapting to climate impacts on the supply and demand of water. *Climatic Change, 37*, 141–156.
Goklany, I. M. (1995). Strategies to enhance adaptability: Technological change, sustainable growth and free trade. *Climatic Change, 30*, 427–449.
Hardin, G. (1968). The tragedy of the commons. *Science, 162*, 1243–1248.
Hirsch, R. M. (2011). A perspective on nonstationarity and water management. *Journal of the American Water Resources Association, 47*(3), 436–446.
Hoekstra, A. Y. (2011). The global dimension of water governance: Why the river basin approach is no longer sufficient and why cooperative action at global level is needed. *Water, 3*(1), 21–46.
Hoekstra, A. Y., & Hung, P. Q. (2002). Virtual water trade. A quantification of virtual water flows between nations in relation to international crop trade. *Value of Water Research Report 11*, IHE Delft.
Jeffrey, P., & Gearey, M. (2006). Integrated water resources management: Lost on the road from ambition to realisation? *Water Science and Technology, 53*(1), 1–8.
Kingdom, B., Liemberger, R., & Marin, P. (2006, December). *The challenge of reducing non-renewable water (NRW) in developing countries. How the private sector can help: A look at performance-based service contracting*. Washington, DC. The World Bank Group, Water Supply and Sanitation Board Discussion Papers Series, Paper No. 8.
Lasage, R., Aerts, J., Mutiso, G.-C. M., & de Vries, A. (2008). Potential for community based adaptation to drought: Sand dams in Kitui, Kenya. *Physics and Chemistry of the Earth, 33*, 67–73.
Maliva, R. G., & Missimer, T. M. (2010). *Aquifer storage and recovery and managed aquifer recharge using wells: Planning, hydrogeology, design, and operation*. Houston, Schlumberger Water Services, Methods in Water Resources Evaluation Series No. 2.
McClurg, S. (1998). Climate change and water: what might the future hold? In *Western Water* (pp. 4–13). (Water Education Foundation), May/June.
Means III, E. G., Laugier, M. C., & Daw. J. A. (2010). Evaluating decision support planning methods for incorporating climate change uncertainties into water planning. *Final report: Prepared for water utility climate alliance*.
Milly, P. C. D., Betancourt, J., Fallenmark, M., Hirsch, R. M., Kundzewicz, Z. W., Lettenmaier, D. P., et al. (2008). Stationary is dead: Whither water management. *Science, 319*, 573–574.
Missimer, T. M., Drewes, J. E., Amy, G., Maliva, R. G., & Keller, S. (2012). Restoration of wadi aquifers by artificial recharge with treated wastewater. *Ground Water.* (in press)
Mutikanga, H. E., Sharma, S., & Vairamoorthy, K. (2009). Water loss management in developing countries: Challenges and prospects. *Journal American Water Works Association, 101*(12), 57–68.
O'Brien, K., Leichenko, R., Kelkar, U., Venema, H., Aandahl, G., Tompkins, H., et al. (2004). Mapping vulnerability to multiple stressors: Climate change and globalization in India. *Global Environmental Change, 14*, 303–313.
Pearce, F. (2006). *When rivers run dry*. Boston, MA: Beacon Press.
Pearce, F. (2010). *The coming population crash and our planet's surprising future* (p. 289). Boston, MA: Beacon Press.
SKM. (2009). Adaptation options for climate change impacts on groundwater resources. *Report to World Bank*, Washington.
Smit, B., & Wandel, J. (2006). Adaptation, adaptive capacity and vulnerability. *Global Environmental Change, 16*, 282–292.
Sukhija, B. S. (2008). Adaptation to climate change: Strategies for sustaining groundwater resources during droughts. In W. Dragoni, & B. S. Sukhija (Eds.), *Climate change and groundwater* (pp. 169–181). Geological Society of London Special Publication 288.
Tarlock, A. D. (2000). How well can international water allocation regimes adapt to global climate change? *Journal of Land Use & International Law, 15*, 423–449.

Tompkins, E. K., & Adger, W. N. (2004). Does adaptive management of natural resources enhance resilience to climate change? *Ecology and Society, 9*(2), p. 14, Article 10.
Turton, A. R. (1999). Water scarcity and social adaptive capacity: Towards an understanding of the social dynamics of water demand management in developing countries. *MEWRE Occasional Paper No. 9*, School of Oriental and African Studies, University of London, England.
United Nations. (2004). *World population to 2300*. New York: United Nations.
Vogel, R. M., Yaindl, C., & Walter, M. (2011). Nonstationarity: Flood magnification and recurrence reduction factors in the United States. *Journal of the American Water Resources Association, 47*(3), 464–474.
Vrba, J., & Verhagen, B. T. (2006). Groundwater for emergency situations. A framework document (p. 94). *International Hydrological Program (IHP) VI, Series on Groundwater No. 12*. Paris: UNESCO.
Waage, M. D., & Kaatz, L. (2011). Nonstationarity water planning: An overview of several promising methods. *Journal of the American Water Resources Association, 47*(3), 553–540. doi:10.1111/j.1752-1688.2011.00547.x.
Yohe, G., & Tol, R. S. J. (2002). Indicators for social and economic coping capacity—moving toward a working definition of adaptive capacity. *Global Environmental Change, 12*, 25–40.

Part X
Conclusions

Chapter 40
Conclusions

40.1 Introduction

Water management in arid and semiarid lands touches upon a wide variety of different technical, economic, political, governmental, and social issues. The challenge of developing a reliable and safe supply of water that meets the needs of societies is ever increasing because of a combination of population growth and, in some areas, increasing per capita demands associated with growing affluence. Superimposed on the trend of growing demands is the prospect of significant changes in water supply caused by global climate change.

It is abundantly clear that there is no one universal solution to the water supply challenges in arid and semiarid lands. Local solutions will have to be developed based on available local water supplies and economic and socio-cultural realities. The latter are important because water management policies must have widespread local (stakeholder) support and acceptance in order to be successful. Nevertheless, there are a number of themes that are widely applicable in arid and semiarid lands water management.

40.2 Importance of Groundwater

Management of groundwater resources is a critical issue as most arid and semiarid lands either lack significant surface-water resources or, where perennial rivers are present (e.g., Nile, Tigres-Euphrates, Colorado, Rio Grande Rivers), the resources are usually already fully utilized (or close to full utilization). Groundwater has the great advantage in that where it is not over-exploited, it is a reliable source of water that can buffer variations in precipitation and surface water flows. Groundwater can also serve as a strategic resource in the event of disruption of other water supplies, such as desalination facilities. Groundwater is, therefore,

R. Maliva and T. Missimer, *Arid Lands Water Evaluation and Management*,
Environmental Science and Engineering, DOI: 10.1007/978-3-642-29104-3_40,

the lynchpin of conjunctive water use and integrated water resources management. Aquifers can also provide enormous storage space for water that can be utilized in MAR systems. A key point is groundwater may have a much greater value as an element of a conjunctive water use system, than through stand alone use.

A fundamental challenge is utilizing groundwater in a sustainable manner and obtaining the greatest societal value from the resource. Unsustainable groundwater use can ultimately deplete or degrade the resource. Nonrenewable groundwater resources are, by definition, a finite (although in some areas a very large) resource that will inherently be depleted with use. The economic concept of "discounting" is central to management of non-renewable resources, in that people place a much greater value on current use of water than on future uses. The course of least resistance is often to continue current unsustainable water use patterns rather than making painful changes now, especially if the main impacts of current water use (including aquifer depletion) will occur at some time in the future.

A principal water management threat in some arid and semiarid lands is that current groundwater use is not sustainable and there is no real exit strategy in place concerning how to cope with declining groundwater resources. With respect to the use of nonrenewable groundwater, its use is often justified as its being an element of a transitional stage in economic development. It is assumed or anticipated that over time, the national or local economies will evolve away from large-scale irrigated agriculture or that alternative water supplies will be developed. Ultimately, over-exploitation of groundwater resources is self limiting, as declining water levels and quality will force a curtailment of pumping. The critical water management issue is whether over-exploitation is controlled through active management or allowed to go on unchecked until the aquifers are severely depleted or degraded to the point that they can no longer be economically used.

40.3 Water as an Undervalued Resource

Water is vital for life, but is often provided to users at such a low cost (or at no cost at all) that it is perceived as having little value. With respect to groundwater, the cost of water to users is commonly only the pumping and conveyance costs. The water itself is considered to be free. The low value placed on water creates dis-incentives to conserve water and invest in measures to increase water efficiency. Water users tend to respond to the price of water in a rational economic manner. Where water is very inexpensive, the business case typically cannot be made to invest in water saving technology (i.e., costs exceed the financial benefits).

Charging users more for water is a politically difficult issue particularly where users have long taken cheap water for granted or are unable to pay more. It is also important to recognize that social considerations may have greater importance than the economic efficiency of water use. The challenge is addressing both economic efficiency and socio-cultural and political realities. For example, the cost of water provided to farmers is subsidized or provided for free. These subsidies indirectly support the entire rural community. Societal decisions have to be made as to whether or not rural agricultural communities are worth saving and if the consequences of their demise (e.g., more urban migration, increased rural poverty, and less local food production) are acceptable. Water policy decisions tend to be much more complex than simple economic analyses.

On the domestic side, per capita water use rates in many arid regions of developed countries are quite high. The cost of domestic water supply is typically a very small fraction of household budgets, and therefore, little economic incentives exist for people to educate themselves about water conservation and implement conservation measures that might involve minor lifestyle changes. Undercharging for water (or failure to collect fees) in developing countries denies utilities needed revenues to finance the operation, maintenance, and expansion of their water-supply systems.

A fundamental issue in water management in arid lands is that the costs of increasing water use efficiency are typically much less than the costs of developing new water sources. Water users rarely are charged the marginal cost of new water supplies. There is growing consensus that water should be treated as an economic good in the sense that it is recognized that water has an economic value that should be considered in allocation decisions and that the cost of its provision should be recovered to the extent practically possible from users. However, recognizing that water has an economic value does not equate to its being treated as a pure economic commodity and traded in a free market.

40.4 Groundwater Quality

The focus of this book is on water quantity, not quality. The omission of detailed discussion of water quality issues is not a statement on their importance, but rather is due to the fact that it is a highly complex subject that cannot be adequately addressed in one or several chapters. Indeed, where water is scare, protection of the quality of existing sparse resources has much greater importance.

Groundwater resources in arid and semiarid lands, in general, are not especially more vulnerable to contamination than resources in more humid regions. However, some differences in vulnerability related to climatic and hydrologic conditions may exist between arid and humid climates. The greater depth to water (and thus thickness of the vadose zone), provides a buffer between surficial contamination and underlying unconfined aquifers and may be more conducive for biodegradation of organic compounds, particularly volatile compounds. Lower recharge rates may result in lesser dilution of any contaminants that do enter underlying aquifers.

Recharge in arid lands tends to be predominantly indirect (focused) with the bulk of it occurring in relatively small geographic areas (e.g., wadi channel). It is therefore important to identify the main recharge areas and control land use in them in order to reduce the potential for groundwater contamination. Fortunately, upland recharge areas in arid lands often have relatively low population densities.

Wadis have been historically used for wastewater disposal, which results in poorer groundwater quality in wadi aquifers downstream of populated areas.

Coastal aquifers tend to have a high vulnerability to saline-water intrusion induced by high pumping rates and low recharge rates. Saline-water intrusion can be controlled to some degree by managed aquifer recharge systems, such as wadi dams (which increase freshwater recharge) and salinity intrusion barriers.

40.5 Groundwater Resource Assessments

Assessing the volume of usable groundwater resources and impacts from their utilization is critical for effective groundwater management. Much progress has been made in recent years evaluating groundwater resources in arid and semiarid lands, but much work still needs to be done. Developing countries in arid and semiarid lands that have perilous water situations, commonly lack the economic and human resources (i.e., technical expertise and staff) to assess their groundwater resources, monitor groundwater use, monitor changes in aquifer water levels and quality, and develop and implement an effective water policy.

Groundwater resources evaluation is an on-going process in which models need to be refined as new data become available. As reviewed in this volume, there are numerous tools available for assessing groundwater resources. Professionals involved in groundwater management need to have an understanding of what tools are available, the types of information that can be provided, their limitations, and costs. An important technical challenge is obtaining the maximum value from the employment of technology and choosing the right tools that most cost-effectively provides the needed information.

The most critical data for water management is water level (pressure) and quality monitoring data, which are very limited in many arid and semiarid lands. In some areas (e.g., United States), the numbers of wells that are actively monitored has actually decreased dramatically over time. Groundwater monitoring is viewed by some as being mundane and unimportant compared to more high-technology projects. In a time of increasing stress on groundwater resources, and the likely prospect of greater hydrologic stresses from climate change, there may be in some areas less data upon which to assess the state of groundwater resources and to calibrate and verify groundwater models used to make water management decisions.

Another important issue is maximizing the value obtained from hydrologic and hydrogeological data. In well studied areas, an enormous volume of data has been collected over the years. The data includes lithological and geophysical logs, surface geophysical data, water level and quality monitoring data, and meteorological data. Much high-quality data are often obtained or already available for groundwater investigations, but are not fully incorporated into what are ultimately over-simplified groundwater models that are used for resource assessment. There is a need for greater utilization of workflow software that can be used to analyze and up-scale data, and then incorporate it into groundwater flow and solute-transport models. Unfortunately, the water industry is far behind the oil and gas industry in how information is extracted from available data sources and used in simulations. The workflow software exists (e.g., Petrel), but most people in the water field are unaware of its existence and the type of data analyses are now possible.

40.6 Data Management

Collection of hydrologic and hydrogeologic data represents a substantial investment. It has commonly been the case that the storage of the data is fragmented and disorganized. In order to obtain the maximum benefits from the investment in data collection, it is necessary to have the data stored in a manner that is both readily accessible and analyzable by all parties involved in water resources development and management. The future of groundwater data management ultimately lies in GIS-based hydrologic data management systems in which all available data are stored, processed, analyzed, and accessed in real time. Fragmentation of data between government agencies can present a significant hindrance to accurate groundwater resource assessments. Much useful information is also often in the form of "gray" data, which includes the results of studies and reports that had a limited or restricted distribution, and of which often few people are even aware of its existence.

GIS technology offers tremendous opportunities for improved hydrologic information management that is just beginning to be realized. The ultimate goal is to have all available data on the hydrology and hydrogeology of a region georeferenced and incorporated into a single database that could be accessed via the internet by authorized users. Meteorological (temperature and rainfall), monitoring (water level and flow), and major water use (well pumping rates) data should be stored and accessible in real time.

The software for such a hydrologic information system currently exists (e.g., HydroManager and others). The greatest effort, by far, involves entering the available data into the system, much of which may be in hard copy (paper) format. It is also important that the systems be continuously upgraded to include new data as it becomes available. Continued funding for the hydrologic information management system is, therefore, necessary. The investment made in developing the system would be lost if the system is allowed to become obsolete by not continuously incorporating recent data.

The greatest value is obtained from hydrologic information management system when the consolidated data are used for water resources analysis including groundwater modeling and optimization analyses using decision support systems. For example, the hydrologic information management system would provide the input data for a groundwater flow model and calibration targets. A regulatory agency reviewing a well construction or water use permit, would be able to quickly obtain a map of existing wells and the available current and historic water-level monitoring data in the proposed well location, and therefore, be able to make an informed decision as to whether or not to approve the permit. A single centralized database will ultimately result in cost savings, as it would avoid the need to create and compiled new databases for each individual project.

40.7 Demand Management

The most difficult decision facing societies in arid and semiarid lands is how to allocate scare water resources between competing sectors and users within sectors. The societal goal should be to obtain the maximum economic and social benefits from available resources. Reallocation of water is a politically charged issue, because existing water users often have a strong vested interest in maintaining the status quo.

Although there is increasing recognition that water should be treated as an economic good, there is considerable disagreement as to what this really means in terms of practical water management. Arguments for and against a market approach to water management tend to reflect the personal philosophies of advocates with selected anecdotes as evidence, rather than presenting objective, impartial analyses. Nevertheless, it is clear that water has an economic value (and costs associated with it production, treatment and distribution) that needs to be considered in water management decisions.

Demand management through increased water use efficiency for individual tasks is less controversial than reallocation, and is a necessary element of improved water management. Incentives need to exist to encourage the adoption of more efficient water use technologies. Where water is expensive, there is an inherent economic incentive to avoid waste. On the contrary, where users receive water for free or at a nominal cost, there is no incentive to invest in water-saving technology. Water use efficiency needs to be considered on a groundwater (or surface water) basin scale as water apparently lost through inefficient use by individual users (e.g., irrigation return flows) may be later recovered and used by others (form of aquifer recharge).

It makes little economic sense to use scarce water resources for relatively low value uses (e.g., agriculture), if the same water could be reallocated to other sectors of society in which it could result in greater economic returns. In reality, that tradeoff option may not locally exist. There are usually other overriding issues why more water efficient industries are not present in an area, particularly the lack of economic and technical resources. The reason why a developing country does not have, for example, thriving, water-efficient high-technology industries, is invariably not related to the allocation of needed water to agriculture.

40.8 Virtual Water Trade

Virtual water, which has also been referred to as the embedded water, referring to the amount of water use involved in the growing and manufacture of products that are traded (Allan 2003). The term "virtual" refers to the fact the most of the water used to produce a product is not actually present in the product. The basic concept of the virtual water trade is that areas facing water scarcity would benefit by importing food and other products with high virtual water contents, which are grown or manufactured in areas with more abundant water resources.

Virtual water trade makes great sense from a societal water use efficiency perspective. In fact, virtual water trade is a reality in many water scarce countries, which already import a large fraction of their food. However, concerns over food security and vulnerability to global commodity price changes cannot be discounted. Food supply is a national security issue. Countries will need to have assurances that food will always be available to them at an affordable cost, before they will voluntarily embrace the virtual water trade. Most countries in arid lands will have to accept increasing food imports out of necessity, simply because they do not have water resources to be self-sufficient in food supply.

40.9 Integrated Water Resources Management

Integrated Water Resources Management (IWRM) is a concept that is increasing being pursued out of necessity in arid and semiarid lands, although progress has been slow and variable between countries. Where water resources are scarce relative to demands, water cannot be effectively managed in isolation. Instead there is a need to optimize the use of all available water resources, manage demand (as well as supply), and consider how water supply is integrated into other aspects of society. The challenge with implementing IWRM is that on a large scale (country or region) it is too much of an all encompassing concept to fully grasp and handle. The most practical approach is to strive to evolve to more integrated water resources management in a step-by-step manner. It is also critical that IWRM plans be developed with a focus on implementation. Plans without subsequent action are a waste of time and money and currently occupy the shelves of offices with little good use.

A good starting point is the conjunctive use of water resources, i.e., examining how all available water resources (e.g., surface water, groundwater, reclaimed water, and desalted water) can be utilized together in order to achieve the greatest economic and societal benefits. Implementation of IWRM will necessarily be a gradual process in most cases, as it is simply not practically possible to quickly change current water supply and use infrastructure, which represent great societal and/or private investments. Instead, efforts should be made to implement more integrated management of water resources as opportunities arise (e.g., as resources become available or there is a need to replace and expand existing infrastructure).

IWRM also needs to consider social-cultural issues. All stakeholders should be involved to some degree in the decision-making process. Projects that are discordant with local social-cultural and religious values or are perceived (often accurately) as being foisted on local communities by outsiders without their consent, will likely elicit strong local opposition. The human element cannot be ignored if water management is to be successful.

40.10 Wastewater Reuse

The use of freshwater resources in most water scarce areas already approaches or exceeds sustainable levels. Some areas have non-renewable groundwater resources that can be used for current and future demands, but are ultimately, by definition, unsustainable as they will someday become uneconomical to further exploit. Baring the large-scale importation of water, which is unlikely for political reasons, the primary large water-supply options for water scarce countries are desalination and wastewater reuse, with perhaps rainwater harvesting also being viable in some settings.

Wastewater reuse has the great advantages of eliminating environmental impacts associated with its disposal and being a reliable water supply that has a relatively low vulnerability to droughts. Wastewater reuse is already being practiced to varying degrees throughout many arid and semiarid regions, but the degree of planning and control considerably varies. A fundamental challenge for water resources management in arid and semiarid lands is the optimization of the use of all water resources including reclaimed water, while ensuring that public health is protected. The implementation of wastewater reuse can be divided into four primary stages (Fig. 40.1).

In Stage I, wastewater is treated to varying degrees with the primarily purpose being disposal rather than intentionally putting the water to a beneficial use. Some incidental reuse may occur as the result of discharged wastewater recharging underlying unconfined aquifers. Wastewater treatment focuses on meeting environmental regulations for disposal and treated wastewater is viewed as a liability rather than an asset. Unplanned and unmanaged reuse may also occur, which in developing countries often involves use of untreated wastewater with associated health risks.

Stage II is the developing demand phase in which a utility has established a formal wastewater reuse program. The supply of treated wastewater exceeds demands with excess flows going to waste. Wastewater utilities actively seek new customers and reclaimed water is provided at minimal, if any, cost and in unrestricted quantities. The reuse system reduces the disposal requirements, offsets some fresh groundwater (or desalinated water) use, and often satisfies government mandates.

The mature stage (Stage III) of reuse systems is characterized by demand catching up with available supply. Reclaimed water may no longer be available to

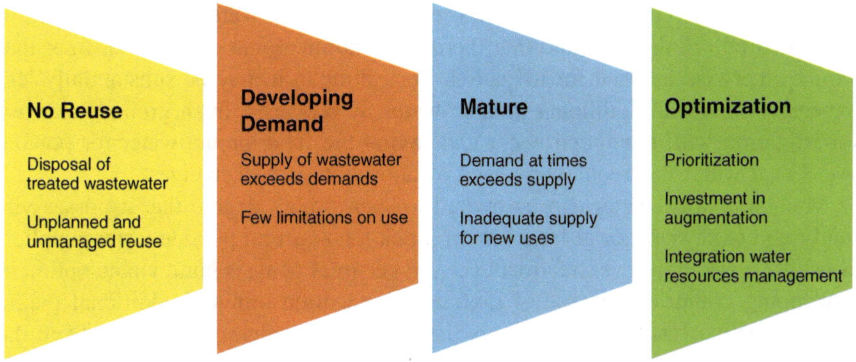

Fig. 40.1 Stages of wastewater reuse implementation

potential new customers, especially during high demand periods. Restrictions on the use of reclaimed water may be emplaced in order to avoid waste (e.g., over irrigation). Utilities may also be able to charge more for the water in order to recoup costs.

The final stage of reuse system development (Stage IV) is the optimization phase. Reclaimed water is recognized to be a valuable resource and the concern becomes how to obtain the maximum societal value from the water. The optimization of reclaimed water use ties into integrated water resources management, which is concerned with the optimization of use of all water resources. Optimization of the use of reclaimed water may include efforts to increase system reliability and total reuse percentage by augmenting the supply during peak demand periods and storing water during periods of low demand (e.g., using ASR) for later use. Optimization may also involve reallocation of reclaimed water to higher value uses.

The time of transition from Stage II to Stage III is quite variable and depends upon the availability of potential customers. Existing urban areas may be stuck in Phase II for a long period of time (many decades) because of low internal demand for water and unaffordable costs for expanding the reclaimed water distribution system (e.g., construction of a dual-piping system in built-out areas or dedicated transmission mains to distant agricultural areas). Where there are abundant potential customers, the transition from Stage II to Stage III may be quite rapid (within a decade). Once initial concerns over the use of reclaimed water are overcome by the experiences of initial users, there may be a rush of other users for what is recognized to be a valuable, often nutrient-rich source of inexpensive water.

A critical issue is the cost to construct and operate reclaimed water systems that can seldom be fully recovered from the sale of water to customers. Where water for irrigation is free or available at low costs, wastewater utilities often cannot make the business case for investing in reuse systems as they will not generate enough cash flow to cover construction and operational costs. Economic analyse of reclaimed water reuse systems need to consider that the water would still have to be treated before disposal and that there is economic value of the freshwater that would otherwise be used for irrigation. For example, it may be substantially less expensive to obtain additional potable water by replacing fresh groundwater use for irrigation with reclaimed water and saving the fresh groundwater for potable use. Local water use regulations need to allow for such exchanges.

Municipal wastewater can be treated to such a high degree that it posses virtually no health risks, or at least less of a risk than current potable water supplies. However, very high level treatment (i.e., hyper-treatment) is not a viable option in developing countries because of high costs and often minimal additional public health benefits. Institutional controls (i.e., means of isolating reclaimed from the public) can be effective in reducing the possibility of unintended exposure. MAR technologies may be a cost effective means to improved reclaimed water quality as an element of a multiple barrier approach to protecting public health (Chap. 31).

In rural agricultural areas, which typically do not have centralized sewerage systems, more work is needed to develop and promote the use of low-cost, small-scale technologies that reduce the health and environmental risks associated with wastewater reuse. Use of graywater for irrigation is an attractive option, because of its much lower risks than "blackwater" that contains fecal matter.

40.11 MAR for Wastewater Treatment

Managed aquifer recharge of reclaimed water can serve three primary purposes:

(1) increase the volume of freshwater stored in an aquifer,
(2) improve the quality of reclaimed water through natural contaminant attenuation processes, and
(3) protect fresh groundwater resources (e.g., salinity barrier systems).

There is now ample laboratory and operational data showing that the quality of wastewater can be substantially improved during aquifer recharge and subsequent storage. The concentrations of pathogens and most chemical contaminants are reduced by a variety of physical, chemical, and biological processes.

MAR of treated wastewater can serve as one element (e.g., a polishing step) of a multiple barrier approach to wastewater treatment and resue. MAR-based wastewater treatment methods are especially attractive for developing countries (compared to advanced engineered treatment systems), because of their lower costs, lesser technical expertise requirements for their operation and maintenance, and high reliability. Natural subsurface contaminant attenuation processes operate continuously without on-going human intervention. For example, a power failure can shut down a conventional wastewater treatment plant, but natural attenuation of pathogens in recharged wastewater within an aquifer would be unaffected.

System performance will depend upon local hydrogeological conditions, which need to be evaluated on a site-by-site basis. Monitoring is also necessary to confirm that contaminant attenuation is occurring at expected rates and that recovered water meets quality targets. A key operational issue is managing system clogging and associated reduced recharge rates.

Although the advantages of MAR of treated wastewater are compelling, progress has been slow in its implementation. The main impediment is concern over the introduction of wastewater into what are considered uncontaminated aquifers. Local regulations or policies may not allow the practice. The reality is that reclaimed water is often of better quality than native groundwater. Systems can also be located and operated so that recharged wastewater will not enter potable water supply wells, and thus they pose no threat to public health. It is expected that as operational data from the initial wastewater MAR systems become available, the technology establishes a safe track record, and the benefits of the technology become more widely known and accepted, there will be an acceleration in the implementation of wastewater MAR.

40.12 Desalination

Desalination can provide a virtually unlimited supply of water to any country with access to a coast or, in some instances, to abundant brackish or saline groundwater resources. Desalination also has the great advantage of being a drought-proof source of water. The principal constraint on the implementation of desalination is cost, which usually limits its use to potable supply and high-value industry uses.

Financing the construction and operation of desalination plants will clearly be a greater challenge in many developing countries with limited economic resources. The greatest barrier to the use of desalination to expand the potable supply in some cases (poorer regions) may be limited by an inadequate distribution system (e.g., poor coverage and high leakages rates or unaccounted for water) and an ineffective means for billing and collecting fees to recoup operational costs. However, citizens in developing countries often already pay a high human and economic price for the absence of a safe, convenient, and reliable potable water supply.

The costs of desalination of seawater and brackish groundwater have been decreasing over time, and it is expected that system construction and operational costs (corrected for inflation) will continue to gradually decrease in the future, albeit at a slower rate. On the other hand, increases in energy costs would increase desalination system operational costs (unless systems are powered by alternative energy sources). Hybrid systems including various combinations of reverse osmosis and thermal desalination, electrical power production, and aquifer storage and recovery can increase overall efficiencies and decrease costs. New desalination technologies, such as adsorption desalination, are low-cost potential solutions to water supply development. These technologies need to be further researched and developed.

There is growing attention being given to the environmental impacts of desalination. Greenhouse gas emissions associated with the production of energy to power plants (i.e., their carbon footprint) is being given growing attention and it is expected that the use of renewable-energy powered systems will gradually increase over time. The transition to renewable-energy powered systems will more likely be prompted by rising conventional (fossil fuel) energy costs and decreasing renewable energy costs rather than from environmental consciousness.

The environmental impacts associated with concentrate disposal are also growing in concern. Point discharge impacts can be reduced through outfall design. Salt loading of receiving water bodies may, in the long-term, be a more intractable problem as economically viable and environmentally sound options for the disposal of the large concentrate flows from large seawater desalination plants may not be available. In light of the crucial role of desalination as the primary source of potable supply for rapidly growing metropolitan areas in the Middle East, it is inevitable that environmental impacts from salt loading will not be of prime importance (and force a discontinuation of desalination), although economically feasible efforts to mitigate impacts may be implemented.

40.13 Managed Aquifer Recharge

Managed aquifer recharge (MAR) consists of a diverse suite of technologies to store and treat water. MAR dovetails with some types of rainwater harvesting, which can be considered small-scale versions of MAR. MAR is already being widely practiced in arid and semiarid lands, although there are differences between regions as to which techniques are most commonly used. For example, wadi dams are widely used for both flood control and groundwater recharge in the Middle East. ASR systems for strategic store freshwater to be used in the event of a shutdown in desalination facilities are under development in the United Arab Emirates. The water resources management benefits of MAR are still far from being fully realized.

Although the main MAR technologies are now well established, there remain substantial opportunities for improved implementation. Perhaps the greatest challenge associated with MAR is that it is increasingly being treated as "commodity engineering" rather than a specialized field that requires technical sophistication to successfully implement. The basic design of a recharge well or infiltration system is fairly simple and does not require much technical skill. The challenge lies in designing MAR systems that actually achieve water resources management goals and safely and reliably operate over the long term. MAR system performance depends upon site-specific hydrogeological conditions, which may be locally unfavorable.

The main issue now is not the invention of new MAR techniques but rather optimization of the existing techniques. Optimization involves finding and implementing means for improving system performance. Some specific opportunities for optimization of MAR include:

- Increasing the percentage of water captured in wadi dams that actually recharge underlying aquifers.
- Improved aquifer characterization and groundwater modeling to better predict the movement and mixing of recharged water. Improved predictive modeling would allow for better system design and operation, and addressing of public health concerns. There would be great value in being able to better predict system performance before incurring the costs of full-scale system construction.
- Improved understanding of subsurface geochemical processes including natural contaminant attenuation processes and adverse fluid-rock interactions such as arsenic and metals leaching.
- Improved management of clogging in wells and surface spreading systems, which is a major operational concern.
- Improved economical analysis procedures for MAR systems. A key issue for the growth in the implementation of MAR is making the business case that the systems are cost effective. The benefits of MAR systems should be greater than costs. MAR system costs should also be less than the costs of other water supply, treatment, or storage options. Economic analyses of MAR systems also need to incorporate the risks associated with dealing with natural systems. Hydrogeological systems inherently have an element of uncertainty, which can be significantly reduced (but not totally eliminated) through better aquifer characterization.
- Greater incorporation of MAR in integrated water resources management. MAR technologies need to be considered as a potential integral element of water resources management. Decision support systems need to include or be specifically developed that could identify how MAR can contribute to optimization of water resources.

40.14 Global Climate Change

There is now ample supporting data that the surface temperatures of the Earth are increasing and that anthropogenic emissions of greenhouse gases (e.g., carbon dioxide and methane) are at least a contributing factor. There is still much uncertainty as to the future rate of temperature increase and its impact on the hydrologic cycle. Global warming is predicted to result in an acceleration of the hydrologic cycle. Higher temperatures on a global scale will result in greater ET rates and corresponding increases in precipitation, as under steady state conditions ET and precipitation are balanced. However, changes in precipitation are predicted to have a high degree of spatial variation. Some regions will benefit from increased rainfall, while others may experience significant reductions in rainfall and, therefore, surface water runoff. Increased temperatures are also expected to result in an increase in the amount of precipitation that falls as rain (rather than snow) and the melting of glaciers, which will ultimately imperil water supplies that are dependent on snowmelt. Snowmelt dependent areas may initially benefit from increased water supply through the release of water stored in glaciers, which is essentially a not a sustainable resource.

As discussed in Chapter 38, climatic models suggest that the future climate will be more variable, with more common extreme events. Temperatures are expected to have a greater range of fluctuations around a gradually increasing mean. Annual rainfall will tend to occur more frequently as short-duration intense events with longer intervening dry periods. Water supply and management systems designed based on historic climatic data may not be able to cope with future climate changes. At the present time, climatic models cannot accurately predict future climatic changes on a local level, although there is increasing agreement of regional trends. Modeling results suggest that an increase in the number of consecutive dry days (i.e., drought conditions) will be most significant in North and Central America, the Caribbean, northeastern and southwestern South America, southern Europe and the Mediterranean, southern Africa and western Australia (Bates et al., 2008).

Societies will vary in their ability to cope with climate change. The wealthier nations will have the financial resources to mitigate many of the impacts of climate change, although it is clear that at the present time that there is a profound lack of understanding and appreciation by many decision-makers and the general public as to how expensive such measures could be. Sea level rise represents perhaps the greatest economic threat because of the large populations and commercial, industrial, and transportation infrastructure occupying low-lying coastal cities. Poorer countries in arid and semiarid lands are particularly vulnerable to the impacts of climate change because they already suffer from water scarcity and face increasing water demands associated with population growth. Many countries are poorly adapted to current climate conditions, much less potential future more challenging climatic conditions.

40.15 Adaptation to Climate Change

At the present time, projections of local hydrologic impacts of climate change are poorly constrained. Water management systems need to have robust and flexible designs that are capable of dealing with a wide range of potential future hydrological conditions. There is great variation in the adaptive capacities of societies to future climate changes, which depend to a large degree on existing available water resources, and financial and technical resources. Poor countries already facing water scarcity will generally have low adaptive capacities. The primary adaptive strategies to water scarcity caused by global climate change (or other factors) are:

- Demand management. Societies can implement measures to increase the efficiency of the use of available water resources. Measures can be taken to increase the efficiency of individual activities (e.g., adopt high efficiency irrigation technologies) and to reallocate water to sectors of the economy that yield the greatest value per volume of water.

- Optimization of existing water resources. Societies can implement techniques such as managed aquifer recharge and rainwater harvesting to capture and store more of available water resources. Conjunctive use schemes may also increase the effectiveness of water management. Optimization of existing water resources also includes increasing the reuse of treated wastewater (reclaimed) water.
- Development of new water resources. New water sources, such as desalination, can be developed to provide addition water resources. New water sources may also include development of new groundwater resources and water importation or "virtual" water trading.

As water resources becomes scarcer, integrated water resources management will become increasing important as societies will be even less able to afford inefficient and ineffective water management. The principal challenge will be overcoming inertia and vested interests in order to make difficult, but necessary, decisions. However, it is critical that socio-cultural considerations become part of (rather than separate from) integrated water resources management. Water management plans that fail to adequately address the human element of water management will ultimately have a low chance for success.

40.16 Closing Thoughts

It has become clear to the authors while researching this book that a very wide range of tools, technologies, and strategies are available to improve the evaluation and management of water resources in arid and semiarid lands. The importance of water in the coming decades is also being increasingly realized by the popular press, politicians, and leaders. The growing importance of water is a stimulus for much public and private sector research and investment, and it is anticipated that more and better (less expensive) water supply and treatment options will be available in the future. Solutions will exist to solve global water supply problems. The critical question is whether the will, courage, and foresight will exist to implement, what may be painful and expensive solutions.

Perhaps the most sobering lesson from ancient failed civilizations is that the decisions that societies make can determine their fate and some choose what turned out to be the wrong strategy for dealing with environmental change (Diamond, 2005) or existing environmental conditions. The focus of this book is on the technical issues related to water management. Scientists and engineers can develop technical solutions to water resources and other problems. Finding solutions to the human element of water resources management will invariably be a much more difficult task.

References

Allan, J. A. (2003). Virtual water—the water, food, and trade nexus. Useful concept or misleading metaphor? *Water International, 28*(1), 4–11.
Bates, B., Kundzewicz, Z.W., Wu, S., & Palutikof, J. (Eds.). (2008). *Climate change and water: Geneva Intergovernmental Panel on Climate Change Technical Paper VI.*
Diamond, J. (2005). *Collapse, how societies choose to fail or succeed.* London: Penguin Books.

Curriculum Vitae

Robert G. Maliva has been a consulting hydrogeologist since 1992 and is currently a Principal Hydrogeologist with Schlumberger Water Services USA Inc. He is currently based in Fort Myers, Florida. Dr. Maliva specializes in alternative water supply projects including managed aquifer recharge and desalination. He has been either the project manager or the technical lead on numerous aquifer storage and recovery (ASR) projects. He has designed raw water supply wellfields for brackish water desalination systems, alternative intakes for seawater desalination systems, and injection well systems for concentrate disposal. He has experience in all aspects of underground injection control (UIC) projects including feasibility assessments, exploratory well programs, system design, permitting, construction supervision and management, and environmental impact assessments.

Dr. Maliva has been investigating and implementing a more technically sophisticated approach to managed aquifer recharge (MAR) projects including the use advanced borehole geophsyical logging technology, more meticulous geological and hydrogeologic analyses, and the computer modeling of the controls of MAR system performance and strategies to optimize system performance. His objective is to learn from the lessons of both failed and successful MAR systems and apply new technologies to improve the performance and obtain the maximum waste resources value of ASR and other managed aquifer recharge projects.

Robert G. Maliva completed his Ph.D. in geology at Harvard University in 1988 on the silicification of marine limestones. He also has a Masters degree from Indiana University (Bloomington) and a BA in geological sciences and biological sciences from the State University of New York at Binghamton. Upon completion of his doctorate degree, Dr. Maliva has held research positions in the Department of Earth Sciences at the University of Cambridge, England, and the Rosenstiel School of Marine and Atmospheric Science of the University of Miami, Florida. He grew up in New York City and attended Stuyvesant High School in Manhattan. Dr. Maliva's university research interests were in sedimentology, sedimentary

R. Maliva and T. Missimer, *Arid Lands Water Evaluation and Management*, Environmental Science and Engineering, DOI: 10.1007/978-3-642-29104-3, petrology and geochemistry, and with a focus on the diagenesis, and controls of porosity and permeability preservation of carbonate rocks.

Dr. Maliva has also managed or performed numerous other types of water resources and hydrologic investigations including contamination assessments, environmental site assessments, water supply investigations, wellfield designs, and alternative water supply investigations. He has maintained his research interests and completed studies on such diverse topics Precambrian silica diagenesis, aquifer heterogeneity, precipitates in landfill leachate systems, carbonate diagenesis, and various aspects of the geology of Florida. Dr. Maliva gives frequent technical presentation and has numerous peer-reviewed papers and conference proceedings publications on ASR and injection well and water-supply issues, hydrogeology, and carbonate geology and diagenesis. He coauthored with Thomas Missimer the book, *Aquifer Storage and Recovery and Managed Aquifer Recharge Using Wells: Planning, Hydrogeology, Design, and Operation,* which was published in 2010.

Thomas M. Missimer is Visiting Professor of Hydrogeology at the King Abdullah University of Science and Technology, Thuwal, Saudi Arabia. He teaches groundwater hydrology and other courses and conducts research at the Water Desalination and Reuse Center. Dr. Missimer received his PhD in marine geology and geophysics from the University of Miami in 1997, an MS in geology from Florida State University in 1973, and a BA in geology from Franklin & Marshall College in 1972. Dr. Missimer began his career with the U. S. Geological Survey (USUS) as a hydrologist in 1973. In 1975, he became a research associate at the Rosenstiel School of Marine and Atmospheric Science, University of Miami, where he stayed one academic year. In May of 1976, he founded Missimer & Associates, Inc., a groundwater consulting firm that specialized in the development of water supplies for municipalities, agriculture, and industry. He served as president of Missimer & Associates, Inc. until May of 1991. In late December of 1991, Missimer & Associates, Inc. became a public corporation traded on NASDAQ exchange. The corporation was subsequently renamed ViroGroup, Inc. in 1992, and he served as Vice Chairman of the company until September of 1993. He cofounded Missimer International, Inc., a groundwater consulting and construction firm in December of 1993, and served as president of that company until March 1999, when it was acquired by Camp, Dresser & McKee, Inc. (CDM). He was vice president and artificial recharge practice leader of CDM until June 2004, when he retired and founded Missimer Groundwater Science, Inc., a groundwater consulting firm. He served as president of that company until it was acquired by Schlumberger Water Services USA Inc. in January 2007. In his time at Schlumberger, he was the water resources global sector leader and a research advisor. He retired from Schlumberger at the end of 2010 and joined the King Abdullah University of Science and Technology as Visiting Professor.

Dr. Missimer is the author or co-author of seven books and about 300 technical publications, including peer-reviewed papers, conference papers, book chapters, educational publications, and field trip guidebooks. He managed and supervised consulting projects in water supply development, groundwater contamination assessment and remediation, and development of water supple plans. His current research includes the design and development of natural systems filtration systems using bank filtration, wells, and galleries as intakes to water treatment plants. He also is doing research on aquifer storage and recovery of treated water as well as aquifer recharge and recovery of domestic wastewater for treatment, storage, and recovery. In 1991, he won the Best Paper Presentation Award at the World Congress on Desalination and Water Reuse. In 2004, he and Robert Maliva were awarded the 2nd Place Thomas A. Philpott Excellence of Presentation Award by the Gulf Coast Association of Geological Societies. In February, 2010, the Houston Section of the Society of Technical Communication presented him with an Award of Excellence in Technical Communication for his book (2009), *Water Supply Development, Aquifer Storage, and Concentration for Membrane Water Treatment Facilities*. In May, 2010, this same book received the highest achievement award of Distinguished Technical Communication at STC International Technical Communications Competition. In February 15, 2011, the Houston Section of the Society of Technical Communication presented Robert Maliva and him with an Award of Excellence in Technical Communication for their book (2010), *Aquifer Storage and Recovery and Managed Aquifer Recharge Using Wells: Planning, Hydrogeology, Design, and Operation*. On May 17, 2011, this same book received an Award of Excellence in Technical Communication at STC International Technical Communications Competition.

Dr. Missimer has served as an appointee to a number of important boards and advisory committees. Florida Governor Lawton Chiles pointed him to the State Board of Professional Geologists for a 4-year term beginning in December 1991. He was appointed and confirmed by the Florida Senate to two terms on this board and served as Vice Chairman and Chairman. Governor Jeb Bush appointed to two terms on the Florida Forever Advisory Council which drafted the goals for the $3 billion environmental land acquisition program in Florida. In 2007, Governor Charlie Crist appointed him to the Strategic Aggregates Task Force, which assessed the future needs for construction materials in Florida.

He was elected Chairman of the Florida Academy of the Earth and Planetary Sciences Section of the Florida Academy of Sciences in 1976 and 1977. He served on the national steering committee on environmental site assessments (contamination) for the Association of Groundwater Scientists and Engineers. He is a Life Member of the American Water Works Association. He is also a member of various other technical associations, including the American Geophysical Union, the Society of Sedimentary Geology, American Association of Petroleum Geologists, the International Desalination Association, the Southeast Desalting Association (former board member), the National Groundwater Association, the American Association for the Advancement of Science and others.

Index

17α-ethylestradiol (EE2), 843
17β-estradiol (E2), 832, 833, 843
1997 Convention on Non-navigation Uses of International Watercourses, 958, 959, 963
$^3H/^3He$ age, 285

A

Accelerator mass spectrometry (AMS), 282
Acoustic-televiewer log (ATV), 426–428
Active pharmaceutical ingredients (APIs), 813
Actual evapotranspiration
  measurement of, 173, 174
    eddy covariance method, 176, 177
    energy balance and micrometeorological methods, 175
    energy balance Bowen ratio method, 175, 176
    geophysical methods, 177, 178
    lysimeters, 174
Actual rate of evapotranspiration (ETa), 22
Actual runoff (RO_v), 258, 303
Adaptation
  basis of, 1002
  climate change
    adaptive capacity, 1002–1005
    greenhouse gas emissions, 1021, 1022
    increased water supply (*see* Water)
    occurrence, 1002
    water management plans, 1005–1007, 1006*f*
    water scarcity causes, 1040, 1041
    water use, decreased (*see* Water)
  definition, 1001
  property rights regime, 1008
  water management strategies, 1007, 1008
Adaptive management, 609–610
Adaptive strategies
  to climate change, 1001–1005
Adsorption desalination. *See also* Desalination, 711–713, 712*f*, 713*f*
  combined with multiple effect distillation, 724, 725
  geothermal energy with, use of, 737
Advanced wastewater treatment (AWT), 779
Aerial photograph interpretation and remote sensing, 438, 439
Age tracers, 265, 266
Agricultural irrigation. *See also* Irrigation
  efficiency of, 686–691
  primer, 682–684
  types of, 682
  wastewater reuse for, 782–786
  water losses associated with, 690
Agriculture
  climate change
    frequency and intensity of flood, 996
    temperate and tropical areas, 995
    virtual water trading, 996
    warmer climate, 995
    water requirements, irrigation, 995
  drought, impact of, 31
  global water use, 4
  lithic mulch, 553, 554
  rainwater harvesting, 545, 546
Agua Fria Recharge Project, 572, 573*f*
Air entrainment, 571
Airborne geophysics, 390, 391

R. Maliva and T. Missimer, *Arid Lands Water Evaluation and Management*,
Environmental Science and Engineering, DOI: 10.1007/978-3-642-29104-3,

A (*cont.*)
arid lands, aquifer mapping in, 391, 392
recharge and salinity mapping, 394
saline-water
contamination plumes, 393, 394
and fresh water interface coastal mapping, 392, 393
Akkadian empire, Mesopotamia, collapse of, 519
Alkylphenol polyethoxylates (APEOs), 833
All-American Canal lining, 618, 968, 969
Allocation, 853
as fundamental water management issue, 883
initial allocation, 897
Alluvial aquifers, 591
Alluvial deposits, 73, 445, 966
in basin-fill aquifers, 105
and wadi, 103
Alluvial fans
arid region versus humid-region, 55, 56
characteristics of, 56
coarse-grained sheetflood deposit, 60, 64*f*
debris-flow and sheetflow process, 57, 58*f*
definition, 54
diagenetic process, 61, 66*f*
factors, 54
lateral relationship of, 61, 67*f*
mountain-basin geomorphic system, 53*f*, 55
parasequence or sequence, 60, 61, 65*f*
primary versus secondary sedimentary, 57*t*
sedimentary facies, 60, 61*f*
sedimentation, 57–59
stages of, 59, 60
subsurface structure, 56
type I and II, characteristics of, 60, 63*t*
types, 55
Alluvial rivers, 321
definition, 72
Alternative energy use, in desalination, 725–736
Alternative water supplies, development of as element of IWRM, 915
Amargosa Desert Basin, 195
American Rule. *See also* Reasonable use doctrine, 866
AMS. *See* Accelerator mass spectrometry (AMS)
Anasazi, Southwestern United States
collapse of, 521, 522
water management in, 510, 511
Ancient climate, collapse of, 517–522
Ancient water management, 501–523
ancient China, 509
ancient Jordan, 502
Jawa, 503
Petra, 503–506
collapse of, 517–518
Akkadian empire, Mesopotamia, 519
Anasazi, Southwestern United States, 521, 522
Angkor, Cambodia, 521
classic Maya civilization, Central America, 519, 520
Mississippian (Cahokian) culture, 520, 521
Neolithic cultures, Central Plain of China, 521
Tiwanaku civilization, 520
Greek civilization, 508, 509
Negev Desert, of Israel, 507, 508
Qanats, 513–517
Roman civilization, 508, 509
Southwestern United States, 510–513
Anasazi, 510, 511
Hohokam, 512, 513
Angkor Empire, Cambodia, collapse of, 521
Antecedent runoff condition", 303
Antelope Valley, California, 359
Anthropogenic organic compounds, 811
APEOs. *See* Alkylphenol polyethoxylates (APEOs)
APIs. *See* Active pharmaceutical ingredients (APIs)
Appropriateness, in water management, 16–19
Aquifer assessment programs, 931
Aquifer characterization, 224, 225
Aquifer hydraulic testing, 229, 230
Aquifer management organizations (AMOs), 934
Aquifer mapping, formation and, 381, 382
Aquifer performance tests (APTs), 230–232, 253
during drilling, 232–233
Aquifer properties
heterogeneity
arid and semiarid land, control recharge, 129
cenozoic carbonate, 127
karstic limestones, 127, 128*f*
sedimentary deposits, 127
solute transport, 128, 129
types, 127
hydraulic
diffusivity, 124

groundwater flow
(*see* Groundwater flow)
heads, 117, 118
potentiometric surface, 117, 118
properties (*see* Hydraulic properties)
porosity
effective, 124
hydraulic conductivity, 126
pore types, 125
primary, 125, 126
secondary, 126
Aquifer recharge, *See also* Groundwater recharge, 155, 659
alluvial, 591, 592
and aquifer heterogeneity, 129
artificial, 600
cost-benefit analysis, 616
as element of IWRM, 915
estimation of, 298, 476
fractures, on groundwater flow, 111
groundwater modeling goal, 492, 493
and infiltration, 269, 336, 589
as injection well application, 563
managed (*see* Managed aquifer recharge (MAR))
playas role in, 204, 205
primary benefits of, 617
rainwater harvesting, 555, 556
spreading basins, 609
stormwater management systems, 538*f*
and transmission losses, 293, 294, 315
unmanaged, 617–619
using wells, 567, 568, 561*t*
wadis role in, 103
and water balance, 659
water management goal, 311
Aquifer recharge and recovery (ARR), 560
Aquifer storage and recovery (ASR), 562, 563–567
advantages of, 564
chemically bounded, 563–565, 564*f*
components of, 563
definition, 563, 564
future innovations of, 739, 740
limitations of, 564
physical storage, 565, 566, 566*f*
using horizontal flow barriers, 600–602, 601*f*
regulatory storage, 566
Aquifer storage transfer and recovery (ASTR), 568, 569, 569*f*
Aquifer water budget-based methods, 254
aquifer water budget, 254–256
Darcy flux in aquifers, 258–260
spatially distributed water budget models, 256–258
stream gauging, 261
baseflow method, 260–261
surface water budget of channels, 261, 262
Aquifers, 158, 159, 187
basin-fill, 105*f*
alluvil, 105
deposition of, 105
flow system, 106, 107
hydraulic energy, 105
hydrologic characteristics, 106, 107*f*
phreatic playas, 106
salinity, 108
sedimentation, 105, 106
thickness, 105, 106
vadose playas
carbonate, 108
categories, 100*t*
confined, 96, 157, 160
flow system, 100–102
fractured aquifer (*see* Fractured rock aquifer)
hydraulic classification
confined (*see* Confined aquifers)
unconfined (*see* Unconfined aquifers)
hydraulic testing, 226, 229, 230
packer tests, 234, 235
performance testing, 230, 232
testing during drilling, 232, 233
intermontane, 108
non-renewable groundwater resources, 112, 113
perched aquifers, 98, 99
recharge (*see* Aquifer recharge)
regional and intermediate flow systems, 112
regional flow orientation mapping, 382, 383
response time, 159
sand dune (erg), 104
stand stone and submontane, 108
unconfined, 160
wadi, 102–104
water budget, 154
water budget-based methods, 254
Darcy flux in aquifers (flow tubes), 258–260
spatially distributed water budget models, 256–258
stream gauging, 260, 261
surface water budget of channels, 261, 262
water budget, 254, 256

A (*cont.*)
Archie's equation, 367, 410
Archie's Law, 377
Arid and semiarid lands
aquifer types in, 95, 100*t*
aquifer hydraulic classification, 95–98
local aquifers, 102–104
perched aquifers, 98–100
desalination, 701
groundwater modeling for, 476, 477
IWRM in, 923, 924, 1033, 1034
rainwater harvesting in, 529–531
concepts, 532–534
and RBF, 643
resource reliability in, 14
Arid environments
recharges estimates in, 271–273
Arid lands, 923, 924
non-renewable groundwater resources, 928
Arid lands, desalination in, 701–744
alternative energy use, 725–736
future innovations of
aquifer storage and recovery, use of, 739, 740
distillation treatment plant, 736, 737
geothermal energy with adsorption desalination, use of, 737
hybrid desalination, 740, 741, 742*f*
seawater RO plants, subsurface intakes use for, 737–739
solar energy, use of, 737
waste heat, use of, 737
geothermal powered desalination, 732, 733
membrane, 702, 714–721
brackish-water reverse osmosis, 717–719
electrodialysis reversal, 714, 719, 720, 720*f*
forward osmosis, 720, 721
seawater reverse osmosis, 714–718*f*
nuclear-powered desalination, 733–736, 735*f*
solar-powered desalination, 727–730
direct distillation, using solar stills, 727, 728
indirect desalination, using solar collectors, 729, 730, 731*f*, 732*f*
indirect desalination, using solar ponds, 728, 729
technologies
costs of, 743, 744
energy use of, comparison of, 741–743, 743*t*
global distribution of, 703*f*
thermal desalination, 702, 704–713
adsorption desalination, 711–713, 712*f*, 713*f*
multiple effect distillation, 708–710, 709*f*
multi-stage flash distillation, 705–708, 706*f*
vapor compression distillation, 710, 711*f*
wind-powered desalination, 731, 732
Arid lands water budgets
main elements of, 152*f*
Arid regions
precipitation in, 164, 165
recharge location, 197, 198
Aridity
causes of, 25
cold ocean currents, 26
continental winds, 25
high pressure, 25
rain shadow effect, 26
classification, 24*t*
definition of, 21–23
Aridity index (AI), 23, 24
ARR. *See* Aquifer recharge and recovery (ARR)
Arroyos, 45, 510
bedrock relief in, 74
definition, 72
ephemeral river systems, 102, 153
valley-bottom gullies, 77
Artificial aquifers. *See also* Aquifers, 600
Artificial recharge, 319, 333
GIS application, 465
as Hueco Bolson aquifer option, 967
in Oman, 593
Arvada, Colorado ASR system, 335, 336
ASR. *See* Aquifer storage and recovery (ASR)
ASTR. *See* Aquifer storage transfer and recovery (ASTR)
Atlantic Multidecadal Oscillation (AMO), 28
Atlantis® Infiltration Tank System, 588, 589
ATV. *See* Acoustic-televiewer log (ATV)
Australian Guidelines for Water Recycling, 805
Australian Managed Aquifer Recharge Guidelines, 606
AWT. *See* Advanced wastewater treatment (AWT)

B
Bacteria
Campylobacter, 837

Cryptosporidium, 837
Desulfuvibrio desulfuricans, 839
Escherichia coli, 835
Pseudomonas dentrificans, 839
Badlands
geology and hydrogeology, 76
gullies, 77
landscape, 74–76
terminology, 74
Bank filtration. *See also* Filtration, 560
Barrier systems
extraction, 663, 664
extractive salinity, 662, 663
Injection, 663, 664
physical, 664
positive salinity, 660–662
Basalt deposits, 87
Baseflow, hydraulic water budgets, elements of, 152*t*
Baseflow method, 260, 261
Baseline monitoring, 807
Basic water requirement (BWR), 3, 4
Basin Characterization Model (BCM), 256, 257
Basin floor recharge (BFR), 199
components of, 200
BAT. *See* Best Available Technology (BAT)
BCM. *See* Basin Characterization Model (BCM)
Bedrock, 200
altitude, 74
as alluvial fan feature, 63*t*
arid lands water budgets, 152*f*
basin fill aquifers, 105*f*, 107*f*
climatic interaction with, 48
groundwater dam, 595*f*
mountainous regions, 101*f*
physical storage ASR system, 601*f*
sand dam, 594*f*
simple bank system, 632*f*
surfaces, low-angle, 50, 51, 43*t*, 53*f*
types, 43
wadi aquifer, 103*f*
Bellagio Draft Treaty, 962
Berlin Rules, 958
Berms, 533, 534*f*, 536
Best Available Technology (BAT), 751
BFR. *See* Basin floor recharge (BFR)
Biodegradation, EDC compounds, removal of, 832
Biosolids, definition, 777*t*
BOD. *See* Biochemical oxygen demand (BOD)
Borehole geophysical logging programs development, 431–433
Borehole geophysical techniques, 401, 402
borehole imaging logs and, 424
acoustic-televiewer log, 426–428
microresistivity imaging, 428
optical televiewer, 426
video survey, 424–426
caliper logs and, 404, 405
elemental capture spectroscopy log and, 430, 431
flowmeter logs and, 417, 418
data interpretation, 421–423
electromagnetic borehole flowmeter (EBF), 419, 421
heat-pulse flowmeter, 421
spinner, 418, 419
logging programs development, 431–433
natural gamma ray log and, 406
nuclear logging and, 413
density log, 413–416
neutron log, 416, 417
nuclear magnetic resonance log and, 428–430
quality assurance and control and, 402–403
resistivity logs and, 409, 410
sonic logs, 410–413
spontaneous potential, 406–408
temperature and fluid resistivity logs and, 423, 424
Borehole imaging logs, 424
acoustic-televiewer log, 426–428
microresistivity imaging, 428
optical televiewer, 426
video survey, 424–426
Borehole televiewer log (BTV), 426
Borehole temperature profiles, 269, 270
Bouguer anomaly, 330
Bouguer slab equation, 331
Bowen ratio method, energy balance, 175, 176
Brackish groundwater reverse osmosis
treatment facilities, physical/chemical properties of, 755*t*
Brackish-water desalination, 1013
Brackish-water reverse osmosis. *See also* Reverse osmosis (RO), 717–719
BWR. *See* Basic water requirement (BWR)

C

Calcretes, 44
Calibration, 403
of EBF tool, 421
of groundwater mounding models, 297

C (*cont.*)
of instrument, 243, 269
in quantitative geophysical log interpretation, 403
of soil moisture probes, 239
California wastewater quality standard, 820
Caliper logs, 404, 405
and gamma ray, 405*f*
Canals, in Hohokam, 512, 513, 513*f*
Canary Islands, fog harvesting in, 552
Capillary pressure
vadose zone hydrology
definition, 131
and radius, inverse relationship between, 132, 133
Young–Laplace equation, 132
Captured discharge and recharge, 154–156
Carbonate rock, 81*f*
chemical process, 48
macroporosity measurements, 428
Catchments, in rainwater harvesting, 165, 532
artificial catchment surface, 533
groundwater catchment, vital elements in, 247
location selection factors, 597
microcatchments, 533, 535*f*
runoff coefficients, 532
surveillance of, 541
traditional technology, 545
versus wadi, 102
CECs. *See* Compounds of emerging concern (CECs)
Cenozoic carbonate, 127
Center for Environment and Development in the Arab Region and Europe (CEDARE), 941
Central Arizona Project, 1016*f*, 1017
Central Plain of China, neolithic culture collapse of, 521
Check dams, *See also* Dams, 532, 533
advantages of, 592
Chemical contaminants
in reclaimed water, 810, 811
removal, in riverbanks, 637–639
risks, associated with MAR system, 615
Chemical hazards, 806
Chile, fog harvesting in, 531
Chilean water law, 899–900
China
ancient, water management in, 509
Loess Plateau
rainwater harvesting in, 545–546
Chloride mass balance (CMB) method, 262–265, 296, 298, 461
Chlorine, 36, 286, 287
Chlorofluorocarbons, 267, 268
Chloroform, 842
Cisterns
in ancient Greek civilization, 508
in ancient Jordan, 502
in ancient Petra, 505
in Jordan, 531
Classic Mayan Civilization, collapse of, 519, 520
Clean Water Act, 751, 752
Climate change
abrupt, 976, 977
adaptation (*see* Adaptation)
agriculture and food supply security
frequency and intensity of flood, 996
temperate and tropical areas, 995
virtual water trading, 996
warmer climate, 995
water requirements, irrigation, 995
in coastal regions, 657, 658
glacial cycles, 975, 976
global change, 1039, 1040
global warming, 979
groundwater system
modeling, stimulated changes, 991, 992*f*
population growth, 991
rainfall variability, 991, 993
regional variability, 993, 995
socioeconomic responses, 993
temperature and precipitation changes, 991, 992
human activities, impact of, 979, 980*f*
human population, affecting, 977, 978
hydrologic cycle, 975
hydrologic systems, effects of
freshwater and uncertainty (*see* Freshwater resources)
GCM prediction, variation changes, 980, 981
modeling (*see* Modeling)
snow- and glacier-dominated regions, 981, 982
water cycle, changes in, 980
surface-water resources impacts
affects, 988
Palmer Drought Index, changes in, 989, 990
regional variability, 993, 994
variations, 975
water management (*see* Water)
Climate Precipitation Center Morphing (CMORPH) technique, 169

Climate variability, water resources, 979
Climatic data, 28
Clogging, 578–581, 639, 640
CMB method, 592
Coastal water management, 647–665
  climate change and, 657, 658
  mitigation of, 658–665
  saline–water interface, location and characterization of, 652, 653
  saline–water intrusion
    causes of, 648–651, 649*f*, 650*f*
    monitoring of, 653–656
    simulation of, 656, 657
Cocksackievirus B3, 835
Cold ocean currents, as aridity cause, 26
Colorado River, 1016*f*, 1017, 1018
Comanchaca project, 548, 549
Committee on Privatization of Water Services in United States (CPWSUS), 904
Common-pool resources, 855–858
Complex Terminal (CT) aquifers, 938
Compounds of emerging concern (CECs), 798, 799, 811–816
Comprehensive approach
  and integrated approach, 913, 914
Comprehensive Assessment of Water Management in Agriculture
  triple water management goals, 7
  water scarcity prediction, 4
Comprehensive Everglades Restoration Plan (CERP), 602, 609
Computer modeling, 475
  groundwater modeling, 270
  MAR system, control, 1043
  MODFLOW, 575
Concentrate disposal issues, in desalination, 752–765
  chemical properties of, 753, 754–755*t*
  crystallization, 763–765
  evaporation ponds, 762, 763
  physical properties of, 753, 754–755*t*
  subsurface disposal methods, 759–762
  surface discharge to tidal waters, 756–759
  zero liquid discharge, 763–765
Concentrate management, 703, 704, 752
Concentrated solar thermal power (CSP), 726
Confined aquifers, 932, 933
  conditions, 111
  as emergency sources, 1009
  hydraulic properties, 96, 97
  semi-confined aquifers, 97, 98
  in transboundary aquifers, 960
  and unconfined aquifers, 96, 97
Conflict resolution, in transboundary water management, 959, 960
Consumer education, 675, 676
Consumptive use permit (CUP), 867
Contaminants
  attenuation processes, 827, 828
  groundwater environment, 843
  redox conditions, 828
Contamination vulnerability mapping, 468
Continental Intercalary (CI, deeper aquifer), 938
Continental winds, as aridity cause, 25
Continuous wave (CW) systems, 374, 391
Convective storm cells, 165
  differentiating arid and semiarid areas, 299
Conventional surveying, 348–349
Conventional vertical electrical sounding (CVES) systems, 372, 373, 379
Cost effectiveness analysis, 795
Cost–benefit analysis, 616, 795
CPWSUS. *See* Committee on Privatization of Water Services in United States (CPWSUS)
CRF. *See* Capital recovery factor (CRF)
Culturally modified recharge. *See* Unmanaged aquifer recharge
Cumulative infiltration, definition, 140
Cumulative recharge, definition, 253
CVES systems. *See* Conventional vertical electrical sounding (CVES) systems
CW systems. *See* Continuous wave (CW) systems

D

DALY. *See* Disability-adjusted life year (DALY)
Dams
  check dams, 532, 533, 535*f*, 592
  flood control, 1036
  groundwater in, 594, 595*f*, 598–600
  in Petra, 505, 506*f*
  operation of, 1016
  sand, 594–597, 594*f*
  wadi, 590–594
Darcy flux in aquifers, 258–260
Darcy's Law, *See also* Darcy–Buckingham law, 119, 136, 137, 592
  and Green-Ampt equation, 140, 141
  from hydraulic conductivity, 592
Darcy–Buckingham law, 136, 137, 241
  methods based on, 251, 252
Data management, hydrologic, 1031, 1032
DBP. *See* Disinfection-byproduct (DBP)

D (*cont.*)
DE. *See* Deuterium excess (DE)
Decision-making process, 1021
stakeholder participation, 863, 868
Deflation flats. *See* Desert flats
DEM. *See* Digital elevation model (DEM)
Demand
management, 1032, 1040
reduction, decreased water supply (*see* Water)
Demand management
as element of IWRM, 915
Denitrification, 829, 830
CEC, reduction of, 844
Density log, 413–416
Desalinated water, post-treatment of, 704
Desalination, *See also* Salinization, 1013, 1014, 1037, 1038
alternative energy use in, 725–736
in arid lands, 701–744
environmental issues in, 749–766
concentrate disposal issues, 752–765
energy consumption issues, 750, 751
intake issues, 751, 752
fundamental aspects of, 703, 704
geothermal powered, 732, 733
hybrid, 721–725
membrane, 702, 714–721
brackish-water reverse osmosis, 717–719
electrodialysis reversal, 714, 719, 720, 720*f*
forward osmosis, 720, 721
seawater reverse osmosis, 714–717
nuclear-powered, 733–736, 735*f*
solar-powered, 727–730
direct distillation, using solar stills, 727, 728
indirect desalination, using solar collectors, 729, 730, 731*f*, 732*f*
indirect desalination, using solar ponds, 728, 729
technologies
costs of, 743, 744
energy use of, comparison of, 741–743, 743*t*
global distribution of, 703*f*
thermal, 702, 704–713
adsorption desalination, 711–713, 712*f*, 713*f*
multiple effect distillation, 708–710, 709*f*
multi-stage flash distillation, 705–708, 706*f*
vapor compression distillation, 710, 711*f*
in urban areas, 31
wind-powered, 731–732
Desert flats
definition, 85
depositional features, 85
inter-mountain valley, 86, 87*f*
Pampa de la Joya, 86*f*
pavement, 86
permeable sediments, 87
Desert landscapes, 22*f*
Desertification, 33, 34
causes and management of, 36, 37
Deserts
and desertification, 33, 34
Great Plains Dust Bowl of 1930s, 34, 35
Desktop feasibility evaluation, 607
Desktop investigations, 225, 226
Deuterium, late spring snowmelt indicator, 193, 194
Deuterium excess (DE), 288, 289
Diagenesis, 44, 126
DIC. *See* Dissolved inorganic carbon (DIC)
Dielectric constant, 449
Dielectric permittivity, 384
Digital elevation model (DEM), 437, 445, 446, 459
DIL. *See* Dual induction laterolog (DIL)
Dimensionless unit hydrographs (DUHs), 306–308
Direct recharge, 189
Disability-adjusted life year (DALY), 818
Disinfection-byproduct (DBP), 781, 840–842
Dissolved inorganic carbon (DIC), 282
Dissolved organic carbon (DOC), 638
Distillation
multiple effect, 708–710, 709*f*
multi-stage flash, 705–708, 706*f*, 754–755*t*
vapor compression, 710, 711
DOC. *See* Dissolved organic carbon (DOC)
Domestic scale rainwater harvesting systems, *See also* Rainwater harvesting, 534–538
Domestic water conservation and saving, *See also* Water conservation and saving, 672–682
implementation of, 674–682
conservation and utilities operation, 679, 680
consumer education, 675, 676
legal mandates, 674, 675
market incentives, 676–679

non-revenue water, 680–682
opportunities of, 672–674
Dominican Republic, fog harvesting in, 552
Dose–response assessment, 817, 818
Double-ring infiltrometer, 236, 237
Drip irrigation method, 1017
Drought indices, 32
normal precipitation, percentage of, 32
Palmer Drought Severity Index, 32
standardized precipitation index, 32, 33
Droughts
definition of, 27–30
impacts of, 30–32
vulnerability to, 31
types of, 30
Dual induction laterolog (DIL), 409, 410
Dual-induction log, 409
DUHs. *See* Dimensionless unit hydrographs (DUHs)
Dust Bowel, of Great Plains, 34, 35*f*

E

Earthwork systems, 530, 534–536
EBBR method. *See* Energy balance Bowen ratio (EBBR) method
EBF. *See* Electromagnetic borehole flowmeter (EBF)
Eclipse, 490, 491
Economic value of water. *See* Water, economic value of
EDCs. *See* Endocrine disrupting compounds (EDCs)
Eddy covariance method, for measuring actual ET rates, 176, 177
Eddy currents, 374
EDM. *See* Electrodialysis metathesis (EDM)
Effective vertical stress, 344
Efficiency
definition, 669
total efficiency, 689
water use, 669, 670, 687–689, 691
irrigation, 691, 692
Effluent, definition, 777*t*
El Nino Southern Oscillation (ENSO), 194, 944
El Paso Water Utilities (EPWU) Kay Bailey 580 Hutchison Desalination Plant, 967
El Paso Water Utilities (Texas, USA) Hueco Bolson Recharge Project, 569
Elastic strain, 347
Electrical resistivity tomography (ERT) program, 379
Electrical resistivity, 368–373
and electromagnetic techniques, 366–368
Electrodialysis, 714
reversal, 714, 719, 720, 720*f*
reversal treatment facilities, physical/ chemical properties of, 755*t*
Electrodialysis metathesis (EDM), 764
Electrodialysis reversal (ER), 714, 719, 720, 720*f*
Electromagnetic borehole flowmeter (EBF), 419, 421
Electromagnetic surveys, 373, 374
frequency-domain electromagnetic soundings for shallow data collection, 374, 375
transient electromagnetic sounding methods for deep data collection, 375–377
Elemental capture spectroscopy log, 430, 431
Endocrine disrupting compounds (EDCs), 798, 832
Endogenous water supplies, 10
Energy balance
latent heat flux, 449
and micrometeorological methods, 175
and water, computation, 318
Energy balance Bowen ratio (EBBR) method, 175, 176
Energy consumption issues, in desalination, 750, 751
English Rule, 865
Entrainment, 751, 752
of marine life, 737
Environmental isotopes. *See* Isotopes
Environmental issues, in desalination, 749–766
concentrate disposal issues, 752–765
energy consumption issues, 750, 751
future innovations of, 765, 766
intake issues, 751, 752
Environmental Protection Agency (EPA)
on contaminants, 758
in power plant intakes, 751
Environmental risk assessment, 806, 807
Environmental tracers, 262, 277, 320, 931
age tracers, 265, 266
chloride mass balance method, 262–265
chlorofluorocarbons, 267, 268
tracer versus water migration, 268
tracers versus mitigation, 268
Environments for Living® Certified Green, 673
EPA. *See* Environmental Protection Agency (EPA)

E (*cont.*)
Ephemeral river system. *See also* Wadis; Arroyos, 102, 153, 202
Ephemeral stream channels. *See also* Ephemeral river system, 72, 77, 189, 198, 202
  assumptions in, 315
  check dams in, 532, 533
  fundamental elements, 315, 316
  near Phoenix, Arizona, 294
  rainfall-runoff relationships in, 318
Ephemeral streams, 153, 202
Equity, 671, 672
ER. *See* Electrodialysis reversal (ER)
Escherichia coli (*E. coli*), 810
Estriol, 832
Estrogenic activity in drinking water, 815
Estrone (E1), 843
Evaporation ponds, discharge of concentrate using, 762, 763
Evapotranspiration (ET), 22, 158, 163, 172, 173, 250
  actual ET, 173, 174
    eddy covariance method, 176, 177
    energy balance and micrometeorological methods, 175
    energy balance Bowen ratio method, 175, 176
    geophysical methods, 177, 178
    lysimeters, 174
  Priestley–Taylor method, 180
    Blaney–Criddle method, 180, 181
    pan evaporation measurements, 181–183
  and recharge rate quantification, 448, 449
  reference ET, 178
    modified Penman–Montieth method, 178–180
ET. *See* Evapotranspiration (ET), 777*t*
"Exit strategy", 930
Exogenous water supplies, 10
Exploratory well construction, 226–228
Extensometers, 349, 350
Extraction barrier systems. *See also* Barrier systems, 663, 664
Extractive salinity barriers. *See also* Barrier systems, 662, 663

F
Fairness, 671
Fecal bacteria, as indicator organism, 809, 810
Fecal coliform concentration, 820, 821
FEFLOW, 490
Field investigations, 607, 608
Filtration
  bank, 560
  infiltration. *See* Infiltration
  microfiltration, 571, 714, 775
  nanofiltration, 714
  riverbank, 631–643
  ultrafiltration, 714
Financial feasibility analysis, 795
Flat lands depressional recharge, 204, 205
Flood control
  as element of IWRM, 915
Flood irrigation. *See* also Irrigation, 683
Flow tubes. *See* Darcy flux in aquifers
Flowmeter logs, 417, 418
  data interpretation, 421–423
  electromagnetic borehole flowmeter (EBF), 419, 421
  heat-pulse flowmeter, 421
  spinner, 418, 419
Flow-tube method, in confined aquifer, 259*f*
Fluid resistivity logs and temperature, 423, 424
Fluvial deposits
  alluvial fan
    horizontal relationship, 64, 67*f*
  facies, 67, 68, 69*t*
  lithofacies and sedimentary structure, 68, 71*t*
  relationship between
    valley profile and stream power, 67, 69*f*
    width and drainage area, 66, 68*f*
  rivers, 70, 71
  sedimentary, characteristic of, 66
  sediment sequences versus humid lands, 64, 68*t*
  vertical sedimentary profile model, 67, 70*f*
Fog harvesting, 531, 547–553, 548*f*
  in Canary Islands, 552
  in Dominican Republic, 552
  in Northern Chile, 547–549
  in Oman, 552
  quality control of, 552
  in Saudi Arabia, 550
  in Southern Peru, 552
  Valencia, Spain, 551, 552
  in Western South Africa, 550
  in Yemen, 550, 551
Food production, and drought, 30
Food security, water, economic value of, 888, 889
Forward osmosis, 720, 721

Fossil water, 927, 928
Fox Canyon Groundwater Management Agency (FCGMA), 622
Fractured rock aquifer
characteristics, 111
development, 111, 112
groundwater flow, 110, 111
hydraulic of, 112
packer-testing program, 111
pumping test, 110
water storage, low volume of, 109
zones, 109, 110
Fractured rock aquifers. *See* Regional aquifer flow orientation
Fractures, tabular or irregular pores, 126
Fred Hervey Water Reclamation Plant, 967
Frequency-domain electromagnetic (FEM) method, 374
for shallow data collection, 374, 375
Freshwater
economic value, 1036
resources, use of, 1034
storage, ASR system for, 1038
Freshwater lenses, 650
mapping of, 380
Freshwater resources
climate change
atmospheric concentration, 982, 984*f*
human activities, impact of, 980*f*, 982
hydrologic cycles and climatic observation, 982, 983*t*
paleoclimate data, 985
water management impact, 985, 986
Furrow systems, 581

G

Ganges River, 877
GCMs. *See* Global climate models
GDEs. *See* Groundwater dependent ecosystems (GDEs)
Gemfibrozil, 833
Geographic information systems (GIS), 457, 458
applications, 459, 460
contamination vulnerability mapping, 468
groundwater model development, 460, 461
recharge investigations, 461–464
site evaluation studies, 464–467
future, in arid lands water management, 471
groundwater applications, 457
hydrologic data management, 468–471
limitation to, 453
principles, 458, 459
Geomorphology
allometric changes, 45
alluvial fans (*see* Alluvial fans)
badlands (*see* Badlands)
bedrock surfaces, 50, 51, 53*f*
characteristics, 45
desert flats (*see* Desert flats)
dominant process, 45
dry watercourses
alluvial rivers, 72
depositional features, 71, 72
wadis and arroyos (*see* Wadis)
dynamic equilibrium, definition of, 45
ergs (*see* Sand seas (ergs))
fluvial deposits (*see* Fluvial deposits)
landscapes, 46*t*
mountains
carbonate rocks, 48, 50*f*, 51*f*
hyperarid and arid regions, 49, 52*f*
narrow and shallow wadis, 48, 52*f*
Oman, 48, 49*f*
physiographic components, 49, 50, 53*f*
sedimentary rocks, 48, 50*f*
tectonic activities, 46, 47
playas (*see* Playas)
volcanic rock deposits, 64*f*, 87, 88, 89*f*
Geophysical methods, 177, 178
Geothermal powered desalination. *See also* Desalination, 732, 733
Ghyben-Herzberg relation, 652
GIS. *See* Geographic information systems (GIS)
Glacial cycles, 975, 976*f*
GLDAS. *See* Global Data Assimilation Systems (GLDAS)
Global climate models (GCMs), 980
Global Data Assimilation Systems (GLDAS), 339
Global Meteoric Water Line (GMWL), 287, 288
Global positioning systems (GPS), 350, 351
system noise sources, 351
Global warming, 979, 980, 985, 986
Global water budget
and water stress, 11–14
indices, 14
Global Water Partnership, 903
IWRM
definition, 911

G (*cont.*)
meaning and implementation, 912
Global water use, in agriculture, 4
GOES Precipitation Index (GPI) algorithm, 461
Governance and water policy. *See* Policy; Water
Governmental water policy, 16
GMRP. *See* Great Man-Made River Project (GMRP)
GMWL. *See* Global Meteoric Water Line (GMWL)
GPR. *See* Ground penetrating radar (GPR)
GPS. *See* Global positioning systems (GPS)
Grain size analysis, estimates based on, 254
Granite, 109
Graphic user interface (GUI), 921
Gravimetric method, 237, 238
Gravity recovery and climate experiment (GRACE), 336–341
artist rendition of, 338*f*
Graywater, definition, 777*t*
Great Artesian Basin (GAB), 943, 944
Great Man-Made River Project (GMRP), 941–943
Greek civilization, water management in, 508, 509
Green-Ampt equation, 140, 141
Greenhouse Earth, 975
Greenhouse gas emission
anthropogenic, 1039
energy production, 1037, 1038
reduction and control of, 1021, 1022
Greenhouse gases
anthropogenic climate change, 979
atmospheric radiation balance, 979
concentration, variations in, 977
impact of, 986
scenarios, 988
Ground penetrating radar (GPR), 383, 384
Groundwater, 916
advantages, 916, 917
economic analysis of, 917
law, 865
monitoring, 581, 582
salinity. *See* Saline–water intrusion
Groundwater abstraction, 11
Groundwater ages, 249
Groundwater and surface water
biosolids concentrated in, 814
budgets, 152
conjunctive use, 623, 692
as element of IWRM, 915
contamination, 893
government policy and water law, 862
resources, 210
Groundwater banking, 620–623
basic requirement for, 620
complications associated with, 622, 623
overall use pattern, 622
prerequisite for, 621
technical and operational challenges of, 621
third party impacts, avoiding, 621
Groundwater dams. *See also* Dams, 595*f*, 598–600
advantages of, 599, 600
construction of, 598
siting criteria for, 598, 599
Groundwater dependent ecosystems (GDEs), 447
Groundwater doctrines, in United States
absolute ownership, 865
correlative rights, 867
prior appropriation, 865, 866
reasonable use, 866, 867
restatements, torts of, 867, 868
Groundwater extraction rate, control of, 935
Groundwater flow
hydraulic conductivity versus permeability, 120
hydraulic gradient, 119
fractured aquifer, 110, 111
Groundwater flow modeling codes
MODFLOW, 488
MicroFEM, 488
for solute transport, 488, 489
ECLIPSE, 490, 491
HST3D, 490
Hydrus, 491
MT3DMS, 489
PHT3D, 491
SEAWAT, 489
SUTRA, 489, 490
Groundwater flow modeling. *See also* Computer modeling
for arid and semiarid lands, 476, 477
codes, 487
MODFLOW, 488
MicroFEM, 488
for solute transport, 488, 491
compaction and land subsidence simulation, 491, 492
development, 478
conceptual model, 479
model calibration, 483–486

model code choice, 480–482
model verification, post audits, and refinement, 486, 487
and population of model grid, 482, 483
predictive simulations, 487
statement of problem and modeling objectives, 478, 479
integrated surface water and, 492–495
limitations, 477, 478
Groundwater management issues, arid and semiarid regions
types of, 6
Groundwater mining, 211, 212, 926–928
Groundwater model development, 478
conceptual model, 479 model calibration, 483–486
model code choice, 480–482
model verification, post audits, and refinement, 486, 487
and population of model grid, 482, 483
predictive simulations, 487
statement of problem and objectives, 478, 479
Groundwater modeling, 18, 270, 271
Groundwater movement, unconfined aquifer, 96
Groundwater pumping-induced land subsidence rates, 354*t*
Groundwater recharge. *See also* Recharge, 129
and dams, 151
and vadose zone, 130
calculations, 462
definition, 187
field moisture-capacity, 139
secondary macroporosity features and, 190
Groundwater recharge movement, 543, 544
Groundwater Replenishment System (GRS), 660
Groundwater resources
advantages of, 1027
assessments, 1030, 1031
brackish water, 1013
conjunctive water use, 1028
development of, 1009, 1010
management of, 156–159
non-renewable, use of, 1028
over-exploitation, 1028
quality, 1029, 1030
Groundwater system
basic water budget, 152
basic elements of, 153*t*
data for analysis, 469*t*
Groundwater use, impact acceptability and sustainability, 214, 215
GSFLOW code, 494

H

Hantush-Walton method, 123
Hard rock aquifer. *See* Fractured rock aquifers
Harmon Doctrine, 956
Hazards, and risk, distinction between, 805, 806
Health-based targets and monitoring, 819, 821–823
Heat transfer, in sediments, 324
Heat-pulse flowmeter, 421
HEC-HMS model code, 317
Helicopter electromagnetic (HEM) investigation, 392, 393*f*
Helium, 285, 286
Helsinki Rules, 957
Article V, 957, 958
Berlin Rules, 957, 958
HEM investigation. *See* Helicopter electromagnetic (HEM) investigation
Heterogeneity
and bedding impact, vadose zone, 142, 143
aquifer
arid and semiarid land, control recharge, 129
cenozoic carbonate, 127
karstic limestones, 127, 128*f*
sedimentary deposits, 127
solute transport, 128, 129
types of, 127
basin-fill aquifers, 105
carbonate aquifers, 108
fractured rock aquifers, 111
perched aquifers, 99
unconfined aquifers, 97
HFBs. *See* Horizontal flow barriers (HFBs)
High Plains Aquifer, of United States Great Plains, 337
High pressure
as aridity cause, 25
Hohokam
multi-story abode building in, 512*f*
water management in, 512, 513*f*
Holistic water management, 911
Horizontal flow barriers (HFBs)
physical storage ASR using, 600–602, 601*f*

H (*cont.*)
Horton's infiltration model, 139
HST3D, 490
Hueco Bolson Aquifer, 966–968
Humid land
versus arid land, 64, 68*t*
regions, 43
Hybrid desalination, 721–725
future innovations of, 740, 741, 742*f*
Hydraulic conductivity
pore size, 126
property of, 119
saturated, 141
semi-unconfined aquifer, 97
unconfined aquifer, 96
unsaturated, 137
Hydraulic diffusivity, 124
Hydraulic gradient, 96
Hydraulic properties
leakance (*see* Leakance coefficient)
storativity, 121, 122
transmissivity, 121
Hydraulic water budgets, 152*t*
Hydrogen isotopes, 280
Hydrogeology, 100
ephemeral streams, 88, 89
groundwater resources, 88
sand dunes (ergs), 89
Hydrograph data
late spring snowmelt indicator, 193, 194
Hydrographs, 304, 305
dimensionless, 306–308
natural, 305, 306
synthetic, 308–310
unit, 306–308
Hydrologic budget method
advantages, 463
Hydrologic climate change, impact of, 1040
Hydrologic cycle, 1039
changes, GCM predictions on, 986
Hydrologic data collection, 1031
Hydrologic data management, 468–471
Hydrologic information management system, 1031, 1032
Hydrologic model systems, 315, 316
HEC-HMS model code, 317
precipitation-runoff modeling system (PRMS) model, 318
soil and water assessment tool (SWAT), 317, 318
SW modeling applications, 318, 319
Hydrus, 491
HYGES scheme, 469, 470

I
IAH. *See* International Association of Hydrogeologists (IAH)
IBWC. *See* International Boundary and Water Commission (IBWC)
Icehouse Earth. *See* Snowball Earth
ILA. *See* International Law Association (ILA)
Impingement, 751, 752
IMS. *See* Inertial Measuring Systems (IMS)
In-channel recharge methods, 589, 590
India, rainwater harvesting in, 543–545
Indicator organisms, 809, 810
Indices of humidity (Ih), 24
Indirect recharge, 189
Industrial wastewater reuse, 799, 800
Inelastic strain, 347
Infiltrated water, migration of, 577, 578
Infiltration. *See also*Filtration, 89
basins, 572*f*
maintenance cost, 580
monitoring requirements for, 581, 582
periodic drying of, 580
cumulative, 140
field moisture capacity, 139
galleries, 587–589
Green and Amp
method, estimation of, 140
various soil types, parameters for, 141
wetting front concept, 140, 141
Horton's model, 139
induced infiltration, 361
percolation, 138
process of, 138
rates, ponding depth effect on, 579
reduction, 139
shafts, 589
trenches, 587–589, 588*f*
Infiltration and recharge rates, direct physical measurement of, 249
Darcy–Buckingham law based methods, 251, 252
lysimeter, 250
seepage meters, 250, 251
Infinitesimal runoff coefficient (IRC), 304
Infrared (IR) data, 169
Injection barrier systems. *See also* Barrier systems, 663, 664
Injection wells. *See also* Wells, 560–571
In-lieu recharge, 620, 622
INS. *See* Inertial Navigation Systems (INS)
Institut de Droit International (Institute of International Law), 957
Intake issues, in desalination, 751, 752
Integrated surface water

groundwater modeling, 492–495
modeling goals, 492, 493
Integrated urban water management (IUWM), 911
Integrated Water Resources Management (IWRM), 886, 911, 1033, 1034
in arid semiarid lands, 923, 924
comprehensive approach, 913
conjunctive use, 916–918
decision support systems, 920–922
elements of, 915
integrated approach, 913, 914
meaning and implementation of, 912–916
multiple criteria decision analysis, 919, 920
optimization, 918, 919
primary value of, 914
in Singapore, 916
and water management planning, 922, 923
workflow, 912*f*
Interferometric synthetic aperture radar (InSAR), 451
Interflow, IF, 130, 200, 253, 261
in natural recharge estimation, 153
in precipitation, estimation, 163
Intergovernmental Panel on Climate Change (IPCC), 981, 1022
aridity criterion by, 23
International (Transboundary) River Water Law, 956, 957
Berlin Rules, 958
Helsinki Rules, 957, 958
International Association of Hydrogeologists (IAH), 963
International Boundary and Water Commission (IBWC), 965
International Court of Justice, 959
International Hydrological Programme (IHP), 963
International Law Association (ILA), 957
International Law Commission of the United Nations, 957
International river basins, legal theories of, 956, 957
International water law, 854
Internationally Shared Aquifer Resource Management (ISARM), 963
IPCC. *See* Intergovernmental Panel on Climate Change (IPCC)
IRC. *See* Infinitesimal runoff coefficient (IRC)
Irrigation
for agriculture, 682–684, 782–786
definition, 682
flood, 683
localized, 683
saline-water, 693, 694
water use efficiency
incentives, 694, 695
techniques, 691, 692
ISARM. *See* Internationally Shared Aquifer Resource Management (ISARM)
Islam
water management
conservation of water, 874
principle, 872, 873
purifying and cleansing agent, 873
wastewater reuse, 874, 875
Isotopes, 277, 278
basics, 278
radioisotopes
chlorine, 36, 286, 287
krypton, 81, 287
radioactive dating concepts, 281, 282
radiocarbon (carbon 14) dating, 282–284
tritium, 284–286
stable, 278–281
of water, 287–290
Israel, 957
Six-Day War, 953
IUWM. *See* Integrated urban water management (IUWM)
IWRM. *See* Integrated Water Resources Management (IWRM)

J
Jawa, water management in, 503
Jordan, 957
cistern use, decline of, 531
rainwater harvesting system in, 536*f*
water management in, 502–506
Jawa, 503
Petra, 503–506

K
Karst systems, 108
Karstic limestones, 127, 128*f*
Kinematic wave theory, 142
Krypton, 81, 287
Kyoto protocols, 1019

L
Lake Belt In-Ground Reservoir Technology Project, 602
Lancaster, California aquifer storage and recovery site, 335

L (*cont.*)
Land subsidence simulation
compaction and, 491, 492
Land subsidence, 343–348
case studies
Antelope Valley, California, 359
Mexico City, 356–358
impacts, 355
measurement, 348
conventional surveying, 348, 349
extensometers, 349, 350
global positioning systems (GPS), 350, 351
measured land subsidence amounts and rates, 353, 354
Satellite Interferometric Synthetic Aperture Radar (InSAR), 352, 353
mitigation, 359–361
Land use maps
IRS-LISS II in, 465
Land use, effects of, 194–196
Landsat 7 Enhanced Thematic Mapper Plus (ETM+) instrument, 437, 466
Large-scale technologies, 10
Las Vegas Valley
subsidence, 349, 358*f*
Law of Transboundary Aquifers, 936, 964, 965
principles of, 964
relevant factors of, 964, 965
Layered heterogeneity, 127
Leadership in Energy and Environmental Design (LEED), 674
Leakance coefficient
definition, 122
field testing method, 123, 124
groundwater microwave, 123
property, 123
Legal and institutional controls, of non-renewable groundwater resources, 934, 935
Legal mandates, for water conservation and saving, 674, 675
Legionella, 810
Libya, 936
LIDAR. *See* Light Detection and Ranging (LIDAR)
Light Detection and Ranging (LIDAR), 437, 438, 442, 443
Limestones, 126
Lineament mapping, 439–443
introduction to, 439, 440
brittle deformation zones, 441
important lineaments, 440
IRS-LISS II in, 465
limitations to, 443–445
Lisse effect, 144, 145
Lithic mulch agriculture (LMA). *See also* Agriculture, 553, 554
Little Ice Age, 978
LMA. *See* Lithic mulch agriculture (LMA)
Local historic rainfall distribution, 29
Local recharge rates
impact factors, 191, 192
Localized irrigation. *See also* Irrigation, 683
Loess Plateau, China
rainwater harvesting in, 545, 546
Lysimeters, 174, 250

M
Macropore recharge, 189
Magnetic resonance sounding (MRS). *See also* Surface nuclear magnetic resonance (Surface NMR), 385–387
Magnetotellurics, 387, 388
Managed aquifer recharge (MAR). *See also* Unmanaged aquifer recharge (UMAR), 228, 334, 380, 381, 559
aquifer recharge
and recovery, 560
using wells, 567, 568
aquifer storage
and recovery, 562–567
transfer and recovery, 568–569, 569*f*
definition, 559
direct versus indirect, 560
groundwater banking, 620–623
hazard risks associated with, 611*t*
assessment of, 612
chemical contaminants, 615
microbiological risks, 613–615
implementation of, 606–610
desktop feasibility evaluation, 607
field investigations and testing of potential system sites, 607, 608
pilot system construction and testing, 609
project review and adaptive management, 609, 610
system design, 608, 609
system expansion, 610
injected water distribution mapping in MAR system, 335
Arvada, Colorado ASR system, 335, 336
Lancaster, California aquifer storage ad recovery site, 335
Weber river ASR site, 336

investigations, 380, 381
by natural recharge enhancement processes, 589–597
groundwater dams, 594, 595*f*
in-channel recharge methods, 589, 590
sand dams, 594–597, 594*f*
wadi dams, 590–594
optimization of, 1036, 1037
projects, economics of, 615–617
redox reaction, 839
stormwater recharge, 602–606, 605*f*
subsurface engineering, 597–602
surface spreading, 571–589
clogging, 578–581
design issues of, 573–577
infiltrated water, migration of, 577, 578
monitoring, 581, 582
recharge galleries, 587–589
recharge shafts, 589
recharge trenches, 587–589, 588*f*
soil-aquifer treatment, 582–587, 583*f*
types of, 560–589, 561*t*
vadose-zone recharge wells, 569–571, 570*f*
wastewater, 790–792, 791*f*
wastewater treatment (*see also* Water quality improvement), 1036, 1037
zone of discharge perimeter, 828
Management, definition, 851
Manual on the Right to Water and Sanitation, 860
MAR. *See* Managed aquifer recharge (MAR)
Marginal water, 774
Market incentives, 676–679
Matrix echarge, 189
Mayan collapse, 518
MBR. *See* Mountain block recharge (MBR)
MCA. *See* Medieval Climate Anomaly (MCA)
Medieval Climate Anomaly (MCA), 985
Medieval Warm Period (MWP), 978
Megadroughts, 31
Membrane desalination. *See also* Desalination, 702, 714–721
brackish-water reverse osmosis, 717–719
electrodialysis reversal, 714, 719, 720, 720*f*
forward osmosis, 720, 721
seawater reverse osmosis, 714–717, 715–718*f*
Metabolites, 814, 815
Mexico City, 356, 357
Arizona, 357, 358
Mexico, groundwater management in, 965
MFR. *See* Mountain front recharge (MFR)
Microbial degradation, 834
Microbiological risks, associated with MAR system, 613–615
Microbiological standards and guidelines, 820, 821
Microcatchments. *See also* Catchments, in rainwater harvesting, 533
capture, 535*f*
MicroFEM, 488
Microfiltration. *See also* Filtration, 571, 714, 775
Microgravity, 329, 330
gravity recovery and climate experiment (GRACE), 336–341
relative gravity surveys, 332–334
use in groundwater investigations, 331, 332
to water management, application of, 334
injected water distribution mapping in MAR system, 335, 336
Microjet irrigation method, 1019
Microorganisms, 808–810
groundwater environment, removal of, 834
Microresistivity imaging, 428
MIKE SHE/MIKE 11 codes, 494
Mississippian Chiefdoms (Cahokia), United States, collapse of, 520, 521
Mitigation
and adaptation, climate change (*see* Adaptation)
and sustainability, 216, 217
Model calibration, 483–486
calibration targets, 483
mean absolute error (MAE), 484
mean error (ME), 484
root mean square error (RSME), 484
weighted least square residual (WLSR), 484
Model code
choice of, 480–482
issues, 480
public domain software, advantages of, 481
Model grid
population and development of, 482, 483
Model verification
post audits, and refinement, 486, 487
Modeling
climate change
GCM prediction, 986
precipitation, 987*f*
Modern carbon (MC), 282
MODFLOW, 232, 488, 493
Monte Carlo simulations, 818
Monturaqui-Negrillar-Tilipozo (MNT), 947
Mounding analysis, 298

M (*cont.*)
Mountain block recharge (MBR), 198, 199, 202
Mountain block, conceptual diagram of, 199*f*
Mountain front recharge (MFR), 198–202
MRS. *See* Magnetic resonance sounding (MRS)
MT3DMS, 489
Multiple criteria decision analysis (MCDA), 919, 920
  weighted scoring systems for, 919
Multiple effect distillation (MED). *See also* Distillation, 708–710, 709*f*
  combined with adsorption desalination, 724, 725
Multiple packer tests, 234
Multi-stage flash distillation (MSF). *See also* Distillation, 705–708, 706*f*
  combined with power generation
    and nanofiltration, 723
    and reverse osmosis, 722, 723, 724*f*
  treatment facilities, physical/chemical properties of, 754–755*t*
Municipal wastewater, defined, 777*t*
Muskkingum-Cunge technique, 318, 319
MWP. *See* Medieval Warm Period (MWP)
Mycobacteria, 810

N
Nanofiltration, *See also* Filtration, 714
  combined with power generation and multi-stage distillation, 723
  membrane treatment process, 829
National Pollution Discharge Elimination System (NPDES) permitting program, 758
National Research Council, 810, 811, 884, 885, 900, 901
National Weather Service (NWS), U.S., 166
Natural biological wastewater treatment, 779
Natural discharge, 153
Natural gamma ray log, 406
Natural hydrographs, 305, 306
Natural recharge, 187, 197
Natural water treatment process
  filtration and containment attenuation processes, 827
  inexpensive, 828
  vadose zone, incorporation of, 828
  water quality improvement (*See* Water quality improvement)
Negev Desert of Israel
  water management in, 507, 508
Neo-environmentalism, 518
Net infiltration, 188
Net virtual water import (NVWI), 891
NETPATH program, 283
Neutron log, 416, 417
Neutron probe measurements, 269
Neutron-moisture probe, 238–240
Next Generation Radar (NEXRAD) weather radar system, 168
Nitrification, 829, 830
Nitrogen
  denitrification, 829
  nitrification, 829
Noble gases, 201, 202
Non-government organization (NGO), 861
Non-polar arid lands, distribution of, 25
Non-renewable groundwater management, examples of, 936
  Great Artesian Basin (GAB), 943, 944
    key issues, 944
  Man-Made River Project (GMRP), 941, 942
    choice between options, 942, 943
  North Western Sahara Aquifer System (NWSAS), 938, 939
  Northern Chile, 945–947
  Nubian Sandstone Aquifer System (NSAS), 939–941
    objectives, 941
  in Saudi Arabia, 936, 937
    goals, 937, 938
Non-renewable groundwater resources, 927
  assessment of, 931–934
  characterization of, 932
  definition of, 927, 928
  examples of, 936
    Great Artesian Basin (GAB), 943, 944
    Great Man-Made River Project (GMRP), 941–943
    North Western Sahara Aquifer System (NWSAS), 938, 939
    Northern Chile, 945–947
    Nubian Sandstone Aquifer System (NSAS), 939–941
    Saudi Arabia, 936–938
  exploitation of, 932
  fundamental question in, 930
  legal and institutional controls of, 934, 935
  transboundary, 935, 936
  use, evaluation of, 932, 933
  utilization of, 928–931
    in planned schemes, 928
    in unplanned schemes, 928
Non-renewable water resources

major tasks, in characterization of, 932
technical factors, 932, 933
Non-revenue water use (NRW), 680–682
Non-stationarity
water resources management, 1005–1007, 1006*f*
Normalized difference vegetation index (NDVI), 446, 465
North Western Sahara Aquifer System (NWSAS), 938, 939
Continental Intercalary (CI, deeper aquifer), 938
Complex Terminal (CT) aquifers, 938
Systeme Aquifere du Sahara Septentrional (SASS), 938
Northern Chile, 945–947
fog harvesting in, 547–549
NRW. *See* Non-revenue water use (NRW)
Nubian Aquifer Regional Information System (NARIS), 941
Nubian Sandstone Aquifer System (NSAS), 939–941
Nubian Sandstone Aquifer (NSA), 939
Post-Nubian Aquifer System (PNAS), 939
Nuclear logging, 238, 413
density log, 413–416
neutron log, 416, 417
Nuclear magnetic resonance log, 428–430
Nuclear-powered desalination, 733–736, 735*f*
Numerical groundwater modeling, 223
NVWI. *See* Net virtual water import (NVWI)

O

Observatoire due Sahara ET du Sahel (OSS), 938
Office of Saline Water (OSW) (United States), 728
Oman, fog harvesting in, 552
Omaruru Delta of Namibia, 591
Operational monitoring, 807, 808, 822
Operational recovery efficiency, *See also* Recovery efficiency (RE), 565
Optical televiewer, 426
Optimization
existing water resources, 1041
MAR, 1038, 1039
wastewater reuse, final stage of, 1035
Organic matter oxidation, 839
Over-treatment, 774
Oxygen isotopes, 279

late spring snowmelt indicator, 193, 194

P

Pacific Decadal Oscillation (PDO), 28, 194
Packer tests, 234, 235
Paleowater, 927
Palestine, 957
Pan. *See* Playas
Parabolic trough collector (PTC), 730
Passive microwave (PMW), 168, 169
Pathogen removal, in riverbanks, 635–637
Pathogens, 808–810
concentration of, 827
removal of, SAT, 830
risk evaluation and management, 836–838
Pediments, 77
Percolation flux, 269
Percolation, 89, 138
Perennial streams, 153
Permittivity, 449
PERSIANN-CCS technique, 169
Persistent scatter INSAR (PSI), 353
PEST program, 485
Petra
treasury (Nabataean city), 504*f*
water management in, 503–506, 506*f*
Petrel workflow system, 482, 483
Pharmaceutical and personal care products (PPCPs), 812, 814
Phoenix, annual rainfall in, 165*f*, 171*f*
PHT3D code, 491
Physical barrier system, *See also* Barrier systems, 664
Piedmont, 198
Pilot system, construction and testing of, 609
Pipeline system, in Petra, 505
Playas
characteristics, 78
classification, 78*f*
closed-basin, 78, 79
definition, 77
depositional pattern, 80, 81*f*
open-basin, 80
salt deposition, 79
sediments, 80, 81
types, 78
water and salts, exchange pathways for, 80*f*
Policy
cultural conservatism, 861, 862

P (*cont.*)
definition, 851
development and implementation, 852, 853
economic, role in, 854
"use it or lose it", 863, 864
and water law
market system, 863
stakeholder management, 863
state control, 863
unregulated use, 862
Policy decision, 851
Poliovirus type 1, 835
Political hydrology, 954
Porosity, 412
effective, 124
hydraulic conductivity, 126
neutron porosity, interpretation of, 417*f*
pore types, 125
primary, 125, 126
secondary, 126
Portable reuse, of wastewater, 786–792
direct, 787, 788
indirect, 788–790, 789*f*
managed aquifer recharge, 790–792, 791*f*
Positive salinity barriers systems, *See also* Barrier systems, 660–662
Post-Nubian Aquifer System (PNAS), 939
Potential evapotranspiration rate (ETp), 22
Potential recharge water
penetration, depth of, 193
Power generation–multistage effect distillation–reverse osmosis desalination, 723, 724
Power generation–multistage flash distillation
and nanofiltration desalination, 723
and reverse osmosis desalination, 722, 723, 724*f*
PMW. *See* Passive microwave (PMW)
Precautionary principle, 774
Precipitation, 163
in arid regions, 164, 165
frequency analysis, 169–172
measurements, 165
radar measurement of rainfall, 167, 168
rain gauges, 166, 167
satellite rainfall measurement, 168, 169
Precipitation-runoff modeling system (PRMS) model, 318
Predictive simulations, 487
Preliminary wastewater treatment, 779
Priestley–Taylor method, 180
Blaney–Criddle method, 180, 181
pan evaporation measurements, 181–183
Primidone, 833
Priorities, in water management, 16–19
Privatization, 904–906
Production well construction, 226–228
Project review, 609, 610
PTC. *See* Parabolic trough collector (PTC)
Public acceptance, of wastewater reuse, 796–799
Public domain software
advantages of, 481
Public Utilities Board (PUD)
in Singapore, 914
Pueblo Bonita, 510, 511*f*
Pumping test
fractured rock aquifer, 110
leaky aquifer, 97, 98*f*
Pumping, optimization of
as Hueco Bolson aquifer option, 967

Q
Qanats, 513–517, 514*f*, 515*f*
attraction of, 516
collapse of, 516
components of, 514
construction of, 514, 515
limitations of, 517
social benefits of, 516, 517
QMRA. *See* Quantitative Microbial Risk Assessment
Quantitative Microbial Risk Assessment (QMRA), 817, 818

R
Radar measurement, of rainfall, 167, 168
Radioactive dating concepts, 281, 282
Radiocarbon (carbon 14) dating, 282–284
Radioisotopes
chlorine, 36, 286, 287
krypton, 81, 287
radioactive dating concepts, 281, 282
radiocarbon (carbon 14) dating, 282–284
tritium, 284–286
Radium, 934
Rain gauges, 165, 166, 167
Rain shadow effect
as aridity cause, 26
Rainfall-runoff relationships, 298–300
geomorphic relationship, 310
hydrographs, 304–310
precipitation, 300, 301
regression methods, 301
soil conservation service curve number method, 301–304

Rainwater harvesting, 501, 502, 529–540
in ancient China, 509
in ancient Jordan, 502
concepts, 532–534
domestic scale, 534–538
earthwork systems, 530, 534–536
storage systems, 530, 536–538
economics of, 554–556
in India, 543–545
in Loess Plateau, China, 545, 546
large-scale type, 542–546
legal issues associated with, 541, 542
limitations of, 531
local prestige, lacking of, 531
roof, 539, 540, 555
types of, 530
use of, 530
Rainwater harvesting agriculture (RHA), *See also* Agriculture, 545, 546
Raw water
development, 704
pretreatment of, 704
Reasonable use doctrine, 866, 867
Recharge, 88, 108, 109*f*, 138, 187–189
arid and semiarid region recharge location, 197, 198
flat lands depressional recharge, 204, 205
mountain block and front recharge, 198–202
wadi recharge, 202–204
captured discharge and, 154–156
direct, indirect, matrix, and macropore, 189
factors controlling infiltration rate and, in arid regions, 190–194
soil moisture gradients effects and changes, 197
temperature effects, 196, 197
vegetation effects and land use and land cover, 194–196
in water budget context, 153, 154
Recharge areas
and discharge areas identification, 446, 447
mapping of, 382
Recharge galleries, 587–589
Recharge measurement, in arid and semiarid regions, 247–249, 270–272
aquifer water budget-based methods, 254
aquifer water budget, 254–256
Darcy flux in aquifers (flow tubes), 258–260
spatially distributed water budget models, 256–258
stream gauging, 260, 261
surface water budget of channels, 261, 262
borehole temperature profiles and, 269
direct physical measurement of infiltration and recharge rates, 249
Darcy–Buckingham law based methods, 251, 252
lysimeter, 250
seepage meters, 250–251
environmental tracers and, 262
age tracers, 265, 266
chloride mass balance method, 262–265
chlorofluorocarbons, 267, 268
tracer versus water migration, 268
groundwater modeling and, 270
neutron probe measurements, 269
water-table fluctuation method, 252–254
Recharge shafts, 589
Recharge trenches, 587–589, 588*f*
Recharge with groundwater catchment
vital elements in, 247
Reclaimed water, definition, 777*t*
Recovery efficiency (RE)
definition, 565
operational, 565
system, 565
Recycled water, defined, 777*t*
Red-Dead Project, 1015
Reduction' droughts, 27
Reference ET rate (ET_o), 173
Reference evapotranspiration, 178
modified Penman–Montieth method, 178–180
Regional aquifer flow orientation
mapping of, 382, 383
Regolith, 109–110*f*
Relative gravity surveys and microgravity, 332–334
Relative water stress index (RWSI), 13
Religious and cultural influence
water management (*see* Water)
Remote sensing (RS), 435–437
aerial photograph interpretation and, 438, 439
application to groundwater management
evapotranspiration and recharge rate quantification, 448, 449

R (*cont.*)
hydrogeological significance, structural and sedimentological features identification, 445, 446
interferometric synthetic aperture radar (InSAR), 451
lineament mapping, 439–445
recharge and discharge areas identification, 446, 447
soil moisture measurement, 449–451
vegetational mapping, 448
and groundwater modeling, 451, 452
light detection and ranging (LIDAR) and, 437, 438
limitations of, 452, 453
radar-based, 165
Remote sensing (RS), in groundwater management
evatranspiration and recharge rate quantification, 448, 449
hydrogeological significance, structural and sedimentological features identification, 445, 446
interferometric synthetic aperture radar (InSAR), 451
lineament mapping, 439–445
recharge and discharge areas identification, 446, 447
soil moisture measurement, 449–451
vegetational mapping, 448
Renewable energy sources (RESs), 725, 726
Renewable water, 12
Reservoir Saturation Tool (RST), 431
Reservoirs
dam and, 1010–1012, 1011*f*
operation of, 1016
Resistivity logs, 409, 410
Resistivity-based geophysical methods, 240, 241
Resistivity-based surface geophysics and applications to groundwater investigations, 377
depth to water table, 380
formation and aquifer mapping, 381, 382
freshwater lenses mapping, 380
managed aquifer recharge investigations, 380, 381
recharge areas mapping, 382
regional aquifer flow orientation mapping, 382, 383
saline-water interface mapping, 377–379
RESs. *See* Renewable energy sources (RESs)
Reuse water, definition, 777*t*
Reverse osmosis (RO), 571, 582, 660, 775
brackish groundwater
treatment facilities, physical/chemical properties of, 755*t*
brackish-water, 717–719
combined with power generation and multistage flash distillation, 722, 723, 724*f*
membrane treatment process, 829
seawater, 714–717, 715–718*f*
treatment facilities, physical/chemical properties of, 754–755*t*
RHA. *See* Rainwater harvesting agriculture (RHA)
Risks, in wastewater reuse, 805–808
assessment of, 806, 818, 819
acceptable or tolerable risk, 818, 819
and hazard, distinction between, 805, 806
management and reduction of, 807
maximal risk, 806
monitoring, 807, 808
residual risk, 806
River
basins, transboundary, 1008
flows, 1011, 1016
Riverbank filtration (RBF), *See also* Filtration, 545, 631–643, 828
advantages of, 632
bank filtered water contribution, modeling of, 641, 642
clogging, 639, 640
design issues of, 633–635
limitations of, 642–643
modeling of, 640, 641
opportunities of, 643
performance evaluation of
chemical contaminant removal, 637–639
pathogen removal, 635–637
technologies of, 631, 632
Roman aqueduct (Segovia, Spain), 509*f*
Roman civilization, water management in, 508, 509
Roof rainwater harvesting (RRWH) system *See also* Rainwater harvesting, 539, 540
characteristics of, 539
health risks associated with, 540
technical issues of, 555
Roofs
gabled, 537*f*
sloped, 537*f*
Roof-tops, 533
RS. *See* Remote sensing (RS)
RTUs. *See* Remote telemetry units (RTUs)

Runoff coefficients, 532
Runoff farming water harvesting (RFWH), 529, 531, 532
Runoff inducement, 532
Runoff, 982, 988, 989
RWSI. *See* Relative water stress index (RWSI)

S

Sabkhas, 77
Safe yield and sustainability, 212–214
  evaluation, 216
Sahelian Solution Foundation (SASOL), 596
Saline-water contamination
  detection of, 393, 394
Saline-water interface
  location and characterization of, 652, 653
  mapping of, 377–379
Saline–water intrusion, 647
  causes of, 648–651, 649*f*
  mitigation of, 658–665
    extraction barrier systems, 663, 664
    extractive salinity barriers, 662, 663
    injection barrier systems, 663, 664
    optimization, 664, 665
    physical barrier systems, 664
    positive salinity barriers systems, 660–662
  monitoring of, 653–656
  simulation of, 656, 657
  vertical type, 650, 650*f*
Saline-water irrigation, *See also* Irrigation, 693, 694
Salinity mapping, and recharge, 394
Salinization. *See also* Desalination
  of soils, 684–686
San Joaquin Valley
  subsidence in, 344*f*, 345*f*
Sand dams, *See also* Dams, 594–597, 594*f*
  as barrier, 595
  construction of, 595, 596
  development, community involvement effect on, 596
  as spillway, 595
Sand dune (erg) aquifers, 104
Sand seas (ergs)
  basin-fill deposits, 83, 84*f*
  bedforms, 82
  definition of, 81, 82
  fluvial deposits, 83, 85*f*
  hydraulic conductivity, 83, 85
  sand, source of, 83
  shapes and size, 83
  thickness, 83, 84*f*
  types of, 82
Sandstone aquifers, 108
Santa Anna River, Orange County, California, 605
  in-channel/off-channel recharge system, 590, 591*f*
SAT. *See* Soil aquifer treatment
Satellite Interferometric Synthetic Aperture Radar (InSAR), 352, 353
  limitations of, 353
Satellite rainfall measurement, 168, 169
Saturated zone. *See* Vadose zone
Saudi Arabia, 936–938
  agricultural development in, 937
  fog harvesting in, 550
  groundwater use in, 937
  non-renewable groundwater management goals, 937, 938
Schlumberger FMI Fullbore Formation MicroImager, 428
Schlumberger sounding curve, 371*f*
SEAWAT, 489
Seawater reverse osmosis. *See also* Reverse osmosis (RO), 714–717, 715–718*f*, 737–739
Secondary wastewater treatment, 779
Seepage loss, 968
Seepage meters, 250, 251
Seismic reflection, 388–390
Self-potential (SP), 373
Semiarid environments
  recharges estimates in, 271–273
Semiarid lands, 923, 924
  IWRM in, 923, 924
  non-renewable groundwater resources, 928
Semiarid region recharge location, 197, 198
  mountain block and front recharge, 198, 199
Semi-confined aquifers. *See* Confined aquifers
Semi-unconfined aquifers. *See* Unconfined aquifers
Sensitivity analysis, 486
Sewage, definition, 777*t*
Shuttle Radar Topographic Mission (SRTM), 445
Silcretes, 44
Siltation, 591
Singapore, IWRM in, 916
Single-pack tests, 234
Single-ring infiltrometer, 235
Site evaluation studies, 464–467
  steps, 464
Six-Day War, 953
  role of water in, 953

S (*cont.*)
Sludge, definition, 777*t*
Small-scale technologies, 10
Snowball Earth, 975
Societal collapse, 518
Sodium adsorption ratio (SAR), 685
Soil
  exchange sodium percentage (ESP), 684, 685
  salinization, 684–686
Soil and water assessment tool (SWAT), 317, 318
Soil aquifer treatment (SAT)
  effectiveness of, 829
  wastewater treatment
    CEC removal, 832, 833
    denitrification and nitrification, 829, 830
    organic carbon, removal of, 830, 831
    pathogen removal, 830
Soil Conservation Service (SCS) Run-Off Number Method, 463
Soil conservation service curve number method, 301–304
Soil matric (water) potential, 241
  tensiometers, 241, 242
  thermocouple psychrometry, 242, 243
Soil moisture, 28
Soil moisture, measurement, 237, 449–451
  gravimetric method, 237, 238
  neutron moisture meter, 238–240
  resistivity-based geophysical methods, 240, 241
  soil moisture blocks, 238
  time-domain reflectometry (TDR), 240
Soil moisture gradients/changes, effects of, 197
Soil-aquifer treatment (SAT) systems, 582–587, 583*f*
  designing of, 582–584, 583*f*
  redox state, pretreatment for controlling, 584
Soil-moisture blocks, 238
Soil–water potential, 133, 134, 135*f*
Solar collectors, indirect desalination using, 729, 730, 731*f*, 732*f*
Solar energy, 737
Solar ponds, indirect desalination using, 728, 729
Solar stills, direct distillation using, 727, 728
Solar-powered desalination. *See also* Desalination, 727–730
  direct distillation, using solar stills, 727, 728
  indirect desalination
    using solar collectors, 729–730, 731*f*, 732*f*
    using solar ponds, 728, 729
Solar-powered hydrologic cycle, 12
Soluble microbial products (SMPs), 811
Sonic logs, 410–413
  example of, 414*f*
Sonoran Desert of Arizona, 672, 673*f*
Sorption
  detection of EDC, 843
  DOC reduction, 840
  pathogen, removal of, 830
  physical, 832
South African Development Community (SADC) Groundwater Grey Literature Archive, 226
Southern Peru, fog harvesting in, 552
Southwestern United States, water management in, 510–513
  Anasazi, 510, 511
  Hohokam, 512, 513
Spain, Valencia, fog harvesting in, 551, 552
Spatially distributed water budget models, 256–258
Specific retention, 133
Spontaneous potential (SP), 406–408
  and dual induction log, 407*f*
Stable isotopes, 278–281
  of water, 287–290
Stakeholders, 1003, 1027, 1034
  decision-making process, 863, 868
  management, 863
Standard fog collector (SFC), 549, 551, 552
Standard Mean Ocean Water (SMOW), 280
Stationarity, water resources management, 1005–1007, 1006*f*
STELLA programs, 920
Steroids, removal of, mechanism of, 832
Storage, as element of IWRM, 915
Storage systems, 530, 536–538
Storativity
  confined aquifers, 122
  definition, 121
  specific storage, 121–122
  unconfined aquifers, 122
Stormwater harvesting, 529, 540–541
  large-scale, 542–546
  resue and recycling of, 530, 531
Stormwater management
  as element of IWRM, 915
Stormwater recharge, 602–606, 605*f*
  Straddle-pack tests, 234
Stream gauging, 260, 261

Stream hydrograph, example of, 304*f*
Stress history, and subsidence, 347
Subsidence, impacts of, 355
Subsidence and Aquifer-System Compaction (SUB) Package, 492
Subsidization, 902, 903
Subsurface disposal methods, 759–762
Subsurface engineering, 597–602
Sulfate reduction, 839
Sunsidence measurement, 348
  conventional surveying, 348, 349
  extensometers, 349, 350
  global positioning systems (GPS), 350, 351
  measured land subsidence amounts and rates, 353, 354
  Satellite Interferometric Synthetic Aperture Radar (InSAR), 352, 353
Surface energy balance
  in latent heat flux calculation, 449
  in reference evapotranspiration rate calculation, 461
Surface geophysical methods, 253
Surface geophysics, 365, 366
  electrical resistivity and electromagnetic techniques and, 366–368
  electrical resistivity method and, 368–373
  electromagnetic surveys and, 373, 374
    frequency-domain electromagnetic soundings for shallow data collection, 374, 375
    transient electromagnetic sounding methods for deep data collection, 375–377
  ground penetrating radar (GPR) and, 383, 384
  to groundwater investigations, 395
  magnetotellurics and, 387, 388
  seismic reflection and, 388–390
  surface nuclear magnetic resonance and, 385–387
Surface geophysics, applications to groundwater investigations, 395
  resistivity-based surface geophysics and, 377
    depth to water table, 380
    formation and aquifer mapping, 381, 382
    freshwater lenses mapping, 380
    managed aquifer recharge investigations, 380, 381
    recharge areas mapping, 382
    regional aquifer flow orientation mapping, 382, 383
    saline-water interface mapping, 377–379
Surface nuclear magnetic resonance (Surface NMR), 385–387
Surface permeability, 190
Surface runoff potential ($RO_{V\text{-}POT}$), 257, 258
Surface spreading, 571–573
  clogging, 578–581
  design issues of, 573–577
  infiltrated water, migration of, 577, 578
  monitoring, 581, 582
  recharge galleries, 587–589
  recharge shafts, 589
  recharge trenches, 587–589, 588*f*
  soil-aquifer treatment, 582–587, 583*f*
Surface water and groundwater
  conjunctive use, as element of IWRM, 915
Surface water budget of channels, 261, 262
Surface water system
  basic water budget, 152
    basic elements of, 153*t*
Surface water, 88, 89
Surficial Aquifer System
  of South Florida, 154
Surprise, definition of, 479
Sustainability, 209–210, 670
  groundwater use impact acceptability and, 214, 215
  mitigation and, 216, 217
  safe yield and, 212–214
    evaluation, 216
  water resources management, 210–212
Sustainability water resources management, 210–212
Sustainable development, 929
  concepts of, 209
  fundamental aspect of, 209, 210
Sustainable pumping, 213
Sustainable yield, definition of, 209
SUTRA, 489, 490
Suzhou City
  subsidence in, 346
SWAT. *See* Soil and water assessment tool (SWAT)
Synthetic aperture radar (SAR), 352, 445
Synthetic hydrographs, 308–310
Syria, Six-Day War, 953
System recovery efficiency. *See also* Recovery efficiency (RE), 565

S (*cont.*)
Systeme Aquifere du Sahara Septentrional (SASS), 938

T
TCPP [Tris-(choloroisopropyl)-phosphate], 833
Tectonic
  active versus inactive regions, 46, 47*f*
  activities, 46
  movements, 46
  role, 44
TEM method. *See* Transient electromagnetic (TEM) method
Temperature
  effects of, 196, 197
  and fluid resistivity logs, 423, 424
  histories of, 323*f*
Temporal and spatial management of soil water (TSMSW), 692
Tensiometers, 241, 242
Terraces, conceptual diagram of, 533*f*
Terrestrial sedimentary process, 43
Tertiary wastewater treatment, 779
Testosterone, 832
Texas Bureau of Economic Geology
  airborne TDEM studies, 392
Texas Manual on Rainwater Harvesting, 534
Texas, absolute ownership, 865
Theory of absolute territorial integrity, 956
Theory of absolute territorial sovereignty. *See also* "Harmon Doctrine", 956
Theory of community or integrated drainage basin development, 956
Theory of limited territorial sovereignty or equitable utilization, 956
Thermal desalination. *See also* Desalination, 702, 704–713
  adsorption desalination, 711–713, 712*f*, 713*f*
  multiple effect distillation, 708–710, 709*f*
  multi-stage flash distillation, 705–708, 706*f*
  vapor compression distillation, 710, 711*f*
Thermocouple psychrometry, 242–243
Time-domain reflectometry, 240
Tiwanaku civilization, collapse of, 520
Total coliform bacteria, 810
Total dissolved solids (TDS), 764
Total efficiency. *See also* Efficiency, 689
Total organic carbon (TOC), 638
Total water management, 911
Tracer versus water migration, 268
Tragedy of the Commons, 854–858
Transboundary aquifers, 960, 961
  management of, 961
  models of, 960
Transboundary non-renewable groundwater resources, 935, 936
Transboundary water management, 953
  1997 Convention on Non-navigation Uses of International Watercourses, 958, 959
  Conflict Resolution, 959, 960
  International (Transboundary) River Water Law, 956, 957
    Berlin Rules, 958
    Helsinki Rules, 957, 958
  six models of, 960
  transboundary aquifers, 960, 961
  Transboundary Groundwater Law, 962
    1997 Convention on Non-Navigation Uses of International Watercourses, 963
    Bellagio Draft Treaty, 962
    Internationally Shared Aquifer Resource Management Programme, 963
    Law of Transboundary Aquifers, 964, 965
  in United States and Mexico, 965, 966
    All-American Canal lining, 968, 969
    Hueco Bolson Aquifer, 966–968
Transient electromagnetic (TEM) method, 374
  assumptions and limitations of, 376, 377
  for deep data collection, 375–377
Transient electromagnetic methods, 374, 381–382
Transmission losses, 261, 293–294, 310–312
  in ephemeral channels, 295*f*
  from experimental watersheds, 312–314
  and recharge, 315
  relationships from experimental watersheds, 312–314
  and tributary flows, 314, 315
Transmissivity
  hydraulic property, 121
Treated sewage effluent (TSE), definition, 777*t*
Tritium, 284–286

U
Ultrafiltration. *See also* Filtration, 714
U.N. Conference on Environment and Development (UNCED), 854
Unconfined aquifers
  characteristics, 95

conditions, 97
and confined aquifers, 96, 97
definition of, 96
as emergency sources, 1009
hydraulic conductivity, 96
in transboundary aquifers, 960
semi-unconfined aquifers, 97
storage capacity, 96
UNCED. *See* U.N. Conference on Environment and Development (UNCED)
UNESCO
airidity classification, 24*f*
Unit hydrographs, 306–308
United Arab Emirates (UAE), 104
Wadi Ham, 147
United Nations Convention to Combat Desertification (UNCCD), 33
United Nations Law of Transboundary Aquifers, 935
United States
Colorado River allocation, 969
groundwater use in, 965, 966
United States and Mexico, transboundary water management in, 965, 966
All-American Canal lining, 968, 969
Hueco Bolson Aquifer, 966–968
United States Environmental Protection Agency (USEPA)
WaterSense®Program, 672
Universal Declaration of Human Right, 859
Unmanaged aquifer recharge (UMAR). *See also* Managed aquifer recharge, 617–619
Unsaturated zone. *See* Vadose zone
Urban areas, drought in, 31
Urban domestic uses, 888
water resources, affects, 979
Urban water management, 5
U.S. Environmental Protection Underground Injection Control, 563
U.S. Geological Survey (USGS)
Basin Characterization Model, 193
desert, definition of, 33
USGS. *See* U.S. Geological Survey (USGS)
U.S. National Weather Service non-recording rain gauge, 166*f*

V

Vadose zone, 95, 188, 189, 203
hydrology
air entrapment and flow, impacts of, 143–145
bedding and heterogeneity impacts, 142
capillary fringe, separated by, 129, 130
capillary pressure (*see* Capillary pressure)
definition, 129
functions, 130
groundwater resources, 131
infiltration (*see* Infiltration)
kinematic wave theory, 142
process and flow, 134, 136–138, 138*t*
water potential, 133–135*f*
water passage, 830
hydraulic testing methods, 235
infiltration tests, 235–237
soil matric potential measurement, 241–243
soil moisture measurement, 237–241
Vadose-zone processes, 236
infiltration tests, 236, 237
Vadose-zone recharge wells. *See also* Wells, 569–571, 570*f*
Valencia, Spain, fog harvesting in, 551, 552
Validation monitoring, 807, 822
Values, in water management, 16–19
Vapor compression distillation (VCD). *See also* Distillation, 710, 711*f*
Variable density log (VDL), 412, 415*f*
VCD. *See* Vapor compression distillation (VCD)
VDL. *See* Variable density log (VDL)
Vegetation
control, 580
effects of, 43, 194–196
ergs, types of, 82
erosional process, role in, 64
Vegetational mapping, 448
Verification monitoring, 808, 822
Vertical electrical sounding (VES) analyses, 368, 372, 379
Vertical extensometers, 350
Vertical land subsidence
methods of measurement, 348*t*
Veterinary chemicals, 812
Video survey, 424–426
Vienna Standard Mean Ocean Water (VSMOW), 280
Viral decay rates, 835
Virtual water
adverse impacts, 893
blue water, 892, 893
concept, 889
green water, 892

V (*cont.*)
optimal solution, 893
scare water resources, 890
trade, 891, 892
water deficits through, 889, 890
Volcanic rock deposits, 64*f*, 87, 88, 89*f*
Volume balance, 253
Vugs, 126
Vulnerability, 1002–1005, 1010, 1028, 1030, 1034

W
Wadi channels, 198
in arid and semiarid regions, 202, 293
Wadi dams. *See also* Dams, 590–594
advantages of, 592
aquifer recharge, optimizing, 593, 594
benefits of, 593
budget estimates for, 592
infiltration, efficiency of, 592
operation and maintenance of, 593
Wadi infiltration, 293
Wadi recharge, 202–204, 293–295
and geomorphic relationship, 310
hydrologic model systems and, 315, 316
HEC-HMS model code, 317
precipitation-runoff modeling system (PRMS) model, 318
soil and water assessment tool (SWAT), 317, 318
SW modeling applications, 318, 319
methods for measuring, 295–298
rainfall-runoff relationships, 298–300
geomorphic relationship, 310
hydrographs, 304–310
precipitation, 300–301
regression methods, 301
soil conservation service curve number method, 301–304
seasonal timing, 319, 320
temperature data and, 320, 321
hydraulic conductivity determination and infiltration rates, 324, 325
temporal and spatial patterns determination, 321–324
transmission losses and, 310–312
and recharge, 315
relationships from experimental watersheds, 312–314
and tributary flows, 314, 315
Wadis
bedrock, 74
flow energy, variation in, 73
location, 72, 73*f*
Schwaib, 73, 75*f*
sediments, 72
stream bed/channel, 72
Walnut Gulch Experimental Watershed (WGEW), 312, 313
Waste heat, 737
Wastewater
definition, 777*t*
municipal, 777*t*
recycling, industrial, 1013
reuse, 773 801
for agricultural irrigation, 782–786
benefits of, 773
economic costs and benefits of, 793–796, 795*f*
environmental concerns over, 800, 801
industrial, 799, 800
portable reuse, 786–792
public acceptance of, 796–799
treatment processes, 777–782, 778*t*
terminology, 777, 777*t*
Wastewater reuse, 15, 523
as element of IWRM, 915
environmental concerns, 800, 801
epidemiological studies on, 816–819
RBF, as element of, 643
Wastewater reuse, health and safety issues
advantages of, 1034
chemical contaminants, 810, 811
compounds of emerging concern (CECs), 811–816
epidemiological studies, 816–819
health-based targets and monitoring, 819, 821–823
implementation, stages of, 1034, 1034*f*
microbiological standards and guidelines, 820, 821
pathogenic microorganisms, 808–810
risks, 805–808
assessment of, 816–819
management and reduction of, 807
monitoring, 807, 808
treatment, MAR for. (*see also* Managed aquifer recharge (MAR)), 1036, 1037
Wastewater stabilization ponds (WSPs), 780, 781
Water
allocation, 853
arid and semiarid lands, management in, 1027
demand reduction
agriculture water use, 1019, 1020

conservation, 1017, 1018
permanent and temporary reallocation, 1018, 1019
population growth, 1017
trading, 1019, 1020
use, 1020, 1021
human right
arguments, 860
concept and implications, 860
non-government organization, 861
and sanitation, 859, 861
law and policy. (*see also* Policy), 862–864
management
Buddhism, 876–877
Christianity and Judaism, 875–876
decisions, 852
definition, 851
Hinduism, 877
Islam (*see* Islam)
Tragedy of the Commons, 854–858
resources planning and management, 1005–1007, 1006*f*
scarcity
causes, 1040, 1041
countries, 1033, 1034
supply, increased
benefit from, 1040
desalination, 1013, 1014
groundwater resources, 1009, 1010
rainwater harvesting, 1014
storage, 1010–1012, 1011*f*
transfer and diversion, 1014–1017, 1016*f*
wastewater, reuse of, 1012, 1013
trade, virtual, 1033
undervalued resource, 858, 859, 1028, 1029
use efficiency, 1032
Water, economic value of
agriculture, 883, 884
allocation of, 881
analysis of, 887
centralized water system, 884, 885
Chilean model, 898–900
consumers of, 887
cost of, 886, 887
Dublin principles, 886
effective value, 898
financial transaction, 886
food production, 883
food security, 888, 889
markets, costs and benefits, 896
property rights, 896, 897
ownership and market-driven solutions, 893–895
public water treatment, privatization of, 904–906
pure market approach, 886
reallocation, 896
scarcity, 897
reliability, 883
self-sufficiency, 888, 889
socioeconomic, 884
subsidies, 902, 903
trading, third party effects of, 8900–902
virtual (*see* Virtual water)
Water budget, 151, 152, 253
and captured discharge and recharge, 154–156
component evaluation, 160
and groundwater resources management, 156–159
recharge within context of, 153, 154
surface water components, 152, 153
Water conservation and saving, 665–699
agricultural irrigation, 682–684
domestic, 672–682
conservation and utilities operation, 679, 680
consumer education, 675, 676
legal mandates, 674, 675
market incentives, 676–679
non-revenue water, 680–682
opportunities of, 672–674
irrigation water use efficiency techniques, 691, 692
saline-water irrigation, 693, 694
Water demand management
reallocation, 9
adaptive capacity, 9, 10
technological changes, implementation, 9
Water dependency index (WD), 892
Water Evaluation and Planning Version 21 IWRM model, 921
Water management
challenges, 954
demand-side, 669
planning, and IWRM, 922, 923
political issues, 15, 16
socioeconomic issues, 15, 16
supply-side, 669
technical issues, 14, 15
Water markets, 863
Water ownership, 955
Water potential, 133, 134
Water quality improvement
groundwater environment

W (*cont.*)
CEC, 842, 843
disinfection byproducts, 840–842
microbial risk, evaluation and management, 836–838
organic carbon removal, 838–840
pathogenic microorganism inactivation, 834–836
during SAT (*see* Soil aquifer treatment (SAT))
Water ranches, 894
Water recycling, definition, 777*t*
Water resources assessment methods, 223–225
aquifer hydraulic testing, 229, 230
aquifer performance testing, 230–232
aquifer testing during drilling, 232, 233
packer tests, 234, 235
data collection during well drilling, 228, 229
desktop investigations, 225, 226
exploratory well and production well construction, 226–228
exploratory well and production well construction, 226–228
vadose-zone hydraulic testing methods, 235
infiltration tests, 235–237
soil matric potential measurement, 241–243
soil moisture measurement, 237–241
Water resources, development phases, 8, 9
Water scarcity, 954
elements of, 9
levels of, 3
primary loci of, 5
Water self-sufficiency index (WSS), 892
Water stress
global water budget, 11–14
indicators, 13*t*
Water table
definition, 129
depth to, 380
Water table aquifers. *See* Unconfined aquifers
Water table fluctuation (WTF) method, 296, 297
Water use efficiency (WUE), *See also* Efficiency, 669, 670, 687–689, 691
effective, 690
irrigation
incentives, 694, 695
techniques, 691, 692
Water use efficiency, 18, 883
Water use permit (WUP), 867
Water-allocation decisions, 16
Watercourse Convention. *See* 1997 Convention on Non-Navigation Uses of International Watercourses
Water-table fluctuation method, 252–254
Weber river ASR site, 336
Weighted goal programming, 918, 919
Weighted spatial probability modeling (WSPM), 467
Well drilling, data collection during, 228, 229
Well efficiency, 227
Well Tool (WET), 462
Wells
aquifer recharge using, 567, 568
injection, 560–571
vadose-zone recharge, 569–571, 570*f*
Wenner and Schlumberger electrical resistivity arrays, 369*f*
West Bank of Palestine, 875
Westbay System, 654, 655*f*
Western South Africa, fog harvesting in, 550
Western Water Law, 865
WET. *See* Well Tool (WET)
WetSpass water budget model, 257
Wetting front, 140
Wilting point, 195
Wind convergence, 164
Wind-powered desalination. *See also* Desalination, 731, 732
World Bank, 1010
World Health Organization, 818–820
World Water Council, 891
WSPM. *See* Weighted spatial probability modeling (WSPM)
WTF method. *See* Water table fluctuation (WTF) method
WUE. *See* Water use efficiency (WUE)
WUP. *See* Water use permit (WUP)

Y
Yemen, fog harvesting in, 550, 551
Young–Laplace equation, 132

Z
Zero liquid discharge (ZLD), 763–765
ZLD. *See* Zero liquid discharge (ZLD)

GPSR Compliance

The European Union's (EU) General Product Safety Regulation (GPSR) is a set of rules that requires consumer products to be safe and our obligations to ensure this.

If you have any concerns about our products, you can contact us on ProductSafety@springernature.com

In case Publisher is established outside the EU, the EU authorized representative is:

Springer Nature Customer Service Center GmbH
Europaplatz 3
69115 Heidelberg, Germany

Batch number: 10584428

Printed by Printforce, the Netherlands